Quality

Ivan Barofsky

Quality

Its Definition and Measurement
As Applied to the Medically Ill

Ivan Barofsky
Quality of Life Institute
East Sandwich, MA 02537
USA
ibarofsky@comcast.net

ISBN 978-1-4419-9818-7   e-ISBN 978-1-4419-9819-4
DOI 10.1007/978-1-4419-9819-4
Springer New York Dordrecht Heidelberg London

Library of Congress Control Number: 2011932170

© Springer Science+Business Media, LLC 2012
All rights reserved. This work may not be translated or copied in whole or in part without the written permission of the publisher (Springer Science+Business Media, LLC, 233 Spring Street, New York, NY 10013, USA), except for brief excerpts in connection with reviews or scholarly analysis. Use in connection with any form of information storage and retrieval, electronic adaptation, computer software, or by similar or dissimilar methodology now known or hereafter developed is forbidden.
The use in this publication of trade names, trademarks, service marks, and similar terms, even if they are not identified as such, is not to be taken as an expression of opinion as to whether or not they are subject to proprietary rights.
While the advice and information in this book are believed to be true and accurate at the date of going to press, neither the authors nor the editors nor the publisher can accept any legal responsibility for any errors or omissions that may be made. The publisher makes no warranty, express or implied, with respect to the material contained herein.

Printed on acid-free paper

Springer is part of Springer Science+Business Media (www.springer.com)

*Dedicated to the Memory
of
Dr. Anna-Lisa Barofsky*

Preface

The purpose of this book is to define and measure quality, in general, and quality-of-life, in particular. I illustrate the issues involved in this task by examining the quality-of-life of the medically ill person, but I could have addressed these same issues by focusing on the quality of working life, the quality of the environment, or even the quality of an object or meal. I believe that how a qualitative[1] judgment is made is the same independent of the content being judged, and because of this, any qualitative assessment needs to simulate how such a judgment is made. I am not saying that the quality of a car is the same as the quality of a life; content does make a difference. What remains true, however, is that the cognitive processes involved in how different content is made into a qualitative statement is the same. As I will explain, the cognitive components of qualitative judgment involves: *description, evaluation* and *valuations*.[2]

In this book, I describe the fundamentals of this approach and then apply it to common indicators of the quality-of-life of the medically ill person. In Part I (Chaps. 1–3), I review the principles of language usage and describe various cognitive processes, while in Part II (Chaps. 4–6) I discuss the purpose (e.g., establishing invariant principles), content (objective or subjective indicators), and unit of analysis (categories or domains) of a qualitative judgment. In Part III (Chaps. 7–11), I apply the analytical approach and principles I have described to the specific content ordinarily included in a health-related qualitative judgment. In Chap. 12, the final chapter, I restate and review support for my working hypothesis, and in the Epilog I discuss future research that needs to be done.

I will address a number of specific issues in this book. For example, what is the best model[3] to use to conceptualize and how to obtain information about *quality*? The current standard relies on linear modeling systems and characteristic of psychometric methods (analysis of variance, regression analysis, and so on), but I will encourage the use of alternative analytic methods and schemes such as nonlinear models, neural networks, and complexity-based evolving cognitive systems. I will argue, for example, that a complexity-based self-organizing system characteristic of the central nervous system underlies how the constant stream of experiences I encounter (i.e., my phenomenology) is transformed into entities that can be processed by the brain. This occurs because of the adaptive capacity of the central nervous system, and it is this characteristic that makes this type of model particularly relevant to any assessment of quality, in general, and quality-of-life, in particular. However, once cognitive entities are formed, they can be linearly or nonlinearly combined. In addition, it is also possible to use linear models to assess outcomes generated by nonlinear means. I feel that it is necessary to expand the available models since the current domination of linear-based psychometric assessment models essentially ignores *how* the responses to a quality assessment occurs, and this creates issues concerning the validity of a qualitative assessment. In addition, psychometric

---

[1] The term "qualitative" has a distinct meaning in the behavioral and social sciences, being applied to set of procedures that differ from psychometric methods. I will be using the term in a more restricted manner to refer to different types of quality assessments that may be assessed by either "qualitative" (e.g., grounded field theory) or psychometric methods.

[2] I use the term "valuation" to refer to such procedures as value elicitation, preference determination, and utility estimation (e.g., standard gamble).

[3] One of the primary objectives of this book is to offer an alternative to the current model that evolved from educational psychology. This new model involves the cognitive sciences and pragmatic aspects of language usage.

assessments are usually dependent on an intact language system for their input,[4] whereas nonlinear and complexity-based models can manage a broader array of analytical models. This will give me the opportunity to discuss alternative, nonverbal dependent models for use when dealing with the compromised person.

A reader will notice a number of distinct characteristics to my approach. For example, I use the term "assessment" almost exclusively, as opposed to the term "measurement" when characterizing quality. I also very often talk about "objectification" as opposed to "objective." In both instances I want to emphasize that the science underlying what I am talking about is something that a person creates, it is a process, not a state or entity that exists. Thus, I avoid the terms "measurement" or "objective" since both imply something mechanical that can exist independent of me. The scientific study of quality, however, starts from something that is fundamentally subjective, and as such, is based on data that has to be assessed and transformed into something objective (i.e., the objectification process). I believe quality data can be made as objective as any other type of data, and like other forms of scientific data, can assume the characteristics of invariant statements. This, however, will require the recognition of neurobiological determinism and its influence on how I think and reason. Thus, although I see my scientific objectives as being the same as a physicist, I differ by acknowledging that I start with subjective information. I will discuss this and related issues in Chap. 12.

A second characteristic of my approach is that I follow Bridgman's (1959) recommendation of making sure that I understand the process of understanding so that I continually try to separate what an investigator does from what a respondent does. For example, I make a distinction between categories and domains, where a "cognitive category" is a summary statement that a person generates, while "domain" refers to what an investigator does to summarize or collect information (e.g., as when using a factor analysis to identify a list of items to label). I try to avoid using the terms interchangeably, eventhough an investigator also generates categories, it is just that the categories an investigator cognitively creates (e.g., an item) maybe very different from what a respondent may do with the same information. The intent here, of course, is to have the investigator avoid confusing their views of existence with the respondent's. At the same time, it recognizes that the investigator is as much a participant in the scientific study of quality as the respondent. Thus, I see the investigator and the respondent as inextricably interwoven in the process of assessing quality, and a complete picture requires separating and identifying the role of each.

Understanding the process of how an investigator understands is also essential to my argument that an optimal quality-of-life or HRQOL assessment has to include a valuation, preference, or utility estimation. In its absence, a quality-of-life or HRQOL assessment will only approximate an ideal assessment. My primary evidence for this argument comes from the observation that investigators have gravitated to this type of model from the onset of the field. Thus, some of the earliest efforts in such qualitative assessments include valuations of one sort or another (e.g., the Health Utilities Index or Quality of Well-being Scale). In Chap. 12 I discuss this issue in greater detail, particularly concerning whether weighting an evaluated descriptor makes a difference. To achieve my goals, I have to provide the appropriate intellectual foundation for the issues I raise. Thus, at times, I discuss material from such diverse areas as philosophy, psychology, and neurocognition. In each case, I will make the connection to issues of quality assessment clear. Part of the justification for wielding such a broad brush is that I want to be able to describe an approach to the definition and assessment of quality that is consistent across applications (e.g., quality-of-life, quality of working life, judging a painting, judging a car, and so on). In order to do this, I have to address some fundamental issues, such as how I come to know that what I report as quality; how I can judge that my quality-of-life or HRQOL is real and true; what specific cognitive processes (such as concept formation) and brain mechanisms might be involved; and how these mechanisms mediate the formation of a quality assessment. I also want to know if a definition of quality is derivable as a mathematical proposition,

---

[4] Of course, nonverbal behavior can also be assessed using linear modeling, but the broader issue here is that the intent of linear modeling is to establish a causal relationship, while nonlinear modeling acknowledges that the input and output of a system are best characterized as probabilistic, and requires an assessment model that acknowledges this.

whether it evolves from a social consensus, or whether it is an experience that everyone has (my operating assumption).

There has been a considerable amount of effort to define quality or quality-of-life that involves several disciplines, each of which has its own unique set of assumptions that often contradict each other. To deal with this I will initially determine the extent to which each of a representative sample of quality-of-life or HRQOL definitions uses literal or figurative language (e.g., metaphors). This will permit me to determine if part of the difficulty in finding a common definition of quality-of-life or HRQOL has to do with differences in how language is used in these definitions (Chap. 2). I will also describe various *cognitive processes* and consider their role in generating a quality-of-life or HRQOL assessment. Quality, as an abstract entity, was considered a "concept" and how they were formed and used, will be extensively discussed (Chap. 3). Both of these tactics will be used throughout this presentation.

In this book, I will make explicit how performing a qualitative assessment is an *ethical act*. Thus, it seems obvious that one of the major reasons for performing a qualitative, and particularly a quality-of-life, assessment is because I want to ensure that the events that befall a person or interventions that have been applied to them do them no harm. The politics involved in ensuring this (e.g., quality control efforts), as well as the more general issues dealing with how we live our lives (i.e., the communities we live in, the crime rate, poverty, and so on) and what impact this has on our life quality, are all relevant here. Thus, a major theme running throughout this effort will be the explication of how a quality assessment can be used to improve the ethical character of social interactions, with a particular emphasis on the relationship between the patient and clinician.

I approach my task as someone whose training as an experimental and clinical psychologist makes it natural to approach the preparation of this task from the perspective of the individual. I have, however, felt free to use and integrate the insights and principles characteristic of other disciplines (e.g., sociology, philosophy or linguistics) when necessary, even though this places me at risk of inadequately representing these disciplines. Still, this risk was worth taking, considering that the object of this inquiry, understanding the nature of a quality and quality-of-life assessment, seems to naturally encompass a broad array of issues that would be inappropriate to not consider.

My interest in the ethical and political intent of quality-of-life or HRQOL research comes, not unexpectedly, from what I have learned from patients, friends, and family members who have had to use the American healthcare system or who have experienced changes in social policies (both good and bad). I have also observed that the role bioethics currently plays is limited to that of an outside critic rather than as a intimate determinant of the character of the healthcare process. I propose, as others have, that the continued promotion and integration of quality assessment into, for example, healthcare activities increases the chances of *preventing* or reducing ethical transgressions. This, I argue, is essential if health care is to make a maximum contribution to a person's existence.

Finally, I hope that the model I am proposing will stimulate debate and active discussion about how best to assess quality and quality-of-life. To achieve this I have adopted the tactic of being quite open about the limitations of current efforts to define and assess quality, including considering whether it should be assessed at all. I also acknowledge that there is an inherent ambiguity and complexity about this concept and the task concerning its assessment, as would be expected for such an important personal and societal indicator. I find that these complexities are balanced by the richness of the intellectual opportunities created by the concept; it is a concept worth struggling to define and assess. The length of its history alone attests to its capacity to deal with a fundamental human concern: the nature of our current existence or how one has lived one's life. Its potential as a source for social innovations is also exciting, and remains an underdeveloped activity. The task remaining, however, is to convincingly capture the joy, pain, and suffering real people experience with the numbers that I will use to characterize their experiences.

East Sandwich, MA   Ivan Barofsky

Acknowledgments

There are many, many people I would like to acknowledge; some have provided intellectual support, others social and emotional support, and some, all three. Drs. Penny Erickson and David Feeny provided me with the support I needed to start and complete a task of this magnitude. I publically thank them. Many other colleagues have contributed to my intellectual development, and it is best that I not attempt to list them, lest I leave a deserving person out. They know who they are and I thank them all. Of course, my son Jeremy and my family and friends have been supportive and they deserve special thanks for their efforts.

A number of people have been helpful in reading portions of the manuscript, including K. Irlen and Drs. W. Adler, J. Boderson, P. Erickson, P. Fayers, D. Feeny, C. Ferrans, S. Hunt, and T. Smith. Each of these people have read selected chapters or the entire manuscript and none should be held responsible for the final version of the manuscript. The Harrison Library of the Johns Hopkins Bayview Medical Center and the Public Library of The Town of Sandwich, MA were very helpful in providing articles or access to books, and I thank them. Mrs. Norma Miller, my long time secretary, needs recognition and a "thank you" for having kept my professional life in order when I was an active clinician. Ms. Jessie Gunnard took on the heroic task of copyediting this book and I thank her for her patience and skill.

Ivan Barofsky

Contents

Preface		vii
Acknowledgments		xi

Part I Defining and Assessing Quality and Quality-of-Life

1 The Difficulty of Assessing Quality or Quality-of-Life 3
   Abbreviations 3
   1 Introduction 3
   2 Should Quality or Quality-of-Life be Assessed? 6
      2.1 Counter Arguments 6
      2.2 Wittgenstein's Language Games 8
      2.3 The Simplest Measure: A Self-Assessment 9
   3 What Does the Term Quality Mean? 10
      3.1 Is Quality a *Quale*? 10
      3.2 Quality as a Conceptual Space 12
      3.3 Types of Quality Assessments 15
   4 What are the Objectives of This Book? 17
   5 Summary of Chapter 20
   Notes 21
   References 22

2 The Role of Language in Assessing Quality or Quality-of-Life 25
   Abbreviations 25
   1 Introduction 25
   2 Language Usage Principles 26
      2.1 The Elements of Language 26
      2.2 The Optimal Expression of Quality-of-Life or HRQOL 27
      2.3 Literal Expression 28
      2.4 Figurative Expression and Its Complexities 30
         2.4.1 Defining a Linguistic Metaphor 32
         2.4.2 The Importance of Conceptual Metaphors 35
         2.4.3 Embodiment and Metaphors 36
   3 Application of Language Usage Principles 38
      3.1 The Metaphoric Basis of Quality-of-Life or HRQOL Definitions ... 38
   4 Summary of Chapter 45
   Notes 46
   References 47

3 What Role Do Cognitive Processes Play in Assessing Quality or Quality-of-Life? 49
   Abbreviations 49
   1 Introduction 49

	2	Cognitive Principles	51
	2.1	The Neural Basis of Cognition	51
	2.2	The Formation of a Cognitive Entity	54
	2.3	The Classification of Cognitive Entities	55
		2.3.1 Hierarchical Classification	56
		2.3.2 Cross-Classification	57
		2.3.3 Applications to a Qualitative Assessment	58
		2.3.4 Hieroglyphics: A Language Based on Classification	60
	2.4	Concept Formation	62
	2.5	Metacognition	64
	3	The Application of Cognitive Principles	66
	3.1	The Fast and Automatic Assessment of Quality	66
		3.1.1 Affordance	67
		3.1.2 Trait (Facial) Perception	68
		3.1.3 Emotion Inference and the Appraisal Process	69
		3.1.4 Summary	71
	3.2	Cognitive Models of Quality-of-Life Assessments	72
		3.2.1 Brunswik, Social Judgment Theory and the SEIQoL	72
		3.2.2 Norman Anderson's Cognitive Algebra and the QWB	74
		3.2.3 Summary	77
	4	Summary of Chapter	77
		Notes	78
		References	79

Summary of Part I 83

Part II Issues in Assessing Quality or Quality-of-Life

4 The Role of Objective or Subjective Indicators 91
   Abbreviations 91
   1 Introduction 91
   2 Self-Reports as Data 93
   3 Positional Objectivity 95
   4 What to Measure: Objective, Subjective, or Some Combination of Indicators? 97
      4.1 Social Indicators: The Subjective Processing of Objective and Subjective Information 98
      4.2 Models that Combine Objective and Subjective Indicators 99
          4.2.1 Hierarchical or Taxonomic Modeling 102
          4.2.2 Linear Modeling 104
          4.2.3 Nonlinear Modeling 107
      4.3 Summary 109
   5 Objective Assessment of Subjective Indicators I: Kahneman's Model of Objective Happiness 109
   6 Objective Assessment of Subjective Indicators II: The Rasch Model 110
   7 Invariance 112
   8 Summary of Chapter 114
   Notes 115
   References 116

5 The Role of Signs or Symptoms 121
   Abbreviations 121
   1 Introduction 121
   2 On the Nature of a Symptom 123

		2.1	Language Usage and Symptoms	124
			2.1.1 Metaphors in Patient Symptom Reporting	125
			2.1.2 Metaphors in Doctor–Patient Symptom Reporting	125
			2.1.3 Symptoms as Metaphors	126
		2.2	Assessing Symptoms	127
			2.2.1 Symptom Assessment and the Qualification Process	128
			2.2.2 The Accuracy of Symptom Reports	129
	3	The Objective Assessment of Signs and Symptoms		131
		3.1	*Truth-to-Nature*: Qualitative Assessment	132
		3.2	From Mechanical to Structural Objectivity: Quantitative Assessment	135
		3.3	Trained Judgment: Expertise	139
	4	Symptoms, Subjectivity and Clinical Medicine		141
		4.1	Defining Adverse Events	143
		4.2	Clinical Management of Symptoms	148
		4.3	Clinimetric and Psychometric Perspectives	149
	5	Summary of Chapter		152
	Notes			154
	References			154

6 Summary Measurement: The Role of Categories or Domains 159

	Abbreviations		159
1	Introduction		160
2	The Formation of Categories		161
	2.1	Introduction	161
	2.2	Categories and Language Usage	162
	2.3	Category Learning	164
	2.4	Kinds of Categories	166
	2.5	Cultural Determinants of Category Formation	168
3	The Construction of Domains		170
	3.1	Introduction	170
	3.2	Domains: Language Usage	171
	3.3	The Role of Cognitive Processes in Domain Formation	171
	3.4	Objectification Methods and Domain Formation	173
	3.5	The Complex Structure of Domains	174
	3.6	The Modular Basis of Domains	175
		3.6.1 Are Symptoms Modular?	177
		3.6.2 Are Health States Modular?	178
		3.6.3 Are Functional States Modular?	180
		3.6.4 Are Neurocognitive Indicators Modular?	181
		3.6.5 Are Subjective Well-Being Indicators Modular?	182
		3.6.6 Sumary	183
4	Applied Classification		183
	4.1	Modularity Applied: Visual HRQOL Assessments	183
	4.2	Classification and Domain Systems	186
5	Summary of Chapter		187
Notes			188
References			189

Summary of Part II 193

Part III The Content of a HRQOL Assessment

7 Symptoms and HRQOL Assessment 199

Abbreviations		199
1	Introduction: Symptom Domain Formation	200

		2	Cognitive Basis of Symptom Domain Formation	204
		2.1	Fatigue as a Prototypical Symptom Domain	207
			2.1.1 The Assessment of Fatigue	208
			2.1.2 Fatigue, Depression, and HRQOL	211
		2.2	Pain as a Prototype Symptom Domain	217
			2.2.1 The Language of Pain	217
			2.2.2 The Assessment of Pain	221
			2.2.3 Acquired and Functional (Pain) Modularity	224
			2.2.4 The Classification of Pain	228
			2.2.5 Assessment of Pain in the Compromised Person	230
		2.3	Summary	231
	3	Symptom Domains and HRQOL Assessment		232
		3.1	Introduction	232
		3.2	Integrating Symptoms into HRQOL Assessments	232
		3.3	The Role of HRQOL in Symptom Management	237
	4	Summary of Chapter		239
	Notes			240
	References			240
8	**Health Status and HRQOL Assessment**			**247**
	Abbreviations			247
	1	Theoretical Foundation		248
		1.1	Introduction	248
			1.1.1 Definitions of Health	250
			1.1.2 Defining a Disorder	255
		1.2	Assessing Health Status	259
			1.2.1 End-Results and Outcomes Movement	259
			1.2.2 Operations Research in Health Care	260
			1.2.3 Donabedian's Definition of Quality (of Care)	262
			1.2.4 Patient Satisfaction as an Indicator of Quality (of Care)	264
		1.3	Summary of Theoretical Foundations	268
	2	Applications of Health Status Assessment		269
		2.1	Assessing HRQOL	269
			2.1.1 The Components of a Quality-of-Life or HRQOL Assessment	270
			2.1.2 Health Status and HRQOL Assessment	276
		2.2	Applications of HRQOL: The "QALY"	282
		2.3	Summary of Applications	284
	3	Summary of Chapter		285
	Notes			286
	References			286
9	**Functional Status and HRQOL**			**293**
	Abbreviations			293
	1	Theoretical Issues		294
		1.1	Introduction	294
		1.2	The Language of Function	295
			1.2.1 Function: Its Definition and Expression as a Mathematical Statement	297
			1.2.2 Function as a Form of Figurative Expression	298
		1.3	The Cognitive Basis of Function Statements	299
		1.4	Function Statements in Health and Disease	300
		1.5	Function as a Teleological Explanations	302

		1.5.1	Causes Without Teleology	304
		1.5.2	Teleology Without Values	305
		1.5.3	Teleology with Values	305
	1.6	Summary		306
	1.7	Functionalism: A Philosophical Tradition		306
		1.7.1	Models of Function and Functionalism	308
		1.7.2	Parsons' Structural–Functional Analysis of Medical Practice	310
	1.8	Summary		313
2	Applications of Functional Assessments			314
	2.1	The Role of Functional Assessment in the Rehabilitation of a Person with Disabilities		314
		2.1.1	The Language of Rehabilitation and Disability	315
		2.1.2	Models of Disability	316
		2.1.3	Functional Assessment, HRQOL, and the International Classification of Function	320
	2.2	Summary		329
3	Summary of Chapter			331
Notes				333
References				334

10 Neurocognition and HRQOL Assessment — 339

Abbreviations — 339
1 Introduction — 340
2 Finding a Common Metric: The Qualification of Objective Indicators — 343
3 The Substitution Hypothesis — 344
4 The Preservation of Neurocognition — 347
    4.1 Introduction — 347
    4.2 Neurocognitive Preservation — 348
    4.3 Neurocognitive Capacity and Degeneration — 352
5 Neurocognitive Degeneration — 353
    5.1 Introduction — 353
    5.2 Progressive Semantic Degeneration — 355
    5.3 The Psychometric Properties of Qualitative Assessments of Persons with Alzheimer's and Related Dementias — 359
    5.4 Aging and Neurodegeneration — 363
    5.5 Summary — 368
6 Neurocognition Recovery — 368
    6.1 Introduction — 368
    6.2 Metaphoric Properties of a Quality-of-Life or HRQOL Assessment — 369
    6.3 The Neurocognitive Basis of Metaphor Expression — 370
    6.4 Stroke and Metaphoric Expression — 371
    6.5 Stroke and the Meaning of Metaphors: The Visual Analog Scale — 373
    6.6 The Right Hemisphere, the Sense of Self- and Qualitative Assessment — 375
    6.7 Neurocognition Recovery — 376
7 Models of the Relationship Between Neurocognitive Capacity and Performance — 376
    7.1 Introduction — 376
    7.2 Neural Network Models and Optimality Theory — 378
    7.3 Embodiment and the Assessment of Quality — 381
    7.4 Summary — 383
8 Summary of Chapter: Neurocognitive Indicators and Quality-of-Life or HRQOL Assessment — 384
Notes — 385
References — 386

11	**Well-Being as an Indicator of Quality or Quality-of-Life**		393
	Abbreviations		393
	1	Introduction	393
	2	The Language of Well-Being	399
		2.1 Well-Being in Ordinary Discourse	400
		2.2 Subjective Well-Being as a Form of Metaphoric Expression	401
		2.3 The Language Used in Subjective Well-Being Assessments	403
		2.4 Summary	405
	3	The Cognitive Basis of Well-Being	405
		3.1 Introduction	405
		3.2 Subjective Well-Being as a Persistent Affective State	407
		3.3 Subjective Well-Being as a Momentary Emotional State	410
	4	Subjective Well-Being and the Emotions	412
		4.1 Introduction	412
		4.2 Is Subjective Well-Being a Discrete Emotion?	413
		4.2.1 The Metaphoric Nature of Emotion Statements	417
		4.2.2 On the Nature of Affect	418
		4.3 Is Subjective Well-Being a Cognitive-Emotional Process?	421
		4.3.1 The Affect Primacy Hypothesis	422
		4.3.2 The Cognitive-Primacy Hypothesis	423
		4.3.3 The Cognitive Structure of Emotions	424
		4.3.4 Appraisal in Well-Being and Quality-of-Life Assessment	426
		4.4 Subjective Well-Being as an Outcome of Cognitive-Emotional Regulation	428
		4.4.1 Emotional Regulation and Subjective Well-Being	430
		4.4.2 Regulatory Models of Subjective Well-Being	433
	5	Modeling of Subjective Well-Being	437
		5.1 The Neurobiology of Subjective Well-Being	438
		5.2 Nonlinear Modeling, Appraisal and Subjective Well-Being	438
		5.3 Embodiment Models of Subjective Well-Being	439
	6	Summary of Chapter	441
	Notes		442
	References		444

Part IV Summary and Future Directions

12	**Summary and Future Directions**		453
	Abbreviations		453
	1	Overview	454
	2	Determinants of Valuation	457
		2.1 Weighting Procedures	457
		2.2 Cognitive Basis of Valuation Methods	466
		2.3 Values and Language-Usage	471
		2.4 Summary: Valuation and Quality Assessment	473
	3	Can Quality be Quantified?	473
	4	Summary and Conclusions	477
	Notes		479
	References		480

Epilogue 485

Index 487

Part I
Defining and Assessing Quality and Quality-of-Life

The Difficulty of Assessing Quality or Quality-of-Life

Abstract

The quality of human existence is becoming an increasingly visible and vocal political and social concern. Yet, many of the approaches to this problem do not include a discussion of how the judgment of quality itself is made. This book addresses this issue describing an approach that relies on examining the language used and the cognitive processes involved in a qualitative judgment and assessment. This chapter provides the background to this effort, and describes issues in the assessment of quality-of-life or health-related quality-of-life that once clarified will lead to a model of how a qualitative judgment occurs.

Any fool can know. The point is to understand.

Albert Einstein.

Abbreviations

EQ-5D   EuroQual-5 Dimensions (Brooks et al. 2003)
HRQOL   Health-related quality-of-life
SAHS   Self assessed health status
TMS   Transcranial magnet stimulation

1 Introduction

My first and primary objective for this book is to contribute to what is known about how to infuse an ethical perspective into the conduct of human affairs, particularly in the conduct of medical care, by expanding the role that assessing quality plays in these activities. This reflects my belief that ethics is the primary criteria that I should use when designing my activities and is consistent with the French philosopher Levinas' (1969) view that ethics is "a first philosophy," preceding all other criteria for organizing human affairs. I introduced my approach to this objective in the Preface, where I claim that how a qualitative judgment is made is a product of a universal but complex cognitive process. How this occurs naturally will generate a host of questions, only some of which I will be able to answer. There are many reasons for this, including that the required data may be lacking or the issues raised are so complex that no simple answer is currently available. I will be left with more questions than I started with, and fittingly, I start with some questions.

How is it that I can judge the quality of a painting, a meal and that a person has lived "a worthwhile life?"

How is it that I can quickly judge when something is beautiful or disgusting, yet require time to decide on the quality of a person's character?

How can asking a single question or a series of questions provide me with the information I need to judge the quality of a medical procedure or the quality of living in a particular community?

What does the term quality mean now that it has found its way into the popular press and is used to promote products and experiences, and is heard as part of political speech and policy declarations?

I know that the term quality has evoked much philosophical comment from ancient to modern times (e.g., Aristotle; McKeon 1947), but the term has stirred discussion by economists (e.g., Nussbaum and Sen 1993) and spokespersons from a variety of intellectual disciplines (anthropologists, sociologists, psychologists and health service researchers; Sirgy et al. 2006).

I also know that assessing quality is a private act that I can make public informally (as when I make a comment) or formally (as when I respond to a questionnaire).

Judgments I make about the quality of my existence are a major vehicle through which I express my values. These judgments will also vary when I am a child, living an adult life, or when I am preparing for life's natural end.

I ordinarily think about my existence in different ways; I can reflect about how I have lived, how I expected to live and how I may have hoped to live. Each of these is a legitimate subject for study.

A wide range of formal qualitative assessments exist, each of which is designed to provide insights into the judgments I make about the quality of my existence. These assessments determine if my, or any other person's, optimal status has been altered because of a new social welfare policy, medical treatment, or procedure. Since these assessments can detect adverse events produced by these interventions they also have the potential to prevent bad things happening to people. As a consequence, the administration of these assessments becomes ethically mandated and appropriate to use in a variety of settings (e.g., clinical trials, surgical treatments and so on).

Assessing quality, however, is also a natural monitor of how people get along with each other (e.g., is someone exploiting, mistreating, or even under treating someone else), and thereby a vehicle through which I can assert my human rights (Barofsky 2003). Assessing quality literally gives me the voice I need to ensure that I can achieve my right to avoid a faulty product or policy, receive the care I need, and in this way, continue to work towards my personal goals.

The current interest in assessing quality is a product of a number of historic forces. This includes the emphasis on quality indicators as an assessment of environmental and social impact (e.g., health care; Fig. 1.2; Albin 1992; Ch. 1, p. 8), an expanded concern about human (and patient) rights (Barofsky 2003) since the end of World War II; and the demonstration that the assessment of subjective indicators, such as the quality of art, a life and so on, can be reliable, valid and responsive to change.

It is frustrating, therefore, to discover that the definition of quality, and in particular, the definition of quality-of-life or health-related quality-of-life (HRQOL)[1], remains controversial (Ferrans 2005; Hunt 1997; Nord et al. 2001; Rapley 2003; Reeves and Bednar 1994; Wolfensberger 1994). The discussion here can get quite extreme, with some investigators recommending not using the term quality-of-life at all (Wolfensberger 1994), and others recommending not using it as it is commonly stated (Hunt 1997). Yet others belittle the term and its complex meaning (e.g., CDC: Measuring Healthy Years 2000; p. 5).

There are many reasons why there is so much disagreement about the usefulness of the phrase "quality-of-life," including the fact that the phrase may mean different things when defined in a particular context. One of the purposes of this book is to bring some order to this confusion, first by examining the available quality-of-life definitions from a language comprehension and cognitive perspective, and then by discussing some of the difficult issues that need to be dealt with when the concept is assessed.

The degree of frustration in defining quality-of-life is palpable. For example, Ferrans (2005) asks:

> But what is quality-of-life? The literature contains a bewildering array of characterizations. The term 'quality-of-life' is commonly used to mean health status, physical functioning, symptoms, psychosocial adjustment, well-being, life satisfaction, or happiness… Because the terms have meaning in everyday language, they are frequently used without explicit definition (p. 16).

Rapley (2003) provides an even longer list of alternative uses of the term and concludes by stating:

> Not only are all of these terms used in the literature in discussions of what constitutes (a) 'quality-of-life' but it is difficult if not impossible to reconcile them (p. 27).

Even Angus Campbell, one of the founding fathers of quality-of-life research, has been quoted as saying that the concept of quality-of-life is a "vague and ethereal entity, something that many people talk about, but which nobody very clearly knows what to do about" (Campbell et al. 1976; p. 471).

If, as Ferrans' (2005) and Rapley's (2003) statements imply, investigators are deliberately using different words or phrases but claim to be assessing the same concept, then confusion concerning a common definition of quality-of-life is unavoidable. Of course, it is also possible that investigators are using the phrase in a very general sense, as a label, so that I should not be surprised to find that different investigators are using the term in different ways. Much of Feinstein's (Feinstein 1987a; b; Gill and Feinstein 1994) criticism of the current state of quality-of-life research would suggest that he would concur with the concerns of Ferran and Rapley, and that terms such as quality-of-life and HRQOL encompass such a wide range of assessments that they are essentially non-descript labels.

It may yet be that Campbell et al. (1976) came the closest to putting their fingers on what it is that makes defining quality or quality-of-life difficult. When they refer to it as a "vague and ethereal entity" they are suggesting that quality-of-life is an inherently abstract concept that has to be made concrete for it to be understood. If this is so then different definitions are to be expected, since how I make an abstract concept concrete will vary by the context of my application. Thus, if I am dealing with people who have different diseases, experience distinct treatments or come from different cultures, differences in their definitions of quality-of-life may be unavoidable.

These differences are usually reflected in the domains included in the assessment. This has prompted some investigators (e.g., Cummins 1996; Haase and Braden 1998) to try and find a set of domains which can be used to define a common content that would make up a quality-of-life assessment. This type of effort implies that it is possible to find a consensus, empirically or politically derived, that can be used to provide a universal definition of say, quality-of-life or HRQOL. Yet when operationally defined, quality-of-life or HRQOL varies depending on an individual's life history, cultural background, cognitive capabilities and so on. Thus, even if a common content could be agreed upon, this content will vary to different degrees and not necessarily represent what an individual would state is true for them.

One strategy to deal with this complexity is to study how language and cognitive processes are used when investigators or individuals attempt to define quality or quality-of-life. Such studies may yet reveal that finding a universal content-based definition may be a false quest. This seems quite likely, especially since studies of how respondents and investigators define quality-of-life or HRQOL may involve a variety of literal and figurative forms (e.g., metaphors, analogies) and involve different cognitive processes (e.g., social comparisons, heuristics, use of different types of information, and so on). However, what appears as quite diverse may simply reflect a particular level of analysis and that at a higher and more complex level a common cognitive structure (e.g., an informational hierarchy, a conceptual metaphor or a meta-cognitive process) can be identified that organizes this diversity. What this cognitive structure consists of (i.e., what I will refer to as a hybrid construct) will be made explicit as I proceed. But I am convinced that it is possible to make a universal statement about the cognitive processes involved with how quality is assessed, whereas I suspect that attempting to define quality on the basis of common content will always be suspect, since it forces people into domains that inevitably violate the uniqueness of human experience.

Another reason why it is so difficult to define quality-of-life is because it can be characterized in both objective and subjective terms (e.g., Cummins 2000). For example, some investigators argue that the assessment of quality-of-life should only involve objective indicators. Others insist that quality-of-life or HRQOL is unavoidably subjective in nature and reject the idea of objective-only assessment, while others argue that both types of assessments should be included in any assessment. I take a fourth position: a quality assessment involves a qualification process that is applied to either subjective or objective data, and that the origin of the data is somewhat secondary. By this I mean, for example, that while it is possible to represent a color in terms of wavelength it will become a qualitative statement, a hue, only when it has been *subjectified* or cognitively processed. Once made "of the person," the assessment of this quality indicator requires that the entity be *objectified* (made to represent a number). This too is something that a person (e.g., investigator) can do, since objectivity in the current context is not so much an external state as it is a process that has made the qualitative entity observable for others (Chap. 5).

Rapley (2003) points out that social statistics can also provide information that can be used in assessing quality. Here, social reporting (Michalos in Sirgy et al. 2006) consists of an organized collection of social indicators that can focus on the individual, community or nation as the unit of analysis. While a social report reveals how things are going for a person or community, a social account summarizes the "cost" of this status, whether this cost is assessed by dollars, energy, personal satisfaction and so on (Michalos in Sirgy et al. 2006).

The social indicator movement takes more of a system-wide, sociological perspective to the definitional and assessment task, encouraging assessment of both objective and subjective indicators. This approach contrasts with the more cognitive linguistic approach that I will take. The difference between these approaches raises questions about how best to define quality and quality-of-life so that they can be simultaneously applied to the individual and the group. How to do this remains an essentially unresolved problem, with each discipline's approach claiming to have solved the problem, but close inspection revealing conceptual or empirical deficits. One approach to solving this problem involves developing a theory that covers several levels of analysis (e.g., the same theory describing events at a molecular and more molar level of analysis), as suggested by the economist, Hayek (1994; Ch. 3, p. 51).

Ferrans (2005) points out that the public image of a quality-of-life is that it is very personal and individual in nature, which creates a challenging definitional and assessment task. Joyce (1988), in an effort to quantify the individual nature of quality-of-life or HRQOL assessment, defines it "as whatever a person says it to be," and has developed an appropriate assessment instrument that sustains this individuality. Ruta et al. (1994) has also developed an individual-based qualitative assessment. Other investigators (e.g., Bernheim 1999; Cantril 1965; Spitzer et al. 1981) have provided respondents with a linear scale and asked them to provide anchors for the scale and rank themselves along the scale. All of these investigators demonstrate that it is possible both to respect and assess the diversity found among people.

Even if there were a common definition, it is not clear what it is that a quality or quality-of-life assessment should assess. Should it reflect the current state of a person or project future actions or decisions?[2] Should it be an assessment of an ideal state or reflect some prototypical outcome? What role does a particularly traumatic event play when assessing a person's existence? Each of these perspectives may generate different assessment outcomes.

Another complex set of issues deals with the relationship between the construct used to conceptualize what is being assessed (e.g., health status, quality-of-life) and what is actually being assessed (e.g., Priebe et al. 1998). For example, investigators (Ware et al. 1980) state that they use the WHO definition of health ("a state of complete physical, mental and social well-being and not merely the absence of disease" [World Health Organization 1948]) as the basis for defining and assessing health status. Yet inspection of an assessment such as the SF-36 reveals that it doesn't include items which directly assess well-being. Rather, some of the items on the SF-36[3] assess the impact or limitation created by the individual's health status, while others are descriptive statements, assessing a person's physical, mental or social status. Well-being itself is not directly assessed, but is conceived of (e.g., by psychometricians) as a hypothetical or latent variable underlying the items in the questionnaire. An alternative response is to conceive of well-being as a metaphor, and I will discuss this approach in detail in Chap. 11. However, the gap between what is being assessed and the construct the assessment is meant to represent will be an important issue to be dealt with, and I will repeatedly address this issue as I proceed through this book.

2 Should Quality or Quality-of-Life be Assessed?

One of the most important questions I need to answer is whether quality, particularly quality-of-life or HRQOL, should be assessed. Some claim that a qualitative concept such as quality-of-life is too individual and complex to be dissected into parts, assessed and reconstructed into something that reflects a real person's views (Wolfensberger 1994). Others claim that it is not possible to assess quality fairly, particularly quality-of-life, since it cannot be successfully applied to persons who are intellectually or physically challenged (Aksoy 2000; Rapley 2003). Some state that to claim that they have assessed a person's sense of quality is an illusion that at a minimum can produce something that is not useful, but at worst can be dangerous, especially if inappropriate public policy decisions are made on the basis of such an assessment. There is also the concern that political and social influences can seep into any definition, conditions that can significantly undermine any effort

at assessing the concept. This is particularly true when "life quality" becomes the object of national policy (see Rapley 2003; p. 121).

Even if all the above concerns are acknowledged, there are still major reasons why I will pursue defining and assessing quality and quality-of-life. The first and foremost reason is that it is ethically mandated that qualitative assessments be used to monitor the adverse consequences of a world constantly changing whether imposed by other individuals, diseases, political decisions, natural disasters and so on. While the focus of the assessment maybe an object, event, person, community or nation, only standardized methods of assessment will be able to monitor and evaluate the aesthetic, social, political or medical changes that or might have occurred. It is indeed mandated, if it is assumed that each of us, as individuals, have as a first right the right to decide what is best for ourselves.[4] Limiting this right, as when someone is mentally or medically compromised, should be seen as an exception that requires a social consensus to ensure that the outcome is fair and just.

Second, while the subject matter and life experiences of people who are making qualitative judgments may differ, the way that individual processes these experiences may be sufficiently alike so that common cognitive mechanisms can be identified and measured. Thus, diversity of outcome per se does not preclude the presence of common cognitive principles. The universality of human experience also includes the motivation to lead an optimal if not good quality of existence. Estimating the extent to which this is true, particularly with variation in the social context, also justifies an assessment.

Third, legislation and social policies are often implemented with the deliberate intent of improving quality, whether through the physical appearance of a person's environment, community events or a person's life. However, passing legislation alone will remain a relatively inefficient means of facilitating social change in the absence of defining the goals and formally assessing the outcome of such social policies. Assessment is therefore required to evaluate the policy outcome. Thus, even if there could be agreement on what constitutes a good policy, assessment would be required to confirm the standard's applicability and monitor its achievements, or deficiencies.

2.1 Counter Arguments

A number of investigators have raised questions or concerns about assessing quality-of-life. For example, Cattorini and Mordacci (1994), state:

> Life appears not as a simply biological fact but as an event in which the possibility for new meaning is always open, and in which the freedom of the individual is needed to search for that

meaning. This call cannot be ignored for the search is inevitable, unless one abandons oneself to an inauthentic life. To this extent, neither life nor its quality seems to be something measurable, because the meaning calls for interpretation rather than for measurement (p. 61).

Catharine and Mordacci's (1994) suggestion that what is needed is interpretation, not assessment, can easily evoke a sympathetic node of agreement, yet when closely examined their argument does not dissuade me from the need to assess quality or quality-of-life. First, "interpreting the interpretation" an individual places on their own life and existence can be seen as a difficult task, so shifting from assessment to interpretation is not a guarantee of greater clarity. Thus, determining what the person means by the "meaning" they attach to their life is a difficult task unto itself, especially since what is meant by meaning is a complex issue (Ogden and Richards 1946). Second, having characterized an individual's interpretation of their life, I am still left with counting the frequency with which a particular interpretation occurred, especially if I want to make a statement about a person over time, or as part of a specific or broader group. Thus, while Catharine and Mordacci's (1994) concerns can help clarify what is to be assessed during a quality or quality-of-life assessment, only its actual assessment will determine the appropriateness and usefulness[5] of a particular approach.

Another concern is whether a quality-of-life or HRQOL assessment can be sufficiently precise. Morreim (1986) summarizes the issues as follows:

> In this sense we routinely can and do appraise and compare qualities of life. At the same time, though, it seems impossible and undesirable to try to measure quality-of-life with precision. How could it be possible to place a numeric value upon the having - or the loss - of hearing or upon the specific pleasure derived from listening to a symphony orchestra? Or how could the value of saving a limb be tallied against that of teaching a retarded child to read? Thus, there seems to be a kind of paradox in the idea of measuring quality-of-life: we must, yet perhaps we really should not (p. 45).

Morreim's (1986) quote reminds me that what I want from any assessment is to be able to apply it to an individual. This is a fair expectation, an expectation that I believe can be fulfilled in time by the iterative process characteristic of the scientific method.

In this regard, it is also interesting to note that investigators who rely on qualitative methods continue to remind more psychometrically oriented investigators that there is a difference in how respondents characterize their quality-of-life when asked, as opposed to when they answer a questionnaire. After interviewing a group of elderly persons Hendry and Mc Vitte (2004) observed:

> The possible disparity between measured perceptions and subjective experience suggest that it is indeed possible, if not likely, that quality-of-life has often come to be a proxy for other constructs, such as health, functional ability, or psychological well-being… (p. 973).

The possibility that measured perceptions may not capture subjective experiences is a critical issue, but an anticipated one, since any assessment would be expected to "characterize" and not duplicate an observation. This normally invokes an iterative process whereby the gap between perception and assessment is reduced. The notion that quality-of-life has become a proxy for other constructs is unfortunate and has contributed to the impression that the phrase can be used as an umbrella term (see Feinstein 1987b; Ch. 1, p. 4). Still, Hendry and Mc Vitte's (2004) statement is an important one, and I will continue to discuss it, at various points in the subsequent chapters.

Ackoff (1976) takes another tack when he states, "Quality-of-life is a matter of aesthetics and aesthetics is neither sufficiently understood nor adequately integrated with other aspects of life" (p. 289). He goes on to explain that aesthetics are determined by the style of life I live and whether I strive to reach some ideal state. He also sees that the assessment of such issues would be most useful in identifying problems, rather than determining what it is that should be achieved to ensure an individual's quality-of-life. Since this is true, he would discourage communal efforts at optimizing quality-of-life and instead encourage individuals to do what they need to do to live as full a life as possible. As Ackoff (1976) states:

> For these reasons the planning problem of social planners should not be how to improve the quality-of-life of others, but how to enable them to improve their own quality-of-life. Solution of this reformulated problem does not require measures of quality-of-life for its solution. It does require a different concept of social planning (p. 299).

The key word in this quote is word "enable." Ackoff goes on to describe what he calls participative planning, a type of social planning meant to enable individuals to plan and live the style of life they want under the ideal conditions they want. He further states: "It is only in a society in which most individuals take responsibility for their quality-of-life, rather than passively receive it, that continuous improvement of it can be realized" (Ackoff 1976; p. 303).

If individuals took responsibility for their own quality-of-life, they would be expected to be able to define and evaluate the degree to which they have achieved their personal objectives. Formal assessments of quality-of-life would therefore not be necessary. While involvement of the individual may be critical, reliance on social planning, no matter how much it evolved from the people, is not a guarantee that an ideal quality-of-life will emerge.

Rapley (2003) takes a somewhat similar approach to Ackoff when he ends his book by stating:

> … the place for citizens to comment on the success or otherwise of governments of raising their personal well-being, or quality-of-life, is at elections. And, as Ludwig Wittgenstein also clearly put it, on those matters which we cannot speak, we must remain silent (p. 225).

Thus Rapley sees quality-of-life as being improved by changes in social policy, and he also would recommend not assessing the subjective expression of an individual's quality-of-life, since its scientific assessment is fraught with difficulties. However, Ackoff and Rapley's reliance on the political process to correct quality-of-life errors or transgressions is filled with risk as well, since there are always some who are vested in maintaining the status quo. The alternative is to prevent such politically and socially unacceptable issues from arising. To do this requires integrating the ethical perspective of a quality or quality-of-life assessment with its monitoring of the subjective state of individuals, at a point prior to the occurrence of the adverse events. To assess after events occur and expect subsequently to correct a system has repeatedly been shown to be deficient. This strategy is the essence of integrating quality control efforts (Albin 1992) into the fabric of some processes, such as medical care, that I will continue to suggest is the optimal way to use a qualitative assessment, especially if the purpose of the assessment is to optimize some outcome (Barofsky and Sugarbaker 1990; see Ch. 1, p. 18).

2.2 Wittgenstein's Language Games

Rapley's (2003) use of Wittgenstein's comment prompts a bit more of an examination of Wittgenstein's philosophy, especially since the quote does not give the reader a full account of its context and purpose. In addition, it is important to understand more about his philosophy since I will refer back to it at other points in this book. First, the statement Rapley refers to occurs both in the preface and as the last of several hundred propositions in his *Tractatus Logico-Philosophicus* (Wittgenstein 1961; translated version). As an end statement, it clearly is meant to communicate to the reader that a discussion had been completed. However, after inspecting the various propositions that make up the *Tractatus* it is also clear that Wittgenstein was making a comment about the limits of language, especially when it comes to considering the issues he considered most important: ethics, aesthetics and religion. The *Tractatus* provides a list of propositions that are meant to present a systematic, logic-driven scheme. His approach here resembles that of Aristotle, since he excludes metaphors, allusions and rhetorical speech as part of his propositional language. However, by limiting himself to the "pure" language of propositions (what would be literally true), Wittgenstein claims that Aristotle ends up not being able to deal with the truly important issues of ethics, aesthetics and religion, not at least in propositional terms.

Wittgenstein's approach to language is particularly relevant, since he sees it as a result of how a language is used, not as a word's definition or the language's grammar or form (as would occur if one avoided figurative language when using propositional language, e.g., logic). What Wittgenstein states is that one uses the same words in different ways in different contexts with the result that the same words have partial or completely different meanings. He refers to the different uses as an example of a language game[6] with the similarity in their use constituting a family resemblance or reflecting a cognitive prototype.[7] Thus dictionary definitions of words are not helpful.

His views are important because they forced me to focus on how language is used, and this, it turns out, is an important tool in the understanding of how a quality assessment occurs. An example of this type of analysis is to consider the language that is used when a quality assessment is operationally defined. I discuss this issue in the next chapter when I consider the role of figurative language, or metaphors, in various definitions of quality-of-life.

In addition, Wittgenstein[8] argues that language should permit me to distinguish between statements that make sense and those that lack sense or are non-sense. If they made sense then they would be either facts or would reflect reality. By non-sense he does not mean silly or absurd statements, but rather statements that are not factual or representative of reality. He actually considers all the propositions in the *Tractatus* to be non-sense statements, even though they are "logical" (Burbules and Peters; retrieved from the Web 2/8/2007). However, he also believes that one has to engage in such efforts, since one can learn something from them. Lewis Carroll's *Alice in Wonderland* would be an example of "non-sense" one can learn from, as are many of the stories in Green RL. (1956). *The Book of Nonsense*. New York NY: Dutton.

Now I can return to Wittgenstein's statement that "on those matters which we cannot speak, we must remain silent," and conclude that the statement does not mean that one does not speak or talk about things. Quite the contrary. One must speak if one is to realize whether what is being said is non-sense, in his use of the term. However, having examined this material I personally may have to accept that I do not know much about what I am speaking and decide to be silent about it, rather than speak as if I know what I am saying. Consistent with this idea is Wittgenstein's belief that when he wrote the *Tractatus* he was actually writing two books: what was written and what was not written. And it is the "second book" that he claims was the important work, one that someone can become aware of only by discussing the first book. Thus he acknowledged that there was something to say, but that propositional language and logic may not permit him to write the second book. What is going on here, of course, is intellectual theater on a grand scale.

In this regard it is interesting to note Michalos' (in Sirgy et al. 2006; p. 363) comment that Wittgenstein recognized that he had built himself a conceptual castle that he climbed into and then pulled the ladder up behind himself. Michalos' (in Sirgy et al. 2006; p. 363) then goes on to state, "At some point, philosophical analysis must end..." (Michalos in Sirgy et al. 2006; p. 363), a statement with which I agree.

Returning to Rapley's comment, I contend that to apply Wittgenstein's approach to a quality assessment is premature, since it requires that his approach be simulated. This involves presenting a series of propositions that would create a logical scheme that "appears" to account for a quality assessment. Since I am not prepared to do this, Rapley's request has to be deferred. Yet aspects of what Wittgenstein was concerned about (e.g., the notion that a family resemblance or prototype characterizes a term, rather than a definition) remain relevant, and I will allude to them as I proceed. Thus, Rapley's interjection of Wittgenstein's concerns is a valuable contribution and underscores that I have embarked on a multi-faceted task.

Clearly, there is much to debate and consider when these issues are reviewed, but I do not find this discouraging, nor do I feel I have found any irresolvable reasons that prevent me from continuing to assess quality or any of its manifestations, such as quality-of-life. To provide support for my effort, I next consider how the ability to self-assess makes concerns about whether the assessment of quality is something of a non-issue.

2.3 The Simplest Measure: A Self-Assessment

In order to support my thesis I have to demonstrate that it is possible to assess quality. To a certain extent the answer to this concern is quite straightforward, since I can ask myself to observe a picture and judge its quality, report on a movie I saw and rate its quality, or describe my existence and rate the quality of my life. In each of these cases I can give an answer that can have a number assigned to it and be counted. Thus the simple reply to my question is "yes, quality can be assessed," especially if the person is capable of self-reports. Since not all respondents are equally capable of providing reliable self-reports or using language, the next question is whether the model of quality is sufficiently robust to include such challenged respondents.[9] This issue is not easily dealt with, but in order to avoid having a two-tier assessment approach to qualitative assessment, namely, those who are capable of responding and those who are not. Of course Rapley (2003) believes that the failure to include the challenged among the persons who can assess quality should lead us to "remain silent," much as Wittgenstein would have us do. But the limitation here is the dependence on language to assess quality (e.g., a respondent has to understand or be able to read a questionnaire to respond), not a limitation inherent to being assessed. I discuss alternatives to the verbally dependent assessment system below (Ch. 1, p. 19).

Most impressive, however, is how useful a single global item[10] can be. A single self-report item asks the question "How would you rate your health[11] excellent, very good, good, fair or poor?" This item is most often referred to as the self-assessed health status (SAHS) item.[12] The SAHS item has been used, for example, in 40 or more studies (Benyamini and Idler 1999; Idler 1999; Idler and Benyamini 1997; Idler et al. 2004) which have demonstrated an association between a low score and a twofold to sevenfold increased mortality risk. This association seems to be stronger for men than women. A similar association has been found for cancer patients (Ganz et al. 1991). The predictive capacity of the SAHS item remains intact even when variables gleaned from medical record data, self-reported medical conditions and a wide range of health risk factors are factored out (Benyamini and Idler 1999). This association with mortality also occurs when a multi-item assessment (The Health Utility Index; Torrance et al. 1995) is the primary qualitative determinant, although to a lesser degree (Kaplan et al. 2007). This association has now been observed over a sufficient number of diseases and study sites to warrant the label of "an established finding." Clearly, the ability to ask a single question and obtain information about the person's mortality risk is extremely efficient and of great clinical and experimental interest. This added value makes the SAHS question a unique contributor to any policy or research task.

In addition, literature exists that demonstrates that the global item, when obtained preoperatively, can be a predictor of postoperative morbidity for colorectal cancer patients (e.g., Anthony et al. 2003; Efficace et al. 2006), but that indicators other than the global item (e.g., physical functioning) appear to be predictors of postoperative morbidity for esophageal or gastric cancer patients (Blazeby et al. 2001, 2005).

Single item global self-reports of quality-of-life are important for another reason: they represent an assessment that the individuals themselves have created. This circumstance provides the opportunity to study how this occurs; for example, what cognitive processes were involved in generating the rating of the SAHS item or some similar global item? In contrast, a multi-item assessment, constructed by the investigator usually with the collaboration of the respondent, does not teach me about the cognitive processes involved in the individual's response to items on the assessment. Instead, it usually is intended to answer some medical or policy question. Cognitive processes elicited by items on the multi-item assessment can also be studied, but these items usually assess specific domains (e.g., physical or mental functioning), and the global sense of quality-of-life is inferred, as a latent variable (Fayers and Machin 2000).[13]

Another way to demonstrate the usefulness of a single global quality-of-life assessment is to use it as a substitute for a multi-item assessment. This research tactic rests on the assumption that aggregating a series of content-specific items should approximate what an individual does when they generate a response to a global assessment. Fortunately, a research literature exists regarding this issue. It has been shown that the self-reported quality-of-life item can be substituted for the multi-item assessment when the sample size is large (e.g., >150 persons), and the effect size is adequate (Barofsky et al. 2004; De Boer et al. 2003; McHorney et al. 1992; Sloan et al. 1998). However, it is also important to demonstrate that the two measures are conceptually equivalent. Conceptual equivalence would be demonstrated if the same interpretation of a study could be made independent of the type of assessment used. Conceptual equivalence can be studied in two ways: qualitatively (using cognitive interviewing; Willis et al. 2005), or quantitatively (using correlation, regression and difference measures). Quantitative analyses have suggested, but do not conclusively demonstrate, substitutability (Barofsky et al. 2004). This outcome was not surprising, since the comparisons were based on assessments with very different intellectual histories, yet it is possible that if the two assessments were developed in concert, it might be possible to clearly demonstrate conceptual equivalence. However, there is another explanation of these findings, which has to do with the observation that individuals have different response patterns to items. Some will respond to an item in a fast and frugal way, while others may take the time to be more contemplative with their answers. Thus, it is possible that the modest correlations reflect evidence of dual processing (see Chap. 6). von Osch and Stiggelbout (2005) have provided evidence that dual processing occurs during a HRQOL assessment.

It also has been shown that what a person thinks about when responding to the single global item (Krause and Kay 1994) maybe quite different from the content considered in standard multi-item assessments. Respondents may also use different referents when rating an item. A study by Zullig et al. (2005) adds to the complexity here when they examined the correlations of the SAHS item with global items dealing with mental or physical health for adolescent respondents, and they found that SAHS was more highly correlated with mental health than physical health for adolescents. The opposite was found for adults. These data illustrate how the active presence of cognitive processing during a qualitative assessment complicates the demonstration of substitutability (Barofsky et al. 2004).

I return now to my original question "Should quality-of-life be assessed?" Yes it should, and can be assessed, especially if legitimate philosophical and research concerns are not prematurely interpreted as fundamental barriers to assessment. Not only should the quality of objects or the quality of a life be assessed, their assessment is a vital part of an ethical process leading to social interactions that contribute to a fair and just society (see Ch. 1, p. 3).

3 What Does the Term Quality Mean?

So far I have discussed some reasons why it makes sense to define and assess quality or quality-of-life, but I have not yet addressed the broader issue of what the term quality means. Inspecting the available literature quickly reveals that the term has the same definitional and assessment complexities as was discussed for the phrase, "quality-of-life." As previously noted, there are some social scientists who believe that it is not possible to provide a unique definition of the term, no less measure it. Consider what Reeves and Bednar (1994) state:

> A search for the definition of quality has yielded inconsistent results. Quality has been variously defined as a value..., conformance to specifications..., fitness to use..., loss avoidance..., and meeting and/or exceeding customers' expectations... Regardless of the time period or context in which quality is examined, the concept has had multiple and often muddled definitions and has been used to describe a wide variety of phenomena (p. 419).

Even the American Society for Quality acknowledges difficulties with the term by stating, on its website, that quality is "a subjective term for which each person has his or her own definition." The American Society for Quality also states that it uses the term in at least two different ways: first, to refer to the attributes or characteristics of a product or service, and second, to refer to the state of excellence of a product or service. These two approaches are of interest, since determining the attributes or characteristics of a product or service seems to involve a description, while determining the state of excellence of a product or service implies an evaluation. Description and evaluation are common methods of assessing any number of indicators (e.g., psychosocial indicators), not just quality. Something else must be present to make a quality assessment unique, and that is that the person indicates the *importance* of being in a particular state.

Quality is a term that has been used in a wide range of contexts. For example, hot and cold or pleasure and pain are often referred to as reflecting different qualities. In what way is this so? Hot and cold differ in terms of temperature, a continuous variable, yet would a difference between 30.0 and 30.5°C be perceived as a different quality? Clearly this involves a judgment that requires several different cognitive processes. A similar judgmental process would be involved in determining the quality of a painful or pleasurable experience. For an investigator to label two different experiences as qualitatively different appears on the surface to be arbitrary and misplaced. But is there something common to these different applications of the term quality? One possibility is that they each possess is composed of a *quale*.

3.1 Is Quality a *Quale*?

Attempts at defining and assessing quality range from the philosophical to physical-based models (e.g., Gärdenfors 2007), with stops in between for various context-specific definitions. Philosophers, for example, would start any discussion about quality by pointing to our sensory and perceptual experiences of smelling a rose, feeling the pain from a cut or seeing the purple color of a flower as proof of the existence of introspectively accessible experiences called *qualia* (or singular, *quale*). The word *qualia* derives from Latin, meaning "what sort of" or "what kind," and has also been defined as the qualities of feelings, not the feelings themselves.[14] A *quale* is not a physical stimulus, so while seeing red is dependent on psychophysiological processes, the hotness or passion of the red would not be and it is this experience which would make it a *quale*. Instead *qualia* are considered epiphenomena (Jackson 1982), and what philosophers have been actively arguing about is whether they really exist.

Qualia have been defined in a variety of ways.[15] For example, in his book, *Mind and the World Order*, C.I. Lewis (1929) defines *qualia* as the "recognizable qualitative characteristics of the given." Jackson (1982) defines the term as "certain features of the bodily sensations especially, but also of certain perceptual experiences, which no amount of purely physical information includes" (p. 273). Koch (2004) defines *qualia* as "the philosophical term for the introspectively accessible phenomenal aspects of our lives, the elemental feelings and sensations that are the building blocks of conscious experiences. *Qualia* are at the very heart of mind-body problem" (p. R496).

Jackson's definition and Koch's comment highlights one of the characteristics of the debate surrounding *qualia*, which is the contention by some philosophers (e.g., Tye 1986) that if a *quale* exists then it is possible to argue that the mind exists as an entity independent of the body. Dennett (1988, 1991), however, takes issue with these arguments. He points out that the term *qualia*, as it is currently used, refers to a person's private experiences which cannot be communicated or really understood by another person. A *quale* also does not seem to change with experience, but is only known from what an individual experiences in his or her consciousness. Dennett (1988, 1991) "tests" the concept by considering a number of practical applications of the term (e.g., "his neurosurgical pranks"), and finds that these applications break down primarily because of the characteristics that have to be assumed when a *quale* is defined. As a consequence, he concludes that *qualia* cannot be shown to exist.

The larger question here, of course, is how the brain generates conscious experiences, including the sense of quality. Those that argue for the existence of a *quale* would contend that it is an epiphenomenon of various biological processes, and on this basis cannot avoid being labeled as real. To illustrate this idea, consider the following examples. Michotte (1963) reports a series of experiments during which he exposed subjects to objects that moved towards or away from each other at a certain speed so that the first object appeared to hit the second object, resulting in a second object appearing to be "bouncing" away. Actually, the just moved away. At the same time, he noted that these events produced a variety of emotional comments. Subjects stated, for example,

that "It was as though B was afraid when A approached, and ran off," or, "A joins B, then they have a fall out, have a quarrel, and B goes off by himself." More recently, Koch (2004) presented a similar demonstration. In both these cases the pattern of stimulation (a physical stimulus) leads to a conscious experience (a cognitive process), which, when accessed by introspection could be called a *quale*.

How can one prove that what is only known by introspection actually exists? The best available method seems to rely on neuroimaging studies which could show a distinctive intensity of response or involvement of a particular group of cortical sites or neural circuits when a person provides introspective self-reports. This "correlation" approach has been labeled as neurophenomenology and as Metzinger and Walde (2000) say: "Neurophenomenology is possible, phenomenology is impossible" (p. 361).

Another approach to this problem is to determine if a subjective experience such as a *quale* can be experimentally induced. If so, it would further confirm the existence of the entity. Kupers et al. (2006) report an interesting experiment during which they used transcranial magnet stimulation (TMS) of the visual cortex to induce a sensation of touch in blind subjects who had previously been trained to detect touch on their tongue. The experiment is based on the fact that the cortical site of an individual who is deprived from birth of a particular sensory capability will be recruited by other sensory modalities, suggesting the existence of cross-modality plasticity. Thus, the auditory cortex of a deaf person is activated when the person learns sign language. Braille reading by touch activates the visual cortex in the blind (Kupers et al. 2006). In this study, early and late age-onset blind and normal control subjects were taught to make tactile discriminations when objects touched the tongues. The findings were that TMS induced tactile sensations in the blind subjects, while the normal controls reported phosphenes. These data were interpreted as demonstrating that the subjective experience, or *quale*, of a blind person who had experienced sensory remapping was tactile, not visual. This suggested to the authors that an experience depends on its sensory input, not on the specific cortical location. This also demonstrates that a subjective experience, a *quale*, can be induced by appropriate experimental manipulation, adding further support to its existence.

Cross-modality experiences are the essence of synaesthesia (Pearce 2007). Thus, some people (the estimated range is from 1 in 200 to 1 in 2,000, with most being female) will experience the same stimulus in more than one modality. For example, some persons report that they see a color when they hear a particular musical note (e.g., the composer Jean Sibelius reported this), or see a color when also viewing a number (Grapheme-color synaesthesia; Pearce 2007; p. 121). These individuals were born with this ability (the phenomenon has a strong probability of being inherited), and the person's abilities seem related to specific neurocognitive processes. For example, Nunn et al. (2002) have demonstrated that areas in the visual cortex that are known to respond to color will be active when a synaesthetic individual hears a particular spoken word and reports "seeing" a color. In contrast, normal controls do not show activity in this same visual cortical area, even after having been trained to respond by naming a color in response to seeing a word.

These studies demonstrate that sensory input alone does not produce what one experiences, but rather that neurocognitive mechanisms actively process inputs (e.g., Michotte 1963). Studies of cross-modality plasticity demonstrate how a particular experience can be experimentally induced, while synaesthesia reveals that experience is not inevitably limited to specific modular domains (e.g., vision, hearing and so on). Each of these types of data demonstrate that conscious experiences are dependent on a variety of specifiable cortical mechanisms, reinforcing the view that *qualia* are generated by material processes.

Damasio (1994), as commented on by Dennett (1995), deals with the qualia as a product of the multiple regulator processes ongoing in the body. Thus the viscera, the neurohormonal system and the neurocognitive system all participate in representing the external world in the body. From these multiple regulatory processes, what is called "subjectivity" emerges, and it is this and not what Dennett (1995) has called "some imaginary dazzle in the eyes of the Cartesian homunculus" (p. 4) that determines what is meant by a *quale*.

Can I say, then, that a quality is a type of *quale*, in that quality is a subjective experience that is a complex product of what I sense, perceive, and then cognitively and metacognitively process? Among these brain processes is the ability to recruit various cortical sites in complex new arrangements, as evident from cross-modality plasticity. Thus, quality is an experience that is constructed, starting with the quality I sense and perceive and then processed in a non-linear manner.[16] Cytowic (2002; p. 22) argues this when he claims that a cognitive continuum exists, each based on principles of neural networks, that involve perception, synaesthesia, metaphoric and other forms of language expression.

3.2 Quality as a Conceptual Space

If quality is an experience that is a product of a cognitive–emotional construct, then the question arises how this might occur. It turns out that there are at least three models that have been proposed to account for the way that quality is mentally represented. These models include one that relies on the symbolic nature of language and communication; another that makes no assumptions except that as a result of the association of events change occurs; and a third that relies on the fact that experience can be represented in spatial and geometric terms (Gärdenfors 2000; p. 1–2).

Of these models, the symbolic approach evolved from assumption that cognitive processes can be thought of as a Turing Machine,[17] with cognition conceived of as a form of computation mediated by symbolic instructions (e.g., an algorithm). *Associationism* refers to the assumption that information is distributed among different information elements (e.g., neurons) and the simultaneous occurrence of events leads to the differentiation of information. Neural networks would be one example of this type of associationism (Ch. 4, p. 107). Finally, Gärdenfors (2000) proposes an alternative approach to representing quality based on the spatial and geometric structures implicit in the indicators being examined. Thus, if a quality is described dimensionally it will have an implicit geometric or spatial character. He sees these three levels of representation as synergistic and not competitive.

Gärdenfors (2000) finds support for his proposal by inspecting the literature on sensory perceptions where he finds that the qualities associated with these stimuli have been represented mentally as spatial and geometric dimensions. Here he sees the hue of a color, the timbre of a sound or the taste of a meal as examples of quality, with the experience of quality differing from that of a physical stimulus in that the stimuli have been cognitively processed. Thus, quality is a reflection of a phenomenological process and is therefore different from what might be determined by physical assessment. As Clark (1993) points out, each of these sensations can be described in psychophysical terms, without making surplus assumptions about the nature of the experience. From this hypothesis, Gärdenfors (2000) goes on to declare that the elements along each dimension of a psychological space are ordered by similarity, and that this similarity is reflected in the distance between points along the dimension (Fig. 1.1). He illustrates this concept by pointing to the many sensations that are currently scaled along geometric dimensions. Thus, the color wheel organizes colors within a circumference that runs from yellow, to green, to blue and so on, back to yellow. Shepard (1962a, b) reports an experiment in which he gave respondents pairs of 14 different hues and asked them to rate, on a scale of 1 to 5, how similar the pairs were to each other. An adaptation of the resultant similarity plot is displayed in Fig. 1.1. The figure can be looked upon as representing a quality or *qualia* space. Henning (1916) provides an example of a quality space for taste.

Gärdenfors (2000) also argues that the similarity between two stimuli can be conceived of as the psychological distance between them. Citing a number of papers in the literature (e.g., Hahn and Chater 1997; Nosofsky 1992; Shepard 1987) he argues that the similarity between stimuli can be conceptualized as an exponentially decaying function of psychological distance. However, Nosofsky (1986) argues that the exponential function should be replaced by a Gaussian function, suggesting that a consensus on how best to characterize the mathematical analog of similarity remains open. This discussion illustrates how a geometric expression can be used to characterize a dimension that would be an important part of any qualitative assessment, namely, whether two experiences would be rated the same or different.

Having established that quality is a dimension of a phenomenal space, Gärdenfors (2000) then suggests that multidimensional scaling may be one of several statistical methods available to characterize these dimensions. This is done, he points out, by determining the degree to which a person's similarity judgments correspond to the distances in the dimensional space. Shepard's (1962a; b) study, alluded to above (Fig. 1.1) would be an example of the type of data referred to here. Applying a variant of multidimensional scaling, Shepard found that these comparisons formed a quasi-circle along two axes (yellow–blue and green–red). More complex stimuli and comparisons would be expected to generate a more complex array of dimensions, making the psychological interpretation of the dimensions problematic. Thus, how the identified dimensions are interpreted is an important part of determining a conceptual space, and this would be expected to also include the conceptual space involved in a quality-of-life judgment. In the next section (Figs. 1.2 and 1.3) I describe two-dimensional space that provides a conceptualization of the descriptive component of a qualitative judgment (the valuation component is not addressed).

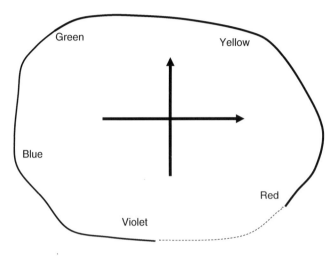

Fig. 1.1 This empirically derived hue space is a rendition of a similar figure presented by Shepard (1962b). The figure can be thought of as representing a *quale* space.

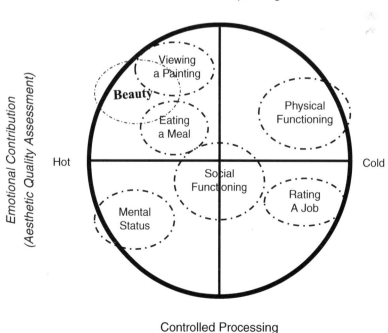

Fig. 1.2 Depicted is the assumption that the assessment of quality consists of both an aesthetic and functional component, although the proportion of each in any assessment will vary. A circumplex is used to illustrate the qualitative conceptual space, with the space dependent on the relative contribution of the two major dimensions. The figure also reflects that cognitive processes are a major determinant of a functional assessment, while emotional responding plays a major role in an aesthetic assessment. Cognitive processes may occur in a fast or automatic manner, or involve a more deliberate and controlled thought processes. Emotions vary from hot (more intense and frequent) to cold (less intense and frequent).

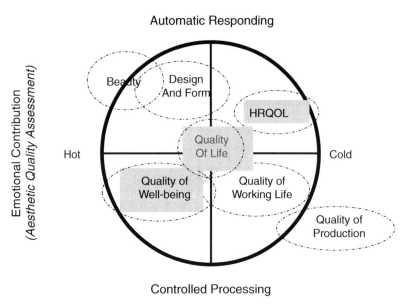

Fig. 1.3 Depicted are various examples of a qualitative assessment, now displayed in a conceptual space. Quality-of-life is placed at the center of the circumplex to establish a reference point around which other examples of a qualitative assessment are placed. Note that what is implied by this arrangement is that each of these assessments contains elements of both components, although they differ in the degree to which each may be present. The *axes* are identical to those illustrated in Fig. 1.2.

To this discussion Gärdenfors (2000) adds the comment that some dimensions of the quality conceptual space have to be conceived of as being innate, while others would be expected to be learned. The issue of what is determined and what is acquired during a qualitative assessment will also be an important one throughout this book (Chap. 6). The point Gärdenfors (2000) is making, however, is that a learned dimension requires the presence of innate dimensions for the learning to occur, otherwise it would not be possible to make sense of the phenomenal world. Other investigators such as Hayek (1952; Ch. 3, p. 51), for example, argue that it would indeed be possible to start with a tabula rasa, and acquire a particular experience, although the structures required for the experience would have to be in place (e.g., a neural network).

In proceeding through this book I will refer to each of the three models I briefly described in the beginning of this section, since any assessment of a person's experience has an implicit model of mental representation and the model being used is a key determinant of how inclusive the assessment will be. Thus, one of the major issues facing current quality-of-life research is how best to assess persons who are verbally and symbolically compromised (e.g., the person with a stroke, or who is demented). The current practice of limiting or excluding these persons seriously compromises the generality of this research and forms the basis of much current criticism of the applicability of current qualitative research (Ch. 1, p. 22). As Figs. 1.2 and 1.3 will reveal (see below), I also conceive of quality from a spatial or geometric perspective, but I do not attempt to present it in terms of a mathematical expression, as Gärdenfors (2000) does. Still, the notion that quality can be characterized as a conceptual space holds promise of avoiding the trap of having to have an intact symbolic system in place in order to assess quality.

One final point, Gärdenfors (2000) not only makes a valuable contribution by conceptualizing quality in terms of spatial relations, he also emphasizes that those stimuli which are perceived have a qualitative dimension. Strongly stated, this implies that all sensory stimuli which I attend to and cognitively process have identifiable qualities. Thus, while there are many stimuli that I may sense (e.g., a wavelength), there are only some that I become aware of, and I will know these stimuli because of their qualities (e.g., its hue).[18] Unfortunately, most quality-of-life researchers rarely use psychophysical scaling methods to place an experience into a qualitative space. Rather, they focus on assessing the consequences of being in a particular qualitative state. They then imply that this descriptive exercise is sufficient to assess quality-of-life, but I have and will continue to argue that this information remains incomplete, if not inaccurate, not only because it lacks a direct assessment of the quality of a person's experiences, but also because it does not include some estimate of the importance that a person places on the assessed consequences. Now to add some complexity here, I also want to point out that it is possible for a statement of quality (e.g., the redness is beautiful, the pain from a medical procedure) to be directly valued, and thus have a cognitive structure that is formally similar to what would be found if some specific consequence were assessed. This explains, as I will discuss in Chaps. 5 and 7, why a person will list experiencing a symptom (e.g., pain, fatigue, depression) as a quality-of-life outcome – because symptoms can have a strong sensory component that also may have a consequence on how a person exists.

3.3 Types of Quality Assessments

Depicted in Table 1.1 are examples of quality assessments that involve an object, an event, a process, or a person's life. What is common to each of these assessments is that they each involve a description, evaluation or qualification of what is being assessed. However, the way a qualitative assessment occurs may vary. Judging the quality of an object may not only rely on an immediate experience and feelings (e.g., the emotional response to a work of art), utilizing an existent qualitative space to evaluate the object, but may also recruit more deliberative cognitive or metacognitive processes. On the other hand, evaluating a concert may rely only on cumulative emotional response or recalling earlier portions of the concert presentation. The admixture of immediate and multiple assessments would become more relevant for assessing the treatment of a person's arthritis or evaluating a person's quality-of-life. These cases would clearly require more complex cognitive and metacognitive processes. Other approaches to organizing qualitative assessments have also been reported.

Pirsig (1974), for example, in his popular book *Zen and the Art of Motorcycle Maintenance* describes what he calls a metaphysics of quality. Quality to him is a state or condition prior to mind or matter, and this may overlap with my previous discussion about whether a *quale* exists. His proof for the prior existence of quality relies on the perceptual fact that in order to know something, actually anything, you have had to have already experienced it. Pirsig (1974) allows Phaedrus,[19] a character in a Plato dialog, to speak for him. Thus Phaedrus is quoted as:

> … speculating about the relationship of Quality to mind and matter and identified Quality as the parent of mind and matter, that event which gives birth to mind and matter. This Copernican inversion of the relationship of Quality to the objective world could sound mysterious if not carefully explained, but he didn't mean it to be mysterious. He simply meant that at the cutting edge of time before an object can be distinguished, there must be a kind of nonintellectual awareness, which he called awareness of Quality. You can't be aware that you've seen a tree until after you've seen the tree, and between the instant of vision and instant of awareness there must be a time lag. We sometimes think of that time lag as unimportant. But there's no justification for thinking that the time lag is unimportant - none whatsoever. The past exists only in our memories, the future only in our plans. The present is our only reality. The tree that you are aware of intellectually, because of that small time lag, is always in the past and therefore is always unreal. Any intellectually conceived object is always in the past and therefore unreal. Reality is always the moment of vision before the intellectualization takes place. There is no other reality. This preintellectual reality is what Phaedrus felt he properly identified as Quality. Since all intellectually identifiable things must emerge from this preintellectual reality, Quality is the parent, the source of all subjects and objects (p. 241).

Pirsig goes on to state that his model of quality or reality can be divided into preintellectual reality and intellectual reality, with intellectual reality further divisible into objective (matter) and subjective (the mind) realities. He describes preintellectual reality as aesthetic quality, while intellectual reality refers to functional quality. Most definitions of quality in use today are probably examples of functional quality, and involve determining the attributes or degree of excellence of an object, event or life history. Determining the quality of medical care is an example of the application of functional quality indicators.

Although Pirsig (1974) defines aesthetic quality as preintellectual, I will use the phrase to refer to attributes or states of excellence of an object, event or life history that also involves an automatic emotional response, involving little or no cognitive processing.[20] This would confirm Pirsig's perspective that judgments can be made which are preintellectual or precognitive, except now they are expressed in information processing terms. However, the notion that emotions or feelings can be expressed independent of cognitions is controversial, and I will discuss this issue in more detail in Chap. 11.

This approach to aesthetics is also common to a number of philosophers. For example, Santayana (1955/1896) claims that "beauty provides pleasure without any reasoning about utility" (quoted by Reber et al. 2004; p. 365). Thomas of Aquinas considers beauty to be a positive experience that one can have without having to think about what it is that one is experiencing. Beauty would be something experienced immediately upon seeing, hearing or experiencing something. Reber et al. (2004) also quotes Read (1972) who denies that "a person with real sensibility ever stands before a picture and, after a long process of analysis, pronounces himself pleased. We either like it at first sight, or not at all" (p. 38). Reber et al. (2004) goes on to make the following distinction:

> Finally, beauty is objectified. For example, the experience of having a cold drink on a hot day is both value positive and intrinsic, but this immediate pleasure lies exclusively in a positive sensation in the body and has little to do with aesthetic appreciation of the object. In contrast, perceivers look at a painting not to

Table 1.1 Prototypical qualitative assessments

Quality assessment of…	Examples
An object	Viewing Michelangelo's "David"
A discrete event	A rendition of Beethoven's Ninth Symphony
An extended process	Chronic treatment of arthritis
A life	A review of Marilyn Monroe's life

please their body, but to enjoy the painting's beauty. Hence, people experience beauty as something that lies in the object. Therefore, beauty is not an "objective," but an "objectified" property (p. 365).

What Reber et al. (2004) meant when he said that beauty is not an objective, but an objectified property is that beauty is not in the object but rather comes from the cognitive processing of the painting, meal or person's life. This is consistent with the notion that a quality assessment involves, along with description and evaluation, a "qualification" process. I will discuss and justify this view as I proceed.

Having clarified what aesthetic quality might mean, I find it interesting that Reeves and Bednar (1994) claim that defining quality in terms of aesthetics "may not be possible…in anything other than abstract terms" (p. 420). Yet, what becomes clear from their discussion of various functional quality definitions is the continued presence of aesthetic issues. Thus they point out that the impact of the Industrial Revolution in Western countries was to shift the definition of quality from producing what was excellent and aesthetic to producing objects or services that satisfied a person's needs. When this happened, price became the determinant of the quality of the object or service. Price or value also had the advantage of providing a quantitative definition, a definition that could be assumed to represent quality (e.g., how much a person was willing to pay for a product or service). However, studies of consumer behaviors soon revealed that purchasing decisions were based not only on convenience, availability or price but also the "quality" of the object or service (Reeves and Bednar 1994). Therefore, even in a predominantly economically driven behavior such as a purchase, aesthetics has a place.

A third determinant of quality other than excellence or value is the degree to which a product or service conforms to specifications. The issue here is to what extent can products, services or parts be sufficiently replicated so that they would be interchangeable. The requirement for interchangeability was stimulated by the military's need for mass production and the desire to decentralize the production of military equipment. If parts are interchangeable then different manufacturers could product different parts and the parts could be assembled to generate a usable piece of equipment. The key to doing this was conformance to specifications (Reeves and Bednar 1994; p. 421) or the precision with which products were produced, and this led to what is now called the quality control movement. Again, however, there was a need for excellence, with its implications concerning aesthetics. Reeves and Bednar (1994) also discuss meeting the expectations of consumers as an important aspect of defining quality, but again this would likely have an aesthetic component. Thus, the aesthetics of any product or service would be a regular component (e.g., an item) of any consumer satisfaction assessment.

If I apply the distinction between functional and aesthetic quality assessment to the types of assessment listed in Table 1.1, then it immediately becomes clear how complex a task there is here. Consider the statues Venus de Milo or Michelangelo's David. These objects epitomize the meaning of "precision production" characteristic of a functional assessment, yet their idealized form and ability to evoke emotions give them an aesthetic that distinguishes them from other objects and types of assessment. In contrast, emphasis on the precision of providing medical care (a functional outcome) will dominate any qualitative assessment, with the aesthetics of this process clearly playing less of a role.

Medical care, however, has distinct characteristics that distinguish it from other types of functional quality assessment (such as ensuring the quality of a car or a television program), and Heinemann et al. (2006) list four reasons why this is so. First, when producing a part for a car, there are quantitative standards that can be uniformly applied. In contrast, medical care is often individualized, and as such the same level of application uniformity is less likely. Second, while in industry one can manage the standards in use (e.g., the precision of the cutting tools used for making a part), this is less likely, but not impossible, for medical care. Thus, it is possible to establish a set of standards for drawing and analyzing blood samples and how this standard should be implemented. Accrediting an individual institutions standards can also be established, but this usually has less of an impact on the impact of the individual providing the service, then if the provider had been properly trained to begin with. Third, car parts manufacturing standards exist so that several manufacturers can produce parts that can be used in the same end-product. Much less uniformity exists in medical care. Just consider the difficulties in producing uniform coronary bypass surgery, or other high-volume procedures. Much more variability of outcome is acceptable, partly due to the greater variability in the biology and psychology of the presenting patients. Finally, car parts manufacturers can influence all aspects of the production process, while a patient who has just had a liver transplant for alcohol-induced cirrhosis can go out and start drinking again. Thus, in medical care there is much less control over the outcome of certain medical procedures and services provided, and this will ultimately affect the apparent quality of such procedures.

The Heinemann et al. (2006) paper makes clear some of the complex issues that have to be considered when the quality of car parts is being compared to the quality of medical care. Finding a place for aesthetic quality in these assessments remains an issue, as it does for a quality-of-life assessment. Therefore it should not be surprising that I have so far come across only one quality-of-life study which explicitly included "aesthetic quality" as a dimension of the assessment. This assessment was designed to measure the quality-of-life of the demented person (Brod et al. 1999). Yet establishing what a beautiful or good life would be is a topic worth considering and will be discussed in Chap. 12.

Sometimes the relationship between aesthetic and functional quality can conflict. For example, some fashion designers (Horyn 2007) go out of their way to make the clothing they design aesthetic but non-functional. They argue that if an outfit is useful its appearance may be ignored, so they make sure that every day they maximize the appeal of their clothing at the expense of its utility. For that reason clothing and apparel are an ideal area in which to investigate the way in which the two dimensions of quality are brought into balance.

So far I have described only one aspect of how quality became defined by some quantity, such as a number. This history dealt with the role that the Industrial Revolution played in shifting the pursuit of quality from one of excellence to one of meeting the needs of the population. As a consequence, the monetary value of the material exchange became an integral part of defining what was meant by quality. Not yet discussed is what role the assessment process plays in determining how a number is assigned to a quality. This starts by the assignment of numbers to events, experiences and qualities (Campbell 1928). The degree to which this approach has succeeded will be discussed later in this book (Chap. 12).

Figure 1.2 is a visual depiction of this discussion. The figure uses the form of a circumplex to illustrate the relationship between aesthetic and functional quality assessment. It is based on a cognitive dimension, which varies in terms of the speed of information processing, and the intensity or frequency of emotional involvement. Of course, emotions may involve the cognitive process of appraisal so when I speak of emotions here it is not devoid of cognitive influence. Also in this figure are examples of objects which can be qualitatively assessed and placed within the grid where they might fit if assessed along these two dimensions. This concept will be used as a guide throughout this book, and I will periodically return to this figure and discuss it in more detail.

What I have illustrated in Fig. 1.2 is a conceptual space consisting of two dimensions, much as Gärdenfors (2000) suggested should be done. The key to understanding the figure, and quality assessment itself, is to see that each of the examples represents some balance between aesthetic and functional quality assessment. How this balance is established will vary as a function of what is being assessed. Thus the quality of a statue, a building, a piece of music, a car, the financial advice I get and so on, will always combine some elements of aesthetic and functional quality.

Before I proceed it is important to acknowledge that defining quality in terms of aesthetics and/or function may not be sufficient, and that a third, new perspective for defining quality may be in order. What this new alternative will be is not clear, but what is clear is that defining quality or the quality-of-life today on the basis of the consequences of the Industrial Revolution is deficient. Illich (1977) and Leiss (1976) made this clear some 40 years ago, and the situation continues to be noted (Layard 2005). What has been repeatedly reported is that people find that basing their satisfaction and estimation of their quality-of-life on things and products do not capture what they mean by quality or the quality of living.

Illich (1977) was an early commentator on these issues. He also observed that the Industrial Revolution modified forms of production, resulting in the detachment of people from the products they used. This condition created a new form of poverty, which represents a poverty of the self and its involvement in determining what a person may mean by quality. When the separation of the individual from the means of production was combined with the marketing of products and services, a combination of forces occurred that could perpetuate this sense of poverty. Concurrently, a new breed of professionals developed who became arbiters of quality further disenfranchising people from themselves. To illustrate this situation, Illich gives the example of a woman who was in the process of giving birth; the nurse tried to push the baby back into the womb because the attending physician was not present (Illich 1977; p. 14).

Illich (1977) clearly has diagnosed a major consequence of modernity, but his suggestions for resolving these adverse outcomes are conservative, offering a return to personal empowerment rather than coming to grips with the need for a major institutional change, such as that occurred following the Industrial Revolution. I suggest that this major institutional change should rely particularly on those assessment procedures that generate invariant qualitative outcomes even in the face of transformation of the primary indicators (e.g., aesthetic or functional quality statements). I discuss what these assessment procedures might be in Chap. 4 (Ch. 4, pages 109–110).

In Chap. 5 (Ch. 5, p. 131) I discuss another approach to assessing quality, after reviewing the evolution of various objectification procedures designed to produce objective outcomes. This approach acknowledges that there has been a shift in how objectification has occurred, from a process that primarily involved classifying natural processes, to the decisions based an expert's opinion. Thus, today the expert physician, architect, composer or chef determines the quality of my health, how I live, what I listen to and what meal I eat. This shift, however, has ethical implications and requires a social consensus to continuation. However, the institutional changes required to replace the psychology of the Industrial Revolution have yet to be articulated the degree that is required to change the quality of my existence.

4 What are the Objectives of This Book?

As I stated in the introduction to this chapter, my first and primary objective for this book is to contribute to what is known about how to infuse an ethical perspective into the

Table 1.2 Models for implementing quality control assessment models

	Quality control	
	Prevention Model[a]	Enhancement Model
Objective of model	"Do No Harm"	Idealized or positive outcome
Methods	Systems analysis	Systems analysis with specific goals
Outcome	Unencumbered development	Prescribed development

[a]The Prevention Model refers to analogs of secondary or tertiary prevention, not primary prevention

conduct of human affairs. As I see it, there are two basic approaches to this task (Table 1.2): one that uses quality assessment to prevent adverse experiences (Prevention Model); and one that attempts to enhance the qualitative state of a person (Enhancement Model). I will discuss each of these models, since each is an area of active research and applied interest. There are many examples of preventing adverse events associated with an object, or enhancing the quality of a product, the environment or even the political process. The unique aspect of this approach, however, is to rely on the medically ill person as the judge of the qualitative consequence of medical care, and not just the clinician. Again, a quality-of-life researcher might argue that this is also currently being done, and I would agree that to a certain extent it is, but not to the extent it optimizes the qualitative outcome of a medical encounter. For this to occur, a more deliberate or directed effort at preventing or enhancing the qualitative state of a person is required (e.g., including a quality assessment at every level of a clinical trial, tempering the "buyer beware" philosophy that characterizes so much of modern industrial society and so on). Before I describe what these activities may involve,[21] I must first convincingly argue that it is possible to define and measure quality or quality-of-life. This will take up most of this book.

A quality assessment is unique in that it permits the optimization of some outcome by monitoring and moderating ongoing processes. It is the integration of these activities into the medical care process that can prevent adverse events or enhance the qualitative state of a medically ill person. However, it is not clear that prevention and enhancement can always occur together, particularly since each method has conflicting goals and objectives. This is particularly true when an adverse event has already occurred and preventing additional adverse events is the goal (secondary prevention). In contrast, in the enhancement model (e.g., the positive psychology movement) a preexisting objective or goal seems to be an implicit aspect of the quality assessment, with its implication that a person's qualitative status can be deliberately augmented. In contrast, the consequence of the prevention model is to maintain the status quo. Each method is meant to enhance the ethical character of a human encounter, which in this case involves providing medical care. And the question now becomes one of how this can be best engineered, considering the complex nature of the medical care process.[22]

I believe that the quality control efforts of the sort I allude to can be implemented without violating the fabric of the scientific process, or my individual freedom to act as I wish. Rather, what is involved here is that an individual, a community or an institution has to recognize that what whatever someone does has a consequence on others or on the environment and that only by individually monitoring the quality of these activities will adverse consequences be moderated or prevented. I certainly do not believe that this is currently being done enough, especially in the process of providing medical care, nor is it often an explicit objective of a patient–clinician encounter.[23]

My views have evolved over 30 years of quality-of-life research, and also from a specific series of studies that occurred early in my career. As so often happens, these studies were formative and have guided my activities in this area of research ever since. The studies I refer to demonstrated that it was possible to identify an optimal treatment for soft-tissue sarcoma while mitigating the adverse consequences of the treatment. The history and successful outcome of these studies, summarized in Barofsky and Sugarbaker (1990), provides a convincing argument for integrating a quality assessment into the fabric of a clinical trial so that the outcome of the evolving treatment regimen can be optimized. As I have indicated, this approach is not unique or innovative, since it has regularly been used when manufacturing products or producing objects.

The optimization of limb-sparing treatment for soft-tissue sarcoma occurred by informally applying an analytic process that originated during the Second World War. This analytical process, referred to as operations research, represents a system analysis of some ongoing activity. It developed when it was found possible to feed back information about the design (i.e., the survivability) of airplanes by examining the conditions that led to their failure. Operations research provides mathematical models or algorithms that simulate the conditions needed to optimize a desired outcome (i.e., a systems analysis). The quality of an object, event or person's life can be one of these desired outcomes, and interestingly enough, operations research was a critical element in the early development of the field of health status and quality-of-life assessment, although it is virtually ignored today. Thus, what I propose is that current quality-of-life or HRQOL researchers return to their intellectual roots to define their current objectives more clearly.

As stated, operations research was an essential element in the early development of health status and quality-of-life assessments. Thus Fanshel, an operations researcher, and Bush (Fanshel and Bush 1970) developed a model of health status, which subsequently evolved into the Quality of Well-being Scale (Kaplan and Anderson 1990). Before he became interested in health status, George Torrance (1976) was an operations research engineer working for a commercial firm. He used his training to develop the Health Utility Index with his colleagues. Also influenced by an operations researcher, was the work of Rachel Rosser, of Rosser and Kind (1978) who developed the EuroQual-5 Dimensions (EQ-5D).[24]

The presence of operations research in the early development of the field makes it clear that health status assessment, and in time quality-of-life or HRQOL assessment, was initially viewed as an "engineering" task, much as is found in many other examples of productive outcomes. Thus, operations research is quite prevalent in the manufacturing of objects (e.g., automobiles) where the repair rate of the product is commonly used as a measure of the quality of the product. As would be expected, this information could then be used to improve the manufacturing process. A product produced in this manner would be "quality controlled" and the manufacturer ethical in that established a method that ensured what they were selling matched the expectations of the buyer.[25]

As I have already mentioned, the quality of medical care can likewise be ethical or non-ethical depending on its consequences (e.g., medication error rate). But how can I characterize the quality of a person's life? What would be considered an error here? One suggestion is to measure the error as the difference between what a medically ill individual expects and what he or she experiences, modified by what the individual considers important in their life. Another measure of error would count the intensity and frequency of adverse events a person experiences. Non-compliance with treatment can also be considered an error, when the person's self-medication is compared to the physician's prescription. Clearly, defining errors that contribute to a person's life quality is and will be complex, but as I will discuss, establishing a definitive definition is somewhat irrelevant, assuming that methods can be put in place that will prevent the occurrence of adverse events.

Still, it would be inappropriate to view the task as limited only to monitoring error rates or adverse consequences. What about the positive consequences of having a car, the medical care I receive or the life I live? Determining what is positive about each of these, however, maybe more difficult to determine than when something has gone wrong. It is more difficult because when something has gone wrong, it usually shows itself by impairing my ability to function, while if something is positive it does not necessarily manifest itself by changes in what I do, it may just make me feel better. This raises philosophical and theological questions about what is important in life. Is feeling better the "end" toward which life should be oriented? I see this in the popularity of the positive psychology movement (e.g., Cloninger 2004), with its promotion of enhanced well-being and better functioning as goals for life. But what specifically are these goals? It is not clear what they are, so most quality-of-life or HRQOL researchers rely on individuals themselves, to define their life objectives. What I have yet to discuss, however, is the information I need to know (Chaps. 4–6), and the methods to use to evaluate quality (Chaps. 7–11) so that I can assess and define the goals and objectives of life.

A major concern of all these activities, however, is that not all persons are equally able to participate in the process of assessing and optimizing quality. There appears to be a group of persons who are challenged by their inability to use language and think about various issues. A second major objective of this book, therefore, is to discuss ways to maximize the participation of the challenged person in the assessment process. This too is an ethical issue, since a policy or a research effort that excludes these persons would not be representative of the entire population, and runs the risk of generating a science that leads to decisions based on inappropriate or limited information (e.g., Aksoy 2000; Faden and Leplege 1992; Rapley 2003; Wolfensberger 1994).

A third major objective involves expanding the nature of the analytic process from the current popular linear model to alternatives including non-linear models. Implicit in the application of linear modeling is the assumption that past events will predict future events, whereas non-linearity allows for unpredictable events, as is characteristic of chaotic systems and emergent phenomena from self-organizing complex systems. Shifting to non-linear modeling will provide a broader range of outcomes than is now possible with linear modeling, and this should help me account for how a quality assessment occurs.

The intent of linear modeling is to help establish structural (causal) laws that describe or predict some phenomena. A classic example is Einstein's statement about the relationship between energy and mass: $E=mc^2$. In this context, the relationship between a toxic effect from chemotherapy and some indicator of quality-of-life would be an example of establishing a linear causal relationship. An alternative, however, is to think of a qualitative assessment as the outcome of a dynamic process, which if non-linear would lead to the emergence of phenomena such as the judgment that an art object was beautiful, a meal delicious or a life meaningful. An example of the application of non-linearity that is relevant is Scherer's (2000) conceptualization of emotions as basically a non-linear process that involves the constant appraisal of a person's emotional status (Ch. 11, p. 438). Specific emotions would emerge from this constant appraisal process, but occur in a catastrophic, chaotic or self-organizing manner. This non-linear model of emotions may be relevant to a discussion of aesthetic quality and may be relevant to the constant monitoring that occurs when I monitor my well-being (Chap. 11). Appraisal as a cognitive process could also be relevant to the assessment of functional quality.

Alternatively, Kahneman and his colleagues report (Kahneman and Krueger 2006) that a person uses the peak and end of a particular experience to globally characterize their experience, and not the linear summation of the experiences (see Ch 4, p. 107 and Ch 11, p. 410). What the respondent is doing, of course, is responding to the most salient feature of an event not its cumulative impact, and this suggests a non-linear process that would contribute to a quality assessment. Thus, both non-linearity and complexity theory will be discussed and found to be useful in conceptualizing quality (see Chaps. 4, 10, and 11).

In Chaps. 2 and 3, I introduce two methods that have not, as yet, found much relevance in a quality assessment: studies of language-usage and the analysis of cognitive processes. In Chap. 2, I demonstrate how the study of language-usage can be used to organize the large number of available definitions of quality-of-life or HRQOL. This will permit organizing the various definitions into a limited number of categories, with each communicating a common meaning. The observation that a limited number of classes of definitions exist will permit definitions to differ, but in a predictable and knowable manner. This information will help me understand what I am communicating when I use a particular type of definition. It will also shift the discussion of whether a particular definition of quality-of-life is appropriate or not, to a question of whether it is appropriate for a particular context.

Chapter 3 illustrates how the study of cognitive process can provide insight into how a quality or quality-of-life assessment occurs. What will be demonstrated is a significant overlap between how information is cognitively organized and various assessment models. I can illustrate my objective by quoting from P. W. Bridgman (1959). He stated:

> I spoke of the two-fold aspects of the problem of understanding - there was the problem of understanding the world around us, and there was the problem of understanding the process of understanding, that is, the problem of understanding the nature of the intellectual tools with which we attempt to understand the world around us (p. 1).

Thus, examining cognitive linguistic or language comprehension issues will help in understanding how figurative language, such as metaphors, became the "intellectual tool" investigators' use to communicate meaning in particular quality or quality-of-life definitions or statements. In the same way, examining cognitive processes will teach me how the information I receive becomes a cognitive *quale* or entity[26] that is classified and organized into a concept. And, studying how I think will provide the intellectual tools needed to understand how a person or an investigator structures the various definitions of quality or quality-of-life.

Tools of this sort will be very helpful in efforts to understand the assessment task. It will become obvious as I proceed the current approaches to assessment in the social and behavioral sciences, with its heavy dependence on linear modeling, will need to be critically examined (Chaps. 4, 10–12). Particularly, since there is the danger that what seems currently appropriate to do when assessing quality or quality-of-life may not capture the complexity of what people experience, this will prompt the development of alternative assessment approaches.

A comment by Mayr (1982) offers an important perspective on the task ahead. He states:

> In the history of biology the phrasing of definitions has often proven rather difficult, and most definitions have been modified repeatedly. This is not surprising since definitions are temporary verbalizations of concepts, and concepts -particularly difficult concepts-are usually revised repeatedly as our knowledge and understanding grows (p. 45)

To paraphrase Mayr's message; Defining a concept is an iterative process that will require continued study, and this will involve examining how the terms quality and quality-of-life are used in ordinary language, and what/how cognitive processes are involved in judging quality and quality-of-life.

5 Summary of Chapter

In this chapter I asked: "Why is it so difficult to define quality or quality-of-life?". Clearly, I could not give a simple answer to this question. Rather, what I attempted to do was to outline some of the issues and approaches that will have to be considered, and I will use the rest of this book to provide a more complete answer to the question. What I also introduced (Figs. 1.2 and 1.3) was the notion that any qualitative assessment has elements of both aesthetic and functional quality dimensions, although the magnitude of the presence of each will vary. In Fig. 1.3 I introduced the types of quality assessments that will be discussed in this book: quality-of-life, HRQOL and quality of well-being. It should also be clear that many of the same issues to be discussed could have been addressed by studying the quality of working life, beauty and so on. I also raised the issue that while aesthetic or functional quality assessments may be necessary, they may not be sufficient to characterize what is meant by quality, particularly the quality of a person's life. What this additional approach to quality might be is not self-evident, but I will consider one option suggested by Daston and Galison (2007); Ch. 5, p. 22). Each of these suggestions justifies excursions into such diverse subject matter areas as philosophy, the neurosciences and assessment theory.

The review of the intellectual history of the concept of health status and HRQOL assessment and of their relationship to operations research made it easier for me to argue that quality assessment should be viewed as a dynamic and non-linear process. Dynamics, of course, is the study of change, and as will be made clear, characterizing the qualitative experiences people have requires methods that can capture this continuously changing process (sometimes referred to as a "stream of consciousness"). I have alluded to using two methods in my quest: first, examining how language is used to describe qualitative issues, and then examining what cognitive processes mediate a qualitative assessment. Finally, I need to emphasize again that the reason I perform a qualitative assessment is because it provides me a means of monitoring the ethical character of events or my and others' actions. However, a qualitative assessment will not tell me what to do, just whether what I did has had an ethical consequence.

Notes

1. The terms, quality, quality-of-life and HRQOL will be distinguished throughout this book. The term "quality," or the phrase "qualitative assessment," will be the most general form of expression, referring to all types of qualitative assessments. As stated in the Preface, I will use the term "qualitative" to refer to those instances involving the judgment of quality, and not limit it to a set of qualitative, as opposed to psychometric, methods. There are phrases, such as "qualitative research" which refer to a particular research orientation, but my use of the term "qualitative" refers to a class of judgments and not these methods. Of course, qualitative judgments can be studied using both qualitative and psychometric research approaches.

   The phrase "quality-of-life" will be differentiated from the phrase "HRQOL" in that one refers to the general population, while the other refers to persons who are medically or psychiatrically ill.
2. The distinction being made here, whether quality-of-life is a reflection of a person's current state or a basis for future action or decisions, is an important one that is often overlooked by investigators. If a quality or quality-of-life assessment is a reflection of a person's past or current state, then it need not lead to action or a decision, but action or a decision will usually imply a projection of a person's future quality-of-life. This, I suggest, is an important difference between a standard quality-of-life assessment and a utilities-based assessment of quality-of-life (see Torrance et al. 1995) that may encompass the past and current state but also project to the future.
3. The SF-36 is a shorter version (36 items) of a questionnaire originally developed at the Rand Corporation, by John Ware and his colleagues as part of the Medical Outcome Study.
4. Clearly, there are medical, political and religious conditions under which a person will have limited or no opportunity to decide what is best for themselves. However, in many of these instances a person will be able to participate in these decisions if given the opportunity. This issue has been previously discussed in a medical care context (Barofsky 2003).
5. Implicit in this statement is the belief that no man is wise enough to capture all of the thoughts, wishes and values of all people and that only empirical observation will provide the reassurance that the relevant data have been approximated by an appropriate measure of quality-of-life or HRQOL. I will address this issue again when I speak about Hayek's contribution to the assessment of quality (Chap. 3).
6. Wittgenstein provides some examples of a language game; for example, giving orders and obeying them, describing an object and giving its assessment, speculating about an event, presenting the results of an experiment in tables, making up a story and reading it, guessing riddles, asking, thanking, cursing, greeting and praying are all examples of language games (Wittgenstein 1953; p. 11–12).
7. Daston and Galison (2007; p. 168) point out that the idea of "family resemblance" came from the early work of Galton who traced the images of members of a family and then generated a composite representation of the entire family. Galton labeled this composite a "family resemblance," which it literally was. Wittgenstein also attended the Vienna group of logical positivists where he was exposed to current psychological research on psychophysics.
8. It is useful to review the relationship between Hume and Wittgenstein, since both contend that "some truths or realities are created by our linguistic practices" (Bloor 1996; p. 356). The extent to which this notion can be extended to a quality assessment, which is based on a "linguistic practice," is something I will be concerned about throughout this book.
9. It is also possible that a culture exists where a person may not have any experience with attaching numbers to feelings or other types of states. This would only be relevant if it could also be demonstrated that the members of this culture could not learn to do this task. If this were so, then these people would join the large group of challenged persons who remain an unresolved task for quality-of-life assessment.
10. This discussion will focus mostly on the single global self-assessed items, including the SAHS or self-assessed quality-of-life items. I do this because the global items come closest to approximating the experiential and cognitive origins of a qualitative assessment. Understanding how an impression of quality emerges will be a key element in the approach to be discussed. I have already hinted at how this might occur as the non-linear emergence of a complex cognitive entity.
11. The SAHS item is sometimes referred to as a quality-of-life self-report, but this would be a clear example of using the phrase "quality-of-life" as a label, since the item is a descriptive statement which lacks the valuation component needed for a complete qualitative indicator. I will review the literature that deals with the SAHS item, since it is particularly complete, but the item itself is most accurately considered a health status indicator.
12. In Chap. 8, I make clear how a health status and HRQOL item differ. The information summarized in this chapter is actually more about the value of a single global health status item, but is not informative about whether a single global quality-of-life item is a good predictor of mortality.
13. Sometimes multi-item assessments include global self-report items along with domain-specific items. In this case, it may be possible to disaggregate the multi-item assessment and study the single global self-report item separately.
14. Accessed at the Wikipedia website (http://en.wikipedia.org/wiki/Qualia; last modified 5/4/2006; accessed 5/31/2006).
15. Accessed at the Stanford Encyclopedia of Philosophy website (http://plato.stanford.edu/entries/qualia/; accessed 5/31/2006).
16. A linear model would state that what I experience when I report seeing something involves having me receive a visual input followed by the activation of the visual cortex leading to some visual experience. A non-linear model could involve the reprocessing of the visual input in such a way that what I experience is not necessarily predicted by the physical characteristics of the stimulus (e.g., Michotte 1968), but involve other cognitive processes.
17. A Turning Machine provides a set of instructions (e.g., an algorithm) presented in symbolic form that relates the input and output of some system. Thus, a Turing Machine output is computation.
18. Clearly there are cognitive processes that I may not be aware of, such as appraisal, which may influence what I am aware of, but in this context I am not dealing with all the factors that might determine what I am aware of, only those qualities that I have already attached to the experience.
19. Phaedrus, a character from Plato's writings, is used by Pirsig (1974) as a "literary devise" and Richardson (2008; p. 11) claims that Pirsig used Phaedrus to express his own views.
20. In Chap. 6, I introduce the notion that people process information in two basic ways: automatically or after due consideration. Dual processing of this sort may yet account for the difference between aesthetic and functional quality. For this to be confirmed will require clarification of the relation of emotions and cognition to

determine which comes first, who influences whom and so on. These issues will be discussed in Chap. 11.
21. An example of a preventative practice that is not widespread and would have a qualitative consequence, is assessing quality-of-life during a Phase I clinical trial.
22. Physicians engage in many activities that are designed to ensure quality control. Medical case conferences, in which specific cases are discussed, would be one example of a quality control effort. Another example is the procedure a department of surgery established to prevent "wrong site, wrong procedure and wrong patient outcomes" (Michaels et al. 2007). Quality control activities, however, does not ensure that the patient has an optimal *qualitative* outcome. Thus, when I discuss quality control there are multiple levels of application.
23. As is often the case, some other author previously stated the same basic idea. Robert Pirsig in his book "Lila" (1991) ends it by stating the following:

> Good is a noun. That was *it*. That was what Phaedrus had been looking for. That was the homer, over the fence, that ended the ball game. Good is a noun rather than an adjective is all that the Metaphysics of Quality is about. Of course, the ultimate Quality isn't a noun or adjective or anything else definable, but if you had to reduce the whole Metaphysics of Quality to a single sentence, that would be it (p. 409).

Each individual, community, or nation doing good – making quality a primary objective of their activities – is what is required for ethical outcomes.
24. Paul Kind (October 2005, personal communication) has indicated that Rachel Rosser's husband was an operations researcher and he felt that his background influenced the model that was developed.
25. Matching the expectations of a buyer is not a simple process, since a person's expectation may not be in concordance with the manufacturer's estimate of the cost of the product. There is obviously an entire topic of interest here, *the ethics of material exchange*.
26. I use the word "entity" to refer to the subjective experience I am assessing, since the *quale* has become infused with the mind–body debate and I want to avoid implying that what will be presented supports a dualistic perspective. In fact, in Chap. 10 I provide several examples of how mind–body or capacity–performance dichotomies can be presented in non-dualistic terms.

References

Ackoff RL. (1976). Does quality of life have be quantified? *Oper Res Q*. 27, 280–303.
Aksoy S. (2000). Can the "quality of life" be used as a criterion in health care services? *Bull Med Ethics*. October 19–22, 2000.
Albin JM. (1992). *Quality Improvement in Employment and Other Human Services: Managing for Quality Through Change*. Baltimore MD: Brookes.
Anthony T, Hyman LS, Rosen D, Kim L, et al. (2003). The association of pretreatment health-related quality of life with surgical complications for patients undergoing open surgical resection for colorectal cancer. *Ann Surg*. 238, 690–696.
Barofsky I. (2003). Cognitive approaches to summary measurement: Its application to the measurement of diversity in health-related quality of life assessments. *Qual Life Res*. 12, 251–260.
Barofsky I, Erickson P, Eberhardt M. (2004). Comparison of multi-item and self-assessed health status (SAHS) indexes among persons with and without diabetes in the U.S. *Qual Life Res*. 13, 1671–1681.
Barofsky I and Sugarbaker PH. (1990). The cancer patient. In, (Ed) B Spilker. *Quality of life Assessment in Clinical Studies*. New York NY: Raven. (p. 419–439).
Benyamini Y, Idler EL. (1999). Community studies reporting association between self-rated health and mortality: Additional studies, 1995 to 1998. *Res Aging*. 21, 392–401.
Bernheim JL. (1999). How to get serious answers to the serious question: "How have you been?": Subjective quality of life (QOL) as an individual experiential emergent construct. *Bioeth*. 13, 272–287.
Blazeby JM, Brookes ST, Alderson D. (2001). The prognostic value of quality of life scores during treatment for oesophageal cancer. *Gut*. 49, 227–230.
Blazeby JM, Metcalfe C, Nicklin J, Barham CP, et al. (2005). Association between quality of life scores and short-term outcome after surgery for cancer of the oesophagus or gastric cardia. *Brit J Surg*. 92, 1502–1507.
Bloor D. (1996). The question of linguistic idealism revisited. In, (Eds) H Sluga, DG Stern. *The Cambridge Companion to Wittgenstein*. New York NY: Cambridge University Press. (p. 354–382).
Bridgman PW. (1959). *The Way Things Are*. New York NY: Viking.
Brod M, Stewart A, Sands L, Walton P. (1999). Conceptualization and measurement of quality of life in dementia: the Dementia Quality of life Instrument (DQOL). *Gerontol*. 39, 25–35.
Brooks R, Rabin R, de Charro F. (2003). *The Measurement and Valuation of Health Status Using the EQ-5D: A European Perspective. Evidence from the EurQol BIOMED Research Programme*. Dordrecht The Netherlands: Kluwer.
Burbules NC, Peters M. (2007). *Tractarian Pedagogies*: Sense and nonsense. Retrieved on the Web, 2/8/2007. http://www.faculty.ed.uiuc.edu/burbules/syllabi/Materials/tip.html.
Campbell, N. R. (1928). *An Account of the Principles of Measurement and Calculation*. London UK: Longmans, Green.
Campbell A, Converse PE, Rodgers WL. (1976). *The Quality of American Life*. New York NY: Russell Sage Foundation- Rutgers University Press.
Cantril H. (1965). *The Pattern of Human Concerns*. New Brunswick NJ: Rutgers University Press.
Cattorini P, Mordacci R. (1994). Happiness, life and quality of life: A commentary on Nordenfelt's "Towards a Theory of Happiness". In, (Ed.) L Nordenfelt. *Concepts and Measurement of Quality of life in Health Care*. Dordrecht The Netherlands: Kluwer. (p. 59–62).
Centers for Disease Control and Prevention (CDC). (2000). *Measuring Healthy Days: Population Assessment of Health Related Quality of life*. Atlanta GA: Center for Disease Control.
Clark A. (1993). *Sensory Qualities*. Oxford UK: Clarendon Press.
Cloninger CR. (2004). *Feeling Good: The Science of Well-Being*. New York NY: Oxford University Press.
Cummins RA. (1996). The domains of life satisfaction: An attempt to order chaos. *Soc Indic Res*. 38, 303–328.
Cummins RA. (2000). Objective and subjective quality of life: An interactive model. *Soc Indics Res*. 52. 55–72.
Cytowic RE. (2002). Touching tastes, seeing smells- and shaking up brain science. *The Dana Forum Brain Sci*. 14, 7–26.
Damasio AR. (1994). *Descartes' Error: Emotion, Reason and the Human Brain*. New York NY: HarperCollins.
Daston L, Galison P. (2007). *Objectivity*. New York NY: Zone Books.
de Boer AGEM, van Lanschot JJB, Stalmeier PFM, van Sandick JW, et al. (2003). Is a single item visual analogue scale as valid, reliable and responsive as multi-item scales in measuring quality of life? *Qual Life Res*. 13, 311–320.
Dennett DC. (1988). Quining *qualia*. In, (Eds.) A Marcel, E Bisiach. *Consciousness in Contemporary Science*. New York NY: Oxford University Press. (p. 42–77).
Dennett DC. (1991). *Consciousness Explained*. Boston MA: Little, Brown and Company. (p. 369–411).
Dennett DC. (1995). Review of Damasio, Descartes' Error. *Times Literary Suppl*. August 25, 1995. (p 3–4).
Efficace F, Bottomley A, Coens C, Van Steen K, et al. (2006). Does a patient's self-reported health-related quality of life predict survival

References

beyond key biomedical data in advanced colorectal cancer? *Eur J Cancer.* 42, 42–49.

Faden R, Leplege A. (1992). Assessing quality of life: Moral implications for clinical practice. *Med Care* 30 Suppl, MS166-MS175.

Fanshel S, Bush, JW. (1970). A Health Status Index and its application to the health services outcomes. *Oper Res.* 18, 1021–1066.

Fayers PM, Machin D. (2000). *Quality of life: Assessment, Analysis and Interpretation.* Chichester UK: Wiley.

Feinstein AR. (1987a). Clinimetric perspective. *J Chronic Dis.* 40, 635–640.

Feinstein AR. (1987b). *Clinimetrics.* New Haven CN: Yale.

Ferrans CE. (2005). Definitions and conceptual models of quality of life. In, (Eds.) J Lipscomb, CC Gotay, C Synder. *Outcomes Assessment in Cancer.* New York NY: Cambridge University Press. (p. 14–30).

Ganz PA, Lee JJ, Siau J. (1991). Quality of life assessment. An independent prognostic variables for survival in lung cancer. *Cancer.* 67, 3131–3135.

Gärdenfors P. (2000). *Conceptual Spaces: The Geometry of Though.* Lexington MA: MIT Press.

Gärdenfors P. (2007). Representing actions and functional properties in conceptual spaces. In, (Eds.) T Ziemke, J Zlatev, RM Frank. *Body, Language and Mind: Volume 1 Embodiment.* Berlin Germany: Mouton de Gruyter. (p. 167–195).

Gill TM, Feinstein AR. (1994). A critical appraisal of the quality of quality of life measurements. *JAMA.* 272, 619–626.

Green RL. (1956). The Book of Nonsense. New York, NY: Dutton.

Haase JE, Braden CJ. (1998). Guidelines for achieving clarity of concepts related to quality of life. In, (Eds) CR King, PS Hinds. *Quality of life: From Nursing and Patient Perspectives. Theory, Research, Practice.* Sudbury MA: Jones and Bartlett. (P. 54–73).

Hahn U, Chater N. (1997). Concepts and similarity. In, (Eds.) K Lambert, D Shanks. *Knowledge, Concepts and Categories.* East Sussex United Kingdom: Psychological Press. (p. 43–93).

Hayek FA. (1952). *The Sensory Order: An Inquiry into the Foundation of Theoretical Psychology.* Chicago Ill: University of Chicago Press.

Hayek FA. (1994). *Hayek on Hayek: An Autobiographic Dialogue.* (editors) Kresge S, Wenar L. Chicago Ill: University of Chicago Press.

Heinemann AW, Fisher WP Jr, Gershon R. (2006). Improving health care quality with outcomes management. *J Prosthet Orthot.* 18, 46–50.

Hendry F, Mc Vitte C. (2004). Is quality of life a healthy concept? Measuring and understanding life experiences of elderly people. *Qual Health Res.* 14, 961–975.

Henning H. (1916). Die qualitätenriehe des geschmacks. *Z Psychol Physiol Sinnesortan.* 74, 203–219.

Horyn C. (2007). Après Rehab: A lovely sobriety. *The New York Times.* August 23, 2007.

Hunt SM. (1997). The problem of quality of life. *Qual Life Res.* 6, 205–212.

Idler EL. (1999). Self-assessment of health: The next stage of studies. *Res Aging.* 21, 387–391.

Idler EL, Benyamini Y. (1997). Self-rated health and mortality: A review of twenty-seven community studies. *J Health Soc Behav.* 38; 21–37

Idler E, Leventhal H, McLaughlin J, Leventhal E. (2004). In sickness but not in health; Self-ratings, identity and mortality. *J Health Soc Behav.* 454, 336–356.

Illich I. (1977). *Toward a History of Needs.* New York NY: Pantheon Books.

Jackson F. (1982). Epiphenomenal *qualia. Philos Q.* 32, 127–136.

Joyce CRB. (1988). Quality of life: The state of the art in clinical assessment. In, (Eds.) SW Walker, RM Rosser. *Quality of life Assessment and Application.* Lancaster PA: MTP Press. (p. 169–179).

Kahneman D, Krueger AB. (2006). Developments in the measurement of subjective well-being. *J Econ Perspect.* 20, 3–24.

Kaplan R, Anderson JP. (1990). The General Health Policy Model: An integrated approach. In, (Ed.) B Spilker. *Quality of life Assessments in Clinical Trials.* New York NY: Ravens.

Kaplan M, Berthelot J-M, Feeny D, McFarland BH, et al. (2007). The predictive validity of health-related quality of life measures; mortality in a longitudinal population-based study. *Qual Life Res.* 16. 1539–1546.

Koch C. (2004). Qualia. *Cur Biol.* 14, R496.

Krause N, Kay G. (1994). What do global self-rated health items measure? *Med Care.* 32, 930–942.

Kupers R, Fumal A, Maertens de Moorhout A, et al. (2006). Transcranial magnetic stimulation of the visual cortex induces somatotopically organized *qualia* in blind subjects. *Proc Nat'l Acad Sci.* 103, 13256–13260.

Layard R. (2005). *Happiness: Lessons from a New Science.* New York, NY: Penguin.

Leiss W. (1976). *The Limits of Satisfaction: An Essay on the Problem of Needs and Commodities.* Toronto CA: University of Toronto Press.

Levinas E. (1969/1961). *Totality and Infinity: An Essay on Exteriority.* (Trans. A, Lingus) Pittsburgh PA: Duquesne University Press.

Lewis CI. (1929). *Mind and the World Order.* New York NY: C. Scribner's Sons.

Mayr E. (1982). *The Growth of Biological Thought: Diversity, Evolution, and Inheritance.* Cambridge MA:Belknap Press/Harvard University Press.

McHorney CA, Ware JE, Rogers W, et al. (1992). The validity and relative precision of MOS short-and long-form health status scales and Dartmouth COOP charts: results from the Medical Outcome Study. *Med Care.* 30, MS253-MS265.

McKeon R. (1947). *Introduction to Aristotle.* New York NY: Random House.

Metzinger T, Walde B. (2000). Commentary of Jakab's "Ineffability of *Qualia*". *Conscious Cogn.* 9, 352–362 doi:1186/1477-7525-3-64.

Michaels RK, Makary MA, Dahab Y, Frassica FJ, et al. (2007). Achieving the Quality Forum's "Never Events:" prevention of wrong site, wrong procedure, and wrong patient operation. *Ann Surg.* 245, 526–532.

Michotte A. (1963). *The Perception of Causality.* New York NY: Basic Books.

Morreim EH. (1986). Computing the quality of life. In, (Eds.)GJ Agich, CE Begley. *The Price of Health.* Dordrecht The Netherlands: Reidel. (p 45–69).

Nord E, Arnesen T, Menzel P, Pinto J-L. 2001 Towards a more restricted use of the term "Quality of life". *News Letter Qual Life*.26, 3–4.

Nosofsky RM. (1986). Attention, similarity and the identification-categorization relationship. *J Exp Psychology: Learn Mem Gen.* 115, 39–57.

Nosofsky RM. (1992). Similarity scaling and cognitive process models. *Annu Rev Psychol.* 43, 25–53.

Nunn JA, Gregory LJ, Brammer M, Williams SCR, et al. (2002). Functional magnetic resonance imaging of synaesthesia activation of V4/V8 by spoken words. *Nat Neurosci.* 5, 371–375.

Nussbaum M, Sen A. (1993). *The Quality of life: Studies in Developmental Economics.* New York NY: Oxford University Press.

Ogden CK, Richards IA. (1946). *The Meaning of Meaning.* New York NY: Harcourt, Brace and World.

Pearce JMS. (2007). Synaesthesia. *Exp Neurol.* 57, 120–124.

Pirsig RM. (1974). *Zen and the Art of Motorcycle Maintenance.* New York NY: Bantam Books.

Pirsig RM. (1991). *Lila: An Inquiry Into Mass Morals.* New York NY: Bantam Books.

Priebe S, Kaiser W, Huxley PJ, Röder-Wanner, Rudolf H. (1998). Do different subjective evaluation criteria reflect distinct constructs? *J Nerv Ment Dis.* 188, 385–393.

Rapley M. (2003). *Quality of life Research.* Thousand Oaks CA: Sage.

Read H. (1972). *The meaning of art.* London UK: Faber & Faber.

Reber R, Schwarz N, Winkielman P. (2004). Process fluency and aesthetic pleasure: Is beauty in the perceived processing experience? *Personal Soc Psychol Rev.* 8, 364–382.

Reeves CA, Bednar DA. (1994). Defining quality: Alternatives and implications. *Acad Manag Rev.* 19, 419–445.

Richardson M. (2008). *Zen and Now: On the Trial of Robert Pirsig and the Art of Motorcycle Maintenance.* New York NY: Knopf.

Rosser R, Kind P. (1978). A scale of valuations of states of illness: Is there a social consensus? *Int J Epidemiol.* 7, 347–358.

Ruta DA, Garratt AM, Leng M, Russell IT, et al. (1994). A new approach to the measurement of quality of life: The Patient Generated Inventory (PGI). *Med Care.* 32, 1109–1126.

Santayana, G. (1955/1896). *The sense of beauty.* New York NY: Dover.

Scherer KR. (2000). Emotions as episodes of subsystem synchronization driven by nonlinear appraisal processes. In, (Eds.) MD Lewis, I Granic. *Emotions, Development, and Self-organization: Dynamic Systems Approaches to Emotional Development.* Cambridge UK: Cambridge University Press. (p. 70–99).

Shepard RN. (1962a). The analysis of proximities: Multidimensional scaling with an unknown distance function: I. *Psychom.* 27, 125–140.

Shepard RN. (1962b). The analysis of proximities: Multidimensional scaling with an unknown distance function: II. *Psychom.* 27, 219–245.

Shepard RN. (1987). Towards a universal law of generalization. *Sci.* 237, 1317–1323.

Sirgy MJ, Michalos AC, Ferriss AL, Easterline RA, et al. (2006). The Quality of life research movement: Past, present and future. *Soc Indics Res.* 76, 343–466.

Sloan JA, Loprinzi CL, Kuross SA, et al. (1998). Randomized comparison of four tools measuring overall quality of life in patients with advanced cancer. *J Clin Oncology.* 16, 3662–3673.

Spitzer WO, Dobson AJ, Hall J, Chesterman E, et al. (1981). Measuring the quality of life of cancer patients. *J Chronic Dis.* 34, 585–597.

Torrance GW. (1976). Health status index models: A unified mathematical view. *Manag Sci.* 22, 990–1001.

Torrance GW, Furlong W, Feeny D, Boyle M. (1995). Multi-attribute preference functions: Health Utilities Index. *PharmEcon.* 7, 503–520.

Tye M. (1986). The subjective qualities of experience. *Mind New Ser.* 95, 1–17.

von Osch SMC, Stiggelbout AM (2005). Understanding VAS valuations: Qualitative data on the cognitive process. *Qual Life Res.* 14, 2171–2175.

Ware JE, Brook RH, Davis-Avery A, Williams KN, et al. (1980). *Conceptualization and Measurement of Health for Adults in the Health Insurance Study: Vol. 1, Model of Health and Methodology.* Santa Monica CA: Rand. Publication No R-1987/1 HEW.

Willis G, Reeve B, Barofsky I. (2005). The Use of Cognitive Interviewing Techniques in Quality of life and Patient-Reported Outcome Assessment. In, (Eds.) J Lipscomb, CC Gotay, C Synder. *Outcomes Assessment in Cancer: Findings and Recommendations of the Cancer Outcomes Measurement Working Group.* Cambridge UK: Cambridge University Press.

Wittgenstein L. (1953). *Philosophical Investigations.* Oxford UK: Blackwell.

Wittgenstein L. (1961). *Tractatus Logico-Philosophicus.* New York NY: Routledge and Kegan Paul.

Wolfensberger W. (1994). Lets hang up "quality of life" as a hopeless term. In, (Ed.) D. Goode. *Quality of life for Persons with Disabilities: International Perspectives and Issues.* Cambridge, Brookline. (p. 285–321).

World Health Organization. (1948). *Constitution of the World Health Organization: Basic documents.* Geneva, Switzerland: World Health Organization.

Zullig KJ, Valois RF, Drane JW. (2005). Adolescent distinctions between quality of life and self-rated health in quality of life research. *Health Qual Life Res.* 3, 1–9, doi: 1186/1477-7525-3-64.

The Role of Language in Assessing Quality or Quality-of-Life

Abstract

The statement below by Skinner is enough to demonstrate that changes in language can result in changes in meaning. This chapter considers the various ways this is true, especially when considering how language is used in a qualitative assessment. This is illustrated by analyzing the language basis of various definitions of quality-of-life or health-related quality-of-life (HRQL).

Once you have formed the noun 'ability' from the adjective 'able' you are in trouble.

BF Skinner (1987)

Abbreviations

AIDS	Acquired immune deficiency syndrome
HRQOL	Health-related quality-of-life
QWB	Quality of Well-being Scale (Kaplan and Bush 1982)
SME	Structure-mapping engine
WHO	World Health Organization

1 Introduction

A defining characteristic of a quality assessment, such as a quality-of-life or HRQOL assessment, is that it is based on the ordinary language that people use to express themselves. As a consequence, it is subject to a full range of uses, from the precision of a logical expression to the confusion of contextual misappropriation. Yet, explicating the role of language in a quality-of-life or HRQOL assessment seems almost mandatory, although it is seldom reported. One approach involves examining a quality assessment in terms of its syntax, semantics and pragmatics, and this should provide insight into how language can be used to express quality. Thus, when a quality assessment is "adjusted" (e.g., made more concrete, uses simpler language, etc.) to facilitate responding by a person who is compromised, then it is appropriate to ask what these changes in linguistic expression do to the meaning being expressed by the different assessments. At the same time, monitoring changes in language expression can also be used to characterize the impact of disease progression, treatment, and aging. This can be done by assessing for semantic errors, changes in speed of responding, impaired facility with abstract concepts (Chap. 10), and any number of additional language-based indicators. Thus, quality researchers can take advantage of an established literature designed to monitor or analyze language changes, and use these tools to gain insight into the nature of a qualitative assessment. To help orient to these tasks, I will first provide a brief, but hopefully useful, review of the grammatical and semantic nature of language, with emphasis on the difference between literal and figurative expression, and between linguistic and conceptual metaphors, but also the role of speech acts. Most qualitative researchers will probably not be familiar with the various concepts and distinctions commonly used in the study of language expression. I will, however, demonstrate the usefulness of this approach in Sect. 3 of this chapter, when I apply a metaphoric analysis to a representative set of quality-of-life or HRQOL definitions.[1] A critical element in this analysis will involve the deconstruction of the available definitions using a method akin to creating a diagram of sentences (e.g., O'Grady et al. 1989).

Fig. 2.1 The tree diagram for the sentence, "The player lost a shoe in the dugout" (O'Grady et al. 1989; p. 135).

"The player lost a shoe in the dugout".

Glossary:
S = Sentence
N = Noun
P = Preposition
V = Verb
Adj = Adjective
Adv = Adverb
NP = Noun Phrase
VP = Verb Phase
PP = Prepositional Phrase
Det = Determiner

In addition, I will keep Wittgenstein's admonishments concerning the limits of language in focus, particularly his notion that it should be expected that the meaning of words or phrases should be derived from how they are used, not defined. His view, that the pragmatics of language usage should determine the role of language in communication, will be reflected in my discussion of both speech acts and the context or surroundings of a qualitative assessment. I will differ from Wittgenstein by regularly including definitions and synonyms of words to serve as fixed points to begin these studies. What will become obvious is that the word "quality" and the phrase "quality-of-life" are subjective terms that can be used metaphorically, and in this way provide different meanings, but they do so by using common cognitive mechanisms (e.g., the mapping of meanings between terms).

2 Language Usage Principles

2.1 The Elements of Language

The study of language (O'Grady et al. 1989; Pinker 1994) reveals how a person uses words and sounds to communicate information. Each verbal or written statement, in effect, is a creative effort that appears to occur in the context of specific rules and constraints, but also in a social and cultural setting. These rules and constraints make up a *grammar* that orders the way I say things and shapes the phrases or sentences I speak or write. This grammar also helps me reliably interpret what I read or hear. All languages have a grammar of varying complexity, and our capacity to use a grammar appears to be biologically determined (Chomsky 1965).

It is literally possible to "see" the grammatical structure of a phrase or sentence by deconstructing it into its component parts, and identifying the rules that are operating when a sentence is formed. For example, the deconstructed form of the sentence, "The player lost a shoe in the dugout" (O'Grady et al. 1989; p. 135) is displayed as a syntactic tree in Fig. 2.1. It can also be written in a linear fashion by combining expressions such as; S → NP VP; NP → (Det) (AdjP) N (PP); VP → V (NP) (PP); PP → P NP (see Fig. 2.1). In either case, the resultant display reveals that for an English declarative sentence the subject or a noun phrase (NP) will always precede the predicate or verb phrase (VP), while a noun phrase will always include a noun (N) but could also include a prepositional phrase (PP), a determinant (Det), or an adjectival phrase (Adj). A similar deconstruction can occur for the verb phrase (VP), and other parts of language. As can be imagined, these trees become more complex as the sentence becomes more complex.

Now what is interesting about this method of illustrating the grammatical structure of a sentence or phrase is that it may also be what I ordinarily do when I read or hear a sentence. Thus, my nervous system may be organized to perceive each part of a sentence and then determine how the parts are related to each other and the sentence as a whole.

This process is referred to as a combinational strategy (e.g., Fodor 1975), and I will refer to it again as I proceed through this and subsequent chapters.

Although a language is governed by a grammar (a syntax), words also have the capacity to change their meaning over time by shifting how they are used, whether it be as a word, or part of a phrase or sentence. Thus, usage-based changes can occur by semantic broadening (or narrowing), or semantic shift (O'Grady et al. 1989). An example of semantic broadening is the use of the word "dog," whose old meaning was limited to a "hunting breed" but now also means "any canine." An example of semantic narrowing is the word "disease," that at one point in linguistic history referred to being in "any unfavorable state" (O'Grady et al. 1989), but now most often refers to "having an illness." Semantic shifting for words can be illustrated with the changes in the meaning of the word "immoral," which originally meant "not customary" but now means "unethical." The word "grasp," as an example of a metaphor, involves shifting the meaning from its original usage of "to hold," to now also mean "to understand." The word "high" that originally meant being "in a position in space" now also refers to "being on drugs."

While the meaning of individual words can narrow, broaden or shift, it is also possible for new meanings to evolve independent of the semantics of the individual words in a sentence or phrase. For example, it would not be possible to know that the phrase "kicked the bucket" meant that a person died, or that "sitting pretty" meant that someone was fortunate, if you just examined the semantics of the individual words in the phrases. These examples are idioms, and they illustrate the fact that language can be figurative, meaning that language can acquire and communicate meaning without the words used being literally true. Idioms, in fact, are a subset of a group of fixed expressions that compose a significant portion of our verbal and written vocabulary. Jackendoff (1995) estimates that nearly half of the estimated 160,000 items that a person remembers are not fixed expressions.

Now while I have emphasized the benefits of deconstructing sentences and phrases to identify their component parts, it is also possible to approach language from a more holistic perspective. Thus, if what is important about some expression of language is to intentionally communicate, then it may be important to establish the role that speech acts play in this process. Speech acts include making statements, asking questions, giving commands, making promises, and similar activities. Searle (1969; p. 17) nicely distinguishes the relationship between rule-governed grammar and speech acts when he states, "It would be as if baseball were studied only as a formal system of rules and not as a game." He also points out that the meaning conveyed by language is not only expressed by the semantics of a phrase or sentence, but also by studying the performance of speech acts. Searle (1969) follows this thought with the insightful comment that Wittgenstein, especially in his early writings, may have been viewing language from the perspective of speech acts – its pragmatics – and on the basis of this finding, the meaning attached to sentences or words may be quite different then their formal literal definitions.

He also distinguishes between what is intentional and what is conventional about a communicative statement, including a speech act (see also Grice 1957). Thus, when information is exchanged between people or when a person fills out a written test, there are certain expectations that exist concerning the content and format of the communication. However, sometimes the communication can violate these conventions and instead communicate another objective or goal. It should be easy to see how an intentional communication could set the stage for figurative expression.

While most assessments of quality-of-life involve written material, the phrasing of the items as questions overlaps with what would be expected if the material were verbally presented. Of course, some assessments require interviews (e.g., the original version of the QWB), and asking questions in this context overlaps with what would occur as a speech act (i.e., a response is expected to the question). Other settings where a person is expected to respond include when they have been asked to, during cognitive interviewing, during doctor–patient discussions, focus group participation, and so on. In each case, assessing the response to the implicit call for performance can be informative. This would be especially true for monitoring the status of the compromised person.

The next question I need to be concerned with is the best way to express what I mean by the terms quality or quality-of-life. Part of my answer will involve inspecting definitions of quality or quality-of-life to determine if they rely on literal or figurative language. I will return to the importance of speech acts in Chap. 11, where their existence is helpful when distinguishing statements of well-being from quality, particularly quality-of-life.

2.2 The Optimal Expression of Quality-of-Life or HRQOL

Mehl and Pennebaker (2004) asked a group of college students to write an essay in response to the question, "How would you describe your quality-of-life?" Following are some of their responses:

> Webster defines quality-of-life as your personal satisfaction (or dissatisfaction) with the cultural or intellectual conditions under which you live (as distinct from material comfort). I define it as living in the moment, having hope for the future, and loving my past. It is being healthy, being loved, and being engrossed in a safe atmosphere.
>
> The thing that I value most in life is that I'm living. I was diagnosed with lupus when I was 13 years old, and I had to go to the doctor every week. Now it's under control, but I'm the youngest to have this disease.
>
> Quality-of-life is the hope one has ....
>
> The things that matter most to me seem to drift in and out of my priority range dynamically. A few years ago, my focus was my ambitions and investments. Then, it became God, and my relationship was centered around Him. While I would love to say that is where it remained, I would be dishonest in doing so. My newest focus has to be either school or my girlfriend.

These students capture the diversity of issues, unique to the individual yet common to them all, that they considered when reflecting on the quality of their existence. For this reason, some would argue that poetry and prose are the best vehicles for expressing this reality, more so than an academic-sounding definition or some attempt at assessment. Still others may wonder if the meaning attached to a person's existence is so complex that it is highly unlikely that it could ever adequately describe it with words, no less assess it. Frank (2001) directly addressed these issues in an interesting article entitled "Can we research suffering?" To quote Frank (2001):

> Suffering involves experiencing yourself on the other side of life as it should be, and nothing, no material resource, can bridge that separation. Suffering lies beyond such help. Suffering is what is unspeakable, as opposed to what can be spoken; it is what remains concealed, impossible to reveal; it remains in darkness, eluding illumination; and it is dread, beyond what is tangible even if hurtful. Suffering is loss.... Suffering resists definition because it is the reality of what is not. Anyone who suffers knows the reality of suffering, but this reality is what you cannot 'come to grips with'.... Suffering is expressed in myth as the wound that does not kill but cannot be healed. (p. 355).

While claiming what can't be done, Frank, of course, is using language to elegantly describe and define suffering. All I need to do is accept his discourse as the content to be assessed and this will contradict his proposition. When he says people can't assess such a complex human experience as suffering, he more than likely means that a questionnaire will not adequately capture this information. This may or may not be true, but methods are available that permit quantification of his essay (e.g., discourse analysis or text analysis), quantification that may yet be shown to adequately characterize what he describes as suffering. Schott (2004) has raised these same definitional and assessment issues about another major subjective indicator, pain. The individual experience of pain, he claims, may be indefinable but he acknowledges that analogies and other forms of figurative language can be used to characterize pain. This is in contrast with Frank's (2001) assertion.

The notion that profound subjective experiences are not easily subject to public examination is not unique to Frank. However, there are a variety of forms of literary expression, especially poetry, which provide appropriate venues for this type of expression.

> "The resident doctor said,
> 'We are not deep in ideas, imagination or enthusiasm-how can we help?'
> I asked.
> 'These days of only poems and depression-
> what can I do with them?
> Will they help me to notice
> what I cannot bear to look at?'"
>
> Robert Lowell (1977)[1]

For example, Harold Schwizer (1995) in an article entitled "To give suffering a language," introduces his article by quoting a poem by Robert Lowell (see below), in which the poet describes his encounter with a psychiatrist asking the psychiatrist how he or she was to free him of his depression. Of course, at the time of this encounter (the 1970s) the treatment options were limited. Schwizer goes on to describe the experience of the medical anthropologist Arthur Kleinman, with a young girl patient who suffered extensive burns. At first, Kleinman did not know what to do to comfort the silent child, except to hold her hand. Soon he felt the child holding his hand, and she started to speak. By extending himself, Kleinman literally gave the child a way to speak of her suffering, and by so doing he also helped the child transform an otherwise abstract indefinable state or concept into a concrete and understandable action. This process of making the abstract concrete will become a central theme in what I will discuss in subsequent chapters. It illustrates Bridgman's (1959) call for understanding how a person understands, and it accounts for why Frank (2001) believes that suffering can't be researched. He is correct if suffering is viewed as only an abstract entity, but it can be studied if it is made concrete, as when concrete language is used to express it. But this paradox is also true for many other abstract concepts, and in each case only by making the abstract concept concrete does it become possible to assess it. This should become clear as I consider other examples of this principle in subsequent chapters.

Clearly, poetry, prose or conversation can, if well crafted, effectively describe what one means by quality or life quality, just as words can be used to describe the beauty of a statue or painting, but these forms of expression are limited in their ability to validate what has been described or predict what may happen in the future. That can only happen if a formal analytic process is applied, such as the scientific method, which would first define what is to be assessed and then attempt to confirm this definition using appropriate assessment procedures. Confirmation (defined probabilistically) of these descriptions would be estimated by determining if the assessment process was reliable and valid, and confidence in the predictability of the assessment process would depend on estimating the responsiveness of the assessment methods that were used. If successful, assessment of this sort should be able to characterize the qualitative nature of our existence.

If diverse types of discourse provided data for some qualitative analysis, then how the language that makes up this discourse was used, becomes a legitimate subject for study. For the same reason, the language used in the assessment process should be examined to determine how it contributes to the objectives of the scientific process. My first challenge, then, is to consider how I would determine the meaning of a word, phrase or sentence, but my ultimate goal is to relate this to what I mean by the words or sentences I use to describe the process of assessment, and the process of assessing quality, in particular.

2.3 Literal Expression

Frege (1892/1966) has argued that someone can understand a sentence no matter who said it, or for whatever reason. This, he suggests, proves that a sentence has a literal meaning, a meaning that is totally independent of the context of its use, and totally dependent on the meaning of the individual words that make up the sentence and the syntactical rules that order these words. The fact that a sentence can be expressed in a variety of settings and yet provide the same meaning suggests that such a sentence, if literally true, represents an invariant statement (Chap. 4, p. 112). Certain sentences, however, may have more than one literal meaning, such as an ambiguous sentence. It may also be true that the literal meaning of a sentence may be defective, as in a nonsense sentence. Also, the literal meaning of a sentence should be distinguished from how a person uses a sentence, as when a sentence is used as if it contained an idiom, metaphor, and so on.

An extension of Frege's (1892/1966) original proposal, the literal meaning hypothesis, suggests that a person will always examine the literal meaning of a sentence first, and only later consider alternative interpretations. Is this so? In addition, is it true that the meaning of literal and figurative sentences differs, and finally, is there a difference between a linguistic interpretation of what a statement may mean and what readers or speakers say these statements mean? Let me examine each of these questions, that were originally asked by Gibbs (1984).

First is the hypothesis that a person will also examine the literal meaning of a sentence before considering other interpretations of the sentence. This implies that all sentences have a literal meaning. Consider these sentences, suggested by Gibbs (1984):

(1) "How about the salt?"
(2) "Take a leak."
(3) "Trip the light fantastic."

Gibbs claims that these sentences do not have a literal meaning, yet they are understandable. This suggests to him that literal meaning is not necessary for comprehension. In addition, Gibbs reports a series of experiments in which he found that people took as long to respond to indirect requests (e.g., "Can you pass me the salt?") as they did to direct requests ("Do not open the window.") when these sentences were asked in an appropriate context. However, when these sentences were taken out of context, then the indirect statements took longer to read and respond to than the direct request questions. Glucksberg (2001) provides additional support for the similarity in the processing time of literal and figurative sentences (see below).

Another question worth asking is if the meaning communicated by literal and figurative sentences is different? Consider the sentence, "A man is not an island unto himself." The literal and figurative meanings of this sentence differ, but both are meaningful. This suggests that the same words or phrases can be interpreted differently, making it unlikely that words alone determine the meaning of a sentence. Gibbs (1984) suggests that one way to resolve the difference between literal and figurative sentences is to assume that they differ in terms of how well they are placed in a context. Thus, a literal sentence would be more isomorphic with its context, while a figurative sentence would be less so. This, according to Gibbs, would make the two classes of sentences equally as understandable, so they would differ more in degree than in kind.

The final question that Gibbs raised was whether a linguistic statement differs from how a reader or speaker would state or use the same sentence. This he describes as the difference between *semantics* and *pragmatics*, a distinction he does not consider important if the primary purpose of the study is to understand how a person understands or uses language.

Gibbs also points out that for *indicative sentences*, the literal meaning of the sentence determines a set of conditions which, if satisfied, makes the sentence a *true* statement. Thus, if a cat is actually on a mat, then the sentence "The cat is on the mat." would be considered true. Some philosophers, therefore, argue that to know the literal meaning of a sentence is to know its truth value (e.g., Carnap 1956), and that this remains so, independent of its context. This issue is an important one for me and I will return to it shortly.

Glucksberg (2001) offers another approach to defining what is meant by literal. He suggests that it is necessary to separate and examine two processes: *linguistic decoding* and *linguistic interpretation*. Linguistic decoding, he states, involves examining an expression in terms of its phonological, lexical, and syntactic characteristics. He also states that the linguistic meaning is very much attached to the results of this decoding process, and as such, is very much dependent on the linguistic theory underlying the postulated decoding process. Thus, he quotes Stern (2000) that the "literal meaning of a simple expression is whatever our best linguistic theory tells us is its semantic interpretation" (p. 23). From this, Glucksberg goes on to say, "Linguistic-literal meanings are thus the product of a particular....theory of semantics and syntax, a theory that does not pretend to describe or explain what people actually do when talking and listening." (p. 11).[2] This view is consistent with Gibbs' (1984) response to the literal meaning hypothesis. An alternative, he suggests, is to rely on the common pattern of usage, or folk theory of language. The folk language would limit the definition of literal to the primary meaning of a word, phrase, or sentence. Thus, the primary dictionary definition of "literal" is, "using or interpreting words in their usual or most basic sense without metaphor or allegory" (Oxford University Press 1996).

Glucksberg (2001) also takes issue with the literal meaning hypothesis view that literal meaning is context independent.

He provides a series of examples to demonstrate that this view is hard to defend. Yet he also points out that while various literal expressions may not be context independent, people treat them as if they are, and this, he says, is what you might expect in a folk theory of language. Glucksberg (2001) summarizes his view in the following quote:

> Perhaps the most useful position is that the concept of *literal* cannot be explicitly defined except in formal linguistic-theory terms. Within our folk theory of language we make a sharp distinction between the literal and nonliteral. However, when we make judgments about specific examples the distinction becomes graded, rather than discrete. People can make reliable judgments about degrees of metaphoricity, for example, suggesting that there is a continuum from the literal to the nonliteral… (p. 14).

If the term literal is best defined as reflecting the ordinary use of language (i.e., a folk theory of language), then it is probably best to assume the same for the language used to express quality or a quality-of-life assessment.

2.4 Figurative Expression and Its Complexities

First, figurative expression can take many forms: as an idiom, hyperbole, indirect request, irony, understatement, metaphor, simile, or rhetorical question. What is common to all these forms of expression is that they can convey meaning in a more efficient manner than an equivalent literal statement, and for this reason are a regular part of discourse. I will follow Glucksberg (2001), Bowdle and Gentner (2005) and others, and focus on the study of the metaphor as a prototypical figurative form, since an extensive literature exists on how meaning is being communicated by some expressive form (e.g., such as the definitions listed in Table 2.1). Glucksberg (2001) states that:

> Theories of metaphor in philosophy, linguistics, and psychology… address one or more aspects of…two senses of metaphor: metaphor as a form of linguistic expression and communication and metaphor as a form of conceptual representation and symbolization. (p. 4).

Table 2.1 List of quality-of-life or HRQOL definitions[a]

Campbell et al. (1976)	Quality-of-life is defined in terms of satisfaction of needs, and "level of satisfaction" can be precisely defined as the perceived *discrepancy* between aspiration and achievement, ranging from the perception of fulfillment to that of deprivation
Calman (1987)	Quality-of-life therefore measures the difference, at a particular period of time, between the hopes and expectations of the individual and the individual's present experience. It is concerned with the *difference* between perceived goals and actual goals
Ferrans (1990)	Quality-of-life is defined as a person's *sense of well-being* that stems from satisfaction or dissatisfaction with the areas of life that are important to him/her
Gotay et al. (1992)	Quality-of-life is *a state of well-being* that is a composite of two components: (1) the ability to perform every day activities that reflect physical, psychological and social well-being and (2) patient satisfaction with levels of functioning and the control of disease and/or treatment-related symptoms
Frisch (1993)	Life satisfaction is equated with quality-of-life and refers to a person's subjective evaluation of the degree to which his or her most important needs, goals and wishes have been *fulfilled*
Patrick and Erickson (1993)	Health-related quality-of-life is *the value assigned to duration of life* as modified by the impairments, functional states, perceptions, and social opportunities that are influenced by disease, injury, treatment, or policy
Osoba (1994)	HRQOL is a multidimensional construct encompassing perceptions of both positive and negative aspects of dimensions, such as physical, emotional, social and cognitive functions, as well as the negative aspects of somatic discomfort and other symptoms *produced by* a disease and its treatment
Cella (1995)	HRQOL refers to the extent to which one's usual or expected physical, emotional, and social *well-being* are affected by a medical condition or its treatment
Ebrahim (1995)	HRQOL may be thought of as those aspects of self-perceived *well-being* that are related to or affected by the presence of disease or treatment
World Health Organization (1995) (WHOQOL)	Quality-of-life was defined, therefore, as an individual's perception of their *position in life* in the context of the culture and value systems in which they live and in relation to their goals, expectations, standards, and concerns
Felice and Perry (1996)	Quality-of-life is defined as an overall *general well-being* that is comprised of objective and subjective evaluations of physical, material, social, and emotional well-being, together with the extent of personal development and purposeful activity, all weighted by a personal set of values
Padilla et al. (1996)	HRQOL is defined as a personal evaluative statement summarizing the positivity and negativity of attributes that characterize one's psychological, physical, social and spiritual *well-being* at a point in time when health, illness and treatment conditions are relevant
Shipper et al. (1996)	Quality-of-life in clinical medicine represents the functional *effect of* an illness and its consequent therapy upon a patient, as perceived by the patient
Revicki et al. (2000)	HRQOL is defined as the subjective assessment of the *impact of* disease and its treatment across the physical, psychological, social, and somatic domains of functioning and well-being

[a] Based on definitions provided by Ferrans (2005). *Italics* added by author

I will consider both usages, but start with describing metaphors as a form of linguistic expression and communication. First, it is important to recognize how prevalent metaphor usage is in our everyday language and especially in scientific discourse. Graesser (Personal Communication, October 1989), for example, claim that one out of every 25 words expressed on television programs is a metaphor. Metaphors are regularly used in the various sciences, where a constant interplay occurs between empirical evidence and the metaphor used as a model. Fiedler (2000) points out that:

> It is not unusual for psychological theories to draw on common physical or statistical metaphors (Gigerenzer 1991; Roediger 1980), such as Lewin's (1951) field theory, Brunswik's (1956) lens model, Thurstone's (1927) law of comparative judgment and the related 'signal detection' approach (1966)…In all these cases, the theoretical innovation gained in the transfer domain does not depend on how novel and influential the imported metaphor is in its home domain. For instance, an optical lens or an arithmetic average do not entail exciting developments in physics or statistics, respectively. What leads to theoretical originality and fertility is often the simplicity of very common metaphors." (p. 659).

Fiedler's (2000) comments suggest that these particular metaphors are not advancing the fundamental basis for the physical or statistical sciences, but rather are efficient means of communicating some meaning that would otherwise be difficult to communicate or not be forthcoming at all. Thus, when Lewin (1951) uses *field theory* to characterize psychological events, he is giving his reader an abstract concept to use when thinking about and understanding a psychological phenomenon. The same process occurs in all the different sciences.

As will become clear, the available definitions of quality-of-life or HRQOL rely, to varying degrees, on various types of metaphors to communicate meaning. But this raises a fundamental question, and that is: if these definitions are not literally true, how can I rely on them when I have to make practical decisions (e.g., selecting a treatment based on quality-of-life or HRQOL data)? Literal language is considered to be real, true, and unambiguous. At least that is what most of us believe. We all assume that what is "out there" is what we all experience. If this is so, what do I believe about figurative expressions, such as metaphors. Are they also real, true, and unambiguous? This is a central issue not only for understanding metaphors, but also for a number of other topics I will discuss.[3]

The implicit question being asked here, however, is how I come to know what is true and real. Philosophically, the question is whether there is an objective reality which will permit me to make true statements. Much of twentieth-century Western Philosophy argued that there was such an objective reality, and this view reached its fullest expression in the doctrine of logical positivism. As Ortony (1993) states:

> A basic notion of positivism was that reality could be precisely described through the medium of language in a manner that was clear, unambiguous, and, in principle, testable – reality could and should be literally described… During the heyday of logical positivism, literal language reigned supreme. (1993, p. 1).

It is, of course, interesting to note how Ortony's comment connects literal language usage and the positivistic approach to reality.[4] An alternative exists, however, that assumes that reality (particularly "scientific reality") is a product of combining immediate experiences with past and contextual knowledge. A cognitive entity is formed based on elements of perception, language, and past knowledge. This *constructionist* view claims that reality, particularly subjective reality, is not directly observable but is constructed and may include the nonliteral use of language, as in the case of a metaphor.

There are actually two major constructionist models (Zuriff 1998), and I will briefly review them so that my approach will be clear. At one extreme is the metaphysical constructionist's view (Gergen 1985; Searle 1995; Zuriff 1998), which, while acknowledging the constructed nature of reality, uses this observation as a stepping stone to engage in an antilogical positivism polemic, and ends up by proposing a dramatic reinterpretation of the nature of science. For example, the above authors claim that since what is known is socially constructed, I cannot transcend my constructs and contact reality directly. Instead, I am limited in my conceptualization of the world to using interpretative categories, concepts, and theories. Observation and theory become indistinguishable since both are constructs. Truth becomes relative since it is determined by the scheme, framework, or language used. The difference between value and fact collapses because the neutrality of fact is a myth. Objectivity is a myth and agreement has to come from social interactions.

Zuriff (1998) claims that metaphysical constructionism is a modern-day version of solipsism. This form of constructionism, he states, makes two errors. The first is the failure to acknowledge that it is possible to distinguish between the social world and the rest of the natural world, and the second is the failure to differentiate a fact from the social acceptance of a belief as a fact. He states that, "These errors gain their plausibility from a version of traditional skepticism which leads ultimately to solipsism and irrelevance." (1998; p. 18).

An alternative model of constructionism, *empirical constructionism*, also acknowledges that my conception of the natural world is constructed and that there may not be a unique description of reality (Zuriff 1998). It differs from metaphysical constructionism, however, by not claiming that my reality is the only form of reality. For example, the empirical constructionism view accepts that human perception of sound may not correspond to the physical sounds presented, and in this sense is a construction of reality, but it also does not deny that physical sound is real. Instead, empirical constructionism proposes that the two forms of reality exist and their relationship should be the subject of study. I alluded to this when I made the distinction between the wavelength (a perception) that I perceive and the hue (a cognition) that I report. Thus, empirical constructionism emphasizes the study of the relationship between people's behavior, their psychological (subjective) world and external reality.

I will engage in just such a task when, to paraphrase Bridgman (1959), I try to understand how I, and others, understand. Thus, I will speak of assessment rather than measurement, and objectification rather than objective in order to emphasize that I am an inseparable part of the scientific process and I have to understand my role in establishing what I consider to be real, if I am going to believe what I have observed. The history of science tells me that my scientific reality has and will change and I think that it is essential that I incorporate this perspective in my approach to such a complex topic as "quality." Again this does not mean that the object I see is not objective, rather that I need to understand what has happened after I have cognitively processing it.

With this review I can now apply what I have learned about the difference between the nonconstructionist/constructionist views to my discussion of literal and figurative language. For example, the logical positivists (i.e., nonconstructionist) treat metaphors as an unimportant, almost deviant, form of expression. As stated above, they are primarily concerned with what is literally true. They would be quite suspicious if a metaphor was used as part of an argument or a definition, or was used to make a decision (e.g., a medical decision). In contrast, the constructionists argue that comprehension and meaning can evolve from both the literal and nonliteral (e.g., from metaphors) use of language. Thus, for a constructionist,[5] truth and reality are acquired, and acquired as a product of the interaction of the person with their physical and social environment.[6] Consistent with this is the demonstration by Asch (1958) and Gardiner and Winner (1978) that children learn the difference between figurative and literal expression over time, and that the duel usage of terms, such as metaphors, is common to different cultures.

As stated above, the concern that definitions that use figurative language may not be "true and real" may account for some of the ambivalence concerning the acceptance of quality or quality-of-life research findings as part of medical decision making or social policy formation. This concern, however, should not last any longer than the observer's willingness to recognize the ubiquitous usage of figurative language, a usage which presumably has not mislead or deceived in other forms of communication or types of decision making. In addition, Glucksberg (2003) and others have pointed out that a metaphor is a very efficient way to communicate a meaning which, if expressed literally, would involve a more extended explanation.

Still, the view that figurative language distorts truth and reality and is not a useful part of communication process persists, as is evident in Sontag's book, *Illness as Metaphor: AIDS and its Metaphors* (1991). Dealing with cancer, tuberculosis and AIDS, Sontag argues that the flagrant use of metaphors in medicine confounds effective treatment, and should be avoided by persons discussing their illness with physicians. She also claims that metaphors distort medicine's task, as when conceiving of a disease and its treatment as "a victory in a battle," a battle that she says sometimes can't be won. Of all of Sontag's admonitions, however, the notion that a person can effectively communicate without the use of figurative language seems hardest to accept. First, because metaphors have such a central role in how I express myself and think about things, and second because it is very hard to communicate meaning without the use of figurative language. A good example of this is Scheper-Hughes and Lock's (1986; in Gwyn [1999]) claim that, "Sickness is a form of communication…through which nature, society and culture speak simultaneously, and it is in that setting that metaphors thrive" (p. 220).

Miller (1993) offers another approach to the issue of the truthfulness of figurative language. To him a metaphor "is an abbreviated simile and that the thought provoked is the kind required to appreciate similarities and analogies. In the nineteenth century that kind of thought was called 'appreception'."[7] (p. 357). Miller goes on to explain that appreception is a class of mental processes which brings "an attended experience…with an already acquired and familiar conceptual system" (1993; p. 357). Thus, a metaphor is compared to what I already know, and the truthfulness of the metaphor would be a product of this comparative process. He states:

> In order to find a compromise between the requirements of the truth assumption and the need to relate the textual concept as closely as possible to general knowledge and belief, the reader must search for resemblances between the textual concept and general knowledge. These resemblances, which are the grounds for the metaphor, can be formulated as comparison statements. Once found and interpreted, the comparison is not added directly to the textual concept, but is used as a basis for imagining a minimally divergent state of affairs in which the metaphorical claim is true. (p. 373).

Thus, to Miller the truthfulness of a metaphor is based on a person's imagining whether what is being claimed in a metaphor matches their past experience. This view is consistent with a constructionist perspective.

2.4.1 Defining a Linguistic Metaphor

My next task is to determine how I can recognize when a word, phrase or sentence is being used metaphorically, and I will follow this with a discussion of how a linguistic metaphor is comprehended. This will involve a discussion of alternative models of metaphor comprehension, a discussion which will introduce a number of conceptual issues. I am doing this because metaphors are an important part of any quality assessment, and this will become evident after I examine the metaphoric basis of various definitions provided by Ferrans (2005).

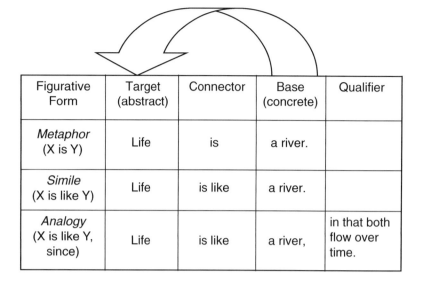

Fig. 2.2 Figurative language mapping. Depicted is the figurative relationship between sentences that reflect an analog, a simile, and a metaphor. This figure is based on a paper by Bowdie and Gentner (2005).

Table 2.2 Characteristics of a linguistic metaphor as defined by Glucksberg (2001)

Metaphoric statements come in the form of X is Y, where X and Y are not literally related. (For example, "alcohol is a crutch.")
Elements of a metaphoric statement are not reversible. (For example, "a crutch is alcohol" is uninterpretable)
Statements vary in their degree of metaphoricity (from purely literal to purely metaphoric)
* The context of a statement (e.g., the use of qualifiers) will affect the degree of metaphoricity
* A prototype member of a metaphoric category is more likely to produce a good metaphoric fit then a nonprototype category member. (For example, "my life is as good as gold," as opposed to, "my life is as good as brass.")
Metaphors come in two forms: *nominal* (a noun is a noun) and *predictive* (e.g., a verb)
Literal statements can function as metaphors (e.g., "people are not sheep")
Metaphors can be understood as readily as literal statements
Metaphors require no more time to process and interpret than a literal statement

Glucksberg (2001) suggests several characteristics which can be used to identify a linguistic metaphor (Table 2.2). Principal among these is that a metaphor takes the form "X is Y statement," and that if reversed, the statement would lose its metaphorical properties. Take, for example, the metaphor, "my boss is a shark." If reversed as "my shark is a boss," it clearly does not evoke the same meaning. Glucksberg (2001) also suggests that metaphors come in two forms: nominal and predictive. The nominal form uses nouns to connect objects, situations, or events, while the predictive form uses verbs to connect certain categories of action. In both instances, a mapping occurs between a concrete (the base) and an abstract (the target) entity. Figure 2.2 illustrates this mapping, but now between several figurative forms. Note that the different meanings associated with the different figurative expressions are due to the connectors and qualifying statements, not the target and base. Also note that the simile and analog, as opposed to the metaphor, would retain meaning if the base and target were reversed.

In addition, data exist which suggest that the ease of understanding and speed of processing metaphors are comparable to those found for literal statements (Glucksberg 2001). In general, the criteria listed in Table 2.2 suggest that functionally, a metaphor is as effective in communicating meaning as a literal statement, yet its truth value remains in dispute.

Aristotle (1996) claimed that metaphors (e.g., "man is a wolf") are basically false, and warned against using them, particularly in definitions. This is an issue that will be relevant for my discussion of the metaphoric basis of quality-of-life or HRQOL definitions. Yet Aristotle was aware that metaphors are an efficient form of communication in ordinary discourse. To explain how metaphors were understood, he suggested that they involved an implicit comparison, with the reader determining what was similar and literally true when comparing the target (e.g., man) to its base (e.g., the wolf). Thus, the way I understand a metaphor is what is literally true about it (Aristotle 1996; p. 38).

Aristotle's view has been reinforced by modern empirical support, starting with the studies of Tversky (1977) and his colleagues. The key element in this research was the demonstration that it was possible to identify specific features of the base which would be mapped to the more abstract target to give the target meaning. This feature-matching process can be visualized graphically by the overlapping area between two neighboring circles (i.e., a Venn diagram).

Evidence to support the model summarized in Table 2.2 includes the observation that the greater the similarity between target and base the greater the aptness and interpretability of the metaphor (e.g., Marschark et al. 1983) and the faster the metaphor is understood (Gentner and Wolff 1997). Criticism of this model includes the fact that not every feature found common to the target and base actually contributes to the meaning of the metaphor (Lakoff and Johnson 1999). Second, because of these "missing" common features, it should be possible to reverse the meaning mapped between a target and base, but if this is done it would violate a defining characteristic of a metaphor (see Table 2.2). Ortony (1979) counters by saying, in effect, that not all features of the target and the base are created equal; some are more salient for the metaphor's meaning than others, with this being more so for the base than the target. Thus, salient features of the base can transfer meaning to the target that would otherwise be ignored, resulting in a situation that would make the statement irreversible. Bowdle and Gentner (2005) also point out that targets and bases are typically from different semantic domains (e.g., what may be communicated in the metaphor "man is a wolf" is predation, but human and animal predation are quite different), and that these differences would ordinarily be sufficient to lead to a failure of information transfer based on a similarity comparison. However, this actually does not occur. Finally, the standard comparison model of metaphors does not explain how metaphors can lead to the creation of new similarities between the base and target.

The *comparison model* of metaphor comprehension is a static model dependent on features that are already present when the base and target are compared. As an alternative, Glucksberg (2001) suggests that metaphor comprehension is based on the creation of abstract classes (a categorization process) that results in the transfer of information from base to target. As Bowdle and Gentner (2005) describe it:

> According to Glucksberg and Keysar (1990), the literal target and base concepts of a metaphor are never placed in direct comparison during metaphor comprehension. Rather, the base concept is used to access or derive an abstract metaphoric category of which it represents a prototypical member, and the concept is assigned to that category. On this view, metaphors differ from literal categorization statements in that metaphors involve *dual reference*: refers simultaneously to a specific literal concept and a general metaphoric category. (p. 195).

They go on to illustrate this model:

> …consider the metaphor '*My job is a jail*'. The base term *jail* literally refers to a building that is used to detain criminals and therefore does not seem immediately applicable to the target term *my job* (assuming the speaker does not actually work inside a jail). On the categorization view, then, comprehension of the statement requires that one uses the base concept to elicit a metaphoric category that it typifies -namely, *any situation that is unpleasant and confining*. If this category is already associated with the base concept, then it is simply assessed during comprehension. if the category is not well established, then it must be abstracted online, much as people may create literal ad hoc categories to achieve certain goals…In either case, once the metaphoric category has been elicited, the target can be understood as a a member of the category. This sets up the kind of inheritance hierarchy that is implicit in all taxonomic relations. Consequently, all properties characterizing the metaphoric category named by *jail* are attributed to the subordinate concept *my job*. (p.195)

To paraphrase, a person reads a metaphor and notes the discordance between the base and the target (e.g., a job is not literally a jail), resulting in a cognitive search that leads to the generation of an abstract class that is true of the base that can now be applied to the target (e.g., a jail's unpleasantness and confining nature might be duplicated in a job). If it is confirmed that the target has these characteristics (e.g., the unpleasantness and confining nature of a job), then it is possible to understand the metaphor. What would also be expected is that if confirmation was not possible, then the metaphor would not be understood, or a question raised about whether the expression was actually a metaphor. If it was understood, it would also demonstrate the more efficient means of the metaphor in communicating meaning, especially when compared to an equivalent literal expression.

There are several aspects of this model worth noting. First is the suggestion that a person abstracts a class or category from the base, or creates one de novo, and uses it to classify the target's characteristics. This immediately places the relationship between the base and the target in an informationally hierarchical relationship, with the hierarchy gravitating from the concrete base to the more abstract target. Second, while both jail and job have many literal references, one uses only a particular set of characteristics found in the base (e.g., the unpleasant aspects of a jail) to classify the information found in the target.[8] Glucksberg, however, recognized that this isn't always the case, since sometimes the informational content of the target can affect the understanding of the metaphor. As Bowdle and Gentner (2005) state, "…metaphor targets provide information about what type of properties they can meaningfully inherit and therefore about what types of categories they can meaningfully belong to" (p. 195). As a result, Glucksberg et al. (1997) have proposed an interactive property attribution model that acknowledges that metaphor mapping is a flexible process, involving both the base and the target.

In contrast to Glucksberg's notion that abstract metaphoric categories are used to map meaning between the metaphor base and target, Bowdle and Gentner (2005) return to feature-matching models and suggest a model that assumes that at least some features of the base are isomorphic to the target. The process they envision as operative involves applying Gentner's (1983) structure-mapping theory to metaphor comprehension. To quote Bowdle and Gentner (2005):

Structure-mapping theory assumes that interpreting a metaphor involves two interrelated mechanisms: alignment and projection. The alignment process operates to create a maximal structurally consistent match between two representations that observes one-to-one mapping and parallel connectivity (Falkenhainer et al. 1989). That is, each element of one representation can be placed in correspondence with at most one element of the other representation, and arguments of aligned relations and other operators are themselves aligned. (p. 196).

Interpreting the metaphor "Socrates was a midwife" illustrates their approach to metaphoric comprehension. They state:

First, the identical predicates in the target and base concepts (i.e., the relations *help* and *produce*) are matched, and the arguments of these predicates are placed in correspondence by parallel connectivity: *midwife* → *Socrates, mother* → *student*, and *child* → *idea*. Next these local matches are coalesced into a global system of matches that is maximally consistent. Finally, predicates that are unique to the base but connected to the aligned structure (i.e., those predicates specifying the gradual development of the child within the mother) are carried over to the target. Thus, the metaphor could be interpreted as meaning something like, 'Socrates did not simply teach his students new ideas but rather helped them realize ideas that have been developing within them all along. (p. 196).

Implicit in this model is the assumption that comprehension can be understood as a linear information-processing model. Consistent with this, Gentner and her colleagues (Falkenhainer et al. 1989; Forbus et al. 1995) have developed a general computer-based model (Structure-Mapping Engine: SME) that can be used to simulate metaphor comprehension. Their description of this computer program is worth reproducing here, since I will revisit elements of their description in different guises, at different times in this book. They state:

In the first stage, SME begins blind and local by matching all identical predicates in the representations being compared. In the second stage, these local matches are coalesced into structurally consistent connected clusters, called *kernels*, by enforcing one-to-one mapping and parallel connectivity. In the third stage, SME gathers these kernels into one or a few global interpretations. This is done using a *greedy merge* algorithm. It begins with the maximal kernel and then adds the largest kernel that is structurally consistent with the first one, continuing until no more kernels can be added without compromising consistency. It then carries out this process beginning with the second largest kernel to produce a second interpretation and so on. At this point, SME produces a structural evaluation of the interpretation of interpretations, using a kind of cascade like algorithm in which evidence is passed down from predicates to their arguments…Up to this point, the mapping process has been nondirectional. Now, however, a directional inferences process takes place. Predicates connected to the common structure in the base, but not initially present in the target, are projected as candidate inferences about the target." (p. 196).

Their description of how kernels are aligned to form a global interpretation is an example of the more general problem of how parts are combined together to form a whole. But they also describe a top-down process, the greedy merge algorithm, which simulates the process whereby information is cognitively organized and projected from one domain to another.

They are clearly dealing with a generic issue that will repeatedly come up in this book, but which is now presented in the context of the transmission of information, as in a metaphor.

The work of another investigator is worth mentioning, since it replaces the literal-metaphor dichotomy with a single central principle: a saliency continuum. Giora (2003) argues that the literal/metaphor dichotomy can't account for the ease of processing, which is characteristic of some literal and metaphorical phrases. She points out, for example, that "conventional metaphors and conventional idioms used unconventionally both behave like highly literal language used innovatively: they all trigger a sequential process."(Giora 1997; p. 26–27). Thus, the speed of information processing due to familiarity and or frequency of exposure may be as informative as dichotomizing expressions as literal or figurative (Giora 1997; 2003).

Many of these issues will be relevant when discussing the next topic, conceptual metaphors. What I will learn is that linguistic and conceptual metaphors raise similar issues, but at different levels of abstraction.

2.4.2 The Importance of Conceptual Metaphors

As Glucksberg (2001) stated, the term "metaphor" is currently used in two ways; as a form of linguistic expression, and as a form of "conceptual representation and symbolization." Conceptual metaphors are important, because they are used to help organize how a person thinks; they are not just used to provide a nonliteral meaning for a word or phrase. Lakoff (1980) and his colleagues have most fully articulated this view. They suggest that a conceptual metaphor plays a central role in managing my "cognitive economy," providing an in-place method whereby new information is efficiently processed. Thus, the presence of a conceptual metaphor helps avoid the time-and energy-consuming process of developing a new cognitive form each time a person is exposed to new information. Conceptual metaphors also facilitate the organization of information into a hierarchy, with the apex of the hierarchy representing the most abstract form of metaphoric expression. This metaphorical hierarchy, when applied in a "top-down manner," provides meaning for abstract concepts and emotional expressions by reference to concrete experiences.

Lakoff (1993) describes how this can be done in the following quote, where he uses the LOVE IS A JOURNEY metaphor as a model *conceptual metaphor*. He says:

What constitutes the LOVE IS A JOURNEY metaphor is not any particular word or expression. It is the ontological mapping across conceptual domains, from source domain of journeys to the target domain of love. The metaphor is not just a matter of language but of thought and reason. The language is secondary. The mapping is primary, in that it sanctions the use of source domain language and inference patterns for target domain concepts. The mapping is conventional; that is, it is a fixed part of our conceptual system, one of our conventional ways of conceptualizing love relationships. (p. 208).

Thus, the mapping of a particular source, or base term, to a particular target is the primary cognitive method whereby information is efficiently processed. McGlone (2001) expands these notions by stating:

> ....Lakoff characterizes the conceptual metaphor that links love and journeys as playing two distinct but related roles: *a representational role* and *a process role*. It plays a representational role in that it structures my knowledge of love. The reasoning behind this claim is that the mind represents abstract concepts (such as love) in an economical fashion, borrowing the semantic structure of more concrete concepts (such as journey) to organize aspects of the abstract concept. One reason for this is that it might be too computationally expensive to represent abstract concepts in a stand-alone fashion. Second, the love-journey metaphor plays a process role in that it mediates expressions pertaining to love…Again, the metaphor's hypothesized process role appears to be economical from a computational standpoint, in that (a) metaphoric meaning may be retrieved rather than constructed de novo and (b) the meaning of any number of metaphoric expressions…may be generated from a single semantic structure (the love-journey conceptual mapping. (p. 91).

Lakoff (1993; p. 220) also discusses the EVENT STRUCTURE METAPHOR. This broad-based metaphor includes such notions as "states, changes, actions, causes, purposes, and means" which are cognitively conceptualized in terms of *space, motion*, and *force*. This type of metaphor will be particularly relevant for my discussion of the applicability of a metaphoric analysis of definitions of quality-of-life or HRQOL. The mappings that occur within this metaphor include (Lakoff 1993; p. 220):

EVENT STRUCTURE METAPHOR
States are location (bounded regions in space).
Changes are movements (into or out of bounded regions).
Causes are forces.
Actions are self-propelled movements.
Purposes are destinations.
Means are paths (to destinations).
Difficulties are impediments to motion.
Expected progress is a travel schedule; a schedule is a virtual traveler, who reaches prearranged destinations at prearranged times.
External events are large, moving objects.
Long-term, purposeful activities are journeys.

According to Lakoff (1993) this metaphor contains a series of metaphors (including the LOVE IS A JOURNEY metaphor) that can be organized into a hierarchy. Thus,

Level 1: The EVENT STRUCTURE METAPHOR,
Level 2: A PURPOSEFUL LIFE IS A JOURNEY,
Level 3: LOVE IS A JOURNEY.

constitute a hierarchy. Lakoff (1993), describes the basis for the hierarchy in the following quote:

> In our culture, life is assumed to be purposeful, that is, we are expected to have goals in life. In the *event structure metaphor*, purposes are destinations and purposeful action is self-propelled motion toward a destination. A purposeful life is a long-term, purposeful activity and hence a journey…Choosing a means to achieve a goal is choosing a path to a destination. Difficulties in life are impediments to motion…In short, the metaphor A PURPOSEFUL LIFE IS A JOURNEY makes use of all the structure of the event structure metaphor, since events in a life conceptualized as purposeful are subclasses of events in general." (p. 223).

The same would be said about the relationship between A PURPOSEFUL LIFE IS A JOURNEY and the LOVE IS A JOURNEY metaphors; the events in a love relationship make use of the structures of a purposeful life as a journey metaphor and can be conceived of as a subclass of A PURPOSEFUL LIFE IS A JOURNEY metaphor.

A critical component of Lakoff's model is his assumption that a conceptual metaphor is not a linguistic phenomenon but instead is a cognitive entity (or type of *conceptual representation*).[9] He also claims that this type of metaphor is the primary method by which the mind represents concepts which are not sensory or perceptual in nature. McGlone (2001) sees this type of metaphor as providing "a way to representationally piggyback our understanding of abstract concepts on the structure of concrete concepts, which presumably are represented in their own terms, that is, in a standalone fashion." (p. 93).

2.4.3 Embodiment and Metaphors

Lakoff and Johnson (1999) also state that metaphors and metaphorical thinking, like my conceptual system in general, is "embodied"'; meaning that it arises from my body's experience with the physical world. Thus, my sense of what is real is embodied, and the metaphor is a central element which gives me this sense of reality. Gibbs (2006) has expanded on these ideas and has applied the notion of embodiment to the several cognitive domains (e.g., perception, concept formation, memory and reasoning, communication, and so on). I discuss Gibbs' (2006) notions in more detail in Chap. 10, where embodiment is presented as a model of the relationship between capacity and performance.

If what is real is embodied, does that mean that truth is "embodied"? Lakoff and Johnson (1999) would say yes. They specifically state:

> What we understand the world to be like is determined by many things: our sensory organs, our ability to move and manipulate objects, the detailed structure of our brain, our culture, and our interactions in our environment, at the very least. *What we take to be true in a situation depends on our embodied understanding of the situation*, which is in turn shaped by all these factors. Truth for us, any truth we can have access to, depends on such embodied understanding. (p. 102).

Truth, they point out, is not the correspondence between words and the world, as would be true if you relied on truth being literal. Rather, truth is a product of the convergence of three levels of activities (neural, phenomenological, and the cognitive unconscious), each of which contributes to what I understand to be true. The neural level helps me understand my experience in scientific terms, while the phenomenological level does so in terms of everyday experience. The cognitive unconscious refers to the vast number of mental processes that I am not aware of, but which govern my language, thoughts and actions. Lakoff and Johnson (1999) ask me to give up the illusion that a unique description exists for each situation. They argue that since multiple levels of embodiment exist there will be multiple forms of truth. And what will be real is what I need to think to be realistically, so that I "function successfully to survive, to achieve ends and to arrive at workable understandings of the situations we are in" (p. 109). Thus, when I say that a "verb," or "energy" or "quality-of-life" are real, I do not expect these words to necessarily have a physical presence, but they are real because they can contribute to *explanation, prediction*, and *understanding*.

Lakoff and Johnson's notions, particularly concerning conceptual metaphors, have not escaped criticism (e.g., McGlone 2001; Murphy 1997). McGlone (2001) points out that Lakoff and Johnson (1999) have limited empirical support for their argument, and their views have the potential for circular reasoning. McGlone's paper (2001) provides a detailed critique, but the essence of it is that he feels that Lakoff and Johnson blur the distinction between literal and metaphorical language and end up with a theory that becomes incoherent. As a rejoinder, data from Meier and his colleagues (2004; 2004; 2005), and what Gibbs (2006) reviews provides empirical support for the Lakoff and Johnson idea that conceptual metaphors are built on a person's experience with the physical world.

Meier and Robinson (2005) illustrate how the development of embodiment may occur, but now relative to the embodiment of affect. Embodiment can be said to start when cognitive development evolves from a child's sensorimotor experiences. This suggests that a connection develops between a child's perception and manipulation of their world (touching and playing with objects) and what the child feels about this world. Also of importance was that Meier and Robinson (2005) recognize that metaphors play a critical role in this linkage. First is the observation that early concrete (sensorimotor) experiences and latter abstract cognitions can be thought of as being linked by the mapping process that is characteristic of conceptual metaphor formation. Second, the physical world biases the relationship between perception and affect. Thus, social activities are more likely to occur during the day – when things are bright –facilitating the impression that positive affect is associated with brightness, while negative affect is associated with inactivity and darkness. Finally, the use of a metaphor when expressing affect is pervasive, which increases the chances that it will be associated with a person's physical environment. Thus, physical metaphors may play a critical linkage role in the formation of abstract concepts (e.g., happiness), and in this way demonstrate how embodiment of affect occurs (Meier et al. 2007).

To test this, Meier and Robinson (2005) proposed three types of experiments. First, they suggest testing the hypothesis that the encoding of affective stimuli will be biased by "metaphor-consistent physical aspects of stimuli (e.g., positive stimulus should be encoded faster if they are white rather than black)" (Meier and Robinson 2005; p. 241). Second was to demonstrate activation of a related perceptual process when an affective experience occurs. Thus, if a negative experience occurs, then the person's visual attention should be downward. Third, if the perceptual-affect link is represented by a metaphor, then this should occur without conscious awareness and occur involuntarily, much as would be expected for automatic behavior (Chap. 6, p. 165).

Meier and Robinson (2005) identify three perceptual dimensions which are regularly associated with metaphors: brightness, vertical position, and distance. I will briefly review the data that Meier and his colleagues present regarding brightness. First, it is quite obvious from inspecting popular culture that brightness is associated with good and darkness with bad. This association of brightness, and good and bad has been found to be true for 20 different countries and cultures (Adams and Osgood 1973), but also is characteristic of children. These studies, and others that Meier and Robinson (2005) cite, support the first hypothesis that affect is structured by a physical metaphor.[10] Meier and Robertson (2005) also found that "the explicit evaluation of positive and negative words led to a metaphor-consistent bias in brightness judgments" (Meier and Robinson 2005; p. 244). This they demonstrated by presenting positive and negative words on various shades of gray background. Respondents were then asked to select the background that the words were presented on, and they found that the respondents selected brighter backgrounds than what was originally presented for positive words, demonstrating the influence of the perceptual experience of brightness on a judgment. Meier et al. (2004b) also report a study which supports the hypothesis that the association between affect and brightness is an unconscious automatic process. They presented respondents with 50 positive words and 50 negative words typed on either a white or black background. They found that negative words were evaluated faster and more accurately when on a black background than on a white background, while positive words were evaluated faster and more accurately when presented on a white, rather than black background. Clearly, the background had an impact on the person's judgment, yet in either case a person would not be expected to be aware of the influence of the background as they responded to their task. These studies support the notion that physical metaphors (e.g., brightness) contribute to the embodiment of affect. They also suggest that these metaphors are part of how a person encodes information, and that they play a far greater role in cognition than just as a means of communication; rather they may actually structure a person's world. I will continue this discussion in Chaps. 10 and 11.

3 Application of Language Usage Principles

3.1 The Metaphoric Basis of Quality-of-Life or HRQOL Definitions

I am now going to apply this background about linguistic and conceptual metaphors to a representative list of quality-of-life or HRQOL definitions. In reviewing this list of definitions (Table 2.1), it is quickly clear that I am not dealing with the type of metaphoric material you might encounter in a poem or even in the example of discourse quoted above from Frank (2001). Rather, what I have here are expressions which vary in their use of figurative language, but regularly rely on conceptual metaphors to communicate meaning (Tables 2.4–2.6, 2.8). Of greatest importance is that by demonstrating the presence of metaphors in these definitions, I will be able to classify them into a limited number of categories.

The first step in this process is to *deconstruct* each definition listed in Table 2.1. As I described earlier (Chap. 2, p. 26), there are several ways to do this, including the use of "syntactic trees" illustrated in Fig. 2.1. These methods, however, would become cumbersome if applied to some of the definitions I have to deal with. So instead, I will use a linear display (Chap. 2, p. 38) that remains sensitive to the syntactic structure of the definition, but presents it in such a way as to make the presence of extensions of meaning from one part of the sentence to another more obvious. Linguists refer to this type of analysis as *anaphoric* (i.e., Tirrell 1989). The components will then be examined to determine if they contain linguistic metaphors, dual usage terms and/or conceptual metaphors. Definitions with a similar figurative structure will then be grouped into categories.

Table 2.3 illustrates the deconstruction process for the Osoba (1994; Table 2.1) definition. Inspection reveals that HRQOL (A) is also described as "a multi-dimensional construct" ($\sum_{1-x} F$) which, if summed, would presumably approximate the global indicator, HRQOL. The definition goes onto state that HRQOL *is produced by* a disease or treatment (DT). Thus, the Osoba (1994) definition can be expressed in a general form as:

(I) A or ($\sum 1-_x F$) is caused by DT.

where A = quality-of-life or HRQOL; ($\sum_{1-x} F$) = the sum of indicators provided; and DT = a disease or treatment. Inspection of Table 2.1 reveals two additional definitions (Revicki et al. 2000; and Shipper et al. 1996) which match this general format (Table 2.4). I have labeled these defini-

Table 2.3 Deconstruction of the Osoba (1994) HRQOL definition

Definition: HRQOL is a multidimensional construct encompassing perceptions of both positive and negative aspects of dimensions, such as physical, emotional, social and cognitive functions, as well as the negative aspects of somatic discomfort and other symptoms *produced by* a disease and its treatment

Symbol	Components
A	HRQOL
($\Sigma_{1-x}F$)	A multidimensional construct encompassing perceptions of both positive and negative aspects of dimensions, such as physical, emotional, social and cognitive functions, as well as the negative aspects of somatic discomfort and other symptoms
⇐	*produced by*
DT	A disease and its treatment

Table 2.4 Deconstruction of various definitions of quality-of-life or HRQOL: *causal models*

Reference	A=($\Sigma_{1-x}F$)		⇐	D
Osoba (1994)	HRQOL	is a multidimensional construct encompassing *perceptions* of both positive and negative aspects of dimensions, such as physical, emotional, social and cognitive *functions*, as well as the negative aspects of somatic discomfort and other *symptoms*	*produced by*	a disease and its treatment
	HRQOL		*produced by*	a disease and its treatment
Shipper et al. (1996)	Quality-of-life in clinical medicine	represents the functional	*effect of*	an illness and its consequences as perceived by the patient
	Quality-of-life		*effect of*	an illness and its consequences
Revicki et al. (2000)	HRQOL is defined as	the *subjective* assessment… across the physical, psychological, social, and somatic domains of functioning and well-being	*impact of*	of the disease and its treatment
	HRQOL		*impact of*	disease and its treatment

Fig. 2.3 Figurative language mapping for causal definitions of quality-of-life or HRQOL. Included are the definitions of Osoba (1994), Shipper et al. (1996), and Revicki et al. (2000).

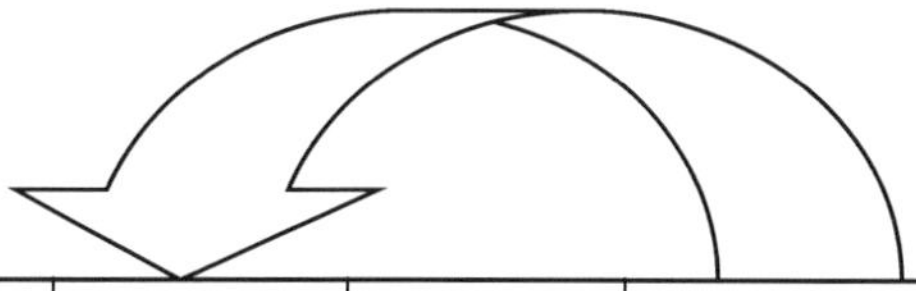

Figurative Form (Author)	Target (abstract)	Connector	Base (concrete)	Qualifier
Metaphor (X is Y)	Life	is	a river.	
Osoba (1994)	HRQOL	produced by	disease or treatment	
Shipper et al, (1996)	Quality of life	effect of	illness or its consequences	
Revicki et al (2000)	HRQOL	impact of	disease and its treatment	

tions as reflecting a "causal" relationship, as between A and DT, and this is based on an analysis of the relationship of the deconstructed components, and the meaning of the phrases *produced by*, *effect of*, or *impact of* (see text below).

First, let me restate these definitions (Table 2.4; leaving out, for the moment, the $[\sum_{1-x} F]$ expression). Thus, the three definitions can be stated as:

(A') "HRQOL ... *produced by* disease or treatment." (Osoba 1994).
(B') "Quality-of-life ... *effect of* an illness and its consequences." (Shipper et al. 1996).
(C') "HRQOL ... *impact of* disease and treatment." (Revicki et al. 2000).

My next task is to apply the rule that Glucksberg (2001) suggested to determine if I am dealing with a linguistic metaphor, which involves determining if the statement retained meaning if it were reversed. Reversing these definitions would read as:

(A'-revised) "Disease or treatment ... produced by HRQOL."
(B'-revised) "Illness or its consequences ... effect of quality-of-life."
(C'-revised) "Disease and its treatment ... impact of HRQOL."

Clearly, these statements have experienced a decrement in meaning, suggesting that I am dealing with linguistic metaphors.

But am I also dealing with conceptual metaphors? Figure 2.3 illustrates these scaled-down definitions, now presented in the format of a conceptual metaphor. Inspection of the figure illustrates that just as the term "river" tells me something about life, so does the phrase "disease or treatment" inform me about the life a person is leading when they are sick and receiving treatment. Thus, meaning is being transported from concrete terms (D) to abstract terms (A = $[\sum_{1-x} F]$), which is characteristic of a conceptual metaphor.

As was illustrated in Fig. 2.3, the *connector* phrases make a unique contribution to the meaning of a phrase. It is also interesting to examine the figurative character of these phrases. For example, *produce* is a verb whose primary definition is "to make, manufacture, or create," and secondary usage is to "cause to happen or exist" (Oxford University Press 1996). The first definition suggests the creation of physical objects, while the second usage suggests the establishment of a nonphysical association between events. Thus, the phrase is a dual usage term, but it is its second usage involving nonliteral (nonprimary) form which appears to be playing a linguistic metaphoric role. This provides further support for the presence of metaphoric properties in the Osoba (1994) definition.

The word *effect*, which has several meanings when used as noun, is used in the Shipper et al. (1996) definition as a verb which has the dictionary definition of, "cause to happen or bring about" (Oxford University Press 1996). The phrase "*effect of*" can easily be seen as having dual usage; alluding to both physical and psychological consequences. However, the phrase in the Shipper et al. (1996) definition appears to be serving a more literal role, since the definition limits what is meant by quality-of-life to the "*functional effect of* an illness and its consequent therapy" (Table 2.4).

Consider the phrase "*impact of*". The word "impact" comes from the Latin *impingere* meaning, "drive something in or at." When used as a verb, the primary literal usage of the word means, "come into forcible contact with another object," but its secondary nonliteral usage is defined as "to have a strong effect on." *Impact of*, therefore, is at a mini-

Table 2.5 Deconstruction of various definitions of quality-of-life or HRQOL: *matching models*

Reference	A	≈	B
Campbell et al. (1976)	Quality-of-life is defined in terms of satisfaction of needs	"level of satisfaction" (*of needs*) can be precisely defined as the perceived *discrepancy* between	aspiration and achievement, ranging from the perception of fulfillment to that of deprivation
	Quality-of-life = satisfaction of needs	*discrepancy* between	aspiration and achievement
Calman (1987)	Quality-of-life therefore measures the	*difference*, at a particular period of time	between the hopes and expectations of the individual and the individual's present experience
	It is concerned with the	*difference* between	perceived goals and actual goals
	Quality-of-life, is concerned with	*difference* between	perceived goals and actual goals
Frisch (1993)	Life satisfaction is equated with *quality-of-life* and refers a person's subjective evaluation	have been *fulfilled*	the degree to which his or her most important needs, goals and wishes
	Quality-of-life = Life satisfaction	have been *fulfilled*	needs, goals and wishes
World Health Organization (1995) (WHOQOL)	Quality-of-life was defined, therefore, as individual's perception of their	*position in life*	in the context of the culture and value systems in which they live and in relation to their goals, expectations, standards, and concerns
	Quality-of-life	*position in life*	their goals, expectations, standards, and concerns

mum a dual usage phrase with an action orientation, and appears to be used in this definition in a nonliteral manner.

Each of the phrases (e.g., "produced by", "effect of", or "impact of"; Table 2.4) is functioning as a verb and communicating actions and consequences. This is also consistent with the phrase's role in describing a causal relationship, as well as in being a predictive metaphor (especially if you accept that the phrase had metaphoric qualities in a particular definition).

Causal definitions of this sort also suggest the presence of a conceptual metaphor independent of the extent linguistic metaphors are present in the definitions. As discussed earlier, Lakoff (1993) discussed THE LOCATION EVENT-STRUCTURE METAPHOR, which included as sub-metaphors Causes Are Forces or Causes Are Movement. They describe a source domain as the domain of forces or motion-in-space (as is in the phrase, "produced by a disease or treatment") and a target as the domain of events (as amongst the components of HRQOL). Including causal metaphors in these definitions (e.g., "produced by," "effect of", or "impact of"; Table 2.4) helps me understand the author's definitions by reference to my personal experiences with the physical consequences of forces or motion in space. Thus, what gives meaning to these definitions is the embodied nature of such phrases as "produced by," "effect of", or "impact of."

Now a concern may be expressed about the value of an analysis of this sort. Thus, it was useful to find that Fernandez-Duque and Johnson (2002)[11] applied this type of analysis in their effort to bring some order to the various models or definitions of attention. Their paper confirmed the presence of conceptual metaphors in both cognitive, psychological and neuroscience models of attention, and illustrated how this type of analysis can bring some conceptual clarity to a relatively diverse and complex subject matter area.

Linguistics use the term *anaphora* (Tirrell 1989) to refer to the extension of meaning from a metaphor to other parts of a sentence or text. Thus, if the phrases "produced by," "effect of", or "impact of" are functioning as linguistic metaphors, then their reference to "disease and treatment" or "physical, psychological, social, and somatic domains of functioning and well-being," as in the Revicki et al. (2000) definition, represents the target of the extension of meaning of the metaphor. This suggests that the structure of a sentence, its syntax, plays an important role in communicating the meaning within a definition or an item of a qualitative assessment (see Chap. 10).

The definitions in Table 2.4 were described as "causal models" since the definitions described forces being applied over space and time. In contrast, the definitions listed in Table 2.5 suggest that people have some sense of the needs or goals that they want to achieve and the extent that they have been achieved, and this determines how satisfied they are with their with their quality-of-life or HRQOL. Each of these definitions capture the cognitive process of a person *matching* what the person ideally *hoped for, expected, aspired to, or set as goals* and with what has been *achieved*. The definitions take the general form of:

$$(\text{II}) \ A \text{ is} \approx B,$$

where A refers to the person's aspirations and B refers to a person's various achievements.

The notion that I may have some sense of what I want my quality-of-life or HRQOL to be raises the interesting

Table 2.6 Deconstruction of the Campbell et al. (1976) definition

Definition: Quality-of-life is defined in terms of satisfaction of needs, and "level of satisfaction" can be precisely defined as the perceived *discrepancy* between aspiration and achievement, ranging from the perception of fulfillment to that of deprivation

Symbol	Components
A	Quality-of-life
≈	"Level of satisfaction" can be precisely defined as the perceived *discrepancy* between aspiration and achievement, ranging from the perception of fulfillment to that of deprivation
B	Satisfaction of needs

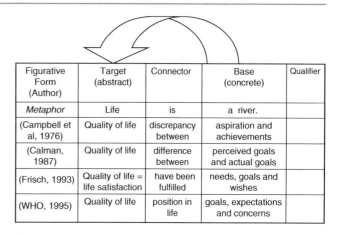

Fig. 2.4 Figurative language (conceptual metaphor) mapping for matching models. Included are the definitions by Campbell et al. (1976), Calman (1987), Frisch (1993), and the WHO (1995) definition of quality-of-life or HRQOL.

question of where this sense comes from. Is it something I was born with? Something that evolved as I experienced life? Or is it something I can be taught to expect? These questions will ultimately have to be answered if Type II (matching) definitions are to play a significant role in quality-of-life or HRQOL research.

Table 2.6 disaggregates and simplifies the Campbell et al. (1976) definition. If I do this for each of the definitions in Table 2.5, then I have the following list:

(IIa) "Quality-of-life…discrepancy between…aspirations and achievements" (Campbell et al. 1976)
(IIb) "Quality-of-life…difference between…perceived goals and actual goals" (Calman 1987)
(IIc) "Life satisfaction is equated with quality-of-life… fulfilled needs goals and wishes" (Frisch 1993)
(IId) "Quality-of-life…position in life…their goals, expectations, standards and concerns" (WHO 1995)

My next task is to determine if these phrases contain *linguistic metaphoric properties*. To do this I will need to reverse the content of each definition and determine if the resultant phrases remain interpretable.

(IIa-revised) "Aspirations and achievements…discrepancy between…quality-of-life."
(IIb-revised) "Perceived goals and actual goals…difference between…quality-of-life."
(IIc-revised) "Needs goals and wishes…fulfilled…life satisfaction is equated with quality-of-life."
(IId-revised) "Their goals, expectations, standards and concerns…position in life…quality-of-life."

Inspecting these phrases suggests that they, as opposed to similes and analogies, do not retain their meanings, suggesting the presence of linguistic metaphoric properties.

Figure 2.4 illustrates the conceptual metaphor mapping that is occurring for each of the definitions. Note that in each case, the base provides concrete information that helps define what is meant by quality-of-life. This is consistent with what would be expected if these definitions have conceptual metaphor properties.

The connector relating the base to the target also plays an important role in these definitions. For example, the literal meaning of the word "discrepancy" (Campbell et al. 1976; Table 2.6) is that "an illogical or surprising lack of compatibility between facts" exists (Oxford Dictionary 1996). In the Campbell et al. (1976) definition, however, the word appears to be being used to describe the extent of fulfillment or level of deprivation (i.e., more like reaching a point as opposed to describing a gap). If so, then the term discrepancy is not being used in its primary literal sense.

A similar analysis can be done of the phrases "difference between hopes and expectations" or "difference between perceived goals and actual goals" (Calman 1987; Table 2.5). A dictionary definition of the word "difference" refers to the "way in which people or things are dissimilar; the state or condition of being dissimilar; a disagreement, quarrel, or dispute; or the remainder left after subtraction of one value from another" (Oxford University Press 1996). Yet the term is being used in the Calman (1987) definition to refer to what is left over from the subtraction of one's "hopes and present experience," or "perceived goals and actual goals." Clearly, a difference can be based on the gap between physical values or subjective experiences. However, the meaning I derive from the second usage comes from my familiarity with the first usage, and this, of course, is characteristic of the role that physical metaphors play in the embodiment of affect (e.g., hope, perceived goals, or desires).

Both phrases ("discrepancy between" or "difference between") function in their respective definitions to relate antecedent and consequential conditions that could not otherwise be reversed (Table 2.5) and remain interpretable. As illustrated above (Sentences IIb and IIb-revised), one can't easily reverse the elements in the Calman (1987) definition and expect to generate a statement that has comparable meaning. Thus, to state that, "hopes and expectations can be measured by differences in quality-of-life," is much less meaningful than saying that "quality-of-life can be measured by differences in hope and expectations."

The Frisch (1993) definition has a form very similar to the Campbell et al. (1976) definition. In both, "quality-of-life" is defined in terms of "satisfaction of needs, goals, and wishes." Frisch is using the word "fulfilled" in much the same manner as the position in life is used in the WHO definition of quality-of-life (Table 2.5); both refer to reaching some condition or state. Both terms also rely on reference to physical as well as psychological applications to provide their meaning.

Lakoff (1993) describes achieving a state of fulfillment or position in life as indicative of the A PURPOSEFUL LIFE IS A JOURNEY METAPHOR. Reference to aspirations, needs, and goals suggests that implicit in these definitions of quality-of-life or HRQOL is a motivational construct that is dependent on the A PURPOSEFUL LIFE conceptual metaphor. Inspection of Maslow's (1968) hierarchy of needs theory will make this association clearer.

Maslow (1968) claims that five basic sets of needs exist, and that satisfaction with life progresses from resolving physiological needs (e.g., adequate food, comfort) to maximizing self-actualization, with the achievement of safety, love and self-esteem as intermediate levels.[12] Presumably, maximizing self-actualization would also include maximizing one's quality-of-life or HRQOL. There is also the assumption that at each level, lower level needs have been satisfied and are included at the next higher level of need. The assumption of a hierarchy implies that past achievement of certain needs or goals are required before self-actualization occurs. This directionality, which is characteristic of the LIFE IS A JOURNEY and EVENT STRUCTURE conceptual metaphor, is present in Maslow's hierarchical motivational model.

Lakoff and Johnson (1999) illustrate how the selective or directional quality of a metaphor can affect the meaning communicated by the metaphor. For example, the statement, "He is a *cold* and unfeeling person" contains a linguistic metaphor, *cold*, which differentiates this person from others and encourages avoidance of the person. Contrast this with the statement, "He is a wide and unfeeling person," which is less likely to evoke an avoidance response. Thus, the presence of the metaphor "*cold*" adds a directional quality to the statement that would not otherwise be present. The same is true for Maslow's model; consider his model as if it were presented without the hierarchy metaphor. The model would then consist of five categories that by themselves would not suggest that a journey is being undertaken. Thus, the meaning Maslow wishes to communicate in his model requires the presence of the conceptual metaphor, as is apparently true for theories of motivation and attribution, in general (Weiner 1991). The same can also be said for the quality-of-life or HRQOL definitions listed in Table 2.5.

The definitions listed in Table 2.7 refer to a person being in a particular *state of well-being*. The term *well-being* is used in several different ways, and I will have an extensive discussion of these usages in Chap. 11. Three of the definitions listed in Table 2.7 (Cella 1995; Ebrahim 1995; Padilla et al. 1996) allude to disease or treatment affecting a person's *well-being*, that is then used as an indicator of HRQOL. These definitions are similar to the *causal* definitions listed in Table 2.4, except that *well-being* has been substituted for an indicator of quality-of-life or HRQOL. The definitions would be formally described as:

$$(III)\ A = C$$

where $C = (\sum_{1-x} G)$; and $C =$ is caused by DT

In this definition A = quality-of-life or HRQOL; C = well-being; $(\sum_{1-x} G) =$ the sum of well-being indicators; and DT = disease or treatment. The Padilla et al. (1996) definition (Table 2.1) states,

> "HRQOL is defined as a personal, evaluative statement summarizing the positivity and negativity of attributes that characterize one's psychological, physical, social and spiritual *well-being* at a point in time when health, illness and treatment conditions are relevant."

The same basic form would characterize the Cella (1995) and Ebrahim (1995) definitions. The remaining three definitions are *descriptive* definitions specifying, in more detail, what is meant by *well-being*. They can formally be described as,

$$(IV)\ A = C;$$

where $C = (\Sigma_{1-x} G)$

The Gotay et al. (1992) definition (Table 2.1) illustrates this type of definition. The deconstructed definition states that,

> (IIIb) "Quality-of-life is *a state of well-being* which is a composite of two components..."

and the definition goes on to describe the elements of these components.

The definitions listed in Table 2.7 function as *metonyms*, where *a metonym* (Chap. 11, p. 400.) is a figurative word or expression that is used to substitute for another well-known word or expression. In these definitions the term well-being plays this role relative to the phrase *quality-of-life*. Whether this is an appropriate substitution is discussed in Chap. 11. Three of the definitions also include examples of the metaphoric content of causal definitions, and since this has already been discussed (see above) I will not discuss it here. Do these definitions contain other examples of metaphoric content?

Linguistically, well-being can be defined as "good existence." Lakoff and Johnson (1999; p. 290–334) point out, however, that the phrase "good existence" is intimately involved in what is meant by "moral." They state that well-being, as a conceptual metaphor, refers to what is physically best for me and how I ought to live. This sets my moral ideals,

Table 2.7 Definitions of quality-of-life: *Well-being models*

Reference	A=C	C=($\sum_{1-x} G$)	DT
Ferrans (1990)	Quality-of-life is defined as a person's *sense of well-being* which stems from satisfaction or dissatisfaction	With the areas of life that are important to him/her	
	Sense of well-being	Areas of life	
Gotay et al. (1992)	Quality-of-life is *a state of well-being* which is a composite of two components	The ability to perform everyday activities which reflect physical, psychological and social well-being, and	
		Patient satisfaction with levels of functioning and the control of disease and/or treatment-related symptoms	
	A state of well-being	The ability to perform everyday activities which reflect physical, psychological and social well-being, and satisfaction	
Cella (1995)	HRQOL refers to the extent to which one's usual or expected physical, emotional and social *well-being*		Affected by a medical condition or its treatment
	Well-being		Medical condition or its treatment
Ebrahim (1995)	HRQOL may be thought of as those aspects of self-perceived *well-being*		That are related to or affected by the presence of disease or treatment
	Well-being		Presence of disease or treatment
Felce and Perry (1996)	Quality-of-life is defined as an overall *general well-being*	That is comprised of objective and subjective evaluations of physical, material, social, and emotional *well-being,* together with the extent of *personal development and purposeful activity,* all weighted by a personal set of values	
	General well-being	Objective and subjective evaluations of physical, material, social, and emotional *well-being*	
		Personal development and purposeful activity	
		All weighted by a personal set of values	
Padilla et al. (1996)	HRQOL is defined	As a personal *evaluative* statement summarizing the positivity and negativity of attributes that characterize one's psychological, physical, social and spiritual *well-being*	At a point in time when health illness and treatment conditions are relevant
		Well-being	Health, illness and treatment conditions

as when I define justice, fairness, virtue, freedom, rights, and so on. Thus, I prefer to be healthy rather than sick, wealthy than poor, free than enslaved, having rights than being denied rights, nurtured than isolated – all of which contributes to my sense of well-being. An increase in well-being is a gain, and a decrease is a loss. If I am sick, as Sontag (1991) indicated, I may be judged to be immoral, since the onset of a disease may be associated with a person's behavior.

A variety of conceptual metaphors can be used to define moral concepts. Lakoff and Johnson (1999; p. 293) discusses moral accounting as an example of a metaphor that can also be applied to characterize the dynamics of well-being. Thus, when Cella (1995) defines HRQOL as referring to the extent "one's usual or expected physical, emotional and social well-being" (Table 2.1) are affected by a medical condition or treatment, he is alluding to an implicit accounting exercise in which a decrease in well-being would imply a less then moral outcome, a less than optimal well-being. The phrase "usual or expected....well being" also implies that some baseline assessment of well-being is possible, which may change in response to a medical condition or treatment. Accounting seems to be an implicit element of each of the definitions in Table 2.7, and in this sense each uses a conceptual metaphor to convey a moral intent.

Diener (2006) describes the term well-being as an umbrella term that can assume a variety of meanings [including those suggested by Lakoff and Johnson (1999)], depending on how it is being used. As I explain in Chap. 11, this conceptualization of the term makes the term more of a state indicator than an evaluative term, as is characteristic of the term quality or the phrase quality-of-life. What appears to be true, however, is that the term is used in common discourse as a quality indicator, and one of the questions I address in Chap. 11 is whether this is appropriate.

The Ferrans (1990) and Felce and Perry (1996) definitions (Table 2.7) also include reference to what a person considers important and what they value. Both definitions acknowledge that more is needed in a definition of quality-of-life or HRQOL then description or even implicit moral accounting. Rather, what is needed is an explicit estimate of the importance or value a person attaches to the various states (e.g., health states) they are in. The Patrick and Erickson (1993) definition of HRQOL, as stated in Table 2.8[13] provides the clearest example of this by making the value statement the central focus of their definition, while the descriptive portion of the definition modifies the value statement. In contrast, the Ferrans (1990) and Felce and Perry (1996) definitions (Table 2.7) focus primarily on the relationship between quality-of-life or HRQOL and well-being.

Inspecting the Patrick and Erickson (1993) definition reveals that it consists of several elements which can be formally defined as follows:

$$(V)\ A = (\sum_{1-x} F)\ \text{caused by DT},$$

where $H = V \times A$

In this definition, V refers to the value assigned; × refers to an operator (e.g., addition, multiplication, and so on); $(\sum_{1-x} F)$ refers to sum of the descriptors (e.g., impairments, social opportunities, and so on) which modify a value statement; and DT refers to disease, injury, treatment and policy. What is interesting about this definition is that it contains both a value statement and a causal definition; two independently derived indicators. To demonstrate this I will deconstruct the definition in stages (Table 2.8).

The key phrase in this definition, from a linguistic perspective, is of course, "the value assigned." Value is another dual usage term, meaning that it has both physical and psychological referents, as well as literal and nonliteral references. Value assigned suggests that some quantity is being attached to an entity, as may occur when a source or base is mapped to a target or a number is assigned to a length of life. The first mapping is characteristic of a metaphor, while the second requires the application of a "known production rule" or an operator (e.g., addition, multiplication) in order to occur. The two differ in that the quantity being assigned to a length of life can't be confirmed by a method that is independent of the person. For example, length of life can be assessed by asking the person how old they are, but also by examining their birth certificate and the current date. In contrast, a person's valuation of their life has no independent basis for con-

Table 2.8 Definitions of quality-of-life: value statements

Reference	$A = (\sum_{1-x} F) \Leftrightarrow DT$		
Patrick and Erickson (1993)	Influenced by		
	Influenced by		
	$V \times A = (\sum_{1-x} F) \Leftrightarrow DT$		
HRQOL is	The value assigned to duration of life	As modified by	*Impairments, functional states, perceptions, and social opportunities influenced by disease, injury, treatment or policy*
HRQOL is	The value assigned	As modified by	*impairments, functional states, perceptions, and social opportunities influenced by disease, injury, treatment or policy*

firmation. Thus, Equation (V) consists of two quite disparate assessments, which makes this definition an example of a hybrid construct (Chaps. 3, 4, 6 and 8). What is of interest now is to ask the linguistic question of what it means to have a number which is based on such disparate sources.

A number is another example of a word whose meaning I have come to understand because of its association with physical objects. Thus, it has an embodied nature which I started acquiring when I was child. Then I learned that three objects can be represented by the number "3," and I also learned to use numbers to represent the consequences of collecting or splitting groups of objects (addition and subtraction), or to simplify my collecting and splitting (multiplication and division).

When someone is asked to value their life, they may be given a piece of paper with a 100-cm line, and be and asked to place a mark on it, or they may be given a set of anchors ("perfect quality-of-life" and "death") and asked to imagine a line that they are to select a position on. In both cases what is being asked of the person is to rely on their experience with a line segment to participate in this task. I am familiar with line segments, since I have used parts of body (my arm, as in arms' length; fingers, as in width between fingers; foot length, as three feet placed in succession to assess a yard, and so on) to provide a unidimensional and continuous assessment. Lakoff and Nunuz (2000; p. 68–71) suggest that these experiences with physical line segments form the basis of THE MEASURING STICK METAPHOR. In its abstract form, they state that this metaphor has been essential to my understanding of Euclidean geometry and irrational numbers. In the current context, it illustrates that my ability to generate a number to represent a value is based on my physical experiences with line segments.

The formal form of the Patrick and Erickson (1993) definition comes closest to demonstrating that it is possible to have a unique definition for quality-of-life or HRQOL. It does this by combining description with individual explicit valuation. In contrast, if you were to substitute the phrase, "psychosocial variables" for quality-of-life or HRQOL in the definitions listed in Tables 2.4 and 2.5, they would still make sense.[14] This would also be true for the three "causal" well-being definitions in Table 2.7, and the remaining three definitions satisfy the requirement for description but they do not include a valuation component.

I took a somewhat different approach to the definitional task (Barofsky 1990). I suggested that since a quality or quality-of-life assessment is a process that a person can perform, it should be studied in its own right, and that any definition should be based on what is learned about how a person performs this task. This experiential-based approach is in contrast with virtually all of the definitions I have considered so far that have been either experimentally (e.g., based on factor analysis of questionnaire results) or arbitrarily defined by investigators. Thus, to define quality-of-life or HRQOL involves, "the process whereby values are quantified, associated with descriptors and consequences and incorporated into critical human decisions" (Barofsky 1990).[15] Besides identifying who defines a quality or quality-of-life assessment, this definition also raises the question of how "operational" the proposed definition should be. Most of the definitions I have considered describe what is essentially a descriptive task: collecting specific types of content such as listing various types of functions, assessments of well-being, and so on. In contrast, the Barofsky definition is content independent, suggesting instead that a unique yet universal cognitive process can be identified (Barofsky 2003; Chap. 12).

Some investigators argue that while they do not directly assess a person's values or preferences, they have done so indirectly by selecting items which implicitly address these issues. Thus, if a questionnaire is developed that addresses "the impact of cancer," this questionnaire is assumed to reflect the person's values or preferences, since "impact" implies a negative valuation or preference. This, of course, is not necessarily true, since the impact of cancer is not uniformly bad, but gives the person the opportunity to adjust to their disease and its consequences can, on occasion, facilitates personal growth and development.

4 Summary of Chapter

To appreciate the role that language plays in a qualitative assessment, all an investigator need do is inspect the language used in the items that make up the questionnaires and surveys that they administer. Gannon and Ostrom (1996) present such a study entitled, "How meaning is given to rating scales: The effects of response language on category activation." Their paper, which addresses response language from a linguistic perspective, also utilizes cognitive concepts that will be introduced in the next chapter. The model they propose:

> …posits that semantic end labels on rating scales activate corresponding cognitive categories. Take for example the case where the researcher needs to measure perceptions of political candidates on an honest versus dishonest continuum. If a rating scale is labeled from *not at all honest* to *very honest*, the category honest is activated. If the same question is asked, but instead a scale is given which is labeled *not at all dishonest* to dishonest, the category dishonest is activated." (Gannon and Ostrom 1996; p.338).

This, they say, suggests that the category dishonest is not the opposite of honest, and that this occurs not just because different words are being used but also because different cognitive categories are involved. Dichotomies of this sort are common, and include such examples as "pain and pleasure" and "happiness and sadness." Their discussion reinforces the notion that language expression and cognition can differ, and this happens to also explain the difference between a linguistic metaphor and a conceptual metaphor. Gannon and Ostrom's (1996) and Ostrom and Gannon's (1996) models also take advantage of Rosch's (1973) notion that a category such as honesty consists of a series of terms, some of which are better examples of the category than others, and they report a series of experiments that support this model.

The first part of this chapter reviewed principles of language usage, but I have now pointed out that language usage represents only one set of tools that will be needed to define and assess qualitative states. That a language usage analysis can be quite useful was demonstrated by my attempt to bringing some order to the confusing array of definitions that have been offered to define quality-of-life or HRQOL. While I did not deal with all known definitions (e.g., Rapkin and Schwartz 2004[16]) those I have reviewed could be classified into one of four categories. These categories can be summarized as follows:

(I) $A = (\sum_{1-x} F)$, is caused by DT (Causative definitions)
(II) $A is \approx B$ (Matching definitions)
(III) $A = (\sum_{1-x} F)$ (Descriptive definitions)
(V) $H = V \times A$ (Valuation definitions)

I also pointed out that not all classes of definitions are equal. Thus, including a valuation statement in a quality-of-life or HRQOL definition makes that class of definitions unique. In the absence of a unique indicator, quality-of-life or HRQOL definitions can be easily confused with other definitions (e.g., definitions that include psychosocial adjustment variables). This distinction is also relevant to the issue discussed in Chap. 1, where I asked if a qualitative assessment is meant to prevent (quality control) or enhance the consequences of events, processes or life. It is also relevant to whether a well-being statement can be used as an expression of quality-of-life; a statement which, I would argue, is not obvious (Chap. 11), especially if the purpose of the assessment is the prevention of adverse events.

Most important was my discussion concerning conceptual metaphors and how they helped me understand how language was being used in these quality-of-life or HRQOL definitions. What appears to be true is that while the linguistic metaphoric content in definitions may vary, they consistently contain conceptual metaphors. For example, Definitions I and II (Table 2.4) use the EVENT STRUCTURE Metaphor. Thus, Causes Are Forces or Causes Are Movement metaphors helped me see the linkage between the phrases "provided by," "effect of," "impact of" and cause–effect relationship being posited in the definitions in Table 2.4. The A PURPOSEFUL LIFE IS A JOURNEY METAPHOR helped me understand that the definitions in Table 2.5 were referring to reaching a goal or objective. I also noted that the term well-being was being substituted for quality-of-life or HRQOL in a number of definitions (Table 2.7), and raised the issue of whether this type of substitution enhanced or complicated the usefulness of the resultant definition. I will discuss this more in Chap. 11, where it will be made clear that well-being is a term that is a state indicator, although it could assume a number of other semantic roles (e.g., as an evaluator and if "qualified" as a contributor to a qualitative assessment). I also noted in this chapter the relationship between well-being and moral accounting, suggesting that if this accounting were "qualified" it could be used to express the inherent moral purpose of a quality or quality-of-life assessment. Finally, I discussed how THE MEASURING STICK METAPHOR helped me understand how I could produce a rating, or imagine a line and select a position on it to reflect my values.

More generally, my discussion about the difference between literal and nonliteral language usage gave me the opportunity to discuss "how I come to know", an issue that was captured in the debate between the nonconstructionists and constructionists. I will return to all of these topics again as I continue to consider how I define and assess quality, quality-of-life, or HRQOL.

Finally, it is important to restate that examining the language of a quality assessment is important because of both the analytical and monitoring opportunities it creates, and that as a result it provides unique research opportunities for an investigator. The next chapter will add another set of conceptual tools for the discussion to follow, by describing some of the cognitive processes that may be involved during a qualitative assessment.

Notes

1. I have not been able to find a similar list of definitions for the term "quality," so I will confine this discussion to different definitions of "quality-of-life" or "HRQOL."
2. Glucksberg (2001) view, of course, has strong resemblance to what Wittgenstein was telling us.
3. For a continuation of this discussion, see Chap. 4's discussion about objective and subjective indicators of quality-of-life. Also, Raphael (1996) provides an overview of some of the philosophical alternatives relevant for quality-of-life or HRQOL research.
4. The relationship between language used and reality is an important issue which I will address at various places in this book. I will argue that the spoken and written language that I know is deficient in characterizing the experience of quality, especially of the compromised person, and that the more precise language of mathematics, and its analog, a *conceptual space*, may offer an alternative means of characterizing quality. I will discuss this most directly in Chap. 10. Also, in Chap. 4 (p. 97) I will discuss details about the relationship between objectivity and subjective assessments. I will also argue in Chap. 4 that the assessment process itself tends to optimize the objective character of any observation, independent of its subjective or objective origin. This, however, is not contradicted by my argument here that my sense of reality is a form of constructionism. What is assessed is not necessarily the same as what I consider reality, although it is also clear that the two may overlap.
5. Although I have distinguished between a metaphysical and empirical forms of constructionism, I will use the term "constructionism" from now on to only mean "empirical constructionism."
6. See the next section and Chap. 3, where this idea of acquired reality is expanded during a discussion of Hayek's contribution to the understanding of cognitive processes.
7. Miller (1993) states that terms such as encoding, mapping, categorization, inference, attribution, and so on are all examples of appreception. Thus, appreception can be thought of as a "superordinate term" referring to several mental processes.
8. Glucksberg (2001) refers to this as "dual referencing," where a term may have a concrete reference, such as a law limiting the freedom of a class of citizens, or as a superordinate category referring to a wide range of restraints on a person's freedom.
9. As an alternative, Steen (1999) describes a *linguistic* approach to conceptual metaphor analysis. Steen would like to be able to determine if a particular metaphoric expression is a conceptual metaphor (so called one-shot metaphors), without alluding to broader issues. Thus, he is interested in *metaphor analysis* as opposed to *metaphoric understanding*. He identified five steps which he suggests are sufficient to do *a linguistic analysis of a conceptual metaphor*.
10. The association between perceptual processes and affect is important and I will discuss it when I discuss the topic of well-being in Chap. 11.
11. I will discuss the Fernandex-Duque and Johnson (2002) paper in more detail in Chap. 7.
12. Koltko-Rivera (2006) has recently suggested that Maslow had another level to his motivational hierarchy in mind, beyond self-actualization, that he referred to as "self-transcendence."
13. Patrick and Chiang (2000) have presented an expanded version of this definition which seems designed for broader policy oriented issues and will be discussed in Chap. 8.
14. The reader is probably aware that many investigators fail to differentiate quality-of-life and psychosocial indicators when they describe what they mean by a quality-of-life or HRQOL assessment. One of the reasons for this has to do with the fact that these investigators feel that descriptive statements are sufficient to define quality-of-life or HRQOL, the consequence of which is that there is no operational consequences to using either class of indicators.
15. Feldstein (1991) pointed out that the quote from Barofsky (1990) was originally reported in 1983 at a *Conference on Methodology in Behavioral and Psychosocial Cancer Research* in response to comments by Wellisch (1984).
16. The Rapkin and Schwartz (2004) definition proposes a complex cognitive structure that will be discussed in Chap. 11.

References

Adams FM, Osgood CE. (1973). A cross-cultural study of the affective meanings of color. *J Cross-Cultural Psychol.* 4, 135–56.
Aristotle. (1996). *Poetics*. Heath M, Translator. New York NY: Penguin Books.
Asch S. (1958). The metaphor: A psychological inquiry. In, (Eds.) R Tagiuri, L Petrullo. *Person Perception and Interpersonal Behavior*. Stanford CA: Stanford University Press. (p. 86–94).
Barofsky I. (1990). Conceptual issues in quality of life assessments. Unpublished paper presented at the *15th International Cancer Congress*, Hamburg, FRG, August 16–22, 1990.
Barofsky I. (2003). Cognitive approaches to summary measurement: Its application to the measurement of diversity in health-related quality of life assessments. *Qual Life Res.* 12, 251–60.
Bowdle BF, Gentner D. (2005). The career of metaphor. *Psychol Rev.* 112, 193–216.
Bridgman PW. (1959). *The Way Things Are*. New York NY: Viking Press.
Brunswik E. (1956). *Perception and the Representative Design of Psychological Experiments*. Berkeley CA: University of California.
Calman KC. (1987). Definitions and dimensions of quality of life. In, (Eds.) NK Aaronson, JH Berkman. *The Quality of life of Cancer Patients*. New York NY: Raven Press. (p. 1–9).
Campbell A, Converse PE, Rodgers WL. (1976). *The Quality of American Life*. New York NY: Russell Sage Foundation, Rutgers University Press.
Carnap R. (1956). *Meaning and Necessity: a Study in Semantics and Modal Logic*. Chicago, Ill: University of Chicago Press.
Cella D. (1995). Measuring quality of life in palliative care. *Sem Oncol.* 22, 73–81.
Chomsky N. (1965). *Aspects of the Theory of Syntax*. Cambridge, MA: The MIT Press.
Diener E. (2006). Guidelines for national indicators of subjective well-being and ill-being. *J Happiness Stud.* 7, 397–404.
Ebrahim S. (1995). Clinical and public health perspectives and applications of health-related quality of life measurement. *Soc Sci Med.* 41, 1383–94.
Falkenhainer B, Fortus KD, Gentner D. (1989). The structure-mapping engine: Algorithm and examples. *Art Intel.* 41, 1–63.
Feldstein ML. (1991). Quality of life-adjusted survival for comparing cancer treatments; A commentary on TwiST and Q-TWiST. *Cancer*. 67, 851–54.
Felice D, Perry J. (1996). Exploring current conceptions of quality of life: A model for people with and without disabilities. In, (Eds.) R Rewick, I Brown, M Nagler. *Quality of life in Health Promotion and Rehabilitation: Conceptual Approaches, Issues, and Applications*. Thousand Oaks CA: Sage. (p. 51–62).
Fernandez-Duque D, Johnson ML. (2002). Cause and effect theories of attention. The role of conceptual metaphors. *Rev Gen Psychol.* 6, 153–165.
Ferrans CE. (1990). Development of a quality-of-life index for patients with cancer. *Oncol Nurs Forum.* 17, 29–38.
Ferrans CE. (2005). Definitions and conceptual models of quality of life. In, (Eds.) J Lipscomb, CC Gotay, C Synder. *Outcomes Assessment in Cancer*. New York NY: Cambridge University Press. (p. 14–30).
Fiedler K. (2000). Beware of samples! A cognitive-ecological sampling approach to judgment biases. *Psychol Rev.* 107, 659–76.
Fodor J. (1975). *The Language of Thought*. New York NY: Thomas Y Crowell.
Forbus KD, Gentner D, Law K. (1995). MAS/FAC: A model of similarity-based retrieval. *Cogn Sci.* 19, 141–205.
Frank AW. (2001). Can we research suffering? *Qual Health Res.* 11, 353–62.
Frege, G. (1892/1966). On Concept and Object. In, (Eds.) P Geach, M Black. *Translations from Philosophical Writings of Gottlob Frege*. Oxford UK: Oxford University Press. (p. 42).
Frisch MB. (1993). The quality of life inventory: A cognitive-behavioral tool for complete problem assessment, treatment planning, and outcome evaluation. *Behav Therapy.* 16, 42–44.
Gannon KM, Ostrom TN. (1996). How meaning is given to rating scales: The effect of response language on language on category activation. *J Exp Psychol Soc Psychol.* 32, 337–360.
Gardner H, Winner E. (1978). The development of metaphoric competence: Implications for humanistic disciplines. In, (Ed.) S Sacks. *On Metaphor*. Chicago IL: University of Chicago Press. (p. 121–140).
Gentner D. (1983). Structure-mapping: A theoretical framework for analogy. *Cogn Sci.* 7, 155–170.
Gentner D, Wolff P. (1997). Alignment in the processing of metaphor. *J Mem Lang.* 37, 331–355.
Gergen MM. (1985). The social constructionist movement in modern psychology. *Am Psychol.* 40, 266–275.
Gibbs RW. Jr. (1984). Literal meaning and psychological theory. *Cogn Sci.* 8, 275–304.
Gibbs RW Jr. (2006). *Embodiment and Cognitive Science*. New York NY: Cambridge University Press.
Gigerenzer G. (1991). From tools to theories: A heuristic of discovery in cognitive psychology. *Psychol Rev.* 98, 254–267.
Giora R. (2003). *On our Mind: Salience, Context, and Figurative Language*. Oxford UK: Oxford University Press.
Giora R. (1997). On the priority of salient meanings: The graded salience hypothesis. *Cogn Ling.* 8, 183–206.
Glucksberg S. (2001). *Understanding Figurative Language: From Metaphor to Idioms*. Oxford UK: Oxford University Press.
Glucksberg S. (2003). The psycholinguistics of metaphor. *Trends Cogn Sci.* 7, 92–96.
Glucksberg S, Keysar B. (1990). Understanding metaphorical comparison: Beyond similarity. *Psychol Rev.* 97, 3–18.
Glucksberg S, McGlone MS, Manfredi D. (1997). Property attribution in metaphor comprehension . *J Mem Lang.* 36, 50–67.
Gotay C, Korn E, McCabe M, Moore TD, Cheson BD. (1992). Quality of life assessment in cancer treatment protocols: Research issues in protocol development. *J Natl Cancer Inst.* 84, 575–79.
Grice HP. (1957). Meaning. *Philosop Rev.* 66, 377–88.
Gwyn R. (1999). "Captain of my own ship" Metaphors and the discourse of chronic illness. In, (Eds.) L Cameron, G Low. *Researching and Applying Metaphor*. Cambridge, UK: Cambridge University Press. (p. 203–220).
Jackendoff R. (1995). The boundaries of the lexicon. In, (Eds.) M Everaert, E van den Linden, A Schenk, R Schreuder. *Idoms: Structural and Psychological Perspectives*. Hillsdale NJ: LEA. (p. 133–166).
Kaplan RM, Bush JW. (1982). Health-related quality of life for evaluation research and policy analysis. *Health Psychol.* 1, 61–80.
Koltka-Rivera ME. (2006). Rediscovering the later version of Maslow's hierarchy of needs; Self-transcendence and opportunities for theory, research, and unification. *Rev Gen Psychol.* 10, 302–317.
Lakoff G. (1980). The metaphoric structure of the human conceptual system. *Cogn Sci.* 4, 195–208.
Lakoff G. (1993). The contemporary theory of metaphor. In, (Ed.) A Ortony. *Metaphor and Thought*. 2nd edition, New York NY: Cambridge Press. (p. 202–251).
Lakoff G, Nuñez RE. (2000). *Where Mathematics Comes From*. New York NY: Basic Books.
Lakoff G, Johnson M. (1999). *Philosophy in the Flesh: The Embodied Mind and its Challenge to Western Thought*. New York NY: Basic Books.
Lewin K. (1951). *Field Theory in Social Sciences: Selected Theoretical Papers*. New York NY: Harper.

Lowell R. (1977). *Day by Day*. New York NY: Farrar Straus & Giroux. (p. 117).

Marschark M, Katz A, Paivio A. (1983). Dimensions of metaphors. *J Psychol Res*. 12, 17–40.

Maslow AH. (1968). *Towards a Psychology of Being*. Princeton NJ: Van Nostrand.

McGlone MS (2001). Concepts as metaphors. In, (Ed.) S Glucksberg. *Understanding Figurative Language: From Metaphor to Idioms*. Oxford UK: Oxford University Press. (p. 90–107).

Mehl M, Pennebaker J. (2004). *Unpublished Research*.

Meier BP, Robinson MD. (2004a). Why the sunny side is up: Association between affect and vertical position. *Psychol Sci*. 15, 243–47.

Meier BP, Robinson MD. (2005). The metaphorical representation of affect. *Metaphor Symbol*. 20, 239–257.

Meier BP, Robinson MD, Clore GL. (2004b). Why good guys wear white: Automatic inferences about stimulus valence based on color. *Psychol Sci*. 15, 82–87.

Meier BP, Robinson MD, Crawford LE, Ahlvers WJ. (2007). When "light" and "dark" thoughts become light and dark responses: Affect biases brightness judgments. *Emotion*. 7, 366–376.

Miller GA. (1993). Images and models, similes and metaphors. In, (Ed.) A Ortony. *Metaphor and Thought*. 2nd edition, New York NY: Cambridge Press. (p. 357–400).

Murphy GL. (1997). Reasons to doubt the present evidence for metaphoric representation. *Cogn*. 62, 99–108.

O'Grady W, Dobrovolsky M, Aronoff M. (1989). *Contemporary Linguistics: An Introduction*. New York NY: St. Martin Press.

Ortony A. (1979). Beyond literal similarity. *Psychol Rev*. 86, 161–180.

Ortony A. (1993). Metaphor, language and thought. In, (Ed.) A Ortony. *Metaphor and Thought*. 2nd edition, New York NY: Cambridge Press. (p. 1–16).

Osoba D. (1994). Lessons learned from measuring health-related quality of life in oncology. *J Clin Oncol*. 12, 508–516.

Ostrom TM, Gannon KM. (1996). Exemplar generation: How respondents give meaning to rating scales. In, (Eds.) S Sudman, N Schwarz. *Answering Questions: Methodology for Determining Cognitive and Communicative Processes in Survey Research*. San Francisco CA: Jossey-Bass. (p. 293–318).

Oxford English Dictionary (1996). Oxford UK: Oxford University Press.

Padilla GV, Grant MM, Ferrell BR, Presant CA.(1996). Quality of life: Cancer. In. (Ed.) B Spilker. *Quality of life and Pharmacoeconomics in Clinical Trials*. 2nd edition. New York NY: Raven Press. (p. 301–309).

Patrick DL, Chiang Y-P. (2000). Measurement of health outcomes in treatment effectiveness evaluations. *Med Care*. 38 (Suppl II), II14- II25.

Patrick DL, Erickson P. (1993). *Health Status and Health Policy: Quality of life in Health Care Evaluation and Resource Allocation*. New York NY: Oxford University Press.

Pinker S. (1994). *The Language Instinct: How the Mind Creates Language*. New York NY: Morrow.

Raphael D. (1996) Defining quality of life: Eleven debates concerning in measurement. In, (Eds.) R Renwick, I Brown, M Nagle. *Quality of life in Health Promotion and Rehabilitation: Conceptual Approaches, Issues, and Applications*. Thousand Oaks CA: Sage.

Rapkin BD, Schwartz CE. (2004). Toward a theoretical model of quality of life appraisal: Implications of findings from studies of response shift. *Health Qual Life Outcomes*. 2: 14. doi:10.1186/1477-7525-2-14. (p. 1–12).

Revicki D, Osoba D, Fairclough D, Barofsky I, et al. (2000). Recommendations on health-related quality of life to support labeling and promotional claims in the United States. *Qual Life Res*. 9, 887–900.

Roediger HL. (1980). Memory metaphors in cognitive psychology. *Mem Cogn*. 8, 231–246.

Rosch E. (1973). On the internal structure of perceptual and semantic categories. In, (Ed.) TE Moore. *Cognitive Development and Acquisition of Language*. New York NY: Academic Press. (p. 111–140).

Scheper-Hughes N, Lock M. (1986). Speaking 'truth' to illness: Metaphors. Reification and a pedagogy for parents. *Med Anthropol Quart*. 17, 137–140.

Schott GD. (2004). Communicating the experience of pain: The role of analogy. *Pain* 108, 209–212.

Schwizer H. (1995). To give suffering a language. *Literat Med*. 14, 210–221.

Searle JR. (1969). *Speech Acts*. Cambridge, UK: Cambridge University Press.

Searle JR. (1995). *The Construction of Social Reality*. New York NY: Free Press.

Shipper H, Clinch J, Olweny C. (1996). Quality of life studies: Definitions and conceptual issues. In, (Ed.) B Spilker. *Quality of life and Pharmacoeconomics in Clinical Trials*. Philadelphia, PA: Lippincott-Ravens. (p. 11–23).

Skinner BF. (1987). Whatever happened to psychology as the science of behavior? *Am Psychol*. 42, 780–786.

Sontag S. (1991). *Illness as Metaphor and AIDS and its Metaphors*. New York NY: Picador USA.

Steen GJ. (1999). From linguistic to conceptual metaphor in five steps. In, (Eds.) RW Gibbs Jr, GJ Steen.. *Metaphors in Cognitive Linguistics*. Amsterdam, The Netherlands: John Benjamins Press. (p. 57–77).

Stern J. (2000). *Metaphor in Context*. Cambridge MA: Bradford/MIT Press.

Thurstone LL. (1927). A law of comparative judgment. *Psychol Rev*. 34,273-286.

Tirrell L. (1989). Extending: The structure of metaphor. *NOÛS*. 23, 1–34.

Tversky A. (1977). Features of similarity. *Psychol Rev*.; 84: 327–52.

Weiner B. (1991). Metaphors in motivation and attribution. *Am Psychol*. 46, 921–930.

Wellish DK. (1984). Work, social, recreation, family and physical states. *Cancer Suppl*; 53, 2290–2302.

World Health Organization. (1995). *The World Health Organization Quality of life Assessment- 100.*, Geneva Switzerland: The WHOQOL Group, Program on Mental Health.

Zuriff G. (1998). Against metaphysical social constructionism in psychology. *Behav Philosop*. 26, 5–28.

3
What Role Do Cognitive Processes Play in Assessing Quality or Quality-of-Life?

Abstract

This chapter addresses the role that cognitive processes play in a qualitative assessment. It provides an introduction to various cognitive principles, and also gives two examples of how these principles are relevant to current quality-of-life research. The overall objective of the chapter is to provide the investigator with a set of analytical tools that can be applied to understanding how a qualitative assessment occurs.

The whole of science is nothing more than a refinement of everyday thinking.

Albert Einstein[1]

Abbreviations

fMRI	Functional magnetic resonance image
HCI	Human–computer interaction
HRQOL	Health-related quality-of-life
PDP	Parallel distributed processes
PGI	Patient-generated Index (Ruta et al. 1994)
QWB	Quality of Well-being Scale (Kaplan and Bush 1982)
SEIQoL	Schedule for the evaluating of individual quality-of-life (O'Boyle et al. 1992)

1 Introduction

In Chap. 1, I described quality as a unique phenomenological experience (a *quala*), but I also introduced Gärdenfors' (2000) proposal that quality can be mentally represented in at least three ways: as a product of a symbolic system; as an association or connection between elements that can provide information; or as a conceptual space. In Chap. 2, I demonstrated how the analysis of language, as a symbolic system, can be used as an analytical tool to organize and understand various quality-of-life or HRQOL definitions. In this chapter, I start to address how associationism and connection of entities can be used to represent quality. Specifically, I will discuss how analyzing how people think can contribute to the task of defining and assessing quality or quality-of-life. What will become evident is that understanding how people describe, evaluate, or qualify (or reflect about) their experiences will provide another useful, yet uncommonly used, set of tools that can contribute to the task set for this book. Note that the emphasis is on the "how" a quality judgment occurs, not "what has occurred." The "what has occurred" question leads to more of an accounting approach, and it and the "how" approach are both needed to understand "why" – in the sense of "why a person thinks a particular meal, painting or life lived is of high quality." I have divided this chapter into two parts: the first discusses cognitive principles, and the second illustrates the presence of these principles in specific, currently available, quality-of-life assessments.

To start this discussion I want to address the somewhat controversial issue of whether putting a number to an experience essentially eliminates the need to consider quality as an experience. First, it certainly seems that I do not have to put a number to an experience for me to know that I have experienced a high-quality event. However, the psychometrician in me says that this last statement is not quite true, since distinguishing one experience from another is enough to create a type of assessment: a nominal scaling. Thus, the very act of discriminating light from dark, or a pleasant experience from an unpleasant one, is an assessment, but not an assessment that demands a number; it is more a linguistic label.

So before I begin to place numbers on these experiences, I will inquire about what makes up these experiences.[2]

The first step in this process is to recognize that I can become aware of the experience of quality in a variety of ways. Thus, when a person is asked a single or series of questions (e.g., "in the last 3 weeks, how often did you…") they may respond spontaneously and give a quick and unreflective answer, or respond deliberately, reflecting on their past history, current status, or sense of future life quality. Each of these self-reports would be a reactive indication (a response to an external probe such as a questionnaire) of their quality-of-life or HRQOL. But I can also become aware of my life quality by self-reflection, as occurs following personal inquiry or conversation. Self-reflection, of course, can be expressed in various ways, from poetry to prose (e.g., diaries or even novels), or the discourse characteristic of a psychotherapy session. Methods exist for assessing this type of material, such as text analysis (Mehl 2006) or discourse analysis, methods which minimize the amount of a priori coding of what is being examined.[3] Finally, a person's awareness of their life quality may emerge spontaneously, as in response to a particular life event (e.g., when a person's child gets married, when I received a diagnosis of a disease, and so on). Here the assessment methods also require minimal prior coding of potential outcomes. Each of these types of experiences would be expected to involve different cognitive processes, requiring different assessment methods. Table 3.1 summarizes these alternative experiences and how they may be differentially assessed. I will discuss the relevance of these distinctions in greater detail as I proceed, although I will mostly be concerned with monitoring the cognitive processes associated with the reactive cognitive processes that are evoked by a questionnaire.

My next question is how I can cognitively conceptualize quality, and particularly, quality-of-life or HRQOL. Figure 3.1 is a scaled-down version of a figure depicted in Barofsky (2003) which assumes that a basic input–output system exists that is a product of a *complex adaptive cognitive process*, a process that is also conditional in nature. What makes these cognitive processes conditional is that each individual brings with them a different set of antecedent conditions, as when a life history consists of unique expectations, or inferences concerning causal relationships (e.g., attribution). Figure 3.1 also assumes that different levels of awareness of these cognitive events exists, with awareness varying from minimal, as when a person gives a quick and automatic response, to maximal, as when they take a more deliberate thoughtful approach and engage in a metacognitive process (Chap. 3, p. 64).

The primary outcome of the model depicted in Fig. 3.1 is information about the quality of objects, events or a person's existence. This information (as impressions, beliefs, or feelings) has an existence and a natural history. I think of this information as kernels of thought whose presence guides my judgment of a painting, or appreciation of a musical presentation, but most importantly, it is what I use when I judge my work, play, and relationships. It is the quality I seek in my existence. I will refer to this kernel as a *cognitive entity*. A cognitive entity can be formed by combining experiences (e.g., by addition or multiplication), although these experiences may be a produced by complex nonlinear processes. Most importantly, a cognitive entity retains its integrity even as it is changed by the continuous nature of experiences. Thus, when I reflect upon, describe my current status, or predict the future quality of my existence, I am creating de novo or modifying an existing cognitive entity. Emotions are also present, providing the "energy" for these changes, which are inherent to a cognitive entity. As I will describe, this cognitive entity has many of the properties of a complex

Table 3.1 The variety of ways a person can become aware of the quality of an event, process, or their quality-of-life. Examples of assessment methods are also provided

Assessment methods	Continuum of experiences		
	External probes: reactive	Self-inquiry: reflective	Spontaneous: emergent
Questionnaires	X		
Text analysis	X▲	X	X
Discourse analysis	X▼	X	X

▲: Text analysis of questionnaire items that encourage extended verbal responses
▼: Discourse analysis of verbal exchanges initiated by an external probe

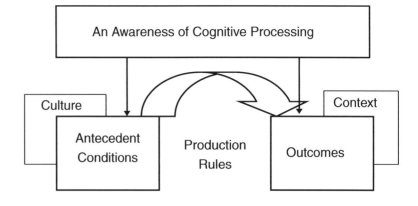

Fig. 3.1 The cognitive components of a quality-of-life or HRQOL assessment are depicted. Included is a basic input–output system that is conditioned by antecedent and consequential conditions. Production rules guide the input–output relationship, while cognitive processes vary across levels of awareness. When awareness is minimal, the input–output relationship occurs quickly and automatically. If awareness is prominent, as during metacognition, then the input–output relationship is slower and considered (Barofsky 2003).

adaptive system, which evolves as a person experiences life, including experiencing the rather artificial event of answering questions on a questionnaire.

A cognitive entity can be considered a *qualia*, if a *qualia* is thought of as an outcome of a neurocognitive process. The modification or formation of the cognitive entity would involve a variety of inputs including inputs from the person's culture and life history (Fig. 3.1). Evidence supporting this includes studies which demonstrate that people from the West and Far East think differently about the form and character of many aspects of ordinary discourse, including conversations (what is said, who says it, when it is said, and so on), and this difference structures how each group perceives and processes information. These groups also differ in how they conceive of their environment (holistically or analytically: Park et al. 1999), what notions they have about poverty, and so on. As the model depicted in Fig. 3.1 implies, this cultural and life information would be cognitively imported into the process, whereby a quality assessment would be formed. What is not clear is whether each ethnic or cultural group would describe, evaluate, set expectations, or infer causal relationships (e.g., attribution) in the same or different ways, as they may do when they generate a qualitative judgment.[4] This remains an empirical but critical unanswered question.

As stated above, Fig. 3.1 assumes that the sequence of events involved in forming a quality-of-life or HRQOL assessment is governed, to varying degrees, by my awareness of ongoing cognitive processes. My ability to cognitively monitor and modulate my own cognitive processes is commonly referred to as a metacognitive process (see Chap. 3, p. 64). This suggests that at some point, I will become aware of and govern my thoughts involved in generating some concept of quality. This implies that at other times I am not aware of these ongoing cognitive processes, which should not be surprising, since most cognitive processes occur below a person's awareness. It also suggests that a cognitive entity can be assumed to exist at an unconscious level of awareness. This would help me account for the unconscious mapping characteristic of a conceptual metaphor, or the unconscious formation of a global assessment in response to a person's life experiences (e.g., their sense of their "good luck," how fortunate they feel, and so on). When the cognitive entity is activated, as when I am asked a series of questions, specific cognitive processes may become involved. My next question, then, is how such a cognitive entity can be formed.

To answer this question, I will have to explain much more about how neurocognitive processes lead to the formation of a cognitive entity. I will start this discussion in this chapter, and continue it at different places in this book (e.g., Chap. 10). But I first need to explain how information is classified, then categorized, and finally, how concepts are formed (such as concepts of quality-of-life or HRQOL). I will also have a more extended discussion on the nature of metacognition, including what it is and how it can be conceived. Thus, this chapter will be divided into two parts: the first will expand the model presented in Fig. 3.1 and provide more details on how a cognitive entity characteristic of a quality assessment can be formed; and the second part will give some examples of the application of these cognitive principles. I will then use these cognitive principles throughout the remainder of the book as I discuss different aspects of the assessment of quality.

2 Cognitive Principles

2.1 The Neural Basis of Cognition

In the Preface, I stated that an adequate presentation concerning the assessment of quality requires that I make an explicit statement about the neurocognitive model underpinning how people come to know and assess quality. Many theorists have struggled with this very issue, and I do not seek to review all their efforts. Rather, I will highlight one author, Friedrich August von Hayek (Caldwell 2004), whose main intellectual contributions were in the area of economics, but who early in his career developed a model of neurocognitive functioning that has become an integral part of modern economic and social theory. His book, *The Sensory Order* (1952), provided the intellectual foundation for an associationistic or connectionist model of neurocognitive functioning, and that model has applications to the current interest in the neurosciences, complexity theory, and its applications to economics. What is most interesting is that he used his model not only to explain an individual's experiences (e.g., how I sense, remember, and so on), but also to account for broader social phenomena (e.g., how various economic processes occur). It seemed worth considering the possibility that he could identify a single principle or idea that could be applied at several levels of analysis, and could also be relevant to my discussion concerning the definition and assessment of quality. Hayek's principle can be stated simply as, "all information is local and distributed."

Let me start this discussion with a brief review of the range of alternative neurocognitive theories that are available. Some of these theories claim that specific areas of the brain mediate specific functions, so that an activity such as cognition is the sum of the coordinated actions within these areas. Other theorists think of the brain as consisting of a series of differentiated modules (e.g., vision, hearing, language, and so on) which function somewhat independent of each other, yet act in a coordinated manner, as when they are part of a network. There are also theorists that view the brain as a computational machine, much as a computer, that processes information in parallel networks (Chap. 4, p. 107), and produces images, perceptions, or thoughts that lead to actions. The required computations may involve specific brain areas, but this is less important than the fact that different modules mediating specific cognitive functions can be demonstrated to exist (Pinker 1997). Some theorists who also promote a computational approach to modeling neurocognitive processes suggest that an "optimality principle" exists that governs these processes (e.g., Smolensky and Legendre 2006). There are also the embodiment theorists (e.g., Gibbs 2006; Lakoff and Johnson 1999) who claim that brain function alone cannot account for what I experience, but that I also need to take into account my physical encounter with the environment. The environment's presence is felt by action-dependent proprioception and kinesthesis that feeds into specific brain areas and thereby contribute to a particular cognitive outcome. These models emphasize the inseparableness of a person's actions from their brain functions in determining their cognitions.

Developmental studies also can be helpful in conceptualizing how cognitive entities are formed. For example, Keil (2006: p. 611) points out that philosophers as ancient as Plato wondered about what capabilities a child had when it was born. Plato believed that children were born with intellectual capabilities that only had to be encouraged to become manifest, and this justified using the "Socratic dialogue" method as an instructional procedure. Thus, it is natural to ask about the initial state of a newborn. Is an infant a tabula rasa, devoid of formative cognitive structures that determine what is experienced, or do babies possess such structural determinants? Of course, the answer to this question is not simple (Keil 2006). It ranges from the demonstration of prenatal conditioning (e.g., Nazzi et al. 1998), includes Chomsky's hypothesis that a person is born with the innate ability to use grammar to organize words into coherent sentences, but also involves learning which presumably occurs at each level of maturation.

Also of interest is whether development proceeds from the concrete to the abstract, as was stated by many classic developmental psychologists (e.g., Bruner 1967; Inhelder and Piaget 1958). This issue is of interest because I am concerned about whether the assessment of quality (or, for that matter, well-being) proceeds from the concrete to the abstract, or vice versa. These early researchers assumed that abstract expressions of language were acquired by clustering together individual words, which formed phrases and then sentences. However, as Keil (2006) states:

> At an early age, children throughout the world seem to learn ordering relations based on abstract syntactic categories such as subject and object and not on number of words or word tokens, simple word order patterns, or sentence size. They appear to have abstract parameters that are 'set' in ways that allow them to unpack the structure of a particular language…there is no doubt that a dominant model of language acquisition is that children start out with abstract skeletal expectations about a grammar of a language that they then fill in with more concrete language-specific details of tense marking, subject–verb agreement, and sentence embedding. (p. 617–618).

Keil's comments illustrate how an abstract capability of a child can contribute to concrete expressions of spoken or written language. I will continue to discuss the relationship between abstract constructs and concrete entities, since it emphasizes the importance of structural hierarchical determinants of cognition, an issue that is relevant when discussing the role of health status (Chap. 8), or well-being (Chap. 11) in a quality assessment.

Hayek developed his original neurocognitive model in response to a series of questions he asked when he was a student; questions which evolved from his efforts to make sense of the difference between the mind and the body. He wanted to understand the relationship between the objective and subjective world, and he framed his approach by asking how order can create itself.[5] He continued to ask this question throughout his career.

After participating in World War I, Hayek returned to Vienna to discover that his university facilities were not sufficiently restored to permit his continued study. Instead, he traveled to Zurich where he studied for a period of time with the brain neuroanatomist von Monakow (Hayek 1994). There he was able to directly examine the structure of the brain and see for himself the large numbers of neurons present. He also observed that the neurons frequently contacted each other. He asked how order could evolve from this apparent disarray and wrote a paper entitled, "Contribution to the Theory of the Development of Consciousness" (Hayek 1919), in which he considered consciousness to be a product of a developmental or evolutionary process, an idea that was to become central to his economic research program. A revised version of this paper was published 33 years later (Hayek 1952). Even though Hayek did not immediately publish his original paper,[6] the idea that evolutionary processes, particularly the notion that chance events set the stage for adaptive change, significantly influenced his subsequent work (Nadeau 1997). Also of interest was that his book was finally published a couple of years after Donald Hebb's (1949) book, *The Organization of Behavior: A Neuropsychological Theory*, which tackled the topic in a similar way.

Looking at the organization of the nervous system, Hayek could have easily concluded that it is impossible to create order out of what appeared to be so disorderly. Instead, he saw in this disorder a means of creating order, as long as the entire central nervous system was conceived of as a network which generated unique patterns of responses in response to different physical stimuli, with each physical stimulus producing a distinct pattern of neural excitation. Each time a person is exposed to the same physical stimulus, a similar pattern of neuronal responses occurs, resulting in repeated classifications of the stimulus. Physical changes in the neurons themselves were assumed to occur, and in this way the pattern of neural response characteristic of a particular experience was retained. This idea was similar to what Hebb

(1949) subsequently postulated when he stated that experience (e.g., learning) leads to physical modification of the neuronal synapse. Hayek does not assume that order exists prior to experience, nor does he assume that specific neurons mediate specific sensations. Rather, he states:

> We do not first have sensations which are then preserved by memory, but it is the result of physiological memory that the physiological impulses are converted into sensations. The connections between the physiological elements are thus the primary phenomena which creates the mental phenomena. (Hayek 1952: p. 53).

Thus, the "external world," meaning everything that is not in the central nervous system, has no meaning independent of the physiological memory (or pattern of neural responses) present in the central nervous system. Hayek, therefore, is a empiricist in the tradition of John Locke (1690), a monist and a constructivist, since his view suggests that no reality exists for a person that cannot be elaborated upon by the central nervous system (Nadeau 1997).

The spontaneous but unintended regularities resulting from the repeated classifications results in a "sensory order," which suggests that the neural network was functioning as a complex adaptive system. As this proceeds, Hayek postulates, increasingly abstract entities evolve, such as consciousness, a sense of quality, or maybe even a sense of quality-of-life or HRQOL.[7] Thus, Hayek's views are reminiscent of the early developmental psychologists cited earlier. Irrespective of this, life experiences would be expected to evolve into a cognitive entity, and it is the natural history of these cognitive entities that is the subject of this book.

Now what is fascinating about these ideas is that Hayek took what is basically a biological idea, applied it to understand cognitive processes, and then applied it to economics. Thus, when he discusses how economic (e.g., price) equilibrium occurs, he talks about how local knowledge or information leads to economic decisions (e.g., someone purchases a product), which are transmitted between people via an information network, leading to decisions that other people make. This information transmission leads to the evolution of a preferred product, price stability, satisfaction with outcome, and so on. Because equilibrium results from this distributed but local knowledge, no one person can know all that is needed to be known, including someone who attempts central planning. Using this cognitive characteristic, Hayek (1944) goes onto claim that any political system that promotes central economic planning is not likely to succeed. Thus, Hayek takes a subjectivist perspective (concern with how people perceive, what their expectations are, what satisfies them), relies on what he understands about how people process information (i.e., use transmitted local knowledge), and uses this to understand how economic systems work.[8]

Next, I have to ask in what way Hayek's views are relevant to a quality assessment. One answer is that he has shown me how it is possible to have an assessment approach, which can be applied at both the individual and group level. His neurocognitive model, which assumes that information is local but distributed, would argue that each person would have a unique view of their assessment of quality (i.e., local knowledge), but when distributed or communicated (as for example, when it is part of a political process) it would somehow emerge, as is true for any complex adaptive system, into a distinct entity (e.g., a preference for a political candidate who supports particular qualitative outcomes).

Contrast this with what is considered normative in quality assessment, where a person is individually assessed and then their data grouped. The grouping, of course, may serve some analytical purpose, as when the person's data is grouped according to a treatment, or for experimental manipulation, but it also leads to an analysis of the variance of these measures as opposed to seeing these differences as the outcome of a complex adaptive process.

I can gain some insight into how a complex adaptive system leads to a particular outcome by inspecting the overlap between Hayek's model and the economic principles of Adam Smith (1776). Adam Smith suggested that if people were allowed to pursue their self-interests, that an "Invisible Hand" (a metaphor for a complex adaptive system) would operate to guarantee that diverse local events would maximize the wealth for all. Thus, the principle being espoused here is that humans, being rational, would maximize their self-interests, which would produce maximum benefit for society as a whole.[9] In addition, if self-interest leads to wealth that enhances well-being, then the pursuit of self-interest becomes morally appropriate.

This model also has significance for how quality-of-life or HRQOL research should be performed. It suggests that if I assume that a person will take a rational approach to maximizing the quality of their existence, then the person should be free to pursue this objective. Like Ackoff (1976), Hayek would discourage the formal assessment of quality-of-life or HRQOL, since these types of assessments tend to artificially structure a person's concept of quality. Instead he would argue for allowing economic and political processes to unfold.

In contrast, what is expected from current quality-of-life or HRQOL research can be summarized by the following quote from Haase and Braden (1998) who state:

> Given the lack of clearly identified essential characteristics of QOL, the problem is to provide a meaningful way to distinguish between concepts related to QOL and indicators that clearly identify the essential characteristics of QOL. QOL will cease to be a primitive term only when consensus is reached about a set of clearly identifiable characteristics or attributes that will be consistently used as the basis for definition and measurement." (p. 55)

The Adam Smith and Hayek models imply that there is no need to find a "consensus …about a set of clearly identifiable characteristics" (Haase and Braden 1998: p. 55), since to do so would be as arbitrary as a central planner determining the output of a particular community. Quality-of-life or HRQOL for Hayek would be a unique local outcome resulting from a networked process that involved a person adaptively (or rationally) pursuing their self-interests.

Also relevant here is Hayek's notions about "spontaneous order," its relationship to complexity theory (Arthur et al. 1997) and nonlinear dynamics. Vaughn and Poulsen (1997) argue that many of Hayek's ideas predate what has been claimed by Arthur et al. (1997) and other "complexity" authors.

The question I have asked here is whether a quality assessment by an individual or a group represents the outcome of a complex adaptive system. For this to occur, I would have to assume that an individual or group of persons would have to have many independent experiences which act in parallel, leading to an emergent outcome that never quite reaches a static equilibrium. The assumption would be that as the outcome continued to evolve, it would be increasingly adaptive, which, in the case of a quality or quality-of-life assessment, would mean that the outcome would be increasingly useful for the person or group. Thus, Hayek would have us characterize an individual or group's quality-of-life as unique, adaptive, and continually open to change.

Another aspect of Hayek's neurocognitive model that may be important for understanding a quality, quality-of-life or HRQOL assessment is his contention that if a person continues to process and classify experiences, an increasingly abstract entity emerges. This suggests that I come to know what health is by my experiences with having a cold, having a sore back, being athletic, and so on. Thus, the presence of a language that differentiates the concrete from the abstract provides indirect support for such a cognitive process. This process may also be consistent with what embodiment theorists would postulate, which is that the experiences I have with my body contributes to my cognitions (e.g., my qualitative status).

However, data are available suggesting that the elements of a hierarchical relationship (concrete experiences and abstract concepts) each may be located in different cortical areas. Thus, Binder et al. (2005), using fMRI, found that abstract concepts and concrete examples involve distinct but partially overlapping brain systems. This suggests, as Keil (2006) implied, that abstract constructs may be centrally in place, acting to govern and organize concrete entities.

In addition, Koechlin et al. (2003) have demonstrated, also using brain imaging techniques, that specific areas of the prefrontal cortex are arranged in a hierarchical manner, with specific sites in the lateral prefrontal cortex responding to past events, another site responding to contextual signals and another to current stimuli. These sites are than activated sequentially and form the basis for cognitive control: "the ability to coordinate thoughts and actions in relation to internal goals" (Koechlin et al. 2003: p. 1181). Both these studies, therefore, suggest that Hayek's contention that abstract concepts literally emerge from concrete experiences has to be qualified to a statement that suggests that the connection of concrete experiences and abstract concepts into a hierarchy "functionally emerges" as a result of continued association of specific cortical areas.

This discussion of Hayek's theory and complexity theory promotes a connectionism approach to quality-of-life or HRQOL assessment, a topic I will continue to discuss in Chaps. 4, 9, and 10. Now I need to introduce what is known about various cognitive processes, especially if they are to help me understand how a quality, quality-of-life, or HRQOL assessment occurs. Overall, I will consider four topics: (1) how cognitive entities are formed; (2) how cognitive entities are classified; (3) how concepts are formed from the information that has been classified; and (4) how metacognitive processes govern cognitive processes.

Before I proceed, I want to summarize a paper by Mareschal et al. (2007), who have proposed a theory of neuroconstructivism that is reminiscent of Hayek's notions concerning the "sensory order" (although they do not cite the overlap). The presence of their theory reinforces the view that the issues that Hayek raised are still very much of interest today. Neuroconstructivism is a developmental theory of cognition, with brain development and cognitive development being correlated. As the authors state, "We see constructivist development as a progressive increase in the complexity of representations, with the consequence that new competences can develop based on earlier, simpler ones…Neuroconstructivism implies the creation of genuinely new cognitive abilities and not just the better use of pre-existing abilities." (p. 5–6) Thus, this theory differs from Plato's notion that intellect is present and only waiting to be revealed, but also Piaget's (1970) notion concerning the hierarchical integration of knowledge.

2.2 The Formation of a Cognitive Entity

One of the operational characteristics of human cognition is that not all the information that is received is processed, rather that the information is selected and used to form a cognitive entity (Barofsky 2003). The cognitive entity represents, in effect, a summary measure or statement of the cognitive processing that has transpired. Hayek's (1952) idea that the repeated classification of physical stimuli (or experiences) via a neural net resulting in a "sensory order" (or global impression) represents one method whereby a summary measure may be neurally formed or represented.[10] However, at this point I would like to shift the discussion to how higher cognitive processes, such as language (e.g., conceptual metaphors) and reasoning, are involved in the formation of a summary statement. What will become evident is that a complex set of production rules participate in the formation of a qualitative cognitive entity. These production rules participate in sensory-perceptual processes that create a conceptual (quality) space, and also a symbolic system, as reflected in grammar and logic, that is than enlisted to form summary statements, or cognitive entities. Sometimes the symbolic production rules are known prior to their application, as when calculating a mean, but sometimes they are not, as when someone provides a self-report of quality or quality-of-life.

Several types of self-report summary measures exist, and each is dependent on varying combinations of production rules. As discussed above, production rules can describe, evaluate, and anticipate a particular end (i.e., expectation); direct an association (i.e., attribution);[11] and also reflect or qualify what has cognitively transpired (that is, a metacognitive process). Descriptive production rules select and organize information that contributes to the formation of a cognitive entity, but it is not sufficient to provide an assessment of quality. Description can occur in an unconscious "fast or fugal manner" (Goldstein and Gigerenzer 1996; Oppenheimer 2003) or in a more deliberate manner. Heuristics (cognitive shortcuts) are examples of unconscious descriptive processes. In one type of heuristic (e.g., the representativeness heuristic; Goldstein and Gigerenzer 1996), a person may select a salient or chronic feature of their health state, or that person may select a recent health experience (the availability heuristic). Retrieval of information may also take advantage of "implicit theories" that a person has developed from past experiences (Ross 1989).

Evaluation during information processing can occur in several ways. For example, descriptive statements can connote an evaluation by the grammatical structure a statement is placed in (see below). Also, information already available to the person can become part of a cognitive entity. Thus, a person may have a prior notion of poverty or happiness, which then becomes part of what they ultimately judge about their own quality-of-life. Evaluation can also take place concurrent with the descriptive task. The classic example of this is when a person is asked to rate how satisfied they are with their medical care. This question prompts both the descriptive activity of identifying what medical care they received, and also the opportunity to evaluate the care. Since the description and evaluation is meant to occur simultaneously, satisfaction indicators are at risk of being confounded. Finally, evaluation and description can occur consecutively, with description preceding evaluation. This is clearly the least biased method of summary measure formation, and, interestingly, is the method investigators use when designing several generic quality-of-life or HRQOL assessments (e.g., Fayers and Machin 2000; Patrick and Erickson 1993; Torrance et al. 1995; Ware et al. 1980).

Expectancy production rules anticipate that the information a person receives will be processed in a particular way. Thus, when a quality judgment is made, I may also be asking whether the quality of an object or experience is what I expected. Expectancy production rules utilize both description and evaluation, but are recruited prior to the formation of a new cognitive entity. What is anticipated, such as an ideal outcome, is projected into the task itself. Thus, what goals are or are not achieved, or what needs are met or not met, can become part of the formation of a cognitive entity (see Chap. 12). "Expectancy," therefore, is a particular cognitive process that reflects the motivational aspects of a qualitative assessment.

Attribution production rules imply that the information a person receives will be used in a particular way. Here the cognitive entity includes a directional character, in which the information is associated with an object, experience, or person. The association created by attribution maps elements involved in a quality or quality-of-life judgment. Thus, the definitions listed in Table 2.5 attribute changes in quality-of-life or HRQOL to a disease or its treatment.

Finally, what has been described and evaluated can also be valued. This usually involves a metacognitive process, and while a person may be aware or unaware of engaging in metacognition, its presence is a critical part of a process whereby a qualitative assessment becomes "of the person." Thus, metacognitive processes are involved in the qualification process, by which I mean the process whereby a person projects their presence (e.g., values their status) onto what they are experiencing. I will also, at times, use the term "subjectification," which involves a person making an objective stimulus subjective. Hedden et al. (2008) illustrated this process when they showed that cultural (Western vs. East Asian) differences influence the neural substrate involved during ordinary attention tasks. These two terms (qualification and subjectification) differ in the origin of their cognitive control (top-down, or bottom-up).

Once an entity is formed, it continues to be cognitively processed. First it is tagged or classified in terms of its characteristics (e.g., using hierarchical or cross-sectional classification methods), and then it is organized into a broader cognitive entity, a concept. Part of the formation of a concept may also involve metacognitive processes. Now, however, I will examine how the formation of a cognitive entity also leads to its classification.

2.3 The Classification of Cognitive Entities

If various production rules are indeed used to form a cognitive entity (e.g., expectancies, attributions, and so on), then the newly formed cognitive entity has characteristics that can be used to classify the entity. Once classified, the cognitive entity has the properties characteristic of a category. However, there are several ways that a classification can result in a category (Table 3.2). In this section I will mostly discuss hierarchical and cross-classification, with a particular emphasis on linguistic classification methods. I will also address the role

Table 3.2 Methods for classifying cognitive entities

Methods of classification
1. Hierarchical
Taxonomic
Abstract or concrete
Linguistic classification:
Subordinate
Basic
Superordinate
2. Cross-classification
Script
Thematic
Goal-derived
3. Contextual (e.g., cultural) classification
Folk classification
Scientific classification

that abstractness or concreteness has in classifying cognitive entities. Additional topics will be discussed elsewhere in this book. It will become clear that the categories and concepts can also be classified (Chap. 6). For example, categories may also be classified into (proto-)typical or exemplar types. If the features or attributes that make up a category or concept "fit together," then they can be said to be coherent. Thus, classification can occur at several levels of cognitive processing.

I would like to define several important terms that I will repeatedly use; they are: classification, categories, and concepts. First, consider classification. It is, as Estes (1994: p. 4) states, "basic to all of our intellectual activities." Estes distinguishes "classification," which involves the collection of objects or experiences into groups, and "categorization," which includes prior knowledge of the category within which objects or experiences are grouped. For example, jogging, walking, and dancing are all physical activities, and together they can be labeled as referring to "physical activity." However, the phrase "physical activity" has a broader meaning than is implied by these three activities alone, and in this sense, the meaning of the phrase must have been established by the labeling person's prior experiences. Of course, if the person is not familiar with the several activities, objects, or expressions, then they may form a new cognitive entity to organize the information they receive and this would also be characteristic of a classification. In addition, an investigator may classify a set of cognitive entities for very different purposes than a person would, so I will distinguish between the formation of a category by a person, and a domain by an investigator (Chap. 6).

Estes (1994) also distinguishes a concept from a category. The difference between them here is that a concept refers to an abstract idea, which in my setting would be used to organize information, whereas a category is usually determined by specific characteristics of the objects and experiences being organized (e.g., similar objects or experiences are grouped together). Categorization is always an important part of a concept, it is just not all of it. What is also true is that there are some instances where a category and concept are indistinguishable: such as, when what is being organized is a taxonomical category such as plants, animals, or nouns. A domain will be subject to the same rules, just that it would reflect the influence and purposes of an investigator.

Having clarified these terms, let me proceed to illustrate what "hierarchical classification" and "cross-classification" mean, follow with a discussion about how different concepts are formed, discuss the nature of metacognitive processes, and then start to apply these approaches to a quality assessment. My purpose in detailing this literature is based on my belief that understanding how categories and concepts are formed will be quite useful in understanding how a quality assessment occurs, and this then can provide the basis for understanding and dealing with the various types of quality assessments (assessing the quality of an object, a meal, a person's life), but also the exceptional problems that are created, as when assessing the quality of the compromised person.

2.3.1 Hierarchical Classification

Rosch and her colleagues (1973; 1975; 1976; 1977; 1978 and Mervis and Rosch 1981) demonstrated that when a concept is formed, the information communicated is naturally organized into a hierarchy. The researchers identified three categories that differ in degree of *informativeness*: a basic level, and categories that are below (subordinate) or above (superordinate) this level. It appears that the basic level provides language information that best matches the physical and biological world, and for this reason reinforces the impression that what is perceived matches the real world. Perceptions above and below this level are more problematic, but for different reasons. Thus, I would have no difficulty discriminating between cats and dogs, but may have more difficulties distinguishing different species of cats (a failure of perception), or understanding the concept of "animal" (a failure of comprehension). Lakoff and Johnson (1999) suggest that the basic level category is biologically determined and is a significant contributor to a person's survival. They state that:

> Basic-level categories are distinguished from superordinate categories by aspects of our bodies, brains, and minds: mental images, gestalt perceptions, motor programs, and knowledge structure. (p. 27: 1999).

The basic level of information organization, according to Berlin et al. (1974) and Rosch et al. (1976), has four defining characteristics. First, the basic level is the highest level of cognitive category which can be represented by a mental image. Thus, while I can close my eyes and image what a "chair" looks like, I would have great difficulty, if I could do it at all, imaging what "furniture" looks like, unless I returned to a basic level and thought about the term "furniture" in more concrete terms, such as chairs, tables, and so on.

Second, the basic level is the highest level of cognitive category where percepts are organized into wholes (gestalt perceptions) and compared. Thus, I can perceive chairs and tables and group them to the extent they are similar and different. There is no generalized image of furniture that I can compare to another image of furniture, unless again I lower my classification level to the basic level.

Third, the basic level is the highest level of cognitive classification that I can expect to interact with an object in the category. Thus, I can sit on a chair and eat at a table, but I cannot usually[12] sit on, or eat at "furniture," unless I want to use the term more concretely.

Fourth, the basic level is also the level where most of my knowledge exists. Here it is clear that I may know a lot more about the items in my home, but know much less about the "furniture" (e.g., you may know the style of furniture, but that would be considered a concrete indicator of the abstract term) in my home. A simple example of a taxonomic hierarchy would assume that an "oak" would be a subordinate level concept, a "tree" a basic level concept, and a "plant" a superordinate concept. The hierarchy exists because the elements follow the "IS-A" relationship (Murphy and Lassaline 1997); that is, an oak is a tree and a tree is a plant. Since each term can be included in a higher level of abstraction, a hierarchy can be confirmed to exist. An alternative explanation is that as I ascend the hierarchy (i.e., it becomes more abstract), I am actually dealing with fewer distinct features, but attending to a common core concept (i.e., an oak is a plant, and a tree is a plant) which can be used to organize information. What also seems to be true is that the speed of responding to these cognitive tasks decreases as one ascends the hierarchy. Thus, people respond more quickly state that an oak is a tree than they would state that a tree is a plant. Murphy and Lassaline (1997) summarize their discussion of this topic by saying that:

> ... people are able to learn and use taxonomic relations in order to draw inferences. Secondly, people are able to reason taxonomically about novel material that have not been previously stored in memory. (p. 99).

The second point that Murphy and Lassaline (1997) make is of particular interest, since it implies that I have the cognitive capacity to organize information into a hierarchy that can be applied any time I deal with novel material. This, it turns out, may also be true for persons from different cultures, or for how specific indicators are organized.

I can illustrate this point by examining a particular linguistic domain, called color terms, that are ordinarily organized into an information hierarchy that can be found across languages. Berlin and Kay (1969) illustrated this when they determined what color terms were being used in 98 different languages. They found, among several interesting observations, that the color words used in all the languages studied were limited to 11 focal colors. In addition, the frequency with which these color terms were used most often followed this pattern: white or black most often used; red, green, or blue next most frequently used; followed by gray, brown, and so on. White and black, of course, are not only the terms most often used; they are also the most physically abstract, white being the exclusion of all color, and black being the inclusion of all colors. Red, green, and blue are primary colors present in the cones of the retina, while gray and brown, and so on, are a product of the mixture of primary colors. This suggests there is a hierarchical pattern of usage embedded in languages that is based on certain physical properties of color; a hierarchy that reflects a type of cognitive organization.[13]

The hierarchical relationships between a category and a concept (that is, subordinate, basic, and superordinate) are essential cognitive structures for deriving meaning from words, phrases, or sentences. I will refer to this relationship when I discuss what is required to be in place if a domain is to be described by an abstract concept, as when a particular set of symptoms are described as reflecting some qualitative state (e.g., HRQOL: Chap. 7). Thus, semantics (the study of meaning) provides the interface between perception and cognition, and this explains why it is important to study the language of quality.

2.3.2 Cross-Classification

Information can also be cross-classified (Ross and Murphy 1999: Nguyen 2007) into a category by a person, or into a domain by an investigator. Consider a list of foods. They can be organized by a person into superordinate or taxonomic categories consisting of breads, beverages, or fruits (Murphy and Lassaline 1997), but they can also be organized in terms of breakfast foods, healthy foods, snack foods, or "movie" foods. These latter categories were labeled by Ross and Murphy (1999) as script categories, since they are time and situation dependent and seem to be telling a story. Another characteristic of these categories is that the elements that make up the script can be changed, but the story of the categories remains the same (e.g., oatmeal can be substituted for cold cereal and the category would still be a breakfast script). The script categories are of interest since they often lead to some action (e.g., eating or rejection of food). Categories, whose components must be present, are called thematic categories. Thus, if a category consists of a spoon and a bowl, or a dog and a bone, removal of one component would disrupt the original meaning of the category. An investigator can also create a cross-classification domain.

Barsalou (1991) has also shown that categories or concepts can be classified by the way they are formed. Thus, some categories are formed because I have specific goals or purposes in mind when I create the category. For example, I may create a category of things to take with me when I go on a camping trip, or when I decide what I want to take out of the house if there is a fire. These categories consist of dissimilar items. What is typically included in a goal-derived category or domain is that it approximates some ideal (i.e., the goal or purpose implicit in the category) rather than some measure of central tendency (e.g., a robin and a chicken may be both birds, but differ in terms of their features). Medin et al. (2000) states that it appears that, "goals can create categories and that these categories are organized in terms of ideals." (p. 132) Goal-derived categories or domains will be quite relevant when I discuss how they are formed in Chap. 6. They are also characteristic of clinimetric indicators (Chap. 5, p. 149).

2.3.3 Applications to a Qualitative Assessment

How is this information about the nature of classification relevant to my understanding of a quality assessment?[14] I will illustrate this by first discussing how domain specific and global quality-of-life assessments have a hierarchical relationship, and then discuss how downward social comparisons are a form of cross-classification. I will then consider how autobiographical material, the substance upon which a quality-of-life or HRQOL assessment is often formed, can be considered a goal-derived concept. However, it is important to acknowledge that much of what I will be discussing is based on studies of objects, with less being known about abstract concepts (e.g., justice and truth) or concepts based on experience (e.g., quality-of-life). Still, applying what is known about objects to more abstract concepts is a logical way to proceed. And it is consistent with the principle espoused in this book that a common set of principles can characterize the quality assessment of objects and experiences.

If I apply the four criteria for defining a cognitive hierarchy listed above to a quality assessment, then I can infer that specific categories (e.g., a symptom or a level of physical activity) would be examples of a subordinate or basic level of information, while global single items (e.g., health, quality-of-life, satisfaction[15]) would be examples of superordinate classification. Evidence for this statement follows.

Consistent with the first criterion (see Chap. 3, p. 56), a person should be able to either directly experience or have a mental image of the elements in a specific category (e.g., it should be possible to imagine a tennis ball or playing tennis or jogging), but this would be much harder, if possible at all, for such superordinate concepts as health status or quality-of-life. This is partly because superordinate categories, as abstract categories, are independent of any fixed set of features or stimulus dimensions (Rehder and Ross 2001).

Based on second criterion, it should be possible to organize different types of physical activity (or similar elements) into a whole (that is, a gestalt), but this would not be possible for categories such as health, quality-of-life, or satisfaction. In this regard, it is interesting to speculate about whether the fairly high inter-correlations found between measures of health, quality-of-life, and satisfaction represent a failure of perception, rather than some common informational characteristic of these concepts.

The third criterion states that the basic-level category is the level of action, and certainly I can play tennis and jog but it is not clear what action "quality-of-life" would evoke, unless it was specified in terms of activities that would fit into a basic-level category.

Finally, the fourth criterion states that the basic-level category is the level where I have most of my knowledge and certainly this is also true when I speak of the general category of health or quality-of-life.

Based on the above, it is interesting to ask whether a person asked to rate the quality of their existence or their health is really rating their existence or health, or creating images of these states (meaning using basic level information) to base their ratings on. There is evidence (Benyamini et al. 2003) which suggests that when given a series of alternatives, persons who rate their health status as poor or fair cite different reasons for their rating than those who rate their health status as excellent or very good. Qualitative studies (Borawski et al. 1996: Groves et al. 1992: Krause and Kay 1994; Schechter 1994: Willis et al. 2005) using various cognitive interviewing methods, including the "think aloud" method, have provided some additional insight into what information respondents spontaneously report when rating their health status. These studies have found that a major proportion of the respondents use their physical health status (either the presence or absence of illness) as the most common criteria to rate their health status. In addition, respondents consider what their body can do, their health-risk behaviors, their social activity, their social relations, and finally their psychological and spiritual condition to rate their health status. Idler et al. (1999) suggests that going from a respondents' physical condition to their psychological and spiritual condition constitutes an expanding continuum that approximates a holistic perception. Idler et al. (1999) also demonstrate that those respondents who overestimate their health status (relative to a rating of their medical status) also use a more holistic description to basis their responses on, possibly compensating for some perceived sense of loss because of poor physical health, while those whose health status estimates are realistic tend to limit the descriptors they use.

These studies support the contention that respondents use basic-level information that they can imagine, perceive, and experience as they respond to items on a quality assessment. I have also learned that these images, etc., vary significantly as a function of the person's rating. As Idler et al. (1999) state:

> Our study supports other findings that show that respondents pick and choose their frames of reference and sources of comparison with respect to health and that they tend to do this in patterned, predictable ways. (p. 474).

Thus, not too surprisingly, how people think structures how they define their quality-of-life. It is also possible to learn what content, frames of references, or comparisons a person or investigators is using because the classification level can be reported (a basic or subordinate level of information). It is also important that I know enough to be able to systematically characterize these efforts and also to understand how people's thoughts vary as a function of the rating they provide. Thus, my task may not be defining quality or quality-of-life per se, but rather describing how individuals or investigators give meaning to these concepts using information they can access.

Much of the current discussion has focused on the self-assessed health status item (SAHS), and not many of the other single global items that exist (e.g., measures of well-being, satisfaction, and so on). The study of multi-item assessments raises some special issues, including the impact that the order of items within the assessment has on the assessment. This can be clearly seen in studies of the SF-36 when the SAHS item is moved from its normal first item position to the last item on the assessment. When this is done, there is a statistically significant change in the rating of the item, suggesting that information garnered by the respondent as they responded to other items affected their assessment of the SAHS item (e.g., David et al. 1999).

Another difference between a multi-item and single-item assessment is that items on the multi-item assessment are more concrete (i.e., more easily imagined and perceived), so that the ratings are more likely to occur rapidly and may involve less cognitive processing. This however, is an empirical question that as far as I know has yet to be studied.[16] What is known is that the concept of quality, quality-of-life, or health status that evolves from a multi-item assessment is most often constructed (e.g., metaphorically arithmetized: Chap. 9, p. 298) by combining the responses of individual items, rather than being created by the respondent in response to a single global item. A person may accomplish this by reference to an abstract construct, while an investigator deals with this by assuming the existence of a latent variable.

As will become evident in Chap. 11, the relationship between the terms "feelings" and "emotions" is another example of the conflict that can develop when I have to bridge the gap between basic and superordinate terms. As Wierzbicka (1999) demonstrated, languages differ in which of these terms has become the principal form of expressing affect. She summarizes her studies by stating, "Thus, while the concept of 'feeling' is universal and can be safely used in the investigation of human experience and human nature…the concept of 'emotion' is culture-bound and cannot be similarly relied on." (p. 4) Her studies suggest, therefore, that if you want to make culturally independent statements you best rely on basic level terms, such as "to feel" when referring to emotions.

The relationship between a basic and superordinate (that is, concrete and abstract) levels of "informativeness" (that is, how much information a term provides) also has a profound influence on the nature of the assessment process. As I will discuss, my belief that it is not possible to directly assess quality (Bowling 2005) and that it is necessary to use "latent variables" to conceptualize what it is that is being assessed is an indirect admission that it is not possible to imagine in concrete terms what is meant by the terms "quality" or "quality-of-life."

One extension of this "reality" is reflected in the efforts by some quality-of-life investigators (e.g., de Boer et al. 1998) to determine the number of latent variables that underlie a disease-specific quality-of-life assessment.[17] De Boer et al. (1998), of course, sees their task as assessing different assessment models, but from the respondent's perspective the issue is whether they would process a single-abstract construct directly from concrete instances, or whether they would utilize intermediate levels of abstraction, as represented by physical or mental dimensions, prior to forming a more inclusive abstract construct. De Boer et al. found that a single latent variable would account for responding to an inflammatory bowel disease and Parkinson's disease disease-specific quality-of-life assessment. Their study, however, differs from what was found when a generic assessment (e.g., the SF-36) was analyzed, and where a two latent variable model matched the data (Hays and Steward 1990: Stewart and Ware 1992). De Boer et al. concluded that since disease-specific qualitative assessments were governed by a single latent variable, it was therefore appropriate to sum the scores of the individual components (e.g., physical, emotional, social, symptoms), thereby generating a single score. However, their study also demonstrated that their conclusion may vary as a function of the disease-specific qualitative assessment being considered, such that a prudent investigator would want to demonstrate what type of model best matches their particular assessment.

While studies of this sort are appropriate for the optimal design of assessment models, it is also possible to study what a person does when faced with the vast number of indicators normally found when assessing some qualitative state. Studies of the cross-classification of information can be particularly informative. For example, given a variety of indicators, a person may very well use an abstract category (taxonomic categories) to generate a summary category (Chap. 6). This category could also use cross-classification, involving some combination of taxonomic, script, thematic, or goal-derived criteria when forming the category. Consider if a person were to generate a category they labeled "social activities." It may consist of such diverse activities as the frequency of family members' visits, vacations, going to movies, playing games, completing puzzles, and so on. This category is a good example of a single-dimension script category. However, if the person also included in the category some indication of the time a person spent away from home with others (a more abstract indicator), then a hierarchical category would also be present and this would make the category an example of cross-classification. Ross and Murphy (1999) give a more straightforward example of cross-classification, as when a bagel is described as both a grain (a taxonomic category such as time away from home) and a breakfast food (a script category such as the list of activities the person engaged in).

Another example of a cross-classification is the proposition that a qualitative assessment consists of description, evaluation. and valuation – the "hybrid construct." The collection of evaluated descriptors is an example of either a script-, thematic-, or goal-derived category which, when combined with the taxonomic (more abstract) cognitive process of a valuation, would reflect a cross-classification. Thus, only some of the quality-of-life or HRQOL definitions listed in Table 2.1 are examples of cross-classification. However, the issue here is not what investigator considers to be an appropriate combination of indicators, but rather what indicators a person describing their own qualitative assessment would combine. Actually determining what classification schema a person uses to create a category related to some qualitative statement remains an important, essentially unexamined, research area.

Parducci (1995) maintains that all human judgments involve comparison and are context dependent. If this is so, then cross-classification would be a highly likely part of the cognitive processing that leads to these judgments. However, Diener and Fujita (1997) point out that when subjective well-being is the focus of a judgment, then a person's personal goals (a goal-derived category), culturally prescribed ideals (a taxonomic category), and physiological drive states may be involved in formulating the judgment. Thus, each of these can be seen as involving a different type of classification, and cross-classification would most likely provide the structure upon which these indicators participate in a qualitative judgment. This type of analysis is, of course, relevant to a qualitative researcher, since they illustrate how the application of classification principles can be used when analyzing how a qualitative judgment occurs.

A classic example is the phenomenon of downward comparisons, an activity often occurring during a quality-of-life assessment. A typical example occurs when a person describes their quality-of-life as being better than another person whom they may see as being more ill, disabled. or economically worse off than they are themselves (Stanton et al. 1999). What is involved in this process? First, it is obvious that the person is attending to a variety of informational indicators that they use to estimate the severity of another person's illness, physical status, or economic well-being. They then will collect or classify this information into either a script or goal-derived category. A value judgment is also implied ("I'm better than that person!") in the social comparison, which suggests a more abstract (a taxonomic) categorization is also involved. Heterogeneity of the classification methods active during the act of comparison suggests a cross-classification of information. Of course, the accuracy of this judgment is another matter, but this example does illustrate how considering the method used in classifying a complex behavior can contribute to understanding the dynamics of the judgment process.

Goal-derived and taxonomic-based categories differ in an important way. "Seaside activities" (including swimming and playing football), for example, is an example of a goal-derived category, while "sports" is an example of a taxonomic category. The difference is that seaside activities imply action, while a sports category does not (e.g., Sudoku is a game and has become a competitive sport). Conway (1990) points out that for taxonomic classification to be valid requires that the category be based on "decontextualized knowledge." That is, knowledge that will remain valid across different experiential conditions. In contrast, goal-derived categories, as an example of cross-classification, remain context- or setting-dependent.

The difference between taxonomic, script, thematic, and goal-derived categories has been examined in studies of traits, autobiographic memory, and emotional expression (Stanton et al. 1999: Parducci 1995: Diener et al. 1999). Conway's (1990) study of how the memory of autobiographical material (material which presumably a person considers when they respond to a quality-of-life or HRQOL assessment) can be affected by different concepts seemed most relevant to my discussion. In a series of studies, he found that goal-derived categories speeded up the recall of autobiographical material relative to taxonomic categories, and that goal-derived categories were closely associated with the long-term memory of specific experiences. Studies of this sort reveal that the recall of personal material will depend on the type of classification that was applied to form the category or concept being used to process new or past experiences. Thus, it is much more likely that I recall and use as part of my quality assessment the fun I experienced while playing sports, rather than the list of the different sports I actually played.

As I previously discussed (Chap. 3, p. 56), a hierarchical classification is a necessary part of the cognitive process of generating an abstract construct from concrete experiences, but what also seems to be true is that abstract processes are structurally determined and present early in the development of a child's life. This suggests that an organism is born with the ability to hierarchically organize information. Thus, Chomsky's (1988) proposal that children are born with the ability to use an abstract grammar rests on the capability of persons to utilize the hierarchical relationship found in information. How a classification system can lead to an abstract thought can be literally visualized by studying a pictographic language (e.g., hieroglyphics), which is my next topic.

2.3.4 Hieroglyphics: A Language Based on Classification

Goldwasser (2005) reports a study that literally visualizes how combining the elements of a form of expression can reflect an underlying hierarchical cognitive structure. This, he suggests, can be observed when the classification system inherent to the written form of a particular language, hieroglyphics, is examined. According to Goldwasser, "the events occurring in the hieroglyphic picture – script – as in a cognitive ultrasound device-render mental processes in the making visible for our inspection" (p. 95).

Hieroglyphics is one of several pictographic languages: others include the language of the Mesoamerican Indians, Chinese, Koreans, and Japanese. Each language consists of highly idealized and prototypical pictographs that are meant to represent different words or phrases, and which can also reflect different levels of abstraction. In hieroglyphics each icon has a distinct meaning and the entire set of terms is meant to provoke a representative meaning that is more conceptual than concrete. Goldwasser (2005) also states that, "the hieroglyphic script, which is formed of hundreds of iconic signs with different semiotic roles, completes almost every word with a 'silent icon' (the so called 'discriminative') that carries no additional phonetic value of its own" (p. 96). While the discriminative is not part of the spoken language, it can give each statement a common conceptual metaphoric meaning. Figure 3.2 illustrates some examples of hieroglyphic words that include a common discriminative term or classifier: the sign for a habitat. The presence of this sign in each word classifies the entire sequence of icons relative to a common concept. Thus, the icon habitat functions as a conceptual metaphor. What is unique here is the visualization of the conceptual metaphor (you can literally see it) whereas in most languages its existence has to be inferred.

Figure 3.2 specifically illustrates the icons that make up the words "tomb" and "fortress." I have also included the English words associated with each icon within each "word." Thus, the term tomb includes a door bolt which locks in a collection of reeds, representing the artifacts of a person. The position of the door bolt suggests that the habitat is being locked up. The hieroglyphic for the word fortress includes icons for "men," "ripple of water," and the "pot" icon. These icons, combined with the quail chick, suggests a collection of men in a place, and the place, as indicated by the discriminative sign, is a HABITAT. Considered literally, these icons do not necessarily imply a fortress, but when combined figuratively it is easy to see how they might. What is also evident is that the discriminative HABITAT is functioning in a similar manner to the CONTAINER conceptual metaphor.

Goldwasser (2005) also notes that the relationship between the various icons and the discriminative icon represents a hierarchical relationship. Thus, the icon for HABITAT is taxonomically related to house, fortress, office, and so on, but it is more abstract. The combination of a series of descriptive icons modified by an abstract icon is to be expected in the construction of a cognitive entity. However, not all pictographic languages are as figurative as hieroglyphics, especially those that were in existence when hieroglyphics was a common language of communication. Interestingly, these languages have not survived or left significant numbers of artifacts. Still, the existence of hieroglyphics for some 3,000 years attests to the fact that forming cognitive (hybrid) constructs is a fundamental pattern of thought that has existed over an extended period of time.[18]

Additional evidence supporting the role for a discriminative in a pictographic languages comes from the demonstration that modern (Mandarin) Chinese, unlike English, has a set of terms that classifies objects, affects their perceived similarity and recallability, and guides a person's judgments (Zhang and Schmitt 1998). Lakoff (1987: p. 110) points out that these classifiers are some of the best examples of conceptual metaphors. Saalbach and Imai (2007) report a series of studies comparing Chinese and German and found that taxonomic or thematic relations provide the best methods for classifying objects. In contrast to Nisbett (2003), they did not find significant evidence for cultural differences in how objects were classified.

To this discussion I need to add the evidence that the alphabet characteristic of most modern spoken languages evolved from hieroglyphics. Hieroglyphics functioned not only as a written language, but also a spoken language. Thus, individual icons had sounds attached to them that could be combined together as a spoken word. However, icons or tokens are not an efficient way of communicating, so a system of attaching a common written symbol for each sound was soon developed. In this way words could be formed in a more efficient manner: in terms of space and amount of information required. This development has been attributed to the Phoenicians, and the Phoenician alphabet became the basis for such languages as Hebrew and Greek (Sass 1988). A critical part of this evolution was the introduction of vowels

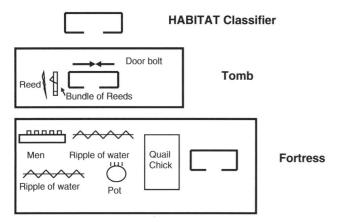

Fig. 3.2 Illustrated are the icons for the words tomb and fortress in hieroglyphics. Common to both terms is the presence of the HABITAT classifier or discriminative. The HABITAT discriminative is found in other words (such as, office, stable, cave, nest, and tent) suggesting it provides a common meaning to a diverse set of terms, and therefore functions as *a conceptual metaphor*, yet remains silent when these words are spoken.

in the Hebrew language, which, when combined with the Phoenician alphabet, evolved into Greek and modern Hebrew. Of course, what I have in this history is another example of a classification process. Thus, the attaching of a symbol, a letter of an alphabet, to a picture or icon creates a new classification system which creates a higher level of abstraction, and in this way contributes to the formation of a cognitive hierarchy. Since discriminative icon or symbols were present early in the evolution of language, these events may also reflect the historical development of the conceptual metaphor.

2.4 Concept Formation

So far I have discussed two characteristics of cognitive processing: first, that the cognitive entities are formed using only some of the information that is received; second, that this information is processed by specific cognitive processes (that is, production rules); and third, that these entities can be classified in different ways. Next I would like to outline what is known about how the summarized information comes to form a concept (that is, a more abstract entity). I am doing this because quality in general, and quality-of-life or HRQOL in particular, can be conceived of as a concept that has evolved from a person's life experiences, not as a series of isolated responses to a questionnaire. I am also convinced that studying the natural history of how a concept is formed will help in the design of a quality assessment. Here again, the available literature will help in this task.

Murphy (2002) and Smith and Medin (1981) describe two major methods whereby concepts are formed. The first method, titled the "classic model," has an extended history dating to Aristotle. It postulates that a concept has to have individually necessary and collectively sufficient defining features. In other words, a concept is determined by its definition. For example, the concept of bachelorhood requires at a minimum that adults be unmarried, but also that these adults express a lack of interest in establishing a permanent relationship. If both these criterion are present then a definition of a bachelor is present and the concept defined. But it is possible for a person to have a permanent relationship and also be unmarried. Is this person still a bachelor? No, in the sense that the second stipulation concerning the concept has not been met. But how many of these violations would be allowed before the definition would be judged to be inapplicable? It is this kind of issue that reinforces Wittgenstein's (Chap. 1, p. 8) view about the limited value of describing a word or concept in terms of its definition. As a result of these and other concerns (Murphy 2002), a second approach to characterizing concepts has evolved. This approach involves various "probabilistic models," which assume that a concept is formed as a representative summary measure, and that this summary measure will vary as a function of the probability that instances of the concept match some ideal outcome. This matching can occur in two ways: in terms of prototypes or exemplars (see below).

Murphy (2002) describes the classic view that a concept is mentally represented by a definition, or as he puts it, "a definition is the concept …" (p. 15). Thus, a concept is acquired by learning which characteristics of some entity define the entity. This can occur by maximizing within-category similarity relative to between-category similarity. For example, if someone rates their quality-of-life as good, then it would be expected that the person considers other experiences and events that would be rated in a similar way. Medin (1989) and Murphy (2002), however, both point out that to assume that a concept is formed by simply combining common features or similar ideas and experiences is flawed. The reasons for this, they state, is that it is very hard to determine the necessary and sufficient features that define what is meant by "similar." In addition, not all ideas or experiences that are similar are good fits of a concept. Thus, a person may provide a list of items or experiences that they consider important indicators of their quality-of-life or HRQOL, yet finding what is common amongst them may be quite difficult.

Murphy (2002: p. 18) used to think that in certain technical domains, the classical view may still prevail, especially where it is possible for a concept to be well defined. Thus, Major League Baseball has a specific definition for defining when a pitch is a strike (that is, the ball crosses a plate [a designated object, on the ground] within a certain space area), yet as any baseball fan will report, there is still great variability in how this definition is applied. Many of the domains considered characteristic of a quality assessment also appear at first glance to be definable as concepts, yet in practice are far more fuzzy, even when a perfectly clear set of rules exist to define membership in the domain (e.g., a symptom domain, or a physical functioning domain, and so on).

Some of the features of the classic model that contribute to its popularity include its ties to traditional logic and ability to account for hierarchical relationships. Thus, the statement "all roses are plants" can be shown to be true by demonstrating that roses are an example of bushes, and bushes are a form of plant. Implicit in this logic is also a hierarchical relationship, where definitions are nested within one other. However, when people are given material to organize into categories, as when they are asked to decide if a set of items fits into another set of items, they do not necessarily follow the logic and hierarchy implicit in the material presented to them. For example, Murphy (2002: p. 27: quoting Hampton 1982) gives an example where subjects had to decide if a car seat was a chair and if this type of chair had the defining features of a piece of furniture. What Hampton (1982) found was that people agreed that a car seat was a chair, but disagreed that it was a piece of furniture. Examples of this sort again raise questions about the usefulness of the classic model, but I will still be able to demonstrate the analytic usefulness of examining some data set from the perspective of a hierarchical relationship (Chap. 7, p. 211).

The notion that "a definition constitutes a concept," to quote Murphy (2002), is a familiar phrase in quality-of-life research, except that I have also heard it as part of the definition of behavioral and social assessment. Thus, I am familiar with the statement that "an operational definition can be used to define an abstract concept, such as quality-of-life or HRQOL." What is interesting is to consider whether the criticisms of the classic model can also be applied to this approach to assessment. What is apparent here is that the act of using an operational definition to define a concept may encounter the problem of determining if such a definition of quality is necessary and sufficient, and whether the operational definition can clearly establish what is required for membership in this concept. If it does not, then it may be necessary to consider other methods whereby concepts are formed to conceptualize how quality is assessed.

Similarity does appear to play a role in concept formation, but more likely after the concept is formed rather than prior to its formation (Medin 1989: Medin et al. 2000: Smith and Medin 1981). Thus, instead of a concept or category being formed on the basis of similar elements, it may be that after a coherent concept is created, elements that make up the concept appear similar. This occurs, as Medin (1989) points out, because people act as if concepts have underlying properties or essences. Thus, I might use a heuristic in which I postulate that things that look alike tend to share deeper properties, such as similarity. I would conclude this, however, only after the concept or category was formed.

What also seems true, is that people develop theories based on what they learn, and they use this knowledge to organize a concept. These concepts are considered knowledge based, and change as new knowledge is acquired. This dynamic is similar to what I earlier postulated is going on when a cognitive entity is formed and changed. The theory-based (or prior knowledge) approach also offers an explanation of how diverse categories are grouped in a coherent manner (that is, cross-classification). Medin (1989) gives the example of having a category consisting of children, money, photo albums, and pets that only makes sense as a category if it is understood that these elements are things someone would want to remove from a house in case of a fire. I could also add personal papers to this list, and I would be consistent with the intent of the category, yet select an item that is different than the other items in the category.

If similarity between ratings, ideas, or experiences is not sufficient to define a concept, then what is the alternative? The alternative, as mentioned above, is probabilistic models, which assume that concepts are "fuzzy," and are organized around a set of correlated but knowable characteristics (Rosch et al. 1976: Medin 1989). This knowable set of characteristics originates from a "basic" level in an information hierarchy (Rosch et al. 1976). The characteristics may function as a cognitive summary measure or prototype. Thus, examples of the concept will have a probability of matching the prototype.

The idea that concepts are "fuzzy" (Wittgenstein 1953: Rosch and Mervis 1975) was popularized by Wittgenstein (1953), who pointed out that the different characteristics of games have a common concept without being limited to a set of similar features; that is, they have a "family resemblance." Thus, *Monopoly* is called a game because it has rules and more than one person is needed to play it, but throwing a ball against a wall is also called a game even though it has no rules and may involve only one person. On the other hand, "debating" has many game-like features, but would not be called a game. So the similarity of features does not appear to be sufficient to define a category or concept.

The diagnostic term "depression" is also an example of a fuzzy concept. Inspection of the *Diagnostic and Statistical Manual IV* (American Psychiatric Association 1994) reveals that for a person to receive the diagnosis of major depression requires that they have five of nine symptoms that have been present over a 2-week period. Yet, two people who have both received the label of "depressed" may have quite different symptom profiles. It is in this sense that depression is a "fuzzy" concept.

Earlier I reviewed the quality-of-life or HRQOL definitions in Table 2.1, and I suggested they each contained linguistic or conceptual metaphors that were used to establish the definition. One such group of definitions included a causal relationship as part of the definition (e.g., a person's disease and treatment impacted or "caused" various changes in their life). Lakoff and Johnson (1980) have suggested that causal statements are actually examples of prototypes in the sense that they have a variety of characteristics, only some of which may be relevant in any one setting. They list 12 characteristics (Lakoff and Johnson 1980: p. 70) that have the common feature of suggesting "cause." Some examples from their list include:

- The agent has a goal that some change of state in the patient.
- The change of state is physical.
- The agent has a "plan" for carrying out this goal.
- The plan requires the agent's use of a motor program.
- They go on to state:

The twelve properties…characterize a prototype of causation in the following sense. They recur together over and over in action after action as we go through our daily lives. We experience them as a gestalt: that is, the complex of properties occurring together is more basic to our experience than their separate occurrence. Through their constant recurrence in our everyday functioning, the category of causation emerges with this complex of properties characterizing prototypical causation. (p. 71).

Smith and Medin (1981) describe a second probabilistic approach: the exemplar model. The exemplar model denies that there is a single cognitive summary representation that functions as a prototype for all comparisons, and instead claims that concepts are represented by examples. Thus, a clinician would be able to diagnosis a depressed person on the basis of his or her past experience with specific persons whose depression was manifest in different ways. They would be able to deal with multiple patients, because they have had examples of these different patients in the past.

The research literature is filled with studies that support one or the other of the two basic models presented by Smith and Medin (1981). Fodor (Cain 2002: p.71), however, challenges the view that concepts are "fuzzy" and prototypical in form. Fodor claims that is not probabilistic, but instead is "compositional," with more primitive linguistic units being sequentially combined in unique ways to form a concept. The sentence tree in Fig. 2.1 illustrates the elements that make up a "composition." Thus, there are an infinite number of concepts, most of which are not prototypes. Consider the example (Cain 2002: p. 71) where someone was labeled as a prototypical "grandmother." This particular grandmother has children who are married to twin dentists. Thus, while the grandmother was prototypical, she had some unique characteristics. The point here is that the concept of "grandmother" can exist, but only some examples of the concept would be prototypical. A second reason why Fodor believes that prototypes "don't compose" is because you can have a complex prototype which bears little resemblance to the prototypes of its component parts. For example, you may claim that a goldfish is a prototypical pet, but a prototypical pet is not likely to be a fish and a prototypical fish is not likely to be a goldfish. This implies that prototypes are dependent on the context within which they are presented, which would not be what would be expected if they were major determinants of how a concept was formed. Finally, arguing that a concept is based on a prototype does not free it from reliance on a definition, since what a prototype is also requires a definition.

This discussion should confirm that a rich literature exists in the area of concept formation and that there are many interesting ideas to apply to the study of the term "quality" and its applications as a concept. Again, the argument here is that a concept is a critical part of a hybrid construct, and knowing how concepts are formed and evolve should help determine what is being assessed when a person is asked to judge the quality of an object, event or their life.

2.5 Metacognition

If a person reflects about a thought they just had they are engaged in a metacognitive process. Figure 3.3 illustrates

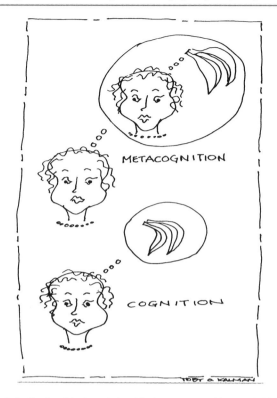

Fig. 3.3 Depicted is the relationship between cognition, and thinking about what the person is thinking about (i.e., metacognition).

this process. Other examples of metacognition include a person thinking about how to recall past memories, how to plan an argument, or how to process unique experiences. I have proposed (Fig. 3.1; Barofsky 2003) that metacognition plays a defining role in a qualitative assessment.[19] Thus, metacognition is involved when a person considers the various attributes of a painting that make it special, decides on what criterion to use decide what is an important life event, how to evaluate the difference between pleasant and unpleasant experiences, how to weight the importance of various components of a qualitative assessment, and so on. What also has to be considered is how to combine the diverse components that make up a quality assessment (e.g., physical or mental components) into a composite.[20] However, it is also appears that I can make qualitative judgments in a way that appears fast and without reflection. I will consider this type of judgment in the next section, and attempt to resolve this apparent discrepancy in what I consider a general principle.

It is also interesting to note that the absence of metacognitive activities, as when a person has become emotionally impaired, may account for variability in qualitative responding, while the encouragement of metacognitive processes may actually increase the reliability of qualitative responding. Support for these possibilities comes from the suggestion that certain forms of psychopathology (e.g., narcissism: Dimaggio et al. 2002) may be associated with impaired metacognitive processing, and interestingly, the person's capacity to judge their qualitative state may also be impaired. In such instances, the encouragement of metacognition (such as with thought stopping) may be therapeutic, and facilitate the ability of a person to report on their quality-of-life.

Koriat (2006) describes metacognition as follows:

> Metacognition concerns the study of what people know about cognition in general, and about their own cognitive and memory processes, in particular, and how they put that knowledge to use in regulating their information processing and behavior…Nelson and Narens (1990) propose a conceptual framework that has been adopted by most researchers. According to them, cognitive processes may be divided into those that occur at the object level and those that occur at the meta-level: The object level includes the basic operations traditionally subsumed under the rubric of information processing-encoding, rehearsing, retrieving, and so on. The meta-level is assumed to oversee object-level operations (monitoring) and return signals to regulate them actively in a top-down fashion (control). The object-level, in contrast, has no control over the meta-level and no access to it. For example, the study of new material involves a variety of basic, object-level operations such as text processing, comprehending, rehearsing, and so on. At the same time, metacognitive processes are engaged in planning how to study, in devising and implementing learning strategies, in monitoring the course and success of object-level processes, in modifying them when necessary and in orchestrating their operation…We should note, however, that the distinction between cognitive and metacognitive processes is not sharp because the same type of cognitive operation may occur at the object level or at the meta level, and in some cases it is unclear to which level a particular operation belongs (Brown 1987: p. 290). (p. 290–291).

Metacognition has been extensively studied by developmental psychologists and memory researchers. There are two fairly common research areas where metacognition has been studied: research dealing with "feelings of knowing" (Hart 1965, 1967) and research dealing with "judgments of learning" (Dunlosky and Nelson 1994). These areas of research have documented the monitoring and control function of metacognition (Koriat 2006).

What is unique about metacognition is its top-down regulation of behavior. Thus, metacognition not only monitors experiences so as to control some outcome, but also can control various cognitive processes themselves. I will invoke this characteristic of metacognition when I discuss how the term well-being might function as a regulatory process in Chap. 11, but also suggest that it could offer an alternative explanation of the experimental findings common to the response-shift phenomena (Schwartz and Sprangers 2000).

Metacognition operates its top-down regulation by mediating between automatic (unconscious) and controlled (conscious) cognitive processes,[21] and also between subjective experience and behavior. To illustrate, while the dominant view is that subjective experience determines behavior (e.g., Bless and Forgas 2000),[22] there is also available evidence suggesting that behavior can feed back and influence what I experience. An example of how this might work is illustrated by the James-Lange theory of emotion. This model postulates that my experience of an emotion can literally be an afterthought, meaning that I will become aware of an experience after I have behaved in a particular way. For example, I became afraid only after I ran away from a bear. Since the behavior of running away occurred automatically (almost reflexively), it is an example of an unconscious process, and as such, fear was experienced after the behavior occurred (see Chap. 6). This sequence of events suggests how a metacognitive process might work, in that the feedback from a person's behavior determines what was experienced.

Evidence of this type will be quite important in any theoretical approach to quality assessment, since a quality assessment may also occur in ways that a person may or may not be totally aware of, or be aware of only after reflection. Thus, a number of social psychologists have postulated that a person's feelings (e.g., sense of well-being), attitudes, and beliefs are based on a person observing their own behavior. A particular dramatic demonstration of this phenomena is when people report feeling happy after they are asked to smile, and angry after they were asked to make an angry face (Strack and Deutsch 2004).

In addition, it might be expected that the greater the effort involved in recalling past experiences and performance, the better the recall; an outcome relevant to any quality assessment. However, studies have shown that if a person is asked to recall 12 examples of their own assertive behavior they would rate themselves as more passive then if they were asked to only recall six examples of assertive behavior (Schwarz et al. 1991). This is a somewhat counter intuitive result. Of course the accuracy of these self-assessments is not known, but the initial interpretation of these results would be that the more I know about myself, the more I would confirm my initial impression about myself (that is, 12 examples would be more likely to confirm what I know about myself than six examples would be). This suggests that the amount of information a person recalls is not simply related to the frequency of experiences or events that they have had.

Still, the ease and accuracy with which I can recall a large number of events reflects how fluent I am with my past, and this would also be expected to be relevant to any quality assessment. In addition, I might believe that the more examples I have had of a particular experience, the easier it would be for me to recall the experience. These beliefs form the basis of naive theories which structure how the informational value of thoughts and metacognitions are determined. Figure 3.4 illustrates Schwarz et al.'s (2007) notion of the relationship between these factors. Most important is their assumption that if the informational value was informative, then this would, in effect, qualify the relationship between metacognitive experiences and thought content. Of course, Schwarz et al.'s (2004; 2007) view that metacognitive experiences qualify thought is similar to my notion (Barofsky

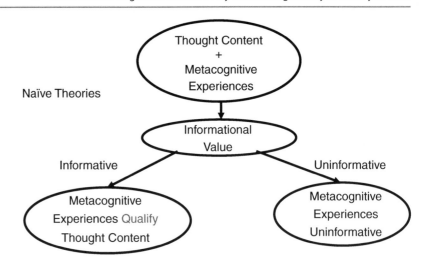

Fig. 3.4 This figure is a simplified version of a figure published by Schwarz et al. (2007); Permission to reproduce the figure was granted by the publisher. In their model, a quality judgment would be the outcome of combining thought content (e.g., descriptive or evaluative elements of a hybrid construct) and accompanying metacognitive processes. The informational value of this cognitive process would then be determined.

2003) that the qualification of terms that have been described and evaluated constitute a quality assessment (Fig. 3.1).

Koriat (2006) also provides an interesting discussion concerning the relationship between metacognition and consciousness, and his discussion may help clarify some of the issues that arise when I claim that a person has to be aware of their cognitive processing (Fig. 3.1) when they perform a quality assessment. Koriat (2006) focuses his discussion within the area of adult memory research and asks a number of questions about this research which can be transposed into questions concerning how a qualitative assessment occurs. Thus, he asks what the basis is for metacognition; that is, "How do I know that I know?" (Koriat 2006: p. 291). He also wants to know how accurate or valid subjective impressions about knowledge are, and what processes underlie the accuracy or lack of accuracy of these judgments. Also of concern is how the output from monitoring processes affect various control processes. He states that, "the study of subjective monitoring knowledge addresses a defining property of consciousness, because consciousness implies that not only that I know something, but also, that I know that I know it." (Koriat 2006: p. 292) This state of knowing that I know can be valid or an illusion, but most importantly, forms the basis of the top-down functions that metacognition promotes. Thus, the study of metacognition implies that the person is an active participant who has an arsenal of cognitive options they can apply at will towards achieving various goals and solving various problems. Most importantly, the selection of these cognitive options seems to be guided by feelings.

Implicit in this model of metacognition are two assumptions. The first is that self-controlled processes affect behavior, and the second assumes that subjective experiences (e.g., feelings affect judgments; Schwarz and Clore 2003) have a causal role in determining behavior. The first proposition reaffirms the statement above that a person is an active participant in regulating various cognitive processes (e.g., memory), even while it is acknowledged that this type of processing can occur automatically, below awareness. The second proposition points out that while a person is not only an active determinant in what they think, what they think can cause changes in behavior. I will discuss the implications of this second assumption directly in Chap. 11, where I review Schwarz's views that feelings inform us about our well-being (Schwarz and Clore 2003).

What does the relationship between metacognition and consciousness tell me about a quality assessment? First, it suggests that one of the consequences of monitoring my qualitative status would be an awareness of this state that could lead to changes in my behavior. Thus, if a breast cancer patient decides to separate from her spouse after a mastectomy, then the awareness of the quality of her existence (e.g., vulnerability) following the diagnosis and treatment may have been sufficient to make this life-altering decision.

3   The Application of Cognitive Principles

3.1   The Fast and Automatic Assessment of Quality

Although I have just argued that a quality assessment requires reflection (that is, a metacognitive process), I also mentioned that fast and automatic cognitive activities could be involved. A prime example of this is the experience of first impressions. Here it seems that I can quite quickly make a judgment with minimal reflection about the quality of what I am seeing, hearing or tasting, or about a person's life (Fig. 3.1). How can this happen and what does it mean about the process of quality assessment? For one, it suggests that there may be two distinct cognitive processes that determine how I experience or judge quality. This notion can be traced to William James (1890) and has been referred to as "dual

Table 3.3 The cognitive task of assessing quality

Type of quality assessment	Cognitive process	Neurocognitive system
Fast/automatic	Affordance	Perceptual
	Trait (facial) perception	Perceptual-cognitive
	Emotion inference and appraisal processes	Cognitive–emotional

process theory," with one process occurring fast and automatically, while the other requiring more time-dependent deliberative (possibly reflective) processing. These two processes also seem to trace phylogenetic history, with the fast system based on the ancient subcortical neural mechanisms that mediate fight and flight, and the slower system reflecting the higher cortical development characteristic of latter phylogenetic development. Automaticity can also be acquired, as when a particular behavior becomes so well learned that it occurs with minimal cognitive processing (Navon and Gopher 1979).

If the experience or judgment of quality occurs automatically, how will I know that it has occurred? Moors and De Houwer (2006) have reviewed the various definitions of automaticity and state that "automaticity should be diagnosed by looking at the presence of features such as unintentional, uncontrolled/uncontrollable, goal independent, autonomous, purely stimulus driven, unconscious, efficient and fast." (p. 297) They point out that several investigators have different models of automaticity, but at its core conceptualizing automaticity is a problem in how best to conceptualize information processing.

Here I will briefly review three different perceptual and cognitive processes that may facilitate a rapid and automatic assessment of quality (Table 3.3). I will then point out, as in Chap. 6, that most often fast and automatic and reflective cognitive activities occur together, even though it is analytically possible to separate them. In fact, it will become obvious that the importance of the information that has been detected, described, and evaluated may vary from being biologically predetermined (why I run when I see a bear) to acquired by experience. Several cognitive mechanisms will emerge as being active in the automatic assessment of quality. These include ecological factors (e.g., the optical array); the fact that I can extract information from the stimulation I receive (e.g., a face); and that cognitive appraisal may determine which emotions I experience. All three of these processes may contribute to the quality of what I experience.

3.1.1 Affordance

Affordance is a perceptual process which indicates how well an object, an animal, or a person interacts with its environment. It was proposed by Gibson (1979), who conceived of perception not as a passive process of receiving and responding to sensory stimulation, but rather as the dynamic process of acting upon, and interacting with, the environment. The extent of the match that occurs between a person's actions and interactions with their environment can then be viewed as an indicator of quality. Gibson, according to Warren (1984), proposed that, "animals visually guide their behavior by perceiving what environmental objects offers or affords for action. An affordance is the functional utility of an object for an animal with certain action capabilities… Affordance are thereby based in material properties of the animal and the environment and exist whether or not they are perceived." (p. 683) Thus, the reason I can walk up four flights of stairs while an infant may not is partly due to differences in muscle capabilities, but also because I and the infant differ in the affordance the stairs has for each of us. Clearly, I have learned something about what I am perceiving that an infant has not yet learned. Eleanor Gibson (2000; J. Gibson's wife and collaborator) describes the process of learning affordance as a process of differentiation and selection, not the addition or construction from smaller units. Differentiation occurs by means of exploratory activity and selection. This can be illustrated by contrasting my ability to detect an object placed in my hand without being allowed to manipulate it, as opposed to being able to turn it and feel it from different sides. Clearly, being able to engage the object (being able to act on the object) will determine the ease and accuracy with which I perceive it. Gibson referred to his approach as an ecological one, and his instance that how I act is determined by the structure of the environment, whether or not I know it, will be an important one in this book, especially when applied to understanding how the structure of the central nervous system determines how I think, reason and feel.

Norman (1988) applied the concept of affordance to human–machine interactions and expanded the term to refer to just those action possibilities which are readily perceivable by someone.[23] This opened the concept to the influence of a person's goals, plans, values, and beliefs. This, Norman felt, was important because Gibson's model did not always predict how a person would act in their environment. Consider the following example. Say that someone came into a room that had a chair and a table. Based on Gibson's version of affordance, it would be just as likely for the person to sit on the table (something that could be done) as he or she would to sit on the chair. By allowing the person to influence with what he or she would interact, Norman opens the concept of affordance to influences from other cognitive processes, including the more time-dependent process of reflectance. Affordance, as Gibson used it, is a descriptive term that implies an evaluation but has no place for reflection. Norman's extension of the term allows for the presence of reflection.

The concept of affordance, as used by Norman, will be helpful when understanding a number of settings where quality is being judged. For example, affordance helps me understand why I like the display panels in some cars and not others; why I appreciate how a suit fits; how a building's architecture is designed; and so on. However, the concept has been expanded to include certain social psychological situations, such as perceiving the intent and appropriateness of the actions of others (Stoggregen et al. 1999) and cooperative tool-grasping activities (Richardson et al. 2007). In both examples, the quality of the encounter may be determined by the degree of affordance of the person acting on its environment. However, the most important application of the concept of affordance may be its potential role in conceptualizing what is meant by the phrase "quality control."

To appreciate the role that affordance can play in quality control, consider how it can be applied in HCI studies, particularly when a medical device is involved. Luo et al. (2006), for example, were interested in analyzing the role of affordance in the design of a mammography teaching program. The software that made up the teaching program provided the student with a set of tools for optimizing displays, scanning images, comparing different projections, marking clinically important regions, and so on. The authors found that the teaching program reached its goals by creating an interactive environment. The quality control of mammography could be increased by reducing the number of errors or discordant interpretations between observers. Xie and Zhang (2006) discuss the application of affordance to electronic health record systems. They were particularly interested in categorizing the type of affordance as correct, false, or hidden. These studies suggest that an ecological perspective maybe a useful way to conceptualize quality control.

3.1.2 Trait (Facial) Perception

A trait is a characteristic of an individual which remains the same over time and situations. In a biological context it refers to an individual's phenotype (such as hair color), but in a psychological context it usually refers to some aspect of a person's personality (such as being an introvert or extrovert). Trait perception, however, refers to one person drawing inferences about another person by inspecting some aspect of that person's appearance, as would occur if I inferred something about a person's personality by inspecting his face. This situation is of interest because it is similar to what I do when I inspect a statue and decide if it is beautiful; when I consider the contents of a meal and decide if I want to eat it; or if I contemplate a particular style of life and decide whether I want to live it. In many instances making these judgments occurs quickly and automatically, much as occurs when inferences concerning facial perceptions are made. Thus, examining the cognitive processes involved in facial trait perception should provide insights into what may occur during some qualitative assessments.

Making judgments about the appearance of a person's face and relating it to the person's personality dates to the ancient Greeks, but was formalized by Lavater's 1797 monograph where he carefully described how different facial components (e.g., eyes, eyebrows. lips, and so on) are related to a person's disposition or temperament. Lavater's monograph, while bordering on phrenology, is still relevant today since there is sufficient empirical evidence to support the supposition that attractive people have a better outcome in life than less attractive people (e.g., Hammermesh and Biddle 1994). However, the evidence that it is possible to accurately associate facial characteristics to specific personality indicators is less clear. Hassin and Trope (2000) reviewed a variety of studies and found that while the judgments a person makes concerning the facial expression–personality association is often reliable, there is little evidence that it is valid. However, repeated exposure to a person's behavior appears to improve the validity of the facial expression–personality association.

Hassin and Trope (2000) also report a series of studies that examined a respondent's tendency to read personality information either from or into a face. They found that, depending on the verbal information provided, both processes occur. For example, they report a study in which the respondents were provided with either ambiguous or unambiguous verbal information about test faces, and found that the more ambiguous the verbal information, the more likely the respondents read information from the faces into their verbal descriptions of the faces. Thus, a person with a cleverlooking face who had an ambiguous achievement record was perceived as more likely to have succeeded than a person who had a less clever-looking face, but the same ambiguous career information. Further, the person whose face was not particularly distinctive but had a record of achievement was perceived as more attractive than the person who was attractive and had a comparable level of achievement.

Hassin and Trope (2000) speculate that a dynamic relationship exists between the perception of the face and its interpretation, and that this is an important part of interpersonal exchanges and relationships. Most interesting is that they speculate that while the dynamics of perception/interpretation can occur serially (that is, linearly), it may also result from a parallel distributed process (PDP; that is, nonlinearly). As they state:[24]

A network in which focal and contextual information are both represented, and in which (the process towards) the end state can affect both kinds of information, will by definition yield dialectical processes. A network of this sort will yield dialectical processes if the flow of excitation between the two nodes (e.g., A and B) is bidirectional and if it works the same way in both directions (either A or B excite each another or they inhibit one another). …Whether in serial or in PDP the implementation, the implication of the above argument is the same: The effect of physiognomy might grow when taking part in a dialectical process. (Hassin and Trope 2000: p. 849)

They also point out that the dynamics here can be influenced by memory and recurrent judgments and decisions. Finally, they claim that the perception/interpretation process can occur in a fast/automatic manner, although they do not provide formal data to support their claim.

The paper by Hassin and Trope (2000) is of interest because their observations seem compatible with what might occur when the quality of an object, event, or life is assessed. In particular, the notion that I both read from, as well as read into, what is observed, suggests the existence of a dynamic that would be expected when a quality assessment occurs, particularly if a reflective process is part of this assessment. Thus, it would be expected that when a person describes their quality-of-life, they not only read from the history of their life, but also they read into their life history, their expectations, and preferences.

If, as Hassin and Trope (2000) contend, these processes occur automatically, then it is of interest to know how fast they occur, especially since a reflective controlled process would be expected to take longer than an automatic response. Studies by Bar et al. (2006) and Willis and Todorov (2006) demonstrated that subjects will respond to faces presented to them within 35–100 ms. Thus, it appears that drawing inferences about another person's face occurs very rapidly. However, these two studies are confounded by the fact that the respondents were primed to detect a particular type of facial expression; it would have to be demonstrated that the reported speeds of detection would also occur naturally.

The perception of an object has been described as involving its detection, perceptual categorization, and recognition. These steps, of course, exemplify a linear model, but an interesting study by Grill-Spector and Kanwisher (2005) suggests that while these steps may be analytically separable, in practice when you know an object is present you also recognize what it is. This suggests that when an object is detected, the significance of the object (an otherwise reflective process) will simultaneously be determine. Consider the situation where a person encounters a threatening animal, or an animal that is known to have been harmful to humans. My prior knowledge concerning the animal will, therefore, be part of my current response to this situation. This sequence of cognitions is also present when a qualitative assessment occurs. Thus, I describe the appearance or disposition of the animal, evaluate whether the animal is threatening, and knowing something about how the animal interacts with humans, I conclude that it is important for me to remove myself from the vicinity of the animal (a reflective process). Responding in this manner, of course, is characteristic of the flight response common to a wide range of organisms; a response pattern that contributes to the survival of the organism and is phylogenetically quite old. What is interesting here is that it is also the essence of a (negative) qualitative assessment. Trait face perception can obviously be the basis of a flight response, a response which can also be characterized in ecological or affordance matching terms. A similar scenario can be written for an approach response, or a positive qualitative assessment.

3.1.3 Emotion Inference and the Appraisal Process

When I look at a painting by van Gogh, listen to Mahler's ninth symphony, or read an essay by Winston Churchill, I find myself experiencing the emotion each author is attempting to communicate. Alternatively, I may have a clinician that is not being particularly emphatic to my plight (e.g., Pollak et al. 2007), and I wonder what emotion he or she might be feeling. In social situations, I may notice that a colleague has made a terrible mistake in his job, and I am wondering how he may be feeling. In each of these instances I am inferring something about the emotions someone else is experiencing. Thus, I might say, "What a beautiful painting – I can feel van Gogh's pain", or, "What a moving piece – Mahler must be greatly distressed", or "I feel pity for my friend for having made such a mistake," and so on.

How is it that I can identify such emotions? Seimer and Reisenzein (2007) have suggested that to infer the emotions being expressed by others requires that I engage in the cognitive process of appraisal, and it is what I appraise that tells me what emotion is being expressed. This appraisal mediation hypothesis of emotional inference, however, is a special case of the more general cognitive appraisal process (Lazarus, 1991: also see Chap. 11, p. 426), a process assumed in defining how I experience particular emotions. Cognitive appraisal is of interest because the emotions engendered by this process are assumed to contribute to the qualitative experiences I have (Fig. 1.2). The same would be assumed concerning the emotion inferences I make. Thus, my judgment concerning the quality of the van Gogh painting would be based on my appraisal of the feelings I had after viewing it, and this process would be repeated each time I judged the quality of some event, object or life.

This suggests that the judgment of quality is based on the cognitive appraisal of feelings, whether inferred or directly experienced. Support for this notion comes from folk psychology where the lay public assumes that one needs to engage in an appraisal process before the quality of an experience becomes evident. Support also comes from the observations that many of the terms commonly used to operationalize appraisal (such as desirable, expected, fair, important, responsible, and so on) are also used to express a qualitative outcome. However, the experimental evidence supporting the role that appraisal plays in emotion expression is not clear cut (e.g., Siemer et al. 2007), as would be its extension to a judgment of quality.

One of the reasons why the experimental evidence supporting appraisal's role in emotional expression is not clear cut is because appraisal is inherently an abstract term that is ordinarily operationalized by more concrete basic-level indicators (see above). These indicators overlap with but do not occupy the exact same semantic space as the abstract term, appraisal. This discrepancy represents a cognitive and linguistic gap that is commonly found when operationalizing abstract concepts. I discussed a related aspect of this issue earlier (Chap. 3, p. 56), when I pointed out that I could not imagine the term "furniture" without referring to its concrete indicators that I could objectively assess, such as chairs, tables, and so forth. The same situation exists for the term "appraisal"; I cannot generate an image of what appraisal would be like without using more concrete terms or indicators. Thus, the collected evidence describes the concrete indicator, with the relevance of these data to appraisal being dependent on an inference, not a direct demonstration. Unfortunately, many investigators interested in studying abstract constructs, such as "appraisal" or "quality," do not acknowledge that a gap exists between the abstract concept and its concrete manifestations. As a consequence they persist in arguing that it is possible to define the abstract concept, when what they are doing, at best, is assessing and defining the "basic" indicators which they assessed.

Siemer et al. (2007) report a study which attempts to override this cognitive–linguistic gap by demonstrating that appraisal is a necessary and sufficient cause of individual emotions. Their hope is that by demonstrating that specific appraisals cause specific emotions, they will, in effect, be demonstrating that the abstract and concrete indicators are one and the same. They also claim that the appraisal-dependent emotions determine the quality of some object, event or life. Thus, they state that, "appraisals are sufficient causes of the quality and intensity of an emotional response" (p. 592). By sufficient, they mean that different appraisals of the same situation would be sufficient to result in different emotional responses (the sufficiency hypothesis). They then cite evidence that supports this hypothesis (e.g., Ellsworth and Scherer 2003: Roseman and Smith 2001). However, Siemer et al. (2007) also point out that many of these studies are confounded by allowing both the type of appraisal and situation where the appraisal occurs to vary. In contrast, a strict test of the sufficiency hypothesis requires that different appraisals would result in different emotions with the situation remaining constant. A second concern expressed by Siemer et al. (2007) is that many of these studies do not directly assess emotions, but rather assess past memories of emotions. In addition, respondents are not very accurate predictors of their reaction to emotion-inducing conditions, as compared to what transpires in real life. Both these issues raise questions about the validity of the results of the reported studies.

The second part of the appraisal hypothesis of emotions states that appraisals are necessary causes of emotions (the necessity hypothesis). Here it is assumed that if the same situation induced different emotions, the situation was appraised differently. Thus, this hypothesis makes the strong claim that only appraisals cause the emotions which occur in a particular situation. The necessity hypothesis also claims that the quality of a particular experience would only be caused by the appraised emotion. Siemer et al. (2007) point out that some investigators (e.g., Berkowitz and Harmon-Jones 2004: Izard 1993) claim that noncognitive (nonappraisal) factors can cause emotions such as pain and hunger. Other investigators, however, propose that appraisal processes could still mediate the emotions caused by these noncognitive factors (e.g., Clore and Centerbar 2004). The investigators acknowledge that it is very difficult to directly prove that appraisals cause emotions. Instead, they propose that if an individual's emotional response to a stressful task was different than others, then these individuals must have had different appraisal profiles.

In the research study that Siemer et al. (2007) performed, respondents were exposed to a stressful task and received negative social reinforcement. Then they answered a questionnaire which asked to rate how guilty, shameful, sad, angry, amused, or pleased they felt, along an 11-point scale, from none at all to extremely. After this, they filled out a five-item questionnaire that addressed different appraisal dimensions: controllability, self-importance, unexpectedness, other-responsibility, and self-responsibilities. Siemer et al. interpreted the results of their study as demonstrating that appraisal predicts the intensity of individual emotions, and in addition, that the profile of emotional responses is compatible with the profile of appraisal dimensions. From this they conclude that their study supports the strong claim that appraisals cause emotions in specific situations.

This conclusion, however, can be contested, and I do so in Chap. 11. For example, there is a strong literature (Zajonc 1968) which suggests that the emotions can be shown to be present prior to onset of cognitive processes (Chap. 11, p. 422), which if true, would raise questions about the necessity of appraisal for emotional expression. It is also interesting to note that a person in cognitive decline can still retain considerable emotional presence, including stating preferences (values) in the absence of short-term or even long-term memory (Mozley et al. 1999). More directly related to the studies by Seimer and his colleagues is a critique offered by Parkinson (1999; 2007) who claims that it is possible for a person to process situational information and reach emotional conclusions without involving the appraisal process. My intent here, however, is not to provide a detailed critique of appraisal theory, but rather to illustrate the usefulness of a cognitive perspective when studying how the assessment of quality might occur. However, I do want to briefly review a study by Schmitz and Johnson (2007), since I believe they provide a meaningful perspective on how to conceive of the role appraisal plays in emotional expression and a qualitative assessment.

Basically, the point that Schmitz and Johnson (2007) make in their paper is that when neurocognitive processes are examined, it appears that supramodal processes exist, and these processes emerge from the coordinate activities of several different neural sites or subsystems. They specifically propose that:

> large scale supramodal processes mediate appraisal of self-relevant content…we distinguish between two top-down sub-systems involved in appraisal of self-relevance, one that orients pre-attentive biasing information (e.g., anticipatory or mnemonic) to salient or explicitly self-relevant phenomena and another that engages introspective processes (self-reflection, evaluation or recollection) either in conjunction with or independent of the former system. (p. 586).

Thus, appraisal from this perspective exists but it is an epiphenomenon, being a product of the interaction of selected neurocognitive sites. This notion provides a way of accounting for the abstract–concrete gap I discussed earlier, but also increases the credibility of the Zajonc (1968) and Parkinson (1999; 2007) arguments that appraisal is not a necessary component of emotional expression. Also of interest is that Schmitz and Johnson (2007) postulate that top-down systems are active, one of which involves reflection and evaluation (e.g., metacognitive). Thus, it would not be surprising to find that appraisal can exist in several formats: as an unconscious autonomic process, or as an integral part of reflection.

This discussion is important not only because it provides an opportunity to demonstrate how a cognitive process might function to produce a qualitative assessment (cognitive appraisal leads to emotions, and emotions determine the quality of an experience), but also because of its emphasis on the role emotions might play in a qualitative assessment (Chap. 11). As illustrated in Figs. 1.2 and 1.3, I acknowledge a role for emotions in a qualitative assessment, but also argue that situations exist which would involve minimal, if any, emotional participation in a qualitative assessment. Finding the limits of each dimension in a qualitative assessment will be a continuing task. However, there is no doubt that determining the role of appraisal in quality-of-life research is relevant, as becomes evident after reviewing the important papers published by Rapkin and Schwartz (2004) and Schwartz and Rapkin (2004). In addition, an empirical study by Padilla et al. (1992) demonstrates that appraisal of "opportunity" or "danger" are significant contributors to the variance of a measure of overall quality-of-life, and provides support for a role for appraisal in a qualitative assessment.

3.1.4 Summary

When I first learned to ride a bicycle, I started out rather crudely, engaging in gross motor movements resulting in scraped knees and elbows. However, with practice I became much more coordinated and stable, so much so that today I probably could get on a bicycle and ride fairly well, even after an extended absence from bicycling. This transformation from a deliberately directed sequence of thoughts to a well-retained automatic motor process is replicated in multiple venues, including skill development (e.g., bicycle riding), knowledge acquisition (memorizing a poem), and a variety of social processes (e.g., stereotyping). It describes one method whereby automaticity can be established. However, as the discussion of affordance, trait perception, and appraisal suggests, automaticity may also have a more biological or modular origin. I will discuss this issue again in Chap. 6, in the context of domain formation.

Is automaticity relevant to quality or quality-of-life research? The answer is obvious to anyone who has ever watched someone respond to a painting they have never seen, or who is participating in a quality-of-life assessment. In both instances it is almost impossible to avoid noting different patterns of responding in these situations, compared to situations that require more deliberate consideration. For example, response to items on a questionnaire reveals that some persons respond in a rapid manner that seems to have not involved much thought, while other persons responses seem to take their time and make a more deliberate response. Von Osch and Stiggelbout (2005) provided empirical support for these observations when they found that respondents selected one of three specific deliberative response patterns, or report minimal consideration when responding (automatic responding).

Next I would like to review a paper that brings together dual processing and metacognition, and identifies a role for metacognition in the process whereby automatic and controlled processes are regulated. Alter et al. (2007) conceptualize dual processing in terms of types of reasoning: intuitive and deliberate. They claim that the available theories on dual processing indicate that deliberative reasoning can override or undo more intuitive and associative responses. Alter et al. (2007) hypothesized that deliberative reasoning would be evoked in the absence of specific cues, while intuitive reasoning would occur when cues are present. One type of cue is the ease or difficulty with which information comes to mind or is processed. The authors argue that the more difficult it is to process information, the more likely the information would evoke deliberative reasoning. They also speculate that the disfluency would evoke a metacognitive process that would prompt more systematic processing. In the experiments, they report disfluency was induced by varying stimulus parameters: for example, the font size of the written material, the facial appearance of an informant (appearance of competence or specific facial characteristics [e.g., furrowing of one's brow]), or some combination of these cues. They found that if a respondent was exposed to what was considered an ambiguous stimulus, then they assumed a more analytical or deliberative approach to responding. They also showed that this effect was not due to a delay in responding or the person's mood.

This and related studies are quite relevant to quality-of-life research, since items on a qualitative assessment may vary in terms of their ambiguity, or concreteness, and these differences may evoke intuitive or deliberative responding. Thus, ambiguously worded items would be expected to evoke more reflective responding, whereas simply worded items may lead to fast and automatic responding, which may be less reliable. I am not aware of studies by quality-of-life researchers who have designed studies to take advantage of these observations. These types of studies could be particularly useful when studying the response pattern of the compromised person. To date, quality-of-life researchers of the compromised person simplifying items on an assessment under the assumption that this would increase the probability of a response, without taking into account that this would also increase the likelihood of automatic responding with the impact this could have on response reliability. This raises the interesting question of whether a person may be considered less competent because they are being asked to respond to items which are liable to increase response variability, rather than items which encourage thoughtfulness when responding.

3.2 Cognitive Models of Quality-of-Life Assessments

This section is designed to provide other examples of the usefulness of examining how cognitive processes contribute to the assessment of quality. Here I will briefly describe two standard quality-of-life assessments: the Schedule for the Evaluation of Individual Quality-of-life (SEIQoL: O'Boyle et al. 1992: Joyce 1988) and the Quality of Well-being Scale (QWB: Kaplan and Anderson 1990). Each of these assessments explicitly uses perceptual/cognitive principles as part of their conceptual foundation. Thus, for the SEIQoL it is the notion of "probabilistic functionalism" of Egon Brunswik as expanded by Hammond's Social Judgment Theory (1975) that was used to design the assessment, while the QWB used Norman Anderson's (1979) cognitive algebra to justify additively aggregating the various components of a qualitative assessment. What is unique about these applications is that by integrating specific cognitive principles, each assessment provides the opportunity to empirically confirm or deny its usefulness. Unfortunately, theory-based assessments are rare in quality-of-life research, with most investigators being content to empirically develop assessments that, while adequate for descriptive purposes, leaves them immune to critical evaluation (e.g., Bullinger 1999) and, not surprisingly, their continued accumulation (as expressed in Chap. 1).

3.2.1 Brunswik, Social Judgment Theory and the SEIQoL

Joyce (1994) has defined quality-of-life as follows: "Quality-of-life is what a person says it is" (p. 1923). This rather terse definition is deceptive, since it masks a sophisticated approach to quality-of-life assessment based on a complex theory of perception, cognition, and judgment that finds its roots in Brunswik's theory of probabilistic functionalism (1956) and Hammond's social judgment theory (1996). Both Brunswik and Hammond suggest that an investigator should be quite explicit about how they conceive of their empirical task, just as Bridgman (1959) (Chap. 1, p. 20) would have us do. Thus, Brunswik proposed that the design of experiments should be representative of ongoing behavior, with the focus of research being ideographic (the study of individuals). This contrasted with the dominant mode of behavioral research found in much of the twentieth century that focused on systematic studies designed to establishing nomothetic (or lawful) relationships. Not too surprisingly, Brunswik's research perspective met with resistance or indifference and has taken a long time to become of interest to investigators; the development of the SEIQoL is one sign of this shift.

There are several aspects of Brunswik's theorizing that are important to examine. First, he saw his psychology as a "psychology in terms of the object" (Leary 1987: p. 119). By this he meant that the task of the individual is to make veridical contact with the environment, and the degree that the person succeeds is reflected in the achievement of their goals. He understood that this involves dealing with a representative sampling (what was sensed) of the environment, which makes knowledge probabilistic. Second is his emphasis on a systems perspective (the Lens Model), with both the organism and the environment as equal components of the system. Since the system involves feedback loops, regulation of the components could occur, thus providing the basis for quality control. Third is Brunswik's focus on ideographic analyses and how this encouraged the development of individual quality-of-life assessments. Finally is his assumption that an individual functions as an intuitive statistician or synthesizer when encountering the perceptual world. To this Hammond (1996) added his Cognitive Continuum Theory (which overlaps with dual process theory) and his restriction of Brunswik's model to the judgment process.

To appreciate the task Brunswik set for himself requires inspecting his early education and career in Europe, a period that spanned the first half of the twentieth century and included the development and application of logical positivism, Gestalt psychology, behaviorism, and new statistical methods being applied to a variety of experimental settings. Each of these traditions, in one way or another, impacted Brunswik's intellectual development, theorizing, and career. He, like Hayek and Parsons (Chap. 9, p. 310), followed the European tradition of forming a Grand Theory,

Fig. 3.5 Presented is a modified version of Hammond's rendition of Brunswik's Lens Model. Here the Lens Model is applied to a clinical decision: a medical diagnosis. Since the model is descriptive, each element is known, with correlations between cues acting as weights reflecting their relative importance in the clinical diagnosis. Permission to reproduce this figure was granted by the publisher.

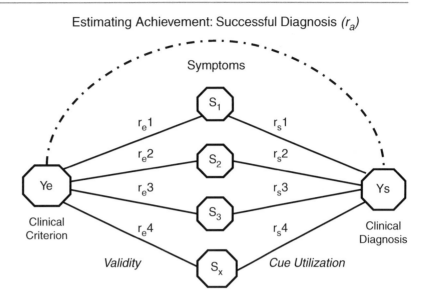

a theory that would offer a comprehensive solution to a well-conceived problem. Stated abstractly, the problem he focused on was how to conceive of the relationship between the subject and object (or environment). More concretely, he saw himself solving the problem of how an organism was going to make its way and survive in its environment, or, as he put it, how the organism was going to "achieve." To do this, he had to take into account activities both within the person (e.g., sensation and perception), and characteristics of the textured environment that an individual normally functioned in. Thus, he thought of his approach as holistic (or gestaltlike), an approach that ensured that his observations would be generalizable. In contrast, observations based on nomothetic studies (studies looking to establish lawful relationships) always have to demonstrate their ability to be generalized.[25]

Leary (1987) traces Brunswik's intellectual development to his teacher, Karl Bühler, who was influenced by Brentano's "act psychology." What Brantano stated was that all psychological activities involve reference to some object. Thus, when I am conscious I am always conscious of something. The same can be said about assessing quality: when I assess quality I first describe something, whether an object, event or experience. This eliminates concern about the dualistic nature of subject and object, or knower and known. Bühler accepted this postulate and interpreted it as a basic statement of the functional or relational nature of the mind. A strong statement of the functional relationship is that every object I encounter is directly related to what I experience. However, Bühler's psychological studies demonstrated that when the background upon which a perceived object varied, the outcome also varied. This suggested that perception was fundamentally an ambiguous process.

Most interesting was that Bühler also applied these views to the study of language. He claimed that no word had a single fixed meaning, rather that words receive their meaning from whole sentences or paragraphs within which they can be found. Thus, a single word can have several meanings depending on the sentences and paragraphs within which they are presented. As a consequence, the person hearing the words has to infer the meaning of the words within the context they are presented. Inferences, therefore, are not absolute but probabilistic. From these observations and his continued research, Bühler concluded that words represent things or thoughts, they are not the things or objects themselves. As Leary (1987) states, "There is, in other words, no invariant one-to-one relationship between representations or signs (whether these be perceptual cues or words) and the things they represent." (p.118) And it is from these notions that Brunswik developed his own views concerning representativeness. Also of note is that both Bühler and Wittgenstein frequented the meetings of the logical positivists in Vienna, so it would not be surprising to find that some of Bühler's ideas concerning language resurfaced in Wittgenstein's philosophical writings.

Brunswik visualized his model (e.g., Brunswik 1956) by using the metaphor of a lens. The lens, which "focuses" the information being exchanged between the person and his or her environment, makes up a dynamic system that is organized to achieve specific goals or purposes. To illustrate the model I will rely on Hammond and his colleagues' application of the model to making judgments (e.g., clinical inferences), especially since the cognitive process underlying a diagnosis will be of interest and discussed in Chap. 5.

Figure 3.5 summarizes an example of the lens model now applied to the making of a clinical decision such as a medical diagnosis. Hammond (1975) describes the task implicit in the Figure as follows:

...judgment is a cognitive process similar to inductive inference, in which the person draws a conclusion, or an inference Y_s about something Y_e, which he cannot see (or otherwise directly perceive), on the basis of data X_i which he can see (or otherwise directly perceive). In other words, judgments are made on palpable data, which serve as cues to impalpable events and circumstances. (p. 73)

Thus, a clinician makes a medical diagnosis by inspecting the symptoms or cues he or she can perceive, and on the basis of this, infers the existence of a disease.

Hammond (1975) specifically recommends correlation and multiple regression statistical methods to establish the degree of relationship between the indicated variables. Thus, the correlation between a clinical diagnosis and the external criteria used to establish the disorder would be a measure of the achievement or accuracy of the diagnosis, while the correlations between the symptoms and the clinical diagnoses act as weights that are part of the clinical decision. Hammond assumes that both the clinical diagnosis and clinical criterion are known, since the model he is applying is descriptive with no intent of establishing lawful relationships. The degree to which the diagnostician used various symptoms or cues would be a measure of cue utilization, while the degree to which the clinical criteria were correlated to the available symptoms or cues would be a measure of ecological validity. Thus, if the variance accounted for was a small proportion of the variance of the clinical criteria, then this would suggest that other symptoms or indicators should be included as part of the lens model.

Hammond (1975: p. 77) also suggests that it is possible to have single system judgment tasks. This case would only involve the right side of Fig. 3.5, with the question being to what extent the cues contribute to the judgment. Since there is an element of selection in terms of cue utilization here, Hammond thinks of the research task as identifying the judgment policy that the clinician or patient has adopted. The elements of the judgment policy that would have to be known include the weights (or importance) attached to each symptom or cue in the diagnosis; the relationship between the symptoms and the criterion or disease (e.g., is a high temperature linearly, curvilinearly, on an inverted U-shaped function of disease status); how the symptoms are organized into a construct (e.g., hierarchically or by cross-classification)[26]; and the reliability with which the judgment is made. Also of interest is that a person's judgment policy may be a product of a metacognitive process.

Interestingly, the instruction manual for the SEIQoL addresses many of the same issues I have just summarized for the single-system judgment task. The reason why the SEIQoL is limited to a one-sided judgment is because there are no known or agreed-to quality-of-life outcomes that can be used as criteria to compare an individual's personal assessment. In the SEIQoL, the cues refer to such qualitative domains as family, work, health, physical capabilities, and so on. Subjects are given the opportunity to select their domains of importance and then asked to rate their current status in each of these domains. For example, they may rate their spiritual status as bad, or their work activities as very good. They are then given a set of case studies and asked to rate these cases along the domains they selected as important to them. So if a person thinks that work, physical activities, and spirituality are important to them, they are asked to read over the cases and rate them in terms of their chosen important domains. From these case studies ratings, a judgment policy is determined. Specifically, the weighting given to each cue or domain is entered into a regression equation, and the R^2 determined. The R^2 is considered an estimate of an individual's quality-of-life. A SEIQoL index based on a summation of level and weights attached to each domain is calculated to permit group comparisons (Hickey et al. 1999).

Along with Ruta et al.'s (1994: The Patient-Generated Index: PGI) assessment, the SEIQoL is increasingly being used to provide an estimate of an individual's quality-of-life or HRQOL. The SEIQoL, in particular, with its intellectual heritage tracing back to Egon Brunswik, offers a unique glimpse into the cognitive underpinnings that are involved in a quality assessment. Noteworthy about the SEIQoL is that the methods used in its generation consist of description (selecting domains or cues), evaluation (determining the good/bad level of each cue), and valuation (calculated indirectly from the relative weights attached to each cue or domain estimated by multiple regressions). The same is true for Ruta's (1994) PGI method of assessment. Thus, the SEIQoL and the PGI are each based on a common cognitive model: a hybrid cognitive model of quality assessment (the same is true for the Patrick and Erickson (1993) definition of HRQOL.) The same will again be true when I examine the QWB (see below). The fact that a series of conceptually independent efforts end up using the same cognitive components to create a quality assessment reinforces my conviction that a general principle is operative here, a principle that is best observed when an investigator asks how a quality assessment occurs, not what content to include in an assessment. I will continue this discussion in Chap. 12 where I will provide additional details about cognitive processes active during a qualitative assessment.

3.2.2 Norman Anderson's Cognitive Algebra and the QWB

As I tried to make clear in the previous section, a method for assessing quality can have a distinctive theoretical foundation, yet its assessment be based on a common cognitive process. This is possible because what is common between these different theoretical statements is the investigator, who generates different theories but does so using a common thought process. Thus when disaggregated, the Quality of Well-being Scale (QWB), which evolved from the General Health Policy Model originally proposed by

Fanshel and Bush (1970), reveals descriptive, evaluative, and valuation components, much as was found for the SEIQoL, and the PGI. What I would like to do in this section is to briefly review the conceptual foundation of the General Health Policy Model, and provide details of Norman Anderson's cognitive algebra. As I do this, I will demonstrate how calculating the QWB also involves description, evaluation, and valuation. From this it will become apparent that the cognitive processes that I have been describing have been used to operationalize and formalize the concept of expected utility.

The General Health Policy Model evolved in the 1970s in response to a number of specific needs. First was the need for a single indicator, a measure that combined mortality and quality-of-life and that could be included in health policy decision making. A second major need was an alternative to the "human capital" perspective for health policy assessment, an approach that placed a dollar value on a person's life. The General Health Policy Model provided a more egalitarian approach and avoided the discriminatory biases of a monetary-based assessment (Kaplan and Bush 1982). A third characteristic of the General Health Policy Model was its integration of utility or preference estimates into a lifetime estimate of health status assessment and its use of lifetime expected utility for evaluating health services or differences in health status. As Kaplan and Bush (1982: p. 63–63) explain, they selected the term "Well-Years" to denote HRQOL, and by so doing utilized the General Health Policy Model to express "the output of health programs in comparable units of life years adjusted for lost 'quality' due to disease or disability." (p. 64)

The General Health Policy Model, as an egalitarian indicator, assumes that each day a person lives is of equal social value, regardless of a person's economic or social status. This assumption permits the direct comparison of all persons in terms of assessing the impact of treatments, health programs, or other policy alternatives. Since the purpose of health policy is to ensure that people live longer with higher quality, the manner in which decisions (or judgments) are made becomes of paramount importance. Kaplan and Bush (1982) claim that decision theory guided them to identify three dimensions that are involved in any assessment.[27] These dimensions are:

1. The states that a person may occupy at any one point in time
2. The probabilities ("risks") of being in the states at different times
3. The relative desirability of occupying the states (p. 65)

The state a person is in is a descriptive statement; the risk of being in a particular state is an evaluative statement; and the desirability of being in a particular state is a valuation statement. Thus, the conceptual foundation of the QWB replicates the components of a cognitive model of quality assessment.

Kaplan and Bush also point out that these cognitive components permit the operationalization of the concept of expected utility, where expected utility is formally defined as:

Expected Utility = Expectancy (Probability) X Value (Utility)

With subjective probability including both description and evaluation (that is, a probability estimate), and utility being a valuation estimate. In this chapter I will also provide an overview of different theoretical conceptualizations of the cognitive processes ongoing during subjective probability estimates (see below).

The QWB itself is composed of the three cognitive processes cited above. Thus, the QWB provides a respondent with a list of symptoms and problems, and asks the respondent to select the ones that were present in the last day, and present in previous days. This is meant to provide the respondent with the opportunity to describe their individual clinical condition. They are also asked to rate their level of mobility, physical activity, and social activity, thereby providing an opportunity to evaluate their qualitative status. Finally, each of the symptoms, problems, and ratings are quantified by preference weights generated from a prior, independent population-based study. The QWB itself is calculated by the additive summation of each of the preference weights, and the question that has yet to be answered is whether preferences can be added together.

The first answer is that they can be added because the preferences assigned to symptoms, ratings and so on, all are based on a common valuation denominator. However, this does not resolve whether the preferences should be added, multiplied, or become part of some complex equation. Kaplan et al. (Unpublished manuscript, 1987) attempt to answer this question by demonstrating that each of the components of the QWB matches the "additivity" assumptions of Norman Anderson's cognitive algebra. Before I review their study, however, I would like to provide some background to Anderson's theoretical approach. He, unlike Brunswick, had an extended research career, but like Brunswik has not received the attention that his work might have expected (Noble and Shanteau 1999). Anderson (1996) felt that what he had done was to create a meta-theory: a theory that would be expected to guide the formation of other more micro theories. Both he and Brunswik required the investigator to think differently about psychological phenomena, which has been difficult for the empirically oriented American psychologists, but which is less challenging for the more theoretically oriented European psychologists who seem to be reporting studies based on his theories (e.g., Juslina et al. 2008).

Anderson refers to this meta-theory as "Information Integration Theory" (IIT: Anderson 1996). IIT has three defining characteristics: it is a functional theory; it is based on the demonstration of a cognitive algebra; and both these characteristics become principle parts of a functional approach to assessment. The starting point for Anderson is the assumption that behavior is purposeful and goal directed. Because of this, behavior, thoughts, and action are conceptualized in terms of their function to achieve some purpose and goal. This view of function is consistent with a lay person's conception of everyday life; I do what I do because it has a function. A second major characteristic of his theory is that behaviors, thoughts, and actions have an approach-avoidance character. This unidimensionality is represented in various ways such as likes–dislikes, success–failure, good–bad, and so on. This, Anderson (1996: p. 3) suggests, makes purposefulness a continuous measurable dimension. Since approach-avoidance may be positive or negative, what has to be measured here are values. Values represent the quantification of sweetness or loudness, the feelings of affection, the unfairness of certain social situations, and other everyday experiences. Fortunately, Anderson says, it is possible to measure each of these dimensions, since each are subjects to the rules of cognitive algebra. Each of the stimulus terms represent one-dimensional values, while the responses all represent the one-dimension of purposefulness. Since I am usually exposed to multiple stimuli, integration has to occur (information integration), leading to some purposeful singular response. Most important is the fact that "cognitive algebra can thus provide an underlying order and unifying framework for the surface complications of innumerable determinants of thought and action" (Anderson 1996: p. 4–5). Thus, the cognitive algebra provides a common metric that can be used both descriptively and analytically.

It is particularly interesting that the values Anderson refers to can easily be seen as quality indicators (e.g., sweetness, loudness, fairness, and so on). I can illustrate this with an example. Say you are trying to decide if you should buy a particular type of car. The car you are interested in has the color you want, and gets the gas mileage you want, but is more expensive than you want to pay and does not have enough room for the dog and the kids – in that order. Each of these indicators says something about the quality of this potential purchase. Even the cost of the car is being considered from a qualitative perspective: is this what you want to pay? In addition, since you are considering each of these indicators in series, the information you consider will be subject to a cognitive algebra: they will be added to each other. Anderson claims that this addition happens intuitively, this combining of cues or issues with the outcome being expressed as a judgment. This process is also capacity limited, and it is in this sense that it is a component of dual processing (e.g., controlled cognitive processing).

The method Anderson used to test the additive nature of judgments involved performing a study in which the cues or issues were presented to a respondent according to a factorial design, with the failure to observe an interaction term evidence that the main effects were additive. Anderson (1996: p. 41–43) provides the following example to illustrate his argument. Subjects were asked to judge how much blame should be assigned to a story of a child who threw a rock that injured another boy. The intention of the harm doer was varied across three levels: definitely to do harm, to scare or to be careless. The harm that was intended to be done was to produce a bruised shin, bloody nose, or black eye. Each subject was asked to make nine judgments, with the judgment task involving estimating how much each case reflected the boy's blame for the consequences. Thus, if the respondent was told the boy was being careless and produced a bruised shin, would the boy be blamed less than if his intention was to harm and a bruised shin resulted? The results were as expected: blame was distributed in a parallel fashion (no interaction) as the consequences varied. What Anderson found, of course, was that this same parallelism occurred over a wide range of stimulus conditions, leading to his belief that this was a general cognitive law. From this experiment, he felt he could write the (cognitive) algebra equation: Blame = Intent + Harm.

Kaplan et al.'s (Unpublished, 1987) study followed Anderson's design but now included different health status scenarios as the cues to be judged. Each scenario was made up of a description of a person's age, mobility, physical activity, and social activity state, plus some example of symptom/problems. The scenarios were varied using an orthogonal design, following Anderson, and community residents (N = 790) participated in the study. Their results indicated that 13 of 20 main effects were statistically significant, while only two of the 30 tests of interactions were significant'; 1.5 interactions would have occurred by chance alone. They state that, "Statistical analysis on our total sample, using all the items in addition to the balanced designs, generally concur with the results of the analyses reported here. Regressions of the health related quality-of-life scores on the well-state dimensions show an overall lack of evidence of interactions." (Kaplan et al., Unpublished, 1987: p. 10).

I have included a discussion of this unpublished paper by Kaplan et al. (1987) because it is such a clear example of applying a cognitive theory to the analysis of a model of quality, particularly quality-of-life. However, the demonstration that the cues or components that make up a quality assessment may be added to leading to an overall judgment is also important in terms of what model to use. Kaplan et al. (Unpublished, 1987) also point out that the results of their study supports the view that the various states they identify are separated by equal intervals. They also raise concerns about generalizing their results to individual respondents.

Next I will briefly review a paper by Juslin et al. (2008), since it provides an additional example of a study demonstrating the presence of a cognitive algebra, but also where the respondents were faced with cues that do not add up. What happens here, apparently, is that a person may adapt the cognitive strategy of using exemplars to "normalize" the apparently nonadditive outcomes. Consider the following example: a person has cancer and yet assumes that they have all the rights and privileges of a person who does not have cancer, including the hopes that they will be able to engage in a certain level of activity. Yet objectively, this person has certain limitations that would preclude such beliefs. Faced with this "nonadditivity," the person may turn to a role model who has had equally as severe cancer and yet has succeeded (e.g., Lance Armstrong). This person acts as an exemplar for the respondent, and is used to support the respondent's belief that they can lead a full normal life. When such a person takes stock of their life quality they will include in their addition the hope that their exemplar provides, and then proceed to combine indicators.

3.2.3 Summary

In this section I have tried to present two examples of how examining cognitive processes can contribute to the study of quality. The first set of examples dealt with how fast and automatic judgments of quality may occur. They gave me the opportunity to discuss how ecological and perceptual factors may contribute to a quality assessment, but also introduced the concept of appraisal and what role it might play in a quality assessment. I also made clear that while a person may make a rapid quality judgment, this judgment may still have been based on a more controlled reflective process. It was also evident that fast and automatic quality assessment may have a strong structural, if not biological (perceptual, emotional), basis.

The second set of examples demonstrated how two different theoretical orientations have been used to provide the cognitive basis of a quality assessment. Interestingly, these seemingly diverse theoretical orientations to quality may actually have a common conceptual base: expected utility theory. In addition, a general principle became more explicit in this discussion. Information is received serially and the capacity of the processing systems has inherent limitations, and it is these two characteristics which significantly contribute to the observed outcomes, including the assessment of quality. It is also evident, after reviewing the Juslin et al. (2008) paper, that a person has the cognitive capacity to transform nonadditive information into a form which can be additively combined with other indicators. This transformation constitutes a reclassification of information, which illustrates how dynamic cognitive processes can be, and how effectively they may be in providing the basis for a cognitive algebra.

Each of the examples I have given can be inspected to identify the presence of specific cognitive processes. Thus, the formation of a cognitive entity, and the impact of classification, concept formation, and metacognition, can be found in all the entities generated. Now I will briefly allude to a substantial research literature that exists dealing with a central concept of prospective quality assessment: subjective

Table 3.4 Cognitive basis of subjective probability

Theory	References
The representative heuristic	Kahneman and Tversky (1972)
Cue-based relative frequency	Estes (1976); Hasher and Zacks (1979); Gigerenzer and Todd (1999); Juslina (1994)
Exemplar memory	Dougherty et al. (1999); Sieck and Yates (2001)

probability. Nilsson et al. (2005) points out that there are three different theoretical approaches to understanding the cognitive basis of subjective probability (Table 3.4). However, this is a bit of a simplification, since some of these categories have multiple approaches to the same theoretical statement. For example, they identified three different approaches to the representative heuristic. Thus, what exists here is a rich empirical literature, very little of which has been applied to the study of quality assessment, yet studies of this sort could be quite informative about the nature of the qualification process.

4 Summary of Chapter

This chapter and the discussion in Chap. 1 concerning a *qualia* introduced a number of concepts about the nature of cognition such that it would be possible to start thinking about a quality assessment as a cognitive process. Two principles were identified. First is the idea that what I sense and perceive is represented by a *qualia*, a cognitive entity that reflects the qualities inherent to experience. Second is that this information is then classified, categorized, conceptualized, and metacognitively processed. The information involved represents a selective summary, not a reproduction, of what a person receives. This information naturally forms three broad types of information (superordinate, basic, and subordinate), and can be classified into a number of categories: hierarchical, cross-sectional, and contextual categories. These structural determinants lead to the formation of concepts based on the similarity, prototypicality, or exemplars. The cognitive entities formed then become part of a metacognitive process that can lead to a number of outcomes, including a quality assessment. Since a person receives information serially, he or she processes this information in either a fast automatic manner or by deliberate processing. The deliberate processing may be viewed as a means of coping with the capacity limitations of any system that receives information. I will continue in this book to identify the role cognitive processes play in the assessment of quality, with particular emphasis on how they may help define and measure the experience of quality.

Notes

1. Hickey et al. (1999) introduces a paper with this quote from Albert Einstein, and I find that it succinctly states what I hope to communicate in this chapter.
2. The notion that different experiences constitute a type of assessment has implications for the relationship between qualitative and quantitative assessment. It suggests that quantitative assessment is built on qualitative assessment. Thus, it has been clear since Stevens' conceptualization of scale of measurement that ordinal, interval, and ratio scaling would not occur if nominal measurement was not possible. Thus, quantification evolves out of a subjective state and, as I discuss in Chap. 4, this forms the basis for viewing subjective and objective assessment as a continuum.
3. What I am referring to here is the practice of having an investigator classify the verbal material created by an exchange between, say, a patient and a clinician, and using this classifying system to order the discourse between individuals. I am stating that these methods, having been created by an investigator, are inherently artificial and that other methods exist (e.g., discourse analysis) which are more natural to the exchange. Still, it is fair to say that while investigator-based classifications are at risk to be biased, the mechanical-based assessments of test analysis are at risk of being uninformative.
4. I will address this issue more directly in Chap. 10 where a number of studies use brain imaging techniques to determine if different cognitive mechanisms are used by different groups when performing the same task.
5. Of course, he communicated part of his answer in how he phrased the problem he wanted to solve. What he seems to be telling us is that he saw order as not being imposed, but created by a natural evolutionary process. Thus, it becomes important to provide a method of how this could occur.
6. Birner (1996) claims that Hayek did not use the ideas that were in his 1919 paper, before, as much as he did after the publication of the "The Sensory Order" in 1952.
7. Here I find an overlap with the notions of Gentner and her colleagues (Bowdle and Gentner 2005) about how a metaphor may be formed (Chap. 2, p. 33).
8. The reader may want to review Ackoff's (1976) paper, since there seems to be some overlap with Hayek's ideas. Papers by Birner (1996), Nadeau (1997) and Smith (1997) provide more details on the relationship between Hayek's neurocognitive model and economic systems.
9. Adam Smith was referring to situations where economic forces can efficiently act, leading to the successful operation of a market. However, in healthcare, market processes often fail, so applying market principles here may not be as successful (David Feeny, Personal communication, 25 February 2008).
10. In Chap. 6 I will introduce the notion that categories of information, or domains, can be biologically determined. Thus, vision and language are domain specific. This will appear to contradict Hayek's notion that cognitive "order" results from indiscriminate sensory input. However, the two views can be seen as compatible if it is assumed that the sensory order occurs within a specific domain. In Chap. 10, I continue and expand my discussion and propose some additional methods whereby a cognitive entity can be formed using neural network processing.
11. Several of these production rules may be combined in a more general cognitive process, such as the appraisal process.
12. Sometimes "furniture" is used to refer to specific items such as chairs or couches so that someone can "sit on furniture." In these cases, the word has been provided sufficient specificity so that it can be understood as a member of a basic category, and not a superordinate category.
13. The Berlin and Kay (1969) study has generated a substantial research literature, since it implied that there may be limits to the influence that a culture has on a language (see Roberson et al. 2004). However, what I am interested in is that in the background of this discussion is the existence of a hierarchy that reflects how I cognitively organize the information in color terms, which should also be true independent of the influence of culture. Thus, the existence of a hierarchical structure underlying some cognitive process should be a universal characteristic, which would be expected to exist independent of the prevalent culture. It should be an invariant characteristic of information processing (Chap. 4, p. 112).
14. The distinction between hierarchical and cross-classification of information will also be important when I discuss the difference between psychometric and clinimetric approaches to HRQOL assessment.
15. I will not include well-being, another term sometimes thought of as a superordinate category or concept, in this list, and will instead leave a discussion concerning its cognitive characteristics to Chap. 11.
16. There is actually a way to test this question, which is to measure the reaction time of responses to domain-specific items vs. single global items. The experimental hypothesis is that responses to the informationally less-demanding items on a multi-item assessment would elicit a quicker response than the single-item global assessments.
17. It is interesting to read in the de Boer et al. (1998: p. 138) paper that the authors feel that by determining the number and type of latent variables underlying some quality-of-life assessment, they are actually defining quality-of-life. Of course, what they are doing is describing whether a one- or two-latent variable model matches their assessment model. However, this linguistic slippage contributes to the confusion concerning what is actually being accomplished by the investigators. All they need do to see the limits of their efforts is to remember that a person may create a very different model of quality-of-life than what is psychometrically determined, although I would admit it is not their concern. Their concern is generating a useful assessment model, not capturing how a person thinks about their qualitative state.
18. The reader might find it interesting to note that the primary icon of ancient Egypt was the pyramid, an artifact that can be considered a physical analog of a cognitive hierarchy. Much of ancient Egyptian culture revolved around ascending to higher levels of abstraction, and the pyramids can be thought of as a physical analog of this ascension process.
19. Metacognition plays critical role in defining a qualitative assessment. It differentiates indicators which are primarily descriptive (e.g., symptoms or psychosocial indicators) from those that are descriptive, evaluative but also reflective (or other types of metacognitive processes). What is of interest, of course, is that any "evaluated descriptor" can become a quality indicator, if a clinician or person considers the consequences and asks a person how important the changes they are experiencing are to them. Thus, quality assessments are a far more common occurrence than is obvious from the study of quality-of-life or HRQOL.
20. The alternative view is that a quality or quality-of-life assessment can be conceived of as a black box, and all that is needed to understand a person's qualitative judgments is to know the conditions of their existence (i.e., the input) and their state (i.e., the outcome). This view, I believe, is potentially misleading, since there is ample evidence that people process the information they receive and use this information in different ways when they make their qualitative judgments. Understanding how this occurs is essential for dealing with a variety of issues, including how you can combine qualitative measures from diverse groups (Barofsky 2003).
21. The dichotomy of automatic and controlled has been labeled in a variety of ways. They are most often referred to as "dual process" theory, but one method which I find particularly useful, suggested

by Strack and Deutsch (2004), describes these two processes as impulsive and reflective.

22. The relationship between experience and behavior is at the basis for any number of causal statements. An example would be if an investigator claimed that feelings of well-being caused the person to do something, such as jump with joy. Another example is if an investigator claimed that the diminished quality-of-life of a woman with breast cancer leads to an increase in suicide or divorce.

23. McGrenere and Ho (2000) have expanded on Norman's definition of affordance by illustrating how the term can be applied to human–computer interactions (HCIs). Hartson (2003) also attempts to define the role affordance in HCI.

24. By "dialectical process" the authors are referring to the fact that both sources of information – reading from faces, and reading into faces – affect and interact with each other. Thus, information exchange occurs in both directions, and this is why the authors feel that a nonlinear model, such as parallel distributed processes, may be more appropriate for this type of phenomena.

25. This argument concerning the importance of generalizability persists today in the debate between Donald Campbell and Lee Cronbach concerning the approaches to validity, with the influence of Brunswik very much in evidence (Albright and Malloy 2000).

26. In Chap. 5, I will discuss the difference between a psychometric and clinimetric approach to medical decision making, and that discussion will give substance to what Hammond is referring to here as the way cues or symptoms are organized.

27. While Kaplan and Bush (1982) argue that decision theory promoted their operationalization of expected utility in terms of specific cognitive processes, it can also be argued that the concept of expected utility evolved from the combination of the specific cognitive processes of description, evaluation, and valuation prior to the articulation of the concept of expected utility. In other words, if you were not able to describe, evaluate, or place values on some experience, you would not have been able to conceive of the concept of expected utility. In that sense, expected utility is a cognitive construction whose nature has been confirmed by its axiomatic formulation.

References

Ackoff RL. (1976). Does quality of life have to be quantified? *Oper Res Q.* 27, 280–303.
Albright L, Malloy TE. (2000). Experimental validity: Brunswik, Campbell, Cronbach and enduring issues. *Rev Gen Psychol.* 4, 337–353.
American Psychiatric Association. (1994). *Diagnostic and Statistical Manual of Mental Disorders: Fourth Edition.* Washington, DC: American Psychiatric Association.
Alter AL, Oppenheimer DM, Epley N, Eyre RN. (2007). Overcoming intuition: Metacognitive difficulty activates analytic reasoning. *J Exp Psychol: Gen.* 136, 569–576.
Anderson NH. (1979). Algebraic rules in psychological measurement. *Am Sci.* 67, 555–563.
Anderson NH. (1996). *A Functional Theory of Cognition.* Mahwah NJ: L Erlbaum.
Arthur BW, Durlauf SN, Lane DA. (1997). *The Economy as an Evolving Complex System* II. Reading MA: Addison-Wesley.
Bar M, Neta M, Linz H. (2006). Very first impressions. *Emotion.* 6, 269–278.
Barofsky I. (2003). Cognitive approaches to summary measurement: Its application to the measurement of diversity in health-related quality of life assessments. *Qual Life Res.* 12, 251–260.
Barsalou LW. (1991). Deriving categories to achieve goals. In, (Ed.) GH Bower. *The Psychology of Learning and Motivation: Advances in Research and Theory.* Volume 27. San Diego CA: Academic Press. (p. 1–64).
Benyamini Y, Leventhal EA, Leventhal H. (2003). Elderly people's ratings of the importance of health-related factors to their self-assessment of health. *Soc Sci Med.* 56, 1661–1667.
Berkowitz L, Harmon-Jones E. (2004). Toward an understanding of the determinants of anger. *Emotion.* 4, 107–130.
Berlin B, Breedlove D, Raven P. (1974). *Principles of Tzeltal Plant Classification.* New York, NY: Academic Press.
Berlin B, Kay P. (1969). *Basic Color Terms.* Berkeley and Los Angeles CA: University of California Press.
Binder JR, Westbury CF, McKiernan KA, Possing ET, et al. (2005). Distinct brain systems for processing concrete and abstract concepts. *J Cogn Neurosci.* 7, 905–917.
Bless H, Forgas JP. (2000). *The Message Within: The Role of Subjective Experience in Social Cognition and Behavior.* Philadelphia PA: Psychology Press.
Borawski E, Kinney J, Kahana E. (1996). The meaning of older adults health appraisals: Congruence with health status and determinants of mortality. *J Gerontol: Soc Sci Sect.* 51B, S157-170.
Bowling A. (2005). Just one question: If one question works, why ask several? *J Epidemiol Community Health.* 59, 342–345.
Bridgman PW. (1959). *The Way Things Are.* New York, NY: Viking.
Birner J. (2004). Mind, market and society: Network structures in the work of F.A. Hayek. *Unpublished Manuscript.* (1996) http://reprints.biblio.unitn.it/achieve/00000033/ (Assessed from the web, 2/15/2004).
Bowdle BF, Gentner D. (2005). The career of metaphor. *Psychol Rev.* 112, 193–216.
Brown HI. (1987). *Observation and Objectivity.* New York NY: Oxford University Press.
Bruner JS. (1967). On cognitive growth. In, (Eds.) JSA Bruner, RR Oliver, PM Greenfield, *Studies in Cognitive Growth: A Collaboration at the Centre of Cognitive Studies.* Volume 1. New York NY: Wiley. (p. 1–67).
Brunswik E. (1956). *Perception and the representative design of psychological experiments.* Berkeley CA: University of California.
Bullinger M. (1999). Cognitive theories and individual quality of life assessment. In, (Eds.) CRB Joyce, HM McGee, CA O'Boyle. *Individual Quality of life: Approaches to Conceptualization and Assessment.* Amsterdam The Netherlands: Harwood. (p. 29–39).
Cain MJ. (2002). *Fodor: Language, Mind and Philosophy.* Cambridge UK: Polity.
Caldwell B. (2004). *Hayek's Challenge: An Intellectual Biography of F.A. Hayek.* Chicago Ill: University of Chicago Press.
Chomsky N. (1988). *Language and Problems of Knowledge.* Cambridge MA: MIT Press.
Clore GL, Centerbar DB. (2004). Analyzing Anger: How to Make People Mad. *Emotion.* 4, 139–144.
Conway MA. (1990). Associations between autobiographical memories and concepts. *J Exp Psychol: Learn Mem and Cognition.* 16, 799–812.
David KM, Ganiats TG, Miller C. (1999). Placement matters: Stability of the SF-36 EVGFP responses with varying placement of the question in the instrument. Abstract presented at the Sixth Annual Meeting of the International Society For Quality of life Research, Barcelona Spain, November 1999. *Qual Life Res.* 8, 623.
de Boer AGEM, Spruijt RJ, Sprangers MAG, de Haes JCJM. (1998). Disease-specific quality of life: Is it one construct? *Qual Life Res.* 7, 135–142.
Diener E, Fujita F. (1997). Social comparison and subjective well-being. In, (Eds.) BP Buunk, FX Gibbins. *Health, Coping and Well-being: Perspective from Social Comparison Theory.* Mahwah NJ: Erlbaum. (p. 329–357).
Diener E, Suh EM, Lucas RE, Smith HL. (1999). Subjective well-being. *Psychol Bull.* 125: 276–302.

Dimaggio G, Semerari A, Falcone M, Nicolo G, et al. (2002). Metacognition, states of mind, cognitive biases, and interpersonal cycles: Proposal for an integrated narcissism model. *J Psychother Integr.* 12, 421–451.

Dougherty MRP, Getty CF, Ogden EE. (1999). MINER VA-DM: A memory process model for judgment of likelihood. *Psychol Rev.* 106, 180–209.

Dunlosky J, Nelson TO. (1994). Does the sensitivity of judgments of learning (JOL) to the effects of various study activities depend on when the JOL occur? *J Mem Lang.* 33, 545–565.

Ellsworth PC, Scherer K R. (2003). Appraisal processes in emotion. In, (Eds.)RJ Davidson, HH Goldsmith, KR Scherer. *Handbook of Affective Sciences*. New York NY: Oxford University Press. (pp. 572–595).

Estes WK. (1976). The cognitive side of probability learning. *Psychol Rev.* 83, 37–64.

Estes WK. (1994). *Classification and Choice*. New York NY: Oxford University Press.

Fanshel S, Bush, JW. (1970). A Health Status Index and its application to the health services outcomes. *Oper Res.* 18, 1021–1066.

Fayers PM, Machin D. (2000). *Quality of life: Assessment, Analysis and Interpretation*. Chichester UK: Wiley.

Gärdenfors P. (2000). *Conceptual Spaces: The Geometry of Though*. Lexington MA: MIT Press.

Gibbs RW Jr. (2006). *Embodiment and Cognitive Science*. New York NY: Cambridge University Press.

Gibson, E. (2000). Perceptual learning in development: Some basic concepts. *Ecol Psychol.* 12, 295–302.

Gibson JJ. (1979). *The Ecological Approach to Visual Perception*. Boston MA: Houghton Mifflin.

Gigerenzer G, Todd PM. (1999). *Simple Heuristics that Make Us Smart*. Cambridge MA: MIT Press.

Goldstein DG, Gigerenzer G. (1996). Reasoning the fast and frugal way: models of bounded rationality. *Psychol Rev.* 103, 650–669.

Goldwasser O. (2005). Where is metaphor?: Conceptual metaphors and alternative classification in the hieroglyphic script. *Metaphor Symb.* 20, 95–113.

Grill-Spector K, Kanwisher N. (2005). Visual Recognition: As Soon as You Know It Is There, You Know What It Is. *Psychol Sci.* 16, 152–160.

Groves R, Fultz N, Martin E. (1992). Direct questioning about comprehension in a survey setting. In, (Ed.) JM Tanur. *Questions about Questions : Inquiries into the Cognitive Bases of Surveys*. New York NY: Russell Sage Foundation. (p. 49–64).

Haase JE, Braden CJ. (1998). Guidelines for achieving clarity of concepts related to quality of life. In, (Eds.) CR King, PS Hinds. *Quality of life: From Nursing and Patient Perspectives. Theory, Research, Practice*. Sudbury MA: Jones and Bartlett. (54–73).

Hammermesh D, Biddle J. (1994). Beauty and the labor market. *Am Econ Rev.* 84, 1174–1194.

Hammond KR. (1975). Social Judgment Theory: Its Use in the study of psychoactive drugs. In, (Eds.) KR Hammond, CRB Joyce. *Psychoactive Drugs and Social Judgment: Theory and Research*. New York NY: Wiley. (p. 69–105).

Hammond KR. (1996). *Human Judgment and Social Policy*. New York NY: Oxford University Press.

Hammond KR, Stewart TR, Brehmer B, Steinmann DO. (1975). Social Judgment Theory. In, (Eds.) MF Kaplan, S Schwartz. *Human Judgement and Decision Processes: Formal and Mathematical Approaches*. New York NY: Academic Press. (p. 272–312).

Hampton JA. (1982). A demonstration of insensitivity in natural categories. *Cognition.* 12, 151–162.

Hart JT. (1965). Memory and the feeling-of-knowing experience. *J Educ Psychol.* 56, 208–216.

Hart JT. (1967). Second-try recall, recognition, and the memory-monitoring process. *J Educ Psychol.* 58, 193–197.

Hartson HR. (2003). Cognitive, physical, sensory and functional affordances in interaction design. *Behav Inf Technol.* 22, 315–338.

Hasher L, Zacks RT. (1979). Automatic and effortful process in memory. *J Exp Psychol: Gen.* 108, 356–388.

Hassin R, Trope Y. (2000). Facing faces: Studies on the cognitive aspects of physiognomy. *J Personal Social Psychol.* 78, 837–852.

Hayek FA. (1919). *Beiträge zur Theorier der Entwicklung des Bewussteins*. Unpublished Manuscript.

Hayek, FA. (1944). *The Road to Serfdom*. Chicago, Ill.: University of Chicago Press.

Hayek FA. (1952). *The Sensory Order: An Inquiry into the Foundation of Theoretical Psychology*. Chicago Ill: University of Chicago Press.

Hayek FA. (1994). In, (Eds.) S Kresge, L Wenar. *Hayek on Hayek: An Autobiographic Dialogue*. Chicago Ill: University of Chicago Press.

Hays RD, Steward AL. (1990). The structure of self-report health in chronic disease patients. *Psychol Assess: J Consult Clin Psychol.* 2, 22–30.

Hebb DO. (1949). *The Organization of Behavior. A Neuropsychological Theory*. New York NY: Wiley.

Hedden T, Ketay S, Aron A, Markus HR, et al. (2008). Cultural influences on neural substrates of attentional control. *Psychol Sci.* 19, 12–17.

Hickey A, O'Boyle CA, McGee HM, Joyce CRB. (1999). The Schedule for the Evaluation of Individual Quality of life. In, (Eds.) CRB Joyce, HM McGee, CA O'Boyle. Individual Quality of life: Approaches to Conceptualization and Assessment. Amsterdam The Netherlands: Harwood. (p. 119–133).

Idler EL, Hudson S, Leventhal H. (1999). The meaning of self-rated health: A qualitative and quantitative approach. *Res Aging.* 21, 458–476.

Inhelder B, Piaget J. (1958). *The Growth of Logical Thinking from Childhood to Adolescence*. New York NY: Basic Books.

Izard CE. (1993). Four systems for emotion activation: Cognitive and noncognitive processes. *Psychol Rev.* 100, 68–90.

Joyce CRB. (1988). Quality of life: The state of the art in clinical assessment. In, (Eds.) SW Walker, RM Rosser. *Quality of life Assessment and Application*. Lancaster, PA: MTP Press, (p. 169–179).

Joyce CRB. (1994). How can we measure individual quality of life? *Schweizer Med Wochenschri.* 124, 1921–1926.

James W. (1890). *Principles of Psychology*. Vol 1. New York NY: Holt.

Juslina P. (1994). The overconfidence phenomena as a consequence of informal experimenter-guided selection of almanac terms. *Organ Behav Hum Decis Process.* 57, 226–246.

Juslina P, Karlsson L, Olsson H. (2008). Information integration in multiple cue judgment: A division of labor hypothesis. *Cognition.* 106: 259–298.

Kahneman D, Tversky D. (1972). Subjective probability: A judgment of representativeness. *Cogn Psychol.* 3: 430–454.

Kaplan RM, Bush JW. (1982). Health-related quality of life for evaluation research and policy analysis. *Health Psychol.* 1, 61–80.

Kaplan RM, Bush JW, Blischke WR. (1987). Additive utility independence in a multidimensional quality of life scale for the general health policy model. Unpublished Manuscript.

Kaplan RM, Anderson JP. (1990). The General Health Policy Model: An integrated approach. In, (Ed.) B Spilker. *Quality of life Assessments in Clinical Trials*. New York NY: Ravens Press. (p. 131–149).

Keil FC. (2006). Cognitive Science and Cognitive Development. In, (Vol. Eds.) W Damon, R Lerner (Series Eds.) D Kuhn, RS Siegler. *Handbook of Child Psychology*. Vol 2: *Cognition, Perception, and Language*. (6th ed.). New York NY: Wiley. (609–635).

Koechlin E, Ody C, Kouneiher F. (2003). The architecture of cognitive control in the human prefrontal cortex. *Science.* 302, 1181–1185.

Koriat A. (2006). Metacognition and consciousness. In, (Eds.) PD Zelazo, M Moscovitch, E. Thompson. *Cambridge Handbook of Consciousness*. New York NY: Cambridge University Press. (p. 289–326).

References

Krause N, Kay G. (1994). What do global self-rated health items measure? *Med Care.* 32, 930–942.

Lakoff G. (1987). *Women, Fire, and Dangerous Things: What Categories Reveal About the Mind.* Chicago Ill: University of Chicago Press.

Lakoff G, Johnson M. (1980). *Metaphors We Live By.* Chicago Ill: University of Chicago Press.

Lakoff G, Johnson M. (1999). *Philosophy in the Flesh: The Embodied Mind and its Challenge to Western Thought.* New York NY: Basic Books.

Lavater JC. (1797). *Essays on Physiognomy.* (C. Moore, Translator) London UK: HD Symonds.

Lazarus RS. (1991). *Emotion and Adaptation.* New York NY: Oxford University Press.

Leary DE. (1987). From act psychology to probabilistic functionalism: The place of Egon Brunswik in the history of psychology. In, (Eds.), MG Ash, WR Woodward. *Psychology in Twentieth-Century Though and Society.* Cambridge UK: Cambridge University Press. (p. 115–142).

Locke J. (1690). *An Essay Concerning Human Understanding.*

Luo PL, Eikman EA, Kealy W, Qian W. (2006). Analysis of mammography teaching program based on an affordance design model. *Acad Radiol.* 13, 1542–1552.

Mareschal D, Johnson M H, Sirois S, Spratling M, et al. (2007). *Neuroconstructivism: How the brain constructs cognition.* Oxford UK: Oxford University Press. (Vol. 1).

McGrenere J, Ho W. (2000). Affordances: Clarifying and evolving a concept. *Proc Graph Interface.* 1, 79–186. Accessed from the web 12/20/2007.

Medin DL. (1989). Concepts and conceptual structure. *Am Psychol.* 44, 1469–1481.

Medin DL, Lynch EB, Solomon KO. (2000). Are there kinds of concepts? *Annu Rev Psychol.* 51, 121–147.

Mehl M. (2006). Quantitative text analysis. In, (Eds.) M Eid, E. Diener. *Handbook of Psychological Measurement: A Multimethod Perspective.* Washington DC: American Psychological Association. (p. 141–156).

Mervis, C, Rosch E. (1981). Categorization of natural objects. *Annu Rev Psychol.* 32, 89–113.

Moors A, de Houwer J. (2006). Automaticity: A theoretical and conceptual analysis. *Psychol Bull.* 132, 297–326.

Mozley CG, Huxley P, Sutcliffe C, Bagley H, et al. (1999). "Not knowing where I am doesn't mean I don't know what I like: Cognitive impairment and quality of life responses in elderly people. *Inter J Geriat Psychiatr.* 14, 776–783.

Murphy GL. (2002). *The Big Book of Concepts.* Cambridge MA: MIT Press.

Murphy GL, Lassaline ME. (1997). Hierarchical structure in concepts and the basic level of categorization. In, (Eds.) K Lamberts, D Shanks. *Knowledge, Concepts and Categories.* Cambridge MA: MIT Press. (p. 93–132).

Nadeau R. (1997). Hayek and the complex affairs of the mind. *Paper presented at the Sixth-seventh Annual Meeting of the Southern Economic Association.* Atlanta, George. November 21–23.

Navon D, Gopher D. (1979). On the economy of the human processing system. *Psychol Rev.* 86, 214–255.

Nazzi T, Bertoncini J, Mehler J. (1998). Language discriminations by newborns: Towards an understanding of the role of rhythm. *J Exp Psychol: Hum Percept Perform.* 24, 756–766.

Nelson TO, Narens L. (1990). Metamemory: A theoretical framework and new findings. In (Ed.) G Bower. *The Psychology of Learning and Motivation: Advances in Research and Theory.* New York NY: Academic Press. (p. 125–173).

Nguyen SP. (2007). Cross-classification and category representation in children's concepts. *Dev Psychol.* 43, 719–731.

Nilsson H, Olsson H, Juslin P. (2005). The cognitive substrate of subjective probability. *J Exp Psychol: Rev Mem Cogn.* 4, 600–620.

Nisbett RE. (2003). *The Geography of Thought.* New York NY: The Free Press.

Noble S, Shanteau J. (1999). Book Review: Information Integration Theory: A Unified Cognitive Theory. *J Math Psychol.* 43, 449–454.

Norman DA. (1988). *The Psychology of Everyday Things.* New York NY: Basic Books.

O'Boyle CA, McGee HM, Hickey A, O'Malley K, Joyce CRB. (1992). Individual quality of life in patients undergoing hip replacement. *Lancet.* 339, 1088–1091.

Oppenheimer DM. (2003). Not so fast! (and not so frugal): rethinking the recognition heuristic. *Cognition.* 90, B1-B9.

Padilla GV, Mishel MH, Grant MM. (1992). Uncertainty, appraisal and quality of life. *Qual Life Res.* 1, 155–165.

Parducci A. (1995). *Happiness, Pleasure and Judgment: The contextual theory and its Applications.* Hillsdale NJ: Erlbaum.

Park DC, Nisbett R, Hedden T. (1999). Aging, culture and cognition. *J Gerontol.* 54B(2), 75–84.

Parkinson B. (1999). Relations and dissociations between appraisal and emotion ratings of reasonable and unreasonable anger and guilt. *Cogn Emot.* 13, 347–385.

Parkinson B. (2007). Situations to emotions: Appraisal and other routes. *Emotion.* 7, 21–25.

Patrick DL, Erickson P. (1993). *Health Status and Health Policy: Quality of life in Health Care Evaluation and Resource Allocation.* New York NY: Oxford University Press.

Piaget J. (1970). *The Origins of Intelligence in the Child.* New York NY: International Universities Press.

Pinker S. (1997). *How the Mind Works.* New York NY: Norton.

Pollak KL, Arnold RM, Jeffreys AS, Alexander SC, et al. (2007). Oncologist communication about emotion during visits with patients with advanced cancer. *J Clin Oncol.* 25, 5748–5752.

Rapkin BD, Schwartz CE. (2004). Toward a theoretical model of quality of life appraisal: Implications of findings from studies of response shift. *Health Qual Life Outcomes.* 2:14. doi:10.1186/1477-7525-2-14. (p. 1–12).

Rehder B, Ross BH. (2001). Abstract coherent categories. *J Exp Psychol: Learn Mem Cogn.* 27, 1261–1275.

Richardson MJ, Marsh KL, Baron RM. (2007). Judging and actualizing interpersonal and interpersonal affordance. *J Exp Psychol: Hum Percept Perform.* 33, 845–859.

Roberson D, Davidoff J, Davies IRL, Shapiro LR. (2004). The Development of Color Categories in Two languages: a longitudinal study. *J Exp Psychol: Gen.* 133, 554–571.

Rosch E. Natural categories. (1973). *Cogn Psychol.* 4, 328–350.

Rosch, E. (1975). Cognitive representation of semantic categories. *J Exp Psychol.* 104, 573–605.

Rosch, E. (1977). Human Categorization. In, (Ed.) N Warren. *Advances in Cross-Cultural Psychology.* New York NY: Academic Press. (p. 1–72).

Rosch E, B Lloyd. (1978). *Cognition and Categorization.* Mathwen NJ: L Erlbaum.

Rosch E, Mervis CB. (1975). Family resemblances: Studies in the internal structure of categories. *Cogn Psychol.* 7, 573–605.

Rosch E, Mervis CB, Gray W, Johnson D, et al. (1976). Basic objects in natural categories. *Cogn Psychol.* 8, 382–439.

Roseman I, Smith C. (2001). Appraisal Theory: Overview, assumptions, varieties, controversies. In, (Eds.) KR Scherer, A Schorr, T Johnson. *Appraisal Processes in Emotion: Theory, Methods, Research.* New York NY: Oxford University Press.

Ross M. (1989). Relation of implicit theories to the construction of personal histories. *Psychol Rev.* 96, 341–347.

Ross BH, Murphy GL. (1999). Food for thought: Cross-classification and category organization in a complex real-world domain. *Cogn Psychol.* 38, 495–553.

Ruta DA, Garratt AM, Leng M, Russell IT, et al. (1994). A new approach to the measurement of quality of life: The Patient Generated Inventory (PGI). *Med Care.* 32, 1109–1126.

Saalbach H, Imai M. (2007). Scope of language influence: Does a classifier system alter object concepts? *J Exp Psychol: Gen.* 136, 485–5501.

Sass B. (1988). *The Genesis of the Alphabet and its Development in the Second Millennium B.C.* Wiesbaden Germany: Kommiission bei Otto Harrassowitz.

Schechter S. (1994). *Proceedings of the 1993 NCHS Conference on the Cognitive Aspects of Self-reported Health Status.* (Cognitive Working Staff Working Paper No. 10). Hyattsville MD: Centers for Disease Control and Prevention/National Center for Health Statistics.

Schmitz TW, Johnson SC. (2007). Relevance to self: A brief review and framework of neural systems underlying appraisal. *NeuroSci BioBehav Rev.* 31, 585–596.

Schwartz CE, Rapkin BD. (2004). Reconsidering the psychometrics of quality of life assessment in light of response shift and appraisal. *Health Qual Life Outcomes.* 2, (16) 1–11.

Schwartz CE, Sprangers MAG. (2000). *Adaptation to Changing Health: Response Shift in Quality of life Research.* Washington DC: American Psychological Association.

Schwarz, N. (2004). Meta-cognitive experiences in consumer judgment and decision making. *J Consum Psychol.* 14, 332–348.

Schwarz N, Bless H, Strack F, Klumpp G, et al. (1991). Ease of retrieval as information: Another look at the availability heuristic. *J Personal Social Psychol.* 61, 195–202.

Schwarz N, Clore GL. (2003). Mood as information: 20 years later. *Psychol Inq.* 14, 294–301.

Schwarz N, Sanna LJ, Skurnik I, Yoon C. (2007). Metacognitive experiences and the intricacies of setting people straight: Implications for debiasing and public information campaigns. *Adv Exp Social Psychol.* 39, 127–161.

Sieck WR, Yates JF. (2001). Overconfidence effects in category learning: A comparison of connectionist and exemplar memory models. *J Exp Psychol: Learn Mem Cogn.* 27, 1003–1021.

Siemer M, Mauss I, Gross JJ. (2007). Same situation- Different emotions: How appraisals shape our emotions. *Emotion.* 7, 592–600.

Siemer M, Reisenzein R. (2007).The process of emotion inference. *Emotion.* 7, 1–20.

Smith A. (1776/1904). *An Inquiry into the Nature and Causes of the Wealth of Nations.* London UK: Metheuen.

Smith B. (1997). The Connectionist Mind: A study of Hayekian Psychology. In, (Ed.) SF Frowen. *Hayek: Economist and Social Philosopher: A Critical Retrospect.* London UK: Macmillan,. p. 9–29. http://maelstrom.stjohns.edu./archival.htlm.

Smith EE, Medin DL. (1981). *Categories and Concepts.* Cambridge MA: Harvard University Press.

Smolensky P, Legendre C. (2006). *The Harmonic Mind From Neural Computation to Optimality-theoretic Grammar. Volume 1: Cognitive Architecture.* Cambridge MA: MIT.

Stanton AL, Danoff-Burg S, Cameron CL, Snider PR, et al. (1999). Social comparison and adjustment to breast cancer: An experimental examination of upward affiliation and downward evaluation. *Health Psychol.* 18, 151–158.

Stewart AL, Ware JE Jr. (1992). *Measuring Functioning and Well-being: The Medical Outcomes Study Approach.* Durham NC: Duke University Press.

Stoggregen TS, Gorday KM, Sheng YY, Flynn SB. (1999). Perceiving affordance for another person's action. *J Exp Psychol: Hum Percept Perform.* 25, 120–136.

Strack F, Deutsch R. (2004). Reflective and impulsive determinants of social behavior. *Personal Social Psychol Rev.* 8, 220–247.

Torrance GW, Furlong W, Feeny D, Boyle M. (1995). Multi-attribute preference functions: Health Utilities Index. *PharmEcon.* 7, 03–520.

von Osch SMC, Stiggelbout AM. (2005). Understanding VAS valuations: Qualitative data on the cognitive process. *Qual Life Res.* 14, 2171–2175.

Vaughn KI, Poulsen JL. (1997). Is Hayek's social theory an example of complexity theory? Paper presented at the Southern Economic Association, November, 1997. http://www.gmu.edu/departments/economics/working/Pages/9807.html.

Ware JE, Brook RH, Davis-Avery A, Williams KN, et al. (1980). *Conceptualization and Measurement of Health for Adults in the Health Insurance Study: Vol. 1, Model of Health and Methodology.* Santa Monica CA: Rand. Publication No R-1987/1 HEW.

Warren WH. (1984). Perceiving affordances: Visual guidance of stair climbing. *J of Exp Psychol: Hum Percept Perform.* 10, 683–703.

Wierzbicka A. (1999). *Emotions Across Languages and Cultures: Diversity and Universals.* Paris France: Cambridge University ress.

Willis G, Reeve B, Barofsky I. (2005). The Use of Cognitive Interviewing Techniques in Quality of life and Patient-Reported Outcome Assessment. In, (Eds.) J Lipscomb, CC Gotay, C Synder. *Outcomes Assessment in Cancer: Findings and Recommendations of the Cancer Outcomes Measurement Working Group.* Cambridge UK: Cambridge University Press.

Willis J, Todorov A. (2006). First Impressions: Making up your mind after a 100 ms exposure to a face. *Psychol Sci.* 17, 592–598.

Wittgenstein L. (1953). *Philosophical Investigations.* Oxford UK: Blackwell.

Xie Z, Zhang J. (2006). Development of a taxonomy of representational affordances for electronic health record system. *Am Med Inf Assoc.* Annual Meeting 1146, (accessed from the web 12/20/2007).

Zajonc, R. B. (1968). Attitudinal Effects of Mere Exposure. *J Personal Social Psychol.* 9, 1–27.

Zhang S, Schmitt B. (1998). Language-dependent classification: The mental representation of classifier in cognition, memory and evaluations. *J Exp Psychol: Appl.* 4, 375–385.

Summary of Part I

One of the first impressions I developed while writing Part I was how often and in how many different ways the term "quality" can be found in the common discourse and scientific reports. In addition, there are a large number of terms used that are assumed to monitor the quality of objects, events or life histories. Thus, terms such as "hue," "blueness," "beauty," "happiness," and even "well-being" are used as a quality indicator. Part of what makes them quality indicators is that they are products of the cognitive processing of physical and physiological stimuli. Hue, for example, can be used to denote quality of the colors a person is observing, whereas wavelength, a physical measure, does not. The emotions evoked by seeing a particularly beautiful face could also indicate quality, but again the correlated physical or physiological stimuli alone do not. Sometimes calling a term a quality indicator may not be appropriate, and I will provide examples of this. It is also clear that a quality assessment is a ubiquitous part of human affairs, reflecting a human being's constant attempt to control the nature of their existence by constantly monitoring and assessing its quality.

The conceptualization of quality and quality-of-life assessment has an extended history, starting with Aristotle and continuing to this day. I found Gärdenfors' (2000, pp 1–2) suggestions that quality can be conceived of in symbolic terms; in terms of an association or connection between entities; or as a conceptual space, to be quite useful. The symbolic perspective, of course, involves the use of language, and in Chap. 2, I outline some of the concepts that must be attended to when analyzing language usage patterns. My objective in Chap. 2 was to learn about how people use language so that I can understand how the concept of quality is expressed. I illustrated how some of these analytical tools (e.g., demonstrating the presence of conceptual metaphors) can be used to analyze and organize the diverse quality-of-life or HRQOL definitions available. I suspect that different settings (e.g., diseases, treatments, symptoms, functional indicators, and so on) and different languages will express quality differently, so I will continue to use these tools.

If a qualitative assessment is modeled using principles from associationism or connectionism, then this gives me opportunity to use *non-linear processes* to characterize how a cognitive entity is formed (Stephen et al. 2009). In addition, I should be able to use complexity theory to account for how a person's constant monitoring of the quality of their existence leads to the development of cognitive entities in a qualitative assessment. If so, a model of quality as an emergent phenomena would be reinforced, suggesting that what a person reports as their life quality is based on a complex adaptive system.

Gärdenfors' (2000) emphasis on conceiving of quality as a conceptual space was also useful to me. Thus, in Chap. 1 I introduced the notion that a quality assessment involves both aesthetics and functional assessments. I used a two-dimensional circumplex to characterize the descriptive and evaluative tasks involved in a quality assessment, leaving the valuation component as a third dimension (not illustrated in Figs. 1.2 or 1.3). Gärdenfors (2000) also points out that each of the three approaches he proposes may have a place in the conceptualization of quality. I concur.

One task I have taken on in is to illustrate how a symbolic rule system, such as language and reasoning, can be related to a complex evolving cognitive system. In this regard, Smolensky (2004) nicely describes the task:

> Precise theories of higher cognitive domains like language and reasoning rely crucially on complex symbolic rule systems like those of grammar and logic. According to traditional cognitive sciences and artificial intelligence, such symbolic systems are the very essence of higher intelligence. Yet intelligence resides in the brain, where computation appears to be numerical, not symbolic; parallel not serial; quite distributed, not as highly localized symbolic systems.

Smolensky sees his own research task as bridging the gap between these two perspectives, and of interest is that he uses a mathematical construct – optimality theory – to do this. This is important for two reasons. First, it approximates what is required in fundamental measurement, as compared to the current approach that has serious conceptual deficiencies (see Chap. 12). Also, it may provide an alternative to a language-based symbolic system approach to assessing quality, a system that can be compromised when a person has had a stroke, or is progressively losing the ability to remember. Under these circumstances the neural network, the brain, may be damaged, or the person may lack adequate neural network functioning. Thus, while I ordinarily use language and reasoning to communicate, I rely on my biology for this to happen, and I will continue to explore the relationship between biological factors and the subjective world of quality-of-life or HRQOL assessment (Chaps. 6–11).

Part I was designed to introduce the analytic tools I that I have used in succeeding chapters. Most importantly, I introduced the notion that a quality assessment is based on a defined set of cognitive processes (description, evaluation, and valuation/reflection), and these same processes occur whether a person is judging a painting, a fine wine, a treatment regimen, or a particular life style. However, I acknowledged and will demonstrate in greater detail that this universal cognitive process (the formation of a hybrid construct) can be influenced by various contextual factors (e.g., culture, disease status, and so on). In addition, I acknowledged that some qualitative judgments occur in a fast and automatic manner (as occurs when a person bases a judgment on first impressions), and I also discussed how these types of cognitive processes may still be accounted for by a hybrid construct (Chap. 3 p. 66).

One of my objectives was to shift the attention of quality researchers away from continuing to catalog indicators of quality to studying *how a quality judgment is made*. The rationale for this shift is really a question of how best to demonstrate the validity of a quality assessment. Currently, there are a variety of psychometric methods (e.g., Nunnally and Bernstein 1994) that are routinely used for establishing validity (e.g., face validity construct validity, criterion validity, and so on), but as I have implied and discussed in a different context (Barofsky 2000), exclusively using psychometric methods to demonstrate the validity of a quality assessment has inherent limitations. For example, criterion validity is routinely established by correlating one set of psychometrically derived indicators with another, with the result that there is limited evidence of external validity. An alternative is to demonstrate the cognitive equivalence between the two assessments. In Part II, I shift my attention to discussing three topics: what is the *purpose* (e.g., establishing invariant principles), *content* (objective or subjective indicators), and *unit of analysis* (categories or domains) of a qualitative assessment.

References

Barofsky I. (2000). The role of cognitive equivalence in studies of health-related quality of life assessments. *Med Care.* 38 (Supp II), 125–129.

Gärdenfors P. (2000). *Conceptual Spaces: The Geometry of Though.* Lexington MA: MIT Press.

Nunnally JC, Bernstein IH. (1994). *Psychometric Theory.* New York NY: McGraw-Hill. Third edition.

Stephen DG, Dixon JA, Isenhower RW. (2009). Dynamics of representational change: Entropy, action, and cognition. *J Exp Psychol Learn, Mem Cog.* 35,1811-1832.

Smolensky P. Unpublished statement of research objectives. (Accessed July, 2004). http://www.cog.jhu.edu/faculty/smolensky/#top.

Part II

Issues in Assessing Quality or Quality-of-Life

Introduction to Part II

Abbreviations

HRQOL   Health-related quality-of-life
HUI   Health Utilities Index (Torrance et al. 1995)
QWB   Quality of Well-being Scale (Kaplan and Bush 1982)

This book starts with the observation that a person is capable of providing self-reports about quality, and that my first priority[1] should be the study of how these self-reports occur. Since a quality statement involves description evaluation and qualification, it is also natural for me to study the judgment process that leads to self-reports. Studying the language used and cognitive processes involved in, multi-item assessments has to become a secondary goal; one that would logically occur only after I have learned how quality-related self-reports occurred.

Although this book starts from a phenomenological observation, it should not be confused with a philosophical phenomenological approach. Phenomenology, as a distinct philosophical tradition, emphasizes the study of immediate experience (Baron 1985). It encourages an investigator to isolate experiences, avoiding any assumptions of cause or consequence. I wish to avoid the debates that follow from this particular tradition (e.g., whether it is actually possible to separate an experience from its context); although others have found it worth including when considering particular aspects of self-reports. Hudak et al. (2004), for example, described a phenomenological approach to embodiment while studying satisfaction with medical care.

In addition, the approach I propose contrasts with the current standard for assessing quality-of-life or health-related quality-of-life (HRQOL), a standard which evolved from the application of psychophysical principles to education issues followed by its extension to any number of venues, including the evaluation of quality indicators. As I will and have discussed, there are very specific reasons justifying this approach, some of which include an investigator's desire to avoid unobservable introspectively based reports and what that might mean about the objectivity and reliability of that being assessed. The development and early success of behaviorism and its influence on the philosophy of science also exercised a strong influence on how the study of quality was to occur. The psychometric methods that evolved, therefore, were limited to measuring input and output relationships using hypothetical constructs, such as latent variables, to account for possible intervening events. Formal efforts to study qualitative experiences, a person's preferences or values, and the cognitive mechanisms that mediates these activities were largely ignored or avoided.

---

[1]Of course, as is so often the case, my approach is not unique. An example can be found in Bernheim (1999), who has promoted using global qualitative assessments since the 1980s.

Also evident is that applying the test and measurement approach to assessing quality has certain limitations. For example, the approach basically samples the issues that make up a person's quality-of-life or HRQOL, and assumes that if the sampling is large and diverse enough, it would be able to capture what is true for the totality of a person's experiences. However, describing a person's status may not be sufficient to capture what is meant by quality, a concept that implicitly or explicitly has an evaluative and value/preference dimension. Of course, quality or quality-of-life researchers have recognized this and have developed procedures that combine the descriptive results of testing with methods designed to attach value statements to these descriptors (e.g., the Health Utilities Index (HUI) or Quality of Well-being Scale (QWB)).

A second set of limitations has to do with the tendency of multi-item assessments to particularize the content being assessed. The argument is that this enhances the analytical value of any assessment, but the limitation is that the outcomes may have less to do with the respondent's interests than the research question posed by the investigator. There is also the problem of the methods used to draw inferences by reaggregating the information collected by the multi-item assessment. Again, this is often done by the investigator rather than the respondent, raising the risk that conclusions drawn by the respondent and the investigator may not concur. This, of course, raises questions about the validity of the assessment. Thus, as indicated in Chap. 1, there is much concern about the outcome of a multi-item quality-of-life or HRQOL assessment and its usefulness, particularly when used to characterize the exceptional person.

A third concern expressed in Chap. 3 was that there are no specific rules that can be used to define how an abstract concept like quality-of-life or HRQOL can be operationalized or made concrete. Of course, it is possible to find empirical support for including a particular set of domains (Cummins 1996; Haase and Braden 1998), or even for developing a consensus among investigators on what concrete domains to use to operationalize the concept of quality-of-life or HRQOL. However, this does not avoid the comment that any list of domains appears arbitrary, no matter how much agreement exists between investigators. In addition, any set of domains would at best reflect a nomothetic perspective, sacrificing the inherent idiopathic characteristic of qualitative data.

However, preserving the idiopathic character of a qualitative assessment leads to a constructionist, rather than a positivistic, approach to a qualitative assessment. Norman (2003) points out that the rationale for this includes the observation that standardized questionnaires often do not capture the individual's unique state, although several quality-of-life or HRQOL assessments exist which give the respondent the opportunity to indicate their individual values or preferences (Joyce et al. 1999; Ruta et al. 1994).

These and other issues have made me cautious about using multi-item assessments as the starting point in studying how to define or measure quality-of-life or HRQOL. As a consequence, when I use the phrase "quality-of-life or HRQOL assessment" in this book, it is appropriate to assume that I am referring to a person's self-report which may not only be global but also could be content-specific. Thus, understanding the "self" in either type of report is my first task.

While the test and measurement approach has been very successful, its intellectual foundations, based on psychophysics and behaviorism, have not been left unchallenged. For example, in the last 50 years or so, a revolution has occurred in the scientific study of cognitive processes (the so-called cognitive revolution), resulting in it being increasingly acceptable to speak of internal processes not as hypothetical constructs, but processes that can be empirically examined either by appropriate experiments or brain imaging studies (e.g., Gardner 1985). As a consequence, a significant research literature exists which can be applied to the assessment of quality or quality-of-life, though to date it has not been done to a great extent. One of the purposes of this book is to bridge this gap.

But how shall I proceed to bridge this gap, especially when it is evident that a vast body of information exists that can be included as part of the definition or assessment of quality or quality-of-life? For example, should I be cataloging specific domains, as would occur when I measure a person's health status, physical functioning, or symptoms? I could also focus on measuring what is important to a person, or what they prefer or value, and use this to define life quality. Should I be tracing the consequences of a disease or treatment, or should I be concerned with broader issues such as whether a person has achieved their life goals, or whether they have lived "the good life"? Should I be measuring all these indicators, and if I did, how would I combine or organize them in some meaningful way? Is there a rational basis for deciding what should or should not be included in either the definition or assessment of quality or quality-of-life? Should I rely on what the public thinks about these issues or should I rely on an investigator's personal intellectual history or aesthetics to decide on what is important here?

Fortunately in Part I, I identified some tools I can use to deal with these issues. Thus, I learned how information is classified, concepts formed, and how metacognitive processes monitor and control the formation and reformation of a cognitive entity. These cognitive processes will, in effect, act as an overlay that structures my subsequent analyses. In addition, I have also discussed how quality or quality-of-life can be viewed as a conceptual metaphor, and indicated that it is part of a conceptual hierarchy.

In Part II, I intend to use these tools to describe the descriptive information that should be included in a qualitative assessment. Thus, in Chap. 4 I will ask if objective, subjective, or some combination of these types of data should be included in a quality assessment. Chapter 5 will be a continuation of this discussion, since signs and symptoms are analogs of the objective/subjective distinction discussed in Chap. 4. In Chap. 6, I discuss how people cognitively form categories and investigators create domains.

How categories or domains are formed is a critical issue, since they may combine a complex array of indicators, for example, both experiential and biological factors. In addition, some of this information may be acquired (e.g., conditioning that results in anticipatory nausea and vomiting) and some predetermined (e.g., Chomsky's notion about grammar). A single construct, such as fatigue, may actually have both types of indicators.

Dual-process theories (e.g., Birnboim 2003; Feldman Barrett et al. 2004; Schneider and Shiffrin 1977; Chap. 6), which combines automatic and attention-dependent controlled information processing will also be important to discuss. Dual-process theories have been successfully used to account for a number of psychological phenomena (e.g., attribution, stereotyping, persuasion, and so on), and should help me explain how the domains of information that result from classification contributes to decisions, ratings, and verbal description of the quality of a person's existence. The notion that judgments, in general, and judgments of quality, in particular, may occur automatically with minimal conscious control (Chap. 6) or by active cognitive or metacognitive processes will provide a range of conceptual tools that will be helpful in accounting for the variability sometimes observed in qualitative assessments (Fig. II.1).

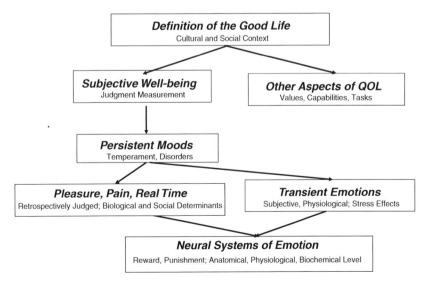

Fig. II.1 Quality-of-life defined as an emotional state associated with the "good life" as conceptualized by Kahneman et al. (1999) (With permission).

Kahneman et al. (1999) provide an example of how some of the concepts I have been discussing can be organized into a model (Fig. II.1) that includes an implicit hierarchy. The model starts with a reference to a "definition of the good life," a phrase suggesting that the Kahneman et al. have adopted the Aristotelian notion that quality should be based on an idealized prototypical outcome (see Chap. 12). They go on to disaggregate this definition into a measure of subjective well-being and what they refer to as other aspects of quality assessment including values, capabilities, and tasks. Interestingly, this formulation remotely resembles the basic valuation definition discussed previously, where descriptors are combined with a valuation statement (Chap. 2). These authors also describe other levels of analysis, including "persistent moods," followed by "pleasures, pains, or transient emotions," all of which are based on neural systems dealing with emotions. The hierarchy implicit in this model suggests a movement from a purely subjective to more objective indicators of affect (from the cultural and social level to the neural). Thus, to these authors, quality is predominately an emotional expression, with minimal emphasis on the cognitive task of making a judgment.

Of course, defining quality or quality-of-life as some aspect of a feeling state seems limited and limiting. Quality, as I reviewed in Chap. 1, has a functional component that deserves to be attended to, yet it is hard to detect its presence in Kahneman et al.'s (1999) model (Fig. II.1). In addition, the model does not clarify whether an assessment of quality should be based on subjective or objective affective indicators (Chap. 4), or as Kahneman et al. (1999) suggests some combination of both. For example, if I limited my inquiry to only objective indicators, I might find it necessary to avoid certain types of self-reports (e.g., those dealing with feelings, preferences, attributions, etc.) that would otherwise be informative, while if I relied exclusively on subjective indicators I might find that the data is highly suspect and susceptible to confounding by respondent biases. I will discuss these issues in detail in Chap. 4.

Chapter 5 continues and expands on the discussion in Chap. 4, now applying it to the subject of interest in this book: the medically ill person. Signs and symptoms at first glance seem to be analogs of objective and subjective indicators, and one of the questions explored in Chap. 5 is to what extent this is true. I start the chapter by discussing how language is used to express symptoms, and also address some of the complex issues involved in assessing symptoms. At this point, I review the history of the concept of objectivity, illustrating that the meaning and methods of studying objectivity and subjectivity, like signs and symptoms, has evolved and continues to evolve. An additional topic I discuss is the role of subjectivity in clinical medicine, how integral a part of medicine is, and what this means about the assessment of quality. I end this chapter by asking whether I should use psychometric or clinimetric assessment approaches to measure signs or symptoms.

Chapter 6 considers two basic methods, category and domain formation, whereby information can be summarized and presented in a format that can be communicated and/or used in statistical analyses. As stated in Chap. 3, classification is the first step in the process that leads to the formation of a category, and categories are required to formulating a construct or concept. Each of these cognitive processes are required for judgments to occur. Domains are also summary statements, but now generated by an investigator for specific purposes that include making data amenable to statistical analyses. Chap. 6 includes an extended discussion of the role of dual processing, different forms of modularity, and other structural determinants of category and domain formation. I will provide examples of this by applying some of the principles identified to specific research tasks.

A recurring issue in this book is whether quality can be adequately characterized by some quantity (Chap. 12). Is this possible, or is something missing when just a number is used to characterize an experience? I raised this issue earlier when discussing how often different definitions of functional quality elicit qualification by statements of aesthetic quality. This is a topic that has generated much philosophical discussion (e.g., Churchland 1985; Jackson 1982; Nagel 1974), and it is not my purpose to digress into it in great depth (see Chap. 1, p. 10, where I introduced some of these issues). Instead, I would like to focus on whether it is possible to measure meaning, with the number that I use when I perform a quality or quality-of-life assessment. Of course, there are various methods I might use to approach this task, and I will review some of them as the book progresses.

References

Baron RJ. (1985). An introduction to medical phenomenology. *Ann Inter Med.* 103, 606–611.

Bernheim JL. (1999). How to get serious answers to the serious question: "How have you been?": Subjective quality of life (QOL) as an individual experiential emergent construct. *Bioethics.* 13, 272–287.

Birnboim S. (2003). The automatic and controlled information-processing dissociation: Is this still relevant? *Neuropsychol Rev.* 13, 19–31.

Churchland PM. (1985). Reduction, Qualia, and the Direct Introspection of Brain states. *J Philos.* 82, 8–28.

Cummins RA. (1996). The domains of life satisfaction: An attempt to order chaos. *Soc Ind Res.* 38, 303–328.

Feldman Barrett L, Tugade MM, Engle RW.(2004). Individual differences in working memory capacity and dual-process theories of the mind. *Psychol Bull.* 130, 55–573.

Gardner H. (1985). *The Mind's New Science: A History of Cognitive Revolution.* New York NY: Basic Books.

Haase JE, Braden CJ. (1998). Guidelines for achieving clarity of concepts related to quality of life. In: (Eds.) CR King, PS Hinds. *Quality of Life: From Nursing and Patient Perspectives. Theory, Research, Practice.* Sudbury MA: Jones and Bartlett.

Hudak PL, McKeever PD, Wright JG. Understanding the meaning of satisfaction with treatment outcomes. *Med Care.* 42; 718–725: 2004.

Jackson F. Epiphenomenal qualia. (1982). *The Philos Q.* 32, 127–136.

Joyce CRB, O'Boyle CA, McGee H. (1999). *Individual Quality of Life: Approaches to Conceptualization and Assessment.* Amsterdam The Netherlands: Harwood.

Kahneman D, Diener E, Schwarz N. (1999). Preface. In: (Eds.) D Kahneman, E Diener, N Schwarz. *Well-being: The Foundation of Hedonic Psychology.* New York NY: Russell Sage Foundation. (p. ix–xii.).

Kaplan RM, Bush JW. (1982). Health-related quality of life for evaluation research and policy analysis. *Health Psychol.* 1, 61–80.

Nagel T. (1974). What is it like to be a bat? *Philos Rev.* 83, 435–450.

Norman G. (2003). Hi! How are you? Response shift, implicit theories and differing epistemologies. *Qual Life Res.* 12, 239–249.

Ruta DA, Garratt AM, Leng M, Russell IT, MacDonald LM. (1994). A new approach to the measurement of quality of life: The Patient Generated Inventory (PGI). *Med Care.* 32, 1109–1126.

Schneider W, Shiffrin RM. (1977). Controlled and automatic human information processing: I. Detection, search and attention. *Psychol Rev.* 84, 1–64.

Torrance GW, Furlong W, Feeny D, Boyle M. (1995). Multi-attribute preference functions: Health Utilities Index. *PharmEcon.* 7, 503–520.

4. The Role of Objective or Subjective Indicators

Abstract

The role objective and/or subjective indicators play in a qualitative assessment is central to the process whereby a quality indicator is transformed into a quantity. This chapter attempts to clarify the issues involved in this transformation, with a particular emphasis on reviewing the various models that have been proposed to deal with this issue. This discussion is continued in Chaps. 5 and 12.

How pathetically scanty my self-knowledge is compared to my knowledge of my room. There is no such observation of the inner world, as of the outer world.

Franz Kafka, *The Third Notebook*[1]

Abbreviations

ADL	Activities of daily living
CHF	Congestive heart failure
COPD	Chronic obstructive pulmonary disease
EQ-5D	EuroQoL -5D (Kind 1996)
HUI	Health Utility Index (Torrance 1976)
ORHQoL	Oral health-related quality-of-life
PGI	Patient Generated Index (Ruta et al. 1994)
QWB	Quality of Well-being Scale (Kaplan and Anderson 1990)
SEIQoL-DW	Schedule for individual quality-of-life-direct weighting (O'Boyle et al. 1993)
WHO	World Health Organization

1 Introduction

One of the more contentious issues in assessing quality is deciding what type of information should be included. Should I, for example, limit myself to objective indicators, especially because the statistical methods I will use to make decisions seem most compatible with this type of data? Or should I be measuring subjective indicators, since self-reports are the best method of gaining information about what a person might mean by quality, especially quality-of-life?

Complicating things however, is the fact that *how* I assess these indicators may vary (Table 4.1). Thus, it is possible to perform a quality assessment of either objective or subjective indicators, just as it is also possible to perform a nonqualitative assessment of objective and subjective indicators (Table 4.1). For example, the medication a person consumes can be objectively assessed (e.g., the number of pills a person consumes can be independently counted), but to determine if an error (pills consumed relative to the prescribed dose) has occurred will require a subjective judgment or evaluation.[2] Thus, it would meet the criteria of a quality assessment of an objective indicator.

Measuring the distance a person walks or the ability to dress oneself are also objective indicators, because they are self-reports that others can confirm. However, there is nothing inherent in these indicators that ordinarily suggests or connotes a qualification (e.g., that they have clinical importance). Thus, it is possible to have a nonqualitative assessment of an objective indicator. This doesn't mean that these objective indicators may not be "qualified," but for that to happen requires at least two distinct cognitive processes: first, the assessment of the objective or subjective indicator (e.g., the scaling of the indicator), and then the "qualification" (see Table 4.1, footnotes) of these indicators.

It is also possible to have subjective indicators, or *self-reports,* which are descriptive and evaluate a state a person is in, but that do not represent a quality assessment. Pain and

Table 4.1 The relationship between type of assessment and indicator

	Qualitative assessments Based on subjective evaluations	Nonqualitative assessment ▲Based on objective evaluations
Objective indicators	Medication errors	Feet walked per minute
	Car repair rate	Self-care: dressing oneself
Subjective indicators	Self-reports that includes qualifications. ▼	Self-report of pain confirmed by X-ray exam; self-report of fatigue confirmed by test for anemia

▲: By "qualification" I mean the implementation of a metacognitive process; that is, *a second order process that requires reflection and is in essence a valuation of the significance of a prior evaluation*. QWB, HUI, EQ-5D, are examples of assessments that can be used to qualify an evaluation. These assessments provide standardized (population-based) person-based preference systems that can be used in this regard, although SEIQoL or PGI provide valuation systems that rely on the direct assessment of an individual

▼: A nonqualitative assessment can be "qualified," but qualification is not an inherent part of the assessment. Thus, soiling of a garment per se is not an indicator of a qualitative state – unless the respondent indicates the importance of being in that state. Obtaining this information is required for the state to be qualified. The same would be required for a report of colonic pain or the time a person spends socializing

fatigue self-reports fit into this category, because these self-reports can be confirmed by objective assessments (e.g., an X-ray can be used to evaluate whether a person has a fracture to confirm the report of pain, or a blood test for anemia would confirm a person's complaint of fatigue), but will remain nonqualitative if the person is not asked to evaluate the importance of the pain or fatigue. When this occurs, of course, then the self-report would be considered a quality-of-life statement, or a "qualified" statement.

Finally, it is possible to have a subjective evaluation of a self-report (that is, a subjective response). These self-reports combine both a descriptive evaluation and a qualification of these evaluations. Sometimes these outcomes are synthetically "constructed," with persons other than the respondent placing an importance rating on the initial evaluation or scaling. Some examples are listed in Table 4.1 (see also Table 2.8). This process of combining several cognitions when performed by a person involves metacognitive processes (Fig. 3.1; Chap. 3, p. 64), but investigators have also modeled these experiences using different conceptual approaches (e.g., Kaplan et al. 1987; Torrance et al. 1995). In Chap. 8 (Table 8.6), I provide more details about these cognitive "constructs." As I have already implied, I will use this type of hybrid construct as a cognitive model for a quality assessment throughout this book.

From this I can conclude that both objective and subjective indicators participate in a qualitative assessment, and because of this, some investigators have suggested that they can be combined into one composite indicator. This can be done in two basic ways; experientially or empirically. What I have just been describing are examples of experientially based cognitive constructions; that is, a person either "makes the objective subjective" (the *subjectification* process) or uses metacognition to insert characteristics of themselves into the experience (the *qualification* process). It will usually be followed by the application of various statistical measurement models. Empirically derived constructs involves an investigator deciding on how objective and subjective indicators should be combined, and I will provide some examples

Table 4.2 Methods for making subjective indicators objective (i.e., quantities)

Methods	Examples
Prescribed cognitive pathways	*Strong models*: Standard gamble; Time-trade off *Weak models*: Visual Analog Scale
Modeling methods	Rasch modeling and item-response theory Neural networks and dynamic system analysis
A priori valuation methods	Methods that establish valuations as part of the process of identifying items to include in a Multi-item Assessment

of these models. Here, statistical measurement models would be directly applied.

However, what also remains possible is to "objectify" the subjective self-reports, and I will also discuss this notion in some detail (e.g., Sen's notion of positional objectivity; Kahneman's ideas about objective happiness; and the Rasch Model; see below). This process of making the subjective, objective is a primary goal in quality and quality-of-life research, since it makes possible the optimal application of statistically based decision-making methods.[3] Table 4.2[4] summarizes examples of the methods currently being used to objectively assess subjective data. I will discuss these methods in more detail, but what characterizes the "prescribed cognitive pathways", for example, is that a respondent is given very explicit instructions on how to generate an indicator (e.g., a utility). Some of these indicators have characteristics that make them compatible with the von Neumann and Morgenstern (1944) mathematically derived theory of making decisions under uncertainty. In agreement, the Rasch model also applies probabilistic concepts, but now to items that are selected to match the model. Both of these approaches illustrate the potential of applying mathematically derived concepts to the task of objectifying subjective indicators. This includes the neural network-based optimality theory (Chap. 10; Smolensky and Legendre 2006). This approach will play an important role in this book.

2   Self-Reports as Data

A general issue that has bedeviled qualitative research is its reliance on self-reports and the inherent limitations of such reports to external verification. Some investigators have expressed considerable concern about the usefulness and inferences that can be drawn from such subjective data, and remain concerned about the extent such data can be used in individual and public health policy decision making. Sen (2001, 2002) is one such investigator and he states his concerns as follows:

> Objectivity is central to scientific knowledge…That objective knowledge can be deeply threatened by the subjectivity of observations has been often discussed and emphasized (in my judgment rightly so). The observer's personal prejudices and leanings can leave a lasting impression on observation and analysis. The fear of 'subjectivity' has bred a deep suspicion of the influence of the observer's position on what is observed. (Sen 2001; p. 115)

Sen's view has had practical consequences. For example, the investigators involved in WHO's World Health Survey (Murray et al. 2002) refer to and agree with Sen, and suggest that health surveys should avoid what they refer to as the "paradoxical findings" found when subjective (e.g., self-rated) assessments are included in measures of health status (Salomon et al. 2004). They state:

> Clinical trials and national surveys rely heavily on self reported measures of health, but interpretation of these measures is complicated by incomparability when different people understand and respond to a given question in different ways. Paradoxical findings have been reported in many analyses of population health surveys, suggesting that self-reported measures may be misleading without adjustment for these differences. Distinguishing between differences in self ratings due to actual health differences and differences due to varying norms or expectations for health is a key challenge in interpreting self reported measures of health. (Salomon et al. 2004; p. 328).

For Salomon et al. (2004; p. 328), the problem was the "incomparability when different people understand and respond to a given question in different ways." They deal with this issue by using culturally specific vignettes to anchor a self-rated health status scale. This, they claim, permits the assessment to reflect cultural differences, yet be sufficiently general (because the same scale was used) so that cross-cultural statements can be made. However, they have not reported, as of yet, the optimization of their assessment by using item response theory, and particular differential item responding, as a method for selecting items in a cross-cultural setting.

Sen (2001) and Salomon et al. (2004) were not the only investigators to raise questions about the value of self-reports. Stone et al. (2000) reviewed many of the issues involved in assessing self-reports, as does Schwarz et al. (2007). Kahneman (1999) was particularly concerned about how retrospective recall may confound a self-report (see Chap. 4, p. 109; Chap. 11, p. 410). Dunning et al. (2004) pointed out that self-assessments are particularly deficient when persons are asked about their skills or knowledge. Broderick et al. (2006) report a study in which they interviewed patients who were asked to self-report pain intensity over the past week, and found that the respondents were using several different aspects of the pain experience upon which to base their judgments. Thomas and Diener (1990) report studies of persons who were asked to recall the intensity and hedonic level of various emotional experiences. They found that respondents overestimated the intensity of the emotion they experienced, and underestimated the number of positive, as opposed to negative, experiences they had. Atkinson et al. (1997) found that persons with an affective disorder (e.g., persons with bipolar disorders) were less able to assess their quality-of-life, as compared to persons with schizophrenia or a physical illness. Barsky (2002) reports that when a person is asked to describe symptoms or illness episodes, they may forget, fabricate, or project recent experiences as happening in the past. And I am sure there are other examples where the quality and value of self-reports as a form of data can be raised. All these types of data reinforce the concern that subjectivity confounds empirical observations, including the interpretation of surveys used in making individual or policy decisions (see Chap. 11, p. 395). But what is the difference between subjectivity and objectivity, and is the influence of subjectivity as corrupting as Sen (2001) seems to imply, or as paradoxical as Salomon et al. (2004) claim? In some ways, the complaint here is that qualitative assessments are "local" phenomena, with the uniqueness of their conditions reflecting the individuality of respondents, much as Hayek would expect. In addition, the problems here are not all due to the nature of subjectivity. Schwarz (1999), for example, has shown that the types and conditions of asking questions "shapes" a person's self-report, suggesting that some of the unreliability of self-reports may be due to the imprecision of the assessment process itself, not something that is inherent to a self-report. What is not obvious is what proportion of the self-report variance would be due to such limitations. But what also remains unresolved is whether the complex empirical situation that exists justifies restricting the type of information that should be collected. I will return to this issue later.

As a first step in addressing these questions, I have provided definitions of the terms *subject* and *object* (Table 4.3) and their role as parts of speech (e.g., adjective or noun). For example, an "objective statement" is considered to be a statement about an object that is free of personal influence. Thus, the statement "an orange is round" would be considered "objective," because a criterion *external* to the object (e.g., a mechanical indicator of "roundness") can be applied. In addition, this information about the object can be communicated to others (i.e., it is public) and can be verified by others.

Table 4.3 Definitions of common terms used in quality or quality-of-life assessments[a]

Term	Definition
Subjective	*Adjective*: (1) Based on or influenced by personal feelings, tastes, or opinions. (2) Dependent on the mind for existence. (3) [Grammar] related to or denoting a case of nouns and pronouns used for the subject of a sentence
Subjectivity	*Noun*: Being in a subjective state
Objective	*Adjective*: (1) Not influenced by personal feelings or opinions. (2) Not dependent on the mind for existence. (3) Grammar relating to a case of nouns and pronouns used for the object of a transitive verb or a preposition. *Noun*: (1) A goal or aim. (2) The lens in a telescope or microscope nearest to the object observed
Objectivity	*Noun*: Being in an objective state
Well-being	*Noun*: The *state* of being comfortable, healthy, or happy
Happiness	*Noun*: The *condition* of being happy
Pleasure	*Noun*: (1) A feeling of happy satisfaction and enjoyment. (2) An event or activity from which one derives enjoyment. (3) [Before another noun] intended for entertainment rather than business: pleasure boats. (4) Sensual gratification
Satisfaction	*Noun*: (1) The state of being satisfied. (2) Law: The payment of a debt or fulfillment of an obligation or claim. (3) What is felt to be due to one to make up for an injustice
Need	*Noun*: (1) Circumstances in which a thing or course of action is required. (2) A thing that is wanted or required. (3) A state of poverty, distress, or misfortune *Verb*: (1) Require (something) because it is essential or very important. (2) Expressing necessity or obligation
Value (s)	*Noun*: (1) A desire for something. (2) Lack or deficiency. (3) Lack of essentials; poverty
Preferences	*Noun*: (1) A greater liking for one alternative over another or others. (2) A thing preferred. (3) A favor shown to one person over another or others
Quality	*Noun*: (1) The degree of excellence of something as measured against other similar things. (2) General excellence. (3) A distinctive attribute or characteristic
Life	*Noun*: (1) The condition that distinguishes animals and plants from inorganic matter, including the capacity for growth and functional activity. (2) The existence of an individual human being or animal. (3) A particular type or aspect of people's existence: school life. (4) Living things and their activity. (5) A biography. (6) Vitality, vigor, or energy. (7) [Informal] a sentence of imprisonment for life. (8) [In various games] each of a specified number of chances each player has before being put out

[a]Based on definitions provided by the *Oxford English Dictionary* (1996)

In contrast, the statement "I am in pain" is a "subjective" statement because it is "of the subject" (i.e., it reflects a person's beliefs or feelings). As a subjective statement, no one other than the person can confirm it; that is, it is a private statement. Because it is a private statement, pain-related behaviors such as not going to work are often viewed with suspicion. Even so, on some occasions it is possible to establish the reliability of what the person said (that is, others can agree that the person said what they said) and it can lead to actions that are consistent with their statement (e.g., taking pain medication that leads to reports of relief and return to work). In addition, external objective indicators (e.g., an X-ray of a broken bone) may reinforce the basis for the subjective response.

Sen's (2001) views about subjectivity can be traced to a very specific study with which he was involved.[5] This study compared self-reported health status and longevity in different regions of India. What was found was that the region with the lowest *mortality* (death) rate also reported the highest *morbidity* (health complaints) rate, whereas the opposite was true for an "comparison" region. It was counterintuitive that the area which reported the lowest *morbidity* and had the highest *mortality* rate was also the area which was objectively least "healthy," in that it had the fewest health facilities, available medical staff, and so on. Sen suggested that differences in education (i.e., literacy levels), available public health facilities, etc., confounded the self-reports, in the sense that the respondents were not aware of the level of health care support possible. What was of particular concern to Sen, however, was how a policy maker could justify decisions and an intervention based on these sorts of data. How could a policy maker justify expanding the health services of a community whose morbidity rates were low?

Salomon et al. (2004), on the other hand, were concerned that people of different cultures think differently about their lives and that this will complicate, if not prevent, any cross-cultural comparisons in surveys such as the World Health Survey (Murray et al. 2002). Thus, these investigators acknowledge that cognitive processes play a role in accounting for the differences in cross-cultural studies, but consider this as a research distraction (something to be avoided) rather than an important research opportunity.

Both Sen and Murray and their colleagues accept that health status (or quality-of-life) is "of the subject," but they are concerned about how to interpret the results of such studies. One approach to this problem, suggested by Sen (1993), is to consider *self-reports as being conditionally dependent on objective indicators*. Thus, if the respondents from the two regions in India, *objectively* differed in their *position* relative to literacy and available health care facilities, then these differences in objective indicators or "objective position" should be taken into account to help understand the counter-intuitive self-report results of his study.

3 Positional Objectivity

Sen's notion of *position-dependent objectivity* (or *positional objectivity*) is important since it helps clarify the relationship between objective and subjective indicators. It does this by defining the conditions under which an objective, but also by extension a subjective, statement may vary because of differences in how observations are made. The classic view of objectivity, Sen (1993) points out, requires that statements be invariant or "*trans-positional*" (see Chap. 4; p. 112 for an extended discussion of invariance). This occurs when an observation remains true independent of the conditions of the observation. Sen believes that while this may occur under certain conditions, it does not occur under all of them, and the conditions under which an objective observation will be dependent on its position will also be applicable to decisions concerning beliefs and actions. Consider the following illustration Sen gives (1993; p. 128–129):

(A). The sun and the moon look similar in size.

This statement is considered *positionally dependent*, since it will vary by where the person views the sun and the moon. Thus, a person can say,

(B). From here, point A, the sun and the moon look similar in size.or they can say,

(C). From here, point B, the sun and the moon look similar in size.

Not only are these statements positionally dependent, but they are also person-invariant, meaning all observers can make statements like (A), (B) or (C).

As an extension of this, consider the following:

(A*) All cultures optimize their members' well-being.

This is a subjective statement (i.e., based on self-reports of well-being) that could also be interpreted as being positionally objective. It is objective because it refers to a measure that is *trans-positional* (applicable to all cultures[6]), is coherent (meaning it is believable), and is a statement that *others can confirm*.

It can also be stated as a *positional statement*. Thus, (A*) can be expressed as:

(B*) From *here* in North America, all cultures optimize their members' well-being. or,

(C*) From *here* in the Far East, all cultures optimize their members' well-being.

(B*) and (C*) are position-dependent terms and including them does not appear to alter the objectivity of the original subjective statement. Thus, it appears possible to make *statements that are position-dependent, but person-invariant* for a subjective indicator.

Sen (1993) acknowledges that positional objectivity and subjectivity can overlap, especially if an effort is made to control for some of the sources of variance in a subjective response. Still, he states (1993) that, "subjectivity and positional objectivity do, in general, remain different; the possibility of overlap does not undermine this distinction" (p. 137).

Bayes' Law (1763) is an example of a subjective (probability) statement that Sen acknowledges can be influenced by objective positions. The term "subjective probability" in Bayes' Law would appear to deny any claim to objectivity, since the Law is often defined in terms of personal beliefs or bets a person is willing to make. However, by responding to the objective consequences of these decisions (e.g., losing money or confirming a decision) the person's (subjective probability or betting) behavior may be modified. As Sen says, "The idea of positional objectivity is precisely what is needed to understand the Bayesian concept." (1993; p. 142)

What makes positional objectivity analysis applicable to subjective probability statements but not to other forms of subjective statements is that Bayes' Law,

> …makes extensive use of demands of reason, rejecting reliance on merely idiosyncratic persuasions and subjective beliefs. In some respects, therefore subjective probabilities are thus required to be objective after all. (Sen 1993: p. 141).

Sen also points to the extensive decision analysis theory which supports the "rational use of positional information" when studying subjective probabilities.

Now we can see what Sen considered to be important about a positional objective analysis of a subjective statement; he wants to avoid those subjective indicators that are based on "idiosyncratic persuasions and subjective beliefs," not all subjective statements. This distinction will be particularly useful when I discuss utility models, such as the Health Utilities Index (Torrance et al. 1995) in Chap. 12, but I will also discover that I can take a "positional" perspective to subjectivity when defining quality-of-life or HRQOL items (see Chaps. 8 and 9).

Sen also states that a culture, as a characteristic of a society, may under certain situations make particular beliefs positionally objective. Sen (1993; p. 138) gives the example of a culture claiming that women are not capable of performing certain tasks. However, if women are not given the opportunity to learn or engage in these tasks, then an objective position exists which places the society's beliefs in perspective.

One of the more interesting applications of Sen's notion of *positional objectivity* is what he calls *objective illusions*, or what Salomon et al. (2004) referred to as *paradoxical findings*. Sen's (2001) description of the region in India where residents self-reported low *morbidity* even though they have a high *mortality* rate was an example of an objective illusion. It is an illusion that a large proportion of the Indian population in that particular region believed.

A somewhat different example of an objective illusion is the double amputee diabetic who claims to have a good quality-of-life. Here the objective position of the person is that both legs have been amputated and the person has severe mobility limitations. The self-report of having a good quality-of-life may be considered contrary to what others would describe. Yet, Sen would not feel comfortable with this application, even though comparing the objective characteristics of the person to their subjective statement makes the presence of an "illusion" obvious. This discomfort exists because while the objective facts make the illusion obvious, they do not change the illusion. Rather, what has transpired here is a change in expectation, so that the person with diabetes now makes a judgment about their HRQOL from the perspective of a person who has lost both legs. An alternative explanation involves the assumption that a common rating on a quality-of-life or HRQOL assessment can occur, but by utilizing different neurocognitive pathways (Chap. 10). As a consequence, while the psychometric ratings may be equivalent, cognitively the persons may produce these ratings in very different ways (Barofsky 2000).

Clearly, determining how objective determinants play a role in subjective statements is a useful analytic strategy. It could be helpful in determining what role objective aspects of symptoms, or physical functioning, play in a quality-of-life or HRQOL assessment. Positional objectivity may also help to understand the definitions listed in Tables 2.4–2.5, and 2.7–2.8, particularly the causal definitions in Table 2.4. Thus, a person's financial status, disease or treatment are objective determinants of a person's subjective status. They are position-dependent and perhaps person invariant.

Brown (1987) takes a somewhat different tack than Sen. He points out that taking an *objective* research approach does not necessarily lead to empirical truth, and as a result he suggests that investigators should think of objectivity *not as an achievement, but as a process* (emphasis mine; Brown 1987; p. 201). In this respect, subjectivity is unavoidably involved, but the question remains to what extent objective observations meet the criteria of an acceptable scientific statement, when these observations are made in a sea of subjectivity. Brown's approach, therefore, is quite helpful since it de-emphasizes the dichotomous, and instead emphasizes the continuous or interactive, nature of the objective–subjective continuum.

The notion that the objective–subjective continuum involves a dynamic process can be confirmed by any number of everyday experiences. For example, I can expect that the person I love (a subjective experience) will provide objective evidence of their affection (someone could count the number of times I am kissed per day). The same can be said when I look at an art object; I can report my emotional response to the object, but when describing it I usually will describe its objective characteristics (e.g., its shape, color, form, and so on) that can be confirmed by others. And I may use this information to make a judgment. Adorno (1997) described some of the issues involved in a paper entitled, *Aesthetic Theory* (p. 163–175).

My discussion to this point has made it clear that characterizing quality or quality-of-life/HRQOL data as purely objective or subjective is simplistic. Rather, it is possible that subjective data may have objective determinants, just as objective data may have subjective determinants. Thus, the statement "50% of people like oranges" is a public statement that can be independently verified, even though "liking oranges" is subjective. The statement "I prefer temple oranges" reflects a person's private feelings (i.e., is subjective) about an object whose existence can be publically verified (i.e., it is object). As Brown (1987) would state, both these statements can be empirically supported and can be considered to have "objective" characteristics, independent of the origin of the original statement.

Table 4.4[7] includes a number of candidate indicators that I might include in a quality-of-life and HRQOL assessment. This Table reveals that both objective and subjective statements and quantitative measurements can be made for health status, symptoms, and functional measures, but obviously not for hedonic variables (limited to a direct measure of *well-being*) or utility estimates. Psychosocial variables were not included in Table 4.4, but if they were it should be possible to identify both objective and subjective indicators and quantitative measures.[8] Thus, hedonic variables or utility

Table 4.4 Objective and subjective indicators and potential quality-of-life or HRQOL measures

Potential quality-of-life indicators	Objective indicators	Quantitative measures of objective indicator	Subjective indicators	Quantitative measures of subjective indicators
Health status (e.g., muscle tone)	Muscle load factor, duration of contraction	Maximum weight lift, duration of lift	Satisfaction with physical capabilities	Rating of satisfaction with muscle tone
Symptoms (e.g., pain)	X-ray of fracture	Extent of fracture as seen on X-ray film	Burning and throbbing sensations	Numerical rating of pain reports
Functions (e.g., mobility)	Going out, driving a car, gardening, etc.	Distance walked in feet	Satisfaction with social activities	Numerical rating of social activities
Hedonic variables			Happiness, satisfaction	Numerical rating of well-being
Utility estimates			Preferences, values	Numerical estimate of preferences or values

estimates may be most compatible with an *exclusively* subjective definition of quality-of-life or HRQOL assessment.

Fuhrer (2000) points out that there is also some confusion in the literature over how the terms subjective and objective are used, and how the terms are generalized and individualized. He says (Fuhrer 2000):

> We may read, for example, that 'patient's' preferred outcomes are highly 'subjective'. The commentator may be simply asserting that people's outcome preferences vary a great deal from individual to another, so that ultimately we must focus on one person at a time. Alternatively, the claim may be that preferred outcomes ultimately reflect people's private experience and that at best, external observers can only crudely surmise those preferences. The components of the distinction are not mutually exclusive. Outcome preferences may be essentially idiosyncratic, but nevertheless be knowable by external observation. Conversely, other preferences may be characteristic of most or all people, but be essentially subjective in nature. (p. 482)

The notion that an objective statement can be idiosyncratic, while a subjective statement can be generally true, adds another dimension to the issues I have been discussing. It reinforces the positional dependence of both objective and subjective measures.

Steadman-Pare et al. (2001) add to this discussion by pointing out that both objective and subjective indicators may have an external or internal reference. For example, observing that a patient has lost a limb would amount to an external objective characteristic of the person, while the money a person paid for some experience they desired would be an example of an objective (the amount of money) indicator that has an internal (what they liked) reference. Subjective indicators may have an external reference, as when a person selects a color of a car, but also an internal reference, as when they refer to how satisfied they are with family life.

The Steadman-Pare et al. (2001) perspective is particularly relevant when the quality of different entities is at issue. Thus, determining the quality of a piece of sculpture or a painting involves making a subjective judgment about an external object; the piece of marble or wood that makes up the object, or the colors and forms displayed on a surface. Judging the quality of a poem or a novel, however, involves making a subjective judgment of an internal experience, an experience that would be cognitively processed before a judgment can be made. The same occurs when judging a person's life.

Clearly, expressing quality or quality-of-life/HRQOL information in purely objective or subjective terms is not sustainable, partially because the objects, events or feelings that are being judged are complex combinations of externally and internally mediated processes. Rather a variety of combinations of information have to be considered, and this will be illustrated in the next topic where I describe a number of different approaches to combining objective and subjective indicators of quality or quality-of-life.

4 What to Measure: Objective, Subjective, or Some Combination of Indicators?

As discussed in Chap. 1, early efforts at describing quality involved determining the beauty or aesthetics of an object, an event, or a person's life. Thus, the initial debate about whether to include objective, subjective, or some combination of indicators as quality indicators focused on aesthetic issues. The introduction of aesthetics[9] into a culture and society helped establish standards or prototypical outcomes that could then be used to guide expectations and behavior within the culture and also provide the basis for the transmission of the culture. By the time of the Industrial Revolution, however, functional quality assessment replaced aesthetics as the primary determinant of quality. Thus, the same standards or prototypes that were used to produce a beautiful canoe, weapon, or sculpture now were expected to create useful objects, and to the extent that they engaged a person's thinking and evoked subjective responses (e.g., emotional responses and self-reports), would also be expected to raise issues about the aesthetics of the object.

Now by asking about the aesthetics of an object, an event, or a person's life, I can characterize these entities in subjective terms. This provides me with one method of describing the external world, but science provides another method. Science teaches me that it is possible to separate the objective from the subjective, a separation that, arguably, is necessary if new knowledge is to be accumulated. This suggests, however, that the person is an important ingredient in defining what is objective and what is subjective. The concern about the role of a person in defining what is objective and what is subjective has resurfaced at different points in intellectual history, most recently during the early development of the behavioral and social sciences. For example, psychology (Boring 1957) was established as a scientific discipline when it separated itself from philosophy. This it did by substituting *empirical (objective) observation* for philosophical speculation, with the consequence that experimental methods were used to establish new knowledge. Many of the earliest "psychologists" were trained as physicists or physiologists (e.g., Helmholtz, Pavlov [Boring 1957]) who applied their native disciplines to the study of psychological phenomena. The subsequent history of psychology continued the debate about "what to measure," with the *subjectivists*, such as Freud or Titchener, using free association or introspection to measure subjective states. At the same time, those committed to exclusive objective measurement, such as Skinner and his colleagues, used operational definitions to describe what is to be measured and rejected any measure that could not be directly observed (e.g., learning without awareness; Bargh and Ferguson 2000). More recently the cognitive sciences have attempted to balance the influence of behaviorism by using a combination of modern brain research, with a computer analog of the mind to provide empirical approaches to the study of subjective or mental phenomena. Thus, the cognitive sciences narrows the gap between objective and subjective indicators, as will be illustrated in Chap. 10 where two specific models designed to bridge this gap are discussed (the *optimization* and *embodiment* models).

In contrast, modern quality-of-life research originated in the political and social policy arena, and thus comes from a fundamentally applied setting.[10] I will briefly review this history, but will also examine its current application to social policy, where the difference between objective and subjective measurement is a central research and political issue. Various researchers have suggested models meant to bridge the gap between these types of indicators, and their suggestions will be summarized here. What I will discuss is that just as an object can evoke a subjective response, as is true when an *aesthetic judgment* is made, so too can a subjective response (such as a report of social activities) be objectively monitored (e.g., when the time dancing or spent with family is measured by a stop watch). This discussion will permit me to clarify the different ways objective and subjective data can be combined (Table 4.1). I will also argue that while there may be political or philosophical reasons for combining objective and subjective data, these reasons may not stand up to scientific examination. Several different approaches to this problem will be examined.

4.1 Social Indicators: The Subjective Processing of Objective and Subjective Information

The collection of social statistics dates to the 1700s, although the data collected at that time was limited to ensuring adequate taxation and/or recruitment for military preparation (Fox 1986). The collection of demographic data can be dated to the 1920s, while the collection of social statistics for the purpose of (United States) social policy analysis appears to have started in 1929 when President Hoover called for a study of social trends. The results of this study (*Presidential's Research Committee on Social Trends* [1933]) was a product of economists, sociologists, and political scientists (Fox 1986). The period from 1926 to 1976 witnessed a revolution in the collection of United States federal statistics, although much of these data focused on economics. It was in the 1960s when the type of data expanded to include the other social sciences. As Barofsky (1986) pointed out, specific political initiatives prompted these developments including the measurement of social indicators and quality-of-life.[11] These initiatives included President Eisenhower's statement of National Goals (*Report of the President's Commission*, 1960); President Johnson's call for *The Great Society*; Bauer's (1966) efforts at evaluating the social impact of the space program and many others (Liu 1975). In most of these applications, social indicator or quality-of-life measurements were expected to "prove" the worthiness of the social programs being promoted. While it was not clear whether such evidence was actually provided, what did occur was that sociologists, psychologists, and economists became increasingly interested in social indicator and quality-of-life measurement (e.g., Prutkin and Feinstein 2002). It was also recognized that measurements of this sort could be a good way to identify social problems and monitor social innovations. From this, the social reporting movement developed, a development that has rapidly spread so that many countries regularly assess their constituents' social condition and presumably base policies on the information generated. Policymaking was facilitated when social reports were combined into *social accounting systems*.

A second major effort at measurement also started prior to World War I, and could be traced to E. A. Codman (Mallon 1999), who was interested in measuring the quality of hospital care.[12] His pioneering efforts helped establish the field of *health outcomes research*, and specifically, health status measurement. Heath status measurement, as will be discussed in Chap. 8, became the progenitor of HRQOL measurement.

Social reporting raises two issues: what to measure (subjective or objective indicators) and how to connect these measures to social policy. As Rapley (2003) describes, two basic "schools" have evolved: the Scandinavian "Level of Living" approach (Erickson 1993) and the American "Quality-of-life" approach. The Scandinavian approach states that *only* "objective" measures of quality-of-life (or welfare) should be collected and that "subjective" concerns should be used to set political priorities, as part of any democratic political process. This contrasts, of course, with much of the literature on social indicators and HRQOL assessment as practiced in the United States, where the measure of interest are usually individual and "subjective" measures. For example, Campbell (1972) said, "Quality-of-life must be in the eye of the beholder." Thus, these two views not only emphasize different measures but also different approaches to achieving social welfare. Noll and Zapf (1994) point out that in Germany, welfare or quality-of-life is defined as "good living conditions which go together with positive subjective well-being."

Table 4.5 (Zapf 1984) illustrates four possible outcomes when objective living conditions are compared to subjective measures of well-being. The "adaption" cell is sometimes called the *disability paradox*. It occurs when a person with clear physical limitations reports that their well-being is good. The "dissonance" cell is referred to as the *dissatisfaction paradox*. Here people are living under "objectively" good

Table 4.5 Categories of individual welfare[a]

Objective Living Conditions	Subjective Well-being	
	Good	Bad
Good	Well-being	Dissonance
Bad	Adaption	Deprivation

[a]Zapf (1984). Permissions to reproduce was granted by author and publisher

Table 4.6 Objective and subjective quality-of-life measures[a]

Objective measures	Subjective measures
Life expectancy	Happiness
Crime rate	Sense of safety
Unemployment rate	Satisfaction with life as a whole
Gross domestic product	Material possessions
Poverty rate	Relationships with family
School attendance	Sense of community
Working hours per week	Job satisfaction
Suicide rate	Sex life

[a]Abstracted from Cummins (1996), Hagerty et al. (2001) and Noll (2000)

conditions but still report being dissatisfied with their life quality. Clearly, having a large proportion of the population living well and feeling good is the most desired outcome (the well-being cell). However, of the various outcomes, persons in the "adaption" cell are the most difficult to theoretically deal with, since adjusting to their living conditions masks deficits.

It is interesting to apply Sen's (1993) notion of positional objectivity to Table 4.5. Both the "dissonance" and the "adaption" cells would be examples of "objective illusions," in that there are inconsistencies between the person's "objective" circumstances and their self-reports (much as Sen found in his study of the health status of different regions of India). These types of "objective illusions," however, occur quite regularly, especially for persons with medical illnesses. For example, Brinkman et al. (1978) found that paraplegics were just about as happy as people who just won the lottery. Sackett and Torrance (1978) found that people on dialysis rated their quality-of-life as 0.56 (on a 0.00–1.0 scale), while healthy persons rated those same people's quality-of-life as 0.39. Similar findings were found by Baron et al. (2003), using persons accessed via the Internet, and by Buick and Petrie (2002) who assessed persons who know someone who has breast cancer. One explanation of these findings is that they represent the ill person's normal psychological adjustment to the circumstances created by their disease and its treatment. Riis et al. (2005), however, have suggested that it is the healthy person's inability to properly estimate the "ill" person's ability to adjust that is producing the discrepant outcomes. They refer to this type of adjustment as "hedonic adaptation." Thus, it would appear that Sen's (2001) suggestion to label these discrepancies as "illusions" while consistent with a commitment to using objective indications when making policy decisions (e.g., a physician's recommendations), also seems premature and a failure to recognize that cognitive processes can be both trans-positional (meaning that all people can think about quality or their quality-of-life in the same manner) and position-dependent (meaning that persons may differ in terms of what specific aspects of their existence they consider important for their quality-of-life or HRQOL).

Table 4.6 provides examples of objective and subjective measures that are reported in the literature (Cummins 1996; Hagerty et al. 2001; Noll 2000; Rapley 2003). These measures are different, in that subjective indicators exclusively rely on self-reports while objective indicators may or may not. A number of quality-of-life researchers (e.g., Rapley 2003, p. 31; Cummins 2000a) have provided empirical support for models that include both types of measures in a single quality-of-life or HRQOL assessment. Concern may be expressed about this strategy, because it involves combining very different types of indicators. Michalos (in Sirgy et al. 2006) lists 15 issues (Table 4.7) that should be considered when combining objective and subjective social indicators. Michalos (in Sirgy et al. 2006) states the following when considering how or what to combine:

> It should not be thought that the selection of the appropriate mix of options from each of the 15 issues is merely a technical problem to be resolved by statisticians or information scientists. On the contrary, the construction of social indicators of the quality-of-life is essentially a political and Table 4.5 as being the typical outcomes associated with combining objective and subjective indicators. philosophical exercise, and its ultimate success or failure depends on the negotiations involved in creating and disseminating the indicators, or the reports or accounts that use those indicators. (p. 349).

Michalos (in Sirgy et al. 2006) goes on to describe the four outcomes summarized in Table 4.5 as being the typical outcomes associated with combining objective and subjective indicators.

4.2 Models that Combine Objective and Subjective Indicators

There are many situations where combining objective and subjective indicators or data becomes an issue. For example, assessing the quality-of-life of the intellectually compromised person often pits objective measures of a person's living conditions with the inability of the individual to properly respond to inquiries concerning their past experiences or current state. Some investigators contend that when assessing the challenged person it is best to separate these assessments and not attempt to combine them, but that both types of assessments are important to collect (e.g., Cummins 1997). Other investigators claim that because the subjective assessment

Table 4.7 Issues to Consider when combining objective and subjective indicators[a]

Issue no.	Issue
1	*Unit of analysis*: How can you aggregate indicators that have very different domain sizes?
2	*Time frames*: How can you aggregate indicators which are based on different time frames?
3	*Population composition*: How homogenous or heterogeneous should your study sample be, in terms of language, gender, age, education and so on?
4	*Domains of interest*: Which domains should be included in a study: health, job, family life, housing, etc.?
5	*Source of data*: How can you combine data that differ in terms of who provided the data? Can you combine data that comes from people living in a neighborhood with that of expert appraisers?
6	*Positive and negative indicators*: How can you combine positive and negative indicators when these indicators can be very different?
7	*Causal and indicator variables*: Should you include variables that cause a particular outcome with those that indicate the outcome?
8	*Benefits and costs*: How do you combine indicators that may differ in terms of inherent value, especially if you are doing a cost/benefit analysis?
9	*Measurement scales*: How do you combine data that derives from different measurement scales (e.g., 0–7; −7 to +7; 1–4; etc.)?
10	*Report writers*: Investigators differ on what to, and how to monitor a particular outcome. How do you combine data with such differences?
11	*Report readers*: The target audience for a study may differ, affecting how the data is analyzed and presented. How do you deal with this situation?
12	*Quality-of-life model*: How can the accumulated data be integrated into a qualitative model?
13	*Distributions*: Measurement scales can easily mask or highlight data extremes. How do you deal with this problem?
14	*Distance impacts*: How do you account for geographic distributions of subjects (distance from home to target site) when accounting for some outcome?
15	*Casual relations*: How can you combine outcomes when you are not aware of the causes of the outcomes?

[a]Adapted from Michalos (in Sirgy et al. 2006; p. 348–349). I mainly retained the labels Michalos used but changed some of the examples

of the challenged person is not sufficiently valid or reliable, it is best not to even attempt to collect subjective data, and instead rely on "objective" measures of quality. The rationale given for this is that the challenged person doesn't understand the concept of "quality-of-life" (e.g., Hatten and Ager 2002), so that any direct assessment of it would have to be considered suspect. Implicit in the Hatten and Ager (2002) view is that the challenged person has no concept of quality, just as they don't understand other aspects of their existence (e.g., why they should not walk into the street, leave their hand on a hot stove, and so on). However, I will take issue with this view, especially if it is applied generally, and instead argue that the cognitive skills the challenged person (e.g., the person with Alzheimer's disease) retains are lost in a progressive manner, so that premature limitations in decision making would underestimate the capacity of the challenged person to participate in decisions concerning their welfare at any particular time. In addition, there is some evidence (Mozley et al. 1999) that the challenged person retains the ability to express preferences (which may be emotion based), even though they may lack particular cognitive skills, such as short-term memory.

Of course, there are many other situations where both objective and subjective information are clinically relevant, especially in situations where *causal relationships* have to be established or the information is being used to make a decision or judgment. In some of these situations both types of data are considered simultaneously, but only one type of data is actually used (that is, the data are not combined). For example, if a family has police reports that a family member with Alzheimer's disease is wandering around the neighborhood and endangering themselves and the person in question denies this, then this is a situation where both types of data are available. However, a physician or a judge who may be adjudicating this information will rely on the statements of the family and the police reports, because they think that the objective information is more reliable than what a Alzheimer's patient may say. In contrast, if the person with Alzheimer's disease is placed in a facility and reports being physically abused and there is objective evidence of this abuse, then their reports of being dissatisfied with where they are living would be considered along with the objective data in order to decide whether or not to move the person. However, the person making the decision may claim that they would not have relied on the patient's self-report alone, so that in effect, the objective data was given priority when the decision was made.

In extreme contrast is the debate ongoing in economics about whether it is appropriate and mathematically feasible to combine subjective and objective utilities. Here investigators take advantage of the probabilistic nature of procedures to elicit utilities to provide mathematical solutions to the combinational task (e.g., Anscombe and Auman 1963; Kaneko 1985). This research area is said to have started with the 1944 publication of von Neumann and Morgenstern's book *Theory of Games and Economic Behavior*. What von Neumann and Morgenstern (1944) demonstrated was that it was possible to account for decision-making under uncertainty using axiomatically derived expected utilities. This was followed 10 years later by Savage's (1954) book, where

Table 4.8 The correlation between objective and subjective quality-of-life measures[a]

	Objective vs. objective (r)	Subjective vs. subjective (r)	Objective vs. subjective (r)
General population			
Number of comparisons	6	7	31
Mean ± SD	0.315 ± 0.051▲	0.380 + 0.145▼	0.120 + 0.082▶
"Threat" to quality-of-life			
Number of comparisons	8	9	35
Mean ± SD	0.325 ± 0.138▲	0.440 + 0.131▼	0.255 + 0.124◀
O/S general population vs. O/S threat; $P < 0.001$			

[a]Based on Cummins (2000a)
▲: O vs. O not statistically significant
▼: B. S vs. S not statistically significant.▲
▶: C. O/O vs. O/S; $P < 0.01$ and S/S vs. O/S; $P < 0.001$
◀: D. O/O vs. O/S; Nonsignificant and S/S vs. O/S; $P < 0.01$

he demonstrated that it was also possible to generate expected utilities using subjective probability estimates.

There are also many examples of research studies where objective and subjective quality indicators are directly compared, and low or modest correlations between these indicators are found (e.g., Campbell et al. 1976). These findings reinforce the view that objective and subjective indicators are measuring fundamentally different phenomena, and that making them quality indicators requires additional cognitive processing (Table 4.1). There are also examples where objective and subjective indicators are combined into an index, and the index is statistically analyzed (Liu 1975). This avoids the combinational task but leaves the interpretation of the data open to concern. There are also investigators (e.g., Cummins 2000a) who claim that both types of indicators should be measured, but that objective indicators become relevant for a qualitative assessment only if a threshold value is reached (see below). More fundamentally, however, are the investigators who claim that all objective indicators are actually subjective. Cummins (2000a) states their argument, "Since there can be no measured reality beyond our capacity to experience the world, the so called 'objective' measures are actually a product of our perceptions and as a consequence, subjective" (p. 56). Cummins (2000a) goes on to quote Andrews and Withey (1976) as an example of investigators who take this position. Andrews and Withey (1976) state:

> We believe...that this classification is neither clear nor very useful. Even birth and death and what defines human life are currently matters of legal, medical, and doctrinaire dispute. Presumably objective indicators of these matters turn out to involve subjective judgments. Conversely, it can be argued that many subjective indicators (such as people's evaluations of their lives) provide rather direct and objective measurements of what they intend to measure. (Andrews and Withey 1976; p. 5)

Cummins, of course, takes issue with this view, pointing out that distinguishing these two types of indicators remains experimentally and philosophically useful. Of interest is that Cummins (2000a) has also shown that objective and subjective indicators can interact or covary with each other, raising the possibility that simply combining these data may be confounded by the lack of independence between the indicators. Thus, Cummins (2000a) has shown (Table 4.8) that for persons whose quality-of-life is "threatened" (for example, by a medical illness) the average correlations between subjective and objective indicators doubles when compared with persons from the general population (from r = 0.120 ± 0.082 to r = 0.255 ± 0.124).[13]

Further evidence for an interaction between objective and subjective indicators comes from a number of studies (e.g., Cummins 2000b, p. 133–134) which examined the relationship between "objective" measures of *wealth* and subjective measures of social well-being. An analysis of these data (Cummins 2000b, p. 133–134) revealed that a threshold exists above which the relationship between increasing wealth and well-being does not hold, but which does hold for persons whose income was below the threshold (Diener et al. 1999). However, Hagerty and Veenhoven (2003) claim that independent of what a person may be aware of, it is still possible to demonstrate that income and well-being, for example, are directly related. To do this requires that data from countries with low gross domestic products or low per capita income be included when calculating the income per country. Thus, the debate continues. These are just some of the issues to consider when combining or comparing objective and subjective indicators. As I proceed, I will continue to provide examples of situations where the combinational task is an issue, but for now I will examine examples from the quality-of-life research literature.

Stated most generally, combining of objective and subjective indicators is a basic problem in the *aggregation* of data, and the related issue of generating *summary measures*. Summary measurement, as I explained in Chap. 3, is one of the basic ways I cognitively deal with the vast amount of information I receive, moment to moment. In contrast, there is not much agreement on how summation or aggregation might occur statistically or as an exercise in the modeling of cognitive processes. Inspecting the research literature reveals three approaches to this task. For example, many of the clinical

Table 4.9 Models that include objective and subjective indicators

Model type	Reference	Depiction
Hierarchical (taxonomic) models	Fries (1991)	(Fig. 4.1)
	Spilker and Revicki (1996)	(Fig. 4.2)
	Torrance (1976)	(Fig. 6.1)
Linear models	Cummins (2000a)	(Fig. 4.4)
	Kahneman et al. (1999)	(Fig. 2.1)
	Ormel et al. (1997)	(Fig. 4.3)
	Patrick and Chang (2000)	(Fig. 9.1)
	Wilson and Cleary (1995)	
Nonlinear models	Smolensky and Legendre (2006)	(Fig. 10.7)
	Scherer (2000)	

examples I discussed reflect a comparative model, in which objective and subjective information were compared, rather than being combined into a composite indicator. A second approach, that of statistical modeling, combines these indicators and justifies it by assuming that "a number is a number" and that combining such diverse entities as objective and subjective indicators would not violate the assumptions of a statistical analysis. What is implicit in this argument is that the application of statistical procedures is different from the interpretation of the results, and it is only when the resulting analysis is interpreted that the origin of the numbers used in the analysis become relevant. What is fairly commonly done is that structural equation modeling is applied to determine if the objective indicator contributes to the variance of the dependent subjective indicator (e.g., a quality-of-life outcome). Another approach involves using the Rasch model (see Chap. 4, p. 110). Finally, there is the common denominator model, where an investigator attempts to find a common dimension to use to combine objective and subjective information.

Of these three approaches, finding a common denominator comes closest to characterizing the cognitive process that a person (as opposed to an investigator) might use to deal with this task, and the question is whether there are any examples of this approach that are applied to quality-of-life data and what the common denominator is. Table 4.9 provides a list of models dealing with quality-of-life and health status issues which have been proposed and which can be used to combine objective and subjective data. Three classification models were identified; models based on a *hierarchical* or taxonomic[14] classification, *linear* models and *nonlinear* models. A review of these models will illustrate how *hierarchical* or *linear* models might provide the common denominator needed to combine data. *Nonlinear* models have not been applied to quality or quality-of-life data so I will be confined to illustrations that use other types of data, illustrating how such modeling might occur.

4.2.1 Hierarchical or Taxonomic Modeling

Hierarchical or taxonomic modeling is a natural product of how a person thinks and organizes his or her experiences.[15] For example, children at an early age are quite adept at distinguishing categories, correctly ordering information from a more concrete basic format to the abstract, and so on. Thus, children are able to use properly such terms as "animal" as opposed to "cat or dog," or "food" as opposed to "cereal or apple" (Carey 1995).

Not surprisingly, there are formal mathematical analogs of this cognitive process, including a variety of statistical modeling procedures that are meant to provide multilevel or *hierarchical linear models* (e.g., Bryk and Raudenbush 1992). *Random coefficient modeling, latent curve modeling,* and *growth curve modeling* are all examples of hierarchical modeling. Each of these models has as an underlying structure where measures are nested within units at a higher level of the hierarchy. The rationale for using hierarchical linear models is so that an investigator can avoid inducing high collinearity among predictors or the large standard errors that can come from the aggregation or disaggregation of information. Most important is that hierarchical modeling permits an investigator to study variation at different levels of the hierarchy.

The hierarchical organization of information can also be found in various scaling procedures. *The Guttman Scale* is an example of a formal hierarchical framework that classifies information in a progressively more inclusive manner, and in this way provides a quantitative scaling of some indicator. Katz's *Index of Activities of Daily Living* Scale (ADL; Katz and Apkom 1976) is an example of this type of scaling (see Brorson and Åsberg 1984 for empirical support) for essentially objective indicators. It consists of a series of ratings by an independent observer concerning whether a person can bathe themselves, dress themselves, toilet properly, move between places, maintain continence, and feed themselves. The person receives a "dependency rating" depending on the number of functions they need assistance with. Thus, the most dependent person will require assistance in all six domains, but the next lower level of dependence requires that the person be dependent in all domains except in either self-feeding or continence. The second and next but higher level of independence assumes the person can transfer themselves, as well as feed and maintain continence. This continues until the person can be classified as being completely independent. What tells me that a hierarchy is present is the fact that if a person required assistance in feeding or maintaining their continence, then they would also need assistance in bathing, dressing, going to the toilet, and transferring. It was this scalability of the Katz's Index of ADL that the Brorrson and Åkpom (1984) study supported. Also of interest is the report by Njegovan et al. (2001) demonstrating that the

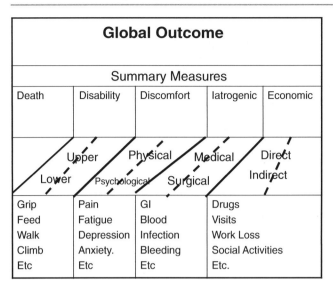

Fig. 4.1 Fries' hierarchical model of patient outcomes. Specific domains are subsumed into particular summary measures, with the aggregation of the summary measures determining the global outcome. Permission to reproduce this figure granted by the publisher.

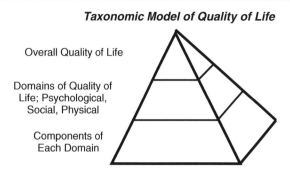

Fig. 4.2 The taxonomic model of quality-of-life presented by Spilker and Revicki (1996) orders analytical units along an abstract to concrete hierarchical dimension. Permission to reproduce this figure was granted by the publisher.

decline in functional independence is related to progressive decline in cognitive functions.

Now while the Katz's scale demonstrates that a hierarchy can be used to collect information about a common dimension, it does not mandate that either an objective or subjective indicator be included in this scaling. To gain some insight into what is possible, I would like to examine the qualitatively oriented hierarchical model suggested by Torrance (1976). He and his colleagues used the hierarchy model (Fig. 6.1; Chap. 6, p. 179) to create a formal multi-attribute assessment system: the *Health Utilities Index* (HUI; Torrance et al. 1995). The utility estimates developed to weight these attributes conformed to the von Neumann and Morgenstern *axiomatic* approach to risk assessment, with the standard gamble or time tradeoff providing the descriptive states, the utility estimates for which are aggregated. Not surprisingly, therefore, the nature of the descriptive state (that is, whether they are objective or subjective) is not critical. The HUI, therefore, would be an example of a common denominator model, with the utility estimates associated with objective and subjective indicators being the common denominator.

Fries (1991; Fig. 4.1) also proposes using a hierarchy, in this case to organize information about "patient outcomes," some of which would be objective and some subjective. Fries and Singh (1996) list seven advantages to using a hierarchy. The first is that knowing that several assessments are addressing a common abstract construct provides an investigator with a wider range of assessments to select from. Second, knowing the hierarchical relationships between indicators allows an investigator to select assessments at a more concrete level of abstractness (e.g., subordinate level), so as to trace the origin of a specific observation. Third, knowing where an assessment fits into the hierarchy provides an investigator the opportunity to trace the influence of more abstract constructs in the particular assessment (e.g., patient satisfaction is influenced by the person's psychological status, but also by whether they are subject to medical iatrogenic effects; see Fig. 4.1). Fourth, the Fries (1991) model provides a minimum number of summary measures, and in this way eases the task of communicating results. Fifth, the hierarchy makes implicit assumptions explicit. Sixth, the presence of the hierarchy and the identification of summary measures make it easier to do longitudinal studies, especially if changes occur in the assessment used. Thus, disability can be assessed by different assessments at different times and still be addressing the same abstract concept. Finally, the hierarchy emphasizes and encourages the completeness of an assessment by explicitly displaying the various components that should be considered as part of the construct.

While Fries and Singh (1996) illustrate some of the advantages of using a hierarchy, their model does not offer any formal method whereby objective and subjective indicators can be combined, other than assuming that the reader will agree that the combination of outcomes can be aggregated into a global outcome indicator and be interpretable.

Spilker and Revicki (1996; Fig. 4.2), in contrast, suggest *a taxonomic model*. This type of classification method organizes information by using a specific set of inclusion criterion. Thus, in the taxonomy that starts with an "oak," progresses to the more abstract "tree," and then to the more abstract category of "plant," information is organized according to the principle of a common kind. Spilker and Revicki (1996) applied this model to qualitative issues when they proposed that separate hierarchies be used to characterize health and nonhealth issues. Each of these hierarchies can include both objective and subjective indicators, but how to explicitly combine these indicators was not addressed.

Lawton (1991) provides another example of a hierarchy now applied to the issue of "behavioral competence." His scheme is two dimensional, varying from the simple to the complex, or along a health dimension (that is, health, functional health, cognition, time use, and finally social behavior).[16] Thus, the simple to complex dimensions of functional health vary from the simple physical ADL to instrumental ADL, to financial management, to paid employment. This, he says, illustrates that both the person (the subjective) and the person's environment (the objective) are essential to have a complete picture of the person's quality-of-life. He sees these dimensions as being in parallel, each providing important information about the other. Lawton (1991), like Sen (1993), believes that it is not possible to understand the classic "dissonance" patient's (Table 4.5) complaints about their quality-of-life, if the objective nature of the person's existence is not taken into account. When Lawton describes this outcome he is, of course, describing the view of an outside observer, not necessarily the respondent. Thus, Lawton (1991) appears, as is true for Cummins (1997), to use the combinational model approach to making judgments or decisions. However, as I will argue later, Lawton can't avoid finding some sort of a common denominator at some point in his decision-making process if his decisions or judgments are to be reliable and valid.[17]

The approach of Torrance and his colleagues is an excellent example of a hierarchical model that provides a solution for how to combine objective and subjective indicators, and they do this by asking a representative group of persons to provide utility estimates for each type of data. Since the utility estimates are determined using a common method (e.g., standard gamble, or time tradeoff), the difference in the type of data the utilities are based upon can be ignored. In addition, there are a variety of statistical models available (see above), including Bayesian and hierarchical modeling methods (Ciarleglio and Makuch 2007; De Leeuw and Kreft 1995; Draper 1995), which can also be used to aggregate data into a summary measure. Aggregating hierarchical data is also a classic problem in engineering and operations research, and much attention has been placed on developing mathematical and statistical methods that can be used to optimize the functioning of a system (e.g., Chen and Hanisch 2000; Rogers et al. 1991).

4.2.2 Linear Modeling

Linear models can also be used to display and map relationships between objective and subjective quality-of-life indicators, and a representative sample of these models is listed in Table 4.9. Linear modeling can be said to have its origin in the study of vectors in a Cartesian space (Chap. 9, p. 297), with a vector representing an entity, and its length the magnitude of the entity. Most linear models listed in Table 4.9 are unidimensional, although a vector space can have as many dimensions as necessary. Linear algebra can be used to characterize the vector space, permitting the description of a higher order construct. Linear modeling, using linear algebra, is particularly useful, leading to solutions of a wide variety of problems.

Each of the models listed in Table 4.9 progress from objective indicators to subjective indicators. For example, in the Wilson and Cleary (1995) model, biological and physiological variables are assumed to lead to symptoms, which lead to certain functional outcomes. Functional outcomes contribute to general health perceptions, which determine overall quality-of-life. Although the model is graphically presented to imply a linear progression, from the objective to the subjective, the authors acknowledge that feedback processes, as well as moderating factors exist (e.g., characteristics of the individual or the environment), making the model more dynamic than visual inspection would suggest.

Ormel et al. (1997) have also suggested a model that includes objective and subjective indicators that is hierarchically related. The foundation of the hierarchy is based on objective indicators and reaches its apex with a subjective measure (psychological well-being). In between these extremes, indicators exist that have objective or subjective characteristics. The model they propose is based on *Social Production Function* theory which "basically asserts that people produce their own well-being by trying to optimize achievement of universal human goals via six instrumental goals within the environmental and functional limitations they face" (Ormel et al. 1997; p. 1051). When they apply their model to understanding health effects, they start with various objective indicators of physical disease or neurophysiological disorders which they expect to lead to physical and mental impairments. Symptoms and functional limitations activate internal and external resources of the person, and are used to combat the short-term adverse quality-of-life or HRQOL consequences created by the physical and mental impairments. The more successful these resources *substitute* for what was lost or impaired, the less severe the long-term quality-of-life or HRQOL effects (see Fig. 4.3). Thus, at a conceptual level, *Social Production Function* theory provides a common denominator that orders the objective and subjective indicators.

Both the Wilson and Cleary (1995) and Ormel et al. (1997) models suggest that any model relating objective and subjective indicators would be expected to be more dynamic than static. Cummins (2000a) is most explicit about this issue, suggesting that the subjective status of an individual is regulated by homeostatic processes, which can be activated when an objective indicator exceeds a threshold level (Fig. 4.4). Cummins (2000a) states:

> It has been proposed that this changing relationship can be understood in light of two linked propositions. One is that subjective quality-of-life is held under the influence of homoeostatic

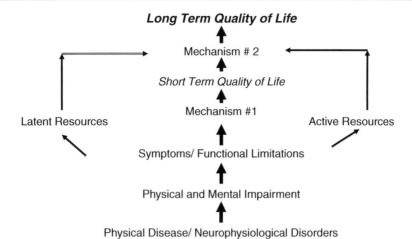

Fig. 4.3 Adopted from Ormel et al. (1997; p. 1058). This figure is presented in reverse order to the original which starts with physical disease and neurophysiological disorders and descends downward to long-term quality-of-life. Mechanism #1 refers to the "impairment of ongoing activities and endowments." This involves the increased cost of achievement of instrumental goals, what highly productive activities and endowments are lost, and what is the worst short-term effects. Mechanism #2 refers to substitution, with the better the possibilities for substitution the less severe the long-term effects. Permission to reproduce this figure was granted by the publisher.

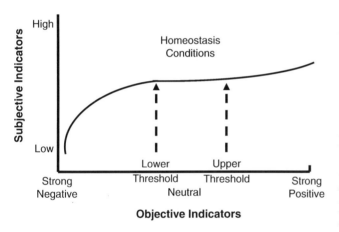

Fig. 4.4 Cummin's homeostatic model of well-being as applied to the relationship between objective and subjective indicators.

control. This means it is normally maintained within a quite narrowly defined range and the processes of adaptation explains the independence of O and S indicators. The second is the existence of a low objective threshold of living which, when experienced, exerts a sufficiently powerful influence to compromise homeostasis. At that level of environmental challenge, O starts to drive S and the new dimensions show increased covariation. (p. 68)

Thus, Cummins resolves the dilemma of how to combine objective and subjective indicators in a quality-of-life or HRQOL assessment by essentially shifting the debate away from the combinational task, and instead taking advantage of Headey and Wearing's (1989) suggestion that a well-being set point exists which a person tries to keep constant. This set-point or homeostatic range (Fig. 4.4), when violated, leads to changes in magnitude of some subjective indicator, such as well-being.

This approach seems to make sense because, for example, I only pay attention to the crime rate when it impacts me, or the gross domestic product when I am concerned about my own standard of living. At other times (e.g., when homeostasis exists), objective indicators can be present and even change, yet not become part of an assessment of our qualitative status. There are numerous examples of this. For example, someone may have a cancer and not know it, but when they do learn about their status they will modify their behavior, as well as their sense of well-being. Of course, the cancer can be "objectively" detected (e.g., with an X-ray) prior to a person being aware of its presence, but objective detection alone has no qualitative importance until the person becomes aware of the presence of the information. This could then lead to a change in the person's sense of well-being, as Cummins et al. (2002) describes, or disrupt the person's efforts at "optimal achievement," as Ormel et al. (1997) describe. Thus, a subjective state is the common denominator that mediates our "Contact" (see preface to Chap. 10) with the objective world and mediates our qualitative status.

Now it is also important to assess how well a linear model can be applied to some set of data. Table 4.10 summarizes a representative series of studies which used the elements of the Wilson and Cleary (1995) linear model to organize a study, and then used regression analyses and/or structural equation modeling to determine the degree the available data fit the linear model. Inspecting the studies in Table 4.10 revealed that it was possible to identify a linear model in each of the these studies,[18] although including individual and/or environmental characteristics would often improve the fit of the model (e.g., Baker et al. 2007; Sousa et al. 1999; Sullivan et al. 2000; Ulvik et al. 2008). In addition, 30–55%

Table 4.10 Research studies utilizing the Wilson and Cleary (1995) linear conceptual model of HRQOL

References	Sample (N)	Comments
Sousa et al. (1999)	AIDS patients (142)	"The regression, model including the environmental and individual characteristics, explained 32% ($R^2=0.3186$) of the variance in overall quality-of-life, while the model excluding the environment and individual characteristics explained 20% of the variance" (p. 183)
Crosby et al. (2000)	AIDS patients (146)	"Although there were relationships among several of the variables in the five dimensions...many of the expected associations were not statistically significant" (p. 277)
Sullivan et al. (2000)	Community-dwelling elders (5279)	"The Linear Model did not satisfactorily account for the observed data...so the saturated Nonlinear Model...is described. When anxiety and depression were added to this Nonlinear Model they best fit in a position mediating the relation between perceived health and overall quality-of-life" (p. 801)
Janz et al. (2001)	Female cardiac disease patients (570)	"...General Health Perceptions explained more of the variation in QOL than any other category (38%), Symptoms Status explained 26%...Functional Status and Biological/Physiological variables explained 19% and 13% respectively... When considered jointly, all model variables explained 46.8% of the variation in the baseline QOL rating" (p. 593)
Arnold et al. (2005)	COPD (95) and CHF (90) patients	Regressions analyses accounted for 54% of the variance in general health perceptions for COPD patients, and 47% for CHF patients. QOL was not measured
Heo et al. (2005)	Heart failure patients (293)	"Subjective variables – symptom status, health perception, and functional status – were consistently related to HRQOL. Objective variables (with the exception of age, which was related to several aspects of HRQOL) – etiology, number of comorbidities, marital status, having a confidant and educational status – were unrelated to or weakly related to HRQOL" (p. 378)
Phaladze et al. (2005)	Community-dwelling HIV/AIDS patients (743)	"...a hierarchical, six-step multiple regression analysis was performed with a total of 10 predictor variables, with 522 complete data sets...The overall model explained 53.2% of the variance in quality-of-life as measured by life satisfaction" (p. 124)
Vidrine et al. (2005)	AIDS patients (348)	"The squared multiple correlations for the three endogenous latent constructs (i.e., symptoms, role-specific functional status and overall HRQOL) were 0.67, 0.23, and 0.74, respectively. Thus, 74% of the variance in overall HRQOL was explained by the model" (p. 927)
		"...results from this study suggested that nadir CD4 counts had only marginally significant effects on symptom status and nonsignificant effects on role-specific functional status and generic HRQOL" (p. 931)
Orfile et al. (2006)	Community-dwelling elders (544)	Using sequential linear regressions, the authors found that either chronic conditions or the sum of chronic conditions, functional capacity, and alcohol consumed accounted for between 41.4 and 42.4% of the variance in an HRQOL assessment (the Nottingham Health Profile)
Sousa and Kwok (2006)	AIDS patients (917)	"Both symptom status and general health perceptions accounted for 38.2% of the variance in overall quality-of-life" (p. 733)
		CD4 counts were added to the model, but the quality-of-life variance accounted for was not specified
Baker et al. (2007)[a]	Xerostomia patients (85)	"...in our model, the measured variables accounted for only 9, 22, 24 and 21% of the variance in symptom status, ORHQoL*, global oral health perceptions, and subjective well-being respectively, suggesting an important role for individual and environmental factors" (p. 305)
Mathisen et al. (2007)	Heart surgery patients	Regressions of health status on global life satisfaction or life satisfaction on global health accounted for comparable amounts of variance, making it difficult to interpret the supposed linear causal relationship implied by the primary model.
Ulvik et al. (2008)	Cardiac catheterization (753)	"Wilson and Cleary (1995) included biological and physiological data and patient-reported symptoms in their model, but assert that traditionally it not been included in conceptualizations of HRQOL. Although they recognize the importance of including emotional and psychological factors, they preferred to avoid it because of its complexity and the possibility that the causal relationship might go in both directions. We included all these factors, and found that 43% of the variance of overall QOL was explained by this model" (p. 10)

CHF congestive heart failure; COPD chronic obstructive pulmonary disease; ORHQoL oral health-related quality-of-life

[a]Baker et al. (2007) argued that a measure of psychological distress, the Hospital Anxiety and Depression Scale, could represent one aspect of well-being or overall quality-of-life

of the variance in overall quality-of-life was accounted for by the independent indicators (e.g., Arnold et al. 2005; Janz et al. 2001; Orfila et al. 2006; Phaladze et al. 2005; Sousa et al. 1999; Sousa and Kwok 2006; Ulvik et al. 2008). The Vidrine et al. (2005) study was an exception, finding that 74% of the variance in overall HRQOL was accounted for by symptoms, role-specific functional status and overall HRQOL. Combining objective indicators (e.g., disease severity, number of diseases) with subjective indicators either decreased the amount of variance in overall quality-of-life, or had no impact on the variance accounted for (e.g., Heo et al. 2005; Sousa and Kwok 2006). Ulvik et al. (2008) found that measures of cardiac function alone did not contribute to variation in HRQOL, but that adding personality characteristics (e.g., such as a measure of anxiety or depression) resulted in the final regression model accounting for 43% of the variance in a quality-of-life measure.

There is a temptation to think that if a linear model fits the data, this provides evidence for a causal relationship between the indicators. Thus, if the Wilson and Cleary model were confirmed it would seem natural to conclude that some disease or physiological condition "caused" changes that eventually caused changes in overall quality-of-life. However, as Sousa and Kwok (2006) and others point out, demonstrating a correlation between indicators does not establish a causal relationship. What is required for establishing a causal relationship is the demonstration that there is a temporal sequence between the indicators of interest, as would be provided by a longitudinal study. Unfortunately, all of the studied reported in Table 4.10 involved the analysis of cross-sectional data.

Inspection of Table 4.10 provides general support for Wilson and Cleary's assumption that the indicators they selected fit a linear model, but what was also evident was that a variety of moderating variables (e.g., anxiety, depression, and so on), when included in the regression analyses, increased the variance accounted for by the regression analyses. This suggests that the relationship between objective and subjective indicators is far more dynamic and context dependent than implied by the Wilson and Cleary model. This justifies raising the question of whether other types of models, such as nonlinear models, would be equally suited to model the combining of objective and subjective data.

4.2.3 Nonlinear Modeling

To answer the question of whether nonlinear modeling can do a better job at combining objective and subjective indicators than linear models can, I first must point out that the application of a model, linear or nonlinear, implies that a particular mathematical construct matches data generated by a particular set of variables. Usually, an investigator has a sense of what model would provide a match, because the original organization of the variables may reflect an implicit or explicit model. For example, the graphic display provided by Wilson and Cleary (1995) made it clear that they felt that a linear model would best fit their data. If, of course, it was found to be inadequate, then other models could be tested. In the same sense, applying a nonlinear model implies that a particular set of variables exists that could not be accounted for by a linear model, and may be best matched by a nonlinear model. Ulvik et al. (2008) actually provided an example of this, when they found that a physiological indicator (left ventricular ejection fraction) was nonlinearly related to measures of self-reports of angina and depression.

One type of nonlinear model evolved from *catastrophe theory* (Zeeman 1976). In this model, it is assumed that a sequence of consecutive events occur until suddenly something unexpected happens which then changes the course of events. This change (e.g., referred to as *hysteresis*) may produce a very different outcome even if the sequence of events that follows returns to what could be described as a linear process. Both chaos and complexity theories exhibit nonlinearities of this sort. For example, Scherer (2000) uses catastrophe theory to account for the sudden onset of emotions (e.g., anger, joy) following a sequence of appraisals.

Nonlinearity, however, can also occur even in the presence of linear modeling procedures (e.g., structural equation modeling). Thus, if the assumption of linearity can't be sustained, as when X and Y are curvilinearly related, or if there is an interaction between variables, then according to Loehlin (2004; p. 111–113) it is still possible to use linear structural equation modeling. Thus, the presence of nonlinearity in some data set does not, per se, determine the model to use to account for some set of variables.

I, however, am interested in sites where nonlinearity is more likely to account for an emergent phenomena. This involves such complex organizations as the central nervous system, or complex processes such as how quality emerges from my incessant experience with the world around me, and also as a product of particularly complex biological or psychological processes (e.g., reproduction, growth, creativity, and so on). It is, as Hayek, Hebb and others (Chap. 3, p. 52.) have proposed, the nature of the organization of these sites that can be used as templates to conceptualize complex social and psychological phenomena. One example of this approach is to apply neural network models (see below) to account for the concept of quality, and I will consider how well this model can help us as I proceed through the book.

The intellectual foundation upon which neural networks are based can be said to have started with Leibniz (1646–1716) who, according to Lettvin (1998; p. 3), stated that "any task which can be described completely and unambiguously in a finite number of words can be done by a logical machine." It was Alan Turing (1936–1937), however, who first mathematically and then practically demonstrated that such a logic machine was possible. Turing, as Abraham (2002) states, pointed out that "The task of computation could be described 'completely and unambiguously' in a finite number of steps, and as such, Turing's machine could be seen as an 'engine of logic'" (p. 18). The key point about this model was Turing's demonstration that "the 'operations' of a logic machine or a human computer can be split into 'simple operations' so elementary they cannot be further divided" (Abraham 2002; p. 18). This is, of course, an example of recursive thinking which is now common to any computer algorithm. Because logic itself is both the study of reasoning and the language used to state propositions, it can be expressed mathematically and symbolized, as it was in Boolean algebra and set theory, and this provides a means for describing the relationship between these "simple operations."

This history is very important, because it makes it possible to posit that "a logic machine" can be developed that will characterize a quality assessment. However, the next major application of this logic involved its application in the development of neural networks, as initially described in the McCulloch and Pitts (1943) paper entitled *A Logical Calculus of the Ideas Imminent in the Nervous Activity*. Basically, what McCulloch and Pitts' paper did was to apply the ideas about the relationship between logic, language and "simple operations" to the nervous system, arguing that just as

> propositions in propositional logic can be 'true' or 'false', neurons can be 'on' or 'off'…This formal equivalence allowed McCulloch and Pitts to argue that the relations among propositions can correspond to the relations among neurons, and that neuronal activity can be represented as a proposition. (Abraham 2002; p. 19)

Thus, the McCulloch and Pitts paper demonstrated that it was possible to apply Boolean logic to the study of the functional relationship between neurons, and most importantly, to solve some of the complex logistic problems that are ongoing in the central nervous system (e.g., modeling how the central nervous system might work).

Although McCulloch and Pitts never contended that their model actually represented how the nervous system works, this did not stop the development of the cognitive sciences with its assumption that a computer may be a reasonable analog of the mind, brain or cognitive processes. Two fields of investigation emerged from these developments, and they remain active areas of research and application: *artificial intelligence* and *connectionism/neural networks*. Artificial intelligence emerged in the 1950s, while the connectionistic or neural network perspective in the neurosciences developed in the 1980s. Artificial intelligence refers to the creation of logical structures, which, with the manipulation of symbols according to rules, allows for the imitation of some desired outcomes.[19] Neural networks consist of a number of individual neurons that interact with each other, with some activating and some inhibiting other neurons, but also with some neurons doing both.

One of the limitations of the McCulloch and Pitts model was that it did not propose connection weights between neurons, so that the network were limited in their ability to self-adjust and solve different problems. As Scarborough and Somers (2006) state, "the McCulloch-Pitt neuron had no *learning rule*" (p. 32). It was here that the notions of Hayek and Hebb (Chap. 3, p. 51) become relevant; that is, the repeated activation of neurons could lead to the modifications of the neuron and these different neurons would function as weights of different magnitudes in some computational system. Alternatively, it could have been assumed that neural networks were predisposed to develop focal points or nodes, which functioned to "teach the network" as it processed information. This would be analogous to the network having a grammar that would order the information being processed. Thus, the nodes in a neural network could model how abstract constructs provide structures to organize concrete entities (Chap. 3, p. 54).

In the 1950s a number of investigators grouped together in an effort to model various biological functions, but by the 1960s it became clear that a series of problems (Minsky and Papert 1969) emerged that could not be solved by the neural networks available at that time (so-called single layer networks), so the field went into a period of minimal research activity. This was reversed in the 1980s (see summary in Scarborough and Somers 2006) by a number of computational innovations, including the introduction of "backpropagation training algorithms." As Scarborough and Somers (2006) state, "backpropagation was hailed as a computationally efficient way to train multilayered networks to represent nonlinearly separable pattern functions" (p. 38). They go on to describe back propagation as an iterative procedure that involves six steps performed as each new case or data point is presented to the network. Today there are many different examples of networks, each of which is classified by its training algorithm. Thus, an ordinary regression analysis, which would be considered a linear network, still has a (linear) algorithm as its basis. Scarborough and Somers (2006), however, identify 12 different nonlinear feed-forward network models.

But what is nonlinear here? It is the interaction of the basic units, whether they be neurons or some computer chips, and all that the network requires is that the information provided occurs in a digital manner (on–off, yes–no). What this means theoretically is that there are no restrictions on the type of data that can be processed by the neural network. But what is also clear is that the optimal use of the network requires large volumes of data, so that the reliability and reproducibility of the data becomes more of an issue than whether it is objective or subjective in nature. One source of such a large volume of data is the stream of consciousness of daily life, or the continuously changing nature of a person's qualitative state. As it turns out, a whole technology has developed which focuses on "real-time data capture" (Stone and Shiffman 2002), including using telemetry to prompt periodic data queries, daily diaries, and so on. Monitoring and recording conversational exchanges is another potential source of a large set of data.

What remains as an important question, however, is how these daily experiences lead to the impression of quality or quality-of-life. How does this happen and how can these experiences be modeled? One approach is Kahneman's (1999) notion of *objective happiness*. Kahneman (1999) utilizes data based on momentary assessments to generate temporal integrals of instant (or experienced) utility over a relevant time period. I will discuss his approach in detail in the next section of this chapter and also in Chap. 11. Relevant to this discussion, however, is that he is modeling momentary data that may be generated in a nonlinear manner, as would occur when any continuous process is sampled.

A more direct example of building a model based on a neural network is "optimality theory," as proposed by Smolensky and Legendre (2006) (I discuss this model in detail in Chap. 10). Optimality theory attempts to relate higher level cortical events, modeled as neural networks, with symbolic structures such as the grammar or structure of a sentence. The model does this by using mathematical procedures to characterize the relationship between these different levels of analysis, and does so without relying on language (Fig. 10.6). Another example is the approach that Scherer's (2000) takes to the study of emotion as an outgrowth of neural network models (see Chap. 11). Each of these examples should increase the credibility of conceiving of quality assessment in terms of nonlinear processes and models.

4.3 Summary

In this section I was concerned about whether objective, subjective, or some combination of these indicators should or could be combined when performing a quality or quality-of-life assessment. Most of my examples came from the quality-of-life or HRQOL literature, because the conflict about what indicator to use seems clearly stated. I provided some background to this controversy, and then discussed various models that have attempted to relate various indicators, only some of which combined both subjective and objective indicators. Also of interest was the description of the history of neural networks, and their relationship to nonlinear modeling. Potential applications of neural network modeling that would be appropriate for quality or quality-of-life studies were also given. Of most importance was the discussion of how digitizing nonlinear (not necessarily random) processes can lead to "logic machines," which in our case would mean modeling how quality assessments emerge from the stream of daily events that make up an experiential, if not existential, life. Recently, Diamond (2007) has suggested that dynamic system modeling, that includes nonlinearity as a component, could be used to account for the onset and changes in same-sex relationships. Applications of this sort are bound to increase with time, considering the historic shift from deterministic to probabilistic nature of causality.

5 Objective Assessment of Subjective Indicators I: Kahneman's Model of Objective Happiness

Kahneman (1999) points out that one of the major limitation of self-reports is their dependence on a person's reconstruction of past events. In fact, evidence is available that suggests that recall is a reconstruction, not reproduction, of past events (Bartlett 1932). Thus, the concern expressed by Sen (1993), Murray (Sadana et al. 2002), and others about the variable and idiosyncratic nature of self-reports could partially be due to the limitations in a person's ability to accurately reconstruct what they have experienced and felt. To avoid this problem Kahneman (1999) suggests that instead of asking a respondent to recall past events, any subjective assessment should rely on the recall of current or "momentary" events or states. Thus, if a person were asked how happy or sad they were, their immediate response would be a subjective one, which when rated as good or bad, would reflect an instant or *experienced* utility. Instant utility, Kahneman (1999; p. 4) states, refers to the probability that someone would continue, or interrupt, their current activities, experiences or feelings. One method that he and his colleagues (Fredrickson and Kahneman 1993) have used to measure instant utility involves a subject moving a sliding knob that controlled a number of colored lights corresponding to different levels of an intensity scale. Remembered utility, in contrast, refers to a global evaluation of events or experiences that have happened in the past. But in what way can instant utilities provide an objective measure? Kahneman (1999; p. 5) states:

> Objective happiness, of course, is ultimately based on subjective data: the Good/Bad experiences of moments of life. It is labeled objective because of the aggregation of instant utility is governed by a logical rule and could in principle be done by an observer with access to the temporal profile of instant utility (Kahneman et al. 1997).[20]

Thus, it is the fact that an independent observer can assess the instant utility profile that makes it possible to speak of an objective assessment. Kahneman et al. (1997) also provided a formal logical defense of their approach.

Kahneman (1999) considers his objective measurement of subjection indicators a bottoms-up approach, in which he starts with a basic datum and integrates it over some time interval. The idea of temporal integration is not new to him, he acknowledges, having been suggested by Edgeworth (1967/1881) and a variety of contemporary investigators, including Parducci (1995). Several experiments were reported which demonstrated the concept of temporal integration, but the classic and most well-known study involves patients reporting the intensity of pain during a colonoscopy (Redelmeier and Kahneman 1996). Patients were asked to rate their pain every 60 seconds on a 0 (no pain at all) to 10 (intolerable pain) scale. Later the patients were asked to provide a global estimate of their pain experience, compare it to other unpleasant experiences, and decide if they would prefer a future colonoscopy or barium enema.

The Redelmeier and Kahneman (1996) study offered the investigators a number of data analysis opportunities, including determining what combination of experimental conditions would be informationally and experientially equivalent. For example, would a person who had a high pain intensity exposure for a short duration judge a less painful experience for a longer duration to be experientially equivalent? To determine the answer the examiners would have to integrate the pain responses over the duration of the study, but because time or duration is measured on a ratio scale of physical units, their assessment has the advantage of being an objective measure (that is, nonsubject-based measurement). But for this to be true, the scales used would also have to support the assumption of a stable and distinct zero point. Thus, there would have to be a condition which was neither good nor bad, painful or not painful, and so on. Even if a zero existed, Kahneman (1999) points out it would still be possible to have different values separating the scale values. Thus, the difference between 5 and 6 on a 7-point scale may not be the same as the difference between 2 and 3. Psychologically, therefore, the scales need not be equivalent.

Also required by the theory is that an observer be knowledgeable about the subject's response, and this observer could indeed be the subject. What the observer does, to quote Kahneman (1999; p. 6), is to "effectively determine the equivalence between the original utility scale and duration." Thus, a transformation of the utility scale into a duration scale is made and this, he feels, can be done if a number of axioms are satisfied. These axioms are (Kahneman 1999; p. 22):

1. The global utility of a utility profile is not affected by concatenation with a neutral utility profile.
2. Increases of instant utility do not decrease the global utility of a utility profile.
3. In a concatenation of two utility profiles, replacing one profile by another with a higher global utility increases the global utility of the concatenation

He goes on to say,

> Peter Wakker has proven the following theorem: 'These three axioms hold if and only if there exists a nondecreasing (value) transformation function of instant utility, assigning value 0 to 0, such that global utility profiles according to the integral of the value of instant utility over time'. (p. 22)

A formal proof is provided in Kahneman et al. (1997), and if met, justifies the rescale of the utility functions into duration. They see what they are doing as being analogous to what happens when Quality-Adjusted Life Years (QALYs) are calculated. They see their formal analysis as clarifying the logic that can be applied to the evaluation of the instant utility profiles, and reiterates that in order for rules of temporal integration to apply, rescaling of the profiles that incorporate judgments which establish the equivalence between intensity of an experience (good-bad, painful-not painful, and so on) must be demonstrated.

Kahneman (1999; p. 6) asks, "What should a concept of instant utility include?" His first candidate is the hedonic qualities of a sensory experience, but anticipating some future outcome or appreciating the past should also play a role in an instant utility. He is particularly interested in including Csikszentmihalyi's (1990) notion of *flow* as part of what is meant by instant utility. Thus, if someone finds herself engaged in the flow of activities and tasks, so much so that she is not attending to the sensory aspects of her experiences, then this flow could be a major determinant of this person's report of instant utility. Other elements may enter the assessment of instant utility, such as the person's position along the good–bad dimension and how this can be affected by the person's mood, and other affective indicators. Kahneman (1999; p. 7) clearly recognizes the complexity of the task of generating a ratio scale of instant utility, but believes that it will be possible to make progress with weaker measures of instant utility.

What has been most important for me is to see how it is possible to work with subjective indicators and yet generate measures whose measurement characteristics are comparable to other types of objective measures. Kahneman's approach should draw the attention of such critics of self-reports as Sen (1993). I will continue to discuss Kahneman's contributions, including criticism of it (Alexandrova 2005) in Chap. 11. In the next section I will discuss another approach to objectifying subjective measures, this time by changing the method of measurement.

6 Objective Assessment of Subjective Indicators II: The Rasch Model

The belief that objective and subjective indicators are fundamentally different has a long history, which is intimately involved with the development of modern science. It is interesting to inspect this history, because some believe that the meaning of the term "objective" has no history, and in this regard transcends human intervention. However, this, as Daston (1992; with Galison, 2007; Chap. 5) demonstrates, is not the case. For example, from Enlightenment to the twelfth century the term *objective* meant "an object of thought" (Daston 1992; p. 300), while today it is understood to mean something "out there in the real world." The term *subjective* has also changed from meaning "existing as a subject; existing in itself" to its use today as "in the mind." From their inception, the objective and subjective have always been paired together. Daston and Galison (2007; p. 30) claim that the modern definitions of objectivity and subjectivity were established by the 1850s.

Daston (1992; p. 599) also pointed out that the term objective can be used in a variety of ways. Thus, "objective ontology" refers to determining the structure of reality; "mechanical objectivity" is about limiting the human tendency to make judgments about what is being observed; and "aperspectival objectivity" is about eliminating the role of the individual (e.g., their idiosyncracies) from what is observed. As Daston (1992) states,

> Although aperspectival objectivity is only one component of our layered concept of objectivity…it dominates current usage. Indeed, it is difficult for us to talk about objectivity without enlisting the metaphor of perspective or variants such as 'point of view', 'centreless', 'stepping back', 'climb[ing] outside of our own mind', or Thomas Nagel's brilliant oxymoron 'view from nowhere'. (p. 599)

To help cope with this "view from nowhere," researchers have learned to use instruments to record observations, and models to either test if the data match the model or if the model can help select items to include in the assessment (e.g., the Rasch Model). These approaches have clearly been successful when applied in the natural sciences, but less so in the social and behavioral sciences.

Not only have changes occurred in the meaning of the term objective, changes have also occurred in what is meant by objective measurement. The great achievement of classical physics (e.g., Newtonian mechanics), for example, was to demonstrate that it is possible to use known laws to predict a particular outcome, and this gave the impression that nature is determined. However, when these laws were applied, they were not always successful in accounting for a particular phenomenon (e.g., Brownian movement, gas dynamics, and so on). As a result, modern physics abandoned the notion that nature was determined and instead viewed it as "indeterminate," using probability theory to account for the irregularities that were observed. This shift from a deterministic to probabilistic statement of physical laws has had a profound effect on all of science, including the assessment of quality. For example, the Guttman (1944) scale, an example of a deterministic scale of measurement that is relevant to a quality-of-life assessment, is rarely reported, whereas the Rasch Model (which has the characteristics of a probabilistic Guttman scale), is regularly reported.

Nagel's "view from nowhere" promoted comment from Sen (1993; p. 126–127), who pointed out that the notion of positional objectivity has implications for objective measurement. For example, Sen (1993; p. 126–127) raises questions about Nagel's characterization of objective observations as being invariant. As he states, "The objectivity of observations (or measurement) must be a position-dependent characteristic; not a 'view from nowhere', but one 'from a delineated somewhere'." (Sen 1993: p. 127)

In contrast to Sen's view that objective measurement is positionally dependent, Rasch's (1960; Andrich 2004) approach starts from the premise that "fundamental measurement" principles can be applied to the social and behavioral sciences. What Rasch (1960; 1977) demonstrated was that data from an educational setting could fit models that reflected fundamental types of measurements. To do this required changes in the assessment instruments (e.g., by including or excluding items) so that the data generated was compatible with the model. This represented a paradigm shift from classical measurement theory, where if the data did not fit the model then the model was changed (Chap. 12).

To place Rasch's contributions in an appropriate perspective, I have to go back and review a bit more of the history of the term "objective" in measurement. Fisher (1992), for example, traced the history of the term to the Pythagoreans' who asserted that "the world is number" and that everything is made up of indivisible elementary spatial units (paraphrased from Fisher, 1992). Thus, the basic units of existence were objective, and this could be confirmed by the application of appropriate mathematical methods. Mathematics did this by demonstrating that a simple, elegant, and parsimonious solution could be applied to any number of practical problems. This made what was objective transsituational, which contributed to the separation of the "figure from meaning," to use Fisher's phrase, or as I might state it, a separation of the objective from its linguistic, cultural, or subjective context.

Fisher (1992) states that Plato took the Phythagoreans' concept of number and idealized it, thereby:

> …extending the notion of *ideality* into the definition of the objects of geometry conceiving of a point as 'an invisible line' and a line as 'length without breath' (Cajori 1985; p. 26). These definitions make it most obvious that what is analyzed, or rather, what reveals itself through the dialectical interplay of theory and data mediated by instruments…is not the actual divisible line that is drawn out, and which is called a point, but instead the conceptual ideality of that figure. (p. 36).

Fisher (1992) goes on to state that the mathematization of sociocultural phenomena, which presumably would also include quality or quality-of-life data, requires the same "conceptual ideality" (e.g., fundamental principles of measurement) independent of whether one is dealing with numbers or metaphors. Fisher (1992), referring to Rasch (1960, p. 11), suggested how this could be achieved:

> …we stand in need of measurement models in which chance plays a decisive role because of the impossibility of deterministically predicting human behavior; only when the problem of objective discourse is formulated probabilistically can we approach the rigor of geometry in social measurement. (p. 37).

However, the absence of "rigor of geometry in social measurement" has had its consequences. Fisher (1992), quotes Marcuse, and goes on to state:

> 'The great gap which separates the new [Galilean] science from its classical [Platonic] original' (Marcuse 1974; p. 230) was that the mathematics of modern science was seen as strictly numerical, devoid of the moral, political and aesthetic implications pursued by Plato. The effectiveness of modern mathematics took on such a force that philosophy came to be irrelevant and unneeded in face of the seeming self-evident way figure separates from meaning. Science will not regain its meaning for life, however, until the contemporary and ancient senses of mathematics are reconnected in the form of a critically constituted domain of human, moral and cultural investigation. (p. 38).

Fisher (1992) goes on to state,

> What this means for contemporary social sciences is that the conception of mathematics as strictly numerical–'the numbers don't remember where they come from' (Lord 1953; p. 751)–prevents the introduction of rigor and continues the alienation of human and natural by preventing the recognition of the need for instruments that will allow things to communicate themselves… Only when the wider, Platonic sense of mathematics is used to calibrate instruments can count on to mediate relations with a relative and probabilistic invariance will the rigor of geometry be introduced to social sciences. (p. 38).

To do this will require that I go beyond the separation of figure from meaning, question from answer, or difficulties with items and abilities to answer questions, and instead consider both simultaneously. The Rasch (1960;1977) Model does this by applying *conjoint measurement,* which has an underlying stochastic (or probabilistic) structure, and can be applied to a variety of practical problems.

Perline et al. (1979) provide a succinct description of *conjoint measurement*. They state:

> The theory of conjoint measurement was motivated by the realization that the theory of physical measurement was too simple to be applied in psychological research. Physical science deals with objects which admit to combining operations. It is easy to show that the weight of two lumps of clay joined into one is equal to the sum of the weights of the individual lumps. Weight is considered as a measurement system based on an empirical combining (concatenating) operation. Such a system allows more than just the comparison between single objects: It is possible to compare X concatenated with Y to the object Z. While not all physical properties admit to concatenation (temperature does not), there are several that do, and these form a fundamental basis from which other forms of measurement are derived. Unfortunately, it does not seem that the attributes of interest to psychologists can be concatenated. From this it was formerly argued that the measurement of psychological properties cannot do better than ordinal scales.
>
> Research in measurement theory, however, has led to a different conclusion. It is now known that an empirical concatenation operation is not necessary for interval scale measurement, and several models have been proposed which yield interval scales (Coombs et al. 1970). Additive conjoint measurement is one such model.
>
> Conjoint measurement is concerned with the way the ordering of a dependent variable varies with the joint effect of two or more independent variables. The situation can be compared to ordinary analysis of variance where the two (or more) independent variables (factors) form a completely crossed factorial design. ANOVA tests whether the dependent variable can be represented as the sum of row and column effects. In the case of additive conjoint measurement, the question is whether or not there exists a monotonic transformation of an ordinal measure of the dependent variable from which an additive representation can be constructed. In effect, can interaction be removed by a monotonic rescaling of the dependent variable? (p. 238).

It was Rasch's demonstration that it was possible to do "monotonic rescaling of the dependent variable" that convinced him that it was also possible to apply fundamental measurement principles to qualitative or subjective data.

Andrich (2004) described the events that lead Rasch to the realization that his approach constituted a paradigm shift, a new way of thinking about measurement in the behavioral and social sciences that was also compatible with the laws of physics. However, other investigators offer different approaches to item response theory, so that the extent of its applicability has remained controversial (Andrich 2004). The Rasch Model, therefore, provides the best available example of how to objectively measure a subjective indicator. By so doing it also introduces the possibility that subjective measures properly objectified can be invariant even if transformed in various ways, just as is found with the various laws of physics.

7 Invariance

One of the cardinal objectives of any scientific enterprise is to be able to demonstrate that what is being observed or stated remains unchanged over various conditions or transformations. In the absence of this, a researcher can only approximate what is meant by a true statement. This is certainly one of the issues that continues to undermine the credibility of self-reports. However, the discussion in this chapter has demonstrated that with appropriate methods, self-reports can meet this criteria; certainly, this was Kahneman's intent and Rasch's accomplishment. Still, the demonstration of a shift in responding over time, the so-called *response shift* (Spranger and Schwartz 1999), and the complexity of measuring responses over time (Chap. 11, p. 407) certainly keeps the question of whether self-reports can be characterized as invariant very much alive.

I am actually raising several issues here, but probably most important is the concern about whether the common discourse that forms the basis of the language of self-reports precludes invariant self-reports. Frege (1892/1966) addressed this issue when he claimed that words or sentences, if true, would retain their meaning when presented in different contexts. Thus, to the extent that a self-report is literally true it could be considered to be invariant. Wittgenstein (1961), of course, would cast doubt on this assumption, arguing instead that when language is used in its varied social settings, and takes full advantage of figurative expression, it is not likely to be invariant, but rather has the "family resemblance" common to different forms of expression.

Demonstrating that a statement is invariant is also important because of what it implies about objectivity. Nozick (2001; p. 75–119) has pointed out that *invariance under various transformations* is one of four criteria involved in defining what is meant by objective. The other three criteria involve demonstrating that an objective statement can be assessed from different directions and times and remain the same (this criteria is not unlike Sen's [1993] notion of positional objectivity); that people can agree about the objectivity of the statement (that is, intersubjectivity can be demonstrated); and that if a statement is objective then it will remain true independent of people's belief, hopes, and so on. Since the degree any observation matches these criteria may vary, it is also clear that the degree to which an observation can be called "objective" will vary.

The association of invariance with the establishment of objectivity suggests that the very process of assessment itself represents an implicit effort to transform indicators to a common objective format. Thus, all data independent of its origin when assessed approximates an objective indicator to varying degrees. This notion has implications concerning the assessment of quality, or quality-of-life. As I have previously stated, an investigator has to start with either subjective data, or objective data that has been qualified when measuring quality. The task then becomes determining the degree to which these data meet the criteria for objectivity. Table 4.2 summarizes three popular approaches to the objectification task.

Demonstrating invariance in quality research is a particularly important goal since it would have broad social and political consequences. Thus, if it could be demonstrated that the quality of a social policy was invariant (meaning that it would remain the same independent of which person or circumstance it was being applied to), then it could be argued that this policy should be applied universally. An example of such a policy might be ensuring the safety of cars and vehicular transportation. In Chap. 1 (p. 13) I suggested that to achieve this goal would require the concurrent measurement of both aesthetic and functional quality (see Fig. 1.2). To date, this is rarely reported in a qualitative assessment.

At this point I would like to clarify the relationship between invariance and objectivity, and in particular, take advantage of what Nozick (2001) has written on this topic. First, the meaning of the phrase invariance under various transformations can be illustrated fairly easily. Consider the situation where I am traveling and I buy a liter of gasoline, but being from the United States, I want to transform this quantity into the more familiar quantity of gallons. This I can do by multiplying the liters by a constant that represents the proportion that a liter occupies in a gallon, and I can do this without concern that the transformation has altered the fundamental character of the quantity. In contrast, if I compare Fahrenheit and centigrade temperature measures, the ratio of these two measures would not be invariant since their value is an artifact of their particular scale. As Nozick, (2001; p. 77) states, "One quantity's being twice another is not invariant under the transformation F= (9/5) + 32. Also, the difference or interval between two temperatures (T_1 -T_2) has no fixed significance but is an artifact of a particular scale." However, the ratio of (T_1 -T_2)/(T_3 -T_4) would be the same for each scale of measurement, and is thereby invariant under a positive linear transformation.

Invariance under various transformations is what makes Einstein's theory of relativity so important, but is also what Gibson (1966) claims can be found in stimulus energy flux that reaches our sensory receptors. In both cases the same principle can be used to explain many different observations. Nozick (2001) states the following about Einstein's theory of relativity:

> Einstein taught us that spatial distance and temporal distance are relative to an observer; their magnitudes will be measured differently by different inertial observers, and spatial and temporal intervals are not invariant under Lorentz transformations. However, inertial observers will agree about another, more complicated interval between events, involving not just spatial separations alone or temporal separations alone but a particular mixture of the two, namely, the square root of the square of the time separation minus the square of the spatial separation. This more complicated interval *is* invariant under Lorentz transformations. The principle of relativity of Einstein's Special Theory holds that all laws of physics are the same for all inertial observers; they are the same in every inertial reference frame and so are invariant under Lorentz transformations. (p. 76).

Can the same kinds of laws be found or approximated by the assessment of quality? Are there universal laws of quality? I believe they exist, especially if both aesthetic and functional quality assessments are performed. What is not clear is how these two dimensions can be combined and then transformed so as to demonstrate invariant outcomes. Proof of the existence of such transformations may involve demonstrating that *a common cognitive process* has occurred (Chap. 12).

What Nozick (2001) makes clear is that there are degrees of *invariance under various transformation*, just as there are degrees of objectivity. The degrees of invariance will be determined by the type of transformation. Nozick (2001) identifies two basic types of transformations. The first involves a mapping from one set of functions to another (that is, if certain things have a property it is possible that other things will also have that property). The second transformation involves dynamic changes or alterations over time, while retaining properties that are invariant and do not change.

In his extended discussion, Nozick (2001) first distinguishes between *a fact* and *a belief* (a subjective statement), noting that it is possible to speak of both in either objective or subjective terms, much as I illustrated in Table 4.1. Consistent with this is his view that there are degrees of objectiveness for facts or beliefs (Nozick; 2001 p. 87). Of most interest in the present context, is what Nozick (2001; p. 94–99) says concerning the objectivity of beliefs. He states that "A judgment or belief is objective when it is reached by a certain sort of process, one that does not involve biasing or distorting factors that lead belief away from the truth." (p. 94) Thus, Nozick would agree that a judgment concerning the quality of an object, experience, event, or life lived can be objective. Biasing diminishes the objectivity of a belief by distorting the truth value of the belief (by making it more irrational), but this bias is not an inherent characteristic of the belief. Biases also adversely impacts the reliability of a belief or other subjective statements. Nozick (2001) goes on to make an interesting related point; "Science could contain values, and even make value assertions, and still be objective *if* values themselves are objective." (p. 95) Again the notion that a value statement is not inherently biased (that is, can be objective) will be an important part of my discussion in Chap. 12.

It should be clear from this discussion that Nozick's (2001) notions concerning invariance; objectivity and subjectivity; facts and beliefs are based on an implicit, unstated additive model of cognition. His basic cognitive unit is an objectively determined cognitive entity that can be influenced by biases that make such a statement irrational or unreliable. This cognitive entity can also be transformed and shown to be invariant, much as when Gibson (1966) speculated about how the optical array is perceived as constant in a continually changing sensory and perceptual world. Sen (1993) would probably be comfortable with Nozick's perspective, since both recognize that what needs to be done is to identify the method used to achieve objective measurement in the presence of subjectivity.

Now, psychometricians have also been active in developing techniques that can be used to characterize invariance. Their approach involves the use of confirmatory factor analytical technique, as it is applied to current quality-of-life and health care data. The invariance demonstrated here is not of the type which claims that a relationship remains valid as the data is transformed from setting to setting, rather the invariance is more of an indication of whether the same statement can be applied across different data groups; that is, *measurement invariance*. For example, Gregorich (2006) describes a series of procedures that demonstrate that invariance can vary in degree, but now in terms of the precision of the inferences that can be drawn from the factor analysis of different groups of data. The least precise inferences of these procedures would involve demonstrating the presence of *dimensional invariance*: the demonstration that the number of factors in two or more groups of data is equal. A second type is referred to as *configural invariance*, where claiming invariance requires that each common factor is associated with identical item sets across groups. A third type of invariance is called *metric invariance*, and here the claim of invariance requires that the factor loadings between groups are equal. Metric invariance assumes that both dimensional and configural invariance can be demonstrated, suggesting that a hierarchy of increasing precise inferences is possible. Gregorich (2006) also considers the potential confounding effect of bias, suggesting that if group comparison statements can be made which are not diluted by cultural or other contextual biasing factors, it would be possible to make a strong factorial invariance claim. He also discusses making *strict factorial invariance* statements, where common, not residual, factor variation across groups is compared. Thus, Gregorich (2006) demonstrates that a variety of statistical procedures are available to determine if some observed factor structure remains invariant across groups of data.

In this regard it was interesting to find a report by Bertoquini and Ribeiro (2006) of a study in which they created an index that combined a subjective measure of well-being with measures of positive and negative affect, but which also determined the degree of measurement invariance. Applying confirmatory factor analytic techniques, they compared the individual components and the index, and were able to demonstrate the presence of *configural* and *metric invariance*. Strict invariance was rejected. They concluded that the partial invariance they demonstrated suggests some measurement equivalence across groups, with the degree of measurement equivalence being an indication of how confident they would be in using their index to infer conclusions across studies following the application of their well-being assessment.

Borsboom (2005), in his book *Measuring the Mind*, has identified three major trends in psychometrics. The first approach he labels as the study of error (that is, a true score is assumed to exist) and is characteristic of the classical approach to psychometrics. The second refers to the existence and assessment of latent variables, which reflects the interest in item-response theory in psychometrics. The third he describes as the axiomatic approach to psychometrics. Of these three models, the axiomatic approach is the most likely to lead to invariant principles (see Chap. 12).

8   Summary of Chapter

One of the objectives of this book is to describe an approach to the assessment of quality that can be simultaneously applied to an object, an event or a person's life. Key to achieving this goal is clarifying in what way objective and subjective indicators are similar or different, as well as describing how objective and subjective indicators can be combined. What has become clear is that rather than a dichotomy, objective and subjective indicators should be seen as points on a continuum that can be context dependent (e.g., the phenomena of positional objectivity). It was also implied in this chapter, and will be demonstrated as I proceed, that objective indicators in a quality or quality-of-life context become "subjectified" or "qualified," and as such, should be viewed as modulating rather than mediating changes in quality. This suggests that those who insist on combining objective and subjective indicators do not take into account how objective indicators change when placed in a subjective context. Saying this does not question whether objects and objective indicators exist, only that they change when combined with value-laden issues, such as judging quality or quality-of-life.

Also discussed was the notion that while subjects normally subjectify or qualify objective indicators, the assessment process itself involves objectification. Two examples of this transformation were presented: Kahneman's concept of *instant utility* and the *Rasch model*. What was also implicitly stated was that other assessment processes vary in the degree to which they succeed in objectifying data. The history of change from a deterministic to probabilistic view of measurement was noted, as was the shift from classical test theory to item response theory prevalent in current quality-of-life measurement (Chap. 12). The next chapter will continue this discussion, but this time in the context of determining the role of signs and symptoms in qualitative research of the medically ill.

I would like to make explicit some of the implications of what was presented in this chapter. For example, a purely objectively defined indicator (e.g., the number and type of bacteria in the ground I walk on) would not normally be a useful part of a qualitative assessment. Consider Lawton's (1991) model for assessing the quality-of-life of the person with dementia. This model includes as a separate dimension what he refers to as the "objective environment" (see footnote 12). Of course, there are many aspects of the objective environment that I may not ordinarily attend to, such as the rocks on the ground, the color of the floor, or even the type of window in a house. Yet each of these factors can become a salient part of a person's environment if they are perceived as affecting some subjective state, such as the person's well-being or orientation in space. Lawton (1991) recognizes this when he considers physical measures of the objective environment, or "measurement by consensus," as determining when a house is well lighted or home-like. What is implicit in his approach is a concern for the person's *comfort* and ability to function, which of course are subjectively based concerns. Thus, as I have discussed earlier and will again, when objective indicators become relevant to a person they inevitably become "subjectified" (Fuhrer 2000), and it is these data that can become objectified via the application of statistical measurement models resulting in objective assessments of quality or quality-of-life.

If, as I contend, objective indicators that have not been qualified have a limited role in a quality-of-life or HRQOL assessment, then this has policy implications. It challenges the "Swedish" model of social welfare (assuming that Rapley's [2003] interpretation of this model is current) that relies on reducing the crime rate or the poverty rate, increasing the gross national product, and so on, to optimize the quality of a person's existence. As with cancer, changes in these indicators could occur and yet have no presence or meaning in a person's life. The argument that I don't need a person's voice to confirm that a social policy has led to an improvement in the quality of a person's existence seems to contradict the very purpose of social policy, which presumably is to positively impact various aspects of a person's life. Pure objective indicators alone simply do not capture this information.

Implicit in the criticism by Sen and Murray, concerning the reliability and validity of self-reports, is a view of "what is real" that reflects a dualistic perspective. Thus, it is not surprising that these investigators reject subjective representations, and see subjectivity as essentially idiosyncratic, confounding policy making and science. There is, however, an alternative approach that I mentioned earlier, but which I will discuss in more detail in Chaps. 5, 10, and 11. This approach assumes that reality is *embodied* (Chap. 2, p. 36), and that the assumptions that objective and subjective entities exist as separate realities is misplaced. Lakoff and Johnson (1999) state the issue clearly:

> The problem with classical disembodied scientific realism is that it takes two intertwined and inseparable dimensions of experience - the awareness of the experiencing organism and the stable entities and structures it encounters - and erects them as separate and distinct entities called subjects and objects. What disembodied realism…misses is that, as embodied, imaginative creatures, *we never were separated or divorced from reality in the first place*. What has always made science possible is our embodiment, not our transcendence of it, and our imagination, not our avoidance of it. (p. 93).

Lakoff and Johnson quotation offers an alternative to the Sen/Murray approach, an approach that will be particularly relevant to a discussion on the role of symptoms in a quality-of-life or HRQOL assessment. It also offers a possible mechanism whereby the continuum underlying objective and subjective indicators may be formed, which is a discussion I will continue in the next chapter.

Notes

1. This quotation comes from a paper by Gawronski (2009).
2. It is possible that a computer monitor the pills a person consumes, and this information could indicate if an "error" has occurred. However, to determine that the "error" was not due to a lost pill or a request by the patent's physician will involve a review and the judgment about whether the drug protocol has or has not been violated. This evaluative process is considered a subjective assessment.
3. The methods here have been described as transforming the qualitative to the quantitative, but making something quantitative doesn't necessarily mean that it has been transformed into something objective.

4. Dr. David Feeny contributed to the creation of this Table.
5. Campbell et al. (1976) reported a similar finding some 25 years prior to Sen's report.
6. I am aware that in some countries the notion that the government is interested in assessing the well-being of its population is an inappropriate assumption. However, I believe my point remains even in the face of this reality.
7. I have not discussed measures such as environmental factors, crime rates, work status, and so on. Each of which's positional dependence can be determined.
8. Psychosocial measures, such as coping with stress, can be monitored objectively by cortisol levels and subjectively by self-reports. Both indicators can be quantitatively measured.
9. According to Averill et al. (1998), the widespread use of the term *aesthetics* occurred after Baumgarten's publication, in 1750, of his monograph entitled *Aesthetica*. Prior to this, topics considered to be aesthetic were referred to as different aspects of *beauty*.
10. I say this even though Aristotle's notion of *Eudaimonia* predated modern applications of quality-of-life research by several thousand years. The difference in these two approaches is reflected in the difference between aesthetic and functional quality.
11. Dr. Carol Ferrans (personal communication; November 2, 2005) pointed out that when discussing social welfare, Pigon (1920) used the phrase "quality-of-life." She suggested that this may have been one of the earliest *modern* usages of the phrase.
12. Note the contributions of Florence Nightingale also in this regard.
13. The reader should note that while the two correlations are significantly different, the higher correlation accounts for only 6.5% of the variance in a study that compares over 30 studies.
14. Simpson (1961) defines a *hierarchy* as a systematic classification framework, and a *taxonomy* as the theoretical study of classification.
15. Please note that hierarchical models simulate the cognitive process of classifying information Chap. 3 (35). Thus, a *superordinate* concept, such as health, can be disaggregated into *basic* concepts, such as symptoms, stamina, well-being, and these basic concepts can be further disaggregated into *subordinate* domains, such as depression, miles run per week, happiness. I will refer to this relationship as I proceed, because it reinforces the statement that I made in Chap. 1 which is that it will be important to understand how I understand, which in this case is reflected by the concordance between a hierarchical model I use to theorize how I think.
16. Lawton's (1991) discussed his conceptualization of *behavioral competence*, but he also included measures of psychological well-being, perceived quality-of-life, and the objective environment in a total description of the quality of a person's existence.
17. For example, if a respondent is asked how many rooms he lives in, and this objective indicator is entered into an equation that includes some subjective qualitative assessment (e.g., how satisfied he is with his life) the resulting statistic would be less likely to be informative then if the person was asked how satisfied he was with the number of rooms he lives in and this was included with their life satisfaction estimate. My point here is that assessing data which has been scaled on a common subjective denominator is more likely to be meaningful and useful then if the data consists of heterogeneous concepts and content.
18. You may have noticed that in the Comments section of Table 4.10, I quote Sullivan et al. (2000) as claiming that, "The Linear Model did not satisfactorily account for the observed data..." (p. 81). This statement should be interpreted as meaning that the Wilson and Cleary model did not fit the data, not that some other linear models would not fit the data.
19. A discerning reader might have noticed that I used the phrase "production rules" in Fig. 3.1 to refer to the relationship between the antecedent and consequential conditions of a cognitive model of a qualitative assessment. The reader who interprets this to mean that I am suggesting that an algorithm underlies how the production rules operate would be correct. However, in this book I am emphasizing the role that a neural network can play in accounting for various phenomenon, one example of which is of interested is the relationship between capacity and performance (Chapter 10).
20. This statement by Kahneman implies that an investigator is capable of simulating the cognitive processes, in this case a logic, that a respondent may also have engaged in when they reconstructed (recalled) their most recent experiences.

References

Abraham TH. (2002). (Physio)logical circuits: The intellectual origins of the McCullock-Pitts neural networks. *J Hist Behav Sci.* 38, 3–25.

Adorno TW. (1997). Aesthetic Theory. In, (Eds.) G Adorno, R Tiedemann; Translator. R Hullot-Kentor. *Theory and History of Literature*. Vol 88. Minneapolis MN: University of Minneapolis.

Alexandrova A. (2005). Subjective well-being and Kahneman's 'Objective Happiness". *J Happiness Stud.* 6, 301–324.

Andrews FM, Withey SB. (1976). *Social Indicators of Well-being: American's Perceptions of Life Quality*. New York, NY: Plenum Press.

Andrich D. (2004). Controversy and the Rasch Model: A characteristic of incompatible paradigms? *Med Care.* 42, 1–7.

Anscombe FJ, Auman RJ. (1963). A definition of subjective probability. *Ann Math Stat.* 34, 199–205.

Arnold R, Ranchor AV, Koeter GH, de Jongste M, Sanderman R. (2005). Consequences of chronic obstructive pulmonary disease and chronic heart failure: The relationship between objective and subjective health. *Soc Sci Med.* 61, 2144–2154.

Atkinson M, Zibin S, Chuang H. (1997).Characterizing quality of life among patients with mental illness: A critical examination of the self-report methodology. *Am J Psychiatr.* 154, 99–105.

Averill JR, Stanat P, More TA. (1998). Aesthetics and the environment. *Rev Gen Psychol.* 2, 153–174.

Baker SR, Pankhurst CL, Robinson PG. (2007). Testing relationships between clinical and non-clinical variables in xerostomia: A structural equation model of oral health-related quality of life. *Qual Life Res.* 16, 297–308.

Bargh JA, Ferguson MJ. (2000). Beyond behaviorism: On the automaticity of higher mental processes. *Psychol Bull.* 126, 925–945.

Barofsky I. (1986). Quality of life assessment: Evolution of a concept. In: V Ventafrida, R Yancik, FSAM van Dam, and M Tamburini. (Eds.). *Quality of life Assessment and Cancer Treatment*. Amsterdam: Elsevier. (p. 11–18).

Barofsky I. (2000). The role of cognitive equivalence in studies of health-related quality of life assessments. *Med Care.* 38 Supp II, 125–129.

Baron J, Asch DA, Fagerlin A, Jepson C etal. (2003). Effect of assessment method on the discrepancy between judgments of health disorders people have and do not have: A web study. *Med Decis Mak.* 23, 422–434

Barsky AJ. (2002). Forgetting, fabricating and telescoping: The instability of the medical history. *Arch Intern Med.* 162, 981–984.

Bartlett FC. (1932). *Remembering: A study in experimental and social psychology*. Cambridge UK: Cambridge University Press.

Bauer RA. (1966). *Social Indicators*. Cambridge, MA: MIT Press.

Bertoquini V, Ribeiro JP. (2006). Measurement invariance of subjective well-being model. International Society for Quality of life Research meeting abstracts; *Quality of life Res Supplement*, 237/Abstract #1056. [www.isoqol.org/2006mtgabstracts.pdf].

Boring EG. (1957). *A History of Experimental Psychology*. New York, NY: Appleton-Century Crofts .

Borsboom D. (2005). *Measuring the Mind: Conceptual Issues in Contemporary Psychometrics*. Cambridge UK: Cambridge University Press

Brinkman P, Coates D, Janoff-Bulman R. (1978). Lottery winners and accident victims: Is happiness relative? *J Personal Soc Psychol.* 36, 917–927.

Broderick JE, Stone AA, Calvanese P, Schwartz JE, Turk DC. (2006). Recalled pain ratings: A complex and poorly defined task. *J Pain.* 7, 142–149.

Brorson B, Åsberg KH. (1984). Katz Index of Independence in ADL: reliability and validity in shor-term care. *Scand J Rehabil Med.* 16, 125–132.

Brown HI. (1987). *Observation and Objectivity*. New York NY: Oxford University Press

Bryk, A. S. & Raudenbush, S. W. (1992). *Hierarchical linear models*. Newbury Park CA: Sage

Buick DL, Petrie KJ. (2002). "I know just how you feel": The validity of healthy women's perceptions of breast cancer patients receiving treatment. *J Appl Soc Psychol.* 32, 110–123.

Cajori F. (1985). *A History of Mathematics*. New York, NY: Chelsea.

Campbell, A. (1972). Aspiration, satisfaction and fulfillment. In: (Eds.) A Campbell, P Converse. *The Human Meaning of Social Change*. New York, NY: Russell Sage. (p. 441–446).

Campbell A, Converse PE, Rodgers WL. (1976). *The Quality of American Life*. New York, NY: Russell Sage Foundation, Rutgers University Press.

Carey S. (1995). On the origin of causal understanding. In: (Eds.), D Sperber, D Premack, AJ Premack. *Causal Cognition: A Multidisciplinary Debate*. Oxford UK: Oxford University Press. (p. 268–302).

Chen H, Hanisch H-M. (2000). Model aggregation for hierarchical control synthesis of discrete event systems. *Proceedings of the 39th IEEEConference on Decision and Control*. Sydney Australia. (p. 418–423). Accessed from the web, March 2007.

Ciarleglio MM, Makuch RW. (2007). Hierarchical linear modeling: An overview. *Child Abus Negl.* 31, 91–98.

Coombs CH, Dawes RM, Tversky (1970). A. *Mathematical Psychology: An Elementary Introduction*. Englewood Cliffs, NJ: Prentice-Hall.

Crosby C, Holzemer WL, Henry SB, Portillo CJ. (2000). Hematological complications and quality of life in hospitalized AIDS patients. *AIDS Patient Care STDS.* 14, 269–279.

Csikszentmihalyi M. (1990). *Flow...The Psychology of Optimal Experience*. New York, NY: Harper and Row.

Cummins RA. (1996). The domains of life satisfaction: An attempt to order chaos. *Soc Ind Res.* 38, 303–328.

Cummins RA. (1997). Self-rated quality of life scales for people with an intellectual disability: A review. *J Appl Res Intell Disabil.* 10, 199–216.

Cummins RA. (2000a). Objective and subjective quality of life: An interactive model. *Soc Indic Res.* 52, 55–72.

Cummins RA. (2000b). Personal income and subjective well-being: A review. *J Happiness Stud.* 1, 133–158.

Cummins RA, Gullone E, Lau ALD. (2002). A model of subjective well-being homeostasis: The role of personality. In: (Eds.), E Gullone and RA Cummins *The Universality of Subjective Well-being Indicators*. Dordrecht NE: Kluwer.

Daston L. (1992). Objectivity and the escape from perspective. *Soc Stud Sci.* 22, 597–618.

Daston L, Galison P. (2007). *Objectivity*. New York, NY: Zone Books.

de LeeuwJ, Kreft IGG. (1995). Questioning multilevel models. *J Educ Behav Stat.* 20, 109–113.

Diamond LM. (2007). A dynamical systems approach to the development and expression of female same-sex sexuality. *Perspect Psychol Sci.* 2, 142–161.

Diener E, Suh EM, Lucas RE, Smith HL. (1999). Subjective well-being. *Psychol Bull.* 125, 276–302.

Draper D. (1995). Inference and hierarchical modeling in the social sciences. *J Educ Behav Stat.* 20, 115–147.

Dunning D, Heath C, Suls JM. (2004). Flawed self-assessment: Implications for health, education, and the workplace. *Psychol Sci Public Interest* 5,69–106.

Edgeworth FY. (1967). *Mathematical psychics: An essay on the application of mathematics to the moral sciences*. New York, NY: Kelly (Originally published in 1881).

Erickson R. (1993). Descriptions of inequality: The Swedish Approach to welfare Research. In, M Nussbaum, A. Sen (Eds.) *The Quality of life*. Oxford, UK: Clarendon Press. (p 67–87).

Fisher WP Jr. (1992). Objectivity in measurement: A philosophical history of Rasch's separability theorem. In, (Ed.) M Wilson. *Objective Measurement: Theory and Practice*. Norwod, NJ: Ablex. (p. 29–58).

Fox KA. (1986). The present status of objective social indicators: A review of theory and measurement. *Am J Agric Econ.* 68. 1113–1120.

Fredrickson BL, Kahneman D. (1993). Duration neglect in retrospective evaluations of affective episodes. *J Personal Soc Psychol.* 65, 45–55

Frege, G. (1892/1966). On Concept and Object. In (Eds.) P Geach, M Black, *Translations from Philosophical Writings of Gottlob Frege*. Oxford UK: Oxford University Press. (p. 42).

Fries JF. (1991). The hierarchy of Quality of life Assessment, the Health Assessment Questionnaire, and issues mandating development of a Toxicity Index. *Control Clin Trials.* 12, 106s–117s.

Fries JF, Singh G. (1996). The Hierarchy of Patient Outcomes. In, (Ed) B Spilker. *Quality of life and Pharmacoeconomics in Clinical Trials*, 2nd edition. New York, NY: Raven Press. (p. 33–40).

Fuhrer MJ. (2000). Subjectifying quality of life as a medical rehabilitation outcome. *Disabil Rehabil.* 22, 481–489.

Gawronski B. (2009). Ten frequently asked questions about implicit measures and their frequently supposed, but not entirely correct answers. *Can Psychol.* 50, 141–150.

Gibson JJ. (1966). *The Senses Considered as a Perceptual Systems*. Boston, MA: Houghton Mifflin.

Gregorich SE. (2006). Do self-report instruments allow meaningful comparisons across diverse populations groups? Testing measurement invariance using the confirmatory factor analysis framework. *Med Care.* 44 Suppl 3, S78-S94.

Guttman L. (1944). A basis for scaling qualitative data. *Am Sociol. Rev.* 9, 139–150.

Hagerty MR, Cummins RA, Ferriss AL, Land K, et al. (2001). Quality of life indexes for national policy: Review and agenda for research. *Soc Ind Res.* 55, 1–96.

Hagerty MR, Veenhoven R. (2003). Wealth and happiness revisited – Growing national income does go with greater happiness. *Soc Indic Res.* 64, 1–27.

Hatten C, Ager A. (2002). Quality of life measurement and people with intellectual disabilities: A rely to Cummins. *J Intell Disabil.* 15: 254–260

Headey B, Wearing A. (1989). Personality, life events and subjective well-being: Toward a dynamic equilibrium model. *J Personal Soc Psychol.* 57, 731–739.

Heo S, Moser DK, Riegel B, Hall LA, Christman N. (2005). Testing a published model of health-related quality of life in heart failure. *J Card Fail.* 11, 372–379.

Janz NK, Janevic MR, DodgeA, Finerlin TE, et al. (2001). Factors influencing quality of life in older women with heart disease. *Med Care.* 39, 588–598.

Kahneman D. (1999). Objective Happiness. In, (Eds.) D Kahneman, E Diener, N Schwarz. *Well-being: The Foundations of Hedonic Psychology*. New York, NY: Russell Sage Foundation. (p. 3–25).

Kahneman D, Diener E, Schwarz N. (1999). Preface. In, (Eds.) D Kahneman,. *Well-being : The Foundations of Hedonic Psychology*. New York, NY: Russell Sage Foundation. (p.i-xii).

Kahneman D, Wakker PP, Sarin R. (1997). Back to Bentham? Explorations of experienced utility. *Q J Econ.* 112, 375–405.

Kaneko M. (1985). An axiomatization of utility and subjective probability based on objective probability. *Cowles Foundation Discussion Paper No. 746.* New Haven CT: Cowles Foundation for Research in Economics at Yale University. Accessed from the web; April 2007.

Kaplan RM, Anderson JP. (1990). The General Health Policy Model: An integrated approach. In, (Ed) B Spilker. *Quality of life Assessments in Clinical Trials.* New York, NY: Ravens.

Kaplan RM, Bush JW, Blischke WR. (1987). Additive utility independence in a multidimensional quality of life scale for the general health policy model. *Unpublished Manuscript.*

Katz S, Apkom CA. (1976). A measure of primary socio-biological functions. *Int J Health Ser.* 6, 493–507.

Kind P. (1996). The EuroQoL instrument: An index of health-related quality of life. In, (ed) B Spilker, *Quality of life and Pharmacoeconomics in Clinical Trials.* Philadelphia PA: Lippincott-Raven. (p. 191–201).

Lakoff G, Johnson M. (1999). *Philosophy in the Flesh: The Embodied Mind and its Challenge to Western Thought.* New York, NY: Basic Books

Lawton MP. (1991). A multidimensional view of quality of life in frail elderly. In, (Eds.) JE Birren, JE Lubben, JC Rowe, DE Deutchman. *The Concept and Measurement of Quality of life in Frail Elderly.* New York, NY: Academic Press. (p. 3–27).

Lettvin JY. (1998). (Interview with JA Anderson and E Rosenfeld) In: (Eds.) JA Anderson and E Rosenfeld. *Talking Nets: An Oral History of Neural Networks.* Baltimore MD: Williams and Wilkins, pp. 1–21.

Liu B-C. Quality of life: Concept, measure and results. *Am J Econ Sociol.* 34; 1–14: 1975.

Loehlin JC. (2004). *Latent Variable Models: An Introduction to Factor, Path and Structural Equation Modeling.* Mahwah, NJ: Lawrence Erlbaum. Fourth Edition.

Lord F. (1953). On the statistical treatment of football numbers. *Am Psychol.* 8, 750–751.

Mallon B. (1999). *Ernst Amory Codman: The End Result of a Life in Medicine.* New York NY: Elsevier.

Marcuse H. (1974). On science and phenomenology. In: A Giddens (Ed.) *Positivism and Sociology.* London, UK: Heinemann. (p. 225–236).

Mathisen L, Andersen MH, Veenstra M, Wahl AK, et al. (2007). Quality of life can both influence and be an outcome of general health perceptions after heart surgery. *Health Qual Life Outcomes.* 5:27. doi: 10.1186/1477-7525-5-27. (1–10).

McCulloch WS, Pitts W. (1943). A logical calculus of the ideas immanent in the nervous activity. *Bull Math Bio.* 5, 115–133.

Minsky M, Papert S. (1969). *Perceptrons.* Cambridge, MA: MIT Press.

Mozley CG, Huxley P, Sutcliffe C, Bagley H, Burns A, Challis D, Cordingley L. (1999). "Not knowing where I am doesn't mean I don't know what I like: Cognitive impairment and quality of life responses in elderly people. *Int J Geriat Psychiatr.* 14, 776–783.

Murray CJL, Salomon JA, Mathers CD, Lopez AD. (2002). *Summary Measures of Population Health: Concepts, Ethics, Measurement and Applications.* Geneva, World Health Organization.

Njegovan V, Man-Son-Hing M, Mitchell SL, Molnar FJ. (2001). The hierarchy of functional loss associated with cognitive decline in older persons. *J Gerontol : Med Sci.* 56A, M638-M643.

Noll, H-H. (2000). Social indicators and social reporting: The international experience. http://www.cccsd.ca/noll1.html (Accessed June 6, 2010).

Noll, H-H, Zapf, W. (1994). Social Indicators Research: Societal Monitoring and Social Reporting. In, (Eds.) I Borg, PP Mohler. *Trends and Perspectives in Empirical Social Research.* Berlin/New York: Walter de Gruyter. (p. 1–16).

Nozick R. (2001). *Invariances: The Structure of the Objective World.* Cambridge, MA: Harvard University Press.

O'Boyle CA, McGee HM, Hickey A, Joyce CRB, Brown J, O'Malley K. (1993). *The Schedule for Individual Quality of life. User Manual.* Dublin IR: Department of Psychology, Royal College of Surgeons in Ireland.

Orfila F, Ferrer M, Lamarca R, Tebe C, et al. (2006). Gender differences in health-related quality of life among the elderly: The role of objective functional capacity and chronic conditions. *Soc Sci Med.* 63, 2367–2380.

Ormel J, Lindenberg S, Steverink N, Vonkorff M. (1997). Quality of life and social production functions: A framework for understanding health effects. *Soc Sci Med.* 45: 1051–1063.

Oxford English Dictionary. (1996). Oxford UK: Oxford University Press.

Parducci A. (1995). *Happiness, Pleasure and Judgment: The contextual theory and its Applications.* Hillsdale, NJ: Erlbaum.

Patrick DL, Chiang Y-P. (2000). Measurement of health outcomes in treatment effectiveness evaluations. *Med Care.* 38 (Suppl II), II-14- II-25.

Perline R, Wright BD, Wainer H. (1979). The Rasch Model as additive conjoint measurement. *Applied Psychol Meas.* 3, 237–256.

Phaladze NA, Human S, Dlamini SB, Hulela EB, et al. (2005). Quality of life and concept of "living well" with HIV/AIDS in sub-saharan Africa. *J Nurs Scholarsh.* 37, 120–126.

Pigon (1920). Presidential's Research Committee on Social Trends. (1933). *Recent Social Trends.* New York, NY: McGraw-Hill.

Prutkin JM, Feinstein AR. (2002). Quality of life measurement: Origin and pathogenesis. *Yale J Biol Med.* 75, 79–93.

Rapley M. (2003). *Quality of life Research.* Thousand Oaks, CA: Sage.

Rasch G. (1960). *Probabilistic Models for Some Intelligence and Achievement Tests.* Copenhagen DK: Danish Institute for Educational Research. (Expanded edition, 1983. Chicago, MESA Press).

Rasch G. (1977). On specific objectivity: An attempt at formalizing the request for generality and validity of scientific statements. *Dan Yearbook Philos.* 14, 58–94.

Redelmeier D, Kahneman D. (1996). Patients memories of painful medical treatments: Real-time and retrospective evaluations of two minimally invasive procedures. *Pain.* 116, 3–8.

Riis J, Baron J, Jepson C, Fagerlin A, Ubel PA. (2005). Ignorance of hedonic adaptation to hemodialysis: A study using ecological momentary assessment. *J Exp Psychol: Gen.* 134, 3–9.

Rogers DF, Plante RD, Wong RT, Evans JR. (1991). Aggregation and disaggregation techniques and methodology in optimzation. *Oper Res.* 39, 553–582.

Ruta DA, Garratt AM, Leng M, Russell IT, MacDonald LM. (1994). A new approach to the measurement of quality of life: The Patient Generated Inventory (PGI). *Med Care.* 32, 1109–1126.

Sackett DL, Torrance GW. (1978). The utiilty of different health states as perceived by the general public. *J Chronic Dis.* 31, 697–704.

Sadana CD, Mathers CD, Lopez AD, Murray CJL, Iburg KM. (2002). Comparative analysis of more than 50 household surveys of health status. In (Eds.) CJL Murray, JA Salomon, CD Mathers, AD Lopez. *Summary Measures of Population Health: Concepts, Ethics, Measurement and Applications.* Geneva: World Health Organization.

Salomon JA, Tandon A, Murray CJL. (2004). Comparability of self rated health: cross sectional multi-country survey using anchoring vignettes. *Brit Med J.* 328, 258–264.

Savage LJ. (1954). *The Foundations of Statistics.* New York, NY: Wiley.

Scarborough D, Somers MJ. (2006). *Neural Networks in Organizational Research: Applying Pattern Recognition to the Analysis of Organizational Behavior.* Washington DC: American Psychological Press.

Scherer KR. (2000). Emotions as episodes of subsystem synchronization driven by nonlinear appraisal processes. In (Eds.) MD Lewis, I.

Granic. *Emotions, Development, and Self-organization: Dynamic Systems Approaches to Emotional Development*. Cambridge, UK: Cambridge University Press. (p. 70–99).

Schwarz, N. (1999). Self-reports: How the questions shape the answers. *Am Psychol.* 54, 93–105.

Schwarz N, Sanna LJ, Skurnik I, Yoon C. (2007). Metacognitive experiences and the intricacies of setting people straight: Implications for debiasing and public information campaigns. *Adv Exp Soc Psychol.* 39, 127–161.

Sen A. (1993). Positional objectivity. *Philos Public Aff.* 22, 126–145.

Sen A. (2002). Health: perception versus observation. *Brit Med J.* 324, 860–861.

Sen A. (2001). Objectivity and position: Assessment of health and well-being. In, (Eds.) J Drèze, A Sen. *India: Development and Participation*. Oxford, UK: Oxford University Press. (p. 115–128).

Simpson GG. (1961). *Principles of Animal Taxonomy*. New York, NY: Columbia University Press.

Sirgy MJ, Michalos AC, Ferriss AL, Easterline RA, Patrick D, Pavot W. (2006). The Quality of life research movement: Past, present and future. *Soc Indic Res.* 76, 343–466.

Smolensky P, Legendre C. (2006). *The Harmonic Mind From Neural Computation to Optimality-theoretic Grammar. Volume 1: Cognitive Architecture*. Cambridge, MA: MIT.

Sousa KH, Holzemer WL, Henry SB, Slaughter R. (1999). Dimensions of health-related quality of life in persons living with HIV disease. *J Adv Nurs.* 29, 178–187.

Sousa KH, Kwok O-M. (2006). Putting Wilson and Cleary to the test: Analysis of a HRQOL conceptual model using structural equation modeling. *Qual Life Res.* 15, 725–737.

Spilker B, Revicki DA. (1996). Taxonomy of quality of life. In, *Quality of life and Pharmacoeconomics in Clinical Trials*. 2nd Edition. New York NY: Raven Press. (p. 1–10).

Spranger MA, Schwartz CE. (1999).Integrating response shift into health-related quality of life research: a theoretical model. *Soc Sci Med.* 49, 1507–1515.

Steadman-Pare D, Colantonio A, Ratcliff G, Chase S, Vernich L. (2001). Factors associated with perceived quality of life many years after traumatic brain injury. *J Head Trauma Rehabil.* 16, 330–342.

Stone AA, Shiffman S. (2002). Capturing momentary, self-report data: A proposal for reporting guidelines. *Ann Behav Med.* 24, 236–243.

Stone AA, Turkkan JS, Bachrach CA, Jobe JJ, Kurtznab HS, Cain VS. (2000). *The Science of Self-report: Implications for Research and Practice*. Mahwah NJ: Erlbaum.

Sullivan M, Kempen GIJM, von Sonderer E, Ormel J. (2000). Models of health-related quality of life in a population of community dwelling Dutch elderly. *Qual Life Res.* 9, 801–810.

Thomas DL, Diener E. (1990). Memory accuracy in the recall of emotions. *J Personal Soc Psychol.* 59, 291–297.

Torrance GW. (1976). Health status index models: A unified mathematical view. *Manag Sci.* 22, 990–1001.

Torrance GW, Furlong W, Feeny D, Boyle M. (1995). Multi-attribute preference functions: Health Utilities Index. *PharmEcon.* 7, 503–520.

Ulvik B, Nygård O, Hanestad BR, Wentzel-Larsen T, Wahl AK. (2008). Associations between disease severity, coping and dimensions of health-related quality of life in patients admitted for elective coronary angiography – a cross sectional study. *Health Qual Life Outcomes.* 6, 38. doi:110.1186/1477-7525-6-38.

Vidrine DJ, Amick BC, Gritz ER, Arduino RC. (2005). Assessing a conceptual framework of health-related quality of life in a HIV/AIDS population. *Qual Life Res.* 14, 923–933.

von Neumann J, Morgenstern O. (1944). *Theory of Games and Economic Behavior*. Princeton, NJ: Princeton University Press.

Wilson IB, Cleary PD. (1995). Linking clinical variables with health-related quality of life: A conceptual model of patient outcomes. *JAMA.* 273, 59–65.

Wittgenstein L. (1961). *Tractatus Logico-Philosophicus*. New York, NY: Routledge and Kegan Paul.

Zapf W. (1984). Individuelle Wohlfahrt: Lebensqualität. In, (Eds.) W Glatzer, W Zapf. *Lebensqualität in der Bundesrepublik*. Germany: Campus Vlg. (p. 13–26).

Zeeman EC. (1976). Catastrophe theory. *Sci Am.* 234, 65–83.

The Role of Signs or Symptoms

Abstract
The focus of this chapter is on the complex relationship between signs and symptoms, as an analog and therefore continuation of my discussion of the relationship between objective and subjective indicators. I discuss the figurative language used to express signs and symptoms, and a history of the development of the concept of objectivity. A final major topic covers the role of subjectivity in clinical medicine, and includes a discussion of the difference between clinimetric and psychometric assessments.

Abbreviations

AE	Adverse events
BFI	Brief fatigue inventory (Mendoza et al. 1999)
BPI	Brief pain inventory (Cleeland 1989)
CES-D	Center for epidemiology scale – depression (Radloff 1977)
CTCAE	Common terminology criteria for adverse events
FACT-G	Functional assessment of cancer treatment – general (Cella et al. 1993)
FACT-P	Functional assessment of cancer treatment – prostate (Esper et al. 1997)
HRQOL	Health-related quality-of-life
HUI	Health Utility Index (Torrance et al. 1995)
IES	Impact of Event Scale (Horowitz et al. 1979)
LC	Local control
LCSS	Lung Cancer Symptom Scale (Hollen et al. 1993)
QLQ-30C	Quality-of-life questionnaire for cancer (Aaronson et al. 1991)
Q-TwiST	Quality-adjusted time without symptoms and toxicity
QWB	Quality of Well-being Scale (Kaplan and Anderson 1990)
REL	Relapse
SF-36	Short Form-36 (Ware et al. 1993)
TOI	Summary measure of physical well-being, functional well-being, and disease-specific overall score from disease-specific FACT assessments
TOX	Toxicity

1 Introduction

If my physician measures my heart rate, blood pressure, or blood chemistries, should these signs of physiological function or dysfunction be accounted for when I consider the quality of my existence? Alternatively, if I complain about chest pain, fatigue, shortness of breath, or heart palpations, should these *symptoms* be what I consider when I assess my HRQOL? While most quality-of-life researchers agree that the "unqualified" signs of a person's physical status are not primary indicators of HRQOL,[1] there is no universal agreement on what role symptoms should play in an HRQOL assessment. This is so even though symptoms are often cited by a person as a major contributor to their diminished quality-of-life, and many HRQOL assessments contain symptom items (Ferrans 2007).[2]

Feinstein (1967) defines a sign as "the name given to an entity objectively observed by the clinician during physical examination of the patient," while he defined a symptom as "the name given to a subjective sensation or other observation that a patient reports about his body or its products" (p. 131). Of course, what immediately becomes an issue is what Feinstein, or any clinician, means by "objectively observed" or "subjective sensation," and this brings to mind the same issues that I discussed in Chap. 4, but now in a clinical setting. For example, I previously argued that the terms objective and subjective represented extreme points on a continuum. Is the same true for signs and symptoms? Are they extremes on a continuum? In addition, does the assessment process, per se, "objectify" signs or symptoms, much as appears to happen when "objective or subjective" indicators are operationalized and attempts are made to combine and assess them? Also, is it possible to combine signs and symptoms, and generate a meaningful outcome? When I previously discussed these issues I suggested that combining objective and subjective indicators could co-occur if the objective indicators were "qualified" (i.e., "made of the person"). This would permit combining these indicators, since they would now be scaled along a common subjective dimension. Thus, when Lawton (1991; Chap. 4, p. 104) labels the number of rooms in a person's house (supposedly an objective indicator) as a quality-of-life indicator, then he is implicitly "qualifying" the number of rooms. If Lawton did not do this and yet insisted on combining these indicators, then he'd be creating a significant poststudy interpretive task for himself, since he would lack a common basis for interpreting the combined indicators.

A quality assessment, of course, is a generic activity that can occur in any number of situations, including eating a meal, viewing a painting, receiving health care, or labeling bodily events as a sign or a symptom. Associating any of these indicators with the quality of a person's existence, however, requires an additional step. I refer to this process as the *qualification* process. To be explicit, the *qualification process* occurs when a person reflects and values the importance of some aspects of their interaction with their internal or external environment. Thus, from a respondent's perspective a quality assessment is a retrospective cognitive process that also involves metacognition. When this higher-order cognitive activity is applied to concrete situations, then the outcome can be said to reflect a "top-down" process. When investigators design a qualitative assessment, they either deliberately or incidentally select assessment tasks (e.g., ratings, utility generation, and so on) and items which are meant to match the respondent's cognitive processing. However, as will become obvious, most HRQOL assessments are only partially qualified, being limited to including descriptive items that are evaluated (e.g., providing scales for ratings) but not usually valued (e.g., assessment of the importance of the particular outcome to the person). However, as I will illustrate in Tables 8.3–8.5 HRQOL assessments do exist that contain all three components of the hybrid construct.

In contrast to how a person processes information about a sign or symptom, an investigator is involved in transforming this information into a quantity. I will refer to this as the *objectification process*.[3] Objectification consists of a number of steps, including the coding, labeling and scaling of an indicator; operationally defining it; and applying appropriate statistical procedures or models to the assessment. Each step is meant to create some order about what is being assessed by identifying and collecting common characteristics of the indicators being studied. Thus, the minimum outcome of objectification is nominal scaling. I refer to this process as an "objectification" since the criterion that define an assessment method becomes progressively more operational; a defining characteristic of objective assessment. Thus, the criteria required for interval scaling are more extensive then the criteria required for ordinal scaling, and so on. This approach places "qualitative" and "quantitative" data on a common footing, but acknowledges that these data may differ in the degree to which they can be objectified.[4] Two general approaches to the assessment task have been identified; in one, the model is adjusted to match the data, while in another, data are selected to match a model that produces objective invariant statements (as occurs in Rasch Modeling). Both approaches represent progressive steps toward a purely objective indicator whose invariant properties can be identified (Chap. 4, p. 112).

To illustrate this, consider how an interview would be assessed and compared; these data are often considered subjective in nature. Thus, I could decide to count the number and type of words in the interviews, list the topics discussed, and also examine the content to determine if figurative language was used. Each of these methods – the counts, lists, and defined content areas – are part of an objectification process, since they are methods which can be replicated by others, yet the subject matter being objectified remains subjective. These procedures, while they differ in degree, do not differ in principle from what Rasch (1977) accomplished with data that originated from an educational setting. Thus, the assessment process itself can define objective characteristics of either objective or subjective data, and it is this process that I have labeled as the "objectification process."

Signs and symptoms are clearly an important part of the information exchanged during a health care encounter, and this setting becomes the natural venue to examine the characteristics of this communication. Both signs and symptoms are best viewed as summary statements, reflecting a broad array of information that may be accumulated over time or by different experiences. Thus, a symptom, such as a person's complaint of pain, may refer to what they have just experienced, or a particular sensation they have had over time. These experiences can be modified by the situation the person is in, who is listening to their complaints, and so on, but all labeled as a symptom called "pain." Since signs and symptoms are usually verbally reported, the language used to facilitate this process also becomes quite important.

I will focus on clarifying what the word "symptom" means, how it is expressed, and what cognitive processes are involved in its expression, since its expression is more difficult to objectively confirm. Of special interest will be elucidating the role that metaphors and figurative language play in symptom expression. In addition, I will continue to track the role that cognitive processes play in formulating symptoms, and also what role they play when a clinician establishes a diagnosis.

To summarize, the primary purpose of this chapter is to continue the discussion in Chap. 4 on the relationship between subjectivity and objectivity, now as applied to signs and symptoms. Also of interest is determining how signs or symptoms are "qualified" and then "objectified" as part of the process of being assessed. After reviewing the complex nature of symptoms, I will discuss some of the issues dealing with their assessment. This will be followed by an extended discussion of the history of the development of the concept of objectivity, with a particular emphasis on its relevance to clinical medicine. Next I will discuss a number of different aspects of how adverse events (AE) are monitored or used in clinical medicine. The final major topic will deal with the application of either clinimetric or psychometric approaches to symptom assessment. As I will make clear, assessments based on clinimetric or psychometric methods each represent different cognitive classification methods and involve different assessment models, including the use of overlapping but different statistics. Interwoven with these various topics will be a discussion of the role of HRQOL assessment in clinical medicine and clinical decision-making. I, however, will leave a more extensive discussion of whether symptoms should be a part of a HRQOL assessment for Chap. 7.

2 On the Nature of a Symptom

To start, I want to consider a number of somewhat philosophical questions about the meaning of the term "symptom." My answers to these questions should help clarify why "a symptom" is such a difficult term to define and conceptualize. My first question is whether a symptom would exist if I were not aware of it. This is similar to the more common question of whether a falling tree would make noise if there were no one around to hear it. The answer I give to both questions is no, since in both instances I have to be aware of the presence of the symptom or hear the sound in order for it to exist for me. However, I can become aware of an experience in a number of ways, and only one of which would I directly report the experience. A classic example

involves the appraisal of emotions, where appraisal can occur at an unconscious level (Lazarus 1991). Another example is when a person implicitly learns something (such as a racial or ethnic prejudice), but only becomes aware of it when explicitly challenged (Murphy and Zajonc 1993). Since implicit associations occur immediately, and explicit associations take time (Ranganath and Nosek 2008), it is also possible for a person to be influenced by conditions associated with a symptom but not be able to label it as a symptom. This then is the first (cognitive) characteristic of a symptom, which is that a symptom exists only at a particular level of awareness.

In addition, there is evidence suggesting that many more people have symptoms than they report or act upon ("the Iceberg of Morbidity"; Verbrugge and Ascione 1987). One reason may be because a person is not always able to label their experience as abnormal; they are not familiar with the language ordinarily used to label what they are experiencing. Another reason may be due to difficulties in defining a particular type of symptom, and this may be difficult for both the person and the investigator. This occurs because of limitations in the,

> ...conceptualization and measurement of pain, depression and fatigue (as examples of symptoms); heterogeneity of conditions or phenomena defined as pain, depression and fatigue; lack of consensus on the criteria to define symptoms individually or in combination; and lack of consensus on the 'best' measure(s) in terms of validity and reliability for each of the symptoms separately or in combination. (National Institutes of Health State of the Science Conference Statement 2004; p. 11)

One way a person can resolve this problem is to rely on the clinician to provide a label for an experienced symptom. And very often the person feels a sense of relief upon being given a name to attach to their complaints. This suggests that characterizing a symptom involves a social process, such as the doctor–patient relationship. One additional product is the labeling of a person's symptoms as indicative of a diagnosis or disorder. This may require another degree of social exchange, with the clinician proposing and the person accepting the label now attached to their complaints. The interpretation of the symptoms as indicative of a disease or disorder reveals a second cognitive characteristic of a symptom, which is that concrete experiences (such as symptoms) are used to justify abstract categorizations, such as the diagnostic terms "pain," "cancer" or "depression."

Another interesting question is whether a symptom would exist if a person felt that what they were experiencing was normal. The answer would depend on what was known about the respondent. Thus, if the person insisted that their hallucinations were normal, then their past history of a diagnosis of schizophrenia would have to be taken into account. Yet, as any clinician who has had to work with a person with schizophrenia knows, part of the "therapy" provided for this disorder involves convincing the person of the nonfunctionality and potential dangers of hallucinating. This suggests that the person with schizophrenia is unaware of the abnormality of their behavior, and therefore, would not recognize their bizarre behavior as a symptom of a disorder. On the other hand, an elderly person may be aware of their reduced activity and account for it as the result of "aging," when in fact it may have a more immediate organic cause. In both these instances the social influence of an outside observer is a vital part of establishing what is meant by a symptom, and in the absence of this input the activity of either person would not be labeled as a symptom. This, then, is another characteristic of a symptom, and that is, that attributions attached to a symptomatic experience often rely on a social consensus.

This raises another question: if a person complains about having a symptom (e.g., back pain, headache, fatigue, hot flashes, and so on), does that mean their complaint should be considered a symptom of some underlying pathology? This too is an age-old question, and again, a clinician may have a difficult time determining the true nature of the person's complaint, since there is an extensive literature suggesting that aspects of a person's personality influence symptom reporting. Still, as Pennebaker (1982) indicates, people in general are reasonably capable of monitoring and appropriately reporting their physiological status independent of the exaggeration of a particular personality type (e.g., neuroticism) and their ability to find an appropriate label for their complaints.

One of the reasons why a person may have difficulties identifying and labeling an experience as a particular symptom is because symptoms interact. Thus, in cancer, it is not uncommon that a patient reports the presence of pain, depression, and fatigue. The question becomes, then, how do these symptom clusters interact and how are they best treated (Fleishman 2004; Paice 2004)? Walsh and Rybicki (2006) report an analysis of 25 symptoms from 922 patients with advanced cancer and confirm that the symptoms clustered into specific domains (e.g., fatigue/anorexia-cachexia, neuropsychological, and so on). Badr et al. (2006) continuously monitored cancer patient mood and physical symptoms (e.g., pain and fatigue) and also found an interaction, suggesting that a symptom can radiate and interact with many aspects of a person's existence.

What also complicates the assessment of symptoms is that a person may have more than one symptom, but only report the most salient one or the symptom that they feel is treatable (e.g., pain). This is often true for patients with major illnesses (e.g., cancer). As Jonas et al. (2001) state, this may result from selective attention or be a product of cognitive impairments secondary to a disease or its treatment. Determining the contribution of each factor is clearly a complex task (see Chap. 10).

Palmer and Fisch (2005) report that particular patterns of symptoms are associated with survival in terminally ill cancer patients. Thus, they found that persons who reported dyspnea, drowsiness, problems with appetite, and nausea were less likely to survive than those with pain, depression, and other symptoms. The notion that the emotional state of a person structures their symptom reporting is not surprising, but adds another degree of complexity to treating the person.

The meaning of symptoms also varies. For example, Jones et al. (1981) gave a group of 131 undergraduates 45 symptoms and asked them to rate these symptoms in terms of severity, usualness, threat to life, whether it requires treatment, disruption of normal functioning, painfulness, whether the person is responsible for the symptom, and embarrassment. Using factor analytic procedures they found that the college students clustered symptoms into three factors: the first consisted of symptoms which were threatening, disruptive, or painful; the second included symptoms that were familiar and which a person could take personal responsibility for; while the third major factor involved symptoms that were embarrassing. The implication of this study is that symptoms will vary in meaning, and of course this raises the interesting question of how these different classes of symptoms would act when part of a cross-classification of information.

Thus, the nature of a symptom turns out to be quite complex, involving issues concerned with how aware the person is of changes in their physiological, psychological or emotional status; who labels these changes as a symptom; and whether explicit criteria or a social consensus exist to label particular experiential reports as symptoms. It was also clear that specific cognitive processes (e.g., awareness, abstraction, and so on; see Chap. 5, p. 124) are required for a symptom to be recognized and accepted as indicative of a diagnostic condition. Not too surprisingly, these processes have developed their own unique language, and understanding how language usage contributes to the experience and assessment of symptoms is my next topic.

2.1 Language Usage and Symptoms

As mentioned earlier, Susan Sontag (1991) was clearly concerned that the meaning attached to symptoms and indications of illness could be distorted by using figurative language, particularly metaphors, to express the meaning attached to the symptoms. She states:

> …the most truthful way of regarding illness – and the healthiest way of thinking of being ill – is one most purified of, most resistant to metaphoric thinking. Yet it is hardly possible to take up one's residence in the kingdom of the ill unprejudiced by the lurid metaphors with which it has been landscaped. (p. 3–4).

Metaphors can distort what a person is experiencing when they report symptoms, and also when a patient and clinician attempt to communicate about symptoms. Metaphoric expressions can also be used to suggest that medical policy (e.g., the "war on cancer") can achieve something that literally cannot occur (actual combat) and is not likely to be achieved (e.g., to defeat cancer). In this section, I will discuss the role of metaphors and other forms of figurative expression in patient reports of symptoms; the role of metaphors in doctor–patient communication; and how symptoms can be considered metaphors for ongoing psychological processes.

2.1.1 Metaphors in Patient Symptom Reporting

The term "symptom" comes from the Greek word *symptoma*, meaning "anything which has befallen one." How can that which has befallen someone become part of a person's considerations concerning their HRQOL? First, symptoms may be viewed not only as direct causes of changes, but also as an indicator of a person's HRQOL. Thus, pain, a symptom, can be correlated with a person's degree of anxiety, which is also a symptom. Of course, what is implicit here is a linear modeling with the objective of establishing a cause and its consequences (see Fig. 8.3). Alternatively, a symptom can be viewed as a type of information that a person cognitively processes, processes that may occur in a nonlinear manner and lead to the formation of a cognitive entity. Here both pain and anxiety are salient parts of the person's experience, and these experiences can be conceived of as part of a non-linear self-regulatory system. Thus, what usually befalls a person is the totality of his or her experience, and it is usually the clinician who imposes a linear cause–effect model on that experience, primarily to justify a treatment regimen.

Using metaphors to describe symptoms is very common, and attaches meaning to experiences that may otherwise not be describable. Jairath (1999), for example, reports a study where a group of patients was asked to describe their chest pain during their heart attacks. The patients expressed the physical experience using such metaphors as, "pain like nerves," "pain like gas," "chest pain all over like a headache," "awful pain…like a hot knife, sharp…," and so on. Interestingly, terms of this sort are also used to formally assess pain.

The McGill Pain Questionnaire (Melzack and Katz 1992) is a good example of where linguistic metaphors are regularly used to characterize a symptom. The assessment consists of 78 descriptors grouped into domains (Table 5.1), which also permits scaling within the domains. Thus, a person can describe their pain as "cool, cold or freezing," but since skin or body temperatures are not likely to vary to the extent implied by these terms, it can be assumed that the descriptor are being used metaphorically. The same would be expected for terms which describe pain as "flickering, quivering, pulsing, throbbing, beating, or pounding." Here the person may objectively experience their pulse as in a migraine headache, and the person's pain may even be intermittent but not likely to physically quiver, throb, or pound. Thus, metaphors are an essential element of a person's subjective reports of pain quality.

Table 5.1 Examples of descriptors from the McGill pain questionnaire[a]

Flickering	Jumping	Pricking
Quivering	Flashing	Pressing
Pulsing	Shooting	Gnawing
Throbbing	Cool	Cramping
Pounding	Cold	Crushing
Hot	Freezing	Dull
Burning		Sore
Scalding		Hurting
Searing		Aching
		Heavy

[a]Adapted from Melzack and Katz (1992). Reproduced with permission of the publisher

Metaphors are frequently used to describe "mental health symptoms." For example, Skårderud (2007) interviewed persons with an established history of anorexia and asked them to describe the relationship between their body and mind. He found that the respondents used fairly concrete metaphors to describe this relationship. He labeled these metaphors "body metaphors," and they included such terms as emptiness/fullness, purity, spatiality, heaviness/lightness, solidity, and removal. These concrete metaphors indicate that the respondents had a reduced reflective capacity (i.e., they were less able to think of their state in abstract terms) and reduced capacity to process the symbolic meaning of their body's appearance. Skårderud also recommended a form of therapy for these persons that involved expanding the linguistic basis, or meaning, these metaphors provided. Of course what is implicit in this research is the assumption that language can control or affect thought (Boroditsky 2001).

Of related interest is a study by Woolrich et al. (2008) who found that the metacognitive control strategies used by persons with anorexia were more likely to make them feel worse; that is, it reinforced negative self-evaluations. Combining these observations with the Skårderud (2007) study suggests that the concrete body metaphors, when used as metacognitions, may contribute to negative self-evaluations. Woolrich et al. (2008) also suggests that persons with anorexia who have poor metacognitive coping turn to more concrete behavioral strategies (e.g., not eating) to deal with their negative thoughts. The notion that the concrete metaphoric character of a symptom can govern metacognitive processes reinforces the view that there is a constant interplay between concrete and abstract processes, which, in the case of the anorexic, may actually be one of the factors which sustains this form of psychopathology.

2.1.2 Metaphors in Doctor–Patient Symptom Reporting

Skelton et al. (2002) report a study in which they analyzed the metaphors used when patients and doctors communicated with each other. They found that patients used metaphors to describe the interface between their physical

Table 5.2 Examples of linguistic metaphors used to identify a conceptual metaphor: THE STUDENT DOCTOR/PATIENT RELATIONSHIPS AS WAR (Rees et al. 2007; p. 730)

Linguistic metaphors
1. "I think it's very easy for people the doctors… To see patients as over the other side them and us and them kind of situation." (Medical educator)
2. "You've got to be able to see other side or understand the other side." (Medical educator)
3. "Subconsciously just puts up that barrier." (Medical student)
4. "Knowledge on both sides of the fence." (Medical educator)
5. "She (patient) is terrified she will be struck off his or her (doctor's) list." (Patient)
6. "The (the students) were quite shocked to hear this experience where she (patient) could pull her punches." (Medical educator)
7. "The cutting way in which he (patient) gave immediate feedback (to the student) on the whole idea of seeing a medical student." (Medical educator)
8. "You give somebody a gun, they play around with it, put a blank in it and they're happy. Put them on a range and put a live round in it, its different kettle of fish yet it's the same gun… You have got to teach them (students) to move from textbook (patients) to real life (real patients)." (Patient)

and psychological states. On the other hand, doctors used mechanical metaphors to describe a disease and their role as "problem-solver" or "controllers of disease." The authors concluded that the different pattern of metaphor-usage reflected different communication agendas for each participant. However, these different patterns of metaphor usage could also be viewed as different types of speech acts. Thus, the "mechanical" metaphors used by doctors may be intended to justify some action, such as establishing a diagnosis and selecting a treatment, while the patients are selecting metaphors that are meant to direct the attention of the doctor to issues of more concern to them. Thus, each participant has a different agenda that is reflected not only in the type of metaphor being used, but also in the type of illocutionary (speech) acts implied by the form of expression used.

Rees et al. (2007) report a study in which metaphors were identified in the transcripts of focus groups following the interaction between individual medical students, persons who were trained to act as patients, and their medical supervisors (i.e., medical educators). The data suggest that all three groups used six different conceptual metaphors to describe their encounter. Thus, the relationship between each group was described as either a war, a hierarchy, doctor-centeredness, a market, or the theater. Table 5.2 illustrates some of the statements made by each group that use various linguistic metaphors, each of which fit under the conceptual metaphor of THE STUDENT DOCTOR/PATIENT RELATIONSHIPS AS WAR. A similar table was provided in the Rees et al. paper for each of the other five conceptual metaphors.

Examining the role of metacognition in the doctor–patient communication process is also of interest. It would be expected, as suggested for the anorexic, that the "top-down" application of conceptual metaphors would determine the nature of this interaction. Thus, conceptualized as war, market or theater, each has implications concerning how the doctor–patient relationship is implemented.

2.1.3 Symptoms as Metaphors

The assumption that a physical symptom may be a metaphoric expression of an underlying psychological conflict dates to the advent of medicine. This assumption has had an upsurge in interest following the development of psychoanalysis and its application to psychosomatic medicine (Alexander 1950). The prototypical psychoanalytic hypothesis whereby psychological conflict and anxiety is transformed into a physical symptom is referred to as "the conversion process." What is metaphoric here is the term conversion, with its implication of a transformation from one state to another (i.e., the psychological into the physical). Implicit in this process is the mind–body separation, although the current interest in embodiment offers an interesting alternative conceptual model of how the conversion process may proceed.

Apparently any symptom may be the focus of a conversion, but pain is the most common. Thus, patients may unconsciously select a particular symptom because it is a metaphor for their psychosocial condition. For example, someone may complain of chest pain after being rejected by a lover; a women may complain of being bloated all the time when she desperately wants to have a baby; someone complains of back pain when they are dealing with an abusive relationship; and so on. There is an extensive literature expounding on the significance of a symptom being a metaphor, especially among psychoanalysts, and even suggestions that examining the metaphoric character of a symptom can be a useful aid in psychotherapy (Koop 1995). Skårderud's (2007) recommendation (see above) of expanding the meaning of concrete metaphors for anorexics is an example of this type of therapeutic strategy.

Various "functional somatic syndromes" (e.g., Barsky and Borus 1999) are often cited as including symptoms that have metaphoric properties. Examples of these syndromes include multiple chemical sensitivity, sick building syndrome, repetitive stress injury, chronic fatigue syndrome, irritable bowel syndrome (IBS), and fibromyalgia. Consider the IBS patient; they present with a primary complaint of abdominal or colonic pain, distention, and usually disturbed stooling patterns (e.g., diarrhea or constipation). Of course pain is a subjective indicator, while bloating can be measured with a tape measure, and stooling pattern can also be objectively assessed (i.e., a metric can be designed for this task and several individuals can use this metric to produce a

Table 5.3 Type of assessment and type of indicator for persons with irritable bowel syndrome (IBS)

	Qualitative Assessments Based on subjective evaluations	Nonqualitative assessment[a] Based on objective evaluations
Objective indicators	Time spent in social activities	Stomach distention measured by the difference between "normal" and bloated girth
	Changes in eating pattern	Stooling pattern
Subjective indicators	Self-reports that include "qualification"[b]	Self-report of pain correlated with colonic distention
		Stooling accidents (When is staining an accident?)

[a]By "qualification" I mean the implementation of a metacognitive process; that is, *a second-order process that requires reflection and is in essence a valuation of the significance of a prior evaluation*. QWB, HUI, and EQ-5D are examples of assessments that can be used to qualify an evaluation. These assessments provide standardized (population-based) person-based preference systems that provide the valuations, although SEIQoL or PGI provide valuation systems that rely on the direct assessment of an individual's preferences

[b]A nonqualitative assessment can be "qualified," but qualification is not an inherent part of the assessment. Thus, soiling of undergarments, *per se*, is not an indicator of a qualitative state – unless the respondent indicates the importance of being in that state. Obtaining this information is required for the state to be qualified. The same would be required for a report of colonic pain or the time a person spends socializing

similar quantitative outcome). In addition, studies of colonic distention reveal that persons with IBS report pain at a lower level of air inflation of the colon than persons who report normal bowel functions (Ritchie 1973; Whitehead et al. 1990), suggesting an enhanced mucosal sensitivity and possible objective physical basis to the reported symptoms. Yet, these same persons will not ordinarily experience this irritability when asleep, suggesting that the source of the symptoms may not be irritated colon per se, but rather something more central. The association of dysmotility with pain reports has suggested that IBS is a functional disorder, meaning that normal functioning of the gastrointestinal tract is somehow disrupted, although evidence of a pathophysiological basis for this dysfunction may be lacking (Schuster et al. 2002). Thus, it can be asked, what is the source of irritation for IBS, and where is it? The same type of question applies about the fatigue in chronic fatigue syndrome, the pain in fibromyalgia, and so on.

Table 5.3, which is similar to Table 4.1, reveals the wide range of indicators normally assessed when the diagnosis of IBS is being established. Note that there are at least four different classes of indicators that can be studied, each of which differs in terms of whether they are based on an objective or subjective evaluation processes and generate objective or subjective outcomes. However, only one would be considered an HRQOL assessment, as defined in this book (i.e., self-reports that include qualifications). Again, each of these types of indicators represent steps toward the objectification that is characteristic of assessment.

If the symptoms of IBS are meant to be metaphors, then what are these metaphors and what do they represent? Zimmerman (2003) has suggested that the symptom of altered motility that is characteristic of IBS can be conceptualized by the metaphor of a river, with the constant flow of the river being its normal pattern and the excessive and slowed motility each being a product of an obstruction in flow reflecting different psychological conflicts. A classic example of a psychoanalytical perspective on a gastrointestinal disorder is the notion of Alexander et al. (1968) that peptic ulcers result from repressed passive wishes that keeps the stomach ready for feeding resulting in excessive stomach secretions that results in an ulcer. Thus, the ulcer would be a metaphor for passivity–activity dynamics that characterizes the psychological conflict ongoing in the patient.

Psychosomatic disorders are an ideal setting in which to consider the issues of subjectivity and objectivity in a clinical setting. To illustrate this, in the next section I will discuss a specific group of patients who report physical symptoms that have no discernable demonstrable physical bases. What will also be of interest is determining how these various indicators become part of a quality assessment. Up until this point, I have indicated that any statement of quality requires that the available data be transformed into a "qualified" (subjective) outcome. First, I need to explore the cognitive and emotional basis of symptom formation, and I will follow with a discussion of how signs and symptoms have historically been objectified.

2.2 Assessing Symptoms

The previous discussion should have made it clear that figurative language is an important part of the description and expression of symptoms. In this section, I am going to first examine the linguistic and cognitive foundation of a symptom assessment. This is because I want to demonstrate that currently available symptom assessments generally characterize symptoms in subjective terms. This, it turns out, is a necessary step in the process of demonstrating that a symptom can be a quality indicator, but it also makes clear that it is the subjective character of a symptom that is then objectified. I will leave the discussion of how symptoms are collected together to form a "symptom domain" for Chaps. 6 and 7.

A second topic I will consider describes some of the cognitive and emotional processes active when a symptom is reported. These data will make it clear that a symptom report is not a simple stimulus response reflex (e.g., a pain, an episode of breathlessness, dizziness, and so on), but that a variety of cognitive and emotional processes intervene to contribute to the symptom report. Of particular interest will be the role that automatic or effortful cognitive processing (see Chap. 6), including biases and heuristics, play in a symptom report. I will also describe some of the neurocognitive determinants of the cognitive mediators of a symptom report. These data offer an "explanation" of why a symptom assessment evokes subjectivity; namely, that a variety of cognitive and emotional factors, including personality and culture, are involved when a person responds to the items on a questionnaire.

2.2.1 Symptom Assessment and the Qualification Process

Symptoms, even when naively observed, appear cognitively complex. Thus, symptoms may be represented in terms of discernable features that are used in their classification, as a collection of entities that make up a category or domain, or as an abstraction that reflects its symbolic base. Detecting features, collecting entities, or applying abstractions to concrete situations are each cognitive processes that quite generally define the cognitive structures upon which categories and concepts are formed, one example of which can be labeled a symptom.

When the available literature is inspected, however, it quickly becomes clear that to characterize a symptom only in terms of its parametric features (e.g., its frequency, intensity or duration) is often empirically and linguistically deficient (Cleeland 2007). Instead, alternative words or phrases are proposed by investigators that imply processes supposedly better able to capture the symptomatic experience of an individual. These shifts in language and indicators are invariably more introspective which is consistent with the presence of a qualification process. Thus, an investigator may claim that he or she is assessing the experience of severity rather than the more objective indicator intensity, but in so doing is also shifting the semantics by attaching a more negative (emotional) valance to the descriptor, at least as compared to the comparatively neutral valance of the term "intensity." In addition, distress (an even more emotive term) may be considered more informative then severity, while a cluster of symptoms may be described as reflecting the person's symptomatic burden. Interference of a person's functions and discomfort may also be useful descriptors. Substituting terms of this sort would be acceptable if the resultant indicators were found to be more informative or shown to be better predictors then parametric indicators, and the available evidence, in general, supports this inference (see below). In so doing, however, it would also support the proposition that the patient *subjectifies* or the investigator *qualifies* an indicator as a condition of its assessment, explicating its emotional and functional dimensions, and thereby defining and clarifying the qualitative nature of a symptom.

The design of cancer symptom assessments illustrates how investigators have approached the problem of objectively (i.e., quantitatively) assessing the subjective consequences of symptoms. For example, the *McCorkle and Young Symptom Distress Scale* (McCorkle and Young 1978) assesses parametric aspects of a symptom and distress. This was done by first constructing an assessment where the symptoms (e.g., nausea, vomiting, pain, and so on) were anchored in terms of frequency, intensity, and duration. Then respondents were asked to rate these item-anchors in terms of their degree of distress. However, it was the respondent's perceived distress (emotional response) associated with frequent vomiting, poor sleep, and so on, that became the primary indicator of the symptom and outcome of the assessment.

Haig et al. (1989) report a similar type of experimental design but this time dealing with "discomfort," another emotion-laden term, as an outcome indicator. They operationalized what they meant by discomfort in terms of its quality, intensity, and duration. The quality of discomfort is monitored in terms of bodily sensations (e.g., fatigue, malaise, and so on), physical sensations (e.g., pain, nausea, and so on), or psychological distress (e.g., worry, fear). The intensity of discomfort was measured in terms of the severity of impact of the symptom on the person's ability to function, while the duration of discomfort was determined by dividing a particular day into three parts and estimating the presence of the symptom during these periods. By including a quality indicator in their discomfort assessment, these investigators were acknowledging that subjective assessments are a necessary part of any characterization of a symptom. But what is not clear is whether this makes any difference in terms of the assessment of a symptom. Data from a number of alternative studies suggest that it may.

Sloan et al. (2001), for example, report a study that summarizes a clinical trial cooperative group's effort to develop a method for measuring hot flashes. What they found to be reliable and "valid" sources of information included daily diaries of hot flashes to estimate not only frequency of occurrence, but also a composite measure of frequency and severity that provided a hot flash score. By including a measure of severity, the authors, of course, were also suggesting that the emotional (negative valance) consequence of the hot flash experience was an important part of its assessment. Supporting this interpretation, Portenoy et al. (1994) found that measures of emotional distress provided more information about symptoms than the frequency or severity of the symptoms alone, but also that combining some measure of distress with the frequency or severity of symptoms was more informative than distress alone. Tishelman et al. (2005) also found that distress was more informative than measures of intensity or frequency. Tishelman et al. (2007) have extended these observations, asking patients to rank the distress associated with specific symptoms (e.g., breathing, pain, and fatigue). The authors comment that "the dimensions of symptom prevalence, intensity and distress are not equivalent, but provide complementary information of clinical importance" (Tishelman et al. 2007; p. 5384). Samarel et al. (1996), on the other hand, found that frequency, intensity or distress were equally able to describe the symptom experience of breast cancer patients. Chang et al. (2003) used item response theory to compare two 5-point rating scales (frequency or intensity) as applied to a 13-item fatigue assessment (FACT-F; Cella 1997). They found that the two scaling methods were significantly correlated ($r=0.86$; $P<0.001$). They recommend using the frequency scores, however, since the resulting scale is better at covering the fatigue continuum. They did not include a more subjective rating scale, such as degree of distress or burden.

An example of a symptom assessment that includes both parametric and qualified components is the *Memorial Symptom Assessment Scale* (MSAS; Portenoy et al. 1994). Respondents are asked to first indicate if they have experienced 1 of 32 symptoms (a type of nominal scaling), then how often (frequency) they experienced the symptom, followed by the severity (some indication of intensity) of the symptom, ending with the more reflective question of whether the symptom was associated with distress. A global index of distress is the primary outcome of the MASA. A somewhat similar assessment is the *MD Anderson Symptom Inventory* (Cleeland et al. 2000). Respondents here are asked to rate the severity of 13 symptoms, but are also asked the more reflective question of whether symptoms interfere in the person's general activity, mood, work, relations with others, walking, and enjoyment of life. Both of these symptom assessments reveal an investigator's efforts at making a subjective indicator (e.g., distress or interference) the primary outcome.

Kirkiva et al. (2006) have summarized the characteristics of 21 (rather broadly defined) assessments used to define cancer symptoms. Inspection of these assessments (Tables 2 and 3; Kirkiva et al. 2006) revealed that 81% of the assessments included either or both *severity* or *distress* as one of their outcome indicators, but that parametric indicators (such as intensity, frequency, or duration) were found in only one-third of the assessments. Clearly, investigators recognize that a symptom assessment is most useful when it has a clear subjective or emotive reference.

When they design a symptom assessment, investigators ordinarily use various methods (focus groups, cognitive interviews, factor analyses, and so on) to identify words and concepts that match a person's symptomatic experiences. This information is then used to design items that presumably ensure the face validity of the (symptom) assessment. Thus, while the terms "intensity" or "severity" may appear to overlap in meaning, their semantic and emotive space is sufficiently different that using one term over the other when assessing symptoms appears to be supported empirically and linguistically. When this practice is expanded to include other terms such as distress, discomfort, and burden, then this suggests that a general principle is operating here, that a symptom report should include an explicit emotional component. This, of course, is consistent with the definition of quality that includes both a cognitive and emotional component (e.g., Figs. 1.2 and 1.3), but most symptom assessments will still lack the reflective metacognitive component necessary to create a complete qualitative assessment.

2.2.2 The Accuracy of Symptom Reports

Having discussed some of the complex issues dealing with the nature of a symptom (Chap. 5, p. 123), I now want to consider some of the determinants of the accuracy of symptom reports. Symptom reports, of course, are an important guide to a diagnosis and treatment. Thus, the *precision* of this self-report is particularly important because over-reporting of symptoms can lead to overtreatment and under-reporting to under treatment. Symptom reporting, however, is also something a perfectly healthy person can do, and in some ways is a more natural place to study the determinants of symptom accuracy, since symptom reports are not confounded by the different disease states a person may have. Pennebaker's (1982) famous book on *The Psychology of Physical Symptoms* (1982) provides an introduction to research on just these issues and I will review parts of it here as an introduction to the topic.

First, Pennebaker (1982) defines a physical symptom or sensation as "a perception, feeling or even belief about the state of our body. The sensation is often – and not always – based on physiological activity" (p. 1). Thus, a symptom report says something about a person's internal state. Pennebaker (1982) also claims that the sensations forming the basis of symptoms can both cause and affect behavior, especially since a symptom can both initiate and signal that an activity has occurred, or describe a state. In addition, he claims that symptom reporting may be subject to a large number of perceptual biases and distortions. These biases and distortions are likely to compromise the accuracy of symptom reports and may involve the process whereby a symptom is *encoded*, issues dealing with the *awareness* of the symptom, and situational factors that determine the *reporting* of a symptom. Examples of the internal states (i.e., symptoms) that were studied include heart rate, nasal congestion, warmness of the hand, pain, and so on.

Pennebaker (1982; p. 153–156) draws the following conclusions from his and his colleagues' research. First, he contends that physical symptoms as percepts are subject to the same perceptual, cognitive, and emotional processes as is true for the perception of external objects or events (e.g., a visual stimulus). There is a continual competition between external and internal stimuli, and since people have a limited capacity to process information, the probability of noting an internal cue is inversely related to the amount of external stimulation. A corollary of this conclusion is that with minimal or no external stimulation, a person will be increasingly aware of their internal state. A second conclusion was that people interpret their internal state by reference to cognitive schemes or beliefs that they have in place. These schemes evolve from developmental (e.g., familial) experiences, personality characteristics, and a person's cultural background.

Pennebaker also found that people are not very accurate about perceiving their internal physiological state, although there is evidence that this can improve with practice (e.g., Ádám 1980) or when internal stimulation is intense. A person's ability to perceive internal sensory information may not be related to the absolute intensity of stimulation, but rather the *relative* intensity of that stimulation. Not only do external stimuli dampen, if not inhibit, the perception of internal states, they may also be used to define the internal state. Thus, if a person finds herself in a threatening environment, she may be more likely to report pain in her stomach or a headache.

The personality and developmental history of the symptom reporter may lead to a particular response style or pattern of symptom reporting. Thus, Pennebaker (1982) reports that "the high symptom reporter tends to be female, from a conflict-ridden home, anxious, socially insecure, oriented toward and/or seeking to please members of the opposite sex, with an unhealthy lifestyle in relation to that of the low symptom reporter" (p. 155).

Pennebaker (1982) also claims that, as far is known, persons from all cultures perceive and report physical symptoms. This suggests that the reporting of symptoms is highly adaptive. He speculates that the organism that can take advantage of the information revealed by a symptom report is more likely to survive. This occurs because symptom reporting can induce self-regulation, but it is also possible for the person to over- or under-report symptoms, confounding survival potential.

An important observation by Pennebaker and others (e.g., Leventhal et al. 1996; Watson 1988; Watson and Pennebaker 1989) is that individuals differ in their affective style when reporting physical symptoms. Thus, persons with high negative affect (NA) experience subjective distress, nervousness, angry, fear, and so on, also report consistent and moderate correlations with their health complaints. In contrast, some people report high levels of positive affect (PA), suggesting that they receive pleasure from engaging with the environment, but show no significant correlation with their health complaints. These associations are found for both state and trait personality indicators, and remain present independent of the time frame (past week, past year, or no specific time; Watson 1988). From this has evolved the *symptom perception model* (Watson 1988; Watson and Pennebaker 1989). The symptom perception model claims that people differ in how they perceive, respond to, and complain about their physical symptoms. As Watson (1988) states:

> In its strongest form, the symptom perception model posits that the association between NA and health complaints is completely spurious and simply reflects the fact that high NA subjects are more likely to attend to or complain about internal physical sensations. A weaker form of this model posits that high NA subjects exaggerate the magnitude of their health problems (p. 1021).

Evidence in support of the symptom perception model comes from studies relating negative affect and self-rated health (McCrae et al. 1976; Tessler and Mechanic 1978). These studies reveal that negative affect is significantly (inversely) correlated to self-rated health, while self-rated health is significantly related to physician ratings of health status. Negative affect status, however, is not correlated with physical rating of health status. This supports the notion that patient reports of physical symptoms are psychologically important, reflecting the influence of negative affect without necessarily reflecting basic biological dysfunction.

Verbrugge (1989) states that "Women are bothered much more by daily physical symptoms than are men" (p. 287). This quotation, which was quoted by Gijsbers van Wijk and Kolk (1997), summarizes the substantial literature supporting Verbrugge's contention. The Gijsbers van Wijk and Kolk (1997) review supports the finding that the greater proportion of women reporting physical symptoms cannot be attributed to greater physical morbidity, but rather by greater negative affect facilitating somatic attention and awareness. A mixed gender study by Gendolla et al. (2005) supports the linkage between these variables, especially during momentary symptom experiences. Wearden (2005) adds another dimension to these associations by demonstrating that a person with an avoidance attachment history is more likely to report physical symptoms. They find that fearful and preoccupied attachment styles are associated with symptom reporting by a negative model of self and others.

These studies with their focus on somatic attention and awareness, fearful preoccupation of attachment and so on, all support the hypothesis that worry mediates the relationship between negative affectivity and physical symptom reporting (e.g., Esterling and Leventhal 1989). A study by Mora et al. (2007) demonstrated that negative affect-motivated persons, via worry about asthma, are more aware of asthma-specific symptoms and use this information to correctly report their symptoms. This suggests that negative affectivity and symptom-specific worry can provide a survival advantage.

Table 5.4 Characteristics of objectification procedures (based on Daston and Galison (2007; p. 371))

Objectification procedures	Truth-to-nature	Mechanical objectivity	Trained judgment
Observer	Experienced observer: idealized type	Noninterfering observer: mechanical devices	Trained observer: expertise
Image	Reasoned: four-eyed sight	Mechanical: blind-sighted	Interpreted: physiognomic sight
Procedure (cognitive processes)	Stimulus selection and synthesis	Suppression of self: ▼automated data collection	Pattern recognition
Object of assessment	Identification of universals	Identification of particulars	Identification of patterns of stimuli

▼: *Metacognitive* processes would be involved here; a person has to deliberately adopt an attitude which minimizes the presence of the self in observations.

The *symptom perception model* and its supporting data are of great interest, since it makes clear that the accuracy of physical symptom reporting is a complex product of perceiving both the external and internal physical environment, as modified by the lens of a person's affective style and developmental (e.g., attachment) history, all leading to the accurate (or inaccurate) detection of a physical symptom. This model can be drawn in terms of the *Brunswik's Lens Model* (Hammond et al. 1975) as illustrated in Figure 3.5. It also provides an explanation of how the qualification process occurs, but now from the perspective of the individual, not the investigator. Critical elements in the qualification process, therefore, would include the competition between internal and external stimulus awareness (i.e., cognitive processes), and the role affect (emotions, such as worry) has in sensitizing a person to detecting the presence, real or imagined, of physical stimuli. Thus, the accuracy of physical symptom reporting has some of the elements of a qualitative assessment (Figs. 1.2 and 1.3).

A physical symptom is or can be an objective indicator. Thus, heart rate can be measured by a clinician, as well as be estimated by a person. The same could be said for a large number of other examples of physical symptoms. Yet, as demonstrated by the discussion in this section, accurate reporting of a physical symptom is quite complex, readily influenced by characteristics of the reporter, the reporting situation and even the audience. This raises again the question of what it means to make an objective statement. In the next section I will review the *intellectual history* of the concept of objectivity to provide some insight into what the term has evolved to mean, and then in the following section apply it to understanding the clinical encounter. While at first glance this may seem like a diversion, in fact, this review will provide the background necessary for an understanding of the different roles that a patient and a clinician assume when they have to agree on the presence and nature of a symptom. What will become clear is that the clinician functions as a *naturalist*, *scientist* or *expert* at different points in the clinical encounter, and it is the co-mingling of these roles that complicates the process whereby a physical symptom is defined and assessed.

3 The Objective Assessment of Signs and Symptoms

In the previous two sections, I reviewed some of the complexities created by a person reporting symptoms. This includes how the patient or the investigator modifies the indicator being assessed, and what determines the accuracy of symptom reports. In this section, I will address the different roles that an investigator can assume during the *objectification process*. In order to help clarify these roles, I will rely on Daston and Galison's (2007) review of the history of different approaches to objectivity. This will provide a necessary background to the subsequent sections, where I will directly consider the nature of the clinical encounter and how these different objectification processes contribute to sign detection or symptom reporting.

The three historically related objectification procedures that Daston and Galison (2007) identify (Table 5.4) remain as currently active forms of representation. Each is considered a means of gaining knowledge, but also as an effort to state what is true. Each of these procedures has been challenged, so no one can claim to be more relevant than any other; rather each assessment process has certain assets and limitations, with the investigator matching the different objectivity procedures to the task and purpose of the assessment. The first set of procedures, which Daston and Galison label as the "truth-to-nature" stage, involves a highly experienced observer (e.g., a naturalist or "sage") who, after making multiple observations (e.g., of a particular bird), uses their knowledge to generate an *idealized* version of their observations (e.g., a particular species of bird). The same may be said to occur when a person, after a great deal of self-observation, declares the presence of a physical symptom.

Daston and Galison (2007) visualize this method by referring to it as form of *four-eyed sight*, which refers to the fact that the naturalist and the illustrator both have to come to see the depicted object in the same way. This consensus is also similar to what happens when two pathologists use a two-person microscope (four eyes) to come to some agreement concerning what they are observing, or when a person and a physician concur concerning the person's physical complaints. By labeling some idealized outcome, the naturalist or clinician is also engaging in a type of nominal scaling; an essential element in any assessment process.

A second set of procedures, particularly relevant to the detection of signs, involves separating the person from the observation by using a mechanical device. As a consequence all observations, including accidents and unpredictable outcomes have to be considered not just those that match some ideal outcome, as in the truth-to-nature procedures. If a mechanical device is not available, then people can learn to separate themselves from what it is they are observing. Daston and Galison (2007) describe this type of seeing as "blind-sighted," and it is the essence of what is called "the scientific method". A variant of this form of objectivity, structural objectivity, shifts emphasis to demonstrating that a relationship remains invariant after multiple transformations. The third set of objectification procedures involves "trained judgment," the development of expertise. Daston and Galison use the metaphor of "physiognomic sight" to characterize the cognitive process involved in this type of assessment. What is "seen" in this setting is akin to seeing a (facial) pattern, as opposed to a creation of a type, which is the intent of the first model. Table 5.4 is a modified version of a table presented in Daston and Galison (2007; p. 371), and is meant to summarize these different objectification procedures. And as I will discuss, clinicians also assume each of these roles.

While each of these historically different procedures can be applied to the study of a sign, a symptom, or of quality for that matter, mechanical objectivity, and the related concept of structural objectivity, are considered the preferred methods for accumulating new knowledge. It is the essence of what scientists are trained to do, this separating the subjective from the objective, with all of the complications this creates in terms of the ethical consequences of their *blind-sightedness*. The notion that mechanical objectivity requires that a person adopt a particular mental set that is then applied to some set of observations suggests the presence of a *metacognitive* or a top-down process. This strategy is most successful when a mechanical device can be inserted into the process of accumulating knowledge. Structural objectivity differs in its emphasis on the identification of abstract relationships that remain invariant after various transformations. However, as Feinstein acknowledges, for the clinician assessing signs and symptoms, this effort at separation is never totally successful.

As stated earlier, all three objectification methods are attempting to represent nature. However, Daston and Galison (2007) end their book by suggesting that another mode of depiction may be possible, and this mode involves the fusion of the natural with the artificial. This combination inevitably leads to new images and new ways of seeing, and thus constitutes a "presentation of what is," rather than a "representation of what was." They illustrate their point by referring to the fact that creating objects at the nano-technological level (atomic level) inevitably creates a situation where an artifact (some human-developed tool) interacts with nature (at the atomic level), leading to the emergence of new complex and sometimes unexpected outcomes. These outcomes will rely on aesthetics for their selection and expression. The person who would do this selection would be an engineer-scientist who would create virtual images that would be made and could be seen at the same time.

Daston and Galison's (2007) suggestion is based on a rapidly expanding technology involving physical principles, but what if their ideas were transferred to a biological or social context? What would be the consequence? Would a situation be created where someone would be asked to select a particular biological or behavioral outcome (hair or coloration, behavior pattern, and so on) over others? Wouldn't this create a host of ethical issues? My concerns are not that far-fetched or far away, since research is progressing to where certain biological organisms are being engineered to perform specific functions (Weiss 2007). In addition, the concern about human involvement in climate change may be viewed as another example of human beings "engineered" what is meant by natural.

What is of interest here is the implication that these activities have on what is meant by *quality*. For example, in the past the quality of an outcome was judged in terms of its aesthetics or function, but what may now be happening is that aesthetic and function criteria may be applied to "construct" an outcome, with the consequence that man is replacing nature as the determinant of quality. This trend will become evident as I review the three examples that Daston and Galison (2007) discuss. Thus, it should not be surprising that the last stage in the development of objectification procedures involved the development of the expert. The physician, the architect, the composer, the sculptor, and so on have always existed, but their role has now evolved to the point where they have a significant input in determining the nature of our health, the places where we live, what we listen too, and what we see. Thus, the next stage in the evolution of what is meant by quality (Chap. 1, p. 10) may require an institutional consensus rather than allowing it to be a product of natural forces. And in that sense, such characterizations as *truth-to-nature* may be replaced by *truth-to-man*.

3.1 *Truth-to-Nature*: Qualitative Assessment

Naturalists have been quite successful in developing procedures that permit the systematic observation of natural phenomena. The methods they use rely upon repeated observations to select and synthesize characteristics of natural objects (e.g., plants or animals). Key to this activity is that the naturalist "knows they self," and as a consequence, assumes a particular cognitive or mental set when making multiple observations. Daston and Galison (2007) describe the cognitive task as follows:

> To see like a naturalist required more than just sharp senses; a capacious memory, the ability to analyze and synthesize impressions, as well as the patience and talent to extract the typical from the storehouse of natural particulars, were all key qualifications. The ideal Enlightenment naturalist…was endowed with an 'expansive mind, master of itself, which never receives a perception without comparing it with a perception; who seeks out what diverse objects have in common and what distinguishes them from one another'… (p. 58).

They go on to quote Johann Wolfgang von Goethe who reflected about his research in morphology and optics and described his task as a,

> …quest for the 'pure phenomena', which could be discerned only in a sequence of observations, never in an isolated instance. 'To depict it, the human mind must fix the empirically variable, exclude the accidental, eliminate the impure, unravel the tangled, discover the unknown.' These were the concrete practices of abstract reason as understood by the Enlightenment naturalist; selecting, comparing, judging, generalizing. (p. 58–59).

One of the consequences of these activities is that the naturalist constructs a cognitive prototype that reflects his or her observations. The prototype, as a summary measure, is meant to represent a wide range of instances, but not be subject to the characteristics of any one instance.

A second part of the naturalist activities involves transforming the cognitive prototype into a format that can be communicated to others. This required that the observer be able to communicate what they considered to be prototypical to an illustrator or person skilled in the communication arts. This was not always accomplished easily; rather there was often a struggle between the naturalist and illustrator to avoid exceptions and keep to the general. This could only be accomplished by the naturalist being enmeshed in the preparation of the representation.

Daston and Galison (2007) claim that science as pursued under the rubric of truth-to-nature was different from a science based on (mechanical) objectivity. They illustrate this by describing Linnaeus' effort at creating a taxonomy of leaf shapes. Linnaeus was interested in the "essential" leaf shape, not particular leaf forms. Thus, Daston and Galison (2007) state:

> A Linnaean botanical description singled out those features common to the entire species (the *descriptio*) as well as those that differentiated this species from all others in the genus (the *differentia*) but at all costs avoided features peculiar to this or that individual member of the species. The Linnaean illustration aspired to generality – a generality that transcended the species or even the genus to reflect a never seen but nevertheless real plant archetype: the reasoned image. …The type was truer to nature – and therefore more real – than the actual specimen (p. 60).

Thus, Linnaean botany was *a science about the rules of nature*, not its exceptions, and in this sense he was seeking to express universal biological principles. To the extent these universal principles were applicable in all situations, he was able to achieve a type of "descriptive invariance." Most interesting is that his descriptive invariance has the same purpose as the invariance demonstrated by mathematical procedures, and used by Rasch (1977) and others. This principle was that the biological form he proposed would retain its character independent of its applications or transformations. This type of invariance differed from mathematically derived invariance, in that descriptive invariance provided by the individual naturalist was less likely to be reproduced. Still, there was no doubt that the naturalists were engaged in a type of scientific enterprise, such that by excluding the exception and by relying on themselves as the primary measuring instrument they limited their analytical opportunities.

The *truth-to-nature* perspective, however, is still very much alive today. An example of a modern-day naturalist is the author Oliver Sacks, who has written a number of books that look for universal neurological principles amongst the intricacies of clinical material. He can truly be called "a naturalist in the field of symptoms." The same is also true of qualitative psychological assessments, where the objective is also to find descriptive invariance. Here, the investigators deal with the universe of verbal material and attempt to find order in it. I will briefly review some of Sacks' writings, and also demonstrate how the truth-to-nature perspective is achieved by examining the procedures used in *grounded theory* (Glaser and Strauss 1967). What will become evident is that there is a striking resemblance between the naturalism of Linnaeus, Sacks' clinical interpretation of symptoms, and those who call themselves grounded field theory researchers. What also seems to be true is that art or music critics also practice the science of extracting universal principles from repeated observations, relying on themselves to provide these principles. This suggests that the truth-to-nature approach is an important part of the assessment of quality, as applied in a number of different venues.

Sacks (1985), in his famous book, *The Man Who Mistook His Wife for a Hat and Other Clinical Tales*, illustrates with a variety of clinical vignettes the art of the neurological naturalist. The subject of the title of the book suffered from *prosopagnosia*, or facial agnosia. Thus, he failed to recognize the faces of pictures of his family, and at the end of a visit to the clinician, he thought his wife's head was his hat and tried to lift her head to put it on (he was dissuaded from continuing). What is of interest is how Sacks established this diagnosis; he created a series of provocative tests that the patient responded to, and in this way generated signs or symptoms which he judged to reflect a particular neurological state. The patient retained a reasonable level of functioning by singing to himself and telling himself what he had to do for daily living (e.g., bathe, dress, and so on). The patient was also a gifted painter, but when Sacks inspected his painting, he noted a progressive change over time. His early paintings were natural and realistic but his more recent paintings were in Sacks' opinion, abstract and chaotic. Of interest is that the patient's wife viewed this progression as artistic development, while Sacks saw the progression as the advancement of what was now his profound visual agnosia. Sacks concluded that the patient had lost his ability to perceive the concrete, and lived in an abstract world; an abstract world that would manifest itself by the inability to recall specific memories necessary for functioning in the world at large.[5]

Sacks was a naturalist in that he used repeated (provocative) observations to establish a "reasoned image" of the patient's clinical state. He differed from Linnaeus in that he based his generalizations on multiple observations of a single subject, as opposed to making limited observations of many objects. Both men were aiming to achieve descriptive invariance; the ability to make a statement that would remain true even if applied across varying conditions, or multiple instances. Each also recognized himself as the measuring instrument, and had to be quite aware of how his own presence could add or subtract from what that they were observing.

Grounded theory is another example of an assessment method where the investigator is the measuring instrument, and where particular attention is placed on the role that the investigator plays in the process of collecting data. It is, like other naturalistic observational methods, basically inductive in nature, where the investigator follows multiple observations and draws an inference that is meant to reflect a higher level of abstraction and generate a theoretical statement. A grounded theory is different from an ordinary theory, in that it is literally and figuratively grounded in data, and is not a product of the imagination of an investigator. Heath and Cowley (2004) state that:

> …grounded theory's aim is to explore basic social processes and to understand the multiplicity of interactions that produces variation in that process. …Fundamental to grounded theory is the belief that knowledge may be increased by generating new theories rather than analyzing data within existing ones. (p. 142).

Implementing this philosophy, however, has generated controversy, mostly about how an investigator should proceed relative to the amount of knowledge acquired before initiating their observations, how they should observe and how to present their outcomes. Interestingly, the two founders of grounded theory have contributed to this controversy by positing different approaches (see Table 1 from Skeat and Perry 2008; p. 5).

Each founder asked an investigator to demonstrate that they have grounded their theory with data. Glaser (1992), for example, recommends that an investigator only have a general notion of what the problem area is before they start interviewing subjects, and allow more specific hypotheses to emerge from the data. Once this is accomplished, they can then take the formulated categories and search the literature to confirm what they have learned. The key for Glasser is that the theory must "emerge" from the data, not be imposed on the data by the investigator. This explains his concern about how much an investigator knows about what they are about to observe. Glaser (2004) is also concerned about efforts to describe grounded theory as a form of qualitative data analysis, which he considers to be misguided and a reflection of the lack of appreciation of the uniqueness of discovering an emergent phenomenon.

In contrast, Strauss (e.g., Strauss and Corbin 1998) suggests that the investigator be more active in preparation for observing the phenomena of interest. They acknowledge that it is very hard to be a truly naive observer, especially when what has to be known to select a problem to study is considered. They also encourage investigators to ask more questions during interviews. Strauss and Corbin (1998) are quite explicit about their expectations for an investigator applying grounded theory, and their concerns remind me of what Daston and Galison (2007) describe when the eighteenth and nineteenth century naturalists interacted with their illustrators. The issue in both instances was about how the assessing instrument, the investigator, conceptualized grounded theory as being more of an iterative process than a purely inductive process, as espoused by Glaser. Starting with a preconceived notion of what to expect, an investigator deduces a hypothesis from the accumulated data. They then follow this by trying to validate their observations and repeat this until "an analytical stance" emerges (Heath and Cowley 2004; Fig. 3; p. 145). The cognitive process implicit in this strategy involves going from the concrete to the abstract and back, until a succinct validated concept emerges.

Both Glaser (2004) and Strauss and Corbin (1998) suggest methods with the objective of establishing descriptive invariance. A key element involved is the method used in coding data. While Glaser and Strauss differ in the details of their coding methods, both recognize the importance of *data saturation*. Data saturation involves an investigator repeating an observation until he or she is convinced that what they are observing will not change with additional observations. Also of interest here is that this is the same problem that Linnaeus had to face when deciding whether he had seen sufficient numbers of the features of the animal or flower to generate an idealized image. Thus, there is a continued presence of the truth-to-nature perspective in current qualitative research.

I started this section reviewing the observations of Daston and Galison (2007) concerning the truth-to-nature perspective used by naturalists in the eighteenth and nineteenth centuries. Several issues emerged from this discussion. First, that the "self" was the observing instrument used to generate an idealized image (resulting in "descriptive invariance," as I described it). Second, that the "self" had to be trained to perform the role of an observer. Third, that the investigator's attempts to communicate observations to either an illustrator or others was a task onto itself, that could be as much of a determinant of the outcome as the care taken in the investigator's observations. I then demonstrated that many of these same objectives are present in the current clinical assessment of symptoms, and in qualitative research. Thus, the clinician's efforts to establish that a diagnosis is invariant across observations for an individual, as opposed to a group, illustrates that the objectification procedures are present in the clinical encounter. In addition, the discussion of grounded theory illustrated how the assessment of qualitative data has many of the elements of the objectification process (e.g., coding, scaling, reliability of categories, and so on). In both these examples, how the clinician or the investigator conducts their affairs is a critical determinant of the outcome. The clinician has a dedicated education to achieve skill in diagnosis, while the qualitative researcher has a protocol to follow, designed to guide them in their task. Each has many of the characteristics of a naturalist, but in different settings.

3.2 From Mechanical to Structural Objectivity: Quantitative Assessment

One of the issues a naturalist had to face was ensuring that the illustrator generated truth; that is, that the illustrator did not artistically embellish the image the naturalist wanted to convey. One way to achieve this is to use a mechanical device to produce the image, not a human illustrator. Daston and Galison (2007) claim that these efforts produced a new kind of objectivity: *mechanical objectivity*. Daston and Galison (2007) first describe what they mean by objectivity:

> What is the nature of objectivity? First and foremost, objectivity is the suppression of some aspect of the self, countering of subjectivity. Objectivity and subjectivity define each other, like left and right, up or down. One cannot be understood, even conceived, without the other. If objectivity was summoned into existence to negate subjectivity, then the emergence of objectivity must tally with the emergence of a certain kind of willful self, one perceived as endangering scientific knowledge. The history of objectivity becomes, *ipso facto*, part of the history of the self. (Daston and Galison 2007; p. 32–33).

Two aspects of this quotation are of interest. First is the notion that objectivity and subjectivity are inseparable parts of one another, not two independent concepts or conditions. Second is that a willful act (e.g., a metacognitive action?) is required to control the impact of subjectivity on some observation. This implies that our perception and characterization of what we experience are fundamentally subjective, and only become objective by a deliberate action which can only be achieved in degree, rarely entirely. This view, of course, is consistent with the notion that qualification of an indicator unavoidably precedes its objective assessment.

Daston and Galison (2007) claim that the development of mechanical objectivity had a specific impact on the investigators themselves. As they state:

> …scientists came to see mechanical registration as a means of reining in their own temptation to impose systems, aesthetic norms, hypotheses, language, even anthropomorphic elements on pictorial representation. What began as a policing of others (artists, printers, engravers, woodcutters) now broadened into a moral injunction for the investigators, directed reflexively at themselves. (p. 174).

This shift made the observer or scientist himself the focus of how objectivity was to be achieved, and it would not be surprising to find that the use of mechanical devices to generate and record observations was encouraged in place of an imperfect human. One advantage of this approach to objectivity was that if the conculsions based on "mechanical" observation were true, then it can be concluded that what was observed was not confounded by the observer's subjectivity.

I can illustrate the relevance of mechanical objectivity to a clinical setting by comparing human- and machine-based decision-making for the task of weaning patients off mechanical ventilation. Mechanical ventilation is not only used during and following surgical procedures as an assist to breathing, but also may be important in sustaining life after a stroke or accident (Epstein 2007). Obviously, it is most desirable and cost-effective to restore a person's ability to breathe on his own as quickly as possible, but how is this decision made? Clearly, reliance on a clinician to make this decision is only one way to approach this task, and it may be confounded by the clinician's subjectivity. Another approach involves the "observation" of a patient by a machine, with the decision based on certain physiological parameters. A number of decision algorithms have been proposed (e.g., Giraldo et al. 2006; Tehrani 2007). Gottschalk et al. (2000), however, report a study in which they trained a neural network (which I will consider a machine) by repeatedly exposing the network to various respiratory parameters, and then compared the ability of the network to successfully predict the outcome of a weaning effort. The neural network's predictions were compared to those of a clinician. The researchers also varied the number of indicators that the neural network and the clinician used when making their decisions. They concluded that, "When both are restricted to the same limited set of patient data, appropriately trained neural networks can be as effective as human experts in predicting whether weaning from mechanical ventilation will be successful" (Gottschalk et al. 2000; p. 160). Gottschalk et al. (2000) go on to discuss the pragmatics of each approach (e.g., that it takes time to train the neural network), demonstrating that the actual decision about how to decide is more complex than simply estimating the "percent correct number of persons weaned."

Although a machine seemed to have performed as accurately as a clinician in this study, the objectivity created by a machine can still be quite variable (e.g., as when judging the photographic plates following an X-ray). To circumvent this problem another form of objectivity has evolved: *structural objectivity*. Mechanical and structural objectivity deal with different aspects of subjectivity. To quote Daston and Galison (2007):

> Mechanical objectivity restrained a scientific self all too prone to impose its own expectations, hypotheses and categories on data – to ventriloquize nature. This was a projective self that overleaped its own boundaries, crossing the line between observer and observed. The metaphors of mechanical objectivity were therefore of manful self-restrain, the will reined in by the will. The metaphors of structural objectivity were rather of a fortress self, locked away from nature and other minds alike. Structural objectivity addressed a claustral, private self, menaced by solipsism. The recommended countermeasures emphasized renunciation rather than restraint; giving up one's own sensations and ideas in favor of formal structures accessible to all thinking beings. (p. 257).

The origin of structural objectivity can be traced to sensory physiology research performed during the late nineteenth and early twentieth century. For example, Helmholtz (1869/1954) demonstrated that the subjective sensation of color was actually not dependent on the perception of a broad array of colors, but rather on some combination of three primary colors. From this he concluded that all sensations were only signs of, not a picture of, external reality. Objectivity could be preserved, however, by the *invariant relationships* between (in this case) the three color primaries. Thus, while mechanical objectivity can be made visible, structural objectivity cannot. Helmholtz and much of twentieth century science involved the pursuit of such nonobservable invariant relationships. For Helmholtz, the structures were law-like sequences of signs. For the mathematician, these invariant relationships were found in differential equations, and for the philosopher logical relationships established the invariant relationship. This version of objectivity was the only way of breaking out of the private mental world of subjectivity. It was the only way of communicating scientific observations, since, as discussed above, mechanical observations (e.g., interpreting EEG records) can still be influenced by subjectivity.

Helmholtz believed that mathematical concepts and logic could be derived from experience. Thus, Euclidean geometry can be derived from experience, with different experiences generating different geometric intuitions. As Daston and Galison (2007) state, quoting from Helmholtz (1878), "There was nothing transcendental about geometric axioms and definitions…rather they were 'empirical knowledge, gained through the accumulation and reinforcement of similar, repeated associationism, not transcendental intuitions given prior to all experience" (p. 263). Even arithmetic can be understood in this manner. The key to Helmholtz's notions was the fact that thought and other mental processes take time and can be measured.

In contrast to Helmholtz, Frege (1884/1966) rejected an empirical approach to logic and science and instead insisted that objective reality existed independent of its psychological manifestations. Thus, Frege argued for a structural objectivity that was true to the nature of things. A necessary part of his approach was that ordinary language could describe this objectivity; that is, that it would be literally true. From these views came Frege's (1892/1966) *literal meaning hypothesis* (Chap. 2, p. 28), and with it, its support of Frege's concept of objectivity. He rejected the notion that the concept of a number had to be based on experiences such as counting pebbles, or other objects. The essence of his argument was that the empiricists were conflating subjective representation and intuitions with objective concepts, and that this was not necessary. Psychological representation and intuition were subjective because they were unavoidably private, whereas objectivity is inherently common to all people. Instead of such subjective constructs, he sought a purely symbolic language to express objectivity: the language of logic.

The conflict between the *empiricists* (e.g., psychophysiologists) and the *mathematical logicians* (e.g., Russell, Carnap, and others) continues to this day with the conflict between objectivity and subjectivity still the central issue. Structural objectivity, with its emphasis on invariance in relationships, found complex support in Einstein's theory of relativity. In fact, Einstein was preoccupied throughout his career with the meaning of objectivity as used in physics (Daston and Galison 2007; p. 302). For example, for Einstein time could only be defined relative to space, shattering the notion that time is a purely objective entity. Daston and Galison (2007) describe Einstein's approach as follows:

> In the case of relativity, Einstein took subjective time to be the beginning of our construction of objective, coordinated time. That subjectivity starting point, alongside what he always insisted was a *conventional* method for coordinating clocks, showed very clearly how inextricable the subjective and objective were within a theory… So was Einstein a structural objectivist? Yes and no. Yes, he was relentless in his hunt for theoretical structures that 'conditioned' our sense impressions. Yes, within the relativity theories he sought invariance – in many ways, this was his life work. But, at the same time, Einstein insisted over and over that an indispensable as objectivity was, physics did not come to it element by element or even symmetry by symmetry. Instead, objectivity issued from the integrity of a theory like relativity taken as a whole, complete with principles, observations and conventions. For Einstein to take invariant structures as objectivity was far too narrow. But to identify mathematical–physical structure per se with objectivity was far too broad; Einstein took each theory, with its peculiar combination of conventional and nonconventional elements, to pick out the objective. (p. 305).

There are several aspects of this quote that are of interest. First is the realization that the issue of the relationship between subjectivity and objectivity is not confined to the assessment of quality, but rather pervades the very foundation of modern science. Second is Einstein's contention that *subjectivity precedes objectivity*; a notion that reinforces the view espoused in this book that the assessment of quality first requires the subjectification or qualification of an indicator in order to be able to establish objective invariant statements. Third is Einstein's view that neither invariant nor mathematical–physical statements are sufficient to account for some phenomena, but rather that both aspects need to be considered in the larger context of a theoretical statement. The same may yet be confirmed when evaluating some approach to quality assessment.

How is this discussion about structural objectivity relevant to the assessment of symptoms, and, more generally, the assessment of quality? First is the realization that the relationship between subjectivity and objectivity is as much an issue for physics as it is in discussing the definition and assessment of a sign or symptom. This suggests that the nature of the problem is not unique to qualitative assessments. Rather, what is different for each problem area is the nature of the solution (see below). In addition, the efforts at defining objectivity have naturally lead to the increasing reliance on abstract expression of invariant relationships, especially since it became clear that more concrete relationships are vulnerable to subjective considerations.

Another issue is the fundamental suspicion of the private, the self, and what it means for the assessment of a symptom or of quality. The pursuit of a solution to this dilemma – objectivity – has lead to what Nagel (1986) has described as "the view from nowhere" (p. 5; Chap. 4, p. 111). Nagel sees objectivity in its purest form as a core circle surrounded by increasingly subjective circles or indicators. In this instance, the core is the most "real." This obviously has implications concerning the assessment of a sign, symptom and quality, each of which has an unavoidable start as a subjective statement. Does this mean that when I say I like a painting I am not making a "real" statement, or if I complain about a pain, that the pain is not real? Nagel would deny it, and would say that they are just not as real as an abstract statement of some relationship that remains invariant over its many transformations.

One approach to the problem is to apply what Bridgman (1959) charged us to do when he said that it was necessary to "understand how we understand" a phenomenon, such as objectivity. Daston and Galison (2007) state what needs to be done fairly clearly (see above) when they said that Einstein's pursuit of objectivity was a "relentless hunt for the 'theoretical structures that' conditioned' our sense impressions" (p. 305). The cognitive equivalents of these "theoretical structures that condition our sense impressions" would then become the object of study. Of course, these cognitive equivalents are more than likely the structures that mediate language, thought and reasoning; the processes that are used to characterize an objective indicator, a subjective statement of quality or a description of a sign or symptom.

Structural objectivity as applied in physics provides a standard to compare current approaches to the assessment of signs, symptoms, and quality. Here, I am assuming that structural objectivity means being able to identify a relationship that remains invariant over its transformations. What Cronbach and Meehl (1955) suggest is that psychological constructs, such as a symptom or the quality of some entity, should be viewed as a theoretical construction that would be used to account for behavioral consistency over varying contexts. If a "theoretical construction" is thought of as a relationship and accounting for "behavioral consistency over contexts" (i.e., invariance over transformation), then there would appear to be an overlap between the Cronbach and Meehl approach to psychological assessment and structural objectivity. The question now becomes determining the extent of this overlap.

Typical examples of symptoms and quality indicators include self-reports of fatigue, depression, life satisfaction, or beauty. These indicators, however, are not directly observable, but psychometric assessment models assume they are, and, in effect, simulate an assessment that would be characteristic of a more objective indicator. Thus, depression is assumed to be a measurable construct, like heart rate, but since it is not directly observable, it is referred to as a "latent variable" that underlies the behavior. What is observable and objective is the behavior that occurs when the construct is assessed. For example, if a five-point rating scale is used to assess life satisfaction, then a person's rating is assumed to reflect the person's true life satisfaction. The reliability and validity of this rating can be established, but the number selected to quantify the rating is often arbitrary. In contrast, the relationship between a physical construct (e.g., length, weight) and the arbitrariness of its numerical expression is much less of an issue. However, it is just this type of difference that raises concerns about the extent to which psychometric approaches to symptom and quality assessment adequately reflect the principles of structural objectivity.

Blanton and Jaccard (2006) discuss the complications that occur as a result of use of "arbitrary metrics," but now when dealing with psychological data. They define a metric as being arbitrary, as follows:

> …when it is not known where a given score locates an individual on the underlying psychological dimension or how a one-unit change of the observed score reflects the magnitude of change on the underlying dimension. This definition of metric arbitrariness makes explicit that an individual's observed score on a response metric provides only an indirect assessment of his or her position on the unobserved hypothetical psychological construct. It is assumed that some response function relates the individual's true score on the latent construct of interest to his or her observed score on the response metric… When a metric is arbitrary, the function describing this relationship and the parameter values of that function are unknown. (p. 28).

Table 5.5 Kazdin's (2006) comments based on the Blanton and Jaccard (2006) paper[a]

Point	Key points
1	Measures reflect arbitrary metrics if the connections between the observed score and the true score on the underlying dimension are not known. In this sense, height, weight or income are not arbitrary, but marital satisfaction, depression, and self-esteem on self report inventories or interviews are arbitrary, if not connected to other referents, are arbitrary
2	Reliability and validity in their many forms, as important as they are, do not themselves address or resolve the arbitrariness of a metric
3	High and low scores on a measure do not necessarily reflect high or low status on the psychological construct of interest; small or large amounts of change on a measure, however operationalized, do not necessarily reflect small or large changes on the underlying construct
4	One should be wary of high, low, slight, moderate, and other designations of psychological characteristics unless these designations can be linked to defensible referents
5	Data transformations, rescaling, conversion of a metric to familiar units (e.g., effect size), use of normative data and more fancy tactics do not necessarily savage the arbitrariness of the metric
6	Solid, sound, clear, and rigorous measures, such as direct counts, latency, or duration, are excellent measures if used as descriptions of behavior but may become arbitrary metrics if they are used to infer some psychological construct
7	Measures of any applied relevance or interest (e.g., attributes, clinical importance) must be linked to observed referents to have meaning
8	Many facets of a measure, such as the number of response alternatives for individual items, the anchors or endpoints of the scale for the items and the order in which the items appear, can influence a person's score on individual items and on the scale overall. This fact is relevant because two "identical measures" of a given construct can yield different results as a function of these facets and other facets of the assessment format
9	A score at a midpoint or zero point of a scale may not reflect a person's neutrality, medium status, or mid-level equivalent on the underlying construct
10	Arbitrariness of a measure may not be of concern in many contexts in which theoretical propositions and hypotheses are tested or generated in relation to some theory

[a]Not all examples Kazdin provided were reproduced and some points were shortened by leaving out portions of the comment

In the remainder of their article, Blanton and Jaccard (2006) spell out the significance of their concerns with the specific reference to the role that *arbitrary metrics* plays in studies of prejudice and clinical significance. In Table 5.5, I summarize the key points that Blanton and Jaccard (2006) make, as summarized and augmented by Kazdin (2006).

The Blanton and Jaccard (2006) paper raises two issues. The first deals with concerns that arbitrary metrics complicates the task of establishing scale intervals. Thus, when a length is measured it is possible to assume that 4 in. is twice as long as 2 in., but an investigator would have much less confidence in assuming that four units of depression were twice as much as two units. Embretson (2006) reviews the history dealing with this problem, not only pointing to Thurstone's (1927, 1928) efforts to establish an interval scale for intelligence, but also to Otis' (1917) introduction of standard scores. She also notes that Townsend and Ashby (1984) have suggested that scales be constructed to meet the criteria of fundamental measurement, although Lord (1953) would argue this is not necessary and the appropriate assessment models are up to the task. The second concern deals with the fact that while arbitrary metrics do not necessarily undermine research findings, they do raise the concern that individual scores and score changes may not be meaningfully interpreted (Embretson 2006). This second concern is particularly important, since a clinician would want to be able to make statements concerning symptoms and quality indications which are relevant to the individual.

Because depression, fatigue, etc., are theoretical constructions, there are no natural metrics that can be assumed to characterize them. Instead it has become necessary to justify the numbers used by making estimates of the latent variable assumed to exist, and to do this requires the development and application of assessment theories. There are two currently active theories: *classical test theory* and *item response theory*. The two models differ in ways that impact the numbers assigned to a construct. As Embretson (2006; p. 51) describes, in classical test theory the psychometric properties of items are considered to be fixed and must be balanced to obtain equivalent measures. On the other hand, in item response theory the psychometric properties of items are included directly in the model, and as a result, comparable trait estimates may be obtained from tests with nonequivalent items. In addition, classical test theory and item response theory transform raw scores into meaningful metrics in qualitatively different ways. From this and additional comments, Embretson (2006) concludes:

> The adequacy of the scaling of scores impacts not only test interpretations but also psychological research findings. What, one might ask, constitutes a nonarbitrary scaling? That, of course, has been the subject of considerable debate. According to some scholars, appropriate applications of Rasch models can achieve the necessary desirable qualities. But, of course, for some psychological measure the Rasch model simply will not fit. For other measures, achieving fit to the Rasch model leads to significant narrowing of the construct.

Achieving nonarbitrary scales requires mapping the observed outcomes to fallible items onto the latent construct. Applying contemporary methods in model-based measurement is one possible solution. (p. 54).

As previously stated, the measurement task for physics, signs, symptoms, quality assessment, or psychological assessments is the same: each case requires that objectification be achieved. This discussion about arbitrary metrics demonstrated that the achievement of objectification is qualitatively different depending on the subject matter being studied. In this sense, I would say no matter how controversial achieving structural objectivity may be in physics, it is even more difficult when dealing with symptoms and quality indicators. The good news is that successful efforts have been extended to develop assessment models (e.g., Rasch model) that are designed to bridge this gap. For example, Borsboom (2005) discusses *measurement invariance*, which occurs when the mathematical function relating the latent variables to the observations are the same in each group (e.g., cultural group) being considered.

3.3 Trained Judgment: Expertise

As so often happens in intellectual history, an idea is rejected only to find itself resurrected some time later in a modified form. This is certainly what appears to have happened with the development of the role of the expert, a role that has historic connections to that of the naturalist. Like the naturalist, the expert is expected to develop a level of knowledge of their topic that permits them to exercise judgments of one sort or another. What both the naturalist and the expert have in common is that both are expected to recognize basic patterns or principles and to use them to either characterize the entity (e.g., depict a species of flower) or make a judgment, such as diagnose some disorder (as a clinician would do). How they go about their respective tasks differs, however. Thus, the health care expert is expected to rely on information generated by mechanical objectivity (e.g., a cardiogram, an EGG, a galvanic skin response), but can also take advantage of conceptual models that were forthcoming from efforts to establish invariant relationships (such as the Rasch Model and item response theory). In contrast, these options were not available to the naturalist of the seventeenth and eighteenth centuries. There was also a shift away from seeing the self as a debasing influence in the process of doing science, to seeing the educated or informed self as an instrument that could make decisions. Part of the reason for this shift, according to Daston and Galison (2007), was the demonstration by Freud and his contemporaries that unconscious processes control our current behavior. The implications of this work were that the self could creatively contribute to the scientific process by generating intuitions and taking advantage of specific intellectual instincts.

As they regularly do, Daston and Galison (2007) conceptualize these changes in terms of how things are seen. Thus they say, "Instead of the four-eyed sight of truth-to-nature or the blind sight of mechanical objectivity, what was needed was the cultivation of a kind of physiognomic sight – a capacity…to synthesize, highlight and grasp relationships in ways that were not reducible to mechanical procedure, as in the recognition of family resemblance" (p. 314). This intellectual history is not to be interpreted as indicating that one form of seeing replaced another. Quite the contrary, all three are active contemporary representatives of the scientific process. For example, the current conflict about the difference between psychometric and clinimetric approaches to symptom and quality assessment should now be seen in the context of the difference between mechanical and structural objectivity and trained judgment or expertise (see below). And as Daston and Galison say; "Sequence matters–history matters" (2007; p. 317). I interpret their statement to mean that since the need for trained judgment evolved from the limitations of mechanical and structural objectivity, trained judgment or clinimetrics has to be viewed as a distinct intellectual tradition (see below).

How does training to be an expert occur? Fortunately, there is a fairly extensive literature on the cognitive processes that underlie the development of expertise. In Chap. 3 (p. 62) I spoke of three basic methods about how information is gathered together to generate a concept (e.g., establishing a diagnosis). In the first method, the classical view, items or experiences that have specific or similar features are grouped together. In a probabilistic approach, items or experiences are grouped together because they match some average example or prototype, or have a family resemblance. Finally, specific examples, or exemplars, can be recalled and used to organize a category or concept. Specific features, matching prototypes or examples can also be nested, to form hierarchies of categories or concepts. Thus, the first question is what features of a symptom are used to determine a diagnosis? Is it the similarity of features, their prototypicality or exemplariness?[6]

One of the ways that an expert clinician generates a diagnosis is by taking advantage of the natural groupings of symptoms and using the similarities in these groupings to form a diagnosis. Kushniruk et al. (1998), in fact, have speculated that clinicians create what they refer to as "small worlds" or theories in the process of collecting signs and symptoms together to form a diagnosis. As they describe, "the 'small world' hypothesis…states that expert physicians organize knowledge on the basis of similarities between disease categories, forming numerous small worlds which consist of subsets of logically related diseases and their distinguishing features" (p. 256). Creating diagnostic categories on the basis of similarity of features is, of course, a defining characteristic of the classic view of concept formation, and can be considered a method that permits the efficient processing of information.

Cantor et al. (1980) used these two basic models (classic or probabilistic models) to conceptualize how a group of psychiatrists organized a list of symptoms of functional psychosis into diagnostic categories (e.g., schizophrenia, manic-depression, and so on). They also gave the psychiatrists case studies of patients whose diagnosis was ambiguous. They found that psychiatrists selected symptoms that matched a prototypical expression of a diagnosis. Thus, the authors concluded that a psychiatric diagnosis is best seen as a type of prototype matching. They state (Cantor et al. 1980):

> The forgoing analysis suggests that it may be useful to study the psychiatric diagnostic system and diagnostic process by analogy to *naturalistic classification systems* used to classify common objects...and types of people. Psychiatric diagnosis and the diagnostic system look reasonably orderly when viewed within the context of these other systems. (Italics added; p. 190).

Of interest was a study by Ratakondar et al. (1998) that confirmed the Cantor et al. (1980) findings by using factor analytic techniques. They found that the same cluster of symptoms that were provided by psychiatrists' direct ratings also formed distinct statistical factors.

One of the determinants of a naturalistic classification system is the prior knowledge a person has about a particular category or concept. This prior knowledge can function as a theory to guide how a category or concept is organized. Norman et al. (2006) have been studying the role of prior knowledge of signs and symptoms when physicians establish diagnostic categories. Their paper reviews the available literature on this topic, with an emphasis on identifying the cognitive processes involved in establishing diagnostic categories.

Norman et al. (2006) describe three forms of prior knowledge – causal, analytical, and exemplar-based – all three of which may be involved in establishing a particular clinical domain. For example, if a physician is dealing with a difficult case, he may decide to go "back to basics" and review his prior knowledge concerning the "cause" of the particular disease or symptoms. Analytical knowledge refers to using a list of signs and symptoms the physician has learned about a particular disease or diagnostic domain. The specific learned list functions as a prototype, if applied when reasoning about a particular diagnosis. Exemplar knowledge comes from experiences with prior cases. As Norman et al. (2006) state:

> The basic notion is that every learned category is accompanied by a number of examples acquired through experience, and that these examples are still individually retrievable and provide support for the categorization of new cases that are similar to at least one prior example. (p. 345).

Norman et al. (2006) described several types of prior knowledge that may also be involved when a physician is making a judgment about a person's quality-of-life or HRQOL. Causal knowledge is of particular interest, since it may help us understand the role symptoms play in an HRQOL assessment; for example, "Did symptoms cause the changes observed in a person's HRQOL?"

But what does the phrase "causal knowledge" mean? Rehber (2003) describes it as referring to what is understood about how things happen or work. Physicians may use factual knowledge to establish causal linkages, while lay people may use culture-specific beliefs (e.g., the hot and cold theory of disease prevalent among certain ethnic groups), superstitions (e.g., letting a black cat run across your path is an omen of bad luck), or purposeful attributions to objects and things (e.g., "That rock had my name on it!," as an explanation of why a rock struck a person), each of which explains why and how things happen. Cognitively, causal knowledge creates expectations, and to the extent that a new experience matches this expectations, then the experience is classified in a particular category. This approach contrasts with an empiricist view claiming that categories are formed on the basis of what is directly observed.

Kim and Ahn (2002) report a study that illustrates the practical consequence of this cognitive (modular) characteristic. They studied clinical psychologists trained to use the atheoretical *Diagnostic and Statistical Manual* when diagnosing psychiatric disorders. The manual asks the clinician to identify specific symptoms, count the number of these symptoms and compare this number to predetermined criteria meant to establish if the person fits a diagnostic domain. The researchers found that in spite of their training, the psychologists used *preconceived theories* concerning the nature of psychiatric disorders to diagnose a particular patient.

Kim and Ahn (2002) suggest there is a reason why this occurs. As they state:

> As philosophers of science have argued...the goal of scientific research is to eventually develop a theory that explains a set of observations, not just to collect more and more observations. It seems to follow, then, that clinicians are also justified in developing theories that make sense of the knowledge they have amassed about mental disorders. (p. 473).

The Kim and Ahn (2002) study also implies that the training or experiences a clinician receives contributes to the formation of distinct cognitive attributes (e.g., implicit theories) that are used when making a diagnosis. This was confirmed in a study by Eva and Cunnington (2006), which demonstrated that physicians with greater experience were more likely to make diagnostic decisions on the basis of first impressions than physicians with less experience. Other studies have demonstrated the important role that memory plays in the skills an expert acquires (see summary in Brauer et al. 2004). Thus, the expert, when compared to the nonexpert, can store more information in short-term memory; is better able to distinguish relevant from nonrelevant information; can process a larger amount of information; is more efficient in allocating attentional efforts, has more differentiated knowledge structures; is better at developing abstract categories; and tends to adopt a forward-thinking rather than a backward-thinking approach to problem solving (Brauer et al. 2004). How the expert engages these skills, however, will vary depending on what they are providing advice on; for example, providing advice or judging chess playing, vs. a medical diagnosis (Brauer et al. 2004).

The purpose of this section was to review the history of the development of the concept of objectivity, and apply it to sign and symptom reporting. Much was revealed, especially about the various efforts to deal with the relationship between subjectivity and objectivity, and their relevance to assessing signs or symptoms. In the next section, I will build upon this and examine the clinical encounter and clinical medicine. For example, a classic struggle within clinical medicine is determining the extent to which the clinician should exercise interpretative (subjective) skills when assessing signs or symptoms. Based on the above discussion, it is clear that this role will involve some combination of each of the three major forms of objectification. Thus, finding the most efficacious balance between subjective impression and objective fact becomes a major task for the clinician. The same concern about the relative role of subjectivity and objectivity can be seen in defining a disorder. Also in the next section I will discuss three topics: first, how investigators define and rate AE; second, how clinicians manage patients who complain of symptoms that have no clear physical basis, and finally, the difference between psychometrics and clinimetrics.

4 Symptoms, Subjectivity and Clinical Medicine

The clinical encounter is where signs and symptoms are diagnosed, defined, and used to make decisions. As Feinstein (1967) states:

> For the clinical encounter to attain their full scientific potential...they must be reliable as basic observational data, and the observations must be performed not by laboratory techniques… but by an irreplaceable human medical apparatus – the clinician. (p. 66).

Thus, the clinician as *the final common path* is expected to make a clinical judgment and use information about a person's signs and symptoms to provide a diagnosis and select a treatment. Since this decision is being made by the clinician and is person-based, the critical decision point is a subjective event, even if objective indicators are available and involved. However, there is far from universal agreement that this should be the situation.

For example, Scriven (1979) states that the idea that a clinician, as an expert, is superior to scientific-based evaluations is, to use his word, "bogus." Referring to the work by Meehl (1954), he points to the data that demonstrates the advantage of using statistics-based decisions for predictions over clinical judgments. Meehl's original observations have been repeatedly confirmed (e.g., Dawes et al. 1989). Goldberg (1991), using a statistical metaphor, describes the problem with a clinician's decision making as being that he or she is an "imperfect, unreliable generator of regression weights." Kleinmuntz (1990) however, points out that most investigators use both "the head and the formula," either consecutively or simultaneously, when making clinical judgments.

Positing this debate in terms of a conflict between clinical and statistical decision-making, has, according to Westen and Weinberger (2004), created the false impression that clinicians are seriously flawed in their "observations, thought processes, and beliefs" (p. 595). Yet when Meehl's (e.g., Grove and Meehl 1996) comments are closely inspected, it appears that he was most concerned about the use of informal aggregation of clinician reports, not clinician reports per se. Westen and Weinberger (2004) suggest that instead of there being two extremes, there is actually a continuum between statistical and clinically based decisions. They suggest that intermediate conditions exist, such as using statistical aggregation methods based on clinician reports or informal aggregation of psychometrically based self-reports.

Meehl, a clinician for his entire professional life, was also aware of the difference between making decisions for groups and individual patients. Thus, he would have appreciated Feinstein's (1987b) view that at the critical moment when a clinician is to advise an individual patient, most would not mechanically decide on the basis of some number, but would rather reflect about a variety of issues relevant to the patient before providing advice.

Feinstein is not only concerned about how clinical judgments were made, he is also concerned about what information was used and how it was collected. Thus, he sees what is being measured in a clinical setting as fundamentally different from what is usually measured and used in a statistical-based decision. He argues that there are two types of assessment relevant in a clinical setting: the first is "mensuration,"[7] which refers to identifying characteristics of a person (such as weight, height, functional status, and so on); while the second type involves *quantification*, which involves collecting data into groups and numerically comparing the groups. Mensuration is what Feinstein (1967) thinks a clinician would use in clinical judgment, even though he recognizes that quantification is useful for prediction. Feinstein (1987b) calls the assessments based on "mensuration" *clinimetrics*, or the assessment of clinical indicators. He defines clinimetric indexes as follows:

> Clinimetric indexes are arbitrary ratings for the diverse phenomena of clinical care that are observed *subjectively* and that cannot be expressed in dimensional numbers. (Italicized for emphasis; Feinstein 1987b; p. 245).

Feinstein (1987b) was acutely aware that there would be resistance to the use of subjective indicators in clinical settings, but he felt this had less to do with the type of assessment than whether it was standardized, and thus, reliable and sensitive to change. He sees observed data obtained from histopathology, radiology, and electron microscopy as no less subjective then the clinician's judgment that a patient is depressed. His views, of course, fit nicely into the historical changes in the meaning of the term "objectivity" I have traced above, although not all investigators are particularly aware or sensitive to this history.

What differentiates a clinimetric index from other types of measures is that it usually consists of a collection of heterogeneous indicators with no clinically relevant common dimension, although the collection of indexes may have a common goal. Thus, the Apgar (1953) score, which is used to rate the viability of a newborn, is an example of a clinimetric measure and includes such diverse indicators as skin color, heart rate, respiration, reflex response to nose catheter, and muscle tone. Each of these individual indicators is rated or measured and the total score represents a summary statement of the index. Many other examples of indexes are summarized in Feinstein (1987b). What should also be obvious at this point is that a clinimetric index is a near perfect example of the *goal-directed* or *cross-classification* of information: a heterogeneous collection of indicators, each of which contributes to a common objective or goal.

By suggesting that clinical decision making is almost always a subjective event, Feinstein was contributing to the growing awareness that subjectivity can play a meaningful role in clinical medicine. When this is combined with influence that studies of health status and HRQOL assessment are having on clinical decision making, then a dramatic rethinking of what medicine is about appears to be occurring. Sullivan (2003) summarizes the changes as follows:

> Facts known only by physicians need to be supplemented by values known only by patients. Outcomes research has pointed to the importance of the patient's view on the goals of medicine in its call to emphasize "patient-centered" outcomes such as quality-of-life. …When combined, these innovations suggest a radical realignment between the objective and subjective elements on clinical medical science. Now many of the most important patient outcomes, like patient choices before them, are valid because they are subjective. …Medicine is thus turning away from the scientific ideal of fully objective or "perspective-less" assessments of disease and health. (p. 1595).

He goes on to state that for these changes to succeed, the practicing physician would have to expand what they consider acceptable medical practice. He states (Sullivan 2003):

> If patient-centered outcomes such as quality-of-life do unseat the defeat of death and disease as the primary goals of medical care, then physician autonomy will be qualified further. Patient values will shape the goals as well as the means of medical care. Patients' lives rather than patients' bodies will be the focus of medical interventions. (p. 1602).

These views, however, are not universally accepted. Kaplan et al. (2000),[8] arguing in another but relevant context, suggests that the assessment of subjective indicators would offer little to the understanding of medical care, the quality of medical care, or other primary clinical outcomes. Kaplan et al. (2000) uses the Wilson and Cleary model (1995) as a way of ordering types of clinical indicators, starting from concrete measures of pathology or physiology (e.g., signs), to symptoms, functions, perceived health status, and finally to the abstract concept of overall quality-of-life. To illustrate their argument, they apply the Wilson and Cleary model to a diabetic, who, as a result of being treated with a hypoglycemic agent experienced positive glycemic control, improved physiological functions, ability to physical functioning, health perceptions and overall quality life. Kaplan et al. (2000) then point out that if each of these variables accounted for 50% of the variance of the next variable in the sequence, then "the percentage of variance of overall quality-of-life accounted for by better use of the hypoglycemic agents would be $(0.50)^6 = 0.0156$ or 1.56%" (Kaplan et al. 2000; p. II-185). Kaplan and her colleagues interpret this example as evidence that subjective indicators, such as overall quality-of-life, would make a rather limited contribution to clinical care, and they suggest that if investigators are interested in evaluating and ensuring the effectiveness of medical care, they should study endpoints that are more likely to be affected (such as some more biological outcome).

Kaplan et al.'s (2000) argument has limitations, since it depends on the number of intervening variables. Thus, if only five variables are in the sequence, then 6.3% of the variance would be accounted for by a quality-of-life measure, and so on. Resistance to the changes that Sullivan (2003) addresses would not be surprising, since clinicians have to practice in different ways, and clinical researchers would have to expand what they are assessing and what is considered acceptable data. Ultimately, the challenge created by "subjective medicine" may require that a different social contract be developed between health care professionals and the patient.

What also has to be recognized is that both subjective and objective data are subject to the same process of being ordered and structured so as to be "made objective," even though there may be differences in the degree and nature of this objectification. The goal of the objectification process is to produce an outcome that has the property of being invariant (Chap. 4, p. 112), and the continued tug-of-war between subjectivity and objectivity so prevalent in science would also be expected to be present in clinical practice. This issue will become clearer after I discuss my next topic, the relationship between toxicity and symptoms.

Before moving on, I would like to "connect a thread"; that is, refer back to some previous discussions and trace them to our current topic. One of the issues I have been concerned about is how to deal with the apparent dualistic nature of medical information. While not directly expressed, Feinstein and Sullivan's approach to clinical assessment attempts to deal with this issue. Both investigators are aware that a physician has to deal with information about a person that is of both internal and external origin, has an objective or subjective nature, or refers to the body or the mind. An optimal approach to this inherent dualism would take advantage of both perspectives. However, Scriven (1979) and many others emphasize the traditional approach and rely on, or prefer, objective indicators or their surrogates to make clinical medical or social policy decisions. They are convinced that clinicians are not good decision makers, and idiosyncratic subjectivity creates unmanageable bias in data that ultimately will distort social policy and other types of decisions making. They would argue that the instances where a decision has to be made on the basis of a subjective indicator could be significantly reduced by the development of objective indicators with greater predictive power. In contrast, Feinstein (1967), Sullivan (2003) and others argue that, independent of the origin of medical information, an ethical clinician would have to apply all the available information and not just rely on a "formula." This would presumably apply to both individual and social policy decisions where the decision maker would also have to decide if the objective information were sufficient.

Lakoff and Johnson's (1999) notion of "embodied realism" (see above) offers an alternative way to think about these issues. They point out that it is not possible to know what is external to a person without the involvement of internal processes; it is not possible to know what is objective if I am not aware of my own subjectivity, and I could not know the mind without its presence in the body. They suggest that the continued separation of these dimensions is more a product of the language I use (i.e., there are separate words for each entity) than my biology. In this regard it is interesting to note (e.g., Singer et al. 2004; Lau et al. 2004) the rapidly accumulating data demonstrating that such characteristically "mental" phenomena as empathy or intention, are, with the aid of functional magnetic resonance imaging (fMRI), being shown to activate different but specific brain areas. Thus, phenomena which historically have been considered "mental" are now being shown to have an irrefutable material basis.

So far I have argued that subjectivity plays a critical role in clinical medicine, independent of the degree to which the clinician relies on objective indicators. Next I am going to discuss how an adverse event can be defined. AE, of course, are a critical determinant of the quality of a clinical encounter, and a natural setting to examine how signs and symptoms can be combined to produce a particular outcome (in this case, adverse). Then I will discuss the complexity of symptom presentation, particularly for symptoms that lack a demonstrable physical origin. The ensuing clinical encounter creates a unique interpretative task for the clinician and patient. How this is coped with becomes an example of how clinical medicine deals with symptoms that provide the rationale for medical intervention. The complexity of this encounter has convinced some investigators (e.g., Feinstein 1987a) that clinical medicine creates a unique empirical environment that justifies a different assessment approach. I will also explore whether or not this is true, by comparing clinimetric and psychometric assessment approaches.

4.1 Defining Adverse Events[9]

Signs and symptoms are usually informative about some bodily dysfunction, but not all signs and symptoms are considered clinically significant. In this section, I will be discussing three different aspects of how the clinical significance of a sign or symptom is established and used for different analytical purposes. First, I will describe and illustrate a system that researchers and clinicians have developed to define when a sign or a symptom has reached an unacceptable level. This type of system is an essential element in any quality control effort, and of course, can be used to monitor the ethical character of treating and caring for patients. Next, I will describe how an adverse event generated in a Phase I clinical trial is used to establish the maximum tolerable dose of a new drug or treatment. Finally, I will describe the Q-TWiST methodology, where survival time is partitioned into symptom-free time, time in a toxic or adverse state, or time in relapse. This type of display is very useful when patients have to make decisions between treatments and need to know what they can expect over an extended period of time.

Let me clarify what is meant by "an adverse event" or "an unacceptable level of clinical significance." I will use the National Cancer Institute's *Common Terminology Criteria for Adverse Events* v30 (CTCAE 2006) to illustrate how the phrase "adverse event" can be operationalized. The CTCAE (2006) defines an adverse event (AE) as follows:

> An AE is any unfavorable and unintended sign (including an abnormal laboratory finding), symptom, or disease temporally associated with the use of a medical treatment or procedure that may or may *not* be considered related to the medical treatment or procedure. An AE is a term that is a unique representation of a specific event used for medical documentation and scientific analyses. (CTCAE 2006; p. 1).

The resultant classification system also allows groupings of signs, symptoms, diagnoses, and disease processes into *superordinate* categories. For example, within the domain of neurology, there is a superordinate domain labeled "Mood Alternation" that refers to different feelings such as agitation, anxiety, depression, or euphoria. Mood alteration is not considered an AE, nor is it reported as such. I mention this because it suggests that the authors of the CTCAE were aware of the potential confusion that could result if they allowed both concrete and abstract (superordinate) domains to be operationalized and combined in the same assessment.

Table 5.6 provides examples of AE extracted from the domain *Neurology* in the CTCAE (2006). Inspection of the table makes clear that a wide range of criteria were used to establish different AEs, requiring that the person doing the ratings be trained to perform these judgments (i.e., an *expert*).

Table 5.6 Examples of adverse events (AE) abstracted from the CTCAE v3.0 (2006)[a]

Adverse event (short name if different)	Grade 1	Grade 2	Grade 3	Grade 4	Grade 5
Ataxia (medically or operatively induced)	Asymptomatic	Symptomatic, not interfering with ADL	Symptomatic, interfering with ADL; mechanical assistance indicated	Disabling	Death
CNS cerebrovascular ischemia		Asymptomatic radiographic findings only	Transient ischemic event or attach (TIA) <24 h duration	Cerebral vascular accident (CVA, stroke) neurologic deficit >24 h	Death
Cognitive disturbances	Mild cognitive disability not interfering with ….	Moderate cognitive disability interfering with ….	Severe cognitive disability; significant impairment of ….	Unable to perform ADL: full time specialized resources or institutionalization	Death
Mental status	–	1–3 points below age and education norm in MMSE	>3 points below age and education norm in MMSE	–	–
Personality	Change, but not adversely affecting patient or family	Change, adversely affecting patient or family	Mental health intervention indicated	Change harmful to others or self, hospitalization indicated	Death
Speech impairment (e.g., dysphasia or aphasis)	–	Awareness of receptive or excessive dysphasia, not impairing ability to communicate	Receptive or expressive dysphasia, impairing ability to communicate	Inability to communicate	–

MMSE mini-mental status exam; Folsten et al. (1975)
[a]The National Cancer Institute Cancer Therapy Evaluation Program (CTCAE v3.0); August 9th 2006 (http://ctep.cancer.gov/reporting/etc. HTML)

Both subjective and objective criteria were used in different AEs. Judging when a personality change has adversely affected a person or his or her family, requires making a clinical (or subjective) judgment, while determining that an ischemic event has occurred can be radiologically confirmed (an example of *mechanical objectivity*). Judging the adversity of ataxia involves establishing that it is symptomatic, while determining the magnitude of the adversity involves estimating the degree the ataxia interferes with the person's activities of daily living. In contrast, change in mental status is defined as occurring following a change in the score of a particular questionnaire (the Mini-Mental Status Exam; MMSE; Folstein et al. 1975). Overall, nearly 80% of the *Neurology* AEs involved interpreting when a symptom interfered with some function (e.g., activities of daily living).

Investigators regularly use the CTCAE in clinical trials to track the consequences of innovative treatments and disorders secondary to these treatments (e.g., fevers secondary to treatment with interferon-like compounds; peripheral neuropathies secondary to cancer treatments). Inspecting the cognitive basis of the grading system reveals that each level consists of descriptive statements that have been scaled (i.e., evaluated), but not directly valued. What does seem to be in place, however, is an implicit form of valuation[10]; that is, there seems to be an *a priori* consensus that at certain grades the AE listed are important and generally unacceptable outcomes of treating a patient. These consequences were obviously known prior to the formation of the CTCAE and were readily integrated into the grading system. Thus, value statements have structured the development of the monitoring system. In addition, although not advertised as such, the CTCAE can be broadly interpreted as a quality assessment, and to the extent that AEs promote the activities that lead to changes in procedures or treatments, they can also be said to exercise quality control functions.

This interpretation is consistent with an examination of the semantics and use of the term "adverse." The Oxford English Dictionary (1996) reveals that etymologically the term originates from the Latin word "adversus," meaning against, or opposite; or from the Latin word, "advertere," to "turn to." Thus, the literal interpretation of an adverse event is that it is an event opposite from what might have been expected or desired. The term also has figurative properties, particularly conceptual metaphoric properties, since its negative connotation is suggestive of a type of emotion. As Kövecses (2000; p. 52) suggests, the EVENT STRUCTURE conceptual metaphor is compatible with various emotional states, including presumably, being in an "adverse" state.

However, the values attached to these states are a product of social construction, meaning that investigators had to agree that the grading system and their own clinical experiences with a disease and its treatment concurred. Sometimes an adverse event occurs that is not anticipated, and it would be expected that such an event would be added to the grading system. For example, rashes seem to be a comparatively unexpected adverse event associated with monoclonic antibodies and small molecular inhibitors, and as such, would find their way into the grading system (Gerber 2008). It was interesting, therefore, to find the report by Pond et al. (2008) describing a nomogram that was meant to estimate the risk a patient would face in experiencing a serious adverse event due to molecularly targeted agents. The nomogram, in a direct sense, is a statement of values, in that as for any quality control effort it is necessary to detect and minimize, if not prevent, an adverse event.

While AE are generally considered undesirable, in the context of most Phase I clinical trials they are literately induced. They are induced, however, for a specific purpose that makes their occurrence empirically understandable but morally complex. The problem is how to expose people to treatments that have never been tested in humans before, and how to "do no harm" but also determine if the new treatment can be tolerated or have some benefit. This fine balance is usually achieved by selecting healthy patients who may get paid for their participation, or by using a small number of patients who have exhausted their therapeutic options and are willing to participate in an explicitly stated (via informed consent) experimental effort. To minimize the risk to the participants, a researcher ordinarily starts at a dose level they expect would have little chance of producing an adverse event and then escalating the dose, over trials, until a "Maximum Tolerable Dose" (MTD) is determined.[11] Sometimes the MTD is determined by the participant's symptoms (e.g., fatigue) or sign (e.g., thrombocytopenia), but it can also be based on the pharmacokinetic and pharmacodynamic properties of the agent being studied. Thus, the MTD sets limits to further dose escalation, but demonstrating therapeutic benefit of the MTD usually involves Phase II and III clinical trials.

Serious questions have also been raised about whether the cancer patients who participate in these studies receive any benefit. This is an important issue because often a participant will expect to receive benefit for participating even though they have been explicitly told not to expect a benefit. This observation, referred to as the "therapeutic misconception phenomenon," has substantial empirical support (e.g., Glannon 2006; Henderson et al. 2006). There have also been a series of studies that have estimated the benefit of participating in a Phase I trial (e.g., delayed recurrence, extended survival, and so on). Estey et al. (1986), for example, found that only 4.2% of 6,447 patients received some benefit from participating in a Phase I study. More recently, Horstmann et al. (2005) report a review of 460 Phase I trials involving 11,935 cancer patients, all of which have been evaluated for toxicity, and 10,402 who have been evaluated for therapeutic benefit (e.g., extended survival, tumor regression, and so on).

They found that for the 20% of the Phase I studies that involved the classic one agent type of design, only 4.4% of the patients received benefit. The benefit increased to 17.8% when the Phase I study combined the new agent with an agent that had an established therapeutic benefit. Also of interest was that only 14.3% of the 3,465 patients who experienced a Grade 4 toxic event had only one toxic event. Thus, as advertised, Phase I clinical trials have minimal therapeutic benefit for the participant, but a reasonable risk of that patient experiencing an adverse event.

As I have stated, exploratory Phase I trials are a necessary step in the effort to establish effective treatments. While this seems to be an important and justifiable activity, one question that remains unanswered is whether a Phase I clinical trial can or should be an appropriate platform for qualitative studies. A concern is that Phase I trials have only a small number of participants (as in single agent studies), each of which may have been treated (e.g., doses) somewhat differently, and for relatively short periods of time (e.g., 1–3 months). In addition, Phase I studies are meant to be performed quickly and inexpensively, so that a more definitive early Phase II study can be initiated.

Barofsky (1993) countered this argument by claiming that the logic of a Phase I trial alone would, in fact, be sufficient to justify providing a qualitative profile of the MTD.[12] The rationale is that just as titrating a dose would be expected to increase the probability of a therapeutic effect, so too would it be increasingly likely to produce toxic effects. The same would be expected for qualitative indicators; continuing to increase the dose of a drug or intensity of a treatment would be more and more likely to produce an adverse impact on physical and mental functioning and related global quality indicators. Thus, the protocol of a Phase I clinical trial provides sufficient reason to justify performing qualitative studies, independent of any empirical support. In addition, persons who participate in these trials (i.e., healthy persons, or persons who have exhausted their therapeutic options) are putting themselves at risk, yet are doing so for reasons other than a reasonable probability that they will experience some therapeutic benefit. The risks that is being taken is an important reason to give the participant an opportunity to report about their qualitative status. This is particularly so, since the CTCAC toxic event grading systems usually do not include qualitative indicators. A final reason for doing qualitative studies in Phase I studies has to do with the possibility that some participants (especially the healthy participants) may experience long-term consequences of exposure to experimental treatments (Shamoo and Resnik 2006), and a qualitative assessment is a convenient method for monitoring these long-term outcomes.

Table 5.7 summarizes a representative list of studies that have included an HRQOL assessment as part of their Phase I study. The table reveals that only recently have investigators started to regularly integrate qualitative assessments into Phase I trials. Most of these trials involve dose escalation and limit assessment frequency to baseline and follow-up. Sample size varies from 19 to 157, with a median sample size of 28.5. In general, collecting qualitative data seem to be feasible, even if studies involve dose escalation, and the results appear to be informative. The advantage of incorporating an HRQOL assessment into a Phase I clinical trial is that at least theoretically, it can provide a means of ensuring the quality with which the trial is conducted, and also to provide vital information for assessing HRQOL in Phase II and III trials.

While monitoring for AE can help prevent unacceptable treatment-induced consequences, in some situations (e.g., many cancer treatments) a patient may be faced with deciding if the potential benefit from a particular therapy matches or exceeds the adverse or toxic outcomes (e.g., fatigue, hair loss, and so on) of the therapies. This is especially true when a patient has to select from two or more treatments. Gelber et al. (e.g., 1986) have developed a statistical model (the Q-TWiST) that is designed to estimate these tradeoffs. Benefits are measured by the time free of symptoms and toxic effects, and the costs are measured by the presence of symptoms and toxic effects and also relapse and progression of disease. The Q-TWiST methodology is an excellent example of the *objectification process*, since it is an investigator who takes subjective and objective data and transforms them into a common metric, with *time in a particular state* (e.g., survival time) as the metric.

Gelber et al. (1989) describe the four major parameters of the QTWiST as follows:

1. The time period during follow-up when an individual patient experiences subjective toxic effects (TOX).
2. The time period following systemic relapse (REL).
3. The time recovering from treatment of local recurrence (LC).
4. The remaining survival time without TOX, REL, and LC.

This time period represents time without symptoms and toxicity (TWiST).

What is particularly interesting is how Gelber and his colleagues transform the TWiST into a measure of quality by multiplying measures of TOX and REL (LC is usually included in estimates of REL) by *utility estimates* (u), as presented in the following equation:

$$Q\text{-}TWiST = u_t \times TOX + TWiST + u_r \times REL.$$

The TWiST duration is not presented as being qualified in the equation, probably because the authors assume that being toxic and symptom-free has a utility of one. If this assumption is correct, then it becomes straightforward for the three components to be added.

Table 5.7 Phase I trials which have included an HRQOL assessment[a]

References	Study design	Dx; Groups; (N =)	HRQOL assessment	Assessment protocol	HRQOL findings only
Melnick et al. (1985; 1992)	7 different single agents (cytotoxic) Dose escalation	Cancer, mixed (N=45) Supportive care (N=10)	14 LASA also PS	Before each treatment session	No change in mean total LASA and PS scores for Tx group; significant reduction for supportive care group in both mean total LASA and PS scores
Cohen et al. (2002a)	Single agent (vaccine) – Dose escalation	Stage III or IV melanoma (N=25)	SF-36; IES	Baseline, at 3 weeks and at 7 week follow-up	Significant improvement over Tx time in physical status of SF-36, but not mental status or IES
Cohen et al. (2002b)	Single agent (vaccine)- Dose escalation	Stage IV renal cell carcinoma (N=36)	SF-36; IES	Baseline, at 3 weeks and at 7 weeks follow-up	Significant improvement over Tx time in physical status of SF-36, but not mental status or IES
LoRusso et al. (2003)	Single agent (biologic) Dose escalation	Cancer, mixed (N=157)	FACT (multi-disease specific) TOI	Baseline,14, 28, day 28 after each dose	No change or improvement in median FACT score as function of disease group
Hollen et al. (2004)	Combination chemotherapy (No dose escalation)	Stage III or IV lung cancer (N=20)	LCSS (6 symptoms; 2 global, 1 global QOL)	Assessed every 2 weeks for 18 months or until death	Single global QOL indicator was used to determine that every 3 week assessment provides optimal data input
Melhem et al. (2005)	7 Oral+4 iv anti-oxidants (No dose escalation)	Chronic hepatitis C (N=50)	SF-36	Baseline and within 2 weeks follow-up	SF-36 scores were inappropriately presented as total score; total scores were found to increase significantly over study
Michael et al. (2005)	Radio and cytotoxic chemotherapy Dose escalation	State III B lung cancer (N=19)	LCSS	Baseline and 3 weeks postradiation	Global QOL did not change postirradiation
Tester et al. (2006)	Combination chemotherapy Dose escalation	Hormone-refractory prostate cancer (N=20)	FACT-P TOI BPI	Baseline, 8 weeks and 24 weeks	Descriptive statistics suggested that only a portion of subjects reported improved HRQOL scores and pain reduction
Isell et al. (2007)	Noni fruit extract Dose escalation	Cancer, mixed (N=29)	QLQ-C30 BFI CES-D	Baseline and 3 month follow-up	Physical functioning, global health, fatigue and depression statistically significantly worsened with increasing dose
Lim et al. (2007)	Cytoreductive surgery and chemotherapy Dose escalation	Soft-tissue sarcomas (N=28)	FACT-G SF-36	Baseline (N=9) 6–8 weeks (N=5) 3–6 months (N=3)	Small sample size precludes drawing inferences, but consistent with other studies that show transient (6–8 weeks) QOL effect, followed by recovery

BFI brief fatigue inventory (Mendoza et al. 1999); *BPI* brief pain inventory; Cleeland (1989); *CES-D* Center for Epidemiology Scale – Depression; Radloff (1977); *FACT-G* functional assessment of cancer treatment – general; Cella et al. (1993); *FACT-P* functional assessment of cancer treatment – prostate; Esper et al. (1997); *IES* impact of event scale; Horowitz et al. (1979); *LCSS* Lung Cancer Symptom Scale; Hollen et al. (1993); *QLQ-30C* quality-of-life questionnaire for cancer; Aaronson et al. (1991); *TOI* summary measure of physical well-being, functional well-being, and disease-specific overall score from disease specific FACT assessments; *SF-36* short form –36; Ware et al. (1993)

[a]This table presents details only of the HRQOL aspects of each study; information about toxic events, symptoms and other indicators will require a review of the appropriate paper

Of particular interest to me was the origin of this approach to creating a qualitative assessment. Where did it come from and what is its intent? It is clear from the authors' publications that they were working in an environment where they would continue to evaluate innovative cancer treatments, and where the primary objective of their activities was extending the survival of the patient. The standard method of measuring survival was in terms of the Kaplan–Meier (1958) equation; that is, *survival duration analyses*. I imagine that Gelber and his colleagues asked how they could integrate qualitative issues into what they consider their basic outcome measure: survival. This they did by first dichotomizing TOX (into good or bad components) and then asking how long the patient remained in the TOX state (meaning that sometimes TOX only lasted for the brief treatment period or period immediately after treatment); in effect, they were performing a survival analysis on TOX. They did the same for REL. As a consequence, Gelber and his colleagues were partitioning survival time into three components. The Q-TWiST, therefore, is not an example of structural objectivity (an equation generated from some set of basic principles, as is the HUI), but rather has to be understood as being based on the judgment of the authors regarding what parameters were important to assess when integrating a qualitative assessment into a survival analysis. From a cognitive perspective then, what the authors of the Q-TWiST have done is to create another example of a hybrid construct (evaluated descriptive statements whose importance has been valued) that I have suggested underlies all efforts at establishing qualitative assessments. Most important is that the Q-TWiST enhances the ethical character of clinical decision making by providing a common metric upon which to make treatment decisions.

In this section, I have reviewed three examples of how the adversity associated with certain toxic events or symptoms can be monitored: by a formal monitoring system (the CTCAE); by integrating HRQOL into a particular research paradigm (Phase I clinical trial); or by providing a method whereby patients can estimate the benefit and cost of their decisions. Each of these can also be seen as contributing to the optimization of medical care: an objective that is consistent with a quality control effort. Next, I want to address the complex issues that surround symptoms when they have no apparent physical basis, but yet have the potential to have major qualitative impact.

4.2 Clinical Management of Symptoms

Inspecting the clinical literature quickly reveals that a series of terms exist that reflect different conceptualizations of the relationship between social, psychological, and biological determinants of a physical symptom. These include terms such as psychosomatic, sociosomatic, and biopsychosocial. Each of these terms has been interpreted as connoting a causal relationship; that is, psychological and/or social factors determine some somatic (physical symptom) outcome. For example, the term "psychosomatics" is often interpreted to mean psychogenic (Lipowski, 1984) that is, psychological factors cause physical symptoms. When this association is considered abnormal (i.e., when there is no evidence of a linkage between the indicators), then terms such as somatization and symptom conversion are used to describe the processes that have been labeled as somatoform disorders, or functional somatic syndromes. Not too surprisingly, distinct fields (psychosomatic medicine, behavioral medicine, consultation psychiatry, etc.) have developed to deal with persons whose physical complaints do not seem to have an organic base.

Functional somatic syndromes are of particular interest, since they encompass such publically visible disorders as multiple chemical sensitivity; sick building syndrome; repetitive stress injury; side effects of breast implants; Gulf War syndrome; chronic whiplash; chronic fatigue syndrome; irritable bowel syndrome and fibromyalgia (Barsky and Borus 1999). Each of these disorders creates a major conceptual and empirical challenge to what constitutes a physical symptom. While a physical determinant may eventually be identified for these disorders, they have certain common features that make their clinical management particularly difficult. As Barsky and Borus (1999) state:

> Patents with functional somatic syndromes have explicit and highly elaborated self-diagnoses, and their symptoms are often refractory to reassurance, explanation, and standard treatment of symptoms. They share similar phenomenologies, high rates of co-occurrence, similar epidemiological characteristics, and higher-than-expected prevalences of psychiatric morbidity…the suffering of these patients is exacerbated by a self-perpetuating, self-validating cycle in which common, endemic, somatic symptoms are incorrectly attributed to serious abnormality, reinforcing the patient's belief that he or she has a serious disease. Four psychosocial factors propel this cycle of symptom amplification: the belief that one has a serious disease; the expectation that the one's condition is likely to worsen; the 'sick role', including the effects of litigation and compensation; and the alarming portrayal of the condition as catastrophic and disabling. (p. 910).

When these syndromes are placed in the social or political contexts that ordinarily accompany them, the complexity of their clinical management becomes evident. For example, there is usually a profound discrepancy between the clinician's and the patient's interpretation of the patient's somatic complaints. Interestingly, the lay public's willingness to accept the notion that the mind and body are separate actually makes it more difficult for the patient to accept the argument that psychological or social factors can somehow contribute to physical symptom reporting. As a result, an "explanatory gap" (Bakal et al. 2008) exists between what a clinician says and what the patient believes.

Efforts to bridge this explanatory gap have involved a number of distinct interventions. For example, Morris et al. (2007) performed a randomized clinical trial where primary care physicians were trained to help their patients reattribute their physical symptoms to emotional determinants. The reattribution training consisted of the physician eliciting patient reports of physical symptoms, psychosocial problems, mood states, beliefs held by patients about their problems, and so on. Next, the physician summarized the patient reports, and linked the patient's physical complaints with psychosocial aspects of their lifestyle using time or physiological factors as determinants of the linkage. This was followed by an effort to negotiate further treatment. The researchers found that reattribution lead to better physician–patient communication, but no change in the patient's belief system. Interestingly, the experimental group reported a significant lowering in their quality-of-life (as measured by a "health thermometer") compared to the nonintervention control group.

Cognitive behavioral therapy has also been used to attempt to modify the behavior of functional somatic syndrome patients (e.g., Escobar et al. 2007). The rational for this type of intervention is that symptom reports of these patients reflect catastrophization, and that modulation of this behavior would improve symptom reporting. This rationale, however, does not recognize the biological basis of the patient's complaints, and not surprisingly, physicians seem reluctant to become trained to use cognitive behavioral interventions (Salmon et al. 2007).

Bakal et al. (2008) suggested a third approach, but did not provide any empirical support for it. They recommended that clinicians take advantage of the Fink et al. (2007) demonstration (see below) that bodily distress is a construct that can account for all the varied manifestations of functional somatic syndromes. This demonstration creates therapeutic opportunities because bodily distress leads to a greater bodily awareness, and even though these patients tend to ignore those parts of their body that are not related to their symptoms, this discrepancy in focus offers the opportunity for beneficial redirection. This, Bakal et al. (2008) suggest, offers a physician the opportunity to engage a person in activities which may actually alleviate distress by distraction, if not by desensitizing their complaints. They also point out that by emphasizing somatic awareness as an intervention point, they avoid the dualism (how can the mind influence the body?) that perplexes the patient.

The recommendations of Bakal et al. (2008) are based on a study by Fink et al. (2007). Fink et al. were interested in determining if the number and type of symptoms reported, that lack a definitive physical cause, cluster into distinct factors and diagnostic entities. To determine this, Fink et al. (2007) interviewed 978 inpatients using the *Schedule for Clinical Assessment in Neuropsychiatry*. They found that women, on average, complained of six symptoms, while men complained of four symptoms. Principal component factor analysis identified three factors: a cardiopulmonary, a musculoskeletal and a gastrointestinal symptom cluster and these factors accounted for 36.9% of the variance. However, inspection of these factors revealed that very few patients totally matched the particular symptom profile; rather, patients reported symptoms from several factors. Using a latent class analysis the researchers were able to demonstrate that symptom clusters were likely to emerge in the same patients, suggesting that it was possible that a common (latent) factor determined the symptom profile for each patient. When five general symptoms (e.g., headache, dizziness, memory impairment, concentration difficulties, and fatigue) were added to the latent class analysis, it was possible to identify a three-class diagnostic system: no bodily distress (Median symptoms = 1); moderate bodily distress (Median symptoms = 9); and severe bodily distress (Median symptoms = 21.5).

The notion that a single concept, such as bodily distress, could account for symptom reporting independent of the particular symptom profile, is very revealing. When this is combined with the notion that what may be going on physiologically is that the central nervous system of these patients may have become sensitized (i.e., the "central sensitization" hypothesis; Sarkar et al. 2001; Woolf and Salter 2000; Yunus 2008), then an otherwise disparate and confusing array of symptom reports can be reduced to a common set of determinants. Still, the clinician is left with the task of regularly dealing with patients whose symptom reports lack a distinctive physical rationale for treatment, and this reinforces the notion that a significant part of clinical medicine deals with subjective factors that requires intervention skills different from standard medical training. It also supports the notion that the assessment of clinical medical phenomena may have unique assessment requirements, a topic I discuss next.

4.3 Clinimetric and Psychometric Perspectives

Feinstein's (1987a) claim that clinimetrically based HRQOL assessments are fundamentally different from psychometrically based HRQOL assessments has lead to an animated and protracted debate (e.g., Bech 2004; De Vet et al. 2003a, b; Fava and Belaisea 2005; Nierenberg and Sonino 2004; Streiner 2003a, b). Investigators who support Feinstein claim that clinimetrics provides the tools required to assess a particular clinical issue, and simultaneously avoids using methods developed for other disciplines (e.g., as applied to psychological and educational issues). They also claim that the product of each approach is fundamentally different, with clinimetrics involved in the construction of an index based on multiple indicators, and psychometrics creating a single construct using multiple items.

DeVet et al. (2003a) and Nierenberg and Sonino (2004), in a medical and psychiatric setting, respectively, promote the continued existence of clinimetrics as an independent field of study. They offer a variety of reasons to justify this position, including that a clinician's statistical concerns are clearly different than those facing the psychometrician. Bland and Altman (1986), for example, point out that clinical assessment is often indirect, and that when a clinician wants to replace one measure with another they have no known quantity with which to compare the new measure. Thus, when two methods are compared there is no unequivocally correct assessment. Instead, an investigator tries to estimate the degree of agreement between the assessments. This, however, is not a statistically straightforward task. For instance, if two measures are compared, it would be possible to have agreement only if the points lie on "the line of equality,"[13] but it is also possible to have a perfect correlation ($r = 1.00$) and the straight line cross the x- or y-axis at a point different from zero. Bland and Altman (1986), therefore, raise concerns about the use of the correlation coefficient and instead recommend simpler statistics including confidence interval estimates and various graphic displays.

Streiner (2003a) rejects the implication of the Bland and Altman (1986) paper, pointing out that the psychometrician would have the same concerns. Streiner's (2003a) basic argument is that clinimetrics is a subfield of psychometrics, and that separate fields of study are not necessary. His reasons for making this claim are based on his view that clinimetrics lacks methods which are unique onto itself. Thus, he refers to his paper (Streiner 2003c) in which he distinguished between *scales* that include homogenous items, and *indices* that include heterogeneous items. He points out that both types of measures are ordinarily found in psychometric assessments, and because of this there is no need to establish a separate field of study such as clinimetrics. He also asserts that the methods used in item selection and for establishing reliability and validity are similar between the two assessments methods. He concludes by stating:

> The conclusion is that clinimetrics is not describing a new family of techniques that should be used with a unique scale, but is simply another word for a portion of what is done in psychometrics. (Streiner 2003a; p. 1144).

De Vet et al. (2003b) responded to Streiner's (2003a) paper by claiming that psychometrics and clinimetrics are "two sides of the same coin"; two independent assessment methods working together. First, they state that in the absence of clinimetrics, it is not likely that medicine would have become involved with assessment, since psychometrics is not a discipline that ordinarily addresses clinical concerns. In addition, they claim that clinimetrics is more content dependent, while psychometrics is more statistically or mathematically driven, suggesting that the two methods use different criteria for establishing an assessment. The main point they make, however, is that two different types of assessment models exist, with one being consistent with a clinimetric approach to HRQOL assessment, and the other, a psychometric approach. The authors base this on the Fayers and Hand (2002) paper which describes two types of variables *indicator* and *causal variables*. The difference between the two, as described by Fayers and Hand (2002), is that indicator variables would not ordinarily affect the underlying construct assumed to model the variable being studied (e.g., depression or anxiety), whereas causal variables have the potential to impact the underlying construct. For De Vet et al. (2003b) the possibility that clinimetric variables cannot affect an underlying construct is sufficient to justify a separate assessment approach.

Streiner (2003b) responds to this argument by pointing out that the Fayers and Hand (2002) paper was based on a prior paper by Bollen and Lennox (1991), where structural equation modeling was used to illustrate the relationship between the underlying construct and the type of variable (indicator or causal). This, Streiner (2003b) claims, supports his view that both types of assessments fit into the theory and methods of modern psychometrics. What was also made clear in the Fayers and Hand (2002) and Bollen and Lennox (1991) papers was that *some* of the statistics ordinarily applied in a psychometric study cannot be used with indices that include causal variables. Fayers and Hand (1997) document the relevance of these concerns by showing that factor analysis (as an example of a statistical procedure) is largely irrelevant as an analytical tool when dealing with multi-item assessments that contain causal variables (e.g., data from the Rotterdam Symptom Checklist). Streiner (2003b) acknowledges that limits exist when applying specific statistical procedures to particular types of assessments, but that these limits do not justify modifying the basic assessment model (as reflected by indicator or causal variables), as clinimetrics represents.

Bilsbury and his colleagues (Singh and Bilsbury 1989; Bilsbury and Richman 2002) have developed additional applications of the clinimetric approach, now involving the assessment of psychological change following psychotherapy. Their assessment is a variant of a procedure originally developed by Shapiro (1961), where each respondent lists a series of problem areas (e.g., symptoms, mood, specific performance or activity issues, relationship concerns, concerns relative to self-esteem, and so on) and then scales these problems along a continuum from "illness to recovery" using the method of paired comparisons. By repeating this assessment regularly, a quantitative assessment is generated which is also focused on an individual's personal definition of their problems and perception of their stage of recovery at any one time. It captures what Feinstein (1967) considered to be the essence of a clinimetric approach: measuring "the cluster of individual clinical manifestations that constitutes the general clinical state" (p. 143).

Bilsbury and his colleagues (Singh and Bilsbury 1989; Bilsbury and Richman 2002) see the history of quantitative clinical outcome assessment as having progressed from the use of generic assessment methods to disease-specific methods to now patient-specific methods. To them, monitoring the course of a disorder, rather than grouping people according to some common characteristic, is the purpose of assessment, and clinimetrics achieves this.

Bech (2004) suggests another approach to these issues that uses modern (as opposed to classical) psychometrics, such as item-response theory, to incorporate clinically relevant issues into the psychometric assessment of depression. He points out that modern improvement in the diagnosis of depression started with the advent of the *Diagnostic and Statistical Manual of Mental Disorders* (DSM-IV) system that, according to Fava et al. (2004), has resulted in the reliable and valid diagnosis of depression. Fava et al. (2004) also claim that a psychometric approach to depression using the Hamilton Depression Scale (HAM-D) has actually been an obstacle to progress in clinimetric research of depression. Bech does not disagree with this assessment, and acknowledges that the HAM-D, although established on the basis of classical psychometrics, lacks item homogeneity. The impact of this lack of item homogeneity was illustrated in a study by Gram et al. (1981), who reported a curvilinear relationship between plasma levels of tricyclic antidepressants and HAM-D. These results were eventually accounted for by the heterogeneity of the items on the HAM-D. Item response theory, on the other hand, provides a method whereby items can be selected on the basis of their having a common dimension. When this is done, the resultant HAM-D6 has been shown to be more sensitive than the full HAM-D in discriminating between antidepressants and placebo (Bech et al. 2000). Bech (2004) ends his paper by stating that "Clinimetrics integrate clinical coherence with statistical coherence within the clinical framework" (p. 137), and modern psychometrics can contribute to this outcome.

Juniper et al. (1997) and Marx et al. (1999a,b) report a number of studies where the difference between a clinimetric and psychometric assessment impacted what was measured. They formally compared health status measures developed on the basis of either clinimetric or psychometric principles. Juniper et al. (1997), for example, (Streiner was a coauthor) report a study in which a clinimetric impact model (where only items a person consider most important are included) was compared to an assessment based on psychometric methods (where factor analysis was used to select items included in the assessment). The impact of an item was calculated by multiplying the importance by the frequency of the item selected. This lead to a 32-item assessment reduced from the original 152-item pool. In comparison, the assessment based on a psychometric method generated a 36-item assessment based on the same original 152-item pool. Twenty items were common to the two assessments, but the highest-impact emotional and environmental items were discarded in the psychometric-based version, and included lower impact items dealing with fatigue. Seven items from the impact model were not included in the psychometric-based assessment, and eight items on the psychometric-based assessment were not included in the impact model. The authors conclude:

> Given these differences is it possible to decide which approach is "better"? ...We believe that all items of functional impairment that are important to patients, irrespective of their association with each other, should be included in a disease-specific quality-of-life instrument and therefore we use the impact model. Those who believe that there should be a mathematical linkage between items within a questionnaire will continue to rely primarily on factor analysis. (Juniper et al. (1997); p. 237).

The Juniper et al. (1997) study demonstrated that designing an HRQOL or health status assessment on the basis of clinimetric or psychometric principles can lead to assessments that differ in terms of the number and type of items. Does this make a difference in an actual study? Marx et al. (1999a) provides an example of such a comparison. They constructed a 30-item HRQOL assessment based on either clinimetric and psychometric assessment principles. The two scales had only 16 items in common. The interclass correlation coefficient was 0.93, which improved to 0.97 after clinicians reviewed the selected items. The authors conclude:

> A scale developed with a clinimetric strategy can measure a complex (so-called heterogeneous) clinical phenomenon (thought to be composed of several patient attributes) but still fulfill psychometric criteria for 'homogeneity'. Thus, these strategies for the development of health measurement scales, which have been considered potentially opposite or conflicting, may be complementary. (Marx et al. 1999a; p. 105).

One would think that after a study of this sort, the controversy concerning the role of the two types of assessments and their capabilities would be resolved, but as the citations discussed above suggest, the discussion rages on.

For example, Ribera et al. (2006) recently reported a study in which they administered a clinimetrically based HRQOL assessment to a group of persons who had had heart attacks, but also recalibrated the assessment in psychometric terms (using factor analysis). Thus, they were able to directly compare the outcome of the same test in either clinimetric or psychometric terms. They found that the clinimetric version of the assessment seemed to account for the data better, although they acknowledged that development of an improved psychometric-based version of their assessment might also account for the original data.

At this point, it is worthwhile to review Gill and Feinstein's (1994) recommendations for improving psychometrically based quality-of-life assessments. They recommend a greater utilization of global indexes, ratings of severity and importance, and that supplemental items be permitted in an assessment. They also state that a vital task remaining for quality-of-life researchers is to develop strategies for including patient preferences in assessments. They end their paper by stating:

> The challenges arise because quality-of-life, rather than being a mere rating of health status, is actually a uniquely personal perception, representing the way that individual patients feel about their health status or nonmedical aspects of their lives. Accordingly, quality-of-life can be suitably measured only by determining the preferences of patients and supplementing (or replacing) the authoritative opinions contained in statistically 'approved' instruments. Unless greater emphasis is placed on the distinctive sentiments of patients, quality-of-life may continue to be measured with a psychometric statistical elegance that is accompanied by unsatisfactory face validity. (Gill and Feinstein 1994; p. 624).

These comments by Gill and Feinstein (1994) remain relevant today and are helpful because they focus attention on what is unique about HRQOL and what may not even now be adequately assessed by standard psychometric-based assessments. To paraphrase, the reason why clinimetrics continues to gain adherents is because it implicitly or explicitly incorporates clinically relevant *value and preference-related* issues into its assessment model, while psychometric assessments tend not to.

There is, of course, no reason why psychometric-based assessments cannot explicitly include value and preference-based components, and there are examples of such assessments in the literature (e.g., the HUI, the QWB; e.g., Ferrans 1990). What seems to be at play here is the influence of behaviorism and its emphasis on objective assessment, and a lack of appreciation of the potential contribution of cognition to assessing quality. I can demonstrate this by figuratively, "stitching a thread," with the thread being the cognitive processes that are active in each type of assessment. Thus, the bringing together of a heterogeneous collection of indicators around a common theme or clinical entity, as is common in a clinimetric assessment, is characteristic of a goal-directed category or the cross-classification of information; while a psychometric assessment with its interest in establishing causal relationship is more characteristic of a hierarchical type of classification. However, when Streiner (2003b) claims that it is not necessary to think of clinimetrics as a separate assessment model, what he may mean is that since the basis of classifying information is not important to psychometrics (e.g., a number is a number), numbers generated from, say, the cross-classification of information will be dealt with in the same manner as numbers generated from hierarchically organized information. If this is what Steiner is saying, then of course he is correct, but at the same time he is also admitting that psychometrics is a relatively blunt analytical instrument, since it fails to take into account the origin of the information that is being assessed.

These observations become important in any critical assessment of the role of latent variable in a qualitative assessment. One tactic that has been used to deal with these issues is to rely on statistical modeling (e.g., structural equation modeling), models that can be used to relate concrete indicators to a more abstract constructs, such as a latent variable. As I proceed, I will also make clear that advances in the capacity to assess HRQOL will depend on understanding how specific cognitive processes lead to different conceptualizations of quality or HRQOL.

5 Summary of Chapter

Much of this chapter has dealt with the complexities of assessing a physical symptom, and I spent much less time discussing signs. One issue I did raise in the introduction was whether signs and symptoms should be considered two extremes of a continuum, as I argued was true for subjective and objective indicators. I also asked if signs and symptoms could be added together, much as has been proposed for subjective and objective indicators. Part of the answer came from reviewing the history of the concept of objectivity, as presented by Daston and Galison (2007). This review make it clear that different methods have evolved over time to define objectivity, but what has remained constant is the struggle to establish objectivity in what is essentially a subjective milieu; how I come to know. Thus, objectivity and subjectivity are inextricably interwoven, and the historically different methods proposed (the naturalist, the scientific method, the expert) each deal with subjectivity in different ways and to varying degrees of success, so that today all three options remain active.

Does this intellectual history apply to signs and symptoms? First, re-reading Feinstein's definitions of a sign and symptom (Chap. 5, p. 123) make it clear that he considered the objective (or externally manifest) nature of a sign and the subjective nature of a symptom as defining features of each type of indicator. In addition, he was aware that a clinician would have to deal with his or her own subjectivity in trying to make objective statements, while a patient would have difficulty being "objective" when assessing their own symptoms. Thus, the struggle between objectivity and subjectivity is alive and well in clinical medicine, irrespective of Frege's claim that it is possible to separate subjective from objective statements and thereby confirm the independent existence of objectivity (Chap. 5, p. 131).

Additional information about this issue can be gained when the relationship between sign and symptom *domains* are examined. In Chap. 6 it will become clear that a domain (e.g., of symptoms, functions, and so on) can have both a biological (e.g., modular) and experiential basis. This implies that some of the components of a sign (e.g., its biological presence) can be found in what otherwise would be considered a subjective indicator, such as a symptom domain, even if the person is not aware of this connection. Thus, anemia may induce fatigue, or a growing tumor can induce pain, and the person might not be aware of the hematocrit or cancerous tumor. There are many examples of this, so many that it seems reasonable to argue that a portion of the symptoms reported will have a biological determinant, which will overlap with what would be considered a sign.

However, as was discussed in this chapter, there are some symptom reports that do not seem to have a demonstrable physical basis that is clinically detectable. The ambiguity surrounding these clinical states, however, has been significantly reduced primarily by the advent of brain imaging studies. What brain imaging has confirmed is that many of these previously considered inaccessible mental processes generate distinctive cortical patterns, with the result that their apparent subjectivity has now become visible and material. Thus, it would be expected that self-reports of fatigue, depression, or pain, which have no detectable physical basis, would still create distinctive cortical imaging patterns. Thus, signs and symptoms can be conceived of as having a common neurocognitive basis, but they differ in how it is manifest and the degree to which each can be externalized. However, making the subjective visible (i.e., objective) is a technological deficit, not an ontological limitation. This approach is a very different way of thinking about a sign and symptom, since they would not now be thought of as separate entities, like the mind and body; dualistic processes that have little or no overlap. Instead, a sign and symptom would be seen as historically distinct with one set of observations more externalized, and in this way, more likely to be characterized as being objective, while the other would in time and with technological innovation also be externalized and made objective. I believe that this is one of the lessons to learn from reviewing the history of objectivity: methods for externalizing and independently confirming the existence of a phenomenon will continue to evolve.

Also discussed in this chapter was the existence of the *subjectification* and *objectification* processes. Subjectification occurs when an indicator is linguistically restated to be "of the person," and this occurs when more affective terms are used to represent what could also be stated parametrically. This process occurs when the purpose of the assessment is to determine the person's quality-of-life. However, there are many situations where a quality assessment can occur which are not quality-of-life assessments; as when a clinician evaluates and indicates the importance of a person's symptoms and makes a diagnosis. Although the clinician's statement is a quality statement (it has the same components as a quality-of-life assessment), it can also become a quality-of-life statement if the person indicates the importance of being in this particular diagnostic and clinical state.

This, of course, may also occur when a person claims that a symptom should be part of an HRQOL assessment. For example, Levine et al. (1988), as part of the process of constructing their Breast Cancer Questionnaire asked patients about the effect of chemotherapy on their lives. The patients were then given a list of 99 items that the patients had previously selected and were asked to rate the importance of each item to themselves. What became evident was that when given the choice, symptoms were part of what a patient considered important to their life quality. What was not done in this study was to trace the meaning these various symptoms had for the person. Thus, it would have been interesting to know if hair loss was important to the person because of the consequences it had on their social and interpersonal life. Then it would have been possible to clarify if the symptoms were being included in this list because they were perceived as causes of changes in life quality.

A second part of the task of assessing quality involves taking a subjective statement (e.g., an idea or thought) and transforming it into an assessable entity. This process I have referred to as the objectification process. I refer to it as a *process* since it implies that some activity is involved, and the history of objectivity (Daston and Galison 2007) documents what activities have been available to describe this process. Thus, objectifying (externalizing) some subjective entity may be expressed by the idealized type of the naturalist, the X-ray of mechanical objectivity, an equation relating variables, or the trained judgment of the expert clinician.

Also relevant here was the discussion of whether the assessment of signs and/or symptoms is better served by clinimetric or psychometric methods. Streiner's (2003a) argument that clinimetrics is a form of psychometrics does not take into account that the most efficient assessment method should match the cognitive structure of the domain being assessed. Thus, psychometrics, with its interest in establishing causal relationships, relies on a hierarchical (e.g., relating concrete indicators to an abstract latent variable) structural analysis to perform its analytical tasks, while the primary objective of clinimetrics, being more descriptive, relies on goal-directed categories or the cross-classification of information. In this regard it is worth referring back to what Bech (2004) said when he ended his paper (Chap. 5, p. 151). To quote him again: "Clinimetrics integrate clinical coherence with statistical coherence within the clinical framework" (p. 137). While modern psychometrics can contribute to this outcome, the two types of activities are worth keeping distinct.

Also discussed in this chapter is the complex nature of a symptom, particularly symptoms forthcoming from AE. The existence of systems that function to identify and monitor

AE (e.g., CTCAE 2006) is an admission that some aspects of medical care can do harm, and that it is ethically appropriate to monitor these outcomes. Establishing these systems, of course, is one element of a quality control process, now set in an experimental medical environment. Also discussed in this section was the collection of HRQOL data during a Phase I clinical trial (Barofsky 1993/1990). As acknowledged, at the time this idea was proposed there were minimal data available to support the feasibility of the task, but in the time since, a sufficient number of studies have been performed to demonstrate that it is technically feasible and informative. Covered also in this section was the Q-TWiST (Gelber et al. 1989) methodology: a method for partitioning survival duration into time without symptoms, time with toxic symptoms, and time following recurrence. The developers of this methodology have since demonstrated that it is a useful tool for patient decision making, as well as a means of characterizing the outcome of an innovative treatment clinical trial.

The next chapter discusses the nature of a category or a domain: how they are formed and what they are made up of. This will set the stage for a discussion in the next five chapters of the role of five specific types of domains (symptoms, health status, functional status, neurocognitive processes, and well-being) in an HRQOL assessment.

Notes

1. Of course, a person's heart rate, blood pressure, or blood chemistries could become part of a quality assessment, these symptoms are qualified (i.e., a person indicates the importance these indicators have for them). Evidence that this occurs naturally can be confirmed when respondents claim that various *signs* are impeding their quality-of-life.
2. I will limit my discussion in this chapter to the assessment of a sign or symptom, and discuss how sign or symptom *domains* are formed in Chaps. 6 and 7.
3. The term "objectification" has another use different from how I am using it. In the alternative usage, objectification refers to the phenomenon that a person views his or her body as an object; something outside themselves (Fredrickson and Roberts 1997). This has been associated with the presence or eating disorders, particularly for women.
4. As will be discussed, "Grounded Field Theory" as an example of a qualitative assessment method, can reliably extract information from available data (e.g., interviews) and generate statements that have some of the properties of an invariant statement. This is so even though invariant statements generated from qualitative data may differ from those generated from quantitative data.
5. In Chap. 10, I provide another example of a neurological (dementia) deterioration and what it does to artistic expression.
6. This same question can be asked when establishing the relationship between symptoms and HRQOL. Is it the similarity, prototypicality, or exemplariness of features of a symptom that determines its relationship to HRQOL?
7. Mensuration refers to assessment, but the part of assessment or geometry concerned with ascertaining lengths, areas, and volumes. It originates from the Latin, "from mensurare" or "to measure."
8. The Kaplan et al. (2000) paper was published prior to the Sullivan paper, but both are responding to a general debate that pervades medical science, to which Feinstein has made significant contributions.
9. Defining and monitoring adverse events is a fairly complicated field involving several alternative definitions of what is adverse; application sites (e.g., clinical trials, postmarketing surveillance); settings (e.g., research or clinical); and type of intervention (e.g., drugs, biological agents, various device, so on). There are also different agencies involved in this type of activity, such as the World Health Organization, the Food and Drug Administration, the European Union, etc. The WHO organization definition of adverse events is "An injury related to medical management, in contrast to complications of disease. Medical management includes all aspects of care, including diagnosis and treatment, failure to diagnose or treat, and the systems and equipment used to deliver care. Adverse events may be preventable or nonpreventable." (WHO 2005; p. 8).
10. The issue of *explicit* and *implicit valuations* is a topic I will continue to discuss in this book. There are many more instances of implicit valuations, as opposed to formal explicit valuations, and the extent of this should reinforce the cognitive model of quality assessment being promoted in this book.
11. A typical dose-escalation protocol would have three subjects being exposed to a particular dose, until at least one person experiences a toxic response. Then three more people are exposed to the dose, and if none of the participants respond to this dose, they are exposed to the next higher dose, but if one of the three did respond, then the MTD would be said to have been reached. Dose-finding, however, is not unique to Phase I trials, since it is also found in Phase II and III, as well as in clinical practice.
12. At the time of presentation of this paper (October 1990), there were only two papers published that supported my argument: Berkel et al. (1988), which I was not aware of and did not cite; and an abstract by Melink et al. (1985), which I cited, and which was eventually published as a full paper in 1992.
13. The "line of equality" is a 45-degree angle with the origin at 0,0.

References

Aaronson NK, Ahmedzai S, Bullinger M, Crabeels D, et al. (1991). The EORTC core quality-of-life questionnaire: Interim results of an international field study. In, (Ed.) D. Osoba, *The Effect of Cancer on Quality-of-life*. Boca Raton FL: CRC Press. (p. 185–203).

Ádám G. (1980). *Perception, Consciousness, Memory; Reflections of a Biologist*. New York NY: Plenum.

Alexander F. (1950). *Psychosomatic Medicine*. New York NY: Norton.

Alexander S, French T, Pollock G. (1968). *Psychosomatic Specificity*. Chicago IL: University of Chicago Press.

Apgar V. (1953). Proposal for a new method evaluation of the newborn infant. *Anesthesiol Analg*. 32, 260–267.

Badr H, Basen-Engquist K, Taylor CLC, de Moor C. (2006). Mood states associated with transitory physical symptoms among breast and ovarian cancer survivors. *J Behav Med*. 29, 461–475.

Bakal D, Coll P, Schaefer J. (2008). Somatic awareness in the clinical care of patients with body distress symptoms. *Bio Psycho Soc Med*. 2:6 doi:10.1186/1571-0759-2-6.

Barofsky I. (1993). Integrating quality-of-life assessments into Phase I clinical trials. In, (Eds.) CD Furberg, JA Schulttinger, SA Shumaker, NK Wenger. *Quality-of-life Assessment: Practice, Problems, and Promise. Proceedings of a Workshop*, October 15–17, 1990. National Institutes of Health. (p. 47–49).

Barsky AJ, Borus JF. (1999). Functional somatic syndromes. *Ann Inter Med*. 30, 910–921.

Bech P. (2004). Modern psychometrics in clinimetrics: Impact on clinical trials of antidepressants. *Psychother Psychosom.* 73: 134–138.

Bech P, Cialdella P, Haugh M, Birkett MA, et al. (2000). A meta-analysis of randomized controlled trials of fluoxetine versus placebo and tricyclic antidepressants in the short-term treatment of major depression. *Brit J Psychiatr.* 176, 421–428.

Berkel WG, Knopf H, Fromm M, et al. (1988). Influence of phase I clinical trials on quality-of-life of cancer patients. *Anticancer Res.* 8, 313–322.

Bilsbury CD, Richman A. (2002). A staging approach to measuring patient-centered subjective outcomes. *Acta Psychiatr Scand.* 106, 5–40.

Bland JM, Altman DG. (1986). Statistical methods for assessing agreement between two methods of clinical measurement. *Lancet.* 1: 307–310.

Blanton H, Jaccard J. (2006). Arbitrary metrics in psychology. *Am Psychol.* 61, 27–41.

Bollen K, Lennox R. (1991). Conventional wisdom on measurement: A structural equation perspective. *Psychol Bull.* 110, 305–314.

Boroditsky L. (2001). Does language shape thought? Mandarin and English speakers conceptions of time. *Cogn Psychol.* 43, 1–22.

Borsboom D. (2005). *Measuring the Mind: Conceptual Issues in Contemporary Psychometrics.* Cambridge UK: Cambridge University Press.

Brauer M, Chambres P, Niedenthal PM, Chatard-Panetier A. (2004). The relationship between expertise and evaluative extremity: The moderating role of experts task characteristics. *J Pers Soc Psychol.* 86, 5–18.

Bridgman PW. (1959). *The Way Things Are.* New York NY: Viking.

Cantor N, Smith EE, de sales French R, Mezzich J. (1980). Psychiatric diagnosis as prototype categorization. *J Abnorm Psychol.* 89, 181–193.

Cella DF. (1997). The Functional Assessment of Cancer Therapy-Anemia (Fact-An) scale: A new tool for the assessment of cancer anemia and fatigue. *Semin Hematol.* 34 (3, Suppl 2), 13–19.

Cella DF, Tulsky DS, Gray G, Sarafian B, et al. (1993). The Functional Assessment of Cancer Therapy (FACT) scale: Development and validation of the general measure. *J Clin Oncol.* 11, 570–579.

Chang CH, Cella DF, Clarke S, Heinemann AW, et al. (2003). Should symptoms be scaled for intensity, frequency or both? *Palliation Support Care* 1, 51–60.

Cleeland CS. (1989). Measurement of pain by subjective report. In, (Eds.) CR Chapman, JD Losser. *Advances in Pain Research and Therapy.* Vol.12 *Issues in Pain Measurement.* New York NY: Raven Press. (p. 391–403).

Cleeland CS. (2007). Symptom burden: multiple symptoms and their impact as patient-reported outcomes: Review. *J Nat'l Cancer Inst Monogr.* 37, 16–21.

Cleeland CS, Mendoza TR, Wang XS, Chou C, et al. (2000). Assessing symptom distress in cancer patients: *The M.D. Anderson Symptom Inventory. Cancer.* 89, 1634–1646.

Cohen L, Parker PA, Sterner J. de Moor C. (2002a). Quality-of-life in patients with malignant melanoma participating in Phase I trial of an autologous turmor-derived vaccine. *Melanoma Res.* 12, 505–511.

Cohen L, de Moor C, Parker PA, Amato RJ. (2002b). Quality-of-life of patients with metastatic renal cell carcinoma participating in a Phase I trial of an autologous tumor derived-vaccine. *Urol Oncol.* 7, 119–124.

Cronbach LJ, Meehl PE. (1955). Construct validity and psychological tests. *Psychol Bull.* 52, 281–302.

CTCAE. (2006). The National Cancer Institute Cancer Therapy Evaluation Program (CTCAE v3.0); August 9th 2006 (http://ctep.cancer.gov/reporting/ctc. HTML).

Daston L, Galison P. (2007). *Objectivity.* New York NY: Zone Books.

Dawes RM, Faust D, Meehl PE. (1989). Clinical versus actuarial judgment. *Science.* 243, 1668–1674.

De Vet HCW, Terwee CB, Bouter LM. (2003a). Current challenges in clinimetrics. *J Clin Epidemiol.* 56, 1137–1141.

De Vet HCW, Terwee CB, Bouter LM. (2003b). Clinimetrics and psychometrics: two sides of the same coin. *J Clin Epidemiol.* 56, 1146–1147.

Embretson SE. (2006). The continued search for nonarbitrary metrics in psychology. *Am Psychol.* 61, 50–55.

Epstein SK. (2007). Weaning the "unweanable": Liberating patients from prolonged mechanical ventilation. *Crit Care Med.* 335, 2640–2641.

Escobar JI, Gara MA, Diaz-Martinex AM, Interian A, et al. (2007). Effectiveness of a time-limited cognitive behavior therapy-type intervention among primary care patients with medically unexplained symptoms. *Annu Fam Med.* 5, 328–335.

Esper P, Mo F, Chodak G, Sinner M, et al. (1997). Measuring quality-of-life in men with prostate cancer using the Functional Assessment of Cancer Therapy-Prostate (FACT-P) instrument. *Urology.* 50, 920–928.

Esterling D, Leventhal H. (1989). Contribution of concrete cognition to emotion neural symptoms as elicitors of worry about cancer. *J Appl Psychol.* 74, 787–796.

Estey E, Hoth D, Simon R, Marsoni S, et al. (1986). Therapeutic response in Phase I trials in antineoplastic agents. *Cancer Treat Rep.* 70, 1105–1115.

Eva KW, Cunnington JP. (2006). The difficulty experience: Does practice increase susceptibility to premature closure. *J Continu Educ Health Prof.* 26, 192–198.

Fava GA, Belaisea C. (2005). A discussion on the role of clinimetrics and the misleading effects of psychometric theory. *J Clin Epidemiol.* 58, 753–756.

Fava GA, Ruini C, Rafanelli C. (2004). Psychometric theory is an obstacle to progress of clinical research. *Psychother Psychosom.* 73, 145–148.

Fayers PM, Hand DJ. (2002). Causal variables, indicator variables and measurement scales: An example from quality-of-life. *J R Soc Stat.* 165, 233–261.

Fayers PM, Hand DJ. (1997). Factor analysis, causal indicators and quality of life. *Qual Life Res.* 6, 139–150.

Feinstein A. (1967). *Clinical Judgment.* Huntington NY: Krieger

Feinstein AR. (1987a). Clinimetric perspective. *J Chronic Dis.* 40, 635–640.

Feinstein AR. (1987b). *Clinimetrics.* New Haven NY: Yale.

Ferrans CE. (1990). Development of a quality-of-life index for patients with cancer. *Oncol Nurs Forum.* 17, 29–38.

Ferrans CE. (2007). Differences in what Quality-of-Life instruments measure. *J Nat'l Cancer Inst Monogr.* 37, 1–9.

Fink P, Toft T, Hansen MS, Ørnbol E, et al. (2007). Symptoms and syndromes of bodily distress: An explanatory study of 987 internal medical neurological, and primary care patients. *Psychosom Med.* 69, 30–39.

Fleishman SB. (2004). The treatment of symptom clusters: Pain, depression and fatigue. *J Nat'l Cancer Inst Monogr.* 32, 119–123.

Folstein MF, Folstein SE, McHugh PR. (1975) Mini-mental state: A practical method for grading the cognitive state of patients for the clinician. *J Psychiatr Res.* 12: 189–198.

Fredrickson BL, Roberts T-A. (1997). Objectification Theory: Towards understanding women's lived experiences and mental health risks. *Psychol Women Q.* 21, 173–206.

Frege, G. (1892/1966). On Concept and Object, in (eds.) P. Geach, M. Black, *Translations from Philosophical Writings of Gottlob Frege.* Oxford UK: Oxford University Press. (p. 42).

Gelber RD, Goldhirsch A, and the Ludwig Breast Cancer Group. (1986). A new endpoint for the assessment of operable breast cancer. *J Clin Oncol.* 4, 1772–1779.

Gelber RD, Goldhirsch A. (1989). Comparison of adjuvant therapies using quality-of-life considerations. *Int J Technol Assess Health Care.* 5(3):401–413.

Gendolla GHE, Abele A, Prurk D, Richter M. (2005). Negative mood, self-focused attention and the experience of physical symptoms: The Joint Impact Hypothesis. *Emotion.* 5, 131–144.

Gerber LR. (2008). Targeted therapies: A new generation of cancer treatments. *Am Fam Physician.* 77, 311–319

Gijsbers van Wijk CMT, Kolk AM. (1997). Sex differences in physical symptoms: The contribution of symptom perception theory. *Soc Sci Med.* 45, 231–246.

Gill TM, Feinstein AR. (1994). A critical appraisal of the quality of quality-of-life measurements. *JAMA.* 272, 619–626.

Giraldo B, Garde A, Benito JS, Diaz J, Ballesteros D. (2006). Support vector machine classification applied on weaning trials patients. *Proceedings of the 28th IEEE EMBS Annual International Conference*, New York City, USA, August 30-September 3, 2006. Accessed from the Web 12/28/2007.

Glannon W. (2006). Phase I oncology trials: Why the therapeutic misconception will not go away. *J Med Ethics.* 32, 252–255.

Glaser BG. (1992). *Emergence vs Forcing: Basics of Grounded Theory Analysis.* Mill Valley, CA: Sociology Press.

Glaser BG, with the assistance of Holton J. (2004). Remodeling grounded theory. *Forum Qual Soc Res.* 5(2), Art. 4. Available at: http://www.qualitative-research.net/fgs-texte/2-04glaser-e.htm. (Assessed; 12/20/2007).

Glaser BG, Strauss A. (1967). *The Discovery of Grounded Theory: Strategies for Qualitative Research.* New York NY: Aldine.

Goldberg LR. (1991). Human mind versus regression equation: Five contrasts. In, (Eds.) D Cicchetti, DE Gerber. Thinking Clearly About Psychology: Essays in Honor of Paul E. Meehl. Minneapolis Minnesota: University of Minnesota Press.

Gottschalk A, Hyzer C, Geer RT. (2000). A comparison of human and machine-based predictions of successful weaning from mechanical ventilation. *Med Decis Mak.* 20, 160–169.

Gram LF, Bech P. Reisby N, Jørgenson OS. (1981). Methodology in studies on plasma level/effect relationships of tricyclics antidepressants. In, (Ed.) E. Usdin *Clinical Pharmacology in Psychiatry.* Amsterdam, North-Holland:Elsevier. (p. 155–171).

Grove WM, Meehl PE. (1996). Comparative efficiency of informal (subjective, impressionistic) and formal (mechanical, algorithmic) prediction procedures: The clinical-statistical controversy. *Psychol Public Policy Law.* 2, 293–323.

Haig TH, Scott DA, Stevens GB. (1989). Measurement of the discomfort component of illness. *Med Care.* 27, 280–287.

Hammond KR, Stewart TR, Brehmer B, Steinmann DO. (1975). Social Judgment Theory. In, (Eds.) MF Kaplan, S Schwartz. *Human Judgment and Decision Processes: Formal and Mathematical Approaches.* New York NY: Academic Press. (p. 272–312).

Heath H, Cowley S. (2004). Developing a grounded theory approach: a comparison of Glaser and Strauss. *Int J Nurs Stud.* 41, 141–150.

Helmholtz H. (1869/1954). *On the Sensation of Tone as a Physiological Basis for the Theory of Music.* Mineola NY: Dover Publishing.

Henderson GE, Easter MM, Zimmer C, King NMP, et al. (2006). Therapeutic misconception in early phase gene transfer trials. *Soc Sci Med.* 62, 239–253.

Hollen PJ, Gralla RJ, Kris MG, Potanovich LM. (1993). Quality-of-life assessment in individuals with lung cancer: Testing the Lung Cancer Symptom Scale (LCSS). *Eur J Cancer.* 29A, S51–S58.

Hollen PJ, Gralla RJ, Rittenberg CN. (2004). Quality-of-life as a clinical trial endpoint: Determining the appropriate interval for repeated assessment in patients with advanced lung disease. *Support Care Cancer.* 12, 767–773.

Horowitz M, Wilner N, Alvarez W. (1979). Impact of events scale: Measure of subjective stress. *Psychosom Med.* 41, 209–218.

Horstmann E, McCabe MS, Grochow L, Yamamoto S, Rubinstein L, et al. (2005). Risks and benefits of Phase I oncology trials 1991 through 2002. *NEJM.* 352, 895–904.

Issell BF, Gotay CC, Pagano IS. (2007). Quality-of-life assessment in a Phase I trial of noni. Poster presentation at the 12th Annual Conference of the International Society for Quality-of-life Research, San Francisco, CA. (October, 2007)

Jairath N. (1999). Myocardial infarction patients' use of metaphors to share meaning and communicate their frame of reference. *J Adv Nurs.* 29, 283–289.

Jonas DL, Kline Leidy N, Siberman C, Margolis MK, Heyes A. (2001). Cognitive factors in symptom assessment: Are we asking too much? Abstract Presented at the Annual Meeting of the International Society of Quality-of-life Research, Amsterdam The Netherlands. 2001. *Qual Life Res.* 10, 274.

Jones RA, Weise HJ, Moore RW, Haley JV. (1981). On the perceived meaning of symptoms. *Med Care.* 19, 710–717.

Juniper EF, Guyatt GH, Streiner DL, King DR. (1997). Clinical impact versus factor analysis for quality-of-life questionnaire construction. *J Clin Epidemiol.* 50, 233–238.

Kaplan EL, Meier P. (1958). Nonparametric estimation from incomplete observations . *J Am Stat Assoc.* 53, 457–481.

Kaplan RM, Anderson JP. (1990). The General Health Policy Model: An integrated approach. In, (Ed.) B Spilker. *Quality-of-life Assessments in Clinical Trials.* New York NY: Ravens. (131–149).

Kaplan SH, Kravitz RL, Greenfield S. (2000). A critique of current uses of health status for the assessment of treatment effectiveness and quality of care. *Med Care.* 38 (Supp. II), 184–191.

Kazdin AE. (2006). Arbitrary metrics: Implications for identifying evidence-based treatments. *Am Psychol.* 61, 42–49.

Kim NS, Ahn W-K. (2002). Clinical psychologists' theory-based representations of mental disorders predict their diagnostic reasoning and memory. *J Exp Psychol: Gen.* 131, 451–476.

Kirkiva J, Davis MP, Walsh D, Tiernan E, et al. (2006). Cancer symptom assessments instruments: A systematic review. *J Clin Oncol.* 24, 1459–1473.

Kleinmuntz B. (1990). Clinical and actuarial judgment. *Sci.* 247, 146.

Koop RR. (1995). *Metaphor Therapy: Using Client Generated Metaphors in Psychotherapy.* Bristol PA: Brunner-Mazel.

Kövecses Z. (2000). *Metaphors and Emotion: Language, Culture and Body in Human Feeling.* Cambridge UK: Cambridge University Press .

Kushniruk AW, Patel VL, Marley AAJ. (1998). Small worlds and medical expertise: Implications for medical cognition and knowledge engineering. *Int J Med Inf.* 49, 255–271.

Lakoff G, Johnson M. (1999). *Philosophy in the Flesh: The Embodied Mind and its Challenge to Western Thought.* New York NY: Basic Books.

Lau HC, Rogers RD, Haggard P, Passingham RE. (2004). Attention to intention. *Science.* 303, 1208–1210.

Lawton MP. (1991). A multidimensional view of quality-of-life in frail elderly. In, (Eds.) JE Birren, JE Lubben, JC Rowe, DE Deutchman. *The Concept and Measurement of Quality-of-life in Frail Elderly.* New York NY: Academic Press. (p. 3–27).

Lazarus RS. (1991). *Emotion and Adaptation.* New York NY: Oxford University Press.

Leventhal EA, Hansell S, Diefenbach M, Leventhal H, et al. (1996). Negative affect and self-report of physical symptoms: Two longitudinal studies of older adults. *Health Psychol.* 15, 193–199.

Levine MN, Guyatt GH, Gent M, De Pauw S, et al. (1988). Quality-of-life in Stage II breast cancer: An instrument for clinical trials. *J Clin Oncol.* 6, 1798-1810.

Lim SJ, Cormier JN, Feig BW, Mansfield PF, et al. (2007). Toxicity and outcomes associated with surgical cytoreduction and hyperthermic intraperitoneal chemotherapy (HIPEC) for patients with sarcomatosis. *Ann Surg Oncol.* 14, 2309–2318.

Lipowski ZJ. (1984). What does the word "psychosomatic" really mean? A historical and semantic inquiry. *Psychosom Med.* 46, 152–171.

Lord F. (1953). On the statistical treatment of football numbers. *Am Psychol.* 8, 750–751.

LoRusso PM, Herbst RS, Rischin D, Ranson M, et al. (2003). Improvements in quality-of-life and disease-related symptoms in Phase I trials of selective oral epidermal growth factor receptor tyrosine kinase inhibitor ZD1839 in non-small cell lung cancer and other solid tumors. *Clin Cancer Res.* 9, 2040–2048.

Marx RG, Bombardier, Hogg-Johnson S, Wright JG. (1999). Clinimetric and psychometric strategies for development of a health measurement scale. *J Clin Epidemiol.* 52, 105–111.

Marx RG, Bombardier C, Hogg-Johnson S, Wright JG. (1999). How should importance and severity ratings be combined for item reduction in the development of health status instruments? *J Clin Epidemiol.* 52, 193–197.

McCorkle R, Young K. (1978). Development of a symptom distress scale. *Cancer Nurs.* 1, 373–378.

McCrae RR, Bartone PT, Costa PT. (1976). Age, anxiety and self-related health. *Aging Hum Dev.* 7, 49–58.

Meehl PE. (1954). *Clinical Versus Statistical Prediction.* Minneapolis MN: University of Minnesota Press.

Mendoza T, Wang XS, Cleeland CS, Morrissey M, et al. (1999). The Rapid Assessment of Fatigue Severity in Cancer Patients: Use of the Brief Fatigue Inventory. *Cancer.* 85, 1186–1196.

Melhem A, Stern M, Shibolet O, Isreali E, et al. (2005). Treatment of chronic Hepatitis C virus vai antioxidants: Results of a Phase I clinical trial. *J Clin Gastroenterol.* 39, 737–742.

Melink TJ, von Hoff DD, Clark GM, Coltman CA. (1985). Impact of Phase I clinical trial trials on the quality-of-life and survival of cancer patients. *Proc Am Soc Clin Oncol.* 4, 251.

Melink TJ, Clark GM, von Hoff DD. (1992). The impact of Phase I clinical trials on the quality-of-life of patients with cancer. *Anti-Cancer Drugs.* 3, 571–576.

Melzack R, Katz J. (1992). The McGill Pain Questionnaire: Appraisal and current status. In, (Eds.) DC Turk, R Melzack. *Handbook of Pain Assessment.* New York NY: Guilford. p. 152–168.

Michael M, Wirth A, Ball DL, MacManus M, et al. (2005). A phase I trial of high-dose palliative radiotherapy plus concurrent weekly Vinorelbine and Cisplatin in patients with locally advanced and metastatic NSCLC. *Brit J Cancer.* 93, 652–661.

Morris R, Dowrick C, Salomon P, Peters S, et al. (2007). Cluster randomized controlled trial of training practices in reattributions for medically unexplained symptoms. *Brit J Psychiatr.* 191, 536–542.

Murphy ST, Zajonc RB. (1993) Affect, cognition and awareness affective priming with optimal and suboptimal stimulus exposures. *J Personal Soc Psychol.* 64, 723–739.

Nagel T. (1986). *The View from No Where.* New York NY: Oxford University Press. (p. 5).

National Institutes of Health: State-of-the-Science Conference Statement. (2004). Symptom management in cancer: Pain, depression and fatigue, July 15–17, 2002. *J Nat'l Cancer Inst Monogr.* 32, 9–16.

Nierenberg AA, Sonino N. (2004). From clinical observations to clinimetrics: A tribute to Alvan R. Feinstein, MD. *Psychother Psychosom.* 73, 131–133.

Norman G, Eva K, Brooks L, Hamstra S. (2006). Expertise in medicine and surgery. In, (Eds.) CN Ericsson, R Hoffman, Feltovich PJ. *Cambridge Handbook of Expertise and Expert Performance.* New York NY: Cambridge University Press. (p. 339–353).

Otis AS. (1917). A criticism of the Yerkes-Bridges Point Scale, with alternative suggestions. *J Educ Psychol.* 8, 129–150.

Oxford English Dictionary. (1996). Oxford, UK: Oxford University Press.

Paice JA. (2004). Assessment of symptom clusters in people with cancer. *J Natl Cancer Inst Monogr.* 32, 98–102.

Palmer JL, Fisch MJ. (2005). Association between symptom distress and survival in outpatients seen in a palliative care cancer center. *J Pain Symptom Manag.* 29, 565–571.

Pennebaker JW. (1982). *The Psychology of Physical Symptoms.* New York NY: Springer-Verlag.

Pond GR, Siu LL, Moore M,Oza A, et al. (2008). Nomograms to predict serious adverse events in Phase II clinical trials of molecularly targeted agents. *J Clin Oncol.* 26, 1324–1330.

Portenoy RK, Thaler HT, Kornblith AB, McCarthy LJ, et al. (1994). The Memorial Symptom Assessment Scale: an instrument for the evaluation of the symptom prevalence, characteristics and distress. *Eur J Cancer.* 30A, 1326–1336.

Radloff LS. (1977). The CES-D Scale: a self-report depression scale for research in the general population. *Appl Psychol Meas.* 3, 385–401.

Ranganath KA, Nosek BA. (2008). Implicit attitude generalization occurs immediately; Explicit attitude generalization takes time. *Psychol Sci.* 19, 249–254.

Rasch G. (1977). On specific objectivity: An attempt at formalizing the request for generality and validity of scientific statements. *Dan Yearbook Philos.* 14, 58–94.

Ratakonda S, Gorman JM, Yale SA, Amador XF. (1998). Characterization of psychotic conditions. *Arch Gen Psychiatr.* 55, 75–81.

Rees CE, Knight LV, Wilkinson CE. (2007). Doctors being up there and we being down here: A metaphorical analysis of talk about student/doctor-patient relationships. *Soc Sci Med.* 65, 725–737L.

Rehber B. (2003). Categorization as causal reasoning. *Cogn Sci.* 27, 709–748.

Ribera A, Permanyer-Miralda G, Alonzo J, Soriano N, Brotons C. (2006). Is psychometric scoring of the Mc New Quality-of-life questionnaire superior to the clinimetric scoring? A comparison of the two approaches. *Qual Life Res.* 15, 357–365.

Ritchie J. (1973). Pain from distention of the pelvic colon by inflating a balloon in the irritable bowel syndrome. *Gut.* 14, 125–132.

Sacks O. (1985). *The Man who Mistook his Wife for a Hat and Other Clinical Tales.* New York NY: Summit Books.

Salmon P, Peters S, Clifford R, Iredale W, et al. (2007). Why do general practitioners decline training to improve management of medically unexplained symptoms? *J Gen Inter Med.* 22, 565–571.

Samarel N, Leddy SK, Greco K, Cooley ME, et al. (1996). Development and testing of the Symptom Experience Scale. *J Pain Symptom Manag.* 12, 221–228.

Sarkar S, Hobson AR, Furlong PL, Woolf CJ, et al. (2001). Central neural mechanisms mediating human visceral hypersensitivity. *Am J Physiol Gastrointest Liver Physiol.* 281, G1196–1202.

Schuster MM, Crowell MD, Koch KL. (2002). *Schuster Atlas of Gastrointestinal Motility in Health and Disease.* Hamilton Canada BC: Decker.

Scriven M. (1979). Clinical judgment. In, (Eds.) HT Engelhardt Jr, SF Spricker, B. Towers, *Clinical Judgment: A Critical Appraisal.* Dordrecht Holland: Reidel.

Shamoo AE, Resnik DB. (2006). Strategies to minimize risks of exploitation in Phase I trials on healthy subjects. *Am J Bioeth.* 6, 1–13.

Shapiro MB. (1961). A method of measuring psychological change specific to the individual psychiatric patient. *Brit J Med Psychol.* 34, 151–155.

Singer T, Seymour B, O'Doherty J, Kaube H, et al. (2004). Empathy for pain involves affective but not sensory components of pain. *Science.* 303,1157-1162.

Singh AC, Bilsbury CD. (1989). Measurement of subjective variables: The Discan method. *Acta Psychiatr Scad.* 79 (Suppl 347), 1–38.

Skårderud F. (2007). Eating one's words, Part I: 'Concretized Metaphors' and reflective function in anorexia nervosa – An interview study. *Eur Eat Disord Rev.* 15, 163–174.

Skelton JR, Wearn AM, Hobbs FD. (2002). A concordance-based study of metaphoric expressions used by general practitioners and patients. *Brit J Gen Pract.* 52, 114–118.

Skeat J, Perry A. (2008). Grounded theory as a method of research in speech and language therapy. *Int J Lang Commun Disord.* 43, 95–107.

Sloan JA, Loprinzi CL, Novotny PJ, Barton DL, et al. (2001). Methodologic lessons learned from hot flash studies. *J Clin Oncol.* 19, 4280–4290.

Sontag S. (1991). *Illness as Metaphor and AIDS and its Metaphors.* New York NY: Picador USA.

Strauss A, Corbin J. (1998). *Basics of Qualitative Research: Techniques and Procedures for Developing Grounded Theory.* Thousand Oaks CA: Sage.

Streiner DL. (2003a). Clinimetrics vs, psychometrics: an unnecessary distinction. *J Clin Epidemiol.* 56, 1142–1145.

Streiner DL. (2003b). Test development: two-sided coin or one-sided Möbius strip? *J Clin Epidemiol.* 56, 1148–1149.

Streiner DL. (2003c). Being inconsistent about inconsistency: when coefficient alpha does and doesn't matter. *J Personal Assess.* 80, 217–222.

Sullivan M. (2003). The new subjective medicine: Taking the patient's point of view on health care and health. *Soc Sci Med.* 56, 1595–1604.

Tehrani FT. (2007). A new decision support system for mechanical ventilation. Proceedings of the 29th *Annual International Conference of the IEEE EMBS Cité Internationale*, Lyon, France. August 23–26, 2007. Accessed from the Web, 12/28/2007.

Tessler R, Mechanic D. (1978). Psychological distress and perceived health status. *J Health Soc Behav.* 19, 254–262.

Tester W, Ackler J, Tijani L, Leighton J. (2006). Phase I/II study of weekly docetaxel and vinblastine in the treatment of metastatic hormone-refractory prostate cancinoma. *Cancer J.* 12, 299–304.

Thurstone LL. (1927). A law of comparative judgment. *Psychol Rev.* 34, 273–286.

Thurstone LL. (1928). The absolute zero in measurement of intelligence. *Psychol Rev.* 35, 175–197.

Tishelman C, Degner LF, Rudman A, Bertilsson K, et al. (2005). Symptoms in patients with lung carcinoma: distinguishing distress from intensity. *Cancer.* 104, 2013–2021.

Tishelman C, Petersson L-M, Degner LF, Sprangers MAG. (2007). Symptom prevalence. intensity, and distress in patients with inoperable lung cancer in relation to time of death. *J Clin Oncol.* 25, 5381–5389.

Torrance GW, Furlong W, Feeny D, Boyle M. (1995). Multi-attribute preference functions: Health Utilities Index. *Pharm Econ.* 7, 503–520.

Townsend JT, Ashby FG. (1984). Measurement scales and statistics: The misconception misconceived. *Psychol Bull.* 96, 394–401.

Verbrugge L. (1989). The twain meet: Empirical explanations of sex differences in health and mortality. *J Health Soc Behav.* 30, 282–304.

Verbrugge L, Ascione FJ. (1987). Exploring the iceberg: Common symptoms and how people care for them. *Med Care.* 25, 539–569.

Walsh D, Rybicki L. (2006). Symptom clustering in advanced cancer. *Support Care Cancer.* 14, 832–836.

Ware JE, Snow KK, Kosinski MA, Gandek B. (1993). *SF-36 Health Survey: Manual and Interpretation Guide.* Boston MA: The Health Institute, New England Medical Center.

Watson D. (1988). Intra-individual and inter-individual analysis of positive and negative affect; Their relation to health complaints, perceived stress and daily activities. *J Personal Soc Psychol.* 54, 1020–1030.

Watson D, Pennebaker JW. (1989). Human complaints, stress, and distress: Exploring the central role of negative affectivity. *Psychol Rev.* 96, 234–254.

Wearden AJ. (2005). Adult attachment, alexithymia, and symptom resporting: An extension of the four category model of attachment. *J Psychosom Res.* 58, 279–288.

Weiss R. (2007). Synthetic DNA on the brink of yielding new life forms. Washington Post December 17, 2007. Accessed on the web; http://www.washingtpost.com/wp-dyn/content/article/2007/12/16/2007. December 17, 2007 at 11:36 AM

Westen D, Weinberger J. (2004). When clinical descriptions become statistical predictions. *Am Psychol.* 59,595-613.

Whitehead WE, Holtkotter B, Enck P, Hoelzl R, et al. (1990). Tolerance for rectosigmoid distention in irritable bowel syndrome. *Gastroenterol.* 98, 1187–1192.

Wilson IB, Cleary PD. (1995). Linking clinical variables with health-related quality-of-life: A conceptual model of patient outcomes. *JAMA.* 273, 59–65.

Woolf CJ, Salter MW. (2000). Neuronal plasticity: Increasing the gain in pain. *Science.* 288, 1765–1769.

Woolrich RA, Cooper MJ, Turner HM. (2008). Metacognition in patients with anorexia nervosa, dieting and non-dieting women: A preliminary study. *Eur Eat Disorders Rev.* 16, 11–26.

World Health Organization Draft Guidelines for Adverse Event Reporting and Learning Systems. (2005). Geneva Switzerland: WHO. (p. 8).

Yunus MB. (2008). Central sensitivity syndromes: A new paradigm and group nosology for fibromyalgia and overlapping conditions, and the related issue of disease versus illness. *Semin Arthritis Rheum.* 37, 339–52.

Zimmerman J. (2003). Cleaning up the river: A metaphor for functional digestive disorders. *Am J Hypn.* 45, 353–359.

Summary Measurement: The Role of Categories or Domains

Abstract

The focus of this chapter is on how the constant stream of information that a person experiences every day is summarized into units that are used to communicate or provide the basis for quantification. Two methods are described: those based on the person using cognitive processes to form categories of information and those that involve the investigator using statistical procedures to create domains. In addition, I describe how the content (modular or nonmodular) of these summary statements can vary and what it implies about the summary measure formed. Finally, I illustrate how differences in the content of a domain and the quality-of-life assessment used (generic or disease specific) contribute to the predictive ability of a quantitative assessment.

Abbreviations

ADVS	Activities of Daily Vision Scale (Manguione et al. 1992)
CAT	Computer adaptive testing
CL	Confidence limits
DSM	Diagnostic and statistical manual (American Psychiatric Association 2000)
EQ-5D	EuroQol group 5 dimensions (Brooks et al. 2003)
Generic	Generic quality-of-life assessment
HADS	Hospital Anxiety and Depressions Scale (Zigmond and Snaith 1983)
HRQWOL	Health-related quality-of-life
IRT	Item response theory
LogMAR	Log of minimal angle of resolution
PROs	Patient-reported outcomes
RSCL	Rotterdam symptom checklist (De Haes et al. 1990)
SF-36	Medical outcome study short form 36 (Ware and Sherbourne 1992)
SF-36 PFS	Physical functioning items of the SF-36
SG	Standard gamble utility measure
SG-VH	Standard gamble-visual health
SIP	Sickness Impact Scale (Bergner et al. 1981)
SVA	Snellen visual acuity
SWB	Subjective well-being
TyPE	The TyPE specification assesses visual functioning in five dimensions: (1) distance vision; (2) near vision; (3) daytime driving; (4) nighttime driving; and (5) glare
VA	Visual acuity
VF-14	Visual functioning 14 items (Steinberg et al. 1994)
Vision-Specific	Vision-specific HRQOL assessment
VR	Verbal utility report
WHO/PBD VF20	World Health Organization Standards for Vision 2003
WMC	Working memory capacity

1 Introduction

One of the distinguishing characteristics of making a judgment about the quality of an object, event, or a life lived is that it is usually based on multiple experiences. Sometimes the same experience occurs repeatedly; sometimes past experiences are used to make a current judgment; and sometimes the judgment is based on a new experience and guided by how a person has made judgments in the past. To illustrate, when I judge the quality of an object, I already have had a lot of experiences with objects to compare it with, or I may have noted something unique about this particular object that sparks my interest and contributes to my judgment. However, it would be very awkward for me to list all these experiences, and it should not be surprising that I would want to somehow summarize this information. Interestingly, the brain has the same problem and also solves it by collecting experiences together to form categories or concepts. In fact, both the person (e.g., categories or concepts) and the investigator (e.g., domains) have developed methods for summarizing the continuous stream of information that characterizes life's experiences. In this chapter, I will describe some of the determinants of these summary statements, and in this way describe the role that a category or a domain plays in a qualitative assessment (see also Chap. 3).

The summary statement can also be viewed as a unit of analysis that a person cognitively processes, or which an investigator statistically assesses. By distinguishing between a person-generated category and an investigator-generated domain, I am implying that there are two somewhat independent sources for summarizing information, and it is possible that the voice of the person and goals of the investigator may not coincide. However, these different classification methods have a common set of cognitive principles that may be affected by the *context* within which they are applied. An example is the study by Medin et al. (1997), where they found that when taxonomists, garden maintenance persons, and landscape gardeners were asked to sort a list of trees, each went about it differently, in a way that reflected their occupation. However, when the context was changed, as when the participants were to assume that a blight affected the trees, then some of the participants (e.g., landscape gardeners) changed their sorting principle and sorted the trees more like the taxonomists.

In creating a domain, a researcher accomplishes several tasks, including initiating the assessment process. This occurs when criteria are applied to select and label content (a type of nominal scaling). When this is done, an "information reduction" procedure has also occurred; while it may wash out the uniqueness of the available data (that is, the sensitivity of the data may be reduced), it also significantly enhances the ability to apply statistical decision-making tools. Thus, when asked if they regularly engage in (the domain of) "physical activity," someone whose principal form of exercise is walking can be compared to someone who regularly swims. Next, reducing the number of indicators by collecting them together into a domain also reduces the number of multiple comparisons thus reducing the chances of making Type I errors (Fairclough 2002). This is particularly important for studies occurring in the context of a clinical trial. Third, forming domains makes it easier to communicate about quality or quality-of-life. This occurs because a domain statement, like any summary measure, is usually more abstract and encompasses a wider range of information. However, communicating information via a domain statement also has risks, since its origin may differ from its applications. In general, however, a domain statement is a helpful way for an investigator to organize information into manageable analytical units.

While it is common for students of quality or quality-of-life research to use a variety of objectification methods when defining a domain (see below), what may be even more important is that the respondent himself has a variety of cognitive methods to use when organizing information into categories. Thus, my first task will be to discuss what is known about the formation of categories. With this background, I will describe different ways that respondents themselves cognitively form categories, and follow this with a discussion of how an investigator, using various formal objectification methods, collects and classifies information into derived domains. Tables 6.1 and 6.2 include examples of each approach.

A person generates summary statements or categories because the nervous system organizes information in terms of the type, place, and temporal pattern of stimulation. For example, the central nervous system is capable of differentiating the light of a particular wavelength, from the pounds of pressure of a particular sound, or pounds of pressure on a particular part of the skin. This information is received via different receptors (e.g., the eyes, ears, or skin), and is processed in different places in the brain (that is, specific subcortical and cortical areas). The frequency or intensity of the neural input establishes a temporal pattern of information that is used to organize, integrate, and summarize information leading to thoughts or behavior. Two broad extremes of what is essentially a continuum have been identified. One represents responding in a direct and *automatic* (more biologically determined or over-learned) manner, while the other reflects the processing of information in a cognitively *controlled* and deliberate way. Each of these methods constitutes a different cognitive process, even though they both may be simultaneously present and functionally combined. However, the existence of these processes reinforces the importance of attending to a cognitive perspective when defining a category.

An investigator also naturally organizes information into cognitive categories. However, when investigators create a domain, they are engaging in an activity different from forming a category; in this case, they are using external information to achieve some purpose or experimental goal. For this reason, I will persist in distinguishing between category and domain formation. Also, since domains are usually arbitrary constructions, it is legitimate for me to ask about the purpose and intent of the formed domain. Thus, I will examine in detail the five candidate domains (symptoms, health status, functions, neurocognitive activities, and well-being; Chaps. 7–11) that are commonly considered to be part of an HRQOL assessment. I will be particularly interested in what cognitive processes are active in the formation of these domains.

Examination of these domains will reveal that they rarely consist of homogeneous content. Rather, what will be more commonly found is that some domains consist of predominately modular (biological) components, while others have primarily experiential components. For example, it will be shown that symptom domains are more dependent on the automatic (more but not exclusively biologically determined) processing of information than a functional activity domain. In contrast, a person naturally uses a variety of methods (biological, cognitive, and learning) to form the categories of information that make up a qualitative judgment.

While I am separating category and domain formation for analytical and rhetorical purposes, in practice both processes occur in the same person at the same time. For example, pretend I am a tour guide and have been asked to examine a statue, such as Michelangelo's David, and discuss it with a group of tourists. When I observe the statue, I might wonder out loud about whether this depiction of a biblical person is appropriate. I could note the sheer size of the object (17 feet tall); its realistic display of the human form; its color or texture; and I might even notice that I momentarily stop breathing as I take in the enormity of what I am seeing. All these elements make up an aesthetic and functional experience that clearly becomes part of my judgment of the statue. Since my role here is as a guide, I will attempt to describe how I came to my judgment, and in doing so will inevitably adjust my experience for my audience, their level of knowledge, experience with art objects, and so on. By so doing, I transform my summary statement into a domain, and I suggest generally that something akin to this occurs when a domain based on a qualitative assessment is formed.

There are a variety of methods an investigator can use to organize information from a conversation, written text, or speech into a domain: these include content analysis, discourse analysis, or text analysis. Classical (e.g., the items on a quality-of-life or HRQOL assessment) or modern psychometrics (e.g., item banks derived from the application of item response theory) can be used to create domains that, when used in an experimental study, produce responses (e.g., self-ratings, or the frequency of a response to an item on a questionnaire) that after statistical examination (e.g., factor analysis) can further define what is meant by the domain. Thus, from an investigator's perspective, there is a range of methods that can be used when forming and defining a domain. I will review some of these methods (Tables 6.1 and 6.2), and continue to discuss these issues when I consider whether specific types of domains (e.g., symptom, health status, functional, neurocognitive, or well-being domains) should be included in a quality-of-life or HRQOL assessment. To start, I will review the general topic of how categories are formed, follow with a discussion of how domains are formed, and end by applying what has been learned to a particular set of data and the analysis of some established classification systems.

2 The Formation of Categories

2.1 Introduction

In this section, I will briefly review the methods for forming a category. I consider a person's summarization of information as forming a *natural category*, while an investigator uses various methods to create an *experimental domain*. Thus, the roles each has is different and this is certainly what psychologists (e.g., Medin et al. 1997) and cultural anthropologists have made clear in their research (Lakoff 1987).

As described in Chap. 3, the first step in cognitive processing is the classification of an entity. There are at least three basic types of classifications (Table 3.2) – hierarchical, cross-classification, and contextual – each of which can contribute to the formation of a natural category or be combined to generate a complex category. Of course, investigators may also use these methods to form domains. But how does classification contribute to the formation of a category? First, classification identifies the features that can be used to include an entity in a category. For example, the statement "having a job that pays well and is satisfying" has features that can be used to classify the job, but if the statement were also qualified (for example, the respondent adds "and these features are important to me"), then this becomes a complete statement of the quality of a work experience. However, it is also possible to use the features of a job to exclude an entity from a category, or learn how to use these features to classify an entity. Thus, the pay level that would be acceptable to a person is something that may require a comparison with others, experience with what is to be expected for a particular type of work, and so on. One of the issues that makes a qualitative assessment so complex is that there are usually many features that can be used in creating the qualitative category. Thus, it would be interesting to learn how various features (e.g., those attributable to aesthetic or emotional factors) add, subtract, or interact with other features (e.g., functional features) to contribute to a quality judgment.

The most common method used when forming a category is to minimize the errors that might occur during the process of classifying entities (Bott et al. 2007).[1] For example, a person who has a job and is trying to determine if he or she is being paid appropriately might ask coworkers what their salaries are, or if not what they think the salary range for their type of work should be. However, as is often the case, a worker will learn a lot more about their employer from their coworkers than the operative salary range. Bott et al. (2007; see also Allen and Brooks, 1991) demonstrated that this is so, and that many more features are learned than just what is needed for classification (e.g., what is a fair salary?). They also speculate that the function of these additional features is that they can be used to make inferences, such as that "this employer is not someone to work for."

It is easy to imagine that something like this occurs when a quality assessment is being made. Consider the momentary evaluation of a person's quality-of-life after they participate in a family outing. The first thing they might do is classify the event as enjoyable (an emotional response), and then recall features associated with the event (a cognitive process), such as the family playing some sport together, preparing food for a meal, and so on. Thus, a range of features becomes apparent that are then used by the person to make the inference reflected in a rating of quality-of-life; that is, in defining the constituents of a particular qualitative category.

Another aspect of the role of classification in category formation is that a person will, if given a variety of information, spontaneously organize it along one or more dimensions (Pothos and Close 2008). For example, if a person is cognitively interviewed after rating their health status as "poor," then they will most often reveal that they based their rating on their physical status (Krause and Kay 1994). This suggests that not all indicators of quality are of equal salience, and that people have preferences about the dimension they use to classify the information they receive and presumably use when forming categories that make up a qualitative assessment. These activities are consistent with Gärtenfors' (2000) view that quality is a dimension in a conceptual space.

This is an important issue since most often studies of category formation in qualitative studies involve "supervised settings." For example, when an investigator is interested in performing a utility estimation, she provides a respondent with a scenario that includes multiple pieces of information. It is easy to see how artificial this setting is compared to one in which a person provides and organizes information on their own. Thus, the difference between spontaneous and supervised classification of information simulates the difference between natural category and experimental domain formation. Of course, an investigator may have a specific experimental question to answer and would want to govern the cognitive processes a respondent engages in by structuring the items the respondent is asked to respond to. Still, I would argue that if an investigator wants to optimally structure the items on an assessment that they first attend to how a person ordinarily organizes information, such as forming categories. This I believe has not been sufficiently attended to.

It is expected that a qualitative category will be a complex cognitive structure, consisting of multiple features and dimensions. In addition, these categories consist of features that have a *family resemblance* (see below), rather than being a homogeneous collection. In this regard, it is interesting that Hoffman and Murphy (2006) found that categories with four or eight pieces of information that made up a family resemblance category were learned equally as well, suggesting that complexity alone does not preclude efficient category formation.

These brief comments are meant to introduce the analytical opportunities that are created when the substantial cognitive literature is applied to the task of classifying entities used in category formation, including a category used when assessing quality. Four topics will be discussed: first, the role category formation plays in language expression and usage; second, how categories (particularly perceptually based categories) are acquired; third, whether "kinds" of categories exist; and fourth, the role that contextual factors, such as a person's culture, plays in the formation of categories.

2.2 Categories and Language Usage

As stated in Chap. 3 (p. 62), a category or concept can be conceived of as consisting of entities that have similar properties (the so-called classical view of categories), where the entities within a category are determined probabilistically (that is, where abstractions such as a prototype define the category); or where recalled exemplars determine the nature of the category (that is, Medin's context model of category formation; Smith and Medin 1981). Lakoff (1987) claims that the historic shift from a similarity- to probabilistic-based determinant of a category has had a profound impact on linguistics, philosophy, and even scientific progress. This shift started with Wittgenstein's rejection of the notion that a category has clear boundaries, so that the entities to include can't be clearly delineated. Instead, he described categories as consisting of entities that have a "family resemblance," using as his example a collection of what is common to a set of games (Chap. 1, p. 8). As noted earlier (Chap. 3, p. 72), Wittgenstein's use of an essentially psychological phenomenon in his philosophical deliberations I suspect came about because of his exposure to the activities of the Vienna logical positivists, who included Bühler and Brunswik. His reformulation challenged the Aristotelian notion that a logic based on discrete categories (e.g., the syllogism) was possible, and also challenged Frege's (1892/1996) argument that literal definitions of words and concepts were invariant across applications. Wittgenstein also suggested that categories were expandable. Lakoff (1987; p. 17) gives the example of the argument among mathematicians about how to define a polyhedron, a mathematical construct that can be defined in a number of different ways (e.g., with or without tetrahedra or cubes). Lakoff (1987; p. 17) also points out that Wittgenstein argued that elements within a category are not equally representative of the category. Thus, in mathematical terms, an integer is more representative of what is meant by the term "number" than, say, a transfinite number. Thus, according to Wittgenstein, items that have a *family* resemblance define the structure of a category, and this structure can be expanded and organized so that some items are more central to the category then others.

Lakoff (1987) also discussed the contribution of the philosopher John Austin (1962), whose concern about the meaning a word communicates resembles Wittgenstein's concern about how to conceptualize the nature of a category. Lakoff quotes a discussion Austin (1961; p. 71) had about the meaning of the adjective "healthy." Austin asked what is common about the concepts "healthy body," "healthy complexion," and "healthy exercise." His answer was that for each use of the word "healthy," there is a primary nuclear sense present. According to Lakoff (1987), this contrasts with the various uses of the word "bank," where in one use it refers to a place to put money, and in another to the side of a river. These two uses of the term would not conjure up the impression of a primary nuclear sense. This primary nuclear sense of Austin's is what Wittgenstein referred to as a "central indicator of a category," and what Lakoff (1987; p. 18–19) claims in modern cognitive sciences is referred to as the abstraction characteristic of a prototypical "sense."

Of course, it was Rosch et al. (1976) who clearly articulated "prototype theory" (Chap. 3, p. 56). Lakoff's (1987) contribution was to demonstrate that prototype cognitive categorization could be extended to linguistic categorization. However, what is of concern is whether the same argument can be extended to qualitative statements, which would ordinarily involve evaluated descriptors that have also been qualified. My first task here is to establish that the evaluated descriptors (e.g., ratings of physical activities, social activities, mental activities, and so on) constitute a category that was formed by the same principles as any other category.

Consider if an evaluated descriptor constituted a prototypical category. For this to happen, members of the category might vary in the degree of membership (that is, the category has fuzzy boundaries); have members that are graded as to their centrality; vary in terms of the extent they are linguistically "basic"; or have elements that can be organized into a system. Then, the question becomes whether these determinants can properly be considered a prototypical evaluated descriptor that would ordinarily be part of a qualitative statement.

A major concern about evaluated descriptors is their preferred level of informativeness. An issue of this sort may appear obscure, but in fact is quite relevant to the concerns about how language is used for effective communication about quality. Consider the issue (see Chap. 11) of the relationship between feeling and well-being. Do these terms have equivalent meaning? Is one more abstract than the other? Can an abstract term be used more concretely, and can a more concrete term be used abstractly? The same questions can be asked of the phrase "health status," as I will illustrate later in this chapter when I discuss the various uses of the phrase by Patrick and Erickson (1993). Thus, the issue here is whether abstractness and concreteness are inherent to the words being used, context specific, or if abstractness and concreteness are cognitive process that a person applies. This is a critical issue for a qualitative researcher since he or she is constantly arguing that the basic level items (e.g., can you walk around a block?) that make up a questionnaire that reflects the superordinate construct referred to as "quality-of-life."

Rosch et al. (1976) has speculated that basic level categories have two advantages over superordinate categories: they are more specific and distinct. These distinguishing characteristics are reflected in the observation that a person will respond more quickly to a basic as opposed to a superordinate term, yet basic terms are less likely to be remembered as compared to superordinate terms, while an object is more likely to be labeled with a basic term as compared to a superordinate term, and so on. In addition, children learn the meaning of superordinate terms *later* than more basic terms (e.g., Horton and Markman 1980).

With basic terms possessing clear advantages over superordinate terms, Murphy and Wisniewski (1989) have experimentally asked what good are superordinate terms – what do they contribute to effective communication? This question was also asked in this book when a variety of authors (Chap. 1) questioned the value of the phrase *quality-of-life* – surely a superordinate concept (Chap. 1, p. 2) – and asked what good is it? Murphy and Wisniewski (1989) suggest that basic and superordinate categories and concepts differ in that:

> Superordinate concepts may contain more *relational* information; information about the relative location of exemplars, information about the functional relations between exemplars, and information about how concepts and exemplars interact with nonexemplars. For example, the superordinate concept *clothing* could contain information about how different items of clothing are spatially arranged when they are worn. If people first learned superordinates as collections, this relational information could be a vestige of the earlier form of representation…What kinds of relational cues might be represented in superordinate concepts? One possibility is that *global shape* cues may indicate that a number of exemplars of a superordinate are present…When eating utensils or silverware are placed on a table, they form a

consistent pattern that is repeated at each place setting. Wheeled vehicles on a street or parking lot may form parallel patterns. Thus, some superordinates may have particular forms associated with configuration that their members (and other objects) take in a scene. A second possibility is that particular category members may have specific, very predictable spatial relations. For example, lamps normally appear on top of tables, shirts above the pants…A third, related possibility is *functional information* that may be apparent in a scene. For example, it is no accident that lamps appear on top of tables rather than under them – the function of both requires this relation. To the degree that people can perceive functional relations of the objects in a scene, these relations might be useful in object identification. (p. 573).

In a series of experiments designed to test their hypothesis, Murphy and Wisniewski (1989) demonstrated that object recognition speed is reduced when the object is placed in a scene, and recognition speed is even slower when the scene is relevant to the object. In contrast, when the basic terms are kept constant and the superordinate categories are placed in incompatible scenes, then the speed of recognition is increased. These data were interpreted as supporting the hypothesis that superordinate terms possess relational capabilities that basic level terms may not. Thus, Murphy and Wisniewski (1989) manipulated the context within which a term was placed, and this facilitated or inhibited cognitive processing (as measured by reaction time).

Another way of studying the role superordinate categories play in cognitive processing is to examine their lexical and semantic space. Is it true, for example, that a superordinate category will evoke fewer terms ordinarily associated with it than basic level terms? This does not seem to be the case, at least as reflected in a study reported by Marques (2007) that involved respondents listing features (words) associated with superordinate and basic categories or concepts. While this type of evocative assessment strategy may not reflect how cognitive processing ordinarily proceeds, delineating the lexical and semantic space surrounding different categories is a legitimate exercise. Marques (2007) found that "in comparison to basic-level concepts, superordinate concepts are not generally less informative and have similar feature distinctiveness and proportion of individual sensory features, but their features are less shared by their members" (p. 891). Thus, the lexical and semantic space surrounding both basic and superordinate categories appears to be more flexible than might seem from the early literature, and this provides support for the notion that concreteness and abstractness are processes that are cognitively applied under certain circumstances.

It is quite easy for me to apply this discussion to a quality or quality-of-life assessment. For example, when Murphy and Wisniewski (1989) talk about superordinate categories or concepts (such as furniture) consisting of collections of objects in a relational space, this suggests that many of the individual experiences I have in my life that I label as contributing to the quality of my existence may also be part of a relational space. This relational space may be defined by the emotions created by these experiences (Fig. 1.2). In addition, when I talk about *a* global quality-of-life assessment, I may mean a fairly concrete relational statement (where good experiences are collected together along a dimension). Thus, abstract statements need not be obscure; they can be as defined as objects and may even have a physical analog (e.g., a written number). Consider the objectification process; it can be seen as literally transforming an abstract category into an object: a number. Of course, much of the assessment process is how well objectification has been achieved. Also, note that the term "global" is an example of a conceptual metaphor – the CONTAINER METAPHOR – suggesting that figurative expressions are being used when making qualitative statements.

Murphy and Wisniewski (1989) also claim that a lamp and table fit together as a natural pair reflecting a relational statement, and I would say that the prevention of aversive events fits naturally with the assessment of quality, especially if this is seen as an effort at quality control. Thus, the cognitive connectedness between lamp and table and quality and control is why I see a qualitative assessment as a distinctly ethical act; they simply go together. Here, of course, I am referring to the functional consequence of a quality assessment. To this has to be added Marques' (2007) study demonstrating that the lexical and semantic space surrounding basic and superordinate terms are broader than anticipated, and this might account for some of the more concrete forms of expression of superordinate categories.

This discussion also has implications for the assessment of the compromised person, who loses the capacity to remember concrete basic level information as their illness progresses, but retains the capacity to recall and use superordinate abstract concepts. This suggests that the capacity for a compromised person to report about the quality of their existence may be retained, although the specifics about what they like or dislike may not be.

Much of this discussion has focused on what might be called "cognitive linguistic issues," rather than language usage per se, but I will correct this discrepancy when I review the several candidate domains.

2.3 Category Learning

As Zeithamova and Maddox (2006) state, "Humans live in a world of categories, rather than unique instances" (p. 387). This divides the world into meaningful pieces, and people use these categories to remember, communicate, and understand, and also to account for and predict what to expect from their continued encounter with their phenomenal life. Zeithamova and Maddox (2006) point out that a number of different processing models have been proposed to account for how people learn to categorize. These models are reflected in different tasks that a person faces, but can be simulated during an experimental study (see below). Support for the notion that multiple methods are used to learn a category comes from neural imaging studies demonstrating two basic systems that compete during learning (Zeithamova and Maddox 2006). This *dual process* system has been described by Ashby et al. (1998) and Ashby and Waldron (2000) as involving an *explicit* hypothesis testing system (that uses logical reasoning and relies on working memory and executive attention) and an *implicit* procedure learning-based system.

Ashby and Maddox (2005) selectively reviewed this literature and described at least four different types of perceptual learning tasks that may be involved in learning about categories.[2] The tasks Ashby and Maddox (2005) discuss include rule-based tasks; information integration tasks; prototype distortion tasks; and probabilistic rule-based tasks (also called the "weather prediction task"). Of course, the assumption here is that the cognitive processes evoked by these tasks reflect how a person learns to form a category.

Rule-based tasks are those that evoke explicit reasoning processes. The rules are usually verbally presented, focused on producing accurate responses, and ask the respondent to think along a specific dimension. Thus, when a person is asked to rate their quality-of-life using a visual analog scale, the instructions provided (which the respondent reads) constitute the rules; the visual analog scale represents the dimension; and the accuracy of complying with the rules will be assessed by the person's success in completing the task. Several theories of how this learning occurs have been developed. In one such theory, a person may be given an example of how to respond as part of their instructions, but in another approach they may learn more incidentally, following successful completion of the task (that is, the reinforcement of a particular response pattern). These instances are actually examples of *exemplar* learning. Another way this learning that may occur involves the application of the classic view of category or concept formation. In this approach, it is the *similarity* between stimuli being presented (e.g., format of the items) that is a critical determinant, as is the pre-established standards that form the basis for a person's response. However, Ashby and Maddox (2005) suggest that natural category learning will combine several such models (e.g., to include both similarity and rule-based models). Thus, they propose that classical theory and other forms of category learning be combined into a single neuropsychologically based model of category learning (Ashby and Ell 2001).

Information-integration tasks were defined by Ashby and Maddox (2005) as those "whose accuracy is maximized only if the information from two or more stimulus components (or dimensions) are integrated at some predecisional state" (p. 153). They go on to point out that perceptual integration could involve a wide range of activities, including computing weights for stimuli in a linear combination, or evoking gestalt perceptual processes. They also claim that the precise strategy in information integration may not be known, and illustrate this by reference to how difficult it is for an expert to verbalize exactly what they base their judgments on. In his book, *Blink*, Malcolm Gladwell (2005) provides an example of this phenomenon. He describes a tennis coach who, when asked how he identified the faults of a tennis player, confessed that he had no specific idea, except that he was confident he could. He admits that he was able to make a judgment after observing the player's performance, and that he compares his sense of adequate performance with the person's actual performance. Thus, as an expert he is making a comparative judgment that seems based on prior information integration. Having made the comparative judgment, he could then go on to advise the tennis player. Gladwell's point was that the expert was using his first impressions to make his comparisons, and that these impressions were often unconscious, yet highly accurate, suggesting that they were based on the process of perceptual integration that Ashby and Maddox (2005) refer to as "information integration."

Judgments based on automatic or first impressions are also commonly found in qualitative assessment. Walking into a newly decorated room and saying how much I like it, drinking wine I have never tasted before and commenting about how good it was, meeting a person for the first time and indicating my appreciation of their appearance, quickly answering a quality-of-life question, and so on, are all examples where a quality judgment is being made that may involve immediate perceptions that are combined with prior perceptual integration leading to information integration. What usually happens in these situations is that a person may be asked to explain their judgment, and a more reflective process is initiated whereby the person actually verbalizes and learns what features of the wine, etc., they may have used to base their first impressions on. Thus, after tasting a new wine, the evaluator may describe the wine's color, odor, and so on, using a variety of descriptive terms to characterize their judgment, but also to learn what they based their judgment on. Of course, what is going on is that the person is generating and testing a hypothesis about what they have implicitly experienced, reflecting again the presence of dual processing, but also creating a hybrid cognitive construct.

A third approach to studying category learning involves *prototype distortion methods*. In these types of studies, a prototype of a category is first created and then systematically distorted. In most studies of this kind, the category consists of visual stimuli so that the distortion is perceptual in nature. However, it would be easy to simulate this experimental paradigm in a qualitative assessment setting. For example, in utility estimation, a person is presented with a *scenario* that consists of a variety of pieces of information, and if a particular scenario is labeled as prototypical, and its content systematically distorted, then it should be relatively easy to observe the consequences this maneuver has on utility estimates. The same type of manipulation could occur to the visual analog scales, scale anchors, and so on. Ashby and Maddox (2005) point out that prototype distortion tasks are particularly important because neuropsychologically compromised persons respond differently when confronted with these tasks than when they need to deal with rule-based or information integration-based tasks. The authors cite literature demonstrating that persons with Parkinson's disease, schizophrenia, or Alzheimer's disease have no difficulty with particular types of prototype distortion tasks, but do with rule-based or information integration-based tasks.

A fourth approach to category learning is based on the concern about whether an entity's membership in a category is *determined* or *probabilistic*. An entity is considered "determined" when it is an exclusive member of a category, while a probabilistic entity may belong to more than one category. The experimental model that Ashby and Maddox (2005) use to illustrate this phenomenon involves a person being asked to predict the weather (will it be sunny or rainy?) after viewing one, two, or four tarot cards. In some trials, the same card is present (that is, the card is *determined*), while in others the presence of the target cards is varied (the stimulus is *probabilistic*). This paradigm has been used to study learning in both normal persons and persons with neuropsychological deficits. Thus, amnesic patients do not have trouble learning the task just recalling what they have learned, while both Parkinson's and Huntington's disease patients show deficits in learning from the outset (Ashby and Maddox 2005).

To summarize, a person responding to the task of making a qualitative judgment will learn something. For example, they will learn how to describe and evaluate an experience, especially a new one. Alternatively, given a set of instructions, they will learn whether they have successfully completed the requested task, which is often facilitated by a questionnaire illustrating an appropriate response. Part of what is learned are the features to attend to when making a judgment. Thus, a person interested in judging a statue, wine, meal, or life will more than likely learn different *features* of what they are judging. Of course, most quality-of-life research involves asking a person a fixed set of questions they may or may not find relevant, but there are sufficient hints in the literature to support the assumption that a qualitative assessment involves some level of learning. This includes the evidence that as a person proceeds through a multi-item questionnaire, they learn something that impacts their subsequent ratings. Thus, for example, placing the self-assessed health status item at the beginning or end of the SF-36 will affect the ratings of the item (e.g., David et al. 1999). The response shift phenomenon (Spranger and Schwartz 1999) is another example of a well-established empirical observation, suggesting that active cognitive processing ongoing during and after a qualitative assessment; this processing implies that something about the experience will or will not be retained.

This discussion implies that the organization of information involves deliberate cognitive processing, such as the application of specific rules, recall of information, and so on. This helps me understand how a person who has never been asked a quality-of-life question can still have a sense of the quality of their existence, as well as what happens as the consequences of making a qualitative judgment (e.g., filling out a questionnaire). As will become evident, the existence of these natural categories or *kinds* also supports a dual process conceptualization of category formation.

2.4 Kinds of Categories

One of the more interesting aspects of a quality assessment is that the terms used for descriptive and evaluative purposes often have an emotional connotation (e.g., good/bad; satisfying/dissatisfying; happiness/unhappiness; etc.). These terms denote emotional clusters that form perceptually and cognitively distinct categories, and since this clustering seems to occur naturally, they are referred to as *natural kinds* (Chap. 11, p. 415). Natural kinds differ from *artificial categories*, in that a person uses external criteria to classify the entities that are part of the category, while *artifactual categories* include items that can be organized along a common dimension (e.g., items that are examples of furniture). Evidence supporting the existence of natural kinds is most obvious in the language used by children, whose ability to use words to designate objects, particularly biological entities, occurs at a fairly young age (e.g., Hirschfeld and Gelman 1994). This suggests that children are capable of organizing information into categories, even though they may not use the same labels adults would for these categories. Since this occurs at such a young age, it also suggests that this type of information organization may require little or no learning (see below). Classification of this sort usually reflects a *basic* level of informativeness, as occurs when a child refers to his pet as "a dog," or "a cat," rather than a Poodle or a Siamese. As Medin et al. (2000) point out, the demonstration of natural kinds has been interpreted by some investigators as having broad implications, including supporting the predisposition to organize information in determined ways. This issue of the relative role of *determined* vs. *acquired* language expression will be recurrent in this chapter, overlapping with the more generic issue of whether the cognitive processes involved in domain formation are determined or acquired.

Whether natural kinds exist or not is controversial, with Weiskopf (2009) arguing that a kind of concept, or category, is context dependent and can be represented in a number of different ways. In contrast, Medin et al. (2000) at least considers the possibility that natural kind can be based on a uniform form of mental representation, and not be context dependent. In support of this view, Medin et al. (2000) evaluated the literature available at that time to determine if the type of category can be distinguished on the basis of structural differences, processing differences, or content-laden principles. For the purposes of their evaluation, features characteristic of a category constitute a structural determinant that can be used to distinguish categories. One such set of features is those characteristics of *nouns* and *verbs*. Gentner and Boroditsky (1999), for example, have demonstrated that nouns and verbs refer to different clusters of correlated properties that create perceptually distinct "chunks." Most important, these distinctions persist following cross-cultural comparisons.

Another example of a lexical distinction that reflects a structural characteristic of a category is the *mass/count distinction*. The word "dog," for example, is a count word, while the words "rice" and "sand" are mass words. Wisniewski et al. (2005) have found that "count superordinate" terms refer to single objects or concepts, while "mass superordinates" refer to many objects or concepts, suggesting that mass superordinates are not true taxonomic categories. Other structural characteristics that Medin et al. (2000) cite include the observation that categories of *objects* or *mental events* behave differently when part or the whole of the category is compared. Another structural characteristic that distinguishes categories is whether the entities are clusters of features or reflect a relationship between features. Each of these structural characteristics provides the qualitative researcher with analytical tools that can be used to assess how a person, particularly a compromised person, forms a category focused on the quality of their existence.

If a category is formed by a process where data lead to an abstract category (that is, a bottom-up process), or an abstract category guides the formation of concrete elements of a category (a top-down process), then these two categories would differ on the basis of the different *processes* involved in their formation. Medin et al. (2000), however, acknowledge that it is often not possible to distinguish structural from process determinants of a category, and that they often occur together. An example where both types may be active is in goal-directed categories. Brasalou (1983) has pointed out that goal-directed categories are organized around some ideal, with the ideal governing the process whereby items are selected to be included in the category. The resultant category is distinctly different from a natural kind. The notion of forming a category that has a purpose or goal is widely applicable, and appears to provide coherence to dispositional categories such as traits (Read et al. 1990). Goal-directed categories are also common in qualitative assessments. For example, a quality-of-life researcher may ask a person what he would want to be true for himself if he had the best possible quality-of-life, and the items he selects may be quite diverse and heterogeneous, yet reflect an idealized outcome.

A third type of category formation are *content-laden principles*. These principles refer to the fact that categories seem to divide themselves into unique content summary statements. Thus, Medin et al. (2000) refer to categories that consist of items that are examples of naive psychology, naive biology, or naive physics. Whether these distinct categories exist has been controversial, with developmental studies providing critical information about the distinction between biological and psychological categories (see below). For example, do children have the ability to understand that animals have categorical features that make them biological or psychological different? Springer and Keil (1989) argue that children can understand animals from a biological perspective, but that this understanding is not as complex as that of adults, while Carey (1985) argues that children learn about animals from a folk language psychology perspective, implying that the distinction between the two categories is not as separable as some would argue.

Medin et al. (2000) trace the history of the notion of *category specificity* or domain specificity to a shift from viewing a human being as a general-purpose computer, to a computational system focused on adaptation with the result that the organism develops specialized computational mechanisms designed to cope with particular environments. A corollary is the notion that learning is constrained and dependent on certain given or determined structural characteristics of the organism. This shift has broad implications about the nature of summary statements and I will discuss it in more detail later.

The notion of category specificity or domain specificity is quite relevant to a quality-of-life researcher, especially when a typical qualitative assessment is examined. It becomes obvious that most assessments are composed of specific (sub)domains, such as physical functioning, social functioning, mental functioning, etc. Most of these assessments rely on respondent input for identifying items, and empirical existence of these domains has been supported by factor analytic studies. Whether these domains would remain as specific when viewed from a folk language perspective is less clear.

The issues raised by Weiskopf (2009) are also relevant here. Basically, he views concepts or categories as not uniform singular forms of mental representation, but rather determined by any number of cognitive processes. Thus, the concept of quality-of-life, or even the category of entities that make a qualitative domain, may be based on prototypes; exemplars; causal relations between qualitative properties; unique properties of the person's life; cultural or ideal properties of a quality-of-life; and so on; all of which are stored in a person's long-term memory and recruited depending on the context of the retrieval. This view challenges the notion of category or domain specificity, and prompts the next topic: the role that cultural differences may play in category formation.

This discussion of how categories are learned should be particularly relevant to a quality-of-life researcher, yet the information provided is rich and requires careful translation into qualitative studies. The question is how qualitative assessments involving different cultures can be combined into composite indexes, as is commonly done now in large multinational clinical trials. This is currently justified by demonstrating the psychometric equivalence of the qualitative assessments, but as I have argued this doesn't ensure that such assessments are cognitively equivalent (that the assessments have the same meaning to the respondents from the different cultures). Establishing the cognitive equivalence of these assessments is critical, however, to sustaining the validity of these assessments. The extent to which contextual factors (such as a person's culture) contribute to category formation will be clear in the next section.

2.5 Cultural Determinants of Category Formation

Murphy (2005), in a paper meant to celebrate the intellectual contributions of Medin, describes some of the important issues during category formation, and how categories are formed in different cultural settings. He points out that most models of category formation are laboratory-based, where the specificity of the experimental design suggests that what is learned during category formation are specific features of the category. These features function as *exemplars* and exemplars are quite useful when building models, since they are often concrete and can be objective assessed (e.g., "walking 100 yards"). In contrast, prototypes tend to be abstract and fuzzy (e.g., "physical activity"). Thus, it is not surprising to find that exemplar-based models of category learning have proliferated (e.g., Nosofsky and Palmeri 1997). These models, Murphy (2005) points out, do a very good job of accounting for what he describes as the "artificial" category learning that occurs in a laboratory setting, but they do not account for the presence of hierarchical structures in categories, the predominance of basic level terms, and what has been learned from categorical and conceptual developmental studies. Murphy also suggests that how categories are actually learned may make a difference in how categories are formed, not only the structure and content of categories (e.g., Weiskopf 2009).

Medin's work, Murphy points out, has led the way in the development of an *exemplar model of category learning* (e.g., Medin and Schwanenflugel 1981), but also led the way in studies of category learning in different social and/or cultural contexts (e.g., Medin et al. 1997). Medin et al. (1997) found that when taxonomists, maintenance workers, and landscapers were asked to sort different types of trees, they created different categories; categories that were not necessarily preserved when the person had to face specific tasks and decisions (e.g., what types to treat when a blight affects one type of tree within a category). Murphy (2005; p. 190) summarized the Medin et al. (1997) study as indicating that: (1) given the same domain of objects, different people may construct different categories; (2) the basis for this categorization may differ; (3) the categories formed may be similar at specific levels but deviate at more abstract levels; (4) prior knowledge and how the categories are to be used will influence how the categories are formed; and (5) the same person may have different reasons for forming a category, and then use these categories in different ways when making inductions.

Murphy's summary is again easily transposed into what might be expected during a qualitative assessment. Thus: (1) it would not be hard to imagine that different people would sort the same set of experiences into different quality categories; (2) they would use different bases for these categorizations (e.g., classifying experiences into good or bad, satisfying or not, or happy or sad); (3) that the categories formed would reflect a basic level of information, but only some could be part of an abstract category (thus, pain and fatigue can be conceived of as basic categories, while fatigue is more likely to be part of the abstract construct of depression); (4) prior knowledge, such as growing up poor, will influence how the respondent structures a category that reflects the quality of the existence of a poor person; and (5) a person may form a category of features found in their cultural group, but abandon these criteria when evaluating their life in the larger culture they live in. Clearly, category formation can be a complex process, significantly influenced by the context or culture of the construction, but studies of this sort should be essential when considering quality-of-life issues.

In Fig. 3.1, cultural factors are depicted as an element of the antecedent conditions leading to a qualitative judgment. This suggests that cultural factors function at an *a priori*, perceptual and/or cognitive level, shaping how the various production rules act when generating a qualitative assessment. Some studies by Nisbett (2003) and Nisbett et al. (2001) suggest how this might occur. Nisbett et al. (2001), who extensively studied East Asian societies and compared them to Western cultures, conclude that considerable social differences exist between these societies, and that these differences influence various cognitive processes. They

describe the link between social organization and cognition as follows:
1. Social organization directs attention to some aspects of the field at the expense of others.
2. What is attended to influences metaphysics, that is, beliefs about the nature of the world and about causality.
3. Metaphysics guides tacit epistemology, that is, beliefs about what it is important to know and how knowledge can be obtained.
4. Epistemology dictates the development and application of some cognitive processes at the expense of others.
5. Social organization and social practices can directly affect the plausibility of metaphysical assumptions, such as, whether causality should be regarded as residing in the field vs. the object.
6. Social organization and social practices can influence directly the development and use of cognitive processes such as dialectical vs. logical ones. (p. 291–292).

So, the social organization a person lives in structures how she sees, what she thinks is important, and how she processes information. Nisbett et al. (2001) provided support for these propositions by comparing ancient Chinese and Greek thought and modern experimental studies with children and adults of these two different cultures. Most important for the present discussion is the researchers' view that social organization affects cognitive processes by indirectly focusing attention on particular parts of the environment, and by making certain kinds of social communication more likely than others. Thus, they found that East Asians perceive the environment as a whole or as a relationship, while Westerners tend to be analytic and see the environment in terms of individual objects or categories. Thus, if shown a picture of a pond with fish in it, an Asian person would first describe a pond, while a Western observer would first report seeing a fish. These perceptual differences affect many aspects of thought, including the possibility of contradictions. The famous Ying and Yang of Chinese symbolism attests to the fact that an opposite or contradiction is possible in Chinese thought, whereas it is not characteristic of Western thought. In the West, a contradiction is considered a refutation of a proposition. This, Nisbett et al. (2001) claim, affected the kind of science each civilization created, with the Chinese being more pragmatic in their inventions, and the Greeks developing scientific methods including geometry that required being able to refute a proposition (as in a proof). Also related to these cognitive differences is the Asian preference for finding a middle way, while the Western preference is for using principle-based decisions.

If, as Nisbett et al. (2001) claim, Asians use relationships to perceive their environment while Westerners tend to particularize what they see, then it is natural to ask if this difference is transposed into qualitative assessments. Chin (1972) reports a study in which American and Chinese students were asked to form categories from provided lists of objects. When the Chinese children were shown pictures of a male and female adult and a child, they clustered the mother with the child and explained their decision in terms of the relationship between a child and parent. In the same situation, American children clustered the two adults, since they both had the feature of being "adult." Similar findings were reported by Ji and Nisbett (2001), but now with adult college students. Studies of this sort can be very useful for understanding how cultural factors impact a qualitative assessment, since they provide an experimental paradigm that does a better job of simulating the experience of forming a qualitative category than the fixed structure of a questionnaire.

Nisbett et al. (2001) also reviewed a number of studies that touched on category learning. For example, Norenzayan and Nisbett (2000) report a study in which a group of East Asians, American Asians, and European Americans were presented with cartoons of animals on a computer screen, and told that some of these animals were from Venus and some from Saturn. Some were given examples of each type of animal (exemplar group) and some were trained to follow a rule using five different features to identify the animal (rule-based group). They found was that Asians and Americans performed equally as well for the exemplar categorization task, but differed for the rule-based group. Thus, as in previous instances, East Asians tended to group objects on the basis of relationships and similarity, whereas Americans were more likely to group objects on the basis of categories and rules.

One of the implications of this research is that culture or language, either together or separately, influences cognition. One statement of this relationship is the Sapir-Whorf (Whorf 1956) hypothesis, which claims that culture, through language, influences how someone thinks. If this is true, then persons from different cultures would be expected to differ in *how* they would make qualitative judgments, even though they may appear equivalent when responding to a structured information format (e.g., a questionnaire, Nisbett 2003). The notion that language influences cognition and cognitive development has an extended history (e.g., Vygotsky 1962). Recent evidence to support the Whorf hypothesis comes from a study by Boroditsky (2001), which demonstrated that Mandarin and English speakers think differently about time; Mandarin speakers describe time using vertical descriptive terms (e.g., references to "front/back"), while English speakers use horizontal descriptive terms (e.g., "there are good times ahead, or hardships behind us"). However, Boroditsky's (2001) study has been challenged (e.g., Chen 2007; January and Kako 2007). These studies failed to replicate the vertical vs. horizontal common time references found in Mandarin newspaper reports. More recently, Boroditsky and her colleagues (Winawer et al. 2007) reported a study in which linguistic categories influenced perceptual decisions (for example, there are two color categories for blue in Russian, but only one in English).

An example of how language, independent of culture, may influence thought is Logan's (1986) suggestion that the difference in writing style between Asian and Western cultures (pictograms vs. the alphabet) influences thinking patterns. Thus, Logan claims that the presence of a phonetic alphabet formed the basis for abstract, logical and systematic thought, which helps explain why science started in the West not the East, even though Chinese technology surpassed the West until the sixteenth century (Ji et al. 2004). Pictograms are based on drawn, concrete characters, and this concreteness is claimed to discourage the development of abstract notions of law, science, and deductive logic.

An area where the role of culture and language is relevant to qualitative assessments is in translation from one language to another. The standard method used to demonstrate the "equivalence" of the translated and original versions of the questionnaire involves "back translation," where the new version is translated back into the original language. If the resultant back translation is semantically equivalent to the original version, then the translated questionnaire is assumed to be equivalent to the original questionnaire. However, this practice does not recognize the independent influence of culture and language on thought processes (Ji et al. 2004). One way to examine the influence of culture independent of language is by studying persons who are bilingual, some of whom may have learned their second language as children and some as adults. The difference between these acquisition histories is that early learners are more likely to have only one cortical representation system for the two languages (compound bilinguals), whereas the adolescent or adult learners have two representation systems (coordinate bilinguals). The existence of these two types has been confirmed by neuroimaging studies (Kim et al. 1997), as has the separate cortical representation of monolingual and bilingualism (e.g., Kovelman et al. 2008).

Ji et al. (2004) were interested in taking advantage of this supposed difference in cortical representation and in using it to determine the separate role that culture and language play in cognitive processes. The cognitive task they selected involved asking compound and coordinate bilinguals and monolingual persons to classify two of three items into categories. They assumed that if cultural beliefs are the driving force underlying any difference between two cultural groups, the categories formed by a compound bilingual person should not vary with their language (e.g., Mandarin or English). However, since the coordinate bilinguals have two representations for the two languages, it would be expected that they would demonstrate a language effect, meaning that they would be more likely to form the category differently in each of the test languages.

The results of this study confirmed that there was a strong cultural effect when comparing bilingual or monolingual English speakers, but that there was much less of a cultural effect in coordinate as opposed to compound bilingual persons. The authors interpret their study as demonstrating that different cultures tend to focus on different things when engaged in the cognitive task of forming categories. Also, that the language development of coordinate bilinguals plays a critical role in category formation, suggesting that language can cue different reasoning styles.

Category generation can take many forms, from the grouping of two or three items into a category, or the clustering of events along an important qualitative dimension in a person's life (e.g., the dimension of contentment). Unfortunately, there is very little known in this regard when quality is the object of the classification, but there is a lot to be learned from the general literature on these topics. I have reviewed a small portion of this literature that should be sufficient to indicate its relevance to a qualitative assessment. My next task is to shift from what a person does to form a summary measure, to consider what an investigator does to form a summary measure, with or without the assistance of a respondent.

3 The Construction of Domains

3.1 Introduction

To start, I will assume that the principles an investigator uses when forming a domain are the same a person uses to classify entities into a category. Investigators generate *supervised* domains for various reasons, one of which is as an aid in clinical decision making. Thus, *The International Classification of Diseases-10* (WHO -ICD-10; 1994) and the *International Classification of Functioning, Disability and Health* (WHO, 2001) are used to establish diagnoses and justify treatment decisions. A constant source of debate and empirical study is determining the features to include in each example of a domain, since domain establishment is based on the acuity and precision of the data that makes it up. Domains have been developed as summary measures because they provide an efficient means of presenting information. To support this, a set of statistical procedures has been developed (e.g., factor analysis) which objectifies these summary measures. As will become evident, some of these domains are quite complicated, I will explore the nature of this complexity. This topic also has implications for the current effort to include "Patient-Reported Outcomes (PROs)" into various clinical trials that include qualitative assessments. I will examine the components of symptom, health status, function, neurocognitive domains, and well-being, and determine the commonalities and differences among them (Chaps. 7–11).

3 The Construction of Domains

Table 6.1 Types of domains

Domain formation methods	Type of domain	Examples of domain
Cognitive (Investigator)-based methods		
Autonomous processes	Domain-specific modularity	Collection of visual images
	Functional modularity	Types of language limitations
	Acquired modularity	Conditioned food aversions
Controlled processes	Implicit theories	Personal histories
	Culturally determined perceptions	Different stereotypes

3.2 Domains: Language Usage

The dictionary (*Oxford English Dictionary* 1996) has several meanings for the word *domain*, including "an area controlled by a ruler or government" or "a sphere of activity or knowledge." In mathematics, a domain is a set of independent variables. The word, however, clearly has metaphoric properties, although more likely to be conceptual than linguistic. Thus, the conceptual metaphor A CATEGORY IS A CONTAINER appears to apply to a "domain." A "domain" is a container in the sense that objects, information, experiences, thoughts, and feelings can be mapped to it, or collected in it.

Lakoff and Johnson (1999) also make clear that the CONTAINER METAPHOR is an essential element of many formal ordering schemas. They illustrate this by reference to Aristotle's "container logic" (Lakoff and Johnson 1999; p. 380). Thus, for a syllogism to work (If all B's are C's and all A's are B's, then all A's are C's), it depends on the proper classification of entities and their containment within categories. Another example is the famous syllogism "if all men are mortal and Socrates is a man, then Socrates is mortal." Here a prediction is also being made on the basis of what is contained within the category. Lakoff and Johnson (1999) point out that prediction on the basis of the CONTAINER METAPHOR was what Aristotle meant by the scientific method, the method whereby new knowledge is acquired. Less obvious, but equally as important for his logic, was Aristotle's view that the elements that made up the syllogism had to be stated as declarative factual sentences not containing metaphoric information. It is not clear what he would have thought of Lakoff and Johnson's (1999) notion of a conceptual metaphor being at the heart of the syllogism.

Another example of a formal ordering schema that utilizes the CONTAINER METAPHOR is Boolean algebra. Boole (1854) noticed there was a relationship between arithmetic and classes (or categories). To quote Lakoff and Núñez (2000):

Boole observed that if you conceptualized classes as numbers, and operations on classes (union and intersection) as operations on numbers (addition and multiplication), then the associative, commutative and distributive laws of arithmetic would hold for classes. In cognitive terms, he constructed a *linking metaphor* between arithmetic and classes, mapping members to classes, arithmetic operations to class operations, and arithmetic laws to 'laws of thought' – that is, the laws governing operations on classes. (p. 124).

Feinstein (1967) made extensive use of Boolean algebra and the graphic displays of Venn diagrams to illustrate how a disease and its characteristics could be conceptualized as classes, thus providing examples of the CONTAINER METAPHOR. I should not be too surprised, therefore, to find the CONTAINER METAPHOR as part of an ordering schema used to construct domains that are part of any quality or quality-of-life assessment.

3.3 The Role of Cognitive Processes in Domain Formation

In Sect. 2, I mostly discussed the *cognitive* processes involved in forming a category, although with allusion to the possibility that various biological factors or modules can also play a role in category formation. Now I will focus on how the various cognitive processes previously discussed combine with specific structurally determined modules (e.g., the visual system) to form a domain. I refer to this summary measure as a *domain* to emphasize the presence of structural constrains that determine the form of the summary statement, but also because it is usually the unit of analysis an investigator will process during a qualitative assessment. I will also discuss the various statistical methods an investigator uses to form a domain in the next section, although Table 6.1 implies there are a variety of domain types to be studied (e.g., functional domains, acquired domains, and so on).

Both a person and an investigator use the same cognitive processes to form a category, categories that have structural or modular constraints.[3] It is when the investigator-based category refers to something external to him or herself that the category they are forming is a domain.

Domains are formed as a result of a classification process, and some emerge early, are reliably, and effortlessly acquired. Different types of living kinds (e.g., animals, plants) or facial expressions fit this type of classification suggesting that they have a built-in, if not, biological (perceptual) origin. The same would be true for domains based on color perception, or phonetic interpretation. Both are again dependent on presence of an intact biological (vision or audition) process.

These types of domains would primarily be dependent on *autonomous processes*, which would occur quickly and unconsciously (Table 6.1).

Hirschfeld and Gelman (1994; p. 23) have argued that the term "domain" should only be applied to these autonomous processes, but practically speaking, whatever role this type of domain plays it is altered by the influence of other cognitive processes so that most functional domains are a product of both autonomous and controlled processes. Smith and De Coster (2000) describe these two processes as follows:

> When people perform tasks as diverse as solving logical problems, evaluating persuasive arguments, and forming impressions of other persons, they can make use of different processing strategies. People can (and in everyday life do) use a sort of "quick and dirty" approach, arriving at usually reasonable answers efficiently and effortlessly. For example, they may agree with an argument because a quick glance reveals that it is presented by an expert and contains statistical data. People also, when adequately motivated and given enough time and freedom from distraction, can try hard to think deeply about these tasks, sometimes arriving at qualitatively different answers. The expert's argument, on close examination, may prove specious, the statistics biased. (p. 108).

They also suggest that these two processes are mediated by two different long-term memory systems.[4] These two aspects of long-term memory reflect the end-product of two relatively distinct cognitive methods for processing information, which can lead to two distinct domains.

The acknowledgement that information processing takes two basic forms is commonly referred to as *dual processing theory*. Most important for the current discussion is to demonstrate that dual processing is also relevant to a qualitative assessment. This was reported in an initial study by van Osch and Stiggelbout (2005) who used cognitive interviewing techniques to determine the cognitive processes at work when subjects respond to a visual analog scale. Sixteen persons were interviewed to reveal more about what they were thinking when they responded to the scaling task, and 57% of the respondents were not able to specify why they placed a mark on a particular point on a line, other than that it "felt right." The remaining respondents considered some aspect of the task as part of their scaling decision. Thus, some said they essentially bisected the scale and decided to place their line above or below the midpoint (29%), but only 8% reported actually referred to a number as part of their valuation decision. Thus, the first group responded without much awareness and deliberation, which is consistent with the *automatic* processing of information; while the others were more deliberate and *controlled*. Clearly, two distinct patterns of responding occurred, suggesting there were two different cognitive processes involved. Not clarified by this study was whether the ratings of respondents who used different cognitive processes also differed, though this might be expected.

Dual-cognitive processing, (Bargh and Ferguson 2000; Birnboim 2003; Chartand and Baugh 1999; Moors and De Houwer 2006) has been found to be involved in a variety of cognitively mediated activities, including working memory capacity (WMC: Baddeley et al. 1997), attribution (Uleman et al. 1996), person perception (Zárate et al. 2000), and emotions (Teasdale 1999). It has also been used as an analytical tool to examine specific neurocognitive disorders such as aphasia, amnesic patients, and persons with Alzheimer's disease (e.g., Birnboim 2003), and has been demonstrated to have a neurophysiological basis (Kelly and Garavan 2005; Jansma et al. 2001; Ramsey et al. 2004). A number of theoretical statements (Birnboim 2003; Hasher and Zacks 1979; Schneider and Shiffrin 1977) have been proposed to account for dual processing. For example, Schneider and Shiffin (1977) argued that automatic and controlled processing are two parts of a continuum, dependent on the amount of training a person has had. Thus, if I were to learn a complex task (e.g., riding a bicycle), at first I would attend to it using controlled processing, but with practice I would be able to respond automatically. The neurophysiological evidence suggests that similar anatomical areas may be involved in both types of processing, but they differ in their efficiency. This suggests that a cortical reorganization occurs as a result of practice, leading to the observed automaticity.

Automatic (or quick and dirty) processing has been described as occurring reflexively, usually initiated by a external stimulus. Thus, in-place knowledge structures such as concepts, scripts, or schemas are activated by attending to cognitive entities or representations which lead to actions, thoughts, or feelings. This automatic processing has also been called "nonconscious," "implicit," or "heuristic," and is considered the default form of cognitive processing (Barrett et al. 2004). It is also the cognitive process underlying "first impressions" (Chap. 3, p. 66). In these models, automatic processing is assumed to be highly biologically determined. For example, Stanovich and West (2003) proposed that automatic processing is a cognitive system that both humans and animals share, which includes instinctive behaviors that are innately programmed.

Automatic processing is a class of cognitive process that can take several forms, each of which can impact an outcome. This can be demonstrated by examining automatic processing's role in judgments and decision making. For example, Gigerenzer and his colleagues (e.g., Gigerenzer and Todd 1999) summarized in their book, *Simple Heuristics That Make Us Smart*, how a number of different heuristics are active in facilitating different patterns of responding for a person asked to make a judgment or a decision. These heuristics occur in an automatic, nondeliberative manner.

A cognitive module (Fodor 1983)[5] would be another example of an automatic process (Fig. 6.1), but now a biologically dependent process that would also operate in a fast, modular-specific manner, having evolved to handle particular types of information (e.g., vision, language). A *modular domain* responds reflexively to predetermined inputs that lead to predetermined outputs; operates out of awareness

Table 6.2 Objectification methods and domain formation

Objectification methods		
Analysis of qualitative data	Content analysis	List of topics
	Discourse analysis	List of themes
	Text analysis	Frequency of words
Quantitative methods	Classical psychometrics and test development	Items selected on the *brief fatigue inventory*
	Applications of item response theory	Item bank for the *FACIT-fatigue assessment*

(meaning it is not possible to know the operations of the module); cannot be influenced by prior knowledge, expectations or beliefs; and is assumed to be mediated by specific neural systems. Thus, a modular domain represents an informationally constrained, structurally determined system and the question is to what extent it can be found in the candidate domains being considered as determinants of a quality or quality-of-life assessment (see Chaps. 7–11).

Barrett et al. (2004) introduced an interesting alternative concept of modularity; *functional modularity*. Here, a system appears modular because of capacity constraints. Thus, if a person has limited ability to attend to some task, then the resulting informational domain may appear as constrained structurally determined as a modular domain. This notion may be particularly relevant for the challenged individual, whose limited capabilities determine how information is organized into domains. But it is also relevant for a disorder such as irritable bowel syndrome (IBS), a regulatory system involving gastric motility whose capacity to self-regulate is limited, resulting in either over- or under-responding (that is, IBS with diarrhea, IBS with constipation, or IBS with intermittent diarrhea or constipation).

A third type of modular domain is "acquired," as may occur in response to a disease or treatment. For example, I was not born with chronic fatigue, yet it can develop and remain as a fixed and recurrent part of my life. Conditioned food aversions, or anticipatory nausea and vomiting, are also associated with characteristic behaviors that can evolve. *Acquired modularity*, as will become evident, is probably the most common determinant of a domain, but these qualities (that is, the cognitive bases of a domain) are seldom considered when domains are examined in an analytical quality or quality-of-life study.

The second component of the *dual process theories* involves *controlled processing*. Controlled processing refers to a goal-directed, top-down, or dependent type of activity that determines the degree to which automatic processes influence thoughts, feelings, and behaviors.[6] Thus, when a respondent deliberately considers her past history when responding to a single quality-of-life item, she is engaging in a *controlled process*. I do the same when I focus my attention on finding a label that captures what is common to a factor derived by a factor analysis. And I do the same when I construct a mathematical or statistical model. This capability appears to be uniquely human, is slow and sequential in nature, and permits abstract hypothetical thinking. I typically call this type of processing conscious, explicit, or systematic.

It also has been demonstrated (Evans 2003) that the two processes function simultaneously, with the controlled process system able to override automatic processing. As a result, the two cognitive systems should be thought of as competing for control over overt behavior, as would occur when responding to items on an assessment, in judgments and decision making, reasoning, or theoretical thinking.

3.4 Objectification Methods and Domain Formation

To paraphrase Estes (1994), classification is basic to all of our intellectual activities . Classification is not only something I can cognitively do, it is also an activity I can apply as when I construct domains for a quality or quality-of-life assessment. There are many labels for such activities, including classifying information into a *trait*, a *domain,* or a *taxon* (Meehl 2004), and this can occur using informal or formal (theory-dependent) methods (Table 6.2). Some of the formal methods for collecting various indicators into domains range from statistical methods (e.g., such as factor analysis, or item response theory) to symbolic logic systems such as Boolean algebra or set theory (e.g., see Feinstein's application to disease or symptom classification; 1967, p. 156). Less formal methods, such as content analysis, discourse analysis, and text analysis, function in the same manner, but classify indicators on the basis of criteria applied by an investigator or by a predetermined dictionary, as in text analysis. Earlier it was mentioned that psychometric methods, both classic and modern, permit the selection and organization of items into domains with common characteristics. In this regard, the items in a questionnaire or the items in an item bank can be thought of as a domain (Table 6.2). Each of these methods will be discussed at various points in this book.

Item response theory (IRT) and computer adaptive testing (CAT), when applied to quality-of-life or HRQOL assessment, illustrate how a formal statistical method can be used to form a domain (or as commonly referred to as a "data bank"). As summarized in Table 6.3, an investigator starts with a large pool of items, usually selected from established health status, quality-of-life or HRQOL assessments. Each of these items is then assessed to determine whether they fit onto a common

Table 6.3 Item bank development framework[a]

Determining appropriate banks for development
1. Determine domains to be covered in bank: (a) literature review; (b) clinical input (including patients)
2. Determine the availability of relevant data source(s). If "yes," go to step 3; if "no," go to step 6

Develop the initial item bank
3. Identify common items and rating scales
4. Data analysis: (a) examine dimensionality; (b) examine item fit; (c) calibrate items on continuum
5. Examine construct deficiency: (a) statistical deficiency (gaps); (b) clinical deficiency (gaps)

Developing an operational item bank
6. Acquire or write new items with clinical input
7. Content validation
8. Field testing: (a) CAT programming; (b) data collection
9. Data analysis: (a) examine dimensionality; (b) examine item fit; (c) calibrate items on continuum
10. Evaluate item parameter equivalence across subgroups
11. Establish an operational item bank: (a) psychometric results; (b) clinical input
12. Implement CAT: (a) establish parameters; (b) simulate across continuum
13. Create short forms

CAT computer adaptive testing
[a]Adapted from Cella et al. (2007). Reproduced with permission of the publisher

dimension (e.g., "Is differential item functioning not present?"). To illustrate, consider Cella et al.'s (2007) effort to develop a *fatigue item bank* (Table 6.3); they first selected the 13 items from the FACT-Fatigue scale (Yellen et al. 1997), but found that only 10 of these items met their criteria for inclusion into an item bank. Upon empirical testing, they found floor and ceiling effects that prompted them to expand their item pool to 92 items, 20 of which were subsequently found to be "misfits," resulting in a final pool of 72 items with minimum (<3%) floor and ceiling effects. They then developed a CAT format to present the items to respondents. Cella (2004) reports that he has repeated this process for a number of different categories (e.g., physical functioning, pain, and so on), and in this way he hopes to develop a comprehensive set of item banks (or domains) to use with cancer patients.

There are several interesting advantages to using item banks for defining a domain. First, the items in the pool will have to fit along a single psychometrically defined dimension, ensuring the homogeneity of the items included in the domain. Second, since the properties of the items are empirically determined it is usually not necessary to use a theoretical construct to select the items, although the items' clinical relevance may be an issue. Of course, a theoretical construct – a latent variable – underpins the statistical model being applied, but a theoretical statement is not used to *a priori* select the items included in the domain. Third, the item bank can be viewed as operationally defining some abstract concept, such as fatigue.

Implicit in this last stated advantage is the assumption that the collection of items making up an item bank or domain represent what is meant by an "abstract concept." Is this true? There are two concerns about this assumption. First is the completeness of any particular collection of items as an abstract concept. Second is how an investigator can be sure that a collection of items approximates an abstract concept without also having some *prior* sense of what the concept being operationalized should mean.

A study reported by Arnold et al. (2004) documents some of these concerns. They administered the Cantril Ladder (Cantril 1965; a global linear quality-of-life assessment), and the SF-20 to 1,851 healthy subjects and 1,457 patients who had one of eight chronic diseases. The researchers were interested in determining the extent to which the physical, social, and psychological domains of the SF-20 were correlated to and contributed to the variance in overall quality-of-life (as measured by the Cantril Ladder). Even though 26/27 of the domains were significantly correlated to overall quality-of-life, the correlations only ranged from 0.16 to 0.54, with a median of 0.34. Regression analyses included demographic variables, and the three SF-20 domains revealed that the psychological functioning domain was a significant contributor to overall quality-of-life variance for each of the nine comparisons (healthy controls and persons with one of the eight diseases). However, the total variance accounted for by these analyses (i.e., the R^2) ranged from 0.20 to 0.44, with 7/9 of the comparisons having R^2's less than 0.30. The authors state, "This study shows that separate QOL domains make a limited contribution to the explanation of overall QOL" (p. 893). The Arnold et al. (2004) study reinforces the concern about whether a collection of items, as in an item bank or a series of domains will be sufficient to capture what is implied by an abstract concept, such as a global quality-of-life measure.

3.5 The Complex Structure of Domains

The previous discussion should have made it clear that investigator-generated domains have complex structures. This has to do partially with the difference in how domains are formed (modular or nonmodular), but also for what purpose they are formed. These domains may also differ in terms of the classification method used (e.g., hierarchical or crossclassification), as well as the number of components that make up the domain.

When these domains are inspected, at least three types of ordering systems can be identified. First, domains may have a single major component with a single function, such as vision, language, or fatigue. Second, domains may have multiple components, all of which contribute to a common function, as would be true for memory. And third, domains may consist of multiple components, each of which has a different function, yet a common attribute or purpose that permits grouping these components together. An example of this type of domain would be a person's mental status.

These ordering systems, of course, also can be described as producing either a hierarchy or crossclassification. Although I speak of three types, each can be included in the next larger class, as is characteristic of a *Guttman Scale* (Guttman 1944). A single modular component dealing with vision or language, for example, can also be part of working memory (see below), and working memory can be part of mental status. Thus, when I speak of a domain, I need to be clear about what level of complexity or abstraction I am dealing with.

In addition, the context or the setting of the various candidate domains that are formed contributes to the structure of a domain. Thus, the content of Chaps. 7–11can be considered different contexts, and as I will demonstrate the domains formed in these contexts will differ. If I consider the domains created in each of these settings as a tool for studying some aspect of a person's quality of existence, then I am obliged to understand how these contexts affects the structure of the domains.

Another source of complexity is related to the fact that a domain is a point on an informational continuum, a continuum that can be modified by how the term is used. For example, health status is sometimes considered a direct analog of an HRQOL assessment, and in this sense functions as a superordinate concept defined in terms of other domains (e.g., functional activities, symptoms, and so on). But it can also be thought of as a domain unto itself, with an emphasis on health-related issues.

Patrick and Erickson (1993) provide a definition of health status illustrating that they conceive of health status as a superordinate concept. They state:

> *Health Status*: Most often defined by the World Health Organization's definition: 'A state of complete physical, mental and social well-being, and not merely the absence of disease or infirmity' (1948). Not included in this definition are physiological phenomena, the probability of health in the future, or the means of determining which states of well-being are more healthy or desirable than others. Nonetheless, WHO's popular definition encompasses most of the usual meanings given to health status, including functional status, morbidity, and well-being. A complete representation of health includes a definition of health status, weights for these states, and prognosis, or the probability of movement to future states based on all the evidence. (p. 420).

This definition indicates what Patrick and Erickson (1993) think should and should not be included in the definition. A critical part of clarifying the role of health status in an HRQOL assessment is defining what is meant by the term *health*, and Patrick and Erickson (1993) touch on these issues:

> Concepts of health often lack clarity. The terms used to define health include positive states – 'wellness' and 'normal' – and negative states -'disability' and 'illness.' These terms do not specify what health is; neither do they indicate where it begins or ends. The ambiguous nature of these concepts, however, is only part of the definition problem. The notion of disability or of illness is distinct from health is a value judgment. Illness definitions usually contain assumptions about what states of physical being are desirable or undesirable, normal or pathological, damaging or benign. What separates the 'healthy' from the 'unhealthy' and the 'sick' from the 'well' is a subjective interpretation of what is 'good' or 'bad', 'desirable' or 'undesirable. (p. 19).

These concerns will prompt an extended discussion in Chap. 8 on what health, disease, or disorder means, but what is also evident is that Patrick and Erickson (1993) accept the WHO definition as workable, since it captures most of the usual meanings attached to the concept of health and health status.

This discussion has made it clear that each domain should be examined to determine if it can be used in a quality or quality-of-life assessment. To a certain extent, I have already started to answer how to do this. For example, I pointed out that the information a person receives is *local (*meaning unique to each individual); selected so as to form a *summary statement* (as in forming a cognitive entity); *networked* (communicated to others); cognitively *classified* (hierarchically or by crossclassification); and organized (into *concepts*). Also true is that cognition can occur automatically or reflect a controlled process. By studying these cognitive dynamics, it should be possible to get an understanding of the information the domain provides (e.g., learning about the interplay between subordinate and basic level information and superordinate concepts), and whether this information can be useful in a qualitative assessment.

Each domain should also be examined in terms of its unique nature. For example, neurocognitive indicators have a variety of natural settings, actually many natural sites, for study. Some of these sites include persons who are intellectually challenged, those who have sustained a neurological injury (e.g., a stroke, the cancer patient), or persons who have abnormal neurological functioning, as is true for the epileptic or a person with schizophrenia. Examining and understanding the cognitive dynamics in each of these settings is a critical element of what is required to properly assess the domains of a quality assessment.

3.6 The Modular Basis of Domains

Hirschfeld and Gelman (1994) offer a definition of a domain, with a strong emphasis on automatic processing. They state:

> A domain is a body of knowledge that identifies and interprets a class of phenomena assumed to share certain properties and to be a distinct and general type. A domain functions as a stable response to a set of recurring and complex problems faced by the organism. This response involves difficult-to-access perceptual, encoding, retrieval, and inferential processes dedicated to that solution. (p. 21).

They go on to say that domains are guides to partitioning the world. Thus:

> ...domains function conceptually to identify phenomena belonging to a single general kind, even when these phenomena fall under several concepts. For example, living things can be classified in a number of different ways, ranging from foodstuffs to zoo animals. The psychological correlates of competing classifications and their internal structures have significant effects on the way many common categories of living things are sorted, recalled, and recognized... Yet in spite of these competing ways of classifying living things, beliefs about living things are typically early emerging, consistent and effortlessly acquired. Domain competency facilitates this by focusing attention on a specific domain rather than general knowledge... (Hirschfeld and Gelman 1994; p. 21).

Hirschfeld and Gelman (1994) also state that the automatic processing underlying a specific domain is independent of will and awareness. To them, domain operations "involve focused, constrained, and involuntary perceptual, conceptual or inferential processes" (Hirschfeld and Gelman 1994; p. 23). They go on to state that not only does a person have the capacity to form domains, they also use these domains to explain things. Although the capacity to form domains from available information may be predetermined, the skill involved in forming domains increases with age and experience.

In contrast to this approach, most investigators think of a domain as the simple collection of items or terms, and make no assumption about the origin of the items and the extent to which the domain may be predetermined and fixed. Instead they use the psychometrician's preferred statistical method, factor analysis, to form domains. The objective of factor analysis is to reduce the number of indicators into a smaller number, but to do so while retaining the information present in the larger set of indicators. This is usually done by correlating each indicator with the other and determining which have a high degree of association. This high degree of association or correlation between indicators is interpreted as evidence that the indicators are assessing the same underlying construct and that they may be grouped together, as when forming a domain. The next stage is the labeling of the factor and requires that an investigator decide what is common among the various indicators in the factor. At this point, a cognitive (controlled) process is initiated that, at its core, requires that the investigators agree that the factor has domain-like qualities, and justifies a particular label. Thus, the two models differ in terms of how indicators are selected, but agree that a nonstatistical but cognitive process is involved in labeling and confirming its appropriateness for a particular domain.

Psychometric methods are applied quite successfully when a domain has a common content, but are less successful for domains heterogeneous in nature. For example, if the defining characteristic of a domain is that it includes *what a person does* (e.g., their activities or roles), it may consist of activities involved with bladder control, eating, or physical activities and socializing. As I will discuss later, bladder control and eating are either reflex- or sphincter dependent, and therefore have the characteristics expected of automatic processing, and dysfunction of these automatic processes may have profound qualitative consequences. When factor analytical procedures are applied to help sort these data, it becomes clear that some of these measures may *cause* change in quality-of-life or HRQOL rather than just *indicate* the status of the concept, and that these differences require sensitivity concerning the appropriate statistical procedure to use (see later for a discussion of the paper by Fayers and Hand 1997).

If part of the argument here is that heterogeneous domains require different statistical and conceptual approaches, then criteria should be identified that can be used to determine when a domain is modular.[7] I start with Chomsky's (1988) work on the nature of language, since it provided a major impetus to the view that mental processes include modular domains. To Chomsky, *the language faculty* was different from the visual or auditory systems, facial recognition, etc. He considered the existence and autonomy of these specific domains as evidence for a *modular* view of mental functioning. As Hirschfeld and Gelman (1994) state:

> The modular claim has three components; first, the principle that determines the properties of the language faculty are unlike the principles that determine the properties of other domains of thought. Second, these principles reflect our unique biological endowment. Third, these peculiar properties of language cannot be attributed to the operation of a general learning mechanism. (p. 7).

Marr's (1982) studies of visual functioning, and Liberman and Mattingly (1989) studies of auditory processing are often cited as supporting the notion that cognitive processes are modular in nature. Marr, for example, was interested in how it is that what I see remains fairly stable even though the colors, shapes, and sizes of visual information constantly change. To explain this, Marr created a computational theory of vision (Marr 1982). He constructed algorithms that detect edges, apparent motion, surface texture, and so on. What is seen, therefore, combines these atomic visual modules into a coherent whole.

Fodor (1983), in his book, *The Modularity of Mind*,[8] suggests that the mind is not a single, homogeneous, general-purpose processing system as most quality or quality-of-life researchers might assume, but rather consists of several task-specific informationally encapsulated subsystems that operate relatively independent of each other. He thinks there are six such systems: one for language and five for each of the senses. I will discuss his ideas in some detail because they not only provide insight into what it means to speak of domain specificity, but also because it is a model that bridges the gap between a computationally based complex adaptive system and a system based on language and reasoning.

3 The Construction of Domains

Sperber (1994) provides a succinct description of Fodor's ideas:

> His target was the then-dominant view according to which there are no important discontinuities between perceptual processes and conceptual processes. Information flowed 'up' and 'down' between these two kinds or processes, and beliefs inform perception as much as they are informed by it. Against this view, Fodor argued that perceptual processes (and also linguistic decoding) are carried out by specialized rather rigid mechanisms. These 'modules' each have their own proprietary data base and do not draw on information produced by conceptual processes. (p. 39).

Fodor (1983) speaks of three basic subsystems: central systems, input and output systems, and transducers. The *central system* is involved with reasoning, problem solving, constructing scientific explanations, and other higher cognitive processes. *Input transducers* translate physical information about the world into symbolic outputs, such as verbal reports on the intensity and color of a visual stimulus. *Output transducers* take symbolic inputs and transform them into muscle and organ activity that function as nonsymbolic outputs. An example would be if neural firing activates a group of muscles leading to a movement. Transducers perform their functions automatically and do not use computation. The role of input systems is to take the symbolic output of input transducers and produce representations of the external world as outputs. These outputs are then fed to the central processing system, where some impression or belief of the external world is generated. These representations of the external world are created by means of computation, and these computations make the external world accessible to thought. The output system takes symbolic output from the central system and delivers it to output transducers. The output transducers then change this information into motor movements and actions.

Of these various components, Fodor only considers the *input and output systems* to be computational in nature. Cain (2002, p. 187) claims that Fodor identified nine properties of the task-specific informationally encapsulated subsystem he defines as "modular" (Table 6.4), and I will use these criteria to examine entities that may be labeled as a domain, including my candidate domains (i.e., symptoms, health status, functions, neurocognitive activities, and well-being).

Before I proceed, let me summarize what I have said so far about the nature of a domain. First, I discussed how a domain can be viewed as an example of a conceptual metaphor; for example, A CATEGORY IS A CONTAINER METAPHOR. This helped me understand that the reason why domains are of interest is because they bring some order to the large set of indicators of quality or quality-of-life. I next considered two ways of studying domain formation; from what was essentially a *cognitive* or *psychometric* perspective (Tables 6.1 and 6.2). I also pointed out that these two models differ in how elements that make up the domain

Table 6.4 Are candidate domains *modular* in nature?[a]

Criteria
1. Domain specific
2. Operationally mandatory
3. Limited central access
4. Input systems are fast
5. Informationally encapsulated
6. "Shallow" outputs
7. Fixed neural architecture
8. Characteristic breakdown
9. Characteristic pace and sequence

[a]Abstracted from Cain (2002)

are selected. The cognitive model proposes that I am born with or acquire modules that order the information I receive into particular domains (e.g., visual or auditory information), but that I also have the capacity to classify information that I gain from experience. Thus, I classify information along an abstractness to concreteness or objective to subjective continuum (as in a hierarchical classification), or by a common goal (as when I am crossclassifying). In contrast, in the *psychometric* model, all elements are assumed to be the same (the only exception seems to be the distinction between causal and indicator variables) and ordered into domains using subject-neutral statistical procedures. My next task is to examine the various candidate domains and determine if each is modular or has modular components, and what this means for the study of quality or quality-of-life.

3.6.1 Are Symptoms Modular?

Symptoms, as a class of indicators, range from pain to fatigue to depression. This immediately suggests that a domain of symptoms will include a variety of components, some of which will be modular and some nonmodular. Suspecting that this may be true, I will select a symptom, for example, pain, that has a strong biological basis and a good chance of being modular or meeting the criteria of *acquired modularity*. Pain, however, is associated with a variety of qualitative experiences that could be considered; at one extreme is the radiologically detectable bone pain common in terminal cancer, and at the other is the abdominal pain reported by young children or adolescents that may defy physical characterization.

Physically based pain, in all of its instances, has modular characteristics. Thus, these types of pain are domain specific (Criterion #1 in Table 6.4), meaning that pain is reported when a specific biologically dependent sensory modality is activated (e.g., free nerve endings that mediate pain; Table 6.4). A second criterion is whether or not the operation of the input system is mandatory. Cain (2002; p. 186) likens this to a reflex, where a movement is mandated by the nature of the neural circuitry underlying the reflex. Clearly, I can think of instances where pain has this characteristic, such as when I withdraw my finger from a hot surface. Cain (2002) describers a third criterion:

> In the course of generating output from input, computational systems typically generate a whole series of intermediary representations. Modules deliver their output to the central system but not the intermediary representations from which that output was computationally generated. (p. 186).

This criterion has an analog in pain physiology, where it has been shown that a great deal of information processing occurs at the level of dorsal horn cells of the spinal cord prior to transmission to the central nervous system (Wall and Melzack 1994; Barofsky 1997). Clearly, pain information is rapidly transmitted (Criterion #4) and encapsulated, in that specific neural circuits mediate pain reception, particularly at spinal cord sites where pain input is processed, much as is found with vision or audition, but at subcortical sites. It is also here where *upregulation* occurs, leading to acquired modularity.[9] However, while it is clear that there is not a "pain center" in the central nervous system, there is ample evidence that specific centrally mediated processes (e.g., the sympathetic nervous system) can be recruited to participate in the experience of pain. Another example of this relationship is captured in the famous Melzack and Wall (1965) model of pain perception that includes activating a feedback loop from the central nervous system, thereby modulating (e.g., inhibits) spinal cord activity. This mechanism is often postulated as the mechanism whereby a person can affect their pain experience. This type of organization would meet Cain's idea of a constrained output (Criterion #6). Fodor illustrates what he means by a constrained output by pointing to the difference between a language output and a language output that also includes the communicative intent of the speaker (Cain 2002; p. 186). In the pain example, this translates into a basic message about pain transmitted from the spinal cord, and its modulation by the central nervous system. Criterion #7 states that the input system be associated with a fixed neural architecture, which is true of pain sensation and perception. Criterion #8 requires that the input system be sensitive to disruption if damage occurred to the brain or genetic makeup of the person. This is clearly true of pain (Wall and Melzack 1994). Finally, Criterion #9 requires that a modular domain have a developmental history independent of a person's individual experiences. Pain, as with the other senses, develops in stages, although a rudimentary level of pain perception is present in the infant at birth (McGrath and Finley 1999).

Thus, physically based pain matches, with a reasonable degree of success, the criteria that Fodor would list as characteristic of a modular domain (particularly, acquired modularity). Whether a broader domain consisting now of an array of symptoms, as is characteristic of a particular disease, would match these criteria is less clear. Psychological pain, depression, or anxiety, for example, are less likely to fit these criteria, since the neural structures underlying these disorders are less localized and may involve multiple brain sites.

In addition, these symptoms do not seem to be informationally encapsulated, but instead are rather diffuse in their impact on ongoing functioning (e.g., depression affects memory, gastrointestinal functioning, and so on). Some symptom domains may actually constitute systems onto themselves and fit a definition of a domain reflecting a higher order of complexity. Thus, examining symptoms in terms of whether or not they are modular has been useful since it has clarified that symptoms as a domain may be quite heterogeneous, including modular and nonmodular components. It also may provide a basis for classifying various domains (Chap. 7, p. 211). Thus, if an assessment task includes an array of indicators, then an investigator would best consider the nature of the domains in the assessment, as well as in the analysis and interpretation of the resultant data.

Pain, as an example of a symptom, can also serve to illustrate the cross-classification of information. What is common about all the components that make up pain (e.g., sensory, behavioral, physiological) is that they are designed to avoid or minimize noxious experiences. This common goal of protecting the organism helps organize the domain. To achieve this goal (e.g., Barsalou 1983), the domain includes multiple components, some of which are modular and some nonmodular. This gives the pain domain both adaptive and nonadaptive characteristics that are well suited to ensure the survival of the species. It also allows the domain to be psychometrically robust and able to detect several types of changes. Examining domains in this manner can be repeated for each symptom, as well as a more global symptom domain.

3.6.2 Are Health States Modular?

Next, I will consider whether *health status* should be considered a modular domain. The term *health status* has been used to estimate the level of health of an individual, group, or population (Patrick and Erickson 1993), but it can also be used more broadly to measure the effectiveness of a health care system. It is considered *a point- in- time* indicator (a state of health) that may include domains consisting of both subjective and objective indicators. Patrick and Chiang (2000) describe health status (including HRQOL) as including symptoms, functional status, and health perceptions, each of which contributes to the opportunity a person has for optimizing their quality-of-life. They acknowledge that their approach to defining health status evolved from Torrance's (1986) efforts to create a multiattribute health state classification system (Fig. 6.1), which leads to a single global outcome measure. To quote Torrance (1986):

> The general idea of a multi-attribute health state classification system…is based on the concept that health status can be defined in terms of a number of attributes, possibly *hierarchically* nested…The lowest level attributes in the hierarchical structure are divided into levels that represent step-wise decrements in function on that particular attribute. For example, attribute 'dressing' might be divided into three levels: (i.) able to dress oneself normally, (ii.) able to dress oneself with difficulty or with the use

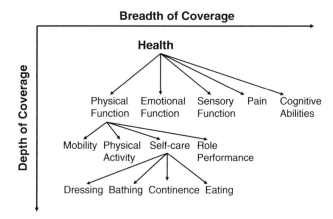

Fig. 6.1 Hierarchical basis of a health indicator system. Permission to reproduce granted by publisher.

of mechanical aids, (iii.) requires assistance of another person in dressing. Within each attribute the function-levels must be mutually exclusive and exhaustive, so that at any point in time each individual can be classified on each attribute into one and only one function-level. (italics added for emphasis; p. 14).

First, it should be noted that Torrance's model (Fig. 6.1) resembles what I earlier discussed as *a hierarchical classification of information*. A superordinate concept, in this case health, is being disaggregated into more concrete units. This, of course, provided Torrance with the units (or items) that he needed to establish a health status classification system, but it does so by defining what is meant by health in terms of other domains. Thus, Patrick and Erickson (1993) could apply this model to defining health status, and do so as Torrance did, by defining the concept in terms of other domains: basic or subordinate units of information (e.g., dressing, bathing, and so on).

The question is at what level of analysis am I operating on when I study a specific health status domain. The answer, at least for Patrick and his colleagues (1993; 2000) and Torrance (1986), seems to be at the level below the concept level. This means that if health status or HRQOL are considered concepts, then they would not ordinarily be classified as a domain, and, therefore, do not need to be assessed as to their modularity. Instead, this question would be best directed to the domains that make up the concept, which in their current usage of the phrase included symptoms, functional status, and health perceptions.

However, Patrick (2003; p. 26) has more recently described symptoms, functional status, perceptions, disadvantages/opportunities, treatment satisfaction, and adherence as *concepts*, and their disaggregated form as examples of *domains* (e.g., fatigue as a domain of symptoms, walking as a domain of a functional state, and so on). The good news about this approach is that he retains the hierarchy classification method (i.e., concept/domain distinction), but does it by shifting the defining categories down the informational continuum. Also

interesting is that there is no clear reason to reject this revised approach, which suggests that health status, in and of itself, does not have the characteristics that would require it to be defined as modular, as occurs for vision, for example.

The changes in Patrick's thoughts over time (Patrick and Elinson 1984; Patrick and Erickson 1993; Patrick and Chiang 2000; Patrick 2003) nicely illustrate that if I rely on defining a domain in terms of its abstractness relative to some comparison, then I am actually left with no specific criterion to use to determine when a *specific* body of information should be considered a domain. Instead, a definition can shift along the informational continuum depending on the particulars of a study, the research question being asked, and so on.

In contrast, to what I might call the "Torrance-Patrick approach," other investigators have proposed that health status refers to a person's *level of wellness vs. illness*. This approach would include biological and physiologic symptoms, but also the degree of illness control. Alternatively, health status domain could include age-adjusted mortality rates; infant mortality; low birth weight; potential years of life lost; smoking rates; and self-rated health status. Even when the domain is defined in this way, it is evident that its definition relies on reference to other domains. Still, alternative approaches to defining health status is more consistent with its intellectual history (see discussion later), and would include instances of what has been referred to as "acquired modularity."

Most quality-of-life or HRQOL investigators do not think it necessary to apply a multiattribute hierarchy or distinguish an abstract concept from a domain. Rather, investigators see the assessment task as essentially a *descriptive* process, in which health status is one of several ways of collecting information about the person. Much more concern has been expressed about the comprehensiveness of any list of descriptors (e.g., Padilla and Kagawa-Singer 1998), than whether the descriptor can appropriately be called a "concept" or a "domain." The obvious outcome of this process is that any assessment may include a diverse set of indicators, reflecting the implicit assumption that combining these indicators in a single assessment has no meaningful impact on the assessment outcome. An example is the Wilson and Cleary (1995) model. As previously stated, quality-of-life/HRQOL researchers use statistical clustering methods to describe the relationship among indicators.

To summarize, what the Torrance-Patrick approach does is shift what is to be assessed to the more concrete indicators of health status (e.g., "walking" instead of "physical activities"), and leaves health status to be inferred from these assessments. If I accept this strategy, I am acknowledging that the cognitive processes, which form a domain, may not always be applicable to health status, and that health status need not be classified as a domain of interest.[10] This, of course, would be less true if I defined health status as a level of wellness relative to illness, and could provide unique indications of both wellness and illness.[11]

Besides varying in terms of abstractness and concreteness, the interpretative information contained in a domain may vary. For example, Torrance (1986) points out that when different investigators were asked what information was contained in a particular domain, he found that responses varied from a few words or phrases, sentences, or paragraphs, to even complete video or audio tape reproductions of interview sessions. However, the available data (Torrance 1986) are mixed as to whether these different descriptors have an impact on the content of a domain.

The content of a domain can also be *framed*, and as Kahneman and Tversky (1982) indicated, this can influence the results of an assessment. There is a substantial literature on the impact of framing effects on health status domains. For example, postmenopausal hot flashes have been labeled as a *medical problem* or as a *coping problem*, depending on how the hot flashes are described to the patient (Hust and Andsager, 2003). These data suggest that a domain (e.g., hot flashes and their collateral symptoms) can be perceived as a health status issue or not, depending on how the symptoms were presented. Framing, therefore, is an example of a "controlled cognitive process"(Table 6.1), where the information a person has is used to structure their thoughts about some experience (e.g., hot flashes) or judgment they are making. Health status, while an important conceptual term, engenders a level of abstractness that at times limits its usefulness as a domain designation (e.g., Patrick and Elinson 1984; Patrick and Erickson 1993; Patrick and Chiang 2000; Patrick 2003). It is most useful when used to characterize wellness relative to illness of some study group. However, when used in this way it usually recruits indicators (e.g., symptoms) from other domains, domains that can also have modular components.

3.6.3 Are Functional States Modular?

As Fig. 6.1 indicates, and as was true for my discussion of pain, functions can be disaggregated into several levels of concreteness. Thus, physical functioning may involve self-care, which includes such activities as dressing, bathing, continence, and eating. While it may be difficult to demonstrate that "self-care" is modular in nature, it would not be difficult for such functions as continence and eating. Both are activities which can be considered *modular* (Criterion #1), which for Fodor (1983) means that the systems mediating these activities respond to a narrow range of stimuli. In both cases, the involved stimuli regulate specific sphincters, which control biological products in the case of continence, and ingestive processes (e.g., swallowing) in the case of eating. Both are operationally mandatory (Criterion #2), in that sphincters are regulated by reflexes, and both are mediated by subcortical sites, which have limited central control (Criterion #3). Thus, psychogenic dysphagia patients, (i.e., patients who report difficulty swallowing in the absence of a physical basis for this difficulty) still retain the reflexive capacity to swallow. The systems that regulate continence and swallowing operate quickly (Criterion #4), reflecting the mandatory nature of the processes that are activated by the input system. These systems are also informationally encapsulated (Criterion #5), meaning that the sequence of events involved in continence and swallowing are computationally regulated at specific brain sites, and are not easily influenced by central processes. Evidence for this comes from the limited success of central (voluntary) efforts to restore either function following a site-specific brain lesion. This seems to be generally true for continence and swallowing. Also true is that the information provided by these subcortical sites is only a limited part of the total functional activity (Criterion #6). Thus, the reflex that governs the continence or swallowing reflex is only part of what is considered good bladder control or adequate ingestion. The subcortical sites regulating these functions have a fixed neural architecture, which meets the requirement of Criterion #7. Criterion #8 requires that if the input system is damaged, the function is disrupted in a characteristic manner. This is generally true for both functions. Finally, Criterion #9 requires that functions have a characteristic developmental history that is not easily disrupted by voluntary efforts, and this seems to be true for both continence (an infant lacks sphincter control at birth) and swallowing (where the ingestive process originates as a sucking response and only develops an adult pattern with time).

From this discussion, I can say that at least some "functions" will fit a modular definition of a domain, although, of course, there are exceptions. However, even "nonmodular" functions examined at a particular level may still reveal a strong biological dependence and justify the label of an acquired module. For example, if "spending time with others" is an item meant to assess social functioning in a quality-of-life or HRQOL assessment, clearly the person requires adequate physical mobility in order to engage in such activities. If the person lacks such mobility, this is a situation akin to the one Sen was concerned about when he raised the issue of *positional objectivity*. In this situation, the respondent with mobility problems is objectively limited, which could confound their ability to spend time with others. This interdependence between so-called independent levels of analysis is not always recognized. In general, however, functions as a general domain are as heterogeneous as was found for symptom domains.

We also must attend to the level of abstractness or concreteness, or objectivity or subjectivity, included in the assessment of a domain. This issue can be illustrated by posing the following series of questions to a respondent:

1. Are you satisfied with your functioning in the last 2 weeks?
2. Are you satisfied with your self-care in the last 2 weeks?
3. Are you satisfied with your continence control in the last 2 weeks?

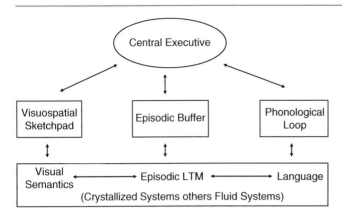

Fig. 6.2 Illustrated is Baddaley's model of working memory.

Which of these questions should I ask? Clearly, all three can and have been asked, depending on the purpose and goals of the study. However, does this mean that a definition of a domain should be broad enough to include all such examples, and if so, how do I decide to exclude an item as part of a domain? Psychometricians would argue that you select the level of inquiry that will lead to least error variance, and accept the informational loss you do experience as unavoidable (Norman 2003).

3.6.4 Are Neurocognitive Indicators Modular?

The question here is whether a neurocognitive indicator can be characterized as a task-specific informationally encapsulated system that operates relatively independently. The answer is a conditional "no," since the ability to perceive, recall, and think is usually dependent on a *system* of modular and nonmodular components, not on a single module. Supporting evidence involves the demonstration that disruption of elements of a particular cognitive activity will not necessarily cause a complete breakdown of the goals of the system, as would be expected for a single modular domain.

An example of this type of domain can be found in Baddeley's (Baddeley 2002) working memory model of short-term memory and its extension to long-term memory. The short-term memory component of the model consists of four components (Fig. 6.2). Both the *phonological loop* (e.g., rehearsing material) and the *visual-spatial sketch pad* (e.g., an image of the experienced material) are devices designed to temporarily retain material so it can be consolidated into long-term memory. The *central executive* has a number of *attention-related roles*, including focusing a person's attention to the task at hand, facilitating divided attention, and possibly switching attention between tasks (Baddeley 2002). A fourth role for the central executive is that of forming an interface between long-term memory and the subsystems (Fig. 6.2). This is the so-called *episodic buffer* role.

The *episodic buffer* is particularly relevant to a quality-of-life or HRQOL assessment. For example, when a person reads an item, he is expected to retain the question in his working memory. During this time, he is expected to connect the content of the item to his past experiences (long-term memory) and uses this information to respond to the item. The time he is able to keep this material in mind, called the "working memory span," may be a major determinant of the outcome of the assessment. This is especially so since the working memory span has been shown to be a surrogate measure of general intelligence (Kyllonen and Christal 1990).

Figure 6.2 makes it clear that working memory, as Baddeley (2002) has proposed, is a multicomponent system that includes various modular (e.g., language or vision) and nonmodular domains. It does not rely on a hierarchical arrangement to define the relationship between the components. Also, there are data indicating that disruption of one component does not disrupt overall working memory. For example, patients with Williams syndrome (Phillips, et al. 2004) who have impaired *spatial* short-term memory are still able to recall nonspatial material.

Describing working memory as a domain may create some questions when compared with, for example, a more familiar domain such as physical functions. However, I argue that the logic and cognitive activities depicted in Fig. 6.2 are as relevant for physical functioning as they are for working memory. In addition, the CONTAINER METAPHOR is as applicable to the phrase "working memory" as it is to "physical functioning." True for both domains is that modular and nonmodular components are collected together into a fairly complex system. A neurocognitive domain, however, is different in that it appears to be organized around a particular goal (e.g., "will a person be able to hold information in place to respond properly to a question?"), and often involves the cross-classification of information. Therefore, in general, domain formation, like summary measure formation, is a necessary and common correlate of many cognitive activities.

The notion of Barrett et al. (2004) of *functional modularity* adds another dimension to what is meant by a modular domain. They postulate that if a person's capacity to attend is limited, this limitation creates the appearance of modularity even though there is no disturbance in the neural structures underlying the module. Differences in the capacity to attend can be assessed by the person's working memory capacity (WMC). Barrett et al. (2004) apply this to language comprehension:

Language comprehension, however, requires some appreciation of context (i.e., the individual must not only parse the syntax of a given sentence but also must track the meaning of a concept across several sentences). To accomplish this, participants must hold information from sentences in working memory as they continue to parse those upcoming. ...Those lower in WMC are unable to use context to disambiguate a syntactically ambiguous sentence, making their language comprehension appear more "modular" and cognitively impenetrable. In contrast, those higher in WMC are able to use context to help understand a syntactically ambiguous sentence, such that their language comprehension does not have the properties of a modular system... (p. 565).

They go on to state that if a person lacks adequate resources to attend, then the ordinary dynamics between automatic and controlled processing might be disrupted, "creating a functional boundary between the two. When this boundary occurs, the phenomenon appears modular, or reflex-like" (p. 565). This, they claim, will vary from one person to another, providing a basis for accounting for individual differences in domain formation. As such, the concept of *functional modularity* may have particular relevance for assessing the quality-of-life or HRQOL of the "challenged person" who may have attention deficits.

3.6.5 Are Subjective Well-Being Indicators Modular?

If subjective well-being (SWB) is a form of emotional expression (see Chap. 11), then determining whether the emotions have a modular base would be one way to determine if SWB indicators have a modular basis. For example, if there are a limited number of cortical sites that mediate happiness, satisfaction, anger or other emotions, then it could be argued that there is sufficient domain specificity to consider if Fodor's other criteria (Table 6.4) can be met. On the other hand, if the nature of the module mediating forms of well-being is dependent on a system of interrelated cortical or subcortical sites and activities, then this is a situation not very different from what was found for characterizing the modular nature of neurocognitive indicators.

Davidson (2004) has summarized the available neural site data demonstrating that a particular neural activity pattern mediates SWB. He and his colleagues demonstrated that the prefrontal cortex, the amygdala, the hippocampus, and related structures are involved in emotional responses. In particular, Davidson (2004) demonstrated that "low basal levels of amygdala responding, effective top-down regulation and rapid recovery, characterize a pattern that is consistent with high levels of well-being" (p. 1400). In addition, high levels of anterior cingulate cortex activity facilitate emotional regulation and maintain positive SWB. In a more recent study, van Reekum et al. (2007) confirmed these earlier findings, demonstrating that individual differences in amygdala and ventromedial prefrontal cortical activity are associated with the speed with which persons respond to negative or neural stimuli. They found that those persons who responded slower to negative (as opposed to neutral) stimuli also scored more positively on the PWB, and this was mediated by less activation of the amygdala and more activation of the anterior cingulate cortex. Thus, there seem to be distinct patterns of neural activity associated with a person's appraisal of the threat created by a negative stimulus, and responses to these threats seem to be dependent on the reactivity of specific subcortical and cortical sites. These differences more than likely have a structural basis, which reinforces the view that a person's genetic history may play a role in the expression of SWB.

A person's personality has also been posited as a possible determinant of SWB (DeNeve and Cooper 1998), but this can occur in at least two ways. It may be that a person is genetically predisposed to either high or low emotional subcortical reactivity, or may have developed a particular personality because of major life events, gender, or family dynamics. Most likely is that both dispositional and experiential factors interact to produce a particular personality. To the extent that personality factors cluster and persist together, these clusters may function as acquired modules. Indirect support for the existence of acquired modules is the evidence that personality factors remain stable over time, although major life events (e.g., death of spouse, loss of job, or onset of a major illness) can alter this pattern (Chap. 11).

Guytiérrez et al. (2005) report a study that typifies many addressing the relationship between personality and SWB. They administered the NEO Five Factor Inventory (Costa and McCrae 1992) and the Affect Balance Scale (Bradburn 1969) to 236 nurses. The Affect Balance Scale was used as an indicator of well-being. They found that positive affect was inversely related to neuroticism, while negative affect was inversely related to extraversion. Many other studies of this sort exist (Chap. 11).

The relationship between SWB, personality, and genetic background has also been studied (e.g., Lykken and Tellegen 1996; Nes et al. 2006; Weiss et al. 2008). Weiss et al. (2008) report a study using 973 sets of twins (including both mono- and dizgotic twins) and found that SWB was accounted for by a unique genetic factor that accounted for all five personality factors in the NEO. They also found that the correlation between SWB and personality factors differed for monozygotic twins and dizygotic twins. The researchers found that for monozygotic twins, low neuroticism, high extraversion, openness, agreeableness, and consciousness scores correlated with SWB in the range from 0.35 to 0.52, while the same correlations for dizygotic twins ranged from 0.11 to 0.23. The difference between the mono- and dizygotic twins supports a contribution of genetic disposition to SWB. The authors state that:

...there were no genetic effects unique to subjective well-being. Instead these findings show that the genetic variance underlying individual differences in happiness is also responsible for differences in Neuroticism, Extraversion, and to a lesser extent Conscientiousness. (p. 208).

They go on to speculate that the observation that a single genetic factor accounts for the variance in all five personality domains and SWB suggests that a higher-order factor may exist that accounts for these results and is possibly related to life experiences.

What do these data tell me about whether well-being meets Fodor's criteria for a module? It does not, but as

I found for several other candidate domains (e.g., symptoms, health status, and so on), it is quite possible that some aspect of SWB would, so that SWB could be characterized as some admixture of both modular and nonmodular components. Part of the difficulty is that not that much is known about the biological basis of well-being (Chap. 11), and until this issue is clarified, characterizing the modular nature of well-being will be limited.

3.6.6 Summary

In this section, I used Fodor's criteria (Table 6.4) to evaluate whether the candidate domains have modular properties. Using Fodor's criteria, some symptoms and functions could clearly be considered modular, but the phrase "health status" is sufficiently abstract to suggest that it not be considered a domain, although I acknowledge it is an important construct to assess. Neurocognitive processes were found to be sufficiently complex and redundant in structure such that it would be simplistic to characterize them in single domain terms, and it is more appropriate to think in terms of systems of modular and nonmodular domains. The same may be true for SWB. These conclusions will be reinforced by the discussion in Chaps. 7–11.

4 Applied Classification

4.1 Modularity Applied: Visual HRQOL Assessments

If a domain has a strong modular component (e.g., vision, hearing, pain, and so on), then it is expected that variation in its biological basis would be more likely to determine the qualitative (or subjective) outcome of the domain than a domain with a predominantly nonmodular base. For example, *visual acuity* is highly biologically determined and easily affected by a number of visual impairments (e.g., glaucoma, macular degeneration, cataracts, etc.). Thus, it can be predicted that diminished visual acuity would be correlated with self-reports of diminished vision-dependent qualitative indicators (e.g., ability to read a paper, ability to drive, depth perception, and so on). In contrast, visual acuity is likely to be less correlated with the domain of social activities, which are more likely to be primarily nonmodular and highly determined by a person's history. Of course, this is a relative statement, especially since social activities are also vision dependent (social activities are aided by a person being able to see someone's facial expressions, being able to see where they are going when dancing, etc.). In addition, a person's capacity to adapt to the impact of their diminished visual acuity will also affect the magnitude of these correlations.

The question now becomes whether it is possible to demonstrate this more likely linkage between a modular vs. nonmodular component and a specific quality-of-life or HRQOL assessment. There are at least two ways to approach this task: first, by determining if generic and vision-specific HRQOL assessments differ in their association with modular domain (e.g., visual acuity); and second, to determine (e.g., by using regression analysis) whether a modular or nonmodular domain accounts for more variation in a *global* or specific quality-of-life or HRQOL indicator. Vision-specific qualitative assessments are expected to include many more items directly linked to vision-related behaviors and the changes in these indicators could be expected to come from diminished visual acuity, or other biological determinants of vision. In contrast, most generic quality-of-life assessments would not expect to have many, if any, items that directly address vision and its contribution to functioning.

To test this hypothesis, I performed a *Medline* literature review ($N=391$; 6/14/2008) to identify papers that included old age–onset cataract patients who filled out both a generic and/or vision-specific HRQOL assessment. There are now quite a few vision-specific qualitative assessments (e.g., de Boer et al. 2004), so to simplify my task, I excluded papers that included patients who had specific visual defects (such as glaucoma, diabetic retinopathy, and so on) and child patients, and limited myself to papers written in English. However, it was clear that many of these elderly persons had a variety of comorbid physical conditions that were not always identified in the study reports. Since I was just interested in papers that reported the desired correlation or regression analyses, this naturally reduced the number of possible papers to include in the table. Table 6.5 summarizes the studies that reported the desired correlation/regression analyses. Several types of measures were included (e.g., Pearson's Product Moment correlations, Spearman correlations, Beta weights from regressions, and percentage differences). In addition, some authors surveyed their patients preoperatively and others postoperatively, while a third group calculated difference scores. Also, some authors reported single eyes and others presumably assessed both eyes. Finally, I included some correlations of self-reported vision indicators and either a generic or vision-specific assessment (e.g., Rosen et al. 2005).

The paper by Lee et al. (2001) is a prototype of the kind of study that can test the first approach listed earlier. Visual acuity, as measured by the Snellen Visual Acuity (SVA) chart scores, was correlated with each of the eight domains of the SF-36 (an example of a generic assessment), and all were nonsignificant ($P>0.05$; $r=-0.000$ to 0.146). In contrast, scores of the SVA (an example of a vision-specific assessment) were significantly correlated with the VF-14 ($r=0.266$; $P=0.004$),. Thus, there was a significant correlation between the vision-specific HRQOL assessment, but not for any of the eight domains of the generic HRQOL assessment and visual acuity. Particularly interesting about this study was that it also included utility estimates, using the standard

Table 6.5 The relationship between visual acuity, generic and disease-specific HRQOL assessments

References	Diagnoses (N)	Visual acuity ↔ generic	Visual acuity ↔ vision-specific	Generic ↔ disease-specific
Manguione et al. (1992)	Cataracts (334) (preoperative)		VA ADVS $r=-0.39$	SF-36 PFS and ADVS $r=0.31$
Manguione et al. (1994)	Cataract (458) (postoperative)	SVA SF-36 PFS 95% vs. 36%	SVA ADVS 95% vs. 80%	
Steinberg et al. (1994)	Cataracts (776) (preoperative)	SVA SIP $r=-0.23$ SVA SIP-VR $r=-0.20$[a]	SVA VF-14 $r=0.23$	
Fletcher et al. (1997)	Cataracts (100) (preoperative)	LogMAR Aravind-QOL $r=0.40$	LogMAR Aravind-QOL VF $r=0.40$	
	(Postoperative)	LogMAR Aravind-QOL $r=0.41$	LogMAR Aravind-QOL VF $r=0.44$	
Lee et al. (2001)	Cataracts (132) (postoperative)	SVA § SF-36 $r=-0.065$ to 0.146 ($P>0.05$) SVA SG-G $r=0.122$ ($P>0.05$) SVA VR-G $r=0.147$ ($P>0.05$)	SVA VT-14 $r=0.266$ ($P=0.004$) SVA SG-VH $r=-0.024$ ($P>0.05$) SVA VR-VH $r=0.202$ ($P=0.031$)	
Lau et al. (2002)	Cataracts (310) (postoperative)	LogMAR QOL[b] $r=0.313$ (0.207–0.412 95% CL)	LogMAR $r=0.420$ (0.322–0.509 95% CI)	
Javitt et al. (2003)	Cataracts (1832) (postoperative)	LogMAR SF-36 PFS $r=-0.10$	LogMAR TyPE[c] $r=-0.32$	SF-36 PFS TyPE $r=0.27$
Rosen et al. (2005)	Cataracts (233) (single eye postoperative)	SVA QWB $r=0.141$ ($P<0.05$) Visual Symptoms QWB $r=0.195$ ($P<0.01$) Trouble with vision QWB $r=0.256$ ($P<0.01$)	SVA VF-14 $r=0.157$ ($P<0.05$) Visual Symptoms VF-14 $r=0.465$ ($P<0.01$) Trouble with vision VF-14 $r=0.520$ ($P<0.01$)	
Polack et al. (2007)[d]	Cataracts (196) (preoperative)	SVA EQ-5D $r=-0.22$ ($P<0.002$)	SVA WHO/PBD VF 20 $r=0.46$ ($P<0.001$)	
Datta et al. (2008)[e]	Cataracts (289 elderly women) (pre-post–operative changes)	LogMAR – EQ-5D Beta=0.07 ($P=0.32$) Stereopsis – EQ-5D Beta=0.04 ($P=0.53$) Contrast sensitivity – EQ-5D Beta=−0.11 ($P=0.10$)	LogMAR – VF-14 Beta=−0.14 ($P=0.08$) Stereopsis – VF-14 Beta=−0.21 ($P=0.01$) Contrast sensitivity – VF-14 Beta=0.16 ($P=0.03$)	

CL confidence limits; *Generic* generic quality-of-life assessment; *LogMAR* log of minimal angle of resolution following administration of ETDRS charts; *SF-36 PFS* physical functioning items of the SF-36; *SG* standard gamble utility measure; *SG-VH* standard gamble-visual health; *SVA* Snellen visual acuity; *TyPE* the TyPE specification assesses visual functioning in five dimensions: (1) distance vision; (2) near vision; (3) daytime driving; (4) nighttime driving; and (5) glare; *VA* visual acuity; *Vision-Specific* vision-specific HRQOL assessment; *VR* verbal utility report

References: *ADVS* Activities of Daily Vision Scale (Manguione et al. 1992); *EQ-5D* Euro Qol group 5 dimensions (Brooks et al. 2003); *SF-36* Medical Outcome Study Short Form 36 (Ware and Sherbourne 1992); *SIP* Sickness Impact Scale (Bergner et al. 1981); *VF-14* visual functioning 14 items (Steinberg et al. 1994); *WHO/PBD VF20* WHO Vision-Related Quality-of-Life Scale (2003)

[a] The SIP-VR were items that a respondent selected from the SIP, and then the respondent was asked if response was based on vision
[b] The quality-of-life assessment used in this paper seems to be a study-specific assessment that includes items asking about self-care, mobility, social activities, and mental health issues
[c] The TyPE is best described as a functional assessment, but will be considered an approximation of an HRQOL assessment for this discussion
[d] Correlations were provided by S. Polack, based on the Polack et al. (2007) study
[e] Data presented in terms of beta weights following regression analyses calculated at follow-up. Data for visual acuity, stereopsis, and contrast sensitivity are presented (see text for discussion)

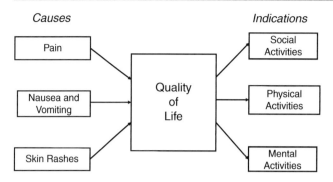

Fig. 6.3 Determinants and outcome of a qualitative assessment

gamble and a verbal rating. Both utility measures were estimated from a general health or visual health perspective, and they were then correlated with SVA scores. While the standard gamble results were not significantly correlated with visual acuity scores, the verbal ratings were, though only for visual health (Table 6.5).

I interpret Table 6.5 as providing the kind of support that would be expected for establishing a *working hypothesis*; the hypothesis being that modular domains are more likely to be correlated with variation in vision-specific HRQOL data than generic quality-of-life data, especially if the assessment were done postoperatively. Additional support for this hypothesis could be garnered by inspecting other visual impairments (e.g., age-related macular degeneration). However, a well-designed prospective study providing both correlation and regression analyses would be particularly helpful. My hypothesis overlaps with the continuing discussion about the relative merits and limits of generic and study-specific quality-of-life assessments (e.g., Patrick and Deyo 1989).

Domains, as CONTAINER metaphors, are neutral relative to the role they play in the research enterprise. What defines their role is the content of the domain, and how this content functions within the domain. For example, an investigator-created modular domain has many of the characteristics of a goal-directed category (Chap. 6, p. 167), since its content can include a wide range of entities collected together for a specific purpose or ideal. Thus, a vision-related modular domain could include the subjective analog of a physical (e.g., wavelength of light) stimulus: hue; the subjective consequences of visual acuity as impaired vision–dependent behaviors; and its social ramifications. A nonmodular domain is usually less determined, and instead is more likely to consist of only indicators of the consequences of events. Thus, modular domains (Fig. 6.3; pain, nausea and vomiting, skin rashes, etc.) should properly be seen as *causing* changes in quality or quality-of-life, while nonmodular domains are more likely to *indicate* the outcome of these causal factors.

This difference has practical statistical significance. This was illustrated by a study of Fayers and Hand (1997) that traced the impact of an exploratory factor analysis on the effort to establish the construct validity of the Hospital and Anxiety and Depression Scale (HADS) and the Rotterdam Symptom Checklist (RSCL). They found that the factor structure of the HADS (Zigmond and Snaith 1983), for example, confirmed that the assessment's investigators achieved their goal of selecting items to represent the concepts of depression and anxiety. A path analysis illustrated that the selected items could indeed be seen as a reflection of, or an indication of, the concept (that is, the analytic path was from the concept to the item; Fig. 6.3). In their exploratory factor analysis of the RSCL (de Haes et al. 1990), Fayers and Hand (1997) found four factors: psychological distress; pain-related issues; nausea and vomiting; and various symptoms. Using the same path analytic model as they had for the HADS, the initial assumption was that each of these factors was an indication of the concept of quality-of-life or HRQOL. However, certain counterintuitive components of the factor structure were found suggesting that instead of these factors being an indication of quality-of-life, they actually caused changes in quality-of-life. This is so because these factors were clear examples of modular domains (e.g., pain, nausea and vomiting, and other symptoms). As a result, Fayers and Hand (1997) specifically recommend that confirmatory factor analytic procedures be used with casual variables. Thus, the statistical procedure that should be used will vary depending on whether the variable being measured has modular or nonmodular characteristics.

In this section, I hoped to illustrate how examining the modular or nonmodular nature of a domain can be applied to understanding the role that a domain performs (as causal or as an indicate of a qualitative change), and also how this difference in the content of a domain has practical statistical significance. It is clear that the data I reported in Table 6.5 could be replicated with any number of other indicators, and doing so may very well demonstrate that most domains have both modular and nonmodular components. What also becomes evident is that the makeup of a domain may be culturally specific (Lee et al. 2001), which adds another set of issues.

If the composition of domains is relatively modular and nonmodular, then this has implications for the Wilson and Cleary (1995) model. A more accurate conceptualization may be that the dimensions of the Wilson and Cleary model may will vary in their degree of modularity, with all the dimensions likely having both modular and nonmodular components. The degree a domain is modular also has implications for the effort to subsume qualitative indicators as a subset of Patient-Reported Outcomes (PROs), when in fact it may be that all PROs vary in their degree of modularity. It is also not surprising to find that investigators create modular domains, since they are interested in explaining some phenomenon, and establishing a causal relationship helps in this process. The alternative is to draw inferences from correlational studies, involving domains that have predominantly nonmodular components. This option clearly would be less informative when the objective is prediction of outcomes.

Table 6.6 Examples of domain systems

Examples
1. Multiattribute Health System (Torrance et al. 1995)
2. Diagnostic and Statistical Manual of Mental Disorders (American Psychiatric Association, 1994 to 2000)
3. International Classification of Diseases and Related Health Problems-10 (WHO 1994)
4. International Classification of Functioning, Disability and Health (WHO 2001)

4.2 Classification and Domain Systems

An investigator-created domain is clearly an essential aspect of many scientific endeavors. One of the more important applications of domain formation is the effort to organize domains into systems. Torrance's (1986; Fig. 6.1) hierarchy of health states is one example, organized so that domains of limited informativeness are assumed to be incorporated into more abstract domains. Thus, the hierarchical relationship governs the relationship between domains in this particular system. There are many additional examples of domain systems that are of great practical value (Table 6.6).

Investigator-generated domain systems are, as Glushko et al. (2008) point out, examples of explicit classification systems. They require effort to create; are designed for specific purposes (e.g., to facilitate diagnoses); permit communication between users of the system (that is, two clinicians should be able to communicate about a particular patient with the same disorder); and are subject to change. They differ from cultural classification systems that are implicitly acquired with little awareness by children, and they also differ from individual classification systems that are explicitly created by a person and have limited generality (that is, they are idiosyncratic). Each of these classification systems is an example of a top-down cognitive process (like metacognition), within which category formation facilitates classification. In contrast, the bottom-up process involves sensory and perceptual information that is first classified, and then categories are formed.

The *Diagnostic and Statistical Manual of Mental Disorders*, or DSM system (DSM; American Psychiatric Association, 2000) is an example of an investigator-generated domain system and I will use it to illustrate some of the complex issues involved in creating and using such systems. The DSM system is an example of an explicit classification system since it is dependent on a consensus of investigators (e.g., social workers, psychologists, and psychiatrists) to reliably classify psychopathology. Since the DSM is meant to characterize psychopathology, there has also been an extended debate about whether psychopathology can be conceived of as a series of discrete domains or whether it is best viewed as a continuum, ranging from some norm to a pathological state. This issue has played itself out during the development of the DSM system. Thus, there have been four major versions (a fifth is on the way), the first two of which were based on psychoanalytical theory. These early versions of the DSM did not generate satisfactory inter-rater reliability scores, were vaguely worded, and reflected a unidimensional view of psychopathology. The third version abandoned psychoanalytical jargon, introduced more behaviorally based categories, and used explicit criteria for including or excluding a person's symptoms as a basis for a diagnosis. Development of the DSM system continues (e.g., the DSM V) as additional research amplifies particular diagnostic domains and demonstrates the reliability of particular diagnostic syndromes. It has shown itself to be a practical system and is used in health care reimbursement activities in the United States.

Schmidt et al. (2004) address some of the complexities involved in generating a classification system such as the DSM, including the concern that the system's authors may have sacrificed validity for the sake of reliability. This, they suggest, has happened because some of the less reliable but essential indicators of a disorder have been eliminated from the criteria used in a diagnostic domain. In addition, they point out that the DSM system is based on presenting symptoms that do not consider the natural history of the disorder in the diagnosis. This has occurred because psychiatric criteria rely on cross-sectional studies, as opposed to longitudinal ones.

The debate concerning whether psychopathology should be viewed dimensionally rather than as a series of discrete domains was a major concern of the psychologist Paul Meehl (Meehl 1992), especially as it was applied to the DSM system.[12] Meehl asked whether it is possible that domains, instead of being made up by a committee,[13] could be natural nonarbitrary entities called *taxons*. He developed a series of statistical procedures to demonstrate the presence of taxons, and Ruscio et al. (2007) have reported an investigation of major depressive disorders based on these procedures. Their results provide evidence supporting the view that psychopathology consists of a series of taxons or domains. However, their analysis was based on a restructuring of the symptoms that make up the major depression domain and, thus, the analysis appears forced.

In addition, Zimmerman et al. (e.g., 2006) have reported a series of 12 studies designed to determine the symptoms to include as criteria for major depression, as described in the DSM-IV. The diagnosis of major depressive disorder requires that five of nine characteristics must be present. However, Zimmerman et al. 2006) showed that either one or two of the nine (low mood or loss of interest) criteria have to be minimally present for the application of the diagnosis. Zimmerman and his colleagues (Zimmerman et al. 2006) have also reported how criteria not listed in the DSM-IV criteria for major depression can be better discriminators for the listed criteria (e.g., diminished drive).

One of the risks of a domain approach is that it can be arbitrary, impeding the study of any underlying disease process (Hayes, et al. 1996). There is also the issue that the DSM system was designed for North American clinicians, and that some of the categories and criteria for inclusion in a category maybe culturally dependent. Add to this the argument that when the DSM system is compared to a biological classification system (e.g., a classification of trees), it is found wanting (Blashfield 1986). Thus, while the mixed DSM system allows for multiple diagnoses, its hierarchical organization does not correspond to what is found in a biological classification. Schmidt et al. (2004) note a number of other deficiencies in the DSM system, including its lack of dependence on an explicit theory of pathogenesis, the presence of political influence to define diagnostic categories, and so on. They are particularly concerned that an atheoretical system may give the appearance of being useful when it is not. They state:

This process is potentially deceiving and may produce the impression of progress when, in actuality, there is an enduring stagnation. A purely descriptive system is, in a sense, nonfalsifiable. If there is no hypothesis, not even about the nature of the categories, the system is nothing more than drawings in the sand. (Schmidt et al. 2004; p. 26).

The DSM system attempts to link specific criteria to specific diagnostic categories (a criterion-referenced approach), so that if a clinician identifies the presence of these criteria, a person can be placed in a particular diagnostic domain. Cronbach and Meehl (1955), however, argue that psychiatric disorders are open concepts with fuzzy boundaries. For this reason, they propose that psychiatric classification should be based on a construct validity approach. Construct validity requires developing and empirically supporting an appropriate theory, an objective they acknowledge that had yet to be achieved.

This discussion should be sufficient to reveal some of the issues involved in organizing a number of domains (e.g., diagnoses) into a classification system. The same task exists for those investigators (e.g., Cummins 1996; King and Hinds 1998) who want to identify a series of quality-of-life domains that can be organized into a system. Thus, the questions asked of the DSM-IV system can also be asked of any prospective quality-of-life classification system. Are the domains based on *a priori* theoretical rationale that can be accepted or rejected, or do they represent statistical clustering dependent on whatever information is placed in the statistical analysis? Do the domains organize themselves into a hierarchy, or some other explicit system? Does the created system have a specific goal or purpose that can be applied to all of its components (e.g., should both physical and mental health be part of one quality-of-life classification system?)? What is now considered typical and "consensual" is that a quality-of-life assessment consists of domains that include items that assess physical, social, and mental functioning; various indicators of health status; measures of cognitive activities; and so on.[14]

As I discussed in Chap. 2, lists of this sort are a regular part of a definition of quality-of-life or HRQOL, but they also constitute a classification system, since they are used to monitor the consequences of disease and treatment. However, this "system" has never been carefully examined and remain deficient when compared to the efforts extended to the systems listed in Table 6.6. Still, this doesn't mean that a series of quality domains can't be organized into a system by an investigator, especially if the system has a specific function or purpose, such as ensuring an optimized outcome. Thus, just as the DSM system permits persons to be classified and labeled, it should be possible to classify and label the quality of a product, meal, or person's life. Once classified, it should also be possible to optimize an outcome, as would be forthcoming in any quality control effort. I will consider aspects of the other systems listed in Table 6.6 in subsequent chapters.

5 Summary of Chapter

There were several important ideas developed in this chapter. First, (Table 6.1) that it is conceptually and practically possible to separate how a summary statement is made. This distinction is useful, even though empirically, both the person and the investigator contribute to the summary measures construction. Second, categories or domains can be formed automatically, reflect a controlled cognitive process, or most likely be a product of the interaction between the two. Third, when analyzed, summary statements reveal that they consist of subtypes (modular and nonmodular); some with one component (pain symptoms include mostly modular components, while depression does not), and others with a combination. For example, the extent to which a modular component accounts for a summary statement potentially could have empirical implications, although I did not document this in this chapter (see Chaps. 7–11). Thus, symptoms that have a greater physical and biological basis are more likely to be salient to a respondent, so that domains including these symptoms might be more likely to be correlated with global indicators, than, say, depression or mood indicators. Certainly, this occurred in the Krause and Jay (1994) study, as was evident by the fact that most respondents identified physical status indicators as determinants of their global quality.

A domain can be thought of as a point on an informational continuum that can be defined relative to a more abstract entity, such as a concept. It can be formed as a result of a collection of a common kind (similar items collected together) or a collection around a common theme (cross-classification of information). Domains can also be created to achieve a specific purpose, or by investigator-based formal methods such as factor analysis, Boolean algebra, and Venn diagrams.

Only domains that are predominantly modular[15] have specific criteria that can be used in their definition. All other investigator-generated summary statements are arbitrarily labeled. In this chapter, I also discussed functional modularity and acquired modularity, both of which may be far more prevalent than currently considered, and which can be used as part of a classification system.

Some have argued (e.g., Sperber 2002) that much, if not all, cognitive processing is modular. This massive modularity view, Sperber argues, reflects the fact that cognition (specifically modular domains) represents an adaptive response to evolutionary pressures. This issue overlaps with the notion that a quality-of-life or HRQOL assessment may also represent an evolving adaptive response reflecting a person's continued assessment of the circumstances of their life and what they want the nature of their existence to be.

Review of studies in this chapter revealed that the modularity of a domain has practical statistical significance. This was illustrated by the study of Fayers and Hand (1997) that clarified when and why particular factor analytic techniques should be used (e.g., exploratory or confirmatory factor analytic techniques). I also raised the possibility that since any domain has varying degrees of both modular and nonmodular components, this will have implications for the statistical procedures that can and should be used in a study. It also provided another perspective on what the Wilson and Cleary (1995) model actually assesses.

I conclude by adding another stitch to the "cognitive" fabric I am constructing: classifying information and determining the role of categories or domains are essential steps in a scientific approach to a qualitative assessment. Classification is the first step, and one of the most important. As revealed in Tables 6.1 and 6.2, there are a variety of methods whereby this classification can occur. My concern in the next five chapters is to determine which of the candidate domains actually function as a domain, and to learn how they contribute to clarifying what is meant by quality or quality-of-life.

Notes

1. Much of the research on category formation that I cite involves studies of objects and perceptual arrays. I recognize that while these studies may be quite relevant when the object of study is the qualitative assessment of a statue or painting, there is a risk when applying these studies to the more complex environment of quality-of-life assessment. However, I find this extension to be acceptable in most instances, especially if it stimulates the required research.
2. Ashby and Maddox (2005), like Estes, distinguish the terms *categories* and *concepts*, although they acknowledge that it is common for them to be used interchangeably. Thus, a category is defined as a collection of similar objects belonging to the same group, while a concept refers to a collection of more abstract entities, such as ideas. Again, the interchange of the terms category and concept is another demonstration of the fact that a person has the cognitive capability to make a concrete entity abstract and vice versa.
3. As expressed here, the difference between a category and a domain constraint is that the constraints are not made explicit when a category is formed, but they usually are (by an investigator) when a domain is formed.
4. These two memory systems do not refer to short- and long-term memory, both systems are long-term memory system, with short-term memory being an activated portion of the information held in long-term memory.
5. A number of authors have proposed different schemes for characterizing modular domains. Geary and Huffman (2002) propose that modules can be conceived of in terms of neural, perceptual, cognitive, and functional modules.
6. In Chap. 2, I described the process of how a conceptual metaphor is formed. I suggested that a structural mapping was occurring between the base and the target (Bowdle and Gentner 2005). In the context of the current discussion, I suggest that this mapping is, at first, a controlled process that can become automatic with continued exposure, as would be expected with a conceptual metaphor.
7. Hirschfeld and Gelman (1994) and others use the term "domain-specific" to refer to entities that are quite prescribed in terms of their neurological and psychological structure and function (e.g., vision). This term may be confusing, especially since I am using the term domain to refer to a wide range of types of domains. I have opted to use the term "modular" instead of "domain-specific". However, the term modular also has a broad reference, but I don't believe this should be as confusing as concurrently using the phrase domain-specific and domain. I will clarify the different usages of the terms when I discuss Fodor's (1993) nine criteria for a module.
8. Fodor has written a more recent book on this topic published in 2000, but I will focus on the 1983 book and leave the issue of Fodor's current thoughts for another time and place.
9. The phenomenon of "upregulation" is an excellent example of acquired modularity. This is so because a specific biologically-based event occurs which produces a change in a person's pain sensitivity that did not exist prior to the event (i.e., it is acquired), and which represents a change in a particular modality (pain perception) that has a physiological basis.
10. Describing health status as "not being a domain of interest" does not mean that I am suggesting it not be measured. Just that it will be measured by the domains that make it up, which if I follow Patrick and Erickson's (1993) recommendations would include symptoms, functional states, and health perceptions.
11. By "unique" measures, I mean measures that do not have to refer to other domains for their definition.
12. Whether and how the content of a domain can be organized into a dimension is a critical issue. This may occur statistically, as during an IRT analysis and the creation of a data bank, but may also involve the interaction between an investigator's own cognitive processing and various statistical procedures (e.g., as when an investigator labels a factor following a factor analysis and orders the elements in the domain along a dimension). However, it is also possible that the entities within a domain spontaneously form a dimension. Thus, you can have a quality-of-life domain and have the dimension of goodness or badness of the qualitative entities within the domain. In fact, it can be argued that you can't have a cognitive dimension without first grouping this information into a category, or a dimension with the domain.
13. The authors of the DSM system make clear that they are aware that the domains of their system contain arbitrary elements (American Psychiatric Association, 1994 to 2000; p. xxii).
14. Some would argue that the Patient Reported Outcomes (PROs) movement is an example of domain formation gone wild, with practically any number of combinations of indicators included in a domain. Actually, PROs are an excellent example of domains that

could benefit from the type of analysis being discussed in this chapter.

15. Although the criteria for defining a modular domain belong to Fodor as interpreted by Cain (2002), his general philosophical orientation of *radical nativism* has been criticized (Laurence and Margolis 2002).

References

Allen SW, Brooks LR. (1991). Specializing the operation of a explicit rule. *J Exp Psychol Gen.* 120, 3–19.

American Psychiatric Association. (1994 to 2000). *Diagnostic and Statistical Manual of Mental Disorders: Fourth Edition, Text Revision*. Washington, DC: American Psychiatric Association.

Arnold R, Ranchor AV, Sanderman R, Kempen GIJM, et al. (2004). The relative contribution of domains of quality-of-life to overall quality-of-life for different chronic diseases. *Qual Life Res.* 13, 883–896.

Ashby FG, Alfronso-Reese LA, Turken AU, Waldron EM. (1998). A neuropsychological theory if multiple systems in category learning. *Psychol Rev.* 105, 442–481.

Ashby FG, Ell SW. (2001). The neurobiology of category learning. *Trends Cogn Sci.* 5, 204–210.

Ashby FG, Maddox WT. (2005). Human category learning. *Annu Rev Psychol.* 56, 149–178.

Ashby FG, Waldron EM. (2000). The neuropsychological bases of category learning. *Curr Trends Psychol Sci.* 9, 10–14.

Austin JL. (1961). *Philosophical Papers*. Oxford, Oxford University Press.

Austin JL. (1962). *How To Do Things With Words*. Cambridge MA, Harvard University Press.

Baddeley AD. (2002). Is working memory still working? *Eur Psychol.* 7, 85–97.

Baddeley AD, Della Sala S, Papagno C, Spinnler H. (1997). Dual-task performance in dysexecutive and nondysexecutive patients with frontal lesions. *NeuroPsychol.* 11, 187–194.

Bargh JA, Ferguson MJ. (2000). Beyond behaviorism: On the automaticity of higher mental processes. *Psychol Bull.* 126, 925–945.

Barofsky I. (1997). Functional gastrointestinal disorders: Chronic pain management. In, (Eds.), R McCallum, G Friedman and ED Jacobson *Gastrointestinal Pharmacology and Therapeutics*. Raven Press, New York, pp. 133–145.

Barrett L, Tugade MM, Engle RW. (2004). Individual differences in working memory capacity and dual-process theories of the mind. *Psychol Bull.* 130,55–573.

Barsalou LW. (1983). Ad hoc categories. *Mem Cogn.* 11, 211–227.

Bergner M, Bobitt RA, Carter WB, Gilson BS. (1981). The Sickness Impact Profile: Development and final revision of a health status measure. *Med Care.* 19, 787–805.

Birnboim S. (2003). The automatic and controlled information-processing dissociation: Is this still relevant? *NeuroPsychol Rev.* 13, 19–31.

Blashfield RK. (1986). Structural approaches to classification. In, (Eds.) T. Millon, GL Klerman. *Contemporary Directions in Psychopathology: Towards the DSM-IV*. New York NY:Guilford. (p. 363–379).

Boole G. (1854/1951). *An Investigation of the Laws of Thought. On Which are Founded the Mathematical Theories of Logic and Probabilities*. New York NY: Reprinted by Dover Publications.

Boroditsky L. (2001). Does language shape thought? Mandarin and English speakers conceptions of time. *Cogn Psychol.* 43, 1–22.

Bott L, Hoffman AB, Murphy GL. (2007). Blocking in category learning. *J Exp Psychol Gen.* 136, 685–699.

Bowdle BF, Gentner D. The career of metaphor. *Psychol Rev.* 112; 193–216, 2005.

Bradburn NM. (1969). *The Structure of Subjective Well-being*. Chicago Il: Aldine.

Brooks R, Rabin R, de Charro F. (2003). *The Measurement and Valuation of Health Status Using the EQ-5D: A European Perspective. Evidence from the EurQol BIOMED Research Programme*. Dordrecht The Netherlands: Kluwer. (p. 254).

Cain MJ. (2002). *Fodor: Language, Mind and Philosophy*. Cambridge UK: Polity.

Cantril H. (1965). *The Pattern of Human Concerns*. New Brunswick NJ: Rutgers University Press.

Carey S. (1985). *Conceptual Change in Childhood*. Cambridge MA: Bradford Press.

Cella D. (2004). The Functional Assessment of Cancer Therapy-Lung and Lung Cancer Subscale assess quality-of-life and meaningful symptom improvement in lung cancer. *Semin Oncol.* 31(3 Suppl 9), 11–15.

Cella D, Gershon R, Lai J-S, Choi S. (2007). The future of outcome measurement: Item banking, tailoring short-forms and computerized adaptive environment. *Qual Life Res.* 16, 133–141.

Chartand TL, Baugh JA. (1999). The chameleon effect: The perception-behavior link and social interaction. *J Personal Soc Psychol.* 76, 893–910.

Chen J-Y. (2007). Do Chinese and English speakers think about time differently? Failure of replicating Boroditsky (2001). *Cognition.* 104, 427–436.

Chin L-H. (1972). A cross-cultural comparison of cognitive styles in Chinese and American children. *Int J Psychol.* 7, 235–242.

Chomsky N. (1988). *Language and Problems of Knowledge*. Cambridge MA: MIT Press.

Costa PT, McCrae RR. (1992). *NEO PI-R Professional Manuel*. Odessa FL: Psychological Assessment Resources.

Cronbach LJ, Meehl PE (1955). Construct validity and psychological tests. *Psychol Bull.* 52: 281–302.

Cummins RA. (1996). The domains of life satisfaction: An attempt to order chaos. *Soc Indic Res.* 38, 303–328.

Datta S, Foss AJ, Grainge MJ, Gregson RM, et al. (2008). The importance of acuity, steropsis, and contrast sensitivity for health-related quality-of-life in elderly women with cataracts. *Investig Ophthalmol Vis Sci.* 49; 1–6.

David KM, Ganiats TG, Miller C. (1999). Placement matters; Stability of the SF-36 EVGFP responses with varying placement of the question in the instrument. Abstract presented at the Sixth Annual Meeting of the International Society For Quality-of-life Research, Barcelona Spain, November 1999. *Qual Life Res.* 8, 623.

Davidson RJ. (2004). Well-being and affective style: Neural substrates and biobehavioral correlates. *Philos Trans R Soc (Biol Sect).* 359, 1395–1411.

de Boer MR, Moll AC, de Vet HCW, Terwee CB, et al. (2004). Psychometric properties of vision-related health quality-of-life questionnaires: a systematic review. *Ophthalmic Physiol Opt.* 24, 257–273.

De Haes JCJH, van Knippenberg FC, Neijt JP. (1990). Measuring psychological and physical distress in cancer patients: Structure and application of the Rotterdam Symptom Checklist. *Br J Cancer.* 62, 1034–1038.

DeNeve KM, Cooper H (1998). The happy personality: A meta-analysis of 137 personality traits and subjective well-being. *Psychol Bull.* 124, 197–229.

Estes WK. (1994). *Classification and Choice*. New York NY: Oxford.

Evans JSBT. (2003). In two minds: dual-process accounts of reasoning. *Trends Cogn Sci.* 7, 454–459.

Fairclough DL. (2002). *Design and Analysis of Quality-of-life Studies in Clinical Trials*. Boca Raton FL: Chapman & Hall/CRC.

Fayers PM, Hand DJ. (1997). Factor analysis, causal indicators and quality-of-life. *Qual of Life Res.* 6, 139–150.

Feinstein A. (1967). *Clinical Judgment*. Huntington NY: Krieger.

Fletcher AE, Ellwein LB, Selvaraj S, Vijayakumar V, et al. (1997). Measurement of vision function and quality-of-life in patients with cataracts in southern India. Report of instrument development. *Arch Ophthal.* 15, 767–774.

Fodor J. (1983). *Modularity of the Mind*. Cambridge MA: MIT Press.
Frege, G. (1892/1966). On Concept and Object. In, (Eds.) P Geach, M Black. *Translations from Philosophical Writings of Gottlob Frege*. Oxford UK: Oxford University Press. (p. 42).
Gärdenfors P. (2000). *Conceptual Spaces: The Geometry of Though*. Lexington MA: MIT Press.
Geary DC, Huffman KJ. (2002). Brain and cognitive evolution forms of modularity and functions of mind. *Psychol Bull.* 128, 667–698.
Gentner D, Boroditsky L. (1999). Individuation, relativity and early word learning. In, (Eds.) M Bowerman, S Levinson, *Language Acquisition and Conceptual Development*. Cambridge UK: Cambridge University Press. (p. 215–256).
Gigerenzer G, Todd PM. (1999). *Simple Heuristics that Make Us Smart*. Cambridge MA: MIT Press.
Gladwell M. (2005). *Blink*. New York NY: Little Brown.
Glushko RJ, Maglio PP, Matlock T, Barsalou LW. (2008). Categorization in the wild. *Trends Cogn Sci.* 12, 129–135.
Gutiérrez JLG, Jiménez BM, Hernández EG, Puente CP. (2005). Personality and subjective well-being: Big five correlates and demographic variables. *Personal Individ Differ.* 38, 1561–1569.
Guttman L. (1944). A basis for scaling qualitative data. *Am Sociol Rev.* 9, 139–150.
Hasher L, Zacks RT. (1979). Automatic and effortful process in memory. *J Exp Psychol Gen.* 108, 356–388.
Hayes SC, Wilson KG, Gifford EV, Follette VM, et al. (1996). Experiential avoidance and behavior disorders: A functional dimensional approach to diagnosis and treatment. *J Consult Clin Psychol.* 64, 1152–1168.
Hirschfeld LA, Gelman SA. (1994). Towards a topography of mind: An introduction to domain specificity. In, (Ed.) LA Hirschfeld, SA Gelman. *Mapping the Mind: Domain Specificity in Cognition and Culture*. New York NY: Cambridge University Press. (p. 3–36).
Hoffman AB, Murphy GL. (2006). Category dimensionality and feature knowledge: When more features are learned as easily as fewer. *J Exp Psychol Learn, Mem Cogn.* 32, 301–315.
Horton MS, Markman EM. (1980). Developmental differences in the acquisition of basic and superordinate categories. *Child Dev.* 51, 708–719.
Hurt SJ, Andsager JL. (2003). Medicalization vs. adaptive models? Sense-making in magazine framing of menopause. *Women Health.* 38, 101–122.
International Classification of Diseases and Related Health Problems-10. (1994). Geneva Switzerland: WHO.
International Classification of Functioning, Disability and Health. (2001). Geneva Switzerland: WHO.
Jansma JM, Ramsey NF, Slagter HA, Kahn RS (2001). Functional anatomical correlates of controlled and automatic processing. *J Cogn NeuroSci.* 13; 730–743.
January D, Kako E. (2007). Re-evaluating the evidence for linguistic relativity: Reply to Boroditsky (2001). *Cognition.* 104, 417–426.
Javitt JC, Jacobson G, Schiffman RM. (2003). Validity and reliability of the Cataract TyPE Spec: an instrument for measuring outcomes of cataract extraction. *Am J Ophthalmol.* 136, 285–290.
Ji L-J, Nisbett RE. (2001). Culture, language and categories. *Unpublished Manuscript*, University of Michigan MI. (Referenced in Nisbett et al, 2001).
Ji L-J, Zhang Z, Nisbett RE. (2004). It is culture or is it language? Examination of language efforts in cross-cultural research in categorization. *J Personal Soc Psychol.* 87, 57–65.
Kahneman D, Tversky D. (1982). The psychology of preferences. *Sci Am.* 246, 160–173.
Kelly AMC, Garavan H. (2005). Human functional neuroimaging of brain changes associated with practice. *Cereb Cort.* 15, 1089–1102.
Kim K, Relkin N, Lee K, Hirsch J. (1997). Distinct cortical areas associated with native and second languages. *Nature.* 388, 171–174.
King CR, Hinds PS. (1998). *Quality-of-life: From Nursing and Patient Perspectives. Theory, Research, Practice*. Sudbury, MA: Jones and Bartlett.
Kovelman I, Baker SA, Petitto LA. (2008). Bilingual and monolingual brains compared: A functional magnetic resonance imaging investigation of syntactic processing and a possible "neural signature" of bilingualism. *J Cogn NeuroSci.* 20, 153–169.
Krause N, Kay G. (1994). What do global self-rated health items measure? *Med Care.* 32, 930–942.
Kyllonen PC, Christal RE. (1990). Reasoning ability is (little more than) working memory capacity. *Intelligence.* 14, 389–433.
Lakoff G. (1987). *Women, Fire, and Dangerous Things: What Categories Reveal About the Mind*. Chicago Il: University of Chicago Press.
Lakoff G, Johnson M. (1999). *Philosophy in the Flesh: The Embodied Mind and its Challenge to Western Thought*. New York NY: Basic Books.
Lakoff G, Nunez RE (2000) *Where Mathematics Comes From*. New York NY: Basic Books.
Lau J, Michon JJ, Chan W-S, Ellwein LB. (2002). Visual acuity and quality-of-life outcomes in cataract surgery patients in Hong Kong. *Brit J Ophthalmol.* 86; 12–17.
Lee JE, Fos PJ, Zuniga MA, Kastl PR, et al. (2001). Assessing health-related quality-of-life in cataract patients: The relationship between utility and health-related quality-of-life treatment. *Qual Life Res.* 9, 1127–1135.
Liberman A, Mattingly I. (1989). A specialization for speech perception. *Science.* 243, 489–494.
Logan RF. (1986) *The Alphabet Effect*. New York NY: Morrow.
Laurence S, Margolis E. (2002). Radical concept nativism. *Cognition.* 86, 25–55.
Lykken D, Tellegen A. (1996). Happiness is a stochastic phenomenon. *Psychol Sci.* 7, 186–189.
Manguione CM, Phillips RS, Seddon JM, Lawrence MG, et al. (1992). Development of the 'Activities of Daily Vision Scale': A measure of visual functional status. *Med Care.* 30, 1111–1126.
Manguione CM, Phillips RS, Lawrence MG, Seddon JM, et al. (1994). Improved visual function and attenuation of declines in health-related quality-of-life after cataract extraction. *Acta Ophthalmol.* 112, 1419–1425.
Marques JF. (2007). The general specific breakdown of semantic memory and the nature of superordinate knowledge: Insights from superordinate and basic-level feature norms. *Cogn NeuroPsychol.* 24, 879–903.
Marr D. (1982). *Vision*. New York NY: Freeman.
McGrath PJ, Finley GA. (1999). *Chronic and Recurrent Pain in Children and Adolescents*. Seattle WA: IASP Press.
Medin DL, Lynch EB, Coley JD, Atran S. (1997). Categorization and reasoning among tree experts: Do all roads lead to Rome? *Cogn Psychol.* 32, 49–96.
Medin DL, Schwanenflugel PJ. (1981). Linear separability in classification learning. *J Exp Psychol Hum Learn Mem.* 7: 355–368.
Medin DL, Lynch EB, Solomon KO. (2000). Are there kinds of concepts? *Annu Rev Psychol.* 51, 121–147.
Meehl PE. (1992). Factors and taxa, traits, and types, differences of degree and differences in kind. *J Personal.* 60, 117–174.
Meehl PE. (2004). What's in a taxa? *J Abnorm Psychol.* 113, 39–43.
Melzack R, Wall PD. (1965). Pain mechanisms: A new theory. *Science.* 150, 971–979.
Moors A, de Houwer J. (2006). Automaticity: A theoretical and conceptual analysis. *Psychol Bull.* 132, 297–326.
Murphy GL. (2005). The study of concepts inside and outside the laboratory "Medin versus Medin." In, (Eds.) W Ahn, RL Goldstone, BC Love, AB Markman, P Wolff. *Categorization Inside and Outside the Laboratory: Essays in Honor of Douglas L. Medin*. Washington DC: American Psychological Association. (p. 179–195).
Murphy GL, Wisniewski EJ. (1989). Categorizing objects in isolation and in scenes: What a superordinate is good for. *J Exp Psychol Learn Mem Cogn.* 15, 572–586.

Nes RB, Røysamb E, Tambs K, Harris JR, et al. (2006). Subjective well-being : Genetic and environmental contributions to stability and change. *Psychol Med.* 36,1–10.

Nisbett RE. (2003). *The Geography of Thought.* New York NY: The Free Press.

Nisbett RE, Peng K, Choi I. (2001). Culture and systems of thought: Holistic versus analytic cognition. *Psychol Rev.* 108,291–310.

Norenzayana A, Nisbett RE. (2000). Culture and causal cognition. *Curr Direct Psychol Sci.* 9, 132–135.

Norman G. (2003). Hi! How are you? Response shift, implicit theories and differing epistemologies. *Qual Life Res.* 12, 239–249.

Nosofsky RM, Palmeri TJ. (1997). An exemplar-based random walk model of speeded classification. *Psychol Rev.* 104, 266–300.

Oxford English Dictionary. (1996). Oxford UK: Oxford University Press.

Padilla GV, Kagawa-Singer M. Quality-of-life and culture. In, (Eds.) CR King, PA Hinds, *Quality-of-life From Nursing and Patient Perspectives: Theory-Research-Practice.* Sudbury MA, Jones and Bartlett, 1998. (p74–92).

Patrick DL. (2003). Concept of health-related quality-of-life and of patient-reported outcomes. In, (Eds.) O Chassany, C Caulin. *Health-Related Quality-of-life and Patient-reported Outcomes: Scientific and Useful Outcome Criteria.* Paris France: Springer. (p. 23–34).

Patrick DL, Chiang Y-P. (2000). Measurement of health outcomes in treatment effectiveness evaluations. *Med Care.*; 38 (Suppl II), II-14- II-25.

Patrick DL, Deyo RA. (1989). Generic and disease-specific measures in assessing health status and quality-of-life. *Med Care.* 27(suppl. 3), S217–232.

Patrick DL, Elinson J. (1984). Sociomedical approaches to disease and treatment outcomes in cardiovascular care. *Qual Life Cardiovasc Care.* 1, 53–62.

Patrick DL, Erickson P. (1993). *Health Status and Health Policy: Quality-of-life in Health Care Evaluation and Resource Allocation.* New York NY: Oxford.

Phillips C, Jarrold C, Baddeley AD, Grant J, et al. (2004). Comprehension of spatial language terms in Williams syndrome: evidence for an interaction between domains of strength and weakness. *Cortex.* 40, 85–101.

Polack S, Kuper H, Mathenge W, Fletcher A, et al. (2007). Cataract vision impairment and quality-of-life in a Kenyan population. *Br J Ophthalmol.* 91,927–932.

Pollak KL, Arnold RM, Jeffreys AS, Alexander SC, et al. (2007). Oncologist communication about emotion during visits with patients with advanced cancer. *J Clin Oncol.* 25, 5748–5752.

Pothos EM, Close J. (2008). One or two dimensions in spontaneous classification: A simplified approach. *Cognition.* 107, 581–602.

Ramsey NF, Jansma JM, Jager G, van Raalten T, et al. (2004). Neurophysiological factors in human information processing capacity. *Brain.* 127, 517–525.

Read SJ, Jones DK, Miller LC. (1990). Traits as goal-based categories: The importance of goals in the coherence of dispositional categories. *J Personal Soc Psychol.* 58, 1048–1061.

Rosch E, Mervis CB, Gray W, Johnson D, et al. (1976). Basic objects in natural categories. *Cogn Psychol.* 8, 382–439.

Rosen PN, Kaplan RM, David K. (2005). Measuring outcomes of cataract surgery using the Quality of Well-being Scale and the VF-14 Visual Function Index. *J Cataract Refract Sur.* 31: 359–378.

Ruscio J, Zimmerman M, McGlinchey JB, Chelminski I. (2007). Diagnosing major depression XI: A taxometric investigation of the structure underlying DSM-IV symptoms. *J Nerv Ment Dis.* 195, 10–19.

Schmidt NB, Kotov R, Joiner TE Jr. (2004). *Taxometrics: Towards a New Diagnostic Scheme for Psychopathology.* Washington DC: American Psychological Association Press.

Schneider W, Shiffrin RM. (1977). Controlled and automatic human information processing: I. Detection, search and attention. *Psychol Rev.* 84, 1–64.

Smith EE, Medin DL. (1981). *Categories and Concepts.* Cambridge MA: Harvard University Press.

Smith ER, DeCoster J. (2000). Dual-process models in social and cognitive psychology: Conceptual integration and links to underlying memory systems. *Personal Soc Psychol Rev.* 4,108–131.

Sperber D. (1994). The modularity of thought and the epidemiology of representation. In, (Eds.) LA Hirschfeld, SA Gelman. *Mapping the Mind: Domain Specificity in Cognition and Culture.* New York NY: Cambridge University Press. (p. 39–67).

Sperber D. (2002). In defense of massive modularity. In, (Ed.) E Dupoux. *Language, Brain, and Cognitive Development: Essays in Honor of Jacques Mehler.* Cambridge MA: MIT Press. (p. 47–57).

Spranger MA, Schwartz CE. (1999). Integrating response shift into health-related quality-of-life research: a theoretical model. *Soc Sci Med.* 49, 1507–1515.

Springer K, Keil FC. (1989). On the development of biologically specific beliefs: the case of inheritance. *Child Dev.* 60, 637–648.

Stanovisch KE, Wast RF. (2003). Evolutionary versus instrumental goals: How evolutionary psychology misconceives human rationality. In, (Ed.) DE Over, *Evolution and the Psychology of Thinking.* Philadelphia PA: Psychology Press. (p 171–230).

Steinberg EP, Tielsch JM, Shein OD, Javitt JC, et al. (1994). The VF-14: An index of functional impairment in patients with cataracts. *Arch Ophthalmol.* 112, 630–638.

Teasdale JD. (1999). Multi-level theories of cognition-emotion relations. In (Eds.) T Dalgleish, MJ Power, *Handbook of Cognition and Emotion.* Chichester UK: Wiley. (p. 665–681).

Torrance GW. (1986). Measurement of health state utilities for economic appraisal: A review. *J Health Econ.* 5, 1–30.

Torrance GW, Furlong W, Feeny D, Boyle M. (1995). Multi-attribute preference functions: Health Utilities Index. *Pharm Econ.* 7, 503–520.

Uleman JS, Newman LS, Moskowitz GB. (1996). People as flexible interpreters: Evidence and issues from spontaneous trait inference. In, (Ed.) MP Zanna, *Advances in Experimental Social Psychology.* San Diego CA: Academic Press.(Vol. 28, p. 211–279).

van Reekum CM, Urry HL, Johnstone T, Thurow ME, et al. (2007) Individual differences in amygdala and ventromedial prefrontal cortex activity are associated with evaluation speed and psychological well-being. *J Cogn NeuroSci.* 19, 237–248.

von Osch SMC, Stiggelbout AM. (2005). Understanding VAS valuations: Qualitative data on the cognitive process. *Qual Life Res.* 14, 2171–2175.

Vygotsky LS. (1962). *Thought and Language.* Cambridge MA: MIT Press.

Wall PD, Melzack R. (1994). *Textbook of Pain.* Edinburgh UK: Churchill Livingstone.

Ware JE, Sherbourne CD. (1992). The MOS 36-Item Short-Form Health Survey (SF-36®): I. conceptual framework and item selection. *Med Care.* 30,473–83.

Weiskopf DA. (2009). The plurality of concepts. *Synthese.* 169. DOI 1007/S11229-008-9340-8.

Weiss A, Bates TC, Luciano M. (2008). Happiness is a personal(ity) thing: The genetics of personality and well-being in a representative sample. *Psychol Sci.* 19, 205–210.

Whorf BL. (1956). *Language, Thought and Reality.* Cambridge MA: MIT Press.

Wilson IB, Cleary PD. (1995). Linking clinical variables with health-related quality-of-life: A conceptual model of patient outcomes. *JAMA.* 273, 59–65.

Winawer J, Witthoft N, Frank MC, Wu L, et al. (2007). Russian blues reveal effects of language on color discrimination. *Proc Natl Acad Sci (USA)* 104; 7780–7785.

Wisniewski EJ, Clancy EJ, Tillman RN. (2005). On different types of categories. In, (Eds.) W Ahn, RL Goldstone, BC Love, AB Markman, P Wolff. *Categorization Inside and Outside the Laboratory: Essays in Honor of Douglas L. Medin*. Washington DC: American Psychological Association. (p. 103–126).

Yellen SB, Cella DF, Webster K, Blendowski C, Kaplan E. (1997). Measuring fatigue and other anemia-related symptoms with the Functional Assessment of Cancer Therapy (FACT) measurement system. *J Pain Symptom Manag.* 13,16–74.

Zárate MA, Sanders JD, Garza AA. (2000). Neurological disassociations of social perception processes. *Soc Cogn.* 18, 223–251.

Zeithamova D, Maddox WT. (2006). Dual-task interference in perceptual category learning. *Mem Cogn.* 34, 387–398.

Zigmond AS, Snaith RP. (1983). The hospital and anxiety and depression scale. *Acta Psychiatr Scand.* 67, 361–370.

Zimmerman M, McGlinchey JB, Young D, Cheiminski I. (2006). Diagnosing major depressive disorder IX: Are patients who deny low mood a distinct subgroup? *J Nerv Ment Dis.* 194, 864–869.

Summary of Part II

The material discussed in Part II was meant to address a number of fundamental issues that were raised in Part I. For example, how can a qualitative assessment be based on a subjective judgment, and how can the respondent, investigator, or policy maker feel confident about the conclusions drawn from such judgments? What is it about a self-report or subjective assessment that creates such doubt? The most likely answer is the fact that subjective indicators appear at times to be non-verifiable, meaning that it is often not possible to know exactly on what basis a person makes a particular comment and decision. Of course this is not absolutely true, since brain imaging, cognitive interviews, momentary behavioral analysis (e.g., Kahneman's notion of experienced utility; Kahneman et al. 1997), and the application of the Rasch methodology provide some information about how a subjective judgment is made. Still, doubts linger about the usefulness of such information, even though subjective indicators are best at revealing the complexities of the human condition (e.g., suffering, pain, depression, and so on), and are the natural subject matter of any qualitative assessment.

What was also revealed in Part II is that there are concerns about the type of information being provided by an objective assessment. This became evident when Daston and Galison (2007) reviewed the history of objectivity and demonstrated that the meaning of the term has evolved and will likely continue to evolve with time. It appears that "to be objective" is very much a set of behaviors or mental attitudes that a person can be trained to perform; a deliberate effort to separate the self from what is being observed or assessed. This makes "being objective" a human construction, and the success of this effort varies. How a person conceives of the "objective," however, is different from the reality of objects, events, and experiences.

Daston and Galison's (2007) historical review identified at least four approaches to objective assessment: the naturalist's; the use of mechanical observations; the identification of structural (e.g., mathematical) determinants of objectivity; and the expert (who may actually take advantage of each of the other approaches to objectivity). They even have speculated about the next form of objective reality: the modification of natural processes by human (aesthetic) intervention, resulting in new physical and biological forms. The idea that the subject and the object interact is consistent with the notion that objectivity and subjectivity are not totally independent states. They coexist. This is because the objective has to be sensed, perceived, and cognitively processed if it is to be made a part of aesthetic, financial, or policy decisions, and in this way is "made of the subject."

If subjective indicators can be objectified and objective indicators qualified, then what is the purpose of assessing these indicators? The answer was nicely described in a discussion Daston and Galison (2007) had about Einstein's approach to science (Chap. 5, p. 136). Einstein, it appears, was acutely aware that his effort at identifying physical principles started with subjective observations, progressed to objectification of these observations, but most importantly strove to *establish invariant principles*. To Einstein, establishing invariant relationships was the primary task for science, and the issue of objectivity and subjectivity was of secondary concern. Consistent with this is my contention that subjectively based qualitative principles can also be shown to be *invariant*, and that once established should eliminate concern about the role these indictors play in public affairs.

Quality-of-life investigators are currently most interested in demonstrating *measurement invariance* (Chap. 4, p. 110). Measurement invariance involves demonstrating that an observation in one set of data can be replicated in other data sets, and this usually involves applying confirmatory factor analytic techniques. However, measurement invariance is not the same as demonstrating conceptual or perhaps more accurately cognitive invariance. Thus, I believe that it should be possible to demonstrate that the method that persons of different cultures, ethnicities, disease states, or treatments use when making a qualitative judgment involves forming a hybrid construct, and that the specific content that makes up these judgments is of secondary interest. Thus, I would not expect that varying the content of a qualitative assessment would alter the basic cognitive processes of description, evaluation, and metacognition (e.g., reflection).

There are, however, a number of settings where the presence of a hybrid construct can be questioned. For example, the observation that qualitative judgments can be made very rapidly suggests that they can occur without reflection. However, when reviewed (Chap. 3, p. 66), this threat can be dealt with by the demonstration that *first impressions*, and other instances, continue to be cognitively processed, sometimes prior to and sometimes concurrent with the judgment task. This suggests that the elements of a hybrid construct need not all occur at the time of the qualitative judgment, but must be available (e.g., memory of importance of a particular outcome) when the judgment is made.

Another issue has to do with the observation that there are many types of quality indicators, ranging from sensory stimuli (such as hues, odors, tastes, and so on); a particular model of a car; a dinner party; a person's life; etc. The question is whether the quality of each of these indicators necessarily involves the elements of a hybrid construct. It is not clear that they do, although I would speculate that they should. These are empirical questions that should be answerable by appropriate studies. For example, Gärdenfors (2000) points out that the difference between a wavelength and hue of a color is that the hue has been cognitively processed. Whether this cognitive processing involves all of the elements of a hybrid construct is not clear, although in some instances it might.

A second major topic discussed in Part II was how information is summarized when a qualitative assessment is performed. This is a natural issue to consider, since these summary statements are the units of analysis needed for efficient cognitive processing, but also for the initiation of any assessment process. Thus, both the person and an investigator summarize information, but usually for very different purposes. To keep these differences clear, I tried to consistently refer to what a person does as forming *categories*, and what an investigator does as forming *domains*. I understand that sometimes the person can be the investigator, but it is clear to me that when an investigator summarizes some data, their objective is very different from what a person does when they form a category of the same data. I am also aware that the two terms are often interchanged and used as synonyms, and my didactic efforts have more of a rhetorical purpose than a call for a language usage change.

The first step in the formation of any category, or domain for that matter, is the *classification* of the information being processed. This can occur in several ways (Table 3.2), including organizing entities in terms of hierarchical relations, the cross-classification of information, or by considering various contextual factors. Hierarchical relationships are reflected in the varying levels of the informativeness of a category, which may range from being concrete to abstract, or alternatively, progressing from a subordinate, basic to superordinate category. The cross-classification of information can be observed in categories that reflect a particular script, theme, or are goal-directed. In addition, the context of the qualitative assessment, such as a person's cultural background, would be expected to influence how information is classified.

I also discussed that when a person forms a category he or she learns something about the category. This presumably is what happens when a person fills out a quality-of-life questionnaire and learns something from the items that the person was not aware of or thought about. Ashby and Maddox (2005) and Ashby and Waldron (2000) have suggested that this learning may occur in either of two ways: as involving either an explicit hypothesis-testing process (that uses logical reasoning, relies on working memory and executive attention) or an implicit procedure-learning process. I briefly reviewed four experimental paradigms that could be used to study this learning (Chap. 6, p. 164).

Although a person can learn something about the categories they form, I also noted that some categories seem to be organized into natural kinds with little or no learning. Evidence for natural kinds comes from the presence of these types of categories in more primitive forms early in the lives of children. Children, for example, are able to distinguish one type of animal from another, or not confuse an animate with an inanimate object, while persons who are increasingly intellectually compromised fail to make these distinctions. This ability seems to have some adaptive or survival significance, and this suggests that some types of information may be classified into categories as *determined cognitive structures*.

The presence of determined cognitive structures seems to originate from two sources: overly learned behaviors that become automatic, and those whose onset is quite early and suggests a biological or genetic origin. The same would be expected to be true for domains, and I spent a lot of time in Chap. 6 discussing dual process theory and its role in domain formation. One observation seemed to become quite clear: that a domain is far more complex a structure then is generally considered. It seems that every domain can be inspected as to whether it has both modular and non-modular components; whether one component dominates relative to the other; and so on. It became clear from the discussion in Chap. 6 that it may be wise to classify domains in terms of their degree of modularity. This would be an alternative to what is now done which is to classify domains in terms of common themes such as physiological determinants, symptoms, health perceptions, or quality-of-life indicators. I will make an attempt to do this in Chap. 7 when I try to organize the various pain syndromes in terms of their degree of modularity.

I can illustrate the complex structure of a domain by examining the modularity of the various domains of the SF-36. Ideally, domains would be designed so that they have as little overlap as possible, but this is rarely achieved. Rather, what is usually found are varying levels of shared variance. Inspecting Table Pt II:1 illustrates this by summarizing the product moment correlations between the eight domains of

Table Pt II:1 Product–moment correlation and reliability coefficients (in *parentheses*), SF-36 scales for the general US population $(N=2,474)$[a]

	PF	RP	BP	GH	VT	SF	RE	MH
PF	(0.93)							
RP	0.65	(0.89)						
BP	0.52	0.61	(0.90)					
GH	0.55	0.55	0.56	(0.81)				
VT	0.44	0.50	0.52	0.58	(0.86)			
SF	0.45	0.52	0.49	0.47	0.51	(0.68)		
RE	0.30	0.42	0.32	0.35	0.44	0.53	(0.82)	
MH	0.28	0.35	0.39	0.46	0.63	0.56	0.54	(0.84)

[a]Ware et al. (1994; p. 3:10)

the SF-36 (Ware et al. 1994; p. 3:10). For example, the correlation between the physical functioning (PF) and role-physical (RP) domains was 0.65 ($R^2 = 0.4225$), which reduced to 0.45 ($R^2 = 0.2025$) when PF and social functioning (SF) were compared, and reduced further when PF and mental health ($r = 0.28$; $R^2 = 0.078$) were compared. Note that the variance accounted for (R^2) decreased as the domains being compared to PF became less likely to be dependent on physical functioning.

Why is there so much shared variance between domains, and why does it decrease over comparisons? My first answer is there are decreasing amounts of modularity implied by the different domains. By this I mean that the degree that biological factors play a role in determining a response decreases as the domains being correlated move from items that are dependent on the physical characteristics of the person to items that reflect the mental health of the person. Thus, while a social activities domain may include items that ask a person about dancing or playing a game, both these activities are dependent on the physical activity of walking, an issue that also would be addressed in items included in a physical functioning domain. Spending time with grandchildren may involve the ability to lift the child, and lifting weights would be addressed in items in a physical functioning domain. On the other hand, the capability to walk or lift would not likely be part of the items addressing mental health (MH), and as expected, these domains have minimal shared variance.

Ware et al. (1994) uses these and similar data sets to perform a principal component factor analysis, and identifies two major factors: physical and mental health. Part of the rationale of Ware et al. (1994) for supporting the presence of two summary measures was the demonstration that factor loadings from the orthogonal factor analysis clustered into two regions with a gap in between. However, this same presentation could also be interpreted as demonstrating that all eight domains were linearly related (see Ware et al. 1994; Fig. 3.2; p. 3:9). Thus,

an alternative label for the two Ware et al. (1994) summary measures might be *predominately modular* or *predominately non-modular*. The advantage of using these labels is that it would acknowledge the role that structural determinants play in domain/summary measure formation.

It is also interesting to note that Ware et al. (1994) present their model in terms of a hierarchy, much like Torrance (1976) does, with items, scales, and the two summary measures making up the increasingly abstract components of a hierarchy. This may be another example where an investigator's cognitive processes are displayed in their efforts to organize the data they accumulated.

Part II begins by asking whether objective or subjective indicators should be assessed when quality is assessed. The questions raised here were used to gain some insight into the nature of signs and symptoms: major determinants of the quality-of-life of the medically ill person. The final chapter deals with defining the unit of analysis in a qualitative analysis. In Part III, I have applied what has been learned in Parts I and II, now to the common domains assessed during an HRQOL assessment.

References

Ashby FG, Maddox WT. (2005). Human category learning. *Annu Rev Psychol.* 56, 149–178.

Ashby FG, Waldron EM. (2000). The neuropsychological bases of category learning. *Curr Trends Psychol Sci.* 9, 10–14.

Daston L, Galison P. (2007). *Objectivity*. New York NY: Zone Books.

Gärdenfors P. (2000). *Conceptual Spaces: The Geometry of Though*. Lexington, MA: MIT Press.

Kahneman D, Wakker PP, Sarin R. (1997). Back to Bentham? Explorations of experienced utility. *Q J Econ.* 112, 375–405.

Torrance GW. (1976). Health status index models: A unified mathematical view. *Manag Sci.* 22, 990–1001.

Ware JE, Kosinski MA, Keller SD. (1994). *SF-36 Physical and Mental Health Summary Scales: A Users Manual*. Boston MA: The Health Institute, New England Medical Center.

Part III

The Content of a HRQOL Assessment

Introduction to Part III: What Content Should Be Included in an HRQOL Assessment?

As I stated in the Preface, there are many venues where the issues of how to define and assess quality can be discussed. Questions about what to assess and what unit of analysis to use are universal and are relevant to assessing the quality of an object, event, or a life lived. Since I have chosen to consider the quality-of-life of the medically ill, the next question becomes what *content* should be included in any assessment. This is the focus of the next five chapters of this book. One advantage I have is that there has already been a fair amount of research on this topic, so I have 30–40 years of material to work with. Thus, symptoms seem a natural topic to consider because they are a very prominent part of a medically ill person's experiences, while health and functional status are also obvious concerns of a medically ill person. Including neurocognitive aspects of a person's existence in an HRQOL assessment is not so obvious. I am going to include it because various investigators include such items in an HRQOL assessment, and also because I realize that neurocognitive factors are a major determinant of a person's symptoms, health status, and functioning. I have also included them because they provide enormous insight into the nature of a qualitative experience, and also into what has been compromised in a compromised person. Thus, discussing these issues are a necessary step in making progress towards including the comprised person as for full assessment participant. A fifth chapter deals with subjective well-being, and the extent that it is or is not an analog of quality-of-life. Finally, these domains are not the only ones I could have included. Patrick and Chang (2000), for example, include *spirituality* and *opportunities* as domains, while others use different but overlapping terms to organize the basic information I will review.

Each chapter in this section was designed to cover a number of related topics. For example, in Chap. 7. I continue my discussion on the role of symptoms in an HRQOL assessment. First, I illustrate how cognitive principles (e.g., the hierarchical or cross-classification of information) can be used analytically to understand the relationship between fatigue, depression, and HRQOL. I follow this with an examination of the assessment of pain, the development of functional and acquired pain, and how pain is assessed in a challenged person. The second part of the chapter deals with the role of symptoms in an HRQOL assessment. In this section, I expand on material introduced in Chap. 5 about whether including symptoms into an HRQOL assessment adds value, and I also continue to try to clarify the *causal* or *indicator* role that symptoms presumably play. I end this discussion by addressing how HRQOL is used in symptom management.

Chapter 8 considers what role health status should have in a qualitative assessment. I review the theoretical basis for the terms *health* and *disorder*. My reasoning is that if the state of health is a quality indicator, then I should be able to clearly define what it is I am proposing to assess. I will follow this by reviewing the health status assessment movement, and this should reinforce the vital role that quality control has and continues to play in HRQOL assessment. The second part of the chapter applies this background in an effort to determine if HRQOL is a viable construct, but also to clarify how current versions of HRQOL assessment match the cognitively based hybrid construct model I am promoting in this book. I end the chapter with a comment on the relationship between health status and HRQOL.

The terms *function, functional,* and *functionalism* represent different aspects of a central determinant of a qualitative experience, event, or life history. Its centrality naturally evokes a broad set of topics, and since its relevance to a qualitative assessment has not been previously reviewed, I take the time in Chap. 9 to provide this background. Finally, I apply what I have been discussing to the functional assessments of the person with disabilities who is being rehabilitated.

Chapter 10 focuses on the relationship between *capacity* and *performance,* or as is sometimes; structure and function. This, as I have stated, is a necessary step in the process of clarifying what is compromised in a compromised person. I start by asking whether neurocognitive reports and subjective indicators can be combined to produce a composite HRQOL index. I acknowledge that neuropsychological reports are primarily determined by an investigator, and ask if a person's self-reports about their cognitive status act as an adequate proxy. The rationale is that if self-reports do a reasonable job capturing what a neurocognitive assessment would reveal, then it would be much easier to combine these and other self-reports so as to form a common qualitative (subjective-based) index. I follow this discussion by expanding on the discussion in Chap. 3 on the neural basis of cognition. In this section I address the use of the phrase *cognitive reserve* to help bridge the conceptual gap between capacity and performance. I follow with a discussion of the cognitive degeneration of language, giving examples from the several disorders. The next major section deals with the effect of acute neurological events on cognitive capacity, using changes in figurative expression as the vehicle to describe these changes. The last part of the chapter discusses different models of the relationship between capacity and performance. Here, I make my best effort to describe the currently available approaches to the qualitative assessment of the compromised person. I conclude that what is most impressive is the wide number of opportunities this literature provides to study the biological basis of a qualitative assessment.

Chapter 11 examines the question of whether the term *well-being* can be included as an indicator of quality, and in particular, quality-of-life. To address this question, I assume that the phrase "subjective well-being" represents various forms of emotional expression, and I will examine the available literature to determine the extent to which this is true. I will ask if subjective well-being is a discrete emotion, a form of cognitive-emotional expression, or a product of cognitive-emotional regulation. I end by discussing various conceptual models of subjective well-being.

The five chapters in this section are meant to illustrate the benefit of applying the principles of language usage and cognition to a specific set of problems. In addition, the content covered illustrates the broad range of issues relevant for the qualitative assessment of the medically ill person. I find it quite useful to have a single cognitive model – the hybrid construct – to account for all the different types of qualitative assessments that occur under a wide variety of conditions.

Reference

Patrick DL, Chiang Y-P. (2000). Measurement of health outcomes in treatment effectiveness evaluations. *Med Care.* 38(Suppl II), II14–II25.

Symptoms and HRQOL Assessment

Abstract

Should symptoms be included in an HRQOL assessment? This chapter addresses this issue using the research literature based on the assessment of fatigue and pain. It will become apparent that it is possible to assess the qualitative consequences of a symptom without having to relate it to HRQOL. Thus, a symptom, if qualified, can be a quality indicator. To achieve this first requires a shift in the semantics of symptoms with the incorporation of emotive determinants of symptoms (e.g., a shift from pain intensity to pain burden), and if this is followed by a valuation, then the components of a qualitative hybrid construct would be in place. I also discuss the role of modularity and how it can help in predicting different outcomes between symptoms, using fatigue, depression, and HRQOL as the model. Also discussed is the issue of how to assess pain in the compromised person.

Abbreviations

BCQ	Breast Cancer Questionnaire (Levine et al. 1988)
BDI	Beck Depression Inventory (Beck et al. 1961)
BPI	Brief Pain Inventory (Mendoza et al. 1999)
CARES	Cancer rehabilitation evaluation system (Ganz et al. 1992)
CART	Classification and regression tree (Breiman et al. 1984)
CES-D	Center for Epidemiological Studies-Depression (Radloff 1977)
CPM	Cognitive processing model (Tourangeau 1984)
CRPS	Complex regional pain syndrome
DSM-IV	Diagnostic and statistical manual of mental disorders
EORTC-QLQ-C30	European Organization for Research and Treatment of Cancer-Quality-of-Life Questionnaire-Cancer specific-30 items (Aaronson et al. 1991)
ESAS	Edmonton Symptom Assessment Scale (Bruera et al. 1991)
FACT	Functional assessment of cancer therapy (*B* breast cancer; *BMT* bone marrow transplantation; *F* fatigue; *G* general; *O* ovarian)
FLIC	Functional Living Index-Cancer (Schipper et al. 1984)
HADS	Hospital Anxiety and Depression Scale (Zigmond and Snaith 1983)
HRQOL	Health-related quality-of-life
HUI	Health Utility Index (Torrance 1976)
IASP	International Association for the Study of Pain
IBS	Irritable bowel syndrome
MAC	Mental adjustment to cancer (Watson et al. 1988)
MAPS	Multidimensional affect and pain survey (Clark 1984)
MDASI-BMT	MD Anderson Symptom Inventory-bone marrow transplantation (Cleeland et al. 2000)
MFI-20 PF	Multidimensional Fatigue Inventory (Smets et al. 1995)
MOS-SF-36	Medical outcome study-short form-36 (Ware et al. 1993)

MPI	West Haven-Yale Multidimensional Pain Inventory (Kerns et al. 1985)
PFS	Piper Fatigue Scale (Piper et al. 1989)
POMS	Profile of mood states (Mc Nair et al. 1992)
PRI	Patient Reported Inventory
PROMIS	Patient reported outcome measurement system
QIR	Qualitative item review (DeWalt et al. 2007)
QWB	Quality of Well-being Scale (Kaplan and Bush 1982)
RSCL	Rotterdam symptom check list (De Haes et al. 1990)
SDS-F	Symptom Distress Scale-Fatigue (Mc Corkle and Young 1978)
SSQ	Social Support Questionnaire (Sommer and Fydrich 1999)
VAS	Visual Analog Scale
VAS-F	Visual Analog Scale-Fatigue

1 Introduction: Symptom Domain Formation

Part II of this book discussed three aspects of a qualitative assessment: its *purpose* (e.g., establishing invariant principles), *content* (objective or subjective indicators), and *unit of analysis* (categories or domains). This chapter continues this discussion, now by asking whether a symptom itself can be considered a quality indicator, and if not, what needs to be in place for this to happen. A second question addressed is whether symptoms can be directly included in an HRQOL assessment, and the assessment be considered an indicator of quality-of-life. Both these questions contain a number of implicit cognitive assumptions; first, that it is possible to map concrete entities (e.g., symptoms) to an abstract concept (e.g., HRQOL), and second that a valuation has occurred. But how can I demonstrate that these cognitive processes have occurred? I need to answer this question in order to answer my primary questions, especially if I want my statements to be more than figurative expressions.

As I discussed in Chap. 3, indicators are cognitively organized into categories in three basic ways: hierarchically, by cross-classification, or by context. A domain, as I am using the term, is formed deliberately by an investigator using statistical manipulation, and is, therefore, cognitively neutral. Thus, there are no *natural domains* as there are *natural kinds*, rather the content and internal structure of a domain is determined either by statistical analysis or arbitrarily by the investigator selecting the items. This contrasts with the formation of a category, which is a cognitive construction, and provides an internal structure to the summary statement. Of course, there are intermediate conditions where the investigator enlists the respondent in an effort to cognitively generate a summary statement, and uses statistical manipulations of what the respondent has provided in order to form a domain (e.g., a collecting of items). To the extent that the domain reflects the respondent's cognition, the domain will also have a cognitive structure. The interaction between the respondent's thought processes and the investigator's modeling of these processes is probably the most common way domains are formed. Thus, there are several methods whereby summary statements are formed, and I will discuss each as I proceed in this chapter.

If indicators in a domain are cognitively organized into *a hierarchy*, then it will be possible to claim that a concrete entity (a symptom) is an indicator of an abstract concept. This occurs because of the logic of a hierarchy, where inclusion rules exist that permits more concrete indicators to be included in more abstract classes. An example where this type of inclusion rule is active is the Guttman scale (Guttman 1944). Alternatively, if, a diverse set of indicators was selected to reach some common goal, or reflect a script or theme, then the domain formed would constitute a *cross-classification* domain. Here the principle that organizes the indicators is the specific goal, script, or theme, and only indicators are included that meet this rule. However, the ability to infer that the concrete entities of the domain refer to an abstract concept is dependent on whether the goal, script, or theme is expressed as an abstract concept. In either instance, however, it is necessary to demonstrate that the indicators in the domain have been valued if the category or domain is to be considered a summary statement of an HRQOL assessment.

A number of alternative approaches to the relationship between symptoms and an HRQOL assessment have been proposed (Table 7.1). There are many options: (1) outright rejection of the notion that an HRQOL assessment needs to be included in a clinical assessment of symptoms (e.g., Hollenberg et al. 2000); (2) claims that qualitative aspects of symptoms (e.g., symptom burdens) are sufficient so that it is not necessary to include HRQOL assessments (Cleeland 2007); (3) claims of a specific role for an HRQOL in a symptom assessment (Osoba 2007); (4) consideration of the causal role of symptoms in a person's HRQOL status (Chap. 6); and finally, (5) the creation of a common qualitative metric that would permit the formal combination of symptoms and HRQOL indicators.

This wide range of options (Chap. 7, p. 200) suggests that the role of symptoms in an HRQOL assessment, or vice versa, remains controversial, but there also may be natural barriers that would ordinarily inhibit the simple integration of symptoms and HRQOL indicators into a common assessment. I can identify at least three issues that would create such a barrier. First, the format of the items found in a symptom and an HRQOL assessment usually differ; symptoms are expressed as discrete units of analysis (e.g., present or absent), while HRQOL items more commonly involve

1 Introduction: Symptom Domain Formation

Table 7.1 Alternative conceptualizations of the relationship between a symptom and HRQOL

Alternative conceptualization	References
Symptom distress is a sufficient measure, so HRQOL indicators are not necessary for clinical decision making	Hollenberg et al. (2000)
Symptom burden, which combines symptom *severity* and the *impact* of symptoms, is a sufficient descriptor, so that HRQOL indicators need not be regularly assessed	Cleeland (2007)
HRQOL indicators have to be assessed so that the *impact* of symptoms is monitored, and thereby included in the symptom management of clinical trials	Osoba (2007)
Symptoms are *causally related* to HRQOL indicators	Chapter 6 of this book
Symptoms and HRQOL differ in content but can evoke *similar cognitive processes* (e.g., evaluation and valuation), and therefore each can be an independent measure of quality. This observation represents an example of the *cognitive equivalence* hypothesis	Chapter 5 of this book

ratings along some dimension. Second, symptoms tend to be expressed concretely, while HRQOL indicators have a wider range of expression, including being construed as an inherently abstract statement. And third, symptom domains are more likely to be modular (biological) in nature, while an HRQOL domain, again, has a wider content basis. Each of these potential barriers will be discussed in this chapter.

If the term HRQOL is being used to refer to an abstract construct, this has implications about which statistical procedures are used, and how. For example, an investigator may want to use a regression analysis to determine if depression predicts an HRQOL outcome. However, if the investigator also wanted to determine if HRQOL predicted depression, then this regression analysis would be uninformative, because it would claim that an abstract concept (e.g., a global quality of assessment) was somehow causing or predicting a concrete entity. How could this be operationalized? The only way would be if some concrete aspects of a person's HRQOL were identified, and these concrete indicators would then be predictive. The issue here is in part linguistic, with some investigators claiming that HRQOL is a causative agent taking advantage of the connotative capacity of an abstract concept.

An alternative approach to this issue is to inspect the neurocognitive representation of abstract concepts and concrete indicators. When this is done (Chap. 10), it is clear that the cortical areas involved in abstract concepts are associative in nature, while those involved with concrete domains are category specific. This reinforces the view that mapping abstract and concrete concepts is accomplished by making the abstract concept concrete (the associative areas are mapped to a specific categorical cortical sites).

The term "symptom" also has a wide range of linguistic uses, referring to any number of functional states. Thus, a symptom may be used to characterize a malfunctioning mechanical device, a particular personality, or an emotional state. These uses of the term symptom reflect that it is context specific. The term symptom can also be misused, as happens when an investigator claims that a particular domain, without a supporting cognitive structure, implies some abstract construct. This unfortunately regularly occurs in the HRQOL literature.

As discussed in Chap. 5, the best approach to measuring a symptom remains an unresolved issue. For example, should symptoms be measured in terms of the frequency of their occurrence, their intensity, or some qualitative indicator, such as the distress they create? Should symptoms be measured individually or in some combination? What is the best way to measure a symptom when it varies in terms of temporal distribution (e.g., does the symptom persist or occurs episodically)?

Depression is a particularly useful symptom to examine, since it raises a number of important issues. For example, depression as a domain has been used to both explain (cause) and describe (indicate) changes in HRQOL (Visser and Smets 1998). How appropriate is this? First, assigning this role to a domain is not unique to depression and occurs for a wide range of other domains (e.g., fatigue, anxiety, social support, lack of intimacy, etc.). Second, it is not clear that indicators of the depression domain that are claimed to cause changes in quality-of-life are the same ones that indicate quality-of-life changes. For example, if the term depression is used as a label, it may connote but not necessarily state what it is that is concretely contributing to a person's mood state. The difference in the two uses of the term depression can be more clearly seen when the modular content of the domain is inspected. For example, weight gain (a modular component) and sadness (a nonmodular component) both are items commonly found on a depression assessment, yet weight gain is more likely to be a cause, and sadness more likely to indicate a depressed state. However, it is also common to hear someone say, " I eat because I am depressed." This raises a third issue, which is whether it is appropriate to use a linear model to conceptualize the relationship between these indicators. The problem is that a linear model limits the dynamics between elements of a domain since the purpose of such an analysis is to establish cause, while a nonlinear model would not be limited in this way (e.g., Scherer 2000).

Part of the complexity here has to do with how domains, including depression, are constructed. Thus, when depression assessments are inspected, a fair amount of variability can be observed in the number (e.g., from 1 to 24 or more items) and content (e.g., feeling suicidal, blue, tired, etc.) of the items that make up the assessment. This suggests that it is far from clear what exactly a depression domain is assessing. This ambiguity can be clearly seen when depression is examined as a diagnostic category. Chochinov et al. (1997), for example, found that the single item "Are you depressed?" was an effective diagnostic tool when dealing with particular class of patients; the terminally ill persons. At the other extreme, an extended semistructured interview based on the DSM-IV has also been used to establish the diagnosis of depression. In this regard, Zimmerman et al. (2006a, b) have published a series of papers evaluating the clinical utility of the diagnostic domain of major depression disorder, and found that certain criteria of the DSM-IV definition (e.g., weight loss, weight gain, indecisiveness) could be eliminated without affecting the usefulness of the diagnostic domain. Some of the criteria that remain (e.g., appetite, loss of energy, insomnia), however, are likely to have a strong modular base, while others (depressed mood, anhedonia, excessive guilt) would be less likely to. So the mapping of these different criteria to some qualitative outcome remains to be clarified. Many of these same issues can be raised when pain expression is the object of study.

The assumption that nonmodular components of a domain are associated with causal changes in an HRQOL assessment deserves special consideration, especially since defining a nonmodular domain remains arbitrary – limited by the items selected to be part of the assessment. Part of the response to this situation, using classical psychometric procedures, has been to include as many items as possible, and thereby create domains that are comprehensive and then subject them to multidimensional statistical analyses. Most often correlation analyses are used to estimate the degree of association between indicators, and regression analyses are used to identify predictors. Both of these procedures can be used to provide a post hoc definition of a domain. However, the attribution of a causal character to these domains remains suspect. Even the recent application of item response theory to increase the precision of the domain defining process by creating item banks does not resolve these concerns.

One way to avoid the complications that modularity creates for an HRQOL assessment is to examine components of a complex domain prior to its assessment. Thus, in the studies of the relationship between depression, fatigue, and HRQOL in cancer patients, Visser and Smets (1998) separated out items that had a somatic or modular-like base from other elements of a depression measure. These authors were aware that a person's disease or treatment may exacerbate the somatic, but not necessarily the nonsomatic, aspects of depression and fatigue. By separating these indicators the authors, of course, were implicitly confirming the importance of considering the modular nature of the elements that make up a domain.

Fatigue (Servaes et al. 2002) is an example of a symptom whose formulation as a domain can be affected by many of the issues I have been discussing. I will examine it in detail in order to illustrate the impact these complexities have on the conceptualization and interpretation of fatigue symptom domains in particular, and domains in general. Cancer patient fatigue will provide the context in which to consider these issues, although much of the discussion will also be relevant to other clinical disorders (e.g., chronic fatigue syndrome, fibromyalgia, arthritis, etc.), and will be referred to in order to complete the discussion of this topic. I will start by examining the literal and figurative nature of the term fatigue, followed by a review of how it is currently defined in the scientific literature. Then I will consider its qualitative nature, review some of the issues involved in establishing a fatigue domain, and consider the relationship of fatigue to depression and HRQOL. This will provide a direct illustration of the analytic opportunity provided by examining the modularity of a domain.

A second major objective of the first part of this chapter involves examining *pain* as a prototypical symptom which is often independently assessed (e.g., Mc Dowell and Newell 1996; p. 335–379), but can also be included as a single item(s) in an HRQOL assessment. As discussed earlier, there are natural barriers to integrating symptoms into an HRQOL assessment, and this is particularly evident when pain is the symptom. This leads to the question of whether the assessment of pain itself can be considered a quality assessment. Thus, there are two questions to answer: first, whether a pain assessment can be a qualitative assessment; second, whether or not pain items can be successfully integrated into an HRQOL assessment.

Investigators interested in assessing pain have recognized that the term pain, like depression, has several linguistic uses. Like depression, pain can be felt (e.g., as would happen if I touched a hot surface), such that pain has a concrete sensory and emotional component. It also can be used to describe an abstract construct, such as a pain disorder or disease state (e.g., complex regional pain syndrome, CRPS). Both forms of expression have cognitive and emotional components, as would be expected for a qualitative statement (Figs. 1.2 and 1.3). Thus, the pain associated with a cut finger can be as much a part of a qualitative statement as is a report of chronic pain, even though both differ in terms of the length and complexity of the associated experiences. Adding to this complexity is the fact that the quality of a pain experience varies between bodily sites, type of tissue involved (e.g., visceral vs. muscular), presenting characteristics of the person (e.g., the person's personality), and so on.

This wide range of qualitative expressions creates definitional and assessment problems, and could also complicate the integration of pain items into health status or HRQOL assessments. Consider the SF-36: it includes two items on bodily pain – one asking if the person senses pain ("How much bodily pain have you had during the past four weeks?"); one monitoring the consequences of the pain ("During the past four weeks, how much did pain interfere with your normal work, including both work inside and outside the home and housework?"). The scoring of the items assumes that the person will have to sense the pain, as indicated by their response to the first item, before its consequences are rated. This is an important maneuver because it creates a conditional relationship between the items that can be interpreted as a simulation of how a person might experience pain. If this is correct, then it suggests that a cognitive structure exists within the pain domain for this particular assessment, but this structure is not sufficient to claim that the domain assesses some aspects of the abstract construct of HRQOL. For this to happen, the investigator has to demonstrate that the components of the domain are a product of cross-classification or that a hierarchical relationship exists between the components. This, of course, is consistent with the point I made earlier that the strongest way to demonstrate that a concrete entity (e.g., a particular symptom) is associated with an abstract construct, such as HRQOL, is to demonstrate that a cognitive structure exists within the domain. Following from this is the question of whether other HRQOL assessments that include pain items attempt to verify the sensing of pain before addressing its functional consequences.

One of the major sources of complexity in pain assessment is that pain can occur acutely, but can also be transformed into a recurring event or become a chronic process. Acute pain, if it occurs reflexively, can be a highly biologically determined event, and thus would function as a modular component in a pain domain. A traumatic event can also induce an acute pain episode, even in the absence of a physical cause for the pain report. In this type of pain, a strong emotional event may be sufficient to induce what appears to be a reflex-like pain event.[1] Chronic pain can develop in a number of ways, ranging from the modification of fundamental neural processes (resulting in up-regulation[2]) to a person's personality characteristics and psychosocial factors contributing to recurrent pain reports. While most people are born with the capacity to perceive pain, they are rarely born with chronic pain. This suggests that chronic pain is acquired and would, therefore, function in any pain domain as a form of *acquired modularity* (Table 6.1). In contrast, the alteration of various physiological processes that are painful (e.g., disturbed gastrointestinal motility as in irritable bowel syndrome, IBS) would be characterized in a domain as a form of *functional modularity* (Table 6.1), even though this type of modularity is also acquired. Thus, the difference between acquired and functional modular components of a domain is that for the former, the onset of pain is more acute but persistent, but not permanently disruptive of a regulatory system, while it is assumed that physiological regulation has been altered for a functional modular.

Functional pain syndromes often are chronic and recur on a regular basis, particularly when stressful conditions precipitate an acute pain episode. Contributing to this chronic state is a person's notion of what pain is (implicit theories; Table 6.1). Implicit theories would be expected to affect the components of a domain, now as a nonmodular component. The same would be expected for a person's culturally determined perceptions. Also, there is a large array of neuropathic conditions that produce recurrent pain episodes that can evolve into a chronic pain syndrome. This type of chronic pain would be highly biologically determined and function as a modular component in any pain domain.

There are a number of examples of functional modularity, including irritable bowel syndrome, fibromyalgia, dyspepsia, and so on. Formal efforts to classify these have also been reported (e.g., Drossman 2006), and I will review several of them to provide an overview of past efforts. I will then follow this summary with an explicit effort to see if it is feasible to classify different types of pain according to their qualitative status. If it is feasible, then pain may best be viewed dimensionally (e.g., Figs. 1.2 or 1.3), and the effort to create discrete pain domains should be left for the more analytical task of mapping specific symptoms to specific etiologies.

A final topic in this section is a review of the current status of assessing pain in the compromised person. My intent is to lay out some of the issues involved in assessing a person whose ability to verbalize or think may be compromised to the extent that they are unable to respond to or understand verbal inquiries concerning their qualitative status. How to assess the compromised person is, of course, a practical clinical problem that demands resolution, but it is also a question of broad theoretical and political significance that must be solved if all voices are to be heard when quality is assessed. Thus, there is an ethical imperative surrounding this issue that justifies a serious intellectual effort.

I will make two suggestions about how to deal with this problem. First, the nature of the communication exchange leads naturally to linear modeling, but the deficits characteristic of the compromised person might best be captured by a nonlinear model. Second, the intellectual deficits of a compromised person develop with time and monitoring this process provides an opportunity to predict a person's preferences as a disease process progresses. Additionally, the severely cognitively impaired person seems to retain the capacity to communicate emotionally, so there may be opportunities available to assess some specific aspects of quality, even though the interpretation of these emotional responses may have to be considered by persons other than the compromised person. I will discuss the first suggestion in this chapter and the second in Chap. 10.

The communication process leads naturally to linear modeling because the normal expectation is that when verbal information is being exchanged, one person initiates (providing the stimulus) and another responds. To the extent that the response is reactive, the exchange takes on the causal appearance, implicit in any linear model. However, when dealing with the compromised person the nature of the verbal exchange can be disrupted such that alternative means of communication are required. This may involve emotional or facial expressions, movements, touch, etc. Communication under these conditions becomes a problem of cross-modality communication. For example, when a nurse asks an aphasic person to roll over so that he or she can fix the patient's bed, the success of this exchange is dependent upon the nurse initiating movement by the patient; usually by lifting the sheet the patient is resting on. When the patient does move, this usually occurs in an all-or-nothing manner, and this situation can be modeled with such nonlinear concepts as chaotic onsets or complexity. The basic argument here, therefore, is that while a linear model is best at capturing the sequential nature of verbal communication, a nonlinear model is best at capturing the binary onset of events that will then be ordered into a sequence. This difference becomes particularly important when dealing with the compromised person.

Many studies have been reported that assess the qualitative status of a person with mild to moderate intellectual deficits, but the severely deficient person remains particularly intractable to at least some form of qualitative assessment. It is these people who are usually not included in a qualitative assessment (see Chap. 10), and they will be the focus of my attention in this section. However, before I discuss the three topics I have listed earlier, I would like to continue the discussion started in Chaps. 5 and 6 on the cognitive basis of symptom domain formation.

2 Cognitive Basis of Symptom Domain Formation

One of the complexities of forming symptom domains is that an investigator is unaware of the categories individuals use when responding to queries. Although the investigator could perform cognitive interviews to get some insight into what a particular category means to a person, he or she usually takes a person's response (e.g., "I am in pain"; "I am depressed") at face value. The extent to which this is a problem can be gauged by examining studies that consider common symptoms in their cultural context. It is clear from my previous discussion (Chap. 6, p. 168) that while there is commonality, there are also differences in Western and Eastern linguistic expression that reflect differences in the meaning of words for members of different cultures. The question is whether these differences migrate into how a domain is formed.

Lee et al. (2007) report an ethnographic study of native Chinese persons whose clinicians indicated that they were depressed. The investigators recorded open-ended in-depth interviews in Chinese, and performed content analysis on the English translation of the person's comments. They found that the patients reported many of the same depressive symptoms reported in Western cultures (e.g., loss of interest/drive, loss of appetite and weight loss, early morning awakenings, psychomotor retardation, concentration problems, uselessness, hopelessness, and suicidal ideas), but also a unique set of domains leading to a different overall conception of what being depressed meant to the Chinese respondents. Thus, the patients used expressions to describe depressive experiences that were recognizable, but not concordant with Western expressions and language. A second difference involved combining affective distress with bodily experiences, specifically, embodied emotional experiences. Thus, the persons would report they had "heart pain," "heart dread," or "heart exhausted" when asked if they were depressed.

Another domain evident in the study was the distress of social disharmony. Some participants felt they were irritable, vexed, shaken, bad tempered, and so on, and although Western clinicians may not consider these symptoms clinically significant, they were in a culture that encourages togetherness. In addition, many of the terms were being used metaphorically, with the reference to physical pain used to communicate the meaning of psychological pain. In addition, the investigators found that not all experiences could be put into words. Thus, Lee et al. (2007) labeled a fifth domain implicit sadness. Again, people would report a variety of symptoms but not speak of them as depression, and in fact were surprised when the interviewer made that suggestion. A final domain dealt with the centrality of sleep, with sleep disturbances being causally linked by the respondent to their depression.

An analysis of these domains suggests that the Chinese mix the modularity of their categories, often expressing their emotions by reference to a bodily part. There was also a liberal use of metaphors in this study. This contrasts with the Western expression of emotions, where a deliberate effort is made to make literally true statements, resulting in more of a differentiation between the two modalities. Another contributor to these distinctions is the nature of the language; Chinese is a pictographic language combining basic units to create new meanings, similar to what was described during the earlier discussion on hieroglyphics (Chap. 3, p. 60). Thus, there are over 100 Chinese lexical constructions to express emotion that contain reference to the heart as a core element. Consistent with these observations is the fact that Chinese medicine conceives of depression as a disorder of *qi* (Chinese word for energy flow), where the life force that flows around the body is disturbed. Thus, *qi* flow is stagnated in the liver, spleen, and lung, and treatment involves disrupting this stagnation.

Lackey (2008) provides another example of the role of culture in the experience of depression. Seven focus groups were used to determine the meaning of the term "depression" for male Mexican immigrants (total $N=39$) to the United States. Again, it appeared that interpersonal problems and affective responses were the most salient qualities used by these respondents to identify depression. Reporting somatic correlates of depression also occurred, as did the expression of emotions through reference to bodily processes. Remedies for depression were expressed as involving altering the person's social relations. Lackey also identified 11 different words that were used by his interviewees to describe depression.

Both these examples illustrate that while different cultures may use a common set of terms to express depression, there are also important differences that give the term a different meaning. Investigators most likely use these context(cultural)-specific categories in the cross-classification of components to form a domain, especially if the domain has a specific purpose or goal. To illustrate how this might occur, I will revisit a study I reviewed in Chap. 3 by Ross and Murphy (1999), which demonstrated the nature of cross-classification for different food items. In the first of their studies, the investigators asked respondents to organize a list of food items into self-chosen categories they deemed appropriate. Thus, 45 food items[3] were placed into 826 categories, 49% of which were superordinate, or taxonomic (hierarchical) categories, while 42% of the responses fit categories that define how food was used (e.g., as breakfast food, junk food, etc.). This second group of food items was referred to as script categories. These data are of interest because they suggest first that cross-classification via script categories is a fairly likely form of classification, and second that this classification resulted from a person's interaction with the items rather than simply judging food items on the basis of how similar the items may be to one another (as would be expected in a taxonomic classification). Also true is that the speed with which a person organized food terms into categories was fastest for hierarchically organized categories than for heterogeneous categories and slowest for totally ad hoc categories. Both types of primary classification methods could be used to make inferences about food properties, and presumably symptoms as well.

Ross and Murphy (1999) also addressed the cognitive basis of the cross-classification they observed. They state:

> It seems clear that there are different kinds of organization, because the two specific sorting solutions (taxonomic and script) deviate in a number of important respects and are only correlated 0.54 overall. Thus, these could not be embedded within a single hierarchy. Rather, it seems necessary to conclude that different clusters of items exist simultaneously, presumably through different category links. For example, bagel may be connected to breads and grains as well as breakfast foods and sandwich foods. There does not seem to be any contradiction in one item being fairly strongly connected to multiple categories... (p. 539)

Thus, the results of the Ross and Murphy (1999) study suggest a kind of nonhierarchical network of categories, in which items are connected to all the domains they exemplify. "Items are related to other items by shared category membership…and/or shared properties…access must be determined in part through goals and contexts" (Ross and Murphy, 1999; p. 540). They go on to state that the outcomes of these classifications can then be used for drawing inferences, making predictions, problem solving, planning, explanation, and communication; or more generally contribute to goal-oriented activities.

These observations can be applied to studies of the classification of symptoms into domains. Studies on symptom clustering (e.g., Miakowski et al. 2004, 2007) are particularly suited for this type of analysis. Much of the research literature on symptom clustering involves the administration of investigator-developed symptom assessments that are then subjected to various statistical procedures. Walsh and Rybicki (2006), for example, report that they were able to identify seven symptom domains using agglomerative hierarchical analysis methods to a group of 922 cancer patients who had entered a palliative care unit and were interviewed to determine the presence of 38 symptoms. They found that 84% of the patients reported pain, 69% reported "easy" fatigue, 66% reported weakness, 66% reported anorexia, 41% reported depression, and so on. Persons reported from 1 to 27 of the 38 symptoms. Walsh and Rybicki's cluster analysis revealed domains that included symptoms of fatigue-anorexia- or cachexia; neuropsychological symptoms; upper gastrointestinal symptoms; nausea and vomiting symptoms; aerodigestive symptoms; debility symptoms; and pain symptoms. The number of symptoms reported together by individual patients ranged from 2 to 8, accounting for 26 of the 38 symptoms presented to the patients. Each of these clusters reflects the fact that the same patient used several of the terms within the cluster to report their symptom. Thus, for the fatigue cluster, there was a high probability that the patient would also report symptoms of "easy fatigue, anorexia, lack of energy, dry mouth, early satiety, weight loss," and "taste changes." The neuropsychological cluster or domain included "sleep problems, anxiety, and depression."

It is interesting how varied symptom reporting was, considering that these patients were in a debilitated clinical state. Several questions come to mind. Would the patient organize these symptoms in the same manner as the statistical cluster analysis did? This would require a study similar to what Ross and Murphy (1999) first reported, in which respondents would be given a list of 38 symptoms and asked to sort them into self-determined groups. Another question has to do with the distinctiveness of these clusters. Walsh and Rybicki (2006) report that gastrointestinal symptoms were found in five of the seven clusters. They also reported that confusion and edema formed a debility cluster, which was not an obvious domain, and they wondered if treating one of the symptoms in the domain had affected the others. Thus, did the successful treatment of depression alleviate symptoms of anxiety and poor sleep?

Chow and his colleagues (e.g., Chow et al. 2007; Fan et al. 2007; Hadi et al. 2008) have reported several studies using principal component analysis with varimax rotation as their method for determining the presence of symptom clusters for different groups of patients, using different assessments. Thus, they used the Edmonton Symptom Assessment Scale (ESAS; Bruera et al. 1991) or the Brief Pain Inventory (BPI; Daut et al. 1983) to assess symptoms in patients with bone metastases (only using the ESAS; Chow et al. 2007); for a mixed group of patients referred for palliative radiation therapy (only using the ESAS; Fan et al. 2007); and for a group of patients who had metastatic bone pain (using the BPI; Hadi et al. 2008). In the two studies using the ESAS, the investigators found that a three-component model best fit their data, but interestingly, the symptoms that made up two of the three components were not the same. For example, the study reported by Fan et al. (2007; $N=1,296$) found that the first component consisted of the symptoms of poor appetite, nausea, and well-being, while in the study involving the bone metastases patients (Chow et al. 2007; $N=518$) they found that fatigue, pain, drowsiness, and well-being made up the first component. In both studies, anxiety and depression made up a separate component, while in the Hadi et al. (2008) study, relations with others, mood, and sleep made up a second component.

A third approach to symptom clustering was reported by Olson et al. (2008). In this study, the ESAS was administered daily to 82 palliative care patients, and structural equation modeling was used to determine whether symptoms collected 1 month and 2 weeks prior to death were causally related. The model generated assumed that so-called endogenous (e.g., appetite, tiredness, depression, and well-being), as opposed to exogenous (e.g., pain, anxiety, nausea, shortness of breath, and drowsiness), symptoms would be differentially predictive of symptom status over time. The investigators found that their model explained between 26 and 83% of the variation between appetite, tiredness, depression, and well-being. In addition, their data indicated that anxiety had a direct effect on well-being and an indirect effect on tiredness 1 month prior to death, with a shift to a direct effect on depression 1 week prior to death. The observation that anxiety's influence shifted from tiredness (e.g., fatigue) to depression was interesting, suggesting that tiredness and depression are hierarchically related (see next section).

These three sets of reports are sufficient to illustrate how various statistical procedures can be used to form symptom domains. The statistical procedures here included hierarchical analysis methods, principal component analysis, and structural equation modeling. All the reported studies dealt with palliative care patients, clearly a group worthy of study, but it is also a group of patients that has a high probability of experiencing multiple symptoms, which raises the question of whether symptom clusters of this sort would be found for patients who were not as seriously ill. Also, it is not clear that respondents would sort their symptoms into the same domains formed by various statistical procedures. Whether it would be ethical to ask a group of palliative care patients to perform this sorting task is also not clear.

A report by Clark et al. (2002), however, provides an example where sorting was actually used as part of the development of a symptom assessment: the Multidimensional Pain and Affect Scale (MAPS; Clark et al. 1995). In this 2002 study, earlier versions of the MAPS ($N=270$ words) were reduced to a set of 189 descriptors that dealt with pain and its emotional consequences, and 104 college students were asked to first sort the descriptors into piles based on the similarity of the meaning of the terms used. The volunteers then were asked to sort these piles into piles of two that were similar. This continued until only two piles remained. The different piles of items were evaluated by an agglomerative hierarchical clustering technique that yielded a hierarchy of clusters of descriptors that were similar in meaning (a dendrogram; Yang et al. 2000). Thus, the internal structure of the domains created by a person's response would be expected to reflect this cognitive structure; a decided advantage when drawing inferences from a data set.

As I reviewed in Chaps. 5 and 6, different classification procedures (i.e., hierarchical, cross-classification, or contextual) can be used to form categories. I have also argued that the strongest argument that can be made that the components of a domain are related to an abstract construct when the indicators in the domain can be organized into a cognitive hierarchy. The weakest argument would involve an investigator labeling the domain as reflecting an abstract construct, without clarifying the relationship between the components. In this section, I tried to determine if there was evidence for the influence of cognitive structures on symptom domain formation using two examples: first, the influence of contextual (cultural) factors on domain formation; second, the demonstration that palliative care patients appear to cluster their symptoms. In both examples, respondents may form a category that statistical procedures can confirm also exists as a domain. In the next two sections, I will expand on these observations by considering different aspects of fatigue and pain domain formation. The section on fatigue will provide an example of how a hierarchy can permit the inference that symptoms and an abstract construct, such as HRQOL, are related, while the section on pain will cover a number of topics including the role cognitive classification can play in symptom classification.

2.1 Fatigue as a Prototypical Symptom Domain

The word *fatigue* originates from the Latin word *fatigure*, meaning to "tire out," or from *ad fatim*, meaning "to bursting, to excess." A dictionary definition of the word *fatigue* (Oxford English Dictionary 1996) includes reference to extreme tiredness resulting from hard physical or mental work, but also to brittleness in metal or other materials resulting from a weakened state caused by repeated stress. A third usage, which requires a modifier, refers to boredom resulting from a person being overexposed to something. Thus, a person may suffer from "museum fatigue" or "baseball fatigue" both resulting from excessive participating in a particular activity.

My previous discussions concerning the *figurative* nature of words suggest that fatigue would be a *dual* usage term with reference to *physical brittleness* due to stress being an important part of the meaning of the word. Examining its usage in the current context suggests that the word can also be part of a conceptual metaphor's mapping. Thus, the "DIFFICULTIES ARE BURDENS" metaphor would be applicable, since fatigue refers to a variety of instances, the sum of which would constitute a domain of impediments towards a person achieving their desired goals. In addition, fatigue is a term common to French and English but is not necessarily found in all languages (Flechtner and Bottomley 2003). Thus, when used in these other languages, the meaning of fatigue has to be defined, and its meaning may not be obvious, since it can be used to refer to a symptom, a syndrome (a collection of symptoms), and even a disease (chronic fatigue syndrome), which also confounds the task of defining the term.

Objective measures of fatigue obtained by monitoring muscle exertion, force, or power are usually associated with the reduced performance of a specific task. Reduced performance alone, however, is not a sufficient condition for defining fatigue, since it can occur for a variety of reasons other than physical limitations (e.g., lack of training, poor vision, etc.). When a person is "objectively" tired, however, they will fairly reliably report being tired (a subjective response), even though they can also report being tired in the apparent absence of objective evidence. For example, the chronic fatigue syndrome patient reports being tired in the absence of physiological measures of muscle fatigue (Wessely 1993) or reduced cognitive functioning (DeLuca, et al. 1995), although increased "central sensitization" (Bell et al. 1998; Staub and Smitherman 2002) has been postulated as a possible neural mediating mechanism. There are also conditions where performance and the feeling of fatigue appear disconnected, as when a person is in the manic phrase of their bipolar disease. Finally, Mallinson et al. (2006) examined the association of fatigue self-reports of cancer patients with observer-based performance measures after a chemotherapy session and found moderately high correlations between these assessments. Studies of this sort are essentially criterion-based validity studies of the self-reports and as such, should be reassuring that self-reports have a reasonable degree of correspondence to what is independently assessed by an observer.

Fatigue for the cancer patient is different than that in a healthy person, since the fatigue is usually dependent on the person's treatment or disease. In a qualitative study of cancer patients receiving radiation therapy, Smets et al. (1997) describe the complex and varied nature of fatigue. Thus, the cancer patient describes fatigue as being present throughout the body, their limbs feeling heavy, and lacking strength. One patient stated, "it is even too much trouble to lift a cup or newspaper" (Smets et al. 1997; p. 38). Weakness, stiffness, and trembling were the patients' most common symptoms. One patient reported that he wobbled so much when walking that he would only walk at night when no one could see him, thus avoiding being mistaken for someone who was intoxicated. The respondents also reported that sleep did not relieve their symptoms. The fatigue was so extreme that one patient reported, "I feel sick with tiredness" (Smets et al. 1997; p. 39). Their fatigue varied within a day and between days, changing as the patient progressed and finished treatment. Respondents also described a wide range of consequences of their treatment on their socializing and other important activities. They also had difficulty convincing others that their fatigue was different from what a person experiences after a day's work.

Paterson et al. (2003) reviewed a series of 34 qualitative studies that focused on the meaning of fatigue during a chronic illness (11 of which dealt with cancer patients), and found methodological deficiencies. Only one of the 34 studies asked if a respondent's fatigue could have been due to reasons other than their chronic illness or treatment. Smets et al. (1997) also found that their patients exclusively attributed fatigue to their illness or treatment. The investigators attributed the absence of concerns by their patients about other sources of fatigue to their unwillingness to acknowledge causes other than their disease even though this may have been possible. Paterson et al. (2003) also point out that many of the studies do not consider the role gender, socioeconomic, or cultural differences can contribute to fatigue assessment variability. They also noted that persons who knew that their fatigue was time limited acted differently than those whose fatigue was chronic. Fatigue was often assessed in isolation, even though there was ample evidence that it could be part of a cluster of symptoms, including depression and pain (Miakowski et al. 2004). In this regard, it was interesting to note that some studies described depression as a *cause* or precursor to fatigue (Richardson and Ream 1996), while in others it was described as an *indication* or outcome of fatigue (Glaus et al. 1996). Visser and Smets (1998) considered both outcomes in their analysis of their data.

Table 7.2 Proposed criteria for a clinical syndrome of cancer-related fatigue[a]

A. Six or more of the following symptoms have been present every day or nearly every day during the same 2-week period in the past month, and at least one of the symptoms is item #1 – significant fatigue

  1. Significant fatigue, diminished energy, or increased need to rest, disproportionate to any recent change in activity level
  2. Complaints of generalized weakness or limb heaviness
  3. Diminished concentration or attention
  4. Decreased motivation or interest in engaging in usual activities
  5. Insomnia or hypersomnia
  6. Experience of sleep as unrefreshing or nonrestorative
  7. Perceived need to struggle to overcome inactivity
  8. Marked emotional reactivity (e.g., sadness, frustration, irritability) to feeling fatigued
  9. Difficulty completing daily tasks attributed to feeling fatigued
  10. Perceived problems with short-term memory
  11. Postexertional malaise lasting several hours

B. The symptoms cause clinically significant distress or impairment in social, occupational, or other important areas of functioning

C. There is no evidence from history, physical examination, or laboratory findings that the symptoms are a consequence of cancer or cancer therapy

D. The symptoms are not primarily a consequence of a comorbid psychiatric disorder such as major depression, somatization disorder, somatoform disorder, or delirium

[a]Cella et al. (1998). Reproduced with permission from the publisher

Finally, Cella and his colleagues (1998) have proposed a definition (Table 7.2) of fatigue that has been accepted by the *International Classification of Diseases 10th Revisions – Clinical Modification* (1992). The definition follows the format of a psychiatric diagnosis based on the *Diagnostic and Statistical Manual* (Diagnostic and Statistical Manual of Mental Disorders: Fourth Edition, 1994) methodology. As such, persons may receive a common diagnosis, although the diagnosis may be based on a different set of indicators. Thus, one person who is classified as being fatigued may match the first six criteria, while another similarly labeled may match the last six criteria. I will use this definition in my discussions.

Common to all persons diagnosed with fatigue is that they have "significant fatigue, diminished energy, or increased need to rest, disproportionate to any recent change in activity level" (Table 7.2). The Cella et al. (1998) definition identifies fatigue as a subjective state that can be measured by a self-report, and it also provides criteria to use to distinguish fatigue from "normal" tiredness (e.g., "not relieved by rest"). Excluded are reports based on physical or psychiatric determinants of fatigue. The usefulness of this approach was demonstrated in a study reported by Sadler et al. (2002), who found that the patients who were identified as fatigued with this definition reported symptoms which were more severe, longer lasting, and interfered more in their life than patients who did not meet the "fatigue" diagnostic criteria (Table 7.2).

Lesage and Portenoy (2002), Flechtner and Bottomley (2003), and Hwang et al. (2003b), however, point out that there are substantial biological determinants of fatigue, including disease and treatment-related factors. To these researchers, fatigue is not an isolated subjective symptom but is best seen as an element of a multidimensional construct. This, of course, raises the question of whether fatigue contains modular, nonmodular, or some combination of both types of elements in any one domain. What remains to be determined is how these multiple determinants find expression in a person's self-reports.

2.1.1 The Assessment of Fatigue

One of my operating assumptions is that a domain is defined not by the number of items in the domain, but rather by who develops the assessment and whether the results of the assessment are formally (i.e., statistically) processed. Thus, when an investigator develops a one-item fatigue assessment, such as a visual analog scale (VAS), then responses to this item would be as much a domain, as say, one consisting of 20 or 30 items (e.g., Hwang et al. 2003a). The rationale for this assumption is that the number generated from the respondent's scaling of the single item is as representative of an outcome as the 20 numbers generated by the 20-item questionnaire. What is interesting is whether there are differences in outcome when using a single or multi-item fatigue assessment. Evidence that helps clarify this question will be displayed in Table 7.5.

Table 7.3 provides a representative list of cancer fatigue assessments (see Christodoulou (2005) who has identified 22 general fatigue assessments, and Mota and Pimenta (2006) who identified 18 assessments), while Table 7.4 illustrates the diversity of the items found in three typical fatigue assessments. Table 7.4 reveals that the assessments not only vary in terms of the number of items (e.g., from 1, to 9, to 13 items), but also in terms of the phrasing of the items; the scaling task involved; and the context for assessing the fatigue item, as when the item was part of a larger heterogeneous symptom assessment, or one item among many dealing with fatigue-related issues. For example, the "lack of energy" item in the Memorial Symptom Assessment Scale (Portenoy et al. 1994) is listed as one of 32 different symptoms, and respondents are asked to indicate if the symptom is present, and if it is, how often it occurs, how intense it is, and how much it bothers the respondent. In contrast, items in the Brief Fatigue Inventory (Mendoza et al. 1999) are clearly focusing on the fatigue state and its consequences, and this focus is obvious to the respondent from the title of the assessment and consistent wording of the items. Finally, the FACT-F, which combines ratings of a person's state of feeling over the last week and whether fatigue interfered with either the initiation or continuation of activities, is usually but not necessarily administered as part of the FACT-G.

2 Cognitive Basis of Symptom Domain Formation

Table 7.3 Examples of fatigue assessments[a]

	Number of items	Time frame	Scale
Unidimensional scales			
Brief Fatigue Inventory (Mendoza et al. 1999)	9	Last 24 h	Severity
Fatigue Severity Scale (Krupp et al. 1989)	9	True, now	Impact and functional outcome related to fatigue
FACT-F (Yellen et al. 1997)	13	Last 7 days	Severity and impact
Pearson-Byars Fatigue Feeling Checklist (Pearson 1957)	13/2	True, right now	Severity
Rhoten Fatigue Scale (Rhoten 1982)	1		Severity
Multidimensional			
Fatigue Questionnaire (Chalder et al. 1993)	11	Last month	Severity
Fatigue Symptom Inventory (Hann et al. 1998)	13	Past week	Severity, impact and duration
Multidimensional assessment of fatigue (Belza et al. 1993)	15/1	Past week	Severity, impact, distress and timing
Multidimensional Fatigue Inventory (Smets et al. 1995)	20	Lately	Experiential: severity and impact
Revised Piper Fatigue Scale (Piper et al. 1998)	22	Experiencing now	Experiential: severity
Schwartz Cancer Fatigue Scale (Schwartz 1998)	28	Past 2 weeks	Experiential: severity
Visual Analog Scale for Fatigue (Lee et al. 1991)	18	Right now	Severity
Multidimensional Fatigue Symptom Inventory-SF (Stein et al. 1998)	30	1 Week period	Experiential: severity
Cancer Fatigue Scale (Okuyama et al. 2000)	15	"First impressions"	Experiential: severity

[a]Adapted from Dittner et al. (2004). Reproduced by permission from the publisher

Table 7.4 The content of fatigue assessments

Item #	Memorial Symptom Assessment Scale-Fatigue Item (Chang et al. 2000)	Brief Fatigue Inventory (Mendoza et al. 1999)	FACT-F (Yellen et al. 1997)
1	Lack of energy	Fatigue right now	I feel fatigued
2		Usual level of fatigue	I feel weak all over
3		Worst level of fatigue	I feel listless ("washed out")
4		Interfered with activity	I feel tired
5		Interfered with mood	I have trouble starting things
6		Interfered with walking ability	I have trouble finishing things
7		Interfered with normal work	I have energy
8		Interfered with relations with others	I am able to do my usual activities
9		Interfered with enjoyment of life	I need to sleep during the day
10			I am too tired to eat
11			I need help doing my usual activities
12			I am frustrated by being too tired to…
13			I have to limit my social activities…

Inspecting the data from Hwang et al. (2003a) study reveals, that while the number of items in the assessment in Table 7.4 may vary the correlations between these assessments are all moderately high and statistically significant. This suggests that there was considerable overlap in the information each assessment provided. However, as is clear from the Smets et al. (1995) study, these correlations are dependent on the content of the multi-item fatigue assessment, with an assessment that reflects mental fatigue less likely to be correlated to the single item. These data suggest that the respondent is playing an active role in the formation of the domain independent of the number of items that make up the domain, and that most nonpsychiatric patients do not perceive mental fatigue as a covariate of fatigue.

There are several issues raised by these data, the first of which is to demonstrate that the Hwang et al. (2003a) data have been replicated. Table 7.5 summarizes 10 studies with 19 correlations of a single fatigue item with a multi-item fatigue assessment (1–20 items). The resulting correlations ranged from 0.23 to 0.82, with six of the 17 data points (modal correlation of 0.67) approximating the Hwang et al. (2003a) studies. Thus, the Hwang et al. (2003a) study should not be considered unique.

Table 7.5 Correlations between a single global item and multi-item fatigue measures

References	Group	Assessment #1 (number of items)	Assessment #2 (number of ITEMS)	N	r
Kobgashi-Shoot et al. (1985)	Cancer	Visual Analog Scale (1)	Physical Fatigue Scale (8)	95	0.60
		Visual Analog Scale (1)	Mental Fatigue Scale (4)		0.54
Smets et al. (1995)	Cancer	Visual Analog Scale (1)	General fatigue (4)	111	0.77
		Visual Analog Scale (1)	Physical fatigue (4)		0.70
		Visual Analog Scale (1)	Reduced activity (4)		0.61
		Visual Analog Scale (1)	Reduced motivation (4)		0.56
		Visual Analog Scale (1)	Mental fatigue (4)		0.23
Brunier & Graydon (1996)	Chronic hemodialysis	Visual Analog Scale (1)	Profile of mood-SF fatigue (5)	43	0.80
Kleinman et al. (2000)	Chronic hepatitis	Visual Analog Scale (1)	Fatigue Severity Scale (9)	1,225	0.75
Okuyama et al. (2000)	Cancer	Visual Analog Scale (1)	Cancer Fatigue Scale (15)	225	0.67
Horemans et al. (2004)	Postpoliomyelitis	Polio problem list: fatigue item (1)	Fatigue Severity Scale (9)	65	0.60
		Polio problem list: fatigue item (1)	Nottingham Health Profile: Fatigue (3)	65	0.43
		Polio problem list: fatigue item (1)	Dutch Short Fatigue Questionnaire (4)	65	0.68
Hwang et al. (2003a)	Cancer	MSAS-Short Form, Lack of Energy (1)	Brief Fatigue Inventory (9)	180	0.72
		MSAS-Short Form, Lack of Energy (1)	FACT-F (13)	180	−0.82
Geinitz et al. (2004)	Breast cancer	Visual Analog Scale (1)	Fatigue Assessment Questionnaire (German) (20)	38	0.59
Rupp et al. (2004)	Rheumatoid arthritis	Visual Analog Scale (1)	MFI-20 general fatigue (4)	652	0.79
		Visual Analog Scale (1)	MFI-20 Physical Fatigue (4)	652	0.72
Wolfe (2004)	Rheumatoid arthritis	Visual Analog Scale (1)	Brief Pain Inventory (1)	7,760	0.76

All correlations included in the table are statistically significant ($P<0.05$)

The results in Table 7.5 could have occurred in a number of different ways. First, a person responding to a single item might recruit information about the subject of the item, and use this information in responding to the items, although a researcher would not know if this occurred without also having performed cognitive interviews. Alternatively, a respondent may adopt a "response set" when responding to the various assessments. Support for this notion comes from inspecting the methods section of the cited papers. It becomes clear that the consecutive administration of assessments was a fairly common practice, which would facilitate the development of a response set. The clearest example of this type of study was reported by De Vires et al. (2003), where six different fatigue assessments were mailed at one time to 351 respondents.

A third way to account for the observed correlations is that a person perceives the various fatigue assessments as informationally equivalent, based on their personal histories of dealing with fatigue-like symptoms. These personal histories, acting as an implicit theory, would provide a cognitive template that would organize the information generated by responding to items in the assessment. A fourth instance where the degree of concordance may be mediated involves the observation that for most medical patients it is the physical dimension of fatigue that is most salient to them, and this may dominate their ratings of a fatigue assessment, independent of the number of items. This is one way to interpret the variability observed in the Smets et al. (1995) study. Reviewing the qualitative data also suggests that different groups may have culture-specific perceptions about the meaning of fatigue, and that this, like physical fatigue, may dominate a fatigue assessment independent of the number of items in the assessment. Finally, studies that compare multi-item fatigue assessments (not shown) also suggests considerable overlap in the observed correlations, independent of the number of items in each assessment.

Identifying the variables that predict fatigue provides another approach to defining the components of a fatigue domain. Hwang et al. (2003b) performed such a study. They dichotomized their fatigue data (a score of >3 on the "Brief Fatigue Inventory") and using logistic regression identified six measures as predictors of fatigue. Figure 7.1 illustrates the results of their regression studies and the variables included in their analysis. The six statistically significant predictors identified included feeling drowsy, dyspnea, pain, lack of appetite, feeling sad, and feeling irritable. These predictors define a fatigue domain and consist of both modular and nonmodular components.

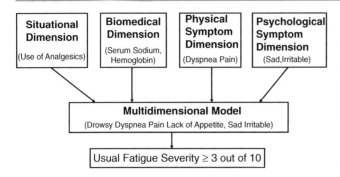

Fig. 7.1 Displayed is an adaptation of Hwang et al. (2003b) multidimensional model of fatigue. Reproduced with permission from the publisher.

Geinitz et al. (2004) report a study of 38 breast cancer patients who received radiation therapy and survived disease free for 2.5 years. These patients were administered the Fatigue Assessment Questionnaire (Glaus 1998) and the VAS-F. They were interested in determining which of their study variables would predict fatigue measures 2.5 years after cessation of treatment. Regression analyses identified that pretreatment fatigue, anxiety, and depression accounted for 60% of the variance in posttreatment fatigue scores as measured by the Fatigue Assessment Questionnaire, and 49% for the VAS-F. They also found that cognitive fatigue was increased posttreatment.

What do the Hwang et al. (2003b) and Geinitz et al. (2004) studies say about how a fatigue domain is formed using psychometric procedures? First, regression analyses can be considered a form of statistical mapping analogous to what occurs cognitively when a conceptual metaphor is applied. Fatigue, as an example of a DIFFICULTIES ARE BURDENS metaphor, would be expected to include various indicators, and the regression analyses in the Hwang et al. (2003b) and Geinitz et al. (2004) studies document what is meant by "burdens." As shown in Fig. 7.1, the predictors (feeling drowsy, dyspnea, pain, lack of appetite, feeling sad, feeling irritable) can constitute the elements of such a domain. Of course the precision of this process could be increased by applying item response theory (Embretson and Reise 2000). The paper by Lai et al. (2003) illustrates how a fatigue-specific item bank can be formed using item response theory, which can then be used to further define an individual's fatigue domain.

This section reviewed some of the methods for assessing fatigue and addressed the issue of whether a single fatigue item can be as useful an indicator as a multi-item assessment. The data reviewed suggests that a moderate but statistically significant correlation exists between these types of assessments. This certainly suggests that a single item can be quite informative, but how might this happen? As discussed earlier, these correlations may be due to a number of factors; (1) study procedural deficiencies (e.g., giving multiple fatigue assessments in a row thereby creating a response set); (2) past experiences with fatigue that have created an implicit theory or schema that structures a person's perceptions; (3) the saliency of the physical component of fatigue that dominates the perception of fatigue; or (4) various cultural determinants. Cognitively, a person may "bring information" to the assessment task. For example, if the single item is an abstract concept (e.g., as the phrase "quality-of-life" is, and "fatigue" can be), then a person may very well think of the term or phrase in concrete terms. As a consequence, the information available to the respondent, when considering the single item, may be comparable, but not likely identical to what is provided by the multi-item assessment. This suggests that the components of a symptom domain and a person's categorical representation of the same material will match to varying degrees, but are unlikely to be exactly the same. A cognitive process of this sort could account for the correlation between the two types of outcome.

One of the tasks that I discussed earlier, that has yet to be fully discussed, was the mapping of the modular and nonmodular elements of a symptom domain to their HRQOL consequences. My next topic gives me an opportunity to do this by examining what is known about the relationship between fatigue, depression, and HRQOL. Here I will discuss how this mapping might be cognitively constructed.

2.1.2 Fatigue, Depression, and HRQOL

Some investigators claim that symptoms such as fatigue and depression are obvious components of an HRQOL assessment (Cella et al. 2002). The logic here is that symptoms, as indicators of disease or treatment, impact and disrupt functioning resulting in increased distress or lost opportunities, states that are presumed to cause changes in HRQOL. Thus, when symptoms are claimed to be part of an HRQOL assessment, an implicit causal assumption is made, although this assumption is seldom specified or supported (see the Wilson and Cleary model).

Part of the reason why these hypothesized causal relationships are not empirically justified is because investigators feel these chains are self-evident and do not require explicit empirical support; they "feel natural and make sense." Thus, a range of symptoms, from pain to lack of social support, may be assigned a causal role relative to observed changes in HRQOL. But, as I have discussed earlier, symptoms as well as other outcomes can function to both cause and indicate some qualitative state. What is natural, of course, is to think in terms of causal chains. For example, I use causal chains to help account for my interactions with others and the environment. These causal chains are usually a product of a "controlled process," an essential element of which is how I categorize or organize information as well as take into account the modularity of the information being processed.

Table 7.6 The Multidimensional Fatigue Inventory-20: items per category (Smets et al. 1995)

General fatigue	Physical fatigue	Reduced activity	Reduced motivation	Mental fatigue
I feel fit	Physically I feel only able to do a little	I feel very active	I am not up to much	I feel like doing all sorts of nice things
Thinking requires effort	I feel tired	Physically I feel exhausted	I think I do a lot on a day	Physically I can take on a lot
I dread having to do things	I think I do very little in a day	When I am doing something, I can keep my thoughts on it	I can concentrate well	I feel weak
I am rested	I tire easily	It takes a lot of effort to concentrate on things	Physically, I feel I am in a bad condition	I have a lot of plans

I can test this causal assumption by examining the structural and modular basis of domains in general, and the domains of fatigue and depression and the abstract concept of HRQOL in particular. To do this I will first engage in a "thought experiment" followed by examining the published literature to determine the extent to which there is support for the hypothesized relationships. The "thought experiment" involves simulating the outcome of a step-wise regression analysis that tests the hypothesis that nonmodular aspects of fatigue, depression, and HRQOL are more likely to be hierarchically related than the modular aspects of these same domains. Support for this hypothesis will come when nonmodular aspects of fatigue and depression are both entered into a regression equation, but only depression is found to significantly contribute to variance in HRQOL scores. This will occur even though entering only nonmodular aspects of fatigue into a regression equation will significantly contribute to the variance of HRQOL scores. This outcome will indicate that nonmodular aspects of depression are capable of accounting for variance in nonmodular fatigue, and in this way are hierarchically related to each other.

Fatigue, although often presented as a specific symptom, can also be conceived of as multidimensional. Thus, the *Multidimensional Fatigue Inventory* (MFI-20), a 20-item fatigue assessment with established psychometric properties (Smets et al. 1995), has been divided with the aid of confirmatory factor analysis into five subdomains of four items each (Table 7.6): general fatigue, physical fatigue, reduced activity, reduced motivation, and mental fatigue. These five subdomains provide the opportunity for separating items in terms of their modularity, although in some cases the modularity of these subdomains may be obvious, while in others it is not. For example, the subdomain of physical fatigue is more likely to consist of modular elements (i.e., physical fatigue is likely to have a somatic reference), and therefore is likely to be related to the equivalent modular components of depression (such as appetite, insomnia, and various performance indicators). In contrast, reduced motivation is more likely related to nonmodular aspects of depression, while general fatigue and reduced activity may have elements of both basic types of modularity. Unfortunately, the type of studies described by Ross and Murphy (1999) has not been performed for the items that make up the fatigue domain, so at this stage I do not know how people themselves would cluster the components of a fatigue domain. Even with these limitations, the MFI-20 provides a unique analytical opportunity to examine the relationship between a domain's modularity and HRQOL.

As I have proposed, it should be possible to first simulate and then empirically determine if a hierarchical or a cross-classification structure exists within an HRQOL domain. Figure 7.2 and Table 7.7 present the basic argument. Figure 7.2 consists of two panels with three overlapping "variance" domains in each. Panel A depicts the relationship between reduced motivation and depression as hierarchical. It is assumed that these fatigue and depression subdomains are predominately nonmodular. Thus, the somatic items in a depression assessment would be expected to be removed. The homogeneity created by this maneuver would be expected to be confirmed by the regression analysis. Of course, the dependent measure, HRQOL, would be expected to be a single global item, or the aggregated score of a multidimensional, multi-item assessment.

Panel B illustrates the potential relationship between these three indicators when information is cross-classified. It includes physical fatigue, but the subdomain of reduced activity could also function in this role. I consider the two relationships (Panels A and B) depicted to be the most likely outcome of an analysis and are the models I will evaluate. The classification of the subdomains of general fatigue and mental fatigue is harder to predict. Mental fatigue, for example, with its heavy dependence on cognition, would be expected to have a strong modular base, yet the distress associated with disturbances in cognition may make the subdomain more likely to covary with a mood-based depression measure.

Inspection of Table 7.7 and Fig. 7.2a reveals that in a hierarchical analysis reduced motivation is assumed to statistically significantly contribute to variance in HRQOL (Regression Analysis #1), if, for the moment, I ignore the contribution of depression. And the same is assumed true for a regression of reduced motivation on depression (Regression Analysis #2), and depression on HRQOL (Regression Analysis #3). When both reduced motivation and depression are entered into an analysis of HRQOL (Regression Analysis #4), however, then only the variance attributable to depression is

2 Cognitive Basis of Symptom Domain Formation

Fig. 7.2 Displayed is the hypothetical relationship between fatigue, depression, and HRQOL based on the hierarchical and cross-classification of information for an HRQOL domain. Nonmodular (reduced motivation) and modular (physical fatigue) aspects of fatigue are included.

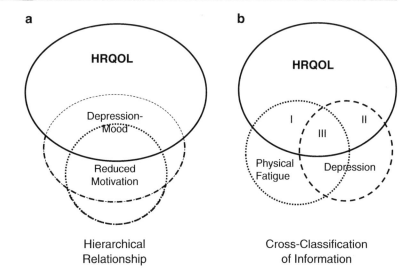

Table 7.7 Hypothetical regression analyses used to demonstrate an informational hierarchy and cross-classification relationship between fatigue, depression, and quality-of-life or HRQOL

Independent variable	Percent variance accounted for ▲	Dependent variable
Hierarchical analysis (regression analyses)		
1. Reduced motivation ($P<0.05$)	0.10	Quality-of-life or HRQOL
2. Reduced motivation ($P<0.05$)	0.25	Depression
3. Depression ($P<0.05$)	0.30	Quality-of-life or HRQOL
4. Reduced motivation (NS▼) + depression ($P<0.05$)	0.35	Quality-of-life or HRQOL
Cross-classification analysis		
5. Physical fatigue ($P<0.05$)	0.15	Quality-of-life or HRQOL
6. Physical fatigue ($P<0.05$)	0.25	Depression
7. Depression ($P<0.05$)	0.30	Quality-of-life or HRQOL
8. Physical fatigue ($P<0.05$) + depression ($P<0.05$)	0.42	Quality-of-life or HRQOL
Complete regression model		
9. Reduced motivation (NS*) + physical fatigue ($P<0.05$) + depression ($P<0.05$)	0.45	Quality-of-life or HRQOL

▲: (NS*) Variance statistically nonsignificant
▼: The values presented here are hypothetical

assumed to be predictive of variation in HRQOL scores. The rationale for this sequence, as stated earlier, is the assumption that the variance attributable to reduced motivation would be "accounted for" by depression and would not therefore directly contribute to the variability of an HRQOL measure. Thus, the variance attributable to reduced motivation and depression would follow the inclusion rule, and their relationship to HRQOL would be described as hierarchical. Empirical support for this model would come from studies that replicate the results of Regression Analysis #4. Graphically (Fig. 7.2a), these relationships can be visualized as overlapping Venn diagrams, with each domain being incorporated into the next more inclusive domain. Cognitively, this relationship can be described as a subordinate level of informativeness (an oak tree) that would be hierarchically incorporated into a basic level (trees), which could then be incorporated into a superordinate category (plants). This sequence requires a third independent variable that might be "mental health," and I would predict that a regression including some measure of fatigue, depression, and mental health would find that only mental health was a significant but moderate predictor of HRQOL.

Information that is *cross-classified* is reflected in regression analyses illustrated in Fig. 7.2b and Table 7.6. Regression Analyses #5, 6, and 7 represent single variable regressions, each of which is assumed to be significant. In contrast with the previous analysis, when both physical fatigue and depression are included in the regression analysis (Regression Analysis #8), both were assumed to be significant predictors of HRQOL scores.

A final regression analysis in Table 7.7 (Regression Analysis #9) includes reduced motivation, physical fatigue, and depression as independent variables. The results of this hypothetical analysis would continue to predict that physical fatigue, but not reduced motivation, or depression as significant predictors of variance in HRQOL scores. The prediction that the variance attributable to reduced motivation would continue to be subsumed by depression illustrates that hierarchical relationships between indicators can exist when domains are also products of cross-classification.

The results of this "thought experiment" may appear paradoxical, so it will be important to determine if there is any evidence to support these hypothetical results. A *Medline* review (last reviewed, January 30, 2010) of the published literature identified at least 347 studies which included measures of fatigue, depression, and HRQOL for cancer patients. I was able to identify at least 35 of these studies that included regression analyses, with seven studies (Table 7.7) providing the elements of the desired analysis plan. I will review most of these studies to provide a sense of the literature at the time of this review. Obviously, this analysis is not meant to *prove* the point I am making, but only demonstrate its possibility when the appropriate data and an analysis plan are in place.

Visser and Smets (1998) report a study in which 250 cancer patients were administered the MFI-20 before completing radiotherapy, immediately after treatment, and 9 months later (Table 7.6). They were also administered a depression measure (Radloff 1977; CES-D), and a single-item global measure of quality-of-life: the Cantril ladder (Cantril 1965). Six items that had a clear somatic reference were removed from the 20 item CES-D, since the authors were concerned that they might be differentially affected by either the person's disease or treatment. The remaining 14 items were referred to as a CES-D mood scale. Thus, their study provides an opportunity to examine the relationship between various dimensions of fatigue, depression, and HRQOL.

First, the authors examined the relationship between the various fatigue dimensions and depression:

> …we investigated by way of regression analyses whether there was any causal relationship between fatigue and depressive mood…neither of these symptoms had much predictive power with respect to the other, suggesting weak causal relationships. Some differences in predictive power between fatigue and depression can be noted, however. When considering the dimensions of general and physical fatigue, it is fatigue that induces depressive mood rather than the other way around. Moreover, a large part of the variance in reduced motivation is predicted by depressive mood, while reduced motivation does not predict depressive mood to the same extent. (p. 106)

These results are, in general, supportive of the directionality intent of my working hypothesis: general and physical fatigue (more modular domains) induce depressive mood, while reduced motivation (a more nonmodular domain) was accounted for by depressive mood. However, more important is the relationship between depression and fatigue and HRQOL. Here the authors report[4] a step-wise regression analysis for each of the time intervals studied, with each of the fatigue subscales and depression being entered in the analysis. They found that only physical fatigue and depressive mood were significant but independent contributors to the variance in quality-of-life or HRQOL (i.e., Fig. 7.2b; areas I and II). This was true when baseline and end of treatment (17% of the variance explained) were compared, and when end of treatment and 9 months posttreatment (20% of the variance explained) were compared. However, when the regression analyses of the interval from baseline to 9 months posttreatment were examined, then only depressive mood was a significant contributor (10% of the variance explained). Still, the total variance accounted for was quite low and this, the authors claim, may be due to their sample not being very depressed, as reflected in their absolute CES-D mood scores. However, the observation that depressed mood was the best and most consistent predictor of HRQOL is consistent with the hypothesis that nonmodular aspects of fatigue are in a hierarchical relationship with depression and would not be a significant predictor of variance in HRQOL scores. In contrast, physical fatigue would be expected to overlap with an assessment of depression that included somatic items (Fig. 7.2b Area III), but would also be expected to include items that would not be ordinarily included in a depression assessment so that one indicator would not be expected to be hierarchically related to the other.

The study of Redeker et al. (2000) provides elements of the desired analysis. They report a series of stepwise regressions, one of which started with scores from the POMS-Depression and Anxiety subscales (Mc Nair et al. 1992), and then added single item measures of fatigue and sleep from the Symptom Distress Scale (Mc Corkle and Young 1978). The dependent HRQOL variable was the total score on the FACT-G (Cella 1997). The investigators reported that depression and anxiety accounted for 43% of the variance in HRQOL scores, and that adding fatigue increased the variance explained by 4%. The authors state:

> Multiple regression analysis revealed that the symptoms and psychological variables explained 47% of the variance in quality-of-life, with the largest proportion of the variance explained by depression. Fatigue and insomnia explained only 4% of the variance in quality-of-life in excess of that contributed by psychological factors. Although overall depression levels were low in this sample, these findings suggest that insomnia and fatigue are related to depression and that depression is more closely associated with quality-of-life than are insomnia and fatigue. (p. 75)

These results are consistent with the hypothesis that fatigue and depression have a hierarchical relationship. Fatigue, however, was not disaggregated, so that the extent to which these results are dependent on modularity remains undetermined.

Holzner et al. (2003) also report a step-wise regression analysis using FACT-Global Score (sum of 27 items) from the FACT-O (39 items) as the dependent HRQOL variable. They initially entered scores related to the MFI-20, general fatigue measure, and found that it explained 33.7% of the variance. They then found that adding clinical variables and sociodemographic variables had no impact on variance explained. Adding the EORTC QLQ-C30 symptom subscales, however, did increase the variance explained by 13.7%. Added next were subscales assessing fighting spirit, anxious preoccupation, and avoidance from the *Mental Adjustment to Cancer* (Watson 1988) assessment, but also the emotional support subscale from the *Social Support Questionnaire* (Sommer and Fydrich 1999). This explained 13.4% of the variance. Finally, total scores for the HADS were added to the final complete regression analysis, and this accounted for 61.1% of the variance.

This study implies that when a variety of psychosocial measures, including depression and anxiety, are added to an estimate of fatigue, being able to predict HRQOL scores nearly doubles. Of course for my purposes, it would have been better had there been an independent measure of depression, that depression had been the first variable entered into the regression analysis, and that the contribution of other subdomains of fatigue would have been reported. It would also have been interesting to examine the consequences of the regression analysis between depression and fatigue measures alone.

Inconomou et al. (2004) report a regression study using the two-item global quality-of-life scale from the EORTC-QLQ-30C scale as the dependent variable. Independent variables included the anxiety and depression component of the HADS, the subdomains of the EORTC-QLQ-30C (physical, role, cognitive, and social functioning, fatigue, pain, nausea, dyspnea, sleep, appetite, and financial impact), and the type of chemotherapy. They performed regression analyses at both baseline ($N=99$) and end of treatment ($N=80$). Not unexpectedly, at baseline they found that depression, physical functioning, and financial impact were significant predictors of global HRQOL scores (adjusted $R^2=62\%$). At end of treatment, they found that only depression was a significant predictor of variance in global HRQOL scores (adjusted $R^2=44\%$).

Ferreira et al. (2008) also provide evidence in support of at least some aspects of the hypothesized relationships. Cancer patients ($N=115$) were administered the EORTC-QLQ-30C scale, the BPI, and the Beck Depression Inventory (BDI). The Karnofsky Performance Status of the patients was also determined. Patients were divided into two groups depending on the severity of their symptoms. The investigators then used a CART (Classification and Regression Tree; Breiman et al. 1984) analysis to determine which co-occurring symptoms would best predict reductions in HRQOL or performance status (as measured by the Karnofsky). The CART analysis is particularly well suited to evaluate the various relationships posited in Table 7.7, since it permits a hierarchical classification of the symptoms based on successive regression analyses. The EORTC-QLQ-30C scale was also disaggregated into physical, role, cognitive, social, and overall domains, and used in separate CART analyses.

What the Ferreira et al. (2008) study lacked, of course, was a fatigue assessment differentiating fatigue into modular and nonmodular components. Thus, the observation that pain was predictive of fatigue and fatigue was predictive of both performance status and physical HRQOL would be expected (Fig. 7.2b), especially if it could be demonstrated that what was being measured was physical fatigue or reduced activity. However, this was not known, so the degree of support for this hypothesis remains unknown. On the other hand, depression, but not fatigue, predicted overall HRQOL, accounting for 26.2% of the variance. Also found was that depression was predictive of pain, which was predictive of role functioning, cognitive and social HRQOL.

Of the studies reviewed, the Visser and Smets (1998), Redeker et al. (2000), and Ferreira et al. (2008) reports were particularly helpful. A number of other studies (e.g., Kim et al. 2008) provide partial support, while others contain the necessary elements for a full test of the proposed hypotheses but have not published the required analyses (e.g., Rossen et al. 2009). Obviously a definitive study has not been reported as yet, but sufficient data are available to encourage the performance of such a study. What was clear from these studies was that fatigue and depression assessments ordinarily contain multicomponents (e.g., modular and nonmodular), and that if the nature of these components (e.g., their modularity) is ignored, then a situation exists which can confound any analysis of the relationship between fatigue, depression, and HRQOL.

Most important about this discussion is that it illustrates the value of examining the cognitive relationship between the elements of a domain. When this is done and if it is possible to demonstrate that a hierarchical relationship exists between these elements, then it is also possible to infer that the elements of the domain represent the abstract label given in the domain. This is equivalent to taking the individual correlations of a factor from a factor analysis, and determining if all the identified elements of the factor are equally as predictive of the label given the factor. In the absence of such a demonstrate referring to indicators, such as fatigue or depression as HRQOL indicators constitutes an arbitrary labeling that lacks empirical support.

In this presentation, I have mostly focused on using regression analyses to demonstrate the presence of a hierarchy (Fig. 7.2; Tables 7.7 and 7.8). However, it is also possible to use other statistical methods to demonstrate the presence of a hierarchy in the elements of a domain. Additional methods

Table 7.8 Fatigue and depression as predictors of HRQOL[a]

References	Patients (sample size)	Fatigue measure (items)	Depression measure (items)	HRQOL measure (items)	Time of assessment
Visser and Smets (1998)	Cancer (250)	MFI-20: PF (4)	CES-D: mood (14)	Cantrill ladder (1)	Pre, post, and 9 months posttreatment
Gaston-Johansson et al. (1999)	Breast cancer (127)	PFS (41) & VAS (1)	BDI (21)	SF-20 (20)	Prior to stem cell transplant
Redeker et al. (2000)	Cancer (263)	SDS-F (1)	POMS-depression-anxiety (24)	Fact-G (47)	During chemotherapy
Holzner et al. (2003)	Ovarian cancer (98)	MFI-20 GF (4)	HADS (14) MAC SSQ	FACT-O (39)	Posttreatment
Iconomou et al. (2004)	Cancer (80)	EORTC-QLQ-C30 – fatigue subscale	HADS (14)	EORTC-QLQ-C30 (30) global QOL	Baseline, end of treatment
Ferreira et al. (2008)	Cancer (113)	EORTC-QLQ-C30 – Fatigue subscale	BDI (21)	EORTC-QLQ-C30 (30)	No active treatment

BDI Beck Depression Inventory (Beck et al. 1961); *CES-D* Center for Epidemiological Studies-Depression (Radloff 1977); *HADS* Hospital Anxiety and Depression Scale (Zigmond and Snaith 1983); *MAC* Mental adjustment to cancer (Watson 1988); *MFI-20 PF* Multidimensional Fatigue Inventory (Smets et al. 1995); *PFS* Piper Fatigue Scale (Piper et al. 1989); *POMS* Profile of mood states (Mc Nair et al. 1992); *SDS-F* Symptom distress scale-fatigue (Mc Corkle and Young 1978); *SSQ* Social Support Questionnaire (Sommer and Fydrich 1999); *VAS* Visual analog scale
[a]See text for details concerning each listed study

(e.g., confirmatory factor analysis, structural equation modeling, and growth modeling; Embretson and Reise 2000) deal with different types of data (e.g., continuous or categorical data) to assess the presence of such associations. These statistical models are designed to determine the relationship between independent and dependent variables, with the assumption that an intervening (latent) variable exists that underlies the supposed association. Since the latent variable is usually stated in abstract terms and the independent or dependent variable in concrete terms, the effort to model these associations can be interpreted as also providing empirical support for the inference that these concrete indicators are examples of the abstract concept.

2.2 Pain as a Prototype Symptom Domain

Pain, its expression and meaning, is a central determinant of a person's existence, and as such, its qualitative consequences (e.g., suffering) should be a central focus of research concerning pain. However, whether pain, and particularly suffering, can be assessed is a question that has been raised both in the scientific and lay literature (e.g., Frank 2001; Chap. 2, p. 27). One of the objectives of the next section is to clarify how pain can and has been assessed as a prototypical example of a symptom domain.

One of the more interesting aspects of studying pain is what the phrase "quality of pain" means. The meaning of the phrase depends on what aspect of the experience of pain is being addressed. Thus, the quality of an acute pain may be limited to the sensory aspects of the painful event (e.g., the stimulus is sharp, burning, throbbing, etc.), while the quality of chronic pain may involve both the sensory aspects of the painful event, but also a variety of pain-related emotions and behaviors that have evolved to help a person successfully cope with their sensory load. In addition, suffering, which also involves the sensory, emotional, and behavioral aspects of a painful experience, has the added element that a person's ability to adjust to a painful experience (particularly the emotional component) is limited. Thus, a hierarchy of experiences exists here, from the concrete or sensory component (e.g., the pain from a cut or a burn) to increasingly pervasive experiences that radiate into all aspects of a person's life and existence. Most important is that each of them makes a distinct contribution to a qualitative outcome.[5]

Pain, like quality-of-life, involves a subjective assessment that has a wide variety of representations. Thus, it can be asked if the assessment of a headache, the pain from a cut, back pain, or stomach pain are all be the same? The answer, of course, is that each type of pain may be different, even though the way they are assessed may be the same; that is, each type of pain may involve the evaluation of descriptors, the importance of which a person can also determine. Thus, what may be common to each of these types of pain is the formation of a common cognitive (hybrid) construct that characterizes its quality. One of the questions I will ask in this section is whether this assessment method can also provide a common denominator that can be used to classify different types of pain.

What is also true of pain is that it is mediated by an elaborate neurological system that is quite ancient in origin. The rapid withdrawal of a finger from a hot surface is a phylogenetic analog of a worm withdrawing in response to a threatening stimulus; both reflect the activation of a neurological alert system, onto which are built the more complex pain mediating systems that are derived from the development of acquired and functional modularity (see below). A key element in this evolution was the development of a system where higher level cortical control could modulate peripheral activation of spinal level pain receptors. This notion is the essence of the classic Melzack and Wall (1965) Gate Controlled Theory of Pain. More recently, Melzack (e.g., 1999, 2005a) has modified his views and now thinks of pain as a centrally mediated phenomena, managed by a neural network. As I will make clear below, this change represents an acknowledgment that a linear, stimulus-response model of pain is not adequate, and that a nonlinear dynamic neural network model is more appropriate to account for the experience of pain. However, before I discuss these topics I want to continue my practice of examining the language used to express pain and identify some of the cognitive processes involved in the experience of pain.

2.2.1 The Language of Pain

The language people use to describe their assessment of the quality of an object, event, or life lived provides a window into what they have previously experienced or are currently experiencing. That which is communicated by language is facilitated by the grammar and semantics of the phrases or sentences used to describe these experiences. In addition, some authors (Halliday 1994, 1998; Lascaratou 2007) suggest that grammar can shape the figurative speech and expressions (i.e., the metaphors) used to communicate complex meaning. This may be particularly relevant when dealing with such difficult-to-understand experiences as pain. In this section, I will examine a variety of self-reported pain statements to determine their grammatical and metaphoric structure. In the section following, I will review and describe the composition of a number of typical pain assessments (Table 7.9), and then compare this material to what is generated from patient self-reports. The question is to what extent these two sources of information concur.

The etymology of the word *pain* can be traced to the old French term *peine*, derived from the Latin *poena*, meaning punishment (*Oxford English Dictionary* [1970] Vol. VII, p. 377–378). Thus, the original meaning of the term *pain* had little to do with pain per se, but instead involved the idea of victimization, penalization, and vengeance (Fabrega and

Table 7.9 A representative list of self-administered pain assessments[a]

Assessment	Scale (number of items)	Developers
Visual Analog Pain Rating Scale	Ratio (1)	Multiple authors (1974)
McGill Pain Questionnaire	Ordinal, Interval (20)	Melzack (1975)
Brief Pain Inventory	Ordinal (20)	Daut et al. (1983)
Medical Outcomes Study Pain Measures	Ordinal (12)	Sherbourne (1992)
Oswertly Low Back Pain Disability Questionnaire	Ordinal (60)	Fairbank et al. (1980)
Back Pain Classification Scale	Interval (13)	Levitt et al. (1978)
Pain and Distress Scale	Ordinal (20)	Zung (1983)
Illness Perception Questionnaire-Revised	Ordinal (52)	Pilowsky and Spence (1983)

[a]Adaption of Table 8.1 from Mc Dowell and Newell (1996). Reproduced with permission from the publisher

Tyma 1976). However, since punishment in Roman times involved physical punishment, it is not hard to see that a semantic shift occurred as the term assumed its current meaning. Hurt, ache, and sore are alternative terms that can be used to describe when a person is in pain, and these terms also reflect the physical dimension of pain.

Studies of the phenomenology of pain (Lascaratou 2007)[6] suggest that pain is construed by a person as an object or entity (Sternbach 1989) that can also act as an agent impacting the person, and that this process is personified as part of a person's coping and adapting to their pain. By thinking of pain as an object or entity, a person separates himself from the pain, and this is reflected in the structure of a sentence, some examples of which are listed in the first two sentences below. Note that the grammatical object of the verb "felt" or "feeling" is pain, and that it is the person who senses the pain. In this way pain becomes a sensation (or an attribute), and thus a possessed entity.

I felt an *unbearable pain* here.
Do you feel *this pain*?
I have a *terrible pain*.

Now the example of the third sentence, "I have a terrible pain," also suggests a possessed entity, but here the sentence relies on a metaphor, the POSSESSING AN OBJECT metaphor, to communicate its meaning. However, the POSSESSING AN OBJECT metaphor evolves from the more general metaphors that form a hierarchy, ending in the EVENT STRUCTURE metaphor (Chap. 2, p. 36). The hierarchical relationship between these metaphors can be displayed as follows:

Level 1: *event structure* metaphor
Level 2: *attributes are possessed objects* metaphor
Level 3: *emotion as a possessed object*
Level 4: (Pain) *a possessed object* metaphor

Lascaratou (2007) describes the relationship between these metaphors as follows:

> To capture the various aspects of the metaphor EXISTENCE OF EMOTION IS POSSESSION OF AN EMOTION/A POSSESSED OBJECT let us trace its emergence with the EVENT STRUCTURE metaphor. Thus, a state, e.g., an emotional, physical, social, etc. state can be viewed as an attribute, i.e., as an attributed state. This understanding of states as attributes motivates their conceptualization as objects that may be given, acquired, owned, possessed, preserved, and lost. Hence, the metaphor ATTRIBUTES ARE POSSESSED OBJECTS. Given the conceptualization of emotions as states, the metaphor EMOTION IS A POSSESSED OBJECT emerges as an instantiation of the higher level metaphor ATTRIBUTES ARE POSSESSED OBJECTS. Pain being defined as an experience of an emotional aspect, it is natural that, like other emotions, it should also be understood as an attributed state and, therefore, as a POSSESSED OBJECT. (p. 144)

The importance of using a grammatical form to make pain a possessed object can be seen in the instructions given to a respondent when they fill out a VAS, or Likert scale; Please rate *your pain* from 0 to 10.

Of course, by making pain a possessed object, a patient and a clinician can communicate about what would otherwise be a private experience. This is one of the components of the objectification process, a step I have previously discussed (Chap. 5, p. 128) as an essential element in the assessment of an indicator. Related to this is the observation that when people are healthy they do not see their bodies as separate from themselves, but when they are ill or in pain, they cognitively separate themselves from themselves. The danger this separation can create is just what Susan Sontag (1991) was concerned about when she warned about using metaphors to describe a disease state and its treatment, and how this can lead to people making inappropriate decisions. However, as Gwyn (1999) points out, the assumption that the literal aspects of an illness can be extracted from its social and cultural context seems difficult to support.

Now if pain is expressed as a possessed object, it will have a variety of manifestations that are extensions of this metaphor. Thus, the *possessed object* metaphor can be an event, as illustrated in the statement:

The pain stopped, but returned again.

But it can also have a force, intensity, or heat. That force, intensity and heat will or will not be contained, which evokes the *container* metaphor. There are many examples of submetaphors that describe some aspect of the *possessed object* metaphor in intensive terms:

PAIN IS DANGER
PAIN IS BURDEN
PAIN IS FEAR
PAIN IS FEELINGS
PAIN IS DARKNESS
PAIN IS AN ILLNESS

The possessed object can also move from one part of the body to another, as illustrated in the following sentence: This pain comes suddenly…sometimes in my leg, sometimes it is in my back.

Again, using these metaphors contributes to a person's sense that they are separating themselves from the experience of pain, but this can also be achieved grammatically, as when the pain is the subject of an intransitive verb. Alternatively, as Lascaratou (2007) points out, a person will personify the possessed object, describing pain as a part of themselves that is an unwelcome visitor, an invader, or as being uninvited. This can also be achieved by the grammatical format, as when pain becomes part of a transitive statement, with the subject acting as an agent. Lascaratou (2007) also points out that there is a strong overlap between the metaphors for emotion and pain, citing Kövecses' (2000) work in this area, especially when the metaphors express exerting a force. Some examples of submetaphors that describe the possessed object as having a force include:

PAIN IS PHYSICAL AGITATION
PAIN IS A PHYSICAL FORCE
PAIN IS A NEEDLE
PAIN IS A SWORD
PAIN IS A SHARP POINT

Most interesting is that these submetaphors resemble the items included in the McGill Pain Questionnaire (Chap. 5, p. 125), which I will discuss in the next section.

Also interesting is that using metaphors to describe a state is actually an admission that the person is having difficulties finding ways to literally describe their state. Sometimes a person can neither literally nor figuratively describe their pain, and this often occurs at the most acute moments of a person's illness, as can happen before death, but can also occur in the less critical situation of chronic pain. This again reflects the complex nature of the phenomenology of pain.

The language used also reflects unique aspects of the physiology of pain. For example, pain does not have a particular stimulus that induces it, whereas the primary sensory modalities do. Thus, it can be demonstrated that a wavelength will induce the experience of a color, a particular frequency a sound, a pressure a touch; each can each be measured, but I can't find an equivalent unique stimulus for pain. Instead multiple types of internal and external stimuli can induce pain and be measured. This suggests that pain is expressed when the stimulation of a particular or combination of sensory modalities is extremely activated and noxious. This contrasts with the sensory modalities, whose language expresses a wide range of states. Thus, while a particular sense can be pleasant or unpleasant, I don't ordinarily talk about "a pleasant pain." Related to this is that pain helps me avoid danger, while the other sensory modalities allow me to either approach or avoid, depending on the conditions.

The language used to express pain also differs from that used to express a sensory modality in that pain is often described as being located in a variety of body parts which may not be the source of the pain, whereas the primary sensory modalities are limited. For example, I might say that "I feel a pain in my stomach" or "I have a pain in my big toe," but I don't usually say "I see a picture in my eyes," "hear the music in my ears," or "I feel the temperature on my face." What is also unique and experientially perplexing about pain, especially if internally initiated, is that I can perceive pain as emanating from one part of my body, and yet its origin can be identified as coming from another. This reflects the phenomenon of referred pain (e.g., Arendt-Nielsen and Svensson 2001). A classic example of a referred pain is the neck, shoulder, or back pain a person feels when having an ischemic episode in the heart, or the chest pain that can be traced to esophageal spasms.

Lay persons consider pain to be a sensation, yet most research lists of sensory experiences do not include pain. For example, Hinton et al. (2008), summarizing a number of authors, suggests that there are 11 types of sensory experiences: the six primary senses that are dependent on external stimulation (tactile sensations were divided into temperature and skin pressure) and five dependent on internal stimulation (e.g., proprioception, vestibular, muscle and tendon tension, gastrointestinal distention, and O_2 and CO_2 level). The absence of pain from such a list certainly reinforces the notion that pain differs from most sensory modalities. In addition, when I say "I don't feel a pain," that is not the same as when I say that I don't see, hear, or touch something. What is different is that I am always hearing, seeing (assuming my eyes are open), or touching, but when I say I am in pain, this implies that some specific neurological event has occurred, a change of state has occurred, I am no longer "normal," as when I have sustained an injury. Thus, pain is akin to having an internal alarm system in the body that is only activated in an emergency, however defined, heralding actions such as getting out of the way, taking a pill, or going to a doctor.

Yet, when pain is inspected more closely it appears to have some important similarities to the other sensory modalities. For example, pain has a threshold for detection; stimuli (even if many different ones) which are reliably capable of inducing pain; and a complex neural system dedicated to its perception. Thus, fast A fibers and slow-responding unmyelinated C fibers express their neural activity through a spinothalamic tract dedicated to the mediation of pain, as would be expected for a sensory modality, but these fibers remain inactive until a noxious stimulus (neural, inflammatory, hormonal, thermal, etc.) activates the neurons. These neurons have been traced to the various nuclei of the limbic system, which is the neural site of attention and emotion. As Clark (2005) describes, "In motivational and emotional terms, the limbic system is a potent place. Pain wraps its tendrils around all parts of it. Unlike most sensory systems, pain has direct access into the innards of our preference functions" (p. 191). Interestingly, smell and taste are equally as intertwined as pain is with the limbic system. Of course pain, smell and taste can easily be seen as the first line of defense for the survival of an organism at all levels of the phylogenetic system.

I can tell that someone is in pain not just by asking them, but also by their body posture, facial expressions, and avoidance of activities. Kirmayer (2008) has suggested that these forms of expression are very culturally sensitive, and often can be described using various metaphors. In addition, the stooped shoulders, furrowed brow, or prolonged inactivity of someone in pain becomes enmeshed in interpersonal relations, social activities, and the political systems (e.g., workers' compensation), all of which adds complexity to assessing the pain experience.

Also true is that people describe themselves as being in pain when the precipitating conditions have an emotional or psychological origin. Examples of emotional pain include the grief following the loss of a loved one, especially if the grief is prolonged (e.g., greater than 1 year); or social rejection, especially if the rejection is profound. But what is interesting about emotional pain is that it usually does not have a peripheral physiological origin (e.g., there is no evidence of a physical injury), even though it may still have an external origin. O'Connor et al. (2008) demonstrated, using functional magnetic resonance imaging (fMRI), that persons suffering from complicated grief vs. noncomplicated grief activate different cortical structures when exposed to pictures of the deceased. In an earlier study, Eisenberger et al. (2003), again using fMRI imaging, demonstrated that emotional pain induced by social rejection activated the same cortical sites that are activated by physical pain. The notion that physical and social/emotional pain have a significant overlap dates to the 1970s when Herman and Panksepp (1978) proposed that social attachment, which is based on various more primitive physiological systems (e.g., place attachment, thermoregulation, and physical pain; Panksepp

1998), activates the endorphin-based pain system when separation occurs. The emotional consequences can be diminished by opiate treatment (Panksepp et al. 1978).

The implication of these and other studies is that social distress and/or an extended emotional state (e.g., sexual or social abuse) may lead to the activation and organization of particular central nervous system sites into a pattern of responding that persists, and which, by overlapping with the sites that mediate pain, are labeled as painful. This, as I will discuss below, is an example of acquired modularity. Related is the demonstration that central sensitization appears to occur for certain types of pain syndromes (e.g., fibromyalgia; Perrot 2008). Fibromyalgia appears to have a peripheral pathophysiological base that suggests it is a painful rheumatological disorder mediated by a heightened central nervous system sensitivity. This sensitivity may also underlie associated symptoms of anxiety, fatigue, depression, irritable bowel, and more (Perrot 2008).

The language of pain and its underling physiological and central nervous system base is obviously complex. But how could I define what pain is? Consider the definition offered by the International Association for the Study of Pain (Merskey and Bogduk 1994; p. 210). It states that pain is "An unpleasant sensory and emotional experience associated with actual or potential tissue damage, or described in terms of such damage."[7] The IASP document goes on to explain that pain is always subjective and unpleasant, that one learns about pain from one's experiences with painful events, and that to be labeled as pain an experience has to have certain sensory qualities. The document also contends that sometimes pain is reported in the absence of tissue damage. This is considered psychological pain which has been accepted as a valid pain report, since there is no way to verify or deny the subjective report. The authors go on to state that this definition avoids tying the definition of pain to a specific stimulus, and acknowledges that pain is always a psychological state that may or may not have a physical cause.

This definition, which is over 20 years old, can now be supplemented by the more recent evidence that pain reports in the absence of injury may still have a specific cortical representation (e.g., Eisenberger and Lieberman 2004; O'Connor et al. 2008). This type of evidence confirms that psychological or emotional/social pain has a material basis, even though a person's report of pain remains subjective and private. This suggests that the next iteration of a definition of pain should acknowledge that pain, whether physical or emotional/social, is embedded and that the different types of pain are best accounted for by the nature of this embodiment.

Before I leave this topic I would like to review one more paper with the provocative title "Is pain every normal?" (Cronje and Williamson 2006). The two authors, a linguist and a physician, point out that many of the pain terms discussed in the IASP manual assume that the designation "normal pain" exists and that deviations from this phrase also exist. *Allodynia*, for example, is defined as "pain due to a stimulus which does not normally provoke pain," whereas *hyperalgesia* is defined as "an increased response to a stimulus which is normally painful" (IASP 2004; p. 210–211). Their paper reviews what the term "normal" means in folk language, and how what is considered normal is approached through evidence-based medicine.

Cronje and Williamson (2006) point out that the prefix *allo-* (from allodynia) means "other" in Greek, and is commonly used in medical terminology to refer to a condition that diverges from expectation. Thus, they argue that "what is expected" is the unspoken operational definition of the term normal when discussing various pain conditions. But whose expectations are involved, and how is "what is expected" established? The answers are pretty straightforward: it is the clinician whose expectations are usually involved in a medical encounter, and a clinician's expectations are established by his or her clinical experiences. Clinicians, like any people, categorize their experiences and create cognitive prototypes that underlie their expectations. The problem with this, of course, is that the prototypical reasoning is always at risk of being inaccurate when applied to an individual, and when it is applied to the pain patient, may lead to under treatment. To support their argument, Cronje and Williamson (2006) cite the large number of studies revealing that patients and clinicians are discordant about their mutual judgments of all sorts of clinical conditions, and that once trained, clinicians are very reluctant to change their prototypical thinking.

This paper illustrates how attending to how language is used and how a clinician thinks can help me understand how clinical decisions are made, some of which may have qualitative implications. It also shows that there are limits to human reasoning that can confound clinical decision making (e.g., deciding on the nature of the pain a patient is experiencing), and that this justifies the development of evidence-based medicine (Chap. 8, p. 249).

2.2.2 The Assessment of Pain

The basic task in assessing pain is to objectify the experience of pain. How can this be done and how successful have past efforts been? First I ask, what it is that I am assessing when I assess pain? Am I assessing pain, or does pain exist independent of its quality? Data suggests that it does. For example, when patients receive a frontal lobectomy, they still report the sensory component pain, but not the aversiveness normally associated with pain (Melzack and Torgerson 1971). The same happens when a soldier or athlete experiences pain during extreme stress, or a person receives an opiate analgesic. There are also people born without the ability to experience pain, but who still have the ability to sense (e.g., pricking, warmth, cold, and pressure; Sternbach 1968). A very different example is the masochist, who finds pleasure in the aversiveness of pain. This suggests that the sensory experience of pain and its aversiveness are neurocognitively separable and manipulable, although they are normally experienced together and represent a hybrid cognitive construct.[8]

If it is the aversiveness of pain that determines its quality, does this mean that painfulness is a *quale*? As discussed in Chap. 1, p. 10, the term *quale* has become enmeshed in a philosophical discussion about whether the mind exists, and whether the existence of a *quale* provides evidence for the nonmaterial existence of the mind. Clark (2005) provides a clear argument about why painfulness is not a quale, and this argument focuses around the fact that painfulness is a cognitive construction, which is not what would be expected to be true of a quale.

Inspecting the structure of some of the available pain assessments (Table 7.9) suggests how pain can be objectified. One approach, the *McGill Pain Questionnaire* (Melzack 1975, 2005b), is almost a textbook example of the objectification process and the role an investigator can play in creating a domain. The questionnaire consists of a number of parts. It provides respondents with an image of the body for them to identify the site of pain, allows the rating of the intensity of pain, and provides a set of descriptive terms to capture the experience a person has with pain. Melzack (2005b) reports that he developed the descriptive part of his questionnaire by first collecting words that patients used to describe their pain, particularly patients with phantom pain (e.g., Melzack 1971). He combined his experience with what he learned from the research literature (e.g., Dallenback (1939) identified 44 commonly used terms that described pain qualities) to form a list of 102 pain descriptors (Melzack and Torgerson 1971). Melzack and Torgerson (1971) describe their interesting decision: "In the course of bring the words together, it was immediately apparent that the list, arranged in alphabetical order, provided a meaningless jumble" (p. 51). Therefore, they decided to classify the words into classes and subclasses, and created clusters of words that had common meaning. Thus, they used their own capacity to form categories to create domains they believed respondents would acknowledge as appropriate. They then grouped these words into three major classes: words that describe "sensory qualities" (in terms of temporal, pressure, spatial, thermal, and other properties); "affective qualities" (in terms of tension, fear, and autonomic properties); and "evaluative words" that describe the overall intensity of the pain. Each of the elements in these groupings was also rated. Patients, students, and doctors were asked to rate each randomly ordered word in terms of the least to the most pain that it represented. Thus, even though the patients who filled out the McGill Pain Questionnaire only selected words that were descriptive of their pain, they were also generating a rating of their words. These ratings did lack, however, the respondent's own independent description of the importance their painfulness had to them.

What struck me about the origin and development of this classification system was that it was a very deliberate effort to capture the phenomenology of pain. Melzack could have, as is true for many other pain assessments, limited his assessment to just assessing the intensity of pain. He, however, not only includes a pain intensity assessment in his questionnaire, he also gave his respondents the opportunity to comprehensively describe their pain experience (e.g., by selecting appropriate descriptors). This, of course, made his assessment quite unique at the time it was developed. Still, his effort raises a host of questions. First, since the McGill Pain Questionnaire was essentially an ad hoc creation by Melzack and his colleagues, can the three-factor structure of the Patient Reported Inventory (PRI) be empirically confirmed? A second issue deals with the relationship between experimental or clinical assessment of pain (e.g., Gracely 1983). Is it true, for example, that experimentally induced pain can replicate self-reported pain? If it can, then it could be argued that the subjective self-reported pain has an "objective" basis. Finally, there is the issue of how well the McGill Pain Questionnaire provides information when compared with a direct interview. Implicit in this question is the issue of whether an assessment, which unavoidably reduces information, is still able to reliably and validly describe the experience of pain.

There have been a large number of studies attempting to empirically support the three-factor structure of the PRI (e.g., Holroyd et al. 1992, 1996). These studies differ in terms of the clinical characteristics of the respondents, but also whether acute or chronic pain is being assessed. In general, the data supports the presence of Melzack's original three factors, but often adds a fourth factor. In addition, investigators regularly recommend that the total score of the PRI be used, rather than being reported in terms of the individual descriptors. This recommendation was made because the mean ratings of each of the descriptors was often highly intercorrelated, suggesting considerable overlap in what each descriptor was measuring.

Melzack and Torgerson (e.g., Torgerson and BenDebba 1983) recognized that their questionnaire was an ad hoc creation that, while clinically useful, still lacked the support of a more formal scaling procedure. To combat this, they applied multidimensional scaling procedures to the verbal descriptors with the aim of quantitatively distributing the descriptors in an abstract space. The scaling process involved subjects rating the degree of similarity between different descriptors and then using the resultant dissimilarities to distribute the descriptors in a conceptual space. The optimum distribution of the descriptor was established by orthogonal rotation. Torgerson and BenDebba (1983) suggested that the domain of pain descriptors consists of a dimensional Euclidean space, since the descriptors vary in intensity, but also of a hyperspherical space that "specifies quality in terms of the great circle distance of a descriptor from each of a number of ideal types" (p. 51). This approach, of course, is consistent with Gärdenfors' (2007; Chap. 1, p. 12) more recent suggestions on the optimal approach to conceptualizing quality as a circumplex.

Alternative approaches to the multidimensional qualitative mapping task have been reported. For example, Clark et al. (2003) report that a factor analysis confirms the empirically derived cluster structure underlying the Multidimensional Affect and Pain Survey (MAPS), thereby challenging the appropriateness of the a priori classification of the McGill Pain Questionnaire. I will discuss the MAPS in more detail below, since it will be useful in any qualitative classification system of pain.

One of the assets of the McGill Pain Questionnaire is that it relies on self-reports for the person to indicate their qualitative state. This same asset, however, can also be seen as a liability in that there is no way for an independent observer to confirm the basis for the person's self-report. This concern has been a continual topic of discussion in this book, as it has been since the advent of behavioral science research. I have suggested that subjective responses can be "objectified" (Chap. 4), making it possible to make invariant statements. One contribution to this end, summarized by Guilford (1954; p. 251), involves using statistical methods to demonstrate objectivity. He points out that Wells (1907) suggested that if the ratio of the variance within an individual was small relative to the variance between individuals, the observation could then be considered objective. Guilford (1954; p. 252) goes on to state that Wells' suggestion roughly resembles what would be true for an analysis of variance, and that calculating the ratio of 1/F (which would ordinarily vary between zero and 1.0) could be used as an objectivity/subjectivity ratio, with a higher ratio indicating greater objectivity. Guilford (1954) also discussed correlation and factor-analysis approaches to objectivity.

Gracely and his colleagues (Heft et al. 1980) report an alternative approach to establishing the objective nature of the verbal descriptors. They used psychophysical methods to calibrate a set of verbal descriptors of intensity (i.e., weak, mild, intense) or unpleasantness (i.e., annoying, unpleasant, distressing) to increasing painful electrical stimulation of the tooth pulp or duration of a cold spray applied to exposed dentin. The verbal descriptors Heft et al. (1980) used overlapped but did not replicate those reported by Melzack (1975), but they were still able to demonstrate that the verbal descriptors they used co-varied with the amount of electrical stimulation (when the electrical stimulation was above threshold), thereby providing an objective metric for the verbal descriptors.

Jensen and his colleagues (Jensen et al. 2006; Victor et al. 2008) have demonstrated that it is feasible to expand the qualitative aspects of an established pain assessment by adding descriptors to the basic assessment. Thus, Jensen et al. (2006) expanded the ten-item Neuropathic Pain Scale (Galer and Jensen 1997) to a 20-item assessment, with each descriptor being rated on a 0–10 scale. They also demonstrated (Victor et al. 2008) that it was possible, by using exploratory factor analysis, to reduce the 20-item assessment into three broad qualitative pain domains (e.g., paroxysmal, surface, and deep), thereby reducing the overall interpretative task the assessment created. This suggests that fewer items on an assessment are not necessarily more informative, and that a balance has to be found between brevity and informativeness. Also demonstrated here was the usefulness of taking a synthetic approach (combining items together for a specific purpose) to the development of a qualitative assessment.

I found it particularly interesting how similar the McGill Pain Questionnaire was to the hybrid construct model of quality that I have been promoting (e.g., Figs. 1.2 and 1.3). The ratings of the descriptors (sensory, affective, or evaluative) are an analog of the evaluation of a descriptor. The inclusion of affective or emotional descriptors was also posited as an important dimension of a quality judgment. Thus, the respondent can be thought of as placing themselves in some two-dimensional qualitative space (e.g., Figs. 1.2 and 1.3), a space that has a descriptive and evaluative meaning as well. What is lacking is the third dimension that involves reflection about the evaluated descriptors that become part of the hybrid construct. An example of a pain assessment that lacks the evaluated descriptors but gets a bit closer to including a reflective statement is the BPI (Daut et al. 1983). This assessment asks respondents to not only evaluate various descriptors of pain, but also asks them to indicate the extent to which these symptoms interfere in their lives. To most quality-of-life researchers, tracking the consequences (the interference) of the events that befall a person is a major indicator of quality-of-life, but as I have pointed out, lots of things happen to people, but only some are important to the person and it is these consequences that must be noted and valued, and this process requires some reflection.

Thus, the McGill Pain Questionnaire appears to have some of the characteristics of an optimally designed quality-of-life assessment. If it also includes some estimate of a person's rating of the importance in being in a particular pain state, then it would have the formal structure of such HRQOL assessments, as the Health Utility Index (HUI), or the QWB scale. This again emphasizes that a qualitative assessment is a cognitive construction and that pain, as a symptom, if completely assessed, would be as much of a quality indicator as would any number of other indicators.[9]

Table 7.9 summarizes a representative selection of pain assessments. The format of the assessments ranges widely, but I will not review them all in detail (there are actually hundreds of pain assessments). Several of the assessments in Table 7.9 are essentially checklists, providing a list of indicators that a respondent selects to characterize their pain (e.g., The McGill Pain Inventory, The Back Pain Classification Scale). Some also include an estimate of how much the pain interferes with daily functioning (e.g., The BPI, the Medical Outcome Study Pain Measures). Some focus on current pain and others ask about pain over an extended period of time. Sometimes pain is just one item in a list of symptoms (e.g., The Illness Behavior Questionnaire, The Pain and Distress Scale). Some assessments ask if a person is currently in pain (e.g., The McGill Pain Inventory, The BPI), prompting the respondent to consider sensory aspect of pain when they rate other aspects of the pain experience. However, few of the assessments listed by Mc Dowell and Newell (1996) approximate an optimally designed qualitative assessment. Not included in Table 7.9 were examples of assessments generated by multidimensional scaling techniques (e.g., Clark 1984; Kerns et al. 1985). I will delay a discussion of these assessments, since they can be useful in the classification of pain, which is the topic of the next section.

A third issue I raise about the McGill Pain Questionnaire is whether it captures a reasonable amount of the information that would be forthcoming if a person was interviewed. There is an extensive literature that suggests that this may not be the case. Consider the comments on the meaning of pain and suffering, found in fiction (e.g., *The Death of Ivan Ilych*, by Tolstoy (1886); *The Magic Mountain*, by Mann (1927), etc.), but also in the anthropological literature (e.g., Morris 1991; Scarry 1985). It is clear that when pain is conceptualized on the basis of the common discourse, it is described in ways which formal assessments such as the McGill Pain Questionnaire have not emulated. Thus, Morris (1991) describes pain as being both a puzzle and a mystery. Pain is a puzzle because of its complex biological, social, psychological, and anthropological nature, and is a mystery because the medical sciences seem to be having difficulties getting their hands around what is essentially a subjective phenomenon. Morris describes a mystery as a truth that is closed off from full understanding, and true mysteries will always be closed off. He also alludes to the potential therapeutic value of using a mystery to conceptualize the existential state a person is in. He notes that current popular descriptions of pain often use the metaphor of war, where pain is combated or conquered, with its implication that the person can do something about their state. This differs from a mystery, which is a state in which a person can dwell, without self-recrimination. As Morris states:

> I would propose that while the doctor typically approaches pain as a puzzle or challenge, the patient typically experiences it as a mystery. Pain takes us out of our normal modes of dealing with the world. It introduces us to a landscape where noting looks

entirely familiar and where even the familiar takes on an uncanny strangeness…Thus, even when we experience pain as a mystery, we continue to think of it – to our infinite perplexity – simply as an unsolved puzzle. …We might say that it is the most earnest wish of almost every patient, ancient or modern, to be released not just from pain but from the requirement of dwelling within its mysteries. (p. 25)

Morris seems to be stating that the consequences of a person developing chronic pain results in a change in his or her state (whether defined by physiological, psychological, or social criteria), and while the person characterizes himself as struggling to escape this state, he still remains captured by it. Morris also asserts that formal pain assessments (e.g., Table 7.9) do not address these issues, and the question remaining is how a questionnaire can characterize a person's existential existence.

In the preface of this book I stated that, "The task which remains, however, is to convincingly capture the joy, pain and suffering real people experience with the numbers that are used to characterize their experiences" (Preface, p. IX). In this section, I suggested that the McGill Pain Questionnaire represents an example of applying numbers to a person's verbal description of their pain experiences. The review I have just completed provides general empirical support for the McGill Pain Questionnaire, although there is concern about how complete this support is. I also summarized Gracely's (Heft et al. 1980) psychophysical approach to pain scaling, and the multidimensional scaling procedure approach of Clark et al. (2003) as alternatives to the procedure Melzack used. Finally, I acknowledged that the process of objectifying subjective pain leads to information reduction that may not capture the verbal expression of the existential dilemma of being a chronic pain patient, although the available assessments provide useful information for a variety of applications. Viewed from this perspective, Rapley's (2003) quotation from Wittgenstein (1961) remain relevant (Chap. 1, p. 8).

2.2.3 Acquired and Functional (Pain) Modularity

I will start this section by briefly reviewing the recent history of ideas concerning the nature of pain, thus demonstrating that the structural basis for the pain experience, although first conceived of in linear terms (e.g., as a stimulus-response-like model), is now thought to be a nonlinear dynamic neural network (Fig. 7.3). This change follows the increasing awareness that pain is not just a peripherally mediated reflex-like event, but rather involves the central nervous system. Next I will discuss the development of acquired and functional modularity, illustrating each with examples from different types of pain experiences. This will give me the opportunity to discuss the full range of pain states, and this will permit me to demonstrate why a detailed study of these pain syndromes (i.e., demonstrating that a pain syndrome contains modular, nonmodular, and conditioned components) is necessary if a researcher is to be able to capture the complexity and quality of existing in these particular pain states.

The history of ideas of what pain is can be traced to Aristotle's view that pain is the opposite of pleasure, and therefore is an emotion. This notion of pain as an emotion changed over time, as is evident in Decartes' (1664) conceptualization of pain as a sensation with a dedicated neural pathway starting in the skin and ending in the brain. This view was augmented in the nineteenth century, when Bell (1811) and Magendie (1822) demonstrated that the peripheral sensory and motor systems were anatomically separable, thereby providing a physical basis for distinguishing between what was sensed and action. By the end of the nineteenth century, sufficient anatomical information was available such that von Frey (1896) could postulate his theory that specific nerve systems existed; he was able to demonstrate that specific nerve receptors existed in the skin that could be associated with specific sensory experiences. Pain was assumed to be mediated by free nerve endings found in the skin and other body organs. However, this still left the problem of how a painful experience was initiated. Sherrington (1906) responded by postulating that the threat or fact of tissue damage was what precipitated the painful or noxious sensation.

von Frey's (1986) theory of specific nerve energies postulated that there was a direct connection between a peripheral sensory input and its central manifestation. This encouraged the belief that a linear model could account for a pain experience, but clinical observations, such as phantom pain, low back pain, or fibromyalgia, suggested alternative views. In response, Melzack (1989, 1999, 2005a) suggested that the central nervous system plays the dominant role in determining the pain experience. For example, Melzack and Wall's (1965) gate control theory assumed that spinal cord activity was modulated by central input, but more recently Melzack (1999, 2005a) has suggested that a central nervous system neural network between the thalamus, cortex, and the limbic system constitutes a neuromatrix that underlies the experience of pain (Fig. 7.3). The neuromatrix consists of genetically determined modules. As Melzack (2005a) states:

> I have labeled the entire network, whose spatial distribution and synaptic links are initially determined genetically and are later sculpted by sensory inputs, as a *neuromatrix*. The loops diverge to permit parallel processing in different components of output products of processing. The repeated cyclic processing and synthesis of nerve impulses through the neuromatrix imparts a pattern; the neurosignature. The neurosignature, which is a continuous outflow from the body-self neuromatrix, is projected to areas in the brain - the *sentient neural hub* - in which the stream of nerve impulses…is converted into a continually changing stream of awareness. (p. 86)

Clearly, Melzack[10] is following in the tradition of Hayek and Hebb (Chap. 3, p. 52) and is formulating a system that

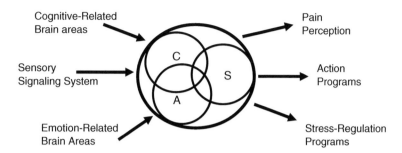

Fig. 7.3 Melzack's (2005a) neuromatrix theory of pain is illustrated. Factors which contribute to the pattern of activity generated by the body-self neuromatrix include sensory (*S*), affective (*A*), and cognitive neuromodules (*C*). The output from the neuromatrix produces pain perception, action programs, as well as activates a variety of bodily and behavioral processes designed to regulate stress. Permission to reproduce the figure was granted by the publisher.

acquires its characteristics by altering some neuromatrix structures with the resulting neurosignatures functioning as relatively fixed output systems. Melzack's description also suggests that a nonlinear process underlies this process, a process that may appear linear from another analytical perspective. This description, of course, can serve as an explanation of how an acquired module is formed.

Acquired modularity is quite common, and represents the fact that much that characterizes a person's experiences evolves from the embodied nature of these experiences. The relatively fixed biologically dependent processes that emerge when a module is formed range from up-regulation (the enhanced sensitivity of spinal cord receptors) to conditioned aversions, but also to language limitations, persistent mood states, and perhaps even personality patterns. Of course, these relatively fixed patterns have been shaped by modulating some genetically determined developmental process, so that what can be modified will vary in degree. However, they are much less modifiable then those physiological processes that have been "conditioned" as a result of associative learning. What occurs during associative learning, of course, is that the probability that one modality will evoke another is increased. The classic example of this occurs during respondent or Pavlovian conditioning. For example, a dog will salivate to a stimulus (e.g., a particular sound) after the stimulus had been repeatedly associated with a reinforcement (e.g., the sight, smell, or taste of food). The modalities being associated (conditioned) are audition and digestion (i.e., salivation in response to food ingestion). Evidence that conditioning has occurred is usually demonstrated by "reversing" the learning process; by exposing the conditioned stimulus repeatedly to the organism without any reinforcement. This results in the reduction of the likelihood that the conditioned response occurs (i.e., the conditioned salivation response is "extinguished"). It is this ease of "reversibility" that distinguishes an acquired module from a conditioned response.

Operant conditioning has also been demonstrated to play a role in sustaining various physiological processes and also has been implicated in maintaining verbal pain reports (Fordyce 1982). More recently, Becker et al. (2008), using operant conditioning techniques, were able to increase a person's sensitivity to a painful stimulus. The authors suggest that this enhanced sensitivity may be a principal means whereby a person's pain experiences become chronic.

The relationships between a module, an acquired module, and an associative construction are complex and require additional explanation. Fodor (1983) described a module to be a form of information encapsulation such that what is sensed or perceived does not flow directly to the central nervous system, but is classified at some subcortical levels into distinct categories, and then sent to higher cortical areas. These modules function at the input (or output) level of the information processing system. All central cognitive processes, Fodor claims, are nonmodular, where various associative processes could occur. A particularly interesting example of an associative process that does not involve conditioning is the formation and use of metaphors to create new knowledge. Thus, the unique mapping (e.g., Fig. 2.2) between a base (a concrete entity) and a target (an abstract entity) would be an example of the creation of a nonmodular entity. In contrast with Fodor (1983), Sperber (2002) has argued for what he calls "massive modularization." In his view both subcortical, cortical input and output systems, perceptual and cognitive processes are all assumed to be modular.

The development and persistence of chronic pain provides an appropriate venue to examine how modularity is acquired. Here I will consider two types of chronic pain: the CRPS and the IBS. IBS provides an example of how functional modularity is acquired. First, however, I need to define chronic pain, and it turns out that there are at least three approaches to this task. For example, the IASP (Merskey and Bogduk 2004) defines chronic pain by determining the duration the pain has been present. Thus, chronic pain is defined as

"…pain which persists past the normal time of healing…In practice this may be less than 1 month, or more often, more than 6 months. With nonmalignant pain 3 months is the most convenient point of division between acute and chronic pain…" (p. xi). In contrast, Olesen et al. (2003) suggest that chronic pain should be defined by using the number of days a person is in pain, while Von Korff and Migliorett (2005) suggest that the phrase "chronic pain" is as much a prognostic statement of future pain as it is a statement of past pain. It turns out that each of these definitions has been used in classifying different types of pain.

The IASP (Merskey and Bogduk 2004) pain classification system describes CRPS as a form of neuropathic pain. CRPS Type I, also known as *Reflex Sympathetic Dystrophy*, was initially thought to lack identifiable neural lesions, but more recent studies suggest that damage occurs to small fiber axons, and these events may be a critical determinant of pain reports (Oaklander 2008).[11] CRPS Type II, historically known as *Causalgia*, evolves after known nerve damage, and therefore has a predominately peripheral (i.e., extremity) origin. The differential diagnosis of these two types of neuropathic pain was thought to be explained by sympathetic nerve blockage, although this procedure remains controversial (Schurmann et al. 2001). Gibbs et al. (2008) provide a useful review of the pathophysiology of CRPS, suggesting that sympathetically mediated pain may be present in both forms of CRPS, with the implication that the designation of two types of CRPS may be less important than the Merskey and Bogduk (1994) classification implies. Harden et al. (2007) provide a current statement on criteria for clinicians to use when diagnosing CRPS.

The signs and symptoms associated with CRPS cover a broad range, including evidence of the involvement of inflammatory, neurological, and sympathetic processes, and evidence of various types of atrophy. Veldman et al. (1993) summarize the signs and symptoms of 829 patients who were diagnosed with reflex sympathetic dystrophy. For example, increased inflammatory responses were assumed to mediate pain reports, skin color differences, swelling, temperature differences between the afflicted and normal limb, limited movement, and increased complaints after exercise. Neurological involvement could be demonstrated by allodynia, hyperpathy (abnormal painful reaction, especially to repetitive stimulation and an elevated threshold), incoordination, tremor, involuntary movements, muscle spasms, paresis (abnormal sensation, whether spontaneous or evoked), and pseudoparalysis. Atrophy of the skin, nails, muscles, or bones is also reported. Evidence of sympathetic nervous system involvement includes the presence of hyperhidrosis and change in hair and nail growth. To this array add the observation that some patients report limb dissociative experiences, reporting feelings that their limb is disconnected from their body (Galter et al. 2000). Clearly this impressive set of indicators has made the diagnosis of CRPS particularly difficult.

Table 7.10 Application of Fodor's criteria, as interpreted by Cain (2002), to determine if particular types of domains have modular components

Criteria	CRPS ▲	IBS ▼
1. Domain specific	CRPS I – Yes CRPS II – Yes	No
2. Operationally mandatory	CRPS I – Not determined CRPS II – Yes	Yes
3. Limited central access	CRPS I – No CRPS II – Yes (?)	No
4. Input systems are fast	CRPS I – Yes CRPS II – Yes	Yes
5. Informationally encapsulated	CRPS I – Yes. but… CRPS II – Yes	Yes
6. "Shallow" outputs	CRPS I – No CRPS II – Yes/No (?)	Yes
7. Fixed neural architecture	CRPS I – Yes CRPS II – Yes	Yes/No
8. Characteristic breakdown	CRPS I – Possibly CRPS II – Possibly	Yes
9. Characteristic pace and sequence	CRPS I – Partially CRPS II – Partially	No

▲ *CRPS* Complex regional pain syndrome
▼ *IBS* Irritable bowel syndrome

Of these indicators I will limit my comments to self-reports of symptoms. I will use the classification criteria suggested by Fodor (1983; Table 6.4) and summarized by Cain (2002) to establish whether the symptoms of CRPS have modular characteristics. First, it is clear that what I have described to be true of CRPS consists of a cluster of different symptoms. Pain, allodynia, hyperpathia, paresis, pseudoparalysis, limited limb movement, complaints after exercise, and limb dissociation all rely on self-reports that an independent observer may not be able to verify. Some components of the symptom cluster are more likely to be considered modular than others. Thus, as I have described (Chap. 6, p. 71), acute pain, if reflex-like, is likely to be modular, whereas chronic pain is best described as acquired.

Applying Fodor's criterion #1[12] to CRPS (Table 7.10), as defined by Merskey and Bogduk (1994) and updated by Oaklander (2008), suggests that CRPS has an identifiable input, or domain specificity, for both CRPS I and II. Considering the complex nature of CRPS I (for example, that it may involve both centrally and peripherally mediated processes), it can't always be assumed that the presence of a stimulus operationally ensures a response, whereas this is more likely to be true for CRPS II (Criterion #2). Jänig and Baron (2002) and others have argued that CRPS I is best conceived of as a disease of the central nervous system, in that a variety of systems (e.g., the somatosensory, sympathetic and somatomotor systems) are altered, This may or may not be true for CRSP II (Criterion #3). Whether the stimuli which induce CRPS I or II are known or not, when a stimulus does occur, the input systems are fast (Criterion #4). Aspects of either CRPS I or II are informationally encapsulated (Criterion #5), in that specific neural circuits are dedicated to mediating these types of pain, although the acquired nature of chronic pain will involve recruiting other portions of the central nervous system that are not ordinarily involved in the pain experience. Cain (2002; p. 188) describes "shallow outputs" (Criterion #6) as occurring when much more information is received as inputs than is forthcoming as outputs in a modular domain. This "constraining of information" is very characteristic of perceptual systems and the same may be true for CRPS. Considering the wide range of systems (sensory, perceptual, and motor systems) involved in CRPS I expression, the assumption of a constrained output does not seem supportable, although this is more likely for the established input/output systems of CRPS II. Both forms of CRPS have an identifiable neural architecture (Criterion #7), but the central portion of this architecture is probably acquired in the manner described by Melzack (1999, 2005a; Fig. 7.3). Criterion #8 is concerned about whether the input system to pain modules can be disrupted by damage to particular brain sites or genetically dependent processes. This is a difficult criterion to evaluate, since CRPS itself can be described as a neural and immunological response to disturbed functions. However, it is highly likely that these responses would not occur if these critical systems were intact. Criterion #9 refers to Fodor's belief that various response systems are genetically dependent and therefore determined (reflecting his nativist position). CRPS clearly has elements that meet this particular criterion, but also appears capable of reorganizing the neural substrate mediating these processes. To summarize, both CRPS I and II have some of the characteristics that Fodor (1983) considered to be evidence of modularity, but differed in a number of ways as would be expected for an "acquired" module. What makes CRPS modular is its biological status, the alteration of which requires active pharmacological intervention.

IBS is another type of "acquired modularity," but here what appears to be acquired is the involvement of the central nervous system in determining the operating characteristics of a peripheral regulatory system (the enteric nervous system that mediates gastric motility). Both biological processes (e.g., bacterial gastroenteritis or alterations in gut microflora) and/or psychosocial factors have been identified as risk factors for these systemic or functional changes. However, the extent to which the disorder relies on physiological and/or psychosocial factors has remained controversial. For example, Nicholl et al. (2008; see also Faresjö et al. 2007) report the first prospective study of a community sample ($N=3,732$) to determine the risk factors for the diagnosis of IBS. They found ($N=2,456$; 71% participation rate) that high levels of illness behaviors, anxiety, sleep problems and somatic symptoms were independent predictors of IBS onset. They concluded that psychosocial factors characteristic of somatization were predictive of the development of IBS (Rasquin et al. 2006). In addition, Dorn et al. (2007) reported that the increased colonic pain sensitivity reported by IBS patients may be more a result of an increased tendency to report pain than a peripherally induced physiological response. This suggests that the episodic abdominal pain relieved by defecation, characteristic of IBS, may be centrally mediated.

From this it would be expected that applying Fodor's criteria (as interpreted by Cain (2002)) to IBS would satisfy fewer of the criteria meant to define a fixed module, especially in contrast to CRPS. However, it is also clear that IBS is not a product of some simply associative process (e.g., conditioning of a sound and salivation), but rather has the permanence of a fixed but episodic module whose precipitating conditions may very well be learned. As I described above, there seems to be a wide range of inputs that initiate the abdominal pain, so it is hard to argue that the onset of IBS symptoms is domain-specific (Criterion #1). What does seem to be true is that once the pain is initiated, the experience of it is unavoidable, and in that sense the operation of the pain system is mandatory (Criterion #2). The observation that the onset of IBS is dependent on psychosocial factors makes it likely that central nervous system processes are involved in mediating pain, so Fodor's Criterion #3 can't be confirmed. The fact that the experience of pain, once initiated, is fast does not reflect the episodic and somewhat unpredictable nature of the precipitating events that lead to the abdominal pain. However, since it takes milliseconds for people to recognize that they are in pain, it is reasonable to conclude that the input system is fast (Criterion #4). Does the enteric nervous system (the neural system that regulates gastric motility) have operating characteristics that are not centrally accessible? Yes, it does (Wood 2008). However, also true is that centrally mediated procedures such as relaxation or behavioral cognitive therapy (e.g., Drossman et al. 2003) have been shown to reduce the frequency of abdominal pain episodes (Criterion #5). In addition, there is also evidence that centrally initiated neurohormones and drugs influence the enteric nervous system, and therefore pain perception. Criterion #6 requires that a module has a constrained output, meaning that it has many fewer outputs than inputs, and this seems to be true for IBS. Criterion #7 requires that a fixed neural architecture exists for the module. To the extent that the enteric nervous system is a system anatomically distinct from the central nervous system, this criterion is met (Wood 2008). However, it is true that the central nervous system plays a critical role in mediating pain for the IBS patient, and in that sense it lacks a fixed neural architecture dedicated to mediating the pain experiences of the IBS patient. Criterion #8 requires that the input to the module have a characteristic

Table 7.11 Representative examples of pain classification systems

References	Title	Classification principle	Classification dimensions
Wiener (1993)	Differential Diagnosis of Acute Pain by Body Region	Hierarchy[a] per body site	Body site → disorders/body site → causes/disorder
Merskey and Bogduk (1994)	Classification of Chronic Pain	6 Axes[b] for each pain disorder	Body site by body system by temporal character of pain by patient report of severity/chronicity by etiology
Spitzer (1987)	Quebec Task Force Classification for Spine Disorders	11 Hierarchies	11 Different spinal pain disorders → duration and work status
Olesen (2004)	The International Classification of Headache Disorders (2nd Ed)	3 Hierarchies	Cause of headache → self-report of pain symptoms (type to subtypes)
Bruera et al. (1989)	Edmonton Classification System for Cancer Pain	5 Axes	Mechanism of pain by incident pain by psychological distress by addictive behavior by cognitive functioning
Muller et al. (2007)	The Classification of the Fibromyalgia Syndrome	1 Hierarchy	Fibromyalgia with or without psychiatric symptoms by depression or somatization
Turk and Rudy (1988)	Multiaxial Assessment of Pain	3 Axes	Dysfunctional profile by interpersonally distressed profile by adaptive coper profile

[a]Here, a hierarchy consists of a progressively more inclusive class. Thus, a pain category may consist of multiple disorders, and a disorder may have multiple causes
[b]Axes create a space consisting of a number of supposedly independent dimensions

breakdown pattern that can be attributed to brain and/or genetic impairment. Again, the response to this requirement is that the complex system (peripheral and central) that mediates abdominal pain has a characteristic breakdown pattern, but this pattern is not nearly as clear for IBS as it is for the impact of a brain lesion on vision or hearing. Finally, the enteric nervous system regulates gastric motility, and in that sense determines its pattern and pace. However, here I am reviewing the pattern and pace of a pain module, and in that sense this criterion is not met (Criterion #9), especially since abdominal pain is usually reported as episodic and is, at times, difficult to predict.

These two examples, CRPS and IBS, should be sufficient to illustrate the complex array of biological and psychosocial processes that contribute to the formation of acquired modularity. This complexity implies that a dynamic nonlinear process underlies the formation of these pain syndromes and would be expected to include various qualitative aspects of the pain experience. My next task, therefore, is to estimate the extent to which this is so, and by so doing, estimate the role qualitative factors might play in the fundamental analytical task of classifying pain.

2.2.4 The Classification of Pain

My objective in this section is to determine if the qualities that are associated with a particular pain state can be used to classify these types of pain. At one level, the answer is straightforward, since it is easy to distinguish headaches from back pain, and back pain from stomach pain, etc. To classify these different pains I will use a variety of criteria, including the location of the pain, the severity or intensity, as well as temporal pattern of the pain. Distinguishing between types of pain, however, is not quite the same as a person being able to use verbal descriptors to describe the phenomenal characteristics of pain, and to gage how important the consequences of living with pain means to them. The issue here is whether folk language, the common discourse, can be used to differentiate and diagnose important types of pain.

My first objective in developing a classification system, therefore, is to determine whether this classification system is best characterized in terms of dimensions (or axes) or summary domains. To answer this question I will review a representative sample of methods currently being used to classify pain, and then discuss multidimensional classification systems. I will also evaluate the extent to which the available assessments approximate a qualitative assessment. What will become clear from these reviews is that much work has been accomplished and that it is feasible to add a specific qualitative component to these multidimensional assessments of pain.

Inspecting the classification literature quickly made it clear that pain classification systems have been developed for more than one purpose (e.g., clinical decision making or scientific purposes), and that this difference impacts the classification process. What I am interested in, however, is how these classifications are organized, what dimensions they include, and whether they also include qualitative (verbal) descriptors. The straightforward answer to this question is that none of them do; however, most of them do rely on self-reports to varying degrees. Table 7.11 summarizes a representative sample of pain classification systems. An example of a classification system that relies on phenomenological reports for distinguishing between types of pain is the *The International Classification of Headache Disorders* (Olesen 2004; p. 10). Thus, a headache can be a migraine headache, and a migraine headache may have a typical aura or not. In a clinical setting this is usually established by the patient's self-reports, and this diagnosis is usually confirmed when a drug treatment alleviates the symptoms. Most classification systems rely on a person's self-report concerning the severity, pattern of the pain symptom, and even the person's work status. Of the classification systems listed, the acute pain classification system seems to minimally rely on self-reports.

Most of these classification systems start with a clinically established diagnosis and then develop axes (i.e., dimensions) or hierarchies that differentiate one clinical entity from another. Turk and Rudy (1988), however, suggest an alternative approach. They recommend that an investigator ignore a priori clinical domains and instead "broadly define, clarify, and synthesize a diversity of variables hypothesized to measure the impact of chronic pain on the patient's physical, psychological, social and behavioral functioning" (Turk and Rudy 1988; p. 233). They applied this more "empirical" approach with the intent of identifying several relatively homogenous groups of chronic pain patients who had common characteristics, and who could conceivably respond more uniformly to treatments. They start by identifying a variety of behavioral and psychosocial indicators using data using the West Haven-Yale Multidimensional Pain Inventory (MPI; Kerns et al. 1985) as their primary multidimensional assessment tool. The MPI is a 60-item assessment that can be clustered into 13 subscales, and has been shown to have good internal reliability and some evidence of validity (Kerns et al. 1985). In their study, Turk and Rudy (1988) identified three psychometrically sound clusters of characteristics that provided the best profile of their sample. The first profile (based on five subscales) comprised 42.6% of the sample and reflected the respondents' sense that their chronic pain disrupted their lives, leading to their being in a dysfunctional state. A second profile (three subscales) reflected the respondents' greater psychological distress, while the third profile (one subscale) indicated the respondents' lower perceived ability to cope and control their lives. These nine subscales can then be used to provide a visual profile of the respondents. When the MPI was administered (Turk and Rudy 1990) to three groups with known types of chronic pain (low back pain, headaches, and temporomandibular disorders), the resultant profiles were able to both differentiate between the diagnostic groups and demonstrate a fair amount of overlap.

Davis et al. (2003) reports an alternative clustering method (a two- rather than three-component model), and their suggestion provides a more useful outcome than the method developed by Turk and Rudy (1988). Sheffer et al. (2007) also report ($N=976$) a re-calibration of the MPI and found three clusters that they labeled as Interpersonal Focused, Stoic, and Adaptively Focused. These studies confirm that the MPI can cluster chronic pain patients into groups, although the exact composition of these clusters has yet to be established. To the extent that the items on the MPI deal with the consequences of being in chronic pain, they match what would be expected for evaluated descriptors of an HRQOL assessment. What MPI lacks in order to be an optimally designed HRQOL assessment, of course, is the valuation of the importance or utility of being in these states. It also does not provide the respondent with the opportunity to select verbal descriptors to describe the phenomenal characteristics of their pain.

Clark et al. (1989a, b) also had a long-term interest in applying multidimensional scaling methods to characterize chronic pain patients. Clark and his coworkers developed a 101-item assessment called the MAPS. These 101 descriptors were initially selected from a list of 270 words provided Clark by Melzack and Torgerson (1971). As previously explained (Chap. 7, p. 222), these words were first sorted by a group of volunteers into words that had similar meanings, followed by a merging of these clusters into piles on the basis of their similarity.[13] This was followed by a hierarchical agglomerative cluster analysis to confirm and further articulate the clusters. The MAPS consists of 57 verbal descriptor items, which Clark labels a "somatosensory pain supercluster." A second supercluster includes 26 items and is labeled "emotional pain," while a third supercluster is labeled "well-being" and consists of 18 items. Each of these descriptors was presented to a respondent as part of a sentence, requiring a 0–5 rating (not at all to very much). Thirty-eight of the 87 descriptors in the McGill Pain Questionnaire are also found in the MAPS, but now their presence is confirmed by objective criteria, not the post hoc clustering of Melzack and his colleagues. Verbal descriptors such as "irritating," "tickling," "tingling," "hot," "stabbing," "numbing," "excruciating," "overwhelming," and so on, were included.

Clark's early work in this field started with an interest in pain sensitivity as approached by sensory decision theory, but in time, lead to an interest in verbal descriptors, and this grew into the development of the MAPS. Much of his research time has been spent in establishing the psychometric properties of the MAPS, with less effort in applying the assessment to clinical settings. Thus, Clark et al. (1989a, b) report a study in which they compared similarity judgments of nine verbal descriptors and found different judgments for cancer patients and healthy volunteers. Cancer patients found pain intensity to be the most important dimension of pain (measured by the degree of similarity), while the healthy volunteers reported that the emotional pain dimension was most important. Most importantly, Clark et al. (1989a, b) displayed their data in terms of individual respondents and found a fair amount of variability between individuals. The similarity of the intent and method displayed in this study and an HRQOL study is striking. In another study, Yang et al. (2000) asked if preoperative responses on the MAPS were used to predict postoperative morphine usage. Subjects were a group of Chinese colorectal cancer patients. It was found that high preoperative MAPS scores on sensory and emotional words predicted higher morphine usage. Preoperative scores on the subclusters for depressed mood (6 descriptors), anger (4 descriptors), anxiety (2 descriptors), and fear (4 descriptors) predicted higher morphine usage, while high preoperative scores in 13 of the 17 subclusters for sensory qualities were positively correlated with a patient's attempt to receive more morphine when they had reached the limit of their prescribed drug in a particular time period. The MAPS has also been used as an analytical tool, as in studies that attempt to determine if VASs (numerical) can provide a reasonable estimate of the respondent's pain state, as compared to more extensive, and time-consuming assessments (e.g., Clark et al. 2002; Huber et al. 2007). Griswold and Clark (2005) have since developed a shortened version of the MAPS.

While the MAPS has not been as extensively used in clinical settings as, for example, the MPI, it has clearly been demonstrated that it is possible to quantitatively and comprehensively characterize the verbal descriptors associated with particular pain experiences, and by so doing, provide a foundation for a quality assessment of the pain. Thus, the MAPS, or some similar assessment, should be able to be used when classifying pain. With this type of information in hand, it will be possible for a symptom, such as pain, to function as a qualitative indicator.

2.2.5 Assessment of Pain in the Compromised Person

An organism's ability to avoid or response to threatening or harmful stimuli is a capability that is retained by all surviving spices. Human beings, however, have expanded this capability by their ability to verbally communicate. This verbal ability is compromised to varying degrees for certain classes of persons. These individuals, however, are still able to detect or avoid painful stimuli. Viewed from this perspective, the deficits of the compromised person can be viewed as a relative not absolute limitation in pain perception and detection.

The people who have limited ability to verbally communicate their pain include the infant, the immediate postsurgical patients in the ICU, the person who is delirious, the stroke person who remains in a vegetative state, the person on a ventilator whose ability to verbally communicate is restricted, people who have a wide variety of cognitive impairments, and even, to some extent, the intact person who has a particular personality type (stoic). Each of these groups is worthy of study, but I will focus my attention on the elderly person who is progressively losing memory and linguistic capabilities.

Not only are there many types of compromised persons, there are also many potential causes of these cognitive impairments, including neurodegenerative, vascular, toxic, anoxic, infections, and social-cultural factors. These causal factors can be exacerbated by aging and ongoing chronic illnesses (e.g., diabetes) limiting a person's ability to detect a painful stimulus, use verbal descriptors, rate these descriptors, or engage in the metacognitive process of applying an importance rating. These factors, of course, are components of the hybrid construct and if they are compromised, then this has implications concerning the person's ability to assess their qualitative state.

There is a extensive literature demonstrating that chronic pain impairs cognitive abilities (e.g., Seminowicz and Davis 2007). However, the literature on the impact of dementia on pain perception is less clear (e.g., Bjoro and Herr 2008). I will be reviewing parts of this material in Chap. 10, so I will minimize my discussion of this topic here. A study by Kunz et al. (2009), however, is worthy of notice, since it clearly demonstrates that dementia differentially impacts the multiple components of the system that mediates pain perception, and offers a way to assess pain in the nonverbal compromised person. In their experimental study, Kunz et al. (2009) compared 35 healthy patients with dementia (i.e., none of the patients had a major illness that might affect their participation in the study) and 46 age-matched health controls, all of which were ≥65 years old. Thirteen compromised patients were diagnosed as having Alzheimer's disease, 14 as having vascular dementia, and 8 as having mixed dementia. Pain was induced by eliciting the "nociceptive flexion reflex" (NFR) by electrically stimulating the calf muscle. Concurrent pain assessments were monitored by self-reports, facial changes, motor reflex parameters (latency, amplitude, and area under the curve for super-threshold responses), and autonomic responses (e.g., skin potentials, heart rate). The results suggested that the threshold to reflexive flexion for demented persons is significantly lower (greater sensitivity) than the healthy controls, but that once the threshold had been established there was no difference between the two groups in the latency, amplitude, or area under the curve (a plot of relationship between intensity of response with increasing intensity of stimulation) of the reflex. The researchers also observed that as the severity of dementia increased the capacity of the respondents to provide usable self-reports diminished $r=0.692$). However, when they focused on those respondents that could provide adequate self-reports, then they found that their interpretation of the intensity of the pain was not different from the control group. There was a clear difference in the frequency and intensity of the facial responses to painful stimuli of the demented persons as compared to the controls. Dementia was also associated with a reduction of autonomic skin responses to painful stimuli, whereas heart rate did not differ between the groups. The authors conclude that the severely demented person retains the ability to respond to painful stimuli, but their pattern of response is different from the healthy controls.

The demonstration by Kunz et al. (2009) that the threshold for response to a painful stimulus is lower for the demented person than it is for healthy controls differs from previous reports where no difference in threshold between groups was reported (e.g., Benedetti et al. 1999; Gibson et al. 2001; Rainero et al. 2000). Kunz et al. (2009) account for this discrepancy by partially attributing it to self-report limitations, but also to the possibility that reflex activation and pain sensation may not be related to one another for the demented person. Their data also suggest that the demented person may be more intensely processing noxious stimulation, possibly because their cerebral mechanism that inhibits or controls their response to the pain is disrupted. Consistent with this hypothesis is the study by Cole et al. (2006) demonstrating, via fMRI studies, that pain-related cerebral activation of demented persons is greater than in controls. Thus, a useful working hypothesis seems to be that because of specific cognitive and central nervous system limitations, the demented person responds intensely to painful events, which further incapacitates them and contributes to their inability to make adaptive responses to their circumstances.

Assuming for the moment that this scenario is accurate, it is easy to see how this information can be used to improve the assessment of the quality of existence of the severely demented person. First, as was confirmed in an earlier Kunz et al. (2007) study, "the facial expression of pain has the potential to serve as an alternative pain assessment tool in the demented patients, even in patients who are verbally compromised" (p. 221). How can this be done? One approach is to tailor the assessment to take into account the specific neuropsychological deficits of the compromised person. For example, Luzzi et al. (2007) tested the ability of 71 compromised persons with Alzheimer's disease to recognize standard drawings of different emotions, and found that 73% of the patients were able to recognize the emotional expressions of the drawings, and that the 27% who were not had right hemispheric deficits. Specifically, they found from neuropsychological testing that those that did not recognize the emotions expressed in the drawings had low scores in constructional praxis and nonverbal memory tests, which are both consistent with right hemispheric deficits. This suggested to Luzzi et al. (2007) that the ability to detect positive and negative facial emotions by the majority of persons with Alzheimer's disease perhaps was persevered, even for persons with severe disease. These findings, and the related literature, also suggest that caregivers should be able to emotionally bond with the compromised person, and that this can be used clinically to detect when the person is in pain. Less known is whether this retained skill can be used to facilitate communication. For example, can the nonverbal Alzheimer's disease patient be taught to use pictures of various emotions to communicate their emotional status? Given the opportunity, would a compromised person point out a face of a particular emotion if they had the faces displayed as pictures in the room they frequented? A procedure like this is done regularly with the verbally limited stroke patient, as when they are provided with a "word book" to use in communicating. These kinds of issues are consistent with the notion that communicating with the nonverbal demented person is a job of cross-modality communication, an area of research that could benefit from an expanded effort.

If communicating with the nonverbal demented person is a task in cross-modality communication, what does this mean when comparing verbal with nonverbal data? First, the assessment task is the same whether I am dealing with a verbal response or nonverbal gestures (movements or facial expressions) in that I need to objectify and put a number to either type of indicator. However, whereas it is possible to use nonverbal information (e.g., behavioral observations) to support the validity of a verbal response, the opposite is not nearly as likely with the verbally limited compromised person. This is so because the verbally compromised person is limited in their ability to understand verbal communication, and to therefore respond to instructions. Thus, the assessment of nonverbal information of the compromised person is more vulnerable to observer biases. The reliability and validity of particular emotional expressions made by the severely compromised person can, however, be enhanced by using a consensus judgment of several external observers. This is reflected in the recent efforts to develop a pain assessment protocol for the demented that relies on specific facial expressions (see Chapman 2008). In addition, there are likely to be differences in the scaling and therefore statistical procedures that can be applied when comparing the two types of data, since the verbal respondent's data are likely to approach interval scaling, while the nonverbal data are more likely to be nominal or ordinal, at best. Still, these complications in how the data is assessed should not prevent the development of cross-modality communication methods. Finally, it is important to keep in mind that the hybrid cognitive construct that I have been promoting offers an excellent analytical template to organize both what is known and what needs to be known about qualitative responding in the compromised person.

2.3 Summary

So far I have discussed two of the three topics I set out to cover in this chapter. The discussion concerning the relationship among fatigue, depression, and HRQOL supported my argument that it is possible to provide empirical evidence to justify making the inference that a particular set of items in a domain are indicators of an abstract construct, such as HRQOL. A second major issue addressed in this section was whether a symptom, such as pain, can be considered a qualitative indicator in its own right. My answer was that it can if it is qualified, meaning that some estimate is obtained concerning the importance, utility, or preference for being in a particular pain state. Particularly impressive was the effort by various pain investigators to quantitate a person's pain phenomenology. This process of starting from what a person experiences and ending with numbers describes just what I would hope would occur in the quality-of-life research. However, most quality-of-life researchers focus on assessing the consequence of adverse events (e.g., disease, medical treatments, family dynamics, etc.), not on first characterizing the qualitative impact of these events. What does occur, however, is that many HRQOL assessments ask respondents to rate various states in terms of satisfaction or dissatisfaction, good or bad, and so on; descriptors that can also be found in a pain assessment. However, the number of these descriptors is more limited than is the number found even in the short form of the McGill Pain Questionnaire.

The third topic deals with the question of whether symptoms can be incorporated into an HRQOL assessment. This is a problem in applying cross-classification methods to the formation of categories, and ultimately to the formation of domains. Domains are usually formed by using statistical methods such as factor analysis to identify entities to include in the domain. Once the items are identified as members of the factor, an investigator has to judge how these items constitute a category before giving the factor a label. Without the interaction between the data and the investigator, the factor would remain a number, and the meaningfulness and usefulness of the domain would be limited.

3 Symptom Domains and HRQOL Assessment

3.1 Introduction

One of the main objectives of this chapter was to determine what role classification principles and modularity play in a symptom domain. Thus, the data reviewed (Table 7.8) on the relationship between reduced motivation → depression → mental health → HRQOL raised the possibility that a hierarchical relationship may exist between these indicators. What I learned from studying the modularity of symptom domains was that some aspects of the biological basis of a symptom may be directly involved in an HRQOL outcome, while other aspects are either indirectly or not involved at all. Unfortunately, most studies of the relationship between symptoms and HRQOL do not acknowledge the role that the cognitive structure and the modular basis of a domain play in organizing the information in the domain. However, based on what I learned so far I can now modestly claim that the broad statement that "symptoms can cause changes in HRQOL" should be replaced by the more precise statement "some entities in a symptom domain may cause changes in HRQOL, while others will indicate the status of HRQOL."

As indicated in Table 3.2, the hierarchical classification of cognitive entities is only one of several ways entities in a category can be classified. Cross-classification is probably a more common method, and illustrating this is the purpose of Sect. 3.2. In this section, I will briefly review the cognitive basis of cross-classification, and then apply it to the task of integrating symptoms into an HRQOL assessment.

3.2 Integrating Symptoms into HRQOL Assessments

Table 7.12 illustrates how often symptoms are found in some of the most popular HRQOL assessments. As I indicated in the introduction to this chapter (Chap. 7, p. 200), assessing a symptom very often requires a different question format compared with dealing with functional status or global assessments. As a consequence, the internal structure of an HRQOL assessment domain can be quite heterogeneous, and include items that differ in scaling and the range of response options. At first glance this heterogeneity would appear to make summarizing the responses to these assessments difficult, but it turns out that five of the seven assessments

Table 7.12 Analysis of items on HRQOL assessments[*]

Assessment (number of items)	References	Descriptive statements: symptoms (percentage of total items)	Description statements: functional consequences	Global: domain specific	Global: universal
EORTC-QLQ C30 ($N=30$)[a]	Aaronson et al. (1991)	20 (66%)	8	1	1
FLIC ($N=22$)	Schipper et al. (1984)	11 (50%)	11	0	0
FACT-B ($N=33$)	Brady et al. (1997)	11 (33%)	14	5	1
RSCL ($N=39$)	de Haes et al. (1983)	30 (77%)	8	0	1
CARES ($N=59$)	Ganz et al. (1992)	10 (17%)	49	0	0
BCQ ($N=30$)	Levine et al. (1988)	20 (66%)	10	0	0
MOS-SF-36 ($N=36$)	Ware et al. (1993)	13 (36%)	20	2	1

[*]Based on HRQOL assessments discussed by Goodwin et al. (2003)
[a]The abbreviation list at the beginning of this chapter contains complete definitions of the assessment names

listed in Table 7.12 are scored as profiles, with only two (the FLIC and the BCQ) being summed and presented as a single number, while the FACT-B can be summed to give a total score or presented as a profile. When given as a profile, however, a researcher or clinician often has the problem of combining this information into a summary statement so that a decision can be made. A researcher would ordinarily rely on statistical procedures, while a clinician would rely on their expertise to aid in making these decisions. Very often both approaches are used, as when a clinician evaluates the clinical significance of some statistically determined outcome. Both participants, at some point, reflect on their information and judge whether the potential inferences that can be drawn make sense. When they do, both evaluators have both cognitively combined information into categories.

The type of items (e.g., symptoms, functional assessments, and global ratings) included in an HRQOL assessment differs in terms of the cognitive demands they create for the respondent. Thus, making a global judgment will evoke different cognitive processes than, say, making a functional assessment or rating a symptom. If an investigator wants to call an assessment an HRQOL assessment, he or she has to combine the results of these different types of assessments into a cognitive (HRQOL) category, usually by using cross-classification. But how does this occur?

Barsalou (1983) suggests that initially, categories of heterogeneous entities are spontaneously formed by a person and usually occur in response to specific circumstances (e.g., which positive indicators a clinician may need to establish a diagnosis or what to take out of the house in case of a fire). He refers to these as "ad hoc categories," which, if they are reused and remembered, are labeled as being "goal-directed." Examples of goal-directed categories a quality-of-life researcher or patient might think of include the social and interpersonal consequences of having a mastectomy, a limb amputation, losing hair due to chemotherapy, and so on. These categories reflect a top-down cognitive process, where the investigator or a person finds a common principle to use to organize the heterogeneous information into a summary statement. Barsalou and his colleagues (Glushko et al. 2008), noting the explicitness of this process, contrast it with what occurs when a child acquires knowledge of cultural norms and practices, which they claim occurs implicitly and without awareness. They consider this to represent a bottom-up approach. They also discuss institutional classification systems that can evolve from collections of goal-directed categories. Thus, individual diagnostic categories for major depression, eating disorders, anxiety, and so on, can be organized into the DSM IV.

One of the advantages of presenting data from heterogeneous domains as profiles is that it permits the administration of statistical procedures to data that have been scaled along a common dimension. Thus, if I summarize the frequency of symptom reports separately from ratings of functional status, I avoid the task of statistically analyzing binary and nonbinary data together. Of course, these data could be combined by transforming the nonbinary into binary, but this would blunt the information the study could provide. The alternative is to leave the task of aggregating the accumulated information to another level of analysis, but this tactic cannot always be applied, especially in clinical situations. Because of this, Feinstein (1987a, b) and others have promoted the development of another type of assessment and decision making: clinimetrics (Feinstein 1987a). As a result of this, a raging debate (Chap. 5, p. 149) has developed about whether psychometric or clinimetric statistical procedures are most appropriate to use in assessing heterogeneous domains.

I can illustrate some of the issues here by reviewing a paper by Levine et al. (1988) that describes the construction of the BCQ. These investigators developed a 99-item questionnaire designed to assess the experiences of receiving chemotherapy for Stage II breast cancer. In their initial study, they directly interviewed 47 patients and asked them whether they had experienced physical, emotional, or social problems as a consequence of receiving chemotherapy. The patients were then asked to select those items among the list of 99 that reflected their experiences while on chemotherapy. They were also asked to rate, using a five-point Likert scale, how important these items were to them. Table 5.9 lists a sample of these ratings for different symptoms. The investigators then multiplied the average importance of the items by the frequency the item was selected, and calculated a total score for each item. They observed that the various items grouped into at least seven topics (e.g., hair loss, fatigue, nausea, positive well-being, etc.), and they selected at least four items from each of these categories to make up a 30-item assessment.

The items in the BCQ are a good example of what constitutes a heterogeneous domain, since they vary from physical indicators (e.g., hair loss has a modular component and can be objectively confirmed) to issues that are not directly observable (e.g., positive well-being). They are also scaled either in terms of duration (all the time to none of the time) or trouble (a great deal of trouble or no trouble). The questionnaire is also an example of a goal-directed assessment, since it was designed for a specific purpose: assessing the consequence of chemotherapy. The questionnaire would have been an example of pure cross-classification if the investigators had not intervened and decided to select four or more items from each of seven categories to make up the 30 items. The alternative would have been to select the first 30 items on the basis of the total score (the importance multiplied by the frequency of the item). Then the questionnaire would have represented what the respondents most often considered important consequences of being a breast cancer patient on chemotherapy.

An alternative, more direct approach of demonstrating cross-classification would involve asking the patients to sort the 99 items into what they would describe as separate categories, as well as asking them to rate the importance of each particular item. The categories formed could then be used for selecting the 30 items on the questionnaire, or the items selected for these categories could be inspected to determine if they form a hierarchy or reflect some specific use or function (i.e., they may be more script-like, to use the Ross and Murphy (1999) label). If both hierarchical and script categories were found, then this would provide direct evidence of cross-classification.

It would be quite interesting to see if the script categories formed were informative about how people connect issues, as would occur if a symptom and its consequences were placed in the same category. As I have previously stated, most quality-of-life assessments are limited to assessing the consequences of adverse events, rather than tracing the sensory and perceptual aspects of an experience to the reported consequences (i.e., the *qualia* of the experience). Yet, tracing this connection could be quite helpful in clarifying the relationship between a symptom and an HRQOL assessment. For example, does a symptom the pressure from limb edema feel, or a prosthesis feel, or a person's fingers feel following chemotherapy, and how does each of these contribute to a women's willingness to participate in social activities? There have been many studies that trace the presence or absence of a symptom, but there have not been many that trace the specific consequences of these experiences. In addition, these studies usually do not include importance ratings, so these studies remain examples of psychosocial descriptions, not qualitative judgments.

Ahmed et al. (2008) report what could be considered a typical study relating a symptom to an HRQOL assessment. In this cross-classification study, three groups of long-time survivors of breast cancer (104 women with self-reported lymphedema; 475 with arm symptoms but not lymphedema; and 708 women without lymphedema or arm complaints) were administered the SF-36. The diagnosis of lymphedema was established during a phone interview in which the respondents were asked a series of questions (Norman et al. 2001), including whether they or others noticed that their arms differed in size during the last 3 months. The primary analysis of the study involved comparing each group's response to each domain of the SF-36, although data was also collected on the percent of respondents who reported swelling, functional limitations, pain, discomfort, and other symptoms. A secondary analysis involved calculating the SF-36 domain scores as a function of number of reported symptoms. However, there was no reported effort to determine the extent to which sensory aspects of the symptom of lymphedema predicted SF-36 responses (i.e., there was no report of a regression analysis with lymphedema symptoms as the independent variable and responses to the SF-36 as the dependent variable).

One of the ways to deal with the heterogeneity of HRQOL domains is to disaggregate the assessment into separate symptom HRQOL domains. The PROMIS (Patient Reported Outcome Measurement Information System) project represents an effort to achieve this goal. As part of this project a large number of descriptive statements from a range of questionnaires were grouped together and subjected to a qualitative item review (QIR; DeWalt et al. 2007). The QIR identifies available items, sorts them, and then selects those items with similar meaning. This pool of items is narrowed by a review by an expert panel. This panel may revise the items if need be, and then they ask a focus group to confirm the domain definitions and identify issues not covered by the pool of items being considered. Finally, cognitive interviews focusing on the meaning of individual items are conducted, a final revision is produced and field tested. The QIR in this study focused on items dealing with emotional distress, fatigue, social functioning, physical functioning, and pain. The review process, for example, found 2,187 emotional distress items, of which, 299 were subject to cognitive interviewing with 224 items making the final item pool. In contrast, 644 pain items were identified, 191 were subject to cognitive interviewing, and 168 made the final item pool. Thus, the role of the respondents was more to confirm the selections made by the expert reviewers than it was to identify and select appropriate terms.

The design of the QIR overlaps with what has become standard for the development of items in the field of quality-of-life assessment. This usually involves "cognitive interviewing" (Willis et al. 2005) to clarify the meaning a person attributes to terms and items being developed, as well as the application of cognitive processing model (CPM; Tourangeau 1984). The CPM asks investigators to determine how well items on an assessment is comprehended, how well the respondent retrieves information necessary to respond to the item, and how well information is recalled to make a judgment. The QIR and the entire PROMIS project most closely resemble the intellectual history behind the development of the SF-36 (Stewart and Ware 1992) which also involved a collation and selection of published items followed by the psychometric characterization of the resultant assessment. It is clearly an investigator-dependent development process.

The emphasis by the PROMIS project on the expert for selecting and constructing the items was, I am sure, considered a practical solution for initially constructing the item banks. The intent was to take this relatively raw set of items and then apply advanced psychometric procedures (Item Response Theory) to construct, analyze, and refine the item banks. The item banks developed could then be used in computerized-adaptive testing or in fixed-length (standard) HRQOL assessments. However, the identification of symptom-specific item banks provides a unique opportunity to design HRQOL assessments with far greater precision than is currently available. One of the consequences of this is the potential to perform HRQOL studies with minimal overlap in content. Currently, there are a number of HRQOL studies where overlapping items in apparently different assessments have led to premature policy decisions (e.g., to limit the use of qualitative data in clinical decision making, drug selection, etc.).

A study by Anderson et al. (2007), which deals with patients about to undergo autologous stem-cell transplantation, illustrates the degree to which the content of assessments used in a study can overlap. The assessment protocol included the MD Anderson Symptom Inventory (Cleeland et al. 2000) adapted for bone marrow transplantation (MDASI-BMT) patients; a 30-item version of the Profile of Mood States (POMS; Mc Nair et al. 1992); the Functional Assessment of Cancer Therapy adapted for bone marrow transplantation (FACT-BMT; Mc Quellon et al. 1997); and the Eastern Cooperative Oncology Group Performance Scale (ECOG PS; Oken et al. 1982). The authors claim that the MDASI-BMT was meant to assess symptoms, while the FACT-BMT was included because "it was an HRQOL assessment." However, upon inspecting the items included in the FACT-BMT, I found that 12 of the 14 symptom items were also included on the MDASI-BMT. In addition, five of seven interference items on the MDASI-BMT could also be mapped to items in the FACT-BMT. The FACT-BMT, like the MDASI-BMT, also addresses issues dealing with the consequences of treatment (e.g., "I am concerned about keeping my job"), but the person is asked to indicate how often they are concerned about these issues, not how important their concerns are. The assessments also differ in that the FACT-BMT consists of 50 items, while the MDASI-BMT has only 21. The overlap between the two assessments, independent of the actual results of the study, suggests that the assumption is suspect that one assessment is measuring symptoms and the other measuring HRQOL.

A study by Goodwin et al. (2003) illustrates the policy or treatment implications of this type of content overlap. The question addressed in this study was whether an HRQOL assessment could contribute to breast cancer clinical decision making. First, Goodwin et al. (2003) defined an HRQOL assessment "as the self-report of the *impact of* breast cancer and its treatment on *some aspect of function* (e.g., physical, role, emotional or social)" (italics added for emphasis; Goodwin et al. 2003; p. 264). They go on to explicitly state that assessments of self-reported symptoms (e.g., pain and nausea) or reported toxicities will not be considered HRQOL measures.

This definition of HRQOL assessment is reminiscent of the causal quality-of-life or HRQOL definitions listed in Table 2.4. It includes the phrase "impact of," but as was true for the definitions in Table 2.4, this definition cannot be distinguished from assessments that include psychosocial indicators. Thus, not surprisingly, the authors (see Table 7.12) include the POMS (Mc Nair et al. 1992) and Hospital Anxiety and Depression Scale (HADS) (Zigmond and Snaith 1983) as HRQOL assessments. However, both assessments are presented as descriptive statements, and as such, lack the valuation component that is required for a unique "HRQOL" assessment.[14]

Although Goodwin et al. (2003) set as a condition for their review that symptom and toxicity assessments were not included, they themselves did not establish that the assessments they used did not, in fact, have symptom items. Thus, upon inspecting the HRQOL assessments they used, it is apparent that the assessments they were considering were not free from symptoms or indicators. Table 7.12 summarizes an analysis of these assessments. Items were classified into five categories[15]: descriptive statements only, descriptive statements that asked about symptoms, descriptive statements that spelled out the functional consequence of being in a particular state, global items that were domain specific, and global items that addressed a universal issue. Since there were very few items that were totally descriptive without any attribution to symptoms or functional consequences, I grouped them with the symptom descriptive statements (Column 3; Table 7.12). An example of an item that is a descriptive statement that spells out the consequence of being in particular state is item #28 on the EORTC-QLQ-30 assessment. This item asks, Has your physical condition or medical treatment caused you financial difficulties? In contrast, item #9 on the EORTC-QLQ-30 asks whether the respondent has pain, and asks them to rate it from not at all to very much over the last week. This item was classified as asking about a symptom.

Inspection of Table 7.12 reveals that none of the HRQOL assessments Goodwin et al. (2003) considered actually meet the criteria of including a homogenous collection of items that only assess either the global or particular functional consequences of cancer and its treatment. The Cancer rehabilitation evaluation system (CARES) comes closest to achieving the author's stated goal (49 of the 59 items appear to ask about functions). In contrast, 31 of the 39 items in the Rotterdam Symptom Checklist are symptoms. This assessment, therefore, deviated the most from the (de Haes et al. 1990) desired goal of considering symptom-free assessments. Thus, the authors appear to have selected a set of assessments that would make it difficult for them to achieve their objective, even though these assessments were considered state-of-the-art HRQOL assessments at the time of their study.

Of interest is determining how Goodwin et al. (2003) used the results of their review to interpret the contribution of HRQOL assessment to breast cancer clinical decision making. The first interpretative question they asked was whether an HRQOL assessment could provide information that could be used in making decisions concerning the primary management of breast cancer. They found that HRQOL assessments could be useful if the survival benefits of the treatments being compared were equivalent, but that this information was not of much use if one treatment made a significantly greater contribution to survival than another.

They also examined HRQOL studies where breast cancer patients received chemotherapy or hormone therapy in an adjuvant setting. These studies, according to Goodwin et al. (2003), failed to contribute to clinical decision making. A study published after the original Goodwin et al. (2003) review by de Haes et al. (2003) did provide such support. de Haes et al. (2003) used the Rotterdam Symptom Checklist as their primary assessment tool, but added items dealing with hot flashes, weight change, social activities, and coping with illness. De Haes et al. (2003) stated the following in their conclusion:

> Goserelin offers improved overall QoL during the first 6 months of therapy compared with CMF chemotherapy in premenopausal and perimenopausal patients with early breast cancer. Coupled with equivalent efficacy in estrogen receptor-positive patients, *these data support the use of goserelin as an alternative to CMF* in premenopausal and perimenopausal patients with estrogen receptor positive, node-positive early breast cancer. (italics added for emphasis; de Haes, et al. 2003; p. 4510)

This conclusion prompted an editorial by Goodwin (2003) where she acknowledged the implications of the de Haes et al. (2003) study, but pointed out that the data provided required that a patient had to make a complex set of trade-offs in order to make a particular decision. Goodwin (2003) suggests that this may be sufficiently complex for a patient such that various decision-making aids (e.g., decision programs described on CD-ROMs) may have to be developed to facilitate the decision process.

Goodwin et al. (2003) also reviewed the various HRQOL studies that were included in clinical trials where patients received chemotherapy or hormone therapy for metastatic breast cancer. According to the authors there was no evidence that HRQOL contributed to clinical decision making in this setting. The same was found for studies dealing with symptom control. But as might be expected, HRQOL assessments were helpful when the issue was deciding about the role of psychosocial treatment factors in clinical decisions.

The review of Goodwin et al. (2003) demonstrates that an HRQOL assessment can contribute to clinical decision making under specific circumstances, but they conclude by stating:

> Our findings in this review…suggest a need for caution in initiating new HRQOL studies in breast cancer using existing general or cancer-specific HRQOL instruments unless treatment equivalency is an expectation…or unless the quality-of-life question targets unique or specific questions that can only be assessed through patient self-report. (p. 279)

Based on their review, Goodwin et al. (2003) see HRQOL assessments as having a limited role in clinical decision making for the breast cancer patient.

It is interesting, however, to examine the logic Goodwin et al. (2003) used to judge these studies, For example, they reviewed a study by Coates et al. (1987), who randomly treated metastatic breast cancer patients with either continuous or intermittent chemotherapy and assessed their HRQOL. Coates et al. (1987) found that continuous, as compared to intermittent, chemotherapy delayed progression of the disease, and concurrently these patients reported improved HRQOL. Goodwin et al. (2003) interpreted these data as demonstrating that HRQOL did not add information to the results of the study, although Coates et al. (1987) state that the HRQOL results should be interpreted by clinicians as supporting their decisions of keeping patients on a continuous chemotherapy program. Thus, what Goodwin et al. (2003) appear to be doing is setting up a situation where extending survival acts as a default decision, preempting any other decisions.

Goodwin et al. (2003) used this logic throughout their paper. For example, they state "in the adjuvant setting, where the goal of treatment is to delay recurrence and/or prolong survival, information obtained using HRQOL measures has had little impact on clinical decision making" (Goodwin et al. 2003; p. 275). If extending survival acts as a default decision precluding other decisions, then evaluating the role of HRQOL in any setting other than treatment equivalence appears moot. Viewed from this perspective, the only legitimate arena to evaluate the contribution HRQOL makes to clinical decision making is when treatment options are equivalent, and under these conditions Goodwin et al. (2003) report a positive contribution from HRQOL data.

But what if patients, not clinicians, were asked to indicate their preference for extending their survival vs. increasing or maintaining their HRQOL? Lenert et al. (2005) report a study in which they asked persons with HIV infections to engage in time trade-offs relative to the length and/or quality of their lives. They found that African-Americans were consistently risk-averse, choosing survival over the quality of their lives, whereas Caucasians were less so. This suggests that assuming that extending survival as the obvious goal of any clinical intervention may be true in some circumstances, but clearly not in others, and that finding the balance between these alternatives remains an empirical question.

One concern with the Goodwin et al. (2003) conclusions is that the studies they considered might not have been optimally designed and implemented. In this regard, Efficace et al. (2003) report a study that first identified and then applied a set of 11 specific criteria to evaluate the design of a group of prostate cancer clinical trials which included the assessment of HRQOL. When these criteria were applied, they found that only eight of the 24 clinical trials they identified had robust experimental designs. Of these eight clinical trials, five included the EORTC QLQ-C30, one used the FLIC, one the FACT-G, and one a SWOG questionnaire. All the trials also included either additional clinically relevant items or questionnaires (e.g., the McGill Pain Questionnaire). Thus, the same concerns I expressed about the Goodwin et al. (2003) study being confounded by overlapping items, or assessments that do not capture the elements of a hybrid construct (e.g., spelling out the consequences of an adverse experience or not including a value statement), were also relevant to the Efficace et al. (2003) review. Overall, only 6 (25%) of 24 randomized trials demonstrated that HRQOL made a contribution to clinical decision making, as compared to 5 (11%) of 46 in the Goodwin et al. (2003) study.

Blazeby et al. (2006) reviewed surgical oncology clinical trials and selected robustly designed studies on the basis of the criteria spelled out by Efficace et al. (2003). They found that HRQOL data made a significant contribution to clinical decision making in 22 (67%) of the 33 randomized trials examined. Many of the studies reviewed by Blazeby et al. (2006) dealt with patients who had more advanced disease. More recently, a study by Efficace et al. (2007a) compared how well HRQOL was assessed in randomized clinical trials performed before and after the year 2000. They found that 39.3% (89/159) of the randomized clinical trials performed before 2000 were judged to robustly assess HRQOL, as compared to 64.3% (70/159) after 2000. This suggested that a learning curve exists and that with time the precision of an HRQOL assessment will increase and be more likely to contribute to clinical decision making. Comparing the Blazeby et al. (2006) study with the Goodwin et al. (2003) and Efficace et al. (2003, 2007b) studies suggests that the potential contribution of an HRQOL assessment to clinical decision making may vary as a function of the procedure or treatment being evaluated, and that investigators are getting increasingly more competent in usefully integrating HRQOL into clinical trials. Of course, many of the questions addressed in this literature should be reevaluated once symptom item banks are developed by the PROMIS project; an effort that should increase the quality of the symptom HRQOL assessments.

Clinical decision making, however, is just one context where an HRQOL assessment can play a role. Barofsky and Sugerbaker (1990) summarized the history and outcome of a clinical trial dealing with treatment of extremity soft-tissue sarcoma. Their initial study demonstrated that patients report certain unexpected dysfunctions that could be attributed to the adverse consequences of treatments. Over a 10-year period, this promoted a series of changes in surgical, chemotherapeutic, and radiation treatment that improved the functional outcome of the medical and surgical treatment. This study demonstrated that an HRQOL assessment can play a role as a quality control agent, and that this role does not require that a choice be made between demonstrating extending survival and reducing morbidity, as was the case in the Goodwin et al. (2003) paper. What it does require is that the investigators acknowledge that the qualitative consequences of treatment are an important indicator that justifies appropriate effort to ensure a patient has minimal or no adverse experience.

The Efficace's et al. (2007a) editorial reviewing the role that qualitative studies play in the assessment of hematologic malignancies illustrates the circumstances where a qualitative assessment could have been helpful. They comment that at first treatments for hematological malignancies were minimally capable of prolonging survival, but created a noticeable adverse qualitative impact. Only now that more effective treatments have become available, has the potential role of an HRQOL assessment come into play in the dynamics that are involved in developing new therapies. The question was whether this was appropriate, and should HRQOL have been assessed throughout the history of treatment development. I suggest that if prevention and quality control had been the objective of a qualitative assessment, then it would have been appropriate to perform these assessments to ensure the ethical character of the clinical trial.

The debate about what role symptoms play in an HRQOL assessment continues. For example, Ferrans (2007) reported an analysis of the presence of symptoms in HRQOL assessments that confirms my concerns, and she recommends that greater effort be made to clarify the consequences of this particular characteristic of HRQOL assessments. So far I have argued that a symptom itself can be a qualitative indicator, if the importance to the person is assessed. Thus, rating the important of being in a particular "pain state" or "level of depression" are qualitative statements. They become quality-of-life statements when the consequence of being in pain or depressed is also spelled out. Symptoms, therefore, can be useful components of an HRQOL if their consequences are traced, but if the assessment just focuses on establishing the presence of the symptom, then its contribution to an HRQOL is limited.

3.3 The Role of HRQOL in Symptom Management

One of the potential applications of an HRQOL assessment is to use it to monitor the consequences of particular symptoms. Thus, it is not uncommon to find an HRQOL assessment included in a clinical trial that contains an evaluation of the consequences of the symptoms associated with a treatment, such as nausea and vomiting that can follow chemotherapy. Yet, some investigators are not convinced that an HRQOL assessment adds much that cannot be determined by direct assessment of symptoms and their consequences. In Table 7.1, I listed several examples of this argument, including a study by Hollenberg et al. (2000). They reported a study where they found that the distress associated with physical symptoms (e.g., slow heart beat, sexual dysfunction, nocturia, light-headedness) was directly related to changes in HRQOL, and on the basis of this observation concluded that it was not necessary to measure HRQOL. However, inspection of their paper reveals a rather limited conceptualization of HRQOL, since the assessments they included focused on a mental health index, a measure of anxiety, loss of control, depression, and general positive affect. They did include a baseline global measure of quality-of-life (the *Cantril's Ladder* (1965), an 11-point scale that ranges from the worst to the best quality-of-life), but only to establish the comparability of the treatment groups. They end their paper by stating that "The distress associated with specific symptoms escapes the ambiguity that is implicit in the assessment of QOL" (Hollenberg et al. 2000; p. 1483).

When Hollenberg et al. (2000) refer to the "ambiguity" of HRQOL, they more than likely have in mind a particular conceptualization of HRQOL, which others (e.g., Campbell 1972) have referred to as "ethereal and abstract." Yet, as I suggested in Table 1.2, quality-of-life or HRQOL can also be conceptualized quite concretely, as implied by a prevention-oriented model of quality assessment. Conceptualizing HRQOL in terms of enhancement (Table 1.2) may very well approximate what Hollenberg et al. (2000) were concerned about, since specifying what is being enhanced (e.g., well-being) may be very difficult, and thus, the HRQOL indicator will remain ambiguous. However, assessing more of the physical and social consequences of their treatment regimen would have given a more complete picture of the impact of their intervention. Distress, of course, can be a quality indicator, but only if the respondents had the opportunity to rate the importance of being distressed. Then distress would have the same formal cognitive structure as any other of the standard HRQOL assessments (e.g., the HUI, the QWB, etc.).

Cleeland (2007) makes another argument. "Symptom reports represent a subset of HRQOL. In most conceptualizations of HRQOL, symptoms are viewed as the patient report most proximal to the disease process, and are thus potentially causative of variation in the more abstract HRQOL components such as well-being, perception of daily functioning, global impressions of the impact of treatment on daily life, satisfaction with treatment, and perception of overall health status" (p. 16). This statement, as Cleeland (2007) acknowledges, is an application of the Wilson and Cleary model (1995), now limited to the relationship between symptoms and HRQOL. Cleeland also claims that symptoms have to be qualified (i.e., their semantics expanded to refer to symptom severity), and because symptoms cluster, it is most appropriate to refer to the qualitative outcome as a symptom burden. Assessing symptom burden for Cleeland (2007) not only includes assessing symptom severity, but also the interference that symptoms create in living. However, by characterizing symptoms in these terms, he implies that it is not necessary to assess the more abstract construct of HRQOL, even though by rating the importance of a symptom burden he would be creating as much of a qualitative assessment as would be found using a standard HRQOL assessment.

In contrast, Osoba (2007; Table 7.1) argues for a direct role for an HRQOL assessment in the symptom management of clinical trials. He points out that various qualitative indicators have to be assessed so that the adverse impact of symptoms is monitored and minimized. To achieve this, he also sees value in adapting the Wilson and Cleary model (1995), but by suggesting that the relationship between indicators should be bidirectional, and therefore more dynamic. To support his argument, Osoba calculated the correlations between each specific domain (e.g., physical functioning and social functioning), and also each of these specific domains and a single-item global quality-of-life assessment of the EORTC-QLQ C30. He found moderately strong correlations following this analysis. The correlation between the physical, social, and global domains suggests the same progressive abstraction postulated by Wilson and Cleary (1995). In addition, Osoba (2007) argues that symptoms (e.g., pain) may function both as indicators and causes, depending on the circumstances of the study. However, he does not address the issue of how a relatively abstract construct such as HRQOL is related to the more concrete indicators that he is dealing with, nor does he recognize that what may have transpired was that the respondent used concrete examples of the abstract construct, and that this is reflected in the observed correlations (see above). This could be demonstrated, of course, by cognitive interviewing that would provide some insight into what the respondents were thinking when they assessed their global quality-of-life. Osoba (2007) also points out that the temporal relationship between indicators may be so rapid that it may not be possible to distinguish between a cause and an effect. He illustrates the usefulness of an HRQOL assessment by demonstrating that an HRQOL assessment can detect outcomes not otherwise expected, and thus provide a more comprehensive view of the impact of the intervention (e.g., chemotherapy) during the symptom management of a clinical trial.

HRQOL assessments are also useful in identifying the role that symptoms play in any causal analysis. In Chap. 6 (Fig. 6.3), I pointed out that modular HRQOL domains (e.g., pain, physical functioning) were more likely to function in a causative role than nonmodular domains (e.g., depression). In this chapter, I discussed the wide range of modular domains that exists (e.g., acquired and functional modularity), and also discussed how they function in a qualitative assessment.

My final argument supporting the role of an HRQOL assessment in a symptom management clinical trial involves demonstrating (Table 7.1) that while the content, questionnaire format, and time frame may differ, the same cognitive processes can be evoked when these indicators are part of a qualitative assessment. The similarity in the hybrid construct for each would be a demonstration of cognitive equivalence, so I can argue that the two types of assessments are formally the same. Unfortunately, most symptom assessments are limited to evaluation of descriptors so that they do not represent a qualitative assessment, and therefore can justifiably be considered different from an HRQOL assessment. They are also more accurately called psychosocial assessments. Still, a number of standard HRQOL assessments (e.g., HUI, QWB), exist which provide importance, preference, or utility estimates for both symptoms and HRQOL outcomes. This reinforces the view that the apparent distinction between a symptom and an HRQOL outcome is more an artifact of how these indicators were assessed than some inherent characteristic of the indicator.

4   Summary of Chapter

As I discussed in this chapter, the content of a domain, symptom or otherwise, is a product of the interaction between the cognitive processes ongoing in an investigator and the statistical procedures the investigator uses to identify the entities to include in the domain. Previously I identified three types of categories: natural kinds, artificial classes, and artifactual classes (Chap. 6, p. 166). The statistical procedures available to an investigator to identify the entities to include in a domain range from arbitrary selecting them, to using factor analysis, item response theory, or grounded field theory. I also pointed out that drawing abstract inferences from concrete entities requires the presence of a particular defined structure in the domain. This defined structure can be a cognitive construction or reflect statistical modeling. For example, if a group of entities are organized into a Guttman scale, then the cognitive structure of inclusion allows the concrete entities to be mapped to a hierarchy that infers an abstract concept. Alternatively, if various latent variable models (e.g., factor analysis, path analysis, and structural equation modeling) are used, then probability statements can be made about the relationship between the concrete entities and an abstract concept. However, by not demonstrating that this inference has cognitive underpinnings the investigator is left with making metaphoric statements, inferring the presence of a relationship that has a probability of occurring but no absolute proof. What is required here, therefore, is that there be a demonstrated cognitive and psychometric (statistical) equivalence in the structural determinants of a domain, in order to ensure the validity of any abstract inference. In its absence, abstract inferences are at risk of remaining metaphoric statements.

Inspection of the HRQOL literature makes it clear that investigators regularly speak of symptoms (e.g., fatigue, pain, depression) as if they were HRQOL indicators (e.g., Cella et al. 2002). It is also common to find that an HRQOL assessment battery includes a separate symptom (e.g., depression) assessment. Table 7.12 certainly suggests that many descriptive HRQOL assessments include specific symptom items within the assessment. When asked to list issues they want included in a quality-of-life assessment, patients also mention symptoms, especially if the respondent has been diagnosed and is being treated for a disease (e.g., Levine et al. 1988). Notwithstanding, it can still be asked whether this confluence is appropriate. If not, what would be a more appropriate way of conceptualizing the relationship between symptoms and HRQOL indicators?

I provided several answers to this question. For example, I argued that it is possible to directly assess the quality of a symptom without it referring to HRQOL. Pain, for example, can be described as flickering, hot, shooting, dull, and so on (Table 5.1); all descriptors which provide a sensory quality assessment. It is even possible to place a value, preference, or utility on this type of descriptor profile. The same would be expected to be true for other symptoms and other categories of indicators. Thus, the quality of a car, health care, working life, and so on, can be described, evaluated, and valued, much as can occur with an HRQOL indicator.

In order for a symptom to be integrated into an HRQOL assessment, the consequences of the symptom have to be spelled out, and all of this information has to be qualified (Chap. 5, p. 128). As previously described, qualifying a symptom involves expanding the meaning associated with the term symptom by explicitly including an emotive component. Thus, instead of describing a symptom in terms of its intensity, frequency, and duration, it would be described in terms of its severity, distress-provoking ability, or burdensomeness. In addition, the consequence of a symptom on different aspects of a person's life has to be included in the assessment.

My discussion of the relationship between fatigue, depression, and HRQOL provided me with the opportunity to discuss how concrete symptoms could be mapped to an abstract construct such as HRQOL. I also illustrated how taking into account the modularity of these symptoms (e.g., motivation aspects of fatigue as opposed to physical aspects of fatigue) leads to the formation of different types of domains. Thus, there was sufficient data in the literature (Table 7.8) to indicate that nonmodular aspects of fatigue are in a hierarchical relationship to depression, and in this manner related to HRQOL, while modular aspects of fatigue lead to the formation of a domain based on cross-classification. Finally, this discussion should reinforce that an investigator should attend to how the elements of a domain are cognitively organized, and that the nature of this organization can provide empirical support for inferring that the label given in the domain is justified.

Pain, as a prototypical symptom, promoted me to discuss a number of important aspects of the process whereby a self-report is transformed into a quantitative indicator. The discussion of the language of pain illustrated the variety of ways that a person expresses their pain experience. The assessment of pain has developed into a complex field of study, with some assessments applying psychophysical methods (e.g., signal detection theory), while others use questionnaires filled with sensory verbal descriptors or items dealing with the consequence of pain (e.g., chronic pain). The rich linguistic base these assessments reveal suggests they may be useful when classifying the different pain experiences. Finally, I addressed the issue of how pain is to be assessed in the compromised person. I focused on the elderly person whose language capacity is degenerating over time. While I offered no unique solution to this complex problem, I did suggest that the task for the investigator was to develop methods of cross-modality communication. The extent to which this is possible remains an empirical question.

Notes

1. I once had a patient who complained about scapula pain. This pain could be evoked when the patient recalled the sexual abuse her daughter experienced. Having noted the association, in time, the patient's pain resolved. By claiming that there was no physical cause, I do not mean that there is no physical analog (e.g., memory, emotion) of the event, just that the cause was not neuropathic.
2. *Up-regulation* refers to process where the threshold to activate a spinal cord neuron is permanently reduced. Thus, a stimulus which would not have been sufficiently intense to activate a neural response now can.
3. Taxonomic categories are similar to what has been referred to as hierarchically organized categories containing homogenous content, while script categories contain heterogeneous content. Ross and Murphy (1999) suggest that if an item is part of both a taxonomic and script category, then *cross-classification* has occurred.
4. In a personal communication, the authors of this paper report that they used a step-wise regression analysis, ultimately including each of the five fatigue components as well as CES-D mood depression measure.
5. There are reports of persons who experience cross-modality sensations (*synaesthesia*; Pearce 2007), which would be considered an exception to the statement I just made, but as far as I have read, no one has reported pain as one of the cross-modality sensations. In addition, it is possible for pain to have a physical analog. Thus, the cold pressor test and ischemic induced pain can be quantified (e.g., Hilgard 1969). What is being quantified here, however, is the response of temperature receptors or pressure receptors, with pain being reported when the stimulation of these receptors reaches an extreme.
6. Much of what I am writing about reflects Lascaratou's (2007) thoughts and examples. She has collected a vast number of interviews between doctors and Greek-speaking patients, and the examples I give are her English translations of these data. Obviously, persons from different cultures may use language to express their experiences with pain differently.
7. Loeser and Treede (2008) recently reaffirmed the appropriateness of this definition.
8. People with leprosy lose the ability to detect pain as their disease progresses. These persons suffer the consequences of this failure by experiencing excessive self-injury, demonstrating the protective and possible survival advantage that is provided by being able to detect a painful event.
9. A reader might ask whether fatigue has the same range of qualitative states as pain apparently does. My answer is that while fatigue can be sensed, the source of the sensation is not an external stimulus (e.g., a surface temperature or the force of an object hitting a person). Thus, fatigue has a more limited number of qualitative states to consider. In contrast, other sensory modalities such as light, sound, and temperature have an external sensory source, but whether these experiences are attended to as part of a person's qualitative state is less clear.
10. Melzack's (1989) theory is an example of an effort to generate a single explanation of a variety of phenomena, and in this sense can be used to classify different pain experiences. Omoigui (2007a, b) offers an alternative approach based on the inflammatory profile. He accounts for a particular pain experience, by determining if one or more of the following inflammatory members of a response profile occur; the inhibition or suppression of particular inhibitory mechanisms, inhibition or suppression of neural affective and motor expression, or the modulation of neural transmission due to treatment interventions.
11. A number of studies (e.g., Hulsman et al. 2009; Oaklander et al. 2006) support the assumption that a physical basis exists that causes the CRPS I, although this literature is confounded by the problem of distinguishing between disuse atrophy and neurogenesis.
12. My presentation here assumes that the reader is familiar with Fodor's criteria for modularity (Chap. 6, p. 177).
13. A paper by Reading and Newton (1978) also reported using a sorting technique of the descriptive adjectives that Melzack used.
14. It should be pointed out that the HADS and the POMS can also be *qualified*, as is true for any assessment.
15. I rated the items in each assessment. Reliability of these judgments was established by repeating these ratings approximately a month later. Table 7.12, therefore, should be viewed as a first approximation. The reader may also want to inspect Table 8.3 in an article by Ferrell and Grant (1998) where these authors also do a count of the number of symptom items of several well-known HRQOL assessments. They also found that symptoms are present in assessments that claim to be HRQOL assessments.

References

Aaronson NK, Ahmedzai S, Bullinger M, Crabeels D, et al. (1991). The EORTC core quality-of-life questionnaire: Interim results of an international field study. In, (Ed.) D Osoba, *The Effect of Cancer on Quality of Life*. Boca Raton, FL: CRC Press. (p. 185–203).

Ahmed RI, DeAnn Lazovich A, Schmitz KH, Folsom AR. (2008). Lymphedema and quality of life in breast cancer survivors: The Iowa Women's Health Study. *J Clin Oncol*. 26, 5689–5696.

Anderson KO, Giralt SA, Mendoza TR, Brown JO, et al. (2007). Symptom burden in patients undergoing autologous stem cell transplantation. *Bone Marrow Transplant*. 39, 759–766.

APA. *Diagnostic and Statistical Manual of Mental Disorders: Fourth Edition*, 1994. Washington: APA

Arendt-Nielsen L, Svensson P. (2001). Referred muscle pain: Basic clinical findings. *Clin J Pain*. 17, 11–19.

Barofsky I and Sugarbaker PH. (1990). The cancer patient. In, (Ed.) B Spilker *Quality of Life Assessment in Clinical Studies*. New York, NY: Raven. (p. 419–439).

Barsalou LW. (1983). Ad hoc categories. *Mem Cogn*. 11, 211–227.

Beck AT, Ward C, Mendelson M. (1961). Beck Depression Inventory (BDI). *Arch Gen Psychiatr*. 4, 561–571.

Becker S, Kleinböhl D, Klossika I, Hölzl R. (2008). Operant conditioning of enhanced pain sensitivity by heat-pain titration. *Pain*. 140, 104–114.

Bell IR, Baldwin CM, Russek LG, Schwartz GE, et al. (1998). Early life stress, negative paternal relationships and chemical intolerance in middle-aged women: support for a neural sensitization model. *J Women's Health*. 7, 1135–1147.

Bell C. (1811). An Idea of a New Anatomy of the Brain; *submitted for the observations of this friends*. London: privately printed pamphlet.

Belza BL, Henke CJ, Yelin EH, Epstein WV, et al. (1993). Correlates of fatigue in older adults with rheumatoid arthritis. *Nurs Res.* 42, 93–99.

Benedetti F, Vighetti S, Ricco C, Lagna E, et al. (1999). Pain threshold and tolerance in Alzheimer's disease. *Pain.* 80, 377–383.

Bjoro K, Herr K. (2008). Assessment of pain in the nonverbal or cognitively impaired older adult. *Clin Geriat Med.* 24, 237–262.

Blazeby JM, Avery K, Sprangers M, Pikhart H, et al. (2006). Health-related quality of life measurement in randomized clinical trials in surgical oncology. *J Clin Oncol.* 24, 31768–3186.

Brady MJ, Cella DF, Mo F, Bonomi AE, et al. (1997). Reliability and validity of the Functional Assessment of Cancer Therapy- Breast quality of life instrument. *J Clin Oncol.* 15, 974–986.

Breiman L, Friedman JH, Olshen RA, Stone CJ. (1984). *Classification and Regression Trees.* New York, NY: Chapman & Hall (Wadsworth).

Bruera E, Kuehn N, Miller MJ, Selmser P, et al. (1991). The Edmonton Symptom Assessment Scale. *Cancer* 88, 2164–2171.

Bruera E, Macmillian K, Hanson J, MacDonald RN. (1989). The Edmonton staging system for cancer pain: Preliminary report. *Pain.* 37, 203–209.

Brunier G, Graydon J. (1996). A comparison of two methods of measuring fatigue in patients on chronic haemodialysis: Visual analogue vs Likert scale. *Int J Nurs Stud.* 33, 338–348.

Cain MJ. (2002). *Fodor: Language, Mind and Philosophy*. Cambridge, UK: Polity.

Campbell, A. (1972). Aspiration, satisfaction and fulfillment. In, (Eds.) A Campbell, P Converse. *The Human Meaning of Social Change.* New York, NY: Russell Sage. (p. 441–446).

Cantril H. (1965). *The Pattern of Human Concerns.* New Brunswick NJ: Rutgers University Press.

Cella D. (1997). The Functional Assessment of Cancer Therapy-Anemia (FACT-An) Scale: A new tool for the assessment of outcomes in cancer anemia and fatigue. *Semin Hematol.* 34:13–19

Cella D, Chang C-H, Lai J-S, Webster K. (2002). Advances in quality of life measurement in oncology patients. *Semin Oncol.* 29, 60–68.

Cella D, Peterman A, Passik S, Jacobsen P, et al. (1998). Progress towards guidelines for the management of fatigue. *Oncology.* 12, 369–377.

Chalder T, Berelowitz G, Pawlikowska T, Watts L, et al. (1993). Development of a fatigue scale. *J Psychosom Res.* 37, 147–153.

Chang VT, Hwang SS, Feuerman M, Kasimis BS, et al. (2000). The Memorial Symptom Assessment scale-short form (MSAS-SF). *Cancer.* 89, 1162–1171.

Chapman CR. (2008). Progress in pain assessment: The cognitively compromised patient. *Curr Opin Anaesthesiol.* 21, 610–615.

Chochinov HM, Wilson KG, Enns M, Lander S. (1997). "Are you depressed?" Screening for depression in the terminally ill. *Am J Psychiatr.* 154, 674–676.

Chow E, Fan G, Hadi S, Filipczak L. (2007). Symptom clusters in cancer patients with bone metastasises. *Support Care Cancer.* 15, 1035–1043.

Christodoulou C. (2005). The assessment and measurement of fatigue. In, (Ed.) J DeLuce, *Fatigue as a Window on the Brain.* Cambridge MA: MIT Press.

Clark A. (2005). Painfulness is not quale. In, (Ed.) M Aydede, *Pain: New Essays on its Nature and the Methodology of its Study.* Cambridge MA: The MIT Press.

Clark WC. (1984). Applications of multidimensional scaling to problems in experimental and clinical pain: An introduction. In, (Ed.) B Bromm, *New Approaches to Pain Measurement in Man.* Amsterdam The Netherlands: Elsevier. (p. 349–369).

Clark WC, Ferrer-Brechner T, Janal MN, Carroll JD, et al. (1989a). The dimensions of pain: A multidimensional scaling comparison of cancer patients and healthy volunteers. *Pain.* 37, 23–32.

Clark WC, Fletcher JD, Janal MN, Carroll JD. (1995). Hierarchical clustering of 270 pain/emotion descriptors: Towards a revision of the McGill Pain Questionnaire. In, (Eds.) B Bromm, J Desmedt. *Pain and the Brain: From Nociception to Sensation.* New York NY: Raven Press. (p. 319–330).

Clark WC, Janal MN, Carroll JD. (1989b). Multidimensional pain requires multidimensional scaling. In, (Eds.) JD Loeser, CR Chapman. *Issues in Pain Measurement.* New York NY: Raven Press. (p. 285–325).

Clark WC, Kuhl JP, Keohan ML, Knotkova H, et al. (2003). Factor analysis validates the cluster structure of the dendrogram underlying the Multidimensional Affect and Pain Survey (MAPS) and challenges the a prior classification of the descriptors in the McGill Pain Questionnaire (MPQ). *Pain.* 106, 357–363.

Clark WC, Yang JC, Tsui S-L, Kwok-Fu N, et al. (2002). Unidimensional pain rating scales: A multidimensional and pain survey (MAPS) analysis of what they really measure. *Pain.* 98, 241–247.

Cleeland CS. (2007). Symptom burden: multiple symptoms and their impact as patient-reported outcomes: Review. *J Natl Cancer Inst Monogr.* 37, 16–21.

Cleeland CS, Mendoza TR, Wang XS, Chou C, et al. (2000). Assessing symptom distress in cancer patients: *The M.D. Anderson Symptom Inventory. Cancer.* 89, 1634–1646.

Coates A, Gebski V, Bishop JF, Jeal PN et al. (1987). Improving the quality of life during chemotherapy for advanced breast cancer. A comparison of intermittent and continuous treatment strategies. *NEJM.* 317, 1490–1495.

Cole LJ, Farrell MJ, Duff EP, Barber JB, et al. (2006). Pain sensitivity and fMDI pain-related brain activity in Alzheimer's disease. *Brain.* 129, 2957–2965.

Cronje RJ, Williamson OD. (2006). Is pain every normal ? *Clin J Pain.* 22, 692–699.

Dallenback KM. (1939). Somesthesis. In, (Eds.) EG Boring, HS Langfeld, HP Weld. *Induction to Psychology* New York NY: Wiley. (p. 608–625).

Davis PJ, Reeves JL, Graff-Radford SB, Hastle BA, et al. (2003). Multidimensional subgroups in migraine: Differential treatment outcome to a pain medicine program. *Pain Med.* 4, 215–222.

Daut RL, Cleeland CS, Flanery RC. (1983). Development of the Wisconsin Brief Pain Questionnaire to assess pain in cancer and other diseases. *Pain.* 17, 197–210.

Descartes R. (1664). *Traite de l"home.* Paris France: C Angot.

De Haes JCJM, Olschwski M, Kaufmann M, Schumacher M, et al. (2003). Quality of life in goserelin-treated versus cyclophosphamide + methotrexate + fluorouracil -treated premenopausal and perimenopausal patients with node-positive, early breast cancer: The Zoladex Early Breast Cancer Resarch Association Trialists Group. *J Clin Oncol.* 24, 4510–4516.

De Haes JCJN, Pruyn JF, van Knippenberg FC. (1983). Klachten lijst voor kankerpatienteen, eerste ervaringen. *Ned Tijdschr Psychologica.* 38, 403–422.

De Haes JCJM, van Knippenberg FCE, Neijt JP. (1990). Measuring psychological and physical distress in cancer patients: Structure and application of the Rotterdam Symptom Checklist. *Br J Cancer.* 62, 1034–1038.

DeLuca J, Johnson SK, Beldowics D, Natelson BH. (1995). Neuropsychological impairments in chronic fatigue syndrome, multiple sclerosis and depression. *J Neurol Neurosurg Psychiatr.* 58, 38–42.

DeVries J, Michielsen HJ, Van Heck GL. (2003). Assessment of fatigue among working people: a comparison of six questionnaires. *Occup Environ Med.* 60 Suppl 1, 10–15.

DeWalt DA, Rothrock N, Yount S, Stone AA. (2007). Evaluation of item candidates: The PROMIS Qualitative Item Review. *Med Care.* 45 Suppl 1, S12–S21.

Dittner AJ, Wessely SC, Brown RG. (2004). The assessment of fatigue: A practical guide for clinicians and researchers. *J Psychosom Res.* 56, 157–170.

Dorn SD, Palsson OS, Thiwan SIM, Kanazawa M, et al. (2007). Increased colonic pain sensitivity in irritable bowel syndrome is the result of an increased tendency to report pain rather than increased neurosecretory sensitivity. *Gut.* 56, 1202–1209.

Drossman DA. (2006). *Rome III Process for Functional Gastrointestinal Disorders.* McLean VA: Degnon Associates.

Drossman DA, Toner BB, Whitehead WE, Diamant NE, et al. (2003). Cognitive-behavioral therapy versus education and desipramine versus placebo for moderate to severe functional bowel disorders *Gastroenterol.* 125, 19–31.

Efficace F, Bottomley A, Osoba D, Gotay C, et al. (2003). Beyond the development of quality of life (HRQOL) measures: A checklist for evaluating HRQOL outcomes in cancer clinical trials- Does HRQOL evaluation in prostate cancer research inform clinical decision making? *J Clin Oncol.* 21, 3502–3511.

Efficace F, Novik A, Vignetti M, Mandelli F, et al (2007a). Health-related quality of life and symptom assessment in clinical research of patients with hematologic malignancies: where are we now and where do we go from here? *Haematology.* 92, 1598–1598.

Efficace F, Osoba D, Gotay C, Sprangers M, et al. (2007b). Has the quality of health-related quality of life reporting in cancer clinical trials improved over time? Towards bridging the gap with clinical decision making. *Ann Oncol.* 18, 775–781.

Eisenberger NI, Lieberman MD. (2004). Why rejection hurts: A common neural alarm system for physically and social pain. *Trends Cogn Sci.* 8, 294–300.

Eisenberger NI, Lieberman MD, Williams KD. (2003). Does rejection hurt? Am fMRI study of social exclusion. *Science.* 302, 290–292.

Embretson SE, Reise SP. (2000). *Item Response Theory for Psychologists.* Mahwah NJ: Lawrence Erlbaum.

Fabrega H, Tyma S. (1976). Culture, language and the shaping of illness: An illustration based on pain. *J Psychosom Res.* 20, 323–337.

Fairbank JCT, Couper J, Davies JB, O'Bryan JP. (1980). The Oswestry Low Back Pain Disability Questionnaire. *Physiotherapy.* 66, 271–272.

Fan G, Hadi S, Chow E. (2007). Symptom clusters in patients with advanced stage cancer referred for palliative radiation therapy in an outpatient setting. *Support Cancer Care.* 4, 157–162.

Faresjö A, Grodzinsky E, Johansson S, Wallander M-A, et al. (2007). Psychosocial factors at work and in every day life are associated with irritable bowel syndrome. *Eur J Epidemiol.* 22, 4734–80.

Feinstein AR. (1987a). Clinimetric perspective. *J Chronic Dis.* 40, 635–640.

Feinstein AR. (1987b). *Clinimetrics.* New Haven CN: Yale.

Ferrans CE. (2007). Differences in what Quality-of-Life instruments measure. *J Natl Cancer Inst Monogr.* 37, 1–9.

Ferreira KASL, Kimura M, Teixeira MJ, Mendoza TR, et al. (2008). Impact of cancer-related symptoms synergisms on health-related quality-of-life and performance status. *J Pain Symptom Manag.* 35, 604–616.

Ferrell BR, Grant MM. (1998).Quality-of-life and symptoms. In, (Eds.) CR King, PS Hinds. *Quality-of-life: From Nursing and Patient Perspectives. Theory, Research, Practice.* Sudbury MA: Jones and Barlett. (p. 140–156).

Fodor J. (1983). *Modularity of the Mind.* Cambridge MA, MIT Press.

Fordyce WE. (1982). A behavioral perspective on chronic pain. *Br J Clin Psychol.* 21,313–320.

Flechtner H, Bottomley A. (2003). Fatigue and quality-of-life: Lesions from the real world. *Oncologist.* 8, 5–9.

Frank AW. (2001). Can we research suffering? *Qual Health Res.* 11, 353–362.

Galter BS, Henderson J, Perander J, Jensen M. (2000). Course of symptoms and quality-of-life measurement in complex regional pain syndrome: A pilot survey. *J Pain Symptom Manag.* 20, 286–292.

Galer BS, Jensen MP. (1997). Development and preliminary validation of a pain measure specific to neuropathic pain: The Neuropathic Pain Scale. *Neurology.* 48, 332–228.

Ganz PA, Schag CA, Lee JJ, Sim MS. (1992). The CARES: a generic measure of health-related quality-of-life for patients with cancer. *Qual Life Res.* 1, 19–29.

Gärdenfors P. (2007). Representing actions and functional properties in conceptual spaces. In, (Eds.) T Ziemke, J Zlatev, RM Frank. *Body, Language and Mind: Volume 1 Embodiment.* Berlin Germany: Mouton de Gruyter. (p. 167–195).

Gaston-Johansson F, Fall-Dickson JM, Bakos AB, Kennedy MJ. (1999). Fatigue, pain, and depression in pre-auto transplant breast cancer patients. *Cancer Practice.* 7, 240–247.

Geinitz H, Zimmerman FB, Thamm R, Keller M, et al. (2004). Fatigue in patients with adjuvant radiation therapy for breast cancer : long-term follow-up. *J Cancer Res Clin Oncol.* 130, 327–333.

Gibbs GF, Drummond PD, Finch PM, Phillips JK. (2008). Unraveling the pathophysiology of complex regional pain syndrome: Focus on sympathetically maintained pain. *Clin Exp Pharm Physiol.* 35,717–724.

Gibson SJ, Vonkelatos X, Ames D, Flicker L, et al. (2001). Am examination of pain perception and cerebral event-related potentials following carbon dioxide laser stimulation in patients with Alzheimer's disease and age-matched control volunteers. *Pain Res Manag.* 6, 126–132.

Glaus A. (1998). Fatigue in patients with cancer: Analysis and assessment. *Recent Results Cancer Res.* 145, 1–172.

Glaus A, Crow R, Hammond S. (1996). A qualitative study to explore the concept of fatigue/tiredness in cancer patients and in healthy individuals. *Eur J Cancer Care.* 5 Suppl 2, 8–23.

Glushko RJ, Maglio PP, Matlock T, Barsalou LW. (2008). Categorization in the wild. *Trends Cogn Sci.* 12, 129–135.

Goodwin PJ. (2003). Reversible ovarian ablation or chemotherapy: Are we ready for quality of life to guide adjuvant treatment decisions in breast cancer? *J Clin Oncol.* 21, 4474–4475.

Goodwin PJ, Black JT, Bordeleau LJ, Ganz PA. (2003). Health-related quality-of-life measurement in randomized clinical trials in breast cancer-Taking stock. *J Natl Cancer Inst.* 95, 263–291.

Gracely RH. (1983). Pain language and ideal pain assessment. In, (Ed.) R Melzack. *Pain Measurement and Assessment.* New York NY: Raven Press. (p. 71–77).

Griswold G, Clark WC. (2005). Item analysis of cancer patient responses to the Multidimensional Affect and Pain Survey (MAPS) determines the content of a short form (SMAPS-CP). *J Pain.* 6, 67–74.

Guilford JP. (1954). *Psychometric Methods.* 2nd Edition, New York NY: McGraw-Hill. (p.251–256).

Guttman L. (1944). A basis for scaling qualitative data. *Am Sociol Rev.* 9: 139–150.

Gwyn R. (1999). "Captain of my ship": Metaphor and the discourse of chronic illness. In, (Eds.) L Cameron, LG Low. *Researching and Applying Metaphor* Cambridge UK: Cambridge Press. (p. 203–220).

Hadi S, Fan G, Hird AE, Kirou-Mauro A, et al. (2008). Symptom clusters in patients with coancer with metastatic bone pain. *J Palliative Care.* 11, 591–600.

Halliday MAK. (1994). *An Introduction to Functional Grammar.* 2nd edn. London UK: Edward Arnold.

Halliday MAK. (1998). On the grammar of pain. *Funct Lang.* 5, 1–32.

Hann DM, Jacobsen PB, Azzarello LM, Martin SC, et al. (1998). Measurement of fatigue in cancer patients: development and validation of the Fatigue Symptom Inventory. *Qual Life Res.* 7, 301–310.

Harden RN, Bruehl S, Straton-Hicks M, Wilson PR. (2007). Proposed new diagnostic criteria for Complex Regional Pain Syndrome. *Pain Med.* 8, 326–331.

Heft MC, Gracely RH, Dubner R, McGrath PA. (1980). A validation model for verbal descriptor scaling for human clinical pain. *Pain.* 9, 363–373.

Herman BH, Panksepp J. (1978). Effects of morphine and naloxone on separation distress and approach attachment: evidence for opiate mediation of social affect. *Pharmacol Biochem Behav.* 9, 213–220.

Hilgard ER. (1969). Pain as a puzzle for psychology and physiology. *Am Psychol.* 24, 103–113.

Hinton DE, Howes D, Kirmayer LJ. (2008).Toward a medical anthropology of sensations: Definitions and research agenda. *Transcult Psychiatr.* 45, 142–162.

Hollenberg NK, Williams GH, Anderson R. (2000). Medical therapy, symptoms, and the distress they cause. *Arch Inter Med.* 160, 1477–1483.

Holroyd KA, Holm JE, Keefe FJ, Turner JA, et al. (1992). A multicenter evaluation of the McGill Pain Questionnaire: Results from more than 1700 chronic pain patients. *Pain.* 48, 301–311.

Holroyd KA, Talbot F, Holm JE, Pingel JD, et al. (1996). Assessing the dimensions of pain: a multitrait-multimethod evaluation of seven measures. *Pain.* 67, 259–265.

Holzner B, Kemmler G, Meraner V, Maislinger A, et al. (2003). Fatigue in ovarian carcinoma patients. A neglected issue? *Cancer.* 97, 1564–1572.

Horemans HL, Nollet F, Beelen A, Lankhorst GJ. (2004). A comparison of 4 questionnaires to measure fatigue in postpoliomylitis syndrome. *Arch Phys Med Rehabil.* 85, 392–398.

Huber A, Suman AL, Rendo CA, Biasi G, et al. (2007). Dimensions of "unidimensional" ratings of pain and emotions in patients with chronic musculoskeletal pain. *Pain* 130, 216–224.

Hulsman NM, Geertzen JHB, Dijkstra PU, van den Dungen JJAM, et al. (2009). Myopathy in CRPS-I: Disuse or neurogenic? *Eur J Pain.* 13, 731–736.

Hwang SS, Chang VT, Kasimis B. (2003a). A comparison of three fatigue measures in a veterans cancer population. *Cancer Investig.* 21, 355–365.

Hwang SS, Chang VT, Rue M, Kasimis B. (2003b). Multidimensional independent predictors of cancer-related fatigue. *J Pain Symptom Manag.* 26, 604–614.

Iconomou MD, Haralabos P, Kalofonos MD. (2004). Prospective assessment of emotional distress, cognitive function, and quality-of-life in patients with cancer treated with chemotherapy. *Cancer.* 101, 404–411.

International Classification of Diseases: 10th Revisions- Clinical Modification. (1992). Geneva Switzerland: WHO.

Jänig W, Baron R. (2002). Complex Regional Pain Syndrome is a disease of the central nervous system. *Clin Auton Res.* 12, 150–164.

Jensen M, Gammaitoni AR, Olaleye DO, Olrka N, et al. (2006). The Pain Quality Assessment Scale: Assessment of pain quality in Carpal Tunnel Syndrome. *J Pain.* 7, 823–832.

Kaplan RM, Bush JW. (1982). Health-related quality-of-life for evaluation research and policy analysis. *Health Psychol.* 1, 61–80.

Kerns RD, Turk DC, Rudy TE. (1985). The West Haven-Yale Multidimensional Pain Inventory (WHYMPI). *Pain.* 23. 345–356.

Kim SH, Son BH, Hwang SY, Han W, et al. (2008). Fatigue and depression in disease-free breast cancer survivors: Prevalence, correlates, and association with quality-of-life. *J Pain Symptom Manag.* 35, 644–655.

Kirmayer LJ. (2008). Culture and metaphoric mediation of pain. *Transcult Psychiatr.* 45, 318–338.

Kleinman L, Zodet MW, Hakim Z, Aledort J, et al. (2000). Psychometric evaluation of the fatigue severity scale for use in chronic Hepatitis C. *Qual Life Res.* 9, 499–508.

Kobgashi-Shoot JAM, Hanewald GJFP, van Dam FSAM, Bruning PF. (1985). Assessment of malaise in cancer patients treated with radiotherapy. *Cancer Nurs.* 8, 306–313.

Kövecses Z. (2000). *Metaphors and Emotion: Language, Culture and Body in Human Feeling.* Paris France: Cambridge University Press.

Krupp LB, LaRocca NG, Muir-Nash J, Steinberg AD. (1989). The fatigue severity scale. Applications to patients with multiple scherosis and systemic lupus erthematosis. *Arch Neurol.* 46, 1121–1123.

Kunz M, Mylius V, Scharmann S, Schepelmann K, et al. (2009). Influence of dementia on multiple components of pain. *Eur J Pain.* 13, 317–325.

Kunz M, Scharmann S, Hemmeter U, Schepelmann K, et al. (2007). The facial expression of pain in patients with dementia. *Pain.* 133, 221–228.

Lackey GF. (2008). "Feeling blue" in Spanish: A qualitative inquiry of depression among Mexican immigrants. *Soc Sci Med.* 67, 228–237.

Lai J-S, Cella D, Chang C-H, Bode RK, et al. (2003). Item bank to improve, shorten and computerize self-reported fatigue: An illustration of steps to create a core item bank from the FACIT-Fatigue Scale. *Qual Life Res.* 12, 485–501.

Lascaratou, C. (2007). *The Language of Pain.* Amsterdam The Netherlands: John Benjamins.

Lee KA, Hicks G, Nino-Murcia G. (1991). Validity and reliability of a scale to assess fatigue. *Psychiatr Res.* 36, 291–298.

Lee DTC, Kleinman J, Kleinman A. (2007). Rethinking depression: An ethnographic study of the experiences of depression among Chinese. *Harvard Rev Psychiatr.* 15, 1–8.

Lenert LA, Gifford A, Bozette S, Lee D. (2005). Race and risk/trade-off aversion in HIV infections: An explanation for disparities in utilization and access. *Qual Life Res.* 14, 2012.

Lesage P, Portenoy RK. (2002). Management of fatigue in cancer patients. *Oncol.* 16, 373–389.

Levine MN, Guyatt GH, Gent M, De Pauw S, et al. (1988). Quality-of-life in Stage II breast cancer: An instrument for clinical trials. *J Clin Oncol.* 6, 1798–1810.

Levitt F, Garron DC, Whisler WW et al. (1978). Affective and sensory dimensions of back pain. *Pain.* 4, 273–276.

Loeser JD, Treede R-D. (2008). The Kyoto protocol of IASP basic pain terminology. *Pain.* 137, 473–477.

Luzzi S, Piccirilli M, Provinciali L. (2007). Perceptions of emotions on Happy/Sad chimeric faces in Alzheimer Disease: Relationship with cognitive functions. *Alzheimer Associated Dis Disord.* 21, 130–135.

Magendie FJ. (1822). Expériences sur les fonctions des racines des nerfs rachidiens. *J Physiol Exp Patholog.* 276–279.

Mallinson T, Cella D, Cashy J, Holzner B. (2006). Giving meaning to measure: Linking self-reported fatigue and function to performance of everyday activities. *J Pain Symptom Manag.* 31, 229–241.

Mann T. (1927/1936). *The Magic Mountain.* London UK: Secker and Warburg.

Mc Corkle R, Young K. (1978). Development of a symptom distress scale. *Cancer Nurs.* 1, 373–378.

Mc Dowell I, Newell C. (1996). *Measuring Health: A Guide to Rating Scales and Questionnaires,* Second Edition. New York NY: Oxford University Press.

Mc Nair DM, Lorr M, Droppelman LF. (1992). *EdITS manual for Profile of Mood States.* San Diego CA: Educational and Industrial Testing Service. (p.1–40).

Mc Quellon RP, Russell GB, Cella DF, Raven BL, et al. (1997). Quality-of-life measurement in bone marrow transplantation: Development of the Functional Assessment of Cancer Therapy-Bone Marrow Transplantation (FACT-BMT) scale. *Bone Marrow Transplant.* 29, 357–368.

Melzack R. (1971). Phantom limb pain: Implications for treatment of pathological pain. *Anesthesiol.* 35, 409–419.

Melzack R. (1975). The McGill Pain Questionnaire: Major properties and scoring method. *Pain.* 1, 277–299.

Melzack R. (1989). Phantom limbs, the self and the brain. *Can Psychol.* 30, 1–16.

Melzack R. (1999). From the gate to the neuromatrix. *Pain.* (Suppl) 6, S121–S126.

Melzack R. (2005a). Evolution of the Neuromatrix Theory of Pain: The Prithvi Raj Lecture: Presented at the Third World Congress of World Institute of Pain, Barcelona 2004. *Pain Pract.* 5, 85–94.

Melzack R. (2005). The Melzack pain questionnaire: From description to measurement. *Anesthesiology.* 103, 199–202.

Melzack R, Torgerson WS. (1971). On the language of pain. *Am Psychol.* 34, 50–59.

Melzack R, Wall PD. (1965). Pain mechanisms: A new theory. *Science.* 150, 971–979.

Mendoza T, Wang XS, Cleeland CS, Morrissey M, et al. (1999). The Rapid Assessment of Fatigue Severity in Cancer Patients: Use of the Brief Fatigue Inventory. *Cancer.* 85, 1186–1196.

Merskey H, Bogduk N. (2004). *Classification of Chronic Pain 2nd Edition.* Seattle WA, International Association of the Study of Pain Press.

Miaknowski C, Aouizerat B, Dodd M, Cooper B. (2007). Conceptual issues in symptom cluster research and their implications for quality-of-life assessment in patients with cancer. *J Natl Cancer Inst Monogr.* 37, 39–46.

Miakowski C, Dodd M, Lee K. (2004). Symptom clusters: The new frontier in symptom management research. *J Natl Cancer Inst Monogr.* 32, 17–21.

Morris DB. (1991). *The Culture of Pain.* Berkeley CA: University of California Press.

Mota DD, Pimenta CA. (2006). Self-report instruments of fatigue assessment: A systematic review. *Res Theor Nurs Pr.* 20, 49–78.

Müller W, Schneider EM, Stratz T. (2007). The classification of fibromyalgia syndrome. *Rheum Int Clin Exp Investig.* 27, 1005–1010.

Nicholl BI, Halder SL, Macfarlane GJ, Thompson DG, et al. (2008). Psychosocial risk markers for new onset irritable bowel syndrome-Results of a large prospective population-based study. *Pain.* 137, 147–155.

Norman SA, Miller LT, Erikson HB, Norman MF, et al. (2001). Development and validation of a telephone questionnaire to characterize lymphedema in women treated with breast cancer. *Phys Ther.* 81, 1192–1205.

Oaklander AL. (2008). RSD/CRPS: The end of the beginning. *Pain.* 139, 239–240.

Oaklander AL, Rissmiller JG, Gelman LB, Zheng L, et al. (2006). Evidence of focal small-fiber axonal degeneration in complex regional pain syndrome-I (reflex sympathetic dystrophy). *Pain.* 120, 235–243.

O'Connor M-F, Wellisch DK, Stanton AL, Eisenberger NI, et al. (2008). Craving Love? Enduring grief activates brain's reward center. *Neuroimage.* 42, 969–972.

Okuyama T, Akechi T, Kugaya A, Okamura H, et al. (2000). Development and validation of the Cancer Fatigue Scale: A brief, three-dimensional self-rating scale for assessment of fatigue in cancer patients. *J Pain Symptom Manag.* 19, 5–14.

Oken MM, Creech RH, Tormey DC, Horton J, et al. (1982). Toxicity and response criteria of the Eastern Cooperative Oncology Group. *Am J Clin Oncol.* 5, 649–655.

Olesen J. (2004). *International Classification of Headache Disorders* 2nd Edition. *Cephalalgia.* 24 Suppl 1, 9–160.

Olesen J. Goadsby P, Steiner T. (2003). The International Classification of Headache Disorders. 2nd Edition. *Lancet Neurol.* 2,720.

Olson K, Hayduk L, Cree M, Cui Y, et al. (2008). The changing causal foundation of cancer-related symptom clustering during the final month of palliative care: A longitudinal study. *BMC Med Res Methods.* 8, 1–11.

Omoigui S. (2007a). The biochemical origin of pain: The origin of all pain is inflammation and the inflammatory response. Part 2 of 3 - Inflammatory profile of pain syndrome. *Med Hypotheses.* 69, 70–82.

Omoigui S. (2007b). The biochemical origin of pain- Proposing a new law of pain: The origin of all pain in inflammation and the inflammatory response. Part 1 of 3- A unifying law of pain. *Med Hypotheses.* 69, 1169–1178.

Osoba D. (2007). Translating the science of patient-reported outcomes assessment into clinical practice. *J Natl Cancer Inst Monogr.* 37, 5–11.

Oxford English Dictionary (1970). Vol VII, London UK: Oxford University Press. (p.377–378)

Oxford English Dictionary. (1996). Oxford UK: Oxford University Press.

Panksepp J. (1998). *Affective neuroscience: The foundations of human and animal emotions.* New York, NY: Oxford University Press. (pp. 466).

Panksepp J, Herman B, Conner R, Bishop P. et al. (1978). The biology of social attachments: Opiates alleviate separation distress. *Biol Psychiat.* 13, 607–618.

Paterson B, Canam C, Joachim G, Thome S. (2003). Embedded assumptions in qualitative studies of fatigue. *West J Nurs Res.* 25, 119–133.

Pearce JMS. (2007). Synaesthesia. *Exp Neurol.* 57, 120–124.

Pearson RG. (1957). Scale analysis of a fatigue checklist. *J Appl Psychol.* 41, 186–191.

Perrot S. (2008). Fibromyalgia syndrome: A relevant recent construction of an ancient condition? *Curr Opin Support Palliative Care.* 2, 122–127.

Pilowsky I, Spence ND. (1983). *Manual for the Illness Behavior Questionnaire* (IBQ). 2nd Ed. Adelaide, Australia: University of Adelaide.

Piper BF, Dibble SL, Dodd MJ, Weiss MC, et al. (1998). The Revised Piper Fatigue Scale: psychometric evaluation in women with breast cancer. *Oncol Nurs Forum.* 25, 677–684.

Piper BF, Lindsey AM. Doff MJ, Ferketich S, et al. (1989). The development of an instrument to measure the subjective dimension of fatigue. In, (Eds.) SG Funk, EM Tornquist, MT Champagne, L Archer Copp, et al. *Key Aspects of Comfort Management of Pain and Nausea.* Philadelphia PA: Springer. (p. 199–208).

Portenoy RK, Thaler HT, Kornblith AB, McCarthy LJ, et al. (1994). The Memorial Symptom Assessment Scale: an instrument for the evaluation of the symptom prevalence, characteristics and distress. *Eur J Cancer.* 30A, 1326–1336.

Radloff LS. (1977). The CES-D Scale: a self-report depression scale for research in the general population. *Appl Psychol Meas* 3, 385–401.

Rainero I, Vighetti S, Bergamasco B, Pinessi L, et al. (2000). Autonomic responses and pain perception in Alzheimer's disease. *Eur J Pain.* 4, 267–274.

Rapley M. (2003). *Quality of life Research.* Thousand Oaks CA: Sage.

Rasquin A, Di Lorenzo C, Forbes D, Guiraides E, et al. (2006). Childhood functional gastrointestinal disorders: Child/Adolescent. *Gastroenterology.* 130, 1527–1537.

Reading AE, Newton JR. (1978). A card sort method of pain assessment. *J Psychosom Res.* 22, 503–512.

Redeker NS, Lev RL, Ruggiero J. (2000). Insomnia, fatigue, anxiety, depression and quality-of-life of cancer patients undergoing chemotherapy. *Sch Inq Nurs Pract.* 14, 275–290.

Rhoten D. (1982). Fatigue in cancer patients. A descriptive study. In, (Ed.) C Norris. *Concept Clarification in Nursing.* Rockville MD: Aspen Systems. (p. 277–300).

Richardson A, Ream E. (1996). The experience of fatigue and other symptoms in patients receiving chemotherapy. *Eur J Cancer Care.* 5 Suppl, 24–30.

Ross BH, Murphy GL. (1999). Food for thought: Cross-classification and category organization in a complex real-world domain. *Cogn Psychol.* 38, 495–553.

Rossen PB, Pedersen AF, Zacharaie R, von der Maase H. (2009). Health-related quality-of-life in long-term survivors of testicular cancer. *J Clin Oncol.* 27, 5993–5999.

References

Rupp I, Boshuizen HC, Jacobi CE, Dinant HJ, et al. (2004).Impact of fatigue on health related quality-of-life in rheumatoid arthritis. *Arthritis Rheum.* 15, 578–585.

Sadler IJ, Jacobsen PB, Booth-Jones M, Belanger H, et al. (2002). Preliminary evaluation of a clinical syndrome approach to assessing cancer-related fatigue. *J Pain Symptom Manag.* 23, 406–416.

Scarry E. (1985). *The Body in Pain: The Making and Unmasking of the Word.* New York NY: Oxford University Press.

Scherer KR. (2000). Emotions as episodes of subsystem synchronization driven by nonlinear appraisal processes. In, (Eds.) MD Lewis, I Granic. *Emotions, Development, and Self-organization: Dynamic Systems Approaches to Emotional Development.* Cambridge UK: Cambridge University Press. (p. 70–99).

Schipper H, Clinch J, McMurray A, Levitt M. (1984). Measuring the quality-of-life of cancer patients: the Functional Living Index-Cancer: development and validation. *J Clin Oncol.* 2, 472–483.

Schurmann M, Gradl G, Wizgal I, et al. (2001). Clinical and physiologic evaluation of stellate ganglion blockade for complex regional pain syndrome type I. *Eur J Pain.* 17, 105–122.

Schwartz AL. (1998). The Schwartz Cancer Fatigue Scale: testing reliability and validity. *Oncol Nurs Forum.* 25, 711–717.

Seminowicz DA, Davis KD. (2007). A re-examination of pain-cognition interactions: Implications for neuroimaging. *Pain.* 130, 8–13.

Servaes P, Verhagen C, Bieijenberg G. (2002). Fatigue in cancer patients during and after treatment. *Eur J Cancer.* 38, 27–43.

Sheffer CE, Deisinger JA, Cassisi JE, Lofland K. (2007). A revised taxonomy of patients with chronic pain. *Pain Med.* 8, 312–325.

Sherbourne CD. (1992). Pain measures. In, (Eds.) ALK Stewart, JE Ware Jr, *Measuring Functioning and Well-being: The Medical Outcome Study Approach.* Durham NC: Duke University Press. (p. 220–234).

Sherrinton CS. (1906). *The Integrative Action of the Nervous System.* New Haven CT: Yale University Press.

Smets EMA., Garssen B, Bonke B, de Haes JCJM. (1995). The multidimensional fatigue inventory (MFI). Psychometric qualities of an instrument to assess fatigue. *J Psychosom Res.* 39, 315–325.

Smets EMA, Garssen B, van Weers W, de Haes JCJM. (1997). Patients' view on fatigue during and after cancer treatment, a qualitative report. In, (Ed.) EMA Smets, *Fatigue in Cancer Patients Undergoing Radiotherapy.* Amsterdam The Netherlands: Printpartners Ipskamp.

Sommer G, Fydrich T. (1999). Social support questionnaire (F-SozU): norms of a representative sample. *Diagnosis.* 4, 212–216.

Sontag S. (1991). *Illness as Metaphor and AIDS and its Metaphors.* New York, NY: Picador USA.

Sperber D. (2002). In defense of massive modularity. In, (Ed.) E Dupoux. *Language, Brain and Cognitive Development: Essays in honor of Jacques Mehler.* Cambridge MA: MIT Press. (p. 47–57).

Spitzer W, and Members of the Quebec Task Force on Spinal Disorders. (1987). Scientific approaches to the assessment and management of activity related to spinal disorders. *Spine.* 12 Suppl, S1–S59.

Staub R, Smitherman ML. (2002). Peripheral and central sensitization in fibromyalgia: pathogenetic role. *Curr Pain Headache Rep.* 6, 259–266.

Stein KD, Martin SC, Hann DM, Jacobsen PB. (1998). A multidimensional measure of fatigue for use with cancer patients. *Cancer Pract.* 6, 143–152.

Sternbach RA. (1968). *Pain: A Psychophysiological Analysis.* New York NY: Academic Press.

Sternbach RA. (1989). Linguistic problems in pain measurement. In, (Eds.) CR Chapman, JD Loeser. *Issues in Pain Measurement.* New York NY: Raven Press. (p. 63–68).

Stewart AL, Ware JE Jr. (1992). *Measuring Functioning and Wellbeing: The Medical Outcomes Study Approach.* Durham NC: Duke University Press.

Tolstoy L. (1886/2006). *The Death of Ivan Ilyich.* New York NY: Penguin.

Torgerson WS, BenDebba M. (1983). The structure of pain descriptors. In: (Ed.) R Melzack. *Pain Measurement and Assessment.* New York NY: Raven. (p. 49–54).

Torrance GW. (1976). Health status index models: A unified mathematical view. *Manag Sci.* 22, 990–1001.

Tourangeau R. (1984). Cognitive sciences and survey methods. In, (Eds.) T Jabine, M Straf, J Tamur, R Tourangeau. *Cognitive Aspects of Survey Methodology: Building a Bridge Between Disciplines.* Washington DC: National Academy Press. (pp. 73–100).

Turk DC, Rudy TE. (1988). Towards an empirically derived taxonomy of chronic pain patients: Integration of psychological assessment data. *J Consult Clin Psychol.* 56, 233–238.

Turk DC, Rudy TE. (1990). The robustness of an empirically derived taxonomy of chronic pain patients. *Pain.* 431, 27–35.

Veldman PH, Reynen HM, Arntz IE, Goris RJ. (1993).Signs and symptoms of reflex sympathetic dystrophy: prospective study of 829 patients. Lancet. 342(8878), 1012–1016.

Victor T, Jensen M, Gammaitoni AR, Gould EM, et al. (2008). The dimensions of pain quality: Factor analysis of the Pain Quality Assessment Scale. *Clin J Pain.* 24, 550–555.

Visser MR, Smets EM. (1998). Fatigue, depression, and quality-of-life in cancer patients: how are they related? *Support Care Cancer.* 6, 101–108.

von Frey M. (1896). Untersuchungen über die Sinnesfunctionen der Menschlichen Haut. Erste Abhandlungen: Druckenpfindung und Schmerz. Berlin sachs *Gess Wissenshaften der mathematisch und physischen Klasse der Königichen.* (KL) 23, 175.

von Korff M, Miglioretti DL. (2005). A prognostic approach to defining chronic pain. *Pain.* 227, 304–313

Walsh D, Rybicki L. (2006). Symptom clustering in advanced cancer. *Support Care Cancer.* 14, 832–836.

Ware JE, Snow KK, Kosinski MA, Gandek B. (1993). *SF-36 Health Survey: Manual and Interpretation Guide.* Boston MA: The Health Institute, New England Medical Center.

Watson D. (1988). Intra-individual and inter-individual analysis of positive and negative affect; Their relation to health complaints, perceived stress and daily activities. *J Personal Soc Psychol.* 54, 1020–1030.

Watson M, Greer S, Yound J, Inayat Q, et al. (1988). Development of a questionnaire measure of adjustment to cancer: the MAC scale. *Psychol Med.* 18, 203–209.

Wells FL. (1907). A statistical study of literary merit. *Arch Psychol.* 1, No.7, 5–30.

Wessely S. (1993). Chronic fatigue. In, (Eds.) R Greenwood, MP Barnes, *Neurological Rehabilitation.* Edinburgh United Kingdom: Churchill Livingstone.

Wiener SL. (1993). *Differential Diagnosis of Acute Pain: By Body Region.* New York NY: McGrall-Hill.

Willis G, Reeve B, Barofsky I. (2005).The Use of Cognitive Interviewing Techniques in Quality of life and Patient-Reported Outcome Assessment. In, (Eds.) J Lipscomb, CC Gotay, C Synder. *Outcomes Assessment in Cancer: Findings and Recommendations of the Cancer Outcomes Measurement Working Group.* Cambridge UK: Cambridge University Press.

Wilson IB, Cleary PD. (1995). Linking clinical variables with health-related quality of life: A conceptual model of patient outcomes. *JAMA.* 273, 59–65.

Wittgenstein L. (1961). *Tractatus Logico-Philosophicus.* New York NY: Routledge and Kegan Paul.

Wolfe F. (2004). Fatigue assessments in rheumatoid arthritis: Comparative performance of visual analog scales and longer fatigue questionnaires in 7760 patients. *J Rheum.* 31, 1896–1902.

Wood JD. (2008). Enteric nervous system: Reflexes, pattern generators and motility. *Curr Opin Gastroenterol.* 24, 149–158.

Yang JC, Clark WC, Tsui SL, Clark SB. (2000). Preoperative Multidimensional Affect and Pain survey (MAPS) scores predict postcolectomy analgesia requirement. *Clin J Pain.* 16, 314–320.

Yellen SB, Cella DF, Webster K, Blendowski C, et al. (1997). Measuring fatigue and other anemia-related symptoms with the Functional Assessment of Cancer Therapy (FACT) measurement system. *J Pain Symptom Manag.* 13, 16–74.

Zigmond AS, Snaith RP. (1983). The hospital and anxiety and depression scale. *Acta Psychiatr Scand.* 67, 361–370.

Zimmerman M, McGlinchey JB, Young D, Cheiminski I. (2006a). Diagnosing major depressive disorder. Introduction: An examination of the DSM-IV diagnostic criteria. *J Nerv Mental Dis.* 194, 151–154.

Zimmerman M, McGlinchey JB, Young D, Cheiminski I. (2006b). Diagnosing major depressive disorder III: Can some symptoms be eliminated from the diagnostic criteria? *J Nerv Mental Dis.* 194, 313–317.

Zung WWK. (1983). A self-rating Pain and Distress Scale. *Psychosomatics.* 24, 887–894.

Health Status and HRQOL Assessment

Abstract

Should indicators of health status be included in a HRQOL assessment? To answer this question I first need to understand what the term health means in the phrase health status. Since the term disorder is also used to characterize a person's health state, I also have to discuss what it means. I discuss both these topics, but also how health status can be assessed, reviewing its historic dependence on operations research approaches. The relationship between health status and quality of care is also discussed, with a particular emphasis on Donabedian's definition of quality (of care). This is followed by a discussion of patient satisfaction and its usefulness as an indicator of quality (of care). A final section deals with using health status as an indicator of HRQOL. In this section, I make explicit the components of a HRQOL assessment, and review examples of its application in different HRQOL assessments. I also review "Quality Adjusted Life Years" estimates, as an application of HRQOL assessments. I end my discussion concluding that health status, especially when expressed abstractly, is of limited usefulness as an element of a HRQOL assessment.

Whenever we have made a word ... to denote a certain group of phenomena, we are prone to suppose a substantive entity beyond the phenomena.

William James (1890; p. 194)[1]

Abbreviations

BCE	Before the Common Era
BCQ	Breast cancer questionnaire (Levine et al. 1988)
BFI	Brief fatigue inventory (Mendoza et al. 1999)
CAHPS	Consumer assessment of health plans
CARES	CAncer Rehabilitation Evaluation System (Ganz et al. 1992)
CCM	Chronic care model (Hung et al. 2008; Wagner et al. 2001)
EORTC-QLQ-C30	European Organization for Research and Treatment of Cancer-Quality of Life Questionnaire-Cancer specific-30 items (Aaronson et al. 1991)
EQ-5D	EuroQol-5 dimensions (Brooks et al. 2003)
FACT	Functional assessment of cancer therapy (B=Breast Cancer; BMT=Bone Marrow Transplantation; F=Fatigue; G=General; O=Ovarian)
FLIC	Functional Living Index-Cancer (Shipper et al. 1984)
HALex	Health and Activity Limitation Index (Erickson 1998)
HRQOL	Health-related quality of life
HUI	Health Utility Index (Torrance 1976)
IASP	International association for the study of pain
JCAHO	Joint Commission on Accreditation of Hospitals Organization
MOS-SF-36	Medical Outcome Study-Short Form-36 (Ware et al. 1993)
QALY	Quality adjusted life years
QOL	Quality of life

QWB	Quality of Well-being Scale (Kaplan and Bush 1982)
RSCL	Rotterdam symptom check list (De Haes et al. 1990)
SEIQoL	Schedule for the evaluation of individual quality of life (Hickey et al. 1999)
SF-6D	Based on the MOS SF-36 or 12 (Brazier et al. 2002)
SIP	Sickness impact profile (Bergner et al. 1981)
WHO	World Health Organization
15D	15 Dimensions (Sintonen 2001)

1 Theoretical Foundation

1.1 Introduction

The act of defining the phrase "health status" is plagued with the problem that William James was alluding to in the above quotation. Evidence in support for this proposition comes from the effort by Patrick and his colleagues to define the phrase health status (Chap. 6, p. 174). Presumably, the phrase is meant to reveal the status of someone along a particular continuum. Thus, the term "status" implies that some entity is in a position (a ranking), while the term "health" acts as an adjective modifying the noun. Together the terms imply that the individual's health is in a particular position (or rank). As a result, if I want to include health status as part of an HRQOL assessment I would need to include items that revealed the individual's particular health level. However, most HRQOL assessments do not contain items that deal with a person's state of health (e.g., red blood count, results of a brain scan, etc.), but rather indications of the consequences of being in a particular health state.

Patrick and his colleagues seem to oscillate between using the phrase as an abstract concept or a concrete domain (Chap. 6, p. 174). What remains unknown is whether the term health will allow for these multiple uses, or whether the term is inherently abstract so that it can only be arbitrarily defined in concrete terms. I will attempt to address this issue by reviewing the various definitions of the term health, and also the term "disorder," which is sometimes used as a surrogate for the term health. A disorder can function as a diagnostic category, and as I will show, it can be helpful when defining what the term health means. This is especially so since health is regularly defined as "the absence of disease," which, unfortunately, does not help when defining health.

Wittenberg et al. (2005) provide an example of how the phrase health status can be operationalized, but also demonstrated some of the empirical complications that can occur when the cognitive underpinnings of these health status domains are not considered. After interviewing eight metastatic breast cancer patients and several clinicians the investigators devised a four state scale that varied from very good, good, moderate, or poor health status. Each of these states involves varying levels of five activity or emotional states. The investigators assume that a respondent will inspect the five descriptors of these four domains and select the descriptors and that best match their personal health status. The question is whether these states indicate something about an individual's health. Is an individual's health determined by whether they are working or not; able to take care of themselves or not; feel well or not; or have a good social support system or not? Clearly, the investigators believed that these response options described some level of health.

In addition, this study failed to reach the investigators objectives of testing Prospect Theory (Kahneman and Tversky 1979). What they found was that while the respondents admitted functional limitation, their global estimate of quality of life was not diminished. One possible explanation of these results was that the investigators did not pay sufficient attention to the process of how a domain was cognitively constructed (e.g., each health state). This was especially important since the intent of the study was to use these domains as hypothetical health states in both a standard gamble utility estimation, and when inferring the existence of an abstract construct (e.g., health status).

There were several ways whereby the results of this study could be altered. First the investigators would cognitively interview their potential respondents after they constructed their health states to determine if the items in the domains were sufficiently relevant or salient to the respondents. They also could give their respondents the opportunity to construct their own domains. The metastatic breast cancer patient is in a unique emotional and social state, yet the descriptors within each health state could be as applicable to any group of respondents. Thus, the entire assessment process could only be partially relevant or of interest to this particular group of respondents.

In addition, the investigators were assuming, but not providing empirical support for, the inference that affirming these concrete options justifies the inference of an abstract concept. While this is a very common practice, I have been arguing that an investigator has to demonstrate (as I did for the relationship between fatigue, depression, and HRQOL [Chap. 7, p. 211]) that the components of the (e.g., health status) domain has an internal cognitive structure that justified such an inference. In the absence of this, any effort at characterizing the data would remain a type of arbitrary labeling.

What makes *health status*, as opposed to *a health state*, an important indicator is that it implies an implicit scale over which various descriptive adjectives can be applied. Thus, both "normal" and "abnormal" health status may exist, each of which can be qualified (i.e., the importance, preference, or utility of a particular health state can be estimated).

1 Theoretical Foundation

It is also important, when discussing what the phrase "normal health status" means, to clarify whether the goal of the qualitative assessment is to prevent adverse consequences of a disease or treatment, or to enhance an individual's health status or well-being (Chap. 1, p. 17). Either type of goal requires interventions that have explicitly stated objectives that would be expected to differ when operationalized.

I started discussing the meaning of the term "normal" when I reviewed a paper by Cronje and Williamson (2006; Chap. 7, p. 221). Their paper questioned the IASP use of pain terminology (Mersky and Bogduk 2004) that includes the claim that a "normal level of pain" exists. At first glance, any level of pain would appear to be abnormal, so the notion that a normal level of pain exists seems counterintuitive. What Cronje and Williamson (2006) were arguing was that this definition was based on what a clinician would expect a patient to say, not what a patient experiences. Allodynia, therefore, was defined as occurring when a respondent reported a level of pain that would not have been expected by the clinician. Hyperalgesia, another pain syndrome, was defined as an increased response to a stimulus that is normally painful. Analgesia occurs when an individual does not respond to a stimulus that would normally be painful. Each of these pain syndromes seems to be using the terms *normal* or *abnormal* differently, which raises serious questions about what the terms actually mean. I would like to examine their meanings a bit more, since they play such an important role in deciding whether a person is healthy or not.

The goal of most political, social, cultural, and medical policies is presumably to encourage an individual to optimally function and continue the process of personal and emotional development. The reality, however, is that most societies are not organized to achieve these goals as much as they are prepared to deal with violations of these goals (e.g., most societies have a police force to prevent crimes, not promote personal and social development). In the area where optimizing and developing health is the goal, the most prevalent approach, the medical model, is based on identifying and treating physical or mental pathology. Interestingly, some investigators raise questions about whether a normal state can ever exist. Thus, Freud (1937; p. 195) said that "A normal ego is like normality in general, an ideal fiction." René Dubos, in his book *The Mirage of Health* (1959), essentially makes the same argument. One of the implications of the term "normal" being an ideal type is that it is also likely to be expressed as an abstract concept, and any effort to operationalize it inevitably leads to multiple concrete examples. Offer and Sabshin (1991; p. 408) recognized this when they claimed that there was probably greater diversity amongst those individuals who were called "normal," then those with a specific physical or mental disorder. Thus, having a diagnosis predicts that a person is likely to have a particular natural history that makes life more defined and knowable, at least in comparison to the individual who does not have this diagnosis.

In the ordinary language, the term *normal* means that which occurs most frequently (Perelman and Olbrechts-Tyteca 1969). This implies that the term is defined quantitatively, yet when considered qualitatively (or by the person), it usually involves a comparison between a standard and the individual's current (e.g., health) status. Perelman and Olbrechts-Tyteca (1969) claim that the standard a person uses is rarely identified or recognized. Of course, implicit here is the assumption that an automatic cognitive comparison has occurred, and on this basis, the individual decides if his or her experience is normal.

Not surprisingly, individuals develop schemas to cognitively represent the experiences they consider to be normal. These schemas are presented cognitively as categories, and this occurs for both the patient and the clinician. Previously (Chap. 3, p. 62), I described several types of categories, including, for example, similarity-based, exemplar, and prototypical. A prototype pain category, for example, will have some defining characteristics, with some of the indicators being more salient to the category than others. Repetition of these indicators gives the clinician and the patient more confidence about what constitutes normal. Violation of this scheme would in time lead to revision of the category. Thus, what constitutes normal pain will vary depending on what the individual reports and also what the clinician reports. Not too surprisingly, therefore, it is common to find that discrepancies exist between these reports (Cronje and Williamson 2006). Both the clinician and the individual will learn more about the individual's pain in time, suggesting that "normal" refers to a dynamic process that continues to change as both learn about the person's symptoms.

Evidence-based medicine represents an alternative, more data-oriented approach to defining the term *normal* (Cronje and Williamson 2006). Sackett et al. (2000) suggest six possible definitions of normal (Table 8.1), some of which will also be used to define health (see below), but some they are critical of. Thus, the Gaussian and percentile definitions assume that the distribution of tests and other results are inherently normal, a condition that is not likely to be found empirically. To use cultural criteria to define what is normal can, at times, conflict with medical information (e.g., at one time smoking was considered normal and acceptable, whereas today medicine has demonstrated that smoking is a health risk). Defining normal as the absence of risk does not mean that a person is not at risk, since risk can be defined in various ways.

Sackett et al. (2000) suggest that criteria relying on the absence of a positive indicator of a disease or disorder, or demonstration of positive therapeutic effects are the most useful ways to define the term *normal*. The reason for this includes the facts that evidence has to be accumulated

Table 8.1 Sackett et al.'s (2000) recommendations on how to define the term normal[a]

Definitions of normal
1. Gaussian: Normal is defined as falling within the statistical area defined by the mean and two standard deviations from the mean
2. Percentile: Normal is defined as an area that falls between the 5 and 95% range
3. Culturally desirable: Normal is defined by what society prefers
4. Risk factor: Normal is defined by the absence of risk
5. Diagnostic: Normal is when the probability of a positive indicator of a disease or disorder is quite low
6. Therapeutic: Normal is defined as the range of results that produce more of a positive than negative outcome

[a]This table was created based on comments in Cronje and Williamson (2006; p. 694)

(usually quantitatively), diagnostic risk has been reduced, and therapeutic benefit demonstrated. This approach replaces the subjective impressions reflected in the definition of the term *normal* found in the common discourse with information that can be reproduced by others. However, both approaches assume that their definitions apply to the population as a whole and is not limited to an individual.

This discussion provides an introduction to some of the issues and approaches I will consider in the rest of this chapter. My first task, however, is to clarify what the term health means in the phrase health status. I will follow this with a discussion of the meaning of the term disorder, since the absence of a disorder or disease is often considered a definition of the term health. Next I will review the "end-results movement"; follow this with a brief description of the purpose and nature of operations research; consider Donabedian's approach and definition of quality of care; and end with a discussion on the nature of patient satisfaction as a quality indicator. Of particular interest will be Donabedian's (1982; p. 3) definition of quality of care, which he posits as being a "valued outcome," and his thoughts will be interpreted as providing additional support for the view that all judgments of quality include a value statement. This review should provide an adequate description of the intellectual history of the phrase health status.

The next major section will focus on the application of health status assessment. I will start with exploring the relationship between health status and HRQOL. I will follow with a discussion of the components of an HRQOL assessment, illustrating which of the available assessments match the hybrid construct model that I claim is applicable to all judgments of quality. I will also consider the question of whether HRQOL is a viable construct, and discuss Beckie and Hayduk (1997), Michalos (2004), and others' concerns. Finally, I will end with the role of HRQOL in economic modeling, addressing how Quality Adjusted Life Years (QALY), a quality indicator, has been successfully applied in economics.

I will draw several conclusions from this discussion. Most important is that the phrase health status has a rather limited role to play when used as a domain, and the term is best used at a conceptual level, similar to how Patrick and Chang (2000) used it in their most recent HRQOL model. This is tantamount to thinking of the phrase as a form of institutional, as opposed to cultural or individual, categorization (Glushko et al. 2008). In addition, I will conclude that a health status assessment differs from an HRQOL assessment in that it exclusively focuses on describing a state a person is, while HRQOL has more of an evaluative connotation. Even so, both types of assessment can be qualified, but if a health status assessment is qualified, then it becomes an HRQOL assessment. I end by suggesting that, in their generic forms, it is best to keep these assessments as separate as feasible, if only to ensure that the outcome of each assessment is properly interpreted (Bowling and Windsor 2001: p. 75).

1.1.1 Definitions of Health

The origin of the word health can be traced to the Anglo-Saxon word *hal*, which gave rise to the words heal, hale, and whole. A dictionary definition describes health as a state of being free from illness or injury, or as a person's mental or physical condition or state (Oxford Dictionary 1996). When used as part of figurative language, health can function as a linguistic metaphor, as when I speak of "the financial health of a nation" or "the emotional health of a relationship." As a conceptual metaphor, Lakoff and Johnson (1999) speak of health in the context of a moral metaphor system, where health (like wealth, strength, protection, and nurturance) is a resource that a person has that contributes to their well-being. Health and health status, therefore, have an implicit quantitative meanings, and accounting for them has moral implications. Lakoff and Johnson (1999) describe moral accounting in the following way:

> The basic idea behind moral accounting is simple: Increasing others' well-being is metaphorically increasing their wealth *(or health)*. Decreasing others' well-being is metaphorically decreasing their wealth *(or health)*. In other words, doing something good for someone is metaphorically giving that person something of value, for example, money *(or for example, extending someone's life)*. Doing something bad for someone is metaphorically taking something of value away from them...Justice *(or good medical care)* is when the moral books are balanced. (p. 292–293; italicized phrases added)

Clearly, it is this moral accounting implicit in the word health that makes performing an HRQOL assessment such an ethical imperative.

Moral accounting also implies that some goal or objective can be attained. Goal attainment, as I will demonstrate, is an integral part of many definitions of health, indicating that the term implies purpose and intention. Not surprisingly, there is a fairly extensive literature dealing with the definition and meaning of the term health. I will start with the René Dubos's book *Mirage of Health* (1959), since he seems most sensitive to the impact language usage has on how health is defined.

Dubos (1959) argues that perfect health is a utopian dream that can never be achieved. Rather, it seems to be true that individuals constantly adapt to their biology, their selves, and their environment. They are best when in some state of equilibrium, which is unavoidably less than an ideal state. Thus, Dubos states that health is "…a *modus vivendi* enabling imperfect man to achieve a rewarding and not too painful existence, while they cope with an imperfect world" (Dubos 1968; p. 69).

Dubos' inability to define perfect or ideal health, I suggest, is another example of the struggle that exists to define an abstract concept using concrete examples. His notion that a dynamic exists to maximize what can be achieved reinforces the view that nonlinear modeling may be appropriate when dealing with what is possible. Of course, his difficulty in defining health is because he literally could not imagine what perfect or ideal health would look like (Chap. 3, p. 57), and it is for this reason that health remains a mirage for him. Dubos (1959), however, was not pessimistic about the impact of this dilemma on advances in medicine. He recognized that what was unique about human nature was that abstract and conceptual thinking can stimulate and direct more concrete efforts that in this case could lead to a treatment or cure of a disease. And it is this constant interplay between the ideal or real with the cognitively abstract or concrete that leads to advancements in medical care and health.

This implies that there will always be cognitive constraints when defining health and it would not be surprising to find that there is not a single definition that lay persons and investigators would agree with.[2] Frances et al. (1991) actually reminds us that the effort of defining health has a very long history, starting with Hippocrates (460–377 BCE) and Galen (131–201 BCE), and continuing unabated until today. A number of scholars have attempted to formally define health, and I would like to briefly present three examples: Parsons (1978), who adopts a role-dependent model characteristic of a sociological approach; Boorse, (1975, 1977) who describes a biologically dependent biostatistical theory; and Nordenfelt (1995), who speaks of health from the perspective of psychology and sociology and describes an action-theoretical approach.

All three conceive of health in *teleological* terms, meaning that health is understood as implying some purpose, goal, or design. If health implies purpose, then this suggests that the activities involved in maintaining health contribute to some end, and these ends are usually described as reproduction and survival. Conceiving of health in teleological terms has cognitive implications, meaning that the lay person, patient, or investigator is making attributions and causal statements to account for what happens to them. Thus, events that would have otherwise been seen as incidental can be causally associated. For example, it is sometimes alleged that depression causes cancer or that women who develop breast cancer must have done something to justify such punishment. Both these cases are instances where incidental events (the association of depression and the onset of cancer, or breast cancer and behavior) have become causally linked with the implication that some purpose is being fulfilled.

It is interesting to examine where these ways of thinking come from. Is it a natural aspect of how we think or is it something we learn to do? Michotte (1963), in his classic study *The Perception of Causality*, demonstrated that causality is a consequence of a biologically determined perceptual and cognitive processes (i.e., it is automatic and modular in nature). In one simple experiment, he demonstrated that if the speed with which one object moves towards another is slowly increased, a point will be reached where the subject will report one object hitting another, "causing" the second object to recoil. Numerous examples of this and other less stimulus-dependent phenomena exist, making it quite clear that purpose and causality is something often biologically determined, and that I can cognitively impose on the information I receive, a process which is critical to my survival. The approach-avoidance reflex is a classic example of a behavior that all species possess and fits the category of being partially biological determined.

Talcott Parsons (1978), borrowing a term that biologist Ernst Mayr used, suggested that health is a form of teleonymy ("as if" teleological). He states, "Teleonymy may be defined as the capacity of the organism, or its propensity, to undertake successful goal oriented courses of behavior" (p. 68). In applying this to health, he defines health as "the state of optimum capacity of an individual for effective performance of the roles and tasks for which he has been socialized" (Parsons 1972; p. 117). Gerhardt (1989) also quotes him: "Health is vital, because the capacity of the human individual to achieve is ultimately the most crucial social resource" (Parsons 1970, p. 281). Thus, health for Parsons is a form of human achievement that occurs in a social context reflecting an individual's capacity to survive. Parsons (1958) has explicitly defined health as the "state of optimal capacity of an individual for the effective performance of the roles and tasks for which he has been socialized" (p. 274).

Boorse (1977) bases his definition of health on a straightforward assumption: he assumes that what is "normal is natural" (p. 554). By this he means that health, but not disease, is a natural part of our lives, and it is in this context that the term should be defined and understood. He uses the *normal probability distribution* (Fig. 8.1) to define what health is and what it is not. Health and disease, therefore, are different points on a biologically determined functional continuum. As such, they can be objectively defined without recourse to value judgments (Boorse 1997). He refers to his approach as naturalism.

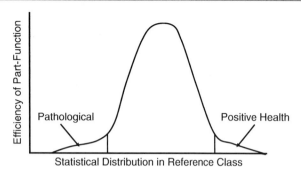

Fig. 8.1 Displayed is Boorse's concept of health and pathology as placed along a normal probability distribution.

Boorse (1977), quoting Temkin (1973), points out that this view is consistent with classical medical tradition, culminating in Galen. Thus, Temkin (1973) states:

> Such a concept of health or disease rests on a teleologically conceived biology. All parts of the body are built and function so as to allow man to lead a good life and to preserve his kind. Health is a state according to Nature; disease is contrary to Nature. (p. 398)

Boorse t goes on to say,

> From our standpoint, then, health and disease belong to a family of topological and teleological notions which are usually associated with Aristotelian biology… (p. 554).

I conclude from this statement that Boorse's biostatistical model of health relies on at least two distinct cognitive processes: the use of *cognitive prototypes* and *causal modeling*, both of which shed light on what he means when he says that health can have a purpose.

Boorse (1997) summarizes his views as follows:

> Theoretical health, I argued, is the absence of disease…the classification of human states as healthy or diseased is an objective matter, to be read off the biological facts of nature without need of value judgments. Let me refer to this general position as "naturalism" – the opposite of normativism, the view that health judgments are to include value judgments. (p. 4)

Nordenfelt (1995) describes his approach to health as a *welfare model*, which permits him to link health and happiness. According to Nordenfelt,

> Health is, roughly, one's ability to fulfill one's goals. Happiness (as an emotion) is a state which arises as a consequence of goal-fulfillment. Thus, health must be an important contributor to happiness. (p. 88)

The goals a person strives for "are necessary and together sufficient for his minimal happiness" (Nordenfelt 1995; p. 90). Vital goals include having food, shelter, and economic security. To the extent that these goals contribute to a person's happiness, they contribute to a person's welfare. Nordenfelt's (1995) model, with its emphasis on happiness as the goal of health, is consistent with a social indicator approach to a quality-of-life assessment.

In contrast to these formal definitions, lay persons provide a more varied list of definitions of health. Hughnet and Kleine (2004) confirmed this following their review of the qualitative literature, especially if they limited their review to studies that assessed the general public's attitude towards health. They found five major themes that characterized the definition of health. These themes seem to focus around the central issue of whether the choices a person can make in their lives will be limited in anyway.

The first, most popular theme, equated health with the absence of illness. A definition that is similar to the dictionary's and a biomedical definition of health. It is a cognitively attractive definition since it only asks the person to indicate what health is not, rather than what it is. To state what health is not can be expressed in concrete terms (e.g., what illnesses, diseases, or treatments a person has experienced), and would not require struggling with an abstract construct.

A second theme equates health with a person's ability to function. If a person can function then their capabilities can be assumed to be intact, leaving the question of whether the person is able to take advantage of their abilities as an additional issue. McKague and Verhoef (2003), however, point out that health is not only about being able to function, but also about being able to function according to a person's expectations. This allows for the discrepancy, often noted, between a person's perceived and actual state of health.

A third theme is the lay person's notion that health consists of achieving a balance, or equilibrium (similar to Dubos's notion) that permits the person to realize their capacity for good health. Herzlich (1973) found that the elements of good health for the general public included achieving happiness, relaxation, feeling strong, and having good relationships with others. The equilibrium theme implies that these goals are balanced by the negative aspects of disease and illness, but that a person is capable of coping with the presence of these factors. This theme is seen as a positive definition and could be achieved by adopting a variety of life style activities (e.g., a good diet, getting proper exercise and rest, maintaining mental stimulation, etc.).

A fourth theme considers health a form of freedom, giving the person the capacity to achieve a satisfactory life and freedom to choose how to live. Thus, the person can eat what they want, do what they want, and there not be any obvious adverse consequences to their health. Health is central to a person because choice then becomes possible.

A fifth theme states that health is a form of constraint. It is a constraint in that a person has to act in certain ways and not others to remain healthy. In this regard, getting sick frees a person from these pressures and allows them to defend themselves against the pressures of society. Here a person is retaining personal control by resisting social expectations about being healthy. Talcott Parsons (1972) notion of the sick role overlaps with this definition of health.

1 Theoretical Foundation

A number of investigators have reviewed the literature and have attempted to organize the various definitions of health into coherent groupings. These efforts seem to overlap with each other, but also overlap with the general public's perceptions. Thus, Offer and Sabshin (1966) and Larson (1999) both include the absence of disease or pathology as a category of health definitions. They also discuss wellness definitions as a category, although Offer and Sabshin (1966) includes wellness definitions within the broader category of a Utopian Model of Health, while Larson (1999) lists the WHO definition of health as a separate category. Offer and Shabshin's (1966) Systems Model and Larson's (1999) Environment Model both consider health to be a product of a person's ability to adapt to their physical, social, and emotional environments. Offer and Sabshin's (1966) model differs from Larson's (1999) primarily in defining what is meant by normal mental health. Thus, they approach the definition of health as a task of separating normal from abnormal or healthy from unhealthy, whereas this does not appear to be Larson's primary objective. To help in this task, Offer and Shabshin (1966) include categories that classify definitions based on statistical criteria, or reflect what pragmatically occurs in clinical practice.

The Absence of Pathology Model or Medical Model defines a persons' health or normality in terms of their symptoms, physical signs, and laboratory abnormalities. Larson (1999) traces the origin of this definition of health to Descartes and his contemporaries who characterized the human body as a mechanical entity. If the body consists of parts then understanding how these parts worked helps understand the whole, or in this case, health.

The Utopia Model or, is its surrogate, The Wellness Model, assumes that some ideal level of healthy functioning exists. The physical wellness movement, humanistic psychology models (Rogers 1959), or the WHO definition of health are examples of utopian models. These models encourage people to strive for a higher state of health and well-being than they might ordinarily attain. Larson (1999) states that "Health is defined in the wellness model as strength and ability to overcome illness, having a 'reserve of health'" (p. 129). He goes on to comment that spirituality and religion also play a role contributing to wellness. In general, the wellness movement by encouraging health promotion and disease prevention would appear to have an important future role in establishing health.

The WHO Model of health, with its emphasis on well-being, has been criticized as defining an unknown, health, with another unknown, well-being. Other criticisms of this model (summarized by Larson 1999; p. 128) include that health or well-being are defined differently in different cultures, that there is no clear statement of which health state is better than another and how this would be decided (e.g., Patrick and Erickson 1993), and that most people would be considered sick if this definition was applied since most people experience at least four symptoms over a 2-week period (Wood 1986). While utopian and difficult to implement, this definition remains popular. In addition, it has received significant empirical support from the RAND Health Insurance study, which used the WHO Model to operationalize the measurement of health (Ware et al. 1980).

Offer and Shabshin's (1991) fourth model, the Systems Model is based on the observation that health is a complex product of the interaction of several independent systems (e.g., biological, psychological, and social). The major outcome of these interactions is adaptation, which reflects the various ways that individuals can ensure their health. Larson's (1999) The Environmental Model also focuses on the individual's ability to adapt to their environment and grow, function, and thrive. Health is also not merely biological attainment but rather a dynamic equilibrium with the environment, so that a person's capacity to live physically, socially, and mentally is maximized.

Offer and Shabshin's (1966) Statistical Model defines health, or normality, as the average level of physical or psychological functioning. Health, therefore, is defined by a cut-off level, above which the person would be healthy and below which they would be unhealthy. Examples of the application of this definition would be a blood pressure or cholesterol level which would define the person as being healthy or unhealthy. A final model of theirs, The Pragmatic Model, defines health and disease in terms of what it is that a clinician treats. Thus, if a set of conditions leads to a treatment then these conditions would define what I mean by unhealthy.

Clearly, formal models, less formal definitions provided by lay persons and investigators attempting to summarize the available definitions demonstrate overlap in how health is conceptualized. Each of the schemas, however, has deficiencies. For example, both lay persons and investigators will define health as the absence of disease. Yet some people have a disease but display no evidence of its adverse effects, at least in terms of its impact on functioning or enjoyment of life (e.g., a person with osteoporosis or a person taking medication for high cholesterol levels). Nordenfelt (1995, p. 105) describes four situations where this can occur. First, it can occur if the progress of the disease is aborted at an early stage and the person is assessed beyond the point of active disease. Second, if the disease is of type that causes disability only at late stages and the person is assessed before this time. Third, if the disease is mild and the effect on the subject is negligible. And fourth, if the person has the ability to compensate so that the disease process does not disturb their ability to function.

In addition, there are situations where people seek out medical care when there is no frank evidence of disease.

To deal with these situations, Parsons (1951) and others distinguish between disease and illness. Feinstein (1967) describes the difference between these terms in the context of a physician caring for a patient. Thus, he states:

> The first type of data describe a *disease* in morphologic, chemical, microbiological, physiologic and other impersonal terms. The second type of data describes the *host* in whom the disease occurs. .... includes both personal properties of the host before the disease began (such as age, race, sex, and education) and also the properties of the host's external surroundings (such as geographic location, occupation, and financial and social status). The third type of data describes the *illness* that occurs in the interaction between the disease and its environmental host. The illness consists of clinical phenomena: the host's subjective sensations, which are called "symptoms", and certain findings, called "signs", which are discerned objectively during the physical examination of the diseased host. (p. 24–25)

So far I have discussed health themes which deal primarily with a person's capacities and abilities, but another dimension has to be added to this list, a person's genetic predisposition to develop a particular disease. Thus, Torres (2002) rhetorically asks if whether a person can be considered healthy, who has been told that they are at risk to develop Huntington's Disease. His answer is that this person should be considered unhealthy since they lack the biological basis for continued health. He does suggest that the person should not be considered ill, until there is evidence that they can't function. This he understands runs the risk of stigmatizing the person, but here he calls on transplant medicine to act to prevent the development of the disease, and in this way prevent the person from becoming stigmatized. In support of Torres (2002) is the evidence that the informed Huntington disease patient experiences significant psychological distress from the time they are informed to the time they develop symptoms of the disease (e.g., Timman et al. 2004). Also Rowe et al. (2010) have demonstrated that it is possible to use deficits in self-paced timing to predict development of Huntington disease systems, thereby providing a basis for monitoring the course of the disease and justifying early preventative interventions.

However, there is a range of genetic backgrounds, not all of which are as determined as the diagnosis for Huntington's disease. Thus, should the person with a strong family history of diabetes who avoids sugar-based foods, and is vigilant about their vision and foot care, be considered "unhealthy"? Or should the women who have had a prophylactic mastectomy because of a history of breast cancer in her family be considered "unhealthy." What is also true is that all of us are bearers of deleterious genes, genes that are rarely expressed except possibly when a person reproduces. If in fact I contribute to the birth of a genetically abnormal child does that make me "unhealthy"? The situation gets even more complicated when it becomes apparent that certain diseases become chronic because of alteration of the *genome* of cells involved in controlling a particular process. A classic example of this is the development of chronic pain following up-regulation of dorsal horn cells of the spinal cord due to excessive and repeated stimulation (Chap. 7, footnote 2, p. 240).

To define health as the capacity, ability, and potential to fulfill one's goals overlaps with many of the other offered definitions and yet allows the definition to be operationalized and supported by positive indicators. However, if health is measured at a particular time, as in a cross-sectional study, then there is no opportunity to prospectively determine if a goal has been reached. Under these conditions, all I can learn is what health state the person is currently in. This is how health status is commonly measured.

Each of the formal models I discussed differs in terms of the context within which it defines health. For Boorse (1977), it is the body, for Nordenfelt (1995) it is the person, and for Parsons (1978) it is society. Each also has its own limitations.[3] For example, Boorse's model ignores the difficulty of using statistical criteria to determine if a state has changed (e.g., from healthy to diseased).[4] Nordenfelt's (1995) definition is based on a person's perceptions and is at risk of the criticism of being a subjective measure. For example, if happiness to a person is to remain intoxicated from drugs, do I call this person "healthy"? Of course, it can be argued that this definition of health can't be sustained, since the addiction that results can, in and of itself, lead to disease and disability (e.g., an inability to work). Still, there are situations where optimizing "health" with drugs over an appropriate period of time appears justified, as when the cancer patient desires opiates for pain relief.

Nordenfelt's (1995) person-based definition can be contrasted with a public health or medical definition of health. Thus, whose definition to use adds another level of complexity to the task of defining health. Still if I adopted a person-based approach to the definition of health and apply it to current health status measures, then I will most often find that it does not match what is ordinarily called a health status measure. In fact, what I currently call a health status measure might be best called a "disease" or "illness" status measures.

In addition, some investigators have suggested that the various definitions of health be combined into a single comprehensive definition. Thus, Johnson (2003) speaks of the *bio-psycho-socio-cultural* model of health (see also Ahmed and Coelho 1979). Johnson claims that this creates a workable model that combines the WHO definition, with Engel's (1977) biopsychosocial model, and suggestions made by the Institute of Medicine (2001). What is important about this model is that it argues that biological, psychological, social, and cultural factors are co-equal determinants of health. Thus, a healthy state would be defined as not smoking, not being overweight, controlling stress, and so on. In the sense that specific behavioral goals are set, this definition overlaps with Nordenfelt's (1995) definition.

Young (1998) cites an additional definition of health, a definition proposed by the European Region of the World Health Organization. It states:

> [Health] is the extent to which an individual or group is able, on the one hand, to realize aspirations and satisfy needs, and on the other hand, to change and cope with the environment. Health is therefore seen as a resource for everyday life, not the objective of living; it is a positive concept emphasizing social and personal resources as well as physical capacities. (p. 435)

Defining health as *a resource* suggests that health is something that is built up, or developed, and must, therefore, be a dynamic and multifaceted process. Health, defined in this context, would not be confined to a single domain, but rather would be a product of the relationship between multiple domains (e.g., biological, psychological, social, and cultural). This "field model" of health has been increasingly used to analyze and intervene on a whole range of health-related issues according to Halfon and Hochstein (2002).

Halfon and Hochstein (2002) offer a definition of health that goes beyond describing a state or achieving a goal, but rather combines each of these criteria as part of a developmental process. They state; "…we define health development as a lifelong adaptive process that builds and maintains optimal functional capacity and disease resistance" (p. 437). Health for them becomes an entity that can be attained and maintained, but since people differ in experience and constitution they each may have unique definitions of health. To achieve health a person has to manage the factors (their capabilities, abilities, and potential) that place them at risk and engage in behaviors that promote their health. The balance of these activities, plus factors that maybe beyond direct control, such as the person's genetic, social, and economic background, contributes to the person's health status. In addition, critical developmental periods (e.g., childhood, adolescence, etc.) exist that will also determine a person's future health status. All these factors provide a person a health trajectory, a life pathway that is influenced by the multiple environments that the person inhabits. Halfon and Hochstein (2002) also see the health trajectory as differing from a life course model that sees development as a linear process, divided into discrete ages and stages bounded by death.

Although I asked a straightforward question in this section ("What does the word *health* mean in a health status measure"?) it is clear that I have no straightforward answer. I have learned that there are several basic approaches to the definition of health: one describes health as a state a person is in; another that health is a goal a person achieves; and a third combines both over time, viewing health as a dynamic, developing, adaptive process. I found that the formal definitions provided by investigators, lay person's views, or investigators provided a wide but overlapping range of ideas about what the term health means. From a practical perspective, this diversity is sufficient to justify that an explanation be provided when the term is used by an investigator. This seldom happens, of course. Instead, there is a tendency to define health as the absence of disease or illness. However, it is apparent that to define health as the absence of a deficit may not succeed, especially since it is quite difficult to define when this state is violated. The alternative is to add additional factors, as reflected in the biopsychosocial model of Engel (1977) or the bio-psycho-social-cultural model of Ahmed and Coelho (1979) or Johnson (2003). This results in definitions of health that consist of multiple levels of analysis and divergent measures (e.g., subjective and objective). An alternative proposed by Halfon and Hochstein (2002) views health as the end result of a developmental process. This approach has the advantage of independently tracking different levels of analysis (e.g., genetic, social, etc.) with health being defined by a particular, more concrete indicator, such as functional status at each time interval. However, I am left with acknowledging that the term health is an inherently ambiguous concept, and I wonder what this means for the phrase "health status."

1.1.2 Defining a Disorder

A disorder, as is also true for the terms disease, illness, or dysfunction, focuses on what it is that is different from normal. These terms refer to clusters of indicators (e.g., symptoms, lab tests, etc.) that can be summarized into a domain. This gives me another opportunity to examine the nature of a domain. It will become clear that the same definitional encumbrances that plagued the definition of health, or HRQOL, also limit the definition of the terms disorder or disease. As such, the effort to shift the definition of health to the supposedly more concrete criterion of "the absence of a disorder or disease" will have significant limitations.

I will start this discussion with a series of papers that were published in the *Journal of Abnormal Psychology* (108; 374–399: 1999a) that dealt with the issue of defining what a disorder is, especially when applied to instances of psychopathology.[5] Two papers, one by Wakefield (1999a) and one by Lilienfeld and Marino (1999), started an extended discussion that included comments by some nine other authors. Wakefield (1999a) defined a disorder as a *harmful dysfunction*, while Lilienfeld and Marino (1999) suggested that a cognitive *Roschian analysis* was sufficient to define a disorder. Their debate raised a number of issues, including:

- How should a disorder domain be constituted?
- What role do value statements[6] play in determining what a disorder is?
- What role do evolutionary processes play in determining what the term disorder means?

I will briefly present each author's approach, and then include selective comments from the nine additional papers responding to the Wakefield (1999a) paper. Note that Clark (1999) states that while Wakefield and Lilienfeld and Marino present themselves as offering opposite views, they also offer suggestions that would permit the integration of their perspectives.

Wakefield (1999a) defines a disorder as a "harmful dysfunction." He acknowledges that a harmful dysfunction is a hybrid cognitive construct, consisting of one term, "harmful," which is used in what he calls a nonscientific evaluative manner, and another, "dysfunction" which can be scientifically established.[7] He states:

> The harmful dysfunction (HD) analysis rejects both the view that disorder is just a value concept referring to undesirable or harmful conditions and the view that disorder is purely a scientific concept. Rather, the HD analysis proposes that a disorder attribution requires both a scientific judgment that there exists a failure of designed function and a value judgment that the design failure harms the individual. (Wakefield 1999a; p. 374)

By "designed function" Wakefield (1999a) is referring to the functions a person is capable of performing as a product of evolutionary processes. Since functions reflect adaptations and are useful, they can be seen as causing certain ends. Wakefield (1999a; p. 375), acknowledging his teleological perspective, gives as an example the heart pumping blood, which is seen as an end of the natural functioning of the heart, and as such, explains why the heart exists and how it is structured. As a consequence, if the heart stops pumping then some internal mechanism is dysfunctional in the sense that the heart cannot fulfill the role developed as a result of evolutionary processes. As Wakefield (1999a) succinctly states, "disorders are failures of mechanisms to perform functions for which they were naturally selected" (Wakefield 1999a; p. 375). Natural selection, therefore, provides a set of "givens," the breakdown of which leads to a disorder.

In contrast, Lilienfeld and Marino (1999) propose a cognitive "Roschian" analysis to account for a disorder. A Roschian model of a concept claims that a disorder is an example of a prototype model that has fuzzy boundaries. As previously discussed (Cronbach and Meehl 1955; Smith and Medin 1981; Wittgenstein 1953), a prototype concept is fuzzy because there is no defined basis to determine if a specific instance belongs to a concept or not, and this leaves placement of an instance in a category as a probabilistic or arbitrary decision. Thus, it will not be possible to absolutely distinguish all cases of a disorder from all cases of a nondisorder. In addition, Lilienfeld and Marino (1999) state that:

> Roschian concepts are organized around an ideal mental prototype that embodies the central features of the category. According to our Roachian analysis (RA) of disorder, judgments regarding whether persons possess a disorder are based on their similarity to an ideal prototype of a "diseased person". (Lilienfeld and Marino 1999; p. 400)

Here, the notion of a "diseased person" comes from experiences with persons who have similar characteristics.

Lilienfeld and Marino (1999), who accept Wakefield's notion that an evaluative/valuation component has to be present in any definition of a disorder, are instead concerned about how he defines dysfunction. They specifically reject Wakefield's conceptualization of dysfunction as being dependent on evolutionary processes. They also reject the idea that physical and mental systems, for example, evolved specifically to perform given functions, and instead argue that the natural selection of systems has incidentally become associated with adaptive tasks. Language, they state, is an example of a system that may have evolved following the natural selection of efficient motor sequencing and timing, with the adaptive significance of language only emerging as a secondarily consequence of these evolutionary changes (Calvin 1983). In addition, since Wakefield (1999a) argues that a dysfunction is the failure of a system to perform as initially designed, his original conceptualization of dysfunction is not able to account for secondary adaptations. Lilienfeld and Marino (1999) do point out that Wakefield's (1999a) conception of dysfunction can account for secondary adaptations if applied to current circumstances.

Where Wakefield's (1999a) ideas break down is when he has to deal with functions that are not a direct product of evolutionary processes but rather reflect specific or general modular domain capacities that have been shown to be adaptive, such as a person's religious beliefs, their athletic and motor skills, their arithmetic or music skills, and so on. His emphasis on evolutionary processes has also led to his not being able to include a wide range of neurological disorders as dysfunctions. According to Lilienfeld and Marino (1999), Wakefield's major conceptual weakness is his assumption that all mental and physical systems are a product of evolutionary processes.

A core difference between these two approaches revolves around what the phrase, "the essence of things," means. Medin (1989) discussed this topic when he considered the difference between a classical and prototype approach to a concept (Smith and Medin 1981). He pointed out that in the classical model, people infer and assume that concepts possess underlying properties or essences (see the quotation from James; Chap. 8, p. 247). Thus, if two objects look alike (i.e., they are similar), then it is assumed that they share some common deeper property. As Lilienfeld and Marino (1999) state:

> Essentialists believe that natural concepts (i.e., concepts referring to entities in nature, rather than to artifacts) correspond to genuine distinctions in the real world. (p. 400)

In contrast, nominalists "believe that natural concepts are mental constructions that bear no direct correspondence to reality and that there are no unambiguous joints in nature to carve" (Lilienfeld and Marino 1999; p. 400).[8] Lilienfeld and Marino (1999) suggest that Wakefield (1999a) is taking an essentialistic view of what a disorder means, suggesting that function and "dysfunction" are parts of a natural separation based on evolutionary processes. They also state that

Wakefield claims that "harm" is a necessary and "dysfunction" sufficient to define a disorder.

Medin and Ortony (1989) propose that the formation of a natural category occurs around an "essence placeholder." An essence placeholder is a set of underlying beliefs that someone may have about, for example, a diseased person who needs medical or psychological care. But Lilienfeld and Marino (1999) go on to state that:

> This placeholder is neither necessary nor sufficient for a disorder but instead provides an approximate anchor that undergirds the prototype for disorder. The inference that there is something wrong with the body or mind is, we hypothesize, itself based on observations regarding several variables, foremost among which is a sudden or marked decrement in functioning of physical or mental systems… (p. 402).

In addition, if someone requires treatment for a disease, then this belief is part of the essence placeholder for the disorder.

Lilienfeld and Marino (1999) claim that someone may have "something wrong" with them and that this need not be evidence that they have a dysfunction or disorder. In contrast, Wakefield (1999a) argues that if there is "something wrong" with a person, they must have a disorder and that this disorder is "rooted in the failure of a naturally selected system to perform its designed function" (Lilienfeld and Marino 1999; p. 403).

Lilienfeld and Marino (1999) also give two examples of where an essentialism perspective has been applied to two totally different problems: how species are defined and the concept of life. In both cases, the attempt to provide necessary and sufficient defining criteria has not succeeded. Thus, Lilienfeld and Marino (1999; p. 409) quote Hull (1976), who states:

> …initially it was thought that the names of all species could be defined by sets of essentialistic traits' but that "no matter how hard they tried, taxonomists could rarely find sets of traits which divided living organisms into neat little packages." Our inability to distinguish most species by sets of necessary and sufficient conditions follows from evolutionary theory. (p. 180)

Lilienfeld and Marino (1999) go on to quote Medawar and Medawar (1983), who state that there is no necessary and sufficient definition for life; the word has no meaning, but rather has a usage that serves a function for biologists.

In contrast, Wakefield (1999a) believes that it is possible to discern a nondisorder from a disorder; that precipitous events mark the onset of a disorder; and that each particular disorder has a unique set of defining characteristics. All of these are characteristics required for establishing a domain. Roschian analysis, in contrast, claims that a disorder is a probabilistic statement based on a mental construction (a prototype), and is not dependent on a precipitous event, such as the breakdown of an evolution-dependent process. Of particular interest to me is that essential elements of each side of this debate reflect one or another of the two major approaches for modeling cognitive concepts (Smith and Medin 1981). This tells me that the decision about how to define a disorder, or even a domain, is very much related to the cognitive model of a concept that an investigator either implicitly or explicitly chooses to use.

A second issue raised by the Wakefield (1999a) paper has to do with his claim that it is possible to combine a value statement (i.e., harmful) with a descriptive statement (i.e., dysfunction) to form a hybrid construct. Whether this is a useful strategy is very relevant to my discussion, since it is essentially the type of definition Patrick and Erickson (1993) used for their definition of HRQOL (Table 2.8). It is also the model used by a number of investigators who have developed generic quality-of-life or HRQOL assessments (e.g., Kaplan and Anderson 1990; Torrance 1976).

Fulford (1999) agrees with Wakefield's (1999a), suggesting that it is possible to create a hybrid definition, but questions whether it is possible for a dysfunction descriptive statement to be value free, especially since the evolutionary processes that are broken down when a dysfunction occurs are conceived of as having a purpose, goal, or design, terms that have evaluative meaning. His own view is that descriptive and evaluative elements are woven together throughout the conceptual fabric of the subject [including of course the concept of dysfunction] (Fulford 1999; p. 420).

In this regard it is interesting that Fulford (2001) has also taken issue with Boorse's (1975) attempt to create a value-free definition of disease. According to Fulford (2001), Boorse (1975) and Wakefield (1999a) are attempting to demonstrate that it is possible to have a value-free scientifically based medical theory by distinguishing between value-laden medical practice and medical theory. Consistent with this, Boorse (1975) distinguishes disease from illness, where illness is thought of as value dependent. Fulford (2001) comments that while Boorse (1975) tries to talk as if a value-free scientifically based medicine exists, in practice the language he uses includes evaluative value-dependent terms. Boorse (1975) succeeds if he confines his examples to physical illnesses, but Fulford (2001) points out that his approach breaks down when applied to mental disorders. While everyone agrees that cancer is a bad condition, I cannot find this agreement about certain desires, emotions, or behaviors that are characteristic of psychiatric conditions. According to Fulford (2001; p. 84), this difference in the role of valuation between physical and mental disorders is not due to some inherent limitation of the definition of mental disorders, but rather to the notion that disorders can be value free. I return to Fulford's arguments in Chaps. 9 and 12, but for now his comments have made me acutely aware of how values can be part of the conceptualization of a disorder and, therefore, a domain.

Others who have weighed in on the Wakefield and Lilienfeld and Marino debate are Klein (1999) and Spitzer

(1999), both of whom are supportive of Wakefield. Spitzer, for example, acknowledges the value of Wakefield's notion of "harmful dysfunction," but says that even if included in the DSM system, it would not likely change the list of categories of mental disorders. Less supportive of Wakefield were Sadler (1999) and Richters and Hinshaw (1999). Richters and Hinshaw (1999) acknowledge that Wakefield's notion of harmful dysfunction comes very close to what most clinicians and lay persons think of when they think of a disorder, but they take issue with him about how applicable his model is for accounting for mental disorders. They state:

> The evolutionary cornerstone of Wakefield's (1999a) proposal is an unexamined premise that the natural functions and dysfunctions of the brain can be defined exclusively with reference to the evolutionary past. We challenge this premise…and argue that it fails to take into account the design openness and experience-based modifiability of higher-level brain processes associated with cognitive, social, emotional and personality functioning. The ability of these recently evolved brain capacities to continue evolving non-genetically in response to environmental contingencies renders the evolutionary past insufficient for conceptualizing their natural functions and dysfunctions. (p. 439)

Cosmides and Toby (1999), while supportive of Wakefield's emphasis on evolutionary processes as a basis for defining dysfunction, point out that the adaptations which evolve are quite complex. Thus, they contend that some evolved functions may cause conditions that are acceptable to a person, while others would not and might require treatment. An example of an evolutionary dependent condition a person would find unacceptable would be the repeated nausea and vomiting which occurs in response to ingesting a foreign substance. This mechanism that protects the person is an example of a positive function that, if it leads to dehydration, may require active treatment. Thus, the breakdown of evolved adaptations can be both harmful and beneficial to a person. The authors go on to sketch "an evolutionary taxonomy of treatable conditions" (Cosmides and Tooby 1999; p. 453).

The final paper in this series by Kirmayer and Young (1999) focuses on the roles that culture and context play in an evolutionary theory of mental disorders. Kirmayer and Young (1999) feel uncomfortable with Wakefield's (1999a) perspective because they believe that it does not correspond to how the term "disorder" is used in a system such as the DSM, and because clinicians do not think of a disorder in term of a monolithic, single category of a disorder, that Wakefield proposes. Instead, they agree with Lilienfeld and Marino (1999) that a disorder is polythetic, with multiple determinants of the category. Some of these determinants are clearly cultural and contextual in nature.

Klein (1999) claims that Wakefield (1999a) successfully enjoins Lilienfeld and Marino's view that a disorder is polythetic. Klein (1999) points out that the classical view of a concept requires that explicit inclusion and exclusion criteria exist that establish the necessary and sufficient conditions for membership in a concept or category. Klein (1999), however, believes that a totally polythetic category may still have a common underlying construct that organizes the various elements in the category. Thus, Klein (1999; p. 427) states that because, "two members of a polythetic illness category may not share a single symptom does not imply that they do not share something even more important: an underlying dysfunction." He gives the example of two persons who meet the DSM system description of having had a manic episode and yet have no common symptoms. Wakefield (1999b) came away from his reading of these various papers concluding that his notion of a disorder as a "harmful dysfunction" remains essentially intact. His work will be of continued interest, since it has implications for how I conceive of "functions" as a part of an HRQOL domain, the topic of the next chapter. In addition, the cognitive basis of Wakefield's notion of harmful dysfunction or evaluation/valuation is regularly replicated in the design of HRQOL assessments.

I reviewed the debate between Wakefield (1999a, b) and Lilienfeld and Marino (1999) because I saw it as a means of gaining insight into how to define a domain. What I learned was that the debate was really a debate about which of two approaches to the cognitive processes of how a category or concept is formed (e.g., classic model or prototype). should be used in defining a disorder. This suggests that the same decision may have to be made when deciding on which model of a domain I should use. Knowing this, I can inspect the available empirical literatures on each of these processes, and use this information to clarify my choice.

I also learned that values have a natural role to play when defining a disorder, and presumably, in a quality-of-life or HRQOL assessment as well. Deciding where this occurs is, of course, a debatable issue, but as I indicated when discussing the various definitions of quality of life or HRQOL (Chap. 1), it is essential for a unique definition of these phrases.

Fulford (2000) raised an additional critical issue when he asked where among the various types of domains value statements should or should not be made. In his response, Fulford, following Boorse (1975) and Wakefield (1999a,b), examined the question in the context of "the naturalization of biology, medicine and psychiatry." By naturalization he meant the effort to characterize these fields in terms of the natural sciences. Thus, Klein (1999), when he discusses the domain of anxiety, does not just talk about some subjective experience, but also considers the neuroendocrine basis of anxiety, and it is this reliance on biological indicators that characterizes the naturalization process. Fulford (2000) points out that function, dysfunction, disease, illness, and disorder are terms whose meanings are linked. He states:

> The details vary, but broadly speaking they are taken in a logical cascade. In this "naturalization", as I will call it, disorder includes disease and illness, illness (experience of illness) is

defined by reference to disease, disease by reference to dysfunction, and dysfunction by reference to function…The importance, therefore, of biological function statements to the naturalization project is that they appear to provide a value-free scientific foundation on the basis of which the other terms in the naturalization cascade can be built up. Most authors realize that values must come in at some point in the cascade: if not with dysfunction, then with disease, if not with disease, then with illness, if not with illness, then with disorder. But provided biological function statements are value free, the naturalization project, it is widely assumed, can at least get underway. (Fulford 2000; p. 78)

The naturalization process, Fulford and others discuss, has also been applied in the Wilson and Cleary (1995) model of an HRQOL assessment. For now what seems clear is that an investigator cannot avoid, as far as Fulford is concerned, finding a place for a value statement in any definition of a disorder, and I would say for a qualitative domain, as well.

In addition, this discussion gave me the opportunity to see how a single conceptual approach (e.g., natural selection and evolutionary processes) provided a basis for differentiating and defining domains: normal and abnormal; functional and dysfunctional; nondiseased and diseased; and nondisordered and disordered. This too was important because it made clear that when I define a domain, I have to be able to distinguish how one domain differs from another.

Finally, what was most informative about Wakefield's (1999a) approach to mental illness was that he used a cognitive model, a complex hybrid construct, to represent how a disorder could be defined, but in so doing he used a cognitive model that I will show reoccurs whenever a qualitative judgment is made (see Chap. 12).

1.2 Assessing Health Status

It is not clear who first used the phrase "health status," but it most likely evolved from a public health setting where methods were being developed to characterize the health of the population.[9] The first efforts in this regard involved surveys of health conditions, progressed to the construction of health indexes, and then developed indicators that described health states. This history can be broadly sketched as progressing from measures based exclusively on mortality statistics to measures (i.e., indexes) that combined both mortality and morbidity statistics, and finally to mortality and morbidity indexes that were utility- or value-adjusted.

The use of mortality statistics to assess the health of a population dates to the 1600s, when Captain John Graunt reported using "bills of mortality" for such calculations (Todhunter 1949). Stouman and Falk (1939; p. 906) refer to Edwin Chadwick's report, *Sanitary Conditions of the Labouring Population of Great Britain* (1842) and Lemuel Shattuck's *Report, Sanitary Commission of Massachusetts* (1850), as examples of health surveys. John Billings (1875) was first to attempt a complete system of health appraisal (Stouman and Falk 1939; p. 907). Stouman and Falk (1939) were asked by the League of Nations to devise numerical indexes of vitality, health and the environment, and public health activities. By the 1970s a variety of such indexes had been created, including the health status ones that were preference (using utility estimates) or value adjusted (e.g., Torrance 1976). The period from the 1970s to the present has seen both the further development and expanded use of these indexes in a variety of settings (e.g., clinical practice, clinical trials, and population assessments).

Much of the impetus for these developments came from the demands of policy makers to monitor the consequences of social programs designed to alleviate social injustices, especially during the 1960s. An example was the effort to establish community health profiles for the purpose of identifying areas with service scarcity. These profiles could then be used to allocate resources. Key to establishing community health profiles was the development of health status assessment (Levine and Yett 1973), which is likely to continue, especially with the applications of computer-based health records increasing the accessibility of such information and the monitoring of the economics of health care becoming an increasingly important public policy objective.

This history, combined with the growing awareness of the cost burden created by medical care, led to the development of methods to monitor the effectiveness and efficiency of such care. One result was the recognition of the importance of patient outcomes to monitor the consequences of disease or treatment. Initial efforts in this regard date to Codman during the beginning of the twentieth century, but required some 70 years before the Joint Commission on Accreditation of Hospitals (JCAHO) formally required health care institutions to include outcome measures in order to qualify for institutional accreditation. A natural product of these activities was attending to the qualitative state reflected in these outcomes, and this led naturally to development and acceptance of HRQOL assessments as an outcome indicator. However, HRQOL assessment is not yet considered a requirement to satisfy Joint Commission accreditation.

1.2.1 End-Results and Outcomes Movement

A major contributor to this effort was the "end-results movement" that was initiated by Ernest Amory Codman (1869–1940), and since it developed at the beginning of the twentieth century, it actually preceded the formal interest in health status assessment. Codman was an orthopedic surgeon who made significant research contributions to shoulder surgery, and who early in his career also became interested in monitoring the consequence of his surgical practice. What separated Codman from his peers was his willingness to publically discuss the outcome of his surgical efforts. In 1910, one year after becoming a full-time staff member of the

Massachusetts General Hospital, he resigned and established his own hospital where he felt he could study and practice medicine as he saw fit, including monitoring and discussing in public his treatment successes and failures (Mallon 2000). He developed a procedure for recording a patient's surgical history on a note card (an innovation at that time) and followed up with each patient after a fixed period of time to determine the outcome of his surgical and medical interventions. A major falling out with his peers occurred in 1915 when, as chairman, Codman organized a meeting of the Surgical Branch of the Suffolk District Medical Society around the topic of hospital efficiency. At this meeting he displayed an eight-foot long cartoon depicting his image of medical practice around the Harvard community. The title was *The Back Bay Golden Goose Ostrich*, and featured an ostrich with its head in the sand generating golden eggs. Present in the illustration were members of the Harvard faculty who were depicted as graciously accepting the eggs, while the trustees of Harvard asked if the ostrich (representing members of the Back Bay community in Boston) would be willing to continue to lay these golden eggs (pay their fees) even if they were informed of the outcome of the medical and surgical care they were getting.[10] Naturally, this presentation was considered an insult, and Codman was asked to resign from his position as chairman.

Codman (1917/1996) made other important contributions to hospital efficiency, including suggesting that the newly formed American College of Surgeons have a committee for the accreditation of hospitals (of which he was initially made chairman). This committee still exists but has now evolved into the Joint Commission for the Accreditation of Healthcare Organizations, an organization whose major function remains accrediting hospitals.

Unfortunately, Codman's contributions were soon forgotten. However, his efforts to improve the quality of medical and surgical care continued through others, some of whom were never aware of his pioneering work, but also had to confront similarly difficult professional and intellectual environments. His activities, like the others, dramatize that the conceptual metaphor "health" carries with it an implicit ethical and moral accounting imperatives that becomes relevant when the consequence of medical treatment is considered. Still, Codman's efforts were not in vain (Donabedian 1989), as evident by the current interest in quality control (Shortell et al. 1998), outcomes research, and evidence-based medicine (Sackett, et al. 1998).

The history that preceded this upsurge in interest is complex and included putting in place a number of building blocks. One was the use of the randomized clinical trial for evaluating alternative therapies, a practice that started in earnest during the 1950s and 1960s. The resulting data provided a more useful standard for use in judging what constituted an adequate treatment program, although clinicians did not fully take advantage of this new higher quality of data (Cochrane 1972). The presence of these standards also made it easier to define when a violation of practice occurred. A second stimulus was the increased interest by consumers in the quality and nature of the care they received, which created pressure for change. A third factor was the economic incentive designed to ensure that the quality of medical care was cost effective and efficient. Thus, while Codman dedicated his life and career to a belief, it has since become clear that an institutional commitment was required if the changes he hoped for were to be accomplished.

An example of such institutional commitment was when the Joint Commission (O'Leary 1987; Roberts et al. 1987) formally required hospitals seeking accreditation to monitor the outcome of their various activities (e.g., mortality rate, medical procedure rates, rehospitalizations, repeat surgical repairs, etc.). This provided the kind of data necessary for policy decisions and the justification for interventions designed to meet various medical needs. Another example of institutional commitment came during the 1970s when further development of health status measurement occurred. At that time, the National Center for Health Services Research and Health Care Technology Assessment sponsored a series of research grants that resulted in the development of several generic health status assessments, as they are known today.[11] As I mentioned in Chap. 1, many of these grants applied techniques and ideas from operations research. Prior to this, the public policy initiatives initiated during the 1960s stimulated the work of a group of sociologists and others to focus on assessing "quality of life," including quality of life in a health care setting. This concordance of events brought the terms "health status" and "quality of life" together, more narrowly stated as "health-related quality of life" (HRQOL).[12] The realm of interest was narrowed because some felt that issues such as income, crime rates, and living conditions would only minimally impact the medical status of a person and were certainly beyond what medicine could be expected to significantly alter (Guyatt 1993, 1995; McCarty 1995). However, there is not universal agreement with this view (e.g., Gill and Feinstein 1994).

Patrick and Erickson (1993) listed five domains to "operationalize" what they meant by HRQOL (see Table 2.8). These domains are resilience, health perceptions, physical function, symptoms, and duration of life (Patrick and Erickson 1993; p. 22). Clearly this is a wide range of domains, although other investigators have suggested different combinations of domains. However, doubt (e.g., Beckie and Hayduk 1997; Michalos 2004) has also been raised about the viability of the HRQOL as a construct, and some have actually suggested that it should be abandoned. I review these papers in Sect. 2 of this chapter.

1.2.2 Operations Research in Health Care

In Chap. 1, I pointed out that many of the early innovators in quality-of-life research were influenced by or trained in operations research. To these investigators, health care assessment offered a unique opportunity to apply principles they had been successfully applying to a variety of environments (e.g., defense applications, product development, etc.). Health status assessment provided the data that would feed an operations research analysis and intervention. In addition, operations research was beginning to be applied to a variety of health-related organizations (e.g., hospitals) and systems (e.g., a health care system; Flagle 1962).

In this section, I hope to provide a general background to what operations research is and can do, leaving a detailed discussion of its various applications to others. I have included this information not just because of its historical significance to quality-of-life research, but because it offers analytical tools important in any successful effort to implement a quality (control) program oriented to preventing adverse outcomes or ensuring that some enhancement program has succeeded (Table 1.2). There are actually very few specific applications of operations research that have quality of life as a primary outcome variable. More commonly, operations research has been integrated into health status assessment (e.g., by Torrance and his colleagues), especially those focused on the economics of health care.

Operations research is a form of applied decision making, the objective of which is to optimize *how* a group (e.g., doctor–patient interaction), institution (e.g., hospital or company), or system (e.g., an entire health care system) functions. Inspection of the history of operations research reveals a linage that can cite significant intellectual contributions from gambling, probability theory, and decision making. Thus, Gass and Assad's (2005) first reference is to Cardano (1501–1576; a physician, mathematician, and gambler), but they also cite Bernoulli (1700–1782), Bayes (1702–1761), Bentham (1748–1832), Frederick Taylor (1856–1915), Pareto (1848–1923), and a host of others. These investigators developed the mathematical models applied to various decision-making settings, and these approaches were also common to economics analyses, especially those concerned with the efficient distribution and utilization of resources (Baumol 1965). When operations research was applied to human welfare, the judgment of how an optimal outcome occurred became an issue, and the principles developed to ensure such an outcome became an issue.

There are three basic ways how a qualitative judgment or decision is made: *normative, descriptive,* and *prescriptive* (Bearden and Rapoport 2005; Bell et al. 1988). In the normative approach, specific models of decision making are postulated and applied. Formal models, such as the von Neumann and Morganstern (1944) game theory, would be an example of a normative approach. A key assumption of these models is that the decision maker will act as a rational agent, an assumption that has been subject to much experimental evaluation and is, in general, found to be suspect (e.g., Stern et al. 2003; Chap. 12). Interestingly, most economists persist in believing that the decision maker is rational, and instead view the discrepant data as an outcome that has to be accounted for, not a basis for reshaping their fundamental assumptions. Descriptive approaches are mostly concerned with how people actually make decisions, not how well they conform to some predetermined model of how decisions are made. Theories evolve from experimental studies of decision making. Prescriptive approaches are concerned with determining how decision makers can be encouraged to make decisions more in conformity with normative decision-making models. Each of these approaches requires additional experimental study, especially if applied to a qualitative judgment.

The overlap in the intellectual foundations of economics and operations research stems from the fact that both fields have models built and developed with mathematical principles (Baumol 1965; Murphy 2001). In addition, operations researchers were often interested in finding the most efficient economic solution to the problems they were dealing with. This resulted in both groups being interested in such topics as optimization theories, inventory control, and game theory. The two fields are also distinct, with economics being primarily concerned with applying models of economic activity to advise policy making, while operations researchers are more interested in engineering decisions at the individual, institutional, or systems level. The operations researcher contributes to policymaking by simulating various outcomes and presenting them to policy makers, as opposed to recommending policy based on specific economic models. Operations research is often seen as providing the basis for making management decisions.

When applied to health care, operations research provides answers to such questions as the best way to assess the quality of medical care, improve medical diagnosis, improve the delivery of health care, identify the therapy that is best for a particular disease, identify medical errors, and so on. Operations research helps model outcomes using simulation techniques, applying stochastic theory, using optimization estimates, and using Monte Carlo statistics. This permitted the development of an adverse events detection system, but also a system that monitors ongoing processes.

An example of an application of operations research intervention that had obvious implications for patient quality of life (Bio-Medical News; 2005) involved a drug (Alimta) that was being developed by the drug manufacturer Eli Lilly. Early trials (phase I) of the antifolate drug showed encouraging antitumor effects, but also that severe toxic side effects occurred. Multivariate regression analysis techniques commonly used by operations research helped determine that lowered foliate and vitamin B_{12} levels most likely caused the observed toxicities. Providing patients with appropriate nutritional supplements prevented the onset of these adverse effects and allowed the continued development of the drug. The drug is now approved by the FDA for treating lung cancer.

Richter (2004) provides some additional examples of how operations research can model issues of interest to the drug development process and its clinical applications. This is especially true in the situation where a technology transfer is the objective; as when a clinician has to rely on the results of a clinical trial to decide whether or not to use a new medication in a clinical setting. Under these conditions, either simple decision trees or complex simulation can be used to model clinical outcomes. Richter (2004) illustrates these options by describing decision tree models for hypertensive medications, Monte Carlo simulations of HIV/AIDS viral load testing frequencies, and Markov models of cancer treatment medications.

The detection of adverse events is an example of a strategy that can minimize the risk or prevent adverse qualitative outcomes. The design of such interventions is another example of how an operations research analysis can be used. The nature of the risk and magnitude of the problem was summarized in a paper by de Vries et al. (2008), who found after an extensive literature review that adverse events in hospitals occurred at a median rate of 9.3% across studies. They also found that 43.5% of these adverse events were preventable, but that the remaining 56% of the patients experienced no or minor disability. Only 7.4% of the 9.3% adverse events were lethal, with 39.6% of the fatalities due to operative events, and 15.1% due to medication errors. These data were sufficient to justify the development of preventative interventions designed to ensure the safety and qualitative experience of the hospitalized person. One example of such an approach (Szekendi et al. 2006) took advantage of 21 electronically obtainable triggers from computerized hospital records to monitor for adverse events. Szekendi et al. (2006) found, following a review of 327 medical records, that 243 of these records had a medical error. Of these errors, 47% required intervention to prevent or ameliorate the potential for harm.

Modeling logistic and resource management issues is another example of an application of operations research with qualitative implications. Xiong et al. (2008) simulated a global scale-up of a treatment process for HIV patients using logistic and operational analysis procedures. They found that it was possible to forecast demand for services, locating and sizing of treatment facilities for maximum efficiency, and determining optimal staffing levels using operations research methods. They were able to make predictions concerning resource utilization, drug consumption, clinical capacity needs, and provide plans for policy and staff needs and evaluation.

Cost–benefit and cost-effectiveness analyses are common operations research methods used in program evaluation, but these approaches have deficiencies when applied to health care (Torrance et al. 1973). Cost–benefit analyses measure the economic, but not necessarily the health, consequences of health care programs, while cost-effectiveness analyses are limited to assessing the health benefits within health programs, precluding comparisons between programs. Both Torrance et al. (1973) and Bush et al. (1973) present alternative approaches to the health care program evaluation task. Torrance et al. (1973) describe his approach as a utility maximization approach, while Bush et al. (1973) describe an alternative method for generating preferences to attach to different health states. These investigators' early efforts eventually matured into two distinct approaches to generic quality-of-life assessments.

Operations research is not only a formal set of analytical procedures (mathematical models), it is also an idea that monitoring and modulating a process (e.g., providing feedback) will increase the productivity and qualitative outcome of some effort. The application of these ideas can be observed in various health care settings (e.g., a hospital or health system) where efforts at quality control are a concern. Donabedian's work (1980) has been particularly important in these applications, providing a theoretical overlay that has guided much research in this area, and by demonstrating the relevance of the concept of quality to these activities. The question of concern to me is whether his model and approach to defining and assessing quality is adequate.

1.2.3 Donabedian's Definition of Quality (of Care)

In his book entitled *The Definition of Quality: Approaches to its Assessment*, Donabedian (1980) provides both a definition of quality and a conceptual model (structure, process, and outcome) of how to assess the quality of medical care. He states (Donabedian 1982), "...an assessment of quality is a judgment concerning the process of care, based on the extent to which that care contributes to valued outcomes" (p. 3). Of course, by reference to valued outcomes, he was also alluding to a cognitive process (i.e., a judgment) that resulted in the outcome that was valued. This definition, while not a precise restatement of the definition I suggested, does acknowledge a role for the key elements of what I claim are involved in the assessment of quality. Donabedian goes on to cast his discussion in terms of the quantity of care: with too little care being evidence of poor quality, excessive care a cause of diminished quality, with an optimum level of care the desired end. Donabedian (1980) states:

> We must therefore conclude that whenever a judgment is made about the necessity or suitability of the quantity of care a judgment of quality is implied. Assessment of the quantity and quality of care are thus inextricably intertwined, and will be so treated in this book. (p. 7)

Donabedian's statement implies that he expects judgments about the services a person received (either in terms of frequency or cost of services) to be valued, and when this occurs a judgment about the quality of this care will have occurred. The alternative, only counting the frequency of care or the amount of money spent on the care, he would reject as an inadequate indicator. Thus, Donabedian recognizes the importance not only of a subjective evaluation but also a valuation of this care when assessing the quantity of medical care.

Donabedian also distinguishes between the "technical management of care" and the "management of the interpersonal processes" involved in determining the quality of care. He expects that his "quantity equals quality" assumption would most likely work when considering technical aspects of care, but would be less likely to when dealing with interpersonal relations. He also refers to the existence of social criteria determining the quality of care, that to him would involve using society-wide or population-based criteria (e.g., what is best for the society as a whole?) to judge quality. Most important is that Donabedian recognizes that each level of analysis provides a legitimate perspective, and the selection of which perspective to take is best determined by the purpose of the assessment.

Just as there are benefits and risk in the management of technical aspects of care, so too there are benefits and risk in the interpersonal management of care. Both generate costs. Thus, the process of care requires that attention be placed on patient's comfort, privacy, and so on, all of which contributes to patient satisfaction. Donabedian acknowledges that it is the patient who is most appropriate to value the care they receive, although he acknowledges that benefits are felt by more people than just the person who uses them. For example, what a person learns to do after developing a disease or illness to prevent what they have experienced may be more useful for persons other than they themselves. The classic example of this is the smoker who advises others to not smoke and who then dies from lung cancer. Thus, there is a place for social valuations as well as individual evaluations that determine the quality of care. Decisions of this sort often pit issues of social justice against political or economic concerns.

Donabedian also points out that traditionally, clinicians judge quality of care with the objective of achieving improved or preserved physical and physiological functioning. Health would, therefore, be defined in terms of physical and physiological functioning. It is possible, however, to add psychosocial concerns and thereby expand the definition of health. Donabedian makes the cautionary statement:

> At the extreme, "health" can be so broadly defined as to become synonymous with the "quality of life." In this case, judgments of quality of care will pertain only to those legitimate medical activities that contribute to the quality of life, and to no more; but also to no less. (Donabedian 1980; p. 18)

This statement suggests that Donabedian recognizes the importance of an HRQOL, as compared to a quality-of-life, assessment. It also implies that an HRQOL assessment may be influenced by quality of care issues.

Donabedian sees a patient's sense of the quality of care they receive reflected in their statements about satisfaction with their care. He also believes that a person's satisfaction is a judgment on the quality or "goodness" of fit of this care. He then makes the following interesting comment:

> It represents the client's assessment of quality in a way the corresponds to a professional's assessment of the quality of the same care, even though the considerations that enter the two judgments may not be the same and the conclusions may differ." (Donabedian 1980; p. 25).

He seems to be saying he realizes that both the client/patient and the clinician think about quality in the same way, even though the content involved in their respective judgments may vary. The patient judging their satisfaction with care may express satisfaction with specific aspects of that care, and may subjectively sum these individual judgments to give an overall indication of satisfaction. Thus, Donabedian seems to be aware that cognitive processes are active when a quality judgment is made, even though he does not use the language of the cognitive sciences to express his views.

Not surprisingly, the feeling of being satisfied or not is a commonly studied indicator of the quality of the medical care process. Donabedian feels it is valuable to the extent that it reflects the patient's values, and as such, it is a major indicator of the relationship between a clinician and a patient. If a person is satisfied with their care then this could be important in ensuring adherence to medical recommendations, and dissatisfaction may prompt intervening in the care process. Satisfaction, therefore, may function both as a mediator of, as well as an indicator, of some outcome.

Some investigators (e.g., Calnan 1988), however, have expressed concern about the usefulness of satisfaction as a quality of care outcome, especially since studies suggest that satisfaction responses vary as a function of assessment method (Ross et al. 1993; Ware and Hays 1988), and are also subject to various biases, such as the socially desirable response set (Ware 1978). There is also data suggesting that respondents will inflate their satisfaction rating when it refers to themselves, as compared with when the reference is the general population (e.g., Hays and Ware 1986). Not surprisingly, some investigators recommend that indicators other than satisfaction be used to assess the quality of medical care. Cleary (1998) recommends that investigators use the indicators found in the Consumer Assessment of Health Plans (CAHPS) program. These assessments do not ask if a patient is satisfied or not; rather they ask respondents to rate the quality of their care from the *best to worst* possible care along a 0–10 scale.

Donabedian (1980; p. 27), however, is "convinced that the balance of health benefits and harm is the essential core of a definition of quality," and that this balance must be valued by the recipient of care. Twenty-five years later Donabedian (2005) defines quality as "nothing more than value judgments that are applied to several aspects, properties, ingredients or dimensions of a process called medical care" (p. 692). His 1982 book has an extensive discussion of how value or importance measures can be used to weight criteria that are used in assessing the quality of care (Donabedian 1982; p. 223–293).

Donabedian also proposes that the assessment of quality can be measured in terms of either the process or the outcome of medical care. The outcome of medical care is typically measured in terms of the extent to which a person recovers from their illness, the extent to which function is restored, the mortality rate posttreatment, and so on. Using outcomes as an indicator of quality, however, does have some limitations. For example, if the treatment's effect requires an extended period of observation, then relying on the outcome as an indicator of quality may not be practical. In addition, some outcomes are hard to define, as revealed by measuring a patient's satisfaction, determining their level of disability, or deciding when they have been rehabilitated. On the other hand, if the process of receiving treatment is detrimental, then the outcome indicator, survival time, may not be totally relevant.

Another approach is to assess the process of care to determine if the care provided has been appropriate, complete, administered with technical competence, and has given the patient the optimal opportunity to reap the benefits of a medical or surgical intervention. Donabedian (2005) points out that process indicators may be less stable than outcome indicators, but may be more relevant in that they provide the information needed to optimize medical care. Process and outcome may appear to be a means-to-end relationship, but Donabedian (2005) suggests it is more appropriate to think of this relationship as "an unbroken chain of antecedent means followed by immediate ends which are themselves the means to still further ends" (p. 694). This suggests that quality assessment should be viewed as a continuous process, and some clinicians have initiated "continuous quality improvement programs" (e.g., Pincus et al. 2007). Donabedian also suggests that health itself may be seen as a continuous means towards some further objective or end.

A third approach to the quality of medical care assessment task is to consider the setting where medical care takes place and the factors that govern this setting. The setting sets the structure around which medical care occurs, and its nature can quickly become the determinant of the process or the outcome of care (e.g., Kunkel et al. 2007). Thus, if a health care setting lacks adequate infrastructure, administrative expertise, qualified practitioners, and so on, then the process and outcome of care may very well be determined. One liability to this argument is that the linkage of these structural determinants with process and outcome indicators may not be clear.

This relatively brief review of Donabedian's contributions to quality of care research shows that he was aware of and writing about the relevance of a value-oriented perspective to quality assessment. Yet, inspection of current research into quality of care assessment seldom reveals deliberate efforts at integrating value or importance indicators into a quality of care assessment. Rather, what seems to have transpired is an emphasis on defining care protocols and using evidence-based medicine for deciding on which treatments or care processes to use (e.g., Feussner et al. 2000; Mc Glynn 2003). This trend, while appearing to be more objective and data-driven, gives the appearance of a flight from "subjectivity," an effort that is not likely to succeed, since, as Fulford (1989) states, values will find a place somewhere in the cascade of information that leads to clinical decisions.

Before I move on to the next section I want to make clear that a huge effort is currently ongoing to improve the quality of medical care. Many countries sponsor agencies dedicated to this task; multiple research journals have come into existence summarizing the high level of research activities; and, of course, considerable resources have been dedicated to these activities. The U.S. Agency for Healthcare Research and Quality, for example, has sponsored a range of activities fostering development of disease-specific treatment protocols, system approaches to preventive medicine activities, development of assessment tools, patient education efforts, and so on. The Chronic Care Model (CCM; Hung et al. 2008; Wagner et al. 2001), for example, specifies a set of activities that a group of investigators can use to start a quality improvement program. The intervention points of the CCM are various aspects of a health care organization. Thus, the assessment and intervention might involve community input, self-management efforts, medical care delivery system designs, decision support, and clinical information systems. Outcomes are measured in terms of the informed activated patients and a prepared proactive care team. Programs of this sort are proliferating, and no doubt will continue to mature. A major indicator of the success of these activities is usually determined by assessing how satisfied a person is with their care. This is my next topic.

1.2.4 Patient Satisfaction as an Indicator of Quality (of Care)

Satisfaction is a term that has been used to indicate a person's health status, quality of care, or HRQOL (e.g., Table 2.5). Respondents are commonly asked to rate their status in terms of degrees of satisfaction or dissatisfaction. There are extensive research literatures dealing with studies of patient satisfaction, life satisfaction, marital satisfaction, work or job satisfaction, consumer satisfaction, and so on. Most important for the present discussion are those studies that grapple

1 Theoretical Foundation

Table 8.2 Theories and definitions of patient satisfaction in healthcare*

Reference	Name	Theory
Fox and Storms (1981)	Discrepancy and transgression theory	If healthcare orientations and care conditions were congruent, then patients were satisfied; if not, then they were dissatisfied
Linder-Pelz (1982)	Expectancy-value theory	Theory postulated that satisfaction was mediated by personal beliefs and values about care as well as prior expectations about care. Theory subsequently expanded by Pascoe (1983) and Strasser et al. (1993)
Ware et al. (1983)	Determinants and components theory	Satisfaction … "a personal evaluation of health care services and providers… Satisfaction ratings are intentionally more subjective [than care reports]; they attempt to capture a personal evaluation that cannot be known by observing care directly… this is their unique strength…" (p. 247)
Fitzpatrick and Hopkins (1983)	Multiple models theory	Satisfaction is based on expectations that are socially mediated, reflecting the health goals of the patient and the extent to which illness and healthcare violates the person's sense of self
Donabedian (1988)	Healthcare quality theory	"Patient satisfaction may be considered to be one of the desired outcomes of care, even an element in health status itself. An expression of satisfaction or dissatisfaction is also the patient's judgment of the quality of care in all its aspects, but particularly as concerns the interpersonal process." (p. 1746)

*Based on Hawthorne et al. (2006; 2006) and Gill and White (2009)

with the problem of defining and assessing patient satisfaction. Hawthorne and his colleagues (2006; 2006) and Gill and White (2009) identify five major theories or definitions of patient satisfaction (Table 8.2). I will discuss aspects of these theories and definitions, but my first task is to consider the etymology, language usage, and cognitive basis of the term satisfaction itself. The reason is well stated by Williams (1994; p. 509): "Patient satisfaction is now considered an important outcome measure for health services…this professed utility rests on a number of implicit assumptions about the nature and meaning of expressions of satisfaction."

The *Unabridged Oxford English Dictionary* (1970) reveals that the Latin for the word "satisfaction" is *satisfacÐre*, meaning to satisfy. The dictionary lists three classes of definitions for satisfaction: including reference to obligations, as an expression of desire or feeling, or use as a designation or label, as in "satisfaction-money." The term was first used in English to refer to an act of penitence, as when someone did something as payment for a punishment or sin.

Consistent with its usage are the definitions that refer to a person fulfilling a requirement (such as paying a debt), atoning for a misbehavior, and so on. It is also used to refer to the action of fulfilling a desire or feeling, and to describe a state a person is in (e.g., ratified or contented).

This review of the etymology of the word "satisfaction" clearly establishes that the term first had a literal usage, but over time assumed a number of other functions characteristic of figurative expression (indeed, this is true for many of the terms discussed in this book). Interestingly, a major impetus that encouraged this expansion was Shakespeare inventive use of metaphors in his prose. One modern example of this expansion, as Hudak et al. (2003, 2004) have pointed out, is the viewing of a patient as "consumer of services." Consumers, of course, literally buy and choose services, and clearly some aspects of health care involve services a person can make a choice about (e.g., the pharmacy at which they can get their prescriptions filled, which doctor to go to, etc.). Regarding their most important concerns (e.g., what treatment they will receive), however, consumers usually have no or limited choice. This makes the reference to a patient as a consumer more metaphoric than literal, and implies that the term satisfaction, when referring to patient satisfaction, is being used in a more metaphoric than literal sense.

What Hudak et al. (2003, 2004) were most concerned about was not whether it is appropriate to use the consumer metaphor when asking about the person's satisfaction with medical care, but whether it is appropriate to apply the consumer metaphor when satisfaction with treatment was being asked about. They were concerned because they do not view the body as a machine, nor do they see the outcome of treatment as something a person can make an objective decision about. However, the debate about whether a person is or is not a consumer of a service illustrates that the term has the characteristics of a dual usage term, in that it has both a figurative (e.g., Donabedian's claim that patient satisfaction can be viewed as a statement of a person's health status; Table 8.5) and a literal (e.g., physically paying of a debt) form of expression.

When the term satisfaction is cognitively examined (e.g., by cognitive interviewing; Williams et al. 1998) it becomes clear that a comparison occurs between a person's expectations and what they actually experience. Most of the definitions of patient satisfaction in Table 8.2 reflect these criteria [the exception being the Ware et al. (1983) definition]. Since a person's expectations and preferences are usually cognitively in place, and act as a standard, they unavoidably color the assessment of any judgment of what the person has actually experienced with their medical care. This is one explanation for why many studies find that patient satisfaction ratings are skewed towards the positive (e.g., Campbell et al. 1976, p. 99; Santuzzi et al. 2009; Williams et al. 1998). This reflects a person's predilection to interpret what they experience in terms of what they want to experience, rather than what they actually have experienced. This tendency has also been called a "social desirability bias" (e.g., Campbell et al. 1976, p. 99). It also seems to be true that the pool of positive words far exceeds the number of negative words in the common lexicon (Zajonc 1968), so that when a word is selected, it is more likely to be positive. Of course, it may also be true that the person does not know what adequate medical care should be or could be, and therefore they exaggerate the value of what they do experience.

Of the various notions of how to define patient satisfaction (Table 8.2), probably the most widely accepted approach is Donabedian's (1980; 1988; Table 8.2). Williams and his colleagues (1994; Williams et al. 1998), however, have raised several questions about the assumptions underlying Donabedian's definition. They point out first that Donabedian assumes that satisfaction is an expression of some conditional utility, by which they mean that Donabedian claims that some prior process existed that reflected the expectations and values of the person, and that these factors influenced the judgment of their satisfaction. However, a review of the available research in this area provides only limited support for the hypothesis that expectations and values account for much of the variance in satisfaction assessments (Linden-Perz 1982; Thompson 1986).

A second assumption implicit in Donabedian's definition of patient satisfaction is that when a person declares they are satisfied, this constitutes an affirmation of the service they have received. This assumption is particularly important because it implies that satisfaction is a form of quality evaluation. However, a number of studies have shown (e.g., Williams et al. 1998) that the evaluation of a service can be different depending on whether a person is satisfied or not with the service. Thus, a patient may indicate they are satisfied with their medical care, but do so because of limitations in the care they expect to receive. These limitations may include their limited ability to pay, or the nature and severity of their illness.

A third assumption, related to the second, is that the patient has in place, explicit expectations and values concerning the medical care they will receive. This is a reflection of the consumer model of medical care, but clearly not all patients are equally as knowledgeable about what good medical care should be. This is particularly true for esoteric and complex medical care, and usually patients make the assumption that the health care professional is giving them good advice. They also may assume the role of "good patient" and uncritically accept their care. However, the notion that a person may be making an independent judgment of the quality of their care is confounded by their prior expectations and values.

All of these issues raise questions about Donabedian's definition of patient satisfaction, but what is actually going on when a person makes a judgment about their satisfaction with the quality of their care? Williams et al. (1998) provided some answers to this question. They report a qualitative analysis of their interviews with patients and determined that the respondents were engaging in two separate cognitive processes: first they evaluated their care, and second they then decided how satisfied they were with their care. However, the judgment of satisfaction also involved a judgment of whether they should have expected a certain quality of care (described in terms of health care professionals fulfilling their "duty"). For example, if a person believes they had a negative experience with a mental health provider and they expected that the services they did not receive should have been provided, then the patient will declare themselves as being dissatisfied. However, if the care provider was not expected to provide the care, then the respondent might indicate they were satisfied with their care, even though they had a negative experience with their medical care. In contrast, if a patient had a positive experience with the mental health services and they felt that the service was not the duty of the care provider, then they might be very satisfied, whereas if the services were expected, they might be only partially or not satisfied, at all.

Edwards et al. (2004) report a patient satisfaction study involving orthopedic patients that expands on the Williams et al. (1998) study. They demonstrated that patient satisfaction judgments were partially dependent on the respondents' relative dependence on the health care options provided, the need to maintain a relationship with a health care professional, and the general attitude of persons to hold a positive attitude as part of their adaptive coping style. Both the Edwards et al. (2004) and the Williams et al. (1998) studies are based on an investigator's interpretation of the qualitative data, but a replication of these studies would be reassuring. Acknowledging these concerns, the Williams et al. (1998) paper is particularly interesting for its suggestion that *patient satisfaction is not a direct evaluation, but rather a comparison between what was expected and what occurred*. This raises concerns about Donabedian's assumption that patient satisfaction can be used as an indicator of quality, since the term quality, whether referring to quality of life or quality of medical care, implies that a direct evaluation has occurred. Rather it seems as if a patient satisfaction judgment is a "judgment about a judgment." These same concerns have to be applied to three of the other suggested definitions of patient satisfaction. Only the Ware et al. (1983) definition seems to cover the role that an evaluation plays in defining patient satisfaction.

As an alternative to only using patient satisfaction indicators, Rapkin et al. (2008) proposed an event-based approach to quality of care assessment. HIV/AIDS patients were interviewed to determine whether a selected series of events occurred, and what this indicated concerning the responsiveness of their providers and their need for medical care services. They assessed both the patient's needs and the likelihood of fulfillment of those needs. The authors concluded that the patient's needs and satisfaction were situation specific, not person specific. They used the two mental and physical health components of the MOS-SF-36 as their indicators of quality of life, and found, after appropriately adjusting for baseline variation, that there were very low correlations between absolute number of services used (as a indicator of need) and quality-of-life indicators.

Both the Williams et al. (1998) and Rapkin et al. (2008) studies demonstrate that there are viable alternatives to using patient satisfaction as an indication of quality of care, and that at a minimum, the judgment of satisfaction occurs after much cognitive processing of prior events. However, examining these intervening events may be quite informative about subsequent judgments, such as the judgment of satisfaction.

When four of the five definitions listed in Table 8.5 are linguistically examined, they take the form A is ≈ B, a linguistic structure similar to the definitions of quality of life or HRQOL listed in Table 2.5 (Chap. 2, p. 40). Interestingly, three of the four definitions listed in Table 2.5 include a reference to satisfaction. In each of these definitions, a mapping characteristic of a metaphoric structure occurs between the abstract target (quality of life) and the concrete base (aspirations and achievements), with the "discrepancy between what was aspired and achieved" connecting the target to the base. The same kind of analysis could be applied to the other four definitions listed in Table 8.2 (but not the Ware et al. 1983 definition), with patient satisfaction functioning as the abstract target and medical care functioning as the concrete base. For these four definitions, "discrepancies" or "fulfillment" act as the connector completing the metaphor.

When assessment of patient satisfaction is considered, an additional set of issues arises that also has implications about whether it can be used as an indicator of quality. For example, it is assumed that it is possible to assess patient satisfaction along a continuum ranging from dissatisfaction to satisfaction, yet when administered, the scale is presented in terms of a series of discrete verbal descriptors (e.g., a Likert scale: very satisfied, slightly satisfied, neither satisfied nor dissatisfied, slightly dissatisfied, totally dissatisfied). This means that the scale is presented to a respondent as an ordinal scale, although it is assumed and may be statistically assessed as if it is an interval scale. Implicit in this scale format is the assumption that the interval between verbal descriptors is equal, thereby justifying the assumption of an interval scale. Hazelrigg and Hardy (2000) have tested this assumption and found it is not sustainable. In addition, they have found that personal characteristics (e.g., the respondent's level of optimism) affected interval size estimates.

They concluded that the intervals "do not closely approximate the unitary or equal intervals assumed when using an OLS estimator" (Hazelrigg and Hardy 2000; p. 175; OLS refers to "ordinary least squares"). These data are consistent with the argument raised by Coyle and Williams (1999) and Collins and O'Cathanin (2003) that dissatisfaction is not the opposite of satisfaction, and that assessing dissatisfaction may be as informative about various determinants of the quality of medical care as are studies of patient satisfaction.

Though paradoxical, most studies of satisfaction persist in presenting response options in terms of verbal descriptors and analyzing the data as if a continuum underlies them. The rationale for this continued approach in health care studies probably has to do with the use of latent variables as an analytical tool in the statistical analyses of these data, but also because using satisfaction indicators appears to be an appropriate response to the demands of consumers to have their voices heard. The fact that satisfaction as an indicator of health care services can be predictably skewed towards the positive may also explain why service providers persist in using this indicator.

These studies of patient satisfaction have implications for the general study of the term satisfaction as an indicator, and I will continue to discuss this topic in Chap. 12. Now I want to take note of a series of papers (e.g., Locke 1970; Mobley and Locke 1970) that stretch back to the late 1960s and 1970s concerning the role of importance weightings in job satisfaction studies. Ross et al. (1993) have raised similar issues in a study of a patient's preferences for different aspects of medical care, and whether this influenced their satisfaction with their care. These researchers did not find that medical care preferences, as weightings, influenced a person's satisfaction judgments. More recently, Trauer and Mackinnon (2001) raised the same issues relative to importance weightings in quality-of-life assessments (e.g., Cummins et al. 1994; Ferrans and Powers 1985). Wu (2008a, b) and his colleagues (2006a, b, 2007, 2009) have also added to this literature. Many of these studies fail to demonstrate any statistical advantage by weighting job satisfaction ratings with some estimate of importance. A number of questions can be raised by these results, including the inference that values are not an important component of a quality assessment. In addition, Locke (1970) and Mobley and Locke (1970) have suggested that a value statement is essentially an emotional statement, and since satisfaction is an indication of emotion, weighting a satisfaction measure with an importance statement has to be considered redundant. However, there is a potential explanation for these results that rests on the cognitive consequences of how the valuations were assessed, and I discuss this in detail in Chap. 12.

One final note. Sometimes an investigator may ask a respondent to rate how satisfied they are with their life, as opposed to a specific domain (e.g., medical care, social relationships, financial status, etc.). Items like this appear to approximate what a quality-of-life assessment would assess, and an investigator will often assume that life satisfaction items are indeed assessing quality of life. Is this an appropriate assumption? There is actually an extensive literature dealing with this issue (see Chap. 11). In this literature, satisfaction with specific domains or life is considered an emotional (or subjective well-being) dimension of quality of life, while quality of life per se is considered a cognitive dimension. However, the data I will review (Chap. 11, p. 398) reveals this assumption is not sustainable. It is not sustainable because evidence reveals that the two indicators are sufficiently correlated that they can't be assumed to be two separate dimensions of a common indicator. These results are consistent with the assumption that I have made (Chap. 1, p. 12) that any qualitative assessment consists of both an emotional and cognitive component, the degree of which varies depending on the assessment context. Thus, a qualitative and satisfaction rating of a diamond may evoke less correlation than a qualitative and satisfaction rating of a life.

In addition, the cognitive tasks involved in each type of assessment are different. When the cognitive basis of the phrases "patient satisfaction" and "quality of life" are compared, both appear as complex cognitive constructions. They differ, however, in that satisfaction ratings usually involves a comparison between what a person has experienced and their expectations, while a quality-of-life assessment will also involve an evaluation but also a reflective process. Now complicating things is that a person when he or she is evaluating their state of satisfaction may also reflect about their state is satisfaction. This implies that a satisfaction rating can assume the cognitive structure characteristic of a quality-of-life assessment (Chap. 8, p. 270). However, to prove that this has occurred requires that the investigator demonstrate that the satisfaction rating was valued by the respondent (see Chap. 12). In the absence of these data inferring that a satisfaction rating is an indicator of quality of life is inappropriate.

1.3 Summary of Theoretical Foundations

So far I have traced the history and conceptual development of the phrase "health status," and laid the foundation for the subsequent discussion concerning the origin and meaning of the phrase "HRQOL." The topics reviewed ranged from the heroics of the End-results Movement, the development and application of Operations Research principles, Donabedian's definition of quality in the context of quality of care assessments, and finally a discussion of what is being assessed when patient satisfaction is the outcome indicator. This review suggested that the issues previously raised about assessing quality of life were relevant when assessing the quality of care, and that where differences existed, they were mostly due to the greater dependence of quality of care on system-related issues. Also suggested was that the phrase "health status" evolved as a concept from efforts to characterize the outcome and quality of medical care.

I was particularly interested in the meaning of the term "health," since as a modifier of the term "status," it was the semantically less-clear part of the phrase. A review of the theoretical approaches to health revealed a range of models that included Boorse's (1977) emphasis on the natural processes ongoing in the body, Nordenfelt's (1995) focus on the welfare of the person, and Parsons' (1977) sociological model. To this has to be added the lay person's views of what health is or can be, creating a wide range of possible definitions. My task was not to resolve, but rather illuminate, the ongoing debate concerning the definition of the term health. At a minimum, an investigator should feel obliged to describe their approach to the definition, since to leave it unstated leaves the audience to define how they are using the term. These various definitions also evoke again Wittgenstein's argument that a specific definition for a term is not likely to exist, but rather that several uses of the term exist, all of which have a "family resemblance" and some of which are more prototypical than others. The context in which the terms are being used is also an important determinant of the meaning associated with the term. For example, does the term health mean the same when applied to a person with schizophrenia, a cancer patient, or to persons who are aware they are at risk but do not now have manifestations of some disease? I believe there are subtle but real differences in the use of the terms health and disorder in these contexts, and investigators need to acknowledge so in their assessment instruments.

One striking observation, following the historical review of the health status movement, is the relatively slow pace with which innovation has occurred. This was partially due to the lack of institutional commitment, and it was only until the middle to late twentieth century when such commitment occurred. One of the themes facilitating this change was the view that health and medical care could be properly viewed as providing a commodity with a cost and benefit. From this grew the view that the consumer of these services should be listened to, and it was this emphasis that led to the common practice of monitoring patient satisfaction after health and medical care.

This transformation of the patient, from a relatively passive recipient of a service to someone who can select how they want to be treated, has enhanced the decision-making activities of both the person and the provider in the health care process. Concurrent with this was the increasing emphasis on the quality received, both in terms of the care provided and consequences to the patient. Now while quality of care has become an acknowledged objective necessitating institutional commitment, the same cannot be said concerning monitoring the consequences (costs or benefits) of the care received. For this to occur, institutions would have to acknowledge that quality-of-life or HRQOL issues are as important to a health care institutions as providing quality medical care, and that this justifies making them an institutional objective and commitment. Some (e.g., Treurniet et al. 1997) have argued that HRQOL outcomes are not able to indicate the quality of care, but I argued (Barofsky 2003) for just such an approach, pointing out that another subjective indicator – pain management – has received this kind of institutional commitment, and I suggest that the same should be forthcoming for HRQOL outcomes following hospitalization, and in related medical and health care encounters.

2 Applications of Health Status Assessment

2.1 Assessing HRQOL

The relationship between health status and HRQOL has both historic and conceptual components. If I start this history with Codman's work then his initial efforts were focused on the efficiency and outcome of a clinician's activities. The shift to concerns about the patient occurred when the development of health status indicators matured during the 1970s and expanded to include a wider range of psychosocial issues that are characteristic of HRQOL. What also seems to be historically true is that there does not appear to be any definitive theoretical basis for distinguishing HRQOL from health status. Rather there seems to be a bit of selectiveness, if not arbitrariness, about how the two concepts are described by different investigators.

Kaplan and Anderson (1996) provide a thoughtful discussion of these issues. In the section of their 1996 paper entitled "Is quality of life different from health status?" they state:

> Some authors define quality of life as health outcomes that are different from traditional health outcomes. Using these definitions, quality of life measures are typically limited to psychosocial and social attributes...By contrast, our definition of health-related quality of life focuses on the qualitative dimension of functioning. It also incorporates duration of stay in various health states. (Kaplan and Anderson 1996; p. 310)

To this they add a value dimension as a necessary part of any qualitative assessment. Further on, they say, "We prefer the term *health-related quality of life* to refer to the impact of health conditions on function" (p. 313). From this I conclude that for Kaplan and his colleagues, HRQOL is a special case of health status limited to assessing the consequences of a person's health condition on their functioning over time. Of course, by referring to the person's health as having an impact on their functioning, Kaplan and his colleagues also suggest that the person has some disease or disorder.

Other investigators insist on distinguishing health status and quality of life in terms of the content area each indicator assesses. Thus, Follich et al. (1988) limit their definition of quality of life to psychological and social health, and do not consider the direct consequences of a health state (e.g., the presence of a symptom). Studies by Smith et al. (1999) and Zullig et al. (2005) reinforce the notion that the content of quality of life, and presumably HRQOL, differ from health status indicators (e.g., the self-assessed health status item). Others insist that the definition of HRQOL should follow the WHO definition of health. I have and will continue to discuss my concerns about the wisdom of this approach (Chap. 8, p. 277).

Patrick and Erickson (1993; p. 19) also have an extended discussion of the meaning of the terms health status, quality of life, and HRQOL. They describe health status as including functional status, morbidity, well-being, and various physiological processes. They also acknowledge the presence of value-related issues whether they are dealing with health states, quality-of-life or HRQOL issues. What their approach lacks is sufficient attention to the cognitive-linguistic process underlying these concepts. Thus, they alluded to the role of connotation in such terms as health and illness, but don't address the implications this has on how a qualitative assessment is optimally designed. In general, they see their task as generating a statement of how both the quantity and quality of a person's life can be objectified.

Bowling (2001; p. 5–13) also joins the conversation about the relationship between health, quality of life, and HRQOL. She argues that health is a "valued" component of quality of life, and this component is ordinarily referred to as HRQOL. She feels that HRQOL must rely on a concept of health as well as quality of life. If displayed as a Venn diagram, HRQOL encompasses the area between two overlapping circles. Bowling defines HRQOL as:

> ...optimal levels of mental, physical, role (e.g., work, part, career, etc.) and social functioning, including relationships, and perceptions of health, fitness, life satisfaction and well-being. It should also include some assessment of the patient's level of satisfaction with treatment, outcome, and health status and with future prospects. (p. 6)

Bowling recognized the role that values play when she identified health as an aspect of quality of life, but sees the assessment of values, preferences, or utilities as a separate part of an HRQOL assessment. Her concerns about including utility values in an HRQOL assessment are based on the practice of using general population-based utility values to calculate the HRQOL of individuals or persons in different patient groups. Not only is Bowling concerned about the ethical implications of whose values to use when estimating a respondent's HRQOL, she is also concerned about how stable and reproducible utility estimates are over time (Kaplan et al. 1993).

Gold et al. (1996; p. 82–134) brought a number of HRQOL researcher together to provide a consensus definition of health status and HRQOL. They defined a health state as:

> The health of an individual at any particular point in time. A health state may be modified by the impairments, functional states, perceptions, and social opportunities that are influenced by disease, injury, treatment or health policy. (Gold et al. 1996; p. 399)

Gold et al. (1996) defined HRQOL as follows:

> As a *construct*, health related quality of life (HRQOL) refers to the impact of the health aspects of an individual's life on the person's *quality of life*, or overall well-being. Also used to refer to the value of a *health state* to an individual. (Gold et al. 1996; p. 399)

These definitions of HRQOL have logic to them, but have little empirical support so remain arbitrary. That empirical support is a necessary part of defining a concept was well expressed by Mayr (1982; p. 45) who, to repeat the quote from Chap. 1, stated:

> In the history of biology the phrasing of definitions has often proven rather difficult, and most definitions have been modified repeatedly. This is not surprising since definitions are temporary verbalizations of concepts, and concepts – particularly difficult concepts – are usually revised repeatedly as our knowledge and understanding grows." (Mayr 1982; p. 45).

Whether a definition of HRQOL has empirically evolved seems less clear than that there seems to be substantial number of alternatives that investigators have proposed. My concern with this led me to suggest that a cognitive approach can be most useful, since it helps an investigator to focus on how a quality assessment occurs. It also provides a means of testing the cognitive task implicit in the definition. Thus, when I argue that a quality assessment requires a reflective process, this assumption can be tested. An obvious example of an experimental approach would measure the reaction time it takes for someone to respond to an item on an HRQOL assessment. Reflection would require time for the task to be completed and this could be assessed using a measure of reaction time.

This is increasingly becoming a feasible objective since qualitative assessments are regularly done using a computer, making the quantification of reaction time quite straight forward. Responding to each item also requires processing detailed instructions and considerations of alternatives before a decision is made. How this is accomplished by the respondent, as during a utility estimation, can also be systematically studied, again with the aid of a computer to conveniently manipulate alternative instructions and record responses. Klein Entink et al. (2009) have developed a statistical model that combines reaction time (speed) and score accuracy (ability). They demonstrate how their model can be used to determine the role of each factor in response dynamics. This could be particularly useful when operationalizing what it means to be "compromised" (e.g., taking longer to complete an assessment).[13] Of course, a person might feign completing the task and respond reflexively, but this also could be studied. Fast reaction time with a low score accuracy would provide the kind of data needed to propose that the respondent is not seriously participating in the assessment process. Another approach did just that to obtain running account of what the person was thinking about as they responded to their task. The study by von Osch and Stiggelbout (2005) illustrated the value of using cognitive interviewing to determine how a person was responding to a qualitative item. Neural imaging studies could also be useful here (Chap. 10).

From the various definitions of HRQOL and the subsequent discussion, it is clear that most investigators consider HRQOL to refer to the consequence of some health state on a person's quality of life. It would, therefore, be expected that the assessment of HRQOL reflect this definition.

2.1.1 The Components of a Quality-of-Life or HRQOL Assessment

The model I propose of a qualitative judgment assumes that a hybrid cognitive construct is formed by combining several relatively distinct cognitive processes. This ideally includes the cognitive formation of an evaluated descriptor that also traces its consequences. This entity becomes an object of reflection that may involve valuation, preferences, or utility estimation. This "thinking about what the person is thinking" is the essence of metacognition and an essential element in forming the hybrid construct. I describe this process in phenomenal terms but it should be clear that it can occur below a level of consciousness (Chap. 3, p. 64). A qualitative assessment may contain some or all of these cognitive processes, either in the assessments instructions (e.g., a description of the patient's disease or treatment status), or in the items themselves. Table 8.3 describes four types of quality-of-life or HRQOL items that may be found in currently available assessments. I contend that items type III and IV should be considered qualitative items, but only type IV items should be considered a completely informative quality-of-life or HRQOL statement. A comprehensive HRQOL item would, therefore, consist of an evaluated descriptive state (e.g., moderately intense pain) that disrupts the person's physical activity (i.e., the pain has a consequence or impact on the person), and a valuation by a person of their limited physical activity. However, as will become clear when actual HRQOL assessments are inspected, most items on an HRQOL assessment are type II items (see Table 8.5). A symptom, theoretically, can also be assessed using each of the four formats, but most frequently matches the criteria in a type I item.

The format described in Table 8.3 can be used analytically to determine the extent to which the items in any

HRQOL assessment correspond to the ideal type (type IV). Consider some of the items on the EORTC-QLQ-C30. An item that asks about the consequences of a health state but does not attach a value statement is item #28. It asks, "Has your physical condition or medical treatment caused you financial difficulties?" Item #9, on the other hand, which asks whether the respondent had pain (from not at all to very much over the last week), does not directly refer to any consequences of the pain or place a value on this state, so accordingly would be considered a type I item (Table 8.3). In general, I should be able to classify each item on any supposed HRQOL assessment in terms of one of these four criteria (Table 8.3). Table 8.4 lists representative examples of each type of item from published assessments, and my judgment on the type of items in the currently available quality-of-life or HRQOL assessments is summarized in Table 8.5.

If I accept the criteria listed in Table 8.3, then how can I distinguish a quality-of-life item from an HRQOL item? An HRQOL item requires an explicit reference to the person's health, treatment, or disease status (whether in the title of the assessment, instructions on how to answer the items or in the item itself), whereas this type of reference is not necessary for an item in a quality-of-life assessment. Type III items illustrate that a quality-of-life item can include a valuation statement, but have no description of the consequence of some event. Types II and III items are frequently called HRQOL items, but they would not be considered completely

Table 8.3 Criteria for defining items on an HRQOL assessment

Type I: The item could be a *simple descriptive* statement that has been evaluated, but does not include reference to a consequence or have a valuation attached to it
Type II: The item could be a descriptive statement that has been evaluated and does indicates the *consequences* of some event (e.g., becoming unemployed, presence of a disease, or treatment), but does not include a valuation of these consequences
Three III: The item would be an evaluated descriptive statement that only has a *valuation* statement attached to it, and does *not* include a description of the consequences of some event
Type IV: The item includes an evaluated descriptive statement, the *consequences* of which are evaluated, but it also includes a *valuation* statement

Table 8.4 Classification of items in an HRQOL assessment

Type	Item	Reference
Type I *Descriptive statements only*	"In general, would you say your health is excellent, very good, good, fair or poor?" "Did you feel full of pep?"	SF-36, Ware et al. (1993)
	"I feel nervous"	FACT-B, Brady et al. (1997)
	"I feel tired"	BFI, Mendoza et al. (1999)
Type II *Consequence statements but no value statements*	"Has your physical condition or medical treatment caused you financial difficulties?"	EORTC-QLQ-30, Aaronson et al. (1991)
	"Has your physical condition or medical treatment interfered with your family life?"	
	People with an illness have said:	FACT-G, Cella et al. (1993)
	"I am forced to spend time in bed"	
	"I feel close to my friends"	
	"I feel sad"	
	"I become nervous when I get chemotherapy"	CARES, Schag et al. (1991)
	"I feel sick when I think about chemotherapy"	
	"I frequently feel overwhelmed by my emotions and feelings about cancer"	
Type III *Value statements ▲ attached to descriptive statements only*	"In the past four days, if any, did you spend any part of the day or night as a bed patient in a hospital, nursing home, mental institution, home for the retarded or similar place?"	Quality of Well-being Scale, Kaplan and Anderson (1990)
	Selecting symptoms and problems	
	"Which of the following best describes your usual ability to see well enough to read ordinary newsprint?"	Health Utilities Index, Torrance et al. (1992)
	A. Able to see well enough without glasses or contact lenses	
	B. Able to see well enough with glasses or contact lenses	
	C. Unable to see well enough with glasses or contact lenses	
	D. Unable to see at all	

(continued)

Table 8.4 (continued)

Type	Item	Reference
Type VI *Consequences and value statements* ▲	"How *satisfied* are you with your health? How *important* is your health to you?"	Ferrans and Powers Quality of Life Index: Cancer Version, Ferrans and Powers (1985)
	How *satisfied* are you with the health care you are receiving? How *important* is your health care to you?	
	(Six domains with a maximum of six levels per domain, with a *preference* statement attached):	SF-6D, Brazier et al. (2002)
	Physical Functioning:	
	Level #1: "Your health does not limit your vigorous activities"	
	Level #6: "Your health limits you in bathing and dressing"	

▲: The value statements in the form of preferences or utility estimates may be *derived* as for the standard gamble or time trade-off procedures, or *directly assessed*, as in the Ferrans and Powers (1985) assessment. The Ferrans and Powers (1985) assessment is difficult to categorize but is considered an example of a type IV assessment. The difficulty with the Ferrans and Powers (1985) assessment is that it first asks a respondent if they are satisfied with various domains, and this is not a direct assessment of the state a person is in, but a comparison between what they expect or want and what exists. A second task involves asking the respondents to indicate the importance various domains are to the respondent. These two scores are then multiplied and summed to give a composite score (see Chap. 12, p. 458)

Table 8.5 Analysis of item type in health status and HRQOL assessments

Assessment/items	Reference	Type I (No C or V) ▲	Type II (C/No V)	Type III (No C/V)	Type IV (C/V)
EORTC-QLQ30 ($N=30$)	Aaronson et al. (1991)	27	3		
FLIC ($N=22$)	Schipper et al. (1984)	15	7		
FACT-B ($N=33$)	Brady et al. (1997)	26	7		
RSCL ($N=39$)	de Haes et al. 1990	39			
CARES SF ($N=59$)	Schag et al. (1991)	41	18		
BCQ ($N=30$)	Levine et al. (1988)		30		
MOS-SF-36 ($N=36$)	Ware et al. (1993)	18	18		
SIP ($N=136$)	Bergner et al. (1981)				136?
QWB ($N=4$)	Kaplan and Anderson (1996)			(35)+3 ▼	
HUI3 ($N=8$)	Torrance et al. (1995)			2	6
HALex ($N=2$)	Erickson (1998)			1	1
15D ($N=15$)	Simtonen and Pekurinen (1989)			15	
EQ-5D ($N=5$)	Kind (1996)			2	3
SF-6D ($N=6$)	Brazier et al. (2002)			2	4
SEIQoL ($N=5$)	Hickey et al. (1999)			5	

C Consequences; *V* Valuations

▲: Valuations attached to 35 symptoms/problems and three activity dimensions are summed

informative in the current classification; only type IV items would constitute a complete HRQOL item. Often an item may consist of a descriptive statement that is evaluated, but the valuation component is left to a separate assessment. For generic multi-item assessments, such as the SIP, HUI, QWB, EQ-5D, or SF-6D, the valuation may not even involve the respondent, whereas it would for the Ferrans and Powers (1985) assessment. Also true is that this classification system can be used to analyze the items on other qualitative assessments, including the qualitative assessment of an object, a meal, a car, a product, or even a person's working life.

The cognitive model (i.e., a hybrid construct) implicit in types III and IV items can also be applied to a wide range of clinical situations. Consider the example where a person was being attended by a clinician who notes that the patient has aspirated (as evident from the rallies that can be heard when the patient's chest is examined). The clinician might tell the patient that the severity of the symptom (an evaluated descriptive statement) may lead to the development of pneumonia and be life-threatening (a consequence statement), and that it is in the patient's interest to take an antibiotic (a valuation statement). Would this be a qualitative statement? I think that it would, with the full understanding that what is being emphasized here is that it is *how* an observation is made that determines a quality statement, not necessarily the content of what is being described. This greatly broadens the realm of qualitative statements, but also demonstrates that when ongoing cognitive processes are examined, many more examples of the presence of a qualitative assessment can be identified.

I would like to clarify some terms. First is the question of how the phrases "evaluated descriptor" and "valuation" differ, since they are both essential parts of the hybrid construct hypothesis, yet are semantically similar enough to potentially create confusion. I have and will continue to limit the use of the phrase evaluated descriptor to evaluation of a specific object, event, or life lived, while valuation refers to a metacognitive process that is a more time-consuming reflective process. Thus, operationally the two types of cognitive activities can be differentiated by the time it takes to complete the designated task. This, of course, is a testable assumption. In addition, both participate in the objectification process; that is, the process whereby subjective experiences are transformed into numbers.

These two processes also differ in that the evaluation of descriptors leads to inductive inferences (a bottoms-up process), while metacognition involves deductive inferences (a top-down process). Much of science has utilized inductive processing and reasoning for its advancement, but qualitative research is particularly burdened by the overuse, if not misuse, of deductive reasoning. An example is what Spates (1983; p. 42) called "deductive imposition," which occurs when an abstract construct such as well-being or quality of life is assumed to exist and is imposed as a label on a set of indicators. I consider deductive imposition to be another way to explain what I call metaphoric arithmetization. It is a product of the cognitive phenomena that the existence of an abstract concept is totally dependent on its concrete instances. In the absence of concrete instances it would be hard to know what well-being or quality of life mean.

Evaluating a descriptor and evaluating the valuation process differ in that one usually occurs explicitly, while the other can occur either implicitly or explicitly. Thus, I consider rating a descriptor, such as pain, to be an explicit conscious act. In contrast, a person becomes aware of the implicit processing of information (i.e., processes occurring below awareness) only when asked to respond to an item on a questionnaire, and I consider this a valuation of pain. Thus, valuation, as opposed to evaluation of a descriptor, is a continuous process. Of course, this implies that a person is constantly monitoring their qualitative status, and this is consistent with what has been labeled as "respondent" behavior (e.g., reflexive approach-avoidance behavior).

The presence of a continuous below-awareness cognitive process has practical consequences. It implies that when a person is confronted with an evaluation task, such as responding to an item on a questionnaire or expressing an opinion about some issue, they already have a value structure in place. This valuation system may reflect social and cultural norms or the unique experiential history of the person. In addition, when I structure what I say to convey a particular meaning, it may also reflect a value system already in place. Finally, the continuous valuation process greatly facilitates the adaptive process that occurs when the chronically ill adapt to their altered state. As stated, the classification system presented in Table 8.3 can also be used to analyze the cognitive basis of the items in available quality-of-life or HRQOL assessment (Table 8.5). The EORTC-QLQ-C30 is a 30-item assessment consisting of 28 fairly specific descriptive statements, plus two global items. The assessment does not provide a respondent the opportunity to value being in a particular state, thereby limiting the items to type I or II (Table 8.3). Only three of the items specifically ask if a person's physical condition or medical treatment has impacted some aspect of their life. As a consequence, I summarize the EORTC-QLQ-C30 as consisting of 27 type I items and three type II items.

Most interesting about the EORTC-QLQ-C30 (Aaronson et al. 1991) is that its authors claim it is an HRQOL assessment (e.g., the abbreviation QLQ refers to Quality of Life Questionnaire). Yet the analysis of the items summarized in Table 8.5 do not support this, assuming that the criteria I proposed are in fact necessary for defining a qualitative assessment. In what way, then, can the EORTC-QLQ-C30 be called an HRQOL assessment? The developers of the EORTC-QLQ-C30 might argue that inspecting the content of the items provides *prima fascia* evidence that the assessment is covering issues of great importance to a cancer patient, and that valuations are present in the assessment, just not explicitly assessed. This brings the discussion back to deciding whether to give a person a voice in their own particular qualitative assessment and to do this requires a direct assessment. Many other quality-of-life or HRQOL assessments are equally as heterogeneous in their items, and this raises the possibility that they may differ in terms of the meanings they are communicating.

As will become evident, only a limited number of the available items in qualitative assessments approximate type IV items (e.g., the SIP; Table 8.5). However, establishing the meaning being communicated by an assessment is required for establishing an assessment's validity, and this requires careful inspection of the content of its items. For example, one of the items on the MOS-SF-36 (item #6) is as follows: "During the past 4 weeks to what extent has your physical health or emotional problems interfered with your normal social activities with family, friends, neighbors or groups?" This item appears to be asking about the impact of the person's physical health and emotional problems on the person's psychosocial functioning, but actually doesn't specify what this impact might be, so it appears to be a weak example of a type II item (Table 8.6). The items on the Breast Cancer Questionnaire developed by Levine et al. (1988) provide a stronger alternative, since the items include explicit information about a treatment (i.e., receiving chemotherapy) and its consequences. Item #7 from the MOS-SF-36 provides another type of item. It asks, "How much bodily pain have you had during the past 4 weeks?" This item, while it describes a physical state, does not explicitly connect to any consequences (e.g., activities, feelings, etc.) relevant to the person, and as such, qualify as a type I item.

What is common to the two MOS-SF-36 items I just described is that while they do not explicitly connect to particular consequences, they are structured so that the items connote feelings that are common to a wide variety of aversive consequences or states (e.g., a person has arthritis or bone cancer). Structuring items in this way is a common practice among generic qualitative assessment investigators, since it avoids having to provide an item more precisely attuned to the respondent's current health state. This procedure, while creating "efficient" items (items that a large group of respondents can respond to), also raises questions about what precisely is being assessed and what it means. For example, even though each respondent may generate the same numerical rating, is the severe debilitating pain rating of an arthritic the same as the life-threatening pain of the bone cancer patient?

The Sickness Impact Profile (SIP; Bergner et al. 1981) is an example of a generic qualitative assessment that also uses connotative expressions of consequences without specifying a health state. Here, the term "sickness" in the title evokes the negative feelings common to all diseases or treatments, and thereby presumably can avoid having to provide person-specific information in the item. In contrast, the HUI refers to "health," which has a relatively neutral to positive connotation, while the QWB refers to "well-being," which has a positive connotation. Although the investigators use different labels, in practice they may include instructions that use the same or alternative terms. For example, in the SIP instructions, reference is made to both being completely well and being sick.

The SIP consists of 11 dimensions, making up a total of 136 items. The task for the respondent is to select descriptive statements that reflect changes in their behaviors that can be attributed to their health. If someone is perfectly healthy, they actually may not select any of the items. The items describe what a person may not do now, how what they used to do has changed (the duration or pattern of these activities), and so on, all of which can be attributed to the person's state of health. Diseases or treatments are not explicitly stated, although the items clearly reflect the consequences of altered health. For example, a respondent may select, "I spend much of the day lying down in bed in order to rest" or "I am sleeping or dozing most of the day – day and night," and this could be interpreted by the investigators as evidence supporting the impact of a person's sickness. In addition, each of these items is given numeric values, by ratings obtained from both clinicians and patients, that reflect the extent to which sickness impacts a person's normal activities. Thus, each item in the SIP addresses a consequence of an unidentified health state and has been valued. I will consider the items in the SIP as examples of type IV items with a question mark, reflecting the discussed deficiencies.

The HUI has had several iterations, and has led to the development of several additional assessments based on its methodology (e.g., the HALex and the 15D: Table 8.5). The HUI Mark 3, for example, consists of eight health-related issues (e.g., vision, hearing, speech) that respondents are to evaluate using five or six response options. Several of the topics covered clearly do not include consequences as an option. Emotions are assessed along a five-level scale ranging from being happy to being "so unhappy that life is not worthwhile." However, there is no explicit reference to the consequences of being in this state. The same can be said for the item on cognition which has as its most severe option, "unable to remember anything at all, and unable to think or solve day to day problems." Again, however, the consequences of being in this state were not spelled out in any of the other options for this item. In contrast, the item on pain has as its most severe rating the statement of "severe pain that prevents most activities." Although this item does not spell out what specific activities are prevented, it does explicitly allude to this happening. Three of the items (e.g., vision, hearing, and ambulation) have some response options that allude to consequences, but the final most severe statement does not. Thus, the vision question asks respondents if they are "unable to see"; hearing, "unable to hear"; ambulation, "unable to walk." The previous response options, however, for each of these items does allude to consequences (e.g., can't read newsprint, etc.). I listed these items as type IV items (Table 8.5).

The HALex (Erickson 1998) is an example of an assessment that also applies Torrance's multiplicative multiattribute utility theory methodology (Torrance et al. 1995). The two components of the assessment include the five-level self-assessed health status item ("How would you rate your health? Excellent, very good...") and a six-level activity limitation assessment. The activity limitation assessment was based on 13 questions that were reduced to a six-level activity score. Scale values were determined for each level of each item, and then multiplied to give an attribute score that represents a particular composite state. These items are regularly administered as part of national health surveys such as the U.S. National Health Inventory Survey (Chyba and Washington 1993), and these data were used to demonstrate the usefulness of the HALex. The self-assessed health status item was considered a type III item, since it described a state that became part of a valuation but did not report any consequences associated with the state. In contrast, the item dealing with a person's activities was based on questions that described consequences (e.g., "Because of health problems or impairments did you need help from others"?) so that it was possible to consider the resultant classification system as a type IV item. The HALex has been used in a number of studies (e.g., Asada 2005; Livingston and Ko 2002). When the HALex is combined with life expectancy data, it provides an estimate of years of healthy life, which has been used in the Healthy People 2000 Years of Healthy Life reports (Erickson et al. 1995) issued by the National Center for Health Statistics. This has been the HALex's primary application.

Another example of an assessment was developed using the multiplicative multiattribute utility theory methodology is the 15D (Simtonen 2001). This assessment is touted as having the advantage of providing both a profile and a single-index summary statement. The assessment represents an attempt to apply the WHO definition of health as augmented by clinical and empirical input. Using the general public, a three-stage valuation system procedure was implemented to provide utilities or preference weights for the assessment. The items in the assessment indicate that the items provided descriptions of states, but no indication of the consequences of these states. As a result, the items in the assessment were classified as type III items.

The design of the QWB (the interview version) is unique in that it is designed to meet the recall and linguistic needs of the respondents. For example, the QWB tries to simplify what is required of the respondents by asking them to first summarize their current health state by selecting from among 35 different symptoms and/or problems. This list is reduced further, and asks the respondent to identify which one of these states was the most "undesirable" for each of the past 6 days. This procedure was designed to minimize the memory burden and presumably prompt more accurate recall. It also allowed each respondent to reflect about which of the states was least desirable, states which would subsequently be independently valued. This exercise is followed by a series of questions about specific aspects of mobility, physical activities and social activities that would presumably identify adverse consequences. Algorithms are available to map values or weights to these different states generated by a representative sample of the general population. The weights attached to the symptoms/problem complexes and activities are then summed to form a composite estimate assumed to numerically represent the well-being of the respondent.

As was true for the SIP and HUI, the QWB has all the elements of an ideal type IV item, except that health states (symptom/problem complex) and its consequences (different activity levels or impairments) were assessed separately with no effort to connect the two by the individual respondent. This procedure was originally justified on the basis that if an assessment consisted of scenarios that combined these different components, the number of scenarios that would have to be administered to a respondent would not be practical.

Instead, respondents are given the opportunity to affirm what symptom/problem complexes, different activity levels, or impairments are true for them, and then, at least for the QWB, the valuations attached to these components are added together to provide an overall estimate of well-being. Interestingly, scenarios of the sort described were actually given to a sample of the general population to provide the basis for generating valuations. However, no effort was made to capture what the health status of the study population was, but rather gave a representative sample of the general population the opportunity to indicate their judgment of the value of the particular health state described in the scenario. This approach, with its lack of reference to the study sample, has raised questions about the ethics of the resultant indicator (see below). Thus, the SIP, HUI, and QWB have the common feature of synthetically deriving their estimate of quality of life. These assessments become particularly relevant to an HRQOL assessment when they are administered to persons with different diseases or treatments and where it can be assumed that the distinguishing characteristics of being in these states translates into particular pattern of responding to these assessments. Again, what becomes problematic is that the valuation system generated by the general population sample risks being discordant with persons with a particular disease or treatment. Balaban et al. (1986), however, have reported a study in which they asked a group of rheumatoid arthritis patients to generate utilities using scenarios similar to what the general population used and found a high degree of correlation with the general population data ($r=0.937$).

An example of an assessment that comes close to also being able to generate type IV items is the EQ-5D Valuation Questionnaire (Brooks et al. 2003). It does so because it very deliberately asks the respondent to simulate the elements of the hybrid construct, though relative to a theoretical person, not themselves. The instructions read:

> We are trying to find out what people think about health. We are going to *describe* a few health states that people can be in. We want you to indicate *how good or bad* each of these states would be for a person like you. There are no right or wrong answers. Here we are interested in your personal view. (Brooks et al. 2003; p. 261; italics added by author)

The instructions provide the respondent with information on particular health states, and they can select one of three response options. I judge that three (mobility, self-care, and usual activities) of the five items on the assessment include response options that reference the consequences of being in these states. The respondent is also given a visual analog scale to rate how "good or bad" their current overall health status is to them.

The EQ 5D Valuation Questionnaire also includes scenarios (e.g., Brooks et al. 2003; p. 265) that are made up of various combinations of response options from the five items (with one unique scenario referring to being unconscious).

The respondent is asked to rate these health states using a visual analog scale (from best possible to worst). This questionnaire provides data that can be used to generate valuation-based indexes, and be applied to a number of different types of analyses (e.g., economic studies or application in clinical trials).

The development and application of the SF-6D (Brazier et al. 2002) is a particularly sophisticated synthetically generated value or preference-based qualitative assessment. It is based on an assessment that consists exclusively of evaluated descriptors (i.e., items from the SF-36 or SF-12) and a valuation system. I believe that four of the six dimensions of the SF-6D directly allude to response options that include consequence statements, and I have classified them as type IV items.

Examples of assessments that permit a person to select and value the issues that are important to them include the Schedule for the Evaluation of Individual Quality of Life (Hickey et al. 1999; SEIQoL) and the Patient-Generated Index (Ruta et al. 1994). The SEIQOL, for example, does not have prescribed items in the assessment, but rather offers a respondent a limited number of "domains of concern" that they can select from. The respondent is asked to rate their self-selected concerns in terms of the best to the worst possible states. The respondent is also asked to rate their current quality of life using a visual analog scale. The second part of the assessment involves presenting the respondents with 30 hypothetical outcome profiles based on their five areas of concern. Each concern is rated, creating different profiles. The respondent is also asked to rate each of these profiles in terms of the quality of life that they reflect. Regressions analyses are then performed with the quality-of-life rating as the dependent variable and the ratings of the profile as the dependent variable. All 30 hypothetical profiles form the data for this analysis, and since the same individual is involved, the regression analysis provides some insight into the "policy" a person uses when making judgments. The assessment procedure does not allude to the person having a disease or treatment, nor does it specify the consequences of being in a particular state, although the quality-of-life rating may be interpreted as reflecting consequences. For this reason, the five areas of concern, whatever they might be, were classified as type III items (Table 8.5).

What I find fascinating about the last eight assessments in Table 8.5 is that they all implicitly or explicitly use the components of a hybrid construct when they aggregate their data into a single summary statement. The aggregation methods investigators use do differ in their details, with some multiplying and others adding components together, but each combines evaluated descriptors with an independently obtained valuation to form a summary measure. One explanation of this "group think" is that it reflects the investigator's own phenomenology; that is, this pattern reflects how they themselves think and cognitively this takes the form of a creating a cognitive category. The category, which would consist of a hybrid cognitive construct, could then be transformed by appropriate statistical means into a domain. The general observation that an assessment may be designed to reflect how an investigator would make a qualitative judgment should not be surprising, but it does offer a clear opportunity to do an interesting qualitative study (i.e., do cognitive interviews of investigators as they try to solve the problem of how to assess various qualitative states).

My explanation may appear a bit simple, since in each case I am sure the investigator would identify a different intellectual history as the foundation of their assessment. Consider the HUI, which Torrance and his colleagues would argue evolved from their commitment to applying the von Neumann and Morgenstern (1944) equations concerning making decisions under uncertainty to decisions concerning health systems or health status. Or the interest of Bush and Kaplan (e.g., Kaplan et al. 1987) to generate a general health policy modal. In response, I agree that these events contributed to the development of the theoretical basis for these assessments, but would still argue that the actual cognitive design of the assessment involved the investigator either explicitly or implicitly deciding how best to facilitate a response to the assessment task, and this includes how the investigators themselves would make the judgment. The presence of this phenomenon is, to me, some of the strongest evidence supporting the thesis I have been promoting: that there is a common cognitive foundation to all types of qualitative assessments.

2.1.2 Health Status and HRQOL Assessment

What, then, is the relationship between health status and an HRQOL assessment? There are actually several answers to this question, depending on how these terms are linguistically used and how they have been empirically determined. For example, if both phrases are used abstractly, then the meaning communicated is more connotative than denotative, while if one phrase has an abstract and the other a concrete reference, then the statement is likely to have metaphoric properties (i.e., a mapping from the base to the target). In addition, the phrase health status can have a normative or disease-specific reference. A normative reference would imply that the phrase is referring to a full range of states from complete health to death, while a disease-specific use would provide a very different context for establishing the meaning of the phrase. Health status can have a wide range of empirical determinants, including laboratory tests, a clinician's rating (e.g., the diagnosis of a disease), or the self-perceptions of the respondent. But how can this wide range of health state indicators be related to basically a subjective assessment, HRQOL? One answer is to "qualify" the health status indicators. This would cognitively equate a health status indicator so that the components to be combined in an HRQOL assessment would have a common empirical denominator. To illustrate, if a measure of blood chemistry is highly correlated with a person's fatigue, the fatigue impacts the person's social activities, and the person thinks this consequence is important (or unacceptable to them), then that item would be expected to be included in an HRQOL assessment. Of course, most health status assessments do not provide all this information so that purely descriptive health status items (e.g., blood chemistries) would not be expected to be included in an HRQOL assessment.

Based on the data presented in Table 8.5, very few of the supposed HRQOL assessments contain all the information required in an ideal type IV item. The SIP items, which I ambivalently believe to approximate type IV items, come closest to the ideal assessment. As a generic assessment, however, it does not specify a specific disease or relevant disorder, nor does it use valuations based exclusively on the persons being assessed. It has the advantages of individualizing the assessment by permitting the respondent to select examples of how their life has been impacted by their diseases or disorders, but the image generated by the responses still must be synthesized by the investigator.

Wakefield's (1999a) definition of a psychiatric disorder as a "harmful dysfunction" expands the generality of the role that a hybrid construct plays in a qualitative assessment. It also makes explicit that a psychiatric diagnosis is a statement about the quality of a person's existence. Also evident is that the pairing of a concrete "dysfunction" with an abstract value statement "harm" is characteristic of mapping a source to target found in metaphor formation. This raises the interesting possibility that a qualitative assessment will always involve figurative language expression.

I can illustrate this by first acknowledging that while the term "dysfunction" need not be interpreted as referring to good or bad, when it is linked to being "harmful," then it is possible to explicitly infer that there is something bad about the dysfunction. The same could be said following the mapping of the item "a financial crisis due to cancer" (i.e., it is not likely that the cancer literally causes any financial consequences, rather, the cost of drugs, hospitalization, and medical care causes the financial burden). In both instances, I infer meaning that is present because of the mapping characteristic of a conceptual metaphor.

The relationship between health status and HRQOL is a controversial subject, and it has provoked much debate in the literature. A number of investigators have expressed concern about the distinction between health status and HRQOL (e.g., Birren and Dieckmann 1991; Bradley 2001; Carr et al. 1996; Covinsky et al. 1999; Gill and Feinstein 1994; Hunt 1997; Leplege and Hunt 1997; Moons 2004; Nord et al. 2001). For example, Bradley (2001) reviews a series of studies involving diabetic patients in which, in her estimation, authors drew inappropriate conclusions because they confused measures of health status with measures of quality of life. She states:

> Greater precision is needed in the use of the term "quality of life" if clinicians are not to be misled into thinking that findings based on a health status instruments indicate that treatments do not damage quality of life when all the data reveal is that treatments damage perceived health. (Bradley 2001; p. 8)

This interpretation, she claims, is due to the fact that investigators consider some indicators of health status to be quality-of-life/HRQOL assessments. While the investigators may have been imprecise in their usage of the phrases health status and quality of life, it is also possible that patients may see these terms as very different. Consistent with this hypothesis is a report by Covinsky et al. (1999) who reported a study with a group of elderly persons (>80 years of age) who were administered a global quality-of-life question ("How would you rate your quality of life right at present? Would you say it is excellent, very good, good, fair or poor?") and a series of health status questionnaires (e.g., the Duke Activity Scale [Hlatky et al. 1989], the Katz Activities of Daily Living Scale [Katz and Apkom 1976], the Profile of Mood States [McNair, et al. 1992] and a series of questions about pain [Desbiens et al. 1997]). They found modest correlations between measures of health status and the global quality-of-life measure, but also, a good proportion of the respondents whose health status measures were reported as good also rated their global quality of life as fair or poor. The investigators state:

> For example, global quality of life was described as fair or poor by 15% of patients in the highest physical capacity tertile, 21% of patients with the least psychological distress, 25% of patients who were dependent in no activities of daily living, and 30% of the patients with no pain. (Covinsky et al. 1999; p. 437)

The data of Covinsky et al. (1999) suggests that a patient's perception of their health and quality of life is discordant. In a more recent study, Heinonen et al. (2004) demonstrated that responding to a global quality-of-life measure is susceptible to the emotional state of the respondent at the time of the rating. Whether the respondents' emotional state would have accounted for the data in the Covinsky et al. (1999) study is not clear, but the notion that a multi-item assessment may evoke different cognitive and emotional responses than a global item seems quite plausible. Covinsky et al. (1999) goes on to conclude that there are intrinsic limitations to using health status measures as a surrogate measure for quality of life, but that on average, health status is a reasonable indicator of quality of life, especially if used in combination with other measures, such as a global measure of quality of life.

What has remained unsaid, of course, is what the terms "health" and "quality of life" mean when used in the items being asked. My previous discussion should have made clear that health, at least, has a wide range of meanings, only one of which deals with the particular disease or disability status of a person. Thus, one explanation of the Covinsky et al. (1999) results is that when asked to rate some aspect of their global quality of life, the respondents based their response on a totally different "semantic space" than when asked to rate their mood, activities of daily living, and so on. Qualitative methods such as cognitive interviewing (Willis et al. 2005) would help determine what the person was thinking about when they responded to these different items. Thus, this study confirms Bradley's (2001) concerns, in that a reasonable proportion of the patients distinguished between the two concepts, although the investigators may not have.

Zullig et al. (2005) adds to this discussion by reporting an interesting study dealing with adolescents. The investigators were interested in whether adolescents distinguished between self-rated health and quality of life, and in particular, whether they viewed physical health more as of an indicator of self-rated health and mental health more as an indicator of quality of life. They found that the correlations between mental health and quality-of-life ratings were higher than comparable correlations between physical health and quality of life, and that similar correlations with self-rated health did not support their hypothesis that physical health would correlate higher with self-rated health than mental health. A concern about this study is that while the investigators compared correlations to draw their conclusions, they did not provide any statistical basis for their decision. To do this, they would have had to test whether the two correlations they were comparing were statistically significantly different.

Concern about distinguishing between health status and quality of life or HRQOL addresses one set of issues, but there has also been considerable criticism about the concept of HRQOL itself. Beckie and Hayduk (1997), for example, make the following statement in the abstract of a paper:

> This paper considers quality of life (QOL) to be a global, yet unidimensional, subjective assessment of one's satisfaction with life. This conceptualization is consistent with viewing QOL assessments as resulting from the interaction of multiple causal dimensions, but it *is inconsistent with proposals to limit QOL to health-related quality of life (HRQOL).* (Beckie and Hayduk 1997; p. 21; italics added)

The initial impression this quote creates is that Beckie and Hayduk are claiming that those investigators who use HRQOL as a construct are somehow suggesting that all measures of quality of life should be limited to HRQOL. This seems misplaced, especially when it is recognized that most investigators who claim they are studying HRQOL deal with a rather circumspect set of problems and issues (e.g., as when dealing with persons with a specific disease or when comparing the qualitative consequences of two alternative medical treatments), and would be aware of this difference. Certainly, I am aware of this distinction since I persist in writing "quality of life or HRQOL" when describing a qualitative assessment.

Beckie and Hayduk also claim that those who study HRQOL are somehow overwhelmed by the broad range of indicators characteristic of a quality-of-life assessment (including happiness, subjective well-being, and satisfaction), but this view is also not sustainable, as is evident from a review of the definitions of HRQOL (see Tables 2.4, 2.5, 2.7, and 2.8). They make an additional point that sometimes inferences are drawn from HRQOL studies, that are generalized to quality of life, and that this may be inappropriate. Tennant (1995), for example, comments that it may be inappropriate to argue that an assessment developed to rate disability and impairment could somehow be interpreted as measuring quality of life. However, it may be true that the investigator is using the phrase "quality of life" as a label to describe a class of assessments rather than as a construct that can be operationalized in a specific way, and if so, then the investigator may feel justified in the phrasing.

Beckie and Hayduk (1997) divide their paper into two major parts: in the first they present their conceptual model of a quality-of-life assessment, and in the second they evaluate their model in the context of a quality-of-life study involving coronary artery bypass graft (CABG) patients. As will become clear, their empirical study requires that they make some assumptions concerning ongoing cognitive processes in order to complete their argument.

Beckie and Hayduk, (1997) point out that many investigators suggest that quality of life is a multidimensional construct, but that these investigators also confuse the dimensionality with multiple causes of the concept, which, they state, is due to "imprecise thinking about variables in causal networks" (p. 23). Using a regression equation as a template, they argue that any number of variables, or causes, can contribute to the same dependent variable (e.g., a global measure of quality of life). Thus, they claim a more appropriate model of a quality-of-life assessment is that it is unidimensional but has multiple causes. They then ask if it is legitimate to separate health and nonhealth causes and monitor their impact as two separate dependent variables. They answer in the negative, since creating these new dependent variables does not tell anything about the adequacy of the original quality-of-life concept. As they state, "Pointing to the disciplinary slant of the causal variables does not constitute evidence of any kind about the adequacy or inadequacy of the original QOL dependent variable" (Beckie and Hayduk 1997; p. 23). They go on to say that creating and using a health status index as a dependent variable would be particularly misleading since it is possible that this index can actually be involved in several causal networks. Their model avoids this problem since all the indicators that make up their index remain as causal, with only one outcome of interest: a unidimensional quality-of-life measure.

Beckie and Hayduk also postulate that a unidimensional quality-of-life outcome would be global. Here they make a number of cognitive assumptions. First, they assume that a variety of variables can be cognitively combined, claiming that this process should not affect the unidimensional nature of the quality-of-life concept. They state that a unidimensional quality-of-life concept can result from a "global assessment spanning diverse and complex domains" (Beckie and Hayduk 1997; p. 25). A second assumption, which plays a principal role in their argument, is that quality of life is more than the sum of its parts, and as a result, restricting the causal determinants of a quality-of-life measure to health-related issues would result in a distorted quality-of-life concept. As they write, "If the complex and interactive global assessment results in QOL being more than the sum of its parts…this is a major concern for the second strategy suggested above where the causal sources of QOL are at issue" (Beckie and Hayduk 1997; p. 25–26). However, as discussed earlier, the operative cognitive processes present when a person responds to a quality-of-life assessment item are more complex than what Beckie and Hayduk have described. For example, dual process theory tells us that while the basic response to an item may be slow and contemplative, it can also be "fast and frugal," with contemplation occurring subsequent to an initial impression. In addition, a fast and frugal response does not suggest a complex summing of parts to create a whole, but rather the selection of specific salient stimuli leading to a specific response. This difference can be seen when a person with cancer is asked about their quality of life, and in response, describes their fatigue, suggesting that it is the physical symptom with its salient consequences that are causally related to their HRQOL. Thus, not all causes of changes in HRQOL or quality of life are equal.

The second part of the paper tests their notion that quality of life is a unidimensional global concept. They postulate that a single underlying dimension (meaning quality of life) can account for several indicators, but propose that the indicators "purporting to measure that dimension can be potentially rejected if the variances and covariances of the indicators do not behave as if they were created through dependence on a single common causal source, namely QOL" (Beckie and Hayduk 1997; p. 25). To test this model they used confirmatory factor analysis and compared five different but global quality-of-life assessments, each of which was administered to patients who had the CABG procedure. They tested five models, mapping the relationship between these five outcomes. Each mapping model accounted for approximately half of the variance for each item, but the last model fit the available data most completely. These data suggested to Beckie and Hayduk that the respondents were drawing on the same source of information when they responded to each of the different global quality-of-life assessments. It is this universality in the face of similarity when dealing with different quality-of-life items that convinced Beckie and Hayduk that quality of life is a unidimensional global measure that may have multiple causes.

Here again, Beckie and Hayduk appear to be making an assumption about a cognitive process. The assumption is that a person has a *prototype*, but fuzzy (meaning that the different items used in their study may fit a common prototype) concept of quality of life that is unidimensional and global. Statistically, they would refer to this prototype concept as a latent variable, although cognitively this prototype would function as a superordinate concept.

A study of this sort deserves to be repeated under a number of different conditions, including comparing persons with different diseases. This is especially so since it is well known that quality-of-life issues important to a person with a specific disease (e.g., breast cancer) can be different than the issues important for a person in the general population. In fact, the discrepancy has, at times, become an important public policy issue. Do you allocate resources to persons with specific disease or do you consider what is best for the general population? Beckie and Hayduk's argument seems to be that confining a quality-of-life assessment to health issues would be misleading, since it does not take into account that there is a legitimate debate regarding who the audience is. If the audience is a group of people who have a unique experience that deserves to be assessed and dealt with for its own sake, then this group's relationship to the general population becomes secondary. HRQOL assessments, especially disease-specific assessments, should not attempt to generalize the population as a whole (a paper by Erickson (2004) actually attempts to bridge the gap between a population and individual patient assessment of HRQOL).

Beckie and Hayduk, however, may argue that quality-of-life assessments of persons with a disease do not differ that much from the general population, again suggesting that a separate and special assessment would not be necessary. Studies of persons who have had either an amputation or limb-sparing procedure for bone or soft-tissue sarcomas illustrate these issues. It was found that soon after either procedure there was evidence of a diminished quality of life that differed for each groups (Sugarbaker et al. 1982), but with time each group adapted and resumed what was to them an acceptable quality of life (Nagarajan et al. 2004).

What is going on here in this debate, of course, is an example of *positional objectivity* (Sen 1993). This becomes evident when the functioning of the amputee or the limb-spared person is compared to the general population: it becomes obvious that these individuals objectively have a diminished quality of life. Also important is how quality of life is assessed. Thus, if only generic quality-of-life assessments are used, it is possible that the groups may not appear different from the general population, whereas if disease-specific quality-of-life assessments were used, then differences characteristic of this particular disease may become evident.

Beckie and Hayduk (1997) administered the same global quality-of-life item in the beginning and again at the end of their assessment session. They found that the initial assessment accounted for 40% of the variance in response, compared to 71% for the second assessment. They suggest that intervening cognitive events contributed to this change, although they do not provide any measure of these processes. I have previously argued (Barofsky 2000) that using only psychometric methods does not provide a sufficient basis for establishing the validity of a measure, and that cognitive methods, such as cognitive interviewing, should also be used. From this perspective, until the authors demonstrate the cognitive equivalence of these global items (i.e., that people are thinking the same way during the pre and posttest) then their argument remains incomplete.

Many of the issues that Beckie and Hayduk (1997, 2004) raise are consistent with concerns expressed in this book. This includes the centrality of the global estimate of quality of life, and the observation that health, health status, or HRQOL are not equivalent to quality of life. They deal with their concerns by using statistical modeling procedures (e.g., structural equation modeling) to explore the relationship between causal determinants of quality of life and its outcomes (e.g., Beckie and Hayduk 2004). This type of modeling provides some empirical support for the notion that quality of life is a unidimensional construct, and while this demonstration is useful, it is not definitive. An alternative approach would involve taking into account what is known about the *cognitive basis of an abstract concept*, such as quality of life or HRQOL, and integrating this into a research program. A key element involves recognizing the limits that an abstract concept, such as quality of life, places on any assessment process.

Beckie and Hayduk are not the only investigators who have attempted to use a statistical modeling procedure to demonstrate that quality of life is unidimensional. Pagano and Gotay (2006) used item response theory as their model, and provided evidence to support the suggestion that quality-of-life items constitute a unidimensional. They dealt with cancer patients and considered three different assessments for their items. Of the subdomains they considered, only cognition was not a significant contributor to the model.

Michalos (2004) also raises these type of issues about the usefulness of an HRQOL assessment. His comments diverge from Beckie and Hayduk's, as he points out the significant lack of clear thinking amongst investigators who claim to be measuring HRQOL. This confusion takes several forms. First, he is concerned that investigators do not adequately distinguish between health status, quality of life, and HRQOL, and use these terms interchangeably (Michalos cites as an example the paper by Eiser and Tooke 1993). He is also concerned that using disease-specific quality-of-life assessments prevents a comparison to the general population. This he considers an important difference between an HRQOL assessment and a social indicator or quality-of-life study. He states:

> From the beginning social indicator researchers have been focused on measuring the quality of life of average people living in diverse circumstances, and they have been interested in measuring people's health primarily as a determinant of the quality of their lives. Certainly more than social indicator researchers, health related quality of life researchers have been primarily interested in measuring the health of people as something good in itself and, secondarily as a cause, effect, component, or all three, of the quality of their lives. (Michalos 2004; p. 54)

Thus, Michalos acknowledges that the two research groups have different agendas.

Michalos is also concerned about so-called generic quality-of-life assessments, assessments that he considers to be confounded. He grants, however, that if someone were to accept the WHO definition of health and define the physical, mental, and social domains quite broadly, that there may be overlap between the social indicator and HRQOL perspectives. He goes on to say, however, that there is no rulebook to decide what to include in these domains. He actually cautions against broadening the domains to include a wider range of indicators than those that are health related, since this would create a series of confounded outcomes. He writes, "So, broad as the WHO definition of 'health' is, it is still not as broad as the idea of quality of life as understood as happiness and it would be a mistake to try and make the ideas equivalent" (Michalos 2004; p. 55). This is a very important statement by Michalos, since it acknowledges that a fundamental and legitimate difference exists between the two perspectives. To reinforce this, he goes on to quote, and implies acceptance of, a definition of HRQOL by Hanestad (1990). In this definition, she makes it clear that quality-of-life research, if defined in the context of disease and treatment, is a unique field of study, and is an activity that can be justified independent of any other application of the phrase quality of life. Michalos rejoins this argument by asking if separate fields of quality-of-life study could be justified in other contexts, such as in an employment setting or family setting. Why not? Some of the earliest social indicator research dealt with the quality of working life (O'Toole 1974).

Michalos is also concerned about investigators discussing health and quality of life as if they are coequal concepts, especially if what is compared is limited to subjective indicators (e.g., Nord et al. 2001). He presents a sufficient number of examples to justify his concerns, but what specifically are they? Apparently, he objects to the implicit assumption in many of these HRQOL studies that good health is equivalent to good quality of life. This is not so, he contends, and presents data that suggest that health accounts for only a modest amount of the variance (e.g., 20–50%) when the general public is asked about their happiness, with happiness being a surrogate measure of quality of life. But what he doesn't do

Table 8.6 Criticisms of the HRQOL scales as stated in Cummins et al. (2004)

1.	Some HRQOL assessments combine both global and more topic specific components in the same assessment (e.g., Patrick et al. 1999)
2.	Applying the WHO criteria of health to a quality-of-life assessment can lead to inappropriate conclusions (Leplege and Hunt 1997)
3.	There is confusion among HRQOL assessments developers who think that emotion/affect can be assessed separately from cognition (see Figs. 1.2–1.3 in this book)
4.	Satisfaction with health is never the primary determinant of a person's happiness when the general public is assessed
5.	Definitions of HRQOL seem to vary to the extent that there is no consensus concerning what the definition should be
6.	Sometimes objective and subjective indicators of quality of life are mixed in an assessment and this creates a confounded empirical situation (Hagerty et al. 2001)
7.	Many HRQOL assessments confuse causal and indicator variables
8.	There is a general lack of understanding among HRQOL researchers about the homeostatic nature of a qualitative assessment, and this confounds inferring causal relationships
9.	Symptoms can be presented as either a causal or indicator variable
10.	There is little justification provided for including symptoms in HRQOL assessments

is report studies that ask these same questions to persons with a particular disease. Instead of equating health and quality of life, Michalos suggests that health status should be the most appropriate indicator of health.

What is also clear, but not explicitly acknowledged by Michalos, is that the two traditions are interested in different types of outcome measures. The social indicator movement describes its primary outcomes in terms of happiness, satisfaction, pleasure, and well-being. In contrast, HRQOL assessments are mostly interested in cataloging the consequences of diseases and treatments, with their impact on well-being and happiness being only one possible way of indicating these events. Thus, HRQOL and social indicators will have overlapping but not concordant outcomes of interest.

Michalos (2004) ends his discussion with examples of investigators who are confused about the meaning and proper usage of the terms or phrases health, health status, quality of life, and HRQOL. An independent inspection of the literature confirms his impression. But confusion surrounding the usage of a relatively new concept cannot be used to impugn the value of the concept, only its applications.[14] Instead, as Michalos himself acknowledges, HRQOL assessment is a legitimate activity when presented in a specific context. This becomes especially clear when studies have been reported (e.g., Barofsky and Sugarbaker 1990) that demonstrate that the narrowing of a qualitative assessment to a HRQOL assessment has been shown to be an effective means of altering either specific treatments, and sensitizing the medical profession to the "voice of the patient." It is not at all clear that the social indicator movement has had an equivalent impact on public policy, which is apparently their target audience.

Both the Beckie and Hayduk (1997, 2004) and Michalos (2004) papers raise legitimate concerns about HRQOL as a measure and as a concept, but neither has administered a fatal blow. Rather, both papers should stimulate additional research and increased efforts designed to refine the definition of the concept and its application. They have failed to acknowledge that HRQOL assessment differs from a social indicator study in two important ways: first, in terms of the domains being studied, and second, in terms of the need for a valuation statement in an HRQOL assessment. An HRQOL assessment considers domains that are not ordinarily included in a social indicator study, many of which are disease and treatment specific, but also items which monitor the consequences of a person's disease and treatment. Also, the literature Michalos (2004) cites is particularly lacking in terms of a respondent's assessment of the significance of what has happened to them. The failure of including explicit value statements when assessing social indicators (e.g., "How important is it to you that you are less happy because of the hair you lost due to chemotherapy?") limits their ability to be considered a qualitative statement.

I want to reiterate that when an investigator tries to operationalize an abstract concept like quality of life, they invariably have to face the cognitive constraints that are inherent to assessing an abstract concept. One way to deal with this is to narrow the informational domain of what is being assessed, thereby making it more concrete; this is essentially what an HRQOL assessment does. One could make the task even simpler by confining an assessment to various functional states (e.g., activities of daily living). But as Michalos would say, each of these maneuvers involves interpretive and statistical risks when referred back to the concept of quality of life. These risks are more an acknowledgment of the limits of language and cognition, and not a basis for rejecting a certain type of assessment. Instead, as Dubos (1959) states, it should be seen as an impetus to the continued study of the relationship between these two levels of cognition (Chap. 8, p. 251).

I will review one more paper critical of the concept of HRQOL. Cummins et al. (2004) have several concerns with current practices used to develop HRQOL assessments, and I have listed ten of them in Table 8.6. Many I agree with and have noted myself and have provided alternative approaches to in previous chapters. I address Criticism #3, for example, in Chap. 1 where I point out that any measure of affect will also have a cognitive component, and I also refer to using a circumplex to conceptualize the relationship between emotions and cognition, as Cummins et al. (2004) suggests. In addition, I depicted an emotional dimension in qualitative assessments of various sorts in Fig. 1.2–1.3. Many of the criticisms listed in Table 8.6 reflect a lack of appreciation by the developers of HRQOL assessments of the role that language plays in the construction of items, but Cummins et al. (2004) also do not acknowledge that persons with a particular disease or disorder will regularly identify symptoms as a major determinant of their quality of life. This reflects the reality that being ill, having a disease, or receiving a treatment is something someone learns to live with, and is a state that persists even if a person is successfully treated. This statement does not deny that there may be some concepts, such as Cummins' notion of affective homeostasis (Chap. 4, p. 104), which can be used to characterize how healthy persons and persons with a disease or disorder respond to their altered states, it's just that these people will have unique experiences that persons who are healthy will not. Cummins et al. (2004) also does not acknowledge that sometimes HRQOL assessments are designed to inform an investigator about the natural history of a disease or disorder or clarify the nature of a new treatment, and that to use a generic quality-of-life assessment may not be able to provide the relevant information. Finally, Cummins et al. (2004) seem to be proverbially throwing the baby out with the bath water, when all that is needed is cleaner water, or as I might argue, the design of HRQOL assessments that recognize the presence of language usage and cognitive principles.

HRQOL has been used in a number of research venues, but most prominently as part of cost-effectiveness analyses of health and medical care. These applications include using the HRQOL assessed over time as an estimate of QALY. Empirical and ethical concerns about the applications of the QALY have been raised, and I will review some of the ethical concerns in the next section.

2.2 Applications of HRQOL: The "QALY"

Examining the intellectual history that forms the foundation of qualitative assessments reveals that major contributions have come from economics and psychology (Hammond et al. 1980). Each of these traditions has contributed to an understanding of how qualitative judgments and decisions are made. A seminal contribution that greatly influenced economics was the mathematical formulation by von Neumann and Morgenstern (1944) of how "decisions are made under uncertainty." This model, and subsequent theorizing, was based on the assumption that the person making a decision would do so based on reason and logic.

The rationality hypothesis, however, has been questioned by the psychological studies of Kahneman and Tversky (1982) and others. This debate continues with classical economic theoreticians seeing any violation of the rationality principle as another variable to take into account, but not requiring rejection of the fundamental principle. Alternatively, behavioral economics has developed as a field of study that acknowledges the fragility of the rationality principle and instead examines the complexity of decision making.

Economic modeling has been particularly influential in operations research, and as described (Chap. 8, p. 260), operations research-based models have been used to describe the quality of health and medical care. These activities led to the development of a number of summary measures that characterize the operation of the health and medical care systems (e.g., Gold et al. 2002). Included amongst these summary measures are estimates of cost-effectiveness and cost–benefit. Cost-effectiveness estimates the incremental cost to achieve a unit of health effect following some health or medical intervention (e.g., preventative program or a new drug). When some measure of effectiveness is replaced by an estimate of quality, then the summary measure is referred to as a cost-utility estimate. One example of a cost-utility summary measure is the QALYs. The term "utility" is used here to refer to value, preference, or utility estimates.

The QALY, as a utility estimate, evolved from welfare economics and expected utility theory (Pliskin et al. 1980). It reflects the philosophical view that a person will be best at describing what is best for him or herself. It can be used with what Torrance (1986) referred to as an "economic appraisal" of a health care system. Torrance (1986; see his Fig. 1) conceptualizes this appraisal process as starting with a policy decision to implement a health care intervention (e.g., administering a new drug or a new way of financing medical care). This decision naturally will incur cost, including direct, indirect, and/or intangible costs. The assumption is that the new health program would lead to improvement in the health of the study population. This would then be expected to be reflected in positive health effects in terms of reduced morbidity and mortality, economic benefits as indicated by financial improvement, and enhanced valuations as determined by various valuations procedures (including value, preference, and utility estimates). These enhanced valuations are an indication of the qualitative changes that resulted from the health care innovation. These valuation estimates are then used to calculate a QALY. This is done by multiplying some estimate of the values attached to particular health states (i.e., an estimate of HRQOL) at any particular time period. If this is repeated over time, such as over the length of a person's life, then the calculation provides an estimate of the qualitative changes over a "life-year." By multiplying the quality of a person's existence by the length of their life, the QALY also provides an estimate of quality and quantity. The resultant calculation is then used as the numerator, to divide the cost of a health care intervention providing an estimate of cost-effectiveness.

A series of papers published in the journal *Value in Health* (Vol 12, Suppl #1, 2009) provides an overview of the current status of the QALY applications. It quickly becomes clear that there are a number of ways of calculating QALY depending on what concept of health is assessed; whose values or preferences are used (those of patients or the general population); which reference group is used within the general population; what valuation technique is used (standard gamble, time-trade off, or visual analog scale, etc.); what health outcomes are monitored (complete health profiles or particular health states); and how changes in valuations over time are integrated into an estimate of QALY (i.e., the issue of time discounting; Weinstein et al. 2009; Table 1).

Nord et al. (2009), however, raise a number of ethical concerns. Among these issues is that the standard methods used to provide values (or utilities) (such as the standard gamble, time trade-off, or rating using the visual analog scale) generate different values, raising the question of which method to use to insure the maximum sensitivity and specificity of the assessment process. These valuation methods have been used to develop specific assessments (e.g., the EQ-5D, the HUI, etc.), and this brings the issue of which of these assessments would provide the most representative data. Fryback et al. (2007) reports a study where he asked 3,844 people to respond to each of six of these assessments (the EQ-5D, the HUI2, HUI3, SF-6D, QWB, and HALex). They analyzed the data in terms of age and gender, and found that HRQOL scores were higher in older respondents and males, but that these differences also varied between assessments (Fryback et al. 2007; Fig. 1). In Chap. 12 (Table 12.3), I will review a series of studies similar to the Fryback et al. (2007) study. Nord suggests that the problem confirmed by Fryback et al. (2007) may be alleviated by the development of an algorithm that would permit translating the utilities of one assessment to correspond to the utility estimates of another assessment. However, even if "crosswalking" could be demonstrated to be reliable, it will still be possible to raise issues of the validity of resultant data.

Nord et al. (2009) express a second concern that respondents who have experienced certain outcomes (e.g., a disabled person, a surviving cancer patient, or a person with recurrent schizophrenia), when asked to participate in the valuation task, approach it differently than the person who has not experienced the consequences of a disease or treatment. When asked to participate in a valuation procedure, such as the time trade-off, the experienced responder will often refuse to sacrifice life expectancy to be relieved of their health problem. There are several explanations of why this happens, including that the time frame respondents are asked to consider may be too long and beyond what they expect to live. Another possibility is that because of their altered health state, the experienced responder may discount their potential opportunities, which would narrow their perception of what is possible for them. Evoking this kind of response will depend on how the valuation task is framed; with a broader frame, a person is more likely to reach their limits in their willingness to trade-off.

The third ethical concern of Nord et al. (2009) deals with the issue of fairness. This is important because in its current usage, the QALYs represent the aggregation of individual utilities, and this practice does not recognize that the preintervention utility levels of some participants may differ significantly from the general population. Thus, the preintervention utilities levels of the severely ill may not be weighted fairly in the aggregated QALY estimate. A second concern about fairness deals with assumption, implicit in the QALY model, that the value of an intervention is proportional to the person's capacity to receive its benefits. Thus, it can be asked whether the person with a limited life expectancy should be given the same access to resources as someone whose potential for extended survival is greater. The answer may vary with different societies, but the empirical data available permits Nord et al. (2009) to conclude that the general public still believes that the most needy deserve the most services, whether they can benefit or not. A third fairness concern focuses on the issue of whether the benefits of a health intervention are the same for a healthy person as for the disabled. They claim that the increment in health values because of an intervention would be greater for the healthy person compared to the disabled, primarily because the healthy person has a greater capacity to show an increment than the disabled. So when the increments for each group are aggregated in a QALY estimate, the net result is that the disabled will be less likely to be represented. This difference may be reflected in the relative valuations attached to particular interventions, such that for policy-making decisions, the healthier person is given a greater societal preference than the disabled person. Part of the problem here, Nord et al. (2009) states, comes from the fact that the cost per QALY ratio is a productivity indicator; that productivity does not encompass the social issues also involved when making health care policy decisions, including allocating resources.

Finally, Nord et al. (2009) discuss a fourth issue: whether the QALY can be an adequate indicator of the gains in health value when it is basically an indicator of a health state. They cast their argument in the context of the persons who have limitations but different potentials for rehabilitation. For example, can the improvement in health of someone who is in a wheelchair but able to improve their health so that they can go from crutches to walking be compared to someone who progresses from a wheelchair to crutches, but can't make further progress? From a societal perspective, but not an individual perspective, the two persons would differ in health gain. The question now becomes how an investigator incorporates this difference into their cost-effectiveness estimation. Merely adding the two estimates of QALYs together may not reflect the fact that the improvement for each was the maximum possible.

An investigator may ask if this is fair, considering that what may be at issue here is the allocation of resources on the basis of potential for health improvement. In response to this dilemma, Nord et al. (2009) report an ongoing study where they asked groups of subjects in focus groups whether they preferred going to a hospital that treated everyone equally, independent of their potential for full recovery, as compared to a hospital that gave priority to those who had a greater potential for recovery. Their data suggests that three out of five respondents would select the hospital that treated everyone equally, independent of the potential for improvement.

Nord et al. (2009) go on to discuss how the QALY indicator can become "equity adjusted," and in this manner deal with some of the ethical dilemmas that the current usage of the QALY may produce. They describe this process as transforming fairness issues into a numerical evaluation. What is familiar about this discussion, and why I was particularly interested in going into it in some detail, is that it represents another example of the qualification/subjectification followed by objectification process that I discussed in Chap. 5. By raising questions about fairness, Nord et al. (2009) express a subjective concern about the meaning or validity of the QALY, especially when used as part of a cost-effectiveness analysis. By offering ideas on how this concern can be translated into a numerical equivalent, they describe how this subjective concern can be objectified. This constant interplay between the subjective and its objectification is the essential method whereby science progresses and can also be a most useful method for how policy decisions are made.

2.3 Summary of Applications

This section focused on two major topics: the components of an HRQOL assessment and the application of an HRQOL assessment as part of a QALY estimate. Tables 8.3–8.5 provided a cognitive analysis of the items on HRQOL assessments; examples of individual items on various assessments of the classification system were provided in Table 8.3, but I also analyzed the components of currently available qualitative assessments. The last eight assessments in Table 8.5 were examined to determine the extent to which they consisted of the elements of a hybrid construct. Inspection of the assessments revealed that each assessment consisted of evaluated descriptors and an independent valuation assessment. All of the assessments can be summarized with a single number, although some could also be presented as profiles.

Differences exist between these assessments in terms of whether they included explicit reference to the consequences of being in a particular health state, what states were selected to study, the response format (e.g., behavioral checklists, items on a questionnaires, or identification of a judgment policy), and so on.

As I noted earlier (Chap. 8, p. 272), inspection of the last eight assessments listed in Table 8.5 suggests that the investigators all had the same conceptual format in mind when they structured their assessment, even though their assessments differed in their details. Each investigator seems to have appreciated that their assessment was meant to simulate how a person made a judgment, and that this required identification of what was to be judged, and what value was to be assigned to what was being judged. I contend that this was occurring whether the investigator created their model on the basis of a person acting like a "rational man" when making their judgments or not. It could also be argued that the same conceptual format was used in the first seven assessments listed in Table 8.5, just that the valuation task was left for alternative, less formal methods. For example, by creating items that asked about the aversive consequences of a disease or treatment (e.g., BCQ), it could be argued that these items were value statements, reflecting the views of either or both the investigators and/or the respondents. Also true was that the items making up the final questionnaire were rated by the respondents in terms of how important they were to them, so in that sense, the content of the questionnaire was a value statement. Even using these assessments in an ethically ambiguous environment, such as a phase I clinical trial, might be seen as a value statement, but now to monitor the process of care, not necessarily the outcome of care. This brings me back to discussing Fulford's (1999) view that it is very hard to find a place in the communication process (e.g., because of the presence of connotation) where values are not present. Thus, the issue of whether the explicit assessment of values in HRQOL assessments can be useful depends in part on what is being assessed (the process or outcome of care), who is doing the assessment (the respondent, the general public, or the investigator), and how this information is being used (descriptively or as an aid to policy decisions). The virtue of a formal assessment, of course, is that it satisfies the ethical requirement that a respondent or the investigator be aware of the consequences of their decisions.

Inspecting the available literature on the evaluation of health care interventions confirms that integrating an HRQOL assessment into a cost-effectiveness assessment via a QALY has become a popular activity, especially in the area of pharmacoeconomics. However, as the Nord et al. (2009) paper indicated, there are also ethical concerns about this research that should be addressed. For example, they point out that estimating QALYs is very much dependent on the method used to generate the values. Thus, the values generated by a standard gamble, time trade-off, or visual analog scale will differ, and the authors ask which set of values should be used in a given situation. Fryback et al. (2007) suggested that it may be possible to standardize each method relative to the other, and in this way solve the problem of how two or more studies can be compared. An alternative is to ask how these judgments are being made, and in what way they differ or are similar. It may even be possible to decide that one method is more informative than the other, a decision that cannot be made now, since what is known about how a person makes a value judgment has not been applied to these valuation methods. Nord et al. (2009) also raise the issue of fairness, especially relative to whose data should be used to make decisions. Should it be the person who already experienced the disease, its treatment and its consequences, or some member of the general public? Dolan (2003), for example, states that if there is to be rationing of health benefits, does it not make sense to rely on the persons who have already had the illness and treatments to provide input into these decisions? Wouldn't they be able to provide the most valid information? Nord et al. (2009) also raise the issue of fairness relative to the disabled. They acknowledge that the disabled may not be able to demonstrate as great a quantitative benefit from a health care intervention than a healthy person, but does this mean that they should be denied adequate access to health care interventions? If only the quantitative changes in QALYs were used as an estimate of the cost-effectiveness then just such an outcome could occur. These issues reflect the relatively defined nature of a QALY when the environment being studied is far more complex than the statistic may allow. Nord et al. (2009), however, feel that these limitations will be dealt with as the indicator continues to be developed.

The cost of care, when divided by a QALY estimate, provides an estimate of the efficacy of a health care intervention, and policy makers can use this to make resource allocation decisions. An analog of this process is the calculation of the "clinically significant difference" or "minimally important difference" (MID) that clinicians make to determine if the change following an intervention is sufficient to justify subsequent clinical decisions (Norman et al. 2003; Wyrwich et al. 2005). An example of this type of algorithm is Cohen's (1988) suggestion that differences in effect size can be graded into three categories. His grading system, he admitted, was based on "convention," meaning it was arbitrary. Norman et al. (2003) point out that a number of investigators have made alternative suggestions to Cohen's grading system, but that inspecting 38 available studies that provided 62 estimates of effect size revealed that the MID can fairly consistently be described as falling close to 0.5 standard deviation on a 7-point scale. Norman et al. (2003) also rhetorically ask why there is so much commonality in these findings. Their answer, illustrating the usefulness of attending to the perceptual/ cognitive processes that may be evoked by an assessment task, is that the 0.5 unit change describes the limits of human information processing. In other words, people are only able to consistently identify a 0.5 unit change when the categories (bits) involved are limited to seven, plus or minus two (Miller 1956). Norman et al. (2003) go on to discuss how this observation about human information processing may be applicable to MID, although they also acknowledge exceptions. What is so interesting about their argument to me is that by referring to the operating characteristics of perceptual/ cognitive systems they were able to provide support for the convergent validity of their decision model. It is this type of opportunity that I expect will be found for other types of HRQOL indicators, including the continued development of the QALY indicator.

3 Summary of Chapter

This chapter addressed a number of important topics, ranging from how health and disorders can be defined, to what the components are of a quality-of-life or HRQOL assessment (Table 8.5). I felt it was necessary to discuss the meaning of the term "health" to clarify what the phrase "health status" means, and I discussed the term "disorder," since health is regularly defined as the absence of disease or a disorder. The chapter opened by asking if a health status domain can legitimately be included in an HRQOL assessment; the conclusion reached was that if the health status items were exclusively descriptive in form, then it would not be considered part of an HRQOL assessment. Obviously, limiting HRQOL items only to those that combined description of the consequences of being in a particular state and the valuation of being in this state will significantly reduce the number of available items that qualify as HRQOL items. However, this emphasis on *how the assessment is made*, rather than what content is assessed, also broadens the potential application sites. Thus, the diagnosis of a mental illness as a *harmful dysfunction* could now be considered a qualitative, if not HRQOL statement, especially if what is meant by harmful is quantified (i.e., a valuation). The comments made by a clinician about a patient, if they covered the criteria listed in Table 8.3 (Criterion #4), would also be considered an HRQOL assessment. In fact, by using the criteria in Table 8.3 it should be possible to classify and compare all items considered quality-of-life or HRQOL items, but it also greatly expands the range of qualitative judgments.

The history of the development of the HRQOL indicator and its differentiation from a health status indicator was also traced, starting with the end-results and outcome movement, and continuing with the development and application of operations research. Donabedian's definition of the term "quality" was also reviewed, and I noted in particular that he seems to have adapted the same conceptual hybrid construct format I have proposed to be operating in all types of qualitative assessments. Donabedian, of course, used the qualitative indicator of patient satisfaction as his primary monitor of quality of care, so I took the opportunity to examine the cognitive basis satisfaction assessments. The qualitative and quantitative data suggested that satisfaction is a "judgment of a judgment," so that it will have limited utility in any qualitative assessment. Also considered were some of the arguments suggesting that HRQOL is not a viable construct. The papers reviewed did point out legitimate concerns about the application of HRQOL assessments, but I argued that these concerns should not preclude the continued development of the indicator. Also discussed was the QALY, seen as an application of an HRQOL assessment that has played a very useful role in assessing the efficacy of health or medical care intervention.

In general, this chapter illustrated again the importance of a cognitive perspective to understanding the definitional and measurement task in a qualitative assessment. The presence of the hybrid construct in some currently available HRQOL assessments was made explicit and this gives me the opportunity to stitch another cognitive thread into the fabric that supports the conceptual foundation required for making qualitative judgments. The next chapter takes my discussion to another level of specificity as I examine the role of functions in a qualitative assessment.

Notes

1. Lillienfeld and Marino (1999; p. 400) included this quote at the beginning of their paper, and it seemed appropriate to use at the start of this chapter, since the issue of the inferences we draw from the terms we use will repeatedly come up here.
2. Of course, this same argument can be made about why there can never be a universal definition of quality of life or HRQOL.
3. I will discuss Parson's notions about health in greater detail in Chapter 9, which deals with whether functional status is an appropriate domain of a health domain.
4. One approach is to use fuzzy logic to decide what state a person is in.
5. To get a sense of the richness of the issues being discussed here, the reader is encouraged to read the papers published in the *Journal of Abnormal Psychology*, Volume 108, No. 3, 1999 (pages 371–472). This issue includes the two papers discussed here and nine comments that deal with how well each of these perspectives has dealt with the topic of what is meant by a disorder (with a particular emphasis on the meaning of "mental disorder" or "mental disease"). I will review only a limited amount of this material here, but it should also be noted that many of the issues discussed in these papers are relevant for my discussion concerning the definition and measurement of HRQOL.
6. "A value statement" is being and will be used as a general category that includes measures of *values, preferences,* and *utilities.*
7. The hybrid construct that Wakefield (1999a) uses is the same basic model previously cited as providing a unique definition of a quality-of-life assessment. There are other examples of this paradigm, each of which attests to the universality of this model.
8. The distinction being made here should remind the reader of our previous discussion of the difference between *logical positivism* and *empirical constructionalism* (Chap. 2, p. 31).
9. See a book by Wegner, *The Illusion of Conscious Will* (MIT Press, 2002), which addresses issues similar to the discussion here.
10. Dr. Penny Erickson in a personal communication suggested this association (December 3, 2004).
11. The reader is encouraged to read the Mallon (2000) book for a complete description of the cartoon and its consequences on Codman.
12. These assessments include, for example, the Bergner et al.'s (1981) *Sickness Impact Profile*; Kaplan and Anderson's (1990) *Index of Well-being*; and Torrance's (1995) *Health Utility Index*.
13. This is a particularly relevant issue for HRQOL assessment, since many assessments are not administered with a fixed time for completion. Thus, someone who is partially compromised may take longer to complete the task than someone who is not, and yet both may have similar response patterns. The question is whether these two individuals are providing the same information, and how this can be judged.
14. See Michalos (2004, p. 59), who also gives examples where this is not so.

References

Aaronson NK, Ahmedzai S, Bullinger M, Crabeels D, et al. (1991). The EORTC core quality of life questionnaire: Interim results of an international field study. In, (Ed.) D.Osoba, *The Effect of Cancer on Quality of Life*. Boca Raton FL: CRC Press. (p. 185–203).

Ahmed PI, Coelho GV. (1979). *Towards a New Definition of Health: Psychosocial Dimensions*. New York NY: Plenum.

Asada Y. (2005). Assessment of the health of Americans: The average health-related quality of life and its inequality across individuals and groups. *Popul Health Metr.* 3, 7. doi:10.1186/1478-7954-3-7.

Balaban DJ, Sagi PC, Goldfarb NI, Nettler S. (1986). Weights for scoring the Quality of Well-being instrument among rheumatoid arthritics. *Med Care*. 24, 973–980.

Barofsky I. (2000). The role of cognitive equivalence in studies of health-related quality of life assessments. *Med Care*. 38 (Supp II), 125–129.

Barofsky I. (2003). Patients right's, quality of life, and health system performance. *Qual Life Res*. 12, 273–284.

Barofsky I and Sugarbaker PH. (1990). The cancer patient. In, (Ed.) B Spilker *Quality of Life Assessment in Clinical Studies*. New York NY: Raven. (p. 419–439).

Baumol, WJ. (1965). *Economic Theory and Operations Analysis*. Englewood Cliffs NJ: Prentice-Hall.

Bearden JN, Rapoport, A. (2005). Operations research in experimental psychology. *Tutorials in Operations Res*. Hanover MD: INFORMS. (Chapter 1; p. 1–25).

Beckie TM, Hayduk LA. (1997). Measuring quality of life. *Soc Indic Res*. 42, 21–39.

Beckie TM, Hayduk LA. (2004). Using perceived health to test the construct related validity of global quality of life. *Soc Indic Res*. 65, 279–298.

Bell DE, Raiffa H, Tversky A. (1988). Descriptive, normative and prescriptive interactions in decision making. In, (Eds.) DE Bell, H Raiffa, A Tversky Decision Making : Descriptive, Normative and Prescriptive Interactions. Cambridge UK: Cambridge University Press.

Bergner M, Bobbitt RA, Carter WB, Gilson BS. (1981). The Sickness Impact Profile: Development and final revision of a health status measure. *Med Care.*. 19, 787–805.

References

Bio-Medical News. (2005). Identifying and Neutralizing the cause of deadly side effects of Eli Lilly & Co. Anticancer drug ALIMTA. (Accessed on the web at http://www.biomedical-market-news.com/) Originally published at http://www.scienceofbetter.org/can_do/success_stories/iantcodseoelcada.htm. (Accessed from the web -7/22/2011).

Birren JE, Dieckmann L. (1991). Concepts and content of quality of life in the later years: An overview. In, (Eds.) JE Birren, JC Rowe, JE Lubben, DE Deutchman *The Concept and Measurement of Quality of Life in the Frail Elderly*. New York NY: Academic Press. (p. 344–360).

Boorse C. (1975). On the distinction between disease and illness. *Philos Public Affa.* 5, 49–68.

Boorse C. (1977). Health as a theoretical concept. *Philos Sci.* 44, 542–573.

Boorse C. (1997). A rebuttal on health. In, (Eds.) JM Huber, RF Almeder. *What is Disease?* Totowa NJ: Humana.

Bowling A. (2001). *Measuring Disease: A Review of Disease-Specific Quality of Life Measurement Scales*. 2nd Edition. Philadelphia PA: Open University Press.

Bowling A, Windsor J. (2001) Towards the good life: A population survey of dimensions of quality of life. *J Happiness Res.* 2, 55–81.

Bradley C. (2001). Importance of differentiating health status from quality of life. *Lancet.* 357(9249), 7–8.

Brady MJ, Cella DF, Mo F, Bonomi AE, et al. (1997). Reliability and validity of the Functional Assessment of Cancer Therapy- Breast quality of life instrument. *J Clin Oncol.* 15, 974–986.

Brazier J, Roberts J, Deverill M. (2002). The estimation of a preference-based measure of health from the SF-36. *J Health Economics* 21, 271–292.

Brooks R, Rabin R, de Charro F. (2003). *The Measurement and Valuation of Health Status Using the EQ-5D: A European Perspective. Evidence from the EurQol BIOMED Research Programme*. Dordrecht The Netherlands: Kluwer. (p. 254).

Bush JW, Chen MM, Patrick DL. (1973). Health status index in cost-effectiveness: Analysis of PKU program. (Ed.) RL Berg in *Health Status Indexes: Proceedings of a Conference Conducted by Health Services Research*. Chicago Il: Hospital Research and Education Trust. (p. 172–194).

Calnan M. (1988). Towards a conceptual framework of lay evaluation of health care. *Soc Sci Med* 27, 927–933.

Calvin WH. (1983). A stone's throw and its launch window: Timing precision and its implications for language and hominid brains. *J Theor Biol.* 104, 121–135.

Campbell A, Converse PE, Rodgers WL. (1976). *The Quality of American Life*. New York NY: Russell Sage Foundation. Rutgers University Press.

Carr AJ, Thompson PW, Kirwan JR. (1996). Quality of life measures. *Br J Rheum.* 35, 275–281.

Cella DF, Tulsky DS, Gray G, Sarafian B, et al. (1993).The Functional Assessment of Cancer Therapy (FACT) scale: Development and validation of the general measure. *J Clin Oncol.* 11, 570–579.

Chyba MM, Washington LR. (1993). *Questionnaires from the National Health Interview Survey, 1985–1989. Vital Statistics Series 1, Number 31*. Hyattsville, MD: National Center of Health Statistics.

Clark LA. (1999). Introduction to the special section on the concept of disorder. *J Abnorm Psychol.* 108, 371–373.

Cleary PD. (1998). Satisfaction may not suffice! A commentary on 'A patients perspective'. *Int J Technol Assess Health Care..* 14: 35–37.

Cochrane A. (1972). *Effectiveness and Efficiency: Random reflections on Health Services*. Leeds: Nuffield Provincial Hospital Trusts.

Codman EA. (1917/1996). *A Study of Hospital Efficiency*. (Reprinted 1996). Oakbrook Terrace Ill: Joint Commission on Accreditation of Healthcare Organizations Press.

Cohen J. (1988). Statistical power analysis for the behavioral sciences. (2nd Ed.). Hillsdale, NJ: Lawrence Earlbaum Associates.

Collins K, O'Cathain A. (2003). The continuum of patient satisfaction-from satisfied to very satisfied. *Soc Sci Med.* 57, 2465–2470.

Cosmides L, Tooby J. (1999). Towards a evolutionary taxonomy of treatable conditions. *J Abnorm Psychol.* 108, 453–464.

Covinsky KE, Wu AW, Landefeld CS, Connors AF Jr, et al. (1999). Health status versus quality of life in older patients: Does the distinction matter? *Am J Med.* 206, 435–440.

Coyle J, Williams B. (1999). See the wood for trees: Defining the forgotten concept of patient dissatisfaction in the light of patient satisfaction research. *Leadersh Health Serv.* 12, 1–9.

Cronbach LJ, Meehl PE. (1955). Construct validity and psychological tests. *Psychol Bull.* 52, 281–302.

Cronje RJ, Williamson OD. (2006). Is pain ever normal? *Clin J Pain.* 22, 692–699.

Cummins RA, Lau ALD, Stokes M. (2004). HRQOL and subjective well-being: Noncomplementary forms of outcome measurement. *Expert Rev Pharmecon Outcomes Res.* 4, 413–420.

Cummins RA, McCabe MP, Romeo Y, Gullone E. (1994). The Comprehensive quality of life scale (ComQoL): Instrument development and psychometric evaluation on college staff and students. *Educ Psychol Meas.* 54, 372–382.

De Haes JCJM, van Knippenberg FCE, Neijt JP. (1990). Measuring psychological and physical distress in cancer patients: Structure and application of the Rotterdam Symptom Checklist. *Br J Cancer.* 62, 1034–1038.

de Vries EN, Ramrattan MA, Smorenburg SM, Gouma DJ, et al. (2008). The incidence and nature of in-hospital adverse events: A systematic review. *Qual Saf Health Care.* 17, 216–223.

Desbiens NA, Mueller-Rizer N, Connors AF, Hamel MB et al. (1997). Pain in the oldest-old during hospitalization and up to one year later. *J Am Geriat Soc.* 45, 1167–1172.

Dolan P. (2003). Developing methods that really do value the 'Q' in QALY. *Health Econ Policy Law.* 3, 69–77.

Donabedian A. (1980). *Explorations in Quality Assessment and Monitoring Volume I. The Definition of Quality and Approaches to its Assessment*. Ann Arbor MI: Health Administration Press.

Donabedian A. (1982). *Explorations in Quality Assessment and Monitoring Volume II. The Criteria and Standards of Quality*. Ann Arbor MI: Health Administration Press.

Donabedian A. (1988). The quality of care: How can it be assessed? *J AMA.* 260, 1743–1748.

Donabedian A. (1989). The end results of health care: Ernest Codman's contribution to quality assessment and beyond. *Milbank Q.* 67, 233–267.

Donabedian A. (2005). Evaluating the quality of medical care. *Milbank Q.* 83, 691–729.

Dubos R. (1959). *Mirage of Health: Utopias, Progress, and Biological Change*. New Brunswick NJ: Rutgers University Press.

Dubos R. (1968). *So Human an Animal: How we are Shaped by Surroundings and Events*. New York NY: Scribner.

Edwards C, Staniszweska S, Crichton N. (2004). Investigation of the ways in which patients' reports of their satisfaction with healthcare are constructed. *Sociol Health Illn.* 26, 159–183.

Engel GL. (1977). The need for a new medical model: A challenge for biomedicine. *Science.* 196, 129–136.

Eiser C, Tooke JE. (1993). Quality of life evaluation in diabetes. Pharmacoeconomics. 4, 85–91.

Erickson P. (1998). Evaluation of a population-based measure of quality of life: The Health and Activity Index (HALex). *Qual Life Res.* 7, 101–114.

Erickson P. (2004). A health outcomes framework for assessing health status and quality of life: enhanced data for decision making. *J Natl Cancer Inst Monogr.* 33, 168–177.

Erickson P, Wilson R, Shannon I. (1995). Years of healthy life. *Healthy People 2000: Statistical Notes Number 7*. NCHS. Washington DC: Public Health Service.

Feinstein A. (1967). *Clinical Judgment*. Huntington NY: Krieger.

Ferrans CE, Powers MJ. (1985). Quality of life index: development and psychometric properties. *Adv Nurs Sci.* 8, 15–24.

Feussner JR, Kizer KW, Demakis JG. (2000). The quality enhancement research initiative (QUERI): From evidence to action. *Med Care.* 38, 1–1 - 1–6.

Fitzpatrick R, Hopkins A. (1983). Problems in the conceptual framework of patient satisfaction research: An empirical definition. *Sociol Health Illn.* 5, 297–311.

Flagle CD. (1962). Operations research in the health services. *Oper Res.* 10, 591–603.

Follich MJ, Gorkin L, Smith T, Capone RJ, et al. (1988). Quality of life post-myocardial infarction: The effect of a transtelephonic coronary intervention system. *Health Psychol.* 7, 169–182.

Fox JG, Sorms DM. (1981). A different approach to sociodemographic predictors of satisfaction with health care. *Soc Sci Med A.* 15, 557–564.

Frances AJ, Widiger TA, Sabshin M. (1991). Psychiatric diagnosis and normality. In, (Eds.) D Offer, M Sabshin. *The Diversity of Normal Behavior.* New York NY: Basic Books. (p. 3–38).

Freud S. (1937/1957). *Analysis Terminable and Interminable.* (Ed.) JL Strachey. Standard Edition. Vol 23. London UK: Hogarth Press. (p. 209–254).

Fulford KWM. (1989). *Moral Theory and Medical Practice.* New York NY: Cambridge University Press.

Fulford KWM. (1999). Nine variations and a coda on the theme of an evolutionary definition of dysfunction. *J Abnorm Psychol.* 108, 412–420.

Fulford KWM. (2000). Teleology without tears: Naturalism, neo-naturalism, and evaluationism in the analysis of function statements in biology (and a bet on the Twenty-first Century). *Philos Psychiatr Psychol.* 7,77–94.

Fulford KWM. (2001). What is (mental) disease?: An open letter to Christopher Boorse. *J Med Ethics.* 27, 80–85.

Fryback DG, Sunham NC, Palta M, Hanmer J, et al. (2007). US Norms for six generic health-related quality of life indexes from the National Health Measurement Study. *Med Care.* 45, 1162–1170.

Ganz PA, Schag CA, Lee JJ, Sim MS. (1992). The CARES: a generic measure of health-related quality of life for patients with cancer. *Qual Life Res.* 1, 19–29.

Gass SI, Assad AA. (2005). *An Annotated Timeline of Operations Research: An Informal History.* New York NY: Kluwer.

Gerhardt U. (1989). *Ideas About Illness, An Intellectual and Political History of Medical Sociology.* New York NY: New York University Press.

Gill TM, Feinstein AR. (1994). A critical appraisal of the quality of quality of life measurements. *JAMA* 272, 619–626.

Gill L, White L. (2009). A critical review of patient satisfaction. *Leadersh Health Serv.* 22, 8–19.

Glushko RJ, Maglio PP, Matlock T, Barsalou LW. (2008). Categorization in the wild. *Trends Cogn Sci.* 12, 129–135.

Gold MR, Siegel JE, Russell LB, Weinstein MC. (1996). *Dose-Effectiveness in Health and Medicine.* New York NY: Oxford University Press.

Gold MR, Stevenson D, Fryback DG. (2002). HALYs and QALYs and DALYs OH My: Similarities and differences in summary measures of population health. *Annu Rev Public Health.* 23, 115–134.

Guyatt G. (1993). Measurement of health-related quality of life in heart failure. *J Am Coll Cardiol.* 22 (Suppl A), 186A-191A.

A.Guyatt G. (1995). A taxonomy of health status instruments. *J Rheum.* 22, 1188–1190

Hagerty MR, Cummins RA, Ferris AL, Land K, et al. (2001). Quality of life indexes for national policy: Review of agenda for research. *Soc Indic Res.* 55,1–91.

Halfon N, Hochstein M. (2002). Life course health development: An integrated framework for developing health policy, and research. *Milbank Q.* 80, 433–479.

Hammond KR, McClelland GH, Mumpower J. (1980). *Human Judgment and Decision Making.* New York NY: Praeger.

Hanestad BR. (1990). Errors of measurement affecting the reliability and validity of data acquired from self-assessed quality of life. *Scand J Caring Sci* 4, 29–34.

Hawthorne G. (2006). *Review of Patient Satisfaction Measures.* Melbourne Australia: Australian Government Department of Health and Aging.

Hawthorne G, Sansoni J, Hayes LM, Marosszeky N, et al. (2006). *Measuring Patient Satisfaction with Incontinence Treatment.* Center for Health Service Development, University of Wollongong and Department of Psychiatry, University of Melbourne.

Hays RD, Ware JE. (1986). My medical care is better than yours: Social desirability and patient satisfaction ratings. *Med Care.* 24, 519–525.

Hazelrigg LE, Hardy MS. (2000). Scaling the semantics of satisfaction. *Soc Indic Res.* 49, 147–180.

Heinonen H, Aro AR, Aalto A-M, Uutela A. (2004). Is the evaluation of the global quality of life determined by emotional status? *Qual Life Res.* 13, 1347–1356.

Herzlich C. (1973). *Health and Illness: A Social Psychological Analysis.* London UK: Academic Press.

Hickey A, O'Boyle CA, McGee HM, Joyce CRB. (1999). The Schedule for the Evaluation of Individual Quality of Life. In, (Eds.) CRB Joyce, HM McGee, CA O'Boyle. Individual Quality of Life: Approaches to Conceptualization and Assessment. Amsterdam The Netherlands: Harwood. (p. 119–133).

Hlatky MA, Boineau RE, Higginbotham MB, Lee KL, et al. (1989). A brief self-administered questionnaire to determine functional capacity (the Duke Activity Status Index). *Am J Cardiol.* 64, 651–654.

Hudak PL, McKeever PD, Wright JG. (2003). The metaphor of patients as customers: Implications for measuring satisfaction. *J Clin Epidemiol.* 56, 103–108.

Hudak PL, McKeever PD, Wright JG. (2004). Understanding the meaning of satisfaction with treatment outcomes. *Med Care.* 42, 718–725.

Hughnet RS, Kleine SS. (2004). Views of health in the lay sector: a compilation and review of how individuals think about health. Health: An Interdisciplinary *J Soc Study Health Illn Med.* 8, 395–422.

Hull DL. (1976). Are species really individuals? *Syst Zool.* 25, 174–191.

Hung DY, Glasgow RE, Dickinson LM, Forshaug DB, et al. (2008). The Chronic Care Model and relationship to patient health status and health-related quality of life. *Am J Prev Med.* 35 Suppl #1, S398-S406.

Hunt SM. (1997). The problem of quality of life. *Qual Life Res.* 6, 205–212.

Institute of Medicine. (2001). *Informing the Future: Critical Issues in Health.* Washington DC: The National Academies Press.

James W. (1890). *Principles of Psychology.* Vol 1. New York NY: Holt.

Johnson NG. (2003). Psychology and health: Research, practice and policy. *Am Psychol.* 58, 670–677.

Kahneman D, Tversky A. (1979). Prospect theory: An analysis of decisions under risk. *Econometrica.* 47, 263–291.

Kahneman D, Tversky D. (1982). The psychology of preferences. *Sci Am.* 246, 160–173.

Kaplan R, Anderson JP. (1990). The General Health Policy Model: An integrated approach. In, (Ed.) B Spilker. *Quality of Life Assessments in Clinical Trials.* New York NY: Ravens.

Kaplan R, Anderson JP. (1996). The General Health Policy Model: An integrated approach. In, (Ed.) B Spilker. *Quality of Life and Pharmacoeconomics in Clinical Trials.* New York NY: Ravens. (p. 309–322).

Kaplan RM, Bush JW. (1982). Health-related quality of life for evaluation research and policy analysis. *Health Psychol.* 1, 61–80.

Kaplan RM, Bush JW, Blischke WR. (1987). Additive utility independence in a multidimensional quality of life scale for the general health policy model. Unpublished Manuscript.

Kaplan RM, Feeny D, Reviciki DA. (1993). Methods for assessing relative importance in preference based outcome measures. *Qual Life Res.* 2, 467–475.

Katz S, Apkom CA. (1976). A measure of primary socio-biological functions. *Int J Health Serv.* 6, 493–507.

Kind P. (1996). The EuroQoL instrument: An index of health-related quality of life. In, (Ed.) B Spilker. *Quality of Life and Pharmacoeconomics in Clinical Trials.* Philadelphia PA: Lippincott-Raven. (p. 191–201).

Kirmayer LJ, Young A. (1999). Culture and context in the evolutionary concept of mental disorder. *J Abnorm Psychol.* 108, 446–452.

Klein DF. (1999). Harmful dysfunction, disorder, disease, illness and evolution. *J Abnorm Psychol.* 108, 421–429.

Klein Entink RHK, Kuhn J-T, Hornke LF, Fox J-P. (2009). Evaluating cognitive theory: A joint modeling approach using responses and response times. *Psychol Method.* 14, 54–75.

Kunkel S, Rosenqvist U, Westerling R (2007). The structure of quality systems is important to the process and outcome, an empirical study of 386 hospital departments in Sweden. *BMC Health Serv Res.* 7, 104 (p. 1–8).

Lakoff G, Johnson M. (1999). *Philosophy in the Flesh: The Embodied Mind and its Challenge to Western Thought.* New York NY: Basic Books.

Larson JS. (1999). The conceptualization of health. *Med Care Res Rev.* 56, 123–136.

Leplege A, Hunt S. (1997). The problem of quality of life in medicine. *JAMA* 278, 47–50.

Levine D, Yett DE. (1973). *A Method of Constructing Proxy Measures of Health Status. In.* (Ed.) RL Berg, *Health Status Indexes: A Proceedings of a Conference Conducted by Health Services Research.* Chicago Ill: Hospital Research and Educational Trust Fund.

Levine MN, Guyatt GH, Gent M, De Pauw S, et al. (1988). Quality of life in Stage II breast cancer: An instrument for clinical trials. *J Clin Oncol.* 6, 1798–1810.

Lilienfeld SO, Marino L. (1999). Essentialism revisited: Evolutionary theory and the concept of mental disorder. *J Abnorm Psychol.* 108, 400–411.

Linden-Perz S. (1982). Social psychological determinants of patient satisfaction: A test of five hypotheses. *Soc Sci Med.* 16, 583–589.

Livingston EH, Ko CL. (2002). Use of the Health and Activities Limitation Index as a measure of quality of life in obesity. *Obesity.* 10, 825–832.

Locke EA. (1970). Job satisfaction and job performance, A theoretical analysis. *Organ Behav Human Perform.* 5, 484–500.

Mallon W. (2000). *Ernest Amory Codman: The End Results of a Life in Medicine.* Philadelphia PA: Saunders.

Mayr E. (1982). *The Growth of Biological Thought: Diversity, Evolution, and Inheritance.* Cambridge MA: Belknap Press/Harvard University Press.

McCarty DM. (1995). Quality of life: A critical assessment. *Scand J Gastroenterol.* 208 (Suppl), 141–146.

Mc Glynn EA. (2003). Introduction and overview of the conceptual framework for a National Quality Measurement and Reporting System. *Med Care.* 41, 1–1 - 1–7.

McKague M, Verhoef M. (2003). Understanding health and its determinants among clients and providers at an urban community health center. *Qual Health Res.* 42, 414–437.

McNair DM, Lorr M, Droppelman LF. (1992). *EdITS manual for Profile of Mood States.* San Diego CA, Educational and Industrial Testing Service. (p.1–40).

Medawar PB, Medawar JS. (1983). *Aristotle to Zoos: A Philosophical Dictionary of Biology.* Cambridge MA: Harvard University Press.

Medin DL. (1989). Concepts and conceptual structure. *Am Psychol.* 44, 1469–1481.

Medin DL, Ortony A. (1989). Psychological essentialism. In, (Eds.) S Vosniadou, A Ortony. *Similarity and Analogical Reasoning.* New York NY: Cambridge. (p. 179–195).

Mendoza T, Wang XS, Cleeland CS, Morrissey M, et al. (1999). The Rapid Assessment of Fatigue Severity in Cancer Patients: Use of the Brief Fatigue Inventory. *Cancer.* 85, 1186–1196.

Mersky H, Bogduk N. (2004). *Classification of Chronic Pain 2nd Edition.* Seattle WA, International Association of the Study of Pain Press.

Michalos AC. (2004). Social indicators research and health-related quality of life research. *Soc Indic Res.* 65, 2 7–72.

Michotte A. (1963). *The Perception of Causality.* New York NY: Basic Books.

Miller GA. (1956). The magical number seven, plus or minus two: Some limits on our capacity for processing information. *Psychol Rev.* 63, 81–97.

Mobley WH, Locke EA. (1970). The relationship of value importance to satisfaction. *Organ Behav Human Perform.* 5, 463–483.

Moons P. (2004). Why call it health-related quality of life when you mean perceived health status? *Eur J Cardiovasc Nurs.* 3, 275–277.

Murphy FH. (2001). Economics and operations research. In, (Eds.) SI Gass, CM Harris. *Encyclopedia of Operations Research and Management Sciences.* 2nd Edition. p. 225–231.

Nagarajan R, Clohisy DR, Neglia JP, Yasu Y, et al. (2004). Function and quality of life of survivors of pelvic and lower extremity osteosarcoma and Ewing's sarcoma: The Childhood Cancer Survivor Study. *Br J Cancer.* 91, 1858–1865.

Nord E, Arnesen T, Menzel P, Pinto J-L. (2001). Towards a more restricted use of the term "Quality of Life". *News Letter Qual Life.* 26, 3–4.

Nord E, Daniels N, Kamlett M. (2009). QALYs: Some Challenges. *Values Health.* 12, Issue S1 S10-S15.

Nordenfelt L. (1995). *On the Nature of Health: An Action-Theoretical Approach.* Dordrecht Netherlands: Kluwer.

Norman GR, Sloan JA, Wyrwich KW. (2003). Interpretation of changes in health-related quality of life: The remarkable universality of half a standard deviation. *Med Care.* 41, 582–592.

Offer D, Sabshin M. (1966). *Normality. Theoretical and Clinical Concepts of Mental Health.* New York NY: Basic Books.

Offer D, Sabshin M. (1991). *The Diversity of Normal Behavior: Further Contributions to Normatology.* New York NY: Basic Books.

O'Leary DS. (1987). The Joint Commission looks to the future. JAMA. 258, 951–952.

O'Toole J. 1974. *Work and the Quality of Life. Resource Papers for Work in America.* Cambridge MA: MIT Press.

Oxford English Dictionary. (1970). London UK: Oxford University Press.

Oxford English Dictionary. (1996). Oxford UK: Oxford University Press.

Pagano IS, Gotay CG. (2006). Modeling quality of life in cancer patients as a unidimensional construct. *Hawaii Med J.* 65, 74–80.

Parsons T. (1951). Illness and the role of the physician. *Am J Orthopsychiatr.* 21, 452–460.

Parsons T. (1958). Definitions of health and illness in light of the American value and social structures. In, (Ed.) EG Jaco. *Patients, Physicians and Illness.* New York NY: Free Press.

Parsons T. (1970). *Structure and Process in Modern Society.* New York NY: Free Press.

Parsons T. (1972). Definitions of health and illness in light of American values and social structure. In, (Ed.) EG Jaco, *Patients, Physicians and Illness.* 2nd edition. New York NY: Free Press.

Parsons T. (1978). *Action Theory and the Human Condition.* New York NY: Free Press.

Pascoe GC. (1983). Patient satisfaction in primary health care: A literature review and analysis. *Eval Program Planning.* 6, 185–210.

Patrick DL, Chang Y-P. (2000). Measurement of health outcomes in treatment effectiveness evaluations. *Med Care.* 38 (Suppl II), II-14- II-25.

Patrick DL, Erickson P. (1993). *Health Status and Health Policy: Quality of Life in Health Care Evaluation and Resource Allocation.* New York NY: Oxford.

Patrick DL, Martin ML, Bushnell DM, Yalcin I, et al. (1999). Quality of life for women with urinary incontinence: Further development of the incontinence quality of life instrument (I-QOL). *Urol.* 53, 71–76.

Perelman C, Olbrechts-Tyteca L. (1969). *The New Rhetoric: A Treatise on Argumentation* Notre Dame IN: University of Notre Dame Press.

Pincus T, Yazici Y, Bergman M, Maclean R, et al. (2007). A proposed continuous quality improvement approach to assessment and management of patients with rheumatoid arthritis without formal joint counts, based on quantitative Routine Assessment of Patient Index Data (RAPID) scores on a Multidimensional Health Assessment Questionnaire. *Best Pr Res Clin Rheum.* 21, 789-80-4.

Pliskin JS, Shepard D, Weinstein MC. (1980). Utility functions for life years and health status. *Oper Res.* 28, 206–224.

Rapkin B, Weiss E, Chabra R, Ryniker L, et al. (2008). Beyond satisfaction: Using the Dynamics of Care assessment to better understand patient's experiences in care. *Health Qual Life Outcomes.* 6:20. doi: 10.1186/1477-7525-6-20. (p. 1–20).

Richters JE, Hinshaw SP. (1999). The abduction of disorder in psychiatry. *J Abnorm Psychol.* 108, 438–445.

Richter A. (2004). Duct tape for decision makers: The use of or models in pharmacoeconomics. In, (Eds.) ML Brandeau, F Sainfort, WP Pierskalla. *Operations Research and Health Care: A Handbook of Methods and Applications.* Norwell MA: Kluwer. (p. 276–296).

Roberts JS, Coale JG, Redman RR. (1987). A history of the Joint Commission on Accreditation of Hospitals. *JAMA.* 258, 936–940.

Rogers CR. (1959). A theory of therapy, personality and interpersonal relationships as developed in the client-centered framework. In, (Ed.) S Koch, *Psychology: A Study of a Science. Vol. III. Formulations of the Person and the Social Context.* New York NY: McGraw Hill.

Ross CK, Steward CA, Sinacore JM. (1993). The importance of patient preferences in the measurement of health care satisfaction. *Med Care.* 31, 1138–1149.

Rowe KC, Paulsen JS, Langbehn DR, Duff K, et al. (2010). Self-paced timing deficits and tracks changes in prodromal Huntington disease. Neuropsychol. 24,435–442.

Ruta DA, Garratt AM, Leng M, Russell IT, et al. (1994). A new approach to the measurement of quality of life: The Patient Generated Inventory (PGI). *Med Care.* 32, 1109–1126.

Sackett DL, Richardson SW, Rosenberg W, Haynes RB. (1998). *Evidence-based Medicine: How to Practice and Teach EBM.* Edinburgh United Kingdom: Churchill Livingstone.

Sackett DL, Richardson SW, Rosenberg W, Haynes RB. (2000). *Evidence-based Medicine: How to Practice and Teach EBM.* Edinburgh United Kingdom: Churchill Livingstone. 2nd Edition.

Sadler JZ. (1999). Horse feathers: A commentary on "Evolutionary versus prototype analysis of the concept of disorder". *J Abnorm Psychol.* 108,433–437.

Santuzzi NR, Brodnik MS, Rinehart-Thompson L, Llatt M. (2009). Patient satisfaction: How do qualitative comments relate to quantitative scores on a satisfaction survey? *Qual Manag Health Care.* 18, 3–18.

Sen A. (1993). Positional objectivity. *Philos Public Affa.* 22, 126–145.

Schag CA, Ganz PA, Heinrich RL. (1991). Cancer Rehabilitation Evaluation System – Short Form (Cares- SF). A cancer specific rehabilitation and quality of life instrument. *Cancer.* 68, 1406–1413.

Schipper H, Clinch J, McMurray A, Levitt M.(1984). Measuring the quality of life of cancer patients: the Functional Living Index-Cancer: development and validation. *J Clin Oncol.* 2, 472–483.

Shortell SM, Bennett CT, Byck GK. (1998). Assessing the impact of continuous quality improvement on clinical practice: What it will take to accelerate progress. *Milbank Q.* 76, 593- 624.

Simtonen H. The 15D instrument of health-related quality of life: Properties and applications. *Ann Med.* 33, 328–336. 2001.

Simtonen H, Pekurinen M. (1989). A generic 15 dimensional measure of health-related quality of life (!%D). *J Soc Med.* 26, 85–96.

Smith EE, Medin DL. (1981). *Categories and Concepts.* Cambridge MA: Harvard University Press.

Smith K W, Avis N E, Assmann SF. (1999). Distinguishing between quality of life and health status in quality of life research: A meta-analysis. *Qual Life Res.* 8, 479–459.

Spate JL. (1983). The sociology of values. *Ann Rev Sociol.* 9, 27–40.

Spitzer RL. (1999). Harmful dysfunction and the DSM definition of mental disorder. *J Abnorm Psychol* 108, 430–432.

Stern WE, Seale DA, Rapoport A. (2003). Analysis of heuristic solutions to the best choice problem. *Eur J Oper Res.* 51, 140–152.

Stouman K, Falk IS. (1939). Health Indexes: A study of objective indices of health in relation to environment and sanitation. *Bull Health League Nations.* 8, 63–906.

Strasser S, Aharony L, Greenberger D. (1993). The patient satisfaction process: Moving toward a comprehensive model. *Med Care Rev.* 50, 219–248.

Sugarbaker PH, Barofsky I, Rosenberg SA, Gianola FJ. (1982). Quality of life assessment of patients in extremity sarcoma clinical trials. *Surgery.* 91, 17–23.

Szekendi MK, Sullivan C, Bobb A, Feinglass J et al. (2006). Active surveillance using electronic triggers to detect adverse events in hospitalized patients. *Qual Saf Health Care.* 15, 184–190.

Temkin O. (1973). Health and disease. In, (Ed.) PPW Wiener. *Dictionary of the History of Ideas. Volume 2.* New York NY: Schribner's. (p. 395–407).

Tennant A. (1995). Quality of life – a measure too far? *Ann Rheum Dis.* 54, 439–440.

Timman R, Roos R, Maat-Kievit A, Tibben A. (2004). Adverse effects of predictive testing for Huntington Disease underestimated: Long-term effects 7–10 years after the test. *Health Psychol.* 23, 189–197.

Thompson AGH, (1986). The soft approach to quality hospital care. *Int J Qual Reliabiilty Manag.* 3, 59–67.

Todhunter I. (1949). *A History of the Mathematical Theory of Probability.* New York NY: Chelsea.

Torrance GW. (1976). Health status index models: A unified mathematical view. *Manag Sci.* 22, 990–1001.

Torrance GW. (1986). Measurement of health status utilities for economic appraisal. *J Health Econ.* 5, 1–30.

Torrance GW, Furlong W, Feeny D, Boyle M. (1992). *Provisional Health Status Index for the Ontario Health Survey.* Final Report of Project No. 44400900187. Hamilton Ont: McMaster University.

Torrance GW, Furlong W, Feeny D, Boyle M. (1995). Multi-attribute preference functions: Health Utilities Index. *Pharmacoeconomics.* 7. 503–520.

Torrance GW, Sackett DL, Thomas WH. (1973). Utility maximization model for program evaluation: A demonstration application. In, (Ed.) RL Berg. *Health Status Indexes: Proceedings of a Conference Conducted by Health Services Research.* Chicago Il: Hospital Research and Education Trust. (p. 156–165).

Torres JM. (2002). The importance of genetic services for the theory of health: A basis for an integrating view of health. *Med Health Care Philos.* 5, 43–51.

Trauer T, Mackinnon A. (2001). Why are we weighting? The role of importance ratings in quality of life measurement. *Qual Life Res.* 10, 579–85.

Treurniet HF, Essink-Bot ML, Mackenbach JP, van der Maas PJ. (1997). Health-related quality of life: An indicator of quality of care? Qual Life Res. 6, 363–369.

von Neumann J, Morgenstern O. (1944). *Theory of Games and Economic Behavior.* Princeton, NJ: Princeton University Press.

von Osch SMC, Stiggelbout AM. (2005). Understanding VAS valuations: Qualitative data on the cognitive process. *Qual Life Res.* 14, 2171–2175.

References

Wagner EH, Glasgow RE, Davis C, Bonomi AE, et al. (2001). Quality Improvement in Chronic Illness Care: A collaborative approach. *J Qual Improv.* 27, 63–80.

Wakefield JC. (1999a). Evolutionary versus prototype analyses of the concept of disorder. *J Abnorm Psychol.* 108, 374–399.

Wakefield JC. (1999b). Mental disorder as a black box essentialist concept. *J Abnorm Psychol.* 108, 465–472.

Ware JE. (1978). Effects of acquiescent response set on patient satisfaction ratings. *Med Care.* 16, 327–336.

Ware JE, Brook RH, Davis-Avery A, Williams KN et al. (1980). *Conceptualization and Measurement of Health for Adults in the Health Insurance Study: Model of Health and Methodology. Vol. 1.* Santa Monica CA: Rand. Publication. No R-1987/1 HEW.

Ware JE, Hays RD. (1988). Methods for measuring patient satisfaction with specific medical encounters. *Med Care.* 26, 393–402.

Ware JE, Snow KK, Kosinski MA, Gandek B. (1993). *SF-36 Health Survey: Manual and Interpretation Guide.* Boston MA: The Health Institute, New England Medical Center.

Ware JE, Snyder MK, Wright WR, Davies AR. (1983). Defining and measuring patient satisfaction with medical care. *Evaluation Pr Planning.* 6, 247–263.

Wegner DM. (2002). *The Illusion of Conscious Will.* Cambridge MA: The MIT Press.

Weinstein MC, Torrance GW, McGuire A. (2009). QALY: The basics. *Value Health.* 12 Suppl 1, S5-S9.

Williams B. (1994). Patient Satisfaction: A valid concept? *Soc Sci Med.* 38, 509–516.

Williams B, Coyle J, Healy D. (1998). The meaning of patient satisfaction: An explanation of high reported levels. *Soc Sci Med.* 47, 1351–1359.

Willis G, Reeve B, Barofsky I. (2005). The Use of Cognitive Interviewing Techniques in Quality of Life and Patient-Reported Outcome Assessment. In, (Eds.) J Lipscomb, CC Gotay, & C Synder. *Outcomes Assessment in Cancer: Findings and Recommendations of the Cancer Outcomes Measurement Working Group.* Cambridge UK: Cambridge University Press.

Wilson IB, Cleary PD. (1995). Linking clinical variables with health-related quality of life: A conceptual model of patient outcomes. *JAMA.* 273, 59–65.

Wittenberg E., Winer EP, Weeks JC. (2005). Patient utilities for advanced cancer: Effect of current health on values. *Med Care.* 43, 173–181.

Wittgenstein L. (1953). *Philosophical Investigations.* Oxford UK: Blackwell.

Wood, PHN. (1986). Health and disease and its importance in models relevant to health research. In, (Eds.) B Nitzetic, HG Pauli, PG Svenson. *Scientific Approaches to Health and Health Care.* Copenhagern Denmark: World Health Organization.

Wu C-h. (2008a). Examining the appropriateness of importance weighting on satisfaction score from -of-affect hypothesis: Hierarchical linear modeling for within-subjects data. *Soc Indic Res.* 86, 101–111.

Wu C-h. (2008b). Can we weight satisfaction score with importance ranks across life domains? *Soc Indic Res.* 86, 469–480.

Wu C-h, Chen LH, Tsai Y-M. (2009). Investigating importance weights of satisfaction scores from a formative model with Partial Least Squares analysis. *Soc Indic Res.* 90, 351–363.

Wu C-h, Yeo G. (2006a). Do we need to weight satisfaction scores with importance ratings in measuring quality of life? *Soc Indic Res.* 78, 305–326.

Wu C-h, Yeo G. (2006b). Do we need to weight item satisfaction scores by item importance ? A perspective from Locke's range-of-affect hypothesis. *Soc Indic Res.* 79, 485–502.

Wu C-h, Yeo G. (2007). Examining the relationship between global and domain measures of quality of life by three factor structure models. *Soc Indic Res.* 84, 189–202.

Wyrwich KW, Bullinger M, Aaronson N, Hays RD, et al. (2005). Estimating clinically significant differences in quality of life outcomes. *Qual Life Res.* 14, 285–295.

Xiong W, Hupert N, Hollingsworth EB, et al. (2008). Can modeling of HIV treatment processes improve outcomes? Capitalizing on an operations research approach to the global pandemic. *BMC Health Serv Res.* 8, 166 (p. 1–10).

Young TK. (1998). *Population Health: Concepts and Methods.* New York NY: Oxford University Press.

Zajonc, R. B. (1968). Attitudinal Effects of Mere Exposure. *J Personal Soc Psychol.* 9, 1–27.

Zullig KJ, Valois RF, Drane JW. (2005). Adolescent distinctions between quality of life and self-rated health in quality of research. *Health Qual Life Outcome.* 3, 64.

Functional Status and HRQOL

Abstract

This chapter reviews the meanings and uses of the term function, and as was found for the term health, documents that what was once a term with a literal meaning has evolved into a predominately figurative form of expression. This history is documented, as is the impact of these changes on the use of the term function as part of the phrase "functional status". A second part of this chapter examines these issues, but now applied to persons with disabilities.

To teach how to live without certainty and yet without being paralyzed by hesitation, is perhaps the chief thing that philosophy, in our age, can still do for those who study it

(Russell 1946; p. 111)

Abbreviations

AAP	Adelaide activities profile (Clark and Bond 1995)
BI	Barthel Index (Mahoney and Barthel 1965)
BPS	Biopsychosocial Model
CCS	Canadian Cardiovascular Society Angina Scale
DASI	Duke Activity Status Index (Hlatky et al. 1989)
DSU	Disease-specific utilities
EORTC-QLQ-C30	European organization for research and treatment of cancer-quality-of-life questionnaire-cancer specific-30 items (Aaronson et al. 1991)
FAI	Frenchay Activities Index (Segal and Schall 1994)
FIM™'s	Functional independence measure (Hamilton et al. 1987)
FLP	Functional limitation profile (Pollard and Johnston 2001)
GH	General health
GOS	Glasgow outcome score
HAQ	Health assessment questionnaire (Fries et al. 1980)
HIPE	History intention, physical environment, events (Barsalou et al. 2005)
HRQOL	Health-related quality-of-life
HUI	Health Utility Index (Torrance 1976)
ICD-10	International classification of diseases-10
ICF	International classification of functioning disabilities and health (WHO 2001)
ICIDH	International classification of impairments disabilities and handicaps (WHO 1980)
IOM	Institute of Medicine
LISREL	Linear structural relations
NHP	Nottingham health profile (Hunt et al. 1985a, b)
NYHA	New York heart association classification system
PGI	Patient Generated Index (Ruta et al. 1994)
PI-HAQ	Personal impact health assessment questionnaire (Hewlett et al. 2001)
QLI	Ferrans and Powers Quality-of-life Index (Ferrans and Powers 1985)
QLI-Stroke	Ferrans and powers quality of life-stroke (Ferrans 1990)

QWB	Quality of Well-being Scale (Kaplan and Bush 1982)
(MOS) SF-36	Medical Outcome Study-Short Form-36 (Ware et al. 1993)
SEM	Structural equation modeling
SG	Standard gamble
SR-FIM™'s	Self-report functional independence measure
SR-BI	Self-Report-Barthel Index
SWLS	Satisfaction with Life Scale (Diener et al. 1985)
TTO	Time trade-off
VAS	Visual Analog Scale
VRS	Verbal Rating Scale
WHO	World Health Organization
WHOQOL-BREF	WHO Quality of life Group-Brief (WHOQOL Group 1996)
WHODASH	WHO disability assessment schedule (WHO 2000)
WTP	Willingness to pay

1 Theoretical Issues

1.1 Introduction

This chapter is divided into two parts. The first clarifies the meaning of the term "function" and also the use of the term as a modifier in the phrase "functional status." The second part of this chapter examines the role the term "function" plays in studies of persons with disabilities. My first task, however, is similar to what I had to face in Chap. 8 where I felt I needed to review the various meanings of the term health in order to properly understand the phrase "health status." My review will also confirm that the term "function" has multiple meanings, and this will confound its usefulness as a modifier of some state indicator. The reason why such terms as health, function, and well-being (Chap. 11) have large semantic spaces is because all three are common to the ordinary discourse and this makes it difficult to establish a specialized language for them, as is possible for biology and physics. In addition, while it is easy to see how symptoms and health can be states (a static concept), it is less clear that a function, which is a statement about relationships (a dynamic concept), can be a state.

I find that the term function has a history, and this history is intimately involved with the evolution of modern science. The term has its origin as a literal linguistic expression in mathematical applications and only in time evolved to include a variety of figurative expressions. One example is the phenomenon of "metaphoric arithmetization", which I suggest is a fairly common phenomenon in qualitative assessments. Metaphoric arithmetization involves the combination (by adding, multiplying, or dividing) of otherwise diverse indicators under the rubric of a figurative expression as a common denominator. For example, it is not uncommon to find a qualitative assessment that consists of a diverse set of items (e.g., symptoms, functional states, feelings, and so on), and yet are summed to give a composite score. The justification for adding these items together is that they all reflect some common qualitative property, such as a person's quality-of-life. However, this common property may not be literally true, but rather be a figurative expression of an abstract qualitative concept. I have previously discussed this as an example of "deductive imposition" (Chap. 8, p. 273). In addition, the presence of such a phenomenon raises serious questions about the nature of assessment in qualitative research. I will discuss this in more detail in Chap. 12, where I raise the issue of what role axiomatic approaches to measurement can play in qualitative assessment.

I will also describe what is known about the cognitive basis of the term function, and what will become clear is that the linguistic role of the term is to establish relationships between an entity and its goals or consequences. There are several potential explanations of how this attribution occurs (Barsalou et al. 2005). One such explanations is that it involves an automatic cognitive process, not unlike the perceptual-cognitive phenomena of affordance (Gibson 1979), that I discussed in Chap. 3 (p. 67). I also discussed the role that the metacognitive process of intentions and a person's personal history may be involved in facilitating a qualitative first impression. Thus, adding a cognitive perspective should enhance my ability to understand the phrase "functional status," but will also provide another example of the value of "understanding how I understand."

Two of the most important function statements are those that establish causal relationships and those that attempt to capture the relationship between capacity and performance. Thus, walking and talking are a "function" of a person's physical capabilities, as are carrying groceries, social dancing, or cleaning dishes. As a functional relationship, mapping a person's physical capabilities to their expression can cause enhanced or diminished quality-of-life or HRQOL, such that these two uses of function statements merge. A function can also refer to a process, such as physiological homeostasis, but can also indicate purpose or design, as when a genetic-based capacity contributes to the survival of the organism.

Once I have described what the term function means I will be able to clarify its relationship to the word "quality," as in "quality-of-life or HRQOL." "Quality," as I alluded to earlier, is a cognitive construction and a product of a dual processing (e.g., the interaction of automatic and deliberate reflective processing of information). Dual processing can be

1 Theoretical Issues

Table 9.1 Definitions of "functional status" published in the literature

Bowling (1991)

"Functional status can be defined as the degree to which an individual is able to perform socially allocated roles free of physically (or mentally in the case of mental illness) limitations. There is a clear distinction from general health status. Functional status is directly related to the ability to perform social roles, which a measure of health status need not take into account. Functional status is just one component of health – it is a measure of the effects of disease rather than the disease itself." (Bowling 1991; p. 6)

Meyboom-De Jong and Smith (1992)

"Level of actual performance or capacity to perform, both in the sense of self-care and in the sense of being able to fulfill a task or role in a given moment or during a given period." Meyboom-De Jong and Smith (1992; p. 128)

Patrick and Erickson (1993)

"An individual's effective performance or ability to perform those roles, tasks, or activities that are *valued*, e.g., going to work, playing sports, maintaining the house. Most often functional status is divided into psychological, emotional, mental and social domains, although much finer distinctions are possible. Deviations from usual performance or ability indicate dysfunction." (Patrick and Erickson 1993; p. 418)

Ware et al. (1993)

"Functional status: the extent to which individuals currently perform their normal or usual behaviors and activities without limitations due to health problems; often used to refer to a variety of concepts of behavioral functioning and well-being." (Ware et al. 1993; Glossary 3)

Leidy (1994)

"…it is proposed that functional status be defined as a multidimensional concept characterizing one's ability to provide for the necessities or life; that is, those activities people do in the normal course of their lives to meet basic needs, fills usual roles, and maintain their health and well-being." (Leidy 1994; p. 197)

observed when I inspect the lines and colors of a great painting, or the notes of a great musical piece. In each case, my sense of quality comes from combining, in an almost automatic sense, the aesthetics of what I see and hear with some contemplation of the meaning of my experience. When I apply this process to the various functions, activities, or roles a person performs, I can again ask about how a person reflects about the aesthetics, as well as the burdensome aspect of these indicators, leading to a global impression of a person's quality of existence.

I am also interested in determining how the meaning of the term function varies with the context of its use. For example, does the term have the same meaning when I am healthy as when I am ill, disabled, or mad? Do these different arenas of human existence affect the meaning of the term? As part of my answer, I will review Boorse's (1977a) notion that health and disease are just different points on the same continuum, and how this relates to what is meant by "context." Also relevant is the notion of fuzzy logic (Sadegh-Zedeh 2001), which can be used to decide when a person is healthy or with disease.

I also include an extended discussion concerning the use of the term function as a process, and as an indication of the purpose or goals of a process. This will lead to a discussion about when and where the term assumes teleological properties. I will also briefly review the history of philosophical functionalism, as well as Parson's theory of social action, as an example of that tradition. Philosophical functionalism is of interest because it models an entire social systems based on the self-regulatory capabilities of biological systems, and this can potentially be a model of quality-of-life or HRQOL outcomes (e.g., Cummins 1996). For example, what would a "grand" theory of quality-of-life or HRQOL be like? Parson's notions concerning the sick role and its relevance to quality-of-life or HRQOL research will also be examined. This background should be sufficient to assess the influence of philosophical functionalism on a public health perspective, and what role a public health perspective plays in a quality-of-life or HRQOL assessment (Patrick and Erickson 1993; p. 60).

The final major topic in this Chapter will consider issues of function and functional status in the context of disability and rehabilitation assessment. This will give me the opportunity to apply what I have learned about functional status domains to a practical problem and contrast this with how a functional perspective is used in a quality-of-life or HRQOL assessment. Finally, I will summarize what I have learned about whether functional status is an appropriate domain to include in a quality-of-life or HRQOL assessment.

1.2 The Language of Function

Table 9.1 provides some examples of the definition of "functional status." Each of the definitions refers to the activities, roles, or behaviors a person engages in, and each defines functional status in terms of the degree to which a person is free of limitations. Bowling's definition differentiates functional status from health status (Table 9.1), but Patrick and Erickson (1993)[1] do not, although Patrick and Chiang (2000) state that, "Within health status, 'functional status' measures usually refer to limitations in the performance of social roles or restrictions in activity (i.e., doing those activities agreed to be important for society or important to the individual)" (Patrick and Chiang 2000; p. II-16). Clearly,

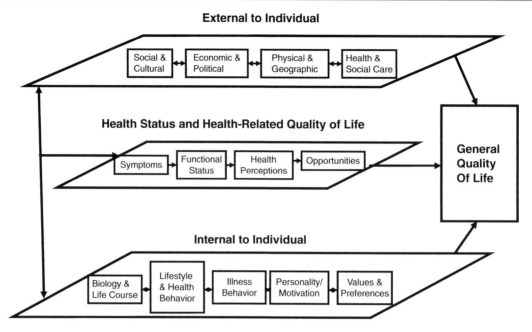

Fig. 9.1 An adaptation of the Patrick and Chiang (2000) Model of Quality-of-life. While the components of health status and health-related quality-of-life are causally related, factors external or internal to the individual are reciprocally related.

Patrick and Chiang (2000) are using the term health status at a conceptual level, one element of which is functional status (see Fig. 9.1). Meyboom-De Jong and Smith (1992) and Ware et al. (1993), like Bowling (1991), define function in behavioral terms, without direct reference to health status.

Leidy (1994) takes a broader approach, first noting that the phrases functional status, health status, and HRQOL have not been clearly differentiated, and also pointing out that functional status itself can be defined in terms of four levels of analysis: functional capacity, functional reserve, functional performance, and functional capacity utilization. Leidy's (1994) approach is attractive because of its comprehensiveness and its ability to systematically consider the role of different concepts in functional relationships. For example (Leidy 1994; Fig. 2, p. 199), if performance is assumed to remain constant (e.g., blocks walked) and the person's capacity diminishes as the person transitions from a jogger to a healthy sedentary adult, to being chronically ill, it can be predicted that the person's functional reserve also diminishes. Alternatively, if the functional reserve is assumed to remain constant, then either the person's capacity or performance, or both, may vary. If the person's functional reserve diminishes because of reduced functional capacity, then it would be predicted that the person's functional capacity utilization would also diminish.

Table 9.2 provides representative examples of domains that might be found in an assessment claiming to measure functional status. These domains, found in the SF-36 (Ware et al. 1993), were established using factor analytic techniques. Note that for both the physical and social functioning domains, the items ask if health limits the person's activities. This suggests that the investigator wants the respondent to take note of the consequences of being in a particular health state. The "vitality" domain consists of items that measure an affective state, with no direct reference to activities or information about health consequences in either the item response instructions or in the text of the item itself. Finally, the bodily pain domain consists of two items, one that measures an affective state (i.e., the intensity of pain) and the other the impact of pain on an activity. Clearly, the domains in the SF-36 are heterogeneous (Tables 9.1 and 9.2), with only some reflecting the consequences of health-related issues. However, most functional or quality-of-life/HRQOL assessments (see Tables 4.7, 8.5, 9.2) consist of several types of items and domains. This diversity has probably been one of the major contributors to the failure to observe some obvious relationships between a disease or treatment effect and some measure of quality-of-life or HRQOL. The discussion (Chap. 7, p. 235) of the Goodwin (2003) paper and a paper by Barofsky (1996) illustrate this point.

I start by considering the role the term function plays in mathematics, which will illustrate that the meaning attached to a function term evolved from concrete experiences to an abstract expression. Subsequent topics will consider other meanings and uses of the term function, each of which may be available to the respondent when they respond to an item on a questionnaire or an open-ended questionnaire. Thus, while the items summarized in Table 9.2 are meant to direct and limit the attention of the respondent to specific issues, the success of this effort is partially dependent of the semantic stability of the terms used.

1 Theoretical Issues

Table 9.2 Examples of items in domains from the SF-36

Physical functioning (does health limit your...?)	Bodily pain	Social functioning
3a. Vigorous activities, such as running, lifting	7. Intensity of bodily pain	6. Extend health problems interfered with normal social activities
3b. Moderate activities, such as moving a table	8. Extent pain interfered with normal work	10. Frequency health problems interfered with social activities
3c. Lifting or carrying groceries	*Vitality*	
3d. Climbing several flights of stairs	9a. Feel full of pep and energy	
3e. Climbing one flight of stairs	9e. Have a lot of energy	
3f. Bending, kneeling, or stooping	9g. Feel worn out	
3g. Walking more than a mile	9i. Feel tired	
3h. Walking several blocks		
3i. Walking one block		
3j. Bathing or dressing		

It remains somewhat odd that functional status was defined in Table 9.1 in terms of the absence of limitations, suggesting that what was "not absence" defined this concept. Defining a concept with the use of a double negative (e.g., if the person is not physically or socially limited, then they are not functionally limited) does not seem to be the optimum way to operationalize a concept, but this is a familiar tactic that I previously discussed when considering the various definitions of the term health. I will return to this topic shortly.

1.2.1 Function: Its Definition and Expression as a Mathematical Statement

The etymology of the term "function" reveals that it is derived from the French word *fonction*, based on the Latin *function* or *functio*, meaning "performance." A dictionary definition (Oxford University Press 1996) includes the following:

As a *noun* it
1. Refers to an activity that is natural to or the purpose of a person or thing.
2. Is a large or formal social event or ceremony.
3. Is a computer operation corresponding to a single instruction from the user.

In mathematics it
4. Refers to a relation or expression involving one or more variables.
5. Is a variable quantity regarded as depending on another variable; a consequence.

As a verb it
6. "Refers to how an entity works or operates in a proper or particular way, and when used in the form "function as," it refers to fulfilling the purpose or task of (Oxford University Press 1996).

Mathematics provides a clear example of the literal usage of the term function, but it also illustrates the progressive changes in its meaning as mathematics itself developed. For example, Ponte (1992) points out that prior to its initial applications in mathematics, elements of a definition of function were present in ancient mathematical practices. For example, the mapping characteristic of a function can be seen in counting, with its linking of a set of objects to a sequence of numbers, with adding, subtracting, multiplying, and dividing; linking a pair of numbers with a product; or in the Babylonian tables of reciprocals, squares, square roots, cubits, and cubic roots which also links a number with a product. Ponte (1992) describes how Oresme (1323–1382) developed a geometric theory of latitudes that included reference to independent and dependent variables (Ponte 1992; p. 3), but the idea of seeing that a relationship was created between these variables had to wait until the seventeenth century and the emergence of the calculus.

Descartes (1596–1650), in his *Monograph on Geometry* (1638/1952), describes how he was able to use mechanical devices to plot curves, and from these curves derive equations. Interestingly, he never abstracted an equation from these plotted points (Dennis and Confrey 1995). However, he did clearly state that the equation resulting from representing any two variables indicated dependence between these variables. This dependence, when expressed by a mathematical equation, constituted a general statement that could be applied to multiple settings. He did not, however, label this dependence a "function" or describe the resulting equation as a "functional relationship." But he did introduce the notion of a derivative, which came from the repeated effort of finding a tangent to a point on the curve.

Both Newton (1642–1727) and Leibniz (1646–1716) have been credited with inventing the calculus; Newton from his study of physical principles, and Leibniz from his study of geometry. But it was Leibniz who first used the term "function," which he did in 1673 when he also introduced the terms "constant," "variable," and "parameter" (Ponte 1992). The first widely dispersed use of the term function, however, had to wait until 1718, when Jean Bernoulli defined the term (Ponte 1992).

Caraça (1951) claims that the development of the term "function" did not arise by accident, but as a necessary tool in the quantitative study of natural phenomena. It evolved following the development of algebraic notions and analytic geometry and became part of the language that permitted the assessment of quantities, identification of physical regularities, and simplification of mathematical expressions (Ponte 1992). It did this by establishing linkages and mappings that permitted prediction and provided a dramatic alternative to medieval (verbal) scholastic thinking and contributed significantly to the development of modern science.[2]

Fourier's (1768–1830) study of heat flow in material bodies provides an example of how a functional analysis could be applied to the quantitative study of a natural phenomenon. Fourier considered temperature to be a function of time and space, two variables that could be related by a trigonometric series to temperature. Fourier, however, did not provide a mathematical proof for this assertion. Instead, Dirichlet (1805–1859) provided the required proof and it was he who proposed expressing the term function as involving a correspondence between two variables, so that for any one independent variable there is one and only one dependent variable. His was the first formal definition of a function in mathematics.

What does this history reveal about the origin of the term function in mathematics? It indicates that what originally involved the linking of pairs of numbers in the context of practical physics and geometry problems was conceptualized in abstract terms as an equation, and then further abstracted and characterized as a relationship. Each stage of this history results in an increasingly abstract concept suggestive of the formation of an informational hierarchy that evolved over decades, if not centuries. The mapping, correspondence, or relationship described in this history has universal applications, in contrast to a taxonomy (e.g., an oak is a tree and a tree is a plant) that is usually unidimensional. My question now is to what extent this history and usage of the term "function" is reflected in current quality-of-life or HRQOL research, and to what extent is the term used in a figurative sense?

1.2.2 Function as a Form of Figurative Expression

As I discussed In Chap. 2, figurative language can be expressed in a variety of ways (e.g., simile, metaphor, irony, and so on). As a metaphor, the term function can be examined either linguistically or conceptually. The conceptual metaphor, the CONTAINER METAPHOR (Lakoff and Johnson 1999), is particularly useful especially when the term function is referring to a variety of entities. For example, when I attend a "function," it is implied that a variety of specific activities may be included. Thus, a wedding function involves a ceremony, a party, the receipt of gifts, and so on. When I measure role function, as in the SF-36, a variety of specific activities represented by specific items are included in the domain. From a mathematical perspective, both these cases would be describing a multivalued function or mappings (one item in a domain, to many items in a codomain).

What is also of interest is that functions can be metaphorically arithmetized; a process that happens often in quality-of-life and HRQOL research. Lakoff and Núñez (2000) describe this situation:

> Literally, functions are not numbers. Addition and multiplication tables do not include functions. But once we conceptualize functions as ordered pairs of points in the Cartesian plane, we can create an extremely useful metaphor. The operations of arithmetic can be metaphorically extended from numbers to functions, so that functions can be metaphorically added, subtracted, multiplied and divided in a way that is consistent with arithmetic. (p. 386).

They go on to illustrate this notion with the following:

> For example, the sum $(f+g)(x)$ of two functions $f(x)$ and $g(x)$ is the sum of the values of these functions at x: $f(x) + g(x)$. Note that the '+' in '$f(x) + g(x)$' is the ordinary literal '+' that is used in the addition of two numbers, while the '+' in '$f + g$' is metaphorical, a product of a metaphor. (p. 386).

Are there examples in quality-of-life or HRQOL research where functions are metaphorically added, multiplied, etc.? I believe there are and suggest that when I speak of the consequence of a treatment or disease being combined with a value statement to create an HRQOL statement, I am metaphorically combining two otherwise independent cognitive processes. Combining these statements is metaphoric because a meaning is communicated that would not have otherwise occurred. Thus, when a patient says that the nausea and vomiting from chemotherapy are not just bad, but are also something they don't want to live with, then they are saying something more than would have been communicated by each statement alone. This type of "meaning construction" has implications for the entire process of qualitative assessment (e.g., formation of a hybrid construct). I will discuss this issue in more detail as I proceed and provide some examples.

Still, the mechanics of combining statements remain worthy of a more detailed study, especially since functional statements are often relational statements, and this justifies examining the available theoretical models that consider how diverse sets of information are cognitively combined. One such approach is referred to as "conceptual blending theory" (e.g., Fauconnier and Turner 1994, 2002). Conceptual blending theory differs from conceptual mapping theory (Lakoff and Johnson 1999) in that the former involves the combining of two or more mental spaces, while the latter is usually conceptualized as involving a base domain being mapped to a target domain. The two approaches also differ in that conceptual blending is more amenable to nonlinear modeling, while conceptual mapping is useful for linear (causal) modeling. Consider the metaphor, "the surgeon is a butcher"

(Grady et al. 1999; p. 105). The meaning communicated by this metaphor can be expressed by either the mapping model (the meaning of being a butcher is mapped to the target, surgeons) or by the blending model. Blending would occur when information is integrated from a generic mental space that identifies the components of two or more input spaces that are then integrated into a blended space. For this example, the generic space identifies the components that are common to both input spaces of the metaphor (the surgeon and the butcher), such as that they both use sharp knives (e.g., a scalpel or cleaver), follow procedures, have distinct but different goals (healing vs. dismembering), and so on. The consequence of combining these differences is that the meaning communicated is discordant, resulting in the inference of incompetence on the part of surgeons.

According to Alexander (2009) and Lakoff and Núñez (2000), blending is common in mathematics. Mathematics, of course, involves generating formal models of relationships. Fauconnier and Turner (2002; p. 11) point out, however, that the resultant formal mathematical models are constantly interacting with meaning systems such that, practically, they are inseparable. To quote them: "meaning systems and formal systems are inseparable. They coevolve in the species, the culture and the individual" (Fauconnier and Turner, 2002; p. 11). The idea that a meaning and a formal mathematical system coevolve is familiar to me as a researcher, since when I engage in the objectification process I am transforming some meaning (e.g., the nature of suffering, or a person's quality-of-life) into numbers that I can use in various ways, including statistical tests of various research hypotheses. Yet, if I only considered these numbers just as numbers, they would be of little use in characterizing a qualitative experience. I can also describe this process as transforming a quality into a quantity. Probably most important is that this process has implications for qualitative assessments in general.

In this section, I have illustrated the role that figurative language can play in mathematical function statements, but I have also raised the issue of how information is cognitively combined. My next topic will continue this discussion by describing cognitive determinants of function statements.

1.3 The Cognitive Basis of Function Statements

Function statements, such as a person's functional status, are involved with several distinct cognitive processes. For one, function statements play a critical role in the classification of entities. For example, in cross-classification the common features of otherwise diverse entities are identified and used to classify the entities (e.g., Ross and Murphy 1999; Chap. 3, p. 57).[3] Thus, the diverse elements in the domain have the common feature of contributing to some goal, and achievement of this goal in essence establishes a (functional) relationship between these entities that justifies their being in the domain.

Children as young as 2 years old use the common functions of the diverse objects they encounter to classify them (Kemler-Nelson et al. 2000). It's not surprising, therefore, to find that quality-of-life assessments can also be used to classify persons on the basis of the functional consequences of their actions, behaviors, or feelings. Thus, the pattern of physical role, or social functioning for persons with physical or mental illnesses is commonly found to differ.

Classification, of course, is a critical step in category and concept formation, and function statements contribute to these processes by integrating various types of information into complex relational statements. Barsalou et al. (2005; HIPE) posit that a person's history (H), intentions (I), physical environment (P), and the sequence of events (E) that they experience all contribute of the formation of function statements. A function statement, therefore, is not a single property, such as some motor activity (e.g., walking a fixed number of feet), but rather a complex cognitive construction based on the relationship between cognitive entities. My description of the common qualitative indicator satisfaction as a "judgment of a judgment" would be a good example of a relational statement (Chap. 8, p. 264). Basically, it is not possible to make a judgment of satisfaction that is not a comparative, therefore relational, statement. The items used in the SF-36 to characterize a person's physical functioning (Table 9.2) are good examples of relational statements (e.g., the relationship of health to…).

The HIPE theory proposes to account for a person's knowledge about function by the integration of the HIPE components. Barsalou and his colleagues also assume that causal chains underlie function statements, with causal relations producing transitions between function statements. Since function statements are second-order cognitions, it is not surprising to find that a number of primary cognitive processes have been proposed to account for function statements. One example is the role that functional affordance has been proposed to play in category formation, especially by children. Thus, following Gibson (1979; Chap. 3, p. 67), it has been argued that the perceptual fit, as reflected by an object or action's function, determines how these experiences are organized into categories. The functional affordance of sitting, for example, is a product of the perception of the physical structure of a chair and the degree it fits a person's capabilities or preferences to sit in a particular chair. The functional affordance of a person's quality-of-life may also be based on the perceived fit, but now on whether a person's physical and emotional conditions fit what it is they want or expect out of life.

A second proposed determinant of a functional statement is the intentions that are reflected in the actions of a person relative to the objects they encounter and experiences they have with these various objects. Chaigneuau et al. (2004) give the example of a teapot, which is ordinarily used to dispense tea, but can also be used to water plants. They point out that even if the teapot is used as a watering can, it still remains a teapot. Thus, what a person's intentions are relative to an object or experience determines its function, not just the perceptual fit of the object or experience. Barsalou et al. (2005) also posit that the exercise of an intention is a metacognitive process. A metacognition (Chap. 3, p. 64) that regularly occurs in a qualitative assessment is the act of valuing an evaluated descriptor, as when a hybrid qualitative construct is formed. The act of valuing, stating a preference, or providing a utility estimate can each be interpreted as intentional cognitive acts.

Basically, Barsalou et al. (2005) argue that affordance, intentions, a person's history, and the events that befall them all participate in forming the function statements that contribute to the formation of categories. They have also provided some empirical support for their model (e.g., Chaigneau et al. 2004). What I wish to explore is their contention that an essential part of the cognitive basis of a function statement is its involvement in causal attributions.

Causal attribution, as when an inference is made, is one of those unique human abilities that has profound survival significance. It demonstrates that our behavior and actions are not just products of associative processes (that is, stimulus-response conditioning cannot completely account for our behavior and decisions), but that various cognitive processes become involved in causally associating various events. White (1988), summarizing the literature available at the time, and points out that infants as young as 3 months are capable of causal processing. This processing, he claims, occurs automatically and is dependent on the ability of the infant to momentarily store perceptual information. As the infant matures to child- and adulthood, more deliberative, controlled cognitive processes become involved. Some of the other contributors to causal processing include concrete familiar event sequences, intentional actions, the observation of regularity and covariation of events, and the perception of generative relations.

Are these various determinants of causal processing relevant for a qualitative assessment? I say they are and that they are used concretely when accounting for various clinical qualitative outcomes, and more abstractly, in such exercises as establishing a definition of quality-of-life or HRQOL (Table 2.4). Thus, when a patient complains that it was the nurse who injected the chemotherapy that made them discontinue their therapy, then a generative attribution is being made, however inaccurate it may be. When Osoba (1994) defines HRQOL as a "multidimensional construct...produced by a disease or treatment," he is making a causal attribution. Even though there is much causal processing that occurs during a qualitative assessment, most quality-of-life or HRQOL researchers do not systematically address the cause of events they are chronicling. Yet, causal processing is a significant component of function statements, something that will become more evident in my next topic.

1.4 Function Statements in Health and Disease

One of my major concerns is whether function statements will vary with the context of use. Let me start by examining Boorse's (1977a) biostatistical theory. This theory gives me the opportunity to explore the role function statements play in health and disease and to do so in the context of a mathematical construct, the normal probability curve (Fig. 8.1). The biostatistical theory implies that health and disease can be defined as two different points on the same continuum.

Boorse (1977a) defines a function as the "causal contribution to a goal" (p. 8). A normal function is defined as a "statistically typical contribution by the individual to survival and reproduction" (Boorse 1977a; p. 7), while:

> a *disease* is a type of internal state which is either an impairment of normal functional ability, i.e., a reduction of one or more functional abilities below typical efficiency, or a limitation on functional ability caused by environmental agents. (Boorse 1977a; p. 7–8; italicized added).

Also as previously stated, Boorse (1977a, b) defines health as the absence of disease, meaning that the portion of the continuum designated as health does not overlap with a disease (Fig. 8.1).

Boorse's (1977a, b) original presentation of the biostatistical theory met with resistance, and as a result, he reformulated his position (Boorse 1977a). As he states, "I realized that this broad usage of 'disease' which so upsets many readers, can be avoided if one switches from 'health vs. disease' to 'normal vs. pathology'" (Boorse 1977a; p. 7). The broad use of the term "disease" he was referring to included such conditions as injuries, poisonings, environmental trauma, growth disorders, functional impairments, and so on; conditions which expanded the meaning of the term disease to the point where perfect health would not appear to be the appropriate alternative outcome. Boorse (1977a) claims that by shifting his views to a discussion of "normal vs. pathological," his theory has come closer to the basic theoretical tenets underlying modern Western medicine. Key to this conceptual shift was his notion that what is called normal is actually *natural* and that what he means by health is "conformity to a species design" (Boorse 1977a; p. 7). What is natural, therefore, is a product of evolutionary processes, and as such, can be considered a "given." He goes on to state:

> In modern terms, species design is the internal *functional* organization typical of species members, which (as regards somatic medicine) forms the subject matter of physiology: the interlocking hierarchy of *functional* processes, at every level from organelle to cell to tissue to organ to gross behavior, by which organisms

of a given species maintain and renew their life. The common feature of all conditions called pathological by ordinary medicine seems to be disrupted *part-function* at some level of this hierarchy. (Boorse 1977a; p. 7; italics added).

Thus, a disease is a partial disruption of an otherwise normal functioning subsystem of the organism.

Returning to my original question, does the use of the term "function" differ when Boorse's (1977a) biostatistical theory speaks to health and disease? I am interested in this question, of course, because ultimately I have to determine whether the functioning of the disabled or challenged individual[4] is normal or pathological; important differences when similar questions are asked of quality-of-life or HRQOL data. Boorse (1977a) would state that a disabled person is capable of many normal functions, but would be different to the extent that they also had examples of abnormal or "part-functioning." What is not clear from Boorse's description is how many abnormal or parts that are partially functioning would be required to reach the threshold that would justify labeling that person's level of functioning as pathological.

Torres (2002; Chap. 8, p. 254), for example, considered Huntington's disease patients unhealthy since they lacked the biological basis for continued health, even though at the time of the assessment they appeared to be functioning normally. A study by Marshall et al. (2007), provides some support for Torres (2002) argument. In this study Marshall et al. (2007) demonstrated that preclinical Huntington disease patients manifest "psychiatric" symptoms that differ from a control group. Still, of concern here is that by giving a person a definitive diagnosis, the meaning of the person's functioning changes for the person and society in general. This, of course, has been observed for the Huntington's disease patient (e.g. Timman et al. 2004), but is also common for genetically predisposed breast cancer patients who opt to remove their breasts to reduce their risk of cancer.

Boorse (1977a) makes it clear that the continuum he is using to distinguish normal from pathological is a quantitative one. He states:

> The function of a physiological process is its contribution to a physiological goals. By 'deficiency' of function then, I mean simply less function, less contribution to the goals, than average. This is an arithmetic, not an evaluative, concept. (Boorse 1977a; p. 21).

Here Boorse articulates one of the issues I am concerned about, which is whether the phrases "normal functioning" and "pathological functioning" can be completely expressed quantitatively, or whether a qualitative change has occurred when a person passes through a threshold to another point on the continuum. For example, does declaring some level of functioning as pathological include an implicit value statement? Certainly, this is what Wakefield (1999) and a variety of other authors would argue, but Boorse denies that this is necessary.

In order to illustrate how this problem might be dealt with, let me reexamine for a moment how functional status is commonly measured in a currently available assessment: the SF-36. Table 9.2 lists a series of items that a respondent may be asked to assess if they have health limitations in their physical activities. The healthy outcome expected would be that the person reports no disruption in physical activities or varying degrees of disruption. But what does a set of items of this sort tell me about what is meant by normal and pathological? Can I assume that no disruption means normal, and if I do, at what point does disrupted normal activities constitute a pathological state?

Ware et al. (1993) do not answer this question. Instead, they provide an investigator with the statistical data needed to determine what would be considered the normal distribution of physical functioning for a sample of the general population or groups of persons with different known diseases (Ware et al. 1993). Defining when a "pathological" condition exists (see Fig. 8.1) is left to the investigator who uses either an arbitrary cutoff point using a statistical criterion, such as the probability of <5% of some measure of physical functioning, or compares scores on a particular domain for a study group and a sample of the general population.

Thus, Ware and many other quality-of-life or HRQOL investigators adopt Boorse's approach and rely on quantitative criteria to describe a person's functional status and leave the interpretation of the significance of the particular level of functioning for another discussion. Clearly, an investigator can define functional status as "some quantitative level of functioning." However, limiting the definition to a quantitative statement becomes harder when these measures are used in clinical or social settings for individual and policy decisions.[5] Also not evident is what these quantitative statements tell us about the quality of a person's existence. Under these conditions, the additional meanings attached to the term function become relevant, as will be evident in the next section.

Boorse's (1977a) model represents one of several alternative philosophical approaches to functions. Thus, Davis (2001) would describe Boorse's naturalism as representing an "historical approach," since it assumes that a prior history, as reflected by the evolution of an organism's traits, leads to the development of a normative set of functions, disruption of which constitutes pathology or disease. Davis, however, points out that alternatives exist to this historic selective function approach; these include the theory of systemic functions and some combination of these two theories. He describes the difference between the two basic approaches to functions as follows:

> The theory of *systemic functions* thus appears to contrast sharply with the theory of *selected functions* in at least three ways. (a) The theories appear to have *distinct explanatory aims*. While selected functions explain the persistence or proliferation of a trait in the population, systemic functions explain how a system exercises some capacity…(b) The theories differ in their *account of the origins of functional properties*. The theory of selected functions asserts selected functions are offices or roles produced

by the natural selective success of ancestral tokens of the trait. The theory of systemic functions, by contrast, makes no specific requirements on the history of the functional trait. It asserts that *systemic functions are specific system capacities* - interactive or structural -within a given system, whether or not the system has been affected by selection. (c) Finally, the theories differ over *the attribution of malfunction*. The attribution of a selected function involves the attribution of a norm of performance that persists when the requested physical capacity is lost. This is thought to underwrite the possibility of malfunctions. The attribution of systemic function, by contrast…is usually taken to mean that if the item has lost the requisite physical capacities it has also lost the associated systemic function, in which case malfunction can't occur. (Davies 2001; p. 28–29; italics added).

The contrast in these models is reminiscent of the difference between linear and nonlinear modeling. These models and their different combinations (Davies 2001; p.29) will provide a helpful perspective as I discuss different aspects of functions and functional status.

1.5 Function as a Teleological Explanations

While the term "function" can be used to describe a category or specify a relationship, it also can assume the meaning of "an activity that is natural to or the purpose of a person or thing" (Oxford University Press 1996). This attribution of a purpose or a goal to a function statement expands the meaning of the term to where it can be used as a causal statement that then can be used as an explanation. For example, the capacity to resist infection, a function of the body, can be said to be one of the factors that "explains" the survival of an organism. The moon illusion,[6] a function of our visual system, can "explain" how a person maintains perceptual constancy in the face of changing visual stimuli. Assuming the sick role, a function of the social system, can "explain" how someone can retain their social standing while ill.

Not only can an individual function be interpreted as having this expanded meaning, but a functional status domain can be constructed that includes functions with a common theme. By acknowledging that the meaning of the term function includes the potential to explain suggests that individual function statements and a function domain can have teleological properties. What does that mean? Teleology explains why things happen, or what such-and-such a thing is for. Behind statements of this sort is the broader question of whether all things, or functions for that matter, have a purpose. Theists believe this, of course, arguing that everything from a rock on the ground to a butterfly flapping its wings has a purpose, reflecting the will of a God. Depressed people, on the other hand, may not believe this. They may feel that nothing has a purpose, and that everything is pointless.

Historically, the establishment of modern science involved rejecting the cosmic teleological explanations of the theists, since too often, empirical evidence contradicted theological claims (e.g., the debate about whether the world was round or flat). So why should I be interested in teleology and how is it relevant to a quality-of-life or HRQOL assessment? First, many types of teleological explanations exist, not just the cosmic teleology of theists. These alternative explanations are regularly used in scientific expositions, as was illustrated when I considered various definitions of health (Chap. 8, p. 250). Less obvious is whether a quality-of-life or HRQOL assessment implies a purpose, and therefore, has teleological properties. One approach is to assume that a quality-of-life or HRQOL assessment is an outcome of a sequence of events that can be interpreted as purposeful. Consider the following. A person's lifestyle augments their ability to engage in physical activities, which enhances their performance, and this affects their sense of satisfaction and well-being which will be reflected in their quality-of-life or HRQOL assessment. If these associations were established, then the sequence of events might be interpreted as purposefully producing changes in a person's quality-of-life or HRQOL, and this could be considered evidence of a teleological outcome.

But does a quality-of-life or HRQOL statement itself have teleological properties? For example, is the purpose of why I live to achieve a certain quality? If so, then I might have some "sense" of what this quality is, and if I do, will this "sense" determine what quality-of-life goals I strive for? If I don't have any idea of what I want out of life, then where do my notions of quality-of-life or HRQOL come from? Thus, I may want to know how this "sense" of what quality I strive for becomes evident. Is it something that emerges from within me, or is it something that comes from outside to influence my expectations or goals (Woodfield 1976)?

Part of the answer to this question overlaps with two major philosophical traditions, Platonic and Aristotelian philosophy, each of which has a distinctive approach to teleological explanation. As Lennox (1992) states, external teleological explanations are derived from Plato, who argued that a goal is imposed by a rational agent or knowing mind with good intentions and purpose. Thus, a "Senate" of selected persons is required to govern a society and a commander is required if an army is to function properly. Another example is the views of theists who believe that an all-knowing God creates order between causal chains and their interactions.

In contrast, Aristotle spoke of internal teleological explanations that derived from an analysis of the function of things. Thus, for Aristotle, a *teleos,* or final cause, can be identified from an analysis of the complex interaction of the various parts of an object or system. And as Lennox (1992, p. 326) paraphrases Aristotle, "Not only do these changes and attributes contribute to an end – they take place and exist in part because they contribute to an end." It is this second assertion, that something exists because it contributes to an end, that constitutes an Aristotelian teleological explanation.

1 Theoretical Issues

For Aristotle, the presence of teleology becomes obvious when examining nature and its internal workings. For example, he would say that the arrangement of teeth in an animal's mouth reflects a purpose or design; sharp teeth are in the fount of the mouth for grabbing and biting, and flat teeth are in the back for chewing, and both are critical parts of the ingestive process (Meyer 1992). Plants have leaves to cover their fruit and seed covers to protect their genome. In each case, functions are there for the "good" of the organism. Thus, not only do these characteristics reflect a purpose or goal, they also have some value to the organism. This type of construct – a state and a valuation of it – is again reminiscent of a hybrid construct.

Secular teleologists, particularly biologists, continue the Aristotelian interest in explaining why things happen by examining the innards of functions, but reject the need to derive a final cause or purpose from a functional analysis. Instead, they examine biological systems from an historical perspective, which is more compatible with Aristotle's formal or efficient causes.[7] Darwin's (1859) theory of evolution provided the evidence for this historical perspective. He demonstrated that the evolution of animals and plants occurred not by external design, but by internal chance events (i.e., mutations), with natural selection determining which changes survived. Thus, if I ask why a heart pumps blood, an Aristotelian teleological explanation would be "in order to survive," but a Darwinian historical explanation would say it's because organisms with hearts which were "less adaptive" did not survive (Wilkins 1997).

Mayr (1982) points out that what distinguished Darwin's approach to teleology from Aristotle's was that Darwin explained how functions or functional organizations could be accounted for by forces of natural selection "pushing" from the past (i.e., the consequences of history), while Aristotle's final causes "pulled" natural events to the future. For Mayr (1982), therefore, Darwin's teleology is not the same as Aristotle's, but is more of an "as *if*" teleology, he called *teleonomy*. Talcott Parsons (1978) borrowed this term from Mayr and used it to define health. The phrase is probably how most biologists think of purpose and design in biological systems.

Mayr (1982; p. 48) has suggested that it is possible to have goal-directed and purposeful biological processes that are not in conflict with a strictly physiochemical explanation. In fact, he says that part of the confusion concerning the term teleology can be attributed to the fact that it can be used in at least four different ways.[8] These include telenomic activities (goal seeking, as in Aristotle's final cause); teleomatic processes (law-like behavior that is not goal seeking); adaptive systems (which are not goal seeking, but exist because they have survived); and cosmic teleology (which refer to end-directed systems).

The discovery of the genetic code, with its program controlling biological expression, would be an example of a telenomic activity. What makes the genetic code telenomic (that is, the genetic code appears purposeful) is the fact that the program and the stimuli that activate the program exist prior to the seemingly purposive behavior: the expression of the code. Thus, the code mediates mechanisms that initiate or cause goal-directed behaviors (Mayr 1982; p. 48).

Is this what happens during a quality-of-life or HRQOL assessment? Is quality-of-life or HRQOL telenomic, in the sense that I have an idea of what my quality-of-life should be, and this idea directs my subsequent (purposeful) behaviors?

Mayr's (1982; p. 49) second of his four examples defines teleomatic processes as "Any process, particularly one relating to inanimate objects, in which a definitive end is reached which is strictly a consequence of a physical law may be designated as 'teleomatic'." An adaptive system can also reflect purpose. According to Mayr (1982; p. 49), he studies adaptive systems to answer the question "why?"; why does a physiological process work the way it does? He gives the example of a reflex, which can't be completely understood unless the investigator asks what is the goal or aim of the behavior.

Complexity theorists, as secular teleologists, also speak of the purpose of random events, but now leading to "emergent complexities" or "self-organization." They, like theists, philosophers, or biologists, claim that universal laws (independent of the source of these laws) exist that govern random events, so that a life history or nature itself is more directed than would appear from an analysis of the components of these interactions alone. They have also added some unique ideas concerning the nature of teleological explanation.

For example, while complexity theorists accept that universal laws exist, they argue that these laws need not have had a prior existence, nor do they need to exist outside of a particular system (Alexander 2002). Instead, these laws emerge from dynamically stable systems, and while they have concrete effects (e.g., death due to dehydration), they do not have a material existence. Interestingly, complexity theorists do accept the view that these laws may exist as mathematical statements prior to their application to a dynamic system.

In addition, complexity theorists also claim that the "emergent complexity" or "self-organization" is irreducible and therefore unpredictable. This is so because new functions or changes in context not accounted for in an initial presentation can emerge, leading to new outcomes that would not have been predicted (Alexander 2002). As a result, any system or process can increase in complexity, a complexity that by way of feedback will be constrained. Again, Davis' (2001) distinction is relevant in that complexity theory appears to be an example of a systems theory of functions.

These ideas of how change occurs are quite interesting and help me understand macrolevel processes as described by John Stuart Mill's "Invisible Hand" or microlevel processes as reflected in Hayak's ideas about how consciousness emerges. In both instances, *order emerges from disorder*. But can an "emergent complexity" or a "self-organized" entity account for the thoughts, experiences, or impressions that give me my "sense" of quality? Certainly, I need a way of accounting for how the collection of experiences that I call my life become "sensible," and once sensed, become the basis of my responses to a quality-of-life or HRQOL assessment. Later I will discuss several mechanisms concerning how this can occur, one of which is self-organization.

For now, I have described a quality-of-life or HRQOL assessment as being based on my phenomenology. Most investigators think of a quality-of-life or HRQOL assessment as a number derived from an elaborately constructed questionnaire, and they make minimal or no assumptions concerning the cognitive equivalent of what they have psychometrically generated. They also may not be interested in asking if the numbers they have produced actually project purpose or intent; if they should attach a value statement to these numbers; or if they should be concerned if the numbers reflect an implicit valuation. If they do not, then, as I have previously suggested, they are missing out on the richness that assessing quality-of-life or HRQOL offers.

Table 2.5 (Chap. 2, p. 40) provides a set of quality-of-life or HRQOL definitions that have teleological properties. Here the definitions listed explain a person's quality-of-life or HRQOL by reporting the degree to which a person's goals "match" what they have achieved, or if they have reached a certain state. The implication here is that some quality-of-life or HRQOL needs, goals, or states exist, and that what is being measured is the degree to which the person has achieved these ends. I have also suggested that the definitions in Table 2.5 (Chap. 2, p. 40) were compatible with the conceptual metaphor LIFE IS A PURPOSEFUL JOURNEY. This too suggests that these definitions have teleological properties. Maslow's (1968) hierarchy of needs is another example of a need-based teleological system, since the goal of satisfying my more basic-level needs (e.g., our physiological needs) "explains" how I achieve my highest need, "self-actualization."

Teleological properties are less evident in other definitions. This is particularly so for the *causal* definitions listed in Table 2.4 (Chap. 2, p. 38) where quality-of-life or HRQOL is determined by the extent to which a disease or a treatment has an adverse impact on the person. Again, implicit in these definitions is the existence of some level of quality that can or can't be achieved. The definitions listed in Table 2.7 (Chap. 2, p. 43) consist of either restatements of a quality-of-life or HRQOL definition in terms of well-being, or causal references made relative to changes to well-being. Finally, Table 2.8 (Chap. 2, p. 44) includes the Patrick and Erickson (1993) definition as an example of a hybrid construct and definition evaluates the consequence of past events and places a value on the length of life a person has or can live.

Their definition, as well as other types of hybrid constructs, such as the HUI (Torrance et al. 1995) and the QWB (Kaplan and Anderson 1990), forces me to carefully consider my next topic; the relationship between teleological explanations and value statements.

To help me in this task I would like to review several papers by Fulford (1989; 2000) that consider the role *causes, teleology,* and *values* play in function statements. First, however, consider the following statement he makes, in the Preface of his book.

> Medicine it seems, rests on a paradox. It gives the appearance of being a science, yet its key defining concepts, illness, disease and dysfunction, look like evaluative concepts - *ill*ness, *dis*ease, *dys*function. (Fulford 1989; p. xii).

It is interesting that I could substitute the words "quality-of-life" or "HRQOL" for medicine and come away with a comparable statement: Quality-of-life or HRQOL assessment, it seems, rests on a paradox. It gives the appearance of being a science, yet its key defining concepts (e.g., well-being, satisfaction, happiness, and so on) look like evaluative concepts.

Fulford suggests that considering the inherent evaluative nature of medicine, and now by extension, quality-of-life or HRQOL, an investigator should aim to accommodate both the ethical and scientific perspectives implicit in these subject matter areas, rather than try to limit their efforts to one perspective or another. This notion, of course, is quite consistent with the thesis of this book. He then goes on to support his views by reviewing a number of alternative models: causes without teleology; teleology without valuations; and teleology with valuations.

1.5.1 Causes Without Teleology

I have already discussed naturalism in reference to the philosophical project of developing a scientific medicine, and psychiatry based on value-free function statements. This is the essence of what Boorse (1977a, b) and others have proposed. It is also the model that some investigators use to justify calling a functional assessment a quality-of-life or HRQOL assessment.

Wakefield (1999) and others took a somewhat different tack, separating causal (dys)function statements from value statements (i.e., they separate fact from value), but then aggregating them at some other point. Wakefield argued that this still leaves him being consistent with the naturalism project, with its emphasis on defining a state without reference to a valuation. The same might be said about the HUI and the QWB, both separate function and value/preference statements, but aggregate them at some point in their decision models. These examples are consistent with Fulford's (2000) notion that value statements have to be made at some point in the information cascade stretching from function, to dysfunction, and on to disease, illness, and disorder.

However, as he points out, there is more to this than just finding a place for values.

Fulford's major objection to the argument that a value-free function statement is possible is linguistic.[9] He claims that words with multiple meanings are being used in such a manner that only part of their meaning is being communicated at any one time. He states, in reference to Wakefield (1999):

> What he presents to us when he sets out his definition of dysfunction is the factual/causal side of the meanings of the terms of which his definition is made up. He presents them, in the conjuring metaphor, factual/causal side up. Hence, "blinded" as we are by the importance of science, we take the terms in his definition to be exclusively factual/causal in meaning. But this is where the illusion comes in. For while our attention is held by the factual/causal meaning of these terms, their real (logical) work in Wakefield's definition is being done, …by the evaluative/teleological sides of their meaning. (Fulford 2000; p. 80).

I believe this same statement could be made when investigators present quality-of-life or HRQOL data. In their effort to be scientific, they ignore that the terms they use often have evaluative meanings, are value statements, and this is part of the meaning communicated, even if not formally acknowledged. If this is so, then the notion that functions can have causes without also having purposes seems unlikely.

It is worth commenting on how this happens, since the analytic process inherent to a scientific endeavor inevitably has me cognitively compartmentalizing how I think. Thus, I learn to separate fact from value,[10] observation from meaning, and so on. All investigators have learned to do this in order to avoid confounding observations. I speak of this as being objective as opposed to being subjective, or nonbiased as opposed to being biased. All investigators have also learned to engage in a series of instrumental activities (e.g., using a recording device, or observing a change in the color of a chemical mixture, and so on) that are meant to confirm or test observations independent of any direct involvement in the observation. These activities are the essence of the "objectification process."

Clearly, I can successfully engage in these activities, but I am left to ponder whether these activities constitute a convenient cognitive ability or whether they reflect some reality that has profound philosophical implications. I have suggested (Chap. 2, p. 31) that being able to engage in these cognitive processes does not necessarily establish what is real (as would be argued by the logical positivists), and I have instead accepted the idea that what I know is a construction, partly based on the nature of my body (how I contact the world); partly based on my past experiences; and partly based on how external phenomena present themselves.

1.5.2 Teleology Without Values

Next, Fulford (2000; p. 81) asks if a function statement can have an implied purpose without also having some value statement attached. He points out that the purpose of things are often their practical aspects, but what I consider practical is itself a value judgment. Thus, when medicine accounts for how a function becomes a dysfunction, it is not possible to do this without involving the evaluative element in the meaning of these terms. Boorse (1977a) would disagree with this argument, and instead claim that it is possible to distinguish states as being "different." But, as Fulford (2000; p. 82) states, how "different is different from dysfunction?" The same argument can be transposed to a quality-of-life or HRQOL context.

1.5.3 Teleology with Values

As Fulford (2000) stated, some notion of values will be introduced somewhere along the informational cascade. He also points out that exactly where this will occur varies by the type of material being considered. For example, since psychiatry is inherently more value-laden than medicine, you would expect value-related issues to be more prominent and visible than for medicine. Thus, in psychiatry you can ask if the patient is mad or bad. This question should be viewed as an evaluation that relies on the presence of past valuations. The same would hold for a comparison of medicine to biology, or biology to physics. For me, quality-of-life or HRQOL would be considered more value-laden than functional status; functional status may be more value-laden than health status or symptoms; and so on. This suggests that the type of information being dealt with also contributes to how and where value- and ethics-related issues are displayed.

In addition, for each of these fields of study how I can "explain" without implicitly or explicitly inserting values into my explanation? I need only to remind myself of what I do when I interpret a statistical inference, and I can recognize the presence of values in even the most data (albeit statistically)-based decisions. Another example would be selecting a probability level to reject the null hypothesis; while based on a probability estimate, what level of probability to accept as significant will change depending on a community consensus and also the circumstances of the study and my judgment as an investigator. These examples represent "resubjectification" of the objectification process, and thereby give meaning to the derived numbers.

Fulford's basic message is that if value-related issues are introduced at all, then values are potentially present at all points on the informational cascade. The alternative is to deny their presence altogether (see Note 9). Since denying or excluding values is ethically and philosophically indefensible, he says that the only alternative is to accept the argument that values are potentially an integral part of each point along the informational cascade. This is a fairly radical notion and would ordinarily meet resistance from most quality-of-life or HRQOL investigators, especially since they can successfully demonstrate that by dealing with what appears as objective nonvalue-based data they have been able to establish reproducible functional statements, which permit predictions. But perhaps the resolution of this issue involves recognizing that value-related statements can be applied quite specifically, as when defining quality-of-life or HRQOL, or broadly, as when someone is asked to place a value on a sequence of experiences.

Fulford (2000; p. 89) also points out that the medical ethics of today involves establishing definitive rules of practice, such as the rules designed to ensure patient autonomy in medical decision making. This gives bioethics a substantive role in health care; it tells me what I ought to do. In contrast, value theory tells me how to reason about what I ought to do. As Fulford (2000) states, "The choice between substantive ethics and value theory is thus a straight choice between ethical certainty and ethical uncertainty" (p. 89). But, as he further points out, this dichotomy is pervasive, with many of the greatest discovers during the twentieth century establishing uncertainty principles. Thus, a transformation in the very nature of science has occurred, from the reductionistic mechanical sciences of the seventeenth to twentieth century to the uncertainty of the twenty-first century that has yet to be fully integrated into thinking about the nature of function or quality. Fulford states:

> From classical times right up to the start of the twentieth century, progress, whether in the arts, in philosophy, in mathematics, or in the sciences, was thought to be a journey toward ever more certain foundations. Now, at the start of the twenty-first century, the only certainly is the journey toward certainty, like chasing the rainbow, is a journey that, for reasons not merely of practice but of principle, cannot be completed. In place of certainty, then, we have radical uncertainty; the openness of interpretive possibilities in post-modern literary theory; the under-determination of theory of data in philosophy of science; the relativization of meaning to forms of life in Wittgenstein's later work; the undecidability of mathematics in Goedel's eponymous theorems; and most radical of all, the limits to the accuracy of simultaneous measurements of the properties of physical systems in Heisenberg's uncertainty principle. (p. 89).

Finally, Fulford quotes Bertrand Russell, who says, "To teach how to live without certainty and yet without being paralyzed by hesitation, is perhaps the chief thing that philosophy, in our age, can still do for those who study it" (Russell 1946; p. 111). Fulford also feels that because psychiatry is based on uncertainty and dependent on value theory, it could provide a model for the development of medicine. In the same way, I would say that because quality-of-life or HRQOL assessment is based on value theory, it could provide a basis for the development of policies to ensure the quality of my existence.

1.6 Summary

The discussion to this point has hopefully demonstrated that the term "function" is central to a variety of issues related to quality-of-life or HRQOL assessment and that this centrality comes from inspecting the various meanings associated with the term (I discussed at least five). One of the issues that emerged was whether an investigator could separate the various meanings of the term, without also changing its interpretation. Fulford (2000), of course, argued that it would, since when the term is applied it will not be possible to ignore the additional meanings attached to the term. He was speaking of the term "function," but many elements of this argument can also be applied to the term "quality," as in quality-of-life or HRQOL.

Also introduced in this section was the notion of self-organization as an expression of a purposeful system, and what role this process may play in the formation of a cognitive entity. The notion of self-organization, when combined with some of the Hayek/Hebb (Chap. 3, p. 51) ideas, will lead to the proposal that complexity theory may have a role to play in the understanding of the cognitive foundations underlying a quality-of-life or HRQOL assessment. The rationale for this approach is that it provides a mechanism to account for how a person's experience leads to an impression and knowledge of their existence, independent of the content of their experiences. Evidence to support this hypothesis will be discussed in subsequent sections.

1.7 Functionalism: A Philosophical Tradition

One of the obvious deficiencies in the current quality-of-life or HRQOL literature is the absence of a comprehensive theoretical statement that can be used to guide research efforts. Most statements, such as the various definitions I have considered (Tables 2.4–2.5 and 2.7–2.8), provide limited guidance in this task. There is, however, a history of social theorizing based on philosophical functionalism that approximates the kind of theoretical statements that would be required to have a "grand theory" of quality-of-life or HRQOL.

This history is very much involved with the origin and development of sociology that was prompted by Auguste Comte's (1798–1857) concerns about what to do about the consequences of the French Revolution. He called for the creation of a science dedicated to studying the nature of social structures and how these structures maintained social order. This science was to describe how a society could organize itself and change, without recourse to such radical violence and other disruptive effects, as occurred following the French Revolution. Comte understood that, for this science to succeed, it had to separate itself from moral philosophy, and to achieve this end, he suggested that it be modeled on biological concepts. This approach[11] leads naturally to asking how each of the organisms' structures contribute to this outcome, and this, of course, involves asking a functional question (i.e., how these structures functioned). Comte, however, did not address this question directly (Turner and Maryanski; 1988). This was left to Herbert Spencer (1876) who popularized the analysis of society as an organism, but also made the important contribution of distinguishing between "function and structure." To quote Spencer from Turner and Maryanski (1988; p. 111), "There can be no true conception of a structure without a true conception of its function. To understand how an organization orientated and developed, it is requisite to understand the need subserved at the outset and afterward." Particularly important about this quote is the reference to a "functional need" and its implicit teleological reference. Spencer appears to be suggesting that the need subserved can be used to explain why a social structure emerges and persists within a larger whole (e.g., a society). This suggests that functions could "explain" why a society survives.

Durkheim (1895) was also a contributor to the foundation of sociology and philosophical functionalism. He was interested in how social mechanisms affected social integration and the maintenance of society (such as the social division of labor; ritual and ceremonial activities; role segmentation; etc.). He was concerned with the concepts of normality and deviance; thus, ill health was for him a form of deviance from the ideal of normal well-being.

Durkheim recognized that the reason a social structure exists is often not the same as what satisfies the social needs of this structure (e.g., a family, a firm, a government). This suggests that a functional analysis and a causal analysis should be separate. As Turner and Maryanski (1988; p. 111) state; "For Durkheim, therefore, sociological analysis must ask two questions; 'What are the antecedent causes of a structure and what need of the larger social system does it meet?'" However, he apparently was not able to maintain this distinction in his own work (Turner and Maryanski 1988; p. 111), as is evident from his notion that the division of labor and religious rituals emerged to meet social needs. This lapse, of course, is not unexpected considering the multiple meanings and uses that the terms function or functionalism can assume.

In the twentieth century, anthropologists also made major contributions to the development of philosophical functionalism. Thus, anthropologists such as Malinowski (1922, 1926), Radcliffe-Brown (1935), and Linton (1936) combined a structural (e.g., kinship relationships) with a functional analysis of cultures. A culture, for them, was conceived of as a social system and its components examined to determine how they self-regulated, a process that was assumed to contribute to the adaptation and survival of the society. Radcliffe-Brown's (1935) conception of function was very similar to Durkheim's: the function of "a structure is to be assessed in terms of its consequences for meeting the 'necessary conditions of existence' in a system" (Turner and Maryanski 1988; p. 113). Radcliffe-Brown (1935) also expected that it would be possible to demonstrate that the structures of a society satisfied certain needs that would, therefore, explain why they existed. A particular kinship system exists because it ensured, for example, optimal distribution of wealth or specific roles. For this to occur, the structures of a society had to demonstrate consistency and continuity; that is, a structure had to provide consistent specifications of the rights and duties of members of the society and its possessions and also provide continuity by specifying the rights and duties between persons in the society.

While Radcliffe-Brown (1935) was interested in "function" as an explanatory tool, Malinowski (1944) advocated a functional method of collecting and analyzing data. Thus, ethnographic data were analyzed in terms of system levels: biological, psychological, social, and symbolic. Talcott Parsons (1937) used this same approach when he developed a structural–functional analysis of the healthcare process (see below), but now also using ideas from psychoanalysis and economics.

Also during the nineteenth century, Darwin (1859) presented his theory of natural selection and described its role in the evolution of species. Darwinism also contributed to the belief that examining society as an organism offers unique analytical opportunities. Thus, I can ask how the adaptive responses of an organism to a changing environment contribute to preferential survival. I can find this same question rephrased and asked in biology, anthropology (How does a culture adapt and survive?), sociology (How does a society adapt and survive?), and psychology (How does the individual adapt and survive?). I can also ask this question in the context of quality-of-life or HRQOL research (How does a person's sense of the quality of their life adapt and survive as their life unfolds?).

William James (1890) is considered, along with Charles Peirce and John Dewey, to be the founders of the philosophical and linguistic tradition of Pragmatism. James, however, was the first to integrate a functional perspective into psychology. He considered mental processes to be adaptive functions that were meant to ensure the survival of the individual as they accommodated their needs to their environment. This approach has developed into a distinct subdiscipline: evolutionary psychology (e.g., Cosmides and Tooby 1987, 1999). Cosmides and Tooby (1987), in fact, claim that psychology need only deal with functions or purposes and needs, and need not rely on the neurosciences. They also believe that natural selection is actually a theory of function and that learning is a "Darwinian algorithm" that "organizes experience into adaptively meaningful schemas or frames" (Cosmides and Tooby 1987; p. 286).

Another important event occurred during the first half of the twentieth century, which was the documentation of the principle of homeostasis. Homeostasis (Cannon 1939), as a biological principle, is an example of the self-regulatory mechanism that contributes to an organism's adaptation and survival. It followed, of course, C. Bernard's (1865/1927)

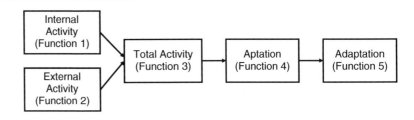

Fig. 9.2 Depicted are five different uses of the term "function" in a biological context and the relationship between these terms. Adapted from Mahner and Bunge 2001. Reproduced with permission of the publisher.

introduction of the concept of *the internal milieu*. Thus, it is a principle which can be used to "explain" or "describe" and has prompted research and theorizing in a number of social sciences.

Also active in the first half of the twentieth century was B.F. Skinner (Moxley 1992), who, under the influence of Ernst Mach, shifted the study of behaviorism from a mechanistic stimulus-response concept to one that took into account the probabilistic nature of behavior change. Skinner (1938/1966) used the term "function" in two senses; first, a respondent is involved in a function in the sense that a correlation between a stimulus and response exists, or alternatively, as an operant function as when the relationship between a prior discriminative stimulus, a response, and its reinforcing consequence exists. Skinner's concept of operant, with its emphasis on the importance of consequences, has been a significant influence on the interest in outcomes research and qualitative assessment.

Out of necessity, this review has been brief, but should be sufficient to appreciate that the developments in philosophical functionalism evolved in response to perceived political deficiencies (What was wrong with society that allowed the French Revolution to occur?) and adopted principles from the biological or physical sciences to model alternative approaches to creating social order. The evolution of quality-of-life assessments in some ways has duplicated this history, since it can also trace it origins to the social policies issues, and its assessment has meant to influence policy decisions. Thus, the interest of Patrick and Erickson (1993) in applying a functional or sociologically based public health approach to quality-of-life or HRQOL assessment is an example of investigators who seem to be aware of the potential influence that a qualitative assessment can have on the political process. Thus, the development of scientific functionalism contributed to the interest in patient-based outcomes research.

1.7.1 Models of Function and Functionalism

Mahner and Bunge (2001) provide a classification system of the different types of functions that they contend brings some order to the confusion surrounding the term. Their approach is essentially a building block approach; starting with a core definition, and then they add various meanings or uses of the term that differentiate and elaborate it, permitting it to be applied to wide variety of settings. Mahner and Bunge (2001; Fig. 9.2) start by applying their approach to biological systems. Thus, they distinguish the internal (function$_1$) and external (function$_2$) activities of the body's subsystems. The liver, for example, has a set of internal metabolic activities that function to ensure its internal milieu (function$_1$). These activities can be distinguished from the liver's external, between subsystems role of cleansing toxic material. Part of the confusion with the term is that investigators tend to use both meanings of the term simultaneously. Thus, when I talk about the function of the legs, not only do I refer to the movement of the legs (function$_2$), but I am also referring to what is happening within the muscles of the leg (function$_1$).

Mahner and Bunge (2001) go on to state:

> Obviously, the function of some organismic subsystems may be valuable to the organism as a whole, i.e., it may favor its survival or reproduction, or it may be indifferent, or even disvaluable. For example, while the function$_3$ of the heart is highly valuable that of the appendix is almost nil, and that of a tumor is disvaluable. If the function$_3$ of the subsystems is valuable to the organism, we call it, or the subsystem in question, an *aptation*... Correspondingly, a disvaluable function$_3$ is a *malaptation* (or *malfunction* or *dysfunction*). And if a function$_3$ should turn out to be neutral, we might call it a *nullaptation*. We call any function1,2,3 that is an aptation a *function4*. Accordingly, the production of a heart sound is not a *function4* of the heart, although it is clearly one of its activities (*function 3*). However, it is clearly an aptation of our noses to support spectacles.
>
> Note that, in principle, the ambiguous word 'function' can be eliminated in favor of the expressions 'internal activity', 'external activity', and 'adaptation'. Any one of these notions may be called 'functions' in a given context, so that we need to watch out which of these different concepts is being referred to in any biological work...Thus, the search for the concept of function in biology is futile." (Mahner and Bunge 2001; p. 78–79)

Mahner and Bunge's (2001) last statement suggests they do not believe there is a unique definition of function in biology, and by inference, unique definition of function in other applications. They illustrate the various uses of the term when only biological functions are considered (Fig. 9.2) and also when psychological, social science, and technological functions are considered (Fig. 9.3). They state that even though social systems are different from biological systems,

1 Theoretical Issues

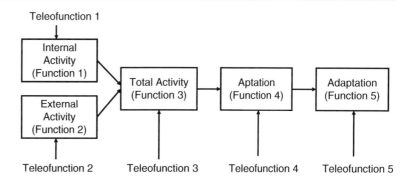

Fig. 9.3 Depicted are five different uses of the term "function" in a psychological, social, scientific, and technological context, and their relationships. Adapted from Mahner and Bunge (2001). Reproduced with permission of the publisher.

they may be subjected to the same analysis as was illustrated in Fig. 9.2. Mahner and Bunge (2001) distinguish these two approaches as follows:

> "…in contradistinctions to biosystems and most animal societies, the function$_{1-5}$ of human individuals and social systems often involve purposive actions. Thus, we arrive at a sixth concept of function (or rather a set of concepts) that is genuinely teleological, involving the notion of intentions, purpose or goal. We call this notion *teleofunction*…teleofunction is a (relational) property of some systems performing any one function$_{1-5}$. That is, a purpose or goal may be attributed to each of the five functions distinguished above, so that social activities, roles, activities cum roles, aptations and adaptations that are intentional are teleofunctions…" (p. 80)

The model that Mahner and Bunge (2001) describe is quite interesting. It suggests that if each category of information (e.g., biological indicators, symptoms, health perceptions, and so on) were inspected, it would be possible to identify a core set of functions, and it would be possible to attach additional indicators to these functions to derive increasingly complex descriptive statements. Thus, adaptive biological functions can be valuable, disvaluable, or have no value; or psychosocial functions can have purpose, intentions, and goals (i.e., they may have teleological properties). Their model clearly suggests that functions are context-specific (meaning that the definition of any indicator will vary as a result of circumstances of how they are studied) and that the terms will have no unique meaning other than as a class indicator (see quote above, Mahner and Bunge 2001; p. 79).

When Mahner and Bunge (2001) apply their model to the social sciences or technology (Fig. 9.3), they also describe how the presence of intentions, purposes, or goals varies with the function level. This may not seem obvious when dealing with technology, but becomes more understandable when considered from the perspective of a technological analog, such as a computer. Thus, the innards of a computer has a function and a purpose that maintains internal processing of electronic activity (function$_1$), while the emails it generates or the papers written on it are external functions (function$_2$). The value I attached to these activities reflect its "aptations" (function$_4$), while its continued usage would reflect its selection and survival (function$_5$). The Wilson and Cleary (1995) model also acknowledges a role for intentions, purposes, and goals, but now as characteristics of the individual and environment, each of which contributes to what becomes perceived as quality-of-life.

By acknowledging that values play a role at the aptation and adaptation level (Fig. 9.3), Mahner and Bunge (2001) have proposed a model that is formally similar to Wakefield's (1999). Both are based on a cognitive hybrid construct; a harmful dysfunction for Wakefield and a disvaluable function (a malaptation, malfunction, or dysfunction) for Mahner and Bunge (2001).

Just as they did for the term function, Mahner and Bunge (2001) clarify the meaning of the term "functionalism" by reference to its various uses. However, they acknowledge that the same complexities that resulted in multiple meanings of the term function will also be true for the term functionalism. First, they point out that for functionalism, what the system does, not what it is made up of, is all-important. To support this view, they quote Putnam (1975; p. 291), who has argued that independent of whether a person was made of Swiss cheese or protoplasms, as long as they functioned in a similar manner, the functions of the system would be considered the same. They apply this to both internal and external functional manifestations of systems. Thus, when the internal activities, or functions, of two different systems are essentially the same, they refer to this as formalistic functionalism (or functionalism$_1$). Following the format in Figs. 9.2 and 9.3, they label functionalism that involves different systems engaged in the same external activities (functions$_2$) as black box functionalism (or functionalism$_2$). They also go on and apply the aptation and adaptation labels to functionalism. When teleological indicators are applied they speak of various forms of teleofunctionalism, which implies that some functions are useful or intend to be useful to somebody or thing.

Some of these ideas are very much part of the justification of a psychometric approach to quality-of-life or HRQOL assessment. Thus, if two people select the same alternative on an item on a questionnaire, the fact that they may have engaged in very different cognitive processes to come to their selection is considered irrelevant. This is the essence of a black box type of functionalism that most quality-of-life or HRQOL researchers have either implicitly or explicitly adopted. However, as the Mahner and Bunge (2001) model implies, this is only one application of the term functionalism.

Mahner and Bunge (2001) have also applied their analytical model to the various examples of how the term functionalism is used in the social sciences. For example, the assumption that a person will make a rational choice when making decisions relies on internal processes (function$_1$) and this makes utility estimations an example of formal functionalism. In contrast, social functionalism is an example of adaptationist functionalism, since the functions of a social system are examined to determine if they contribute to the maintenance or survival of the culture or society. As Mahner and Bunge (2001) state, social functionalism postulates that "all social items (mechanisms, roles, norms, patterns, institutions, etc.) come into being and persist because they are useful to the social system concerned, or even to society at large" (p. 89). Social functionalism, however, offers a description or an account, not an explanation of social processes, although this accounting can still be quite useful when used to analyze specific social processes. Thus, much of what social functionalism provides is a form of taxonomy, identifying and classifying the various elements that make up social processes, with much less of an emphasis on how changes in these systems occur.

The Mahner and Bunge (2001) paper has been helpful since it provides a common analytic model that could be applied to a variety of uses of the term function and functionalism. I will apply their model in a number of settings and in each I compare internal to external functional processes, determine the value of specific functions (such as contributing to the aptation or adaptation of the system), and consider if each involved purpose, intentions, or goals. In this way I could construct a particular meaning and usage for the term function.

The Mahner and Bunge (2001) model should also be helpful when I examine any application of a functional analysis, including Talcott Parsons' approach to illness behavior and the medical care process. It could also be used to assess Patrick and Erickson's (1993) efforts to apply a public health perspective to the assessment of quality-of-life. Each of these topics will lay the foundation for a discussion of the role of function and philosophical functionalism in the rehabilitation process. What will become evident is that functionalism, with its emphasis on homeostasis, actually contributes to some of the difficulties that quality-of-life or HRQOL research get into when dealing with the "compromised" person. These difficulties are related to how a supposedly comprehensive, but closed, system accounts for exceptions (e.g., the disabled, the intellectually compromised, and so on).

1.7.2 Parsons' Structural–Functional Analysis of Medical Practice

To appreciate how distinctive Parsons' structural–functional analysis of medical practice is requires that I first review how other sociologists used the concept at the time Parsons was theorizing and doing research.. There seems to be a consensus (Huaco 1986; Turner and Maryanski 1988) that Merton's 1949 paper (1949/1968) on *Manifest and Latent Functions* and an earlier paper by Kingsley Davis and Wilbert Moore (1945) that dealt with a functionalist theory of social stratification lead to the development of a functional analysis in sociology. These papers are important in the history of sociology because they demonstrated that it was possible to combine a macro- and microstructural analysis of social processes in one theoretical statement. Huaco (1986) summarizes the characteristics of sociological functionalism as follows:

1. It uses a distinctive terminology.
2. It argues for the existence of prerequisites.
3. It tends to ignore causes and attempts to explain in terms of alleged consequences.
4. It makes exclusive use of homeostasis (p. 36).

Consistent with the previous discussion, Merton (1949/1968) also noted that functionalism as a term has a confusing array of uses. Huaco (1986) quotes him as stating:

> From its beginnings, the functional approach in sociology has been caught up in terminological confusion. *Too often, a single term has been used to symbolize different concepts, just as the same concept has been symbolized by different terms.* Clarity of analysis and adequacy of communication are both victims of this frivolous use of works…The large assembly of terms used indifferently and almost synonymously with 'function' presently includes use, utility, purpose, motive, intention, aim, consequences. (Merton, 1968; p. 74,77; italics added).

To resolve this confusion, Merton (1949/1968; p. 78) suggests that the term function, and presumably functionalism, should be defined to mean "observable objective consequences," and that I should avoid using the term to mean "subjective dispositions (aims, motives, purposes)" Huaco (1986) points out, however, that Merton's (1949/1968) effort to objectify the term function did not always succeed, since Merton, at times, used the term to refer to roles, needs, and so on, reflecting the term's[12] "inherent semantic instability" (p. 36).

In contrast to Merton, Parsons (1937) was concerned with action or subjectively meaningful behavior. His interest in subjectivity is reflected in his addition of the category of motivation and personality, along with the social system and the cultural system, as the elements that make up his theory of action.[13] Parsons' subjective, or action-based, orientation to social processes becomes evident when he applies a structural–functional analysis to illness. As Gerhardt (1989) describes and illustrates (Fig. 9.4), Parsons conceives of two models, one representing the structural (i.e., the capacity model) and the other the functional (i.e., deviancy model) component of a

1 Theoretical Issues

Fig. 9.4 Gerhardt's (1989) conceptualization of Parsons' two-component model of illness. Reproduced with permission of the publisher.

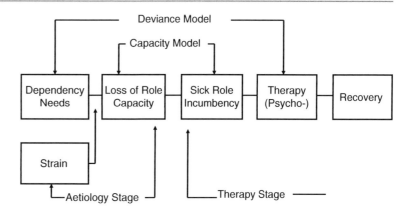

model of illness. Parsons was heavily influenced by psychoanalytic and the economic thought characteristic of the prewar period. This becomes evident from Gerhardt's (1989) description of the relationship between these two models:

> The breakdown of normality is first envisaged as the withdrawal of the libido-like capacity to play roles. It suggests reactivation of this capacity through physician-role/sickness-role medical practice. At the same time, the deviancy model envisages the breakdown of normally repressed dependency needs into active realization (through passivity, as it were). It suggests the repression of these antisocial tendencies in a socialization-like process of therapeutic interaction…The flow of events is as follows: Dependency needs, which normally are neutralized by superego controls, under strain break loose and lead to loss of role capacity. This necessitates sick-role incumbency. This, due to the insulating as well as reintegrative aspects, leads to reestablishing superego controls. These reinstate role capacity and once more instigate abandoning dependency needs in the interest of remaining a respected citizen in the eyes of one's significant others.
>
> The spoils of normality (that is, class, status and power) are thus conditional upon the willingness and, to a certain extent, ability to give up the intermittent indulgence in passivity and irresponsibility. Social inequity necessitates reacquiring the full burden of self-control for modern self-propelled 'economic man'. (Gerhardt 1989; p. 61).

While both elements of this model derive from very different intellectual traditions (e.g., the deviancy notion derives from psychoanalysis, while the capacity model derives from idea of biological homeostasis), they have in common that they contribute to the maintenance and perpetuation of (the social) order. Medical practice and patient behavior illustrate this for a specific case, but for Parsons this subsystem was meant to exemplify the general exchange nature of social action. As Gerhardt (1989) goes on to say:

> In this vein, what matters about illness for Parsons is *how society treats it*; that is, how therapy as a social system goes about eliminating or curtailing it. What matters is the *social order*. Illness becomes a disturbing factor dysfunctional for the upkeep of order in society, lodged with the individual's psychology. To be sure, illness in a psychological sense is of marginal interest to the sociologist; but as an issue of non-functioning in social roles it is at the same time of utmost importance…The issue, then, is that the social order works on the basis of powerful homeostatic mechanisms, and medical practice is meant to constitute one, if not 'the' most important one. (p. 64)

Parsons' ideas have not gone unchallenged. In fact, today much of the theoretical underpinnings of his approach have been questioned. For example, a major concern is his version of Comte's (1875) argument that a society can be studied as if it were an organism and can be expected to operate under the same principles as those for a biological organism. He was also evoking the standard functional assumption that processes ongoing for an individual would be true for society as a whole. This notion has also produced dissenting opinions (e.g., Black 1961). If preservation of the entity, whether the body or the society, is the primary goal of the system, how does organic change in these systems occur? This issue has not been adequately dealt with by Parsons (Gerhardt 1989; Huaco 1986) and remains one of the major limitations of social functionalism.

C. Wright Mills (1959) has been particularly sharp in his criticism of Parsons. After reproducing an excerpt from Parsons' *The Social System* (1951; p. 38–39), he attempts to "translate" Parsons' comments. He states:

> Or in other words: Men act with and against one another. Each takes into account what the others expect. When such mutual expectations are sufficiently defined and durable, we call them standards. We call these expected reactions sanctions. Some of them seem very gratifying, some do not. When men are guided by standards and sanctions, we may say they are playing roles together. It is a convenient metaphor. And as a matter of fact, what we call an institution is probably best defined as a more or less stable set of rules. (Mills 1959; p. 29).

Mills then goes on to comment that while Parsons' ideas are good ones, they are also ideas that can be found in any standard sociological textbook. Of course, I should take note of Mills' suggestion that "playing roles together" may be a convenient metaphor. The notion that Parsons is using metaphors to communicate meaning should not be surprising. Mills goes on to say that one of the consequences of this "grand theory" is that it becomes so abstract that it avoids dealing with the essential issues that it should deal with, such as the power relationships that characterize and determine social processes.

The response to Parsons has not been totally negative, however. For example, it has stimulated interest among European sociologists (c.f., Sciulli and Gerstein 1985), who, while critical, admire his efforts to create a complete theory. Thus, some of the more ambitious theoretical projects that have followed him come from these investigators, such as Habermas's *Theory of Communicative Action* (Habermas 1981), Alexander's *Theoretical Logic in Sociology* (1982–1983), Gidden's *The Constitution of Society* (1984), and Coleman's *Foundations of Social Theory* (1990).

Huaco (1986) describes the ideological implications of social functionalism as conveying the following messages:

1. The highest value is the preservation and survival of the larger system…
2. Since the larger system is like a homeostatic organism, all internal conflicts are evil because they make the system sick.
3. The best social change is no social change…
4. Social problems are not as bad as they seem because regardless of whatever pain, deprivation or injustice they might inflict on specific subgroups, these same social problems contribute to preservation and survival of the larger system. (p. 50).

Considering this, it was not surprising to read Patrick and Erickson's (1993) statement that "In contemporary theoretical sociology, functionalism has been in disrepute because of its inherent conservatism and closed-ended systems approach" (p. 60). Yet, they point out that the definitions of health and illness, function, and dysfunction that Parsons provided have remained useful and have become an essential part of current quality-of-life and HRQOL research. Parsons (1958) defines health, for example, as "the state of optimum capacity for effective performance of valued tasks" (p. 168) which suggests that how well a person performs their various social roles will determine if their behavior becomes a significant issue (i.e., illness) for the social system. Patrick and Erickson (1993) state:

> A person is 'well' if he or she is able to meet the norms or standards for functional behavior that would usually apply to him or her. Thus, quality-of-life indicators, such as life satisfaction, are often based on people's ability to perform the social roles and activities that they want to perform as well as the degree of satisfaction derived from performing them… (p. 61).

This suggests a causal chain exists in which a person's ability to perform social roles leads to good or bad life satisfaction that is then reflected in a persons' rating of their quality-of-life or HRQOL.

Patrick and Erickson (1993) go on to state that the legitimatization of role performance, such as the sick role, is not an important part of how quality-of-life or HRQOL assessments are constructed. Rather, functionalism has influenced how and what types of health indicators are assessed in quality-of-life or HRQOL research. Thus, while Parsons' model of social action includes both the performance and the capacity to perform a role (Fig. 9.4), most models of quality-of-life or HRQOL assessment are limited to measures of performance. As Patrick and Erickson (1993; p. 61) state, "Deviation from the performance of usual activities, rather than capacity, is most commonly incorporated in health indicators…." In addition, since functionalism emphasizes the major social roles, (e.g., work, school, housework, and so on) characteristic of a culture, not surprisingly, they have become the common functional measures of a quality-of-life or HRQOL assessment.

At this point, it seems reasonable to ask what influence Parsons and social functionalism has had on quality-of-life or HRQOL research. First, I find it helpful to think about Parsons' theory as an effort at "grand theory"; an attempt to comprehensively account for major social issues, such as how change and stability occur within a society. There is no comparable theory within quality-of-life or HRQOL research, even though it may be obvious that such a theory would be quite helpful.[14] It is not even clear if the major questions necessary to generate a general theory of quality-of-life or HRQOL have been asked. Certainly, the shift suggested in this book to asking *how* a response to a qualitative assessment occurs will be one element in any "grand theory," but what else is needed?

Parsons' theory is attractive because it captures the extent and depth of human experience, even though criticism of his theory has focused on its static nature and inability to describe how a social system evolves over time. Quality-of-life or HRQOL researchers, however, by discarding many of the mechanisms (e.g., the sick role) that Parsons describes, still end up having to deal with the same issues (e.g., how our sense of quality is maintained, changed, restored, and evolved), but now with many fewer useful conceptual tools. Instead, what has happened is that quality-of-life or HRQOL researchers have invested in cataloging (i.e., counting) functional changes, leaving the interpretation or meaning of these changes to another time or place. This, of course, fills journals with data, data that remain difficult to organize in a theoretically interesting or useful way.

It is interesting to examine Parsons' theory using the function classification system that Davies (2001) proposed (Chap. 9, p. 302). I conclude from this comparison that Parsons' model combines a deviancy (selected functions) and capacity model (systemic model of functions). As Fig. 9.4 illustrates, the sick role and a person's ability to cope with it would reflect their capacity to restore or maintain functioning; an outcome characteristic of a systemic model of functioning. According to Davies, one of the derivatives of a system-dependent model of functioning is that malfunctions would not be possible. They would not be possible because a malfunction implies that a system is not functioning properly, but this is not possible if the system is intact. Thus, change can occur, but a different system would be involved. Instead, he suggests that a systemic model can have nonfunctioning components that would account for its variation in the primary system.

This distinction between malfunctioning and nonfunctioning components of a capacity-based system is a helpful one, especially when I consider the compromised person. For example, is the compromised person a person whose capabilities are malfunctioning, or does he or she lack the *capacity* to function? These alternatives clearly would make a difference in how their quality-of-life or HRQOL should be assessed.

Finally, Parsons' relevance for qualitative research may yet see a revival, especially if interest in the role that genetic factors play in determining the capacity of a person to assess quality becomes of more interest. How this might occur is illustrated in Chap. 10, where I discuss in detail some of the complex issues involved in relating a person's capabilities and their performance of a qualitative assessment.

Now, however, I need to deal with the role that the terms function and functionalism play in assessments used to monitor the rehabilitation of the disabled. This topic will give me another opportunity to examine some of the same issues I will have to consider when deciding if functions or a functional domain are appropriate indicators of quality-of-life or HRQOL.

1.8 Summary

The purpose of the first part of this chapter was to provide the background to the phrase "functional status" by first discussing the various definitions and uses of the term "function." I also discussed how the term has been used in philosophy and sociology to account for social processes and social change. This should provide an adequate background for the next major section, where I discuss the application of the term, but now to persons who have a social as much as a physical limitation that has led them to be labeled as disabled and who requires either physical or mental rehabilitation. I end this chapter by returning to my initial question of whether a functional status assessment is an appropriate indicator of quality or quality-of-life.

From this discussion, I learned that the term "function" has a history that is reflected in the changes in the meaning of the term (e.g., from literal to figurative expressions), but also with the development of science. Function statements are mostly concerned with process issues (e.g., homeostasis), while quality and quality-related issues are mostly concerned with outcome. Also true is that quality-of-life research has generated a vast quantity of data, but has shown very little progress in terms of developing its theoretical foundations. The review in this section was partly intended to identify some of the elements that might be used in such theoretical efforts. At the moment, however, there is no "grand theory of quality-of-life," other than perhaps Aristotle's notion of *eudaimonia.*

The notion that a function statement can be used metaphorically raised some interesting questions. For example, when is a function statement figurative and when is it not? Lakoff and Núñez (2000) suggested that certain mathematical operations could be performed metaphorically, and I ventured that this may be what occurs when a qualitative researcher adds, subtracts, or divides a diverse set of items (e.g., symptoms, physical activities, feelings, and so on) that would not otherwise be mathematically combined, but assumes it's appropriate because the items contain some common underlying qualitative characteristics. Lakoff and Núñez (2000) referred to this as "arithmetic metaphorization."

Cognitive studies of children and adults have shown that the function statements play an important role in category formation. This occurs because function statements facilitate the classification of entities or experiences. Function statements are most often relational in nature, involving the comparison of cognitions, while qualitative statements can occur by direct assessment of objects, events, or experiences and can rely primarily on perceptual processing. This is essentially what occurs when I eat a unique meal for the first time, see a face for a moment and declare it beautiful or handsome, or have a new life experience and declare it important. I discussed these phenomena previously and referred to them as examples of "first impressions" (Chap. 3, p. 66). First impressions are based on perceptual processing that occurs automatically (e.g., affordance), in contrast with a functional statement that will unavoidably be comparative and involve more controlled cognitive processing.

Boorse's (1977a) effort to model health by using the normal probability curve gave me the opportunity to discuss the same set of issues, but now relative to using function status as the indicator. I was primarily interested in whether Boorse's approach could solve the problem that a change in state also changes the meaning attached to terms being used, such as functional status. The importance of this issue will become evident when I discuss the term "disability" and encounter the social image of the disabled, as they attempt to fully participate in society.

Boorse's (1977a) approach involved defining a pathological functional state as resulting from a quantitative change in functioning (e.g., a greater number of dysfunctional states), compared to what would be considered a normal level of functioning. However, he claims that what is normal is natural, not acknowledging that such a change in terms is also a change in the meaning being communicated. The term "natural" has a value connotation suggesting that a qualitative component has entered his argument. Since the value attached to a pathological state is qualitatively different from a normal state, Boorse's assumption that the two states are separate points along a common continuum becomes untenable.

The paper by Torres (2002) illustrates some of the issues that arise when semantic changes of health or functional states occur. In this case, persons labeled as being predisposed to develop Huntington's disease were sufficient to label a person as having a state change (that is, they were not "healthy"). What again was not made clear was that the resultant labeling was also a value judgment that had consequences on the person's perception of themselves and their quality-of-life. In this case, labeling the person as unhealthy meant that he or she was on the track to becoming demented, and that any change in their functioning would be evidence of the progression of this process. Again, a change in the meaning associated with the phrase "functional status" would have occurred.

Of course, when Boorse (1977a) describes a normal or pathological state as "natural," he is also implying that a normal or pathological state has teleological properties. This gave me the opportunity to discuss Fulford's (1989; 2000) concerns about the relationship between causes, values, and teleology. In addition, Fulford's notion that a value statement was unavoidably a part of the cascade of indicators used to assess outcomes was an important reminder of how language and its semantics are intimately involved in discourse, whether lay or scientific. This was followed by a discussion of philosophical functionalism, with an emphasis on models of functionalism and Parsons' two-stage sick role model and its potential contribution to quality-of-life research. I found quite useful Mahner and Bunge's (2001) suggestion that a modular approach be used to account for the various meanings associated with the term function. My review of Parsons' two-component sick role model provided me with the opportunity to discuss the assets and limitations of a grand theory, as opposed to relying on the accumulation of data to provide the basis for a theory. Like Patrick and Erickson (1993), I found it interesting that the assumed homeostatic basis, and the "sick role" itself, remains a useful empirical construct that has persisted.

2 Applications of Functional Assessments

2.1 The Role of Functional Assessment in the Rehabilitation of a Person with Disabilities

The research literature on the rehabilitation of person with a disability quickly reveals that measuring functions, their limitations, and their interactions with a person's environment are its primary indication. Also evident in this literature is a major concern with establishing what is meant by the term "disability." This is understandable, since how disability is defined has important legal and social policy implications,[15] but it also sets assessment objectives. Several models or approaches to defining and classifying disabilities have been proposed over time (e.g., Nagi 1965, 1976; Pope and Tarlov 1991; Brandt and Pope 1997; WHO 2002). These models were meant to coherently organize and relate the various elements (e.g., a person's pathology, impairments, functional limitations, as well as disabilities) that define the "disablement experience" (Verbrugge and Jette 1994). The form and structure of these models (see below) have a striking resemblance to models I have considered in the quality-of-life or HRQOL literature, such as those of Wilson and Cleary (1995) and Patrick and Chiang (2000; Fig. 9.1). Each is an example of a process-outcome analysis with factors internal or external to the individual modulating the overall outcome. Each can also be examined from the perspective of the Mahner and Bunge (2001) model (Figs. 9.2–9.3).

I will discuss three models of disability: first, the Nagi (1976) model; then a model suggested by the two Institute of Medicine (IOM) reports (Pope and Tarlov 1991; Brandt and Pope 1997); and finally, a model based on the WHO (2001; ICF) approach to establishing a common language for functioning, disability, and health. Different from Parsons' social action theory, these models deal with a circumscribed issue – the disablement process – but all attempt to bridge the gap between a micro and macro level of analysis. Parsons, of course, was interested in the relationship between an individual's experience (a micro level of analysis) and social forces (a macro level of analysis), while the disability models emphasize the relationship between individuals and their environment; a "socioecological" perspective. The principal task of these disability models, therefore, is to combine subjective (social) and objective (environmental) indicators. These models also raise additional theoretical issues, such as how the diverse domains are combined into a conceptual whole, what the common denominators are that permit this, and what role quality-of-life or HRQOL assessments play in these models.

A second, equally important set of issues in the disability literature deals with defining deviance and normality as when deciding when a person has an impaired or normal functional status. Thus, it can be asked how many impaired functions define being disabled, or what proportion of normal functions relative to the total number of a person's functions still permits the person to label themselves as normal? I have already considered a variety of approaches to defining these issues, so I have a sense of the complexity here, although I have not considered them from a socioecological perspective. I have also considered mechanisms, such as Parsons' sick role model, that could be used to account for how a person becomes deviant but also returns to normalcy. I will evaluate how well the sick role model applies to the disabled, but also ask if there are alternative social processes that are applicable to the disabled person. Also relevant is the issue of what role rehabilitation plays in the restoration of "normal" functioning, and whether restoring the functional status of the disabled person redefines what is meant by normal.

A third major concern is that the Nagi (1965, 1976) and the IOM reports (1991;1997) involve concepts that are worded in the negative and are value-laden; for example: *pathology, impairment, functional limitation, and disability*. The ICF (WHO 2001), however, claims to emphasize health rather than disability, and this represents a step away from the negativity of previous approaches to disability. Do these conceptual differences make a difference in what the term function means in a rehabilitation setting? This is especially important since I now have learned that the term function is context-specific and very much dependent on the linguistic entailments associated with the term.

We will find that the literature on the rehabilitation of the disabled person is burdened with many of the same complexities I have encountered in the quality-of-life or HRQOL literature. But by focusing on the person–environment interaction, I add an additional level of complexity to my task. The research literature provides no new ideas on how to redefine normality for the disabled, nor does it provide simple answers to the question of how to keep a disabled person's functioning stable. To start, however, let me examine the etymology and language usage of the words "rehabilitation" and "disability." Rehabilitation as a process (e.g., as a system of care) will provide a basis for the study of functioning, and this will guide my interest in the environmental changes required to redefine what is meant by "normal" for a disabled person.[16] Disability, on the other hand, will serve as a state indicator: when a person can be labeled as "not able." I will then return to my original interest and discuss how the term function is used in either context.

2.1.1 The Language of Rehabilitation and Disability

The word "rehabilitate" comes from the Latin *rehabilitare*, or *habilitare*, meaning "to make able" (Oxford University Press 1996). As a verb, it is defined as the process of restoring a person's health or life to a former condition. This could be accomplished by retraining a person or by use of therapy after an illness, imprisonment, or even an addictive experience.[17]

The word "ability," when examined separately from the negative prefix- *dis*, comes from the Latin *habilitas*, or from *habilis*, meaning "able" or "handy" (Oxford University Press 1996). Grammatically, the word functions as a noun and is defined as "the power or capacity to do something," or "refers to a person's skill or talent." Thus, the terms disability and rehabilitation have overlapping etymologies: one describing a person's *capacity* and the other the process of *restoring* a person's *capacity*.

There are many examples of how to make a person "able." A classic one, retold by Gellman (1973; p. 3), is based on the story of Adam in the Garden of Eden, a story that Gellman believes illustrates that the term rehabilitation has metaphoric properties. In the story Adam ate an apple and discovered he now had a "disability"; he gained knowledge of the world around him. He rehabilitated himself by assuming a socially acceptable role as a farmer and agriculturalist, but unfortunately, this was not sufficient to get him back into the Garden of Eden. As a result, Adam might not have thought of himself as having literally been restored and rehabilitated, but rather as having been "accommodated" to a new physical, emotional, and social status (he now had to work to survive) and environment (he was not living in the Garden of Eden anymore!), and that, therefore, he had been only figuratively restored and rehabilitated.

The notion that the term rehabilitation can have metaphoric properties is of great interest, since this could help me understand how a functional term is used in the language of rehabilitation. To review what I discussed earlier (Chap. 2, p. 30), metaphors are a form of figurative language with the unique capacity to generate meaning that would not otherwise be present if it wasn't for the mapping of information from a base to a target. Figure 2.2 illustrates this by depicting Bowdle and Gentner's (2005) suggestion that a metaphor evolves from an analogy. Note that each has the same target and base, but that they differ in terms of the kind of connector phrase and qualifier. Of these three forms of figurative language, only the metaphor would experience a significant decrement in meaning if the target and base were reversed.

Metaphors also have cognitive, affective, or somatic dimensions. For example, Lakoff and Johnson (1999) described a conceptual metaphor as being "embodied," implying that what is learned from the body anchors the meaning being communicated by the conceptual metaphor. This characteristic of metaphors is particularly important, since I will be dealing with people whose "embodied self" or physical status has been altered by injury, disease, or treatments. How these alterations find expression, if at all, during functional assessments is an important issue for study.

Metaphors can also be expressed nonverbally, by gestures and actions. Thus, when a child uses a truck as if it were a phone, or flies a block of wood as if it were a plane, he or she is creating an "enactive metaphor." Inactive metaphors are common forms of nonverbal expression in severe stroke and Alzheimer's patients, and if properly interpreted, can be a helpful mode of communication and important part of the rehabilitation process.

It is also important to recall that linguistic and conceptual metaphors differ in that a linguistic metaphor can provide nonliteral meaning for a word or phrase, while a conceptual metaphor can help organize how I think and reason. Thus, conceptual metaphors are an essential component of cognitive processing, while a linguistic metaphor plays a more restricted communicative role. In both cases, however, a mapping occurs.

In what way is the term "rehabilitation" a conceptual metaphor? To appreciate this, I need to refer again to the EVENT-STRUCTURE METAPHOR (Chap. 2, p. 35), one component of which is the A PURPOSEFUL LIFE IS A JOURNEY metaphor. The PURPOSEFUL LIFE metaphor can be detected in the ordinary discourse, or folk language, as people speak about the purposes or goals they have for themselves. Thus, people have destinations in their lives and they are supposed to do what is required to reach these destinations. Rehabilitation is one such activity, and it is in this sense that the term is an example of a conceptual metaphor.

To what extent is the term "(dis)-ability" a conceptual metaphor? Referring again to the discussion of the EVENT-STRUCTURE METAPHOR (Chap. 2, p. 35), it can be seen that ability, as a state term, implies a location. What a person is able to do or not does reflect their capacity, and that capacity is a state that the person is in. Placing the term in the EVENT-STRUCTURE METAPHOR, however, opens it to a number of other related metaphors, such as "changes are movements (into and out of bounded regions)" or "difficulties are impediments to motion" (Lakoff 1993; p. 220).

Interestingly, when the disability research literature is inspected, it is clear that the term disability is used in a variety of ways, not just as a status or capacity measure. For example, Verbrugge and Jette (1994) define disability as "experienced difficulty doing activities in any domain of life...due to health or physical problems" (p. 4). This definition refers to a full range of possible activities that are specific to the person's life circumstances. The authors then go on to distinguish "disability" from "functional limitations." Here it becomes clear that they interpret the available research literature and their own preferences is to view functional limitations as referring to an individual's capacity and to view disability as a social process.[18] To quote them:

> The words 'action' and 'activity' are simple devices to distinguish the concepts of functional limitations and disability. They help convey the generic (situation-free) features of one and the social (situational) features of the other. The words 'task' vs. 'role' also help distinguish the concepts...Functional limitations and disability refer to different behaviors, not to different aspects or ways of measuring the same behavior. Explicating this point: (1) Some researchers characterize functional limitations as 'can do' and disability as 'do do'. The words separate a person's capabilities from his/her ultimate pattern of behavior. (Verbrugge and Jette 1994: p. 5).

The arbitrariness of their approach can be seen when the definitions are reversed. Thus, functional limitations could indicate what a person "does do" (an activity) and be understandable, while disability could indicate what a person "can do" (a capacity or action statement) and be understandable. Clearly, both terms are, to use Merton's phrase, "semantically unstable," and this makes it more likely that they are at times being used figuratively.

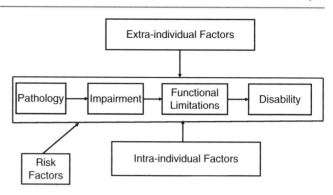

Fig. 9.5 Nagi (1965, 1976) model of disability as modified by Verbrugge and Jette (1994). Reproduced with permission of the publisher.

2.1.2 Models of Disability

Not too surprisingly, the various models of disability have evolved in concert with changes in the meaning attached to the term disability, and this has occurred in response to changes in the political and social status of the disabled person. The nature of this interaction can be illustrated by the following quote from Jette and Badley (2000):

> The 'Social Security Act' defines disability as the 'inability to engage in any substantial gainful activity by reason of a medically determined physical or mental impairment which can be expected to result in death or can be expected to last for a continuous period of not less than 12 months'...This definition in the Social Security Act is at odds with most contemporary thought about the concept of disability and is itself a barrier to the SSA's work disability revision process. (p. 2–3).

Jette and Badley (2000) were suggesting that depending on the definition of disability that was used, a work rehabilitation program could either facilitate or impede achieving its objectives. A major impetus to the politicalization of the term disability came with passage of the Americans with Disabilities Act (1990). This law, along with similar laws in other countries, affirmed the rights of the disabled person, and this significantly contributed to shifting the term disability from a medical to a social phenomenon. In this regard, it is interesting to note that Nagi's[19] original conceptualization of disability had to wait to the early 1990s (see the Institute of Medicine report; Pope and Tarlov 1991) for its full consideration and acceptance.

For Nagi (1965, 1976), disability represented a gap between a person's capabilities and the demands created by their social and physical environment. This gap is created by the degree to which a person's socially defined roles are disrupted. This disruption can occur because of the presence of some pathology that produces impairments, impairments that set functional limitations, and functional limitations that lead to disability. Verbrugge and Jette (1994) elaborated Nagi's model by including references to intra- and extraindividual characteristics, as well as predisposing risk factors that could influence the disablement process. A simplified version of their model is illustrated in Fig. 9.5.

Fig. 9.6 A characterization of the Enabling-Disabling Process. Adapted from the Institute of Medicine, 1997. Reproduced with permission of the publisher.

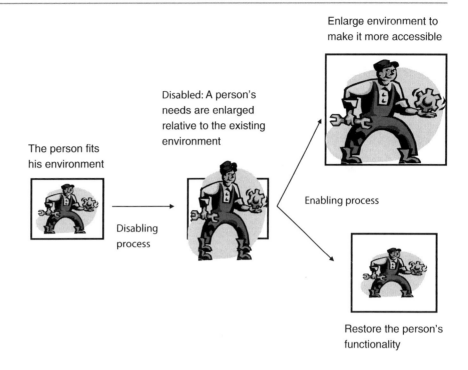

The Institute of Medicine convened working groups in 1991 and 1997, and each meeting resulted in a summary report (Pope and Tarlov 1991; Brandt and Pope 1997). The 1991 report adopted Nagi's original model and added two important concepts to the disablement model: the concepts of secondary conditions and quality-of-life. The 1997 report, responding to Nagi's view that disability was not inherent to the individual, presented a model (Fig. 9.6) which describes the person with disabilities as essentially not fitting into his or her environment. Moving from disablement to enablement would involve either expanding the environmental opportunities for the disabled person or restoring a person's ability to function. This model, therefore, depicts a socioecological approach to the enablement-disablement process.

In 2001, the WHO published a significantly modified version of their *International Classification of Impairments, Disabilities and Handicaps* (WHO 1980), now known as the ICF or *International Classification of Functioning, Disabilities and Health* (WHO 2001). The ICF was designed as a classification system to be used along with the *International Classification of Diseases and Related Health Problems* (the ICD-10; WHO 1994). The ICF has been described (WHO 2002) as follows:

> It is a classification of health and health-related domains - domains that help us to describe body changes in body function and structure, what a person with a health condition can do in a standard environment (their level of capacity), as well as what they actually do in their usual environment (their level of performance)…In ICF, the term *functioning* refers to all body functions, activities, and participation, while *disability* is similarly an umbrella term for impairments, activity limitations, and participation restrictions. (p. 2).

The report goes on to state:

> The ICF is named as it is because of its stress on health and functioning, rather than on disability. Previously, disability began when health ended; once you were disabled you were in a separate category. We want to get away from his kind of thinking. We want to make the ICF a tool for measuring functioning in society, no matter what the reasons for one's impairments… ICF puts the notions of 'health' and 'disability' in a new light. It acknowledges that every human being can experience a decrement in health and thereby experience some disability…ICF mainstreams the experience of disability and recognizes it as a universal human experience. (WHO 2002; p. 3).

As Imrie (2004) makes clear, what is key and unique about the ICF is that it, too, makes disability a sociopolitical issue. It does so by rejecting the notion that if an impairment is presented it has to indicate the presence of a disease, and instead, sees it as a product of social or environmental influences. It also conceives of disability as a relational phenomenon, being the outcome of an individual's interaction with their health condition and personal environmental context. In addition, the ICF adopts the biopsychosocial (BPS) model as an essential element in its conceptual framework. In so doing, it acknowledges that a person's biological, psychological, and social statuses are coequal determinants of a person's status. Finally, the ICF sees itself as being applicable to all persons, not just those with a disability. The universality implied by the ICF comes from its assertion that all people are at risk of acquiring impairments and chronic illnesses, and as such, policies and practices should be oriented to satisfying the needs of all persons. Thus, placing a ramp to permit access to a building doesn't just meet the needs of a person in a wheelchair, it should also be

Fig. 9.7 The Jette and Badley (2000) conceptualization of the relationship between the Disablement Process and Quality-of-life.

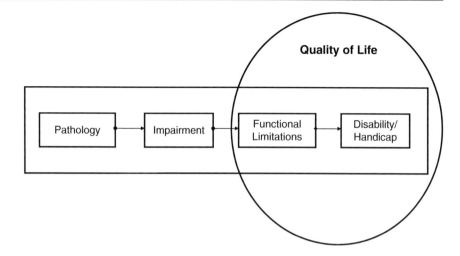

seen as meeting the needs of all persons seeking access to the building.

Imrie (2004) stitches an important thread by pointing out that the conceptual origin of BPS can be traced to Parsons' model of structural functionalism. He states:

> While the ICF does not identify the intellectual origin of BPS, it is derived from structural functionalism, or a conception of society which…exists on different and distinct levels of organization (Parsons 1951). Society comprises 'interlocking systems' in which the four domains of the physiological, personality, social and cultural are seen to operate at one level of a more general hierarchy of interrelated levels…
>
> As Parsons (1951) notes, the body is the foundational level of the social system, in the sense that it is the container for the impulses, desires and motivations that comprise individuals' personalities and actions. For Parsons (1951), however, the nature of personality and action is not to be understood wholly in terms of organic or biological processes; rather it has its own emergent properties." (p. 297–298).

Imrie (2004) goes on to point out that considering the dependence of the ICF on structural functionalism, it should not be surprising that it would be subject to the same criticisms leveled against structural functionalism. Thus, with the ICF, there will be an "over-emphasis on systems harmony, interaction and consensus, and less recognition of systems rupture, tensions and conflict" (Imrie 2004; p. 298).

A review of the disability literature reveals a number of attempts to include quality-of-life issues into a disability model. For example, in the 1991 Institute of Medicine report, quality-of-life was described as follows:

> ..quality-of-life affects and is affected by the outcome of each stage of the disabling process. Within the disabling process, each stage interacts with an individual's quality-of-life. There is no universal threshold - no particular level of impairment or functional limitation - at which people perceive themselves as having lost their personal autonomy and diminished the quality of their lives. Yet perceptions of personal independence and quality-of-life are clearly important in determining how individuals respond to challenges at each of the four stages of the disabling process. (Institute of Medicine, 1991; p. 8).

This suggests that quality-of-life was viewed not as a specific outcome, but rather as reciprocally related to each aspect of the disablement process. Jette and Badley (2000) consider this definition as inconsistent with published definitions of quality-of-life, and instead, suggest that quality-of-life partially overlaps with those aspects of the disablement process which include functional limitations and disabilities (Fig. 9.7). They also acknowledge that there are aspects of quality-of-life that do not overlap with the disablement process, as is depicted in the figure.

Fuhrer (2000) also discussed the role of quality-of-life or HRQOL assessment in the disablement process. First, he points out that the development of subjective measures of quality-of-life or HRQOL and objective measures of medical rehabilitation (the Barthel Index, the FIM™) has proceeded along separate lines until quite recently. Yet, he is interested in the development of subjective models of disablement as a basis for monitoring rehabilitation outcomes. His argument is that to exclude the subjective will narrow the portrayal of a person's experience to the point where it could misrepresent the rehabilitation outcome. To remedy this, he suggests the inclusion of measures of subjective well being as a rehabilitation outcome. How to bridge the gap between objective rehabilitation measures and subjective well-being measures then becomes a question of experimental design. To facilitate this, he recommends that investigators structure questions such that they subjectively assess the indicators of interest; that they need not be concerned with the stability of measures of subjective well-being unless a person's affective state is modified; and that life events, not what specifically happens, will have a time-dependent impact on a person's reports of well-being.

He illustrates the usefulness of a subjective well-being or a HRQOL analysis of the disablement process by examining the literature on the life satisfaction of spinal cord-injured and brain-injured persons. He found for the spinal cord patient that life satisfaction was not related to the extent of the paralytic impairment, but was instead related to the degree of handicap in terms of the limits created in social interactions, mobility, and occupational pursuits. This was confirmed by a meta-analysis (Dijkers 1997) that revealed that correlations between subjective measures of quality-of-life and measures of impairment ($r=-0.05$; $P>0.05$), disability ($r=-0.21$; $P<0.05$), and handicap ($r=-0.34$; $P<0.001$) for spinal cord injury patients progressively increased. In contrast, studies of persons who had a brain injury do suggest that the extent of injury is related to life satisfaction, especially since the extent of the injuries may affect cognition and social relationships.

Up to now, I have been considering conceptual models of disability, where an investigator reviews the available data and organizes this information in a meaningful and efficient manner. Ideally, the conceptual model would lead to the identification and testing of new relationships, but most often it simply acts as a descriptive summary statement. An alternative is to use structural equation modeling (SEM), which provides an empirical method for establishing a relationship between various indicators. SEM (e.g., Loehlin 2004) consists of a series of multivariate statistical methods (e.g., confirmatory factor analysis, regression analysis, discriminant analysis, and so on) that can be used to test a set of hypothesized relationships that can be visualized by a path diagram. For example, both Figs. 9.6 and 9.8, combined with a series of outcome measures for each domain, could be considered examples of path diagrams where the goodness-of-fit to empirical data could be tested using SEM. Thus, what is unique about SEM is that it gives an investigator the opportunity to test if the pieces of a general model consisting of multiple hypothesized relationships literally fit together.

Martz et al. (2005) used SEM to determine if a model consisting of 4 latent variables and 14 observed variables was capable of predicting the psychosocial adaptation of people with spinal cord-based disorders or injuries. To perform their analysis, they conceptualized disability in terms of its impact, rather than viewing it as a state a person was in. Thus, disability's impact was considered a latent variable that was related to a negative affectivity latent variable, while negative affectivity was assumed to be related to both psychosocial adaptation and disengagement coping latent variables. Quality-of-life was considered an observed indicator reflecting a person's psychosocial adaptation. In their final model, they were able to drop two indicators (duration since injury or disorder onset, and denial of death), illustrating another use for SEM. They found that five of the six standard goodness-of-fit methods revealed that they had postulated a "moderately good" model. They also found, in a direct assessment, that psychosocial adaptation and quality-of-life were significantly correlated, with $r=0.456$. Thus, a series of hypothetical relationships were organized into a coherent statement. Since this arrangement also goes from the concrete to the abstract, it reminds me of what is involved in forming an informational hierarchy, but now by using formal mathematical methods.

Sullivan et al. (2000) provide a second example[20] of the application of SEM by testing if the Wilson and Cleary (1995) linear model is capable of characterizing the relationship between measures of physical symptoms, physical functioning, and psychological symptoms to one another and to overall quality-of-life. Their data came from the Groningen Longitudinal Aging Study and involved 5,279 persons. The linear model that they tested expanded the Wilson and Cleary (1995) model from 5 to 7 latent variables[21] and included measures of biological and physiological factors (measured by the number of chronic medical conditions), symptom status, physical function status, daily activity interference, social function status, general health perceptions, and overall quality-of-life. Overall quality-of-life was measured by the Cantril ladder (Cantril, 1965), and each domain was measured by an appropriate assessment or component of a larger assessment (e.g., questions on pain in the SF-20 were used to estimate symptom status).

A LISREL analysis was performed (Joreskog and Sorbom 1993) and provided beta coefficients from the regression analysis to evaluate the contribution of each latent variable. Most interesting for us was the observation that only the beta weight for perceived health was a significant contributor to the variance in overall quality-of-life when the linear model alone was tested. This suggests that if the Wilson and Cleary (1995) model is conceived of as a simple unidirectional model, it would then be limited to the relationship between perceived health and quality-of-life. Wilson and Cleary (1995) made it clear, however, that they see the primary domains of their model as feeding back and influencing prior domains of their model. Sullivan et al. (2000) evaluated this assumption by using a nonlinear model that included hypothesizes concerning the relationship between nonadjacent domains. Analysis of the beta coefficients revealed that physical symptoms had the strongest influence on perceived health (beta=-0.39) and social functioning (beta=0.27), while the number of chronic medical conditions a person had influenced physical functioning (beta=-0.21). This suggested to Sullivan et al. (2000) that functional disability is not the only influence on perceived health.

Wilson and Cleary (1995) also acknowledge that characteristics of a person and their environment, acting as modulating variables, influence a person's perceived health or quality-of-life. Sullivan et al. (2000) tested this by introducing anxiety and depression symptom measures into their model. The resultant nonlinear model was most related to perceived health (beta=-0.46) and quality-of-life (beta=-0.30). Sullivan et al. (2000) state:

Consistent with the predictions of Wilson and Cleary, mental health variables appear to have diverse and widespread effects on health status and quality-of-life. The position in the model for anxiety and depressive symptoms that is most consistent with the data is that between perceived health and quality-of-life. This implies that mental health state affects quality-of-life most potently in its shaping of one's appraisal of his or her overall health status. (p. 808).

Sullivan et al. (2000) point out that this reinforces the Gill and Feinstein (1994) proposition that health status is not a direct measure of quality-of-life, but rather an indicator of how people respond to their health status. This, of course, is consistent with the model presented in Table 8.5 that illustrated the components of an HRQOL item.

These two applications of SEM make it clear that the resultant models are much more data-driven and limited in their scope than the conceptual models presented earlier. Each has its place, with the disability models providing a link between research and policy and the models based on SEM testing specific hypotheses concerning the content of the domains that make up the disability models.

2.1.3 Functional Assessment, HRQOL, and the International Classification of Function

Throughout this chapter, I have been discussing the various meanings and uses of the term function and functionalism, both as an introduction to topics that need continued discussion, but also because I want to be able to answer the question of whether a functional domain can or should be part of a quality-of-life or HRQOL assessment. Now I have to ask if these different meanings make a difference, especially when associated with an estimate of the quality-of-life or HRQOL of the person with disabilities.

Implicit in this, of course, is the assumption that the various meanings of a function statement differ from what is meant by a quality-of-life or HRQOL assessment. However, Bergner (1985) and others have pointed out that function assessments or even health status assessments are sometimes presented as if they are quality-of-life or HRQOL assessments, even though the available data formally relating these different measures are sparse and equivocal (e.g., Fabian 1991; Renwick and Friefeld 1996). Thus, I need to ask what evidence there is that distinguishes a functional assessment from a quality-of-life or HRQOL assessment, and how, if I need to, change a function assessment so that it can assess quality-of-life or HRQOL.

In order to examine these questions, I will adopt the building block approach suggested by Mahner and Bunge (2001) for establishing different types of function statements. I will also summarize a paper by Cieza and Stucki (2005), who offered an alternative methodology that used the ICF to classify the items on a representative group of health status and HRQOL assessments. In the Mahner and Bunge (2001) study, a basic unit of analysis (Figs. 9.2 and 9.3) was identified that when "qualified" led to statements with different meanings and potential uses. In my situation, I selected as the basic unit a descriptive statement (see examples of function statements in Table's 8.8 and 9.2) devoid of any reference to consequences or values. What I will test is whether adding additional information to this basic descriptive (function) statement enhances its correlation with a quality-of-life or HRQOL item or assessment.

My first analysis will compare assessments composed of purely descriptive function statements and available multi-item quality-of-life or HRQOL assessments (Table 9.3, Comparison A). A descriptive functional assessment, however, can occur in more than one way. For example, a health professional can observe the behavior of a person who is disabled and judge their functioning by visual observation or in response to a task the person is asked to perform (Comparison A: independent observer). Or, a person can be interviewed or fill out a questionnaire concerning their functional status (Comparison B: self-report). In addition, the self-reports themselves may vary from being limited to descriptive statements only, to those which assess the consequences of being in a particular state, to those that place values for being in a particular functional or qualitative state (Comparison C). This diversity, however, offers some analytical opportunities. As Table 9.3 indicates, there are at least three comparisons that can be made, and, if there is sufficient data, I should be able to examine the relationship between these comparisons. For example, if I consider Comparison A with B, I would be assessing whether the source of the assessment of the descriptive function statement made a difference in the association with a quality-of-life or HRQOL assessment. Thus, I can hypothesize and test whether a self-reported descriptive function statement will be more highly correlated with various types of self-reported quality-of-life or HRQOL assessments than an equivalent comparison based on an independent observer.

There is also something that can be learned by examining the different types of quality-of-life or HRQOL assessments, again depending on the number of studies available. By arranging the assessments from multi-item to weighed multi-item to those based on derived preferences to a global assessment, a continuum can be assumed to exist. The continuum implied is one of increasing subjective qualification, with the multi-item assessment being primarily descriptive and the global assessment being most likely to reflect subjective processes and the assumed presence of valuations. The question is, therefore, whether the qualitative differences of the various assessment classes are reflected in the correlations between assessments.

Another analytic opportunity comes from inspecting Comparison C. Here self-reported descriptive function statements that may include description of the consequences

Table 9.3 Comparison between different types of function statements and quality-of-life or HRQOL assessments

Comparison	Function statement	Type of quality-of-life or HRQOL assessment
A.	Descriptive statements, assessed by an *independent observer*	Multi-item (e.g., SF-36)
		Weighted multi-item
		Derived preferences (e.g., TTO, SG)
		Global (e.g., VAS)
B.	Descriptive statements, assessed by *self-report*	Multi-item
		Weighted multi-item
		Derived preferences
		Global
C.	*Value-weighted* descriptive statements, *assessed by self-report*	Multi-item
		Weighted multi-item (e.g., Ferrans and Powers Quality-of-life Index-QLI)
		Derived preferences
		Global

VAS Visual Analog Scale; *QLI* Ferrans and Powers Quality-of-life Index; *TTO* time trade-off; *SG* standard gamble

of being in a particular state and a valuation of being in a particular state are compared to different types of quality-of-life or HRQOL assessments. This type of function statement, of course, resembles the hybrid construct I have described as characteristic of an optimal quality-of-life or HRQOL assessment, but now applied to a function statement. It is anticipated that this comparison should have the highest correlation and account for the most variance between assessments, especially if the quality-of-life or HRQOL statement are optimally "qualified." Table 9.3 summarizes the nine comparisons being considered.

The classification system in Table 9.3 implies that progressing from Comparison A to C creates a convergence of meaning between function and quality-of-life assessment, with the final comparison expected to have the highest degree of concordance. Thus, while a function statement and a quality-of-life or HRQOL item may differ in content, they can still be formed by identical cognitive methods. A function statement, like a symptom, can be an evaluated descriptor that is independently valued forming a hybrid construct. For these reasons, a function statement theoretically could serve as an item, domain, and/or an assessment in a quality-of-life or HRQOL study, since it would be cognitively and psychometrically equivalent.[22]

There are several assumptions implicit in this approach that deserve attention. First is the assumption that a task given is a task completed. For example, a person may be given a purely descriptive statement to assess, and yet use information about its consequence or the values placed on it as part of their response. This is, of course, not possible to detect, unless cognitive interviewing (Willis et al. 2005) occurs simultaneously with the person's response. I have discussed the presence of this issue in other situations (e.g., when I addressed teleology and the place where values are expressed; Fulford 2000), so it should not be surprising that it is a concern in the current analysis.

A second assumption concerns who performs the rating, and whether this makes a difference in the outcome of the function assessment. As stated here, this assumes that assessments by an independent observer or a person with disabilities will statistically differ. Is this true? The available data actually suggest this may not be the case, and instead there may be considerable overlap in the response pattern of independent observers and persons with disabilities. Part of the reason for suspecting this comes from the substantial literature suggesting that an independent observer can "substitute" for a respondent, especially when both are assessing the same material. Substitutions commonly occur when data is missing, as in a clinical trial, where there is a need to complete a statistical analysis, but also when a surrogate response is needed because a person is not able to communicate, is considered too young to comprehend, or is intellectually comprised. There is, however, an extensive literature that has established the conditions under which effective substitution is likely or not likely to occur (e.g., Sneeuw et al. 2002; von Essen 2004). Several important observations come from this literature. First, it has been shown that proxies adopt different evaluative criteria than respondents when judging quality-of-life of HRQOL issues, but are more accurate when reporting on objective indicators (Todorov and Kirchner 2000; Epstein et al. 1989). Second, if the assessment task involves health perceptions, then the independent observer is more likely to misreport a respondent's judgment. This suggests that while substitution may be possible, the nature and quality of this substitution will be very much dependent on contextual factors, and this has implications concerning what to expect when Comparisons A and B are examined.

Finally, there is the issue alluded to earlier (Fuhrer 2000) that the type of injury or event a person experienced may also contribute to variation in functional and HRQOL responding. Thus, I should also consider the nature of the disabling event when assessing the results of our comparisons, although the extent to which I will be able to do this will depend on the number and type of available studies.

Each of these assumptions or issues underscore that I will be dealing with a fairly complex comparative environment, so I may end with more of a sketch than a definitive statement supporting a particular hypothesis. Still, I will learn from these analyses that the ultimate goals of this effort are achievable, and this alone provides support for the hypothesis that a properly qualified function statement can be equivalent in meaning to a quality-of-life or HRQOL item, even if the data necessary to prove this may not currently be as available as I would like.

My next task is to determine whether there are examples of purely descriptive functional assessments that could be assessed by an independent observer. It turns out there are many such assessments (Mc Dowell and Newell 1996), with some addressing functional status independent of disease and others reflecting disease severity. Katz's Activities of Daily Living Index (Katz and Apkom 1976) is an example of a function assessment by an independent observer that can be applied to all persons across a wide range of clinical conditions. The same is true for the New York Heart Association (NYHA) Functional Classification (New York Heart Association 1964), which rates the impact of a disease on performance and functioning.

An example of a descriptive functional assessment is the Functional Independence Measure or FIM™ (Hamilton et al. 1987). The FIM™ is a popular measure of activities of daily living, regularly administered in rehabilitation settings by a clinician. There is also a growing literature that includes self-reported versions of the FIM™ (Jensen et al. 2005) and even examples where a preference or value system has been generated for the various FIM™ components (e.g., Stineman et al. 2003). It has also been used with studies that include quality-of-life or HRQOL assessments as a covariate. One weakness of the FIM™ is that it can suffer from ceiling effects, especially for persons with mild disabilities. Thus, the FIM™ is a good example of the type of assessment that can be applied to all three comparison items in Table 9.3. This, however, is not possible for many other functional assessments, which will become evident from the following analyses.

Table 9.4 consists of a series of studies, each of which includes a functional assessment performed by an independent observer, classified in terms of the type of quality-of-life or HRQOL assessment. The table includes studies that only provided correlations of the data, while Table 9.5 summarizes studies that included regression analyses. Table 9.4 revealed that statistically significant correlations ranged from 0.26 to 0.70. If I inspect the correlations between the FIM™ and the eight domains of the SF-36, then the only consistently statistically significant correlation, as might be expected, is with the Physical Functioning domain. The magnitude of the correlation coefficients remained modest independent of the type of HRQOL assessment. The data in the table also reveal that there was no difference between persons who sustained a brain injury (e.g., stroke, multiple sclerosis, traumatic brain injury, etc.) and those who did not (e.g., angina, heart attacks, spinal cord injury, and so on). There were not enough studies of spinal cord injury patients to report them as a separate analysis.

Table 9.5 summarizes the regression analyses found when various indicators, including measures of functional status, were used to predict variation in quality-of-life or HRQOL. Inspecting the table reveals that with the exception of two, all the reported studies dealt with persons with brain injury. It was evident that only some of the FIM™ components contributed to variance in quality-of-life scores, and that this did not vary as the type of assessment changed. I interpreted the table as not supporting the hypothesis that functional status, when assessed by an independent observer, can be used as a surrogate for a quality-of-life or HRQOL assessment. It should also be noted that similar conclusions have been previously reported (Béthoux et al. 1999; May and Warren 2002; Reuben et al. 1995), although they were based on reviews of fewer studies.

My next challenge is to determine if self-reports of both functional status and quality-of-life or HRQOL increase the probability that the two measures will be correlated, and whether the magnitude of this correlation would be higher than expected for an independent observer's assessment of functional status. Also of interest is whether self-reported functional status would be a significant contributor to the variance in quality-of-life or HRQOL outcomes. Examples of self-reported functional status assessments include the self-reported version of the FIM™ (SR-FIM™), the Frenchay Activity Index (FAI), and the Duke Activity Status Index (DASI). A self-report is defined as consisting of data collected by interview or in response to a self-administered assessment. Inspecting the data displayed in Table 9.6, however, does not suggest that self-reports of functional status have an incremental effect on the association with some measure of quality-of-life or HRQOL. Thus, the magnitude of the correlations overlapped with those reported in Table 9.4, and self-reported functional status, per se, did not increase the chances that they would predict variation of quality-of-life or HRQOL scores (only one of six regressions was significant).

The report by Wolfe et al. (2005) is interesting because by comparing a function-based VAS to a number of other VASs, they eliminated the potential confounding that a difference in the type of a quality-of-life or HRQOL assessment could have on the observed correlation. They found that function based VAS was most highly correlated with a VAS for pain (0.774) and progressively less so for the VAS rating for disease severity ($r=0.759$), fatigue ($r=0.672$), sleep ($r=0.605$), and finally the VAS for QOL ($r=0.515$). These data suggest that it was not the type of quality-of-life or HRQOL assessment being compared that contributed to the magnitude of

Table 9.4 Association between an independent observer's function assessment and a quality-of-life or HRQOL assessment (comparison A): correlational analyses*

Type of quality-of-life/ HRQOL assessment	References Patient group (N)	Functional status assessment	Quality-of-life or HRQOL assessment	Results
Multi-item	Webb et al. (1995) Traumatic brain injury (116)	FIM™	Life Satisfaction Index (LSI-A)	$r=0.343$; FIM™/LSI-A[a]
	Indredavik et al. (1998) Stroke (62)	BI	NHP	$r=0.48$; BI/NHP[a], for Stroke Unit
	Sharrack et al. (1999) Multiple sclerosis (50)	FIM™s	General health (GH) from the SF-36	$r=0.69$; FIM™/VAS[a]
	Bezner and Hunter (2001) Traumatic brain injury (34)	FIM™s	Perceived wellness survey (PWS)	$r=0.101$; FIM™/PWS
	Snead and Davis (2002) Traumatic brain injury (40)	FIM™s[b]	SF-36	FIM™ not significantly correlated with all 8 domains of SF-36
	Romberg et al. (2005) Multiple sclerosis (98)	FIM™s	MSQOL-54[c]	$r=0.12$; FIM™/MSQOL-54 (Baseline) $r=0.10$; FIM™/MSQOL-54 (6 months)
	Aprile et al. (2006) Stroke (68)	BI FIM™s	SF-36	$r=0.60$ BI/PF[a]; $r=0.60$ BI/RE[a]; $r=0.30$ BI/MH[a]; $r=0.50$ FIM™/PF[a]; $r=0.30$ FIM™/RF[a]; $r=0.70$ FIM™/RE
	Rabini et al. (2007) Low back pain (108)	BI	SF-36	$r=0.20$ BI/PF[a]; $r=0.20$ BI/RP[a]
	Madden et al. (2006) Stroke (116)	FIM™s	SF-36	$r=0.184$; Total FIM™/PF domain[a] $r=0.051$; total FIM™/PCS
Weighted multi-item	May and Warren (2002) Spinal cord injury (98)	FIM™s	Ferrans and Powers quality of life-spinal cord injury	$r=0.202$; FIM™/QLI-spinal cord injury
Derived preferences	Tsevat et al. (1991) MI survivors (80)	Karnofsky Index	TTO	$r=0.27$; Karnofsky Index/TTO[a]
		NYHA	TTO	$r=-0.32$; NYHA/TTO[a]
		SAS	TTO	$r=-0.03$; SAS/TTO
	Tijhuis et al. (2000) Rheumatoid arthritis (184)	Walk Test	TTO	$r=-0.07$; walk test/TTO
	Havranek et al. (2004) Heart failure (153)	NYHA	TTO	$r=-0.26$; NYHA/TTO[a]
	Pickard et al. (2005) Stroke (98)	BI and modified rankin score	EQ-5D	$r=0.57$; BI-mod Rankin/EQ-Index[a]
		BI and Modified Rankin Score	SF-6D	$r=0.32$; BI-mod Rankin/SF-6D[a]
		BI and modified rankin score	HUI2	$r=0.47$; BI-mod Rankin/HUI2[a]
		BI and modified rankin score	HUI3	$r=0.41$; BI-mod Rankin/HUI3[a]
		BI and modified rankin score	HUI3	$r=0.41$; BI-mod Rankin/HUI3[a]

(continued)

Table 9.4 (continued)

Type of quality-of-life/HRQOL assessment	References Patient group (N)	Functional status assessment	Quality-of-life or HRQOL assessment	Results
Global	Tsevat et al. (1991) MI Survivors (80)	Karnofsky Index	10 cm VAS	$r=0.44$; Karnofsky Index/VAS[a]
		NYHA	10 cm VAS	$r=-0.33$; NYHA/VAS[a]
		SAS	10 cm VAS	$r=-0.34$; SAS/VAS[a]
		Karnofsky Index	VRS	$r=0.46$; Karnofsky Index/VRS[a]
		NYHA	VRS	$r=-0.37$; NYHA/VRS[a]
		SAS	VRS	$r=-0.35$; SAS/VRS[a]
	Granger et al. (1993) Stroke (19)	FIM™s	Single item: Life satisfaction	FIM™ (Total) not significantly correlated with single item
	Indredavik et al. (1998) Stroke (62)	BI	VAS	$r=0.46$; BI/VAS[a] for Stroke Unit
	Kim et al. (1999) Stroke (50)	FIM™s	SAHS	$r=0.48$; FIM™s/SAHS[a]
	Sharrack et al. (1999) Multiple sclerosis (50)	FIM™s	EuroQol VAS	$r=0.69$; FIM™/VAS[a]
	Pickard et al. (2005) Stroke (98)	BI and modified rankin score	EQ-VAS	$r=0.22$; BI-mod Rankin/EQ-VAS
	Witham et al. (2008) Elderly day hospital patients (75)	BI	PGI	$r=0.09$; BI/PGI

* Glossary: *BI* Barthel Index; *HUI* Health Utility Incex; *NHP* Nottingham health profile; *NYHA* New York Heart Association classification system; *PGI* Patient Generated Index; *SAS* Specific Activity Scale; *TTO* time trade-off; *VAS* Visual Analog Scale; *VRS* Verbal Rating Scale; *PF* the physical functioning domain of the SF-36; *RP* role physical; *RE* role emotional domain; *MH* mental health domain; *PCS* physical component summary; *MCS* mental component summary

[a]A statistically significant level was set at $P<0.05$ and applied to all studies

[b]Twelve additional items were added to the FIM™ assessment

[c]The MSQOL-54 consists of the SF-36 plus 18 items specific for multiple sclerosis patients

Table 9.5 Association between an independent observer's functional assessment and a quality-of-life or HRQOL assessment (comparison A): Studies that included regression analyses

Type of quality-of-life/ HRQOL assessment	References Patient group (N)	Functional status assessment	Quality-of-life or HRQOL assessment	Results
Multi-item	Warren et al. (1996) Traumatic brain injury (137)	FIM™	Life Satisfaction Index (LSI-A)	Bowel and memory from the FIM™, and family satisfaction, job status, marital status, and self-blame account for 46% of variance in life satisfaction
	Warren et al. (1996) Spinal cord injury (38)	FIM™	Life Satisfaction Index (LSI-A)	Closeness to family, family activities, and self-blame account for 58% of variance in life satisfaction but FIM™ does not
	Cromes et al. (2002) Burn patients (110)	FIM™	Burn-Specific Health Scale	FIM™ did not predict HRQOL scores at baseline, 2 or 6 months follow-up
	Chen et al. (2008) Spinal cord injury (184)	FIM™	SWSL	FIM™ domain of self-care and total motor predicted life satisfaction at initial assessment
Weighted multi-item	King (1996) Stroke (86)	FIM™	QLI-Stroke	Log FIM™ ($P<0.031$), depression, and social support are significant predictors of HRQOL, accounting for 38% of the variance
	Kim et al. (1999) Stroke (50)	FIM™	QLI-Stroke	$r=0.46$; FIM™'s/QLI-Stroke was significant correlation but not a significant predictor of QLI-Stroke. Depression, marital status, social support, and instrumental activities of daily living accounted for 60.1% of QLI-Stroke variance
	Robinson-Smith et al. (2000) Stroke (63)	FIM™	QLI-stroke	At discharge, FIM™ was not a significant predictor of QLI-Stroke, but coping was. At 6 months, FIM™ accounts for 20.2% and coping with 47.&% of variance in QLI-Stroke scores
Derived preferences	King et al. (2005)[a] Cerebral aneurysms (176)	Glasgow outcome score	SG; TTO; WTP	Age, sex, race, education, history of hemorrhage, number and treatment of aneurysm, GOS, did not predict SG, TTO, and WTP scores. No correlations reported
	King et al. (2005)[a] Cerebral aneurysms (176)	Rankin score	SG; TTO; WTP	Age, sex, race, education, history of hemorrhage, number and treatment of aneurysm. Rankin Score, did not predict SG, TTO and WTP scores. No correlations reported
	Kaambwa et al. (2008) Intermediate care (2,253)	BI	EQ-5D	BI at admission predicted change in EQ-5D over course of stay in intermediate care
Global	Granger et al. (1993) Stroke (19)	FIM™	General life satisfaction item	A multiple step-wise regression revealed that depression (54%), dressing lower body from the FIM™ (21%), visual ability (8%), social cognition (composite of 3 domains from the FIM™) (8%), and Hostility (4%) account for 95% of variation in general life satisfaction
	King et al. (2005) Cerebral aneurysms (176)	Glasgow outcome score	VAS	Age, sex, race, education, history of hemorrhage, number and treatment of aneurysm, GOS, did not predict VAS score
	King et al. (2005) Cerebral aneurysms (176)	Rankin score	VAS	Rankin Score and number of comorbid diseases accounted for 11% of variation in VAS score[b]

FIM™ Functional Independence Measure; *GOS* Glasgow outcome score; *QLI-Stroke* Ferrans and Powers Quality-of-life – (Ferrans and Powers 1985); *SG* standard gamble; *SWLS* satisfaction with life scale (Diener et al. 1985); *TTO* time trade-off; *VAS* Visual Analog Scale; *VRS* Verbal Rating Scale; *WTP* willingness to pay

[a] The results were the same for each analysis and were presented together to minimize the size of the table. However, each study will be counted separately when summarizing the data, for a total of 14 comparisons

[b] A statistically significant level was set at $P<0.01$ and applied to all studies

the association, but rather the type of information being communicated by the rating task. Thus, if a VAS rating of pain is informationally concrete, then it would be expected to be more highly correlated than a similar comparison to a more abstract concept, as when a VAS is assessed for quality-of-life or HRQOL. This study confirms again the role that an informational hierarchy plays in qualitative judgments.

The last part of the analysis plan examined whether a value system has been attached to a function statement, such as the FIM™. This, of course, would qualify the function statement and this should increase the chance that it would be correlated with a quality-of-life or HRQOL assessment. There are actually very few examples of such qualification, but two are available and they provide a first approximation of the desired assessment. The first one was reported by Stineman and her colleagues (1998, 2003). Ideally it would, although there are actually very few published examples of such an assessment. Two examples will be presented here, starting with the work of Stineman et al. (1998, 2003).

Stineman et al. (1998, 2003) report developing a game called the Features-Resources Trade-off game that can be used to calculate utility weights for the various components of the FIM™. This game was used by Stineman et al. (1998, 2003) to compare the weighting system generated by persons with disabilities and rehabilitation clinicians. The game can be played with two or more people, and the task for the players is to come up with a consensus as to what trade-offs between the various components of the FIM™ are acceptable. It was found that both persons with disabilities and clinicians valued communication and cognitive independence more than physical activity independence. This, the authors suggest, requires that the unweighted FIM™ should be adjusted to take these observations into account. For example, a person who is severely physically disabled but cognitively intact (e.g., a spinal cord patient) will have an unweighted FIM™ total score which will underestimate their functional status if it was not qualified (e.g., different physical activities weighted as to importance), while the person who is cognitively impaired and has lower levels of independence may have an unweighted total FIM™ score which overestimates their functional status if it was not qualified. In either case, weighting impacted the interpretation of the total FIM™ score, as predicted.

The Stineman et al. (1998, 2003) report offers a means of generating weights for different groups, and the resultant weighted FIM™ total score could be compared to a value-weighted quality-of-life or HRQOL assessment (Comparison C, Table 9.3). Unfortunately, these data are not yet available. What is available is another example, which combines value statements with a functional assessment and compares it to a derived measure of preference, the TTO. In these studies, Hewlett et al. (2001, 2002) first demonstrated that patients, clinicians, and members of the general public differed in terms of the importance they each placed on having various disabilities. In their second study, they reported developing an eight-item assessment based on the Health Assessment Questionnaire (HAQ) which asks people to rate the importance of having various abilities (Personal Impact Health Assessment Questionnaire; PI-HAQ). Their emphasis on the rating of abilities, rather than disabilities, suggests that they recognized that the term dis-ability already contains a value statement. At first, they considered weighting items in the HAQ in terms of upset rather than importance. They state:

> In developing a measure of personal impact of disability the aim was to create a questionnaire using a language for values with which patients could identify. Patients understood the concept of values and used both positive and negative phrases…Although patient preference was for the negatively phrased 'upset' version, it was decided to use the more positive 'importance' version to capture values. This was decided not only because the 'importance' version had a stronger validation (the 'upset' version was abnormally distributed, associated with personality, and did not correlate with the gold standard measure) but also for conceptual reasons. Conceptually, importance and upset are not necessarily opposite ends of the same continuum and may measure different things, while 'upset' may itself reflect an emotional impact of disability, rather than the value of activities.
>
> (Hewlett et al. 2002; p. 991).

A typical assessment task might be stated as follows; "Rate, on a four point scale, how important it is for you to 'Carry out the tasks involved in dressing and grooming, including tying shoelaces, and so on?'" This item, the other items, and the instructions for filling out the questionnaire did not take into consideration of being in a particular state, but instead seem to be asking about how important it is to be able to carry out a particular task. This suggests that the PI-HAQ is actually asking about a person's capabilities and the importance of these capabilities to that person.

As part of their effort to establish the criterion validity of their assessment, the authors also collected data using the TTO (Torrance 1986). When they correlated PI-HAQ scores with TTO values, they found an $r=0.333; P<0.01$. The magnitude of this correlation, of course, is well within the range reported in Tables 9.4 and 9.6, suggesting that comparing two value-based assessments that differ in format did not hold much of an advantage over correlating a descriptive functional assessment and a value-based quality-of-life or HRQOL assessment (e.g., Table 9.4). They do report, however, that respondents raised concerns about speculating about having to consider trading off when their current treatment regimen was able to keep their disease at bay. Thus, some of the respondents did not feel that the trading off process referred to them, making the data more of an academic exercise than reflecting real life decisions by these respondents. This, of course, raises questions about what is being compared when a derived preference (e.g., the SG, TTO, or WTP) is compared to a more direct importance rating, as was true for the PI-HAQ.

Table 9.6 Association between a self-reported functional assessment and a quality-of-life or HRQOL assessment[a]

Type of quality-of-life/ HRQOL assessment	References Patient group (N)	Functional status assessment	Quality-of-life or HRQOL assessment	Results
Multi-item	Fuhrer et al. (1992) Spinal cord injury (140)	SR-FIM™	Life Satisfaction Index (LSI-A)	LSI-A not significantly correlated with SR-FIM™
	Anderson et al. (1996) Stroke (90)	AAP	General health domain from the SF-36	$r=0.47$; General health/AAP[b]
	Anderson and Vogel (2003) Spinal Cord Injury (216)	SR-FIM™	Satisfaction with Life Scale	SR-FIM™ not a predictor of life satisfaction domains
	Jensen et al. (2005) Neuromuscular disorders (143)	SR-FIM™	SAHS item in SF-36	$r=0.40$; SAHS/SR-FIM™[b]
	Riazi et al. (2006) Ataxia (56)	Barthel Index (SR)	SF-36	$r=0.55$; PF²/BI(SR)?
Weighted Multi-item	Kim et al. (1999) Stroke (50)	FAI	QLI-Stroke	$r=0.41$ QLI-stroke/FAI[b]
Derived preferences	Nichol et al. (1996) Angina (41)	DASI	DSU	$r=0.28$; DASI/ derived DSU utilities
		Self-report-based CCS Angina Scale	Derived utility	$r=-0.35$; CCS blocks walked/derived utility[b]
		Self-report-based CCS Angina Scale	Derived utility	$r=-0.25$; CCS stairs climbed/derived utility
	Tijhuis et al. (2000) Rheumatoid arthritis (184)	HAQ	VAS; TTO	$r=-0.29$; HAQ/TTO[b]
	Havranek et al. (2004) Heart failure (153)	DASI	TTO	$r=0.17$; DASI/TTO
	Sama and Matchar (2004) Stroke (329)	Barthel Index (SR)	TTO	$r=0.24$; BI/TTO[b]
	King et al. (2005) Cerebral aneurysms (176)	Barthel Index (SR)	SG; TTO; WTP	Age, sex, race, education, history of hemorrhage, number and treatment of aneurysm, Barthel Index, did not predict SG, TTO, and WTP scores. Three regressions and no correlations reported
	Riazi et al. (2006) Ataxia (58)	Barthel Index (SR)	EQ-5D Thermometer health state	$r=0.08$; T/BI(SR) $r=0.70$; HS/BI(SR)?
Global	Dorman et al. (1997) Stroke (92)	FAI	VAS from the EuroQOL	FAI, a depression measure, and VAS-pain are significant predictors of VAS-EuroQOL
	Samsa and Matchar (2004) Stroke (329)	Barthel Index (SR)	Verbal global assessment	$r=0.21$; BI/VGA[b]
	King et al. (2005) Cerebral aneurysms (176)	Barthel Index (SR)	VAS	Age, sex, race, education, history of hemorrhage, number and treatment of aneurysm, Barthel Index, did not predict VAS scores. No correlations reported

(continued)

Table 9.6 (continued)

Type of quality-of-life/HRQOL assessment	References / Patient group (N)	Functional status assessment	Quality-of-life or HRQOL assessment	Results
	Wolfe et al. (2005) Rheumatoid arthritis (394)	HAQ VAS-function	VAS-QOL	$r=0.41$; HAQ/VAS-QOL[b]; $r=0.52$: VAS-function/VAS-QOL[b]
	Witham et al. (2008) Day hospital patients (75)	FLP	PGI	$r=0.51$; FLP/PGI[b]

BI Barthel Index; *CCS* Canadian Cardiovascular Society Angina Scale; *DSU* disease-specific utilities; *FAI* Frenchay Activities Index; *PF* physical function domain of the SF-36; *SR* self-report; *SR-FIM™ S'* self-report functional independence measure; *TTO* Time trade-off; *VRS* Verbal Rating Scale; *VAS* Visual Analog Scale; *SG* standard gamble; *WTP* willingness to pay

Self-report Assessments: *AAP* adelaide activities profile (Clark and Bond 1995); *CCS* Canadian Cardiovascular Society Angina Scale (Self-report: Nichol et al. 1996); *DASI* Duke Activity Status Index (Hlatky et al. 1989); *QLI-Stroke* Ferrans and Powers quality of life-stroke (Ferrans 1990); *FAI* Frenchay Activities Index (Segal and Schall 1994); *FLP* functional limitation profile (Pollard and Johnston 2001); *HAQ* health assessment questionnaire; *NHP* Nottingham health profile (Hunt et al. 1985a, b); *PGI* Patient Generated Index (Ruta et al. 1994)

VAS was most highly correlated with a VAS for pain ($r=0.774$), and progressively

[a]A self-report was defined as consisting of data collected by interview or in response to a self-administered questionnaire. It is assumed that the data from an interview represent the respondent's responses, and that the interviewer has not interfered in providing this report. It is only if this is true that it can be classified as self-reports. However, it is often not clear precisely how the data were collected in an individual study, so this remains a risk to the above analysis

A statistically significant level was set at $P<0.01$ and applied to all studies

[b]Probability level of correlation not reported

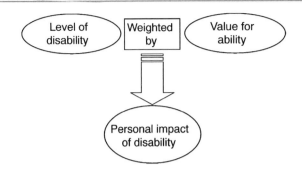

Fig. 9.8 A Hybrid Construct Model of Disability Impact. Reproduced from Hewlett et al. (2001). Reproduced by permission of publisher.

Hewlett et al. (2001) present a model of their approach (Fig. 9.8) that raises an interesting question: is the PI-HAQ a quality-of-life assessment, or more generally a quality indicator? As depicted in Fig. 9.8, the PI-HAQ is a hybrid construct consisting of a series of descriptive statements weighted by some value system. Referring to the classification system summarized in Table 8.5, their model would be an example of a Type III assessment (descriptive statements metaphorically arithmetized by a value system). The PI-HAQ, as an example of a quality indicator, reinforces the point (see Note #22) that a number of specific cognitive processes (e.g., description, establishing judgment consequences, and valuating outcomes) are what makes a quality judgment unique, not a particular display of content or domains of interest (e.g., Cummins 1996) as has been suggested by various quality-of-life or HRQOL researchers. Thus, the PI-HAQ, based on its cognitive characteristics, should be considered a quality indicator, just as an optimally designed quality-of-life or HRQOL assessment would be.

To summarize, adding a value-weighting system to a function statement can make a difference in the meaning of that function statement, and this has clinical and interpretative significance. It may even lead to changes in clinical decision making. This is most clear in the Stineman et al. (2003) study, where applying the weights generated from the Trade-off Game to components of the FIM™ resulted in a reinterpretation of the functional status of a person with disabilities. Less clear was what would be an optimal test of Comparison C (Table 9.3), especially since the Hewlett et al. (2002) study did not demonstrate any increased correlation when the two value-based assessments were compared. Complicating the Hewlett et al. (2002) study was the comparison of a direct assessment of importance vs. a derived preference measure (TTO). Still, it will be worthwhile to continue to study how the addition or subtraction of information to or from a basic descriptive function statement would affect the meaning projected by items in a quality-of-life or HRQOL assessment. Now all we have are published studies with very few examples of studies that deliberately manipulate the meaning being communicated.

Finally, I want to review a study by Cieza and Stucki (2005) that takes a different approach to relating functional and HRQOL assessments. The methodology they use takes advantage of the ICF classification system and asks to what extent individual HRQOL assessments contain components of the ICF. Presumably, if there was significant overlap, this would confirm that the HRQOL assessments contained many function statements. As a test of this hypothesis, Cieza and Stucki (2005) mapped items from the SF-36, the Nottingham Health Profile (NHP; Hunt et al. 1985a, b), the Spitzer Quality-of-life Index (Spitzer et al. 1981), the WHOQOL-BREF (WHOQOL Group, 1996), the WHODASH (WHO 2000), and the EQ-5D to components of the ICF. The 146 items on the assessments were found to address 226 concepts. These concepts were linked to 91 different ICF categories; only 14 of the concepts addressed environmental issues, a topic relevant for any assessment of persons with disabilities, while 12 concepts were not linked to the ICF at all. The authors claim that the degree of overlap suggests that the ICF can serve as a useful framework to compare HRQOL assessments. They also suggest that reference to the ICF can help in the process of selecting items for a new assessment. Cieza and his colleagues have published a series of papers repeating the basic comparison in this study, but now for a variety of clinical conditions ranging from head and neck cancer (Tschiesner et al. 2008) to stroke (Geyh et al. 2007) and osteoarthritis of the hand (Stamm et al. 2006). Overall, these studies support the notion that while HRQOL may vary in degree, there is a fair amount of overlap between the content in the items in these assessments and the classification which occurs with the ICF.

2.2 Summary

My major interest in entering the rehabilitation and disability literature was to gain some insight into how the term "function" was being used in an applied setting. My overall objective was to examine the meaning and use of the term from a system-analytic (e.g., social functionalism) and individual data perspective (that is, from a macro and micro perspective), but also to examine its use as part of a hybrid construct where qualified function statements are metaphorically arithmetized (for example, the term "dis-able"). However, I could have also approached this task in a number of other ways, especially when dealing with persons with disabilities. For example, would the function statements be different and have different meanings if they were defined via a participatory research approach (i.e., where the subjects participates in establishing the research project) as opposed to the traditional investigator-dependent approach used in most of the studies reported here? It is also interesting to speculate about whether a quality-of-life or HRQOL assessment would have different content and meanings if it relied on an analysis on the discourse of disability. I have already raised the possibility that proxies may only provide a limited perspective on the functional status of a person with disabilities. With this background, I can summarize my discussion on the use of the term function and functionalism in a rehabilitation setting.

It seems clear that the term function in the disability and rehabilitation literature is being used as part of a person–environment interaction or as part of a socioecological model. This contrasts with a subjective quality-of-life assessment, which is usually intrapersonal, where a person makes a judgment relative to him or herself. For example, I have discussed matching models of quality-of-life or HRQOL (Table 2.5), where a person's quality-of-life was defined by the discrepancy between their hopes and aspirations and what they felt they were able to achieve, both of which were assessed by the respondent. The socioecological model refers to what is considered an objective quality-of-life or HRQOL model, where external factors such as the crime rate or housing conditions are assumed to influence the qualitative state of the person. One of the difficulties of the socioecological model is that it does not find an easy place for the more micro issues related to linguistic, cognitive, and psychological processes.

Yet when a linguistic perspective was applied, it became apparent that the term rehabilitation was often not being used in its literal sense of restoring someone to prior functioning, but rather to figuratively rehabilitate a person, as when they are accommodated to their functional limitations and impairments. This suggested that expansion of the meaning of the term has communicative and practical consequences. For example, the difference between restoration and accommodation often becomes a source of conflict between clinicians and patients and is usually experienced by the patient as an exaggeration of the benefits of a procedure without sufficient description of its liabilities. Thus, breast reconstruction is sometimes touted for its ability to restore a woman's sense of self, but less attention may be placed on the risk of creating unease because of the reconstructive material implanted in the woman's body. When this becomes obvious, the woman will accommodate to the fact that she has traded off a benefit by assuming a liability, which is not what she may have originally expected restoration to provide. There are many examples of this clinical situation, and their existence can significantly impact the doctor-patient relationship.

An alternative approach to defining rehabilitation was suggested by Fuhrer (1987), who proposed that it be defined in terms of the services provided. Thus, rehabilitation "… may be defined as an individualized array of coordinated services aimed primarily at forestalling, minimizing, or reversing the occurrence of handicap, disability, and impairment" (Fuhrer 1987; p. 2). The services provided would presumably operationalize what is meant by "rehabilitation," which at its extreme could actually differ for each person. Fuhrer (1987) goes on to distinguish rehabilitation and habilitation, where rehabilitation refers to restoring lost ability, while habilitation refers to acquiring abilities not previously possessed. Thus, Furhrer solves the semantic and linguistic problems created by the term rehabilitation by essentially redefining the term, now as a form of habitation.

These changes in language and semantics are also reflected in the continuing debate surrounding what should be considered an appropriate model of disability (e.g., the ICIDH, the ICF, and so on). The ICF, for example, has challenged the notion that a person with disabilities be viewed as abnormal, suggesting that disability is not a state change, but rather a place on a continuum, a continuum that any person can be on. This reconceptualization has been quite helpful in altering the stigmatization of persons with disabilities.

Also true is that the rehabilitation and disability literature have historically used the term "function" to refer to both a person's capabilities as well as actual performance. This reflects the fact that the term bridges the gap between objective (e.g., muscle strength measured by a dynometer) and subjective measures (e.g., asking a person to report their performance). This contrasts with the quality-of-life or HRQOL literature that focuses exclusively on performance-based measures of functions. Does this difference make a difference? It clearly could, especially if a study was limited to one or the other measure. For example, if a person who has limited muscle strength (a capability measure) is able to find a way to perform a task, then knowing the outcome of only one of these measures may result in imprecise, or perhaps better, incomplete inferences.

The debate about whether to use capacity or performance measures of functions also has affected how quality-of-life or HRQOL is modeled. For Wilson and Cleary (1995), functional status is assumed to influence general health perceptions (a subjective measure), but it is not clear whether a measure of functional capacity or performance, or both, is meant to influence health perceptions, and how this would occur. As with the Wilson and Cleary (1995) model, the various disability models gravitate not just from the objective to the subjective, but also from the biological to the social to the psychological, as well as from the concrete to the abstract. Thus, these models are best seen as summary statements that highlight issues to be considered rather than relationships clearly established. The alternative is to consider models that adopt a developmental perspective (e.g., Halfon and Hochstein 2002) or use statistical equation modeling (SEM) to create some order among these indicators.[23]

Verbrugge and Jette (1994; Fig. 2; p. 4) describe the main pathway of the disablement process as involving pathology, impairments, functional limitations, and disability. Functional limitations are meant to refer to restrictions in a person's capabilities, while disability refers to difficulties a person has performing certain activities. Both refer to behaviors, but one refers to what people can do, a capacity indicator, while the other refers to what people do, a performance indicator. What is particularly confusing here is that if I go back and reexamine the etymology of the term "ability," I discover that it refers to the capabilities of a person, while the term disability refers to a disturbance in a person's abilities, as when a physical or mental condition limits a person's ability to move, sense, or behave. This suggests that the term disability should refer to the restrictions in a person's abilities, not difficulties they may have in performing certain activities. So it seems that the term disability, as used by Verbrugge and Jette (1994), represents an expansion of its meaning.[24]

I was only partially successful in my attempt to demonstrate that it was possible to start with a descriptive functional statement and add information to it so that the resultant statement would more closely resemble a quality-of-life or HRQOL statement. If I had completely succeeded, then I would have been able to state that a respondent had processed information in such a way (e.g., establish the consequences of events, and place a value on these events) so as to provide the content that made the resultant comparisons possible. This would have also permitted me to claim that a collection of descriptive function statements, by themselves, does not constitute a quality-of-life or HRQOL assessment.

Still, I did learn that it is possible to alter the meaning of a descriptive function statement by adding value statements or weightings to it. The game that Stineman et al. (1998, 2003) developed provided the weightings that when added to FIM™ statements resulted in a reinterpretation of the person's functional status. I also learned that the type of assessment being compared may not play a decisive role in determining the magnitude of the observed associations. Thus, the Hewlett et al. (2002) study demonstrated that by comparing two valuations-based assessments was not sufficient to dramatically affect the magnitude of the association between these assessments. For example, in the Hewlett et al. (2001) study, importance-based and derived preference-based value systems were compared, but still resulted in a relatively modest correlation ($r=0.333$; $P<0.01$).

The Wolfe et al. (2005) study was also helpful since it estimated the degree of association between assessments that were based on the same exact assessment format (e.g., both were VAS's). They found that the magnitude of the correlations was related to where along the informational hierarchy the two VAS's being compared fell and that using the same format did not significantly alter the magnitude of the correlations. Thus, the VAS for function was more highly correlated with VAS's for pain or disease severity than the VAS for quality-of-life, with the correlation between the VAS for function and the VAS for quality-of-life ($r=0.515$; $P<0.01$) within the range observed with comparisons involving more diverse assessment formats (Tables 9.4 and 9.6).

The Wolfe et al. (2005) study stitches a cognitive thread, since it demonstrated how cognitive processes play a role in what was empirically observed. In this case, the cognitive process dealt with the presence of an information hierarchy that clearly contributed to the observed correlations. Thus, the more concrete and physical the scaling, the higher the VAS was correlated with a VAS for function. Just to extend the generality here, it is interesting to note that Sullivan et al. (2000) found that the four-item general health domain from the SF-20 and the Cantril ladder correlated at the 0.52 level, suggesting that when two relatively abstract constructs were related there was a reasonable level of association. But what does the presence of such cognitive determinants do to my stated goal of transforming a function statement into a quality-of-life or HRQOL statement? It should not prevent it, since the methods proposed simulate the cognitive processes involved in generating a optimal quality-of-life or HRQOL assessment, and would not be content-dependent. These cognitive processes include establishing the consequences of events and valuation of these events.

3 Summary of Chapter

The question this chapter addressed was whether a functional assessment should be included as a domain in a quality-of-life or HRQOL assessment. A wide range of content was covered to provide the appropriate background to answer this question, as well as to prepare for subsequent discussions that will involve different aspects of the terms of "function" and "functionalism." The answer that emerged is that a purely descriptive functional assessment was not consistently correlated with a quality-of-life or HRQOL assessment (Tables 9.4–9.6), and that when it was significantly correlated, these correlations accounted for less than 20% of the variance of a quality-of-life or HRQOL assessment. In addition, data from structural equation modeling (e.g., Sullivan et al. 2000) demonstrated that linear correlations between functional and HRQOL domains were also quite small. These data suggest, therefore, that a functional domain consisting of purely descriptive items, as is found in many functional assessments, should not be considered a quality-of-life or HRQOL assessment.

An alternative approach involves performing a cognitive analysis of the items in the assessment. When this was done, as was for the relationship between fatigue, depression, and HRQOL (Chap. 7, p. 211), then it was possible to determine that a particular group of items (e.g., reduced motivation vs. physical fatigue items) either were or were not part of an informational hierarchy. Those items involved in the hierarchy (e.g., mental fatigue) were less likely to be a significant predictor of variance in HRQOL when a more abstract term, such as depression, was part of the analysis. Data of this sort, however, are not as yet available for analyzing functional assessments.

An alternative analytical tactic suggests that the correlation and the variance accounted for between a functional and quality-of-life or HRQOL assessment would be increased by adding subjective and value-related information to the functional items. The impact of qualifying the functional statements was to be evaluated by reviewing the published literature. Probably the clearest example of the application of this process was provided by Stineman et al. (1998, 2003), who demonstrated that adding weights to an independently based observer assessment of functional status altered the qualitative interpretation of the functional assessments. In addition, structural equation modeling studies and covariance analyses using LISERL demonstrated that adding modulating variables (e.g., anxiety or depression) to a regression of quality-of-life or HRQOL significantly increased the predictiveness of the model (Sullivan et al. 2000). From this it can be concluded that function statements as part of a domain can be qualitatively enhanced (e.g., the consequences of being in a particular functional state were made explicit, and a weighting or valuation system was added), and in this way the cognitive equivalence of such function statements to a quality-of-life or HRQOL assessment increased.

Much of what I have discussed so far has dealt with items on questionnaires, but this is only one source of information about functioning that is informative about life quality. I could learn a lot about what is meant by functioning by examining a person's diaries, conversations, and various forms of written expression (e.g., novels, poetry). This remains a task to be accomplished at some future date.

I also reviewed the history of how the term function evolved in the field of mathematics. For example, at first it was recognized that our ability to add, subtract, and multiply was based on the physical linking of objects to numbers. This mapping leads to the generation of equations that provided general mapping statements that could be applied across situations. But it was for Dirichlet (1805–1859) to suggest the first formal definition of the term function in mathematics when he suggested it be used to express what is implicit in an equation, which is the correspondence between an independent and dependent variable. Today, the term function is used in mathematics and science to describe relationships, as in "a functional relationship."

This progression from physical mapping to an abstract concept, implicit in the history of mathematics, is also what I would expect if the concept I considered was an embedded conceptual metaphor. Thus, it was natural to suggest that the many examples of hybrid constructs (e.g., harmful dysfunction, weighted quality-of-life assessments, derived preferences, and so on) that I have considered have been metaphorically arithmetized. By this, I mean that the combining of terms into conjoint statements only had meaning because of my ability to map information from a concrete base to a more abstract target. This was an important observation, since it suggested that investigators coped with the complex task of defining quality-of-life or HRQOL by engaging in a type of figurative mathematics: adding, subtracting, and multiplying entities that may not have a literal basis.[25] This is certainly what is being done in this book (Table 9.3), as illustrated by my effort to evaluate whether functional statements have been qualified, resulting in a change of the meaning of the term. Alternatively, when the consequences of being in a particular state were added to a valuation statement, then this could lead to generation of a quality-of-life or HRQOL statement.

Most often, the hybrid constructs I dealt with included a value statement. Therefore, it became important for me to consider when and where value statements can become part of a conceptual model of quality. Fulford (1989, 2000) was helpful in this regard, especially when he pointed out that value statements can be inserted at any place along the informational cascade stretching from function, to dysfunction, and on to disease, illness, and disorder; but as he points out, there is more to this than just finding a place for values. There is also the issue of whether the language I use can be sufficiently analytical such that it is possible to make a nonevaluative statement. The approach taken in this book is that it can be possible, although as with Fulford, the applicability of this approach will vary depending on the context (e.g., whether medical or psychiatric information is being considered).

The cognitive-linguistic model of a quality-of-life or HRQOL assessment that has emerged involves, as stated above, the figurative adding of the description of a state a person is in, the consequences of being in a particular state, and a valuation of being in that state. It is content independent, as opposed to many other approaches to quality-of-life or HRQOL assessment. Thus, two people can describe very different states, but do so by using the same cognitive-linguistic methods, and thus, both be assumed to be providing quality-of-life or HRQOL statements.

An example that at first glance may not appear obvious is Wakefield's (1999) definition of mental illness as a "harmful dysfunction." Here, too, is a statement that has the appropriate cognitive-linguistic form, especially if the diagnostic category articulates the consequences of being in a particular state (e.g., the social consequences of being depressed), but which would not ordinarily be thought of as a quality-of-life or HRQOL statement. What makes a psychiatric diagnosis at least an implicit quality-of-life statement is that it is meant to facilitate clinical decision making that has as its objective the improvement of the quality of the person's existence. This is particularly true for the many psychiatric conditions that do not place the person's life at risk. There are actually many examples where the cognitive-linguistic model I have discussed is used, and where the quality-of-life or HRQOL of a person is either an implicit or explicit issue.

This chapter also contained an extended discussion about how function statements can be structured so that they have teleological properties. I reviewed several types of teleological statements, ranging from the cosmic to secular, and discussed Mayr's (1982) four models of teleology. What was most interesting was the discussion about how some entity could assume a purpose without it also having to have theological properties (meaning that it reflected some external design). Darwin and the complexity theorists were most helpful in this regard since they spoke of how random events could lead to species to evolve or to form "emergent complexities" or form "self-organizations." These notions, when combined with Hayek/Hebb's notion that information is local and distributed, gave a sense of what type of a model of how the flow of life experiences can emerge into a coherent experience, the experience of quality.

Complexity theorists, like theists, philosophers, or biologists, claim that universal laws exist that govern random events, but also argue that these laws need not have had a prior existence, nor do they need to exist outside of a particular system (Alexander 2002). Instead, these laws emerge from dynamically stable systems, and while they have concrete effects, these effects are irreducible and therefore unpredictable. This is so because new functions or changes in context not accounted for in an initial presentation can emerge, leading to new outcomes that would not have been predicted (Alexander 2002). As a result, any system or process can increase in complexity that by way of feedback will be constrained.

An additional idea that emerges from complexity theory is that a cognitive entity may be thought of as a complex adaptive system that gives a person their sense of quality, but also keeps this sense stable. How this might occur is a topic that must be left for a future discussion. For now I will be content with the observation that some specific images are beginning to appear in the intellectual fabric I have been weaving. These images include the cognitive-linguistic model of a quality-of-life or HRQOL assessment, and the possibility that it be integrated with an application of complexity theory. Both of these are consistent with the objectives set for this book, which were to bridge the gap between a complex symbolic rule system and a complex evolving cognitive system. What has been described so far, while not a grand theory in Parsons' sense, has at least the conceptual framework of what will be needed for such a task.

Notes

1. The Patrick and Erickson (1993) definition of functional status as "efficient performance" may be interpreted as referring to what a person is able to achieve, rather than what limitations affect their activities. In the Patrick and Chiang (2000), paper they clearly state that functional status usually refers to limitations in performance. Using the later paper as a reference would eliminate any differences in the five definitions presented in Table 9.1.

2. The reader may have recognized that the mapping between entities or concepts that I discussed earlier as being essential for the formation of a conceptual metaphor also played a central role in the development of mathematical concepts. Mapping, therefore, is an essential element of creative activities.

3. In Chap. 3, I used the example of what I would take with me if my home were about to be damaged (e.g., it was on fire). Naturally, the items I would select would be diverse, but they all would presumably function to reassure my continued ability to provide for myself and my family.

4. My concerns here are straightforward; I believe that I have to have a theory comprehensive enough to include exceptions or atypical cases. If I do not, and just have a theory for the "normal," then I run the risk of not fairly treating these exceptions. This is the concern of many who currently criticize quality-of-life or HRQOL research efforts (Wolfensberger 1994).

5. An example of this would be how I determine what is a clinically significant difference, which is an example of a clinically relevant, assessment-dependent topic.

6. The moon illusion refers to the observation that the moon at the horizon appears larger than at its zenith. This illusion is usually attributed to the fact that when seen at the horizon, the perceived image includes elements of the horizon (e.g., trees, buildings, mountains, etc.), which are not present when the moon is viewed at its zenith.

7. Aristotle actually proposed four types of causes: material causes, which are indicative of what things are made of; formal causes, which refer to the form or structure of things; efficient causes, which refer to the ability of things to achieve ends; and final causes, which refer to the purpose or end for which a thing exists (Wilkins 1997).

8. Other authors have suggested different types of teleology. Alexander (2002), for example, suggested five broad categories: Aristotelian, analogical determinism, determined fortuity, pragmatism, and self-organization.

9. While Fulford speaks of values being implicit to statements all along the informational cascade, most definitions of quality-of-life or HRQOL that are value-based include explicit value statements. This difference is important and will permit differentiating the function statements from quality-of-life or HRQOL statements. However, as will be shown later in this paper, it is also possible to add a valuation to a function statement (e.g., Stineman et al. 2003).

10. The fact versus value dichotomy has been the subject of much philosophical speculation. Putnam (2002) has argued that the dichotomy is not sustainable.

11. The reader might recognize how much Comte's suggested science resembles Davies (2001) theory of systemic functions.

12. The Huaco (1986) article discusses a variety of related issues that the interested reader may want to review.

13. The elements that make up the action system have changed over time. First Bales introduced a four-system homeostatic model based on his work with small group dynamics. This model in its original form included adaptation, instrumental, expression of emotion and tension, and integration. Subsequently, Parsons (Parsons and Bales 1955) modified this model to consist of adaptation, goal-directed, latency, and integration.

14. Some might argue that Aristotle's notion of *Eudaimonia,* living the good life, is an example of a comprehensive statement that has implications about a theory of quality-of-life or HRQOL. Chapter 12. will present this example of a grand theory of quality-of-life. In addition, several of the methods used to assess quality-of-life or HRQOL are based on explicit theories (e.g., the Schedule for the Evaluation of Individual Quality-of-life derived from Social Judgment Theory; Hickey et al. 1999).

15. An example of this is the current discordance in the definition of disability used to determine eligibility for Social Security Disability Insurance and other attempts to define disability (Jette and Badley

2000). In addition, once disability and rehabilitation was defined as a part of the *Americans with Disabilities Act* (1990), it became difficult to change these definitions even when thoughtful alternatives were presented.

16. Other factors that may contribute to this end include the political process (e.g., legislation) and a variety of economic and financial issues.
17. While rehabilitation can be applied to a range of states or conditions, I will use it to refer to the physically limited (e.g., medical, accident) person.
18. A search of the web revealed many different definitions for the word *disability*. Reviewing these definitions confirms that in most instances the word was defined to describe *a state* a person was in, and that this state indicated the person's *capacity*, although a variety of words or phrases were used to describe this state. These words or phrases included ability, incapacity, restrictions, impairments, normal abilities, limitations, reduction in a person's capacity, incapable, inability, handicap, loss or impairments, substantially limits, lack of an ability to perform, unable to perform, impedes the completion of daily tasks, and so on. These definitions document that Verbrugge and Jette (1994) were taking a unique approach to the task of defining disability.
19. To simplify this discussion and avoid an extended one concerning alternative definitions and approaches, early versions of the *International Classification of Impairments, Disabilities and Handicaps* (WHO 1980) will not be presented. The reader should refer to a paper by Jette and Badley (2000) for a review of these classification systems. Papers by Shaar and McCarthy (1994) and Grönvik (2007) provide historical background for the discussion on how best to define disability.
20. The Sullivan et al. (2000) study does not directly apply SEM to disability measurement, but will still provide us with an example of how SEM can be used to provide useful information.
21. Sullivan et al. (2000) indicate that the expansion of the Wilson and Cleary model involved disaggregating a person's functional status so as to include measures of a person's physical functioning, daily activity interference, and social functioning.
22. This statement implies that *any descriptive statement can be made into a statement of quality* if it reflects the consequence of some event and includes a valuation of the event.
23. It is, of course, quite possible that structural equation modeling will confirm the elements of the Wilson and Cleary (1995) model, as the Sullivan et al. (2000) study partially does. This would shift the discussion away from what is now mostly competing rhetoric and move towards an empirical basis for deciding about the adequacy of the model.
24. I believe it is accurate to say that a good portion of the efforts to deal with the social, political, and economic consequences of disability has been linguistic and conceptual. Thus, the suggestion by the ICF that any person can be on the same continuum as a person with disabilities represents a conceptual change that has social and political consequences. Key to this is what words were used and how they were defined.
25. The extent to which this is true will be discussed in Chap. 12.

References

Aaronson NK, Ahmedzai S, Bullinger M, Crabeels D, et al. (1991). The EORTC core quality-of-life questionnaire: Interim results of an international field study. In, (Ed.) D.Osoba. *The Effect of Cancer on Quality of life*. Boca Raton FL: CRC Press. (p. 185–203).

Alexander VN. (2002). Non-linearity and teleology. Paper presented at the *Society of Literature and Science*. October 11, 2002. Accessed from the Web, January 2005.Victoria N. Alexander <http://www.dactyl.org/directors/vna/cv.html>.

Alexander JC. (1982–1983). *Theoretical Logic in Sociology*. 4 Vols. Berkeley CA: University of California Press.

Alexander JC. (2009). *Mathematical Blending*. Accessed on the web, April 15, 2009. http:// markturner.org/blending. html.

Americans with Disabilities Act. (1990). 42 USC ‡ 12101 et seq.

Anderson CJ, Laubscher S, Burns R. (1996).Validation of the Short Form 36 (SF-36) health survey questionnaire among stroke patients. *Stroke* 27, 1812–1816.

Anderson CJ, Vogel LC. (2003). Domain-specific satisfaction in adults with pediatric onset spinal cord injuries. *Spinal Cord*. 41, 684–691.

Aprile I, Piazzini DB, Bertolini C, Caliandro P, et al. (2006). Predictive variables on disability and quality of life in stroke outpatients undergoing rehabilitation. *Neurol Sci*. 27,40–46.

Barofsky I. (1996). Cognitive aspects of quality of life assessment. In, (Ed.) B Spilker. *Quality of life and Pharmacoeconomics in Clinical Trials*. 2nd edition. New York NY: Raven Press. (p. 107–115).

Barsalou L, Sloman S, Chaigneau S. (2005). The HIPE theory of function. In, (Eds.) L Carlson, E van der Zee. *Representing Functional Features for Language and Space: Insights From Perception, Categorization and Development*. Oxford UK: Oxford University Press. (p. 131–148).

Bergner M. (1985). Measurement of health status. *Med Care*. 23, 696–704.

Bernard, C. (1865/ 1927). *An Introduction to the Study of Experimental Medicine*. First English translation by Henry Copley Greene. London UK: Macmillan & Co. Ltd.

Béthoux F, Calmels P, Gautheron V. (1999). Change in the quality of life of hemiplegic stroke patients over time. *Arch Phys Med Rehabil*. 78, 19–23.

Bezner JR, Hunter DL. (2001). Wellness perceptions in persons with traumatic brain injury and its relation to functional independence. *Arch Phys Med Rehabil*. 82, 787–792.

Black M. (1961). Some questions about Parsons theories. In, (Ed.) Black M. *The Social Theories of Talcott Parsons*. Englewood Cliffs NJ: Prentice-Hall. (p. 268–288).

Boorse C. A (1977a). Rebuttal on health. In, (Eds.) JM Huber, RF Almeder. *What is Disease?* Totowa NJ: Humana.

Boorse C. (1977b). Health as a theoretical concept. *Philos Sci*. 44, 542–573.

Bowdle BF, Gentner D. (2005). The career of metaphor. *Psychol Rev*. 112 , 193–216.

Bowling A. (1991). *Measuring Health: A Review of Quality of life Measurement Scales*. Philadelphia PA: Open University Press.

Brandt EN, Pope AM. (1997). *Enabling America: Assessing the Role of Rehabilitation Science and Engineering*. Washington DC: National Academy Press.

Cannon W.B. (1939). *The Wisdom of the Body*. New York NY: Norton. 2nd ed.

Caraça BJ. (1951). *Conceitos Fundamentals da Mathmatica*. (1st joint edition, Parts I, II, and III). Lisbon Portugal: Sá da Costa.

Chaigneau SE, Barsalou LW, Sloman SA. (2004). Assessing the causal structure of function. *J Exper Psychol Gen*. 133, 601–625.

Chen Y, Anderson CJ, Vogel LC, Chlan BA, et al. (2008). Change in life satisfaction of adults with pediatric-onset spinal cord injury. *Arch Phys Med Rehabil*. 89, 2285–2292.

Cieza A, Stucki G. (2005). Content comparison of health-related quality of life (HRQOL) instruments based in the International Classification of Functioning, Disability, and Health (ICF). *Qual Life Res*. 14, 1225–1237.

Clark M, Bond M. (1995). The Adelaide Activities Profile: a measure of lifestyle activities of elderly people. *Aging Clin Exp Res*. 7, 174–184.

Coleman JS. (1990). *Foundations of Social Theory*. Cambridge MA: Harvard University Press.

Cosmides L, Tooby J. (1987). From evolution to behavior: Evolutionary psychology as the missing link. In, (Ed.) J Dupré. *The Latest on the Best: Essays on Evolution and Optimality*. Cambridge MA: MIT Press. (p. 277–306).

Cosmides L, Tooby J. (1999). Towards a evolutionary taxonomy of treatable conditions. *J Abnorm Psychol*. 108, 453–464.

Cromes GF, Holavanahalli R, Kowalske K, Helm P. (2002). Predictors of quality of life as measured by the Burn Specific Health Scale in persons with major burn injury. *J Burn Care Rehabil*. 23, 229–234.

Cummins RA. (1996). The domains of life satisfaction: An attempt to order chaos. *Soc Indic Res*. 38, 303–328.

Darwin C. (1859). *On the Origin of Species by Means of Natural Selection or the Preservation of Favored Races in the Struggle for life*. London UK: John Murray.

Davies PS. (2001). *Norms of Nature: Naturalism and the Nature of Functions*. Cambridge MA: MIT Press.

Davis K, Moore WE. (1945). Some principles of stratification. *Amer Sociol Rev*. 10, 242–249.

Dennis D, Confrey J. (1995). Functions of a curve; Leibniz's original notion of function and its meaning for the parabola. *Coll Math J*. 26, 124–131.

Descartes R. (1638/1952). *The Geometry*. LaSalle Il: Open Court.

Diener E, Emmons RA, Larsen RJ. (1985). The satisfaction with life scale. *J Personal Assess*. 49, 71–75.

Dijkers M. (1997). Quality of life after spinal cord injury: a meta analysis of the effects of disablement components. *Spinal Cord*. 35, 829–840.

Dorman PJ, Waddell F, Slattery J, Dennis M, et al. (1997). Is the EuroQOL a valid measure of health-related quality of life after stroke? *Stroke*. 28, 1876–1882.

Epstein AM, Hall JA, Tognetti J, Son LH, Conant L Jr. (1989). Using proxies to evaluate quality of life. Can they provide valid information about patients; health status and satisfaction with medical care? *Med Care*. 27 (Suppl), S91–S98.

Fabian E. (1991). Using quality of life indicators in rehabilitation program evaluation. *Rehabil Coun Bull*. 34, 344–356.

Fauconnier G, Turner M. (2002). *The Way we Think: Conceptual Blending and the Mind's Hidden Complexities*. New York NY: Basic Books.

Fauconnier G, Turner M. (2004). *Conceptual Projection and Middle Spaces*. UCSD Cognitive Science Technical Report.

Ferrans CE, Powers MJ. (1985). Quality of life index: development and psychometric properties. *Adv Nurs Sci*. 8, 15–24.

Ferrans CE. (1990). Development of a quality of life index for patients with cancer. *Oncol Nur Forum*. 17, 29–38.

Fries JF, Spitz P, Kraines RG, Holman HR. (1980). Measurement of patient outcome in arthritis. *Arthritis Rheum*. 23, 137–145.

Fuhrer MJ. (1987). Overview of outcome analysis of rehabilitation. In, (Ed.) MJ Fuhrer. *Rehabilitation Outcomes: Analysis and Measurement*. Baltimore MD: Brooks.

Fuhrer MJ. (2000). Subjectifying quality of life as a medical rehabilitation outcome. *Disabil Rehabil*. 22, 481–489.

Fuhrer MJ, Rintala DH, Hart KA, Clearman R, Young ME. (1992). Relationship of life satisfaction to impairment, disability and handicap among persons with spinal cord disability living in the community. *Arch Phys Med Rehabil*. 73, 552–557.

Fulford KWM. (1989). *Moral Theory and Medical Practice*. New York NY: Cambridge University Press.

Fulford KWM. (2000). Teleology without tears: Naturalism, neo-naturalism, and evaluationism in the analysis of function statements in biology (and a bet on the Twenty-first Century). *Philoso Psychiatr Psychol*. 7, 77–94.

Gellman W. (1973). Fundamentals of rehabilitation. In, (Eds.) JF Garrett, ES Levine. *Rehabilitation Practices with the Physically Disabled*. New York NY: Columbia University Press.

Gerhardt U. (1989). *Ideas About Illness: An Intellectual and Political History of Medical Sociology*. New York NY: New York University Press.

Geyh S, Cieza S, Kolleritis B, Grimby G, Stucki G. (2007). Content comparison of health-related quality of life measures used in stroke based on the International Classification of Functioning, Disability, and Health (ICF). *Qual Life Res*. 16, 833–851.

Gibson, JJ. (1979). *The Ecological Approach to Visual Perception*. Boston, MA: Houghton-Mifflin.

Giddens A. (1984). *The Constitution of Society*. Berkeley CA: University of California Press.

Gill TM, Feinstein AR. (1994). A critical appraisal of the quality of quality of life measurements. *JAMA* 272, 619–626.

Goodwin PJ. (2003). Reversible ovarian ablation or chemotherapy: Are we ready for quality of life to guide adjuvant treatment decisions in breast cancer? *J Clin Oncol*. 24, 4474–4475.

Grady J, Oakley T, Coulson S. (1999). Blending and metaphor. In, (Eds.) RW Gibbs, GJ Steen. *Metaphor in Cognitive Linguistics: Selected Papers from the Fifth International Cognitive Linguistic Conference*. July 1997. Amsterdam Netherlands: John Benjamins. (p. 101–124).

Granger CV, Cotter AC, Hamilton BB, Fiedler RC. (1993). Functional assessment scales: A study of persons after stroke. *Arch Phys Med Rehabil*. 74, 133–138.

Grönvik L. (2007). The fuzzy buzz word: Conceptualizations of disability in disability research classics. *Sociol Health Illn*. 29, 750–766.

Habermas J. (1981). *The Theory of Communicative Action*. (Translated by McCarthy T). Boston MA: Beacon Press.

Halfon N, Hochstein M. (2002). Life course health development: An integrated framework for developing health policy, and research. *Milbank Q*. 80, 433–479.

Hamilton BB, Granger CV, Sherwin FS, Zielezny M, Tashman JS. (1987). A uniform national data system for medical rehabilitation. In, (Ed.) MJ Fuhrer. *Rehabilitation Outcomes: Analysis and Measurement*. Baltimore MD: Brooks.

Havranek EP, Smith TA, L'italien G, Smitten A, et al. (2004). The relationship between health perception and utility in heart failure patients in a clinical trial: results from an OVERTURE sub-study. *J Card Fail*. 10, 339–343.

Hewlett S, Smith AP, Kirwan JR. (2002). Measuring the meaning of disability in rheumatoid arthritis: Personal Impact Health Assessment Questionnaire (PI HAQ). *Ann Rheum Dis*. 61, 986–993.

Hewlett S, Smith AP, Kirwan JR. (2001). Values for function in rheumatoid arthritis patients, professionals, and public. *Ann Rheum Dis*. 60, 928–933.

Hickey A, O'Boyle CA, McGee HM, Joyce CRB. (1999). The Schedule for the Evaluation of Individual Quality of life. In, (Eds.) CRB Joyce, HM McGee, CA O'Boyle. *Individual Quality of life: Approaches to Conceptualization and Assessment*. Amsterdam The Netherlands: Harwood. (p. 119–133).

Hlatky MA, Boineau RE, Higginbotham MB, Lee KL, et al. (1989). A brief self-administered questionnaire to determine functional capacity (the Duke Activity Status Index). *Am J Cardiol*. 64, 651–654.

Huaco GA. (1986). Ideology and General Theory: The case of Sociological Functionalism. *Comp Stud Soc Hist*. 28, 34–54.

Hunt SM, McEwen J, Mc Kenna SP. (1985). Measuring health status: a new tool for clinicians and epidemiologists. *J R Coll Gen Pr*. 35, 185–188.

Hunt SM, McEwen J, Mc Kenna SP. (1985). Measuring health status: a new tool for clinicians and epidemiologists. *J R Coll Gen Pr*. 35, 185–188.

Imrie R. (2004). Demystifying disability: a review of the International Classification of Functioning, Disability and Health. *Sociol Health Illn*. 26, 287–305.

International Classification of Diseases and Related Health Problems-10. (1994). Geneva Switzerland: WHO.

Indredavik B, Bakke F, Slordahl SA, Rolseth R, Haheim LL. (1998). Stroke unit treatment improves long-term quality of life: A randomized controlled trial. *Stroke.* 29, 895–899.

James W. (1890). *Principles of Psychology.* Vol 1. New York NY: Holt.

Jensen MP, Abresch RT, Carter GT. (2005). The reliability and validity of self-report version of the FIM instrument in persons with neuromuscular disease and chronic pain. *Arch Phys Med Rehabil.* 86, 116–122.

Jette AM, Badley E. (2000). Conceptual issues in the measurement of disability. In, (Eds.) N Mathiowetz, GS Wunderlich. *Survey Measurement of Work Disability: Summary of a Workshop.* Washington DC: National Academy Press. (p. 4–27).

Joreskog KG, Sorbom D. (1993). *LISREL VIII. Structural Equation Modeling with SIMPLIS Command Language.* Chicago Ill: Scientific Software International.

Kaambwa B, Bryan S, Barton P, Parker H, et al. (2008). Costs and health outcomes of immediate care: Results from five UK case study sites. *Health Soc Care Comm.* 16, 573–81.

Kaplan RM, Anderson JP. (1990). The General Health Policy Model: An integrated approach. In, B Spilker (Ed.). *Quality of life Assessments in Clinical Trials.* New York NY: Ravens.

Kaplan RM, Bush JW. (1982). Health-related quality of life for evaluation research and policy analysis. *Health Psychol.* 1, 61–80.

Katz S, Apkom CA. (1976). A measure of primary socio-biological functions. *Int J Health Serv.* 6, 493–507.

Kemler-Nelson DG, Russell R, Duke N , Jones K. (2000). Two-year olds name artifacts by their functions. *Child Dev.* 71, 1271–1288.

Kim P, Warren S, Madill H, Hadley M. (1999). Quality of life of stroke survivors. *Qual Life Res.* 8, 293–301.

King JT Jr, Tsevat J, Roberts MS. (2005). Preference-based quality of life in patients with cerebral aneurysms. *Stroke.* 36, 303–309.

King RB. (1996). Quality of life after stroke. *Stroke.* 27, 1467–1472.

Lakoff G. (1993). The contemporary theory of metaphor. In, (Ed.) A. Ortony. *Metaphor and Thought.* 2nd edition. New York NY: Cambridge Press. (p. 202–251).

Lakoff G, Johnson M. (1999). *Philosophy in the Flesh: The Embodied Mind and its Challenge to Western Thought.* New York NY: Basic Books.

Lakoff G, Núñez RE. (2000). *Where Mathematics Comes From.* New York NY: Basic Books.

Leidy NK. (1994). Functional status and the forward progress of merry-go-rounds: Towards a coherent analytical framework. *Nurs Res.* 43, 186–202

Lennox JG. (1992). Teleology. In, (Eds.) EF Keller, EA Lloyd. *Keywords in Evolutionary Biology.* Cambridge MA: Harvard University Press. (p. 324–333).

Linton R. (1936). *The Structure of Society.* New York NY: Appleton-Century.

Loehlin JC. (2004). *Latent Variable Models: An Introduction to Factor, Path and Structural Equation Modeling.* Mahwah NJ: Lawrence Erlbaum. Fourth Edition.

Madden S, Hopman W, Bagg , Verner J, O'Callaghan C. (2006). Functional status and health-related quality of life during inpatient stroke rehabilitation. *Arch Phys Med Rehabil.* 85, 831–838.

Mahner M, Bunge M. (2001). Function and functionalism: A synthetic perspective. *Philos Sci.* 68, 75–94.

Mahoney FI, Barthel DW. (1965). Functional evaluation: the Barthel index. *MD State Med J.* 14, 61–65.

Malinowski N. (1922). Argonauts of Western Pacific. London UK: Routledge.

Malinowski N. (1944). *A Scientific Theory of Culture and Other Essays.* Chapel Hill NC: University of North Carolina Press.

Malinowski N. (1926). *Crime and Custom in Savage Society.* London UK: K. Paul, Trench, Trubner,

Marshall J, White K, Weaver M, Flury Wetherill L, et al. (2007). Specific psychiatric manifestations among preclinical Huntington disease mutation carriers. *Arch Neurol.* 64 ,116–121.

Martz E, Livneh H, Priebe M, Wuermser LA, Ottmanelli L. (2005). Predictors of psychosocial adaptation among people with spinal cord injury or disorder. *Arch Phys Med Rehabil.* 86, 1182–1192.

Maslow AH. (1968). *Towards a Psychology of Being.* Princeton NJ: Van Nostrand.

May LA, Warren S. (2002). Measuring quality of life of persons with spinal cord injury: external and structural validity. *Spinal Cord.* 40, 341–350.

Mayboom-De Jong B, Smith R. (1992). How do we classify functional status? *Fam Med.* 24, 128–133

Mayr E. (1982). *The Growth of Biological Thought: Diversity, Evolution, and Inheritance.* Cambridge MA: Belknap Press/Harvard University Press.

Mc Dowell I, Newell C. (1996). *Measuring Health: A Guide to Rating Scales and Questionnaires.* Second Edition. New York NY: Oxford.

ME Segal and RR Schall (1994) . Determining functional/health status and its relation to disability in stroke survivors. *Stroke.* 25, 2391–2397

Merton RK. (1949/ 1968). *Social Theory and Social Structure.* New York NY: Fress Press. (p. 73–138).

Meyer SS. (1992). Aristotle, teleology and reduction. *Philos Rev.* 101, 791–825.

Mills CW. (1959). *The Sociological Imagination.* New York, NY: Oxford University Press.

Moxley RA. (1992). From mechanistic to functional behaviorism. *Am Psychol.* 47, 1300–1311.

Nagi S. (1976). An epidemiology of disability among adults in the United States. *Milbank Q.* 54, 439–468.

Nagi S. (1965). Some conceptual issues in disability and rehabilitation. In, (Ed.) MB Sussman. Sociology and Rehabilitation. Washington DC: American Sociological Association. (p. 100–113).

New York Heart Association. (1964). *Diseases of the Heart and Blood Vessels: Nomenclature and Criteria for Diagnosis.* Boston MA: Little Brown.

Nichol G, Llewellyn-Thomas HA, Thiel EC, Naylor CD. (1996). The relationship between cardiac functional capacity and patients' symptom-specific -utilities for angina: some findings and methodologic lessons. *Med Dec Mak.* 16, 78–85.

Osoba D. (1994). Lessons learned from measuring health-related quality of life in oncology. *J Clin Oncol.* 12, 508–616.

Oxford English Dictionary. (1996). Oxford UK: Oxford University Press.

Parsons T. (1937). *The Structure of Social Action.* New York NY: McGraw-Hill.

Parsons T. (1951). *The Social System.* Glencoe Ill: Free Press.

Parsons T. (1958). Definitions of health and illness in light of the American values and social structure. In, (Ed.) EG Jaco. *Patients, Physicians and Illness.* New York NY: Free Press.

Parsons T. (1978). *Action Theory and the Human Condition.* New York NY: Free Press.

Parsons T, Bales R. (1955). *Family, Socialization and Interaction Process.* Glencoe Il: Free Press.

Patrick DL, Chiang Y-P. (2000). Measurement of health outcomes in treatment effectiveness evaluations. *Med Care.* 38 (Suppl II), II-14- II-25.

Patrick DL , Erickson P. (1993). *Health Status and Health Policy: Quality of life in Health Care Evaluation and Resource Allocation.* New York NY: Oxford.

Pickard AS, Johnson JA, Feeny DH. (2005). Responsiveness of generic health-related quality of life measures in stroke. *Qual Life Res.* 14, 207–219.

Pollard B, Johnston M. (2001). Problems with the Sickness Impact Profile: A theoretical based analysis and a proposal for a new method of implementation and scoring. *Soc Sci Med.* 52, 921–934.

Ponte JP. (1992). The history of the concept of function and some educational implications. *Math Educ.* 3, 3–8.

Pope AM, Tarlov AR. (1991). *Disability in America: Towards a National Agenda for Prevention.* Washington DC: National Academy Press.

Putnam H. (2002). *The Collapse of the Fact/Value Dichotomy and Other Essays.* Cambridge MA: Harvard.

Putnam H. (1975). *Philosophical Papers.* Vol 2. New York NY: Cambridge University Press.

Rabini A, Aprile I, Padua L, Piazzini DB, et al. (2007). Assessment and correlation between clinical patterns, disability and health-related quality of life in patients with low back pain. *Eur Medicophys.* 43, 49–54.

Radcliffe-Brown AR. (1935). On the concept of function in social sciences. *Am Anthropol.* 37, 394–402.

Renwick R, Friefeld S. (1996). Quality of life and rehabilitation. In, (Eds.) R Renwick, I Brown, M Nagler. *Quality of life in Health Promotion and Rehabilitation: Conceptual Approaches, Issues and Applications.* Thousand Oaks CA: Sage. (p. 26–36).

Reuben DB, Valle LA, Hays RD, Siu AL. (1995). Measuring physical function in community-dwelling older persons: a comparison of self-administered, interviewer-administered, and performance-based measured. *J Am Geriat Soc.* 43, 17–23.

Riazi A, Cano S, Cooper JM, Bradley JL, et al. (2006). Coordinating outcomes measurement in ataxia research: Do some widely used generic scales tick the boxes? *Move Disord.* 21, 1396–1403.

Robinson-Smith G, Johnson MV, Allen J. (2000). Self-care self-efficacy, quality of life and depression after stroke. *Arch Phys Med Rehabil.* 81, 460–464.

Romberg A, Virtanen A, Ruutiainen J. (2005). Long-term exercise improves functional impairment but not quality of life in multiple sclerosis. *J Neruol.* 252, 839–845.

Ross BH, Murphy GL. (1999). Food for thought: Cross-classification and category organization in a complex real-world domain. *Cogn Psychol.* 38, 495–553.

Russell B. (1946). *The History of Western Philosophy.* London UK: George Allen and Unwin.

Ruta DA, Garratt AM, Leng M, Russell IT, MacDonald LM. (1994). A new approach to the measurement of quality of life: The Patient Generated Inventory (PGI). *Med Care.* 32, 1109–1126.

Sadegh-Zedeh K. (2001). The fuzzy revolution: Goodbye to the Aristotelian Waltanschauung. *Artif Intel Med.* 21, 1–25.

Sama GP, Matchar DB. (2004). How strong is the relationship between functional status and quality of life among persons with stroke. *J Rehabil Res Dev.* 41, 279–282.

Sciulli D, Gerstein D. (1985). Social theory and Talcott Parsons in the 1980s. *Ann Rev Sociol.* 11, 369–387.

Shaar K, McCarthy M. (1994). Definitions and determinants of handicap in people with disabilities. *Epidemiol Rev.* 18, 228–242.

Sharrack B, Hughes RAC, Soudaim S, Dunn G. (1999). The psychometric properties of clinical rating scales used in multiple sclerosis. *Brain.* 122, 141–159.

Skinner BF. (1938/1996). The Behavior of Organisms Englewood NJ: Prentice-Hall.

Snead SL, Davis JR. (2002). Attitudes of individuals with acquired brain injury towards disability. *Brain Inj.* 16, 947–953.

Sneeuw KC, Sprangers MA, Aaronson NK. (2002). The role of health care providers and significant others in evaluating the quality of life patients with chronic diseases. *J Clin Epidemiol.* 55, 1130–1143.

Spitzer WO, Dobson AJ, Hall J, Chesterman E, et al. (1981). Catchlove BR. Measuring the quality of life of cancer patients. *J Chron Dis.* 34, 585–597.

Stamm T, Geyh S, Cieza A, Machold K, et al. (2006). Measuring function in patients with hand osteoarthritis – Content comparison of questionnaires based on the International Classification of Functioning, Disability, and Health (ICF). *Rheum.* 45, 1534–1541.

Stineman MG, Maislin G, Nosek M, Fiedler R, Granger CV. (1998). Functional status: Application of a new feature trade-off consensus building tool. *Arch Phys Med Rehabil.* 79, 1522–1529.

Stineman MG, Wechsler B, Ross R, Maislin G. (2003). A method for measuring quality of life through subjective weighting of functional status. *Arch Phys Med Rehabil.* 84 Suppl 2, S15–S22.

Sullivan M, Kempen GIJM, von Sonderer E, Ormel J. (2000). Models of health-related quality of life in a population of community dwelling Dutch elderly. *Qual Life Res.* 9, 801-810.

Tijhuis GJ, Jansen SJT, Stiggelbout AM, Zwinderman AH, et al. (2000). Value of the time trade off method for measuring utilities in patients with rheumatoid arthritis. *Ann Rheum Dis.* 59, 892–897.

Timman R, Roos R, Maat-Kievit A, Tibben A. (2004). Adverse effects of predictive testing for Huntington Disease underestimated: Long-term effects 7–10 years after the test. *Health Psychol.* 23, 189–197.

Todorov A. Kirchner C. (2000). Bias in proxies' reports of disability: Data from the National Health Interview Survey on Disability. *Am J Pub Health.* 90, 1248–1253.

Torrance GW, Furlong W, Feeny D, Boyle M. (1995). Multi-attribute preference functions: Health Utilities Index. *PharmEcon.* 7, 503–520.

Torrance GW. (1976). Health status index models: A unified mathematical view. *Manag Sci.* 22, 990–1001.

Torrance GW. (1986). Measurement of health state utilities for economic appraisal: A review. *J Health Econ.* 5, 1–30.

Torres JM. (2002).The importance of genetic services for the theory of health: A basis for an integrating view of health. *Med Health Care Philos.* 5, 43–51.

Tschiesner U, Rogers SN, Harréus U, Berghaus A, Cieza A. (2008). Content comparison of quality of life questionnaires used in head and neck cancer based on the International Classification of Functioning, Disability, and Health (ICF): A systematic review. *Eur Arch Oto-Rhino-Laryn Head Neck.* 265, 627–637.

Tsevat J, Goldman L, Lamas GA, Pfeffer MA, et al. (1991). Functional status versus utilities in survivors of myocardial infarction. *Med Care.* 29, 1153–1159.

Turner JH, Maryanski AK. (1988). Is neofunctionalism really functional? *Sociol Theory.* 6, 110–121.

Verbrugge L, Jette A. (1994). The disablement process. *Soc Sci Med.* 38, 1–14.

von Essen L. (2004). Proxy ratings of quality of life. *Acta Oncologica.* 43, 229–234.

Wakefield JC. (1999). Evolutionary versus prototype analyses of the concept of disorder. *J Abnorm Psychol.* 108, 374–399.

Ware JE, Snow KK, Kosinski MA, Gandek B. (1993). *SF-36 Health Survey: Manual and Interpretation Guide.* Boston MA: The Health Institute, New England Medical Center.

Warren L, Wigley JM, Yoels WC, Fine PR. (1996). Factors associated with life satisfaction among a sample of persons with neurotrauma. *J Rehabil Res Dev.* 33, 404–408.

Webb CR, Wrigley M, Yoels W, Fine PR. (1995). Explaining quality of life for persons with traumatic brain injuries 2 years after injury. *Arch Phys Med Rehabil.* 76, 1113–1119.

White PA. (1988).Causal Processing: Origins and Development. *Psychol Bull.* 104, 36–52.

Wilkins JS. (1997). Is there progress and direction in evolution? Accessed on the Web, January 18, 2005. http://www.talkorigins.org/faqs/evolphil.html.

Willis G, Reeve B, Barofsky I. (2005). The Use of Cognitive Interviewing Techniques in Quality of life and Patient-Reported Outcome Assessment. In, (Eds.), J Lipscomb, CC Gotay, C Synder. *Outcomes Assessment in Cancer: Findings and Recommendations of the Cancer Outcomes Measurement Working Group.* Cambridge MA: Cambridge University Press.

Wilson IB, Cleary PD. (1995). Linking clinical variables with health-related quality of life: A conceptual model of patient outcomes. *JAMA.* 273, 59–65.

Witham MD, Fulton RL, Wilson L, Leslie CA, McMurdo MET. (2008). Validation of an individualized quality of life measure in older day hospital patients. *Health Qual Life Outcomes.* 6:27. doi: 10.1186/1477-7525-6-27. (p. 1–27).

Wolfe F, Michaud K, Pincus T. (2005). Preliminary evaluation of a visual analog function scale for use in Rheumatoid Arthritis. *J Rheum.* 32, 1261–1266.

Wolfensberger W. (1994). Lets hang up "quality of life" as a hopeless term. In, (Ed.) D. Goode. *Quality of life for Persons with Disabilities: International Perspectives and Issues.* Cambridge MA: Brookline. (p. 285–321).

Woodfield A. (1976). *Teleology.* London UK: Cambridge University Press.

World Health Organization. (2001). ICF: International Classification of Functioning, Disability and Health. Geneva Switzerland: WHO.

World Health Organization. (2002). *Towards a Common Language for Functioning, Disability and Health.* ICF. Geneva Switzerland: WHO.

World Health Organization. (1980). *International Classification of Impairments, Disabilities and Handicaps: A Manual of Classification Relating to the Consequences of Disease.* Geneva Switzerland: WHO.

World Health Organization. (1996). The World Health Organization Quality of life Assessment- Abbreviated Version (WHOQOL-Brèf). Geneva Switzerland:

The WHOQOL Group, Program on Mental Health. (2000). World Health Organization Disability Assessment Schedule (WHODAS). Geneva Switzerland: WHO.

Neurocognition and HRQOL Assessment

Abstract

Can a neurocognitive assessment be included in a qualitative assessment, and if not, how can it be made useful? This issue is addressed by focusing on some of the factors that impact performing a qualitative assessment on the neurocognitively compromised person. Thus, a number of topics are reviewed including what is known about neurocognitive preservation, degeneration, and recovery following an acute neurological event. The chapter ends with a summary of two examples that attempt to model the relationship between neurocognitive capacity and performance.

In the science fiction book Contact *(Sagan 1997), a young astrophysicist takes a journey to a civilization on a distant planet. One of her major concerns is how to communicate with those who represent this new civilization, but she soon learns that they are wise enough to know that her capacity to communicate will not extend beyond what the structure of her body will permit. Thus, "contact" occurs, but in a format that she can sense, perceive and respond to.*

Abbreviations

ADL	Activities of daily living
AEDs	Antiepileptic drugs
AHP	Analytical hierarchy process (Saaty 1980)
AML/MDS	Acute myelogenous leukemia or myeleodisplastic syndrome
AQOL	Assessment of quality-of-life (Hawthorne et al. 1999)
BASQID	Bath assessment of subjective quality-of-life in dementia (Trigg et al. 2007b)
BCQ	Breast chemotherapy questionnaire (Levine et al. 1988)
Ca	Cronbach's alpha
CC	Canonical correlation
DEMQOL	Dementia quality-of-life (Smith et al. 2005)
DQOL	Dementia quality-of-life instrument (Brod et al. 1999)
EORTC QLQ-C30	EORTC quality-of-life questionnaire (Aaronson et al. 1991)
EORTC-CF	Sum of the three items dealing with fatigue in the EORTC QLQ-C30
EQ-5D	European questionnaire – five dimensions (Brooks et al. 2003)
F-IA	Agreement between family and institution proxies
FACT-BR	Functional assessment of cancer therapy-brain (Weitzner et al. 1995)
FACT-Cog	Functional assessment of cancer therapy-cognitive (Jacobs et al. 2007)
FACT-G	Functional assessment of cancer therapy-general (Cella et al. 1993)
FAM	Family as proxy and patient agreement
FLIC	Functional Living Index-Cancer (Shipper et al. 1984)
GDS	Geriatric Depression Scale (Yesavage et al. 1983)
HRQOL	Health-related quality-of-life

HUI2	Health Utility Index 2 (Feeny et al. 1992)
HUI3	Health Utility Index 3 (Torrance et al. 1995)
IAP	Institution as proxy (e.g., physician or nurse) and patient agreement
IQ	Intelligence quotient
IR	Independent rater
KAP	Kappa statistic
MMSE	Mini mental status exam (Folstein et al. 1975)
MOS	Medical Outcome Study (Hays et al. 1993)
NADL	Nishimuria Activities of Daily Living Scale (Nishimura et al. 1993)
NHP	Nottingham health profile (Hunt et al. 1985)
NM	Nishimuria Mental State Scale (Nishimura et al. 1993)
NPT	Neuropsychological testing
PES-AD	The pleasant event schedule-AD short form (Logsdon and Teri 1997)
PET	Positron emission tomography
PROCOG	Patient-reported outcomes in cognitive impairment (Frank et al. 2006)
QLI	Quality-of-life Index (Ferrans and Powers 1985)
QOL-AD	Quality-of-life-Alzheimer's disease (Logsdon et al. 1999, 2002)
QOL-D	Quality-of-life-dementia (Terada et al. 2002)
QOLAS	Quality-of-life assessment schedule (Selai et al. 2001)
QOLIE-89	Quality-of-life in epilepsy inventory-89 (Devinsky et al. 1995)
QWB	Quality of well-being Scale (Kaplan and Anderson 1990)
RMBPC	Revised memory and behavior problem checklist (Teri et al. 1992)
RSQ	Rotterdam symptom questionnaire (De Haes et al. 1990)
SAHS	Self-assessed health status item
SF-12	The Short Form 12. (Ware et al. 1995)
SIP	Sickness impact profile (Bergner et al. 1981)
SIP-64	Sickness impact profile-68 (Nanda et al. 2003)
TBI	Traumatic brain injury
TRT	Test-retest
VAS	Visual Analog Scale
WHOQOL-BREF	World Health Organization Quality-of-life Assessment-BREF (WHO 1996)
WHOQOL100	WHO Quality-of-life Assessment-100 items (WHO 1995)

1 Introduction

How I communicate, whether with an alien being or the person next to me, is dependent on the structure of my body. The same may be said about how I know the quality of an object, event, or life lived. Thus, the relationship between the capacity of my body and how I actually make judgments is a central determinant of any qualitative assessment. It is reassuring that there is so much similarity among how people perform these tasks, suggesting that there is a common structural basis for these qualitative judgments. In this chapter, I will document the relationship between these two levels of analysis and also answer a series of more specific questions. As I have done throughout this book, I will approach my task by using the analytical tools of language expression and cognitive changes.

Figure 10.1 depicts how I will discuss the relationship between a person's capacity, as measured by their neuropsychological test (NPT) results, and their performance on a quality-of-life or HRQOL assessment.[1] Neurocognitive capacity and quality-of-life or HRQOL are, of course, both abstract concepts that I can operationalize by making concrete (that is, I deal with them as "basic" or "subordinate" information-level terms). Thus, I describe what I mean by neurocognitive capacity from the results of NPT (Fig. 10.1a) and performance by the responses to a quality-of-life or HRQOL assessment (Fig. 10.1b). This transforms my task into one where I am now comparing and relating NPT results to responses to a quality-of-life or HRQOL assessment (Fig. 10.1c). In addition, I often use metaphors to depict latent variables, such as cognitive reserve or capacity (Fig. 10.1), which are assumed to model the relationship between these indicators.

But metaphors can also complicate my task, since each time I operationalize these abstract concepts I may do so differently, depending on the context of my study. Figure 10.1, therefore, illustrates the practical consequences of the hierarchical organization of information.

When neuropsychological data and quality-of-life or HRQOL data are examined, it quickly becomes clear that while they are both behavioral outcomes (e.g., responses to specific tasks or questionnaires), they differ in significant ways. For example, NPT's are usually administered and interpreted by a clinician and the results measured in terms of accuracy or speed of task completion (e.g., reaction time). Testing can include a wide range of sensory and motor activities (e.g., clock drawings, block completion tasks). The results of NPT, when combined with brain imaging studies or postmortem biopsies, provide information on the neurological (i.e., brain structures) basis for NPT results. Collectively, these data provide a picture of the person's capacity that can then be related to the person's performance.

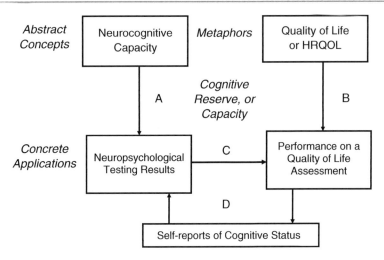

Fig. 10.1 Neurocognitive capacity will be assessed by a person's performance on a qualitative assessment (**A**), the concept of quality-of-life or HRQOL will be monitored by performance on a qualitative assessment (**B**), neurocognitive test results will be reflected in performance on quality-of-life or HRQOL assessments (**C**), while performance on a qualitative assessment will be reflected in self-reports (**D**), and self-reports will impact neuropsychological test results (**D**).

The Dementia Rating Scale (Mattis 1976) is a typical example of a NPT, and it consists of a series of subtests that determine the ability of patients to, for example, recall an expanding list of numbers, follow commands, recall all the things a person can buy at a supermarket, repeat specific phrases accurately, draw particular images or patterns, indicate the difference between objects, and so on. These assessments are then organized into domains (e.g., attention, construction, memory, etc.) and presented as a profile or summary measure. Critically, the clinical significance of these results is left up to the clinician and what he or she has learned about the person, their familial and environmental circumstances, and so on.

In contrast, quality-of-life or HRQOL assessments are usually self-administered, that involve ratings of descriptive statements (e.g., how often can you…; how satisfied are you relative to…; how much impact has "x" had on you…), and can involve the respondents themselves placing a value on the descriptive states. These assessments are often administered by paper and pencil questionnaires, although they can also be assessed with open-ended questions or include complex cognitive tasks designed to provide value statements (e.g., utility estimation using the time trade-off procedures or standard gambles). Most quality-of-life or HRQOL assessments have not been linked to neurocognitive processes, but this is probably due to a lack of empirical data (e.g., brain imaging studies) rather than some inherent barrier to this type of study.

My major concern in this chapter is how can I combine such diverse outcomes. To do this, I will have to identify some of the barriers to this integration (e.g., changes in brain structure with aging, acute neurological events, and so on) and show how these two levels of analysis can be integrated. One approach is to see if there is a common evaluative metric (see below) that can be applied to both types of data. If such a metric exists, then it would be possible to make statements encompassing both types of data. This method, of course, does not literally combine these data as much as transform them into a common dimension that allows me to make general statements about both. An alternative approach has me assuming that a person's self-reports concerning their cognitive status can substitute for NPT results (Fig. 10.1d). If this were true, then I should be able to combine self-reports of both a person's cognitive status and quality-of-life or HRQOL data, since both would be subjective or "self-assessments." A third approach has me acknowledging that it is not possible to formally combine these data, and instead I should shift my attention to studying how these two outcomes can be related. This I can do by considering various models of the relationship between a person's capacity and their performance.

Each of these three approaches will be discussed in different sections of the chapter. Determining if a common evaluative metric can be found in a set of data has already been attempted in Chap. 9 and involved determining if a study contains the elements of a hybrid construct. Thus, I will ask if the items of an assessment spell out the consequences of being in a particular clinical state, and whether the respondent was given the opportunity to value their status. I regularly examine a quality-of-life or HRQOL assessment to determine if it has the elements of a hybrid construct, but I rarely do this for NPT, or for that matter, for many other "objective" indicators, although there is no obvious reason why an objective measure shouldn't be examined from this perspective. What I do know is that NPT, when presented as part of a clinical report, is often expressed in subjective terms, with the clinician spelling out the consequences of being in a particular state and attaching some importance or value statement to the reported outcomes. Thus, a clinical report about NPT data often qualifies the data.

To illustrate the issues here, consider a person who has done quite poorly on a series of tests. The results suggest that he or she has short-term memory deficits (a descriptive statement) that the neuropsychologist interprets (by comparing to established norms) as being sufficient to adversely affect the person's ability to function independently (a statement of consequence). The clinician may then suggest that the person live in a supervised environment. This, of course, is a value statement. Alternatively, the patient may recognize their deficits and judge that they need "prompting aids" to help their memory, but not a major change in their living arrangements. This, too, is a value statement, but now from the perspective of the person. These kinds of exchanges, of course, occur on a regular basis when NPT results are interpreted, but what should be clear here is that, by making value statements about the neuropsychological data, the clinical interpretation of the test results acquires the formal structure of a hybrid construct, much as is true for an optimally designed quality-of-life or HRQOL assessment. And as I tried to illustrate, this could be done from either the perspective of the patient or an independent observer. As a result, both indicators are qualitative statements and have the same evaluative characteristics. This should permit me to compare these statements or measures.

Ideally, I would be able to find examples of qualified NPT studies in the research literature. If I did so and correlated the NPT studies and HRQOL studies that were both qualified, then I would expect them to be more highly correlated, while I would not expect this for two similar assessments that were purely descriptive. This enhanced degree of concordance would suggest, but not prove, that some commonality exists between the two assessments, and this should increase the usefulness of each type of qualified assessment.

To a certain extent, the results of such a comparison can be anticipated. For example, if NPT indicates that the person is cognitively intact, then it would be expected that the person would report a similar state. Critical tests of the relationship between NPT results and cognitive self-reports come from those persons who are determined by NPT to be neurocognitively deficient, but do not perceive themselves as being deficient, or from those persons who are neurocognitively adequate but do not perceive themselves as normal.

My second approach to combining these data has me evaluating whether self-reports of cognitive status can substitute for NPT results. If so, then I would not have to rely on NPT to answer questions about a person's cognitive status. I would then be able to combine self-reports of cognition with those derived from a quality-of-life or HRQOL assessment in some composite index, since both would be subjective. However, it is also possible that I will find that self-reports and NPT are not related, and that I can't simply assume that self-reports can substitute for NPT results. This will set me off on the task of explaining why this is so. What will become evident is that the body has a variety of means of stabilizing and responding to the impact of aging and disease progression that impacts cognitive functioning, and this could account for these findings.

My third approach acknowledges that it may not be possible to combine these data and my attention shifts to understanding the linkages between neurocognition and qualitative outcomes. Several models have been proposed for this task.[2] For example, the embodiment hypothesis states that what I experience is linked to what I have physically experienced, and that while I have separate words to describe each, my body cannot separate them. Second, neural network modeling makes no a priori assumptions about the relationships among each level of analysis, nor does it use figurative language to speculate about the nature of the relationship between these indicators. Rather, it assumes that the parallel numerical processing that occurs in the central nervous system can also be applied at the level of language and reasoning.

Models based on neural networking principles offer two distinct advantages. First, they can be used as statistical tools (e.g., to perform regression-like analyses) to determine if common predictors can be identified for indicators of capacity and performance. And second, they can be used to model brain processes to determine if each type of assessment is dependent on similar neural processes (such as the brain sites involved, level of activation of brain areas, and so on). Of course, if it is found that quality-of-life or HRQOL responses have a definable neural analog, this would reinforce the notion that these qualitative responses are embodied, not just linguistic representations (such as abstract numbers). It will also raise the issue of how embodiment and neural networking can be integrated.

To clarify these issues, I need to determine if data are available for modeling the relationship between neurocognition and quality-of-life or HRQOL. But there is a second related issue that focuses on the role that emotions play in these assessments. I will delay this topic until Chap. 11 where I will primarily discuss hedonic approaches to quality-of-life or HRQOL assessment and their emotive foundation.

As a general operating principle, I assume that neurocognitive indicators refer to a person's capabilities or capacities, and that these indicators have a strong biological foundation that can be altered by changes in the physiological status of the person (that is, they are often modular in nature).[3] I ordinarily use language to describe a person's capacities, but I have learned this is not without its ambiguities, since words that refer to both cognitive or behavioral indicators can be used to describe what a person is capable of, and also their performance (e.g., I can talk about "performance capacity"). I had to deal with this type of confusion when discussing the term "function" in the previous chapter, where the same term was used to refer to what I "can do" and "do do" (Jette and Badley 2000; p. 387). Patrick and Erickson (1993; p. 378) dealt with this semantic instability and potential empirical confounding by suggesting that a research limit quality-of-life or HRQOL assessments to performance indicators of various functions. But I will not be able to do that, since I have

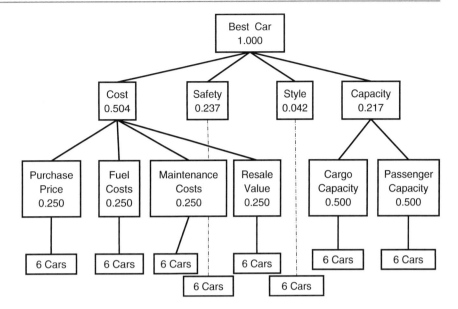

Fig. 10.2 An example of an Analytical Hierarchical Process as applied to the decision to buy a car.

2 Finding a Common Metric: The Qualification of Objective Indicators

taken on the task of clarifying the relationship between neurocognitive capacity and performance on a quality-of-life or HRQOL assessment as part of determining how these types of data can be combined in a single assessment.

Clearly, there is much to discuss. I will consider each approach to the combination task in order. First, I ask if data exist that can illustrate that objective data have been qualified, providing a basis for combining similar valued-based subjective data.

2 Finding a Common Metric: The Qualification of Objective Indicators

The qualification of any objective indicator is really no different than the qualification of any subjective indicator. Consider how I can judge which car to buy. To answer this problem, I will again rely on techniques developed in operations research, but now apply the methodology developed by Saaty in the 1970s called the *analytical hierarchy process* (Saaty 1977, 1980, 1983, 2009; AHP). AHP is considered a form of multicriteria decision analysis and relies on the paired comparison methodology developed by Thurstone (1927). AHP provides me with a method of qualifying an object. It does so by first identifying and then comparing the relative importance of a number of characteristics of the objective indication. Most important, this procedure provides a subjective analog of the objective indicator that I can use to compare to another subjective indicator.

Say I am trying to decide which one of six cars I should buy. AHP has me applying a specific set of criteria (Fig. 10.2) to each of the six cars.[4] My first step is to identify the specific criteria I think I should use to decide which car is best. In Fig. 10.2, these include cost, safety, style, and capacity. Next I would consider what about cost is important to me, and I can decide that purchase price, fuel cost, maintenance, and resale value would be important. The issues of safety and style don't seem to have subcriteria, but my concern about capacity does, and I identify the cargo and passenger capacity as separate issues that might affect my decision.

Having laid out the elements that will go into my decision, I now have to make my decision. According to AHP, I can first do this by comparing the importance of each criterion relative to the other. At this point, I am free to consider what is known about each of these criteria. Thus, the cost of each car should be known, but the resale value has to be estimated. In addition, some variables may rely on my preferences (e.g., the style I prefer, handling, and so on). With this range of information, I will next consider what is more important to me: the cost vs. safety, cost vs. style, cost vs. capacity, and so on. Next I will make the same importance judgment for each of the subcriteria, again asking what is more important to me: purchase price or fuel costs, purchase price or maintenance costs, and purchase cost or resale cost. I also will decide if cargo capacity is more important to me than passenger capacity.

Now I am ready to apply my importance ratings to the cars I am considering.

As is evident from Fig. 10.2, I have eight subcriteria to apply to each of the six cars. Since I want to decide which of two subcriteria is more important for a particular car, I will generate 15 decisions per car. Fortunately, there are simple mechanical devices to record the data and software to summarize these data. However, some of the judgments involved here may not be amenable to a yes/no decision. For this reason, I will be given a scale ranging from 1 to 9 to decide upon the degree of importance for each of these judgments. For example, deciding between cost and safety for a particular car may be quite difficult, so I will use a graded decision system.

I will not get involved in the mathematics used here, but a well-validated set of procedures exists to transform these data into a specific decision (Saaty 2009). The AHP has been used in a variety of situations, from deciding on a leader of an organization, to what characteristics of the environment are important to a person, to any situation where multiple factors determine an outcome. I have included it here because it so obviously illustrates that decisions about objects, as well as subjective indicators, can involve qualification. In other words, qualification refers to the fact that any indicator can be "made of the person," and this can be formally assessed as when a utility estimation occurs (e.g., using standard gamble methodology), or informally as when a person places an importance rating on one of two characteristics of an object.

Thus, the AHP methodology takes what appears to be a subjective assessment of an object and quantitates it. By assigning numbers to the judgments, this makes the judgments amenable to mathematical and statistical manipulation. The existence of procedures of this sort provides an alternative to the efforts of the social indicator movement of combining objective and subjective indicators, without qualifying the indicators that determine quality-of-life (e.g., crime rate, poverty index). In addition, it may provide a basis for a comparison with a self-reported cognitive assessment.

Say, for example, that I am a neuropsychologist who used NPT to assess a person with an intellectual deficit. I have a number of options in presenting my data. I can present it as a series of numbers that reflect the degree to which the person conforms or deviates from some population norms. Or I could take each of these neuropsychological tests, use the AHP methodology to compare pairs of these tests, and end up with ratings that reflect what I consider to be clinically most important for the quality-of-life of the person in question. Naturally, I would have to identify criteria and subcriteria that I think would be clinically important, but once established, I could proceed to estimate these qualities for each of the neuropsychological tests. By so doing, I will have weighted each of the neuropsychological tests by my subjective assessment of the clinical significance of the test. These estimates would, therefore, provide me with a subjective, but qualitative, metric for the test battery of neuropsychological tests I am using, and now I can compare my subjective estimates with what was rated as important (e.g., as assessed by the frequency, the item is identified by a person's self-reports) by the subject.

I have examined the literature for instances where neuropsychological tests had been formally qualified in some manner, but I was not able to find any. I was also not able to find examples of value-weighted descriptive statements about a person's neurocognitive status based on NPT, or formal ratings of the person's behavior by an independent observer which have been qualified. Rather, what seems to be the case is that most comparisons involved comparing objective measures of NPT results and self-reports (only some of which dealt with cognitive issues). In my review, I examined the literature for persons with epilepsy or cancer, persons who experienced a TBI or stroke, and persons who were diagnosed as being schizophrenic and found that correlations between quality-of-life and NPT assessments regularly revealed either no statistically significant correlations, or correlations that accounted for a small proportion of the variance, but were statistically significant. The only exceptions were for persons with brain cancer or schizophrenia. Based on this review and these data, I concluded that while it is theoretically possible that objective indicators, such as NPT results, can be qualified, there were no examples of such data in the literature. I concluded from this review that the current state of knowledge does not permit the conclusion that it is possible to find a common metric that can combine NPT results and self-reports, cognitive or otherwise. Next, I consider a second approach where I ask if self-reports of a person's cognitive status can substitute for NPT results, sidestepping the need to find a common metric.

3 The Substitution Hypothesis

Ideally, substituting a self-report for NPT results would involve demonstrating that self-reports can generate the same information as NPT results. However, people don't ordinarily use the language of a neuropsychologist to describe their cognitive status. People don't discuss their short- or long-term memory capabilities in terms of the number of words they remember, or their executive functioning in terms of their completing complex cognitive tasks, rather they express themselves in more general and experiential terms. To get an idea of the language people might use, Table 10.1 summarizes some typical cognitive items found in several HRQOL assessments. These items should be considered a first approximation of how people express themselves, as in a direct interview. Hopefully, the cognitive process being assessed, in each of these items, was sufficiently clear so that it is possible to map these items to what is being assessed in specific NPTs. For example, if a person claims that they are thinking more slowly, then I can use NPT to assess a person's reaction time, ability to attend, speed of recall of familiar and newly learned material, and so on. Each of these indicators should give me some sense of the person's "thinking speed." Other items may be more difficult to study because of the nonspecificity of the time referral or the lack of appropriate data. For example, if a person claims they are making more mistakes than usual, then I would have to know what their error rate was at some earlier time to get the significance of their complaint. This is usually not possible, so the person is compared to some other group, such as their peers.

In a unpublished study (Barofsky in preparation), I address the substitution issue in studies that reported NPT,

Table 10.1 Examples of Items found in available assessments that deal with cognitive issues

Assessment	Items
SIP[a] 5 of 10 items domain	I have difficulty reasoning and solving problems, for example: making plans, making decisions, learning new things. (Item # 8.4)
	I forget a lot, for example: things that happened recently, where I put things, appointments. (Item # 7.8)
	I do not keep my attention on any activity for long. (Item # 6.7)
	I make more mistakes than usual. (Item # 6.4)
	I have difficulty doing activities involving concentration and thinking. (Item # 8.0)
FACT-Cog (Version 3)[b]	I have had trouble forming thoughts
	My thinking has been slow
	I have had trouble concentrating
	I have had trouble finding my way to a familiar place
	I have had trouble remembering where I put things, like my keys or my wallet
Stroke Impact Scale[c]	In the past week, how difficult was it for you to remember things that people just told you?
	In the past week, how difficult was it for you to remember things that happened the day before?
	In the past week, how difficult was it for you to remember to do things (e.g., keep scheduled appointments or take medication)?
	In the past week, how difficult was it for you to remember the day of the week?
	In the past week, how difficult was it for you to concentrate?

[a] The SIP is a behavioral checklist (Bergner et al. 1981). The respondent is asked to indicate which of the items is true for them, and then the ratings (based on ratings by clinicians and other patients) listed at the end of the item are summed to provide an estimate of the impact of the respondent's illness on their cognitive functioning

[b] The FACT-Cog (Jacobs et al. 2007) is a 37-item questionnaire that can be added to the FACT-G. Of the 37 items, 20 address perceived cognitive impairments; four items address concerns of others about the person's cognitive status; 9 items refer to a person's cognitive abilities; and 4 address the impact of cognitive functioning on quality-of-life. I have selected the first 5 items from the perceived cognitive impairments to illustrate how cognitive functioning is assessed

[c] The Stroke Impact Scale Version 3 (Duncan et al. 2001) is a 60-item assessment covering 6 topics. The items displayed are from the section dealing with memory and thinking

quality-of-life, or HRQOL assessments for persons with epilepsy, cancer, traumatic brain injury, stroke, or schizophrenia. An informative example of this type of study was reported by Perrine et al. (1995), who studied persons with epilepsy. Their study involved a battery of ten neuropsychological tests and a measure of mood that was factor-analyzed, leading to the identification of six factors. These factors were then correlated with specific subscales (that included a varying number of items) of the QOLIE-89 that focused on attention/concentration, memory, and language. They found that the mood factor was correlated with self-reports of attention/concentration ($r=-0.54$; $P<0.05$), memory ($r=-0.42$; $P<0.05$), and language ($r=-0.37$; $P<0.05$). In contrast, only 4 of the 15 correlations between factors based on neuropsychological tests (five factors) and neurocognitive self-reports (three types) were statistically significant. Regression analyses revealed that mood accounted for 28.7% of the variance of attention/concentration cognitive self-reports, while neurocognitive factors accounted for noticeably less variance (verbal memory, 3.6%; psychomotor speed, 2.9%; and cognitive inhibition, 2.5% explained variance). Mood, therefore, accounted for 75.5% (28.7%/38.0%) of the total variance. Mood also accounted for 68.5% of the variance in self-reports of memory (17.2%/25.1%) and 53.5% of the variance in self-reports of language (14.6%/27.3%). Not reported were regression analyses that excluded mood, although the partial correlations suggested that the factors that were statistically significant would remain so. Still, in order to assess the clinical significance of the overlap between the indicators, it would be important to know the magnitude of the variance accounted for by the indicators without the influence of mood. To answer whether a cognitive self-report could predict NPT results would require a different experimental design. However, the data that were reviewed were sufficient so that I was able to conclude that it did not provide support for the substitutability hypothesis.

The only groups that appeared to provide modest support for a relationship between NPT results and self-reported cognitive items on a quality-of-life assessment were brain cancer patients and persons with schizophrenia. A review of the history of these studies revealed that, at first, investigators argued that some composite of the NPT results could substitute for a qualitative assessment (e.g., Dennis et al. 1996; Johnson et al. 1994). Alternatively, some studies tried to map individual neuropsychological tests to a qualitative indicator (e.g., Li et al. 2008), while some investigators (e.g., Giovagnoli et al. 2005, 2009) adopted the Perrine et al. (1995) approach of performing factor analyses to reduce the number of multiple comparisons for both the NPT results and the qualitative assessments.

The Li et al. (2008) study, which dealt with monitoring the consequences of radiation treatment of brain metastasizes,

Table 10.2 Neuropsychological testing and qualitative assessment of persons with primary brain cancer[a]

Reference	Type of cancer (N)	Treatment (surgery in all cases)	Qualitative assessment	Statistical analysis
Choucair[b] et al. (1997)	Astrocytomas (126)	Radiotherapy	ADLs	Total ADLs/MMSE=−0.46[c]
Jason et al. (1997)	Astrocytomas (28)	Radiotherapy	FLIC	Total FLIC/Total NPT=0.32; $P>0.10$
Giovagnoli (1999)	Brain tumors (57)	Chemotherapy/radiotherapy	FLIC	Total FLIC/TMT-A=$r=-0.39$[c]; TMT-B=$r=-0.38$[c]; attentional matrices $r=0.36$[c]
Klein et al. (2001)	Glioma (68) two control groups	Anti-epileptic drugs/corticosteroids	SF-36	SF-36 MCS/information processing $r=0.33$[c]
Fliessbach et al. (2005)	CNS lymphoma	Chemotherapy	EORTC-QLQ-C30	GFS of the EORTC-QLQ-C30/composite NPT $r=0.632$[c]
Giovagnoli et al. (2005)	Glioma Astrocytoma (94)	Chemotherapy/radiotherapy	FLIC	FLIC total score/composite NPT $r=0.26$; $R^2=0.59$[c]
McCarter et al. (2006)	Brain tumors (100)	Various types of treatment	HUI 2/HUI 3	HUI 2/MMSE; $r=0.36$[c] HUI 3/MMSE; $r=0.30$[c]

All abbreviated qualitative assessments are defined in the abbreviation list
[a] All cited studies include NPT. Studies cited present correlational or regression analyses
[b] The Choucair et al. (1997) study was included even though it used the MMSE, a screening tool, to assess neurocognitive status, and ADLs to assess quality-of-life. It did report correlations between these measures
[c] Indicates statistical significance; $P<0.05$

provided supportive evidence of the substitutability hypothesis. Here it was demonstrated that while the correlation between self-reports (FACT-BR) and NPT results remained modest, ranging from −0.429 to 0.327, the frequency with which these correlations were statistically significant was much higher than previous reports. For example, when the ten brain-specific items on the FACT-BR were correlated with eight neuropsychological domains at baseline, six of these domains were statistically significant, and this reduced to five and four of the eight domains at 4 and 6 months postradiation, respectively. This suggests that when the target of the treatment and the expression of the disease involve the brain, the chances increased that a self-report could be a useful and efficient alternative indicator to a battery of neuropsychological assessments. However, as these data indicated, this association appears to diminish as a person recovers from their initial condition. Overall, this method did not appear to account for much of the variance in the data.

An alternative approach to the assessment task (Giovagnoli et al. 2005, 2009) involved reducing the number of multiple comparisons necessary to draw conclusions by using a factor analysis of either the neuropsychological test data or the data generated by the qualitative and psychosocial assessment. For example, Govagnoli et al. (2009) studied a group of persons whose disease expressed itself in the brain (only 15 of 72 were primary brain cancer patients). The researchers did a factor analysis on the data of 15 neuropsychological tests and found that three factors (control functions, cognition, and memory) accounted for at least 70% of the variance in responding. A factor analysis based on the multiple quality-of-life and psychosocial assessments resulted in four factors and accounted for 66.4% of the variance. Correlating these three factors on the NPT with the total score from the WHOQOL-100 and its subscales (eight data points) revealed that five of the 24 potential comparisons were statistically significant ($r=$ranged from 0.39 to 0.51). This suggests that approximately 21% of the comparisons were statistically significant.

Table 10.2 summarizes all seven studies that included either correlations or regression analyses for persons with primary brain cancer.[5] Not shown in Table 10.2 were the seven studies that mentioned the association between NPT and a qualitative analysis, but did not provide a correlation. A review of these studies revealed that most often no data were presented because the investigator did not find statistically significant correlations. If only 30 of 81 studies were eligible for the analysis, and only 14 focused on the association between the indicators of interest, then 43% of the 14 studies were supportive of the substitutability hypothesis.

What was evident was that the issues raised by the substitutability hypothesis were not of great research interest to investigators, especially since half of the eligible studies did not report an analysis of the association. What could account for this? One reason may be that investigators are concerned, if not convinced, that self-reports from persons whose disease or treatment is in the brain may be intellectually compromised, and that this would raise questions about the validity of responses to any qualitative assessment. Consider Meyers et al. (2000) who report a study dealing with patients ($N=80$) with recurrent malignant glioma. The investigators were interested in whether cognitive status or HRQOL predicted survival. They found that cognitive status as assessed by NPT, but not HRQOL, were independent predictors of survival. The authors state:

> The assessment of subjective QOL in patients with neurological deterioration is difficult. Many patients have a diminished

appreciation for their current circumstances and lack insight into their deficits. Thus, the patient may experience a good QOL even though observers might infer that QOL should be poor…It takes grossly intact cognitive functioning for a patient to adequately assess QOL issues, so it is not surprising that QOL scores are not strongly related to outcome in this population. (Meyer et al. 2000; p. 649).

These comments are of interest because they provide a potential explanation for why substitutability has been so difficult to demonstrate. The data suggested that if a person has become neurologically degraded, they would be expected to be intellectually limited and therefore less able to discriminate their qualitative state. This would lead to greater response variability or lack of responsiveness when a person's qualitative state was monitored over time. The extent to which this is so and how a qualitative researcher deals with these compromising issues is a major concern and will occupy the remaining portion of this chapter.

Complicating the Meyers et al. proposition is that neurocognitive intactness is a necessary, but not sufficient, condition to account for the relationship between neurocognitive capacity and self-reports. For example, it is possible that a person's neurocognitive status, as assessed by NPT, could be in the normal range, yet the person could still report a decrement in their cognitive status. A number of studies support this contention. For example, Saykin et al. (2006) report a study with older adults in which they found that persons who complained about memory problems had evidence of cortical deficits, as confirmed by neuroimaging, even though their NPT performance provided no evidence of these deficits. Smith et al. (2007) and Stewart et al. (2008) provide additional empirical support for this contention. One explanation of these findings is that brain function has been preserved in the face of structural deficits. I will discuss how this is possible in the next section.

Meyer and her colleagues were concerned about the situation where a person had demonstrable neurocognitive deficits, by NPT, but do not report such deficits or do they report deficits in quality-of-life as normal. Of course, there are two separate issues here, since a person may or may not be acutely aware of their neurocognitive deficits, yet still value the circumstances and conditions of their life. This simply reflects the capacity of human beings to adapt to their circumstances. Complicating the interpretation of these data is the observation that aging and specific disease processes (e.g., Alzheimer's disease) may lead to a degradation of linguistic and cognitive functioning, while recovery of function is possible following an acute neurocognitive event. Thus, a person's neurocognitive status is a dynamic process reflecting the brain's capacity to preserve, systematically change, and recover functions. Each of these topics will be reviewed in the next section to help understand some of the circumstances that might exist when a person is asked to participate in a qualitative assessment. The next section will also help clarify the relationship between a person's neurocognitive capacity and their response on a qualitative assessment.

4 The Preservation of Neurocognition

4.1 Introduction

As previously stated, a quality-of-life or HRQOL assessment is an experience that is mediated by various cognitive processes (Fig. 3.1), and as such, is intimately dependent upon the integrity of the person's underlying neural structure. But how do these neural structures facilitate the judgment (e.g., forming the hybrid construct) that underlies a qualitative judgment? To start addressing this question, I will assume that, at minimum, I must have some idea of how various cognitive processes are represented at a neural level. Posner et al. (1988) provided such a model when they hypothesized that:

…elementary operations forming the basis of cognitive analyses of human tasks are strictly localized. Many such local operations are involved in any cognitive task. A set of distributed brain areas must be orchestrated in the performance of even simple cognitive tasks. The task itself is not performed by any single area of the brain, but the operations that underlie the performance are strictly localized. (p. 1627).

Thus, cognition is a product of the interaction of multiple but distributed local sites, not unlike what Hayek proposed (Chap. 3, p. 51). How efficiently these multiple sites are organized into a cognitive system is reflected in the potential of the system to perform a variety of tasks. The magnitude of this potential, and therefore the potential to make qualitative judgments, is reflected in a person's cognitive capacity. As I will illustrate, the available evidence supports the existence of cognitive capacity and helps clarify the relationship between capacity and performance.

First, however, I need to provide some background on the literature that deals with cognitive capacity or reserve, as well as cognitive preservation. This will involve learning about some of the operating characteristics of the central nervous system. For example, a number of studies (e.g., Just and Carpenter 1992) have demonstrated that the brain normally has the capacity to perform only a limited number of tasks, either simultaneously or consecutively. This suggests that the brain acts as if it is an "economic system," where the constraints distribute a limited resource (or cognitive reserve) and this is reflected in how well the brain performs its various functions.

Handy (2000) describes the concept of cognitive capacity as follows:

The central tenet underlying capacity theory is that processes can proceed only if sufficient 'capacity' remains available… Translated in the neural domain, when the load of a cognitive process is increased – and the proportion of capacity 'consumed' by that process rises – there is a corresponding increase in activation intensity at the cortical loci where the loaded process is implemented…Within this context, the consequences of capacity limits can be observed as either a ceiling on the activation of a loaded process…or as a reduction in the activation of a second process that is in competition for capacity with the loaded process…Predictions of function-related covariation directly follow from this relationship between load and activity intensity: when the load on a process is manipulated, positive covariation between cortical loci suggests a sharing of processing capacity, negative covariation suggests a competition for capacity, and no covariation suggests an independence of capacities. (p. 1066).

Thus, the neurological analog of capacity for Handy (2000) is the activation of neural structures, the load created by this activation, and the relationship between cortical loci. A critical component of this capacity model is the existence of a limit beyond which cognitive functioning would not be expected. Combining the Posner et al. (1988) model of how neural sites are organized into a cognitive system and Handy's (2000) description of the dynamics of this system provides a working model of the neurological basis for cognition.

Exactly how neural events lead to what I linguistically refer to as "cognitive capacity" brings me back to the discussion about how to bridge the gap between symbolic systems, such as language and reasoning, and its representation in the brain, as a parallel numerical processing system (e.g., Smolensky 2004). Handy was suggesting that the elements of neuronal interaction operationalizes what he meant by capacity (a linear model), but Smolensky and others (e.g., Long et al. 1998; Prince and Smolensky 1997; Rumelhart et al. 1986) would state that a neural network (a nonlinear model) best models neuronal interactions. In addition, the nonlinear interaction has the additional characteristic of being able to optimize the outcome of such a system. For Prince and Smolensky (1997), therefore, cognitive capacity would actually reflect the optimized functioning of a cognitive system. This notion that optimality emerges from the interaction of the elements of a system is characteristic of a self-organizing complex adaptive system.

Rumelhart et al. (1986), Prince and Smolensky (1997), and others claim that a neural network model can also be used to account for symbolic and reasoning systems. Before I discuss this, however, I would like to provide some background as to why an alternative way of thinking about these issues would be in order. As I proceed, I need to be alert to the role figurative language plays as part of my discussion. For example, when Handy (2000) speaks about cognitive capacity, he implies that some quantity exists that is "generated" or "consumed" as cognition proceeds. Is this something that he actually thinks happens, or is he speaking figuratively? And if he is speaking figuratively, does this help or hinder my effort to understand the relationship between capacity and performance? When I described the brain as an "economy" with a bank reserve that can be depleted, do I think this is literally true? These are the types of questions I ultimately have to answer in order to decide on the model that offers the most feasible perspective. I will start my discussion by first describing what is known about cognitive capacity and its preservation.

4.2 Neurocognitive Preservation

A variety of clinical findings suggest that a person can preserve their cognitive functioning even under conditions where their capacity to perform may be impaired from aging or brain damage. For example, Katzman et al. (1989) described 10 cases of persons whose cognitive performance appeared to be normal, yet who showed evidence of advanced Alzheimer's disease following postmortem brain autopsies. In addition, clinicians report seeing patients who have had strokes of comparable severity, yet differ in their degree of functional impairment.

These and other clinical manifestations have prompted investigators to speculate about how this preservation could occur. In summarizing this literature, Stern (2002, 2003, 2007) has proposed that both passive and active neurocognitive mechanisms are involved (Table 10.3). For example, the so-called passive models (Satz 1993) rely on a threshold being passed before a change of state or clinical condition, such as Alzheimer's disease, can be declared. These models are quantitative in that a certain number of neurons have to become nonfunctional or synapses disrupted before a threshold is reached and impacts a person's performance. The speed at which this threshold is reached may depend on the size of the person's available neuronal pool, and this would be indicated by the person's head size or brain volume.[6]

Passive models rely on the number of neurons available as an indicator of a person's cognitive capacity (e.g., head size or brain volume); this is an application of the CONTAINER METAPHOR. In contrast, the active or cognitive capabilities models can occur in two forms, each of which relies on how the resources of the brain are used (qualitative methods), but also is depend on whether the person is or is not brain damaged. Thus for Stern (2003), cognitive reserve refers to the variability seen in noninjured brain function, while cognitive compensation involves the use of alternative brain structures or recruitment of alternative neural networks to preserve performance as a consequence of brain damage. Here the conceptual metaphor TIME IS MOTION is being used, since cognitive preservation occurs over time, and involves the development of a different neural pattern of responding. Thus, Stern (2002) uses two different conceptual metaphors to account for the passive and active models depicted in Table 10.

The role of cognitive preservation can be illustrated with a hypothetical quality-of-life or HRQOL study. Say I was

Table 10.3 Definitions of Cognitive capacity and preservation[a]

Cognitive capacity: passive (quantitative) models	Cognitive preservation: active (qualitative) models
Threshold Models	Cognitive reserve
Neuron depletion	Differential recruitment of neural networks
Brain size hypothesis	Efficient performance of neural networks
	Compensation
	Alternative neural networks

[a]Modified from Stern (2002). Reproduced with permission of the publisher

studying a group of young and older respondents and found that each group rated their quality-of-life or HRQOL about the same. A simple interpretation of these data would suggest to me that quality-of-life or HRQOL remained stable as a person aged. However, I also know that the older respondents have lived physically and emotionally demanding lives that would be expected to impact their capacity, as well as their capability to rate their actual life quality. I might also be concerned that the stability of qualitative responding with aging may actually be artifactual, with the older person's rating being actively preserved by unknown cognitive mechanisms, social desirability effects, personality characteristics, and so on.[7] To help deal with this situation, I may want to know whether the two groups were basing their ratings on the same thought processes (that is, were they cognitively equivalent), or whether they were using the same or different neural circuits (i.e., compensating) to accomplish their ratings. The answers to these questions are important because the validity of the assessment process may depend on them. As an investigator, I hope that what I demonstrate psychometrically (that is, respondents are making the same ratings) would also be true cognitively (the same thought processes would underlie these ratings).

Stern et al. (2005) provided a partial answer to these questions when they reported the results of a study that used positron emission tomography (PET) to scan 17 older (mean age = 70.9) healthy and 20 young (mean age = 23.4) adults performing a visual recognition task. Stern et al. (2005) defined active cognitive preservation (reserve or compensation) as the sum of a person's years of education, their performance on the National Adult Reading Test (NART; Nelson 1982), and the verbal score from the Wechsler Adult Intelligence Scale-Revised (Wechsler 1981). The two groups were not significantly different in terms of their cognitive reserve scores.

An important part of Stein's et al. (2005) procedure was to invite the respondents into the laboratory a day before the PET scan in order to determine each subject's range of accuracy in performing both a low- (one-on-one target and lure matching) and high-demand (one target and on multiple lure matching) visual task. They then selected the 75th percentile of this range to optimize the difficulty of the task for each person. Thus, the difficulty of the task was individually calibrated so that it should not be surprising that the study list size varied significantly between the young ($N = 13.9$) and elderly ($N = 7.5$) adults. However, when given their tasks, each participant was judged within the range that one was capable of, and the investigators found that the percent correct judgments were the same for the low and high visual demand tasks. They also found, as might be expected, that the elderly were not as able as the young adults to remember material presented to them 24 h before.

What Stern et al. (2005) were interested was determining if the PET scans they were inspecting would reveal that each group recruited the same cortical areas, in the same way, as they completed their tasks much as Posner et al. (1988) suggested would happen. They found, to their surprise, that the young responded to the increasing demands of the recognition tasks by efficiently utilizing their available neural circuits, while the older persons engaged in neural compensation. Thus, those young adults with higher cortical reserve increased the activation of cortical areas that were positively loaded to the task and decreased the activation of the areas where the loading to the task was decreased, suggesting that they were using intracortical dynamics to regulate their cortical response. In contrast, those older respondents with higher cortical reserve did the opposite: they decreased their response to the cortical areas that were more activated by the load created by the task and increased their response for areas where the load was less. This suggested to Stern et al. (2005) that the older respondents were taking advantage of an altered or alternative cortical pathway to perform their task, and in this way, they were considered to be neurally compensating.

The design of this study is particularly interesting, since by calibrating each subject, the investigators created a situation where the subjects were psychometrically equivalent (that is, their scores were equally as accurate within the person's individually defined range), so that if variation in cortical function occurred, it could not be attributed to performance demands of the task. This permits the conclusion that these two groups were not cognitively equivalent. Instead, it seems that two different patterns of neurocognitive optimization occurred under conditions where the accuracy of performance, as defined for each individual, was preserved.

This study is of interest because it helped to answer the questions I raised earlier with my hypothetical study (see above). In that study, I speculated about whether the two groups were thinking differently about their rating tasks and may have accomplished them using different neural pathways, even though their ratings were the same. The Stern et al. (2005) data suggest that something like this could happen. A major difference between these two studies, of course, is that the hypothetical study focused on a quality-of-life or HRQOL assessment, while Stern et al. (2005) used a visual recognition task. If it turns out that this difference makes no difference, then drawing the same conclusion seems warranted, including raising concerns about the validity of such a qualitative assessment process. It also raises the possibility that the qualitative outcomes observed were optimized.

Cannon et al. (2005) extend the findings of Stern et al. (2005) to persons with a specific disorder. They used fMRI to study working memory for persons with schizophrenia ($N=11$) and healthy controls ($N=12$). Participants were asked to recall a visual stimulus after a 6-s delay (maintenance phase) or after the stimulus had been rotated (maintenance and manipulation phrase) and delayed for 6 s. In a second study, progressively more complex stimuli were presented by increasing the number of sets presented before the recall test. Participants were given practice sessions and informed about how to perform the task correctly. Cannon et al. (2005) described the purpose of the study as "…to determine the nature of any difference in regional brain activation when comparing patients and controls on conditions in that the two groups show equivalent behavioral accuracy" (p. 1072). Both accuracy and reaction time data were collected. Examining the results of the study revealed that both performance on the memory tasks (not reaction time) and the activation of specific brain areas differed between the two groups. However, when a subsample from each group (Task I, eight patients and seven controls; Task 2, ten patients and ten controls) was selected so that they were matched in terms of behavioral performance, then the difference in the pattern of neural activation between groups persisted. Thus, as was found with the Stein et al. (2005) study, two different groups of people can accomplish tasks at a comparable level of accuracy, yet cognitively achieve this in different ways.

A third example was reported by Cader et al. (2006), who compared a group of 21 persons with relapsing-remitting multiple sclerosis who did not have clinical evidence of memory deficits with a group of 16 age- and sex-matched controls. Each subject was given a progressively more difficult visual memory task concurrent with fMFI brain scanning. The two groups performed the visual memory tasks equally as well, and the initial criterion of psychometric equivalence was established. Brain imaging studies revealed that the task activated similar cortical areas in each group. However, as the tasks became more difficult, the persons with multiple sclerosis showed relatively reduced activation in the superior frontal and anterior cingulate areas, with less cortical areas being recruited then found for controls. Also observed was that the persons with multiple sclerosis activated both right and left prefrontal cortices; this did not happen for the control subjects. The authors interpret their data as suggesting that changes in interhemisphere dynamics constituted an adaptive mechanism that could limit the clinical expression of a person's disease. This occurred in addition to the differential activation of the cortical processing areas usually recruited during the performance of a particular task. I will come across this phenomenon again when I discuss the impact of a stroke on metaphoric expression (Sect. 6.4). For now, however, the Cader et al. (2006) study is another demonstration of the fact that the performance by two groups on a task that may be psychometrically equivalent can also be neurocognitively different.

What factors contribute to cognitive reserve or compensation (e.g., Scarmeas et al. 2004; Stern et al. 2005) for an individual? Principal among these factors are years of education (e.g., Letenneur et al. 1994; Snowdon et al. 1996; Stern et al. 1992); participating in cognitively stimulating activities (Wilson et al. 2003); physical activity (Dik et al. 2003), and other lifestyle activities (e.g., Scarmeas and Stern 2003). However, there are also reports of studies based in India (Chandra et al. 2001) and West Africa (Hendrie 2001) of people who report quite low levels of education and literacy, yet who have some of the lowest rates of dementia observed. These studies suggested that significant contribution to cognitive reserve or compensation may occur that are not dependent on educational attainment. In addition, relying on education level has been found to be suspect, since years of education do not take into account differences in the quality of education a person receives. Thus, Manly et al. (2003) have suggested that literacy level would be a more appropriate indicator of cognitive reserve or compensation. While the debate continues regarding what specific factors determine resistance to developing dementia, the question has been raised of how this resistance would neurophysiologically occur.

Liao et al. (2005), for example, reported a study using SPECT imaging and scanning of a group of 132 Alzheimer's patients to determine if cerebral perfusion varied as a function of the person's level of education. The assumption here is that years of education would be represented by changes in brain physiology. Cognitive status was measured by a number of neuropsychological tests. They found that years of schooling were negatively correlated with cerebral perfusion, but positively correlated with neuropsychological test scores. This was interpreted to mean that the more educated person was more likely to use alternative neural circuits. The alternative interpretation that education impacted synaptic connectivity (a passive form of cognitive preservation) would have required a positive association with cerebral perfusion.

Le Carret et al. (2003) report a study where they administered a battery of neuropsychological tests to 1,022 persons of age 66 years and older. They were interested in determining whether education, age, gender, depression, occupation, and leisure activities were predictive of various measures of neuropsychological functioning (12 measures from 7 tests). They found that age followed by education and gender were the most reliable predictors of neuropsychological outcomes.

Occupations, especially those jobs that were more intellectually demanding, were more likely to predict neuropsychological functioning. In general, education was predictive of a cluster of neuropsychological outcomes that involved controlled processing and conceptualization. For example, in the Digit Symbol Substitution Test, a person has to as quickly as possible enter a symbol that matches a model presented to them. This task involves controlling head and eye coordination as well as being able to scan and keep an image in mind as the response is executed. The deliberative nature of the cognitive task here is similar to the controlled processes I discussed earlier as part of dual process theory (see Chap. 6). Thus, when investigators speak about the relationship between education and cortical reserve, they, as revealed by the results of this study, are also addressing the capability of a person to engage in controlled cognitive processing.

Several longitudinal studies have been reported that provide indirect support for the role of active methods of cognitive preservation. For example, in the famous Nuns Study (Snowdon 2003), 678 women of a Catholic religious order agreed to be neuropsychologically tested as they aged and to have postmortem brain examinations. While only a portion of the participants have passed away to date, those that have reinforce the basic findings found in cross-sectional studies. Riley et al. (2005) have abstracted the ideas expressed from autobiographic material written when the respondents were, on average, 22 years old and related this to their adult cognitive functioning and neuropathological studies. The researchers found that those persons with low idea density when young had significantly lower brain weight, higher degree of cerebral atrophy, more severe neurofibrillary pathology, and were more likely to meet the neuropathological criteria of Alzheimer's disease. Thus, they found that early life idea density was related to latter life cognitive impairment.

Staff et al. (2004) also report a longitudinal study where the cognitive functioning of a group of 92 persons born in 1921 was assessed when they were 11 and 79 years of age. These persons also received an MRI at age 79. The investigators found that education and occupational achievement were significant contributors to variation in cognitive functioning at 79 years of age, but that total intracranial volume was not. Education accounted for 5–6% of variation in memory functioning, but was not a significant contributor to reasoning abilities. Occupational achievement accounted for 5% of variation in old age memory function, and 6–8% of variation in reasoning abilities. These data also suggest that intellectual challenges experienced during life optimize cognitive functioning in old age.

Maruyama et al. (2004) report a 2-year longitudinal study of persons who all entered the study having met the criteria of having a memory disorder. Of the 57 persons selected for the study, 16 patients remained stable over the 2-year period and 17 of the remaining 40 progressed to MMSE scores indicative of Alzheimer's disease. The remaining 24 patients declined, but not at the same rate as the patients who met the criteria of Alzheimer's disease. Regressions analyses revealed that age was the only indicator, among the indicators considered, that predicted progression to dementia. The study, however, did not include information on education and occupation that might have also predicted the rate of cognitive decline. In addition, these authors monitored cerebrospinal-tau levels that permitted them to accurately (87.7%) predict disease progression. This study illustrates that specific physiological indicators can predict the progressive decline of cognitive function.

The studies reviewed here provide insight into the means whereby cognitive preservation occurs in response to changes in the brain due to either aging or disease. When faced with a demanding task, the brain has the ability to alter either the magnitude of a neural response, or the pattern of neural organization. These studies also have implications concerning how to study changes in a person's ability to generate the hybrid construct that underlies a qualitative judgment.

Figure 10.3 summarizes the various issues I have discussed here, but now in one composite display.[8] The figure has as its primary outcome the cognitive construction of the hybrid construct, leading to a qualitative assessment. The integrity of the hybrid construct is dependent on past and current determinants of the person's neurocognitive status.

This includes a set of antecedent conditions including the person's genetic makeup, early childhood environment, educational experiences, occupation, physical health, etc., which may influence their cognitive capabilities or reserve (Fig. 10.3c), but also modulate the impact of CNS lesions (Fig. 10.3d). The person's cognitive reserve is presented as a modulating variable that alters the impact of CNS lesions (Fig. 10.3a) on the hybrid construct (Fig. 10.3b). This model also assumes that these various influences impact the performance outcomes (Fig. 10.3f), that in this case is responding to a qualitative assessment. Thus, the model conceptualizes the impact of CNS lesions on the formation of the hybrid construct not as a direct causal relationship (that is, a person's capacity causes their performance), but rather as a modulated relationship, reflecting the influence of a variety of neurocognitive indicators. This difference, if true, has important theoretical implications, and this justifies a closer inspection of alternative data sets to determine if this type of relationship can be verified.

Barnett et al. (2006) offer a model of neurocognitive reserve as having a modulating role in the relationship between some pathological event and its clinical outcome. They also suggest that this modulating role will vary as a function of the type of pathology. Thus, the loss in neurocognitive capacity as a result of a traumatic brain injury will be directly determined by the person's capability to preserve their cognitive functioning. In contrast, various environmental factors may influence the neuro-degeneration process

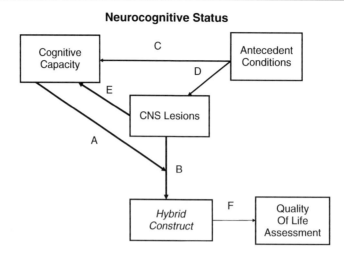

Fig. 10.3 Cognitive capacity (**A**) and CNS lesions (**B**) both contribute to the status of the hybrid construct. Antecedent conditions (e.g., stroke, socioeconomic status, and so on) can contribute to the neuroanatomical status of the CNS (**D**), as well as the person's cognitive capacity (**C**). As a result CNS status could contribute (deplete or augment) a person's cognitive capacity (**E**). Finally, the state of the hybrid construct should be reflected in the outcome of a quality-of-life assessment.

characteristic of Alzheimer's disease, and this will influence what is available for cognitive preservation. In addition, the neuro-development process that characterizes the onset and expression of schizophrenia may significantly alter the nature of a person's cognitive reserve and compensation abilities.

I have not been able to find specific studies that address the relationship between neurocognitive capacity or preservation and quality-of-life or HRQOL assessment. Yet, a variety of determinants (e.g., resiliency, implicit theories, working memory span, and so on) exist that also take advantage of conceptual metaphors (e.g., the CONTAINER METAPHOR) and that may play a role in the relationship between neurocognitive capacity and a quality-of-life assessment. Thus, a person with Alzheimer's disease whose "cognitive reserve" persists may rate their quality-of-life or HRQOL differently than the person whose disease progresses more rapidly. Some of the reported "paradoxical" outcomes (e.g., a diabetic amputee who still rates their quality-of-life as adequate) found in quality-of-life or HRQOL studies may yet be accounted for by differences in a person's cognitive preservation capabilities. Many studies can be cited that illustrate the research opportunities here. For example, a study by Salmond et al. (2006) reports that the degree to which a person becomes depressed following moderate-to-severe head injury varies with the person's premorbid intelligence (as a marker of cognitive reserve). Thus, a researcher may be able to demonstrate that the variance in the relationship between depression and quality-of-life or HRQOL may be related to the role indicators of cognitive preservation play in a person's qualitative ratings. Cader et al. (2006) report a study demonstrating that persons with early-stage multiple sclerosis vary in terms of their "functional connectivity" (as demonstrated by fMRI) as they adapt to disease progression. How would a person's rating of their HRQOL vary with such concurrent structural changes and neurocognitive adaptation?

4.3 Neurocognitive Capacity and Degeneration

A number of disorders exist that are associated with a gradual loss in neurocognitive functioning. Examples include Alzheimer's disease, amyotrophic lateral sclerosis (ALS), Huntington's disease, Parkinson's disease, multiple sclerosis, and so on. NPT results and brain imaging studies are regularly used to characterize these gradual changes that are assumed to reflect alterations in a person's neurocognitive capacity. The degeneration of a person's neurocognitive capacity and its relationship to performance can also be traced in a number of specific ways, including its impact on decision making, functional status (e.g., performance of roles), and changes in communication and language usage. Tracing changes in language usage is of particular interest, since it has the added value of providing the information needed to adjust the quality-of-life or HRQOL interrogative process to the capabilities of a person progressing towards dementia.

Monitoring these changes in performance, however, will not be a simple task, especially since it is often difficult to determine when neuro-degeneration is occurring, or whether the observed changes in performance are occurring because of normal aging. In addition, it would be expected that the pattern of degeneration would vary between individuals as a function of a person's capacity to preserve their cognitive functioning (e.g., their education and occupation), but also because of differences in how dementia occurs in response to different pathologies (e.g., vascular dementia, Alzheimer's disease, Parkinson's disease, and so on). For example, Mesulam (2003) found that primary progressive aphasia results in a language-based dementia, even though the person's memory remained relatively intact. This suggests that memory deficits alone are not a sufficient condition for the diagnosis of dementia. Acknowledging these complexities, it is still both practically and theoretically interesting to determine if language usage can be a helpful analytical tool in the study of the relationship between neurocognitive capacity and performance.

5 Neurocognitive Degeneration

5.1 Introduction

Next I would like to expand the discussion started in the last section of relating a person's cognitive capacity to the consequences of a degenerative neurological process, such as occurs during Alzheimer's disease. Here the person's cognitive capacity is being gradually altered over time, but as demonstrated here, the extent or pattern of this alteration is dependent on the size and nature of the person's initial "neural economy" (e.g., brain volume, education level, etc.). Tracing the impact of these processes on the formation of a hybrid construct will help demystify what is compromised about the "compromised person."

First, I must ask why investigators interested in qualitative assessment would be interested in studying the degeneration of neurocognition, particularly changes in language expression? They would be if they wanted to rationalize the assessment of a person who may be dementing. Thus, if it is assumed that the process of dementing resulted in impaired functioning, then this would provide an opportunity to design studies that identify what specific components of the hybrid construct were impaired and whether the adverse consequences of these events can be identified (e.g., by neuroimaging studies). Once the specific cognitive deficits are identified, an investigator could then develop alternative methods that may supplement or enhance cognitive processing by the person who is dementing. Tracking changes over time would also be helpful, both in terms of assessing any remediation efforts and in assessing the rate of change over time. At a minimum, the investigators would be able to provide empirical support to any contention that these qualitative assessments, while possibly psychometrically equivalent, were not cognitively equivalent. This approach has a distinct advantage over the current practice of degrading the meaning of items to facilitate responding to the assessment, or by using proxies as a means of estimating the quality of a dementing person's existence. In addition, this approach may be a useful method for approaching the task of assessing the severely demented person. An essential step in rationalizing the assessment of the dementing person, therefore, is for the investigator to understand how language degenerates.

Brian Butterworth in the 4 November 1984 edition of the English newspaper the *Sunday Times* illustrated the political implications of linguistic degeneration when he claimed to be able to detect subtle changes in linguistic skills in the speeches of Ronald Reagan some 10 years prior to his receiving the diagnosis of Alzheimer's disease. Such a retrospective approach is potentially confounded, but more recently, Garrard et al. (2005) described the changes in writings of a well-known author, Iris Murdoch, who also eventually received the diagnosis of Alzheimer's disease following a postmortem biopsy. Murdoch published 26 books, the last of which was written in the shadow of her diagnosis without her being aware of her clinical state.[9] Garrard et al. (2005) selected three of her books to study: her first published work; a work selected at the height of her publishing career; and her last book. For each, they performed an extensive textual analysis (including word count, word length, and so on). They found that, as is true for Alzheimer's patients in general, the author was able to retain her capacity to produce well-formed grammatically correct sentences late into her disease, even though she used fewer words and used them more frequently. The investigators interpreted this outcome as being consistent with the biopsy-confirmed temporal lobe presence of her disorder. They also consider that their data provided support for the modularity of language, as was evident from preservation of syntactic but not semantic processes.

The observation that for the person progressing towards Alzheimer's disease semantic, rather than syntactic, abilities are more likely to be limited suggests that a person's capacity to understand the meaning of the words or instructions in a quality-of-life or HRQOL assessment task may be disrupted before the person's ability to construct a sentence, or even fill out a questionnaire is disrupted. This raises questions about the meaning the person with dementia is communicating when they respond to a quality-of-life or HRQOL assessment, and whether the same "informational" weight should be given to these assessments when compared to a person who does not show such neurocognitive differences. It also raises concerns about efforts to ask persons with different degrees of dementia to participate in cognitively complex assessment tasks (e.g., using time trade-offs, or standard gamble procedure for estimating utilities).

These changes in semantic vs. syntactic processes as the person with Alzheimer's disease progresses would also be expected to interact with changes that occur with normal aging. For example, in normal aging, semantic abilities (e.g., comprehension and vocabulary skills) are not disrupted, while a decline with age occurs regarding the speed of mental operations; working memory; the ability to stay focused; the ability to inhibit attending to irrelevant information; and visual and auditory acuity (Park and Gutchess 2000). There is little difference in the young and old in picture recognition, suggesting that some aspects of memory do not decline with aging. Also, narrative speech of the young and old differ in terms of quantity, content, and cohesion (Juncos-Rabadán et al. 2005). Thus, the disturbance in language that the person with Alzheimer's disease is experiencing is occurring in the context of a neurocognitive system where only some aspects are changing as the natural order of aging.

These issues of the changes in cognitive status with aging and disease progression were impressively documented in a study by Auriacombe et al. (2006), who over a 10-year period monitored the cognitive status of 3,777 adults by administering a series of NPT, including verbal fluency tests (e.g., "name all the animals you can think of in 1 min"). During this period, 52 persons received the diagnosis of Alzheimer's disease, and the report summarizes changes in their cognitive status by comparing them to 104 age- and education-matched controls. The investigators examined both the quantity of words produced during the verbal fluency test and whether the quality of these responses declined (quality was measured by the number of repetitions of words or intrusions of nonrelevant words during response to the task).

The researchers found that compared to five years prior, the persons who were subsequently diagnosed with dementia were already statistically significantly different from controls in terms of the number of words they produced. Interestingly, both groups showed age-related declines in verbal fluency over the 10-year period, but the diagnosed person's rate of decline was markedly greater. Persons with Alzheimer's disease also were statistically significantly more likely to repeat a word in response to the fluency task than controls. Inspection of changes in the fluency task revealed that there was no difference than controls in repetition rate 5 year prior to the diagnosis of Alzheimer's disease, but there was 2 years prior the time of diagnosis.

The verbal fluency task, while simple to administer, is a relatively complex cognitive task, evoking several specific cognitive processes, including the executive control mechanism required for cognitive flexibility, working memory, effortful and strategic recall, and inhibition of recalled material. As such, evidence that decrements in performance of the task significantly predates the diagnosis of Alzheimer's is impressive and was consistent with the Garrard et al. (2005) report concerning the natural history of changes in Iris Murdoch's productivity. This study also has relevance to my concerns about what a person with incipient and demonstrated dementia means when they respond to a quality-of-life or HRQOL assessment.

A study by Godbolt et al. (2004) provides further information on these issues. In their study, the neurocognitive status of a cohort of 19 persons with familial Alzheimer's disease,[10] 8 of which had undergone presymptomatic assessments, were assessed over a 1–10-year period (mean=5 years). The investigators used three types of markers to monitor disease onset and progression, including the time of onset of symptoms; when MMSE scores dropped <24; and an "impaired" score on a variety of neuropsychological tests. Using their data, they proposed a temporal sequence of events that would be associated with the progression to the diagnosis of Alzheimer's disease. Included in their model were disruptions in memory and intellectual functioning, which would precede a drop in an MMSE score of <24, and this was followed by disruptions in a person's capacity to calculate, navigate visual space, perceive visual events, be able to name objects, and spell, in that order. This sequence suggests that concrete motor-dependent processes are more likely to be preserved than abstract verbally dependent processes. As the authors state, "The very late preservation of spelling is remarkable, presenting an island of preserved cognitive function when subjects were moderately or severely affected" (Godboit et al. 2004; p. 1746). There was also some evidence of neurological disturbances (e.g., myoclonus in 11 of 18 subjects) that usually preceded early impairment in episodic memory and general intelligence.

The Godbolt model provides a first approximation of the natural history of the degeneration of cognition associated with the progression of Alzheimer's disease. These observations permit several conclusions. First, as was also suggested by the Garrard et al. (2005) study, semantic, rather than syntactic, processes (e.g., recalling the meaning of words) are more likely to be disrupted by the initial progression of the disease. Second, concrete motor-dependent processes (e.g., visuospatial, visuoperceptual, naming, and spelling) are more likely to be preserved than abstract verbally dependent processes (e.g., verbal memory, performance IQ).

These data also raise the interesting question of whether language is dependent on two functionally independent systems (e.g., modular and nonmodular) that could be conceived of as either two dissociable parts of a single system, or two genetically separable and structurally based modules (Pinker 1994). A third alternative, as Bates et al. (1995) has proposed, would be based on a parallel distributed processing model (Mc Clelland et al. 1989) that integrates semantic and syntactic aspects of sentence comprehension into the connections of the same neural network. The Bates et al. (1995) approach, therefore, claims that it would not be necessary to represent semantic and syntactic processes as separate entities.

Evidence that semantic and syntactic aspects of language are represented by two structurally distinct modules includes the observations that persons who have a stroke located in Wernicke's area of the left hemisphere are more likely to report disruption in word meaning than sentence structure, while lesions centered in Broca's area of the left hemisphere are more likely reflected in disturbed sentence structure than word meaning (Garrard et al. 2004). Interestingly, much less is known about the anatomical basis for the pattern of language degeneration for persons with Alzheimer's disease. What does appear clear is that neurocognitive degeneration of the Alzheimer's type follows a distinct anatomical path, starting in the hippocampus, an area of the brain known to be involved in memory, and progressing to the temporal and frontal lobes before it involves the anterior and posterior association areas of both hemispheres (Braak and Braak 1991; Butters et al. 1994). Brain imaging studies provide support for the anatomical specificity of the linguistic breakdown, but the specific cortical sites involved appear to vary depending on the testing method used (Dapretto and Bookheimer 1999; Prince 1998). In addition, other types of dementia (e.g., vascular dementia) may follow a different neurodegeneration pathway.

According to Garrard et al. (2004), case studies provide the best available data supporting the notion that syntactic processing is preserved while semantic processing degenerates (e.g., Breedin et al. 1994; Breedin and Saffran 1999; Garrard et al. 2004). The patient in the Garrard et al. (2004) study, for example, was found to have profound progressive bilateral temporal lobe atrophy, particularly in the left hemisphere, with sparing of frontal and parietal regions. In two experimental studies of this patient and normal controls, it was possible for Garrard et al. (2004) to provide additional evidence to support the separability of the two cognitive systems. However, they also cite a study by Bates et al. (1995) of persons with Alzheimer's disease that reported concurrent decline in grammatical complexity and word finding in spontaneous utterances, a finding that was interpreted as supporting a nonmodular neural network model. Thus, at this point, it is still not clear whether modular degeneration or nonmodular disturbances in neurocognitive capacity, or both, will account for the observed data.

What also appears to be true is that while degeneration may be occurring at specific cortical sites, other sites and functions of the brain may remain intact, and some of these sites may be relevant to a quality-of-life or HRQOL assessment. This point was suggested in a paper by Mozley et al. (1999) entitled; "Not knowing where I am doesn't mean I don't know what I like'; Cognitive impairment and quality-of-life responses in elderly people." As this paper and the previous discussion on cognitive preservation implies, the presence of neurocognitive deficits does not preclude that a person with Alzheimer's disease will retain sufficient capacity to communicate what they like or do not like. Again, if responding to a qualitative assessment is viewed as a judgment, then some aspects of the decision-making apparatus (forming the hybrid construct) may remain intact, and these data suggest that the emotional contribution to cognitive processing remains present.

While quite limited, this review should be sufficient to illustrate that the development of language deficits in the person with Alzheimer's disease has a clear structural basis and occurs in a generally orderly manner, even though there are noticeable individual differences. The existence of this process, however, raises at least three broad issues that are relevant to a person performing a quality-of-life or HRQOL assessment, or to an investigator designing such assessments.

First, if the disruption of semantic recall and processing precedes other types of neurocognitive degenerations, then how does this impact a person's responses to a quality-of-life or HRQOL assessment?

Second, what impact does neurocognitive (semantic or spatial-motor) degeneration have on the reliability and validity of a quality-of-life or HRQOL assessment?

Third, what can be learned from observing cognitive degeneration processes about the relationship between the capacity of the person with Alzheimer's disease and his or her performance on a quality-of-life or HRQOL assessment?

The first two questions are important if I am going to use psychometrically developed qualitative assessments to assess the person with Alzheimer's disease, while the third question is important because of what it tells me about how I can model the relationship between a person's neurocognitive capacity and performance (e.g., their decision making).

5.2 Progressive Semantic Degeneration

Having established that one of the initial changes associated with neurocognitive degeneration is the loss of a person's ability to understand the meaning of verbal or visual material, I can now ask if the meaning associated with quality-of-life or HRQOL assessments is also altered by progressive dementia. Fortunately, there is a distinct group of persons who meet the diagnostic criteria of being semantically demented, but who via cerebral biopsy would not be considered to have Alzheimer's disease. These persons, according to Hodges et al. (1992, 1995), all have difficulty remembering the names of people and objects; are impaired in their ability to comprehend single words; have trouble sorting and matching words to pictures as well as drawing and copying; and in general lack an understanding of the meaning of what they experience. On the other hand, these persons usually have no trouble pronouncing words, creating well-formed sentences, and retaining their day-to-day memories and visual-spatial abilities. These skills can be demonstrated by their normal ability to remain oriented in time and space, their normal visual perception and short-term memory capabilities (that is, they maintain their digit span capabilities; Patterson and Hodges 2000). In addition, brain imaging studies have shown that these persons have focal atrophies that involve the polar and inferolateral regions of one or both temporal lobes. Thus, persons with semantic dementia comprise a unique group of patients whose relatively specific cognitive limitations can act as a model of how a person's neurocognitive status affects their performance, and by extension, their performance on a quality-of-life or HRQOL assessment.

Before I review this material, however, I must place my task in context, especially since the focus of my interest – a quality-of-life or HRQOL assessment – is a more cognitively complex task than most of the stimuli presented in the reports dealing with semantic dementia. I will start by identifying the determinants of a qualitative hybrid construct and follow by considering what is available in the neurocognitive research literature concerning semantic dementia. A knowledge gap will emerge that will have to be bridged, and I will attempt to do this by tracing what the person with semantic dementia can teach me about responding to a qualitative assessment.

Figure 3.1 suggested that a quality-of-life or HRQOL assessment consists of both metacognitive and specific cognitions that involve various production rules. To this can be added a third component, affective or emotional issues, that can also contribute to the formation of a hybrid construct. This is particularly true, since a person's feelings are known to interact with cognitive processes, and may be a primary determinant of valuations (see Chaps. 11 and 12). While I contend that the formation of a quality-of-life or HRQOL judgment follows a universal format (a descriptive statement with its consequences and some valuation statement), I also recognize that the specific content of any qualitative assessment will vary between persons. This variation in what results from a hybrid construction occurs because of individual differences in the content that is important to the person. From a cognitive perspective, a person may construct a qualitative outcome in a fast and frugal (e.g., well-learned and automatic) manner, or take time to consider their response. Their affective state may also vary via the intensity or nature of their feelings (e.g., being highly depressed or manic), and their awareness of their status will also guide their judgments.

What is of interest, of course, is what happens to these various determinants and their interactions as neurocognitive degeneration proceeds. I have already learned a little about what can be expected, including the persistence of the capacity to emotionally express preferences in the presence of dementia (e.g., Mozley et al. 1999). I also learned that following brain injury a person's capacity to detect their own emotions and the emotions of others is diminished, as are their ratings of their quality-of-life or HRQOL (e.g., Henry et al. 2006). I have also described data demonstrating that brain lesions can affect a person's self-awareness, which supports the role of metacognition in a qualitative assessment. The influence of these determinants of a hybrid construct suggests that changes in a person's semantic status alone may not be a sufficient explanation of the changes observed in a qualitative assessment as neurocognitive degeneration proceeds.

Recognizing this complexity, I still need to understand how a person derives meaning from their interaction with the world, and how this is altered by various neurocognitive disorders. I will start by discussing Wernicke's famous studies on the neuroanatomical basis of language disorders, which he used to formulate a theory of semantic knowledge (Wernicke [1900] as quoted by Eggert [1977] and reproduced by Rogers et al. [2004; p. 205–206]). He states:

> The concept of a rose is composed of a 'tactile memory image'- 'an image of touch' – in the central projection field of the somesthetic cortex. It is also composed of a visual memory image located in the visual projection field of the cortex. The continuous repetition of similar sensory impressions results in such a firm association between those different memory images that the mere stimulation of one sensory avenue by means of the object is adequate to call up the concept of the object. In some cases, many memory images of different sensory areas, and in others only a few correspond to a single concept. However, by the very nature of the object, a firmly associated constellation of such memory images that form the anatomical substrate of each concept is established. This sum total of closely associated memory images

Table 10.4 Types of semantic errors[a]

	Examples	Concrete terms	Abstract terms
Associative errors	Antique → Vase Merry → Christmas	10	75
Shared semantic feature errors			
Superordinate errors	Cattle → Animal Hours → Time	15	35
Coordinate errors	Niece → Aunt May → January		

[a]Classification system based on Coltheart (1980). Data here interpolated from Fig. 1c in Crutch (2006)

> can 'be aroused into consciousness' for perception, not merely of sounds of the corresponding words but also for comprehension of their meaning. Following our anatomic mode of interpretation we also postulate for this process the existence of anatomic tracts, fibers, connections or association tracts between the sensory speech center of word-sound-comprehension and those projection fields that participate in the formation of the concept.[11]

Thus, Wernicke (1900; Eggert 1977) appears to be claiming that communication of information via specific transcortical areas leads to the formation of concepts, and that damage to these association areas leads to semantic dysfunctions. This has now been confirmed by modern neuroanatomical and neurocognitive studies of persons with a variety of neurological disorders (e.g., persons with Alzheimer's disease, stroke, head injuries, and herpes simplex virus; Rogers et al. 2004). These data confirm that persons with semantic dementia can provide a unique clinical group with which to study the interplay between a person's neurocognitive capacity and their performance on a variety of assessments, including a quality-of-life or HRQOL assessment.

Hodges and Patterson (2007) summarized the history of the development of the clinical syndrome of semantic dementia, which was first identified by Pick (1892), although it took 75 years for the disease to become the focus of active study. More recently, studies by Warrington (1975), Garrard and Hodges (2000), and Hodges et al. (1998) have been done.

What is common for persons with semantic dementia is that they make systematic errors when asked to read out loud a set of common words (Crutch 2006). The errors can be analyzed using a semantic error classification system developed by Coltheart (1980; Table 10.4). Two basic types of errors occur: those based on an association between the presented word (or the target) and the verbal response or error; and those based on the similarity between the target and error (that is, shared semantic features between the target and error). Shared feature errors could occur in two ways: those targets that elicit superordinate errors and those that elicit so-called coordinate errors. Crutch (2006) also points out that some semantic errors are related to a target in both associative and similarity terms (e.g., savage → cannibals). Most interesting is the following comment by Crutch (2006).

Furthermore, a range of terms has been used to capture a distinction that corresponds closely to that described here with the labels 'association' and 'similarity', including 'metonymy' (a figure of speech consisting of the use of the name of one thing for that of another of which it is an attribute or with which it is associated) and 'metaphor' (a figure of speech in which words or phrases literally denoting one kind of object or idea are used in place of another to suggest a likeness or analogy between them; Jakobseon, 1971). (p. 92).

Thus, Crutch (2006) is acknowledging that the analysis he is performing reflects the same type of mapping I have previously discussed when considering the base and target relationship found for a conceptual metaphor.

The data that Crutch (2006) reanalyzed were based on the errors of a group of four deep dyslexic[12] patients (Cothheart 1980). Ten healthy control subjects classified the target-error pairs according to the system summarized in Table 10.4. The target words were also classified in terms of their concreteness or abstractness using an available database (Crutch 2006). Chi-square test of the error data revealed a significant ($P<0.008$) difference between the number of associative and similarity errors. Table 10.4 suggests that abstract terms were more likely to produce errors than concrete terms, and that when an abstract term resulted in an error, this error was more likely to be associative than based on shared semantic features. In contrast, when concrete terms lead to errors, they were more likely to be based on shared semantic features. From this, Crutch (2006) concludes that different principles of cortical organization underlie abstract and concrete concepts. He believes that these and other data suggest that abstract concepts are predominantly represented in an associative network, while concrete concepts are primarily represented in a categorical network.

In summarizing the available literature, Rogers and Patterson (2007) point out that the progressive decline characteristic of semantic dementia follows an established path, starting with failures of knowledge requiring subordinate or basic terms but with the preservation of knowledge expressed as superordinate concepts. Thus, if a semantically dementing person is shown a picture of a robin, they will not report the picture as a robin, but refer to it as a "bird," and as the disease progresses, will refer to the same picture as an "animal." These data confirm that an informational hierarchy exists that degenerates over time, in a characteristic manner.

The progressive nature of the semantic decay raises intriguing questions and opportunities for a quality-of-life researcher. For example, what if a qualitative researcher could monitor the quality-of-life of a group of persons who were semantically dementing? Would the same retention of superordinate concepts be observed? Would a dementing person be able to respond to a global question, but not a multi-item assessment? And if this were the case, what would this mean about the validity of the qualitative assessment? Would a patient provide coherent responses to a generic but not a disease-specific qualitative assessment?

The notion that abstract concepts and concrete terms may be represented differently in the brain is also of interest, since it reinforces the view that any functional differences between these indicators have a structural basis. Part of the argument that concrete terms and abstract concepts differ comes from the observation that it is not as easy to imagine an abstract concept (other than in arbitrarily selected concrete terms) as it is a set of concrete terms. I have discussed this important observation previously (Chap. 3), but how could it come about? It turns out that there is an extended history dealing with this issue dating to James (1890) and Wernicke (1874), who speculated that knowledge about the meaning of concrete words involves sensory and motor "images" learned through perceptual experiences. Thus, the word "apple" has associated with it a color, taste, smell, appearance, and even a feeling that comes from holding it in one's hand that is stored in memory and recruited when an apple is imagined. In contrast, the word "quality" spontaneously elicits no such broad array of associations, other than the terms that are assigned to it, and thus has not nearly as much of a neurophysiological base to be imagined.

The notion that the difference in imageability of concrete and abstract terms has a neurophysiological basis was supported by a study by Sabsevitz et al. (2005). They used event-related fMRI recordings of persons who were asked to make "similarity in meaning" judgments on a large list of noun triads. The study demonstrated that judgments in response to concrete or abstract noun words activated different cortical areas. The authors concluded that "These data provide critical support for the hypothesis that concrete imaginable concepts activate perceptually based representations not available to abstract concepts" (Sabsevitz et al. 2005; p. 188). They specifically found that concrete terms tended to activate a variety of bilateral association areas, including the ventral and medial temporal, parietal, prefrontal, and posterior cingulate cortical areas; while abstract terms activated almost exclusively the left hemisphere, including the superior temporal and inferior frontal cortex.

Naturally, these differences in the cortical representation of concrete and abstract concepts have practical implications, especially for the person whose neurocognitive capacities are diminished. But Crutch's (2006) comment that the memory for concrete terms is stored in specific categories (e.g., living or nonliving objects are represented in different cortical areas) is also worth following up on, especially since the selective disruption of a person's capacity to recall a specific category may be relevant to how a person assesses their life quality. Capitani et al. (2003) summarize this literature and point out that the most reliable category-specific deficits involve biological objects, with deficits in animate (e.g., animals) and inanimate biological objects (e.g., fruits and vegetables) impacted by neurocognitive deficits that were independent of one another. The same is true for body parts, relative to animals and fruits and vegetables. The changes in the meaning that body parts develop for a person after a surgical event or treatment may be particularly important if that person is also having their quality-of-life or HRQOL assessed.

A study by Grossman et al. (2006) is also of interest, not only because it used fMRI measures to monitor the cortical activity of healthy young adults ($N=9$) responding to several types of visually displayed nouns, but because it also provided an opportunity to see if a major (affective) determinant of a quality-of-life or HRQOL assessment impacts cortical activation patterns during qualitatively different judgment tasks. Thus, subjects were told that they would receive an example of a noun from a particular class of nouns (e.g., natural kinds, manufactured artifacts, and abstract terms) and were asked to judge if the presented noun was typical for the particular class of nouns, but also if it was pleasant or unpleasant. Grossman et al. (2006) summarize the results of the study as follows:

> We observed partially distinct patterns of neural activation associated with each category of knowledge, depending on the nature of the judgment used to probe the category. Typicality judgments elicited a consistent pattern of temporal-occipital activation, emphasizing the role of this anatomic region regardless of the semantic category being judged. By comparison, pleasantness judgments recruited different anatomic regions, for each semantic category, suggesting that the domain of knowledge under consideration contributes to the activation pattern. These observations are consistent with the hypothesis that semantic memory involves at least two components, including semantic knowledge representations and the process to interpret this knowledge. (p. 1006–1007).

The observation that asking a person to make a judgment about the pleasantness of a word is sufficient to create a distinct cortical activation pattern reinforces the previous point that contextual issues, such as the person's affective or emotional state, may modify a judgment such as a quality-of-life or HRQOL assessment. The conclusion of Grossman et al. (2006) that both knowledge (with its cortical representation) and its interpretation (or valuation) are ongoing during a judgment about pleasantness suggests that the components of a hybrid construct may be present.

The research literature dealing with semantic dementia is very rich and offers a quality-of-life researcher all sorts of opportunities to gain an insight into how at least one class of compromised persons responds to a qualitative assessment. And although I have reviewed only a limited portion of this literature, I have learned some important things. For example, the observation that abstract and concrete concepts are processed at different cortical sites suggests that they may be differently affected by an adverse neurological event (e.g., a stroke or brain injury). Thus, if a person has a lesion in the associative areas of the cortex, then it would be expected that they would have difficulty assessing abstract quality-of-life or HRQOL concepts, but retain their capacity to report the meaning of concrete concepts. This is easily tested by asking a person with semantic dementia to fill out a Likert scale, where the anchors refer to a good or bad quality-of-life (fairly abstract anchors) and compare this to a similar visual analog scale that uses more concrete anchors, such as specific physical activities or social activities.

In addition, the site and system specificity of abstract and concrete concepts have implications concerning the neurocognitive basis of a definition. For example, a number of definitions (e.g., Table 2.1) assume that mapping concrete entities to an abstract concept is sufficient to define it. Implicit in these definitions is the assumption that the concrete entities cause the emergence of the abstract concept. Metaphoric arithmetization would be one example of how this emergence could occur. The WHO definition of health ("a state of complete physical, mental, and social well-being and not merely the absence of disease"; World Health Organization, 1948) is also an example of a definition that implies that concrete states lead to an abstract concept: health. Osoba's (1994) definition of HRQOL (Table 2.1) suggests the same process. However, the neuroanatomical studies tell me that the combined concrete entities are not necessarily causally related to an abstract concept, such as quality-of-life or HRQOL, but requires the recruitment of different cortical area as occurs when a valuation occurs.

If this difference can be supported during neuroimaging studies of qualitative assessments, then it would have implications for several issues in quality-of-life or HRQOL research. For example, I would be able to successfully argue that a global, and presumably, abstract, quality-of-life or HRQOL assessment is fundamentally (that is, neurocognitively) different than a multi-item assessment.[13] This might help account for why correlations between these indicators have mostly been moderate (r 0.50; Chap. 1) under all but the most concrete comparisons.

Another persistent issue in quality-of-life or HRQOL research is whether an assessment consisting of a number of different subdomains should be summed into a composite score or not. The notion that abstract concepts have their own neurocognitive representation suggests that this kind of combination may be inappropriate. The mistake here is to assume that if an abstract concept emerges from the combination of a group of indicators, it is therefore acceptable to aggregate a multi-item assessment. Thus, assessments that profile their outcomes, such as the SF-36 (Ware et al. 1994) or the Nottingham Health Profile (Bureau-Chalot et al. 2002), may be providing a more *neurocognitively* appropriate summary of their data than those such as the World Health Organization Quality-of-life Assessment-BREF (WHO 1996) that sum the scores of all items into a composite score.

The Sabsevitz et al. (2005) study was also important, since it affirmed the distinction Rosch (1973) makes between levels of information. Thus, their demonstration that concrete, but not abstract, concepts are imaginable confirms the model that Rosch (1973) proposed that distinguishes between basic and superordinate levels of information. Their study also has implications concerning what to expect when attempting to define as abstract a concept as quality-of-life or HRQOL. It suggests that there will be no universal definition, and that any definition will be context-dependent.

What was also covered here was that persons with semantic dementia reported category-specific deficits. Thus, a demented person may be shown a picture of an apple and call it a fruit, rather than an apple. Errors of this sort appear systematic in semantic dementia and have to be a concern when designing any qualitative assessment. In fact, as will be described in the next section, it is not uncommon for dementia-specific quality-of-life or HRQOL assessments to be "informationally degraded," possibly taking advantage of the propensity of a person with dementia to make semantic errors (Table 10.4). I am not aware, however, of any study that takes advantage of what is known about semantic errors in persons with dementia when designing a quality-of-life or HRQOL assessment. Of course, many of the items on a qualitative assessment are "broadly stroked" (e.g., asking "Can you lift heavy weights?" without specifying what was heavy; "Are you very active?" without specifying what caused the person to be active, and so on).

This discussion provides me the opportunity to refer back to Bridgman's (1959; p. 7) admonishment that in order to do good science, I need to "understand how I understand." Here I am using the neural basis of cognitive representation to guide how I should approach definitions, in general, and the definition of quality-of-life or HRQOL, in particular. I could have this same discussion again if I were to speak about latent variables and their role in modeling quality-of-life or HRQOL data. But in either case, I am illustrating one of the basic tenets of this book: that the principles of my science should be based, as much as I understand, on what my body teaches me about how I process information.

5.3 The Psychometric Properties of Qualitative Assessments of Persons with Alzheimer's and Related Dementias

One of the implications of my discussion concerning semantic dementia is that errors that persons with dementia make may be sufficient so as to jeopardize the reliability and validity of quality-of-life or HRQOL assessments. A number of reviews have been published (e.g., Ettema et al. 2005; Ready and Ott 2003; Salak et al. 1998; Walker et al. 1998) with this research question in mind, especially as applied to the person with Alzheimer's disease. Overall, four types of qualitative assessments have been reported: (1) multi-item assessments (e.g., structured interviews and self-administered; Brod et al. 1999); (2) single-item self-reports (e.g., James et al. 2005); (3) proxy-based assessments (e.g., Logsdon et al. 1999; Rabins et al. 1999); and (4) observational studies (e.g., Bradford Dementia Group as developed by Kitwood and Bredin 1992). Since I am primarily interested in self-reports, I will limit my review to reliability and validity studies of community-dwelling persons with dementia.

The Ankri et al. (2003) paper illustrates the type of data reported in Table 10.5. Their study was designed to determine the acceptability, feasibility, reliability, and validity of the EQ-5D when administered to a group of 142 persons with dementia. Acceptability was demonstrated when 96.5% of the patients completed the assessment. However, only 14 of these respondents were able to complete the assessment themselves; 123 persons required the help of a trained interviewer. The distribution of responses within and between items was within expectation; there was no evidence of a ceiling effect or skewed responding. TRT reliability estimates revealed moderate correlations (in the range of 0.34–0.59) for each of the five items of the EQ-5D, while the correlation of the total score ($r=0.74$) was significant at the $P<0.0001$ level. Validity was established with multiple studies: for example, patient and proxy concordance; covariation of responses to the EQ-5D items; and severity of disease or level of Activities of Daily Living (ADLs), and so on. Significant correlations were found between ADLs and self-reports of mobility, self-care, usual activities, and pain. The authors conclude that it is possible to administer the EQ-5D to persons with dementia, although they acknowledge that respondents had difficulty with the visual analog scale portion of the assessment.

In general, the data in Table 10.5 indicate that the reliability of the response of a person with dementia to disease-specific qualitative assessments was at a acceptable test-retest level, but that this was less so for generic assessments. The table also illustrates the complexities involved in establishing the validity of these assessments. For example, a number of studies used correlations between a proxy and a person with dementia as an indicator of the validity of the patient's responses. Proxy measures, as noted previously (Sneenv et al. 2002; von Essen 2004), have unique features that limit their usage as validity indicators.

Correlations between different types of qualitative assessments, however, do provide some support for the validity of the dementia-specific assessments (Table 10.5), although the number of relevant studies used to make this inference needs to increase. There are a number of reasons why there aren't more studies, including the problem of finding assessments that can simultaneously assess the qualitative state of mild, moderate, or severely demented persons with Alzheimer's disease.

One of the techniques adopted to deal with this issue is to reduce the informational content of the items on dementia-specific quality-of-life or HRQOL assessments. This, it is assumed, would create a lower common denominator that

Table 10.5 The reliability and validity of self-reported quality-of-life or HRQOL assessments for community-dwelling persons with dementia

Reference (sample size)	Quality-of-life assessments	Reliability[a]	Validity[a]
Generic assessments			
Struttmann et al. (1999) ($N=57$)	WHOQOL-100	Physical shape[b]; $r=0.095$; Psychology $r=0.725$; Autonomy $r=0.698$; Social Interactions $r=-0.196$	
Bureau-Chalot et al. (2002) ($N=145$; reliability sample=52)	NHP[c]	C$\alpha=0.54$–0.85 TRT $r=0.45$–0.83	(Proxy-based) FAM $r=0.30$–0.57 IAP $r=0.22$–0.48 F-IA $r=0.20$–0.75
Ankri et al. (2003) ($N=142$; reliability sample=45)	EQ-5D	TRT $r=0.74$ VAS $r=0.44$ KAP $r=(0.34$–0.59)	FAM $r=0.41$ IAP $r=0.42$ F-IA $r=0.54$
James et al. (2005) ($N=193$; 181 completed SAHS)	SAHS and QOL-AD		SAHS/QOL-AD Spearman's$=0.64$
Naglie et al. (2006) ($N=60$)	EQ-5D QWB HUI3 VAS	TRT EQ-5D $r=0.79$ QWB $r=0.70$ HUI3 $r=0.47$ VAS $r=0.38$	
Karlawish et al. (2008) ($N=93$)	EQ-5D HUI2 QOL-AD SF-12	TRT EQ-5D $r=0.58$ HUI2 $r=0.66$	EQ-5D/QOL-AD $r=0.660$–0.906 EQ-5D/SF-12 MCS $r=0.739$–0.892 EQ-5D/SF-12 PCS $r=0.708$–0.934 HUI2/QOL-AD $r=0.767$–0.962 HUI2/SF-12 MCS $r=0.822$–0.913 HUI2/SF-12 PCS $r=0.790$–0.961
Kunz (2010) ($N=333$)	EQ-5D		FAM $r=0.48$
Disease-Specific Assessments			
Brod et al. (1999) ($N=99$)	DQOL	Six domains: C$\alpha=0.67$–0.90 TRT$=0.64$–0.90 Single items$=0.54$–0.75	Four well-being domains: GDS$=(-0.43)$ to (-0.64)
Selai et al. (2001) ($N=2$)	QOLAS	C$\alpha=0.78$	
Logsdon et al. (2002) ($N=177$; 155 completed the study)	QOL-AD	C$\alpha=0.84$ TRT; $r=0.76$	QOL-AD and,[d] ADL $r=-0.31$ GDS $r=0.51$ RMBPC $r=-0.22$ PES-AD $r=-0.30$ MOS: PCS $r=0.22$
Terada et al. (2002) ($N=264$; 10 patients were used for TRT)	DQOL	Six domains: C$\alpha=0.79$–0.81 TRT; $r=0.63$–0.93	Six domains of[e] QOL-D and NM $r=0.00$–0.75 NADL $r=0.04$–0.64
Thorgrimsen et al. (2003) ($N=60, 201$)	QOL-AD	C$\alpha=0.82$ TRT$=0.60$	DQOL$=0.69$ EQ-5D$=0.54$ EQ-5D-VAS$=0.50$

(continued)

Table 10.5 (continued)

Reference (sample size)	Quality-of-life assessments	Reliability[a]	Validity[a]
Frank et al. (2006) (N=153)	PROCOG	Cα=0.82	QOL-AD r=0.53
Matsui et al. (2006) (N=140)	QOL-AD	Cα=0.84 TRT; r=0.84	FAM r=0.60
Patterson et al. (2006) (N=644)	QOL-AD	TRT; r=0.774	QOL-AD (Partner) r=0.486 QOL-AD r=0.348 SF-36 PCS r=0.408 SF-36 MCS r=0.547
Smith et al. (2006) (N=79)	DEMQOL	Cα=0.87 (N=75) TRT; r=0.84 (N=17)	Six domains of DQOL r=−0.40–0.55
Trigg et al. (2007a, b) (N=60)	BASQID	Cα=0.91 TRT; r=0.82 (N=30)	
Yap et al. (2008) (N=67)	QOL-AD	Cα=0.90 TRT; r=0.70	GHQ; r=−0.02 SF-36 physical functioning r=0.13 SF-36 physical role limitations r=−0.06

ADL activities of daily living; *Ca* Cronbach's alpha; *FAM* family as proxy and patient agreement; *F-IA* agreement between family and institution proxies; *IAP* institution as proxy (e.g., physician or nurse) and patient agreement; *KAP* kappa statistic; *MCS* Mental Health Summary Scale; *PCS* Physical Health Summary Scale; *TRT* test-retest; *VAS* Visual Analog Scale

Assessments: BASQID bath assessment of subjective quality of life in dementia (Trigg et al. 2007a, b); *DEMQOL* dementia quality of life (Smith et al. 2005); *DQOL* dementia quality of life instrument (Brod et al. 1999); *EQ-5D* European questionnaire-five dimensions (Brooks et al. 2003); *EQ-5D-VAS* Visual Analog Scale of EQ-5D (Brooks et al. 2003); *GDS* geriatric depression Scale (Yesavage et al. 1983); *HUI2* or *3* Health Utility Index (Torrance et al. 1995); *MOS* Medical Outcome Study (Hays et al. 1993); *NADL* Nishimuria Activities of Daily Living Scale (Nishimura et al. 1993); *NM* Nishimuria Mental State Scale (Nishimura et al. 1993); *NHP* Nottingham Health Profile (Hunt et al. 1985); *PES-AD* The Pleasant Event Schedule -AD Short Form (Logsdon and Teri 1997); *PROCOG* patient-reported outcomes in cognitive impairment (Frank et al. 2006); *QOL-AD* quality of life-Alzheimer's disease (Logsdon et al. 1999, 2002); *QOL-D* quality of life-dementia (Terada et al. 2002); *QOLAS* Quality of life assessment schedule (Selai et al. 2001); *QWB* Quality of Well-being Scale (Kaplan and Anderson 1990); *RMBPC* revised memory and behavior problem checklist (Teri et al. 1992); *SAHS* the self-assessed health status item; *SF-12* the short form 12. (Ware et al. 1995); *WHOQOL-100* World Health Organization Quality of Life assessment; 100 items (WHO 1995); *WHOQOL-BREF* World Health Organization Quality of Life Assessment-BREF (WHO 1996)

[a] Not all the reported correlations are statistically significant
[b] Investigator-selected domains of the WHOQOL-100
[c] Nottingham Health Profile is reported in six dimensions (physical mobility, social isolation, emotional reactions, pain, sleep, and energy). Data presented represents range of correlations
[d] Correlations were reported between the QOL-AD and ADLs (Activities of Daily Living), GDS (Geriatric Depression Scale; Yesavage et al. 1983), RMBPC (Revised Memory and Behavior Problem Checklist; Teri et al. 1992), FEP (Frequency of Personal Events; Logsdon and Teri 1997), and MOS (Stewart et al. 1988). Each reported value is statistically significant
[e] Correlations were reported between the six domains of the QOL-D and the Nishimuria Mental Status Scale (NM; Nishimura et al. 1993) and Nishimuria Activities of Daily Living Scale (NADL). Only 42% of the participants were community based

would increase the number of people who would be able to respond to the assessment. The adjustments made include limiting the reading level of each item; simplifying the item's grammatical structure; using concrete words or phrases as opposed to more abstract terms; focusing on current as opposed to past events; simplifying the response required; and limiting the number of questions asked. These "adjustments" also recognize that the person with dementia may have working memory limitations, lack insight into their own current status, and may not be properly oriented in terms of time and place; each of which could lead to confabulated or overly optimistic responding.

What is of concern here is whether this practice creates an assessment that provides a different kind of validity than what would be found for an assessment of persons who were not demented. Thus, it was surprising to find that none of the studies cited reported the reading level of the respondents, nor did they provide any estimate of how well the respondents comprehended the content of the assessment, relying instead on the presence or absence of response variability to determine if it is an issue to be concerned about.

It is also important to remember that altering the cognitive demand of an item or an entire assessment may shift the basic cognitive process mediating a person's response from a reflective, controlled process to automatic responding. If this occurs, it would help account for what has been observed to be true of memory functions for persons progressing to dementia. Birnboim (2003), for example, following a summary of the literature dealing with dual processing of memory functioning in persons with Alzheimer's disease, concluded that controlled or effortful cognitive processing declined early in the onset of the disease, with automatic responding being preserved until late in the disease (see Starr et al. 2005). Thus, the success in shifting the cognitive demand of items on a qualitative assessment may inadvertently be taking advantage of this shift in neurocognitive capacity, and it is for this reason that the effort at reducing the cognitive demand of items may still result in the person with Alzheimer's disease being able to complete a quality-of-life or HRQOL assessment.

What may also confound the interpretation of the data in Table 10.5 is the observation by Sharp et al. (2006) that while a decline in cognitive processing occurs with normal aging, there is also evidence that aspects of cognitive control can be enhanced when an older person is asked to respond to a cognitively demanding task (see below). As a consequence, the performance of the older normal respondent may be comparable to that of a younger respondent; the difference is that the performance is based on different neurocognitive pathways. The same would be expected for the person with Alzheimer's disease, except that the neurocognitive basis for responding may very well be different from what is found for age-matched nondemented controls or younger persons.

It remains unresolved what this shifting of the cognitive basis of a response does to establishing the validity of a qualitative assessment for these various groups, including the person with Alzheimer's disease. It returns me to being concerned about the difference between psychometric and cognitive equivalence, and what this difference means for assessing the validity of any assessment. One way to interpret the downgrading of the semantic structure of items on qualitative assessments for demented persons is to see it as an attempt to keep the available Alzheimer's-specific assessments psychometrically equivalent to assessments of persons who are not demented or dementing. Of course, this is being done at the cost of the cognitive equivalence between the assessments, a distortion that is seldom acknowledged by psychometricians.[14]

Cognitive demand is also an issue that arises when persons with dementia are asked to participate in complex cognitive tasks, such as making value judgments or generating utilities. The ability of the person with Alzheimer's disease to make value judgments was illustrated in a study using the Schedule for the Evaluation of Individual Quality-of-life (SEIQoL; O'Boyle et al. 1993). The SEIQoL requires that respondents first select five areas of their lives that they most valued or considered important. Coen et al. (1993) administered the SEIQoL to 20 persons with mild dementia and found that only 6 of the participants were able to complete all aspects of the assessment protocol, but a further 11 required prompting to complete the task (the investigator read a list of values to the respondents). Only 10 of the 20 patients were able to divide a circle into five parts representing the relative importance or weightings of the valued aspects of their quality-of-life (e.g., family, housework) and estimate the importance they placed on their selected values. The authors conclude that even with the difficulties reviewed, the SEIQoL can be successfully administered to persons with dementia, although the more the person was suffering from dementia, the less likely they were able to perform different aspects of the assessment task.

The Coen et al. (1993) study suggests that persons with dementia were able to tell the investigator what aspects of life they valued the most, much as Mozley et al. (1999) reported. Respondents were interviewed in both these studies. Were persons with Alzheimer's disease or dementia capable of generating utilities using such techniques as the time-trade off or standard gamble methods (Patrick and Erickson 1993)? Interestingly enough, there do not seem to be any published examples where preferences were successfully estimated directly from persons with Alzheimer's or dementia.[15] Instead, what appears to be available are studies that use commonly available assessment methods (e.g., content specific questions, interviews, visual analog scales, etc.) that have population-based preference systems mapped to the possible outcomes. Examples include the HUI (Torrance 1986), the QWB (Kearner et al. 1998), and the EQ-5D (Brooks et al. 2003). When applied in a specific study (e.g., Ankri et al. 2003; Neumann et al. 1999), the population based preferences were used to provide an estimate of the preferences of persons with dementia. Is this appropriate? A study by Silberfeld et al. (2002), which examined the content validity of each of these assessments for persons with dementia, suggested that it might not be. They found that only the QWB comes close to including content that approximated the interest of persons with dementia. Thus, what is left is a situation where doubt has to be expressed about the descriptors used in these assessments, and this should raise concern about any inference concerning the respondents' preferences based on these generic assessments.

Earlier I mentioned that concern about the impact of loss of insight of persons with Alzheimer's disease justified the use of proxies when their quality-of-life was being assessed. Two studies (Berwig et al. 2009; Ready et al. 2006) have been reported that attempt to quantitate the degree to which impaired and intact insight have on the internal consistency (Cronbach's alpha) of qualitative assessments of persons with Alzheimer's disease. Respondents were classified as retaining or not retaining insight on the basis of the responses to four questions (awareness of reason for visit, awareness of memory deficit, awareness of functional deficit, and awareness of the progression of the illness). One study by Karlawish et al. (2008) used the degree to which the respondent and their caregiver agreed with their judge about the degree of

Table 10.6 Internal consistency (Cronbach's alpha) as an Indicator of the reliability of self-reports of insight intact and impaired insight persons with Alzheimer's disease

Assessment	Domain	Insight intact	Impaired insight
Ready et al. (2006) ($N=67$)			
DqoL	Aesthetics	0.70 fair	0.55 unacceptable
	Self-esteem	0.75 fair	0.65 unacceptable
	Positive effect	0.85 good	0.61 unacceptable
	Negative effect	0.88 good	0.76 fair
	Feeling of belonging	0.58 unacceptable	0.63 unacceptable
Berwig et al. (2009) ($N=27$)			
DEMQoL	Emotions	0.88 good	0.79 fair
	Memory	0.83 good	0.67 unacceptable
	ADL	0.73 fair	0.58 unacceptable
Karlawish et al. (2008)			
Cognitive	Insight score: 7–10	Insight score: 4–6	Insight score: 0–3
HUI2	0.838 ($N=26$)	0.885 ($N=31$)	0.924 ($N=29$)
EQ-5D	0.912 ($N=26$)	0.848 ($N=31$)	0.855 ($N=29$)

insight that the respondent with Alzheimer's disease possessed. Table 10.6 summarizes these data.

The data in the table reveal that there was a difference in the degree to which lack of insight impacted a disease-specific vs. a generic HRQOL assessment. Thus, in the Karlawish et al. (2008) study there was no difference in the correlation between either assessment and the degree to which the respondent and caregiver agreed, whereas there was a difference in the internal consistency of the Cronbach alpha scores of different components of the remaining studies. At a minimum, these data suggest that not all Alzheimer's patients lack insight and those who do are not necessarily impaired in all aspects of a qualitative assessment. As Berwig et al. (2009) state:

> Given these results it thus appears justified to use dementia patients' self ratings QOL as an outcome criterion. However, it will be necessary to develop guideline that specify the conditions under which dementia patients' self-rated QOL can be safely used or needs to be interpreted with caution." (p. 230).

Ecklund-Johnson and Torres (2005) claim that the dementing person's lack of awareness is due to the disruption of specific neural sites, not a language usage deficit, faulty memory, or the consequence of some psychological process (e.g., denial). They state:

> Memory performance was rarely exclusively predictive of unawareness in studies in that this relationship was examined, and there is evidence from the study with circumscribed medial temporal lobe dysfunction that it is possible to have severe amnesic deficit and yet be fully aware of memory impairment (Schacter 1990). (Ecklund-Johnson and Torres 2005; p. 163).

The evidence that unawareness is site-specific comes from two brain imaging studies (Reed et al. 1993; Starkstein et al. 1996) that demonstrated right frontal lobe hypoperfusion. Ecklund-Johnson and Torres (2005) also cite evidence that persons with Alzheimer's disease are more likely to demonstrate a site-specific neurocognitive basis of unawareness than other forms of dementia (e.g., vascular dementia or Parkinson's disease). Longitudinal studies have demonstrated that unawareness becomes progressively worse over time, although the rate at which this occurs varies between persons. The association of unawareness and depression remains unresolved (Ecklund-Johnson and Torres 2005; p. 164), but could be a significant modulator of behavior. Finally, Starkstein et al. (1996) reported data suggesting a separation between awareness of cognitive and emotional deficits. In their study, different sets of covariates predicted emotional (as assessed by behavioral disinhibition) and cognitive unawareness.

These studies are quite informative, since they suggest that a site-specific neurocognitive deficit could compromise the ability of the person with Alzheimer's disease to perform a quality-of-life or HRQOL assessment. It also suggests that the analysis of language usage per se may not be sufficient to characterize this deficit, since the cognitive processes underlying language performance may be sufficiently diverse (e.g., when comparing the younger with the older respondent, or the older respondent with the person with Alzheimer's disease), and that these groups may not be comparable. And if this is so, then this raises the issue of what would be a suitable alternative for modeling this relationship. Also of importance was the demonstration that emotional and cognitive unawareness are not necessarily associated. This suggests that each may have a specific neurocognitive basis that differs as neurological processes degenerate.

5.4 Aging and Neurodegeneration

It is now appropriate to ask what these studies of semantic dementia and neurodegeneration tell us about the relationship between capacity and performance. For example, are the changes in the relationship between a person's capacity and their performance causally related or does the degeneration of neurocognitive capacity modulate quality-of-life or HRQOL outcomes? If this relationship is causal, then it may make sense to include NPT and brain imaging studies as part of a qualitative assessment. If capacity modulates performance, including these measures becomes a more discretionary decision.

I have already reviewed some evidence to clarify these issues. For example, examples of specific deficits exist that can be related to specific changes in performance (e.g., what I have just discussed about the neural basis of awareness). Also, neurocognitive systems as a whole are sufficiently redundant that performance can be preserved in spite of massive

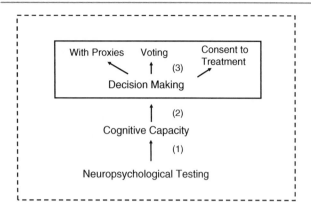

Fig. 10.4 Depicted in the figure is the relationship between neuropsychological testing and cognitive capacity (1), the influence of cognitive capacity on decision making (2), and the manifestation of decision making in the person's consent to be treated and voting behavior (3).

structural changes in capacity. This implies that the degeneration of neurocognitive processes may both cause and modulate performance: a far more complex state of affairs than I anticipated when I originally asked the question. If this is true, then I will need to clarify the circumstances under which I would not include NPT or brain imaging studies as part of a quality-of-life or HRQOL assessment. It also suggests that I will have to include both types of indicators in any model of the relationship between capacity, performance, and a quality-of-life or HRQOL assessment. This is essentially what Wilson and Cleary (1995) attempted to do in their model, with its implied effort to capture the dynamics of a system that is far more complex than a linear model implies.

I propose to examine the neurocognitive basis of decision making as an appropriate venue to clarify these issues. I will then follow this discussion by describing some of the changes that occur with aging. A review of the research literature reveals that several types of studies have examined the cognitive basis of decision making (Fig. 10.4). I am particularly interested in the relationship between NPT (Fig. 10.4, [1]), capacity (Fig. 10.4, [2]), and decision making (Fig. 10.4, [3]), but I will not review the extensive literature dealing with the determinants of decisions (e.g., consenting to treatment, voting, and so on). All of these topics are, of course, related to the broader legal and ethical issue of a person's competence and autonomy when making decisions.

As discussed previously, the decisions a person with Alzheimer's disease makes involve varying levels of cognitive demand. Thus, some decisions require a great deal of processing and are cognitively quite demanding (e.g., whether to sign a consent form, who to vote for, or selecting someone to marry), while others are much less demanding (e.g., selecting a flavor of ice cream, agreeing to physical therapy, or taking a drug). These decisions require a full range of neurocognitive processes, including comprehending what is written or told to them; recalling, retaining, or following instructions; and being able to interpret complex messages. The capacity to make these decisions would also be expected to change as a person ages (Park and Gutchess 2000), raising the question of whether cognitive degeneration represents an exaggeration of normal aging, a unique process, or a combination of both.

The optimal way to examine these issues would be to select a particular recurrent decision (say for example, a person's rating of their health status or HRQOL) and carefully track changes in the cognitive processes that underlie the person's decisions as degeneration proceeds. Selai and Trimble (1999) suggest four cognitive dimensions (based on a model of Lezak 1995) that should be examined: receptive (perceptual) functions (abilities to select, acquire, classify, and integrate information); memory and learning (information storage and retrieval); thinking (mental organization and reorganization of information); and expressive functions (means whereby information is communicated or acted upon). Almkvist (2000) provides a summary of the research literature where brain imaging procedures were used to determine the association between specific cortical sites and task-specific manipulation of cognitive processes such as perception, attention, priming, working memory, and semantic and episodic memory for normal controls vs. persons with Alzheimer's disease (Table 10.7). His paper provides a first approximation of what Selai and Trimble (1999) were proposing, and I have also added some more recent research papers to make this summary more up to date.

Inspection of Table 10.7 reveals the great diversity of cortical sites involved when a person engages in different tasks, and how the sites involved varies depending on whether the respondent was a person with Alzheimer's disease or a normal elderly person. Thus, the person with Alzheimer's disease either increased or decreased the activation of particular cortical sites or recruited additional sites as compared to a normal controls of comparable age, and when each group performed tasks at a comparable level. These differences in the pattern of cortical activity were interpreted as reflecting the capacity of the person with Alzheimer's disease to compensate for their developing cortical deficits. However, Sharp et al. (2004, 2006) have demonstrated that compensation is not unique to the person with Alzheimer's disease, but is a cognitive characteristic common to the aging process as well.

Sharp et al. (2004, 2006) were specifically interested in the neurocognitive changes associated with age-related decline in decision-making capabilities under conditions of receptive or perceptual degradation. Their respondents were cognitively normal, so their studies provide a baseline estimate of an aging effect that might be exacerbated by a neurocognitive degenerative process such as Alzheimer's or Parkinson's disease. The investigators simulated perceptual degradation by verbally masking a set of words or clearly presenting them for a control group. This created a qualitative difference in the material presented, as respondents were

Table 10.7 Brain imaging studies of cortical sites involved in various cognitive processes[a]

Cognitive processes	Normal-aged persons[b]	Persons with Alzheimer's disease
Attention: Sustained and Divided Attention: Frequency Detection Task. Detect changes in a visual oscillating target and a vibrotactile stimulus applied to the right hand fingers. Activation pattern for AD and controls were similar for sustained but not divided attention	Aggregated results: Activation of the right middle frontal gyrus (BA 46, 10), right inferior parietal lobe (BA 40), and right medial frontal gyrus (BA 6)	Aggregated results: Deactivation in the medial frontal regions (BA 11), left orbital (B 11), right lingula (BA 17), and posterior cingulate cortex (BA 23)
Perception: Visual; Faces Matching Task. Performance, accuracy, and reaction time were comparable for each group. Difference in neural activation was interpreted as evidence of compensation by ADs	Normal elderly uniquely activated the ventral occipito-temporal region (fusiform gyrus) bilaterally. Elderly activated larger area than younger normal controls	ADs activated the lateral and medial occipital and occipito-temporal, bilaterally. Persons with AD had greater activation than elderly controls of the occipital pole and frontal region involved in visual fields. Right hemisphere greater than left for ADs than controls
Priming: Word Completion Tasks. Word fragment completions were primed by implicit memory, or explicit word exposure prior to the fragment completion task. Neural activation was interpreted as evidence of compensation	Both young and older respondents demonstrate a decreased blood flow on PET in the right occipital cortex (BA 19)	Mildly demented ADs show an increase in activity in the right occipital cortex (BA 19), but poorer performance than controls. Neural activation varied by magnitude of priming
Working Memory:[c] Detecting Visual Stimuli Task Six colored objects were briefly presented in sequence followed by two test probes. The task of the subject was to determine if the probes were previously presented fMRI provided functional brain images	Normal elderly controls and ADs activated 23 cortical areas in common. Controls-specific activation occurred in the right angular gyrus	Activation of AD-specific cortical areas included left medial frontal gyrus, bilateral amygdale, left cuneus, left putamen, left BA 20, and left insula. Sixteen cortical areas were more activated for ADs than controls. Three cortical areas were less active than controls
Semantic memory and episodic memory:[d] Interpreting the Meaning of a Task, or Recalling a Presented Stimulus Recognizing a single number, semantic task-adding two numbers but not vocalizing them; episodic task – recalling previous single number. Difficulty of task was varied by time between presentations	Cortical areas recruited that were unique to normal controls and more active for episodic memory than semantic memory involved the right parahippocampal gyrus. Cortical areas recruited when semantic memory was more active than episodic memory involved the right and left anterior cingulate. Cortical areas recruited when longer vs. shorter stimulus exposure involved the right superior frontal gyrus	Cortical areas recruited that were unique to ADs and were more active for episodic memory than semantic memory involved the right superior frontal gyrus, left uncus. Cortical areas recruited when semantic memory was more active than episodic memory involved the left medial and middle frontal cortex, right middle occipital gyrus. Cortical areas recruited when longer vs. short stimulus exposure involved the right anterior cingulate, left uncus

AD persons with Alzheimer's disease; *BA* Brodermann's area
[a] Specific references for the information provided can be found in the Almkvist (2000) report. Data presented in this table were limited to PET and fMRI studies
[b] Persons included as normal-aged controls may have had illnesses (e.g., arthritis), but not illnesses that directly affected the brain
[c] A study by Yetkin et al. (2006) was substituted for the working memory studies summarized by Almkvist (2000)
[d] A study by Starr et al. (2005) provided information on both semantic and episodic memory for the same group of respondents and was substituted for the studies presented by Almkvist (2000)

asked to decide whether two of the three words were matched in terms of their meaning (semantics) or number of syllables. The investigators expected that the two types of tasks would involve different neurocognitive pathways or differ in the magnitude of responding for a particular pathway, or both (e.g., Sharp et al. 2004). They also postulated that younger and older persons would respond to these tasks differently, with the older persons exercising greater "cognitive control" than the younger respondents (Sharp et al. 2006).

The hypothesis of Sharp et al. (2006) that as a person ages they exercise greater cognitive control was based on the observation that older persons were more likely to monitor their performance and correct errors then younger persons (Band and Kok 2000; Park and Schwarz 2000). This was supported by evidence demonstrating that a specific cortical network existed that mediated this cognitive control: the anterior cingulate cortex (ACC) and lateral prefrontal cortex (PFC; Shallice 2002). Cognitive control was assumed to occur by modulating the activity within this neural network (McDonald et al. 2000; Shallice 2002). The ACC has been shown to perform an evaluative function, detecting errors, and the presence of conflict between potential responses; while the lateral PFC assumes a more strategic role supporting task-specific processes and high-level monitoring of ongoing behavior. As Sharp et al. (2006; p. 1740) state: "Hence, the ACC is thought to signal the optimal level of cognitive control, with changes in the level of control implemented through an interaction between the ACC and the lateral PFC."

Also of interest was the observation that activity within the ACC and lateral PFC declines with practice (e.g., Jansma et al. 2001). This change in activity was interpreted as resulting from a reduction in the requirement of cognitive control, leading to a shift from controlled to automatic information processing. Thus, my previous discussion (Birnboim 2003) relating a decline in cognitive control early in the onset of Alzheimer's disease concurrent with the preservation of automatic responding until late in the disorder can now be understood as involving a disruption in the ACC and lateral PFC relationship. What may also be true is that this may occur for both the normal-aged person and the person with Alzheimer's disease, although for different reasons.

Sixteen normal persons aged 37–83 participated in the Sharp et al. (2006) study, and PET scans were used to monitor cortical activity. Pilot data (not involving the study subjects) established that responses to all the triplets occurred at a 90% level of accuracy. The subjects themselves were trained in the task by first being exposed to 54 words that were masked to make them difficult to recognize. This period of training was adjusted so that all subjects were correctly identifying the masked words with >50% accuracy. This training also established that the masked words were indeed perceptually degraded (that is, qualitatively different) and were more difficult to comprehend than the clearly spoken words. PET scans data were collected concurrent with the decision-making task.

Sharp et al. (2006) found that when the task involved deciding which two of three clearly heard words had the same meaning, the young and older respondents were equally accurate (that is, the task was age-independent), while matching syllables or matching masked-semantic or -syllable material varied with age (that is, the task was age-dependent). These findings were interpreted as being consistent with the observation that the controlled processing of syllables involved a degree of focused attention to the sound structure of words that is not ordinarily a feature of automatic speech perception and comprehension (Scott and Johnsrude 2003). In contrast, this would occur when the meaning of the presented words is at issue. The perception of masked words would also be expected to differentially involve elements of dual processing as the respondent ages. Thus, Sharp et al. (2004, 2006) have provided a model of how a particular type of performance can be related to a specific neurocognitive system, and the question now is whether the system will retain its capacity to maintain performance when the system is varied (as, for example, by masking the material presented or varying the respondent's age).

Sharp et al. (2006) expected that the rate of decline in activity in the ACC and the lateral PFC would be slower for the older respondents than for the younger, indicating greater cognitive control. This was what they found. They confirmed that performance of semantic-based tasks were age-independent, as were syllable-based task, and semantic or syllable-masked tasks. They then compared the neural activity pattern for these two different performance outcomes and found that declines in performance were related to increase in activity in the rostral ACC and portions of the lateral PFC.

These data make it clear that the relationship between capacity and performance is not a simple input–output system, but is instead a multifactored process where not one element causes a particular outcome, but rather continually (e.g., cognitive control) modulates a system that leads to the final common path of performance. Starr et al. (2005) expand on these ideas by stating that:

> The pattern of progressive recruitment of brain regions would support a 'memory reserve' hypothesis in that compensatory functional reorganization can occur in individuals with high brain reserve in response to brain disease and so limit the level of functional decline that is clinically evident. It is unclear whether similar processes occur with normative cognitive aging. However, my data suggest that this may be the case as the differential loss of function in brain regions associated with episodic memory in AD is mirrored, to some degree by increasing task difficulty by older subjects with no clinically evident pathology…The hypothesis fits with the idea of a spectrum of neuronal dysfunction between 'healthy' old age and clinical AD indicated by neuropathologic surveys…as evidenced by increased recruitment of brain regions in older people carrying the APOE ε4 allele… (p. 267).

Starr et al. (2005) go on to discuss a paper by Allen et al. (2002) where the authors applied structural equation modeling to results of a numerical semantic and episodic memory task for persons younger and persons older than 40 years of age. Allen et al. (2002) found that three factors emerged: episodic, semantic, and a joint episodic-semantic factor. Starr et al. (2005) speculate that as cortical degeneration proceeds, areas that represent overlap in cortical function are recruited to maintain performance, and at some point, this response breaks down. It is at that point where the normal controls and persons with Alzheimer's disease clinically diverge.

Let me assume for a moment that the speculations of Starr et al. (2005) are true (or something close to them). What does this tell me about the relationship between a person's capacity and their performance on a qualitative assessment?[16] Clearly, it raises the question of whether a common response on a quality-of-life or HRQOL assessment for men and women, young and old, and demented and healthy elderly are, in fact, equivalent. These issues will only become clear if both the psychometric and cognitive equivalence (Barofsky 2000) of responding to an assessment are studied. Of course, one of the advantages of studying cognitive processes is that we can use objective brain imaging procedures to provide some insight into how a subjective qualitative assessment may occur.

As I have tried to make clear, psychometric equivalence is a product of controlled processing and automatic responding, both of which are shown to have active roles in governing responding as cognitive degeneration proceeds. Controlled processing is actively involved in cognitive compensation, only to break down as degeneration proceeds, while automatic responding continues well on into the more terminal phases of the illness. Using cognitive interviewing, von Osch and Stiggelbout (2005) demonstrated that dual processing was also present during responding to a visual analog scale, a metric common to quality-of-life or HRQOL assessment. Thus, in time it should be possible to demonstrate that dual processing plays an important role in quality-of-life or HRQOL assessments, as well.

It is also worth noting that in the research reviewed, the word "quality" was used in at least two distinct ways. First, Sharpe (2004, 2006) claimed that distorting a stimulus by masking the clarity of its perception affected its quality, and as Sharpe's data showed, this was reflected in the differences in the neurocognitive processes associated with particular cognitive tasks. Auriacombe et al. (2006) also refer to a qualitative difference, but not in terms of the responses that their participants made. Here the repetition of words or intrusions of inappropriate words during the verbal fluency task (that is, errors) were considered to be indications of quality.

I would now like to describe two rather dramatic examples (Crutch et al. 2001; Sacks 1985) of the progressive nature of the cognitive degeneration process. In each case, pictorial methods were used to depict these changes, and this also has been used with patients with cancer, pain, and a number of other patient groups. The artist William Utermohlen drew a series of self-portraits during the time when his MMSE declined from 22 to 10 of the 30 items. There are many ways of describing the changes illustrated in his paintings, but probably most relevant to me is the loss of personhood and the meaning that life can have. His early paintings depicted a person with an intense, almost piercing expression, and after several years, these pictures transformed into abstract expressions that resembled a nonperson. I could not help but wonder if this was what the person in the last painting was seeing: a world filled with formless objects.

What does a sequence of portraits like this teach us about the nature of quality and its dependence on an intact central nervous system? One way to describe it is to use the Sharpe et al.'s (2004, 2006) definition of quality and assume that what has deteriorated is the artist's perceptual capabilities. Not only may the artist have lost the ability to see himself, he may have also lost the ability to see what he is seeing – a metacognitive loss. Crutch et al. (2001) interviewed the patient and found that he recognized that at times his drawings/paintings were not at the same level of style or technical skill as they had been when he was younger, but that he did not know how to correct them. This is what would be expected for someone with metacognitive deficits: the inability to reflect upon what they were doing. Crutch et al. (2001) also point out that there is an extensive literature on the impact of cognitive degeneration on ability to draw, and NPT suggesting progressive impairment in visuoperceptual and visuospatial skills.

Sacks (1985) provided a second example of this process (Chap. 5) in his famous book, *The Man Who Mistook His Wife for a Hat and Other Clinical Tales*. The subject of the title of the book suffered from prosopagnosia (or facial agnosia), which results in a failure to recognize the faces in pictures of family and friends. The patient demonstrated this limitation when he thought that his wife's head was his hat and tried to lift her head to put it on (he was dissuaded from continuing). The patient was also a gifted painter, but when Sacks inspected his works, he noted a progressive change over time from natural and realistic works of art to his most recent paintings which, in Sacks' opinion, became abstract and chaotic. The patient's wife interpreted these changes as evidence of artistic maturity, but Sacks saw the progression as the advancement of his profound visual agnosia. Thus, the patient had lost his ability to perceive the concrete, and lived in an abstract world, with all that this implied about his ability to function.

Both these examples illustrate the progressive nature of neurological decay and raise interesting questions about the nature of quality, art, and its physiological bases. There are other examples of artistic expression that are products of incipient neurocognitive processes, and I will refer to them as I continue.

Table 10.8 An analysis of the metaphoric content of selected items from the WHOQOL-BREF

WHOQOL-BREF item	Scale[a]	Rephrasing	Deconstruction of items
1. How would you rate your quality-of-life?	Very poor to very good	"Would you rate your quality-of-life as very poor…to very good?"	Your quality-of-life → Poor to very good
2. How satisfied are you with your health?	Very dissatisfied to very satisfied	"Are you very dissatisfied …satisfied with your health?"	Your health → Very dissatisfied to satisfied
3. To what extent do you feel that physical pain prevents you from doing what you used to do?	Not any at all to an extreme amount	"Does pain prevent you not at all or an extreme amount from doing what you used to do?"	You used to do → Not any at all to an extreme amount of pain
4. How much do you use any medical treatment to function in your daily life?	Not any at all to an extreme amount	"Do you not need any at all or an extreme amount of medical treatment to function?"	You need to function → Not any at all to an extreme amount of medical treatment
5. How much do you enjoy life?	Not any at all to an extreme amount	"Do you enjoy life not at all or an extreme amount?"	Your life → Not any enjoy at all or an extreme amount
6. To what extent do you feel your life to be meaningful?	Not any at all to an extreme amount	"Do you feel your life is not at all or extremely meaningful?"	Your life → Not any at all meaningful to an extreme amount
7. How well are you able to concentrate?	Not any at all to extremely	"Do you concentrate not at all or extremely well?"	You concentrate → not any at all or extremely well
10. Do you have enough energy for everyday life?	Not any at all to completely	"Do you have not at all or completely enough energy for everyday life?"	Your everyday life → not any at all or completely enough energy
19. How satisfied are you with yourself?	Very dissatisfied to very satisfied	"Are you very dissatisfied to very satisfied with yourself?"	Yourself → very dissatisfied to very satisfied
26. How often do you have negative feelings, such as blue mood, despair, anxiety, depression?	Never to always	"Do you never to always have negative feelings?"	You → never to always have negative feelings (blue mood, despair, anxiety, depression)

[a] All items were scaled using a five-point Likert scale

5.5 Summary

The topic of this section continued the discussion of Chap. 3 of the neural basis of cognition, but now with a focus on the neurocognitive processes that mediate the adaptive capacity of an organism. The notion of cognitive reserve was given as a model of how this adaptive capacity can be conceptualized and used as a metaphor to design research and test hypotheses. Much of this section dealt with describing the normal aging and pathological processes related to changes in central nervous system functioning. We learned that these processes follow a generally orderly series of steps, although the rate and pattern whereby these changes occurred may vary between individuals. What became obvious to me from this review was that a quality-of-life or HRQOL researcher needs to be acutely aware of whether these changes are ongoing in the persons they are assessing and to design assessments that take advantage of the elements of the hybrid construct these persons still retain. The alternative of degrading the meaning of a qualitative assessment so that its psychometrics are comparable to noncompromised persons perpetuates the impression that if a person is compromised, they will not be able to provide reliable and valid information. The data I reviewed on the reliability and validity of Alzheimer's patients' responses to qualitative assessments (Tables 10.7 and 10.8), especially those who appear to be self-aware, should be sufficient to reject this strategy. What remains to be done is to design assessments that facilitate the assessments of those who are not self-aware or who have suffered profound neurocognitive loss. What little is known about these persons does suggest that they are still able to express their feelings about their state (Mozley et al. 1999), although they may not remember their feelings from one assessment moment to another. I continue this discussion in next section, now dealing with acute traumatic neurological events and their consequences, particularly on metaphoric and figurative expression.

6 Neurocognition Recovery

6.1 Introduction

In the previous section, I traced the path of neurodegeneration and learned some important things about the relationship between capacity and performance and what that relationship means about including NPT results in a qualitative assessment. In this section, I will focus on assessing the recovery from an acute neurological event and its relationship to performance. These neurological events may be due to physical trauma, stroke, or some surgical procedure such as a coronary bypass operation. Ideally, I want to examine the impact these acute neurological events have on a person's rating of their quality-of-life or HRQOL, but also what impact recovery from these events has on these ratings. Obviously, I am also interested in understanding the impact these events have on the meaning a respondent attaches to their responses. Studies of this sort may require some combination of NPT, experimental manipulation of the material presented, and cognitive interviewing following the acute neurological event. However, the quality-of-life or HRQOL research literature I have already reviewed makes it unlikely that I will find such studies. As a consequence, I will take another tack and examine an existing literature that involves studies relating changes in the usage and interpretation of metaphors and other forms of figurative language following an acute neurological event. The rationale for doing so is that since quality-of-life or HRQOL statements have metaphoric or other figurative properties, it should be possible to trace the impact of an acute neurological event on metaphoric expression, and in this way, infer what might happen to a quality-of-life or HRQOL assessment following such an event. The use and interpretation of the visual analog scale will provide a context for illustrating this logic.

In order to do this, I will first have to demonstrate that the items in a quality-of-life or HRQOL assessment also have metaphoric properties, and I must describe them. Then I need to demonstrate that acute neurological lesion can adversely impact metaphoric expression. To do this, I will describe what is known about the neurocognitive basis of metaphoric expression. Using a stroke as my model of an acute neurological event, I will then describe the impact of such an event on metaphoric expression. This will be followed with a discussion about how a stroke also impacts the use of a visual analog scale (VAS), where the VAS, as a ruler, is seemed as a metaphoric expression. The point here is that a stroke may adversely impact not just the metaphoric properties of the verbal material in an item, but also the rating process itself. One of the potential consequences of a stroke is for a person to lose their sense of self, sometimes temporarily and sometimes permanently, and this can impact the reliability and validity of any assessment (e.g., Table 10.6). Applying this approach, however, will not be simple, primarily because the cortical representation of a metaphor is complex (see below) and because an acute neurological event (e.g., a stroke) does not necessarily produce a discrete brain lesion. These issues will complicate any effort to establish a direct association between a particular brain area and metaphoric expression, as will the limited knowledge concerning the neurological basis of a conceptual metaphor. However, what is first needed is a demonstration that the items in a typical quality-of-life or HRQOL assessment have figurative (e.g., metaphoric) properties, something that I have not clearly demonstrated so far.

6.2 Metaphoric Properties of a Quality-of-Life or HRQOL Assessment

As demonstrated in Chap. 2, a metaphoric analysis can be useful when examining the content of quality-of-life or HRQOL definitions. Now I am proposing that it can also be useful when analyzing the meaning being communicated in an item or a scaling method. This implies that figurative language can be used to communicate meaning that a literal expression may not do as efficiently or effectively. Shakespeare was a master at using metaphor to do this, and he even was able to communicate new meanings not otherwise obvious to the listener and reader. Most items on a quality-of-life or HRQOL assessment do not attempt to emulate Shakespearian prose, but they do regularly alter the meaning of the items when dealing with specific patient groups or persons of different ages and intellectual abilities. The investigator will use whatever grammatical form of expression he or she can devise to achieve the objective of a usable assessment. The argument, however, that two such assessments, if demonstrated to have comparable reliability and validity scores, may therefore be assumed to be equivalent defies the obvious fact that each assessment may communicate different meanings. The temptation to aggregate these assessments, especially when making policy decisions, is profound.

Table 10.8 summarizes the items from the WHOQOL-BREF (The WHOQOL Group, 1995), but also attempts to identify whether the items on the assessment have metaphoric properties. I also try to determine if the items have linguistic or conceptual metaphoric properties. The WHOQOL-BREF assessment consists of 2 global items plus 24 items representing each of the 24 domains covered in the WHOQOL-100. I have selected 10 items to deconstruct, and the first 2 are standard global items found in many assessments. What is different about these items, as opposed to a typical quality-of-life definition, is that they have an evaluative, as opposed to declarative, function, and as such, the scaling task (the answer to the question) should be read as an integral part of the item. For this reason, each item was rephrased, for analytic purposes, to include both elements of the item. Thus, when rephrased, the first item reads, "Would you rate your quality-of-life as very poor…to very good?" By doing this, I have eliminated the visual display with its allusion to a linear scale, but at this point I am interested in the semantics of the sentence, and this rearrangement helps me get a fuller sense of the information being communicated to the respondent. I could also have presented these items in a tree diagram (see Fig. 2.1).

Inspecting the list of items in Table 10.8 makes it clear that linguistic metaphors were rarely used in these items, much as was found when I deconstructed the definitions of quality-of-life or HRQOL (Chap. 2). An exception might be the reference to a "blue mood" used to illustrate negative feelings in item #26. All these items have a common reference to the status of a person's self. All 26 items use the word "you" or "your" to direct the responder's attention to some state within themselves that they are to evaluate. While these items do not appear to use linguistic metaphors, they do use a conceptual metaphor to communicate meaning – a metaphor that deals with a person's inner self. The conceptual metaphor I am interested in here is the SUBJECT-SELF METAPHOR. Lakoff and Johnson (1999) describe this metaphor as follows:

> In the general Subject-Self metaphor, a person is divided into a Subject and one or more Selves. The Subject is the target domain of the metaphor. The Subject is that aspect of a person that is the experiencing consciousness and the locus of reason, will and judgment, that, by its nature, exists only in the present. This is what the Subject is in most cases; however, there is a subsystem that is different in an important way. In this subsystem, the Subject is also the locus of a person's Essence – that enduring thing that makes me who I am. Metaphorically, the Subject is always conceptualized as a person. The Self is that part of a person that is not picked out by the Subject. This includes the body, social roles, past states, and actions in the world. There can be more than one Self. And each Self is conceptualized metaphorically as either a person, an object or a location. (p. 269)

If I apply the language that was used in Fig. 2.2, then I can understand that the SUBJECT-SELF METAPHOR maps base or concrete indicators, such as a person's various Selves (that is, their body, social roles, past states, and actions in the world), to the abstract target, the Subject.

Returning to Table 10.8, I can now examine whether this particular conceptual metaphor is compatible with the items on the WHOQOL-BREF. Consider the item "Would you rate your quality-of-life as very poor…to very good?" This item is clearly asking about the Subject, and the various Selves that are involved are left implicit. A second item states "Does pain prevent you not at all…or an extreme amount from doing what you used to do?" Here again there is a clear reference to the person's status (the Subject), and the question is whether the person is able to do what they "used to do" ("used to do" refers, of course, to the roles that the person performs). A third item, "To what extent do you feel your life to be not at all…or completely meaningful?", again clearly is about the Subject, with reference to the person's various Selves left implicit. "Meaningfulness" is a fairly abstract term, but if made concrete, it would be specified in terms of the activities the person finds important in life.

Although many of the base terms of the items in the WHOQOL-BREF are fairly abstract, it is still clear that a base-target relationship exists in each of these items, making them compatible with the SUBJECT-SELF METAPHOR. The data in Table 10.8 show that nearly every item includes the pronoun "you" or possessive determinant "your." The question now is what happens to a person's ability to use this type of a conceptual metaphor following a stroke, or other acute neurological event? Does the person still use the conceptual metaphor (the base-target relationship) contained in the qualitative item, or do they shift their rating on some concrete element in the item, or shift to a more literal interpretation of the item? Alternatively, is the person's capacity to self-reflect disturbed by the stroke, and how does this affect their assessment of their quality-of-life? Which, if any, of these issues occur, and what do they tell me about. whether an acute neurological event alters the meaning of a quality-of-life or HRQOL assessment?

It should be noted that virtually all quality-of-life or HRQOL assessments also use the SUBJECT-SELF conceptual metaphor. This should not be surprising, since these assessments are usually designed to assess the respondent's qualitative state, and this would necessitate that the items refer to the self. An example is the FACT series of assessments, where nearly all the items in the assessments in the series start with the noun "I" (Cella et al. 1993), while all 22 items in the Functional Living Index-Cancer (Shipper et al. 1984) contain the words "you" or "your," much as was found for the WHOQOL-BREF. It is also true for all 30 items in the EORTC-QLQ-C30 (Aaronson et al. 1991). Thus, the observations noted for the WHOQOL-BREF appear to be true for many quality-of-life or HRQOL assessments.

6.3 The Neurocognitive Basis of Metaphor Expression

The next step in applying my logic requires a description of the cortical areas of the brain that have been proposed to represent or mediate metaphoric expression. I need to know this if I am going to be able to attribute the brain lesions that result from a stoke to any change in the ability of a person to understand qualitative items or use scales that contain metaphoric expressions.

It is generally agreed that language is processed in the left hemisphere, with the right hemisphere involved when the language is humorous, ambiguous, unfamiliar, or figurative (Schmidt et al. 2007). This suggests that the two hemispheres process semantic information differently (Beeman and Chiarello 1998). Schmidt et al. (2007) summarized this hypothesized difference as follows:

> In a particular linguistic context, a word or phrase has many semantic features (definitive properties and associative information), some of which is more relevant than others. In the left hemisphere, bottom-up semantic activation is selective, only activating very closely related semantic features, but doing so strongly, resulting in fine semantic coding within a small semantic field. The right hemisphere, on the other hand, weakly activates a broader set of semantic features to create a coarse semantic coding within a larger, but less strong semantic field." (p. 127)

In Chap. 2, I described the Glucksberg (2001) and Bowdie and Gentner (2005) models of metaphor comprehension, and in each case, they proposed a top-down model, in that a cognitive construction (e.g., forming a category, or aligning the base and target) was assumed to be the basis for metaphoric comprehension. These cognitive constructions may be involved when coarse coding occurs in the right hemisphere and may be the basis for online meaning construction that occurs when a person comes to understand a metaphor, particularly an unfamiliar metaphor (Schmidt et al. 2007).

To the bottoms-up and top-down models, I will add a third model that was also first mentioned in Chap. 2. This was Giora's (1997; 2003) graded saliency model (Chap. 2). This model is unique in that it does not assign different literal or figurative roles to the left and right hemispheres when processing semantic information. Instead, it hypothesizes that the saliency of the discourse determines the relative role within each hemisphere, with saliency being determined by the familiarity, frequency of presentation, or the conventionality of the presented material. Thus, the distinction between literal and figurative is seen as a continuum dependent on contextual factors that determine the meaning of the material. As a consequence, this model rejects the view that the literal form of an expression is first perceived before its figurative meaning becomes available (e.g., Grice 1975).

A number of studies by Giora's colleagues have provided information on the neural representation that underlies graded saliency. For example, Mashal et al. (2005, 2007) used fMRI to determine the neural sites involved in comprehending novel nonsalient metaphors as opposed to conventional salient metaphors. They predicted that the right hemisphere would be involved when the meaning associated with metaphor was novel but not salient, and this was what they found. There was increased activity in the right homologue of Wernicke's area, including the right and left premotor areas, right and left insula, and Broca's area; but these areas were not recruited for conventional metaphors, whose presence was expressed by an different set of cortical sites.

Schmidt et al. (2007) suggest that the graded saliency model can be extended, and in this way, be compatible with the bottom-up or fine-coarse coding model discussed by Beeman and Chiarello (1998). This could be done by assuming that nonsalient material was primarily processed in the right hemisphere, while salient material was predominantly processed in the left hemisphere. Schmidt et al. (2007) reported data that supported this interpretation.[17]

These models seem to be describing the cortical representation of linguistic metaphors and do not address conceptual metaphors. However, Lakoff (1993) would argue that every linguistic metaphor has a conceptual counterpart, and as such, these models should also be informative about the cortical representation of conceptual metaphors. It is interesting to note that the linguist Steen (1999) has devised a five-step process whereby the conceptual foundation of a linguistic metaphor can be determined. Still, the relationship between a linguistic and conceptual metaphor remains somewhat controversial (Engstrøm 1999). I, however, will assume that what I learn about the neurocognitive basis of linguistic metaphors will also be relevant to conceptual metaphors.[18]

Having established that metaphoric expression has a neurocognitive basis, I can now ask what happens when a stroke occurs. Does a stroke impact the conceptual metaphoric basis of items and scales on a quality-of-life or HRQOL assessment, and if so, how?

6.4 Stroke and Metaphoric Expression

A paper by Winner and Gardner (1977) is often cited as initiating interest in the study of the cortical representation of metaphors following a stroke. The study compared groups of left and right hemisphere-injured patients to normal controls. Respondents were presented with 18 syntactically comparable sentences that included a metaphoric expression and were asked to listen to the sentence (verbal presentation) and then select one of four pictures (one depicted an object that was literally accurate, one an image that depicted an object metaphorically, and two background matches) that matched the meaning of the sentence. The investigators found that over 70% of the 18 metaphoric sentences presented were matched with metaphoric pictures by the controls, while only about 12% were matched to literal pictures. This demonstrated that the respondents were fairly accurate, but clearly not perfect, when identifying the sentences. Also found was that persons with left hemisphere lesions selected more metaphoric pictures, while the persons with right hemisphere lesions selected more literal pictures. When compared to controls, the right hemisphere-injured persons selected a statistically significant lower percentage of metaphoric than literal pictures, but when persons with left hemisphere lesions were compared to controls, differences in percent metaphoric (approximately 68% for left anterior hemisphere lesions and about 45% for left posterior lesions) and literal matches were found but they were not statistically significant.

These data were interpreted as supporting the notion that, in general, literal and metaphoric expression involves separate neurological domains. Winner and Gardner (1977) claim that this was demonstrated by applying the logic of double dissociation to their data (Tauber 1955). Double dissociation can be demonstrated by showing that a particular outcome was present when hemisphere A was damaged but hemisphere B was intact, while the outcome was not present when hemisphere A was intact and hemisphere B was damaged. However, their demonstration of double dissociation was best described in relative, not absolute terms, since their data indicated that sentences including metaphoric expression were not universally perceived as such and that lesions in the left or right hemisphere did not lead to complete loss of either literal or figurative function.[19]

Winner and Gardner (1977) also discussed the role that each hemisphere plays in the meaning being projected by different verbal expressions. Thus, the left hemisphere, as the center of literal expression, is involved in denotative expression, while the right hemisphere, mediating metaphoric expression, is involved in connotative expression. Lesions in these brain areas, therefore, would be expected to change the meaning of what is being verbally expressed; I cannot help but ask whether this would include the meaning being communicated by a person's responses to a quality-of-life or HRQOL assessment.

Winner and Gardner's (1977) assertion that the logic of double dissociation can be applied to the cognitive processes underlying metaphoric and nonmetaphoric expression was challenged by a study by Gagnon et al. (2003). In this study, persons with right or left hemisphere lesion and normal controls were asked to respond to two different lexical tasks. In one experiment, the subjects were presented with three words and asked to match either a metaphoric or nonmetaphoric word to a target word (i.e., the word triad task), and in a second experiment, they were asked to decide if two words had a common meaning (i.e., yes/no or word dyad task). The word triad task was envisioned as a selection task in that respondents were forced to respond to the material presented, but that responses to these tasks could not be guaranteed as evidence that the respondent had attended to the meaning of the words. In contrast, the word-dyad task required the respondents to indicate whether the two words had the same meaning – a detection task.

The investigators were interested in determining whether persons with right hemispheric lesions would have difficulty processing specific metaphoric material at the word meaning level (that is, whether performance on the word-triad and -dyad tasks would be different), as compared to normal controls or persons with left hemisphere lesions. They found that there was a general deficit in the processing of metaphoric material in right hemisphere-lesioned persons, independent of the experimental task, but this was also true for persons with left hemisphere lesions. These observations were interpreted as not supporting the double dissociation hypothesis. Gagnon et al. (2003), however, pointed out that while their data did not support the hypothesis that the right hemisphere had a specific role in determining metaphoric meaning, their study compared words and it may be true that when sentences or phrases are considered, evidence for double dissociation would be found (see Bottini et al. 1994).

One possible explanation of the Gagnon et al. (2003) data is that lesions in the right hemisphere reduced a respondent's capacity to process alternative metaphoric meaning (that is, it is an example of reduced cognitive reserve). To test this, Monetta et al. (2006) asked normal respondents to perform the triad-task from the Gagnon et al.'s (2003) study under noninterference or interference conditions (e.g., concurrent counting down from 100 in two steps: 100, 98, 96, and so on). They hypothesized that they would find that the accuracy of selecting the matching metaphor and non-metaphor words for the interfered task would be significantly lower than for subjects who performed the task without interference and might even be comparable to right hemispheric-lesioned persons, and this was essentially what they found. Monetta et al. (2006) conclude that an intact right hemisphere was needed to ensure that the attentional capacity used to process alternative metaphoric meanings was in place, as would be expected for a task as cognitively demanding as the triad task.

An alternative to the cognitive reserve hypothesis suggests that what was lost or diminished as a result of a right hemisphere lesion was the capacity to suppress alternative meanings when a person was presented with ambiguous material (Klepousniotou and Baum 2005). This hypothesis assumes that a disturbed cognitive process was present, as opposed to diminished capacity. Kacinik and Chiarello (2007) also claim that lesions to the right or left hemisphere disturb the cognitive process mediating metaphoric comprehension, but they also state that:

…processes in both hemispheres can support metaphor comprehension, although not identical mechanisms. The LH may utilize sentence constraints to select and integrate only contextually relevant literal and metaphoric meanings, whereas the RH may be less sensitive to sentence context and can maintain the activation of some alternative interpretations. (Kacinik and Chiarello 2007; p. 188).

The literature cited should be sufficient to provide a sense of the issues that have to be considered if the intent is to map changes in cortical representation of metaphors as a result of an acute neurological event to a quality-of-life or HRQOL assessment, and to then use this to design studies of the neurocognitive basis of a quality-of-life or HRQOL assessment. Clearly, there is much complexity here, but also sufficient evidence to support an effort to determine how a brain lesion affects the metaphoric character of quality-of-life or HRQOL assessment.

So far I have demonstrated that nearly all quality of-life or HRQOL assessments use a conceptual metaphor. Next, I will address the fact that scales used to rate items also use metaphors. For example, why is it that a respondent given the face scale sees the various faces not just as illustrating different expressions, but also as reflecting some gradual progression from one emotional state to another (e.g., pain to no pain; happiness to sadness)? The same can be asked of the SAHS. Why are the terms "excellent," "very good," "good," "fair," and "poor" perceived not just as words, but as a graded continuum? One explanation is that they reflect the respondent's capacity to conceptualize the faces or the words into a linear scale, even though a literal interpretation of these faces and words would not allow such an inference. The next question, then, is what does a stroke do to this capacity, and how might it impact a qualitative assessment?

6.5 Stroke and the Meaning of Metaphors: The Visual Analog Scale

Winner and Gardner (1977) reported an interesting observation. They asked subjects to describe the meaning of metaphors that were originally presented to them as words or pictures. They found that 40% of the persons with right hemisphere lesions interpreted the meaning of metaphoric *pictures* literally, while these same persons had no such difficulty describing the meaning of metaphors in the *verbally* presented sentences. Rinaldi et al. (2004) repeated the Winner and Gardner (1977) experiment and basically confirmed the dissociation of meaning for metaphors presented verbally or visually for persons with right hemisphere lesions. They attribute this dissociation to a disturbance of cortical integration between visual- (e.g., the picture) and verbal-mediating cortical areas and suggest that this may be true for right brain-damaged persons, independent of the type of material presented.

This dissociation is of interest because it may have significance for the scaling and valuation tools commonly used in quality-of-life or HRQOL studies. For example, if I present a subject with a right hemisphere lesion a visual analog scale, which is usually presented in the form of a picture (e.g., a line with modifiers at each end of the line), and provide this person with verbal descriptors, am I not simulating what Winner and Gardner (1977) and Rinaldi et al. (2004) did? And should I not also expect that the brain-lesioned person would respond to this task in a literal and nonfigurative manner? What if I presented these persons with the Cantril's ladder, a series of faces that represent different levels of pain, or the various Dartmouth COOP stick drawing charts (Nelson et al. 1996)? Should I not expect that the right hemisphere-lesioned person would have difficulty responding to these tasks, especially since they may not perceive the linearity implicit in these assessments of fatigue, pain, or quality-of-life? Is there any data to indicate that the person who has had a right hemisphere stroke has difficulty responding to these assessments?

Before I answer this question I want to make clear how a visual analog scale, and its surrogates – the faces or verbal rating scales, are a type of figurative, if not metaphoric, expression. They are because none of them are literally rulers, and instead, use various perceptual or verbal devices to conjure up the impression of linearity. This can be constructed by presenting these indicators in several ways: along a vertical or horizontal plane; contrasting extremes in verbal anchors (worst possible…or best possible); generating extreme emotions as in different faces (fear vs. happiness); using colors that represent the visual spectrum; or numbers that descend or ascend. However, while some of these indicators may be necessary, none are sufficient to generate this sense of linearity. Linearity, for example, can also be conjured up by any temporal order or even the temporality that is generated by a sentence. At a minimum, therefore, these scales are a figurative expression of a ruler, and they are considered metaphoric in that a mapping is occurring from a concrete basis to an abstract target.

As is the case with some of my other analytical exercises, the available research in this instance only approximates the information I need to answer my question. A related, but somewhat more general, question asks if persons who have had a stroke can use a visual analog scale independent of the location of the stroke. A study by Price et al. (1999) provides some answers to this question. They asked 96 persons with a stroke and 48 controls (i.e., persons without cardiovascular disease) to indicate when they could feel the inflation of a standard sphygmomanometer cuff and then rate this initial tightness using five randomly ordered rating scales. These rating scales included a four-point verbal rating scale (none, mild, moderate, and severe); a numerical rating scale; a mechanical visual analog scale; a horizontal visual analog scale; and a vertical visual analog scale. The cuff was then inflated to twice the initial value and the respondents were again asked to rate the magnitude of the inflation using the five rating scales. Finally, the cuff was deflated to zero and the respondent was again asked to rate the pressure using the five rating scales.

The accuracy of responding was judged by how well the respondent was able to monitor the direction of pressure change and reflect this in their ratings on the five visual analog scales. Thus, if a cuff inflation physically represented a reduction in pressure from the previous value, then the scaled values were expected to decrease to be labeled as accurate. To determine that the respondents had the capacity to make these judgments independent of their use of the visual analog scales, they were also asked, after the second and third change, if the pressure was the same, more or less than the previous pressure they experienced. They were also administered a series of neuropsychological tests designed to assess the extent of the respondent's cognitive or visuomotor impairment following their stroke. Hospital records were used to describe the site of the stroke, although the researchers did not use fMRIs or PET scans to determine the exact site of central nervous system lesions.

They found that 95% of the stroke patients were able to accurately detect the change in inflation of the cuff (i.e., the same, more or less than the previous inflation), as compared to 100% of the controls, but that the stroke patients were less accurate in using each of the ratings scales than the controls, independent of the type of scale. The four-point verbal rating scale produced the highest accuracy (65%) compared with the other types of scales (accuracy range: 47–53%). This greater accuracy was also found for some types of stroke patients. For example, patients who experienced a posterior circulation stroke were 100% accurate when using the verbal rating scale, but was reduced for the four other scales (accuracy varied from 75 to 92%). This effect was also true for the patients who had a lacunar circulation stroke, with 86% accuracy usage for the verbal rating scale, but a range of 59–64% for the other four rating scales. This difference in accuracy persisted for patients with a partial anterior circulation stroke (verbal rating scale=65%; other four scales=41–60%). However, there was no difference in the accuracy of use of all five scales (range 4–7%) for the most seriously impaired patients who experienced a total anterior circulation stroke.

NPT was used to determine the extent to which neurocognitive deficits could account for the inaccurate use of the different rating tools. A measure of aphasia indicated that it did not contribute significantly to the accuracy of usage for any of the rating scales, and that was consistent with having limited the sample to persons who could follow simple commands. A measure of cognition, visuospatial impairment, and tactile inattention or hemineglect each contributed significantly to the inaccuracy on the verbal rating scale. In contrast, only the cognitive measures contributed to variance in the numerical rating scale, while cognition and tactile inattention or hemineglect contributed to variance in the mechanical visual analog scale. Also of interest was that the measure of cognition did not significantly contribute to variation in accuracy of use of the vertical visual analog scale. Unfortunately, this study did not have a sufficient number of participants to examine whether the type of lesion and type of rating scale impacted the use accuracy of the rating scales.

A study by Benaim et al. (2007) involves left and right hemisphere stroke patients rating their pain using a visual analog scale, verbal rating scale, and face scale. Adjustments were made in the scales to take into account that many stroke patients, particularly right hemisphere stroke patients, report unilateral neglect. For this reason, the face scale was aligned vertically, the visual analog scale was colored, numbered (e.g., 0–10), and anchored by the labels "no pain" or "worst imaginable pain." In addition, the verbal scale added numbers (e.g., "0," "1," "2," "3" and "4") to the verbal descriptors ("none," "slight," "moderate," "severe," and "intense"). All 127 stroke patients and 21 controls completed all three scales. After these ratings, subjects were divided into three groups: a validity study, a reliability study, and a preference study. In the validity study, subjects were asked to rate the degree to which each of the faces matched a particular emotion (e.g., pain, sourness, sadness, anger, boredom, and sleepiness) as a measure of content validity and to rearrange a random presentation of the faces into a rank order. In addition, responding to the face scale was correlated with the visual and verbal rating scales and these correlations could be used as an indicator of validity. Reliability was assessed by repeating the ratings of the three scales within 24–48 h to obtain interrater reliability and between each of the scales to obtain intrarater reliability. The reliability groups were asked to rate their preference for each scale.

The results of the study, particularly for the ratings of the face pain scale, were consistent with what would be expected for a person who had a right hemisphere stroke, especially since they would be less likely to recognize the conceptual metaphor implicit in the face pain scale. Thus, the controls and left hemisphere stroke patients ranked pain as most representative of the emotions being expressed in the faces, while the right hemisphere stroke patients rated sadness as the emotion most representative of the faces. In addition, none of the right hemisphere stroke patients were able to successfully rank the set of faces, while 43% and 42% of the controls and left hemisphere stroke patients, respectively, were able to complete the task. While the different stroke groups differed in their responding to the scales, they were highly consistent in their responding between the scales. The visual analog scale was the most reliable (both inter- and intrareliability) for both groups of patients. The left stroke patients often preferred the face pain scale, while the right stroke patients preferred the visual analog scale.

These results are consistent with the Winner and Gardner (1977) observation that right hemisphere-lesioned stroke patients interpreted the meaning of metaphoric pictures literally and thus would lack the conceptual skills to recognize the continuum implicit in the series of faces. In contrast, they had less difficulty describing the meaning of metaphors in the verbally presented sentences, which is consistent with their greater reliability when responding to either the visual analog and verbal rating scales, as compared to the faces pain scale.

These studies clearly have only some of the elements I require to answer whether the stroke patient's ability to use various scales is related to their capacity to operationalize the scaling task. Price et al. (1999) concluded their study by recommending that the visual analog scale not be used with stroke patients. In contrast, Benaim et al. (2007) recommend that the faces pain scale be used with left hemisphere stroke patients, while any of the three scales used in their study would do equally as well for right hemisphere stroke patients. In addition, Dorman et al. (1997) and Korner-Bitensky et al. (2006) both recommend using the visual analog scale for stroke patients. Benaim et al. (2010), however, demonstrated that a verbal scale was better at characterizing changes in depression over time than a visual analog scale for depression.

Admittedly, I have asked a complex question (what meaning a respondent attaches to their ratings) under conditions that are inherently variable (e.g., strokes do not produce uniform damage and the extent to which neurocognition will be affected varies between individuals). The linking concept here is the notion that a quality-of-life or HRQOL assessment contains metaphoric properties whose cortical representation can be affected by an acute neurological event. Stroke patients were my preferred study group, since they were most likely to have lesions in the brain area that mediates metaphoric expressions. The Benaim et al. (2007) study, however, did not provide evidence to support my working hypothesis.

6.6 The Right Hemisphere, the Sense of Self- and Qualitative Assessment

The previous discussion concerning the cortical representation of metaphoric expression raises an interesting question: Is my ability to judge the quality of a painting, a good meal, or my life critically dependent on an intact right hemisphere? Certainly, clinical observations suggest that this may be so. For example, the right hemisphere is usually referred to as mediating creative abilities, processing information in an intuitive way while simultaneously attending to various stimuli. Thus, the creative person first looks at the whole picture and then its details. In contrast, the left hemisphere is involved in more analytical tasks, processing the pieces first, and then bringing them together as a whole. An artist, for example, is encouraged to visualize what she finally wants to produce and then execute her work of art (e.g., Edwards 1979). Is this what a person does when they are asked to rate their own quality-of-life or HRQOL?

Persons with right hemisphere damage have also been found to be impaired in their ability to engage in basic speech acts (Grice 1975). That is, they have difficulties responding to questions, assertions, requests, and basic commands (Soroler et al. 2005). Rehak and Kaplan (1992) studied the ability of persons with right hemisphere damage to discern the presence of deviation in a conversation and found that they were more accepting than normal controls of such deviations. Thus, when engaging such a subject in a quality-of-life or HRQOL assessment, an investigator should be aware of these conversational complexities when communicating task instructions.

The person with right hemisphere damage also tends to be more literal in their interpretation of discourse, as they are less likely to respond to jokes, irony, sarcasm, and as expected, metaphors. These problems with figurative language may be due to an inability to integrate verbal expression with its context. The right hemisphere-damaged person also lacks the ability to distinguish important from unimportant information, interpret body language, and are known to have flat affect, be impulsive, and to confabulate. Also, as alluded to in the Price et al. (1999) paper, these people have a variety of difficulties orienting in time and space, as is evident from their left-sided neglect, and inability to recognize the severity of their own symptoms (that is, they are anosognosic).

Of particular interest is the observation that persons with right hemisphere damage tend not to recognize the faces of others, nor their own faces; that is, they also may be prosopagnosic. The inability to recognize one's own face has suggested to investigators that these persons have difficulty with self-recognition and also self-awareness. Keenan et al. (2005), in fact, summarizes evidence supporting this hypothesis and specifically suggests that the right prefrontal cortex is a critical mediator of a person's sense of self. Johnson et al. (2002) performed fMRI studies asking persons to respond "yes" or "no" to questions requiring self-reflection or factual answers. They found that the anterior medial prefrontal and posterior cingulate areas of the brain were activated when a person was being self-reflective, but not when responding to factual questions. These observations are particularly interesting, especially since I proposed that the SUBJECT-SELF METAPHOR could be used to analyze the items in the WHO-BREF, as well as in other quality-of-life or HRQOL assessments. Putting these two observations together prompts me to ask several questions. If I lack a sense of self, will I be able to estimate my quality-of-life or HRQOL? In other words, if I lack a sense of self, will I be using the SUBJECT-SELF conceptual METAPHOR when responding to items on a qualitative assessment? Is there any data that addresses these issues?

First, let me point out that persons with right hemisphere stroke, autism syndrome, schizophrenia, and Alzheimer's disease have been characterized as having an impaired sense of self, including sensing how others may feel (Frith 1992; Happe et al. 1999; Zakzabme and Leach 2002). Evidence supporting this includes the inability to maintain social relations and participate in normal social exchanges requiring conversation with others. A key element in being able to maintain a conversation is the ability to interpret the facial expressions of others. This is particularly true for the person with autism. Thus, Lahaie et al. (2006) report that even though high-functioning autistic adults have more than adequate intelligence and can process face parts quite well, they have difficulties putting these parts together as a whole face. This presumably hinders their ability to interpret the information being communicated by different facial expressions and body language, much as has been found for the person with a right hemisphere stroke or Alzheimer's disease.

The inability to integrate facial parts into a picture of a whole face raises the interesting question of whether persons with these disorders suffer from a general inability to deal with abstract and concrete items, concepts, or experiences – particularly those tinged with emotional connotations. Thus, it can be asked if these persons would be able to integrate their life experiences into a global estimate of their qualitative state by recruiting abstract concepts. If they are not, then investigators could reduce the informational demands of a multi-item qualitative assessment to its lowest common denominator, with literal expressions being used to express otherwise concrete issues (e.g., asking about the particulars of a person's physical activities).

Another interesting proposal concerning the role of the right hemisphere comes from Mitchell and Crow (2005), who suggest that some of the clinical manifestations of schizophrenia involve deficiencies in the lateralization of language function to the right hemisphere. This, they claim, has consequences for social communication and forms the basis for psychotic thinking. Support for their proposal comes from neuropsychological test results suggesting that schizophrenics are deficient in conversation planning and comprehension; understanding humor, sarcasm and metaphors; and lack the ability to understand emotional prosody. The schizophrenic is also known to have deficiencies in language functions that are dependent on an intact left hemisphere, as is evident from their disorganized speech and difficulties in processing the semantic aspects of language. Mitchell and Crow (2005) propose a bihemisphere theory of the neural basis of language function that, if out of balance, leads to the symptoms characteristic of schizophrenia.

When I combine this information about the relationship between the left and right hemispheres, it seems obvious to ask if many of the conditions that lead to deficiencies in quality-of-life or HRQOL involve functions within each hemisphere, but are also dependent on the relationship between these hemispheres. However, any answer to this question requires recognition that the relationship between these hemispheres may involve several levels of activities. Thus, in Table 6.1 I list three basic sources of cognitive-based modular domains, each that could be dependent on the relationship between the two hemispheres. These modular domains could have a structural basis; be acquired, but function autonomously; or be acquired and involve being actively controlled by the person. It remains to be clarified what role each of these modular domains plays in the interaction between a person's health status and their assessment of their HRQOL.

6.7 Neurocognition Recovery

At this point, it seems appropriate to draw some conclusions. First, in this section I raised the question of whether the items in a quality-of-life or HRQOL assessment have conceptual metaphoric properties. The answer is yes for most of the items on most of the qualitative assessments. Next, I reviewed the complex literature that described the neurocognitive basis for metaphoric expression. With this information, I described what is known about the impact a stroke has on metaphoric expression and followed this by discussing what impact a stroke has on the meaning of a metaphor, such as the visual analog scale. The topic that flowed from this discussion was determining what role an impaired right hemisphere had on a person's sense of self. However, I did not take advantages of the literature on split-brain subjects or longitudinal studies, nor did I extensively review the literature on recovery of function studies; each of which could provide additional insight into the impact of a stroke on a qualitative assessment.

Although the data reviewed here were often insufficient to answer the specific question I asked, there was enough to support the logic of the approach I described. One of the more intriguing questions raised was whether an intact right hemisphere is required for judging the quality of a piece of art, a musical rendition, a new car, or a person's life experiences. A significant part of the answer revolves around whether a person has or lacks a sense of self or awareness, and what impact this has on qualitative assessments. An altered sense of self and awareness seems to be a central determinant of what family members and care givers experience with a wide range of patient groups (e.g., Alzheimer's disease patients, brain cancer patients, schizophrenics, stroke patients, Parkinson's disease patients, and so on) and this has become a central focus of burden of caring for these persons.

The paper cited by Mitchell and Crow (2005) proposes that the balance in the relationship between the two hemispheres may account for the symptoms of schizophrenia. This same lack of balance between the hemispheres may also account for deficits in a person's quality-of-life or HRQOL, but how it impacts the creation of the hybrid construct would be of great interest.

Central to this discussion is the demonstration that an intact central nervous system is a necessary precondition for the self-reflection, recognition of the figurative nature of the language, and performance on a qualitative assessment. NPT results provide some insight into this relationship, but neural imaging studies could also contribute to this information, although both remain most informative about the capabilities of a respondent, and only secondarily about the meaning attached to a qualitative response. Thus, it is logical to ask why it is so important to consider the meaning a person who has had a stroke attaches to their response to a qualitative assessment? Why is it more important than simply deciding that such persons are different from persons who have not had a stroke? Clearly, this is a common level of decision making after a quality-of-life or HRQOL study, but this approach runs the risk of misinterpreting what a patient is communicating with their responses, and this brings me back to asking whether the psychometric or cognitive characterization of a qualitative assessment should be my objective.

7 Models of the Relationship Between Neurocognitive Capacity and Performance

7.1 Introduction

One of the central concerns that motivated me to write this book is my belief that the qualitative status of the compromised person is not adequately assessed. What is evident from the discussion to this point is that the blanket assumption that a person with failing memory, distorted thought processes, stroke-induced impaired visuospatial relations, traumatic or radiation-induced brain injury can't assess the quality of their existence is patently false. Rather, the situation seems to be much more nuanced. Thus, mild or moderate dysfunctional persons, who are optimally pharmacologically treated, exist in an appropriately supporting environment, and are given time to recover from their acute event, have been shown to be as capable of providing data as reliable and valid as age, gender, and culturally matched controls. Remaining problematic are those persons with severe dysfunction; they are the ones creating the most social and emotional impact on the health care system, families, and society. It is these persons who create the impression that the compromised person is not amenable to assessment, remediation, or intervention. Two solutions can be proposed regarding how to deal with this more limited group of subjects, both of which require additional experimental support.

The first is to use the elements of the hybrid construct (e.g., the generation of descriptors, the process of rating these descriptors, and the process of valuation) as an analytical tool to design studies to pinpoint the specific deficits that the severely impaired person may have when making a qualitative judgment. Current practice does not focus on identifying these specific deficits. This greater level of specificity, combined with taking advantage of the time-dependent nature of the evolution of these deficits, should provide an investigator with the opportunity to design intervention programs that may slow the onset of these deficits. This approach differs from cognitive rehabilitation, where the therapist usually waits until the person's intellectual and functional status stabilizes before an intervention is initiated. The rational for doing so is that otherwise a therapist would not know if their intervention had the desired outcome. However, if prior research has established that elements of the hybrid construct have been impaired, then the clinician has a head start on determining where to intervene.

A second approach acknowledges that it is not likely that any approach will succeed that relies on the use of language to analyze, design, or implement an intervention. Rather, what is needed is a totally different conceptual model upon which to base such a research program. In this section, I consider two such alternative models: first, I consider the role that neural networks and optimality theory can play as conceptual models; and then I discuss embodiment theory and its potential applications in qualitative research.

As stated before, NPT can be informative about a person's capabilities. Implicit in this relationship is the assumption of an isomorphic relationship between test performance and brain functioning, and that examining impaired performance will reveal something about the brain systems that have been disrupted (Springer and Deutsch 1993; p. 149). In addition, it is also assumed that performance following a brain injury reflects what was cognitively present before the impairment and whether the mature brain is sufficiently plastic to produce new cognitive modules. Thus, central to the brain's response to an acute neurological event or degenerative process is its capacity to recruit and organize what remains structurally intact into new functional networks. However, while NPT can be informative about these new neural networks, they tell little about ongoing cognitions. Yet, as proposed here, what someone is thinking about when they assess their life quality and what the specifics of these cognitions are is central to understanding how a qualitative assessment occurs. Still, the conditional nature of the relationship between performance and neural processes raises questions about how these two levels of analysis can be related.

Having to relate different levels of analysis to one another is a classic issue in science, and an outgrowth of its analytic and reductionistic nature. Thus, I speak of the "mind and body" or "capacity and performance." Historically, solutions designed to resolve this problem have ranged from assuming that the two or more levels of analysis covary, to identifying some common denominator, or to postulate some intermediate notion. The model implicit in the studies relating NPT results to brain capacity rests on the assumption that the two levels of analysis covary, but remain independent of each other. In contrast, Smolensky and Legendre (2006) assume that the mind and the brain are a single entity; both are computers functioning at two levels of abstraction. At a biophysical level, parallel processing characteristic of brain functioning forms the basis of a numerical connectionist computer, while at a more abstract level characteristic of cognition and language, a serial symbolic computer exists. Both levels of analysis are computational and both generate numerical optimized outcome. The presence of optimality at both levels of analysis has led to the development of an optimality principle that has been usefully applied to syntax, with extensions to semantics and pragmatics (Blutner 2000; Hendriks and Hoop 2001; Prince and Smolensky 1991, 1993).

A third model, the embodiment model, which evolved from a cognitive perspective (e.g., Gibbs 2006), also rejects the implicit dualism present in the capacity/performance dichotomy and postulates a nonlinear interactionist view of the relationship between the body and various cognitive and psychological processes. I alluded to this model when discussing Lakoff and Johnson's (1999) approach to reality and truth and will rely on Gibbs' (2006) book to present a more complete description of the model. The Lakoff and Johnson (1999) approach assumes that embodiment is dependent on three different levels of analysis: the neural, the phenomenological (conscious experiences), and the cognitive unconscious. The role of each in embodiment will be discussed.

The embodiment models are of interest for several reasons. In addition to representing an alternative to dualism, they offer an alternative to Lawton's (1991) insistence on separating the subjective person from the "objective" environment and also the social indicator movement's emphasis on ecological determinants of well-being. Embodiment models do this by essentially cognitizing the environment. A person's early and continued encounter with the environment, the musculoskeletal consequences of their activities, and their perception and sensing of this environment all become embodied. The cognitive-environment template that results becomes part of that person's thought processes.[20]

Both optimality theory and embodiment offer ways of thinking about capacity and performance without adopting a dualistic perspective. However, it remains to be demonstrated whether either or both can be usefully extended to account for quality-of-life or HRQOL assessment.

7.2 Neural Network Models and Optimality Theory

Neural network models have successfully simulated a number of complex cognitive processes (e.g., Long et al. 1998; Rumelhart et al. 1986), and this suggests that they may be able to characterize the relationship between capacity and some measure of performance, such as a quality-of-life or HRQOL assessment. It may even help clarify the role NPT plays in the relationship between a person's neurocognitive capacity and performance.

From their inception, neural networking models were conceived of as analogs of the brain and assumed to have features common to brain structure and function (Rumelhart et al. 1986). Thus, a neural network was to consist of many processing units that were connected in parallel and were able to "learn" by interacting with each other and their environment. In order for learning to occur, the processing units were assumed to be in some state of differential activation that would be propagated throughout the network, modifying the patterns of connectivity. It was also assumed that this process of continually modifying the pattern of connectivity would be retained and lead to the emergence of the new "learned" state, much as postulated by Hayek (1952) and Hebb (1949).

Neural network models have an extensive intellectual history that fits into the general class of connectionist models, including Aristotle's notions about the association of mental ideas; the British associationists; Thorndike's learning theory (1932, 1949); and so on (Medler 1998). This type of modeling has been applied to artificial intelligence, the cognitive neurosciences, economics, linguistics, psychology, philosophy, and physics. It also has been used as a statistical tool (e.g., Warner and Misra 1996), particularly when applied to decision making, prediction, or classification. Neural network models have been found to be useful when a regression analysis is called for.

If I want to use neural networks to simulate the relationship between capacity and performance, then I may want to model such neurocognitive outcomes as preservation; the degeneration a person with Alzheimer's disease experiences; the reorganization of cortical pathways in response to a stroke; and so on. To do this, I would need a model where I could vary the number of functional units; the degree to which these units interact; how quickly they interact; how well they develop new patterns of connectivity following a disruption in the system; etc. I would then "teach" the network the outcomes I want it to simulate by providing large amounts of data and then test this model with a new set of data to determine how well it predicts an expected outcome. Having developed and tested the neural network model, I may also want to ensure if the predicted findings can be confirmed by examining changes in brain structure or function using brain imaging studies. This interplay between available data and a computational model creates an iterative process (Hapgood 2005) that offers a unique method for characterizing complex relationships, including the relationship between capacity and performance.

Neural network models also assume that the interaction of its components occurs in a nonlinear fashion via parallel processing. In contrast, much of the research reviewed in this chapter has often assumed that a linear relationship exists between capacity and performance. For example, a linear model would predict that brain lesions presumably affecting a person's capacity would directly lead to decrements in cognitive functioning. Since this is not regularly found, investigators have postulated the existence of a cognitive reserve, or a cognitive preservation process, to account for the nonlinear outcomes. This logic, of course, is very reminiscent of the economist who has to face some instance of a failure of "rationality". As we have learned, education level and occupation optimize intellectual functioning in old age, and it seems that these factors contribute to a person's ability to use alternative neural circuits to accomplish various cognitive tasks. Also, the adult nervous system may be sufficiently robust (that is, capable of using alternative brain structures and pathways) so as to restore initially lost functions following an acute neurological event. Each of these observations suggests that a nonlinear model may be a more appropriate model than a linear model for fitting the data between capacity and performance.

A number of studies have formally compared linear neural network statistical procedures (e.g., linear regression analyses, path analyses, etc.) to nonlinear ones, demonstrating that neural networks have both advantages and limitations (e.g., Kattan 2002). For example, Buscema et al. (2004) were interested in identifying the statistical procedure that would best predict pathological findings following autopsies of persons with Alzheimer's disease. The authors understood, from anatomical studies, that the number of neurofibrillary tangles was highly predictive of memory impairment, but that this relationship was complex and varied with brain site. They

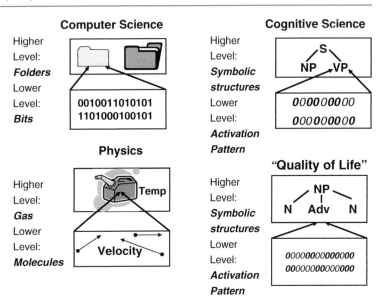

Fig. 10.5 The relationship between higher and lower levels of abstraction with examples from the computer and cognitive sciences. This figure is a modification of Fig. 2 from Smolensky and Legendre (2006; Vol. 1, p. 32). *S* sentence; *NP* noun phrase; *VP* verb phrase; *N* noun; *Ad* adverb.

used NPT data and various measures of function from 117 persons from the Nun study (Snowdon 2003) to predict the extent to which neurofibrillary tangles were present in the brain at autopsy. They found that neural network models were better at predicting the histological outcome then linear discriminant analysis. This study is of interest not only because it demonstrated that a neural network can function as a useful statistical tool, but also because it illustrates that NPT results can provide information about a person's neural capacity. Neural network models have also been used to predict the changes in cognitive functioning that occur over time for the person with Alzheimer's disease (Tandon et al. 2006).

Krongrad et al. (1997) also reported a study that compared statistical procedures based on neural networks or linear regression analyses; this time to determine predictors of the emotional component of a quality-of-life assessment. Scores on the 17-item Rand Mental Health Index (Stewart and Ware 1992), the Functional Living Index-Cancer (Shipper et al. 1984), and additional individual items served as independent variables in the models. Computer programs were used to classify data using feed-forward, back propagation neural network methods. Sixty-three persons with either benign prostatic hypertrophy or prostatic cancer were used to train the neural network to determine if a subject fell into the high or low quality-of-life group, and additional subjects were used to test the ability of the network to predict outcomes. For example, the neural network model was able to correctly classify 89% of the subjects 1 month after their initial assessment, as compared to 84% for linear regression analyses. These basic results were replicated when the 6 month data were collected. The authors also reported using linear and quadratic discriminant functional analyses that were only able to correctly classify high and low quality-of-life outcomes in 57 and 43% of the cases, respectively. The authors concluded by stating a preference for using linear regression analyses, but their data suggest that, while more laborious, neural network modeling can also be a useful statistical tool.

Neural networks may be uniquely suited to model complex high-volume qualitative data that current quality-of-life researchers have tended to avoid. These may include the continuous data streams that come from such procedures as the "thinking aloud" cognitive interviewing procedure; recurrent verbal self-reports; daily diary reports; cued momentary emotional status or physical activity reports; and so on. Monitoring and quantifying these data streams is an important goal, since it will provide insight into the ongoing cognitions that give rise to the experience of quality and a response to a quality-of-life or HRQOL assessment.

Having reviewed some of the ways in which a neural network can be useful, I need to directly address whether they can model the conditional relationship between capacity and performance. Smolensky and Legendre (2006) argued that the "mind and brain" can be conceived of as two parts of a single computer, with the brain being a numerically based parallel processing computer and the mind a symbolically based computer. Their "solution" offers an alternative to a dualistic approach, by assuming that the two outcomes represent one system functioning at two different levels of abstraction. Figure 10.5 illustrates their argument. Consider the area labeled "Computer Sciences" and imagine that at the lower level, bits of information are rapidly flipping between "0" and "1," while at a higher more abstract level folders are slowly opening or closing. As Smolensky and Legendre (2006) state, "One virtual machine, a folder manipulator, is running on top of a different lower-level machine, a bit manipulator. Put differently, the virtual machine is a higher-level abstraction implemented in the primitive bit-manipulator" (p. 32). Each of the examples in Fig. 10.5 illustrates how the atomic and molar level of analysis are related. Smolensky and Legendre's (2006) solution to this problem is to apply the continuous mathematics of mathematical physics to permit brain-type connectionistic computations and also the symbolic computation of higher levels. They describe a new computational architecture (the Integrated Connectionist/Symbolic Cognitive Architecture) to relate these two levels of analysis.

This solution (the architecture they have developed) has the potential to bridge the gap between capacity and performance. Here I am assuming that what is referred to in Fig. 10.5 as "Activation Pattern" for the "Cognitive Science" model refers to processes occurring at the level of the brain, while the "Symbolic Structures" refers to grammar and its related functions. I would extend this by claiming that the term "capacity" is a label that characterizes some set of neural activities, while "performance" represents the outcome governed by the symbolic structures mediated by grammar. Thus, just as a "folder" appears as a symbolic representation of a "bit oscillator," so too would performance be expected to be a symbolic representation of brain-based capacity.

Quality-of-life, as a noun phrase (Fig. 2.1), would involve an even lower level of analysis than a sentence, and this is illustrated in the figure by the units at the activation level. This titrating of the activation pattern upwards as one goes from representing a word, to a phrase, to a sentence illustrates Fodor's (1975) combinational strategy, discussed earlier.

Smolensky and Legendre (2006) have also found that their integrated connectionist/symbolic cognitive architecture can be used to demonstrate that well-formed grammatical representations are optimal symbolic structures. This optimized grammar exists at both the neural and symbolic levels of analysis. Thus, a harmonic grammar evolves from the connectionist neural level and employs numerical computation, while Optimality Theory is nonnumerical and is applied at the symbolic level. Smolensky and Legendre (2006) describe optimality[21] in the following way:

> ...optimization amounts to this: each possible symbolic linguistic structure can be evaluated by a set of well-formed constraints, each of that defines one desirable aspect of an ideal linguistic representation (e.g., a sentence must have a subject). These constraints are highly general and, as a result, highly conflicting. Typically, no structure meets all such constraints, and a means is needed for deciding which constraints are most important: the well-formed or grammatical structures are the ones that optimally satisfy the constraints, taking into account the differing strength or priority of constraints. (p. 39).

In other words, if I were to examine a language, I would find variation in expression and limits to this variation that are dependent on the language's grammar. The constraints that the grammar creates (e.g., a sentence should have a subject) lead to "linguistic conflicts": conflicts between a perfect or less than perfect application of the grammar. The resolutions of these conflicts lead to a hierarchy of outcomes, ending in an optimal outcome.

For example, consider that the standard format for linguistic expression consists of a sentence with a subject, verb, and object. It is true, however, that meaningful sentences can be constructed when the object is stated first, as in "What did John see?," or when the sentence has no subject, as in "It rains." These sentences illustrate that linguistic expressions vary from complete conformity to varying degrees of violation of their grammar. This creates a hierarchy, with the sentence that is the most grammatically correct having the highest optimality rank, while those sentences which are less constrained having a lower ranking. Thus, "Jim said what?." is more grammatically constrained then the sentence "Who said what?." In addition, the former sentence is also more focused and communicates a more specific type of information than the second. This suggests that the more constrained the sentence, the more specific the meaning it conveys. Thus, grammatical constraints lead to ranking and re-ranking of sentences, phrases, or words, in turn leading to an iterative process that determines the optimal outcome of what is being spoken and written. The existence of optimization has obvious adaptive significance, since without it, effective communication would be limited. The universality of these constraints also implies that they play a critical role in such language-dependent processes, such as logic, reasoning, and discourse.

Is the principle of optimality relevant to the task of assessing quality or quality-of-life/HRQOL? It could be, since quality or quality-of-life/HRQOL items, like all sentences, are syntactically dependent, so that they can be more or less "optimally" constructed. But how often are items on a quality-of-life or HRQOL assessment syntactically complex? To answer this question, I applied the QUAID (Question Understanding Aid; Graesser et al. 2000) to the first five items of five standard cancer-specific HRQOL assessments. The QUAID permits an investigator to type in individual items and responses and have them compared to a standard based on 11 Census Bureau surveys consisting of some 500 questionnaire items which were assessed by a panel of experts who classified them in at least five dimensions. Thus, the QUAID identifies words, phrases, or sentences that may be technical and unfamiliar or that consist of vague predicates (verbs, adjectives, adverbs); ambiguous noun phrases; sentences with complex syntax; and questions that overwork memory. Table 10.9 summarizes this analysis for the five questionnaires.

Table 10.9 reveals that many of the five questionnaire items included terms that QUAID considered unfamiliar, vague, or imprecise. This might have been expected since the assessed questionnaires addressed the consequences of a specific disease in contrast to the more general socioeconomic content covered by Census Bureau surveys. Questionnaires that had complex syntax according to the QUAID include the FACT-G and the BCQ. The source of the

Table 10.9 Application of the QUAID (question understanding aid) to the first five items of selected examples of HRQOL assessments: percentage hits per five items

HRQOL assessment	Unfamiliar technical terms (%)	Vague or imprecise relative terms (%)	Vague or ambiguous noun phrase (%)	Complex syntax (%)	Working memory overload (%)
EORTC QLQ-C30*	40	60	0	0	0
FLIC*	80	60	40	20	0
FACT-G	60	100	40	100	0
RSQ*	100	100	20	0	0
BCQ*	100	100	20	100	40
Average	76	84	24	44	8

**EORTC QLQ-C30* EORTC quality-of-life questionnaire (Aaronson et al. 1991); *FLIC* Functional Living Index-Cancer (Shipper et al. 1984); *FACT-G* functional assessment of cancer therapy-general (Cella et al. 1993); *RSQ* Rotterdam symptom questionnaire (de Haes et al. 1990); *BCQ* breast chemotherapy questionnaire (Levine et al. 1988)

inappropriate syntax was located in the phrases used for the ratings the FACT-G and BCQ. Apparently, the QUAID "reads" the rating options as if they are one long sentence or set of phrases, so that it is not surprising that these two questionnaires failed to meet the criteria for being syntactically appropriate. This suggests that the format used to design these questionnaires was not syntactically optimal. For example, most of the questionnaires (e.g., the FACT-G) in Table 10.9 were organized such that they consist of a general statement for context (e.g., "During the past 7 days…"); particular reference (e.g., "I have lack of energy"); and a list of rating options (from not at all to very much). This fragmented presentation creates a cognitive challenge for the respondent who has to integrate the various elements of the item into a comprehensible statement that they can act upon. Consider Item #1 from the FACT-G, with the first rating option: "During the past 7 days I have a lack of energy not at all." Alternatively, consider the Rotterdam Symptom Checklist Item #1 (with rating option "not at all"): "Have you been bothered, during the past week by lack of appetite not at all." Obviously, the investigator expects that the respondent will rephrase these sentences so that they are syntactically appropriate and semantically useful.

However, what may be most important here is how the respondent interprets the information they are being asked to process. This suggests that when I consider the linguistic nature of questionnaires, I am most likely dealing with a problem in interpretation or pragmatics,[22] rather than a problem of production (involving syntax and semantics). Can optimality theory be applied to pragmatics? Blutner (2000) argues that it can (see also Blutner and Zeevat 2004). He proposes a model where syntax and semantics are assumed to be bidirectionally related, as is the relationship between interpretation and semantics. Optimality theory would be active in both relationships. His formal proof (Blutner 2000; p. 197–213) involves applying Grice's (1975) conversational principles that posit a reciprocal relationship between optimal language production and optimal interpretation. For example, in a conversation, it would be expected that both participants would optimize their presentation so that the exchange of information would be as efficient as possible. The same kind of optimized efficient outcome would be expected when a respondent encounters a questionnaire.

I have been particularly brief in my description of this complex material, but it is clear that it provides some interesting ways of examining the nature of a qualitative assessment. The material reviewed here has also been useful in that it provides a perspective on how to relate different levels of analysis. Thus, I discussed how neural networks might be able to model the complex relationship between capacity and performance without lapsing into some form of dualism. NPT results would be expected to provide the data needed to train the neural network. The optimality theory of Smolensky and Legendre (2006) that grows out of neural network modeling has been successfully applied to phonology and grammar, and more recently, to semantics and pragmatics. The extension of optimality theory from language production (syntax and semantics) to language interpretation is particularly useful, since the material commonly found in a quality-of-life or HRQOL assessment (namely, questionnaires or global items) is more likely to generate issues concerning interpretation of meaning than of grammar.

7.3 Embodiment and the Assessment of Quality

Throughout this book I have addressed issues that reflect either the implicit or explicit influence of dualism. Thus, the many discussions about whether *qualia* exist, whether logical positivism or constructionism provides a better view of reality and truth, whether objective data are preferred over subjective data, whether capacity and performance can be conceived of in terms of a neural network model, and so on, each reflects some aspect of the dilemma created by dualism. It is easy enough to understand why this is so; my phenomenology and language taught me about the separateness of the mind and body. In addition, my capacity to attribute purpose to actions and events to feelings reinforced my sense that a separateness exists in various areas of my existence. The standard solution to this dilemma has been reductionistic, reducing complex human experience to apparently simpler levels of analyses, for example physiochemical terms. However, alternatives exist, including the effort of Smolensky and Legendre (2006) to account for several levels of analysis, essentially by using appropriate mathematical principles.

Now I will discuss a second approach to these issues: embodiment. Embodiment will show us that a common process exists that permits the mind and the body, or subjective and objective, to be seen as being inextricably interwoven. As a consequence, embodiment offers another approach to the relationship between capacity and performance.

Lakoff and Johnson (1999) posit that processes that underlie embodiment involve three levels of analysis: the neural, the phenomenological, and the cognitive unconscious. They state that the neural level helped them understand their experiences in scientific terms, the phenomenological level in terms of everyday experience, while the cognitive unconscious refers to the vast number of mental processes I am not aware of but that govern my language, thoughts, and actions.

Gibbs (2006), however, describes the embodiment process as follows:

> People's subjective, felt experiences of their bodies in action provide part of the fundamental grounding for language and thought. Cognition is what occurs when the body engages the physical, cultural world and must be studied in terms of the dynamical interactions between people and their environment. Human language and thought emerge from recurring patterns of embodied activity that constrain ongoing intellectual behavior. I must not assume cognition to be purely internal, symbolic, computational, and disembodied, but seek out the gross and detailed ways that language and thought are inextricably shaped by embodied action. (p. 9).

Thus, embodiment is not just muscle action, nor is it just a cognitive process; rather it involves muscle action interacting with a person's physical and cultural world. Consistent with this, and following the philosophical pragmatists of James and Dewey (see below), was Kelso's (1995; p. 268) point that the brain did not evolve just to cortically represent the external world, but to permit adaptive actions and behaviors. As he states:

> Musculoskeletal structures coevolved with appropriate brain structures so that the entire unit functions together in an adaptive fashion…it is the entire system of muscles, joints, and proprioceptive and kinesthetic functions and appropriate parts of the brain that evolve and function together in a unitary way. (Kelso 1995; p. 268; quoted from Gibbs 2006; p. 9).

This process, set in a person's physical and cultural world, is the essence of what is meant by "embodiment."

Dynamical systems theory (Kelso 1995) is also a critical part of Gibbs' (2006; p.9) approach to embodiment. The theory postulates that complex behaviors, such as language and thought, are higher-order products of self-organization. This occurs because self-assembling occurs in virtually all living organisms. Kelso (1995) describes self-organization as "emergent consequences of nonlinear interaction among active components" (p. 67). Thus, dynamical systems theory has many characteristics of what has previously been described as complexity theory.

But how does dynamical systems theory describe how stable higher-order behaviors emerge from the nonlinear interactions between the brain, body, and environment? Gibbs (2006) states:

> …dynamical systems theory is a set of mathematical tools that can be applied to characterize different states of the systems as these evolve over time. In this way, a dynamical view aims to describe how the body's continuous interactions with the world provide for coordinated patterns of adaptive behavior, rather than focusing on how the external world becomes represented in the inner mind." (p. 10)

Here is a potentially useful model of the relationship between capacity and performance. This model is similar to what Smolensky and Legendre (2006) described in that mathematical tools, a common denominator, are used to bridge the gap between levels of analysis, and this is done without having to rely on a "purely internal, symbolic, computational, and disembodied" model (Gibbs 2006; p. 9), or without focusing "on how the external world becomes represented in the inner mind" (Gibbs 2006; p. 10). Thus, this approach appears to effectively avoid the conceptual separateness that leads to dualistic models or theories.[23]

A second application of embodiment is its role in cognitive linguistics, particularly metaphor expression. Here, Gibbs takes the position, also expressed by Lakoff and Johnson (1999), that the mapping of bodily experiences to abstract ideas is fundamental to how metaphors are formed and interpreted, but also more generally to how people speak and think. Of course, one outcome of speaking and thinking is assessing the quality of an object or experience, or more generally, the quality of one's life. Embodiment would be involved, since metaphors, particularly conceptual metaphors, are an integral part of the qualitative assessment process.

Embodiment clearly has a place in determining a person's quality-of-life or HRQOL as a consequence of normal development and aging (e.g., gender differences); in response to acute medical (e.g., a spinal cord injury) or emotional (e.g., sexual abuse) events; or as part of what appears as a failed adaptive response (e.g., hearing voices). For example, the hormonal environment of the brain, the physiological changes that occur with aging, and the world's responses to these events make it clear that embodiment processes differ for men and women. These differences are also reflected in a person's estimation of their well-being and quality-of-life (e.g., Ryff and Keyes 1995). If a person experiences a spinal cord injury, it is not uncommon for them to require an extended period of time to accommodate to their new physiological and emotional state (Fuhrer 1987). The same is true for the person who has to use a prosthesis (e.g., Murray 2004). Of course, the amputee who experiences a phantom limb or pain is a classic example of where the brain is not able to effectively accommodate so that the embodiment process finds its limits (e.g., Halligan 2002).[24] The same seems to be true for the schizophrenic who hears voices. In each of these cases, reports of quality-of-life or HRQOL would be expected to track the embodiment process.

Social indicator research and embodiment have a natural affinity for one another, since both attempt to include measures of the world outside of the person as part of their conceptualization. They differ in that social indicator research insists that the person's social, cultural, and physical environment can be known objectively, while embodiment considers this distinction a nonsustainable or false dichotomy. Embodiment follows the tradition of the philosophical pragmatists (William James and John Dewey) who claim that an organism's interaction with its environment is functional, focused on adaptive responses so as to survive, reproduce, and flourish. This interaction ties the mind and body inextricably together. Johnson and Rohrer (2007) quote James and then state:

> From the very beginning of life, the problem of knowledge is not how so-called internal ideas can re-present external realities. Instead, the problem of knowledge is to explain how structures and patterns of organism-environment interaction can be adapted and transformed to help deal constructively with changing circumstances…On this view, the mind is never separate from the body, for it is always a series of bodily activities immersed in the ongoing flow of organism-environment interactions that constitutes experience. (p. 21–22).

Embodiment models can be useful to quality-of-life or HRQOL researchers in a number of ways. First, it provides a model that relates capacity and performance (that is, the relationship between brain, body, and the world) and a model that can be used to generate various testable hypotheses. It also can expand the number of areas that can be addressed in qualitative research. For example, body image studies (e.g., Skårderud et al. 2007); the process of emotional appraisal (e.g., Scherer 2000); or embodied experiences (Gibbs 2006) each offer unique qualitative research opportunities. In Chap. 11 (p. 430), I will illustrate the usefulness of embodiment when applied to accounting for how emotions are expressed.

7.4 Summary

In contrast to the two models presented above, the popular models of Wilson and Cleary (1995) and Patrick and Chang (2000; Fig. 9.1) are linear models that perpetuate the inner/outer, mind/body distinctions found in dualistic approaches to complex systems. They both do so by postulating that objective indicators (e.g., biological or physiological variables) precede consideration of subjective indicators such as health perceptions or quality-of-life, and while feedback is assumed to lead to interactions, these indicators remain linearly related. In contrast, I have been promoting the opposite view: subjectification of objects and their qualification precedes objectification, which becomes obvious when the role of the investigator is included in the assessment process.

Also evident was that the data reviewed in this chapter were often not compatible with linear modeling. In response to this issue, investigators have used hypothetical constructs (often phrased metaphorically; e.g., cognitive reserve or latent variables) to communicate a meaning that was not forthcoming from the linear modeling itself. This adds complexity to any analytical process as well as a sense of incompleteness, in contrast to simply acknowledging the nonlinear nature of the data and using a neural network or some equivalent nonlinear model to perform the appropriate statistical analysis.

However, neural network models with optimality theory and embodiment theory lack concrete examples of applications to qualitative data. Consider the following hypothetical study utilizing a neural network model. An investigator interested in comparing the HRQOL of two groups of cancer patients may have preliminary evidence suggesting that the chemotherapy regimens may have a long-term impact on neuropsychological functioning. The investigator is concerned that these neuropsychological effects may impact the respondents' ability to perform a HRQOL assessment (that is, their capacity is impaired), especially since they will be using a visual analog scale to record the respondents' ratings. Fortunately, the investigator has the opportunity to study these patients before they start receiving their chemotherapy, so he or she can administer a series of neuropsychological tests and an HRQOL assessment. The investigator repeat this after the patient's first and last, or sixth cycle of chemotherapy cycle. The investigator also has the ability to ensure that the study sample size, especially since the investigator expects some attrition and can estimate the sample size they need to provide an adequate amount of baseline data to "teach" the neural network to model. Once the network is functioning optimally, the investigator can apply the resultant model to predict what the outcome would be following the first and sixth chemotherapy cycle. This would permit the investigators to determine the accuracy or precision with which patients can use the visual analog scale, their primary outcome measure, and how well the respondents understood the instructions provided and the questions being asked on the NPTs. To find out what a person understands about the instructions and items, you may have to include some cognitive interviewing during baseline assessment and also feed this information into the neural network model (e.g., as a "testing" comprehension score). You expect that those persons who were not able to follow instructions, understand the questions, or use the rating scale will display greater response variability and be less likely to differentiate their HRQOL responding as a function of change in their life quality. If you ask these patients to provide a value statement, you might also find less variation in their ratings over the time interval of the study. Of course, what remains of interest is whether the ratings under different chemotherapy conditions mean the same as at a baseline assessment (e.g., does a rating of "two" at baseline mean the same as a rating of "two" after six cycles of chemotherapy)? This too may require some cognitive interviewing.

What if you wanted to design a quality-of-life or HRQOL study that incorporated an embodiment perspective? How would you do it? Hudak et al. (2004a, b), with a comment by Hargraves (2004), provide an example of this when they monitored the HRQOL of group of persons who had hand surgery. What the investigators were interested in was whether the group's sense of connectedness, that is, their embodiment, with their hands would be associated with their satisfaction with the medical care they received. The researchers first performed a qualitative study (Hudak et al. 2004a) and followed this with a quantitative (statistical) analysis of their data (Hudak et al. 2004b). They adopted a model of embodiment suggested by Gadow (1980), who conceptualized embodiment as consisting of a person viewing their body "as object" (that is, disunity between the affected hand and the self) or as "lived body" (that is, harmony between the hand and the self). Cognitive interviewing provided insight into each patient's attitude towards their hand and their satisfaction with their medical care. The qualitative study found a general relationship between satisfaction (or dissatisfaction) with patients' hands and satisfaction with medical care. The quantitative study (Hudak et al. 2004b) evaluated seven hypotheses, and the investigators concluded that embodiment, along with a person's prior expectations for surgical outcome and issues concerning workmen's compensation, affected satisfaction with surgical care.

In an editorial accompanying these papers, Hargraves (2004) raises questions about the usefulness of an embodiment approach, asking how a phenomenological approach to understanding satisfaction can be applied in medical practice. He encourages physicians to be aware of a patient's expectations and "workers'" compensation issues, but he finds it difficult to conceive of how a physician could make a patient feel more connected to their body, or hand. In addition, both Hargraves (2004) and Hudak et al. (2004a) acknowledge that embodiment is only one of several determinants of satisfaction.

The Hudak et al. (2004a, b) studies nicely illustrate how embodiment can play a role in understanding a particular outcome: satisfaction with medical care. However, I have also discussed embodiment as a critical element in neurocognition, with the implication that it can play a broader role in understanding the relationship between capacity and performance. Support for this role comes from studies of central nervous system functions that demonstrate the intimate interaction of sensory, perceptual, and cognitive brain areas with areas of the brain that mediates motor functions.

8 Summary of Chapter: Neurocognitive Indicators and Quality-of-Life or HRQOL Assessment

The question I asked at the beginning of this chapter was whether it made sense to include the results of NPT directly into a quality-of-life or HRQOL assessment. The answer seems to be no, if the results of NPT that an independent observer has collected are simply to be combined with the results from a qualitative assessment. The alternative of relying on a person's self-reports of cognitive status as a surrogate for neuropsychological test results also did not turn out to be a viable option (see below), except perhaps for the brain cancer patient or the person with schizophrenia. It is, of course, possible for a person to reflect, and thereby value, the results of their NPT results, but this type of information is vulnerable to both underestimating and overestimating the significance of these results, since most lay persons are not familiar with NPT results and their interpretation. However, I have not been able to find data of this type that would place a quantitative face on these issues.

If a clinician describes the results of the NPT as indicative of mild, moderate, or severe cognitive impairment; spells out the consequences of being in this state; and places an importance valuation on this state (e.g., "I am seriously concerned about the welfare of this person living alone, and recommend…"); then this information has the formal structure of a hybrid construct and can be considered cognitively equivalent to a similar combination of cognitions that a patient generates as part of a quality assessment. However, these statements are not both quality-of-life assessments, since a quality-of-life assessment requires input from the person in question. This observation raises the interesting issue of whether adhering to the principle of cognitive equivalence provides a basis for comparing data with rather diverse origins.

To have directly combined NPT results with a qualitative assessment would also resurrect the issue of whether objective and subjective indicators can be combined together into a single qualitative assessment. I have already discussed this issue in a number of different ways (Chaps. 4 and 5), and I take the position that since a quality-of-life or HRQOL assessment is the outcome of a cognitive process, all the components that become part of this process have to become "of the subject," and as such, objective information by itself would not be an appropriate component of a qualitative assessment until it is "subjectified" or "qualified."

Also extensively discussed were different ways of studying the relationship between neurocognition and quality-of-life or HRQOL. In the first part of this chapter, I considered a number of ways that NPT results could be used in a qualitative assessment. I asked if NPT results were ever formally qualified and found no evidence of this, even though most NPT results are valued when presented in a clinical context. Supportive of this was Fulford's (2000) view that connotative value statements pervade clinical activities, and the absence of a formal assessment does not preclude the presence of valuations. I also tested for substitutability by examining those studies that correlated NPT results with cognitive items included in a quality-of-life or HRQOL assessment. Results of these studies did not provide sufficient support to state that substitution was possible in most cases, although the data from the brain cancer patient and the schizophrenic were more likely to support this statement.

This chapter also continued the discussion initiated in Chap. 3 on the neural basis of cognition, with the objective of determining the relationship between neurocognitive capacity and performance. Several topics were discussed, including the capacity of the central nervous system to preserve function in response to intellectual and language usage declines, both age-related or in response to a disease process. Two models were discussed, including the decline with onset of dementia, and the consequences of an acute neurological event. Finally, two models of the relationship between capacity and performance were presented in this chapter: neural network modeling and embodiment.

Many issues were raised in this chapter, but probably none more important than determining how quality-of-life or HRQOL can be assessed in the compromised patient. While no definitive answers were presented herein, some of the issues that must be considered were raised. For example, if I want to assess the quality-of-life of autistic persons, how would I do this? The difficultly here is that these persons lack what may be an essential part of being able to assess quality-of-life; that is, the ability to integrate information, especially if it contains an emotional component? How could I assume that their responses were the same when compared to a non-autistic person? This same question can be asked when comparing the person with dementia to those persons without dementia, or the person with a stroke to those persons without a stroke, and so on. What makes life difficult here are the in-between cases; persons who retain some degree of the essential capacity, yet still show some deficiencies .

I did suggest that the elements of the hybrid construct can be used to design studies to determine what specific cognitive deficits a person may have that would impair their ability to perform a qualitative assessment. Thus, can a person correctly identify descriptors? Are they capable of rating these descriptors? Are they aware of the consequences implied by these descriptors? Are they sufficiently metacognitively intact that they can value their status? In addition, it would be interesting to know if a schizophrenic is able to acknowledge that they know their behavior is risky and even be able to value it as a negative state, and yet proceed to act inappropriately. The same may be asked of the Alzheimer's patient who wanders around and places him or herself in danger. Is it possible to link specific neurocognitive deficits to a specific cognitive component of a person's quality-of-life or HRQOL?

Knowing what specific deficits impair a compromised person's ability to assess their qualitative status requires that these assessments start early in the process of cognitive impairment, with continued monitoring over time. This information would be very useful for family members and clinicians, since it would provide a basis for systematic remediation and adjustment efforts. The current practice of using the MMSE as an indicator of cognitive status, while useful, is minimally informative about the specific cognitive deficits that may impair a quality-of-life assessment. It also indirectly contributes to the impression that the natural history of dementia is determined.

Applying this approach to persons who are recovering from an acute neurological event requires a different tactic. Here, the recovery or stabilization process is to be monitored over time, and the question becomes whether each element of the hybrid construct is equally capable of recovery. If so, at what rate and to what extent? If not, can remedial procedures be developed to provide this capability for the person? Thus, I end this chapter with more questions than answers, as is appropriate for any scientific endeavor.

Notes

1. The terminology here can get a bit confusing, since I could talk about "performance" and "performance capacity." I will use the term "performance capacity" to refer to a characteristic of a person's capacity, as in, for example, "the performance capacity of the brain."
2. Here I will deliberately limit my discussion to two models, even though the literature on this issue is far more extensive and complex. For example, I will not discuss the efforts to reduce quality-of-life or HRQOL responses to specific brain processes. Thus, what I will be discussing should be seen as my estimate of what are reasonable alternatives to modeling the relationship between capacity and performance.
3. The term Capabilities, with its action orientation, and Capacity, with its reference to a stored entity, will be used as related terms. In general, I will assume that the term Capacity (a person's potential) refers to a person's Capabilities (a person's potential to act). A person's Performance will refer to what a person has done. A person's Performance Capacity will refer to what potential a person reveals by acting in a particular way, while a person's Performance Capabilities refers to how well they have achieved their potential to act.
4. Table 10.2 is extracted from an article on Wikipedia on Analytical Hierarchy Process (http://en.wikipedia.org/wiki/Analytic_Hierarchy_Process). 3/2/2010.
5. This table was based on a PubMed search (3/8/10) that initially combined the phrases brain cancer and quality-of-life ($N=1,600$), followed by a search that included neuropsychological testing ($N=81$). Of these 81 citations, only 30 were found to have all three indicators (brain cancer, a quality-of-life or HRQOL assessment, and NPT). Of these 30 studies, 16 did not calculate or discuss the association between NPT and quality-of-life. The table includes the seven remaining studies that presented some data on the association between a composite or an individual NPT result and a qualitative assessment. Seven studies mentioned the indicators of interest, but did not provide any data on the association.

6. Allen et al. (2005) discuss the evolution of brain size and its relationship to longevity, as well as what role passive and active cognitive preservation play in these processes.
7. Although the study described is hypothetical, I have already discussed a number of instances where authors have implied that preservation of quality-of-life or HRQOL responding is somehow considered abnormal. For example, Sen (1993, 2001) spoke of "objective illusions", and Zapf (1984) discussed the "disability paradox when objective and subjective data do not concur". I will show in this section that these apparent contradictions are not statistical artifacts, but rather useful cognitive adaptive mechanisms. The apparent complexity this brings to the scientific process will also have to be considered.
8. Richards and Deary (2005) provided a similar conceptualization, but now applied to a developmental model of cognitive reserve.
9. Murdoch, according to Garrard et al. (2005), was well known for refusing to allow any editorial changes to her manuscripts so that the texts as published can be considered direct evidence of her intellectual functioning, not versions that have been inspected and edited by others.
10. Persons with a familial history of Alzheimer's disease, of course, are ideal to study since they are at risk for developing the disorder with time. Persons with other types of causes for their dementia (e.g., vascular dementia) would be less likely to be available for baseline NPT. However, their pattern of degeneration may be unique, so I need to be cautious about generalizing from the data being presented.
11. I have included this somewhat lengthy quote for several reasons. First, because it represents an early example of a connectionist theory of neurocognitive function, and second, because it probably preceded and formed the basis of Hayek's notions of how new knowledge is formed. Hayek's contributions were, however, to see how this principle can be applied on a broader social and economic level of analysis. This application of the same principle to account for both the microscopic and the molar is no different than Einstein's theory of relativity being able to account for both subatomic and planetary phenomena. It is for this reason that Hayek has made such a unique contribution to social thought.
12. Deep dyslexia is an acquired form of dyslexia that describes a symptom complex in that semantic and visual reading errors are regularly observed, and although there seems to be some arguments about the syndromes status, what remains consistent are the semantic errors that occur (Crutch 2006).
13. The views expressed here differ from what I have previously argued (Barofsky et al. 2004).
14. The same argument can be extended to the difference between generic and disease-specific qualitative assessments. Disease-specific HRQOL assessments portray different life histories from generic assessments, and the effort to make them psychometrically equivalent (e.g., meeting some standard of reliability and validity) masks the adverse consequences of combining these data. The alternative is to acknowledge that these differences are inevitable, whether assessments, groups, or individuals are being assessed, and that the only way to combine these data requires that an investigator demonstrate that they were cognitively generated and combined in the same way: that is, to demonstrate their cognitive equivalence. Demonstrating cognitive equivalence is the objective of the cognitive interview process that is regularly used to assess the meaning of items on an assessment. However, its usage has not progressed to the point where cognitive interviewing is regularly used following an assessment for persons with Alzheimer's disease.
15. A review of the literature, as well as consultation with a number of experts in the field (e.g., Feeny, Neumann, et al.), confirmed this statement.
16. I am making an assumption here when I shift from observations based on visual or auditory perception tasks (Sharp et al. 2006) to a qualitative assessment. I will discuss brain imaging studies dealing with "happiness" and "well-being" in the next chapter and that should demonstrate that there is some continuity between perceptual and language-based tasks, and tasks that have a affective and qualitative component.
17. The literature I have reviewed is rapidly changing with alterative studies regularly reported that support or refute aspects of this review (e.g., Lee and Diapdretto 2006; Rapp et al. 2004; Stringaris et al. 2007). Studies vary by the method used to record cortical activity (e.g., event-related potentials, fMRIs, or PET scans); the type of stimulus material (e.g., single words, comparison between words, review of sentences, and so on); the task asked of the respondent (e.g., reaction time, selection between alternative stimuli, etc.); and the familiarity or novelty of the material presented to the respondent. I have tried to provide a reasonable summary at the time of my review (e.g., I based my review only on studies that used either fMRIs or PET scans), but changes in the details of what I have written may occur in time.
18. Rohrer (2001) discusses some of the research issues involved in studying the neurocognitive basis of conceptual metaphors.
19. The nature of the lesions produced by a stroke are rarely confined to a specific site even in the same hemisphere, so the inability to demonstrate complete dissociation might be expected. Studies following severing of the corpus callosum are probably the best way to determine the role of each hemisphere, but tell little about the impact of one hemisphere on the other.
20. I grew up living in a "railroad" flat in New York City. This image of living in a series of rooms, one following another, has remained with me all my life and is part of how I judge where I live today.
21. The term "optimality," independent of the current context, is usually defined as referring to the efficient solution of problems; solutions that can be expressed mathematically. Thus, optimality in a neural network would represent the shortest and fastest path through the network leading to a particular numerical outcome.
22. "Pragmatics," simply defined, refers to the difference between what a sentence means (literally or figuratively) and what a speaker means when they state that sentence (Wikipedia, accessed 5/29/2006). The *context* of conversations or questionnaires is a major determinant of the meaning conveyed.
23. I say "appears" because avoiding the presence of dualism remains an issue, particularly among philosophers. For example, Thomas et al. (2004) point out that cognitive scientists still speak of the mind as a *thing* that is different from a body, even if this thing is a computer. So it seems that the problem of dualism persists, just now not as a form of ontological separation of mind and body, but rather epistemological separation. Smolensky and Legendre (2006), for example, regularly use the term "mind/brain," suggesting that they too may be subject to the same comment.
24. The reader is encouraged to examine the Halligan (2002) paper since it includes a picture of a patient who has lost his arm and who aided the investigator in creating a picture that depicts the phantom limb. This visualization of the phantom limb is striking!

References

Aaronson NK, Ahmedzai S, Bullinger M, Crabeels D, et al. (1991). The EORTC core quality-of-life questionnaire: Interim results of an international field study. In, (Ed.) D.Osoba. *The Effect of Cancer on Quality of Life*. Boca Raton FL: CRC Press. (p. 185–203).

Allen JS, Bruss J, Damasio H. (2005).The aging brain: The Cognitive Reserve Hypothesis and hominid evolution. *Am J Hum Biol*. 17, 673–689.

Allen PA, Sliwinski M. Bowie T, Madden DJ. (2002). Differential age effects in semantic and episodic memory, *J Gerontol*. 57B, P173–P186.

References

Almkvist O. (2000). Functional brain imaging as a looking-glass into the degraded brain: Reviewing evidence from Alzheimer's disease in relation to normal aging. *Acta Psychol.* 105, 255–277.

Ankri J, Beaufils B, Novella J-L, Morrone I, et al. (2003). Use of the EQ-5D among patients suffering from dementia. *J Clin Epidemiol.* 56, 1055–1063.

Auriacombe S, Lechevallier N, Amieva H, Harston S, et al. (2006). A longitudinal study of quantitative and qualitative features of category verbal fluency in incident Alzheimer's disease subjects: results from the PAQUID study. *Dement Geriatr Cogn Disord.* 21, 260–266.

Band GP, Kok A. (2000). Age effects on response monitoring in mental rotation task. *Biol Psychiatr.* 51, 201–221.

Barnett JH, Salmond CH, Jones PB, Sahakian BJ. (2006). Cognitive reserve in neuropsychiatry. *Psychol Med.* 36, 1053–1064.

Barofsky I. (in preparation). Neurocognitive Indicators and Quality of Life or HRQOL Assessments.

Barofsky I. (2000). The role of cognitive equivalence in studies of health-related quality of life assessments. *Med Care.* 38 Supp II, 125–129.

Barofsky I, Erickson P, Eberhardt M. (2004). Comparison of multi-item and self-assessed health status (SAHS) indexes among persons with and without diabetes in the U.S. *Qual Life Res.* 13. 1671–1681.

Bates E, Harris C, Marchman V, Wulfeck B, et al. (1995). Production of complex syntax in normal aging and Alzheimer's disease. *Lang Cogn.* 10, 487–544.

Beeman M and Chiarello C. (1998). *Right Hemisphere Language Comprehension: Perspective from Cognitive Neuroscience.* Mahwah NJ: Erlbaum.

Benaim C, Decavel P, Bentabet M, Froger J, et al. (2010). Sensitivity to change of two depression rating scales for stroke patients. *Clin Rehabil.* 24, 251–257.

Benaim C, Froger J, Cazottes C, Gueben D, et al. (2007).Use of the Faces Pain Scale by left and right hemispheric stroke patients. *Pain.* 128, 52–58.

Bergner M, Bobitt RA, Carter WB, Gilson BS. (1981). The Sickness Impact Profile: Development and final revision of a health status measure. *Med Care.* 19, 787–805.

Berwig M, Leicht H, Gertz HJ. (2009). Critical evaluation of self-rated quality of life in mild cognitive impairment and Alzheimer's disease further evidence for the impact of anosognosia and global cognitive impairment. *J Nut Health Aging.* 13, 226–230.

Birnboim S. (2003). The automatic and controlled information-processing dissociation: Is this still relevant? *Neuro Exp Rev.* 13, 19–31.

Blutner R. (2000). Some aspects of optimality in natural language interpretation. *J Semant.* 17, 189–216.

Blutner R, Zeevat H. (2004). *Optimality Theory and Pragmatics.* Hempshire UK: Palgrave Macmillan.

Bottini G, Corcoran R, Sterzi R, Paulesu E, et al. (1994). The role of the right hemisphere in the interpretation of figurative aspects of language: A positron emission tomography activation study. *Brain.* 117, 1241–1293.

Bowdle BF, Gentner D. (2005). The career of metaphor. *Psychol Rev.* 112, 193–216.

Braak H, Braak E. (1991). Neuropathological staging of Alzheimer-related changes (Review). *Acta NeuroPathol.* 82, 239–259.

Breedin SD, Saffran EM. (1999). Sentence processing in the face of semantic loss: A case study. *J Exp Psychol Gen.* 128, 547–562.

Breedin SD, Saffran EM, Coslett HB. (1994). Reversal of the concreteness effect in a patient with semantic dementia. *Cogn NeuroPsychol.* 11, 817–660.

Bridgman PW. (1959). *The Way Things Are.* New York, NY: Viking.

Brod M, Stewart A, Sands L, Walton P. (1999). Conceptualization and measurement of quality of life in dementia: the Dementia Quality of Life Instrument (DQOL). *Gerontologist.* 39, 25–35.

Brooks R, Rabin R, de Charro F. (2003). *The Measurement and Valuation of Health Status Using the EQ-5D: A European Perspective. Evidence from the EurQol BIOMED Research Programme.* Dordrecht The Netherlands: Kluwer.

Bureau-Chalot F, Novella J-L, Jolly D, Ankri J, et al. (2002). Feasibility, acceptability and internal consistency reliability of the Nottingham Health profile in dementia patients. *Gerontol.* 48, 220–225.

Buscema M, Grossi E, Snowdon D, Antuono P, et al. (2004).Artificial neural networks and artificial organisms can predict Alzheimer pathology in individual patients only on the basis of cognitive and functional status. *Neuroinformatics.* 2, 399–416.

Butters M, Salmon D, Butters N. (1994). Neuropsychological assessment of dementia. In, (Eds.) M. Storandt, G. VanderBos, *Neuropsychological Assessment of Dementia and Depression in Older Adults: A Clinician's Guide.* Washington DC: American Psychological Association. (p. 3–59).

Cader S, Cifelli A, Abu-Omar Y, Place J, et al. (2006). Reduced brain functional reserve and altered functional connectivity in patients with multiple sclearosis. *Brain.* 129, 527–537.

Cannon TD, Glahn DC, Kim J, Van Erp TGM, et al. (2005). Dorsolateral prefrontal cortex activity during maintenance and manipulation of information in working memory in patients with schizophrenia. *Arch Gen Psychiatr.* 62, 1971–1080.

Capitani E, Laicona M, Mahon B, Caramazza A. (2003). What are the facts of semantic category-specific deficits? A critical review of the clinical evidence. *Cogn NeuroSci.* 20, 213–261.

Cella DF, Tulsky DS, Gray G, Sarafian B, et al. (1993). The Functional Assessment of Cancer Therapy (FACT) scale: Development and validation of the general measure. *J Clin Oncol.* 11, 570–579.

Chandra V, Pandav R, Dodge HH, Johnson JM, et al. (2001). Incidence of Alzheimer's disease in a rural community in India: The Indo-US Study. *Neurology.* 57, 985–989.

Choucair AK, Scott C, Urstasun R, Nelson D, et al. (1997). Quality of life and neuropsychological evaluation for patients with malignant astrocytomas: RTOG- 91-14. *Int J Radiation Oncol Biol Phys.* 38, 9–20.

Coen R, O'Mahony D, O'Boyle C, Joyce CRB, et al. (1993). Measuring the quality of life of dementia patients usind the Schedule for the Evaluation of Individual Quality of Life. *Irish J Psychol.* 14, 154–163.

Cothheart M. (1980). The semantic error : Types and theories. In, (Eds.) M Coltheart, K Patterson, JC Marshall. *Deep Dyslexia.* London UK: Routledge Kegan Paul. (p. 22–47).

Crutch SJ, Isaacs R, Rosser MN. (2001). Some workman can blame their tools: Artistic change in an individual with Alzheimer's disease. *Lancet.* 357, 2129–2133.

Crutch SJ. (2006). Qualitatively different semantic representations of abstract and concrete words: Further evidence from the semantic reading errors of deep dyslexic patients. *Neurocase.* 12, 91–97.

Dapretto M, Bookheimer SY. (1999). Form and content: Dissociating syntax and semantics in sentence comprehension. *Neuron.* 24, 427–432.

De Haes JCJH, van Knippenberg FC, Neijt JP. (1990). Measuring psychological and physical distress in cancer patients: Structure and application of the Rotterdam Symptom Checklist. *Br J Cancer.* 62, 1034–1038.

Dennis, Spiegler BJ, Hetherington CR, Greenberg ML. (1996). Neuropsychological sequelae of the treatment of children with medulloblastoma. *J Neuroonol.* 29, 91–101.

Devinsky O, Vickery BG, Cramer JA, Perrine K, et al. (1995). Development of the quality of life in epilepsy inventory. *Epilepsia.* 36, 1089–1104.

Dik MG, Deeg DJH, Visser M, Jonker C. (2003). Early life physical activity and cognition in old age. *J Clin Exp NeuroPsychol.* 25, 643–653.

Dorman PJ, Waddell F, Slattery J, Dennis M, et al. (1997). Is the EuroQol a valid measure of health-related quality of life after stroke? *Stroke.* 28, 1876–1882.

Duncan PW, Wallace D, Studenski S, Lai S-M. (2001). Conceptualization of a new stroke-specific outcome measure: The Stroke Impact Scale. *Topics Stroke Rehabil.* 8, 19–33.

Ecklund-Johnson E, Torres I. (2005). Unawareness of deficits in Alzheimer's disease and other dementias: Operational definitions and empirical findings. *NeuroPsychol Revs.* 15, 147–166.

Edwards B. (1979). *Drawing on the Right Side of the Brain.* Los Angeles CA: Tarcher.

Eggert GH. (1977). *Wernicke's Works on Aphasia: A Source Book and Review* (Vol 1) The Hague The Netherlands: Mouton.

Engstrøm A. (1999). The contemporary theory of metaphor revisited. *Metaphor Symb.* 14, 53–61.

Ettema TP, Dröes R-M, de Lange J, Mellenbergh GJ, et al. (2005). A review of quality of life instruments in dementia. *Qual Life Res.* 14, 675–686.

Feeny D, Furlong W, Barr RD, Torrance GW, et al. (1992). A comprehensive multi-attitude system for classifying the health status of survivors of childhood cancer. *J Clin Oncol.* 10, 923–928.

Ferrans CE, Powers MJ. (1985). Quality of life index: development and psychometric properties. *Adv Nurs Sci.* 8, 15–24.

Fliessbach K, Helmstaedter C, Urbach H, Althaus A, et al (2005). Neuropsychological outcome after chemotherapy for patients with CNS lymphoma: A prospective study. *Neurology.* 64, 1184–1188.

Fodor J. (1975). *The Language of Thought.* New York NY: Thomas Y Crowell.

Folstein MF, Folstein SE, McHugh PR. (1975). Mini-mental state: A practical method for grading the cognitive state of patients for the clinician. *J Psychiatr Res.* 12, 189–198.

Frank L, Flynn JA, Kleinman L, Margolis MK, et al. (2006). Validation of a new symptom impact questionnaire for mild to moderate cognitive impairment. *Int Psychogeriatr.* 18, 135–149.

Frith CD. (1992). *The Cognitive Neuropsychology of Schizophrenia.* Sussex UK: Erlbaum.

Fuhrer MJ. (1987). Overview of outcome analysis of rehabilitation. In, (Ed.) MJ Fuhrer. *Rehabilitation Outcomes: Analysis and Measurement.* Baltimore MD: Brooks.

Fulford KWM. (2000). Teleology without tears: Naturalism, neo-naturalism, and evaluationism in the analysis of function statements in biology (and a bet on the Twenty-first Century). *Philos Psychiatr Psychol.* 7, 77–94.

Gadow S. (1980). Body and self: A dialectic. *J Med Philos.* 5, 172–185.

Gagnon L, Goulet P, Giroux F, Joanette Y. (2003). Processing of metaphoric and non-metaphoric alternative meanings of words after right- and left-hemispheric lesion. *Brain Lang.* 87, 217–226.

Garrard P, Carroll E, Vinson D, Vigliocco G. (2004). Dissociation of lexical syntax and semantics: Evidence from focal cortical degeneration. *Neurocase.* 10, 353–362.

Garrard P, Hodges JR. (2000). Semantic dementia: Clinical, radiological and pathological perspectives. *J Neurol.* 247, 409–422.

Garrard P, Maloney LM, Hodges JR, Patterson K. (2005). The effects of very early Alzheimer's on the characteristics of writing by a renowned author. *Brain.* 128, 250–260.

Gibbs RW Jr. (2006). *Embodiment and Cognitive Science.* New York NY: Cambridge University Press.

Giora R. (1997). On the priority of salient meanings: The graded salience hypothesis. *Cogn Ling.* 8, 183–206.

Giora R. (2003). *On our Mind: Salience, Context, and Figurative Language.* Oxford UK: Oxford University Press.

Giovagnoli AR. (1999). Quality of life in patients with stable disease after surgery, radiotherapy and chemotherapy for malignant brain tumor. *J Neurol Neurosurg Psychiatr.* 67, 358–368.

Giovagnoli AR, Martins da Silva A, Federicao A, Cornelio F. (2009). On the personal facets of quality of life in chronic neurological disorders. *Behav Neurol.* 95, 247–257.

Giovagnoli AR, Silvai A, Colombo E, Boiardi A. (2005). Facets and determinants of quality of life in patients with recurrent high grade glioma. *J Neurol Neurosurg Psychiatr.* 76, 562–568.

Glucksberg S. (2001). *Understanding Figurative Language: From Metaphor to Idioms.* Oxford UK: Oxford University Press.

Godboit AK, Cipolotti L, Watt H, Fox NC, et al. (2004). The natural history of Alzheimer's disease: A longitudinal presymptomatic and symptomatic study of a familial cohort. *Arch Neurol.* 61, 1743–1748.

Graesser AC, Weimer-Hastings K, Kreuz R, Weimer-Hastings P, Marquis K. (2000). QUAID: A questionnaire evaluation aid for survey methodologists. *Behav Res Method Instr Comput.* 32, 254–262. Available at (http://141.225.14.26/quaid/quaid.htm).

Grice HP. (1975). Logic and conversation. In, (Eds.) P Cole, JL Morgan. *Syntax and Semantics 3: Speech Acts.* New York NY: Academic. (p 41–58).

Grossman M, Koenig P, Kounios J, McMillan C, et al. (2006). Category-semantic effects in semantic memory; Category-task interactions suggested by MRI. *NeuroImage.* 30, 1003–1009.

Halligan PW. (2002). Phantom limbs: The body in mind. *Cogn NeuroPsychiatr* 7, 251–268.

Handy TC. (2000). Capacity theory as a model of cortical behavior. *J Cogn NeuroSci.* 12, 1066–1069.

Hapgood F. (2005). Fearful symmetry: Probing the limits of brain modeling. *Cerebrum.* 7, 7–18.

Happe F, Brownell H, Winner E. (1999). Acquired "theory of mind" impairments following stroke. *Cognition.* 70, 211–240.

Hargraves JL. (2004). Satisfaction with surgical outcomes and the phenomenology of embodiment. *Med Care.* 42, 715–717.

Hawthorne G, Richardson J, Osborne R. (1999). The Assessment of Quality of Life (AqoL) instrument; a psychometric measure of health-related quality of life. *Qual Life Res.* 8, 209–224.

Hayek FA. (1952). *The Sensory Order: An Inquiry into the Foundation of Theoretical Psychology.* Chicago Il: University of Chicago Press.

Hays RD, Sherbourne CD, Mazel RM. (1993). The RAND 36-Item Health Survey 1.0. *Health Econ.* 2, 217–227.

Hebb DO. (1949). *The Organization of Behavior. A Neuropsychological Theory.* New York NY: Wiley.

Hendrie HC. (2001). Exploration of environmental and genetic risk factors for Alzheimer's disease: The value of cross cultural studies. *Curr Direct Psychol Sci.* 10, 98–101.

Hendriks P, de Hoop H. (2001). Optimality theoretic semantics. *Ling Philos.* 14, 1–32.

Henry JD, Phillips LH, Crawford JR, Theodorou G, et al. (2006). Cognitive and psychosocial correlates of alexithymia following traumatic brain injury. *NeuroPsychologia.* 44, 62–72.

Hodges JR, Garrard P, Patterson K. (1998). Semantic Dementia and Pick complex. In, (Eds.) A Kertesz, D Munoz. *Pick Disease and Pick Complex.* New York NY: Wiley Liss.

Hodges JR, Graham N, Patterson K. (1995). Charting the progression in semantic dementia implications for the organization of semantic memory. *Memory.* 3, 463–495.

Hodges JR, Patterson K, Oxbury S, Funnell E. (1992). Semantic dementia: Progressive fluent aphasia with temporal lobe atrophy. *Brain.* 115, 1783–1806.

Hodges JR, Patterson K. (2007). Semantic dementia: A unique clinico-pathological syndrome. *Lancet Neurol.* 6, 1004–1014.

Hudak PL, Hogg-Johnson S, Bombardier C, McKeever PD, Wright JG. (2004b). Testing a new theory of patient satisfaction with treatment outcomes. *Med Care.* 42, 726–739.

Hudak PL, McKeever PD, Wright JG. (2004a). Understanding the meaning of satisfaction with treatment outcomes. *Med Care.* 42, 718–725.

Hunt SM, McEwen J, Mc Kenna SP. (1985). Measuring health status: a new tool for clinicians and epidemiologists. *J R Coll Gen Pr.* 35, 185–188.

Jacobs SR, Jacobsen PB, Booth-Jones M, Wagner LI, et al. (2007). Evaluation of the Functional Assessment of Cancer Therapy Cognitive Scale with hematopoetic stem cell transplant patients. *J Pain Symptom Manag.* 33, 13–23.

Jakobson R. (1971). Recognition reading in paralexia. *Cortex.* 14, 439–443.

James W. (1890). *Principles of Psychology.* Vol 1. New York NY: Holt.

References

James BD, Xie SX, Karlawish HT. (2005). How do patient with Alzheimer's disease rate their overall quality of life? *Am J Geriat Psychiatr.* 13, 484–490.

Jansma JM, Ramsey NF, Slagter HA, Kahn RS. (2001). Functional anatomical correlates of controlled and automatic processing. *J Cogn NeuroSci.* 13, 730–743.

Jason GW, Pajurkova EM, Taenzer PA, Bultz BD, (1997). Acute effects on neuropsychological function and quality of life by high-dose multiple daily fractionated radiotherapy for malignant astrocytomas: Assessing the tolerability of a new radiotherapy regimen. *Psycho-Oncol.* 6, 151–157.

Jette AM, Badley E. (2000). Conceptual issues in the measurement of disability. In, (Eds.) N Mathiowetz, GS Wunderlich. *Survey Measurement of Work Disability: Summary of a Workshop.* Washington DC: National Academy Press. (p. 4–27).

Johnson SC, Baxter LC, Wilder LS, Pipe JG, et al. (2002). Neural correlates of self-awareness. *Brain.* 125, 1808–1814.

Johnson DL, McCabe MA, Nicholson HS, Joseph AL, et al. (1994). Quality of long term survival in young children with medulloblastoma. *J Neurosurg.* 80, 1004–1010.

Johnson M, Rohrer T. (2007). We are live creatures: Embodiment, American pragmatism and the cognitive organism. In, (Eds.) J Zlatev, T Ziemke, R Frank, R Dirven. *Body Language and Mind.* Volume 1. Berlin Germany: Mouton de Gruyter. (p. 17–54).

Juncos-Rabadán O, Pereiro AX, Soledad Rodriguez M. (2005). Narrative speech in aging: Quality, information content, and cohesion. *Brain Lang.* 95, 423–434.

Just MA, Carpenter PA. (1992) A capacity theory of comprehension: Individual differences in working memory. *Psychol Rev.* 99, 122–149.

Kacinik NA, Chiarello C. (2007). Understanding metaphors: Is the right hemisphere uniquely involved? *Brain Lang.* 100, 188–207.

Kaplan RM, Anderson JP. (1990). The General Health Policy Model: An integrated approach. In, B Spilker (Ed.). *Quality of Life Assessments in Clinical Trials.* New York NY: Ravens. (p. 309–322).

Karlawish JH, Zbrozek A, Kinosian B, Gregory A, et al. (2008). Preference-based quality of life in patients with Alzheimer's disease. *Alzheimer's Dis.* 4, 193–202.

Kattan M. (2002). Statistical prediction models, artificial neural networks, and the sopism "I am a patient, not a statistic". *J Clin Oncol.* 20, 885–887.

Katzman R, Aronson M, Fuld P, Kawas C, et al. (1989). Development of dementing illnesses in an 80-year old volunteer cohort. *Ann Neurol.* 25, 317–324.

Kearner DN, Patterson TL, Grant I, Kaplan RM. (1998). Validity of the of the Quality of Well-Being Scale for patient with Alzheimer's disease. *J Aging Health.* 10, 44–61.

Keenan JC, Rubio J, Racioppi C, Johnson A, et al. (2005). The right hemisphere and the dark side of consciousness. *Cortex.* 41, 695–704.

Kelso J. (1995). *Dynamic Patterns: The Self-organization of Brain and Behavior.* Cambridge MA: MIT Press.

Kitwood T, Bredin K. (1992). Towards a theory of dementia care: Personhood and well-being. *Aging Soc.* 12, 269–287.

Klein M, Taphoorn JB, Heimans JJ, van der Ploeg HM, et al. (2001). Neurobehavioral status and health-related quality of life in newly diagnosed high- grade glioma patients. *J Clin Oncol.* 19, 4037–4047.

Klepousniotou E, Blaum SR. (2005). Processing homonym and polysemy: Effects of sentential context and time-course following unilateral brain damage. *Brain Lang.* 93, 308–326.

Korner-Bitensky NK, Kehayla E, Trembly N, Mazer B, et al. (2006). Eliciting information on differential sensation of heat in those with or without post-stroke aphasia using a visual analog scale. *Stroke.* 37, 471–475.

Krongrad A, Granville LJ, Burke MA, Golden RM, et al. (1997). Predictors of general quality of life in patients with benign prostate hyperplasia or prostate cancer. *J Urol.* 157, 534–538.

Kunz S. (2010). Psychometric properties of the EQ-5D in a study of people with mild to moderate dementia. *Qual Life Res.* 19, 425–434.

Lahaie A, Mottron M, Arguin M, Berthiaume C, et al. (2006). Face perception in high-functioning autistic adults: Evidence for superior processing of face parts, not for configural face-processing deficit. *NeuroPsychol.* 20: 30–41.

Lakoff G. (1993).The contemporary theory of metaphor. In, A. Ortony. *Metaphor and Thought. 2nd edition*, New York NY: Cambridge University Press. (p. 202–251).

Lakoff G, Johnson M. (1999). *Philosophy in the Flesh: The Embodied Mind and its Challenge to Western Thought.* New York NY: Basic Books.

Lawton MP. (1991). A multidimensional view of quality of life in frail elderly. In, (Eds.) JE Birren, JE Lubben, JC Rowe, DE Deutchman. *The Concept and Measurement of Quality of Life in Frail Elderly.* New York NY: Academic Press. (p. 3–27).

Le Carret N, Lafont S, Mayo W, Fabrigoule C. (2003). The effect of education on cognitive performance and its implications for the constitution of the cognitive reserve. *Dev Ment NeuroPsychol.* 23, 317–337.

Lee S. Diapdretto M. (2006). Metaphoric vs literal word meanings: fMRI evidence against a selective role of the right hemisphere. *NeuroImage.* 29, 536–544.

Letenneur L, Commenges D, Dartigues JF, Bargerger-Gateau P. (1994). Incidence of dementia and Alzheimer's disease in elderly community residents of south-western France. *Int J Epidemiol.* 23, 1256–1261.

Levine MN, Guyatt GH, Gent M, De Pauw S, et al (1988). Quality of life in Stage II breast cancer: An instrument for clinical trials. *J Clin Oncol.* 6, 1798–1810.

Lezak MD. (1995). *Neuropsychological Assessment.* New York NY: Oxford University Press. (3rd Edition).

Li J, Bentzen SM, Li J, Renschler M, et al. (2008). Relationship between neurocognitive function and quality of life after whole-brain radiotherapy in patients with brain metastasis. *Int J Radiat Oncol Biol Phys.* 71, 64–70.

Liao Y-C, Liu R-S, Teng EL, Lee Y-C, et al. (2005). Cognitive reserve: A SPECT study of 132 Alzheimer's disease patients with and education range of 0–19 years. *Dement Geriatr Cogn Disord.* 20, 8–14.

Logsdon RG, Gibbons LE, McCurry SM, Teri L. (1999). Quality of life in Alzheimer's disease: Patient and caregivers reports. *J Ment Health Aging.* 5, 21–32.

Logsdon RG, Gibbons LE, McCurry SM, Teri L. (2002). Assessing quality of life in older adults with cognitive impairments. *Psychosom Med.* 64, 510–519.

Logsdon RG, Teri L. (1997). The Pleasant Events Schedule-AD: psychometric properties of long and short forms and an investigation of its association to relationship to depression and cognition in Alzheimer's disease patients. *The Gerontologist.* 37, 40–45.

Long DL, Parks RW, Levine DS. (1998). An introduction to neural network modeling: merits, limitations, and controversies. In, (Eds.) RW Parks, DS Levine, DL Long. *Fundamentals of Neural Network Modeling: Neuropsychology and Cognitive Neuroscience.* Cambridge MA: MIT Press.(p. 3–31).

Manly JJ, Touradji P, Tang M-X, Stern Y. (2003). Literacy and memory decline among ethically diverse elders. *J Clin Exp NeuroPsychol.* 25, 680–690.

Maruyama M, Matsui T, Tanji H, Nemoto M, et al. (2004). Cerebrospinal fluid Tau protein and periventricular white matter lesions inpatient with mild cognitive impairment. *Arch Neurol.* 61, 716–720.

Mashal N, Faust M, Hendler T. (2005). The role of the right hemisphere in processing nonsalient metaphorical meanings: Application of principle component analysis to fMRI data. *NeuroPsychol.* 43, 2084–2100.

Mashal N, Faust M, Hendler T, Jung-Beeman M. (2007). A fMRI investigation of the neural correlates underlying the processing of novel metaphoric expressions. *Brain Lang.* 100, 225–226.

Matsui T, Nakaaki S, Murata Y, Sato J, et al. (2006). Determinants of quality of life in Alzheimer's disease as assessed by the Japanese version of the Quality of Life Alzheimer's Disease Scale. *Dement Geriat Cogn Disord.* 21, 182–191.

Mattis S. (1976). Mental status examination for organic disease syndrome in the elderly patient. In, (Ed.) L Bellak. *Geriatric Psychiatry: A Handbook for Psychiatrists and Primary Care Physicians.* New York NY: Grune and Statton. (p. 77–107).

Mc Carter H, Furlong W, Whitton A, Feeny D, et al. (2006). Health status measurement in diagnosis as predictors of survival among adults with brain tumors. *J Clin Oncol.* 24, 3636–3643.

Mc Clelland JL, St. John M, Tarabam R. (1989). Sentence comprehension: A parallel distributed processing approach. *Lang Cogn Process.* 4, SI287–SI335.

Mc Donald AW 3rd, Cohen JD, Stenger VA, Carter CS. (2000). Dissociating the role of the dorsolateral prefrontal and anterior cingulate cortex in cognitive control. *Science.* 288, 1835–1838.

Medler DA. (1998). A brief history of connectionism. *Neural Comput Surv.* 1, 61–101.

Mesulam M-M. (2003). Primary progressive aphasia- A language-based dementia. *NEJM.* 349, 1535–1542.

Meyer CA, Hess KR, Yung WKA, Levin VA. (2000). Cognitive function as predictor of survival of patients with recurrent malignant glioma. *J Clin Oncol.* 18, 646–650.

Mitchell RLC, Crow TJ. (2005). Right hemisphere language functions and schizophrenia the forgotten hemisphere? *Brain.* 128, 963–978.

Monetta L, Ouellet-Plamondon C, Joanette Y. (2006). Simulating the pattern of right-hemisphere-damaged patients for the processing of the alternative metaphoric meanings of words: Evidence in favor of a cognitive resources hypothesis. *Brain Lang.* 96, 171–177.

Mozley CG, Huxley P, Sutcliffe C, Bagley H, et al. (1999). "Not knowing where I am doesn't mean I don't know what I like: Cognitive impairment and quality of life responses in elderly people. *Int J Geriat Psychiatr.* 14, 776–783.

Murray CD. (2004). An interpretative phenomenological analysis of embodiment of artificial limbs. *Disabil Rehabil.* 26, 963–973.

Naglie G, Tomlinson G, Tansey C, Irvine J, et al. (2006). Utility-based quality of life measures in Alzheimer's disease. *Qual Life Res.* 15, 631–643.

Nelson EC, Wasson JH, Johnson DJ, Hays RD. (1996). Dartmouth COOP functional health assessment charts: Brief measures for clinical practice. In, (Ed.) B Spilker. *Quality of Life and Pharacoeconomics in Clinical Trials.* 2nd Edition. Philadelphia PA: Lippincott-Raven, (p. 161–168).

Nanda U, McLendon PM, Andresen EM, Armbrecht E. (2003). The SIP68: an abbreviated sickness impact profile for disability outcomes research. *Qual Life Res.* 12, 583–595.

Nelson HE. (1982). *The National Adult Reading Test (NART): test manual.* Windsor CT: NFER-Nelson.

Neumann PF, Hermann RC, Kuntz KM, Araki SS, et al. (1999). Cost-effectiveness of Donepezil in the treatment of mild or moderate Alzheimer's disease. *Neurology.* 52, 1138–1145.

Nishimura T, Kobayashi T, Hariguchi S, Takeda M, et al. (1993). Scales for mental state and daily living activities for the elderly: clinical behavioral scales for assessing demented patients. *Int Psychogeriat.* 5, 117–134.

O'Boyle CA, McGee HM, Hickey A, Joyce CRB, et al. (1993). *The Schedule for Individual Quality of Life. User Manual.* Dublin Ireland: Department of Psychology, Royal College of Surgeons in Ireland.

Osoba D. (1994). Lessons learned from measuring health-related quality of life in oncology. *J Clin Oncol.* 12: 508–616.

Park DC, Gutchess AH. (2000). Cognitive aging and everyday life. In, (Eds.) DC Park, N Schwarz. *Cognitive Aging: A Primer.* Philadelphia PA: Taylor and Francis. (p. 217–232).

Park DC, N Schwarz. (2000). *Cognitive Aging: A Primer.* Philadelphia PA: Taylor and Francis.

Patrick DL, Chiang Y-P. (2000). Measurement of health outcomes in treatment effectiveness evaluations. *Med Care.* 38 (Suppl II), II-14- II-25.

Patrick DL, Erickson P. (1993). *Health Status and Health Policy: Quality of Life in Health Care Evaluation and Resource Allocation.* New York NY: Oxford.

Patterson K, Hodges JR. (2000). Semantic dementia: One window on the structure and organisation of semantic memory. In, (Ed.) L Cermak. *Handbook of Neuropsychology: Memory and its Disorders.* Amsterdam The Netherlands: Elsevier Science. (p 313–335).

Patterson MB, Whitehouse PJ, Edladn SD, Sami SA, et al. (2006). ADCS Prevention Instrument Project: Quality of life assessment. *Alzheimer Dis Assoc Disord.* 20 Suppl 3, S179–S190.

Perrine K, Hermann BP, Meador KJ, Vickrey BG, et al. (1995). The relationship of neuropsychological functioning to quality of life in epilepsy. *Arch Neurol.* 52, 997–1003.

Pick A. (1892). Über die Beziehungen der senilen Hirnatrophie zur Aphasie. *Prager med Wochenschr* Prague. 17, 165–167.

Pinker S. (1994). *The Language Instinct: How the Mind Creates Language.* New York NY: Morrow.

Posner MI, Peterson SE, Fox PT, Raichle ME. (1988). Localization of cognitive operations in the human brain. *Science.* 240, 1627–1631.

Price CIM, Curless RH, Rodgers H. (1999). Can stroke patients use visual analog scales? *Stroke.* 30, 1357–1361.

Prince A, Smolensky P. (1991). *Notes on connectionism and Harmony Theory in linguistics. Technical Report CU-CU 533–91.* Department of Computer Science, University of Colorado.

Prince A, Smolensky P. (1993). Ms. Rutgers University, New Brunswick, NJ, and University of Colorado, Boulder. *Technical Report RuCCS TR-2.* Rutgers Center for Cognitive Science.

Prince A, Smolensky P. (1997). Optimality: From neural networks to universal grammar. *Science.* 275, 1604–1610.

Prince CJ. (1998). The functional anatomy of word comprehension and production. *Trends Cogn Sci.* 2, 281–288.

Rabins PV, Kasper JD, Kleinman L, Black BS, et al. (1999). Concepts and methods in the development of the ADRQL: An instrument for assessing health-related quality of life in persons with Alzheimer's disease. *J Ment Health Aging.* 5, 33–48.

Rapp AM, Leube DT, Erb M, Grodd W, et al. (2004). Neural correlates of metaphor processing. *Cogn Brain Res.* 20, 395–402.

Ready RE, Ott BR. (2003). Quality of life for dementia. *Health Qual Life Outcome.* 1, 11.

Ready RE, Ott BR, Grace J. (2006). Insight and cognitive impairment: Effect on quality of life reports from mild cognitive impairment and Alzheimer's disease patients. *Am J Alzheimer Dis Other Dement.* 21, 242–248.

Reed BR, Jagust WJ, Coulter L. (1993). Anosognosia in Alzheimer's disease: Relationship to depression, cognitive function, and cerebral perfusion. *J Clin Exp NeuroPsychol.* 15, 231–244.

Rehak A, Kaplan JA. (1992). Sensitivity to conversation deviance in right-hemisphere-damaged patients. *Brain Lang.* 42, 203–217.

Richards M, Deary IJ. (2005). A life course approach to cognitive reserve: A model for cognitive aging and development? *Ann Neurol.* 58, 617–622.

Riley KP, Snowdon DA, Desrosiers MF, Markesbery WR. (2005). Early life linguistic ability, late life cognitive function and neuropathology: Findings from the Nun Study. *Neurobio Aging.* 26, 341–347.

Rinaldi MC, Marangolo P, Baldassarri F. (2004). Metaphor comprehension in right brain-damaged patients with visuo-verbal and verbal material: A dissociation (re)considered. *Cortex.* 40, 479–490.

Rogers TR, Lambon Ralph MA, Garrard P, Mc Celland JL, et al. (2004). Structure and Deterioration of semantic memory; A neuropsychological and computational investigation. *Psychol Rev.* 111, 205–235.

Rogers TT, Patterson K. (2007). Object categorization: reversals and explanations of the basic-level advantage. *J Exp Psychol Gen.* 136, 451–469.

Rohrer T. (2001). The cognitive science of metaphor: From philosophy and neuroscience. Theoria Historia Sci. 6, 27–42.

Rosch E. (1973). Natural categories. *Cog Psychol.* 4, 328–350.

Rumelhart DE, McCelland JL, the PDP Research Group. (1986). *Parallel Distributed Processing. Volume 1: Foundations.* Cambridge MA: MIT Press.

Ryff CD, Keyes CLM. (1995). The structure of psychological well-being revisited. *J Personal Soc Psychol.* 69, 719–727.

Saaty TL. (1977). A scaling method for priorities in hierarchical structures. *J Math Psychol.* 5, 234–281.

Saaty TL. (1980). The analytic hierarchy process. New York NY: McGraw-Hill.

Saaty TL. (1983). *Foundations of the Logical Choice to Harness Complexity.* Unpublished manuscript. School of Business, The University of Pittsburgh.

Saaty TL. (2009). *Theory and Application of the Analytic Network Process: Decision Making with Benefits, Opportunities, Costs and Risks.* Pittsburg PA: RWS Publications.

Sabsevitz DS, Medler DA, Seidenberg M, Binder JR. (2005). Modulation of the semantic system by work imageability. *NeuroImage.* 27, 188–200.

Sacks O. (1985). *The Man who Mistook his Wife for a Hat and Other Clinical Tales.* New York NY: Summit Books.

Sagan C. (1997). *Contact.* New York NY: Pocket Books.

Salek SS, Walker MD, Bayer AJ. (1998). A review of quality of life in Alzheimer's disease. Part 2: Issues in assessing drug effects. *PharmacoEcon.* 14, 613–627.

Salmond CH, Menon DK, Chatfield DA, Pickard JD, et al. (2006). Cognitive reserve as a resilient factor against depression after moderate/severe head injury. *J Neurotrauma.* 23, 1049–1058.

Satz P. (1993). Brain reserve capacity on symptom onset after brain injury: A formulation and review of evidence for threshold theory. *NeuroPsychol.* 7, 273–295.

Saykin AJ, Wishart HA, Rabin LA, Santulli RB, et al. (2006). Older adults with cognitive complaints show brain atrophy similar to that of amnestic. *Neurology.* 67, 834–842.

Scarmeas N, Stern Y. (2003). Cognitive reserve and lifestyle. *J Clin Exp NeuroPsychol.* 25, 625–633.

Scarmeas N, Zarahan E, Anderson KE, Honig LS, et al. (2004). Cognitive reserve-mediated modulation of positron emission tomographic activations during memory tasks in Alzheimer's disease. *Arch Neurol.* 61, 73–78.

Schacter DL. (1990). Towards a neuropsychology of awareness: Implicit knowledge and anosognosia. *J Clin Exp NeuroPsychol.* 12, 155–178.

Scherer KR. (2000). Emotions as episodes of subsystem synchronization driven by nonlinear appraisal processes. In, (Eds.) MD Lewis, I Granic. *Emotions, Development, and Self-organization: Dynamic Systems Approaches to Emotional Development.* Cambridge UK: Cambridge University Press. (p. 70–99).

Schmidt GL, DeBuse C, Seger CA. (2007). Right hemisphere metaphor processing? Characterizing the lateralization of semantic processes. *Brain Lang.* 100, 127–141.

Scott SK, Johnsrude IS. (2003).The neuroanatomical and functional organization of speech perception. *Trends Neurosci.* 36, 309–322.

Selai C, Trimble MR. (1999). Assessing quality of life in dementia. *Aging Ment Health.* 3, 101–111.

Selai CE, Trimble MR, Rossor MN, Harvey RJ. (2001). Assessing quality of life (QOL) in dementia: Preliminary psychometric testing of the Quality of Life Assessment Schedule (QOLAS). *Neuropsychol Rehabil.* 11, 219–243.

Sen A. (1993). Positional objectivity. *Philos Public Aff.* 22, 126–145.

Sen A. (2001). Objectivity and position: Assessment of health and well-being. In, (Eds.) J Drèze, A Sen. *India: Development and Participation.* Oxford UK: Oxford University Press. (p. 115–128).

Shallice T.(2002). Fractionation of the supervisory system. In, (Eds.) D Struss, RT Knight. *Frontal Lobe Function.* Oxford UK: Oxford University Press. (p. 261–277).

Sharp DJ, Scott SK, Mehta MA, Wise RSJ. (2006) The neural correlates of declining performance with age: Evidence for age-related changes in cognitive control. *Cereb Cort.* 16, 1737–1749.

Sharp DJ, Scott SK, Wise RSJ. (2004). Monitoring and the controlled process of meaning; distinct prefrontal systems. *Cereb Cort.* 14, 1–10.

Shipper H, Clinch J, McMurray A, Levitt M. (1984). Measuring the quality of life of cancer patients: The Functional Living Index-Cancer: Development and validation. *J Clin Oncol.* 2, 472–483.

Silberfeld M, Rueda S, Krahn M, Naglie G. (2002). Content validity for dementia of three generic preference based health related quality of life instruments. *Qual Life Res.* 11, 71–79.

Skårderud F. (2007). Eating one's words, Part I: 'Concretized Metaphors' and reflective function in anorexia nervosa – An interview study. *Eur Eat Disord Rev.* 15, 163–174.

Skårderud F, Nygren P, Edlund B. (2005). "Bad Boys" bodies: The embodiment of troubled lives. Body image and disordered eating among adolescents in residential childcare institutions. *Clin Child Psychol Psychiat.* 10, 395–411.

Smith CD, Chebrolu H, Wekstein DR, Schmitt FA, et al. (2007). Brain structural alterations before mild cognitive impairment. *Neurology.* l68, 1268–1273.

Smith SC, Lamping DL, Banerjee S, Harwood R, et al. (2005). Measurement of health-related quality of life for people with dementia: Development of a new instrument (DEMQOL) and an evaluation of current methodology. *Health Technol Assess.* 9, 1–93.

Smolensky P. Unpublished statement of research objectives. (Accessed July, 2004). http://www.cog.jhu.edu/faculty/smolensky/#top.

Smolensky P, Legendre C. (2006). *The Harmonic Mind From Neural Computation to Optimality-theoretic Grammar. Volume 1: Cognitive Architecture.* Cambridge MA: MIT Press.

Sneeuw KC, Sprangers MA, Aaronson NK. (2002). The role of health care providers and significant others in evaluating the quality of life patients with chronic diseases. *J Clin Epidemiol.* 55, 1130–1143.

Snowdon DA. (2003). Health aging and dementia: Findings from the Nun Study. *Ann Inter Med.* 139, 450–454.

Snowdon DA, Kemper SJ, Mortimer JA, Greiner LH. (1996). Linguistic ability in early life and cognitive function and Alzheimer's disease in late life. Finding from the Nun Study. *JAMA.* 275, 528–532.

Soroler N, Kasher A, Giora R, Batori G, et al. (2005). Processing of basic speech acts following localized brain damage: A new light on the neuroanatomy of language. *Brain Cogn.* 57, 214–217.

Springer SP, Deutsch G. (1993). *Left Brain, Right Brain.* New York NY: Freeman.

Staff RT, Murray AD, Deary IJ, Whalley LJ. (2004). What is cerebral reserve? *Brain.* 127, 1191–1199.

Starkstein SE, Sabe L, Chemerinski E, Jason L, et al. (1996). Two domains of anosognosia in Alzheimer's disease. *J Neurol Neurosurg Psychiatr.* 61, 485–490.

Starr JM, Loeffler B, Abousleiman Y, Simonotto E, et al. (2005). Episodic and semantic memory tasks activate different brain regions in Alzheimer's disease. *Neurology.* 65, 266–269.

Steen GJ. (1999). From linguistic to conceptual metaphor in five steps. In, (Eds.) RW Gibbs Jr, GJ Steen. *Metaphors in Cognitive Linguistics.* Amsterdam Netherlands: John Benjamins. (p.57–77).

Stern Y. (2002). What is cognitive reserve? Theory and research applications of the reserve concept. *J Int NeuroPsychol Soc.* 8, 448–460.

Stern Y. (2003). The concept of cognitive reserve: A catalyst for research. *J Clin Exp NeuroExp.* 25, 589–593.

Stern Y. (2007). *Cognitive Reserve: Theory and Applications.* New York NY: Taylor & Francis.

Stern Y, Alexander S, Prohovnik I, Masyeux R. (1992). Inverse relationship between education and parietotemporal perfusion deficit in Alzheimer's disease pathology. *Ann Neurol.* 32, 371–375.

Stern Y, Habeck C, Moeller J, Scarmeas N, et al. (2005). Brain networks associated with cognitive reserve in healthy young and old adults. *Cereb Cort.* 15, 394–402.

Stewart R, Dufouil C, Godin O, Ritchie K, et al. (2008). Neuroimaging correlates of subjective memory deficits in a community population. *Neurology.* 70, 1601–1607.

Stewart AL, Hays R, Ware JE, Jr. (1988). The MOS Short-Form General Health Survey (SF-36): Reliability and validity in a patient population. *Med Care.* 26, 724–732.

Stewart AL, Ware JE Jr. (1992). *Measuring Functioning and Well-being: The Medical Outcomes Study Approach.* Durham NC: Duke University Press.

Stringaris AK, Medford NC, Giampietro V, Brammer MJ, et al. (2007). Deriving meaning: Distinct neural mechanisms for metaphoric, literal and non-meaningful sentences. *Brain Lang.* 100, 150–162.

Struttmann T, Fabro M, Romieu G, de Roquefeuil G, et al. (1999). Quality of life assessment in the old using the WHOQOL 100: Differences between patients with senile dementia and patients with cancer. *Int Psychogeriat.* 11, 273–279.

Tandon R, Adak S, Keye JA. (2006). Neural networks for longitudinal studies in Alzheimer's disease. *Artif Intell Med.* 36, 245–255.

Tauber H-L. (1955). Physiological psychology. *Ann Rev Psychol.* 6, 267–296.

Teri L, Truax P, Logsdon RG, Uomoto J, et al. (1992). Assessments of behavioral problems in dementia: the Revised Memory and Behavior Problems Checklist. *Psychol Aging.* 7, 622–631.

Terada S, Ishizu H, Fujisawa Y, Fujita D, et al. (2002). Development and evaluation of a health-related quality of life questionnaire for the elderly with dementia in Japan. *Int J Geriat Psychiatr.* 17, 851–858.

Thomas P, Bracken P, Leudar I. (2004). Hearing voices: A phenomenologic-hermeneutic approach. *Cogn NeuroPsychiatr.* 9, 12–23.

Thorndike EL. (1932). *The Foundations of Learning.* New York NY: Columbia University Press.

Thorndike EL. (1949). *Selected Writings from a Connectionist Psychology.* New York NY: Greenwood Press.

Thorgrimsen L, Selwood A, Spector A, Royan L, et al. (2003). Whose quality of life is it anyway? The validity and reliability of the Quality of Life- Alzheimer's Disease (QOL-AD) scale. *Alzheimer Dis Assoc. Disord.* 17, 201–208.

Thurstone LL. (1927). A law of comparative judgment. *Psychol Rev.* 34, 278–286.

Torrance GW (1986). Measurement of health state utilities for economic appraisal: A review. *J Health Econ.* 5, 1–30.

Torrance GW, Furlong W, Feeny D, Boyle M. (1995). Multi-attribute preference functions: Health Utilities Index. *Pharmacoeconomics.* 7, 503–520.

Trigg R, Jones RW, Skevington SM. (2007a). Can people with mild to moderate dementia provide reliable answers about their quality of life? *Age Aging.* 36, 663–669.

Trigg R, Skevington SM, Jones RW. (2007b). How can we best assess the quality of life of people with dementia? The Bath Assessment of Subjective Quality of Life in Dementia (BASQID). *Gerontol.* 47, 789–797.

von Essen L. (2004). Proxy ratings of quality of life. *Acta Oncol.* 43, 229–234.

von Osch SMC, Stiggelbout AM. (2005). Understanding VAS valuations: Qualitative data on the cognitive process. *Qual Life Res.* 14, 2171–2175.

Walker MD. Salek SS, Bayer AJ. (1998). A review of quality of life in Alzheimer's disease Part I: Issues in assessing disease impact. *Pharmacoeconomics.* 14, 499–530.

Ware JE, Kosinski MA, Keller SD. (1994). *SF-36 Physical and Mental Health Summary Scales: A Users Manual.* Boston MA: The Health Institute, New England Medical Center.

Ware JE, Kosinski MA, Keller SD. (1995). *SF-12: How to Score the SF-12 Physical and Mental Health Summary Scales.* Boston MA: The Health Institute, New England Medical Center.

Warner B, Misra M. (1996). Understanding neural networks as statistical tools. *Am Statist.* 50, 284–293.

Warrington EK. (1975). Selective impairment of semantic memory. *Q J Exp Psychol.* 27, 635–657.

Wechsler D. (1981). *WAIS-R Manual.* New York NY: Psychological Corporation.

Weitzner MA, Meyers CA, Gelke CK, Bryne KS, et al. (1995). The functional assessment of cancer therapy (FACT) scale: Development of a brain subscale and revalidation of the FACT-G in the brain tumor population. *Cancer.* 75, 1151–1161.

Wernicke, C. (1874). *Der Aphasische Symptomencomplex.* Breslau Germany: Cohn and Weigert.

Wilson IB, Cleary PD. (1995). Linking clinical variables with health-related quality of life: A conceptual model of patient outcomes. *JAMA.* 273, 59–65.

Wilson RS, Barnes LL, Bennett DA. (2003). Assessment of lifetime participation in cognitively stimulating activities. *J Clin Exp NeuroPsychol.* 25, 634–642.

Winner E, Gardner H. (1977). The comprehension of metaphor in brain-damaged patients. *Brain.* 100, 717–729.

World Health Organization. (1948). *Constitution of the World Health Organization: Basic documents.* Geneva, Switzerland: World Health Organization.

World Health Organization. (1995). *The World Health Organization Quality of Life Assessment- 100 (WHOQOL-100).* Geneva Switzerland: The WHOQOL Group Program on Mental Health.

World Health Organization. (1996). *The World Health Organization Quality of Life Assessment- Abbreviated Version (WHOQOL-BREF).* Geneva Switzerland: The WHOQOL Group, Program on Mental Health.

Yap PLK, Goh JYN, Henderson LM, Han PM, et al. (2008). How do Chinese patients with dementia rate their own quality of life. *Int Psychogeriat.* 20, 482–493.

Yesavage JA, Brink TL, Rose TL, Lum O, et al. (1983). Development and validation of the Geriatric Depression Screening Scale: a preliminary report. *J Psychiatr Res.* 17, 37–49.

Yetkin FZ, Rosenberg RN, Weiner MF, Purdy PD, et al. (2006). fMRI of working memory in patients with mild cognitive impairment and probably Alzheimer's disease. *Eur Radiol.* 16, 193–206.

Zakzanis KK, Leach L. (2002). Evidence for a shrinking span of personal and present existence in dementia of the Alzheimer's type. *Brain Cogn.* 49, 249–260.

Zapf W. (1984). Individuelle Wohlfahrt: Lebensqualität. In, (Eds.) W Glatzer, W Zapf. *Lebensqualität in der Bundesrepublik.* (p. 13–26).

Well-Being as an Indicator of Quality or Quality-of-Life

Abstract

This chapter addresses the issue of whether the term well-being, particularly the phrase subjective well-being, can be used as an indicator of quality, and quality-of-life. It examines both the way language is used and the cognitive basis of these expressions. Based on the assumption that subjective well-being (SWB), in particular, is a form of emotional expression, the chapter examines whether the phrase is a discrete emotion, a product of cognitive-emotional processes, or cognitive-emotional regulation. It ends with a discussion of how subjective well-being can be modeled. Neurobiological, nonlinear, and simulated embodiment models are considered. The chapter concludes that SWB and quality and quality-of-life, while overlapping concepts, are best keep distinct, since to use them interchangeably dilutes the unique meaning of each. For this reason it is not recommended to consider SWB, in and of its self, as an indicator of quality-of-life or health-related quality-of-life (HRQOL).

We hold these truths to be self-evident, that all men are created equal, that they are endowed by their Creator with certain unalienable Rights, that among these are Life, Liberty and the pursuit of Happiness.

United States Declaration of Independence; July 4th 1776

Abbreviations

ABS	Affect Balance Scale (Bradburn 1969)
Cantril Ladder	Cantril (1965)
CER	Cognitive-emotional regulation
CPM	Component process model
DT	Dizygotic twins
GWB	General Well-being Scale (Dupuy 1984)
HRQOL	Health-related Quality-of-life
IT	Inferior temporal cortex
MPQ	Multidimensional personality questionnaire (Tellegen 1982)
MT	Monozygotic twins
NA	Negative affect
NHANES	National Health and Nutrition Examination Survey (1973)
NNM	Neural network model
PA	Positive affect
PANAS	Positive and negative affect schedules (Watson et al. 1988)
QWB	Quality of Well-being Scale (Kaplan and Anderson 1990)
SEC	Stimulus evaluation checks
SWB	Subjective well-being
SWLS	Satisfaction with Life Scale (Pavot and Diener 1993)
WB	Well-being Scale
WHO	World Health Organization

1 Introduction

This chapter considers whether the term well-being, particularly the phrase subjective well-being (SWB), can be used as an indicator of quality or quality-of-life. In Chap. 2 I noted and expressed doubt about the wisdom of substituting the term SWB for the term quality in a quality-of-life or HRQOL definition (e.g., Table 2.7). In this chapter, I hope to document this argument primarily by addressing the larger issue of what role the emotions play in a qualitative assessment.

Again, I had started discussing this topic, but now in Chap. 1 (Figs. 1.2 and 1.3), where I acknowledged that to varying degrees the emotions are an integral part of every qualitative judgment. In this chapter I hope to expand on this discussion. I will also follow my pattern of examining the meaning of the term well-being (and by extension the term emotion) by describing how language and cognitive processes are used to express the term with particular emphasis on its various forms of expression (e.g., happiness, joy, sadness, fear, anger, and so on). What I will not do is attempt a review of the extensive SWB literature.

What I learnt from this chapter is that while the term well-being or phrase SWB are abstract expressions when they are used, they will be expressed in both an abstract an concrete way, not unlike what was found for phrase "health status." Thus, when I state that specific feelings lead to a sense of well-being, I am going from the concrete to the abstract, while if I describe well-being as being the normative expression of an "good or end," then I am conceiving of well-being in abstract terms and will eventually discuss its concrete manifestations. This complexity leads to the impression that the term well-being and SWB are semantically unstable with their "family" of uses confounding their meaning, in any particular application. Diener, (2006) has implicitly acknowledged this by referring to well-being or SWB as "umbrella" expressions, implying that any or all uses of the term could fit under the umbrella. This view, I believe is misplaced, since it contributes to the impression that the expressions are examples of literary conveniences when there is ample evidence that they are substantive concepts (see further).

What will also become clear is that the phrase SWB is best understood as a (e.g., emotional) *state* a person is in, in contrast to the term quality which implies presence of a *instrumental process* that involves an evaluation and a valuation of a descriptive indicator. In addition, my ability to define the state of well-being will be directly related to how clearly emotional terms can be defined. This, as my discussion will reveal, is a source of great controversy with much debate on how best to define an emotion and various emotional states (e.g., happiness). As a result, confusion and semantic instability of derivative terms, such as SWB, would be expected.

Also contributing to this complexity is the large number of ways that these expressions have been conceptualized. For example, the hedonic approach to well-being assumes that a person attempts to maximize their pleasure and minimize their pain. Alternatively, the eudaemonic approach to well-being, a term Aristotle (e.g., 1985) used, refers to how well a person or society has achieved some ideal (prototypical) state, while the welfare model, asks how well a person has fared considering all aspects of their life living in a particular society. Sen (e.g., 1985, 1993) has proposed a capacity approach to well-being, an approach he recognizes originates from an Aristotelian perspective, while Ryff (1989a) describes efforts at defining psychological well-being that she sees as an extension of the eudaemonic approach. I will primarily evaluate the hedonic approach to SWB, since it is the dominant approach. After this review I will conclude that the concepts of SWB and quality are best thought of as separate but overlapping concepts or terms, with each having a different pattern of cognitive-emotional determinants.

Part of the difficulty about generating a science of well-being, or SWB, has to do with how widely cited these expressions are. Thus, it is not surprising to find the expressions spoken in reference to political, economic, psychological, social, emotional, and even health-related issues (Sirgy et al. 2006). In addition, the term has found its way into public policy statements, philosophical treatises, behavioral and social sciences texts, and selected areas of biological research. This massive intellectual effort is consistent with the term having multiple purposes, including as an indicator of the moral and ethical achievements of an individual, community, or society.

To illustrate the importance and confusion surrounding the specific term – well-being – consider its role in the WHO definition of health (see also Chap. 8). The definition, which is in the preamble of the Constitution of the WHO (1948), states:

> The States parties to this Constitution declare, in conformity with the chapter of the United Nations, that the following principles are basic to the happiness harmonious relations and security of all people: Health is a state of complete physical, mental, and social well-being and not merely the absence of disease or infirmity. The enjoyment of the highest attainable standard of health is one of the fundamental rights of human being without distinction of race, religion, political belief, economic or social condition.

The obvious implications of this definition is that health is a state of well-being, a state that can be made up of different types of well-being, but will still basically be a state, a person is in. But what does the term well-being mean in this declaration? Is it referring to just SWB, or does it also refer to the amount of resources a person has (e.g., do they own a home, their salary), how long and how well a person lives, or all of these meanings.

As would be expected, this definition has engendered a great deal of controversy (summarized in Bok 2004), so much so that it is now not considered a theoretically viable definition of health. Bok (2004), in summarizing her review, concludes that while the definition remains a historically accurate statement of the aspirations of the international community in 1948, today it "cannot serve operational purposes for assessment, and policy-making purposes" (p. 16). Still, the presence of the term in this document affirms its importance as a method of at least political expression, although its usefulness as a term in scientific discourse remains to be determined.

1 Introduction

What is most important here, however, is whether the term and its characteristics can be used to indicate quality, or quality-of-life. For example, while a person's sense of well-being evolves from interacting with certain objects, experiencing certain events, or living a particular type of life, it also appears that these feelings occur after these events, as is true for emotions in general. In addition, it is not at all clear that my continued appraisal of my experiences will always evoke feelings of well-being, or be relevant to moral and ethical issues. This suggests that a threshold exists, above which issues of well-being become obvious and below which they are not.

In contrast, a quality assessment can occur with minimal input from emotional factors, as during a functional quality assessment, although emotional factors may play a determining role during an aesthetic qualitative assessment. What is not known is the boundaries between these two types of qualitative assessment; how distinct they are and to what extent they can be assumed to be independent indicators of quality.

Subjective well-being, viewed from the perspective of the individual, refers to a feeling a person has about themselves or events and their life history, while the affect associated with quality (especially for functional quality) is often "cold." Goel and Dolan (2003) have provided support for this distinction; they report a neuroimaging study where "hot" or emotion laden reasoning vs. "cold" reasoning activates different neural circuits. This study provides general support for the statement that it is possible for a quality assessment to be predominately "cold," and be different from a well-being statement.

Well-being, in contrast, is often posited as an outcome of a welfare program, whereas this is less likely to occur for quality or quality-of-life assessments. This is so, even though investigators recognize that the quality of housing, the environment, or interpersonal relationships, to give a few examples, are important determinants of a person's welfare status and well-being. In addition, when I discuss SWB I am most often referring to some positive state a person is in (that is, the person is "well"), and not a negative state (e.g., ill-being), although some formal assessments will monitor both states (for overview see, Diener 2006). In contrast, I am quite willing to talk about positive or negative, good or bad quality, or quality-of-life.

What will become evident from this chapter is that quality is predominately a cognitive process that has an affective component that varies from "hot" to "cold," depending on the dominant theme (aesthetic or functional) or object of the qualitative assessment (e.g., a car or a life lived). While, SWB has more of the characteristic of an emotion in that it becomes conscious after a particular behavior or physiological processes.[1] In addition, SWB, as an after-effect of an emotional event, has defining features that distinguish it from other forms of affective expression (e.g., a mood), which I will clarify as I proceed (see Tables 11.1 and 11.6).

Supportive of these distinctions are the dictionary definitions of these expressions; with the term well-being[2] (Oxford English Dictionary, 1996; Table 4.1) being referred to a state that a person is in (e.g., the state of being comfortable, healthy, or happy), whereas the definition of the term quality directly involves an evaluation (Table 4.1). Thus, for the term well-being to function in an equivalent role requires that the well-being statement be qualified or "valued," giving it characteristics of a hybrid cognitive construct. This, of course, is possible and a good example of this type of construct is the "Quality of Well-being" (QWB) scale developed by Bush and Kaplan (e.g., Kaplan and Anderson 1990). This assessment conflates various personal but evaluated descriptors with a weighting or valuation system that reflects the perceptions of a representative sample of the general population.

Diener et al. (1999) also discusses the quality of psychological well-being (e.g., happiness), but now as influenced by mood intensity, emotionality, and mood variability. Thus, two persons may report being equally as happy but may experience substantial differences in their mood swings, hence the quality of their well-being will vary (e.g., "I am happy and everything is going smoothly;" "I am happy, but life is sure rocky"). This also suggests that a well-being statement can be qualified.

While I have been discussing SWB as a personal feeling or type of affect, philosophers talk about it as a good, or as an end unto itself, reflecting how a person should live their life, or what a society should achieve. Cognitively, this approach is an example of going from an abstract construct (e.g., a prototype virtue) to its concrete manifestations (e.g., an individual's specific behaviors). Crisp (2005) states: "A person's well-being is what is 'good for' them" (p. 1). He also recommends that the term "self-interest" be used as a correlate of the term well-being; thus, "my self-interest is what is in the interest of myself, and not others" (p. 1). He also considers the term welfare to be closely linked to the term well-being with its reference to how well a person's self-interests are being achieved. Crisp (2005) also points out that the term well-being can be characterized as a kind of value, a prudential value, as opposed to an aesthetic or moral value. A prudential value refers to what is best for the person (or "good for" the person), as opposed to moral values which refers to what is good for others.

Philosophers talk about happiness not as a momentary event in a person's life, but rather as encompassing a person's whole life. In addition, they tend to avoid using the word happiness, as opposed to well-being, since they often want to talk about the well-being of animals and plants, and talking about a "happy plant or animal" seems awkward. Of course, this awkwardness also reflects the fundamental differences in how these terms are being used.

The happiness approach to well-being, has a long intellectual history dating from Epicurus, continuing in Plato's dialogs, and includes Jeremy Bentham's (1789/1996) famous dictum that any person is under the control of two "masters," pain and pleasure. It is the resolution of the conflict between these masters that determines the nature of any person's existence. Thus, for Bentham, happiness is the difference between pain and pleasure, as modified by the intensity and duration of these indicators. This approach, however, assumes that all types of hedonic outcomes are equal. It does not differentiate the pleasure of eating a meal from the pleasure of reading a great novel, or hearing a particular musical performance. To deal with this problem, John Stuart Mill (1863/1969) proposed that "the quality of hedonic states" also be considered. Thus, for Mill the quality of pleasure is determined by the value a pleasure has for a person. For example, the quality of reading a great book or having a meal would depend on which experience the person considered more valuable. Adopting this approach, Mill was able to not only describe various hedonic states but also to order them, in terms of their value or importance to the person. Interestingly, critics suggested that by doing this Mill changed the nature of what was originally meant by hedonism (Crisp 2005).

Mill's maneuver, however, while in response to a very different set of circumstances, resembles what I have previously discussed concerning the formation of a hybrid cognitive construct, with the description of a hedonic state being scaled in terms of intensity or duration, but modified by an independent value or importance based statement. Consider what Mill (1863/1969) states:

> If I am asked, what I mean by difference of quality in pleasures, or what makes one pleasure more valuable than another, merely as a pleasure, except its being greater in amount, there is but one possible answer. Of two pleasures, if there be one to which all or almost all who have experience of both give a decided preference, irrespective of any feeling of moral obligation to prefer it, that is the more desirable pleasure. If one of the two is, by those who are competently acquainted with both, placed so far above the other that they prefer it, even though knowing it to be attended with a greater amount of discontent, and would not resign it for any quantity of the other pleasure which their nature is capable of, are justified in ascribing to the preferred enjoyment a superiority in quality, so far outweighing quantity as to render it, in comparison, of small account. (paragraph 4).

What Mill seems to be saying is that the quality of my pleasurable experiences is determined by my preferences, not the quantity of the experiences, per se. This, of course, is quite relevant to my discussion of what I mean by quality-of-life, and in particular the relationship between the quantity and quality-of-life (see also, Schmidt-Petri 2003).

In Chap. 1 (p. 13) I described how the shift in emphasis from aesthetic to functional quality assessment occurred around the time of the industrial revolution. Concurrent with these events was the maturation of utilitarianism, a philosophical tradition that was compatible with a functional quality assessment, since it emphasized the use of an object, event or experience, and discouraged "observation for its own sake," as may occur in an aesthetic quality assessment. Philosophical utilitarianism has developed along two dominant paths; egotistic hedonism and universal hedonism. Sidgwick (1884) suggested that egotistic hedonism, being concerned with the happiness of the individual, leads naturally to various ethical statements, while universal hedonism, focusing on what determines universal hedonism and happiness for all, evolves into political expression. Philosophical hedonism claims that pleasure (such as, the satisfaction of desires) is the highest good and should be the aim of our lives. Crisp (2005), however, points out that hedonism and well-being are not quite the same concept. He states that when I have certain pleasurable experiences I may ask "does it feel good?" (a reactive process), whereas when I ask about a person's well-being I am asking if what has transpired was good for the person (a retrospective or reflective cognitive process). Of course, what pleasures I considered good for me may also be those that I enjoyed, and that I may have also desired. Here I am talking about the pleasures I anticipated (a prospective process), when I reflected about what was good for me.

Griffin (1986) and Crisp (2005) claim that desire can be presented in several forms. In its simplest form, if a person's current desires are satisfied then the assumption is that the person is better off (that is, their well-being is enhanced). However, not all current desires are in the best interest of a person. An alternative is to assume that if all the desires a person has over their life are satisfied, then this would maximize their well-being. However, this approach is also not a guarantee of a maximized well-being since at different stages of my life I may have desired things that in retrospect were not of my best interest. The alternative that Crisp (2005) and Griffin (1986) suggest is that I have to qualify my desires by how much I know about them, and only then can I be reasonably sure that what I desire will maximize my well-being. Informed desires, therefore, are most likely to lead to an optimal sense of well-being.

What is interesting about this discussion is how often the descriptive terms being considered (e.g., pleasurable experiences), have had to be "valued" before they became useful. Thus, I needed to be informed about the desires I have if they are to maximize my well-being, just as Mill suggested I needed to know about the quality of my hedonic states to clarify the relationship between these states and well-being. In each case, it can be argued that the authors' thoughts involved forming a hybrid cognitive construct, while formally related, still have distinct meanings. Unfortunately, these terms (e.g., desires, hedonism, and well-being) are often used interchangeably, resulting in a great deal of confusion concerning their definition and conceptual base. This type of confusion also extends to the terms well-being and quality (particularly when applied to quality-of-life assessments), which are also regularly substituted for each other (e.g., Veenhoven 1984; Wheeler 1991).

1 Introduction

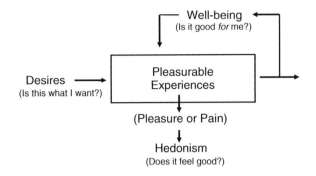

Fig. 11.1 Model of the relationship between hedonism and well-being. Included in the figure is the relationship between current pleasurable experiences, the anticipation of these experiences as desires, and the consequences of these pleasurable experiences as revealed by the person's sense of well-being. The model, as drawn, assumes that the person's well-being needs feedback to reinforce what is considered to be a pleasurable experience.

And, as noted earlier, this occurs even though these terms have different definitions, and as will be demonstrated (Chap. 11, p. 395) have overlapping but distinct synonyms.

This type of confusion makes it easier to refer to both terms as labels or umbrella terms. This is the essence of the review of quality-of-life assessments by Gill and Feinstein (1994), who found that HRQOL includes a group of qualitative assessments each of which may differ in form, content, or purpose. While Diener (2006) directly states that; "... subjective well-being is an umbrella term for the different valuations people make regarding their lives, the events happening to them, their body and minds, and the circumstances in which they live" (p. 400).

In contrast, Gill and Feinstein (1994) were concerned about combining diverse assessments under the same label, (e.g., HRQOL), creating a situation where the broadness of the resulting domain statement makes the concept minimally useful when applied analytically. As a result, a domain term, such as HRQOL, may not be very accurate or useful. However, it is also clear that these assessments can be ordered in terms of their cognitive characteristics (e.g., whether the assessment consists of only evaluated descriptors, or if it includes a "valuation" method). This was essentially what was done in Table 8.4, and this resulted in a relatively homogenous collection of assessments. Classifying HRQOL assessments in this way would remove the need of applying the umbrella label to HRQOL assessments.

In some ways, an abstract term or concept cannot avoid being seen as an umbrella, since characterizing an abstract concept requires the generation of a variety of concrete context-specific terms. Thus, the "umbrella" Diener (2006) speaks of may very well be a product of a particular cognitive process; the repeated use of basic or subordinate level of information to characterize a superordinate concept. However, not all abstract terms lead to as large a number of diverse concrete indicators as the term well-being, just as not all abstract terms function as diffusely as implied by an "umbrella." In fact, it is possible that the category of SWB could avoid the label of an umbrella term, especially if it were organized around some cognitive principle (e.g., whether the statements were valued or not). Unfortunately this has not occurred, so that the term remains a label for a heterogeneous set of descriptors.[3]

As a corollary, an abstract concept carries with it a particular meaning, a meaning which can guide the selection of concrete terms associated with the abstract concept. This would be expected when several terms are organized into an informational hierarchy. Thus, if the term well-being was just a label, it could be referred to as "category #1," or some other nominal indicator. Rather the term well-being has a distinct, although broad, phenomenal meaning that is actually diluted by referring to it as an umbrella term so that what "well-being" means to a person deserves to be clarified.

To think of the term well-being as a "linguistic" umbrella, however, also has a practical consequence. For example, Diener and others (Diener 2006; Diener and Seligman 2004) have proposed that guidelines for national well-being or ill-being be established, yet this effort is at great risk of failing if the indicator of interest is considered a label representing a variety of otherwise, nonspecified, terms. If a specific indicator is identified (e.g., happiness), then there is no need to use the term well-being. Designing national indicators around an outcome whose content may vary from situation to situation will not be very reassuring to policy makers, since it will not be clear what is being achieved by a policy when applied in different settings. They may want to know exactly what is being assessed at any particular time. Kahneman (1999) appears particularly aware of this (e.g., the role of retrospective assessment) when he proposed that an "objective" happiness approach to well-being (Chap. 4, p. 109), and his effort at introducing objective measures can be seen as an attempt to rationalize the "umbrella label."

While the term well-being is a uniquely emotional term, the term quality, particularly functional quality (Chap. 1, p. 12), can be articulated without evoking much if any, emotional correlates. For example, judging the quality of a golf course or a diamond can occur without much attention to the emotional component of such assessments, even though aesthetic aspects of the object may be present. In fact, making "objective" judgments about some object, event, or life lived encourages the minimal intrusion of emotions into the judgment process, since it may "bias" a person's responses or an investigator's interpretation of the results of a study. As I discussed in Chap. 5, the ability of an investigator to be "objective" is a task that requires training (e.g., Truth-to-Nature; Structural Objectivity, or Trained Judgment-Expertise). In contrast, an aesthetic quality judgment is critically dependent on the emotions evoked by the judgment task. Thus, the two types of qualitative assessments differ in that variability in functional quality assessment is not so much a product of the emotional component of a judgment, as it is the precision of production or reproduction of particular outcomes, while in aesthetic quality assessment an analytical process follows the feelings aroused by what is being judged.

The relationship between functional and aesthetic quality assessments raises the more fundamental issue of what role the emotions play in various cognitive processes. For example, Andrews and Withey (1976) noted that the assessment of a person's life involves "both a cognitive evaluation and some degree of positive and/or negative feelings, i.e., affect" (p. 18).This statement places the discussion concerning the role of emotions in cognition squarely in the center of the debate concerning the nature of well-being. Consistent with this, is the proposal of Diener et al. (1999) that SWB consists of an assessment of positive affect, an absence of negative affect, and a person's cognitive evaluation or satisfaction with their life circumstances.

However, the relationship between emotion and cognition is complex, as I illustrate with reference to a paper by Andrews and Mc Kennell (1980). This paper demonstrated that the cognitive form of SWB statements were correlated, to varying degrees, with affective indicators (see also Crooker and Near 1998). These data suggested that cognitive and affective expressions were rarely, if ever, truly independent classes and instead differed only in the degree that emotional factors contributed to the response of an item, such as a measure of satisfaction or a global assessment (Lent 2004). As a consequence it would not be surprising for there to be a continuing debate on the nature of the relationship between emotion and cognition (Duncan and Barrett 2007; Lazarus 1991a; Zajonc 1984).

In addition to the complexities of analytically and empirically separating emotions and cognition, there is also the difficulty of separating the various manifestations of affective expression. Clore and Colcombe (2003), for example, have suggested (Table 11.1; also see Table 11.6) that emotions are current and attached to salient objects (e.g., a person I like or dislike, an animal that is scaring me, and so on), while mood is not attached to an object, but is also current. In contrast, a person's attitudes are not only chronic but also are assumed to be attached to salient objects (e.g., a personal prejudice), while a person's temperament does not change much but is usually not attached to salient objects. Well-being was not included in the original Clore and Colcombe (2003) table, and as will become clear, the concept of well-being is assumed to have evolved from attitude research (Chap. 11, p. 398); yet it appears to differ from attitudes (my likes or dislikes) and has some of the characteristics of a person's temperament (as related to their personality). Thus, most often what contributes to a person's sense of well-being is not dependent on an object.[4]

Table 11.1 Clore and Colcombe's (2003) model of affective information

"Object specificity and temporal duration as constraints on the meaning of experiential information"

Sources of felt affective information

	Current	Chronic
Salient object	Emotions	Attitudes
		Well-being[a]
No salient object	Mood	Temperament

[a]Not included in the original Clore and Colcombe (2003) table. Reproduced with permission of publisher

The primary focus of this chapter, however, is to study the interaction of emotion and cognition, and how this is reflected in the assessment of quality in general, and quality-of-life, well-being, and hedonic indicators. This will also give me the opportunity to continue the discussion of the neurocognitive basis of a qualitative assessment (e.g., Chap. 10), since emotions, as with various other cognitive processes, appear to be differentially represented in the brain. Thus, when I speak about the interaction of cognitions and emotions I assume this interaction will have a cortical spatial representation, involving cortical systems which communicate with each other.

There is also an important theoretical issue here since there are different ways of conceiving of the neurocognitive relationship between emotions and cognition. For example, I assume that cognitions and emotions can exist as relatively separate systems, elements of which can be added or multiplied together, but can also be teased apart to the extent that they are analytically independent of each other. This model would be supported if two somewhat separate cortical sites existed that interacted. This view is also consistent with Zajonc's (1980) view that "affect and cognition are under the control of separate and partially independent systems that can influence each other in a variety of ways and that both constitute independent sources of effects in information processing" (p. 151). Others, such as Damasio (1999) or Duncan and Barrett (2007), think of the two entities as inextricably interwoven so that it would not be possible to separate them without fundamentally changing the nature of the other.

The difference in these two approaches appears not only to be more than an issue of emphasis, but also a difference of substance. The story of Phineas Gage (e.g., MacMillion 2000; see also Chap. 11, p. 428) is commonly cited as evidence of the interwoven nature of the two systems, but demonstrations based on pathology are seldom definitive. At first glance, this discussion appears to raise questions about the nature of a hybrid construct, but a hybrid construct can be formed with cognitive material that is purely cognitive or emotional, or entities that have elements of both processes.

A corollary of the interaction between emotions and cognitions is the observations that measures of well-being remain

Table 11.2 Modeling the semantics of "emotion words"

Type	Definition
The *Label* View	The term emotion is a label for emotion terms (e.g., anger, happiness, fear) each of which may be based on a specific physiological process and/or behavior
The *Core Meaning* View	The meaning of an emotion word comes from its core (or denotative) as opposed to its peripheral (or connotative) meaning
The *Dimensional* View	A number of dimensions (core meanings) define all emotion words and any one word uses some but not all of these dimensions for their meaning
The *Implicational* View	This view is based on the assumption that the connotative (e.g., cultural or situational) meaning of emotion words determines its meaning
The *Prototype* View	Just as some words are prototypical emotion words, so are their meanings which are prototypical for emotion words
The *Social-Constructionist* View	The meanings of emotion words are social-cultural constructions. These meanings are also prototypical for a society

stable overtime. Thus, just as it is important to determine what produces changes in the outcome of hedonic states, such as well-being, it will also be important to determine what keeps such measures stable over time. Thus, the factors (e.g., genetic background, personality) that contribute to the preservation of well-being will be discussed.

With this introduction I will proceed to my first task which is to clarify how the language of well-being, or SWB, is used. I will then follow this with a discussion of the cognitive basis of SWB, and finally with an extended discussion of the role that emotions play in SWB.

2 The Language of Well-Being

From what has been said so far, it is clear that SWB statements are primarily meant to describe the outcome of various emotional states. This would be true whether the term was referring to an indication of a "personal state" or a "societal good." The emotions, however, can be expressed verbally or nonverbally. As a form of verbal expression, the term "well-being" uses ordinary (folk) language to communicate its meaning, and how this occurs can be studied in at least three ways; first by examining different ways, SWB is expressed in ordinary discourse, second by examining the literal and figurative nature of the term, and third by considering how SWB is addressed in formal (e.g., paper-and-pencil) assessments.

First, however, I would like to add to the discussion started in Chap. 2 on the nature of language, by examining various types of "emotion words"[5] and examine their relationship to SWB to illustrate various principles. For example, there are two basic types of emotion words; *expressive* and *descriptive*. A word such as "wow" would be considered an expressive word, while "well-being" would be a descriptive emotion word. Previously discussed emotion words can be organized into an information hierarchy ranging from the concrete subordinate level to basic level words followed by words that have a superordinate level of abstractness. Thus, the sequence of terms starting with "smiling," followed by "happy" and ending with "well-being," would be one example of an informational hierarchy. There are, of course, many subordinate and basic level emotion terms that may be associated with SWB (such as, happy, sad, or angry), and others that are not (e.g., Shaver et al. 1987). Those that are most often associated with SWB constitute a prototypical definition, although the number and type of emotion categories and prototypes remains a topic of continued discussion. The metaphoric and metonymic functions of emotion words, and their relevance to SWB, will also be examined, since their use in figurative expressions, has been reported to occur at a high rate (Fainsilber and Ortony 1987).

Table 11.2 summarizes some of the different approaches to defining the meaning of emotion words as originally described by Kövecses (2000). Clarifying the meaning of emotion words should help in understanding the role of emotions in such terms as, quality, quality-of-life, and SWB. The core meaning approach to emotion words, for example, states that each term has a core meaning (denotative meaning) which consists of a small number of terms that are common to a number of other words that appear to have similar meanings (see further). In contrast, peripheral (connotative) meaning is more likely to be figurative in nature, and more likely to vary between cultures or social situations (Kövecses 2000). Osgood (1964), however, claims that certain connotations are universal (e.g., bad or good; fast or slow; strong or weak), while Wierzbicka (e.g., 1997) claims that a limited number of core or "semantic primitives," as she refers to them, exist that make up the conceptual content of emotion words in a variety of languages. Davitz (1969) has provided some empirical support for the existence of core emotion dimensions in a study he performed in which respondents were asked to match sentences which were associated with a selected number of emotion words. Using cluster analysis, for example, he found that anger was composed of terms which referred to hyperactivity, moving against, tension, and inadequacy. I will return to these semantic models as I proceed.

2.1 Well-Being in Ordinary Discourse

As previously noted, the term "well-being" has gained much visibility as an outcome indicator, partially as a result of its use in the WHO definition of health, but also as part of political speech and policy formation. Part of the reason why this has occurred is not only because the term plays a prominent role in our ordinary conversation, but also because of the cognitive "opportunities" it provides for a respondent. For example, the term well-being, as opposed to quality-of-life, is often used to express a speech act, but is also likely to function as a metonymy. The term, being a descriptive statement, also requires less cognitive processing compared to the qualifying task implied by the hybrid concept characteristic of the phrase "quality-of-life."

A speech act[6] (Searle 1969) occurs when I make a statement which not only communicates a meaning, but also communicates the need for an action. Often there is a one to one relationship between the syntax of a sentence and the action implied. Thus, the declarative sentence, "Run! There is a bear behind you!," would be an example of a speech act since it communicates the need for an action. Likewise the interrogative sentence, "Could you please open the window?", is not just asking a question but it is also expecting someone to act. Metaphors are examples of indirect speech acts, since they "ask" the reader or listener to cognitively transfer information from one word or phrase to another.

The questions; "How are you?", or "How have you been?", or "How's life?"can be interpreted as inquiries about a person's well-being, and as such, act as a common form of greeting. But they can also be interpreted as a verbal emotional communication which should provoke a response. Support for this interpretation comes from a lexicographic analysis of these phrases, which revealed that historically the term was first used to ask about a person's health or welfare (both of which can be construed to be aspects of well-being), but by the sixteen century had expanded so as to be used as a greeting (Szalai 1980; p. 10). Phrases of this sort are also found in all Indo-European languages, even ancient Babylonian inscriptions, and in each case the greeting involves a direct expression of concern about a person's well-being (Szalai 1980). Of course, used in this manner, these phrases can lead to a reciprocal greeting without any discussion of the person's well-being, or to frank responses such as, "I feel great" to "Life is a mess." Thus, part of the reason why the term "well-being" plays such a significant part in our daily language, is because it can be used in such a wide variety of ways.

A quality-of-life statement, as previously noted (Chap. 2), has metaphoric properties, as does a well-being statement (see below). However, a well-being statement differs from a quality-of-life statement in that it is more likely to function as a metonym. A metonym is a linguistic devise whereby a well understood term is used to stand for another well understood term, with the intent of mapping information from the target to the source. For example in the sentence, "Don't let Iraq become another Vietnam," the meanings associated with Vietnam is being attached to Iraq. Lakoff and Johnson (1980) clarify the distinction between a metonym and a metaphor in the following way:

> Metaphor and metonymy are different kinds or processes. Metaphor is principally a way of conceiving of one thing in terms of another, and its primary function is understanding. Metonymy, on the other hand, has primarily a referential function, that is, it allows us to use one entity to stand for another. (p. 36–37)

Metonymy not only can be used to make a reference, it can also function to help me understand. Thus, metaphors and metonyms serve the same purposes, but they differ in that the metonymic concepts are more a part of ordinary discourse reflecting the way I ordinarily think act and talk.

Consider the following definitions of HRQOL; "Quality-of-life is defined as a person's sense of well-being which stems from satisfaction or dissatisfaction…" (Ferrans 1990) or, "Quality-of-life is a state of well-being …" (Gotay et al. 1992). In both these definitions, the meaning associated with the term "well-being" is being attached to the phrase "quality-of-life," as would be expected if the statement was functioning as a metonym. However, it is harder to retain the metonym function if the sentence elements were reversed and restated (e.g., "well-being is defined as a person's sense of quality-of-life" or "well-being is a state of quality-of-life"[7]). Part of the reason why this occurs is because the term quality or quality-of-life connotes an instrumental intent that is less prominent in the term well-being. Thus, well-being, being a descriptive emotion term retains greater linguistic flexibility than the phrase "quality-of-life" which appears to have a distinctive, and more, limited linguistic and cognitive role.

Consistent with its greater linguistic flexibility, is the observation that synonyms for the term well-being cover a wide range of states; including, health, pleasure, contentment, comfort, goodness, and prosperity (Chapman 1992), while synonyms for the term quality have a much narrower focus, referring to essential attributes or a distinctive property of an object, event or experience, or being superior or excellent in value. In addition, there are only a limited number of synonyms for the phrase quality-of-life, two of which overlap with the term well-being; "contentment" and "prosperity" (Roget's Thesaurus of English Words; Chapman 1992).This suggests that contentment and prosperity represent core, as opposed to peripheral, meanings of the two terms (Table 11.2 and Fig. 11.2).

Having argued that the terms well-being and quality or quality-of-life are semantically and functionally separable, I now have to acknowledge Wittgenstein's claim that to

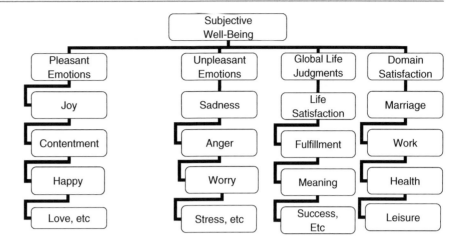

Fig. 11.2 Presented is Diener's (Diener et al. 2003) model of subjective well-being. Permission to reproduce this figure was granted by the publisher.

understand the function of language you should not be as much concerned about its dictionary meaning than how the terms are used. Thus, Wittgenstein seems to be saying that language is totally context-dependent. Applied to the current discussion, what he would say is that if the term well-being is being used to refer to quality-of-life, then the fact that the two terms may have different dictionary meanings is a secondary issue. However, I will persist in distinguishing well-being from quality and quality-of-life, since I believe the failure to do so has contributed significant confusion to the usefulness of both of these terms.[8]

Still, I can ask, what can the study of well-being contribute to the understanding of quality and quality-of-life? First, as I have already stated, if well-being statements are properly "valued" they can be as useful an element of a qualitative assessment, as any other I have considered (e.g., symptoms, functions). If valuations occur then quality-of-life and well-being statements would have the same formal cognitive (hybrid construct) form. What is unique about the study of well-being is its dependence on emotions and feelings, and this offers an investigator the opportunity to study how emotions contribute to the assessment of quality, quality-of-life, or HRQOL. It can also help in the process of understanding the aesthetic dimension of a quality assessment. Next, I will describe the metaphoric nature of the term well-being, and follow this up with a discussion of the metaphoric character of emotion words or phrases. These studies will demonstrate that metaphoric expression, particularly conceptual metaphors, provide common mechanisms whereby the large array of supposedly "well-being" statements can still communicate a common meaning.

2.2 Subjective Well-Being as a Form of Metaphoric Expression

Expressions such as, "I feel on top of the world"; "I feel full of energy"; "Things are going great"; or "I am pleased because of what I have accomplished,"[9] all appear to be addressing a person's well-being, yet there is no direct statement or reference to well-being. How can this occur? One way is if each sentence contained metaphors which communicated meanings that duplicated the meaning implied by the statement about a person's well-being. So for example, the sentence "I feel on top of the world" implies that the person is high or up, and this meaning not only is characterized of positive well-being, it also is consistent with conceptual metaphors that refers to verticality. In contrast, the meaning communicated by the sentence – "I feel full of energy" – refers to quantities of energy which is characteristic of the CONTAINER METAPHOR, a conceptual metaphor that can also be used in various other statements of well-being. "Things are going great" and what a person has "accomplished" reflect quantities that can be "contained" and they are examples of the CONTAINER METAPHOR. Inspecting these sentences suggests that what I have here are conceptual metaphors which are facilitating the communication of a common meaning even though each of the sentences involved are using different words.

There are other methods (e.g., statistical modeling) whereby the items listed earlier might be shown to be representing a common well-being factor, but in each case the investigator will be asked to use language to label the factors or constructs formed. This brings me back to asking what is it about the language used which facilitates the communication of a particular meaning. If I were to ask a group of persons to describe various emotions, as Davitz (1969) did, then what was found was that persons frequently used metaphors to characterize the emotion. But how do metaphors and other figurative expressions facilitate meaning? Fainsilber and Ortony (1987) point out that what metaphors do is to enhance the quality of emotional expression by permitting the expression of ideas that a literal statement may not be able to do, and also enhance the vividness and efficiency of the emotional expression. This is often accomplished by the metaphor characterizing some internal bodily process that a literal interpretation of the term would not. When these linguistic qualitative processes are combined with the framing of information by various heuristics, the mapping of information between a source and target, and the blending of information from different domains (Coulson 2006), then the ingredients are in place for the communication of a common meaning even when different terms and phrases are being used.

In Chap. 2 (p. 43) I pointed out that the term well-being can be defined as "good existence" (Lakoff and Johnson 1999), with the implication that the use of the term represents a summing of the good things that happen to people. This use of the term is consistent with its role as an indicator of a societal good. Inspecting the definitions listed in Table 2.7 (Chap. 2, p. 43), as well as the expressions listed in the beginning of this section, indicates that an implicit accounting was occurring characteristic of a retrospective or reflective process.

The term well-being, with its reference to a "well," also refers to a deep hole, a cavity, a well-spring, a fountainhead, an open shaft, or even an enclosed compartment, as in a ship (WordNet 2005). These references, of course, are also consistent with the CONTAINER METAPHOR, since each refers to a void or space that is characteristic of a container. Conceptual metaphors, of this type, are also found in various definitions of quality-of-life or HRQOL. For example, in the Ferrans (1990; Table 2.7) definition the phrase "with the areas of life" implies the presence of a void which is filled with a collection of things, such as, aspects of life. Gotay et al.'s (1992) definition clearly states that well-being is a state that contains the ability to perform various activities and the person's satisfaction with these various aspects of their life. Activities and satisfactions, therefore, fill the space that make up the person's state. The Felce and Perry (1996) and Padilla et al. (1996) definitions also refer to a collection of things contained within the person's state of well-being. It is evidence of this sort which illustrates how metaphors are used to communicate meaning when defining quality-of-life or HRQOL.

Probably the easiest way to demonstrate the role of a conceptual metaphor in the term well-being is to consider the metaphoric status of one of its primary indicators, such as, happiness. Kövecses (1991) provides an example, in his attempt to define "the concept of happiness as it is understood for everyday purposes through an analysis of the language related to happiness" (p. 29). He summarizes his overall strategy as follows:

> I have proposed that emotion concepts are best viewed as a category, a set of cognitive models with one or more models in the center. Furthermore, I have suggested that the cognitive models in the center – the prototypes – emerge from a system of conceptual metaphors, a system of conceptual metonymies, and a set of inherent concepts. (p. 30).

Thus, for Kövecses a term like happiness would be defined by involving several types of figurative expressions, each of which provides the cognitive mechanism whereby a prototype would be formed. The mappings which are characteristic of a conceptual metaphor or metonym would be an analog of the neurocognitive mechanism[10] that would define the prototype. Implicit in this approach is the idea that the language I use can organize my thoughts, an idea which has its origin in the Sapir-Whorf hypothesis (O'Grady et al. 1989; p. 184–185), and of which Kövecses (1991) is an advocate.

Kövecses (1991) examined ordinary language to identify the metaphors and metonymies commonly used to characterize the term happiness. He provided many examples of sentences which contained these metaphors or metonymies and I will rely on his material for this discussion. Thus, he illustrates the Light and Up metaphors for happiness with the following examples:

> Happiness is Light
> Look on the bright side.
> When she heard the news she lit up.
> He radiates joy.
> Her face was bright with happiness.
> She has a sunny smile.
> Happy is Up
> They were in high spirits.
> Lighten up!
> We had to cheer him up.

Kövecses (1991), states that physical activities or behaviors, as an expression of happiness, are often associated with expressions of Vitality, Energy, and Agitation. Examples of sentences that include these metaphors or are expressed as metonymies are:

> Happiness is Vitality
> He was alive with joy.
> I am feeling spry.
> He's in a lively mood today.
> He was full of pep.

What I also have in these expressions are examples of the hierarchical organization of information; with sentences like "He is full of pep" metaphorically characterizing happiness, and happiness being one of the factors contributing to what is meant by well-being. Thus, an investigator could include an item, such as, "I am full of pep" in their assessment, and interpret the response to the item (e.g., affirmation or denial) as an indication of the person's well-being.

THE CONTAINER METAPHOR is applicable when happiness is thought of as a quantity[11] or as an emotion; with joy being the most common expression of the amount of happiness. The "container" as a metaphor usually refers to the person's body. Thus,

> Happiness is a Fluid in a Container:
> We were full of joy.
> The sight filled them with joy.
> I brimmed over with joy when I saw her.
> Joy welled up inside her.
> He was overflowing with joy.

Kövecses (1991) also points out that the biblical expression "My cup runneth over" is an example of the "Happiness is a Fluid in a Container" metaphor, with the expression referring to a container of happiness.

When the term happiness refers to happiness as a quantity than metaphors, such as, Light and Up carry a positive quantity associated with the term. They refer to an upward orientation as that is found when something is "good." This upward positive connotation is consistent with the folk language expression of such terms as life, health, control, consciousness, and so on. The positive quantity associated with happiness is also something I desire, and the classic expression of this desire, as stated in the United States Declaration of Independence, is "The pursuit of happiness …." Here, happiness is depicted as something hidden which I desire, strive for, and try to achieve. This is consistent with how desires were displayed in Fig. 11.1, where desires anticipate the occurrence of a hedonic event that results, in this case, in happiness.

Kövecses (1991) summarizes the folk language conceptualization of happiness, as follows:

> Happiness is a state that lasts a long time.
> It is associated with a positive value.
> It is a desired state.
> It is pleasurable.
> It give you a feeling of harmony with the world.
> It is something that you can 'spread' to others.
> It exists separately from you and is outside you.
> It is not readily available: It is either requires an effort to achieve or comes to you from external sources.
> It takes a long time to achieve it.
> It is just as difficult to maintain as it is to attain it. (p. 39).

These set of properties constitute a folk language-based prototype definition of happiness, but it might also be considered a prototypical definition of the term well-being. What is confusing here is to conceive of happiness as a chronic state, when as a type of emotion, it would be expected to occur acutely (Table 11.1). Of course, happiness may have more than one type of representation.

What I have presented here is just a glimpse of what Kövecses (1991) and others have discussed, but it should be sufficient to illustrate how the interplay of figurative language and various cognitive processes give life to the term well-being. This flexibility and ease of application in various settings, however, contributes to the impression surrounding the term well-being that it suffers from semantic instability. My next task, however, is to determine whether the language of well-being is captured in the items used in various formal well-being assessments.

2.3 The Language Used in Subjective Well-Being Assessments

The assessment of SWB is a hugely active research area with literally thousands of citations. This research is supported by a large number of paper-and-pencil assessments, each of which consists of a series of sentences structured as declarative or interrogative statements. These statements are composed of words and phrases a person would use in ordinary conversation, yet this material is being used to assess such complex concepts such as "happiness" or "well-being." This is different than the biological and physical sciences where the investigator may have a highly specialized language to assess and communicate their science. The consequence of this, however, is that the behavioral sciences are subject to the rules and regulations that govern ordinary conversation. For example, when a person reads a question on an assessment, he or she is expected to respond to the item just as they would if someone spoke to them directly. It would also be expected that metaphors and use of metonymies would occur in paper-and-pencil assessments of well-being, much as these figurative expressions are present in ordinary conversation or folk language. The following examples will demonstrate that indeed figurative language expression regularly occurs in well-being assessments, particularly assessments dealing with psychological or subjective assessments of well-being.

An example of an assessment that would be considered a SWB assessment is the "Satisfaction with Life Scale" (Table 11.3; Pavot and Diener 1993). Note that the term well-being is not used in any of the items in the assessment. The task for the respondent is to either agree or disagree, using a seven point rating scale, with each of the five statements. Also note that while the assessment is assumed to be indicating life satisfaction, only one of the items directly refers to that issue. Each of the items seems to be asking if the person reached their personal goals. The term "life" is present in each item, and in each case the respondent is encouraged to consider what has happened to them over time. Thus, these items are consistent with the A PURPOSEFUL LIFE IS A JOURNEY METAPHOR (Lakoff and Johnson 1999; p. 60–64). Lakoff and Johnson (1999) present the conceptual metaphor and its submetaphors, as follows:

Table 11.3 The satisfaction with Life Scale[a]

Instructions	"Below are five statement that you may agree or disagree with. Using the 1-7 scale below indicate your agreement with each item by placing the appropriate number on the line preceding the item. Please by open and honest in your responding"
Scale	7-Strongly agree; 6-agree; 5-slightly agree; 4-neither agree or disagree; 3-slightly disagree; 2-disagree; 1-strongly disagree
Items	*In most ways my life is close to my ideal*
	The conditions of my life are excellent
	I am satisfied with my life
	So far I have gotten the important things I want in life
	If I could live my life over, I would change almost nothing

[a]Adapted from Pavot and Diener 1993

A PURPOSEFUL LIFE IS A JOURNEY METAPHOR
A Purposeful Life is a Journey.
A Person Living a Life is a Traveler.
Life Goals are Destinations.
A Life Plan is an Itinerary.

Reading these different metaphors and the items in the assessment suggests a particular type of overlap. For example, if an item asks if a person has reached their ideal life, if life is excellent, if the person is satisfied with their life, if they have gotten what they wanted out of life, and if they would not want to change anything in their life, then this suggests that the person has reached a destination and achieved their goals. Reaching a destination is, of course, the end of a journey, and to the extent that the person judges their journey, then they are being asked to decide if they achieved their purpose in life.

The items in the assessment also seem to be addressing existential issues, alluding to the meaning and purpose of life. A natural question to ask is if these items are an indication of the person's quality-of-life. I would argue that the assessment as presented does not, even though I would agree that it provides a major descriptor of the quality of a person's existence. I can imagine that a person would agree with each item, but still indicate that what determines their quality-of-life has not been asked (e.g., the spiritual nature of their existence, the burden of receiving treatment, and so on). If so, then the affirmative of the life they have just described may not be concordant with their preferred response pattern, demonstrating again the importance of an explicit, independent valuation of their described state.

Another example of a well-being assessment is the "Psychological General Well-being Index" (Dupuy 1984). In its original form it consisted of a 22 item assessment that covered areas such as, anxiety, depressed mood, positive well-being, self-control, general health, and vitality. It has recently been shortened and validated (Grossi et al. 2006) as a six item assessment (Table 11.4). I will use this shortened form for my analysis. The six items included were in the original assessment.

The first impression that comes from reading these items is that they all deal with emotions or feelings, so that what is meant by psychological well-being, in this context, involves assessing the emotional status of the respondent. It is also evident that both "well-being" and "ill-being" are being assessed. There is also evidence of the presence of metaphors in these items. For example, the Happiness is Light or Happiness is Up metaphors discussed earlier can be seen in items #20 as a positive statement (e.g., cheerful, or lighthearted) or in item #7 as a negative statement (e.g., downhearted). The Happiness is Vitality metaphors are present in items #6 and 21 as positive or negative statements. The remaining two items, #5 and 18, have control, or a lack of control, as their central theme. Control, of course, is an important element in the regulation of emotion, and is thereby an entailment of the EMOTIONS IS A FORCE conceptual metaphor (Kövecses 1991; Lakoff 1987). This analysis suggests that happiness (or unhappiness) is the target that the concrete items are mapping to, with the items themselves functioning as the source of the information being mapped.

Table 11.4 The psychological general well-being Scale-short version[a]

Old Item	Dimension	Content
05	Anxiety	"Have you been bothered by nervousness or your 'nerves' during the past month?"
06	Vitality	"How much energy, pep, or vitality did you have or feel during the past month?"
07	Depressed mood	"I felt downhearted and blue during the past month"
18	Self-control	"I was emotionally stable and sure of myself in the last month"
20	Positive well-being	"I felt cheerful, lighthearted during the last month"
21	Vitality	"I felt tired, worn out, used up, or exhausted during the last month"

[a]Based on Table 2, Grossi et al. (2006)

The "Psychological General Well-being Index" (Dupuy 1984) is one example of a well-being scale that assesses various emotional states, while another example is the "WHO Well-being Index" (WHO-5: Bonsignore et al. 2001; Heun et al. 1999). This assessment was originally a 28 assessment that was reduced to a 20 item assessment and then to the five item assessment referenced. The assessment, according to Bonsignore et al. (2001) has been successfully used in a variety of medical and psychiatric settings. The items on the assessment have the same figurative structure as the items in the Dupuy's index, each item starts with a person reference of either "I" or "my," with four of the five items alluding to how the respondent felt or feels about some state they may be in. These items suggest the presence of the SUBJECT-SELF METAPHOR. As I have previously discussed (Chap. 10, p. 368; Table 10.8), what the SUBJECT-SELF METAPHOR does is to map concrete indicators, such as a person's various "Selves" (that is, their body, social roles, past states, and actions in the world) to the abstract target, the "Subject." In this case, the items on the WHO-5 all focus on the person's body and its various manifestations. This same conceptual metaphor also appears to be present in the items in Tables 11.3 and 11.4.

To identify a fourth example, I reviewed the Bowling's (1991) and McDowell and Newell's (1996) compilations of health assessments, and found that in both cases the authors assumed that the assessment of well-being involved psychological assessments. Thus, Bowling (1991) included measures of depression, anxiety, and mental status as a measure of well-being. McDowell and Newell's (1996), on the other hand, included a broader range of psychological assessments,

Table 11.5 The health opinion survey[a]

Item	Content
1.	Do you have loss of appetite?
2.	How often are you bothered by having an upset stomach?
3.	Has any ill health affected the amount of work you do?
4.	Have you ever felt you are going to have a nervous breakdown?
5.	Are you ever troubled by your hands sweating so that they feel damp and clammy?

[a](Macmillan 1957) Reproduced with permission of the author and publisher © Southern University Press

assessments which were not just limited to psychiatric screening tools. Instead they used Campbell et al.'s (1976) method of classifying the various SWB assessments and included representative examples of cognitive based assessments, such as life satisfaction measures (e.g., Table 11.3), assessments which measure affective or feeling states (e.g., Table 11.4), and those that dealt with psychological distress. They also included behavioral check lists, as well as, questionnaires.

Using McDowell and Newell's (1996) lists, I selected a distress related behavior checklist, the 20 item "Health Opinion Survey" (Macmillan 1957), for study. Table 11.5 summarizes the first five items, which I will use for my analysis of the presence of figurative language. Note first, that these items are all addressing ill-being, as opposed to well-being. They all also use the SUBJECT-SELF METAPHOR. The phrase, "nervous breakdown" would be a good example of a figurative expression, since it is unlikely that nerves literally get broken. Other key terms used (such as, appetite, stomach, work, sweaty or clammy, and aliments) have more of a dual-purpose usage (both literal and figurative).

Another set of "well-being" assessments emanate from economic assessments, and they include items dealing with a person's income and work activities, savings, leisure time, purchasing activities, and so on. These items are much less likely to use figurative language to communicate meaning. And as can be shown, these data are usually poorly correlated with measures of SWB.

2.4 Summary

The purpose of this section was to determine how language was used to primarily communicate the meaning of SWB. This task evoked a series of related questions. For example, how is it possible to have an assessment, with varying number of items, which claim to be measuring SWB, when none of the items make direct reference to the phrase SWB? Second, if SWB is an umbrella term or label what does this imply about the usefulness of the term? Can you have a single definition of SWB, which is applicable to all its applications? And if not, how could you use such a concept in policy decisions? Also of concern is how you can provide an explanation of a complex construct, such as SWB, using ordinary discourse? Is it possible to have a unique definition of a term when the method of defining the term is potentially confounded by culture and context? Also relevant is how the phrase SWB differs from a quality or quality-of-life statement? If SWB is an umbrella term or label, how does this help when it is used as part of a definition of quality-of-life or HRQOL (e.g., Table 2.7)?

The issues I have reviewed should have made it clear that one of the advantages of the phrase SWB is that, as a descriptive phrase, it possesses a considerable degree of cognitive and linguistic flexibility, at least when compared to the terms "quality" or "quality-of-life." Evidence of its flexibility comes from its participation in speech acts, or as a metonym. The diversity and extent of its semantic space can be demonstrated by inspecting its synonyms. The term also has multiple references, whether it is expressed at an individual or at global level of analysis. It also engages a large number of emotion terms (e.g., happiness, satisfaction, and the many emotions that can fit under the umbrella) which are part of ordinary discourse, but are not necessarily conjured up in response to a quality statement. The term, or as a phrase, also has a complex metaphoric structure which also provides it with semantic flexibility, if not instability. Thus, the term evokes the meaning of a container which can indicate the quantity of events, feelings, or experiences that a person has had, and this quantity can then be used as an indication of whether the person or society has reached their personal goals or purpose in life. The term also has a positive valence,[12] so that what is found in the container is what a person or a society desires. Metaphorically, therefore, the term well-being has the potential of being an important descriptor of a person's or society's quality, but only if the importance of what has been described is also established. More generally, however, I feel it is best to think of the term well-being as an abstract term, rather than an umbrella term, and as an abstract term it would be subject to all the cognitive characteristics of abstract terms (see further).

3 The Cognitive Basis of Well-Being

3.1 Introduction

Next I shift my attention from discussing how language expresses well-being, particularly SWB, and to how a SWB assessment may occur. I will assume that well-being is something I feel or experience.[13] I am making this assumption because if I am asking how a SWB statement cognitively occurs I need to know that I am not just dealing with a linguistic devise, as is implied by referring to the term as an

Table 11.6 Characteristics of emotional episodes, mood, subjective well-being, and qualitative assessment*

Characteristics	Emotional episodes	Mood: an emotional state	Subjective well-being*	Qualitative assessment*
Duration	Short (seconds/minutes)	Longer (hours/days)	Lengthy (years)	Short (re-object) or long (e.g., a person's entire life)
Speed of onset	Very fast	Slow	Probably very slow	Can be very fast or slow
Intensity	Great	Low but persistent	Low but persistent	From virtually zero to some, depending on what is being assessed
Visibility	Distinctive (e.g., facial signals)	Inferred from behavior or inquiry	Requires inquiry	Can reflect facial signals or requires inquiry
Physiology	Distinctive pattern of CNS activity	Dependent on neurohumeral status*	No known distinctive pattern, but suspected[a]	No known distinctive pattern
Function	Distinctive functions (e.g., fight/flight)	Regulatory indicator*	Regulatory indicator*	Both distinctive functions and regulatory role
Antecedents	Often identifiable	Less specificity	Little known	Often identifiable
Consequences	Specific behaviors (e.g., fight/flight)	General pattern of behaviors but also impact on cognition	Activation of homeostatic processes	Specific behaviors or activation of homeostatic processes
Variability	Fluctuates sharply	Less variable	Stable over time	Fluctuates from variable to stable

*Adapted from Bramston (2002) but modified so as to separate his category of subjective quality-of-life into SWB and a quality or quality-of-life assessment. Also (*) indicated where I have changed the terms that Bramston used
[a]There are some neuroimaging studies, that I will cite herein, that suggest that well-being does generate a distinct set of cortical activities

umbrella, but that instead it can be said that the term has a biological analog (that is, it literally exists). To help in this task, I will take the rhetorical position, and test the assumption, that well-being is one of several types of affective expressions. What becomes quickly clear, however, is that there are many other terms which overlap or are used as substitute for the term SWB (emotion, mood, and quality-of-life; Table 11.6). Thus, my first task is to clarify and distinguish the meaning of these terms. I follow this by asking a series of specific questions in an effort to establish in what ways SWB is an affective expression. In following section I will examine if SWB is a discrete emotion.

Table 11.6 lists some of the terms that are often interchanged when discussing the term well-being. What I ultimately want to do for each of these terms is to apply the same critical analysis that has been used to determine if the different emotions are distinct categories (see below). While this seems like an obvious exercise, it is not something that has been reported that often in the happiness and well-being research literature.[14] Rather, the field appears to assume that the phrase SWB, happiness, positive and negative affect, and so on, are sufficiently distinct so as to justify the accumulation of vast bodies of empirical data. However, when the available literature is inspected (that is, from both a folk language or scientific literature perspective), it will become obvious that demonstrating that emotions are distinct categories is far more complex than what would seem to be true from inspecting the language I ordinarily use, and this will raise questions about how best to conceptualize well-being. What I will finally conclude is that SWB is not an emotion, in the classic sense that it is a brief intense expression that is usually attached to some event or object (for example, the loud noise that scares you, the winning lottery ticket that gives you joy, the good grades that made you happy), rather that SWB is a distinct phenomena, reflecting the constant monitoring of a person's affective status, that becomes conscious only under certain circumstances. Thus, I think of SWB as an experiential affective thermometer, and like the temperature of my body, I would not be aware of it until it reaches a threshold (e.g., my head gets hot and symptoms become obvious) that would indicate a change in my body or affective state.

Bramston (2002; p. 56; Table 11.6) has made an effort to distinguish various affective terms. In his original table, Bramston (2002) compared emotions, mood, and what he called "subjective quality-of-life," but I have modified the table so as to differentiate SWB from a quality or quality-of-life (Table 11.6) assessment. This table illustrates the general similarity between mood and SWB, but it also gave me the opportunity to describe, in more detail, the difference between SWB and a quality or quality-of-life assessment. This should reinforce the differences, noted in the introduction to this chapter, between a SWB and qualitative statement.

In Bramston's model (Table 11.6), SWB was assumed to be a persistent affective state. In contrast, Schwarz and Strack (1999) proposed that well-being is a momentary affective state. To quote them:

> Reports of subjective well-being (SWB) do not reflect a stable inner state of well-being. Rather, they are judgments that individuals form on the spot, based on information that is chronically or temporarily accessible at that point in time, resulting in pronounced context effects. (p. 61).

These context effects can lead to assimilation or contrast effects depending on the information involved, and each can alter the person's perception of their lives. They also point out that people compare themselves to others when deciding on the quality of their own lives. They also claim that mood is likely to affect judgments of general well-being, rather than the judgment of a specific life domain. Finally, they claim that public reports of SWB are often inflated owing to self-presentation concerns.

The Schwarz and Strack (1999) model is clearly an attempt to incorporate cognitive processes (e.g., assimilation and contrast effects) into what I would call qualitative judgments, but they label as "subjective well-being." What they have provided has the characteristics of the evaluated descriptor portion of a hybrid construct, and they appear to rely on the connotation of the term well-being to imply valuation, justifying the use of term well-being as a qualitative statement. However, they do not require a formal valuation assessment, and instead infer the presence of a valuation. This difference between inferring or "declaring" the presence of values by its assessment will be discussed in greater detail in Chap. 12.

The emphasis of the Schwarz and Strack (1999) model that a well-being statement is a product of a series of momentary determinants contrasts with the Bramston model (Table 11.6), where SWB is described as a relatively stable slow changing phenomena (e.g., a trait). Of course, as an umbrella term, well-being may semantically assume either use, but my task now is to determine which of these two approaches to well-being has supporting data. What I will learn is that both processes may be active, most likely together, as occurs during dual processing (Chap. 6). Thus, what Schwarz and Strack (1999) were describing would involve an automatic nonconscious process that with repetition would be transformed into a conscious established feeling called well-being. This cognitive model contrasts with the current most popular model of SWB that assumes that the stable part of a person's sense of well-being is primarily due to dispositional characteristics (e.g., genetics, personality).

3.2 Subjective Well-Being as a Persistent Affective State

Bramston (2002; Table 11.6), by assuming that SWB is a persistent affective state, was implicitly suggesting that the term has some of the temporal characteristics of a mood. However, inspection of the phenomenology of the terms reveals that well-being is less variable and appears to last for a longer duration, while a mood state is more likely to be cyclic. There is also minimal evidence that well-being can be systematically enhanced by pharmacological interventions, whereas this has been shown to be true for specific moods (e.g., depression or anxiety). SWB is not known to interfere with memory and decision making, but mood is known to have such effects (e.g., Kensinger and Schacter 2008). In addition, mood, via its impact on cognitive processes, is known to effect the expression of SWB (e.g., Schwarz and Clore 1983; Schwarz 1990), but the opposite does not seem to be true. In addition, it is usually necessary to ask a person about their state of SWB, while it is often possible to determine a person's mood by observing their facial expressions, body language, or behavior, as well as by asking them. Yet, SWB and mood are alike in that both can be evoked by particular stimuli and while a mood can change over time, a person's sense of SWB appears to remain stable.

One of the most frequently used methods to estimate the stability or persistence of an indicator involves correlating two measures of the indicator over a particular period of time. Table 11.7 summarizes a representative sample of these studies for different SWB assessments (positive affect, negative affect and satisfaction) assessed over varying lengths of time. Inspections of the table revealed moderate correlations that accounted for approximately 25% of the variance. To ask if a change in some indicator has occurred over a time period appears to be a straight forward question; however, its assessment turns out to be a fairly complex task. This is so because it is necessary to assume that the assessing instrument (e.g., a sensor, a person, and so on) remains stable between the baseline and follow-up assessment, so that just the effect of the experimental condition can be observed. Thus, a chemotherapy agent is expected to kill cancer cells, not the brain cells that a person need to respond to items about their life quality. It thus becomes important to determine if the "assessment instrument" was compromised, especially in longitudinal studies. Several other factors may compromise the assessment process (e.g., Hertzog and Nesselroade 2003); such as, the physical aging of the respondent, or changes in a person's emotional or psychological development. Life events, such as experiencing the Great Depression, or being in Vietnam could create a "cohort effect"; that is, a particular group may have had specific experience that gives them information that other respondents may lack. In addition, personality or genetic factors (Table 11.7; Lykken and Tellegen 1996; Nes et al. 2006) may account for some of the variance in well-being ratings over time.

Both cross-sectional and longitudinal studies have been reported examining the relationship between chronological age and subjective well-being (e.g., Schaie and Lawton 1998). A few studies have compared early to late adulthood (e.g., Diener and Suh 1998; Gross et al. 1997; Stacey and Gatz 1991), and while the results that have emerged are not consistent, they do seem to have established a number of findings. For example, the frequency of experiencing negative affect from early to late adulthood does not seem to vary. More

Table 11.7 The stability of subjective well-being indicators: representative studies

Reference	Sample (duration)	Indicator of SWB	Method and summary of results
Atkinson (1982)	2,162 (2 years)	3 items; including Cantril Ladder*	Satisfaction $r=0.41$ Cantril Ladder $r=0.40$ Happiness $r=0.39$
Ormel (1983)	296 (7 years)	ABS*	Stability; based on three time measures; ABS=73%; NA*=73%; PA*=53%
Costa et al. (1987)	4,942 (9 years) General population	10 of 18 items GWB*	NHANES*; baseline to follow-up; total $r=0.48$; stability; NA $r=0.43$; PA $r=0.44$
Lykken and Tellegen (1996)	127 pairs of twins (10 years)	WB scale from the MPQ*	79 MT*; 48 DT*; WB; $r=0.50$ 44–52% of variance accounted for by genetic variation
Suh et al. (1996)	115; 20–21 years old students (2 years)	SWLS*	Stability; NA $r=0.61$; PA $r=0.56$
Kunzmann et al. (2000)	201 (2–5 years)	PANAS*	Stability; NA $r=0.72$; PA $r=0.70$
Maitland et al. (2001)	678 (3 years)	ABS	Stability; NA $r=0.44$; PA $r=0.45$
Fujita and Diener (2005)	136 (17 years)	One item; life satisfaction	Satisfaction; $r=0.42$
Koivumaa-Honkanen et al. (2005)	9,679 (15 years)	Four Item Life Satisfaction Scale	Satisfaction; $r=0.30$
Nes et al. (2006)	4,322 (6 years)	Four Item Life Satisfaction Index	Males; $r=0.51$ Females; $r=0.49$ 80% of variance accounted for by genetic factors
Röcke and Lachman (2008)	3,631 (9 years)	One item; life satisfaction	Summary Statement: "the majority of individuals appraised their life satisfaction to be (high) and rather stable as opposed to increasing" (p. 838)

*Glossary: *ABS* Affect Balance Scale (Bradburn 1969); *Cantril Ladder* Cantril (1965); *DT* dizygotic twins; *GWB* General Well-being Scale (Dupuy 1984); *MPQ* multidimensional personality questionnaire (Tellegen 1982); *MT* monozygotic twins; *NA* negative affect; *NHANES* National Health and Nutrition Examination Survey (1973); *PA* positive affect; *PANAS* positive and negative affect schedules (Watson et al. 1988); *SWLS* Satisfaction With Life Scale (Pavot and Diener 1993); *WB* Well-being Scale

specifically, some studies find no or a small relationship between age and the frequency of negative affect (e.g., Diener and Suh 1998; Malatesta and Kalnok 1984; Shmotkin 1990; Stacey and Gatz 1991), while others have even shown that older adults may experience less negative affect than younger people (e.g., Charles et al. 2001; Gross et al. 1997). Also, Argyle (1999) and Diener and Suh (1998) find that age is associated with a reduced frequency of positive emotional experiences, especially if confounding factors are not partialed out from the data. Finally, life satisfaction is either nonsignificantly or only weakly positively related to chronological age (e.g., Diener et al. 1999; Myers and Diener 1995).

A number of different statistical models have been developed to deal with some of the complexities involved in characterizing the changes in well-being with aging. Charles et al. (2001), for example, report a study that monitored the changes in the Affect Balance Scale (ABS; Bradburn 1969) over a 23-year period. Their study involved several generations of persons, so that part of their assessment task was to determine how to take into account differences in initial age, even though the persons may have all been observed for the same study period. Thus, the initial age for one group may have been 23 and other 46, so that at the end of the observation period the respondents were 46 or 69. They used a linear trend analysis, specifically a growth curve analysis, to model their data. What they found was that; "Examining positive and negative affect separately revealed that age differences in well-being reflected both developmental and historical influences, but these influences vary according to the two types of affect (positive or negative) that comprise the overall measure of well-being" (Charles et al. 2001; p. 148).

They found that negative affect continued to decrease over time, and linear growth trends indicate a consistent decrease for younger and older adults, although the attenuation for older adults was markedly reduced. In contrast, positive affect was stable over aging, but started to decrease for older (60 years) adults. Thus, what seems to be happening here is an age related type of regulation of affect, where the rate of reduction in negative affect and stability of positive affect with aging were both being attenuated resulting in a more moderate or "realistic" world view. Consistent with this, it was not surprising to find that older adults were less likely to affirm items such as; "feeling on top of the world," or "that things were going my way."

Maitland et al. (2001) offer another perspective to the modeling task, again using the ABS. They were interested in whether the meaning of the term well-being remained the same for comparisons that varied across gender and age groups over time. To determine if assessment invariance existed they used methods developed from structural equation modeling. Thus, they state:

> measurement invariance of the ABS may be evaluated using methods described by Alvin and Jackson 1979, Horn, McArdle and Mason 1983, and Meredith 1993, among others. The sequence of testing measurement invariance involves comparisons of nested models of increasingly stringent invariance assumptions. Measurement models employed in the current paper include tests of configural invariance and metric invariance for the factor loadings. Horn and colleagues 1983, suggested that configural invariance requires only that the same variables load on the same number and pattern of factors for different groups. A model with configural invariance provides evidence of qualitative but not quantitative similarity across groups. In contrast, metric invariance involves equivalence of metric (unstandardized) factor loadings between groups or across occasions. Metric invariance establishes comparable units of measurement for the variables and the factors. It therefore allows for meaningful quantitative comparisons of groups or across time … (p. 71).

The results of their study supported the two factor (positive and negative affect) theory of psychological well-being, but they also found that the correlations between these two factors were statistically significant. They also noted differences in these factors when gender was considered. The stability of well-being assessments, by comparing Pearson correlations and factor correlations, were also determined. They found that the factor correlations, that provide a more definitive type of information, were only moderately stabile over the 3 years period of the study.

Overall, they found that their study demonstrated partial, but not complete, invariance over the 3 years of their study, but that only a limited number of items on the ABS contributed to this partial invariance. As a result they recommend caution in the use of the ABS, including interpretation of past research, due to contamination by age or longitudinal changes in item assessment properties in some but not all items on the ABS. Implicit in their interpretation of their data is the suggestion that if these issues are attended to, the ABS would remain a useful method of assessing well-being.

Eid and Diener (2004) have also developed a statistical model for dealing with longitudinal data, a model which assumes that each person (e.g., their personality) has a fairly stable pattern of responding, and that this trait (T) would be expected to be present at any assessment of a person's state. Factors specific to the assessment occasion (O) or the context of the assessment would also be expected to contribute variance to the assessment, as would errors from the assessment process (E). Thus, any measure of the state of well-being or mood would involve determining what part of the variance is attributable to stable characteristics of the person, those that are specific to the assessment occasion, and errors from the assessment process ($S = T + O + E$). They used structural equation modeling to develop a multistate-multitrait model, and used this model to estimate the magnitude of these sources of variance.

Using the SWLS (see Table 11.3) as their indicator of well-being, they applied the multistate-multitrait model to an analysis of a study that included three (every 4 weeks) measures of mood and personality. Unfortunately, they did not report a comparison between time intervals, but what they did find was interesting; they found that SWLS was more strongly related to mood at the trait level than on the occasion-specific level, suggesting that the interaction between well-being and mood was more likely to be a function of stable characteristics of an individual (e.g., their personality) then it was due to what may be going on at the time of assessing well-being (i.e., occasion-specific issues). This has relevance to Schwarz's "judgment model of subjective well-being" (see earlier).

Luhmann and Eid (2009) added another innovation to the tools available to assess the stability of well-being now focused on describing the impact of repeated life events on well-being. They took advantage of data from a yearly social indicator (e.g., job status, income, marital status) survey of German households that started in 1984 and continues today. Their study covered 24 waves of data that included the single question, "How satisfied are you with your life, all things considered?" Participants rated this question on a 10-point scale from completely satisfied to completely dissatisfied. The respondents were also administered a six item, seven point scale designed to assess personality (neuroticism and extraversion). Luhmann and Eid were interested in assessing how repeated experiences of unemployment, divorce and, as a positive indicator, marriage had on the respondents' assessment of well-being, and also examined the role of age, gender, and duration of the first life event.

They extended a statistical model first developed by Lucus et al. (2003) and used it to determine whether changes in a person's response to repeated life events became worse (labeled, sensitization), better (labeled, adaptation), or remained the same with repeated events. What they found was that repeated unemployment leads to sensitization (larger drops in well-being between events), repeated divorces leads to adaptation (less change in well-being), while repeated marriages had no appreciable effect on well-being ratings. Neuroticism, extraversion, and gender also contributed to interindividual variance, while age at the first event and the time between the first and second event were also significant contributors to variance in well-being ratings. In general, this study demonstrated the complex array of determinants of well-being stability.

An alternative to monitoring the stability of well-being in terms of social indicators is to conceptualize change in well-being as a result of various developmental processes. Ryff and her colleagues (e.g., Ryff 1989a, 1989b; Ryff and Keyes 1995; Keyes et al. 2002) have been developing just such an approach, in their effort to generate a model of positive aging.[15] Ryff and Keyes (1995) report the development of a questionnaire that they contend was measuring psychological well-being. A factor analysis of the results of this study reveals six factors: autonomy, environmental mastery, personal growth, positive relations with others, purpose in life, and self-acceptance.

Implicit in Ryff's efforts is the assumption that a psychological approach to well-being can be more, or as, informative as a social indicator approach. This argument, however, rests on the assumption that the six-factor model of psychological well-being was empirically reproducible. However, this does not seem to be completely supported by the current state of the research literature. A series of eight research reports (see Table 1 in Abbott et al. 2006; Kafka and Kozma 2002; Ryff and Singer 2006; Springer and Hauser 2006a, b) have examined the factorial structure of the Psychological Well-being Scale (PWBS) and all, with the exception of Ryff's studies, failed to confirm the original six factors. In addition, Abbott et al. (2006) has suggested that the four factors they confirmed could constitute a second-order factor; a "motivation" factor.

Many of the individual factors that Ryff included in her model can be used to monitor changes that occur to a person's sense of well-being and mood, as they age. The study by Maitland et al. (2001) is a good example of this type of study. For example, they found that there were age/time-related changes in the meaning of certain, not all, items on the ABS, which suggests that items of this sort should be removed from the assessment before a definitive answer can be given relative to question of the stability of well-being, or whether all that is being observed is a change in the meaning of the item with age (that is, a cognitive-linguistic effect).

This review has demonstrated that multiple factors determine well-being responding, some contributing while others distracting from the stability estimates, but when viewed from a total perspective could easily give the impression of stability over time. What remains unclear is to what extent these demonstrated changes in well-being are attributable to, or independent of mood changes. What has also not been clarified is the role the metacognitive process of reflection plays in determining the changes reported.

3.3 Subjective Well-Being as a Momentary Emotional State

A variety of techniques have been developed to monitor momentary behavioral and emotional states. Repeated application of these methods provides one approach to characterizing a complex system such as the emotions or well-being. Csikszentmihalyi and Larson (1987), Kahneman et al. (2004), Levin and Levin (2006), Mehl et al. (2001), and Stone and Shiffman (1994) have each developed procedures which perform this task using modern recording and prompting systems (e.g., a preprogrammed pager or wrist watch). After being prompted, respondents either fill out a questionnaire or verbally report their status on a recorder. This may occur several times a day. These techniques have the potential of reducing the confounding effect that errors in the recall of memories play in determining a person's current affective state.

However, memory limitations are not the only indicator that may confound a SWB assessment. As Schwarz and Clore (2007) point out, question comprehension, response formatting, or social desirability can also differentially affect momentary and retrospective reports. Schwarz and Clore (2007) illustrated the role that "question comprehension" plays by not only demonstrating how questions about being angry could evoke occasions that both produced momentary anger, but also recall of persistent feelings of anger, depending on the context created in the question. Thus, when someone was asked whether they were angry at any time over the last 6 months, what was found was that they gave answers that revealed major occasions of being angry and relatively minor annoyances, that may have occurred recently, were not referred to. Whereas, if they were asked to indicate what angered them yesterday, they were more likely to respond with the annoying little things in their life rather than the big things which occurred with less frequency. Thus, asking the same question in different contexts led the respondents to respond differently creating data that was essentially non-comparable (Bless et al. 2000; Winkielman et al. 1998).

As a corollary to the earlier demonstration, Schwarz and Clore (2007) point out that when a respondent is asked to describe a current or retrospective episode of angry, they may use different kinds of comparisons when making their ratings. For example, since an intense anger episode is more likely to be remembered it will more likely be what a respondent considers when retrospectively rating a particular past event. In contrast, when rating current episodes of anger, a respondent may still be aware of relatively moderate anger episodes and use these experiences when completing their rating task. Also important in this regard is what anchors are used in the rating items. This too can contribute to differences between concurrent and retrospective assessments.

The influence of social desirability implicit in an item also varies depending on whether the item refers to a current or retrospective event. However, social disability will be expected to have less of an influence on responding for current assessments than for retrospective assessments. If this is so, then Schwarz and Clore (2007) ask, whether social desirability increases with repeated assessment?

Again, the issues that Schwarz and Clore (2007) raise suggest that momentary and retrospective assessments do not differ just in terms of the memory-demand they create, but also because of how items are structured and what a respondent brings to the assessment task. This suggests that whatever a momentary valuation assessment provides it is different from a retrospective valuation.

Notwithstanding these complexities, momentary measures may be used to objectify a subjective indicator, such as, SWB, happiness, and so on. Kahneman (1999, with Krueger, 2006) has been particularly interested in this approach, and I introduced some of his ideas earlier (Chap. 4, p. 119). Here, my task is to see how the process of making subjective indicators objective can be applied to the concept of SWB or any of its analogs (e.g., happiness) and used to characterize SWB as a dynamic system. You may recall that Kahneman described the process of objectifying a subjective indicator as involving the integration over time of momentary responses to some specific situation. Applying Kahneman's objectification methodology to SWB includes the explicit assumptions that the brain continuously monitors a person's affective or hedonic state, and that a single indicator can express the value of this state. The first of these assumptions places him in the camp of those who feel that emotions and cognition are functionally separable processes (Zajonc 1980), while the second assumption implies that a single summary measure can adequately characterize a person's affective state. Let me briefly review Kahneman's (1999) arguments that support each of these assumptions.

First, Kahneman notes that the good–bad dimension is an empirically well-established (e.g., Osgood et al. 1957) evaluative dimension of human cognition. He also notes that, since Wundt (1904) and Titchener (1908), it has been recognized that feelings are a basic unit of a person's subjectivity which would ordinarily not be claimed to be a property of an object, and that one aspect of this subjectivity is that it occurs continuously and automatically. While these characteristics of subjectivity are akin to what has been referred to as the cognitive process of appraisal, Kahneman instead takes the position of those who argue that specific stimuli can evoke emotive responses independent of activating cognitive processes. He also cites neurocognitive evidence (e.g., LeDoux and Armony 1999) that supports the anatomical independence of emotional and cognitive functioning. Included in this evidence is a series of studies by Davidson (1992) which showed that positive affective states are associated with activation of the left anterior cortical regions, while negative affective states are activated by right anterior cortical regions. After reviewing a variety of studies Kahneman (1999) states:

> All these lines of evidence, from the introspective to the biochemical, point to the existence of a continuous evaluative process, which manifests itself in physiological responses at several levels, in expression of affect and in an immediate propensity to approach or to avoid. The continuous Good/Bad commentary is not necessarily conscious. When it is conscious, it is experienced as pleasure or distress, with a corresponding acceptance or rejection of the stimulus. The notions of acceptance or rejection imply that the GB commentary is associated with a disposition to respond both emotionally and instrumentally to unexpected interruption of an experience…The GB commentary has multiple physiological and behavioral manifestations that are potentially available for continuous measurement. (p. 8).

Kahneman's argument that a continuous physiologically-based evaluative process exists, provides the foundation for his argument that a single summary measure – a final common path – is able to capture a person's affective (e.g., SWB) state. However, his assumption that this final common path involves an evaluation brings me back to my previous discussion of whether it is logically possible to have an evaluation without first having to posit a description of a state. Thus, even if I accept the notion that an emotional process can be elicited independent of a cognitive process, I would still have to assume that the essence of a perceptual process (that is, a description) and the identification of a stimulus' properties would have to have occurred before the comparative process characteristic of an evaluation. What is common to both an emotive and cognitive process is that they have the same structural determinants; that is, they are both cortically mediated, although each may differ in the degree they are responsive to systemic neurohormonal influences. What makes them different is their anatomical (cortical) distribution, and degree of functional independence. Even so, it is hard to argue that each process is totally anatomically independent of the other, especially since each system may actively inhibit the other[16] (see Chap. 11, p. 438).

Alexandrova (2005) raises some additional questions about applying Kahneman's objectification approach to well-being. She points out that assessing well-being using measures of momentary affective states, precludes relevant pieces of information that can become available only upon reflection. This she says has the consequence of separating the person from what is being assessed, and therefore removes subjectivity from what is being assessed. To quote her (Alexandrova 2005);

> …this approach violates the main assumption of the subjective approach- that when judging happiness, the authority of how to weigh different aspect of our lives and experiences belongs to the subject. (p. 302).

Kahneman, of course, could argue that separating the person from what is being assessed is exactly what he wants to do, if the data he is gathering is to truly be called "objective." Alexandrova (2005), however, does go on to state that the objectification process has a place in the assessment of SWB, just that it should not be considered the first approximation of what is meant by subjective. However, if well-being requires reflection then this sets clear limits to the approach suggested by Kahneman.

In Chap. 3 (p. 66), I presented three different approaches (affordance, trait perception, and appraisal) to how fast and automatic cognitive processes may influence a qualitative assessment. They differ from Kahneman's approach in that they emphasize how a response may occur, whereas he postulates a continuous physiological process without specifying how this process leads to a response. However, neuroimaging studies may provide confirmatory information.

The argument that well-being is a continuous physiologically-based evaluative process that a person becomes conscious of when a certain threshold is reached is consistent with the formation of the evaluated descriptor portion of a hybrid construct, and Kahneman's argument that this is required for an objective assessment of well-being may be correct if all that he wants to accomplish is to assess SWB. However, if the investigator is interested in inferring that the momentary assessment is a qualitative statement, then they would be expected to include some indication of the significance of what has been described and evaluated. This, of course, requires a valuation.

The relevance of my concerns was confirmed by a study reported by White and Dolan (2009). They collected daily activities data on 625 persons who reported at least four activity episodes via a web-based assessment process. What they found was that the activities reported included activities such as driving to work, or playing with children, activities that may evoke feelings, but also could reflect purposeful activities, activities that may be absorbing, represent efforts to reach a personal goal, or involve connecting with others. They interpret these more complex activities as having "reward" value, even though they may generate relatively low levels of moment-to-moment affect. What they also did was to ask each participant to rate each episode of activity in terms of affective terms (happy, nervous/anxious, sad/depressed, and so on), but also whether they felt focused and engaged in these activities, if they felt competent performing these activities, and if these activities were worthwhile and meaningful. They also asked the respondents to indicate how satisfied they were with their activities, and the authors considered this question as reflecting the person's view of the importance of their feelings and thoughts relative to the activities they reported.

Limiting their analysis to single activity episodes, they subjected their 12 ratings to a principal-component factor analysis that resulted in a two (six items each) factor solution. They labeled their factors as reflecting more or less pleasure, or as more or less rewarding. They then plotted the individual activities on this two dimensional display either weighted or not by the duration of the activity. By weighting the activities, they were able to reduce the variability of the display, verifying the importance of Kahneman's notion of integration over time. Limiting my discussion just to the duration-weighted display, it appears that work, activities, exercise, and socializing were the most rewarding or pleasurable to persons, while watching television, resting or relaxing, self-care, or commuting were the least rewarding or least pleasurable. The authors conclude that "more evaluative, or thought based, assessments" (White and Dolan 2009; p. 1006) have a role to play in any assessment of SWB.

White and Dolan (2009) are actually aware that research on SWB has policy implications. They raise the question of how policy makers are going to make decisions when what is pleasurable and rewarding are very different. They see this difference playing out in two ways: first, a policy maker may prefer encouraging those aspects of SWB that contributes to economic or social progress, or second, discouraging those behaviors that lead to feelings of displeasure or are less rewarding.

What I found most interesting about this study was its demonstration that asking questions that encourage reflection or thought processes, may lead to viewing SWB from a different perspective. This is just what would be expected if SWB was qualified, which I believe was approximated by the additional questions that were asked. The notion that a pure hedonic approach to SWB may not be adequate has also been noted by Ryff (1989a) who raised the issue of whether a eudaemonic approach maybe a more adequate perspective. Clearly, there is more to learn here.

What needs to be done next is to clarify in what way a SWB statement is an affect. Is it, for example, a specific emotion, is it a product of a cognitive-emotional process, or is it the outcome of a cognitive-emotional regulation? I will address of each of these topics over the next three sections.

4 Subjective Well-Being and the Emotions

4.1 Introduction

As I previously stated, the term emotion can include several different types of information. Wierzbicka (1999; p. 2), for example, suggested that the term emotion can be characterized as a feeling, a thought, and a bodily experience. Others have divided this landscape differently, with Frijda (1999) speaking of subjective experiences, behavior, and physiological reactions as components of an emotion. Whatever way it is done, it is clear that an emotion would be expected to have several levels of representations, and that if a phenomena did not have evidence of these elements, then it is not likely to be an emotion. Thus, it can be asked whether there is a unique facial expression or a bodily response that corresponds to SWB.

A second set of issues has to do with Wierzbicka's (1999) claim that the presence of the word "emotion" in a language is culturally specific. For example, she states that while the term emotion is found in English and various romance languages, it is not found in German, Russian, or Samoan. Rather, what seems to be true for each of these languages is that certain basic level terms, such as, feelings are used instead of the more abstract term "emotion." Thus, the word for emotion in German is Gefüehl or to feel, and similar words can be found in Russian and Samoan (Wierzbicka 1999).[17] Of course, what can also happen is that the equivalent English word can be incorporated into a particular language, and in this way expand how the terms are expressed in that language.

Terms for well-being may also differ within, as well as between languages. English, for example, uses only one word for well-being, but this is not so for German. Well-being in German can be expressed as, Wohlergehen or, Wohlgefüehl (Cassell's New German Dictionary 1936) or, as part of scientific discourse as Wohlbefinden.[18] The word Wohlfahrt refers to social or public welfare. The prefix "Wohl..." refers to "good" so that, in general, the word Wohlergehen means "good welfare, Wohlfahrt means "good journey" or "good passage," Wohlgefühle means "good or pleasant feelings," while Wohlbefinden means "good health."

Another issue to consider deals with the pattern of language usage for emotion terms. Thus, the term "to feel" (e.g., Gefühle) may allude to "feeling happy" or, as a thought to "feeling ashamed," without referring to bodily processes. Alternatively, the phrase "feeling hungry" may have a clear physiological reference. Thus, carefully determining the reference of the emotion, as was demonstrated earlier for the term well-being, will be an important part of understanding the meaning being communicated.

Wierzbicka (1999) summarized her concerns by stating; "Thus, while the concept of 'feeling' is universal and can be safely used in the investigation of human experience and human nature...the concept of 'emotion' is culture-bound and cannot be similarly relied on"(p. 4). This also suggests that if I want to make culturally independent statements then I best rely on basic level terms, such as "to feel" when referring to emotions.[19]

Applying this to my task I would have to ask what these basic level terms (Chap. 3, p. 56) might be for the phrase SWB? It is not obvious that SWB has a unique basic level reference; many basic level terms can be used to express SWB. In addition, SWB, not unlike what was found for the term health status (Chap. 8), may function at both an abstract and more concrete (or basic) level of information, depending on how the investigator or respondent uses the term. This, of course, has implications concerning the usefulness of viewing SWB as a unique emotion, especially since specific emotions (e.g., anger) usually have a limited number of associated terms.

In the subsequent portions of this section I will consider three different approaches to conceptualizing emotions and by extension, SWB. This will include continuing my discussion concerning whether it makes sense to think of SWB as a discrete emotion, or whether a dimensional approach to emotion is a more appropriate way of conceptualizing emotions and SWB. In addition, I will review a system approach to emotions and SWB, in which emotions are viewed as a product of a continuous and recursive appraisal process. I will also review several different empirical approaches to forming emotion categories, as well as discuss various models of emotions that can be used to conceptualize SWB. I will then discuss the interaction between emotions and cognitions, and what impact this has on the evaluative role of SWB statements within a quality assessment. This discussion should also be helpful in clarifying the contribution of emotions to the hybrid construct. I will end that section by discussing the role that appraisal plays in the assessment of SWB, quality, or quality-of-life.

4.2 Is Subjective Well-Being a Discrete Emotion?

Is SWB a discrete emotion much like happiness, anger, and fear appears to be? I ask this question, since if SWB is a discrete emotion, then it would consist of a unique set of indicators that do not overlap with other emotion types.[20] Some investigators, (e.g., Izard 1977; Johnson-Laid and Oatley 1989) have even suggested that a small set of irreducible emotional categories, or *qualia* exist, that presumably would include the phrase SWB. Evidence in support for these categories or *qualia* comes from the universal perception of facial expression and certain language phenomena.

Part of the reason why emotion and SWB terms appear to be discrete categories or domains is because the denotative meaning associated with each term permits the clustering of these terms into categories or domains. For example, Berlin and Kay (1969) demonstrated that it was possible to take words associated with different colors (Chap. 3, p. 56) from different cultures and organize them into a limited number of prototypical domains that have a physiological reference (reference to the color spectrum). In addition, the research that suggested the existence of universal facial expressions of emotion (e.g., anger, fear, happiness, and so on), and its associated word sets, reinforce the impression that emotion terms are physiologically-based prototypes (Ekman 1992).

The syntactic form of emotion words also contributed to the view that emotions are discrete categories. Thus, in English, most emotion words are nouns, as is true for the term well-being. This reinforces the view that emotions are things as opposed to processes, with emotion words acting as labels for these things. Interestingly, in Japanese, emotion words syntactically are adjectives, while White (2000) points to a group of Solomon Island natives whose emotion words function as verbs.[21] These differences in syntax do not seem to affect the meaning associated with the terms.

It also should be noted that in ordinary discourse and scientific studies, well-being and emotion-based terms, such as happiness, are regularly interchanged, and that this interchange still leads to effective communication; much as Wittgenstein would have predicted. This does not occur, however, by a simple substitution of one term by the other. Rather, SWB statements differ from happiness statements by being more abstract than concrete, and by functioning as a domain indicator (that is, label) rather than as a concrete indicator. Still, it is easy to find instances where the differences between these indicators are confounded. Part of the reason this occurs, of course, is because descriptive emotion or SWB statements can assume different semantic roles. Thus, when emotion or SWB statements are part of an interrogative sentence, or when the connotative meanings associated with these terms are elicited, then both terms may assume an evaluative role, as expressed figuratively.

Examining the etymology of the word emotion reveals that the word evolved from the Latin, e+movere, which means "a moving out," "migration," "transference from one place to another" (The Complete Oxford English Dictionary 1970; Vol. IV, p. 124). These early definitions have a clear physical reference, but by the middle of the eighteenth century, the term emotion had assumed various figurative forms, such as, any disturbance of the mind, feeling, passion, or any vehement or excited mental state (The Complete Oxford English Dictionary 1970; Vol. IV, p. 124). For most of Western intellectual history, however, the term emotion has been defined from three major perspectives; feelings, aesthesis,[22] or passion. Feelings, as discussed later, remain the major synonym for emotion, but the term passion (with its assumption of irrational responses that I have little control over) has also been a major form for the expression of emotion. The term passion comes from the Greek word *pathe* which means to suffer, leading to such emotional terms as "pathetic," "empathy," and "antipathy." Averill (1990) comments that in its original Greek and Latin forms, the term "passion" had very broad connotations, referring to objects, animate or inanimate, that were suffering because of some external action. The connotation of passion, or *pathe*, has had a major impact on the development of emotion metaphors, especially those dealing with sensations, feelings, and diseases.

To determine if SWB is a discrete emotion, I will examine two issues; does SWB occupy the same semantic space as emotion terms, and second, does SWB and emotion terms have the same cognitive characteristics. I will approach my first task by inspecting various dictionaries or lists that define emotions (e.g., Fehr and Russell 1984; Kleinginna and Kleinginna 1981; Plutchik 2001; Shaver et al. 1987; Sweeney and Whissell 1984) to determine if well-being is one of the referents. These lists vary in length with the list identified by Sweeney and Whissell (1984) consisting of as many as 4,500 English words. When these dictionaries or lists were inspected, however, the term well-being was rarely found, and if found was not listed as a primary emotion (e.g., when a group of persons were asked to list emotion terms they rarely cited "well-being" as a type of emotion; e.g., Fehr and Russell 1984).

Clore et al. (1987) report a study in which they asked a group of students to rate a list of 585 emotion words in terms of whether these word referred to feelings or a state of being (e.g., feeling angry or being angry). Well-being was again not one of the words in their list of emotion terms, but the word "well" was listed as a term. However, when they graphed different clusters of emotion terms, they did so along the dimensions of "feeling" and "being."

Brandstätter (1991) reports a series of studies where subjects freely identified adjectives that described their moment-to-moment emotional experience. Hierarchical cluster analysis was used to organize these adjectives, but well-being was not listed as a separate category, nor was it found among the items used to describe eight categories (e.g., joy, relaxation, activation, and so on), plus a "residual" category.

A second method involves determining if a unique semantic space could be constructed for well-being. This approach utilized multidimensional scaling methods to determine where happiness, anger, and possibly well-being, fit in such a space (see also Kövecses 1991) with the intent of determining if discrete emotions and SWB formed a hierarchy. Davitz (1969) reports a study in which he assumed that it would be possible to use semantically simpler terms to determine the category of meaning of a particular emotion. To test this assumption he first collected 556 emotion-based statements. I found that at least 11 of these 556 statements appeared to address some aspects of well-being (including the statement; "a sense of well-being"). He then asked a sample of persons to match these statements with 50 particular emotions (well-being was not one of the 50 emotion terms used in the matching exercise). What he did find was that the statement, "a sense of well-being," was associated with 15 of the 50 separate emotions. Davitz listed all of the 556 statements that were associated with a particular emotion, and reported the frequency with which they were associated with a particular emotion. Seven of the 15 emotions that listed the statement "a sense of well-being" reported the statement as the second or third most frequent statement attached to the emotion. These emotions were: confidence, contentment, delight, enjoyment, gratitude, happiness, and serenity. Interestingly, these terms could also function as synonymies of SWB.

The matching procedure Davitz used provided him with a first approximation of the meaning associated with a particular emotion word. Using cluster analysis he found that the 556 statements could be organized into 12 clusters types. These clusters were named; activation, hypoactivation, hyperactivation, moving toward, moving away, moving against, comfort, discomfort, tension, enhancement, incompetence or dissatisfaction, and inadequacy. Individual emotion could now be defined by reference to these clusters. For example, anger was defined as hyperactivation, moving against, tension, and inadequacy. Davitz also statistically demonstrated (using factor analysis) that larger dimensions could be identified from these 12 clusters. He identified four such dimensions; activation, relatedness, hedonic tone (or valence), and competence. Applying these dimensions to the term anger, the "hyperactivation" cluster would be placed on the activation scale, "moving against" on the relatedness scale, "tension" on the hedonic tone scale, and "inadequacy" on the competence scale. Thus, these four dimensions not only represent a meaning space, they also represent a space that can distinguish between emotions.

Of particular interest is the hedonic tone dimension. This dimension consists of clusters dealing with comfort, discomfort, and tension. Inspecting each of these clusters reveals that the second most popular indicator in the comfort cluster was the statement, "a sense of well-being." This study, therefore, makes it clear that while people do not consider SWB a primary emotion, they clearly use the phrase to describe an emotional state. These data also demonstrate that it is possible to arrange the emotions into a hierarchy of increasing abstraction.

A third approach considers the nature of the cognitive foundation of the term well-being. As previously mentioned, the process of forming a concept also involves the classification of the elements that make up the concept, and this class of elements can define a category. A concept, such as SWB, can potentially be organized in several different ways; for example, it can be a natural kind, an artificial class, or a class of artifacts. I will next consider if the term well-being is *a natural-kind*, since happiness, anger, joy, and so on, have been assumed to be natural kinds. And if SWB is an emotion, or is composed of various emotions, then you would expect it to also be a natural kind (or natural class).

First, some definitions; *a natural kind* differs from an artificial class or artifactual class, in that it involves a classification that occurs naturally,[23] without applying some formal classification system. An artificial class differs from a class of artifacts, in that an artificial class involves applying an external classification system onto a set of entities (as when I label a factor following a factor analysis), while a class of artifacts represents a grouping of entities that have a common property (an example would be the grouping of a chair, table, sofa together as furniture). Now what I would like to do is determine whether the umbrella that Diener (2006) uses to describe SWB is formed naturally, artificially, or consists of a class of artifacts. I will do this by evaluating the extent emotion terms fits one of these classes, and infer that what I find to be true for emotion terms must also be true for SWB statements, especially if SWB is a discrete emotion.

Barrett (2006) describes emotion natural kinds, as follows:

> Certain emotions (at least those referred to in Western cultures by the words 'anger', 'sadness', 'fear', 'disgust', and 'happiness') are given to us by nature. That is, they are natural kinds, or phenomena that exist independent of the perception of them. Each emotion is thought to produce coordinated changes in sensory, perceptual, and motor, and psychophysiological functions that, when measured, provide evidence of the emotion's existence. (p. 28).

She also states:

> In everyday terms, a natural kind is a collection or category of things that are all the same as one another, but different from some other set of things. These things may (or may not) look the same on the surface, but they are equivalent in some deep natural way. In the most straightforward philosophical sense, a natural kind is a nonarbitrary grouping of instances that occur in the world. This grouping, or category, is given by nature and is discovered not created, by the human mind. In a natural kind category instances cluster together in a meaningful way because they have something real in common. (p. 29).

Of course, the claim that a natural kind is discovered not created is another way of expressing the difference between a category and a domain.

There are many examples of natural kinds. Giraffes, tigers, pine trees, and so on, and they represent classes of objects that are unique and not easily confused with each other.[24] These natural kinds consist of fairly concrete entities, and it may not be surprising that these entities form natural groupings. But I am dealing with more abstract emotion-based material that can be represented at several different levels of analysis (e.g., facial expressions, autonomic nervous system responses, behaviors, feelings), so that it is appropriate to wonder if these SWB and emotion terms can be grouped together to form natural kinds. What I need to know is how and on what basis these entities form different classes. As Barrett (2006; p. 29) states, "In a natural-kind category instances cluster together in a meaningful way because they have something real in common." But what is this "meaningful way" and what is "real" that they have in common? Keil (1995) helps with this issue by stating:

> Natural kinds may cohere as meaningful classes of things just because they reflect mutually supportive networks of causal relations among their properties. Properties are highly correlated because, either directly or indirectly through a causal chain, they tend to support the presence of each other. (p. 236).

Keil (1995; p. 237) provides five examples of causal interactions, one of which is maintained by the presence of homeostatic mechanisms. He goes on to state that the combination of causal patterning and homeostasis may be what is required to create a cognitive category, but he also mentions that the category formed has an essence that gives it its character. But what is an essence, particularly a psychological essence? This is a question which I have previously discussed (see, Chap. 3, p. 69; Chap. 8, p. 256), but let me quote Medin and Ortony (1989):

The point about psychological essentialism is not that it postulates metaphysical essentialism but rather that it postulates that human cognition may be affected by the fact that people believe in it. In other words, we are claiming only that people find it natural to assume, or act as though, concepts have essences. (p. 184).

The key phrase here is that "people find it natural to assume, or act as though, concepts have essences" and if I apply this statement to my task, I would have to say that "people find it natural to assume, or act as though, subjective well-being has an essence," an essence that distinguishes it from other terms and sustains its usefulness as a cognitive category, or statistical domain.

Returning to Barrett's (2006) paper, she anticipates her conclusions by stating:

> … however, I suggest that the natural-kind view of emotion may be the result of an error of arbitrary aggregation … That is, our perceptual processes lead us to aggregate emotional processing into categories that do not necessarily reveal the causal structure of the emotional processing. I suggest that, as a result, the natural-kind view has outlived its scientific value and now presents a major obstacle to understanding what emotions are and how they work. (p. 29).

Barrett (2006)[25] based her conclusions following an examination of the available research literature. One set of data tested the assumption that emotions exist as natural kinds by determining if a series of distinctive responses occur, coordinated in time and correlate to intensity. Thus, if an emotion, such as anger, was a natural kind then a feeling, a thought, and a bodily experience (e.g., a facial expression, autonomic responses, subjective reports) would all be expected to be highly intercorrelated. A review of the appropriate literature, and here I rely on Barrett's (2006; p. 33) summary, reveals moderate to small correlations between these indicators, but there were also few instances where more than two of these indicators were simultaneously reported. When these correlations were high they were usually associated with broad affective dimensions, such as valence or intensity (Lang et al. 1993). After reviewing this literature Barrett states:

> Taken together, enough evidence has accumulated for some theorists to conclude that lack of response coherence within each category of emotion is empirically the rule than the exception (Bradley and Lang 2000; Russell 2003; Shweder 1993, 1994). Despite this evidence, however, the science of emotion proceeds as if facial behaviors, autonomic activity, and the like configure into signature response clusters that distinguish particular kinds of emotions from one another. …an equally plausible explanation is that scientists have failed to observe stable and reliable response clusters because they are not really there. Projectable property clusters may not exist because emotions may not be natural kinds. (p. 34).

Izard (2007) criticizes Barrett's (2006) views pointing out that it is important to keep separate the difference between basic emotions and emotion schemes. Izard distinguishes basic emotions from emotional schemes by pointing out that basic emotions can be characterized as having a strong evolutionary background, whereas emotional schemas are a product of developmental and ongoing cognitive (e.g., appraisal) processes. The basic emotions would be examples of natural kinds, but Izard's distinction is of little value in the current context, since both well-being and quality-of-life are predominantly a product of cognitive processing. Even if I accept Izard's model, once the presence of basic emotions have been processed by life circumstances, they more closely resemble emotional schemas than basic emotions.

If emotions are not likely to be natural kinds what could they be? Barrett (2006) suggests that they may be emergent phenomena, emerging from the unique manifestations of an individual's experience of emotion. I would add that they may also represent an investigator's willingness to attribute an essence to these emergent phenomena, giving the term the appearance of a substantive entity, which may be more figurative than literally true. However, I would expect that what has been found for emotion terms would also be true for SWB. This suggests that emotion terms are more likely to be either artificial or artifactual classes.

If SWB is an artificial class, a class or domain of entities that are based on some specifiable inclusion or exclusion rule, then I should be able to identify these rules. The most obvious "rule" or defining characteristics would involve including all emotion terms in the class. An alternative approach is to identify a unique set of facial expressions, autonomic nervous system responses or behaviors that would define a particular emotion term, and to use these patterns as inclusion or exclusion rules. What the data suggest, say for facial expressions, is that a limited range of facial expressions for a particular emotion exists, but that individuals vary within this range (e.g., Ekman and Friesen 1975). Thus, two people may be perceived as smiling, but one has a broad full-facial expression and another has a smile that is closer to a sneer. Thus, if these data are interpreted as establishing a rule, then it is a "noisy" rule.

An alternative is to establish that each emotion term is cognitively equivalent to the other. This would be appropriate if these terms were to cluster into a hierarchy characteristic for an abstract concept, such as SWB. If happiness and satisfaction were assumed to be generated by the same cognitive processes, then a common criterion would be identified. As I have previously discussed, satisfaction is a complex cognitive product (Chap. 8, p. 264), while happiness is best characterized as an evaluated descriptor. From this, I conclude that it is hard to argue that SWB is an artificial class. However, this does not mean that SWB as an abstract construct does not have a neurocognitive representation, as I will discuss shortly (e.g., Davidson 2004).

What I am left with than is to consider whether SWB is either a class of artifacts, or some other yet to be described type of category. To consider SWB a class of artifacts would still be consistent with Diener's (2006) view that SWB is an umbrella term (Fig. 11.2), but would also lay itself open to the criticism that the resultant domain runs the risk of being meaningless.

Diener et al. (2003) have suggested that this class of artifacts would consist of (Fig. 11.2) four major components: pleasant emotions, unpleasant emotions, global life judgments, and domain satisfaction. Assessments of each of these components involve different items or assessment methods, and their outcome would be expected to be expressed as a profile. Correlational studies suggest these indicators are moderately related, and it assumed that they will have a common conceptual foundation (Diener et al. 2003; p. 191). What remains unclear is how these indicators can be combined, or how they can be organized to infer the abstract construct of SWB. Of course, investigators may not see the need to account for how an abstract construct is formed. There are two approaches to this issue: a top-down or bottom-up perspective.

In the top-down perspective, each combination of indicators would be considered equally as valid. The bottoms-up approach, however, would require a set of rules about what indicators should be combined to give an adequate picture of what abstract construct, SWB precisely means, and this would depend on what is currently being assessed.

As illustrated in Fig. 11.2, Diener et al. (2003) seems to be taking the perspective that a series of entities exist that are related by a common characteristic or denominator which becomes progressively more abstract, and inclusive. Thus, joy would be considered an example of a pleasant emotion, while positive emotions would be considered one manifestation of SWB. The common denominator between each level of analysis refers to an affective state; a feeling about an emotion not the emotion directly. The same would apply to the sadness ➡ unpleasant emotions ➡ subjective well-being hierarchy, but the global life judgments and satisfaction domain hierarchies do not seen to have the same common denominator as the affective domains. Both the "global life judgments" and the "domain satisfaction" columns differ from the affective domains in that they focus on judgments, but also differ in that they include objective as well as subjective indicators (e.g., fulfillment measured by the model car the person owns, success measured by the person's salary, domain satisfaction measured by years of marriage, doctors visits, money spent on leisure activities, and so on, each of which can function as an objective indicator). Of course, these indicators can be "subjectified" or valued, but there was no indication that this was considered a necessary component for membership in the domains that made up the hierarchy. For Fig. 11.2 to depict a hierarchy it would have to have a common cognitive determinant within and between each of the dimensions, and since this is clearly not the case I am left to conclude that the model presented is most likely a cross-classification of information based on the common goal of characterizing the different domains of subjective well-being. This is consistent with a domain of heterogeneous artifacts.

One alternative to viewing SWB as a discrete emotion is to characterize SWB in terms of affective dimensions (e.g., valence and arousal). Before I do this, however, I want to consider another possibility, and that is, that SWB is a figurative form of emotional expression, rather than literally being a term for a discrete emotion. This would be consistent with the observation that SWB can be applied in multiple settings, each of which has a distinct meaning. For example, economic well-being has a very different meaning than psychological well-being, yet both allude to the CONTAINER conceptual metaphor with its use referring to a quantity of some entity. In order for SWB to be a metaphoric emotional expression, I have to first demonstrate that emotion statements have metaphoric properties.

4.2.1 The Metaphoric Nature of Emotion Statements

Averill (1990) claims that there are five major metaphors for the term emotion, each of which may evoke the image of motion, or e-motion. These metaphors deal with inner feelings; bodily responses, especially the gut; the animal in human nature; diseases of the mind; and driving forces. The listings that follows below provides examples of basic level metaphors found in ordinary discourse for each of these five abstract emotion metaphors (Averill 1990; see Table 3). These sentences may also be viewed as examples of the type of items that could be used in a SWB assessment.

Emotions are Inner Feelings
He "felt" his anger rising.
Her kindness "touched" him deeply.
He "listened" to his hear, not his head.
Emotions are Physiological Responses:
He had a "gut reaction".
Anger made his "blood boil."
She got "cold feet".

Emotions are the Animal in Human Nature:

Don't be a "brute."
She acted like "an animal."
He "subdued" his fear.
Emotions are Diseases of the Mind:
He was "insane" with rage.
She fell "madly" in love.
He was "paralyzed" with fear.
Emotions are a Driving Force or Vital Energy:
He was "driven" by fear.
Love "makes the world go'round."
He could not "reign in his anger."

According to Averill (1990), these classes of metaphors can also represent different approaches to the study of emotion. Thus, considering emotions to be inner feelings would lead to phenomenological studies, while characterizing emotions as a physiological response would lead to psychophysiological studies. In addition, studies of an animal's nature can lead to studied using ethological methods, diseases of the mind might use psychodynamic methods, while emotions as driving forces or vitality could use models of drive reduction.

Averill (1990) views these five emotion metaphors as abstractions that guide, but are not found in ordinary discourse, while the sentences used to illustrate these categories constitute a basic level of information that is commonly exchanged. He considers metaphors to have a number of specific functions; description (or elaboration), explanation, and evaluation. This suggests that the metaphors allow a term to assume explanatory and evaluative functions. Since this has implications for what I have described as a hybrid construct, I would like to discuss it in a bit more detail and then relate this to the role emotional metaphors play in SWB statements. As Averill (1990) states:

> Metaphors serve two main functions beyond mere description or elaboration. Those functions are explanation and evaluation. Explanatory metaphors are concerned primarily with the transfer of knowledge from the target to the source domain. Evaluative metaphors, by contrast, are intended to convey attitude or mood. Phenomena that call for both explanation and evaluation are especially likely to become a source and/or target of metaphor.
>
> At the core of any emotion is an evaluative judgment. For someone to be angry, frightened, sad, in love, disgusted, proud, and so forth requires that the situation be evaluated in a certain way as good or bad, as beneficial or harmful, as just or unjust, as beautiful or ugly, and so forth. Because of this, emotions are a rich source of evaluative metaphors. However, emotions are also the object of value judgments, and hence the target of evaluative metaphors. Consider, for example, the characterization of emotions as "diseases of the mind." It has an explanatory function (e.g., emotions can disturb orderly thought processes, just as diseases can disturb orderly physiological processes). However the metaphor is clearly evaluative. Emotions are "unhealthy." (p. 106).

In reviewing these comments it appears that Averill seems to be taking advantage of the connotative[26] function of emotion terms. Thus, a happiness statement, besides describing the degree of happiness the person felt, can also be evaluative, in that happiness connotes good, beautiful, just, and so on (Osgood 1969). What may be less obvious is that these evaluations cannot occur without first implicitly or explicitly describing what is being evaluated; that is, it is logically impossible to evaluate something that has not been previously described.[27] This suggests that evaluations or explanations are cognitive add-ons to what is, at first, a fundamentally descriptive perceptual-cognitive process.

Averill goes on to point out that just as emotion statements can be mapped to form an evaluative metaphor (e.g., "diseases of the mind"), the resultant metaphors can feedback and impact emotion statements (e.g., "emotions are unhealthy"). This imposing of a valuation on an emotion is similar to the cognitive processes that lead to the construction of a hybrid construct, which would also require application of a separate cognitive process. However, the type of mapping Averill described differs from what would be expected for a quality or quality-of-life assessment in that the mapping is dependent of the emotions being assessed, and this is only occasionally true for a qualitative assessment. For example, to be angry implies that the person is in an undesirable state, but this evaluation would become a quality-of-life statement only if the person views this statement in terms of whether their anger was of importance to them. To paraphrase; "It is possible to evaluate a situation and be angry about it, but it is also possible that the person recognizes that the source or nature of their anger is not sufficiently important so that it does not affect their quality-of-life." Again it is the application of this independent valuation criterion (that is, a metacognitive process) that creates the qualitative hybrid construct.

What is also not clear from Averill's (1990) presentation is whether his notions have any empirical support, other than his argument that since these various connotations are part of folk language they, therefore, must occur in ordinary discourse. He does go on to state that the evaluative function is usually less evident at the abstract or superordinate level than it is at the more basic level of information. From this discussion I conclude that since SWB may very well be an emotional expression, and emotional expressions can assume metaphoric characteristics, that, therefore, SWB statements may have metaphoric properties.

Before I discuss the next topic I would like to briefly review a paper by Meier and Robinson (2005), that summarizes more recent research on the metaphoric properties of affective terms. What they have found was that evidence exists that supports a role for brightness and vertical position, but less empirical support for distance conceptual metaphors. Next I would like to consider whether SWB can be characterized in terms of affective dimensions (e.g., valence and arousal), as opposed to the categorical approach of viewing SWB as a discrete emotion.

4.2.2 On the Nature of Affect

To consider the dimensional nature of SWB I first have to first discuss what the term affect means. Some investigators use affect and emotion interchangeably, others use it to refer to the experiential or behavioral components of emotion, while others think of affect as a superordinate category including emotions, emotional episodes, dispositional states, moods (such as, depression), and a person's emotion-based personality traits (Clore et al. 1987; Gross 1998; p. 273). Russell (2003), however, provides a model and definition that shifts the definition of emotion to a dimensional rather than a categorical approach. His first task was to define what he calls "core affect," since he wanted to identify basic units that would not be confounded by cognitive processes. Based on the suggestions of Oatley and Johnson-Laird (1987), he identifies two basic units or primitives: pleasure or displeasure (labeled as pleasure or valence) and activation and deactivation (labeled as arousal or energy). Thus, valence and arousal define core affect. Russell (2003) states:

> Core affect is that neurophysiological state consciously accessible as simplest raw (nonreflective) feelings evident in moods and emotions. It is similar to what Thayer (1989) called activation, what Watson and Tellegen (1985) called affect, what Morris (1989) called mood, and what is commonly called a feeling. At any given moment, the conscious experience (the raw feelings) is a single integral blend of two dimensions, hence describable as a single point… The horizontal dimension, pleasure-displeasure, ranges from an extreme (e.g., agony) though a neutral point (adaptation level) to its opposite extreme (e.g., ecstasy). …The vertical dimension, arousal, ranges from sleep, then drowsiness, through various stages of alertness to frenetic excitement. (p. 148).

Russell (2003) visualizes these two dimensions as being part of a circumplex (Fig. 11.3),[28] with terms that represent different combinations of valence and arousal filling the circumference of the circle. Some of the terms are emotional (e.g., elated, upset, depressed), while others are not (tense, calm, serene, contented, fatigue, and so on). The terms serene and contented are of particular interest, since they come close to what SWB might mean (e.g., the synonyms of SWB include comfort and contentment). These terms are situated on the circumference as having positive valence (that is, a form of pleasure), but are below the neutral point along the arousal dimension (Fig. 11.3). In addition, happy and elated are also situated in the positive valence dimension, but involves a level of arousal that is above the neutral well-being and ill-being (Fig. 11.3). One of the conclusions that can be drawn from this way of conceptualizing SWB is that since it consists of several different types of affect, SWB should be considered a cognitive construction and this is consistent with its classification as a category of artifacts, with each artifact being a different measure of affect.

Core affect, as Russell has indicated, are states which are experienced as feelings of good or bad, with more or less arousal. Core affect influences reflexes, perceptions, cognitions, and behaviors. Their influence occurs via various internal and external processes that people cannot directly access. It is only when core affect is attached to some object or experience that its presence becomes known.[29] The list of cognitive processes that associate core affect with objects or experiences is summarized in Fig. 11.4. Since their involvement appears to be required to become aware of a specific emotion, I have listed these associations as causative.

Of the list of cognitive processes depicted in Fig. 11.4, "affective quality" is of particular interest. Russell claims that

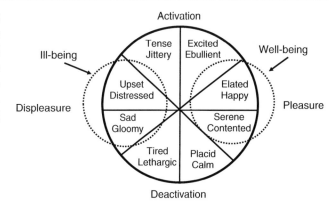

Fig. 11.3 Core affect is displayed as a circumplex along the dimensions of pleasure–displeasure (valence) and activation–deactivation (arousal). "Secondary" affective concepts of well-being and ill-being are proposed. Based on a figure in Russell (2003).

all objects or events that are experienced have qualities attached to them, just that some have affective and others nonaffective qualities. He considers the perception of affective quality to be a ubiquitous and elemental process that other investigators have described with such terms as; evaluation, automatic evaluation, affective judgments, appraisal, affective reaction, and primitive emotions (Bargh 1997; Cacioppo et al. 1999; Zajonc 1980). Affective qualities are properties of the stimuli I perceive, and remain attached to these objects or events. Russell suggests that affective quality and core affect should be viewed as two primitive processes that can change independent of the other. Thus, core affect can change without the influence of external stimuli, and I can perceive affective quality without a change in core affect. This, of course, is a very interesting statement, since Russell sees these primitives as independent entities that can be operated on, much as has been postulated for a hybrid construct. Also of interest is his statement that "To perceive affective quality is to represent rather than to experience core affect" (Russell 2003; p. 149). Claiming that a qualitative indicator "represents" rather than expresses core affect, gives affective quality a level of abstraction that distinguishes it from core affect. He concludes by stating that core affect and the perception of affective quality, whether combined or not combined with nonemotional processes, accounts for most of what a person experiences.

Much of what Russell has described is helpful if I want to describe how an aesthetic quality judgment occurs. For example, Russell describes the construct of "attributed affect" as occurring when a change in core affect is linked to its perceived causes (such as, a person, place, event, physical object or life history). As he states:

> Attributed affect is thus defined by three necessary and, when together sufficient features: (a) a change in core affect, (b) an Object, and (c) attribution of the core affect to the Object. The Object potentially includes the full meaning and future

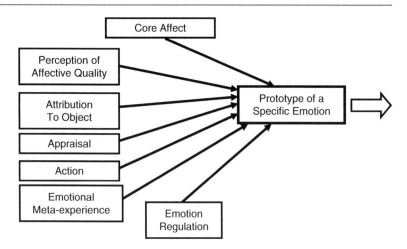

Fig. 11.4 Multiple determinants of a specific emotion as conceptualized by Russell (2003). The figure is a modification of Fig. 3 from Russell (2003).

consequences of that event and has a perceived affective quality. Attribution is the perception of causal links between events and allows room for individual and cultural differences. Attributions usually seem correct to the attributor, but research has demonstrated misattributions. Attributed affect has two functions beyond those of core affect. First, it guides attention to that functions beyond those of core affect. First, it guides attention to and behavior directed at the Object. Second, it is the main route to the affective quality of the Object.

Despite this complex definition, attributed affect is phenomenologically, simple and very common: afraid of the bear, feeling sad at a loss, liking a new tune, feeling uncomfortable from the heat, feeling sympathetic to a friend's woes, and so on. Put more generally, attributed affect covers many topics, including affective reactions, liking, displeasure motives and empathy. (p. 149–150)

Clearly, attributed affect would be one element of a cognitive-affective process (see further) that would help account for how aesthetic quality judgments occur. Thus, looking at a painting a viewer may say; "I like this painting" and this would be considered an example of attributed affect, in that the emotion of liking was attached to the object, a painting, or alternatively the object could be a meal, an event, an experience, or a life. In addition, attributed affect may also be helpful in understanding what has been described as autonomous or automatic processing (Table 6.1); responding that presumably occurs unconsciously. In folk language this type of responding would be called "a gut response," with its implications of being fast and coming from inside the person. Assuming that core affect is a propellant would be consistent with the notion that a gut response is a type of automatic responding. In contrast, controlled processing, that is a more time dependent process, would be expected to involve more cognitive, as opposed to affective, processing.

Summarizing the discussion to this point, I believe I can say that it is inappropriate to argue that SWB is, in and of itself, a discrete emotional expression. The reason for this includes the fact that the concept does not occupy either the same or a unique semantic space, a space that would ordinarily be occupied by emotion terms. In addition, when SWB is considered a cognitive category then it was clear that SWB, as commonly operationalized, is not a natural kind, or a artificial class, but most likely as a category of artifacts. This is consistent with viewing SWB as an umbrella term.

There is, however, another perspective that can be taken to answering the question of whether SWB is a discrete emotion, and this involves considering SWB to be an abstract concepts that is subject to all the rules that are used to operationalize abstract concepts. As previously discussed, abstract terms or concepts can never be completely operationalized only approximated by various concrete terms. These concrete terms would constitute a family of terms, much as Wittgenstein claims, and thereby define a term. They would also share the same semantic space, however wide, and be operationalized by using a common cognitive principle. For example, ample evidence exists that reasonably discrete facial expressions exist for particular affective terms (e.g., happiness, joy, fear, anger, etc). Having a facial analog of an emotion term could therefore be a rule used to operationalize SWB. Thus, if SWB is an abstract concept alluding to discrete emotions then the concrete manifestations of the concept would have to consist of discrete emotions. Applying this criteria to what is considered common well-being terms, such as happiness, joy, and even anger (ill-being), would be acceptable concrete manifestations, but the term satisfaction would not. It would not because it is a derived, second-order cognition that involves a comparison to some implicit or explicit standard. In addition, there is, as far as I know, no primary facial expression or physiological analog for the term satisfaction that would make it comparable to such terms as happiness. Thus, the term satisfaction is not cognitively equivalent to such terms as happiness, fear, joy, or anger, and to combine it with other emotion terms unavoidably make the collection of such terms heterogeneous. On the other hand, if the collection of concrete manifestations of SWB were cognitively equivalent, then the resultant category would be considered an artificial category and thereby avoid labeling SWB as an umbrella term. However, to achieve this goal the category has to restrict the number and type of terms that are currently used to operationalize SWB.

4.3 Is Subjective Well-Being a Cognitive-Emotional Process?

A central theme running throughout the SWB research literature is that there are both cognitive and affective dimensions to the phrase (Andrews and Withey 1976). This tradition appears to be a continuation of a similar distinction found in social psychological attitude research,[30] suggesting that measures of SWB "are fundamentally measures of attitudes" (Andrews and McKennell 1980; p. 127). Andrews and Mc Kennell, (1980) also state that different measures of SWB would be expected to reflect different combinations of cognition and affect. They also claim, following Campbell et al. (1976) that cognition is ordinarily measured by satisfaction and affect by happiness. Diener et al. (1999), basically concur, further claiming that well-being is a multidimensional construct consisting of positive and negative affect, combined with some measure of cognition. Other investigators may tweak this schema, but accept its basic tenants (e.g., Cummins 2005; Kahneman 1999). However, there is a bit of confusion in the literature since some investigators persist in stating that SWB consists of separate cognitive and affective components (e.g., Larsen and Prizmic 2008; p. 258) when there is ample evidence that each component of SWB consists of varying degrees of affect and cognition.

De Haes et al. (1987) also applied the cognition/affect distinction but now in a study of postmastectomy patients who were administered separate seven point satisfaction and happiness scales, 11 months and one and a half years after surgery ($N=34$). They found that the patient's satisfaction remained relatively stable over time but that their happiness improved with time. De Haes et al. (1987) also comment that a satisfaction response involves a comparison to some other group, while the basis of a happiness response may involve a self-rating. Thus, they also acknowledged that the two assessments were not cognitively equivalent. This and related studies (e.g., Michalos 1980) demonstrate that the distinction between affective and cognitive processes can be applied to the medically ill.

Andrews and his colleagues, by conceptualizing well-being as involving both cognitive and affective processes, were also placing the term well-being into the middle of an extended philosophical, psychological, and neurocognitive debate. Some philosophers, for example, argue that an emotion has, by necessity, cognitive content, so that thinking of the two processes as independent is not technically and philosophically possible. Solomon (2000), for example, states this issue as follows:

> What remains at the core of all such theories, however, is an awareness that all emotions presuppose or have as their preconditions certain sorts of cognitions – an awareness of danger in fear, recognition of an offense in anger, appreciation of someone or something loveable in love. Even the most hard-headed neurological or behavioral theory must take account of the fact that no matter what the neurology or the behavior, if a person is demonstrably ignorant of a certain state of affairs or facts, he or she cannot have emotions. (p. 11).

This is also so because emotions create an intention towards an object (Dennett 1987; Searle 1983; Solomon 2000). Thus, if you are angry, you are angry about someone or thing; if you love, you love someone or thing. In each case the emotion is directed at you or an object, and in this sense is intentional.

The assumption by Andrews and other investigators that happiness is indicative of an emotional state devoid of cognitive components, while satisfaction is a cognitive indicator devoid of affect is overly simplistic, and probably wrong. But the exact relationship between emotion and cognition is highly controversial and has been the subject of active debate among psychologists. Some psychologists argue that affective processes precede cognitive processes (James 1890; Zajonc 1980), others that cognitive processes (e.g., appraisal) lead to affective responses (e.g., Lazarus 1991a; Ortony et al. 1988; Strobeck et al. 2006), or that the two processes are inseparable, occurring together (e.g., Duncan and Barrett 2007). Also relevant here are the studies of Schwarz and his colleagues that have demonstrated how cognitive processes determine responding to well-being assessments. I will review each of these perspectives and end this section with a discussion of the role of affective and cognitive processes in the appraisal of well-being and qualitative assessments.

Before I do this, I would like to review some papers by Cummins and his colleagues (Cummins et al. 2007; Davern et al. 2007), since they bridge the previous discussion with the current topic. Davern et al. (2007) report two studies; the first tests the circumplex model of affect, and the second compares three different models of affect (personality driven, discrepancy driven and cognitive-affective model of well-being). In the first study, Davern et al. (2007) traced the initial effort at classifying emotion terms by using a circumplex to a study by Schlosberg (1952). It was he who classified emotion terms along the dimensions of unpleasantness–pleasantness or attention–rejection. In their first study, Davern et al. (2007) asked 478 adults of 2,000 who were surveyed by phone to rate, on a 0–10 scale, how satisfied they were with their life as a whole, and then to rate, again on a 0–10 scale, 31 affective terms relative to their feeling towards their life as a whole. They then used multiple regression analyses to determine which of the 31 affective terms were most effective in explaining variance in SWB (as assessed by satisfaction with one's life). They found that six terms (energized, happy, content, satisfied, stressed, pleased) accounted for 64% of the variance in satisfaction with life. These and the remaining emotion terms tended to cluster along the unpleasant–pleasant dimension, rather than reflect the activation–deactivation dimension. The second study used structural equation modeling to compare the six core affect terms with personality variables, as assessed by a personality inventory, or seven items from Micholas' (1985) discrepancy questionnaire. What was found was that the core affect terms and discrepancy terms (a cognitive-affective model) accounted for 90% of the variance in SWB.

These data confirm that cognitive-emotional processes are major determinants of SWB, but they do not explain how this occurs. To understand how this might occur requires a review of different approaches to the relationship between emotion and cognition, and this is the subject of the next several sections.

4.3.1 The Affect Primacy Hypothesis

William James (1884) in his famous paper entitled "What is an emotion?" initiated what has now evolved into an extended debate on the sequence of events that leads to an emotional experience. The James-Lange theory (Lange made the same general proposal at about the same time) argues that bodily responses (e.g., core affect) or behaviors precede our awareness (a cognitive process) of some emotion. Thus, the common understanding that I run away from a threat because I am afraid is wrong, and that instead I actually become afraid after I run away. Thus, in the James-Lange theory unconscious bodily responses and behaviors precede various actions (e.g., running away) which as feedback, act to make me aware of my feelings (e.g., fear of a threat).[31] One of the implications of the James-Lange theory is that well-being may actually be an afterthought: an experience that a person becomes aware of, following some behavior or activation of some bodily process. Koriat et al. (2006) discusses the role of metacognition in mediating these outcomes.

Consistent with the James-Lange theory, Zajonc (1984) states: "that affect and cognition are separate and partially independent systems and that although they ordinarily function conjointly, affect could be generated without a prior cognitive process. It could, therefore, at times precede cognition in a behavioral chain" (Zajonc 1984; p. 117). Murphy and Zajonc (1993) report a series of studies that were designed to clarify the relationship between affective and cognitive processes. In these studies, priming stimuli (e.g., faces) were presented at either a level below (e.g., 4 ms) or above awareness (1,000 ms), followed by exposure to a above awareness target (2,000 ms), usually a neutral stimulus (e.g., Chinese ideograms). Respondents then were asked to state a preference for (like or dislike) or evaluate (good or bad) the target stimuli. The question they were asking was whether priming the person with a presumably affective-laden stimulus (e.g., a smiling face, or a sad facial expression) or neutral stimulus (pictures of objects) either above or below awareness would somehow conditioned the person's response to the Chinese ideograms.

What they found was that suboptimal affective primes in the form of facial expressions generated significant shifts in subjects' preferences for Chinese ideograms, whereas optimal exposures did not. The same results were found in a second experiment where respondents were asked to rate the ideograms as bad or good. Murphy and Zajonc (1993) concluded that the magnitude of the priming effect was inversely related to the length of exposure to the prime. They then discussed various theoretical models that describe the temporal relationship between affective and cognitive appraisal processes (Murphy and Zajonc 1993; Figs. 2 and 3). These models were based on the assumption, as reflected in a paper by Öhman et al. (1989), that conscious and unconscious processes represent different points on a "consciousness" continuum. Emotions are assumed to enter the information processing system early in the chain, before the encoding of more complex perceptual stimuli.

Murphy (1990) also reports a study in which she presented respondents with six different facial expressions at suboptimal levels of exposure. What she found was that while subjects were able to discriminate at a better than chance level faces that expressed happiness, and faces that expressed fear, disgust, and sadness, respondents were not able to distinguish between negative affect stimuli. This suggests that if a person is to distinguish between the negative emotions they would have to utilize various cognitive processes, and from this they concluded that the priming effect was not specific to a particular emotion.

The independence of cognitive and affective processes can also be studied by examining persons with specific brain lesions. Thus, prosopagnosia (the inability to recognize faces), a fairly rare condition, offers an interesting example of a group of persons with brain lesions who are not able to recognize faces, even of persons they would be expected to be quite familiar with. Interestingly, they remain able to respond to simulated facial expressions. Also of interest was Bauer's (1984) demonstration that persons with prosopagnosia experience elevated galvanic skin responses when shown familiar faces even though they were not able to name the person they were looking at. This suggested to Bauer (1984) that these persons retained a preliminary or "preattentive" ability to detect affective stimuli. Thus, these persons had a positive affective reaction to familiarity without recognition of this familiarity. Murphy and Zajonc (1993) also refer to data from persons with prosopo-affective agnosia, and amygdala damage (especially animal studies) that provides further support for the proposition that cognition and affect are separate processes that can regularly interact.

Murphy and Zajonc (1993) believe that they have demonstrated the independence of affect from cognition, by demonstrating that when stimuli are presented in a nonoptimal manner they elicit responding that reflects the presence of these stimuli. This they felt permitted them to speak of the interaction of two, otherwise independent, entities. Their view, however, has been challenged by those who take a cognitive-primacy perspective.

4.3.2 The Cognitive-Primacy Hypothesis

Lazarus (1991a, b, c, 1995) is probably the most prominent investigator who has promoted the cognitive primacy hypothesis. His theory of emotion combines cognition (in particular, appraisal), motivation and person-environment interactions, and evolved from the social psychological and personality research literature. His primary focus is on examining the role of emotion from a phenomenological, psychoanalytical, and motivational perspective. He uses neurocognitive and developmental data mostly to confirm, not necessarily to adjust, the cognitive primacy hypothesis. He describes the relationship between cognition and emotion as follows:

> The functional relationship between cognition and emotion are bidirectional. As an effect or dependent variable, emotion is the result of appraisals of significance of what has happened for personal well-being. It is also a response to cognitive activity, which generates meaning regardless of how this meaning is achieved. I have taken the strongest position possible, and the most controversial, on the causal role of cognition in emotion, namely, that it is both a necessary and sufficient condition. Sufficient means that thoughts are capable of producing emotions; necessary means that emotions cannot occur without some kind of thought. Many writers who accept comfortably the idea that cognition is sufficient reject that it is necessary. (Lazarus 1991a; p. 353).

Thus, for Lazarus emotions always contain cognitive appraisals, although he also acknowledges that not all cognitions contain emotions. In addition, he also considers the relationship between emotions and cognition as similar to the relationship between an abstract and concrete construct, with cognition, physiological reactions, and behaviors acting as examples of concrete forms of expressing the abstract construct of emotion. In discussing the relationship between knowledge and appraisal, Lazarus (1991a) states:

> Appraisal, on the other hand, is an evaluation of the significance of knowledge about what is happening for our personal well-being. Only the recognition that we have something to gain or lose, that is, that the outcome of a transaction is relevant to goals and well-being, generates an emotion. Because motivation principles states that without knowledge should be viewed as a necessary but no sufficient condition of emotion, whereas appraisal is both necessary and sufficient. (p. 354).

His statement that appraisal evaluates the significance of knowledge which determines well-being is, of course, reminiscent of what was diagramed in Fig. 11.1, but now I would consider appraisal as the cognitive mechanism that mediates between hedonic states (e.g., pleasures, pains, and so on) and well-being.

Claiming that appraisal constitutes an evaluation of knowledge does not tell me much about how the cognitive process of appraisal occurs; it just substitutes one concept for another. Lazarus could just as well have claimed that emotions always contain evaluations. Instead, what Lazarus does is combine a number of different themes to create a hypothetical cognitive mechanism that would account for how appraisals determine emotions. First, he notes the existence of dual processing (Table 6.1; p. 171); where cognitive processes, or behaviors occur in either or both an automatic unconscious manner, or alternatively by a conscious more deliberate controlled process. Thus, Lazarus (1991a; p. 357–358) argues that appraisal is an unconscious automatic process. A second theme focuses on where along the information processing continuum (e.g., sensation, perception, cognition, metacognition) appraisal is most likely to occur. Lazarus proposes that appraisal is essentially a perceptual process, and his argument is as follows:

> Baron's (1988) suggestion that modern approaches to cognition have emphasized complex reasoning and inference at the expense of immediate recognition by human beings and subhuman animals of affordances in the environment, as described by Gibson (1966, 1976) seems also to recognize that much of the time we know instantaneously about what is good for us or bad for us without complex and time-consuming inference processes. We know it automatically on the basis of the optic stimulus array, which also makes his perceptual analysis, almost like radical behaviorism, quite different from a cognitive-mediational outlook (cf., Neisser). (Lazarus 1991a; p. 357–358).

Since I know instantaneously what is good or bad for me, it is obvious that perceptual affordance[32] is a concept that is performing the same task Lazarus assumes appraisal is performing at a cognitive level. Thus, Lazarus is adopting Gibson's notion of affordance as one of the operating characteristics of appraisal. Thinking of appraisal as an unconscious automatic perceptual process makes it clear that Lazarus (1991a, b, c) has adopted a rather innovative model of how appraisal occurs.

In addition, the notion that I can do without complex time-consuming inferences also suggests that I can do without what would be considered a cognitive-meditational approach. Thus, Lazarus's notion of how appraisal works represent a challenge to the conceptual foundation that underlies a hybrid construct. The theoretical papers by Rapkin and Schwartz (2004) and Schwartz and Rapkin (2004), used appraisal as a mechanism to account for the response shift in quality-of-life or HRQOL assessments, and provides an example of how Lazarus's views can be relevant to the current discussion.

Storbeck and colleagues (2004, 2006) by reviewing the available neurocognitive literature helped to clarify the relative importance of affective or cognitive primacy, and offers an alternative to Lazarus's approach. To simplify their task, the authors examined the neurocognitive literature that focused on the perception and processing of visual objects or words which would ordinarily evoke an emotional response (e.g., a face). First, there are two ways affective and cognitive (e.g., semantic) processes may be neurocognitively related; causally or temporally. Thus, an affective response may occur prior to and cause a semantic categorization and identification process or alternatively affective and semantic processes maybe independent of each other, with affective processing occurring more quickly than semantic processing.

Storbeck et al. (2004, 2006) believe that the neurocognitive data they reviewed supports the primacy of semantic processing, whether causally or temporally determined. In addition, their data were supportive of a cognitive-meditational or additive cognitive model, characteristic of a hybrid construct. They demonstrate this by describing the research literature that indicates that specific brain sites perform specific functions, which occur in a sequential manner. Thus, the inferior temporal (IT) cortex appears to categorize visual stimuli and do so without any input from the valence of the stimulus (Rolls 1999). Specific areas of the visual cortex mediate different types of visual stimuli (objects or words and letter strings). Support for this comes from neuroimaging and brain lesion studies in both animals and humans (Storbeck et al. 2004; 2006). Thus, categorization occurs as an unconscious process, but apparently can also occur with minimal affective activation. After describing the data that suggests that the cognitive process of identification also occurs without active retrieval of affect, Storbeck et al. (2006) summarize what they have found:

> Categorization and identification are crucial semantic tasks. The evidence (as reviewed above) suggests that these processes occur in the higher areas of the visual cortex particularly in area IT. Does this mean that we are conscious of such neural activity? No, but semantic analysis should not be equated with consciousness (Lazarus 1995). A good deal of, if not most, semantic analysis is unconscious… Does this mean that area IT "knows" what an object is? Yes. There is a distinct and invariant neural representation for that object. This means that the object has been identified. Does this mean that area IT can represent the psychological significance of the object? No. Area IT creates an invariant code for each object. It then sends the code forward within the brain to elicit memory and emotion-based associations. Although area IT knows what an object is, it does not know the object's goal-related significance. However, identifying an object is a necessary prerequisite for evaluating its significance. (p. 45).

Storbeck et al. (2006) goes on to review the relevant literature on a number of related topics and concludes their paper with the statement "that cognition is primary to affect both in causal and temporal terms" (p. 51). Their presentation also supports a cognitive-mediation model, and rejects Lazarus's Gibsonian "affordance" model of appraisal by requiring separate neurocognitive involvement corresponding to each element (e.g., description, evaluation, and so on) of the process that leads to cognitive primacy.

So far I have reviewed some of the data related to the affective and cognitive primacy hypotheses. Next I would like to present two different approaches that propose that the distinction between cognition and affect is not sustainable. Duncan and Barrett (2007) does this by claiming that optimum cognitive processing would not occur in the absence of affective processes, while Ortony et al. (1988) takes a theoretical approach demonstrating how emotions can be generated by different cognitive processes. Thus, both approaches claim that cognition and affect are analytically inseparable.

4.3.3 The Cognitive Structure of Emotions

Duncan and Barrett (2007) claim that affect is a type of cognition, and that to insist on viewing affect and cognition as separate processes is more a function of language (I have separate words for each concept) and phenomenology, than what is empirically and neurocognitively defendable. They state fairly dramatically, that "There is no such thing as a 'nonaffective thought'" (p. 1185). Affect (that is, core affect) is defined by them as "any state that represents how an object or situation impacts a person" (p. 1185). Affect is experienced as pleasant or unpleasant (valence) and alternatively, as activation or no activation (arousal). Their definition of cognition follows Neisser's (1967, p. 4) who stated that cognition refers to all the processes that transform, reduce, elaborate, store, recover, or use sensory information.[33] They speak of the two concepts as separate terms, as when a cognition may activate an affect or an affect may influence sensory or cognitive processes, but acknowledge that the distinction between the two is not given by nature and instead represents a phenomenological distinction that at times may be useful.

Evidence that supports the inseparable relationship between affect and cognition comes from studies that demonstrate the importance of affective processes in consciousness, how I use and understand language, and how and what I remember. Neurocognitive and neuroimaging studies have succeeded in mapping the neural circuits of affective and cognitive processes, and these studies according to Duncan and Barrett (2007), have failed to demonstrate that separate affective and cognitive brain areas exist. They present a neurocognitive model that has as its focus the various functions of the amygdala and related cortical and subcortical areas (see Fig. 2, in Duncan and Barrett 2007).

Affect, according to Duncan and Barrett, drives neural circuits, but this can be disrupted leading to psychopathological conditions. Among these conditions they list schizophrenia, the delusions or hallucinations which can be due to irregular burst of neural activity in the thalamus. They also describe the Capgras syndrome, which is a disorder in which a person does not attribute an affective value to an object or event when it would be appropriate to do so. There are, of course, a wide variety of other situations where affect impacts a particular cognitive process, such as, decision making (Bechara 2004), or even a quality-of-life or HRQOL assessment (e.g., Heinonen et al. 2004). They, however, do not address the role of affective processes in estimation of well-being.

4 Subjective Well-Being and the Emotions

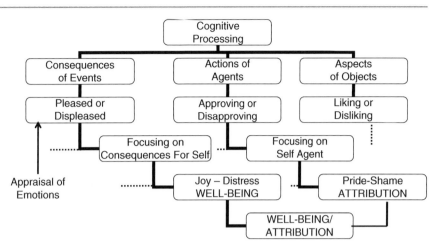

Fig. 11.5 Presented is a modified version of Fig. 2.1 from Ortony et al. (1988) summarizing a theoretical model of how cognitive processes can mediate different emotions. Illustrated is the model for well-being/attribution. Not shown are a variety of other types of emotions (e.g., love/hate as an outcome of liking or disliking objects). Reproduced with permission of the publisher.

Duncan and Barrett (2007) are proposing that affect is a constant in our lives, although its experienced presence may vary. Thus, under certain circumstances affect may function as background and I am not conscious of its presence, while at other times and under other conditions it may be directly experienced. When it is background I experience the world as full of facts, and I have confidence in judgments that use these facts. When I experience affect, I may report pleasure, distress, happiness, and so on. Finally, they acknowledge that since I do report thoughts and feelings as separate experiences it is necessary to investigate why this occurs.

Ortony et al. (1988) have proposed (Fig. 11.5), a theoretical "cognitive model of emotions." Their model assumes that a cognitive arousal mechanism exists that concurrently generates valence, that becomes attached to objects leading to a specific emotion. Thus, this model conceives of emotions as being evoked by combining arousal and valence into one cognitive process. They state, "Our working characterization views emotions as valenced reactions to events, agents, or objects, with their particular nature being determined by the way in which the eliciting situation is construed" (Ortony et al. 1988; p. 13). Emotions, therefore, emanate from prior cognitive processes, which when influenced by the consequences of events, agents, or objects leads to specific categories of emotion.

Each of the three primary branches of their model (reflecting the consequences of events, actions of agents or aspects of objects; see Fig. 11.5) give rise to different basic classes of emotions; pleased or displeased, approving vs. disapproving, or liking or disliking. For example, pleasure or displeasure is associated with the consequences of events that may impact a person or others. If the consequences are irrelevant to a person, then the emotions of hope or fear can be confirmed or disconfirmed. If hope is confirmed then the emotion of satisfaction results, whereas if fear is not confirmed then the emotion of relief occurs. In contrast, if the consequences of events are relevant to a person then the emotions of joy or distress maybe evoked, either of which can lead to a sense of well-being. Combining well-being with attribution leads to the emotions of gratification, remorse, or anger. Thus, for Ortony et al. (1988) well-being is a product of the consequences of events, which when combined with the action of agents leads to a variety of emotions that reflect the combination of well-being with attribution.

The intensity of an emotion is determined by the cognitive process of appraisal, with appraisal assessing an emotion in terms of its desirability, praiseworthiness, and degree of appeal. For example, the appraisal of events leading to well-being can be desirable or undesirable, with the desirable events leading to joy and the undesirable events leading to distress. Thus, Ortony et al. (1988) state that emotions associated with well-being "are the result of reacting to events that are positively or negatively evaluated in terms of their implications for a person's goals..." (p. 86). The type of well-being emotion that represents joy include, being contented, cheerful, delighted, feeling good, happy, jubilant, pleased, and so on. In contrast, well-being (ill-being) emotions that represent distress include being depressed, displeased, dissatisfied, distraught, feeling bad, feeling uncomfortable, grief, homesick, lonely, regret, sad unhappy, and so on.

Ortony et al. (1988) also provides examples of how a sequence of cognitive processes can be used to model the structure of emotions. Thus, the first response to an event, an agent or object is the appraisal of the intensity of the event and this leads to the liking or disliking of the event, which results in joy or distress and from this a sense of well-being develops. This sequence can also be characterized in terms of a formal system of rules and representations, that then can be used to create an algorithm common to artificial intelligence programs, but now designed to instruct computers to "reason about emotions." As discussed, the investigators use well-being to model their system of rules and representations.

While the primary purpose of Ortony's et al. (1988) theory was to outline a research model that uses cognitive processes to characterize the various emotions, it incidentally can do the same for the term well-being. Their theory reinforced the notion that well-being is not a "primary" emotion, in the sense of happiness, fear or anger, and is best described as a cognitive construction or abstraction.

4.3.4 Appraisal in Well-Being and Quality-of-Life Assessment

As discussed in the section on cognitive primacy (Lazarus 1991c; Fig. 11.6) and in Ortony's et al. (1988; Fig. 11.5), model appraisal appears to play a critical role as a cognitive determinant of emotions. Basically, the cognitive primacy theory claims that as appraisal changes emotions change. As Ellsworth and Tong (2006) state:

> Appraisal theories were developed in part to account for this abundant variety of emotions, as an alternative to categorical theories. Emotional experience is like a river, not like a collection of wells. Whenever an appraisal changes, the feeling changes with it, by a little or a lot. Potentially, there is an infinity of emotional states ... (p. 584).

They also state:

> Words have crisper edges than experience, although when it comes to emotions, even the words are somewhat fuzzy ... Words imply categories and suggest that emotions are states. But as Tolsti remarked, 'labels ... have only the slightest connection with the event itself' (1989/1968, p. 733). Emotions are processes changing over time, and emotion labels cannot capture the complexity of this psychological notion. (p. 585).

Appraisal was first given empirical and theoretical prominence by Arnold (1960), who argued that organisms constantly monitor and evaluate the relevance of environmental changes for their own well-being. Lazarus (1966, 1991c; p. 133) postulated that appraisals played an important part in a person's coping response to stress. It has also been suggested that appraisals play a role in the study of psychosocial processes, and quality-of-life or HRQOL research (e.g., Rapkin and Schwartz 2004). Thus, it is a construct that deserves a place in any conceptual or theoretical discussion of emotions, and by association, well-being and quality assessment.

As I usually do, I will start by defining the term I am interested in, since this will give me a fixed point to compare to other usages. When I do this I discovered that the term appraisal has two broad references; as a type of classification or as an instance of assessment, as when a formal assessment is made of a person's performance (WordNet 2.1, 2005). Thus, the term "appraisal" is a type of assessment and inspecting its synonyms supports this; "with judgment, valuation" (Roget's II Thesaurus; Chapman, 1995; p. 46) listed as synonyms. The etymology of the term *apprise* reveals that it originated from the Latin word, appretiare and not only refers to "value, estimate," but also from *adpretium* meaning "to price." The original English spelling of the term was "apprize," but this was altered by the influence of the word, "praise," leading to its current spelling.

Lazarus (e.g., 1991c), as a result of his early work on stress and coping, used the term appraisal in two ways; with primary appraisal referring to when something of relevance to a person's well-being had occurred, and secondary appraisal referring to a person's coping options (e.g., when an action might prevent harm, ameliorate, or produce additional harm or benefit). Primary appraisal has three components; goal relevance, goal congruence, and type of ego-involvement, each of which could be considered to be an outcome of an assessment process. Secondary appraisal, on the other hand, required blame or credit, coping potential, and future expectations to distinguish between the emotions. Primary and secondary appraisals are constantly changing leading to evaluation and reappraisal. In this way, appraisal retains its meaning as an assessment term, but now expanded to include various determinants (e.g., a person's goals) or targets (e.g., blame or harm).

The emotions, as a continuous process, will interact with the environment over time, so that the type of appraisal will change as the person adapts to their circumstances. While theorists differ somewhat, (Ellsworth and Scherer 2003; Table 1), most acknowledge that the appraisal of the novelty of a situation will differ from appraisal of a familiar stimulus, as will appraisal differ depending on the valence associated with the stimulus, goals or needs implicit in the situation, the persons or agents involved, and also the norms or values related.

Since appraisal is an evaluation of what is happening to the well-being of an individual, appraisal can be viewed as a process that evolves into a style – an appraisal style – that can reflect the person's personality. As depicted in Fig. 11.6, appraisal leads to action tendencies, subjective experience affect, and physiological responses.

Presented in this way, appraisal has assumed the role of a category, filled with different types of indicators, and by so doing has also assumed a figurative form of expression.

The relationship between appraisal and well-being (Lazarus 1991c), if taken literally, associates two types of measures, one measuring a process and the other an outcome (e.g., "has something of relevance to a person's well-being occurred?). Used in this manner, appraisal provides a measure of the determinants of well-being, some of which were depicted in the Ortony et al.'s (1988) cognitive model of emotions. Thus, measuring a person's goals and their involvement in achieving these goals will determine their well-being.

Solomon (2000), however, has pointed out that the appraisal process is complex, often involving multiple types of appraisal for a particular emotion. For example, the expression of anger, may involve not only an appraisal of the person or thing expressing anger towards a person, but may also be an appraisal of the person themselves, especially

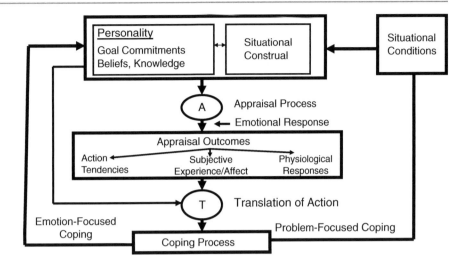

Fig. 11.6 A modified version of Lazarus's (1991c) model of appraisal (Fig. 5.3; p. 210). Reproduced with permission of the publisher.

concerning whether they deserve the anger being expressed. Who this "other" may be will also alter the nature of the appraisal. Thus, if the "other" is a person's supervisor, God, or a child then the appraisal will vary. Solomon also claims there are what he calls, meta-appraisals (or reflective evaluation of emotions) that raise issues about the moral significance of a particular emotional expression. Thus, he suggests that multidimensional appraisals, rather than a single appraisal, are required for characterizing emotion (see also Scherer 1999). What he is describing, of course, resembles the formation of a hybrid construct.

A number of investigators have posited a role for appraisal in a quality or quality-of-life assessment. Smith et al. (1999) have defined quality-of-life as follows; "QOL is the subjective appraisal of one's current life based primarily on psychological functioning and to a lesser degree on physical functioning" (p. 457). Rapkin and Schwartz (2004), also claim that appraisal plays an important role in a quality-of-life or HRQOL assessment, and state that; "Appraisal is a hidden facet in all assessment of QOL, and all studies involving self-reported QOL are influenced by appraisal" (Rapkin and Schwartz 2004; p. 6). These statements imply that appraisal adds something to the assessment of quality or quality-of-life that would not otherwise be present, but it also suggests that appraisal, as a form of assessment, has special features that distinguish it from other types of assessment. What are these special features? To demonstrate this Rapkin and Schwartz (2004) compare Tourangeau et al's. (2000) four process cognitive model (comprehension, retrieval, judgment and response) of self-report responding to their conceptualization of the role of appraisal in a quality-of-life or HRQOL assessment. Thus, they state:

> We have adopted our operational definitions of appraisal processes to correspond to psychological aspects of coping and adjustment intrinsically related to QOL appraisal. This is an important distinction: Tourangeau and colleagues emphasize psychological processes that arise in the survey situation... Alternatively, we believe that cognitive aspects of quality-of-life appraisal are not merely a measurement issue, and may themselves become the focus of clinical interventions to help patients understand and think about their QOL in more adaptive ways. (p. 6).

Thus, the "hidden facet" that is apparently unique to appraisal, appears to be the psychological processes of coping and adjustment that would be monitored during responding to a qualitative assessment. This suggests that Rapkin and Schwartz (2004) have expanded the meaning of the term appraisal from just a measure, to an assessment of a process, a process that can also become the object of an intervention (See quotations from Ellsworth and Tong 2006).

Rapkin and Schwartz (2004) transpose Tourangeau et al's. (2000) four process cognitive model (comprehension, retrieval, judgment and response) model as follows:

> Induction: First, they claim that any quality-of-life or HRQOL assessment induces a unique frame of reference that reflects the personal meaning that a respondent attaches to the items that have been asked to respond to. This they claim corresponds to Tourangeau et al's. (2000) principle of comprehension.
>
> Identification: Second, while Tourangeau's et al. (2000) speak of the retrieval of autobiographic information that a person uses in their responses to an item, Rapkin and Schwartz (2004) think of this process as a sampling that a person does from within the individual's personal frame of reference.
>
> Evaluation: Third, each sample of experiences a person makes involves a comparison to some subjective standard. This process describes what Tourangeau et al. (2000) referred to as the person making a judgment, which, usually involved evaluating how well they retrieved and integrated information, into a response.
>
> Combination: Fourth, for a person to arrive at a qualitative score they would be expected to apply some sort of combinatory algorithm, or, as represented in this book, creation of a hybrid construct. The end result of this process is what Tourangeau et al. (2000) referred as generating a response.

Each of these elements (that is, Induction, Identification, Evaluation and Combination) can also be seen as elements in an assessment process, as well as steps in the appraisal process. They go on to describe these four elements in formal terms. For example, a quality-of-life response (Qt) at time "t" would be some combination of true and error variance, or, $Qt = qt + et$. Unpacking Qt leads to a series of equations reflecting each of the major components in the model. Of particular interest is equation #4, the combinational equation; $qt+[Wt]*[At]$. Rapkin and Schwartz describe this equation as follows:

> ...the QOL true score is the point product of the vector or appraisals [At] premultiplied by a (transposed) vector of the same dimension [Wt]*. [Wt]* consists of the weights needed to combine appraisals across experiences. In other words, these weights represent a combinatory algorithm that dictates the relative impact or importance of specific experiences on QOL by time t. Equation 4 shows that any QOL rating is based on an amalgam of appraisals of different experiences that depends upon weights by their importance at a given time. ... In sum, this measurement model addresses the fact that QOL ratings are not intrinsically meaningful and can only be accurately understood through an underlying appraisal process. (p.7).

The notion that a quality-of-life or HRQOL rating requires an appraisal in order to be meaningful requires an illustration. Rapkin and Schwartz used the role of appraisal in the response shift to demonstrate the role of appraisal in a quality assessment. First, it should be noted, that earlier models of the response shift (e.g., Sprangers and Schwartz 1999) did not include direct reference to the appraisal process. What the current model does (Rapkin and Schwartz 2004: Fig. 2) is to place appraisal in an intermediate position between the various determinants (e.g., antecedent conditions, catalysts, mechanisms) of a response shift and its involvement in determining a quality-of-life outcome. Thus, appraisal is conceptualized as an intervening variable.

The rationale for including appraisal as an intervening process is that it would help account for various, otherwise difficult to understand, quality-of-life or HRQOL outcomes. For example, people with severe chronic illnesses often report equal or better quality-of-life or HRQOL outcomes than persons with relatively mild disorders. This could now be understood if it was assumed that the person engaged in an appraisal that lead to say, a change in values, so that the person's expectations would have shifted leading to the apparently paradoxical outcome. Other examples include the response shift itself, which can be considered an aberrant phenomenon, in that, shifts in values and preferences appear to be confounding the nature of some post-test self-report. Finally, appraisal offers an opportunity to account for individual differences in the outcome of a qualitative assessment. In each of these cases, appraisal has assumed the role of a cognitive mechanism which if active, accounts for some qualitative outcome.

Rapkin and Schwartz (2004; Fig. 2) use a linear regression model to depict the place of appraisal in characterizing a response shift Inspecting their model makes it clear that the role that appraisal plays is exactly the same as various other indicators (e.g., coping, social comparison, goal reordering, reframing expectations, spiritual practices) that were postulated to be active in determining the response shift. Thus, it would be possible to remove the term appraisal without adversely impacting their model of the response shift. I am left therefore wondering what the "hidden facet" is, that the appraisal process offers to a qualitative assessment that has not already been identified as relevant to the response shift or any other kinds of qualitative assessments. Thus, I ask, is it not an appraisal implicit when coping, social comparisons, or goal reordering occurs? I am also left wondering about the utility of linear modeling, in general, since so often the relationships depicted in the model are meant to be figuratively, not literately related. I will present an alternative model that utilizes appraisal (see further), when I discuss Scherer's (2000) nonlinear model systems approach to the emotions.

One possibility is that Rapkin and Schwartz recognize that the term appraisal implies an action (as opposed to a form of assessment) and that they wish to insert a "regulatory mechanism" in what would otherwise be a static model of the response shift. This would make response shift appear as an active, as opposed to passive process, and provide a basis for modeling how change in a response, or its meaning, occurs. In addition, this usage would be consistent with Lazarus's use of the term. Another advantage is that introducing the term provides a bridge between relevant personality and social psychological research literature and the assessment of quality or quality-of-life. This would open quality or quality-of-life research to a variety of concepts ranging from the existential, to the psychoanalytic and psychopathologic.

4.4 Subjective Well-Being as an Outcome of Cognitive-Emotional Regulation

In 1848, Phineas Gage, in what is now a well told story (e.g., MacMillion 2000), experienced a severe brain injury while preparing a bore hole with explosives designed to displace a boulder along a rail road track. What happened was that a tamping rod (3.5 feet long, 2 inches in diameter and weighing 13 pounds) meant to compact the explosives prematurely discharged shooting the rod through Gage's left cheek bone, left eye socket, up into his frontal lobes, and out the top of his skull. The rod landed some 30 m beyond him. Most interestingly, Gage was able to get up and walk, fairly soon after the accident. In time, however, it became evident that while he retained many physical and cognitive abilities, significant changes occurred in his decision making, personality, and emotion demeanor. He became rude, profane, and ended up losing his job as the construction foreman of the rail-laying work crew.

This clinical sequella has fascinated clinicians and researchers (e.g., Damasio 1994), since it suggested that the frontal lobes were ordinarily involved in controlling emotional expression, and that when this control was disinhibited, changes occurred in a selective range of human functions. A more recent case (Cato et al. 2004) of a person with a similar frontal lobe lesion confirmed these findings; here it was found that the person sacrificed accuracy for speed of decision making, and adopted response strategies that demonstrated cognitive inflexibility, impulsivity, and disinhibition. Clinical observations of this sort suggest that emotional expression is a product of a structurally dependent complex regulatory process that if disturbed can lead to various dysfunctions.

Many of the ordinary biological and psychological functions I engage in (e.g., walking, talking, swallowing, expressing my feelings, and so on) are an outcome of a regulated process (e.g., sensory-motor coordination; cognitive-emotional interaction). For example, the fine movement of a hand, results from the balance between the excitation and inhibition of separate muscle groups. If an imbalance occurs, a person may develop a tremor or a flaccid limb. The Parkinson's patient, for example, experiences a tremor at rest as a result of deficits in neurochemicals (e.g., dopamine) at a particular brain site (the Substantia Nigra), while a person who has had their trigeminal nerve severed due to facial pain will report that one side of their face is flaccid and lacks sensation. What is of interest here is that the outcome of this regulation can also be used to judge aspects of a qualitative assessment. Thus, the quality of neuromuscular regulation will reflect how well I can shake the hand of someone else, feed myself, express anger, and so on. Both functional and aesthetic aspects of quality can be assessed. For example, being able to walk does not mean that I can dance the tango, yet being able to dance the tango says something about the quality with which I can move. Broadly speaking, therefore, regulation involves a system coming into compliance with some implicit or explicit functional standard, and the degree this is achieved provides information that can be used in the assessment of quality.

What is also of interest is that the regulatory process, itself, is not static, but can change with repetitions or time, or by resetting a set-point. Thus, when I change the set-point, I am also changing the regulation process itself. Emotional regulation is an example of such a system, since it not only has immediate consequences, but also can alter how the regulation occurs. A classic example of these types of changes is the empirical observation referred to as the "hedonic treadmill." The hedonic treadmill refers to the observation that people adjust to their changing circumstances, adapting to the point where what was at one point attractive or aversive now becomes emotionally neutral. How these alterations may occur may also change. Thus, if a person has an adverse experience, as when they are injured and lose their ability to function, with time the resultant negativity wanes and the person may appear accepting of conditions that an outside observer would find unacceptable. These changes may not only involve a change in a set-point (e.g., what is an acceptable way to live), but may also involve a narrowing of the person's expectation. Changes of this sort are examples of Helson's (1948) adaptation level theory.

An early experimental demonstration of the hedonic treadmill was reported by Brickman et al. (1978) who found that lottery winners were not happier than nonwinners, and that paraplegics were not substantially less happy than those who could walk. In addition, economists have suggested that the hedonic treadmill helps account for the observation that increases in a person's financial wealth does not necessarily lead to increases in perceived happiness, although this is less so for person's whose income was at the lower end of the income range (Layard 2005). What is important to recognize about the hedonic treadmill is the empirical observation that prior events affect the probability of future events. Thus, increases in a person's salary dampens the person's emotional response to subsequent increases in salary, having had an operation dampens a person's response to a second operation, and so on. And not surprisingly, the notion of a hedonic treadmill has become a useful way to account for why economic progress is not necessarily related to perceived increases in welfare.

Diener et al. (2006), however, point out that the current usage of the hedonic treadmill concept underestimates the complexity of its assessment. They suggest that five issues should be considered in any effort to measure the "treadmill." For example, the original treadmill theory suggested that after exposure to an emotionally significant event, the person's affect returns to a neutral level. Empirical studies, however, report that the balance restored is usually above a neutral value. They also point out that individual differences in a person's well-being set-point exists, much of which can be attributed to differences in the person's personality. Diener et al. (2006) also claim that multiple set-points exist, corresponding to the three dimensions of well-being; positive and negative happiness and satisfaction. A fourth point they make is that happiness will vary depending on the circumstances under which a person lives, and their final point is that persons differ in their ability to adapt.

The title of the Diener et al. (2006) article is; "Beyond the Hedonic Treadmill," but their subtitle is "Revising the Adaptation Theory of Well-being." This suggests that the authors recognize that well-being can be thought of as the outcome of a regulated process, but what remains unclear is whether well-being itself can contribute to the regulation of other processes, including emotions. Thus, would the continued downregulation (or adaptation) characteristic of well-being contribute to a person's emotional state? I have already discussed some of the issues involved in cognitive regulation (Chap. 10), but now I will briefly review what is known about emotional regulation, as a step to understanding well-being's regulatory role. This will be followed by a summary of some of the well-being regulatory models that have been proposed. In that section I will review psychologically determined self-regulatory systems that can provide models of well-being (e.g., Higgins et al. 1999), and then I will examine what is known about the neurobiology of well-being regulation.

I am interested in several questions in the next section, for example, what do neuroimaging studies teach me about the cognitive-emotional basis of well-being; are well-being and ill-being two parts of the same neurocognitive system and what can neuroimaging studies tell me if two systems exist? This discussion will then be applied to how a hybrid construct may be formed. This will include a discussion of the neurocognitive role of metacognition in the formation of the hybrid construct. Finally, I will discuss various nonlinear models of emotions and well-being. This discussion will continue the discussion started in Chap. 10 on embodiment and optimality and consider if these approaches can also be used to account for well-being. I will also consider if Kahneman's (1999) approach to objective happiness and how Scherer's approach to emotional appraisal can be modeled in nonlinear terms. The information considered in these sections should provide a greater appreciation of how emotions can be a product of a regulatory process, where well-being fits in such processes, and how emotional factors and well-being statements can be integrated into a qualitative assessment.

4.4.1 Emotional Regulation and Subjective Well-Being

Defining cognitive-emotional regulation (CER), it turns out, is as complex a task as I found for defining the term emotion (e.g., Chap. 11, p. 412). Thus, just as an emotion can be studied as consisting of a subjective experience, a behavior or a physiological reaction, so can CER. In addition, some investigators consider emotion and emotional regulation as inseparable terms, whereas others see them as separate terms and concepts. Cole et al. (2004), for example, argues for separating emotions and emotional regulation stating:

> Emotions are appraisal-action readiness stances, a fluid and complex progression of orienting towards the ongoing stream of experience. Emotions are moving targets that are usually unseen (and unfelt). Emotions must be inferred from evidence of the individual's relation to surrounding events. … The term stance connotes that the individual is evaluating a situation (appraising) and inclining toward a particular class of actions (action readiness)" (p. 320).

Continuing they state:

> Emotional regulation refers to changes associated with activated emotions. These include changes in the emotion itself (e.g., changes in intensity, duration; Thompson 1994) or in other psychological processes (e.g., memory, social interaction). Emotional regulation is not defined by which emotions are activated but by systematic changes associated with activated emotions. …

> The term emotional regulation can denote two types of regulatory phenomena; emotion as regulating and emotion as regulated. (p. 320).

The article by Cole et al. (2004) was followed by a series of comments by other authors (e.g., Campos et al. 2004; Eisenberg and Spinrad 2004) discussing different aspects of the definitional task. Campos et al. (2004), for example, claim that the assumption that emotion and emotional regulation can be thought of as separate is more a product of what our language permits, then reflecting reality; with our language distinguishing a more abstract term, such as, process from a more concrete term, such as, outcome. However, when the process or the outcomes are examined then the dependence of each on the other becomes clear. For this and other stated reasons, Campos et al. (2004) propose that a single-factor process exists relating emotion and emotion regulation. They state:

> … analytically and conceptually, there can be differences between emotion and emotion regulation. …However, such a conceptual distinction does not imply ontological distinctiveness – that each process has a separate, real existence that corresponds to the mental distinction. …We do not accept the assumption that in real life first comes one and then comes the other. Rather, emotion and emotion regulation are conjoined from the beginning in as one observable process-a process that reflects the attempts by the person to adapt to the problems he or she encounters in the world. (p. 379).

They go on to define emotions and emotion regulation by first pointing out that humans are constantly adapting to the social and nonsocial demands of the world that they encounter, and that this process of adaptation determines the events that generate emotions and emotion regulation. They then offer the following definitions of emotion and emotion regulation:

> Emotion is the process of registering the significance of a physical or mental event, as the individual construes that significance. The nature of the significance (perceived insult, threat to life, depreciation by another, relinquishment of a desired state, avoidance or resolution of a problem, etc.) determines the quality of the emotion. … Emotional regulation is the modification of any process in the system that generates emotion or its manifestation in behavior. …Regulation takes place at all levels of the emotion process, at all times the emotion is activated, and is evident even before an emotion is manifest. (p. 379–380).

While the debate about a single or double-factor model of cognitive-emotional regulation (CER) will probably rage on for some time, what is of interest is how this debate can illuminate the nature of SWB. Thus, I can ask if SWB itself is a system, that can be regulated (that is, it can be an outcome) or alternatively can it regulate, as in effecting other processes? Alternatively, SWB may be determined by the system that generates emotion or its manifestations in behavior. And in either case I have to return to the question of whether these distinctions are more a function of the language I use than the reality of how the "well-being" system works?

As discussed in Chap. 10, many of the examples (e.g., loss of language skills following a stroke, or onset of dementia) I considered actually involved the disruption of complex regulated processes. This raises the question of whether there is ever a real "outcome," or whether describing something as an outcome is again an analytic convenience made available by the punctuate nature of language. This can be illustrated for emotional processes by reviewing some of the cortical underpinning of emotional expression. Lewis and Stieben (2004) neatly summarize this vast literature:

> When people report emotional experiences, imaging studies demonstrate changes in blood flow across a large number of sites, ranging from the brain stem to the cortex (Damasio et al. 2000). In contrast, a relatively small number of sites are activated during conscious cognitive activity, and these are specifically cortical … To make sense of these findings, some neuroscientists define emotion as the reciprocal recruitment of subsystems up and down the neural hierarchy, accompanied by endocrine and muscular changes, in a process of self-organization… These subsystems include the brain stem, mediating arousal and behavioral activation; the limbic system, mediating coarse perception, memory learning and affective feeling; and the cerebral cortex, subserving higher order perceptual processes, attention, working memory, and voluntary control. Subsystem synchronization is thought to produce a coherent neural gestalt – a unified brain- whose psychological features include a specific action readiness, a restricted attentional focus, a stable cognitive appraisal, and a distinct emotional feeling …
>
> The problem with defining emotion regulation as an independent system is that synchronization of systems underlying emotion includes regulation of each system by the others. … These regulatory processes do not occur in discrete temporal stages. Rather they coevolve rapidly as attention, perception, emotion, and motor output become spontaneously coordinated … From a neural perspective, then, regulatory processes are intrinsic to the cascade of neural changes underlying emotion. (p. 372).

The argument Lewis and Steiben (2004) present seems convincing; thinking of emotion and CER as functionally separate is just not supported by what is known about how the neuroemotional system works. In the same sense, thinking of SWB as an outcome that is not a product of a regulate process seems difficult to defend.

One of the ways to clarify the issues here is to study the development of emotional regulation and determine the sequence with which events of the regulatory process develop. This can be done, to an extent, by studying the changes that occur during the social and emotional development of a child. An infant, of course, is quite capable of expressing emotions, the modulation of which at first requires the intervention of a parent or caretaker. Thus, following the old prescription that a crying infant is either hungry, wet, or tired many an appropriate parental intervention has occurred. In time, however, the need for external control of a child's emotional expression is modulated by the capacity of the child to toilet, feed itself, or sleep through the night. Bodily control, therefore, develops concurrent with behavioral and cognitive control, leading to the evolution of self-control (Fox and Calkins 2003) and social and interpersonal emotional regulation. Those aspects of cognitive control that are involved with emotional control include regulation of attention, inhibition, and the evolution of various executive functions. Executive functions may include being able to plan, control, reflect, or demonstrate competence in dealing with particular problems (Paris and Newman 1990).

Emotional regulation development, therefore, starts with uncontrolled emotional expression, and leads to a regulated outcome following biological and cognitive maturation and the acquiring of psychological and social skills. The demonstration of the shift from outer to inner regulation includes the development of a sense of what is "right" or "wrong." The question then becomes: does this inner sense feel like something real; like an entity? Less obvious but true is what a child or adult considers to be right or wrong (that is, moral) may represent a developmental determinant of the term well-being, and to the extent that this moral sense has some "phenomenal" existence, it may be evidence that well-being acts on other processes. If you recall, I cited (Chap. 2, p. 43) Lakoff and Johnson's (1999; p. 290–334) comments that linguistically the phrase well-being can be defined as "good existence," and that good existence may be what is meant when I speak of knowing the difference between right and wrong, or being moral. Thus, studying the development of inner regulation may be an effective way of gaining insight into how the meaning attached to the term well-being evolves, especially if individual differences in the use of the term well-being are of concern.

Kochanska and Aksan (2006) suggest that morality, or a person's conscience, can be conceptualized as consisting of three components; moral emotions, moral conduct, and moral cognitions. Their research, and the research of other investigators, suggests that the very young, who would not normally be assumed to be capable, demonstrate "rich consciences" (Kochanska and Aksan 2006; p. 1588). Guilt, experienced or anticipated, appears to be the emotion which infuses particular behaviors with its negative moral valance. The relationship between moral emotions and conduct becomes evident by 2 years old or younger (e.g., Askan and Kochanska 2005). Studies by these investigators reveal that a child's moral emotions and rule-compatible behaviors are moderately correlated, expressed consistently across situations, and have a moderate degree of longitudinal stability. These general statements appear true, even in the face of a fair amount of individual differences in the child's emerging personality.

The demonstration that children develop a sense of what is right or wrong suggests that they are engaged in evaluating themselves, as a step in regulating their emotions and that they do this at a fairly young age. This suggests that something

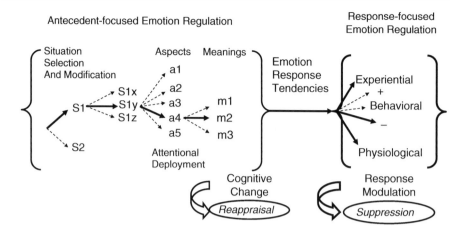

Fig. 11.7 Depicted is a modified version Gross's (1998, 2001) model of emotional regulation. Reproduced with permission of the publisher.

is in place which influences other behaviors and feelings, including what it is that determines a person's sense of well-being. Emotional regulation, however, has many manifestations each of which may influence SWB. For example, experiences such as regret (e.g., Wrosch et al. 2005) for past action and their potential for resolution, forgiveness (e.g., Worthington et al. 2007) as a step in dealing with chronic distress, or acceptance as a requirement of living with discord can all be considered manifestations of emotional regulation and contributors to a person's emotional well-being. Need satisfaction (e.g., Patrick et al. 2007) can also fit into this category, since it is also an effort to restore balance to a regulated process, one element of which may have an emotional component. Needs differ from regret, acceptance, and so on, in that, they may also be used to refer to balancing the physical or financial needs of a person. Need satisfaction, as an example of emotional regulation, plays a prominent role in not only Maslow's (1968) theory of motivation, but also in Murray's (Murray 1938) psychological need theory.

The development of SWB can also occur in specific situations or contexts (e.g., when a person marries into a family, or goes to college), and this would involve repeated positive experiences that in time become automatic and spontaneous. What I am describing here, of course, is dual processing, the process whereby a deliberate thought process becomes automatic (Chap. 6, p. 170). This process has assumed a prominent role in social cognition, accounting for the development of implicit or explicit forms of stereotyping, social comparisons, and attitudes. Since each of these has an emotional base or component, it does not seem farfetched to assume that each of them is also influencing emotional regulation. Thus, one of the consequences of stereotyping of certain ethnic or racial groups is that it helps a person manage some feelings that they might otherwise find hard to deal with (e.g., anger, or a sense of inferiority). Social comparisons, by providing a means of coping with distress of say personal loss (e.g., a loss of a limb) may also prevent other feelings, such as depression or anxiety, and thereby indirectly regulate these emotions.

The model that Schwarz and Strack (1999), Kahneman (1999), and their colleagues present can also be integrated into a dual process model. They have proposed that SWB is a product of momentary events that over time leads to the development of a person's feelings of SWB. For example, the many individual experiences that I have with my significant other ("an object") contribute to my feelings of SWB, but in time I will express my SWB without reference to my significant other. This generalization of affect seems to be true for other emotions as well. Thus, the person who is angry with his or her neighbor often is perceived as "angry with the world." This generalization of affect could also account for the persistence of SWB independent of momentary events.

Figure 11.7 provides a model of CER suggested by Gross (1998, 2001). His model consists of two major components; one which deals with the mechanisms (e.g., reappraisal) a person uses to moderate emotions prior to or concurrent with their occurrence, and those that exist after the onset of emotional response tendencies. Emotional expression is usually short lived (Table 11.7) preparing the individual to engage in a variety of adaptive behaviors, including rapid motor responses, decision making, or providing the information required for change. A person's emotional response tendencies (e.g., approach or avoidance; S1→S1y) are modified by the situation a person finds themselves in, but is also altered by attending to different aspects of the emotion (S1y→a4) being felt. Attending to the emotion gives the emotion a meaning that evolves from the cognitive change induced by a reappraisal process (a4→m2). The resultant response tendencies can be experienced, leading to particular behaviors or having a physiological effect. Also indicated in the model is that these emotional response tendencies can be modified by active suppression. Gross and his colleagues (e.g., John and Gross 2007) go on to use these five elements of emotional

Table 11.8 Regulatory models of subjective well-being

Reference	Subjective well-being model	Regulatory mechanism
Michalos (1985)	Multiple discrepancy theory	Goal matching
Headey and Wearing (1989)	Dynamic equilibrium model	Maintenance of a set-point
Ormel et al. (1997)	Social production function	Optimized goal achievement
Cummins et al. (2002)	A homeostatic model of subjective well-being	Homeostasis
Kim-Prieto et al. (2005)	Time-sequential framework of subjective well-being	Network model[a]

[a]Kim-Prieto et al. (2005) do not label their model as a network model, but I feel what they describe approaches such a model (see text)

regulation to account for how individuals differ in their emotional expression.

A number of neuroimaging studies have confirmed that the cognitive reappraisal contribution to CER is mediated by "at least two independent cortical-subcortical networks" (e.g., Wager et al. 2008; p. 1044). As Wager et al. (2008) states:

> Our findings both replicated prior work and provided information that brain regions implicated in reappraisal are organized into at least two independent cortical-subcortical networks. Replicated findings include: (1) reappraisal-related increases in multiple frontal, parietal, and temporal regions (reviewed in Ochsner and Gross, 2008), (2) decreases in amygdala correlated with self-reported negative affect, and (3) correlations of prefrontal activation with reappraisal success. (p. 1044–1045).

These results confirm Wager's et al. (2008) contention that a mediation model, as opposed to a linear model, best describes the neurocognitive basis of cognitive reappraisal.

The Ochsner and Gross model provide a reasonable approach to how CER occurs. Now I need to relate these processes to the experience of SWB. Three questions can be asked; first, what evidence is available to indicate that momentary fluctuations in CER result in changes in SWB, second how does a stable sense of SWB develop, and third, what evidence is available to suggest that a person's SWB can effect emotional regulation? The first two questions imply that I am distinguishing between more situational and context-dependent momentary determinants of SWB from CER that reflects a stable affective state. This distinction, of course, underlies the different approaches that Schwarz and Strack (1999) and Bramston (2002) have taken to defining SWB.

The data that Schwarz and Strack (1999) provide can be interpreted as supporting the notion that momentary fluctuations in CER result in changes in SWB. I have already partially answered the second question, by reviewing how the feelings of SWB develop (that is, how the person physically and psychologically matures), are enhanced by certain types of training procedures (e.g., learning a sport) or modified by psychotherapy. Each of these interventions usually result in reduced emotional volatility, and this suggests that enhanced CER has occurred. The classic example of this transformation occurs when an adolescent becomes more self-confident and this increases their SWB on the way to adulthood. What is difficult here, of course, is establishing a causal chain. The third question is probably the most difficult question to answer, since SWB is ordinarily considered an abstract construct that can be expressed using figurative language, and is not usually considered a causal determinant.

Gross's (1998, 2001) model is a reasonable approach to how CER occurs, and by inference a reasonable model of how CER contributes to the experience of SWB. However, a number of alternative SWB models have been proposed that do not rely on the assumption that SWB is a product of CER. I discuss these models in the next section.

4.4.2 Regulatory Models of Subjective Well-Being

In this section I will discuss two approaches to the regulation of SWB: investigator-generated models and models based on self-regulation. Table 11.8 summarizes examples of investigator-generated models of SWB, some of which have regulatory functions. I have discussed some of these models in Chap. 4, but now I would like to provide more details about them. What will become clear is that most of these models are descriptive with very little effort being extended to empirically confirm the components of the models. By saying this, however, I do not mean that there are not supporting data for these models, there is data, rather that there are few examples of studies that demonstrate how the various determinants of SWB (e.g., context, personality, genetics) actually impact a SWB assessment. Thus, the elegant studies of Schwarz and Stack (1999) confirming that mood impacts SWB judgment still does not tell me exactly how this happens. For example, do persons who score high on neuroticism subjectively redefine the scale provided to indicate their satisfaction? If they are more emotionally volatile, how does this volatility effect the SWB assessment? In the absence of information of this sort, the models remain descriptive.

Early models of SWB emphasized the role of single major factors, including adaptive processes (Brickman and Campbell 1971; Brickman et al. 1978), personality (Costa and Mc Crae 1980), and major life events (e.g., Abbey and Andrews 1985). These models have now evolved to the point where attempts have been made to integrate these indicators and various parametric determinants (e.g., momentary events, long term processes, regulatory processes, and so on) into a comprehensive model of SWB. Michalos' (1985) offers an early attempt at this effort.

Michalos's (1985) model is formally similar to Campbell's (1972) model who first suggested that when a person's aspiration level is compared to their current situation, an estimation of the person's satisfaction results. Multiple Discrepancy Theory assumes that this cognitive process occurs for each of the domains (marital status, job satisfaction, health status, and so on) important to a person. To the extent that a discrepancy exists, a regulatory process may be activated that is intended to resolve this difference leading to a match between the person's aspiration and achievements.

Headey and his colleagues (e.g., 1989, 2008), have promoted a set-point theory approach to SWB for the last 20 years, or so. The model assumes that each person has had a "normal" set of life events that has or is impacting them and that a equilibrium in life satisfaction has developed that has resulted in the person having a stable sense of SWB. If no change occurs in this pattern, then no change will occur in the person's SWB. What keeps the person's affective state stable in the face of adverse events is their genetic background, personality, and, as I will explain further, their personal goals.

Both Michalos and Headey provide support for their models by the analysis of longitudinal data and in each case they find correlations that they interpret as providing support. Headey (2008), however, noticed that life satisfaction scores of 5–6% of his sample were 1.5 standard deviations above their baseline assessments (a 15-year gap), and that 10–15% of his sample reported scores that were 1.5 standard deviations below their baseline. These changes were sufficient to suggest to him that the set-point theory had to be modified. His response to these data was to examine the changes in life satisfaction but now in terms of the domains (financial status, family, health) of interest to the person. What he found helped him account for some of the changes in his original study, but to do this required a unique way of looking at his data.

Headey (2008), in his paper, cites two papers (Esterlin 2005; Huppert 2005) that directly challenged his model. Headey was particularly interested in Esterlin's (2005) paper, since Esterlin argued that people quickly adapt to improved economic status and show no net change in life satisfaction, while complete adaptation does not occur for noneconomic changes. This suggested to Headey that he apply the game theory approach of distinguishing between zero-sum or nonzero sum decisions to the judgment that people make in assessing their life satisfaction. Thus, the person who pursues material things is bound to end with a net no change or a loss in life satisfaction, while the person whose goals involve having a full family life or engaging in altruistic activities will be more likely to experience a gain in life satisfaction. On the basis of this, Headey (2008) generated a number of hypotheses that he tested in his paper. To quote:

1.1 The greater the importance/priority individuals attach to non zero-sum goals the higher their life satisfaction.
1.2 The greater the importance/priority individuals attach to zero-sum goals the lower their life satisfaction
2.0. On average, satisfaction scores are higher in nonzero sum life domains than in zero-sum domains (indicating that more people can hope to achieve high satisfaction in these domains).
3.1 Individuals who persistently attach high importance/priority to non zero-sum goals are more likely than average to record gains in life satisfaction over time.
3.2 Individuals who persistently attach high importance/priority to zero-sum goals are more likely than average to record losses in life satisfaction over time (p. 217).

Headey used data from the German Socio-Economic Panel Survey to test Hypothesis 1 and 3, and data from Australian Household Income and Labour Dynamics study to test Hypothesis 2. He found supporting data for Hypothesis 1 and 2, but less support for Hypothesis 3. He examined three types of personal goals: zero-sum goals that lead to financial or personal success, and nonzero sum goals of family life or altruism. For Hypothesis 3 he found significant and positive correlations for altruistic and success goals and rating of life satisfaction, but no substantial correlation between family life goals and life satisfaction. He interpreted his data as supporting the view that personal goals are a major predictor of life satisfaction, and, therefore, a contributor to SWB. Thus, the deviations from the set-point model that Headey noted previously could be accountable for by differences in the personal goals of the respondents. These goals, like the person's genetic disposition and personality provide the "content" that is used when the person cognitively and emotionally regulates their sense of life satisfaction. To the extent that they have settled into a particular regulated state than a set-point can be said to exist around which they judge their SWB at any particular moment. How important this regulated state is to the person, and where this fits into their estimate of the quality of their existence remains to be determined.

Like the set-point model of SWB, the Social Production Function (Ormel et al. 1997, 1999) model incorporates a regulatory component. It "asserts that people produce their own SWB by trying to optimize achievement of universal human goals via six instrumental goals within the environmental and functional limitations they are facing." (Ormel et al. 1997; p. 1051). What is optimized are the two universal goals of physical and social well-being. These goals are achieved by maintaining an optimal level of arousal, living in a pleasant environment, satisfying physiological needs (adequate food, medical care), being able to manage personal resources, receiving the positive consequences of appropriate behavior, and benefitting from the involvement with other persons.

Lindenberg, his colleagues (e.g., Ormel et al. 1999), describe Social Production Function theory as integrating psychological and economic perspectives, with a particular emphasis on the tradeoffs that result from limited resources and personal and environmental constraints. Psychological theories have helped specify what goals can be achieved and needs satisfied that lead to feelings of SWB, while economic theories have clarified that people not only consume but also rationally choose cost-effective ways to produce well-being. SWB is produced by a person substituting one activity or goal for another, especially if they have experienced a loss of a previously valued activity or state. Alternatively, a person may note the cost of producing a particular goal and buffer themselves from future losses by altering their means of production (the issue of marginal returns). A number of adaptive mechanisms exist that can be recruited when a life event has changed the person's SWB (Fig. 3; Ormel et al. 1999). Thus, processes such as substitution and buffering constitute regulating mechanisms that stabilize and maintain a person's sense of SWB.

Social Production Function Theory has been strongly influenced by behavioral economic models (e.g., Becker 1976). The economic perspective is reflected in the assumption that each person has a limited set of resources (e.g., genetic background, personality, life events) that they manage so as to optimize their emotional (SWB) status. Economic principles of pricing and marginal costs govern how people made decisions about their priorities and goals. A person achieves their goals through a process of production (Becker 1976; Lindenberg 1996) that may involve substitution and/or buffering. Ormel et al. (1999), however, point out that the economic perspective does not identify universal goals, so that the analysis of SWB is subject to the whims of individual differences, and the relationship between behavior and SWB is not well established. For example, the evidence that greater wealth does not necessarily leads to greater (e.g., Veenhoven 1994) suggests such a disconnect. Still the theory retains many of the principles of behavioral economics and in this regard is unique.

Cummins's et al. (2002) homeostatic model of SWB incorporates the set-point notion of Headey (e.g., 2008), and the multidiscrepancy model of Michalos (1985). The model, as illustrated in Fig. 4.4, describes the relationship between extrinsic indicators (e.g., living situation, life events, or pain) and its relationship to SWB. As depicted in the figure, as the severity of the extrinsic indicators increases, the intensity of SWB increases until a threshold is reached and further changes in SWB stabilize until an upper threshold is reached and then SWB continues to increase. This conceptualization allows for a number of predictions. First, it predicts that there will be a place where increases in extrinsic indicators will have no impact SWB. In addition, it predicts that there will be levels of extrinsic indicators that will be positively correlated with changes in SWB. Cummins et al. (2002) claims that there are data to support both these assumptions.

Cummins et al. (2002), as well as a number of other investigators (e.g., Headey and Wearing 1989; Ormel 1983), assume that SWB is maintained by some brain mechanism, but provide no information on what may be going on in the brain. Cummins et al. (2002; p. 15) provides a model where the dynamic equilibrium model and multiple discrepancy theory are combined to provide some hypothetical processes that could occur in response to changes in extrinsic conditions. For example, if a person perceives that the pain they have in their back is excessive, they might try various adaptive techniques (e.g., rest, medication), but finding that these failed, perceive that they have a need that has been unmet. This would then activate emotional responses, various personality variables, or cognitive buffers (e.g., distraction) that are designed to maintain the person's SWB. Each of these components would be part of the homeostatic mechanism that stabilizes SWB, but also are parts of a regulatory system.

I have described the time-sequential framework of SWB that Kim-Prieto et al. (2005) have proposed as approximating what would be expected for a network model. This comes from my impression that the complex interrelations that the model projects permits such an interpretation. The investigators themselves encourage this interpretation when they say; "We propose that while SWB is a unitary construct, it changes through the passage of time. As such, the different components that make-up the time-sequential framework of SWB are related to each other in systematic ways" (Kim-Prieto et al. 2005; p. 266). Subjective well-being consists, to them, of four major stages that changes over time. These stages include major life events and circumstances, affective reactions to these events and circumstances, recall of past reactions to life events and circumstances, and global judgments that a person has about their life. They also assume that new factors can influence SWB at each stage, with each stage correlating with the stage before and after. They also insist that no one stage gives a complete picture as compared to others, all are needed. They also assume that return loops exist whereby latter stages influence new sets of events. Kim-Prieto et al. (2005) illustrate the complex relationships between the various factors graphically (their Fig. 2), showing large number of possible relationships. If major life events, emotional reactions, recalled feelings, and global judgments are considered nodes then the elements are in place to think of this model as a network. To the extent that the model describes the interaction of various stages of SWB it is also an example of cognitive-emotional regulation.

The five models that I have reviewed are examples of investigator-generated models, but a person can also regulate by self-management. Thus, a person can regulate internal (e.g., the feelings of anger while waiting to be served in a store) or external (e.g., being fired from their job) processes or events, although neither regulatory site is totally independent of the other. Only one of the investigator-generated models has been applied to the self-management of external events. Lindenberg, and his colleagues report a "self-management of well-being" theory (Schuurmans et al. 2005; Steverink and Lindenberg 2008; Steverink et al. 2005). Their theory (Steverink et al. 2005), assessed by questionnaire, assumes that people have six self-management abilities (e.g., self-efficacy beliefs, positive frame of mind, taking the initiative, investment behavior, multifunctionality of resources, and variety of resources) that they apply when they manage their psychological, social, or economic resources. They also assume that these self-management abilities impact various dimensions of well-being (e.g., comfort, stimulation, affection, behavioral confirmation and status). With this emphasis on resource management, as broadly defined, the investigators see their model as an extension of their Social Production Function theory.

Higgins and his coworkers (1999; 2000) have also made self-regulation the focus of a theory, and have applied this theory to quality-of-life assessment. Like White and Dolan (2009), they challenge the usefulness of a purely hedonic approach to assessing well-being, and instead emphasize the role that regulation itself can provide when managing a person's hedonic state. Higgins argues that when a person has a "regulatory focus," as opposed to just seeking pleasure or avoiding pain, then the outcome is different. He points out that Thorndike's (1911) law of effect, and Skinner's (1938) operant principles reflect the purely hedonic perspective that more pleasure results in more positive consequences, and that more punishment may result in more negative consequences. In contrast, "regulatory focus" concentrates on self-regulation towards some desired goals. Thus, his theory, with its focus on attaining goals is a motivational theory: an effort to reduce the difference between the person's current state and their desired goals. Also ongoing, is the effort to avoid undesired outcomes, Higgins assumes that a self-regulatory focus will alter hedonic principles depending on what resources are involved (e.g., nurturance needs, survival needs, security needs, and so on). This can happen in two ways; by a promotion focus, or by a prevention focus. Factors that initiate the prevention focus include nurturance needs, strong ideals, situations which involve gains, while factors that initiate the prevention focus include security needs, what ought to be done, and avoidance of loss situations. The promotion focus leads to certain consequences, such as, an enhanced sensitivity concerning the presence or absence of positive outcomes, cheerfulness in the face of dejection, insuring that decisions are appropriate and are not subject to errors of omission, and also using approach strategies that are designed to insure the person's well-being. In contrast, the prevention focus leads to sensitivity to the absence of presence of negative outcomes, feeling relaxed in the face of agitating emotions, doing what needs to be done to insure correct rejections and avoiding errors of commission, and using avoidance strategies to minimize any impact on the person's well-being.

To this Higgins adds the notion of *regulatory fit*. He describes regulatory fit as occurring when the goals a person wants to achieve and the way they reach these goals are concordant. This is usually achieved by a promotion focus. If the person's pursuit of their goals is considered "nonfit," then the way they pursued their goals may actually disrupt their efforts to achieve their goals. This is considered a prevention focus. He gives the example of two students, one of whom hopes to get an "A" in a course, while another feels that they must get an "A" in the course. To achieve their goals the first student will read additional material beyond the course requirements, while the second student will limit their reading to just what the course requires. Thus, the "mind set" the two students take to achieve their goals will determine the strategies they use, and whether their strategy "fits" their objectives.

Higgins et al. (1999) describes his theory as "a self-regulatory model of emotional experiences." (p. 253). The conceptual components of this motivational model are the promotion focus and prevention focus, as modified by the degree regulatory fit has occurred. This model is relevant to SWB in that it offers an alternative to a purely hedonic model of SWB. Certainly, Kahneman and to a less degree Diener, appear to be advocates of hedonism=SWB model, but Higgins et al. (1999) points out:

> … the hedonic principle does not capture the fact that people's life experiences have as much to do with how they regulate pleasure and pain as it does with the simple fact that they do so. Individuals who experience the pleasure of joy and the pain of disappointment do not have the same life experiences as those who experience the pleasure of relaxation and the pain of nervousness. …Motivational experiences of strategic states, such as feelings eager or cautious, are an important part of life as well. For example, happiness from knowing that one has accomplished one's life goals might not be enough for subjective well-being. …Thus, social policy directed solely at maximizing emotional pleasures might actually undermine quality-of-life. … As has often been noted in classic literature, what makes life worth living is working towards a goal, not simply knowing that it was attainted. (p. 263).

While Higgins is implying that his modification of hedonic theory is a model of quality-of-life assessment, I would not. I would not because it still lacks the explicit assessment of a reflective or qualifying component necessary for generating a hybrid construct. Still his model has some of the characteristics of a eudaemonic approach to SWB, and thus offers an important step in bridging the gap between the various approaches to SWB.

Higgins's model has also received support from a number of neurocognitive studies that deserve to be briefly reviewed (Amodio et al. 2004; Cunningham et al. 2005; Eddington et al. 2007, 2009). Amodio et al. (2004), for example, found greater left vs. right resting EEG's activity when promotion activities were compared to prevention activities. Cunningham et al. (2005) used functional magnetic resonance imaging (fMRI) scanning to compare persons engaged in promotion or prevention focused activities. They found that the promotion focus was associated with greater activity in the amygdala, anterior cingulate, and extrastriate cortex when judging whether positive stimuli were perceived as good/bad (but not abstract or concrete) and they found the same pattern of brain activity when the prevention focus was activated by negative stimuli, and the respondent was asked to judge if the stimuli were good/bad. Eddington et al. (2007) examined the role that a behavioral activation and inhibition motivational systems played in either the promotion or prevention motivation system. They report that they found "preliminary evidence that even while participants were engaged in a judgment task with no relevance to personal goals or approach/avoidance motivation, there was evidence for promotion goal-specific activation following incidental priming of such goals." (p. 1161) Their study suggests that there is a neurocognitive analog of what has been previously referred to as social-cognitive regulatory system.

The last example of a self-regulatory model I will briefly present, of the many available models, is based on the work of Howard Leventhal and his coworkers. Leventhal describes his model as an example of a self-regulation model of adaptation to health threats (e.g., Leventhal et al. 2001). He calls his model "a common sense model of self-regulation," and while he does not claim to be studying SWB, the model he proposes can easily be seen as occupying the same conceptual space. Thus, the model assumes that both a perceptual and conceptual processing systems exist that initiate cognitive and emotional processes that evoke coping strategies and appraisals leading to feedback to the processing system. It is this feedback system that mediates the self-regulatory process. Leventhal et al. (2001) applies his model to account for how a person responds to bodily symptoms and perceived health threats.

Leventhal's model is important because it illustrates that cognitive-emotional self-regulation not only occurs for persons in the general population, but can also occurs to the medically ill. For example, Detweiler-Bedell et al. (2008) apply the model to persons who have both a chronic disease and suffer from major depression. They hypothesize that the difficulty in managing persons with comorbid conditions is confounded by failures in self-management. They speculate that this may occur in several ways: under regulation of the chronic disease because of the person's depressed mood, underregulation of mood because of the person's chronic disease, misregulation of disease management because of how the person's mood was managed, and poor management of the person's mood because of the need to manage the chronic disease. Evidence is reviewed that is supportive of these claims. For example, it has been proposed that various physiological factors link depressed mood and chronic diseases. Evidence in support of this assumption includes the observation that the presence of a chronic disease predicts the onset of depression (e.g., Aneshensel et al. 1984; Bruce and Huff 1994; Cole and Dendukuri 2003), while the presence of depression along with a chronic disease predicts poor outcome (e.g., Billings and Moos 1985). What has also been observed is that psycho-educational programs have been helpful in disease management. Evidence of this sort supports the view that self-regulation has a role to play in the management of comorbid conditions, and ultimately in assessing SWB.

This review should be sufficient to demonstrate that SWB may very well be a product of cognitive-emotional regulation. Previously, I have described SWB as having properties of a conceptual metaphor, but that it is also an abstract construct that is represented by a variety of concrete indicators. This collection of descriptors immediately makes SWB a fairly complex construct, and this complexity will become evident when I discuss the next topic, the neurobiological basis of SWB.

5 Modeling of Subjective Well-Being

Subjective well-being, as is true for any abstract construct, has a cortical representation, and in this sense can be said to exist. If it exists, then it should be possible to demonstrate some specific cortical activity pattern that can be uniquely labeled as SWB. I would also expect that this cortical pattern to reflect the characteristics of SWB that have been identified in this chapter. For example, SWB, more than likely has conceptual metaphoric properties, being the object of a mapping from concrete emotions (e.g., happiness, joy). However, as I have discussed, SWB is not a discrete emotion, and if anything it is best seen as a product of some complex cognitive-emotional regulatory process. Thus, studies of the neurobiological basis of emotions (e.g., Davidson 2004) would not per se be informative about the neurobiological basis of SWB, unless SWB was seen as a linguistic label that could be attached to a particular emotion-based cortical pattern. I, will review this literature, but also present in this section alternative approaches that postulates that the cortical representation of SWB can be conceptualized in terms of neural networks, or alternatively as involving the simulation of centrally mediated emotion-based processes. Since this third approach evolved from the literature on embodiment (e.g., Barsalou 1999, 2008; Niedenthal 2007; Wilkowski et al. 2009), it can be referred to as a embodied simulation model of SWB. As I will show, these models to varying degrees provide different approaches to how the cortical pattern underlying SWB developed. First, I will review some of the research Davidson and his colleagues have reported, since they seem to have done the most work outlining the neurobiology of SWB.

5.1 The Neurobiology of Subjective Well-Being

Urry et al. (2004) report a study in which 84 subjects filled out questionnaires designed to assess positive affect, but also hedonic or eudaemonic forms of SWB before a electroencephalographic (EEG) study, while at rest. Based on previous research (e.g., Davidson et al. 2000) it was predicted that the left prefrontal cortex would be differentially activated for persons with a positive affective disposition, while right prefrontal area would more likely activate a negative affect disposition. Positive or negative affect was determined by the person's responses to the Positive and Negative Affect Schedule (PANAS; Watson et al. 1988). The PANAS consists of 10 items addressing positive and 10 items addressing negative affect. Hedonic well-being was assessed by the Satisfaction With Life Scale (SWLS; Diener et al. 1985), while eudaemonic well-being was assessed by the Scales of Psychological Well-being (PSW; Ryff 1989a). Correlations were calculated between right-left asymmetry, left hemisphere and right hemisphere activity alone, and the two PANAS summary indicators, the total SWLS, and the total and six subscales of the PWB. Significant correlations were found for six of the seven PWS scales for the asymmetry indicator, four of the seven for the left hemisphere but none for the right hemisphere activity. A significant correlation was found for the asymmetry indicator and the SWLS, but not for the left or right hemisphere activity. Only the PA dimension of the PANAS was statistically significantly correlated with asymmetry indicator, but not for the left or right hemisphere activity indicators.

These data, plus a variety of other studies reported by this group (e.g., Tomarken et al. 1990, 1992), were interpreted as providing evidence that the left more than the right hemisphere was associated with a person's disposition to be positive. They claim that the left prefrontal cortex is active in response to appetitive stimuli and approach to the source of stimulation. They interpreted the correlations between scores on the eudaemonic assessment of well-being and hemispheric asymmetry and left hemispheric EEG activity as evidence that this particular pattern of cortical activity also mediates positive well-being.

Davidson (2004), has summarized the available neural site data that demonstrated that a particular neural activity pattern mediates SWB. His basic argument is that SWB is a form of emotional expression and the neural sites that mediate emotions (the prefrontal cortex, the amygdala, the hippocampus, and so on) would be major determinants of SWB. For example, Davidson (2004) states that "low basal levels of amygdala responding, effective top-down regulation and rapid recovery, characterize a pattern that is consistent with high levels of well-being." (p. 1400). In contrast, high levels of anterior cingulate cortex activity would facilitate emotional regulation and maintain positive SWB. In a more recent study, van Reekum et al. (2007) confirmed these earlier findings demonstrating that individual differences in amygdala and ventromedial prefrontal cortical activity were associated with the speed with which persons responded to negative or neural stimuli. What they found was that those persons who responded slower to negative vs. neutral stimuli also scored more positively on the PWB, and that this was mediated by less activation of the amygdala and more activation of the anterior cingulate cortex. Thus, there seems to be distinct patterns of neural activity associated with a person's appraisal of the threat created by a negative stimulus, and responses to these threats seems to be dependent on the person's positive disposition, which presumably refers to the person's personality or genetic background.

The neurobiological model that Davidson and his colleagues have identified tells me what areas in the brain SWB may be dependent on, but it does not tell me how these areas became involved, unless of course, it is assumed that the person was predisposed by personality and genetics to project a positive or negative sense of well-being. This does seem to be part of Davidson's argument, but he and others also argue that SWB can be enhanced and modified (e.g., Ekman et al. 2005) so that how this could occur in the face of a person's disposition that has yet to be explained. An alternative approach to modeling SWB is offered by the literature on *cognitive embodied simulation*.

5.2 Nonlinear Modeling, Appraisal and Subjective Well-Being

Scherer and his colleagues (e.g., 1984, 2001; Sander et al. 2005) have, for a number of years, been interested in developing a neural network-based computational model of the emotions. Their model conceives of emotions as being a product of a continuous process of appraisal and reappraisal of objects, events, and experiences. Thus, the appraisal process functions as the central cognitive mechanism that drives the type and expression of emotions. Scherer also assumes that the evoked emotions are capable of feeding back and

6 Summary of Chapter

The question asked in the beginning of this chapter was whether well-being could be used as an indicator of quality or quality-of-life. To answer this question I have tried to identify where the two indicators linguistically and cognitively overlapped, but I have also addressed what the term well-being, itself, means. I noted that well-being was used in a number of different ways, evoking a number of different meanings, but the one that was of most relevance to this discussion was the subjective assessment of well-being. To determine the nature of SWB, I examined whether it is a discrete emotion, a product of a cognitive-emotional processes, or a product of a cognitive-emotional regulation. I also considered whether SWB is a linguistic label, such as an "umbrella term," or whether it is an abstract concept that is a product of metaphoric mapping. As a category, I considered whether SWB is a natural kind, an artificial category, or a category of artifacts. I also considered various models of emotions, and by inference SWB. Included among these were the models based on neurobiological studies of SWB, neural networks, and embodied simulation.

This review has permitted me to conclude that the assessment of hedonic well-being is less likely to approximate a qualitative assessment, than an estimate of well-being based on a eudaemonic approach. Both models of well-being can provide evaluated descriptors that if formally qualified could be part of a quality-of-life or HRQOL assessment. However, an eudaemonic approach to well-being considers a broader array of indicators and thus could be more easily qualified. In this regard, it is interesting to read what Diener et al. (1998) consider to be the difference between their and Ryff and Singer's (1998) approach. They quote:

> In contrast to Ryff and Singer's approach, the study of subjective well-being pays more attention to people's values, emotions, and evaluations, and does not grant complete hegemony to the external judgments of behavioral experts. We further believe that the characteristics that are functional and lead to well-being, …vary substantially between cultures and in different life circumstances. (Diener et al. 1998; p. 33–37).

What Diener et al. seems to be objecting too is that Ryff and her colleagues are excluding the person from participating in deciding how well they have lived their life, and this, combined with the likelihood that what may be ideal in one community may not generalize to another, is sufficient for them to seriously question the Ryff approach. As I read their debate, it seems to be pitting a self-reported judgment against some idealized version of a lived life, but also a more selected vs. broader realm of indicators. Both speak in terms of the importance of a reflective process, but neither directly access it, with Diener neither indicating how or when the person's assessment has occurred, while Ryff relies on some external standard (e.g., an investigator or philosopher) to judge its presence.

The current practice of assuming that SWB is an indication of quality-of-life derives from the connotative capacity of the phrase. Specifically, the phrase connotes a value orientation – well-being is a desirable and valued state – and this makes the substitution of the phrase for quality-of-life or HRQOL seem natural. There are several reasons why this connotative capacity is in place. First, it is quite likely that qualification is an ongoing process so that when a person has to make a judgment they may already have a relevant value system in place concerning their lived life. However, the extent this is true is usually unknown so that a formal assessment is required, as opposed to be being inferred to be present. An alternative, is that the site of the judgment is semantically complex (or at least vague), which would encourage some semantic drift in the terms being used. I can provide some evidence to support this last point by examining a study by Cher et al. (2009) where the semantics of a qualitative judgment is simplified by the presence of a more concrete task.

Cher et al. (2009) report a study where they attempted to map the semantic space surrounding odors. The investigators point out that an extensive research literature has demonstrated that odors elicit emotional responses, so that the semantic space generated would also be informative of the semantic space surrounding the emotions elicited. Initially a group of 480 terms (147 representing emotion terms experienced in everyday life with no reference to odors, and 333 terms that were common to odors) were identified, and subjects were asked to rate each of these terms as to their relevance to the experience of smelling an odor. From these ratings, a list of 124 terms was identified, with each term being affirmed by 66% or more of the respondents. This list was than inspected by a panel of 10 investigators who found that the list could be split into terms that reflected affect ($N=68$) and those that reflected some quality ($N=54$) of the odors (overlapping items were excluded). This list was then used in an experiment during which respondents were asked to smell odors, and rate the intensity the odors evoked in response to affective or qualitative terms. A factor analysis of the responses to the affective terms revealed five principle factors (happiness -well-being, awe-sensuality, disgust-irritation, soothing-peacefulness, energizing -refreshing), while a factor analysis on the responses to qualitative items revealed four principle factors (delicacy feature, heaviness feature, healthiness feature, sweetness feature).

This study was of interest because it so clearly revealed that in response to concrete sensory stimuli (odors) it was possible to separate the semantic space into affective and qualitative domains, and that this separation was empirically supportable (e.g., the factor structure revealed some internal consistency in the investigator-based division of terms). Also observed was that the terms happiness and well-being were considered affective but not qualitative terms. This study is one of the few that explicitly included well-being in a study of the semantic space surrounding affective terms, and the results provide further support for considering SWB and quality-of-life as separate, but overlapping terms or concepts. An experiment of this sort could be repeated with a group of wine tasters, describing the differences between different wines.

One explanation of these results is that by increasing the concreteness of the stimuli the investigator made it easier for a respondent to distinguish between the underlying dimensions involved in the judgment. Thus, it is easier to declare that a poodle (a concrete object) is a dog, and not a stone, then it is to declare that a dog is an animal, especially since a dog may be a stuffed animal, a statue, or some other form of representation. What the Cher et al. (2009) study implies, therefore, is that the current practice of substituting SWB for quality-of-life or HRQOL is an artifact of the larger semantic space that surrounds one concept as opposed to the other.

Another distinction between SWB and a quality-of-life assessment is that SWB primarily describes the state a person is in, while quality-of-life has more of an instrumental imperative. This connotative difference becomes easier to see if I remember that quality-of-life or HRQOL assessments are subclasses of a boarder domain of quality assessments. For example, if I was to assess the quality of a car it might involve me taking the car for a test run, reviewing its breakdown history, and so on, while I am not nearly as likely to do anything like this when evaluating my SWB. This difference should be sufficient to distinguish the two concepts. Operationally, therefore, the two concepts are quite distinct and I on the basis of these and other observations recommended that the two concepts be kept distinct and not interchanged.

A second objective of this chapter was to clarify the role that emotions and emotion terms play in a qualitative assessment. It was clear from the literature reviewed that the feelings a person has about an object, event, or life lived can affect the assessment of these experiences (e.g., Schwarz and Strack 1999). What remained to be determined is the relative contribution of these emotions to the cognitive processes that mediate a qualitative judgment. Certainly relevant here is the issue of whether cognitive processes (e.g., appraisal) precede, follow, or occur together with emotions. Information of this sort is not available when a person is asked to generate a valuation, preferences, or utilities. For example, it would be very interesting to learn if a person who makes a fast and automatic response during a utility estimation does so by having a more immediate emotional response to the task, than say someone who is more delayed in their judgment. Thus, experimentally examining the interaction and dynamics of the emotions and cognitions, particularly the appraisal process, could be quite informative about the nature of a qualitative assessment. It would also be expected to further differentiate SWB from a qualitative assessment.

Enhancing SWB has also become the object of many clinical interventions and policy initiatives, but as the quotation that starts this chapter indicates, in the United States it is the pursuit of happiness that is declared a right not the achievement of happiness. This difference is an important one, since it shifts the responsibility that exists for individuals, clinicians, and politicians from creating a state (e.g., a state of SWB) to the instrumental one of proving the opportunity to achieve happiness.

Notes

1. The statement I made here is consistent with the James-Lange theory, which posits that my phenomenology is determined by prior bodily processes, such as an emotional response. I recommend an article by Koriat et al. (2006) who provides an interesting discussion about this issue but now discussed in the context of metacognition. Their point is that, at least at the metacognitive level, behavior (or bodily responses) may determine subjective experience, but that subjective experience may also determine behavior.
2. Dictionaries differ, somewhat, in how they define well-being. Many of these definitions refer to a person being prosperous, as well as other mentioned features. Well-being and welfare are associated, with welfare presumably functioning as a general category that includes well-being. The term welfare, of course, can refer to individual or societal issues, and I will be interpreting this term, at least as a first approximation, as referring to an individual's well-being.
3. Some of you might argue that all the indicators included in the phrase SWB have a common origin in emotion or mood terms, and that this can function as a common organizing principle. However, as I will discuss, the measures commonly considered to be SWB indicators do not easily fit this type of classification approach. For example, a satisfaction indicator is generated by a distinctly different cognitive process than a descriptive affect statement. A satisfaction assessment involves two or more cognitive processes: it is a comparison. While, a happiness assessment can be a direct affective response: "Are you happy? Yes."
4. Some may argue that money is an object that like an emotion may determine our feelings of well-being. However, our financial status may be only one of the factors correlated with our sense of well-being, so that it would not be appropriate to consider well-being a primary emotion.
5. I will use the phrase "emotion words" since this is common in the literature, but also realize it tends to defeat the purpose of using such words more precisely. As used here, "emotion words" will also refer to terms used to express well-being, moods, feelings, and the wide variety of terms used to express particular emotions.
6. There are various types of *speech acts* (Bach 2006); constatives (e.g., affirming, alleging, and so on), directives (advising, excusing, warning, and so on), commissives (agreeing, inviting, volunteering and so on), and acknowledgments (apologizing, greeting, accepting, and so on).
7. The reader may point out that reversing the sentence, "Vietnam is another Iraq" would also not retain the metonym properties of the original sentence. However, both terms in the sentence are highly salient with many concrete examples of the terms which can contribute to the meaning they engender. This is much less so for the

terms "well-being and quality-of-life", since both are considerably more abstract, and dependent on their context-dependent definitions and cognitive role to distinguish between them.

8. Diener et al. (1998) state, "Subjective well-being is a person's evaluation of his or her life. The evaluation can be in terms of cognitive states such as satisfaction with one's marriage, work and life, and it can be in terms of ongoing affect (i.e., the presence of positive emotions and mood, and the absence of unpleasant affect). I believe that subjective well-being results from people having a feeling of mastery and making progress towards their goals, from one's temperament, immersion in interesting and pleasurable activities, and positive social relationships" (p. 34). This quotation is representative of the type of statement which contributes to the confusion between the terms quality-of-life and well-being, since it claims an evaluation has occurred without assessing it independent of a description of the person's state. For example, making progress towards one's personal goals is not evidence, in and of itself, of a particular quality-of-life. Of course, it could be declared a favored outcome but that would reflect the investigator's perspective; not necessarily what a person would prefer. As previously mentioned, what this statement lacks is evidence of a valuation of the evaluation; an independent cognitive assessment or preference statement by the respondent for being in a particular valued state; and an act characteristic of a hybrid construct. I will discuss this in more detail as I proceed.

9. These items represent modifications of the positive affect items from the Bradburn Affect Balance Scale (Bradburn 1969).

10. What would be depicted neurocognitively, for example, is the spatial display characteristic of a container, or the movement across space of the motion metaphor implicit in the term e-motion. The content associated with these representations would be of secondary interest. What I am describing here, of course, is the same idea I spoke about in Chap. 7 when I suggested that what was common to all qualitative assessments was not their content but rather a common cognitive mechanism which leads to the formation of a hybrid construct.

11. Kövecses (1991) in his article uses the word "value" to refer to the quantity of happiness, but I will use the term "quantity" to prevent any confusion that may arise in what is already a fairly complex corpus associated with the word "value."

12. The term valence is a good example of a duel usage term, with its literal meaning referring to the number of electrons which are shared by atoms that combine to form a molecule, as opposed to its figurative form which alludes to the force characteristic of an entity.

13. By claiming that SWB has substance does not conflict with SWB being an abstract concept, since brain imaging studies have demonstrated that abstract concepts can generate distinctive cortical patterns.

14. An exception is Bramston's (2002) paper which considers many of the same issues discussed here, although his assessment of the literature is somewhat dated compared to the current review.

15. The distinction between subjective and psychological well-being has also been characterized in terms of the difference between hedonic and eudaemonic conceptualizations of SWB (Ryan and Deci 2001).

16. The classic example of Phinneas Gage (Chap. 11, p. 264), and his change in personality following a frontal lobe lesion, is an example of the consequence of disrupting the functional inhibition system mediating emotional processes.

17. Averill (1990) suggests, as an alternative, that the German word, *Gemütsbewegung*, is used by ordinary people to refer to the term emotion. If *Gemütsbewegung* is disaggregated, then *Gemüts* can be defined as referring to feelings, while *bewegung* would be defined as referring to emotion or agitation. Together these terms maybe read as "feelings of emotion or agitation".

18. When a professional German-English translator was consulted (Personal Communication: E. Russon, December 7, 2006), the word *wohlbefinden* was suggested as an optimal translation for the term well-being, with the noun, *Befinden* used to refer to health. Translated in this way the term well-being would mean "good health".

19. The term *feeling* has a number of definitions (The Complete Oxford English Dictionary 1970), including being a physical sensation, a sensation perceived by touch, an emotion, the capacity or readiness to feel, what one feels in regard to something, the general tone or a building or work of art, and so on. An analysis of the *synonyms* (Roget's et al. 1992) associated with the terms *emotion* and *feeling*, reveals that three terms are considered synonyms of emotion (e.g., feeling, excitement, and attitude), while 12 terms are associated with the term feeling. Of these, the terms, feeling/emotion, and attitude were found to be common to both terms and can be considered to provide "core meaning" (Table 11.2). Also of interest, is the observation that the more abstract term, emotion, had significantly fewer synonyms than the more concrete term, feeling. This is consistent with the existence of an informational hierarchy, and suggests the hypothesis that a large *semantic domain* is associated with a term's presence at a basic level of information which may also be associated with its presence in different languages.

20. The assumption that SWB is an emotion is quite evident in the positive psychology literature which directly associates SWB with happiness, but rarely if ever defines how the terms are similar or different. At best, SWB functions as a label that refers to a variety of positive emotional terms, such as joy, happiness, and so on. A recent newspaper article by Max (2007) entitled "Happiness 101" makes the association between SWB and happiness quite clear, but now in the popular press, although the author refers to the existence of the association in the academic literature.

21. The observation that emotion words have different syntactic roles in different cultures raises questions about whether the translation between languages can capture the same meaning as communicated by the reference language. The same question can be asked of whether the meaning associated with well-being can be accurately translated between languages.

22. Aesthesis is defined as *the sensation of touch*.

23. One way a natural kind may occur is if it occurs automatically, unconsciously, much as occurs when a category is formed.

24. It is interesting to briefly review the history of the notion of a *natural-kind*, since it evolved from a practical problem (Hacking 2002). In the mid-nineteen century, European biologists and explorers were returning from expeditions with various types of plants and animals. The diversity of these items was so great that it became obvious that this material had to be organized into various classes (a problem in taxonomy) to make some sense of them. The question, then became could these items fit into either *natural* categories or *artificially* (arbitrarily) created domains? Thus, all of Darwin's finches fit into a unique category, in that all had a particular type of beak, but their coloration, size, and so on were different requiring a more arbitrary classification.

25. I encourage the reader to read the Barrett's (2006) paper, since I will only be able to summarize some of the content she presents. For an alternative view, see the paper by Panksepp (2007).

26. In semantics and literary theory, connotation and denotation reflect the figurative or literal meaning of words. Thus, if you saw an image of a rose, its denotation would be the physical state of a flower on a stem, but its connotation would be as a symbol of passion and love.

27. What I am claiming is that in order to perform an evaluation a referent has to be unconsciously (e.g., by activating a specific cortical site) or consciously acknowledged.

28. Watson and Tellegen (1985) and Davern et al. (2007) discuss the history and background to the circumplex approach to representing affect.

29. Thus, Russell (2003) and Clore and Colcombe (2003; Table 11.1) have similar views.
30. Tesser and Martin (1996) provide a bit of history here that is useful. They point out that originally an attitude was thought to reflect what people liked or disliked about objects, persons, or events, but that in the 1950s a number of theories were introduced that suggested that attitudes could also be effected by various cognitive processes. For example, Heider's (1958) balance theory, Festinger's (1957) dissonance theory, and Rosenberg's (1956) consistency theory all suggested a role for cognition and behavior in attitudes.
31. Out of necessity, I will not be able to review the history of the development of the concept of emotion. Instead, I recommend Joseph Le Doux's book, "The Emotional Brain…" (1996; Chap. 30) for a very readable description of this history.
32. Gibson believes that it is inappropriate to study vision as an isolated system, independent of its environment. He claims that I perceive at the level of flat, rigid or extended surfaces, and substances rather than at the level of particles or atoms. He states: "… the affordance of the environment are what it offers the animal, what it provides or furnishes, either for good or ill" (Gibson 1979; p. 127). Thus, an affordance exists relative to the capabilities of a person, yet is independent of the person's ability to perceive it, and does not change as the person's needs and wants change. It also should be noted that Gibson's notion of affordance is invariant, reflecting the constancy in our perception in the face of an ever changing world. While Lazarus does not state this directly, the implication here is that the appraisal process may also have invariant characteristics.
33. If you asked me to define what I mean by the term description, in the three components of a hybrid construct (that is, *description, evaluation* and *qualification*) I would also refer to Neisser's (1967; p. 4) description of a cognition.
34. The model that Sanders et al. (2005) present has a distinct similarity to both the Kim-Prieto et al. (2005) and Wilson and Cleary (1995) models, in that it attempts to trace the relationship of a number of indicators. It differs from the Smolensky and Legendre (2006) model in that it can't, at this point, offer a mathematical description of the relationship between the components of the model.

References

Abbey A, Andrews FM. (1985). Modeling the psychological determinants of life quality. *Soc Indic Res*. 16, 1–16.
Abbott RA, Ploubidis GB, Huppert FA, Kuh D, et al. (2006). Psychometric evaluation and predictive validty of Ryff's psychological well-being items in a UK birth cohort sample of women. *Health Qual Life Outcomes*. 4:76. doi. 10.1186/1477-7525-4-67. (p. 1–17).
Alexandrova A. (2005). Subjective well-being and Kahneman's 'Objective Happiness". *J Happiness Stud*. 6, 301–324.
Alvin DF, Jackson DJ. (1979). Measurement models for response errors in surveys: Issues and applications. In, (Ed.) KF Schuessler. *Sociological Methodology 1980*. San Francisco CA: Jossey-Bass. (p. 68–119).
Amodio DM, Shad JY, Sigelman J, Brazy PC, Harmon-Jones E. (2004). Implicit regulatory focus associated with asymmetrical frontal cortical activity. *J Exp Soc Psychol*. 40, 225–232.
Andrews FM, Withey SB. (1976). *Social Indicators of Well-being: American's Perceptions of Life Quality*. New York NY: Plenum Press.
Andrews FM, McKennell AC. (1980). Measures of self-reported well-being: Their affective, cognitive, and other components. *Soc Indic Res*. 8, 127–155.
Aneshensel CS, Frerichs KK, Clouse RE, Lustman P. (1984). Depression and physical illness: A multiwave, nonrecursive causal model. *J Health Soc Behav*. 25, 350–371.
Argyle, M. (1999). Causes and correlates of happiness. In, (Eds.) D Kahneman, E Diener, N Schwarz. *Well-being: The Foundations of Hedonic Psychology*. New York, NY: Russell Sage Foundation. (p. 353–373).
Aristotle. (1985) . *Nicomachean Ethics*. Translated by Terence Irwin. Indianapolis IN: Hackett Publishing Co.
Arnold MB. (1960). *Emotion and Personality: Psychological Aspects*. New York NY: Columbia University Press. Vol. 1.
Askan N, Kochanska G. (2005). Conscience in childhood: Old question, new anwers. *Dev Psychol*. 41, 506–516.
Atkinson T. (1982). The stability and validity of quality of life measures. *Soc Indicator Res*. 10, 113–132.
Averill JR. (1990). Inner feelings, works of the flesh, the beast within, diseases of the mind, driving force, and putting on a show: Six metaphors of emotion and their theoretical extensions. In, (Ed.) DE Leary. *Metaphors in the History of Psychology*. New York NY: Cambridge University Press. (p. 104–132).
Bach K. (2006). *Speech Acts. Routledge Encyclopedia of Philosophy*. On the Web; http://online.sfsu.edu/kbach/spchacts.html. Accessed December 13, 2006.
Bargh JA. (1997). The automaticity of everyday life. In, (Eds.) S Chaiken, Y Trope. *The Automaticity of Everyday Life: Advances in Social Cognition* (Vol. 10). Mahwah NJ: Erlbaum. (p. 1–61).
Baron RM. (1988). An ecological framework for establishing a dual-mode theory of social knowing. In, (Eds.) D Bar-Tal, AW Kruglanski. *The Social Psychology of Knowing*. New York NY: Cambridge University Press. (p. 48–82).
Barrett LF. (2006). Are emotions natural kinds? *Perspect Psychol Sci*. 1, 28–58.
Barsalou LW. (1999). Perceptual symbol systems. *Behav Brain Sci*. 22, 557–660.
Barsalou LW. (2003). Situated simulation in the human conceptual system. *Lang Cogn Process*. 18, 513–562.
Barsalou LW. (2008). Grounded cognition. *Annu Rev Psychol*. 59, 617–645.
Bauer RM. (1984). Autonomic recognition of names and faces in prosopagnosia: A neuropsychological application of the guilty knowledge test. *Neuropsychologia*. 22, 457–469.
Bechara A. (2004). Disturbances of emotion regulation after focal brain lesions. *Int Rev Neurobiol*. 62, 159–193.
Becker GS. (1976). *The Economic Approach to Human Behavior*. Chicago Ill: University of Chicago Press.
Bentham J. (1789/1996). *An Introduction to the Principles of Morals and Legislation*. (Eds.) J Burns, HLA Hart. Oxford UK: Clarendon Press.
Berlin B, Kay P. (1969). *Basic Color Terms : Their Universality and Evolution*. Berkeley and Los Angeles CA: University of California Press.
Billings AG, Moos RH. (1985). Psychosocial processes of remission in unipolar depression: Comparing depressed patients with matched community controls. *J Consult Clin Psychol*. 53, 314–325.
Bless H, Igou ER, Schwarz N, Wänke M. (2000). Reducing context effects by adding context information: The direction and size of context effects in political judgment. *Personal Soc Psychol Bull*. 26, 1036–1045.
Bok S. (2004). *Rethinking the WHO Definition of Health*. Harvard Center for Population and Developmental Studies: Working Paper Series. Volume 14 (7). (p. 1–21).
Bonsignore M, Barkiow K, Jessen F, Heun R. (2001). Validity of the five-tem WHO well-being in an elderly population. *Eur Arch Psychiatr Clin Neurosci*. 251 Suppl 2, II27–II31.
Bowling A. (1991). *Measuring Health: A Review of Quality of life Measurement Scales*. Philadelphia PA: Open University Press.
Bradburn NM. (1969). *The Structure of psychological Well-being*. Chicago Il: Aldine.
Bradley MM, Lang PJ. (2000). Measuring emotion: Behavior, feeling, and physiology. In, (Eds.) RD Lang, L Nadel. *Cognitive Neurocognition of Emotion*. New York NY: Oxford University Press.

Bramston P. (2002). Subjective quality of life: the affective dimension. In, (Eds.) E Gullone, RA Cummins. *The Universality of Subjective Wellbeing Indicators.* Dordrecht The Netherlands: Kluwer. (p. 47–62).

Brandstätter H. (1991). Emotions in everyday life situations. Time sampling of subjective experience. In, (Eds), F Strack, M Argyle, N Schwarz. *Subjective Well-being: An Interdisciplinary Perspective.* Oxford UK: Pergamon Press.(p. 173–192).

Brickman PD, Campbell DT. (1971) Hedonic relativism and planning the good society. In, (Ed.) MH Appley *Adaptation Level Theory.* New York, Academic Press.

Brickman PD, Coates D, Janoff-Bulmann R. (1978). Lottery winners and accident victims: Is happiness relative? *J Personal Soc Psychol.* 36, 917–927.

Bruce ML, Huff RA. (1994). Social and physical health risks for frist-onset major depressive disorder in a community sample. *Soc Psychiatr Psychiatr Epidemiol.* 29, 165–71.

Cacioppo JT, Gardner WL, Berntson GG. (1999). The affect system has parallel and integrative processing components: Form follows function. *J Personal Soc Psychol.* 76, 839–855.

Campbell, A. (1972). Aspiration, satisfaction and fulfillment. In, (Eds.) A Campbell, P Converse. *The Human Meaning of Social Change.* New York NY: Russell Sage (p. 441–446).

Campbell A, Converse PE, Rodgers WL. (1976). *The Quality of American Life.* New York NY: Russell Sage Foundation, Rutgers University Press.

Campos JJ, Frankel CB, Camras L. (2004). On the nature of emotion regulation. *Child Dev.* 75, 377–394.

Cantril H. (1965). *The Pattern of Human Concerns.* New Brunswick NJ: Rutgers University Press.

Casasanto D. (2009). Embodiment of abstract concepts: Good and bad in right- and left handers. *J Exp Psychol Gen.* 138, 351–367.

Cassell's *New German and English Dictionary.* (1936). New York NY: Funk and Wagnalls.

Cato MA, Delis DC, Abildskov TJ, Bigler E. (2004). Assessing the elusive cognitive deficits associated with ventromedial prefrontal damage: A case of a modern Phineas Gage. *J Int Neuropsychol Soc.* 10, 453–465.

Chapman RL. (1992). *Rogert's International Thesaurus.* New York NY: Harper-Collins.

Charles ST, Reynolds CA, Gatz M. (2001). Age-related differences and changes in positive and negative affect over 23 years. *J Personal Soc Psychol.* 80, 136–151.

Chrea C, Grandjean D, Delpanque S, Cayeux I, et al. (2009). Mapping the semantic space for the subjective experience of emotional responses to odors. *Chem Senses.* 34, 49–62.

Clore GL, Colcombe S. (2003). The parallel worlds of affective concepts and feelings. In, (Eds.) J Musch, KC Klauer. *The Psychology of Evaluation: Affective Processes in Cognition and Emotion.* Mahwah NJ: Erlbaum. (p. 335–369).

Clore GL, Ortony A, Foss MA. (1987). The psychological foundations of the affective lelxicon. *J Personal Soc Psychol.* 53, 751–766.

Clore GL, Schnall S. (2008). Affective coherence: Affect as embodied evidence in attitude, advertising and art. In, (Eds.) GR Semin, E Smith. *Embodied Cognition: Social, Cognitive, Affective, and Neuroscientific Approaches.* New York NJ: Cambridge University Press. (p. 211–2360)

Cole MG, Dendukuri N. (2003). Risk factors for depression among elderly community subjects: A systematic review and meta-analysis. *Am J Psychiatr.* 160, 1147–1156.

Cole PM, Martin SE, Dennis TA. (2004). Emotion regulation as a scientific construct: Methodological challenges and directions for child development research. *Child Dev.* 75, 317–333.

Costa PT, Mc Crae RR. (1980). Influences of extravision and neuroticism on subjective well-being. *J Personal Soc Psychol.* 38, 668–678.

Costa PT, Mc Crae RR, Zonderman AB. (1987). Environmental and dispositional influences on well-being: Longitudinal follow-up of an American national sample. *Br J Psychol.* 78, 299–306.

Coulson S. (2006). Conceptual Blending in Thought, Rhetoric, and Ideology. In, (Eds.) G. Kristiansen, R. *Cognitive Linguistics: Current Applications and Future Perspectives.* Amsterdam Netherlands: John H. Benjamins. (p. 187–210).

Crisp R. (2005). *Well-being*. Stanford Encyclopedia of Philosophy. Accessed on the Web (10/27/2006). http://plato.stanford.edu/entries/well-being.

Crooker KJ, Near JP. (1998) Happiness and satisfaction: Measures of affect and cognition? *Soc Indic Res.* 44, 195–224.

Csikszentmihalyi M, Larson R. (1987). Validity and reliability of the experience sampling method. *J Nervous Dis.* 175, 526–537.

Cummins RA. (2005). Moving from the quality of life concept to a theory. *J Intell Disabil Res.* 49, 669–706.

Cummins RA, Gullone E, Lau ALD. (2002). A model of subjective well-being homeostasis: The role of personality. In, (Eds.) E Gullone, RA Cummins. *The Universality of Subjective Well-being Indicators.* Dordrecht Netherlands: Kluwer. (p. 7–46).

Cummins RA, Stokes MA, Davern MT. (2007). Core affect and subjective well-being: A rebuttal to Moum and Land. *J Happiness.* 8, 457–466.

Cunningham WA, Raye CL, Johnson MK. (2005). Neural correlates of evaluation associated with promotion and prevention regulatory focus. *Cogn Affect Behav NeuroSci.* 5, 202–211.

Damasio AR. (1994). *Descartes' Error: Emotion, Reason and the Human Brain.* New York NY: Harper Collins.

Damasio A. (1999). *The Feelings of What Happens: The Body and Emotions in the Making of Consciousness.* San Diego CA: Harcourt.

Damasio A, Grabowski TJ, Bechara A, Damasio BH, et al. (2000). Subcortical and cortical brain activity during the feeling of self-generated emotions. *Nat NeuroSci.* 3, 1049–1056.

Davern MT, Cummins RA, Stokes MA. (2007). Subjective well-being as an affective-cognitive construct. *J Happiness.* 8, 429–449.

Davidson RJ. (1992) Emotion and affective style: Hemispheric substrates. *Psychol Sci.* 3, 39–43.

Davidson RJ. (2004). Well-being and affective style: Neural substrates and biobehavioral correlates. *Philos Trans R Soc (Biol Sect).* 359, 1395–1411.

Davidson RJ, Jackson DC, Kalin NH. (2000). Emotion, plasticity, context, and regulation: Perspectives from affective neuroscience. *Psychol Bull.* 126, 890–909.

Davitz J. (1969). *The Language of Emotion.* New York, NY: Academic Press.

De Haes JCJM, Pennink BJW, Welvaart K. (1987). The distinction between affect and cognition. *Soc Indic Res.* 19, 367–378.

Dennett DC. (1987). *The intentional stance.* Cambridge, MA: MIT Press.

Detweiler-Bedell JB, Friedman MA, Leventhal H, Miller IW, Leventhal EA. (2008). Integrating co-morbid depression and chronic physical disease management: Identifying and resolving failures in self-regulation. *Clin Psychol Rev.* 28, 1426–1446.

Diener E. (2006). Guidelines for national indicators of subjective well-being and ill-being. *J Happiness Stud.* 7, 397–404.

Diener E, Emmons RA, Larson RJ, Griffin S. (1985). The Satisfaction with Life Scale. *J Personal Assess.* 49, 71–75.

Diener E, Lucas RE, Napa Scallon C. (2006). Beyond the hedonic treadmill; Revising the adaptation theory of well-being. *Am Psychol.* 61, 305–314:

Diener E, Scallon Napa C, Lucas RE. (2003). The evolving concept of subjective well-being: The multifaceted nature of happiness. *Adv Cell Aging Gerontol.* 15, 187–220.

Diener E, Sapyta JJ, Suh E. (1998). Subjective well-being is essential to well-being. *Psychol Inq.* 9, 33–37.

Diener E, Seligman MEP. (2004). Beyond money: Toward an economy of well-being. *Psychol Sci Public Interest.* 5, 1–31.

Diener E, Suh EM. (1998). Age and subjective well-being: An international analysis. *Annu Rev Gerontol Geriat.* 17, 304–324.

Diener E, Suh EM, Lucas RE, Smith HL. (1999). Subjective well-being. *Psychol Bull.* 125, 276–302.

Duncan S, Barrett LF. (2007). Affect is a form of cognition: A neurobiological analysis. *Cogn Emot.* 21, 1184–1211.

Dupuy HJ. (1984). The Psychological General Well-being (PGWB) Index. In, (Eds.) NK Wenger, ME Mattson, CD Furberg, J Ellison. *Assessment of Quality of life in Clinical Trials of Cardiovascular Therapies.* Shelton, CO: Le Jacq. (p. 170–183).

Eddington KM, Dolcos F, Cabeza R, Ranga Krishman K, et al. (2007). Neural correlates of promotion and prevention goal activation: An fMDI study using an idiopathic approach. *J Cogn Sci.* 19, 1152–1162.

Eddington KM, Dolcos F, McLean AN, Ranga Krishman K, et al. (2009). Neural correlates of idographic goal priming in depression: Goal-specific dysfunctions in the orbitofrontal cortex. *SCAN.* 4, 238–246.

Eid M, Diener E.(2004). Global judgments of subjective well-being: Situational variability and long-term stability. *Soc Indic Res.* 65, 245–277.

Eisenberg N, Spinrad TL. (2004). Emotion-related regulation: Sharpening the definition. *Child Dev.* 75, 334–339.

Ekman P. (1992). An argument for basis emotions. *Cogn Emot.* 6, 169–200.

Ekman P, Davidson RJ, Ricard M, Wallace BA. (2005). Buddist and psychological perspectives on emotion and well-being. *Curr Dir Psychol Sci.* 14, 59–63.

Ekman P, Friesen WV. (1975). *Unmasking the face.* Englewood Cliffs, NJ: Prentice-Hall.

Ellsworth PC, Scherer K R. (2003). Appraisal processes in emotion. In, R J Davidson, HH Goldsmith, KR Scherer (Eds.), *Handbook of affective sciences.* New York/Oxford UK: Oxford University Press. (pp. 572–595).

Ellsworth PC, Tong EMW. (2006). What does it mean to angry at yourself? ?Categories, appraisals and the problem of language. *Emot.* 6, 572–586.

Esterlin RA. (2005). Building a better theory of well-being. In, (Eds.) L Bruni, P Porta. *Economics and Happiness: Framing the Analysis.* Oxford UK: Oxford University Press.

Fainsilber L, Ortony A. (1987). Metaphorical uses of language in the expression of emotions. *Metaphor Symb Act.* 2, 239–250.

Fehr B, Russell JA. (1984). Concept of emotion viewed from a prototype perspective. *J Exp Psychol Gen.* 113, 464–486.

Felice D, Perry J. (1996). Exploring Curr conceptions of quality of life: A model for people with and without disabilities. In, (Eds.) R Rewick, I Brown, M Nagler. *Quality of life in Health Promotion and Rehabilitation: Conceptual Approaches, Issues, and Applications.* Thousand Oaks CA: Sage. (p. 51–62).

Ferrans CE. (1990). Development of a quality of life index for patients with cancer. *Oncol Nurs Forum.* 17, 29–38.

Festinger L. (1957). *A Theory of Cognitive Dissonance.* Evanston Il: Row, Peterson.

Fox NA, Calkins SD. (2003).The development of self-control of emotions: Intrinsic and extrinsic influences. *Motivat Emot.* 27, 7–26.

Frijda NH. (1999). Emotions and hedonic experience. In, (Eds.) D Kahneman E Diener, N Schwarz. *Well-being: Foundations of Hedonic Psychology.* New York NY: Russell Sage Foundation. (p. 190–210).

Fujita F, Diener E. (2005). Life satisfaction set point stability and change. *J Personal Soc Psychol.* 88, 158–164.

Gallese VS, Lakoff G. (2005). The brain's concepts: The role of the sensory-motor system in conceptual knowledge. *Cogn Neuro Psychol.* 22, 455–457.

Gibbs RW Jr. (2006). *Embodiment and Cognitive Science.* New York NY: Cambridge Press.

Gibson JJ. (1966). *The Senses Considered as a Perceptual Systems.* Boston MA: Houghton Mifflin.

Gibson, JJ. (1979). *The Ecological Approach to Visual Perception.* Boston, MA: Houghton, Mifflin and Company.

Gill TM, Feinstein AR. (1994). A critical appraisal of the quality of quality of life measurements. *JAMA* 272; 619–626.

Goel V, Dolan RJ (2003). Reciprocal neural response within lateral and ventral medial prefrontal cortex during hot and cold reasoning. *Neuroimage.* 20, 2314–2321.

Gotay C, Korn E, McCabe M, Moore TD, et al. (1992). Quality of life assessment in cancer treatment protocols: Research issues in protocol development. *J Nat'l Cancer Inst.* 84, 575–579.

Griffin J. (1986). *Well-being.* Oxford UK: Clarendon Press.

Gross JJ. (1998). The emerging field of emotional regulation: An integrative review. *Rev Gen Psychol.* 2, 271–299.

Gross JJ. (2001). Emotion regulation in adulthood: Timing is everything. *Curr Dir Psychol Sci.* 10, 214–219.

Gross JJ, Cartensen LC, Pasupathi M, Tsai J, et al. (1997). Emotion and aging: Experience, expression and control. *Psychol Aging.* 12, 590–599.

Grossi E, Groth N, Mosconi P, Cerutli R, et al. (2006). Development and validation of the short version of the Psychological General Well-being Index (PGWB-S). *Health Qual Life Outcomes.*4:88. 10.1186/1477-7525-4-88. (p. 1–7).

Hacking I. (2002). How "natural" are "kinds" of sexual orientation? *Law Philos.* 21, 95–107.

Headey B, Wearing A. (1989). Personality, life events and subjective well-being: Toward a dynamic equilibrium model. *J Personal Soc Psychol.* 57, 731–739.

Headey B. (2008). Life goals matter to happiness: A revision of set-point theory. *Soc Indic Res.* 86, 213–231.

Heider F. (1958). *The Psychology of Interpersonal Relations.* New York NY: Wiley.

Heinonen H, Aro AR, Aalto AM, Uutela A. (2004). Is the evaluation of the global quality of life determined by emotional status? *Qual Life Res.* 13, 1347–1356.

Helson H. (1948). Adaptation-level as a basis for quantitative theory of frames of reference. *Psychol Rev.*55,297–313.

Hertzog C, Nesselroade JR. (2003). Assessing psychological change in adulthood: An overview of methodological issues. *Psychol Aging.* 18, 639–657.

Heun R, Burkardt M, Maier W, Bech P. (1999). Internal and external validity of the WHO Well-being Scale in the elderly general population. *Acta Psychiatr Scand.* 99, 171–178.

Higgins ET. (2000). Making a good decision: Value from fit. *Am Psychol.* 55, 1217–1230.

Higgins ET, Grant H, Sah J. (1999). Self-regulation and quality of life: Emotional and Non-emotional life experiences. In, (Eds.) D Kahneman, E Diener, N Schwarz. *Well-being: The Foundations of Hedonic Psychology.* New York NY: Russell Sage Foundation. (p. 244–266).

Horn JL, McArdle JJ, Mason R (1983). When is invariance not invariance: A practical scientist's look at the ethereal concept of factorial invariance. *South Psychol.* 1, 179–188.

Huppert F. (2005). Positive mental health in individuals and populationss. In, (Eds.) F Huppert, N Baylis, B Keverne. *The Science of Well-being.* Oxford UK: Oxford University Press. (p. 307–340).

Izard CE. (1977). *Human Emotions.* New York NY: Academic Press.

Izard CE. (2007). Basic emotions, natural kinds, emotion schemas, and a new paradigm. *Perspect Psychol Sci.* 2, 260–280.

James W. (1890). *Principles of Psychology.* Vol 1. New York NY: Holt.

James W. (1884). What is an emotion? *Mind.* 9, 188–205.

John OP, Gross JJ. (2007). Individual differences in emotion regulation. In, (Ed.) JJ Gross. *Handbook of Emotional Regulation.* New York NY: Guilford Press. (p. 351–372).

References

Johnson-Laird PN, Oatley K. (1989). The language of emotions: An analysis of a semantic field. *Cogn Emot.* 3, 81–123.

Kafka GJ, Kozma A. (2002). The construct validity of Ryff's Scales of Psychological Well-being (SPWB) and their relationship to measures of subjective well-being. *Soc Indic Res.* 57, 171–190.

Kahneman D. (1999). Objective Happiness. In, (Eds.) D Kahneman, E Diener, N Schwarz. *Well-being: The Foundations of Hedonic Psychology.* New York NY: Russell Sage Foundation. (p. 3–25).

Kahneman D, Krueger AB, Schkade D, Schwarz N, et al. (2004). A survey method for characterizing daily life experience: The day reconstruction method. *Science.* 306, 1776–1780.

Kahneman D, Krueger AB. (2006). Developments in the measurement of subjective well-being. *J Econ Perspect.* 20, 3–24.

Kaplan RM, Anderson JP. (1990). The General Health Policy Model: An integrated approach. In, (Ed.) B Spilker. *Quality of life Assessments in Clinical Trials.* New York NY: Ravens.

Keil FC. (1995). The growth of causal understanding of natural kinds. In, (Eds.) D Sperber, D Premack, AJ Premack. *Causal Cognition: A Multidisciplinary Debate.* Oxford UK: Oxford University Press. (p. 234–267).

Kensinger EA, Schacter DL. (2008) Memory and emotion. In, (Eds.) M Lewis, JM Haviland-Jones, LF Barrett. *Handbook of Emotions.* Third Edition. New York NY: Guildford Press. (p. 601–617).

Keyes CL, Shmotkin D, Ryff CD. (2002). Optimizing well-being: The empirical encounter of two traditions. *J Personal Soc Psychol.* 82,1007–1022.

Kim-Prieto C, Diener E, Tamir M, Scollon C, et al. (2005). Integrating the diverse definitions of happiness: A time sequential framework of subjective well-being. *J Happiness Stud.* 6, 261–300.

Kleinginna PR, Kleinginna AM. (1981). A categorized list of emotion definitions, with suggestions for a consensual definition. *Motiv Emot.* 5, 345–379.

Kochanska G, Aksam N. (2006). Children's conscience and self-regulation. *J Personal.* 74, 1587–1617.

Koivumaa-Honkanen H, Kaprio J, Honkanen RJ, Vilinanäki H, et al. (2005). The stability of life satisfaction in a 15-year follow-up of adult Finns healthy at baseline. *BMC Psychiatr.* 5,4.

Koriat A, Ma'ayan H, Nussinson R. (2006). The intricate relationship between monitoring and control in metacognition; Lessons for the cause-and-effect relation between subjective experience and behavior. *J Exp Psychol Gen.* 135, 36–69.

Kövecses Z. (2000). *Metaphors and Emotion: Language, Culture and Body in Human Feeling.* Cambridge UK: Cambridge University Press.

Kövecses Z. (1991). Happiness: A definitional effort. *Metaphor Symbolic Act.* 6, 29–46.

Kunzmann U, Little TD, Smith J. (2000). Is age-related sability of subjective well-being a paradox? Cross-sectional and longitudinal evidence from the Berlin Aging Study. *Psychol Aging.* 15, 511–526.

Lakoff G, Johnson M. (1999). *Philosophy in the Flesh: The Embodied Mind and its Challenge to Western Thought.* New York NY: Basic Books.

Lakoff G. (1987). *Women, Fire, and Dangerous Things.: What Categories Reveal About the Mind.* Chicago Il: University of Chicago Press.

Lakoff G, Johnson M. (1980). *Metaphors We Live By.* Chicago Il: University of Chicago Press.

Lang PJ, Greenwald MK, Bradley MM, Hamm AO. (1993). Looking at pictures: Affective, facial, visceral and behavioral reactions. *Psychophysiol.* 30, 262–273.

Larsen RJ, Prizmic Z. (2008). Regulation of emotional well-being. In, (Eds.) M Eid, RJ Larsen. *The Science of Subjective Well-being.* New York NY: Guilford. (p. 258–289).

Layard R. (2005). *Happiness: Lessons from a New Science.* New York NY: Penguin.

Lazarus RS. (1966). *Psychological Stress and the coping process.* New York NY: McGraw-Hill.

Lazarus RS. (1991a). *Emotion and Adaptation.* New York NY: Oxford University Press.

Lazarus RS. (1991b). Progress on a cognitive-motivational-relational theory of emotion. *Am Psychol.* 46, 819–834.

Lazarus RS. (1991c). *Emotion and Adaptation.* Oxford UK: Oxford University Press. (p. 171–212)

Lazarus RS. (1995). Vexing research problems inherent in cognitive-motivational theories of emotion – and some solutions. *Psychol Inj.* 6, 183–196.

LeDoux J. (1996). *The Emotional Brain.* New York NY: Simon &Schuster.

LeDoux JE, Armony J. (1999). Can neurobiology tell us anything about human feelings? In, (Eds.) D. Kahneman, E. Diener, and N. Schwarz. *Foundations of Hedonic Psychology: Scientific Perspectives on Enjoyment and Suffering.* New York NY: Russell Sage Foundation.

Lent RW. (2004). Toward a unifying theoretical and practical perspective on well-being and psychosocial adjustment. *J Consum Psychol.* 51, 482–509.

Leventhal H, Leventhal EA, Cameron L. (2001). Representation, procedures, and affect in illness self-regulation: A perceptual-cognitive model. In, (Eds.) A Baum, TA Revenson, JE Singer. *Handbook of Health Psychology.* Mahwah NJ: Erlbaum. (p.19–47).

Levin E, Levin A. (2006). Evaluation of spoken dialogue technology for real-time health data collection. *J MedInternet Res.* 8,e29.

Lewis MD, Stieben J. (2004). Emotional regulation in the brain: Conceptual issues and directions for developmental research. *Child Dev.* 75, 371–376.

Lindenberg S. (1996). Continuities is the theory of social production functions. In, (Eds.) H Ganzeboom, Lindenberg S. *Verklarende Sociologic : Opstellen voor Reinhart Wippler.* Amsterdam Netherland: Thesis Publications.

Lucas RE, Clark AR, Georgellis Y, Diener E. (2003). Reexamining adaptation and the set point model of happiness: Reactions to changes in martial status. *J Personal Soc Psychol.* 84, 527–539.

Luhmann M, Eid M. (2009). Does it really feel the same? Changes in life satisfaction following repeated life events. *J Personal Soc Psychol.* 97, 363–381.

Lykken D, Tellegen A. (1996). Happiness is a stochastic phenomenon. *Psychol Sci.* 7, 186–189.

Macmillan AM. (1957). The Health Opinion Survey: technique for estimating prevalence of psychoneurotic and related types of disorder in communities. *Psychol Rep.* 3 (Monogr Suppl. 7), 325–329.

MacMillion M. (2000). *An Odd Kind of Fame: Stories of Phineas Gage.* Cambridge UK: Cambridge University Press.

Maitland SC, Dixon RA, Hultsch DF, Ertoz C. (2001). Well-being as a moving target: Measurement equivalence of the Bradburn Affect Balance Scale. *J Gerontol Ser B: Psychol Sci Soc Sci.* 56, P69–P77.

Malatesta CZ, Kalnok M. (1984). Emotional experience in younger and older adults. *J Gerontol.* 39, 301–308.

Maslow AH. (1968). *Towards a Psychology of Being.* Princeton NJ: Van Nostrand.

Max DT. (2007). *Happiness 101.* New York Times January 7, 2007.

Mc Dowell I, Newell C. (1996). *Measuring Health: A Guide to Rating Scales and Questionnaires.* New York NY: Oxford. Second Edition.

Medin DL, Ortony A. (1989). Psychological essentialism. In, (Eds.) S Vosniadou, A Ortony. *Similarity and Analogical Reasoning.* New York NY: Cambridge University Press. (p. 179–195).

Mehl M, Pennebaker JW, Crow D, Michael DJ, et al. (2001). The Electronically Activated Recorded (EAR): A device for sampling naturalistic daily activities and conversations. *Behav Res Method Instrum Comput.* 33, 517–523.

Meier BP, Robinson MD. (2005). The metaphoric representation of affect. *Metaphor Symbol.* 20, 239–257.

Meredith W. (1993). Measurement invariance, factor analysis and factorial invariance. *Psychom.* 58, 525–543.

Michalos AC. (1980). Satisfaction and happiness. *Soc Indic Res.* 8, 385–422.

Michalos AC. Multiple discrepancies theory (MDT). *Soc Indic Res.* 16, 347–413, 1985.

Mill JS. (1969/1863). *Utilitarianism.* In, (Ed.) J Robson. Collected Works. Toronto, Ontario CA: University of Toronto Press.

Morris WN. (1989). *Mood: The Frame of Mind.* New York NY: Springer-Verlag.

Murphy ST. (1990). *The Primacy of Affect: Evidence and Extension.* Unpublished Doctoral Dissertation. University of Michigan.

Murphy ST, Zajonc RB. (1993). Affect, cognition and awareness affective priming with optimal and suboptimal stimulus exposures. *J Personal Soc Psychol.* 64, 723–739.

Murray HA. (1938). *Explorations in Personality.* New York NY: Oxford University Press.

Myers DG, Diener E(1995). Who is happy? *Psychol Sci.* 6,10–17.

National Center for Health Statistics (1973). *Plan and Operation of the National and Nutrition Examination Survey: United States 1971–1973.* DHEW Publication No. PHS 78–1310. Washington DC: US government Printing Office.

Neisser U. (1967). *Cognitive Psychology.* New York NY: Appleton-Century-Crofts.

Nes RB, Røysamb E, Tambs K, Harris JR, et al. (2006). Subjective well-being : Genetic and environmental contributions to stability and change. *Psychol Med.* 36, 1–10.

Niedenthal PM. (2007). Embodying emotion. *Science.* 316, 1002–1005.

Niedenthal PM. (2008). Emotion concepts. In (Eds.) M Lewis, JM Haviland-Jones, L Feldman Barrett. *Handbook of Emotions.* New York NY: Guildford. (p. 587–600).

Niedenthal PM, Barsalou L, Winkielman P, Krauth-Gruber S, et al. (2005). Embodiment in attitudes, social perception and emotion. *Personal Soc Psychol Rev.* 9, 184–211.

Niedenthal PM. Winkielman P, Mondillon L, Vermeulen N. (2009). Embodiment of emotion concepts. *J Personal Soc Psychol.* 96, 1120–1136.

O'Grady W, Dobrovolsky M, Aronoff M. (1989). *Contemporary Linguistics: An Introduction.* New York NY: St. Martin Press,

Oatley K, Johnson-Laaird PN. (1987). Towards a cognitive theory of emotions. *Cogn Emot.* 1, 29–50.

Ochsner KN, Gross JJ. (2008). Cognitive emotion regulation: Insights from social cognition and affective neuroscience. *Curr Dir Psychol Sci.* 17,153–158.

Öhman A, Dimberg U, Esteves F. (1989). Preattentive activation of aversive stimuli. In, (Eds.) T Archer, LG Nilsson. *Aversion, Avoidance, and Anxiety: Perspectives on Aversively Motivated Behavior.* Hillsdale NJ: Erlbaum.

Ormel J. (1983). Neuroticism and well-being inventories: Measure traits or states? *Psychol Med.* 13, 165–176.

Ormel J, Lindenberg S, Steverink N, Von Korff M. (1997). Quality of life and Social Production Functions: A framework for understanding health effects. *Soc Sci Med.* 45, 1051–1063.

Ormel J, Lindenberg S, Steverink N, Vergrugge LM. (1999). Subjective well-being and Social Production Functions. *Soc Indic Res.* 46,61–90.

Ortony A, Clore GL, Collins A. (1988). *The Cognitive Structure of Emotions.* Cambridge UK: Cambridge University Press.

Osgood CE. (1964). Semantic differential technique in the comparative study of cultures. *Am Anthropol.* 66, 171–200.

Osgood CE. (1969). On the whys and wherefores of E, P, and A. *J Personal Soc Psychol.* 12, 194–199

Osgood CE, Suci GJ, Tannenbaum PH. (1957). *The Measurement of Meaning.* Urbana Ill, University of Illinois Press.

Oxford English Dictionary (1996). Oxford UK: Oxford University Press.

Padilla GV, Grant MM, Ferrell BR, Presant CA. (1996). Quality of life: Cancer. In, B. Spilker (ed), *Quality of life and Pharmacoeconomics in Clinical Trials. 2nd edition.* New York NY: Raven Press. (p. 301–398).

Panksepp J. (2007). Neurologizing the psychology of affects: How appraisal-based constructivism and brain emotion theory can coexist. *Perspect Psychol Sci.* 2, 281–296.

Paris SG, Newman RS. (1990). Developmental aspects of self-regulated learning. *Educ Psychol.* 25, 87–102.

Pavot W, Diener E. (1993). Review of the satisfaction with life scale. *Psychol Assess.* 5, 164–172.

Patrick H, Kness CR, Canevello A, Lonsbary C. (2007). The role of need fulfillment in relationship functioning and well-being: A self-determination theory perspective. *J Personal Soc Psychol.* 92, 434–457.

Plutchik R. (2001). The nature of emotions. *Am Sci.* 89, 344–350.

Rapkin BD, Schwartz CE. (2004). Toward a theoretical model of quality of life appraisal: Implications of findings from studies of response shift. *Health Qual Life Outcomes.* 2 (14) 1–12.

Röcke C, Lachman ME. (2008). Perceived trajectories of life satisfaction across past, present and future: Profiles and correlates of subjective change in young, middle-aged and older adults. *Psychol Aging.* 23, 833–847.

Rolls ET. (1999). *The Brain and Emotion.* Oxford UK: Oxford University Press.

Rosenberg MJ. (1956). Cognitive structure and attitudinal affect. *J Abnorm Soc Psychol.* 53, 367–372.

Russell JA. (2003). Core affect and the psychological construction of emotion. *Psychol Rev.* 110, 145–172.

Ryan RM, Deci EL. (2001). On happiness and human potentials: A review of research on Hedonic and eudaemonic well-being. *Annu Rev Pyshcol.* 52, 141–166.

Ryff CD. (1989a). Happiness is everything, or is it? Exploration on the meaning psychological well-being. *J Personal Soc Psychol.* 57, 1069–1081.

Ryff CD. (1989b). In the eye of the beholder: Views of psychological well-being in middle and old-aged adults. *Psychol Aging.* 4, 195–210.

Ryff CD, Keyes CLM. (1995). The structure of psychological well-being revisited. *J Personal Soc Psychol.* 69, 719–727,

Ryff CD, Singer B. (1998). The contours of positive human health. *Psychol Inq.* 9, 1–18.

Ryff CD, Singer B. (2006). Best news yet on the six-factor model of well-being. *Soc Sci Res.* 35, 1103–1119.

Sander D, Granjean D, Scherer KR. (2005). A systems approach to appraisal mechanisms in emotion. *Neural Netw.* 18, 317–352.

Schaie K, Lawton MP. (1998). Focus on Emotion and Adult Development. *Annu Rev Gerontol Geriat.* Vol. 17 New York NY: Springer Publishing.

Scherer KR. (1984). On the nature and function of emotion: A component process approach. In, (Eds.) KR Scherer, P Ekman. *Approaches to Emotion.* Hillsdale, NJ: Elbaum. (p. 293–317).

Scherer KR. (1999). On the sequential nature of the appraisal processes: Indirect evidence from a recognition task. *Cogn Emot.* 13,763–793.

Scherer KR. (2000). Emotions as episodes of subsystem synchronization driven by nonlinear appraisal processes. In, (Eds.) MD Lewis, I Granic. *Emotions, Development, and Self-organization: Dynamic Systems Approaches to Emotional Development.* Cambridge UK: Cambridge University Press. (p. 70–99).

Scherer KR. (2001). Appraisal considered as a process of multilevel sequential checking. In, (Eds.) K Scherer, A Schorr, T Johnson. *Appraisal Processes in Emotion: Theory, Methods, Research.* New York NY: Oxford University Press. (p. 92–120).

Schlosberg H. (1952). The description of facial expressions in terms of two dimensions. *J Exp Psychol.* 44, 229–237.

Schmidt-Petri C. (2003). Mill on quality and quantity. *Philos Q.* 53, 102–104.

Schuurmans H, Steverink N, Frieswijk N, Buunk BP, Slaets JPJ.(2005). How to measure elf-management abilities in older people by self-report. The development of the SMAC-30. *Qual Life Res.* 14, 2215–2228.

Schwartz CE, Rapkin BD. (2004). Reconsidering the psychometrics of quality of life assessment in light of response shift and appraisal. *Health Qual of Life Outcomes.* 2 (16), 1–11.

Schwarz N. (1990). Feelings as Information: Informational and motivational functions of affective states. In, (Eds.) ET Higgins, RM Sorrentino. *Handbook of Motivation and Cognition: Foundations of Social Behavior.* New York NY: Guilford Press. Vol 2. (p. 61–84).

Schwarz N, Clore GL. (1983). Mood, misattribution, and judgments of well-being: Informative and directive functions of affective states. *J Personal Soc Psychol.* 45, 513–523.

Schwarz N, Clore GL. (2007). Feelings in phenomenal experiences. In, (Eds.) A Kruglanski, ET Higgins. *Social Psychology. Handbook of Basic Principles.* New York NY: Guildford Press. 2nd Edition. (p. 385–407).

Schwarz N, Strack F. (1999). Reports of subjective well-being: Judgmental processes and their methodological implications. In, (eds.) D Kahneman, E Diener, N Schwarz. *Well-being : The Foundations of Hedonic Psychology.* New York NY: Russell Sage Foundation. (p. 61–84).

Searle JR. (1969). *Speech Acts.* Cambridge UK: Cambridge University Press.

Searle JR. (1983). *Intentionality: An Essay on the Philosophy of the Mind.* Cambridge UK: Cambridge University Press.

Sen A. (1985). *Commodities and Capabilities.* Amsterdam Netherlands: North-Holland.

Sen A. (1993). Positional objectivity. *Philos Public Affa.* 22, 126–145.

Shaver P, Schwartz J, Kirson D, O'Connor C. (1987). Emotion knowledge; further exploration of a prototype approach. *J Personal Soc Psychol.* 52, 1061–1086.

Shmotkin D. (1990). Subjective well-being as a function of age and gender: A multivariate look for differentiated trends. *Soc Indic Res.* 23, 201–230

Shweder RA. (1993). The cultural psychology of emotions. In, (Eds.) M Lewis, JM Haviland. *Handbook of Emotions.* New York NY: Guilford. (p. 417–431)

Shweder RA. (1994). "You're not sick, you're just in love": Emotion as an interpretative system. In (Eds.) P Ekman, RJ Davidson. *The Nature of Emotion: Fundamental Questions.* New York NY: Oxford University Press. (p. 32–44).

Sidgwick H. *The Methods of Ethics.* London, Macmillan and Co, 1884.

Sirgy MJ, Michalos AC, Ferriss AL, Easterline RA, et al. (2006). The Quality of life research movement: Past, present and future. *Soc Indic Res.* 76, 343–466.

Skinner BF. (1938). *The Behavior or Organisms: An Experimental Analysis.* New York NY: Appleton-Century-Crofts.

Smith KW, Avis NE, Assman SF. (1999). Distinguishing between quality of life and health status in quality of life research: A meta-analysis. *Qual Life Res.* 8, 447–459.

Smith EE, Semin GR. (2007). Situated social cognition. *Curr Dir Psychol Sci.* 16, 132–135.

Smolensky P, Legendre C. (2006). *The Harmonic Mind From Neural Computation to Optimality-theoretic Grammar.* Volume 1: Cognitive Architecture. Cambridge MA: MIT.

Solomon RC. (2000). The philosophy of emotions. In, (Eds.) M Lewis, JM Haviland-Jones. *Handbook of Emotions.* 2nd ed. New York, Guilford. (p. 3–15)

Sprangers MA, Schwartz CE. (1999). Integrating response shift into health-related quality of life research: a theoretical model. *Soc Sci Med.* 49, 1507–1515.

Springer KW, Hauser RM. (2006). An assessment of the construct validity of Ryff's Scales of Psychological Well-being: Method, mode and measurement effects. *Soc Sci Res.* 35, 1180–1102.

Springer KW, Hauser RM. (2006). Bad news indeed for Ryff's six-factor model of well-being. *Soc Sci Res.* 35, 1120–1131.

Stacey CA, Gatz M. (1991). Cross-sectional age differences and longitudinal change on the Bradburn Affect Balance Scale. *J Gerontol Psychol Sci.* 46, 76–78.

Steverink N, Lindenberg S. (2008). Do good self-mangers have less physical and social resource deficits and more well-being in latter life? *Eur J Aging.* 5, 181–190.

Steverink N, Lindenberg S, Linderberg S, Slaets JPJ. (2005). How to understand and improve older people's well-being. *Eur J Aging.* 2, 235–244.

Stone A, Shiffman S. (1994). Ecological momentary assessment (EMA) in behavioral medicine. *Ann Behav Med.* 16,199–202.

Strobeck J, Robinson MD. (2004). Preferences and inferences in encoded visual objects: A systematic comparison of semantic and affective priming. *Personal Soc Psychol Bull.* 30, 81–93.

Strobeck J, Robinson MD, McCourt ME. (2006). Semantic processing precedes affect retrieval: The neurological case for cognitive primary in visual processing. *Rev Gen Psychol.* 10,41–55.

Suh E, Diener E, Fujita F. (1996) Events and subjective well-being: Only recent events matter. *J Personal Soc Psychol.* 70, 1091–1102.

Sweeney K, Whissell CM. (1984). A dictionary of affect in language: I. Establishment and preliminary validation. *Percept Motor Skills.* 59, 695–698.

Szalai A. (1980). The meaning of comparative research on the quality of life. In, (Eds.) A Szalai, FM Andrews. *The Quality of life: Comparative Studies.* Beverly Hills CA: Sage Publishing Co. (p. 7–21).

Tellegen A. (1982). *Brief Manual for the Multidimensional Personality Questionnaire.* Unpublished Manuscript : University of Minnesota, Minneapolis MN.

Tesser A, Martin, L. (1996). The psychology of evaluation. In, (Eds.) ET Higgins, Edward AW Kruglanski. *Social Psychology: Handbook of Basic Principles.* New York NY: Guilford Press. (p. 400–432).

Titchener EB. (1908). *Lectures on the Elementary Psychology of Feelings and Attention.* New York NY: Macmillan.

Thayer RE. (1989). *The Biopsychology of Mood and Activation.* New York NY: Oxford University Press.

The Complete Oxford Dictionary. (1970). Oxford UK, Oxford University Press. Vol XII p. 289.

Thompson RA. (1994). Emotion regulation: In search of definition. In, (Ed) NA Fox. The Development of Emotion Regulation: Biological and Behavioral Considerations. *Monogr Soc Res Child Dev.* 59 (2–3, Serial No. 240), 25–52.

Throndike EL.(1911). *Animal Intelligence.* New York NY: Macmillan.

Tolsti L (1968). *War and Peace.* (Tr: A Dunningan) . New York NY: Signet Classics, (Originally Published 1869).

Tomarken AJ, Davidson RJ, Henrfiguqes JB. (1990). Resting frontal activation asymmetry predicts emotional reactivity to film clips. *J Personal Soc Psychol.* 59, 791–801.

Tomarken AJ, Davidson RJ, Wheeer RE, Doss RC. (1992). Individual differences in anterior brain asymmetry and fundamental dimensions of emotion. *J Personal Soc Psychol.* 62, 676–687.

Tourangeau R, Rips R, Rasinski K. (2000). *The Psychology of Survey Research.* Cambridge UK: Cambridge University Press.

Urry HL, Nitschke JB, Dolski I, Jackson DC, et al. (2004). Making a life worth living: Neural correlates of well-being. *Psychol Sci.* 15, 368–373.

van Reekum CM, Urry HL, Johnstone T, Thurow ME, et al. (2007). Individual differences in amygdala and ventromedial prefrontal cortex activity are associated with evaluation speed and psychological well-being. J Cogn NeuroSci. 19, 237–248.

Veenhoven R. (1984). *Conditions of Happiness.* Dordrecht Netherlands: Kluwer Academic.

Veenhoven, R. (1994). Is happiness a trait? Tests of the theory that a better society does not make us any happier. *Soc Indic Res.* 32, 101–162.

Wager TD, Davidson ML, Hughes BL, Lindquist MA, Ochsner KN. (2008). Prefrontal-subcortical pathways mediating successful emotion regulation. *Neuron.* 59, 1037–1050.

Watson D, Clark LA, Tellegen A. (1988). Development and validation of brief measures of Positive and Negative Affect: The PANAS Scales. *J Personal Soc Psychol.* 54, 1063–1070.

Watson D, Tellegen A. (1985). Toward a consensual structure of mood. *Psychol Bull.* 98, 219–235.

Wheeler RJ. (1991). The theoretical and empirical structure of general well-being. *Soc Indic Res.* 24, 71–79.

White GM. (2000). Representing emotional meaning: category, metaphor, schema, discourse. In, (Eds.) M Lewis, JM Haviland-Jones. *Handbook of Emotions.* New York NY: Guilford Press. 2 nd Edition.

White MP, Dolan P. (2009). Accounting for the richness of daily activities. *Psychol Sci.* 20, 1000–1008.

Wierzbicka A. (1992). *Semantics, Culture, and Cognition: Universal Human Concepts in Culture-Specific Configurations.* New York NY: Oxford University Press.

Wierzbicka A. (1997). *Understanding Cultures Through their Key Words: English, Russian, Polish, German, and Japanese.* New York NY: Oxford University Press.

Wierzbicka A. (1999). *Emotions Across Languages and Cultures: Diversity and Universals.* Paris France: Cambridge University Press.

Wilkowski BM, Meier BP, Robinson MD, Carter MS, et al. (2009). "Hot-headed" is more than an expression: The embodied representation of anger in terms of heat. *Emotion.* 4, 464–477.

Wilson IB, Cleary PD. (1995). Linking clinical variables with health-related quality of life: A conceptual model of patient outcomes. *JAMA.* 273, 59–65,

Winkielman P, Knäuper B, Schwarz N. (1998). Looking back at anger: Reference periods change the interpretation of emotion frequency questions. *J Pers Soc Psychol.* 75, 719–728.

WordNet; Available on the web at, http://wordnet.princeton.edu/.

World Health Organization. (1948). Constitution of the World Health Organization: Basic documents. Geneva, Switzerland: World Health Organization.

Worthington EL, Van Oyen Witvliet C, Pietrini P, et al. (2007). Forgiveness, health, and well-being: A review of evidence for emotional versus decisional forgiveness, dispositional forgivingness and reduced unforgiveness. *J Behav Med.* 30, 291–302.

Wrosch C, Bauer I, Scheier MF. (2005). Regret and quality of life across the adult life span: The influence of disengagement and available future goals. *Psychol Aging.* 20, 657–670.

Wundt W. (1904). *Principles of Physiological Psychology.* I. (Trans.) EB Titchner, London UK: Sonnenschein.

Zajonc RB. (1980). Feelings and thinking, Preference need no inferences. *Am Psychol.* 35, 151–175.

Zajonc RB. (1984). On the primacy of affect. *Am Psychol.* 39, 117–123.

Part IV

Summary and Future Directions

Summary and Future Directions

12

Abstract

In this chapter I review my working hypothesis and the available data that provide support. Central to this hypothesis is process of valuation. I discuss a number of aspects of the valuation process, first comparing statistical vs. cognitive approaches by examining the weighting task, then I discuss the cognitive basis of valuation methods, and finally discuss the literature demonstrating that values are stable parts of a person's qualitative assessments. I also address what I consider to be the central issue in qualitative research: Can a quality be a quantity? This gave me the opportunity to consider the nature of measurement in the behavioral and social sciences, in general, and qualitative research, in particular. I briefly reviewed the history of development of subjective measurement and concluded that a greater role has to be found for axiomatic fundamental measurement in qualitative research. For this to happen, however, requires that investigators acknowledge the limitations of their current research strategy.

.. if the work of inquiry is to be carried on, it must be at once scientific and philosophic, that is, in particular, the scientist is not philosophic, he will fall into confusions, he will rebuff philosophic criticism... he will be carried away by practical interests, by interests in producing something or implementing a programme instead of in finding something out.

Anderson (1962; p. 183).[1]

Abbreviations

AQOL	Australian quality-of-life assessment (Hawthorne et al. 1999)
ATM	Axiomatic theory of measurement
CBI	Cancer behavior inventory (Merluzzi et al. 2001)
Com-QOL Scale	Comprehensive Quality-of-life Scale (Cummins 1997)
EORTC QOQ30C	European Organization for Research and Treatment of Cancer (Aaronson et al. 1991)
EQ-5D	EuroQoL -5D (Kind 1996)
FACT-G	Functional assessment of cancer therapy – general version (Cella et al. 1993)
FACT-P	Functional assessment of cancer therapy – prostate (Esper et al. 1997)
FIM™'s	Functional Independence Measure (Hamilton et al. 1987)
FLIC	Functional Living Index: Cancer (Shipper et al. 1984)
HALex	Health and Activities Limitation Index (Erickson 1998)
HAQ	Health assessment questionnaire (Fries et al. 1980)
HUI	Health Utilities Index (Torrance et al. 1995)
IDUQOL Scale	Injection Drug User Quality-of-life Scale (Russell et al. 2006)
LSIA	Life Satisfaction Index A (Neugarten et al. 1961)

MAUT	Multiattribute utility approach (Torrance et al. 1995)
MC $X2$	Marlow-Crowne Social Desirability Scale (Strahan and Gerbasi 1972)
NHIS	National Health Interview Survey
PAIS	Psychosocial Adjustment to Illness Scale (Derogatis and Derogatis 1990)
PI-HAQ	Personal impact health assessment questionnaire (Hewlett et al. 2002)
QALY	Quality-adjusted life year(s)
QLI	Quality-of-life Index (Ferrans and Powers 1985)
QOLA	Quality-of-life assessment
QWB	Quality of Well-being Scale (Kaplan and Anderson 1990)
QWB-SA	Quality of Well-being Scale-Self-Administered (Kaplan et al. 1997)
RSES	Rosenberg Self-Esteem Scale (Rosenberg 1979)
SAHS	Self-assessed health status
SEIQoL-DW	Schedule for individual quality-of-life-direct weighting (O'Boyle et al. 1995)
SF-12	SF-12 Health survey (Ware et al. 1995)
SF-36	SF-36 Health survey (Ware et al. 1994)
SF-6D	SF-6D Health survey (Brazier et al. 2002)
SG	Standard Gamble
SWLS	Satisfaction With Life Scale (Diener et al. 1985)
TTO	Time Trade-off (Torrance et al. 1995)
UW-QOL	University of Washington Quality-of-life Assessment (Deleyiannis et al. 1997)
VAS	Visual Analog Scale
WHOQOL-100	World Health Organization Quality-of-life-100 (The WHOQOL Group 1998)

1 Overview

I believe the Anderson's (1962) quote adequately characterizes the current state of qualitative studies, particularly quality-of-life research. In this chapter I hope to sketch out some of the problems that remain to be solved if the study of quality is to become a science. What will become apparent is that, in order to do this, I will have to clarify the role that measurement theory plays in qualitative research.

Throughout this book I have avoided using the term "measurement" preferring to use the term "assessment." I did this, as I explained in the Preface, because I wanted to emphasize that the current science underlying what I was talking about was something that a person creates, it is a (cognitive) process, not a state or entity that exists independent of a person (as when a machine measures some phenomena, or a mathematical theory defines a relationship). As such, using the term assessment seemed to be more appropriate, but in this chapter I want to address what would be required to make a quality assessment more like a measurement, particularly physical measurement. This issue, of course, is not limited to quality-of-life research, but is a much broader one that is also relevant to all of the behavioral and social sciences. My first task, however, is to summarize my working hypothesis.

One of the distinct characteristics of this book is my proposal that a common cognitive process, *the formation of a hybrid construct,* that involves the combination of a number of specific cognitive components using a particular cognitive rule underlies all qualitative judgments. This implies that capable persons from different cultures in different medical, mental, or intellectual states, judging different objects, events, or experiences are all making their qualitative judgments in the same way. At one level, this hypothesis is obvious, since most complex thoughts are a product of combining multiple cognitive (and emotional) components, but what distinguishes the qualitative assessment hybrid construct is that it involves combining specific cognitive components (e.g., evaluated descriptors and a metacognitive reflective valuation) and does so in a particular way. Any example of this particular combination of cognitive components should, therefore, be considered an example of a judgment of quality. This permits me to classify a wide range of judgments as qualitative judgments. These would include the person who after observing a painting and noting that the painting made him or her feel a particular way that they felt was important to them, the physician who felt it was important to tell a patient that their symptoms were serious and had the potential for producing adverse consequences, or a person who used the importance of his or her life history to judge the quality of their life: in each of these instances, the person was generating a qualitative hybrid construct. It would also imply that a qualitative assessment, such as a quality-of-life assessment, is independent of the content of the assessment. This would allow any person's particular set of experiences to be as relevant for a qualitative assessment as any other persons. It would make it unnecessary to develop a consensus about what content to include in a quality-of-life assessment, since the particular content or issues of concern would not be the criteria to use to determine what statements were statements of quality. All that need be demonstrated was that a particular set of cognitive activities had occurred.

This hypothesis has several distinct advantages. First, it provides a means of directly establishing the construct validity of any qualitative assessment by demonstrating that the cognitive basis of a theoretical construct and a qualitative assessment were equivalent. This occurs because an investigator will have established that the respondents were "thinking about"[2] their assessment in the same way that an investigator was when they generated the item. What it doesn't do is reassure the investigator what the respondent was "thinking of" when they generated a particular qualitative judgment, especially if they made a global assessment (e.g., "How would you rate your quality-of-life........ Excellent, very good, good, fair, poor?").

The current practice of including "cognitive interviewing" (Willis 2005) as part of the development of a qualitative assessment provides a method for examining the cognitive discrepancies between the investigator and the respondent. Thus, during cognitive interviewing, an investigator would ordinarily ask a respondent to verbalize what they were thinking about as they responded to an item. By so doing the investigator can then decide if the structure of the item was sufficiently clear so as to permit the respondent to "think about" the content in the way the investigator anticipated when designing the item, and if not, to adjust the item accordingly.

Willis (2005; p. 4) gives a relatively straightforward example of this when he asked a child to determine how many more puppies a poodle had, after reading an item that stated that a poodle had nine puppies and a collie had five puppies. The child answered that the poodle had no puppies, stating that "'You told me that she had nine puppies. But then she didn't have any more, so it's none!'" (Willis 2005: p. 4). This response demonstrated that the child was not "thinking about" the information provided by the item in the way the investigator wanted. Specifically, the word "more" was interpreted by the child as reflecting what was to be and so she did not make the comparison intended by the investigator. As a result, the investigator redesigned the item to increase the chances that the child would make the comparison.

Cognitive interviewing has now become a common part of the development of qualitative assessments. Its presence is a tacit admission that the psychometric properties of any qualitative assessment are critically dependent on the cognitive processes evoked by the items on the assessment, and the failure to monitor this aspect of an assessment has the potential of comprising the validity of the item or the assessment. This is particularly true when a qualitative assessment is translated into another language, especially since translators are usually limited in their ability to find words that match the words or concepts of another language. A good translator, however, is capable of reproducing the meaning intended in one language into another, and they do this by using the target language to facilitate a person "thinking about" the translated expressions in the same way as the author in the original language wanted. This occurs even though, as implied, the literal translations of the words used in the original text may not be totally available in the target language . Thus, optimal translations are rarely word-for-word, but rather closer to thought-to-thought.

A second advantage provided by integrating a cognitive perspective into the development of qualitative assessments (Barofsky 1996, 2000) is that it permits moving beyond the current use of classical test theory and the latent variable model to applying fundamental measurement theory (Borsboom 2005). This would occur if it can be demonstrated that a common cognitive-combination rule is present in all instances of a qualitative assessment. This would then have the potential of establishing an invariant principle, where the statement that evaluated descriptors and a metacognitive reflective valuation process are combined to become an established principle. This would follow in the tradition of Kahneman and Rasch whose pursuit of invariant principles marks the path towards establishing a science of qualitative assessment. Yet, both lack generality so the question remains whether an alternative approach exists.[3]

A third advantage of applying a cognitive perspective to a qualitative assessment is that it opens the assessment process to the significant literature and experimental techniques (e.g., reaction time studies, neuroimaging studies) available in the cognitive sciences. But it also expands the intellectual foundation of qualitative assessments from their current exclusive dependence on methods developed in educational psychology to a broader array of ideas and methods. This will also make it easier to shift from establishing causal relations, characteristic of linear modeling, to alternative modeling schemes, including nonlinear models that are analogs of the central nervous system.

A fourth advantage is that the hybrid construct provides an analytical model that can be used to assess the quality of existence of the compromised person. The model identifies specific cognitive components, the presence or absence of which can be used to design an assessment method or remedial effort. That is, instead of a surrogate being used to simulate the quality-of-life of the compromised person, more limited and targeted interventions are possible. For example, what little data are available suggest that persons with dementia are still capable of expressing their likes or dislikes (Mozley et al. 1999), so that while they may not be able to perform a rating using a visual analog scale, they may be able to offer an opinion on some scenario that is presented to them. This format also provides an alternative to the current practice of adjusting an assessment (usually by making the assessment items more concrete) so that a broader range of persons could respond to the items or concepts, even though the meaning generated by these assessments may be more variable. What this practice does is to preserve the psychometric equivalence of these assessments, although the cognitive and semantic equivalence may not be the same. By examining the cognitive components of the qualitative judgment, it will also be easier to examine the ethical issues involved in decision making about the compromised person. Thus, it should be possible to integrate the extensive neurocognitive data about the various compromised persons into assessing the capacity and nature of these persons' judgment of their quality-of-life.

A fifth advantage is that it permits the application of the language use principles (e.g., an item as a "speech act;" Grice's [1989] postulate of what is expected in normal conversation, and so on) into the analysis of the cognitive processes mediating a qualitative assessment. It does this because it has become clear that most qualitative statements use figurative language, including metaphors, for their expression, and that these linguistic processes reflect basic cognitive processes (e.g., conceptual metaphors). Supporting this connection are the data reviewed on embodiment (e.g., Chap. 11, p. 439) that provide an explanation on how conceptual metaphors might be formed. A language-usage analysis will also be useful in determining if the cognitive-combination rule postulated as part of a hybrid construct uses "metaphoric arithmetization" (Chap. 9, p. 298).

It is natural to ask what evidence have I been able to identify that supports my working hypothesis. My own impression is that the evidence I have reviewed, while sparse, is sufficiently suggestive so as to justify the continued investigation of my working hypothesis. However, it is also obvious that the available evidence would be expected to be sparse since the hypothesis has not been previously studied or tested.

Some of the evidence that I would cite actually provides a more general level of support than determining if a specific cognitive production rule has been applied to several selected cognitive components to form a hybrid construct. Thus, I found examples where specific cognitive processes, such as dual processing, modularity, cross-classification, and hierarchical arrangement of information, were involved in determining qualitative outcomes. Cognitive principles were also involved in efforts to organize different types of pain (e.g., into dimensions or categories/domains) and affective aspects of subjective well-being (SWB) into classification systems (Chaps. 7 and 11). The modular and hierarchical relationships between fatigue, depression, and physical activity were found to be critical determinants for accounting for the variance in HRQOL responding (Chap. 7).

I was also able to find examples where the qualification of functional indicators (that is, creation of a hybrid construct) altered the interpretation of an assessment. An example of this was the Stineman et al. (2003; Chap. 9, p. 326) study, where utility weights for various components of the FIM™'s were generated by patients and clinicians. What they found was that both persons with disabilities and clinicians valued communication and cognitive independence more than physical activity independence. When these weights were used as multipliers for the individual components of the FIM™'s, the interpretation of the results of the study was altered as compared to the investigator who simply summed the unweighted components of the FIM™'s.

In a second example, Hewlett et al. (2001, 2002) report two studies, the first of which demonstrated that patients, clinicians, and members of the general public differed in terms of the importance they each placed on having various disabilities; in their second study, they reported developing an eight-item assessment based on the Health Assessment Questionnaire (HAQ) that asked people to rate the importance of having various abilities (Personal Impact Health Assessment Questionnaire; PI-HAQ). As part of their effort to establish the criterion validity of their assessment, the authors also collected utility data using the TTO (Torrance 1986). The PI-HAQ, by consisting of evaluated descriptors that were qualified (that is, importance weightings were obtained for each person), has to be minimally considered an example of a qualitative hybrid construct.

There are, of course, other examples of quality-of-life and HRQOL assessments that are essentially hybrid constructs, and I have identified a partial list of eight of them (Table 8.6; Keeney 1988). All of these are examples of "synthetic" qualitative assessments, although they differ in degree. For example, the HALex (Erickson 1998) and the SF-6D (Brazier et al. 2002) can be totally synthetic if it involves mapping an established valuation system to already available data, but it can also be calculated using newly acquired evaluated descriptors. Most other assessments are synthetic in that they also involve a mapping of a valuation system that does not involve the respondents, with the respondents' role limited to evaluating the descriptors.

What is also interesting about this list is to ask how the investigators came to formulate their approach to qualitative assessment by essentially creating hybrid constructs. What thought processes and history influenced them, so that weighting some evaluated descriptor by a valuation, preference or utility estimate, seemed an appropriate way to monitor the quality-of-life of the respondents? To answer this question requires a more in-depth discussion, which I provide in the next sections. But what remains true is that the basic format of the hybrid construct has been repeatedly used to theorize about the determinants of decision and judgment making. For example, Keeney (1988) states:

> Most models of decision problems can be broken into two parts. The first part relates the various alternatives available to the possible consequences that these alternatives might have. These consequences describe the degree to which objectives are achieved. The second part evaluates the relative desirability of the consequences. (p. 149).

An additional topic that can conceivably raise concerns about the generality of a hybrid construct is the phenomena of first impressions, and I believe I have adequately addressed this topic in Chap. 3. Basically, my argument was that most cognitions involve dual processing; both rapid automatic assessments and slower more reflective processing, that in the context of a qualitative assessment would still have the elements of a hybrid construct. A final section addresses the role of measurement theory in qualitative assessments.

What I am not going to do is to review the content of most of the chapters, especially Chaps. 7–11, since they each dealt with fairly specific topics that have been adequately summarized in each of the chapters.

2 Determinants of Valuation

In this section I will discuss a series of topics that focus on the process of valuation, a key component of the hybrid construct. The first section deals with the issue of weighting as a way of estimating the valuation component of a hybrid construct. What will become clear is that a range of weighting procedures exist, some using formal models, such as the multiattribute utility function, some use sequential estimates of evaluated descriptors and valuations and then use a formal method of combining the data, and some rely on the person's cognitive processes to value the evaluate descriptor and combine the cognitive entities. The next section considers the apparent variability in outcome when the results of these different combinational methods were considered. This leads to a discussion of the standard gamble (SG) and the role that Prospect Theory (Kahneman and Tversky 1979; PT) plays in a qualitative assessment. A final section addresses the social psychological literature on predetermined values and how these established values of a person are determined by cognitive processes and language-usage. Overall, these sections demonstrate the usefulness of a taking a cognitive and language-usage approach to assessing quality.

2.1 Weighting Procedures

Experimental studies of the components of the hybrid construct vary in terms of the source of the information, the method used in the assessment, and the combinational rules used. Valuations can be determined directly from the respondents or indirectly from community representatives, but can also involve representatives of select groups (e.g., persons with a particular disease or set of experiences). Valuations can be calculated directly from questionnaires, or by various assessment methods (e.g., VAS, TTO, SG). The combination of evaluated descriptors and valuations can occur using formal rules such as addition or multiplication or select cognitive processes as when concrete entities are mapped to abstract constructs.

As was true for the relationship between cognitive categories and domains, formed using statistical methods (Table 6.2), cognitive-combination principles and formal weighting procedures can be conceived of as separate activities, although practically they coexist. This is especially true in a context of a qualitative assessment. Thus, just as an investigator might first cognitively form a category before applying statistical procedures to generate a domain, so too might an investigator cognitively simulate generating an index or numerical summary statement before applying some weighting procedure. In both of these applications, the quantities generated can be thought of as just numbers, but in a qualitative context they are usually meant to have a particular meaning. What remains unclear is the empirical and philosophical question of the extent these numbers convey specific meanings. It should also be noted that the precision of attaching a meaning to a number (that is, the number's semantic significance) is at the core of the objectification process, and if inadequate, may generate a summary statement that is based on metaphoric arithmetization (Chap. 9, p. 298). Of course, the use of cognitive interviewing is meant to minimize the discrepancies in perspectives here, but even after assessing the respondents, the investigator is still tasked with the responsibility of transforming this information into valid items on a assessment.

The term "weight" and the act of weighting can have several meanings, one of which involves its use as a metaphor for the term *importance* (e.g., "That idea has weight or importance"). As Jostmann et al. (2009; p. 1169) states:

> Weight is a metaphor for importance in many languages, including English, Dutch, Spanish and Chinese... The metaphoric use of weight suggests that the association between weight and importance has developed from a concrete link to a conceptual relationship on an abstract level (c.f., Lakoff and Johnson 1980).

Jostmann et al. (2009) go on to report four experiments that demonstrate that a person's bodily experience with heavy objects contributes to the development of their abstract concept of importance. This notion of Jostmann's et al. (2009), of course, is an example of embodied cognition (e.g., Barsalou 2008). It implies that weighting procedures (e.g., the slope in the equation for a line, or multiplying a person's utility estimate by some indication of their functional status) mathematically involves arithmetic, but semantically are importance statements. However, weighting can be done either independent of, or reflect to varying degrees, the context of an assessment. Earlier I reported an example of the influence of context, when I described the child who had misconstrued the wording of an item as requiring an independent as opposed to a comparative assessment (Chap. 12, p. 455; Willis 2005; p. 4).

The Stineman et al. (1998, 2003) studies were examples of where the weighting input from respondents was direct, in contrast to most other studies which were based on synthetic weighting procedures (e.g., where respondents from different groups than the study group are used to generate the weights; Table 12.1) or were based on investigator-generated weights. In the Stineman et al. (2003) studies both patients and clinicians decided that not all the components of the FIM™'s were equally important, and that instead of basing

Table 12.1 Weighting methods used in representative qualitative assessments

Assessment	Investigator	Evaluated descriptors		Origin of weighting		
		Generic	Disease-specific	Person-generated	Surrogate generated	Synthetic ▲
SF-36	Ware et al. (1993)	Several Scales 36 Items				
FLIC	Schipper et al. (1984)	Single Scale 22 Items				
EORTC-QLQ30C	Aaronson et al. (1991)		Several Scales 30 Items			
QLI	Ferrans and Powers (1985)		Satisfaction of ... 34 Items	Importance of ... 34 Items		
SEIQoL	Hickey et al. (1999)			Scenario derived/ based on 5 self-selected values		
HUI	Torrance et al. (1995)	Ratings of eight items			Population-based	
QWB	Kaplan and Anderson (1996)	Symptom Problem Complex + Ratings			Population-based	
HALex	Erickson (1998)	SAHS + Activity Limitations				Population-based
SF-6D	Brazier et al. (2002)	6 Items From SF-36				Population-based

▲: In the synthetic assessments, data from multiple sources are combined to derive a composite measure. Weightings are population-based

inferences on the sum of all the components, it was considered more appropriate to give priority to those components that were identified as particularly important to the respondents (e.g., being able to communicate and retain cognitive independence). In this sense, the differential weightings given these components of the FIM™'s also gave them the label of being "important," even though no direct mathematical manipulation occurred. Thus, it is possible that using the numerical analogs of values, preferences or utilities as weights can have semantic significance.

To answer the question of how general this observation is, I will briefly review the different methods that have been developed to weight-evaluated descriptors (Table 12.1) and then follow this with a review of a representative sample of studies that utilize an investigator-based weighting systems (Table 12.2). Table 12.1 reveals that several different weighting schemes exist for assessing health status (e.g., the SF-36), quality-of-life or HRQOL (Table 8.6). The weighting systems range from those that assume that each item is of equal importance and do not differentially weight the items or domains, to those that ask respondents to formally estimate the importance of an item (e.g., the QLI), rate the importance of scenarios that are based on self-selected values (e.g., the SEIQoL), or use an external group's preferences, or utilities, to weight-evaluated descriptors. These preferences or utilities were generated using either visual analog scales (e.g., Kaplan and Anderson 1990), TTO (Torrance et al. 1995), or standard gamble (Brazier et al. 2002) procedures.

Although the mathematical models used to generate these weights differ, they were all combined with evaluated descriptors,[4] so that the resultant outcomes were *cognitively equivalent*. The next question, therefore, is whether these methods empirically generate equivalent results, and if not, why they do not. To answer this question, I will selectively review the research literature dealing with the usefulness of importance weighting of evaluated descriptors. I start by reviewing the QLI (Ferrans and Powers 1985) as an example of the weighting process in a quality-of-life assessment, review representative studies of the role of importance weightings for healthy people, but as well as medically ill persons, and end with a discussion of the complexities of using weights as revealed in the studies of Campbell et al. (1976) and several review articles. What I will have learned is that not all weighting systems are useful, that certain classes of respondents are more likely to demonstrate the value of weighting than others, that the cognitive processes elicited by an item cannot always be predicted (e.g., does, or does not, the item elicit a reflective metacognitive process?), and that assessing the importance of a domain or some aspects of a life of a person is a far more complex than current assessments recognize and incorporate in their designs.

A standard approach to weighting can be illustrated by reviewing the QLI (Ferrans and Powers 1985) assessment. In this assessment, a subject is asked to rate 33 items in terms of their degree of satisfaction with the state described by the item, and a second independent assessment (followed

Table 12.2 Weighting by importance scores used in summary statements of quality-of-life assessments: representative studies

Reference	Sample (N)	Assessments	Weighting procedure	Study outcome
Healthy subjects				
Bowling and Windsor (2001)	Population-based representative sample ($N=2,031$)	Self-selected important domains, Global QOLA	Importance ratings and ranking of self-selected domains	Importance rating of self-selected domains accounted for only 16% of global quality-of-life assessment
Hsieh (2003)	Random recruitment via phone ($N=90$)	Com-QoL Scale-modified, LSIA, and SWLS	7 Different direct importance and domain rankings	No evidence of an increased correlation between global and domain satisfaction with either weighting or ranking procedures
Todman et al. (2003)	Healthy college students ($N=35$)	SEIQoL-DW, SWLS, VAS, Verbal Analog Scale	SEIQoL-DW as an importance indicator	No evidence that weighting satisfaction scores by importance of use modified results
Wu and Yao (2006a)	Healthy college students ($N=130$)	Investigator- based satisfaction scale based on WHOQOL-100, SWLS	4 Different weighting methods used	No evidence that item importance effected the relationship between item satisfaction and global satisfaction
Wu and Yao (2006b)	Healthy college students ($N=332$)	Investigator-based satisfaction scale based on WHOQOL-100, SWLS	Have-Want discrepancy; item importance, item and global satisfaction	Have-want discrepancy covaries with importance, but is only rarely (2/12 domains) a significant contributor satisfaction (SWLS) variance
Wu and Yao (2007)	Healthy college students ($N=40$)	4 Domain-specific scenarios simulated various importance outcomes	Have-want discrepancy; high-low importance, global satisfaction	Satisfaction ratings covaried with importance demonstrating support for Locke's range-of-affect model
Wu (2008)	Healthy College Students ($N=167$)	WHOQOL-100 modified, SWLS	Importance rankings of items; weighting of satisfaction as a predictor of QOL. #	Have-Want Discrepancy covaries with importance. Weighing or not weighting satisfaction scores did not predict global QOL
Wu et al. (2009)	Comparison of three data sets: Life: college students ($N=237$); Self: College Students ($N=269$); Job: Hospital Workers ($N=557$)	Quality of campus life questionnaire; Rosenberg Self-esteem Scale; SWLS.	Partial least squares model of 3 domain satisfaction indicators, domain importance ratings used to predict global satisfaction	"Unweighted domain satisfaction scores have a stronger predictive effect for global satisfaction measure than importance-weighted domain satisfaction scores" (p. 351)
Medically ill subjects				
Ferrans and Powers (1985)	Hemodialysis Patients ($N=349$)	QYI	Satisfaction and importance ratings	Correlations between Satisfaction and Importance ratings for 32 domains ranged from –0.03 to 0.50; Rankings between domains were also reported
Balaban et al. (1986)	Rheumatoid arthritis patients ($N=288$)	QWB	Replication of weighting system for QWB but now for patients	Weights generated for the rheumatoid arthritis patients not different from General Population weighting values
Gorbatenko-Roth et al. (2001)	Medical outpatients (102)	Quality-of-life ratings of 22 scenarios	Comparison of preference weighted and equal weight ratings of role, emotion, and physical domains	Preference weighting accounted for 14% more variance than equal weighting for persons who showed differential weighting of domains, but opposite effect for minimal differential weighting of domains
Rogers et al. (2002)	Oral-pharyngeal cancer ($N=46$)	UW-QOL 3 global quality-of-life questions	Evaluation and importance rating occurred consecutively for each domain	No evidence of a correlation between functional status and importance ratings
Morita et al. (2003)	Nonsmall cell lung cancer patients ($N=377$)	20-item Japanese-specific QOLA; Global Item	Population-averaged statistical weightings of domains	Results showed that domain-specific studies and relative weights provided useful information
Wettergren et al. (2005)	Hodgkin lymphoma survivors ($N=121$)	Generic and disease-specific SEIQoL-DW; SF-12	Domain and global ratings. importance rating	No difference in correlations of weighted or nonweighted generic or disease-specific SEIQoL-DW and PCS and MCS from the SF-12

(continued)

Table 12.2 (continued)

Reference	Sample (N)	Assessments	Weighting procedure	Study outcome
Hagell and Westergren (2006)	Parkinson disease (N=71)	QLI, Global Life Satisfaction Item, disease severity rating	Applied recommended scoring algorithm (SxI)	No evidence of a benefit from multiplying satisfaction ratings by importance rating
Reed Johnson et al. (2006)	Nonsmall cell lung cancer (N=99); breast cancer (N=159); colorectal (N=117)	EORTC QLQ-C30	Preferences obtained for EORTC QLQ-C30 domains	Results questions the assumption that each domain in a QOLA is of equal importance
Russell et al. (2006)	Injection drug users (N=241)	IDUQOL Scale, MC X2, SWLS, RSES	Most important domains self-selected and rated, separate satisfaction rating	Composite correlations between importance and satisfaction=0.05. 6/22 domains significant correlations
Philip et al. (2009)	Cancer patients (various diagnoses) (N=194)	FACT-G, CBI, SWLS, PAIS	Six representative items from 4 FACT domains, worded (+) or (−) = 24 comparisons: Forced choice between different domains	Statistical weighting of importance score. Weighting algorithms did not reveal that incorporating domain preferences in FACT-G led to more accurate predictors than nonweighted assessments. Persons with no preference best predictors of FACT-G domain scores
Stone et al. (2009)	Prostate cancer patients (N=150)	FACT-P, SEIQoL-DW, VAS	The SEIQoL-DW was used to generate weights for the 5 domains of the FACT-P. Domain scores were multiplied by weights	3/5 domains were significantly higher for weighted vs. unweighted domains. Global HRQOL was significantly greater for weighted vs. unweighted estimates

consecutively) is performed to determine how important these same 33 items are to the person. The satisfaction scores are rescaled so that they range from −2.5 to +2.5 and are than multiplied by the importance scores and are than summed to give a composite score. The format of this assessment is commonly found, but it is a format that does not ordinarily elicit the reflective processes expected when a qualitative assessment is obtained, since the two elements of the assessment process are never concurrently juxtaposed, but rather separated by time.

As a result, the method used to generate a summary statement is subject to metaphoric arithmetization. The investigators, by asking subjects to independently rate selected domains in terms of satisfaction and importance, are structuring the assessment process so that minimal confounding between the two types of assessment occurs. By so limiting the context of the assessment, however, they are also encouraging the respondents to interpret their task as involving evaluating descriptors, rather than making the more comparative judgment involved in a reflective process. This is especially true since they have just finished rating 33 items in terms of satisfaction, and their response set was more than likely to continue the same mode of thought when they rated the importance of each item. This immediately raises questions about whether this assessment meets the cognitive criteria required of a hybrid construct.[5]

As stated, I considered the summary statement of the QLI as an example of *metaphoric arithmetization*.[6] Clearly, the summation that is used to generate the composite score is mathematically appropriate (that is, it is perfectly legitimate to add up multiplied numbers), but what is not clear is whether the semantics of the summary statement justifies the label of a quality-of-life assessment. For example, synonyms for the term satisfaction include, "approval, pleasure, happiness, fulfillment, contentment, and liking," while synonyms for the term importance are distinctly different and include "significance, meaning, weight, consequence, magnitude, substance, and value." Thus, the product of multiplying satisfaction by importance generates a fairly heterogeneous semantic index, so that the interpretation of the composite as an indication of quality or quality-of-life must be considered more of a metaphoric inference than a statement that has elements of being literally true.[7]

The model used in the QLI is similar to many studies reported in the job satisfaction literature. Locke (1969) after reviewing the available job satisfaction literature states:

> With respect to weighting, our previous analysis suggests that *importance is already included in and reflected by the satisfaction ratings* (to the extent that they are valid). Since value importance determines the degree of affect produced by a given amount of value-percept discrepancy, multiplying satisfaction scores by importance scores is redundant. (Locke 1969; p. 331).

To claim, as Locke does, that the term importance was somehow included in the satisfaction ratings does not seem to make sense, semantically. For example, synonyms for the term satisfaction and importance I just reviewed (Chap. 12, p. 457) made it clear that the two terms have minimal semantic overlap. Cognitively, Locke speculates that a discrepancy exists between what a person perceives and what they value, and this evokes emotions proportional to the degree of this discrepancy. Interestingly, this hypothetical cognitive process has some empirical backing as is evident from the cognitive interviewing study by Williams et al. (1998) where he found evidence to suggest that satisfaction was a second-order cognition that involved a comparison of two evaluated descriptors. Thus, his model of the implicit cognitive processes underlying satisfaction rating was accurate, although at the time he was not able to cite empirical support for his assumption.

What Locke also claimed was that a person has a set of *a priori* notions that he or she values (referred to as "value importance") and that these values when compared to current reality evokes feelings that lead to behaviors designed to resolve the perceived discrepancy. His model is consistent with, and derives from, Ann Rand's (1964) political philosophy, that postulated that a person's values are the primary determinant of their behavior. A subsequent empirical study provided some initial support for his views (Mobley and Locke 1970).

Locke's notions have certain implications. For example, they imply that if a respondent rates an item or domain to be satisfying, this automatically and unavoidably also brings along with it an importance rating (that is, it connotes importance). Something like this, of course, can occur but it is not the only inference possible, and its presence would require formal assessment and confirmation (such as by using cognitive interviewing). Interestingly, this would transform the indicator into a qualitative statement, since the respondent would be expected to engage in a reflective process to establish the degree of importance. What may also be true is that what is important to one person may not be important to another, so that in any one assessment a proportion of the selected items or domains may be irrelevant, to all but some of the respondents. Rather, what may be most relevant is the *relative importance* of items and domains to a person. Campbell et al. (1976) made this abundantly clear when they performed a detailed study of importance ratings of a national well-being survey (see below). Lurking behind all these issues are the differences between the investigator's and respondent's interests and the conflict that is inevitably created when the content and design of an assessment are selected. Using the QLI as my model, this issue could be "visualized" by asking respondents if their answers to the assessment captured what was most important to them and reflected the quality of their existence. If a discordance existed, then one interpretation would be that the assessment was designed to meet the investigators', not the respondents', interests.

An example of a HRQOL assessment where respondents first identified important experiences (e.g., symptoms) prior to the development of a qualitative assessment was the Breast Cancer Questionnaire (Levine et al. 1988; see also Guyatt et al. 1989). In this study, the investigators initially interviewed a large sample of breast cancer survivors and asked them to indicate what symptoms and other experiences they had from being treated for cancer that were important to them and then to rank these experiences in order of the most to the least important. This was followed by the development of an assessment that included the top 30 most important experiences. The difference between this study and what Locke (1969) was proposing was that this study asks respondents how satisfied they were with already identified important aspects of their lives and it did not assume that the satisfaction ratings of the items were important or that a satisfaction rating established the importance of the item. Thus, the satisfaction ratings were conditional statements, but were not semantically confounded by multiplying satisfaction by importance, or vice versa.

What may also be true is that while the surgery and chemotherapy these persons received may be new experiences, the disruption in their lives that resulted may not have been, especially since other adverse life events may have precipitated similar experiences leading to the person having a set of values in place. Of course, not all people would have had the same prior experiences. In addition, many of the items on the assessment dealt with the unique consequences of being a breast cancer survivor, including hair loss, intimacy issues, extreme fatigue, medication-induced weight gain, and so on. Again, while the disrupted valued functions may have been the same that other life events produced (e.g., not being able to clean the house, feel embarrassed in public, and so on), the attribution here was unique and the threat to mortality probably quite unique, so that the values could very well be qualitatively and quantitatively different from most adverse life events.[8] These reasons alone argue for formal value assessment.

Locke (1969, 1970) also recognized that value diversity existed and that not all aspects of a job were equally satisfying to the person. This led to his *range-of-affect* hypothesis, where he claimed *a priori* values generated different emotions concerning different aspects of a job (e.g., pay, hours worked, benefits, and so on) impacting the employees satisfaction. Tests of this hypothesis have only recently been accumulating (e.g., McFarlin et al. 1995; Rice et al. 1991; Wu 2008; Wu et al. 2009; Wu and Yeo 2006a, 2006b, 2007). Rice et al. (1991), for example, report a study that created scenarios that described the amount of freedom a person would have on their first job (high or low), and how important the job was (high or low). What they found was that if the scenario described someone as having high freedom but an unimportant job, job satisfaction was low, while if they had low freedom and an important job, then their job satisfaction was high. Varying the amount of freedom on a low importance job had a minimal impact on job satisfaction. These data can be interpreted as supporting the notion that prior values, as reflected in the range-of-affect hypothesis, impact satisfaction judgments. The McFarlin et al. (1995) study replicated the Rice et al. (1991) study, but now examining 12 job characteristics with South African workers. The investigators found that 11 of the 12 job characteristics varied with the importance of the job, providing additional support to the notion that the fulfillment of established values is a critical determinant of job satisfaction.

Critical to the range-of-affect hypothesis is the assumption that a discrepancy exists between what a person has experienced and what they ideally want or expect. The observation that the presence of a discrepancy is a major correlate of the degree of satisfaction should not be surprising since satisfaction itself has been operationalized in the same terms (Chap. 8, p. 265; Table 8.2). Second, Locke also views satisfaction as an emotion when it is clear that it is not a primary emotion, but rather a cognitive construction involving a comparison, which may include to varying degrees an emotional component.[9] This should permit an investigator to use satisfaction as an indicator without having to make assumptions about the *a priori* presence of values and the implications this has for a particular political theory (see above).

Interestingly, quality-of-life or HRQOL has also been defined in terms of a discrepancy or gap (e.g., Calman 1987; Table 2.5), and experimental studies have been reported (e.g., Ferrans and Powers 1985; Welham et al. 2001) where the relationship of the gap to satisfaction has been studied. Wu, in a series of studies (Wu 2008; Wu et al. 2009; Wu and Yeo 2006a, 2006b, 2007), applied the range-of-affect hypothesis to quality-of-life data. She, and her colleagues, noted that several quality-of-life researchers weight life satisfaction scores by importance measures to generate a composite index, and the question she empirically asks, is this necessary, especially if importance is *a priori* determinants of satisfaction. Included in this group were Cummins' (1997) Comprehensive Quality-of-life Scale, Ferrans and Powers's (1985) Quality-of-life Index, and Frisch's (1992) Quality-of-life Inventory, to name three. The method commonly used in these assessments to generate the hybrid construct is to multiple the rating of the evaluated descriptor by the separately obtained weighting of the descriptor. One of the weaknesses of this method is that the weighting process does not take into account the prior evaluation of the descriptor, thus the two assessments (evaluating the descriptor, and weighting the descriptor) may not appear as phenomenally distinct from each other, leading to no differential weighting. The best way to avoid this would be to have both pieces of information simultaneously available to the respondent. This is essential what is done when scenarios are used to generate preferences or utilities and the role of various components is systematically varied. The implicit assumption in the evaluated descriptor by weightings model is that it somehow simulates how a person might combine these two indicators themselves, but this is an inference that may not be correct.[10] Thus, a direct assessment is required rather than investigators inserting their models of how people might think.

A review of representative studies showed that studies involving "healthy" persons failed to demonstrate any advantage of weighting(0 of 8), while the results of studies that included medically ill persons were more likely to demonstrate an advantage from weighting(7/11). The question now becomes why this difference. Table 12.2 summarizes these studies. Inspection of the eight healthy subject studies revealed that six of them asked respondents to separately and consecutively rate satisfaction and importance, so that the concern that I expressed about the QLI is applicable to them. The Bowling and Windsor (2001) study differed in that only importance ratings of various health states were obtained, while the Wu and Yao (2007) study asks subjects to rate their satisfaction with various scenarios that incorporated importance statements. These scenarios, therefore, come closest to creating the kind of situation I was alluding to as being able to generate reflection, except that the importance component was not directly assessed, rather they were arbitrarily defined in terms of the various parameters included in the scenarios (e.g., house size, location, body function, thinking ability, academic achievement, extracurricular activities, and so on). I concluded that the failure to observe a benefit from importance ratings reflected limitations in the experimental design used by these studies, and that they, therefore, do not represent an adequate test of the assumption that importance ratings differentially modify satisfaction ratings. Of course, it would be interesting to determine if importance ratings of other indicators, such as happiness or anger, occurred and if these results were different from satisfaction by importance ratings. I was not able to find examples of this type of study in the research literature.

In contrast, studies of the medically ill person revealed that 7 of the 11 studies in Table 12.2 provided some positive support for the assumption that differential weighting could produce a statistical and clinically significance difference. To these studies must be added the study by Osoba et al. (2006) who found that cancer patients differentially selected hypothetical function/symptom pairs in terms of their diagnosis, stage of disease, or treatments. These results reinforced the view that not all items on a qualitative assessment were equally as important to a respondent. The Gorbatenko-Roth et al. (2001) study demonstrated a modest advantage due to weighting following reading of selected, but hypothetical, scenarios that incorporated different qualitative domains. Of particular interest was their demonstration that the persons who most clearly selected particular domains over others showed the highest advantage to weighting, while those persons who revealed minimal differential preferences among the various domains showed the least benefit from weighting.[11] They state:

> These results indicate that for high- and medium-discriminating participants, the PW model was a better fit of intrinsic decision making, accounting for 14% and 4% more decision variance than the EW model, respectively. For the low-discriminating group the opposite was found: The EW model accounted for 4% more variance. (Gorbatenko-Roth et al. 2001; p. 139; PW refers to preference weighting and EW refers to equal weighting).

In contrast to both of these studies, Phillip et al. (2009) failed to demonstrate an advantage to weighting. They used the design of the Osoba et al. (2006) study (that is, forced-choice domain preference selections), but now with cancer patients who were making decisions based on their current functional and symptom status. They also found that respondents were able to state preferences for particular domains (e.g., emotional well-being), but that when they used these preferences as weights with the domains in the FACT-G they were unable to see any predictive advantage to the weighted scores. They also divided their sample into three groups depending on whether they had high (two domains both of which were one standard deviation above the mean), moderate (one domain which was one standard deviation above the mean), or low preferences for various domains and essentially found the opposite of what Gorbatenko-Roth et al. (2001) found: persons with no domain preferences generated weighting models that predicted the greatest amount of variance than the groups that had moderate or high domain preferences.

How can these differences be accounted for? First, the two studies differ in that one used scenarios that created a context for stating a preference, in contrast to the forced-choice method where the respondents were deliberately limited. This difference was bound to have a cognitive impact, by expanding for one and constraining for the other, the context for stating a preference. The extent this accounts for the difference in the two studies is an empirical question, yet something like this could be found to be relevant, since the results of the two studies were so dramatically different.

Besides the Phillip et al. (2009) study, I considered the Balaban et al. (1986), Wettergren et al. (2005), and Hagell and Westergren (2006) studies as not providing supportive data. The Balaban et al. study was not a direct test of the usefulness of weighting, but was included in Table 12.2 since it was one of the first in the area of HRQOL research to question whether the preferences of persons with a specific disease would be different from preferences generated by persons in the general population.

The results of their study clearly demonstrated that the rheumatoid arthritis patients generated preferences that were essentially similar to the general population. One possible explanation of these results was that the patients had already adapted to their circumstances of having a chronic illness and made every effort to appear and function "normally." The failure to demonstrate a difference in their preferences would then be interpreted as evidence of this adaptation. In addition, the results of the study raise the issue of whether generic or disease-specific qualitative assessments are truly that different and whether it is necessary to persist in distinguishing between a quality-of-life and a HRQOL assessment.

Wettergren et al. (2005), another example of a study where the results did not support the usefulness of importance weighting, used the SEIQoL-DW to provide both a global and disease-specific estimate of HRQOL. A disease-specific assessment was produced by changing the interview schedule to be more specific to the adverse consequences of being treated for Hodgkin Disease or surviving the disease. For both types of assessments, respondents were asked to first identify and rate (from very good to very bad in seven steps) those domains of particular interest, thereby generating evaluated descriptors. They were then asked to rate the evaluated descriptors in terms of their importance, relative to each other, so that the total responses to all five assessments summed to 100. These data were then correlated with summary measures from the SF-12 (the Physical Component Summary and the Mental Component Summary). The results indicated that only small, but consistently higher, differences in the correlation were found as a result of importance weighting of the evaluated descriptors from the standard SEIQoL-DW assessment, while these same correlations for the disease index were much less consistent.

The Hagell and Westergren (2006) study was particularly straightforward in that it did not find any benefit from weighting. One observation sparked my interest: weighting not only did not provide any advantage, it actually appeared as a disadvantage, in that the total and subgroup scores were uniformly, but marginally, reduced, compared to satisfaction ratings alone. This suggested to me that the distribution of the satisfaction by importance ratings may have lowered the total scores. This is possible because the Ferrans and Powers (1985) scoring algorithm requires that the satisfaction scores be rated as both positive and negative so that importance scores may have exacerbated negative scores more than they augmented positive satisfaction scores. It would be interesting, therefore, to determine what particular items contributed to this effect. As I make clear from the Campbell et al. (1976) data (see below), importance ratings can modify satisfaction ratings in a number of ways; the impact of this may not be noticeable from total score analysis and may require the study of particular groups of respondents.[12]

Campbell et al. (1976), following the analysis of a large data set from a national survey on well-being, commented on the complexity of interpreting importance ratings. They interviewed respondents and asked them to first rate their satisfaction relative to a series of life domains (e.g., family, marriage, job, health, home, etc.) and then separately asked them to rate the importance of these domains. Health, marriage, and family life were rated as the three most important domains (Campbell et al. 1976; Tables 3–5). However, when they regressed the importance ratings of all 12 domains to their respective satisfaction scores, then family life and marriage generated the highest regression coefficients, while health was listed as the eighth most important. The relative ranking of a number of other domains was also discordant. Overall they found a correlation of 0.41 between the importance ratings and regression coefficients for satisfaction. They interpret this correlation as indicating "a fair degree of correspondence"(Campbell et al. 1976; p. 84). However, when they examined some of the exceptions, or apparent contradictions (such as, health being quite important but not much of a predictor of life satisfaction for this population), they hypothesized that the two indicators provided distinct dimensions of a person's life that could overlap (such as, the high rating of importance and life satisfaction for family and marriage), but could also not overlap. A good example of an exception is the observation that respondents' rating of national government is the fourth most important domain, but the capacity of the domain to predict life satisfaction was 10th of 12 domains. They acknowledged that more than likely the respondents were making their importance rating independent of the satisfaction rating, so that finding discordance was not be surprising.

They then, as so commonly happens, multiplied the satisfaction score by the corresponding importance score, summed these products, and regressed these scores to a composite measure of well-being or life satisfaction. The question they asked was whether weighting the satisfaction scores increased the variance accounted for, and as was found for so many other studies (Table 12.2), they did not find any benefit from the weighting. I would argue that their results were perfectly predictable, since the experimental format they used maximized the semantic and cognitive discordance between the two assessments, as was reflected in their own data. If I only consider the meaning of the terms satisfaction and importance, then combining them and referring to them as an indicator of well-being would have to be considered an example of metaphoric arithmetization. I would argue this, first because there is minimal semantic overlap between the term, and second because a maximum effort was made to ensure that the two assessments were assessed separately and thus did not cognitively overlap.

A critic of my interpretation might ask how then do I account for the Campbell et al. (1976) findings that regression analyses demonstrated that the same predictors (e.g., marriage and family life) accounted for significant proportion of the variance in importance and life satisfaction scores.

Here again I return to the analysis that Campbell et al. (1976) provided and a review of it made it clear that they felt that asking someone to rate the importance of a domain was a fairly abstract task that they described as generating idealized or hypothetical responses, rather than responses that might have occurred if they were directly asked to indicate how important their particular satisfaction rating level was to them. They also noted, after reviewing their interview data, that the importance ratings were more variable than other types of ratings; that is, the standard deviations of the importance ratings were greater than other outcome indicators. In addition, when they examined subgroups, they found that their importance ratings were "quite intelligible" (Campbell et al. 1976; p. 88). For example, they report that men found their job satisfaction more important than women did. On the other hand, they found that men not married and past 30, and male and female widows, do not find being married as particularly important. An interesting example of the complexity of importance ratings was illustrated by their examining the length of time since a person was divorced. What they found was that women who were recently divorced rated being married as being more important than they did some 5 years later, while divorced men consistently rated the importance of being married lower, independent of the length of time since their divorce. They interpreted these data as evidence that these respondents were engaging in psychological denial, and this, with the tendency to idealize and use hypothetical states, contributed to the variability of importance ratings.

I interpreted their results as just what you might expect considering the context of their study (that is, assessing satisfaction and importance independently). Thus, my response to those who point to the consistency of rankings of satisfaction and importance ratings for some domain as support for not including an independent rating of importance in a qualitative assessment, as being misplaced. It is misplaced for several reasons: first, because the appropriate research context was consistently not used when generating the importance ratings[13]; second, the regression coefficients found, at most, accounted for a small amount of variance (only a maximum of 20% for the domain of family life) in global well-being; third, considering the diversity of results revealed by subgroup analysis, it is possible that the rankings may not be replicated; and fourth, that this would be particularly true if the role of importance ratings were performed with different groups (as was found for studies of healthy vs. medically ill persons; Table 12.2). In contrast to my views, Hsieh (2003) claims that rankings were more effective than ratings in measuring life satisfaction.

Before I go on I want to comment on two review papers (Russell and Hubley 2005; Trauer and Mackinnon 2001) that have made useful comments about the problem of weighting. Trauer and Mackinnon (2001), for example, focus on some of the psychometric issues involved in using importance ratings in quality-of-life research, while Russell and Hubley (2005) cover a broader array of research areas and conceptual issues. The Russell and Hubley (2005) paper reviewed the quality-of-life, self-esteem, and job satisfaction research literature and concluded that using importance weightings did not enhance the assessment of the studied constructs. They go on to discuss the theoretical and psychometric concerns about both importance ratings and the multiplicative model of importance weightings. Of particular interest were the comments by Russell and Hubley (2005) about whether satisfaction and importance were assessing the same construct. As they say:

> Based on the limited evidence provided in the literature, it can be concluded that, although there may typically be some positive relation between the constructs measured by primary and importance ratings, these constructs are not identical or even very highly related." (Russell and Hubley 2005; p. 117; by 'primary ratings they are referring to the rating of quality-of-life, self-esteem or job satisfaction).

This view is consistent with my contention that the two constructs are semantically and cognitively distinct. They also comment on the contention of some investigators (e.g., Trauer and Mackinnon 2001) that importance is incorporated in satisfaction ratings, but they fail to find clear support for this view. They do acknowledge that the importance construct is more likely to be affected by contextual factors, such as, the setting where a person is responding to the items, whether they are union membership when rating job satisfaction, the role of social norms, and the concreteness and abstractness of the two constructs being studied. Evidence supporting the role of the concreteness of the construct comes from comparing the job satisfaction and quality-of-life literatures, where the more concrete judgments relative to a job make it easier to rate the importance of aspects of working, compared to rating the importance of aspects of quality-of-life, usually a more abstract construct (Mastekaasa 1984).

Russell and Hubley (2005) raise the very important question of what the term importance itself means. They offer several ways that a person could interpret the meaning of the term, and in so doing, acknowledge that the way an item is worded and set in a context will have cognitive linguistic consequences. Thus, when a person is asked, "How important is this domain to you?", they may interpret the question as asking them to identify and rate the domain most likely to change, or for the person to indicate how the domain affects a respondent's quality-of-life. They encourage investigators to ask their questions more precisely so that confusion concerning the cognitive task implicit in the item is clear. They also raise the fundamental issue about whether importance is the indicator that should be used, and that instead an investigator should consider alternative indicators, such as what is the *impact* that a particular domain has on the person.

The next topic they discuss dealt with their concerns about the multiplicative model (Satisfaction X Importance) itself, and they included many of the concerns that Trauer and Mackinnon (2001) expressed. Their first concern is that multiplying a satisfaction rating by an importance rating may produce uninterpretable results. For example, how would an investigator interpret two equal scores with one based on multiplying a high satisfaction rating with a low importance rating, and the other by multiplying a low satisfaction rating with a high importance rating? They cite a number of alternative mathematical operations that could be done, but conclude that none of the alternatives totally eliminated the risk of uninterpretable results. A good example of the practical consequences of this concern was the Rogers et al. (2002) study where they report that respondents selected extremes of importance ratings for the same domain, depending on their symptomatic and physical status, resulting in summary statements that failed to represent this heterogeneity. However, the scoring algorithm of the QLI provides a means of avoiding these complexities (Ferrans and Frisch 2005).

Trauer and Mackinnon (2001) also were concerned about the measurement characteristics of importance ratings. They point out, for example, that importance ratings do not represent ratio-level data, and therefore, are not suited to be transformed by multiplication. Concern was also expressed by them about the sensitivity to both satisfaction and importance ratings to distortions in their scaling properties raising concerns about the feasibility of multiplying one indicator by the other. In general, these authors express serious concerns about the practice of multiplying satisfaction by importance.

Clearly, this research area remains complex and in flux, but I find the conclusions of Trauer and Mackinnon (2001) acceptable and predictable, especially considering the experimental design commonly used in these studies (that is, the two assessments were performed separate from each other). An alternative approach would have a respondent rate the importance of a particular satisfaction rating immediately after the satisfaction rating occurred. Thus, if I rate my satisfaction with my health as a five on a seven point scale, and then if I immediately rate the importance of this specific satisfaction rating, my outcome will be an importance score. If these outcomes were summed across a series of different domains, then the outcome would constitute a valuation (that is, a summation of importance scores, or values, preferences, or utilities).[14] Of course, a study could be done where the exact opposite procedure was used. This would happen if the importance rating of a domain were followed by a respondent indicating how satisfied they were with this specific importance outcome. Here the outcome would be the summation of the satisfaction scores. Obviously, these assessments would generate two very different meanings. The advantage of this approach, however, is that it avoids the confounding and conceptual confusion that results from multiplying independently assessed satisfaction and importance scores.

This discussion helped me appreciate how useful it was to examine how language is used and what cognitive processes are involved when I perform a qualitative assessment. The next topic will repeat this demonstration by examining the role that language-usage and cognition play in the valuation process.

2.2 Cognitive Basis of Valuation Methods[15]

Much of what I have and will discuss has involved either direct or indirect efforts to simulate how people think, particularly when they are making valuations. For example, most people are capable of making choices or assigning a numeric value to something that they have experienced, and these activities have been modeled using different formal assessment procedures. Thus, the visual analog scale, standard gamble, or the TTO can be considered examples of methods that formally characterize the variety of ways that people potentially think about when making valuations.[16] Person-generated valuations have also been developed, and these alternative procedures also contribute to the diversity of methods available. The first question that I will ask, therefore, is whether this diversity makes a difference; the second question is to determine the extent to which these models capture what is known about how people actually make choices and ratings. What will become clear from this discussion is that much more has to be learned about the cognitive processes and language used during the valuation process, and how these factors contribute to variability in assessment outcomes.

Fryback et al. (2007b), as part of his introduction to a symposium at the October 12th 2007 Annual Meeting of the International Society of Quality-of-life Research, stated:

> The EQ-5D, HUI2, HUI3, QWB-SA, and SF-6D are self-reported summary indexes of generic health-related quality-of-life (HRQoL), preference-scaled to the same anchor points. Yet they systematically disagree, sometimes dramatically, in their assessment of individual HRQoL.

By "systematically disagree," Fryback meant that each of these methods generated different arithmetic values, so that when these numbers were used in practical settings, such as when estimating a person's quality-of-life, quality adjusted life years, or making policy decisions, there was a risk of generating conflicting recommendations.[17] As I will make clear in this section, this situation was perfectly predictable based on the previous discussion where it became clear that the experimental design evoked cognitive processes that were critical determinants of the semantic and quantitative outcome.

2 Determinants of Valuation

Inspecting the references (see below) listed in Tables 8.6 and 12.2 certainly confirms that there is a fair amount of diversity in the methods used to generate the components of the hybrid construct (e.g., assessments #4 to 9 in Table 12.1). Thus, while these assessments were cognitively equivalent (that is, they all involve the combination of an evaluated descriptor and a valuation), they differed in terms of how the evaluated descriptors were generated and weighted and the mathematical operations employed to form a summary statement. What I need to do now is to describe these differences and speculate about how they may lead to different quantitative outcomes. First, however, I would like to provide an estimate of the magnitude of the diversity.

Table 12.3 summarizes a representative set of studies that have included two or more Class III or IV assessments (Table 8.4). Inspection of the data suggests that there is more variability between studies than within a study. This suggests that the experimental conditions, sample studied (e.g., age of respondents; gender, and so on), and the particular assessment may each contribute to the observed variability. Six of the nine studies involved persons who were medically ill. The Garster et al. (2009) study compares persons who have congestive heart failure and were either taking or not taking medications. This gave me the opportunity to determine what impact taking medication has on a qualitative assessment, and what was found was that taking medication uniformly lowered assessment scores. The Hawthorne et al. (2001) study involved a healthy community sample, while the Franks et al. (2006) and the Fryback et al. (2007a) studies involved the same investigators and dealt with population-based estimates that can also be considered a "healthy" control. Comparing these studies with the remaining studies in Table 12.3 suggested that being medically ill, whether taking medications or not, leads to lower scores on these assessments. What it doesn't account for is the diversity in the results between the assessments within a study. What could account for this? To answer this question I am going to have to study the classifying system or domains included in each assessment, the method used to evaluate or scale these domains, and the valuation methods (VAS, TTO, SG), but also examine the linguistic and cognitive consequences of the valuation methods. There are two potential contributors to the observed variance: differences in the assessments themselves and differences due to respondent characteristics.

Table 12.4 provides an overview of the differences in the format of selected examples of valuation-based qualitative assessments, and in so doing, also provides a template to compare the different assessments.[18] The first question that can be asked, therefore, is whether differences in the evaluated descriptors or domains in each assessment could account for the different summary scores. Inspection revealed a fair amount of heterogeneity in the content included in these assessments, and this can be visualized by inspecting Table 3 in Hawthorne et al. (2001), or Tables 1 or 2 in Bryan and Longworth (2005). The evaluated descriptors varied both in terms of the number and type of domains included in an assessment, the number of response options a respondent had to review (e.g., 15 for the EQ-5D, but 77 for the HUI3), the abstractness and concreteness of the items representing the domains, whether the consequences of being in a particular state were articulated or not, the rating task required (Yes/No; Likert, verbal or numeric descriptors), and so on.

The first observation that can be made was the interpretative complexity of this literature; Feeny and his colleagues (Wee et al. 2007), for example, argued that while they found clinically important differences, the impact of these differences on cost-utility analyses was relatively minor, but needed further study. At the same time, they examined the feasibility of combining preferences from different assessments; they also concluded that "the equivalence of health utilities obtained from preference-based instruments cannot be assumed." (Wee et al. 2007; p. 264). Other investigators (e.g., Brazier et al. 2004; Konerding et al. 2009; Søgaard et al. 2009), while acknowledging the summary scores of different assessments may or may not be statistically significantly different, still argued that the assessments should not be used interchangeably, since the meaning these different assessments created (e.g., differences in the items used) could lead to different interpretations of the cost-utility analyses.

In support of this interpretation was a study by Grieve et al. (2009) who reported that the SF-6D leads to greater valuation scores and lower cost estimates when integrated into a cost-utility analyses, than the EQ-5D. They raised the possibility that the reason for this was because the two assessments differed in terms of their content, with the SF-6D including items on vitality and social functioning that the EQ-5D did not. Not only did the content differ, the two assessments also differed in terms of the number of domains (four vs. six domains), so that the SF-6D would be expected to be somewhat more reliable and a better estimate of the variance inherent to any assessment. Finally, the SG takes a respondent's tolerance for risk into account, while the TTO and other assessments may not.

The potential contribution that variation in content may make can be illustrated by comparing items that are meant to address the same domain in different assessments. For example, pain is listed as a domain in the EQ-5D, SF-6D, and the HUI3. In the EQ-5D, respondents are asked to respond to items phrased "I have …." and then indicate whether they have experienced no, moderate, or extreme pain, while in the SF-6D respondents are asked to select one of six options that each starts with the phrase "You have…." The two phrases differ in that one asks a respondent to affirm and select the severity of a symptom, while the other asks the respondent to acknowledge and indicate the severity of the symptom, a

Table 12.3 The mean scores for eight generic preference-based quality-of-life assessments

Sample	Hawthorne et al. (2001) ◄ Population-based	Conner-Spady and Suarez-Alazor (2003) ▼ Disease-specific	Kaplan et al. (2005) ▲ Disease-specific	Marra et al. (2005) Disease-specific	Stavem et al. (2005) ▼ Disease-specific	Franks et al. (2006) Population-based	Fryback et al. (2007a) ▼ Population-based	Wee et al. (2007) ● Primary-care	Garster et al. (2009) ♦ Disease-specific
EQ-5D	0.79	0.65	0.57	0.66	0.77	0.87	0.85	0.82	0.82/0.74
HUI2			0.64	0.71			0.82	0.85	0.80/0.69
HUI3	0.77	0.66	0.44	0.53		0.79	0.77	0.78	0.75/0.56
QWB%							0.63		0.58/0.52
QWB-SA									
SF-6D	0.89	0.70		0.63	0.73	0.82	0.78	0.87	0.75/0.67
HALex									0.68/0.50
15D	0.85				0.86				
AQOL	0.71								

◄: Total means per assessment were not provided, but four age ranges were presented. These data represent the values found for persons aged 66+ for each assessment
▶: Mean baseline data were presented for those persons who did not experience change in their HRQOL over time
▲: Numeric differences between the HUI3 and other assessments were provided, and mean value of HUI3 itself could be interpolated from Fig. 1, so that the absolute values of all the assessments could be estimated
▼: Total means per assessment were not provided, but five age ranges were presented. These data represent the values found for persons aged 65–74 for each assessment
●: Respondents were primary care patients
♦: Data from persons who reported chronic heart disease ($N=265$), but were not taking medications, or were taking medications ($N=218$)

2 Determinants of Valuation

Table 12.4 Representative examples of valuation-based qualitative assessments

Assessment	Investigators (s)	Evaluated descriptors	Valuation method	Method for forming summary statement
Generic assessments				
EQ-5D	Rosser and Kind (1978)	5 Descriptors, 3 levels of evaluation	243 health states were valued using the TTO procedure	Individual health state profile mapped to population-based valuations constitute summary statement
HUI	Torrance (1976)	Mark I, 4 descriptors 4–8 levels of evaluation; Mark II 7 descriptors 3–5 levels of evaluation; Mark III 8 descriptors 5–6 levels of descriptors.	Mark 1 (960), 2 (24,000) and 3 (97,200) health states valued by TTO and VAS (I), or VAS and SG (II, III)	Individual health state profile mapped to population-based valuations and entered into a multiplicative summary statement
QWB ▲	Fanshel and Bush (1970)	36 descriptors (Yes/No; Symptom-problem complex) + 4 descriptors 4 levels	36 Symptom-problem complexes and 100 health states were valued using VAS	Weights of different descriptors summed
Second-order generic assessments				
HALex	Erickson et al. (1995)	2 descriptors, 6 levels of activity limitations, and 5 levels of self-rated health extracted from the NHIS	30 health states valued by modified HUI Mark I values (no specific valuations study performed)	Each valued health state constitutes a summary statement
SF-6D	Brazier et al. (2002)	6 descriptors 4–6 levels of evaluation, extracted from the SF-36	249 of 18,000 health states were valued using the SG. The remaining health states were modeled	Individual health state profile mapped to population-based valuations constitute summary statement
Disease-specific assessments ▼				
FACT-L	Kind and Macran (2005)	10 descriptors 5 levels of evaluation, extracted from the FACT-L	64 health states were valued using VAS	Individual health state profile mapped to population-based valuations constitute summary statement
EORTC-QLQ-C30	Pickard et al. (2009) ▶	5 descriptors 4 levels of evaluation, extracted from the EORTC-QLQ-C30	144 health states were valued using TTO	Health states mapped to lung cancer patients and the resultant valuations constitute a summary statement
Person-generated assessments				
SEIQoL	McGee et al. (1991)	5 person-selected descriptors evaluated by a VAS	30 (20 unique, 10 replicates) hypothetical cases rated on a global VAS. Regression analysis used to generate weights per descriptor	Evaluated descriptor ratings multiplied by weights and summed
PGI	Ruta et al. (1994)	5 person-selected descriptors + one composite descriptor evaluated by a VAS	Respondents asked to estimate relative importance of each of 6 evaluated descriptor	Evaluated descriptor ratings multiplied by importance ratings and summed

▲: Based on Kaplan et al. (1976) paper
▼: There are a number of citations for each assessment, I have selected a representative one
▶: Pickard et al. (2009) provide data for both multiattribute and global models, but this citation is based only on the multiattribute models, MA-3

shift from the active to a passive expression. In contrast, the HUI3 pain domain asks respondents to select a response from five options each of which is a simple descriptive statement that reflects severity by indicating the consequence of being in pain, but makes no reference to the respondent. Whether such differences would lead to differences in a cost-utility analysis is not clear, but what it does do is to create different cognitive tasks for the respondent. Even the recall time requested by the item (immediate recall, a week, a year ago, and so on) could elicit different cognitive processes. These studies are good examples of how investigators acknowledge the presence of cognitive processes (e.g., the meaning induced by different items) in a statistical analysis without studying the impact these differences have on the outcome.

Konerding et al. (2009), using regressions analyses, estimated the extent that various components of the HUI2 and SF-6D predicted the total score of the EQ-5D. What they found was that the HUI2 sensation component did not significantly predict the total EQ-5D score, while the role limitation component of the SF-6D produced a statistically significant R^2 (0.12), but this result was not considered to be "empirically" significant.

Søgaard et al. (2009) reported first calculating differences scores between the SF-6D and the EQ 5D for patients with low back pain, and then analyzing these data in terms of the mobility, self-care, and anxiety/depression components of the EQ 5D and the social function components of the SF-6D. They summarize the results of their analyses, by stating:

> The two measures were found to produce significantly different mean values of, on the average of 0.085. Such differential need not be vital for decision making, but the fact that it masks more severe bidirectional variation makes it potentially larger in other applications. The expected variation for observations in the future studies was estimated at 0.546, and unless such differential is irrelevant to decision making, the SF-6D and the EQ-5D cannot be generally used interchangeably …" (Søgaard et al. 2009; p. 610).

Finally, there are many other indicators that may contribute to variation between valuation-based qualitative assessments, including age, gender, disease status, and cultural differences between respondents (e.g., Brazier et al. 2004; Dolan 2000), but the consensus among investigators remains that the two assessments should not be interchanged. This, of course, has implications about the comparability of these assessments and this may have policy implications. It is for this reason that Fryback et al. (2009) developed a method whereby linear plots between assessments act as "crosswalks," so that one assessment can be transformed to a comparable score for another assessment. An alternative approach is to determine if the weighting systems' (e.g., the VAS, TTO, and SG) resultant scores for the assessments listed in Table 12.3 can be "corrected" and used to eliminate the observed differences (see below).

Table 12.4 provides an overview of the components of various generic, second-order generic, disease-specific, and person-generated preference-based qualitative assessments. As previously stated (Chap. 12, p. 458), and demonstrated by the analysis of the QLI, each of the first three types of assessment domains are basically "synthetic," since none directly asks the respondents to concurrently provide both an evaluated descriptor and valuation of the evaluated descriptor. The assessments that come closest to doing this are the two "person-generated preference methods," but these and the other methods described in Table 12.4 still remain as models of how a person makes a choice or rates an experience, and the question that remains is what actually happens.

What is common to both evaluating a descriptor and valuing the evaluated descriptor is that a decision or choice has to be made. When evaluating a descriptor a person has to decide on a rating, while when valuing the evaluated descriptor the person has to decide on the importance of this particular state. Bell et al. (1988; p. 1–2) have described three ways these decisions or choices may be modeled: *descriptive, normative, and prescriptive*.[19] Descriptive models include those that describe what and how a person actually makes decisions, normative models describe how people should make decisions or choices, while prescriptive models focus on preparing people to make good or better decisions. An example of a normative model is the standard gamble (SG). It is a normative model since it evolved from a series of axiomatic statements, as represented by the von Neumann and Morgenstern (1944) theory of expected utility, providing a definition of rationality when making decisions under uncertainty. This approach, while theoretical, has been applied to the quality-of-life assessment, as witnessed by the development and successful applications of the HUI (e.g., Torrance 1976).

A typical SG study asks a respondent to make a decision or choice between a certain or an uncertain outcome. A respondent first ranks a series of health state scenarios, and then is given two alternatives that include three health states. To paraphrase: "What would you prefer, to remain in state A and be certain where you are, or take the chance of being in state a B or B*?" In this example, B is preferred to A and A is preferred to B*. (The duration of the health states is specified and, for living health states, equal among the states.) The probabilities between B and B* are then varied until the respondent is indifferent between A for sure and the lottery (with probability p of B and (1-p) of B*). The probability associated with the indifference point (B) is then used to calculate utilities.

The respondent is assumed to make these decisions or choices in a rational manner, but research over the last 50 years has provided a number of examples of violation of the procedural invariance expected of the utility model. Three primary factors have been identified as contributing to the violation of the (expected) utilities generated by the SG, including *probability weighting, loss aversion, and scale compatibility* (Bleichrodt 2002; Hershey and Shoemaker 1985; Tversky and Kahneman 1992). Probability weighting refers to the observation that people do not treat different probabilities equally, but rather overweight small and underweight large probabilities. Loss aversion violates the expected utility theory assumption that the relative gain or losses in utility following an intervention would be the same. Rather, what seems to happen is that the losses in utility following a loss (say, $10) is greater than the gains that occur when a comparable gain ($10) occurs. As a consequence, people try to avoid losses, and in fact become risk averse. Scale compatibility refers to the fact that a respondent has an implicit valuation scale in mind when they make a judgment on a prescribed scale. Responses, therefore, can be scale compatible on incompatible, and this will impact probability estimates. This same dynamic occurs when a person selects a reference point to use when engaging in a SG. The presence of many of these processes has been confirmed by a qualitative study, involving the think-aloud procedure during a SG (von Osch and Stiggelbout 2008).

Procedural invariance is a necessary component of a normative model and this very brief review of the available literature makes it clear that violations of this requirement regularly occur. Instead, the presence of such cognitive processes as probability weighting, loss aversion, scale compatibility, and what a person uses as a reference point are also factors that may make the SG deviate from expected utility theory. The question that remains is to what extent can the demonstration of the presence of these cognitive processes help explain differences in summary scores found in Table 12.3? For example, Bleichrodt et al. (2002; 2007) have demonstrated that the cognitive complexities ongoing during a SG lead to inflated scores, while the same processes may both inflate or diminish TTO scores, again when compared to predictions from expected utility theory. von Osch et al. (2004), have described a number of ways that corrections can be made to both types of assessments and in each cases a linear function was found,[20] but these linear functions still appear inflated. Whether recalculating valuation scores for the EQ-5D, HUI, SF-6D, and so on, based on these correction methods, would make a difference in the observed variability found in Table 12.3 is not clear, but they are likely to continue until the different cognitive processes evoked by these different valuation methods are examined, and a common experimental format found.

Abellan-Perpiñan et al. (2009) have pointed out that many of the limitations of expected utility that are apparent when the SG is administered are predictable from Prospect Theory (Kahneman and Tversky 1979). Prospect Theory has been useful not only in finding a place for psychological processes in economic deliberations, but also when investigating determinant of QALYs. It, however, has not been sufficiently integrated into the development of qualitative assessments. Probability weighting, loss aversion, scale compatibility, and what a person uses as a reference point are cognitive factors that evolve from a person having to make a choice or decision. In a qualitative assessment, choices and decisions occur when a respondent is asked to either evaluate a descriptor or value some descriptive outcome, so that these same issues could be relevant to understanding variability in qualitative responding. Differences in loss aversion, for example, could account for differences between various socioeconomic groups and their tendency to report symptoms.[21] In addition, Ross's (1989) work on the role of implicit theories in the construction of personal histories and Schwarz and Strack's (1999) studies of the cognitive processes (e.g., assimilation and contrast effects) involved in the assessment of well-being also have provided leads that qualitative investigators could use when examining the role that cognitive processes play in the assessment of quality. Finally, it is becoming clear that the presence of these violations of expected utility and the complexity of ongoing cognitive processes are sufficient to justify considering alternative models, some of which include non-linear models of QALY (e.g., Bleichrodt and Pinto 2005).

This section has again demonstrated the usefulness of examining how language is used and what cognitive processes may be involved during the assessment of quality. In the next section, I will address the fact that values are relatively stable enduring descriptive states that can also benefit from an examination from the same perspective as was found useful for the previous two sections.

2.3 Values and Language-Usage

One of the tasks I adopted in this book was to follow Bridgman's (1959) charge that a complete understanding of a phenomenon such as a quality assessment requires that I "understand how I understood." When applied to science, I interpreted this to mean that I needed to examine how the language I used helped me express not only what I had observed and used to make inferences, but also to help me examine my reasoning that underlies my decisions or choices. Thus, it was necessary to examine the role that figurative and literal language played in both the expression of science and in the assessment of quality, and this I did in Chaps. 2, 5, 7–11. Language can't be expressed independent of cognitive processes, so it was also necessary to examine their role again in both the expression of science and a qualitative assessment, and this I did in Chaps. 3, 6, 7, 10, and 11.

In this section, I broaden my discussion, since values are not only cognitive entities that evolve and can be applied, as in a valuation, they are also in place and are engaged when a qualitative assessment occurs. They are in place, of course, because valuation is an outcome, if at times unconscious, of a continuous cognitive-emotional regulatory process. The presence of established values can be felt in linguistic expression, often connoted rather than directly valued. An example of connotation is when an investigator labels a series of descriptive evaluations as a quality-of-life assessment, even if an explicit valuation did not occur. As I discussed in Chap. 9, established values can be found at multiple sites of the informational hierarchy (Fulford 2000) attesting to continuous presence of values.

Established values are also assessed during a quality-of-life assessments. For example, when the SEIQoL assessment is administered respondents are first asked to list their five most important values, and then asked to rate how good or bad they are in these valued states. A second example is when a person is asked to respond to a global qualitative question; in this case, respondents will usually recruit from available information to make their judgments, including the extent to which their lives have permitted them to achieve their valued goals.

Rokeach (1973) defined values as a collection of desired states that can be distinguished from other domains by having a motivational component, distinguished from attitudes by being more stable and transsituational, and are different from simple domains by being hierarchically ordered in terms of importance. Early researchers postulated that values are determents of a person's belief system, and beliefs underlie attitudes and behavior (e.g., Allport 1961; Rokeach 1973). The evidence linking values and behavior, however, does not appear that strong (e.g., Kristiansen and Hotte 1996; McClelland 1985), although more recent research has clarified the conditions under which this relationship can be demonstrated (see below).

A number of investigators have attempted to list the established values that people have. Rokeach's (1973), for example, originally developed a value assessment system that asked respondents to rank two groups of 18 items from 1 (most important) to 18 (least important). The first group of 18 values represented potential means to achieving a person's goals (instrumental values), while a second group of 18 values represented the goals in a person's life (terminal values). This value system has achieved great popularity, although Rokeach acknowledges that the domains were arbitrarily selected. In addition, the Rokeach value system's emphasis on assessing individual domains has resulted in its failing to capture the complex interaction of components of a person's value systems (Bardi and Schwartz 2003).

Schwartz (1992), has reported an alternative empirically derived list based on a 10-component value system that has been demonstrated to be present in as many as 65 different cultures. The value system includes such abstract concepts as power, achievement, hedonism, universalism, conformity, security, and several others. Schwartz and his colleagues (e.g., Bardi and Schwartz 2003) do not conceive these values as being totally independent of each other, but rather as being codependent. Thus, if one value is encouraged, then one or more values may conflict or be congruent with the initial value. Bardi and Schwartz (2003; p. 1208) give the example of the situation where some action fosters a security value such as social order, which also fosters the value of conformity or obedience, but conflicts with the values of independence and freedom. These relationships they suggest can be best represented in the form of a circumplex, with the congruent values set next to each other and the conflicting ones on the opposite side. To calculate the relationship between values within the circumplex, they performed a "Smallest Space Analysis" (Borg and Lingoes 1987; Guttman 1968). This is a multidimensional scaling technique that maps the relationship between all values simultaneously in a two-dimensional space using the importance correlations among each pair of values. They describe the resultant circumplex as a motivational continuum, where the further away a value is from another the more dissimilar the motivations being expressed.

A key question this research raised was whether a person's established values determined a person's behavior. To answer this question, Bardi and Schwartz (2003) reported a series of studies, the first of which involved administering questionnaires that asked people to rate the importance of the 10 values and to estimate the frequency they engaged in behavior characteristics of these 10 values. Correlations of the importance ratings and frequency ratings were calculated for each value domain. The data indicated that some, but not all, values were more likely related to the frequency of behaviors than others. For example, stimulation, tradition, and hedonism were value domains most highly correlated, while benevolence, security, achievement, and conformity generated the lowest correlations (Table 3; Bardi and Schwartz 2003; p. 1212). Two subsequent studies, one that involved partners and another that involved peers, were also reported and involved respondents performing the same two assessment tasks, but relative to the target subjects. The results were very similar to the first study, so that the investigators concluded that relationship between values and behaviors is quite general, although limited to some but not all values.

Torelli and Kailati (2009) report a study that also examines the relationship between values and behavior, but now by examining how the respondent cognitively approached the assessment task. Thus, they were interested in evaluating the hypotheses that values, if presented as abstract concepts, would be more highly correlated to behaviors that are also abstractly stated, than if the described behaviors were concretely stated. The results of their six studies supported this hypothesis. To experimentally study these issues, they used methods attributed to the Würzburg school of thought (Gollwitzer 1996) which were described as inducing "mindsets" in the subjects. Mindsets were induced by first telling subjects that they were to engage in a thought experiment. The purpose of the thought experiment was to encourage a person to think about their actions relative to their life goals or values. To facilitate this, they were then given a hypothetical activity and asked to think about how by engaging in this activity they could reach their life goals. They were then asked to state why they felt they would reach their life goals. Respondents were asked to state why four consecutive times, since it has been shown that repeated questioning induces abstract thinking (Freitas et al. 2004; Fujita et al. 2006). A similar procedure was used to induce a concrete mindset, but now respondents were repeatedly asked to state how they were to achieve their life goals.

There are several aspects of this experimental design that are of interest. First is the demonstration that the abstractness or concreteness induced in an assessment task will impact the observed relationship between indicators. This demonstration reinforces the proposition that the cognitive processes induced by how an item is structured should be viewed as a determinant of its outcome and that the content contained in an item is only one determinant of the outcome. Second is the notion that it is possible to induce ways of thinking by using standardized procedures. Adopting this approach has the potential for providing insight into a number of the problems I have discussed. For example, it is quite clear that many of the procedures and methods developed to facilitate valuations (e.g., VAS, SG or TTO) induce different patterns of thinking or mindsets and that these differences contribute to the observed differences in outcomes between assessment. I am not aware of any studies that have examined this issue, nor attempted to demonstrate its relevance, but I would argue that as long as the cognitive characteristic of these qualitative assessments are not known, the observed arithmetical differences found (Table 12.3) will continue to be present.

2.4 Summary: Valuation and Quality Assessment

The purpose of a quality-of-life or HRQOL assessment is to give persons the opportunity to voice their responses to the events that affect them, but to do this requires that these persons have the chance to reflect on the importance of the events that have transpired. To achieve this objective, a person has to be given an assessment that allows the person to consider these issues, and this usually includes the opportunity to engage in a valuation of the experiences that they have had. It also requires that the assessment allow for the expression of the unique nature of human experience. Modern psychometrics has developed techniques that are increasingly approximating this ideal. Data banks, conforming to item-response theory, are being developed that will allow for the selection of specific content to include in an assessment, while disease-specific (e.g., the FACT; Cella et al. 1993) and symptom-specific assessments (e.g., fatigue assessments; Table 7.3) are also being developed. Still, these assessments do not completely capture the idiosyncratic nature of human experience, and this limitation is important in certain situations. For example, low probability events with high qualitative impact may not be detected by assessments that are fixed or limited to a set of known issues, even if disease- or symptom-specific. In addition, most established qualitative assessments are not titrated to monitor and integrate into the assessment the linguistic or cognitive changes that occur as persons progressively loses their ability to remember or verbally express themselves following a stroke or as the clinical signs of dementia become evident. Most important is the inability of an investigator to grasp the meaning that different aspects of life have for the compromised person. I got a sense of this from the artist, *Willaim Untermohlen*, whose series of self-portraits revealed a loss of personhood and presumably the meaning to his life. As I previously stated (Chap. 10, p. 367), his early paintings depicted a person with an intense, almost piercing expression, and after several years, these pictures transformed into abstract expressions that resembled a nonperson. I rhetorically asked if life had became meaningless to him, with the world now filled with formless objects. Certainly an assessment filled with standardized items is not likely to tell me what life means to him, so maybe alternative modalities and forms of expression need to be considered (e.g., paintings). This is especially so since discovering how to "contact" or communicate with the compromised person remains an unresolved task.

3 Can Quality be Quantified?

At the end of the Preface (p. *vi*), I stated: "The task that remains, however, is to convincingly capture the joy, pain and suffering real people experience with the numbers that I will use to characterize their experiences." This statement summarizes the task that any quality assessment has to face, but it is also true of any of the behavioral and social sciences. Thus, what I am dealing with here is a general problem, a problem that pervades the assessment of any subjective phenomena. The question I am asking can be stated quite directly: that is, can a quality be a quantity? I have argued that it can (Chap. 4) and garnered some support for this view by reviewing the history of the term objectivity (Chap. 5). In this section, I will add to this discussion.

I have referred to the process that transforms a quality into a quantity as the objectification process. The history that marks efforts at the quantification of subjective phenomena can be said to have started when psychology separated itself from philosophy at the end of the nineteenth century. It continues today, with a variety of notable events that have intervened. These events include Fechner's (1860) development of psychophysics, Campbell's (1920) and Bridgman's (1927) efforts at defining measurement, the Ferguson Commission Report (1940) questioning whether qualitative measurement is possible, Stevens' (1946) work on operational definitions and scale development, Luce and Turkey's (1964) development of conjoint measurement theory, and the three volumes published on fundamental measurement by Krantz et al. (1971), Suppes et al. (1989), and Luce et al. (1990). Notable reviews and comments on this literature include Borsboom (2005), Cliff (1992), Fraser (1980), Kline (1998), and Michell (1999).

A key element of current approaches to subjective assessment involves providing an operational definition of how the assessment is to be done. However, there does not appear to be a specific set of criteria to use to determine if the operational definition has achieved its objective, and as a result, many different definitions may result. This same situation is true for qualitative research, as was evident by my discussion of the multiple definitions of quality-of-life in Chap. 2. Thus, to gain some insight into these issues, I will review the history of how operational definitions became such an important part of subjective assessment and how such activities fit into the general effort to create a measurement approach for the behavioral and social sciences. What I will learn is that if a person is relied upon to operationally define a subjective phenomena, then this is not science but rather is an example of trying to solve a problem by using an analog of what a scientific definition would be. It is not scientific because the person's operational definitions are basically subjective statements, they are "of the person," and as such, can't be verified prior to a demonstration of its usefulness. What it is most like is the mapping of a concrete example (the operational definition) to an abstract concept (such as quality-of-life, depression, and so on). Thus, it is closer to a metaphor than it is to mathematical statement. The same can be said about what I have referred to as "arithmetic metaphorization," where elements of a qualitative assessment are summed without demonstrating that a mathematic rationale justifies such an operation. There is a solution to this problem and it have been proposed, but not implemented to any great extent. The proposal I alluded to above and is referred to as "fundamental or axiomatic theory of measurement" (ATM). The reason why ATM has not found wide acceptance, according to Borsboom (2005; p. 118–120), is because it has difficulty accounting for the sloppiness of data (i.e., error).

Michell's (e.g., 1997) has written a series of papers and a book summarizing his view of the history of measurement in the psychological and behavioral sciences. He starts his review of this history with the Pythagorean' (570 BC) proposition that the physical world can be mathematically characterized and that numbers are the ultimate reality. Pythagoras' notion has come to dominate all the sciences, so it became natural to ask where qualitative research fit into such a conception of reality. Clearly, it has been difficult for the behavioral and social sciences to solve the problem of using mathematics to characterize subjective phenomena, even though there has been many attempts to do so. For example, Fechner (1860) and Wundt (1874) both felt it was possible to generate a quantitative psychology analogous to the physical sciences. Fechner, in particular, who was originally trained as a physicist, also believed that reality was quantitative. Consistent with this perspective, he demonstrated that the intensity of sensations and the physical intensity of a variety of sensory stimuli had a particular relationship described as a *psychophysical function*.

A corollary of the Pythagorean imperative was what Michell (1997; p. 362) refers to as the "quantity objection." This notion, which emanated from the seventeenth century scientific revolution states, states that if some phenomenon is quantitative, then it should be possible to add components of the phenomena together, as can be done with two lengths. However, a number of eighteenth century (Kant 1786) and nineteenth century (von Kries 1882) critics argued that subjective phenomenon, such as sensations, could not be added, and therefore, that the psychological and behavioral sciences would never develop into a quantitative science. In response to these concerns, Fechner's approach involved using the change in sensation (that is, the just noticeable difference) as an indicator and mapping it to the change in physical intensity. A typical experiment would have a subject increasing the intensity of some sensory stimulus until they noted it was different from a comparison stimulus. This sensation change would then be related to the physical change (e.g., change in weight, change in wavelength). The resultant physical intensity would then constitute a continuum, a quantitative continuum of sensation. However, if this was so, then it should also be possible to add one unit of sensation to another. Fechner, according to Michell (1997; p. 362), did not see the need to demonstrate this. Fechner's approach, however, seemed to stimulate a variety of applications of psychophysical principles. Thus, the Stanford-Binet intelligence test used a series of test items to determine the person's "intellectual threshold," adjusted by age. From these initial efforts, the field of psychometrics evolved finding application in the study of cognition, intelligence, personality, and a variety of social psychological phenomena (Kline 1998). However, this shift to deal with practical issues has very effectively masked the questionable scientific foundation of this psychological and behavioral research, and this presumably would be true as well for qualitative research. As Michell (1997) stated:

> *Practicalism*, the view that science should serve practical ends, when stronger than the spirit of disinterestedness, can corrupt the process of investigation. Science, as the attempt to understand nature's ways of working, knows nothing of practicalism, for scientific knowledge is neither useful or useless, in itself. (p. 365).

If quality is to be expressed as a quantity, then qualitative indicators have to be formally related to what is done in the physical sciences. Thus, what I have called "assessment" now has to be referred to as "measurement," even if the methods used may not simulate physical measurement. This issue came to a head when the Ferguson Committee, using Campbell's (1920) and Bridgman's (1927) definitions of measurement as a criteria, decided that it would be impossible to measure sensations.

What Campbell (1920; p. 267) proposed was that fundamental *physical* measurement involved the assignment of numbers to the properties of objects, objects which can be ordered (arranged in a sequence) and be part of a sequence (concatenation). Bridgman (1927) offered a similar definition. This was considered a form of *representational measurement* and reflected a constructionist philosophical perspective (Borsboom 2005). However, not all physical measurement can be described (e.g., density), so that what is meant by measurement had to be expanded to include derived measures, measures that were defined in terms of established measures. Campbell referred to these two types as *extensive* and *intensive* types of measurement, and that intensive measurement required a relationship to extensive properties to be defined as a type of measurement. Thus, while density may not be directly measured, it can be derived by the ratio of mass over volume.

Stevens (1946) tried to respond to the Ferguson Committee by adopting Campbell's and Bridgman's definitions of physical measurement, stating that this involved assigning numbers to objects or events according to a rule, and that this could also be done with subjective data. He also noted that mapping from the empirical system to the numerical system can vary and each is equally legitimate. For example, you can represent length in terms of inches or centimeters and both numbers invariantly characterize some empirical system. Stevens goes on to suggest that since there are different but legitimate ways to represent numbers, he needed to use statistics that were appropriate to the numbering system he selected. The numbering systems Stevens was referring to were his suggested scales of measurement: *nominal, ordinal, interval, and ratio scales.*

Some investigators, however, did not agree with Stevens and argued that statistical operations can be carried out independent of the origin of the numbers. Proponents, of this operational theory of measurement, point out that statistical operations only make distributional assumptions and do not need to speculate about the origin of the numbers being processed. They would argue that an investigator needed to be more concerned about *the interpretation* of a particular statistical application, than the application per se. But there is more to an operational measurement theory than this. Operationalism avoids assuming an underlying reality and so is fundamentally different than representationalism. What I would be measuring is defined by the operations I am using, no more or no less. Dingle (1950; p .11) defines a measurement as "any precisely specified operation that yields a number." Sometimes investigators may claim to be measuring the same variable, such as quality-of-life or health status, but because of differences in how the variables were operationalized, different conclusions result. In this case, developing a consensus between investigators would require identifying a common set of operations.

A third approach to measurement, the *classical* or *traditional theory of measurement*, raises questions about whether the matching of a numerical and empirical system is appropriate at all. Michell (1986) describes this theory as traditional because, he claims, it can be dated to the writings of Aristotle and Euclid and has remained the predominate approach to scientific measurement until the beginning of this century. The essence of this theory is the proposition that measurement addresses the question of "how much" of a particular attribute an object has. A quantitative attribute should have ordinal and additive properties. To paraphrase Michell (1986; p 405), "different masses can be arranged in a sequence and added together, but different nationalities can't." Since nationalities do not have these properties, they would be considered a nonquantitative attribute. Is this also true of a measure of quality-of-life?

Hand (1996) compared the operational theory of measurement to others in the following quote:

> According to this theory the hypothesis that an attribute is quantitative is a scientific hypothesis just like any other. Measurement then involves the discovery of the relationship between different quantities of the given attribute. The key word here is 'discovery'. Whereas the representational theory *assigns* numbers to objects to model their relationships, and the operational theory assigns numbers according to some consistent measurement procedure, the classical theory discovers preexisting relationships. By definition, any quantitative attribute has an associated variable. (p. 457).

For the traditional or classic theory, what was measured was always a real number and if you had measured it, then it could be subjected to all arithmetic and statistical manipulations. In this regard, classical theory would consider nominal and ordinal scales as numerical coding scales rather than measurements.

Considering Michell's (1999) work as just philosophical ramblings would underestimate its importance. Kline (1998), for example, thought the issues that Michell raised were of sufficient importance to restructure his approach to psychometrics. His "reformulation" of psychometrics illustrates how an investigator might mount a response to Michell's concerns. Kline considers two classic research areas in psychology: intelligence testing and personality research. His general strategy was to use factor analysis to identify variables to be studied and then to identify methods whereby these factors can be measured quantitatively. He referred to the evidence supporting the existence of two factors, *crystallized intelligence and fluid intelligence*, that best characterized intelligence, both of which have a strong biological basis. His review of the personality research literature revealed that only three major factors reliably characterized personality (*extroversion, neuroticism, and obsessionality*), two of which have a strong biological basis. Thus, what Kline suggested was that the available literature, plus the demands for scientific measurement, should make investigators in the behavioral and social sciences (and presumably quality-of-life researchers as well) identify biological or physical measures, measures which have additive and ratio properties. To quote Kline (1998):

> I have argued that the variables to be measured in the new scientific psychometrics are those which have been shown in factor analysis to be replicable and meaningful: the two g factors and the three personality factors. In these cases, given their psychological nature, it makes sense to assume that they are quantitative and to hold this hypothesis until it is refuted. This is because their high heritabilities suggest that they are variables whose variance is determined by a combination of genes and environmental factors. Such variables tend to be quantitative and normally distributed. (p. 195).

What would be an example of a quantitative measure that Kline would feel met the requirements of having additive and ratio properties? Kline suggested the measurement of the "BIP," or *basic period of information processing*. The BIP, as a measure of the rate of information processing, was considered a measure of intelligence, and as a unit of time, has both additive and ratio properties. He and his student Draycott (1994a; b) report a number of studies which included measures of the BIP. They found that the BIP loaded on the crystallized ability factor and memory, but not fluid intelligence. They interpret the results as providing evidence that the BIP is not a general measure of intelligence. Kline also discussed the use of reaction and inspection time as measures of intelligence but their low correlation with measures of intelligence limits their usefulness at this time. In general, he considered the lack of a precise theory of intelligence to be as much a part of the reason why it has not been possible to develop scientific measures of intelligence, as the nature of the measures themselves. His review of the status of biological measures of personality variables reinforced this view.

Thus, Kline's (1998) "new" approach to psychometrics remains very much a developing program. What is particularly valuable is his call for investigators to acknowledge that the current efforts to measure complex concepts, like intelligence or personality, do not meet the criteria of scientific measurement. He was also prompting measurement approaches that relied on physical measures, such as time, or biological measure each of which may have additive and ratio properties. He was not dismayed by the lack of progress to date, since he recognized that it has taken nearly 100 years to develop an acceptable factor structure for intelligence and personality measures.

A question that can be asked is whether Kline's two-stage strategy can be applied to health status or quality-of-life research. Certainly, the factor analytic work of Ware and his associates (1994) that has identified a physical and mental health component of a health status assessment would be consistent with the first stage of Kline's approach. The second stage that involves identifying physical or biological measures of these factors remains to be done. Again, adopting Kline's approach, there have been a number of studies reported in which differences in *information processing* have been related to measures of self-assessed health status (e.g., Benyamini et al. 2000; Williams et al. 2003; SAHS). The question is, are these measures of information processing (e.g., information recruited, response speed) sufficiently well correlated with self-rated health that they could act as a quantitative analog of the measure? Can they provide quantities to use when scaling the SAHS item, in contrast to what is now done that involves using the distributional characteristics of the rating?

A report by Williams et al. (2003) illustrated the relationship between information processing rate and self-reported health. In this study, college students were asked to rate their physical health, but were also given the task of responding to a modified version of the Stroop test (1935). The Stroop test briefly presents words, in this case 28 illness and 28 nonillness words, each in a particular color. The task for the respondent was to identify the color the word was presented in as quickly as possible. The hypothesis tested in the study was whether the illness-related words would lead to a greater delay (an interference effect) in recognizing the color the word was presented in, than the nonillness words, and whether this delay would vary systematically with a respondent's self-rated health assessment. The study results did show that response speed was slower for health-related words and that people who rated their health poor or fair had slower response speeds than those who rated their health as very good or excellent. Thus, the data provided general support for an association between the variables, but were not sufficient (e.g., the sample size was not adequate) to explicate the relationship between the two variables.

There is also a considerable literature in which SAHS has been studied in an attempt to understand the relationship between age and memory (Park 2000; Tyas et al. 2007). In these studies, poorer health status has been shown to be associated with cognitive slowing. Earles et al. (1997) report a correlation of 0.35 (significant at the $p<0.001$ level) between response speed and self-reported health for a sample of 300 individuals aged 20–90. In this same study, structural equation modeling revealed a significant but low correlation ($r=0.15$) indicating the influence of health on cognitive speed, while age was negatively correlated with speed ($r=-0.71$) and speed positively correlated with working memory ($r=0.60$) and long-term memory ($r=0.41$).

Knäuper and Turner (2003) have provided a useful cognitive-based model (Fig. 12.1) that may provide quantitative indicators of the rating categories (e.g., poor, fair, and so on) of SAHS. Inspection of Fig. 12.1 reveals that they list semantic knowledge, episodic knowledge, and information about change each contributing to a information integration process that once socially compared leads to judgments and rating of health status. The question now becomes do these determinants of memory provide sufficient quantitative indicators (e.g., reaction time, error rate, and so on) that can be mapped to the SAHS rating categories so that these subjective ratings would now play the role of a physical analog and provide an approximation of fundamental measurement? If this can be demonstrated to be useful, then it may be possible to adopt a more empirical approach to fundamental measurement, which in the past had relied exclusively on axiomatic (mathematical) statements. An excellent example of how this objective can be achieved is the effort of Torrance (1976) and his colleague to adapt the von Neumann and Morgenstern (1944) model (an example of fundamental measurement) of making decisions under uncertainty to health status and HRQOL judgments. An additional venue to attempt this type of effort is the response shift phenomena (Schwartz and

4 Summary and Conclusions

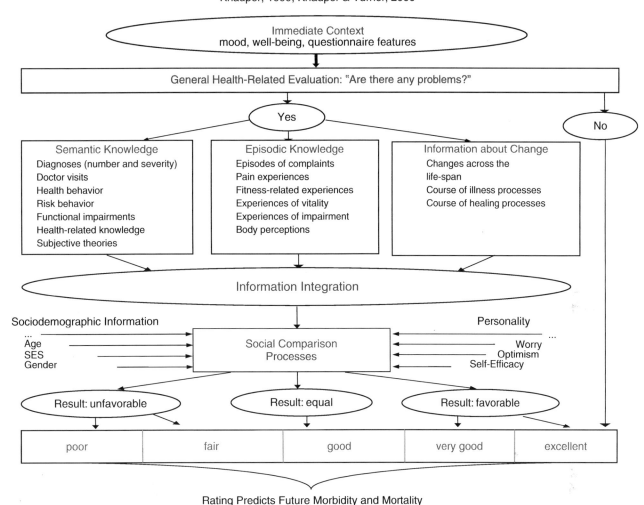

Fig. 12.1 Reproduced is Fig. 3 from Knäuper and Turner (2003). It describes the cognitive processes potentially involved in responding to the self-assessed health status item.

Sprangers 2000), which has been claimed to rely heavily on changes in cognitive processes, yet the role of that these processes have has not been examined. Central to promoting this effort is for the qualitative investigator to recognize and acknowledge that the numbers they are currently dealing with do not meet the requirements for fundamental measurement. This obviously will not occur easily as the intellectual history that I have reviewed has made clear.

4 Summary and Conclusions

The first six chapters of this book accomplished several tasks. First, they demonstrated the usefulness of a language-usage analysis by ordering the various definitions of quality-of-life and HRQOL into a limited number of cognitive linguistic classes. Second, specific cognitive processes were shown to be involved in representative examples of standard qualitative assessments (e.g., the QWB and the SEIQoL: Chap. 3, p. 72). Third, I introduced the concept of dual processing as an analytical tool that could be used in any cognitive analysis of qualitative judgments. I also discussed that most qualitative assessments contain elements of both modular (biological) and nonmodular components, and how this impacts the association between various indicators. Fourth, there was an extended discussion on the role of subjectivity in qualitative assessment and a defense of the value of self-reports as a source of information about qualitative issues. Particularly useful was a discussion on the history of objectivity. What was demonstrated was that objectivity grows out of subjectivity and that there is no one set definition of objectivity, but rather an evolving number of perspectives most of which still are applicable today. This provided a basis for seeing the assessment process as a form of objectification.

There was also a general effort to distinguish what a investigator does and thinks from what a respondent might do and think. The purpose in doing this was to avoid confusing the two sources of information, but to also recognize that the models that were developed to characterize qualitative assessments are based on the interaction between the cognitive processes ongoing in the investigator and what was observed to be true for the respondent. Thus, a distinction was made between a person forming a category and an investigator forming a domain, both of which are summary statements. In addition, the investigator's thought processes were recognized as an important source of information about how a qualitative assessment was formed, and this, in and of itself, became an important source of information about why creating a cognitive hybrid construct is such a popular model of a qualitative assessment.

This last point was also implicit in my examination of the metaphoric basis of various quality-of-life and HRQOL definitions (Chap. 2, p. 38). This analysis also provided some insight into the thought processes of the investigators. Five types of definitions were identified; causative, matching, substitution, descriptive, and valuation definitions. For example, Osoba's (1994) definition of HRQOL implied that he felt that the consequences of a person's disease and treatment caused changes in their qualitative status. Of course, the specifics of how this might occur, especially since several levels of analysis are involved remain to be determined. Matching definitions are a good example of a metaphor, in that the meaning associated with some concrete terms or phrases are being mapped onto the target qualitative term. Thus, terms such as needs, goals, aspiration, and so on have relatively more concrete meanings than is communicated by the phrase health-related quality-of-life. Calman's (1987) definition of HRQOL that is stated in terms of the difference between perceived goals and actual goals is a good example of this. These definitions make me think that the investigators faced with an abstract or ethereal concept, as Campbell et al. (1976; p. 471) put it, coped with their situation by using a literary devise. Also common are definitions that substitute quality-of-life with well-being under the assumption that the term well-being is a more meaningful and concrete than the phrases quality-of-life or HRQOL. My discussion about the nature of well-being (Chap. 11) should raise questions about the wisdom of doing this. Again, however, this effort reflected an investigator's attempts to respond to the ambiguity of the phrase quality-of-life or HRQOL. Descriptive definitions include several aspects of the different definitions I have discussed and in a sense represent a combining of them as a way to define quality-of-life. Included in some of these definitions was references to valuation. Again, these definitions (I only listed one) represent what investigators felt were an appropriate model of quality-of-life or a HRQOL assessment. These definitions were the most comprehensive and were best at simulating a hybrid construct.

Chapters 7–11 covered the subject matter ordinarily found in currently available qualitative assessments. It became clear that each of these areas of study evoked a variety of issues that were relevant to the definition and assessment of quality. For example, symptoms (Chap. 7) have often been included in a qualitative assessment, yet there are some obvious reasons why this can't be done including the fact that any effort to combine symptom ratings with other qualitative indicators runs the risk of the investigator engaging in arithmetic metaphorization. The discussion in the chapter also made it clear that it is possible for a symptom to be a quality indicator if it is qualified, and this would require a shifting of the semantics of the terms used to describe symptoms to include more emotive aspects of the symptoms (e.g., shift from using indicators of intensity to burden), but also requires that valuation of the symptoms occurs. Most important was to recognize that it was not necessary to include a symptom into a HRQOL assessment for it to function as a qualitative indicator. This is true for each of the topics discussed in Chaps. 7–11.

In Chap. 8, I addressed the issue about whether health status should be included in a HRQOL assessment. Again I pointed out that various indicators of health status (e.g., blood levels, radiological scans, as well as feet walked) can be qualified and made into a quality indicator. Whether it should be part of a HRQOL assessment remained less certain, especially since a review of the meaning of the term health revealed a fair amount of ambiguity. I also discussed the term disorder, as an alternative to health, and found some of the same definitional and application difficulties. Wakefield (1999) suggested that a hybrid cognitive construct could adequately capture the meaning of the term disorder. I used the components of the hybrid construct to analyze the different types of HRQOL assessments.

The term function is one of the terms that is an integral part of the dialog concerning what a quality indicator is, and I considered a wide range of issues concerning it in Chap. 9, from its use in language to its cognitive basis. What became clear is that the term is a statement of relationships that has been reshaped into a statement of a person's status. My discussion of the semantic role of the term function in the disability literature helped illustrate the issues surrounding the use of the term. Chapters 10 and 11 dealt with some of the most complex material I had to consider in this book. Not only did I address the appropriateness of including neurocognitive indicators and well-being statements in a qualitative assessment, I also had to examine the many issues generated by assessing each of these terms. For example, in Chap. 10 I addressed the issue of level of analysis by focusing on the relationship between capacity and performance, or stating it in a more classic perspective, between structure and function. I was particularly concerned about what could be learned from this literature about how to assess the physically compromised person. In Chap. 11, I expressed concern that the term well-being had lost its moral imperative and instead was being promoted as a goal onto itself without adequate support for many of its contentions. Much of this I attributed to the ambiguity of the term and its use as a literary device (e.g., an umbrella term). In retrospect, the term well-being is an excellent example (not the only one) of why the presence of fundamental measurement principles needs to be felt more directly when the assessment of subjective phenomena, including quality, is being considered.

Notes

1. This quotation was included in an article by Michell (1997). I have left a portion of the quote out that is available in the article.
2. By "thinking about" I mean both conscious and unconscious cognitive processes.
3. Kahneman's approach has limited generality since it rejects a role for retrospective assessments in a qualitative assessment, yet it is clear that reflection is a regular part of a qualitative assessment, while Rasch's approach is limited to the verbally fluent and breaks down when applied to the compromised person.
4. The summary qualitative statements were calculated by combining weights using either or both additive or multiplicative methods.
5. There are, however, alternative methods that could have been used by the authors of the QLI that would have a greater chance of evoking the reflective process hypothesized as being a critical part of a qualitative assessment. For example, the respondent might be asked to reread their satisfaction rating so that when they made their importance rating it was in the context of an estimate of their satisfaction. This ploy might generate the more delayed thought process characteristic of reflection. It would be even possible to measure the reaction time to the importance question with or without review of the satisfaction question, and in this way, provide some evidence that a reflective process occurred.
6. QLI is not unique in this regard, since I believe many qualitative assessments generate a summary statement using metaphoric arithmetization. This implies that the only "valid" measure of quality is a global assessment produced by the respondent.
7. An example of a study that avoids some of the difficulties that Ferrans and Powers have with their study but which still engages in metaphoric arithmetization is the Laman and Lankhorst (1994) report. In this study, respondents were administered the evaluated descriptor and importance weighting task concurrently, and the summary score involved multiplying these indicators and adding them together. Thus, this study avoids the potential confounding of sequential assessments, but the summary of this assessment still is subject to concerns about the meaning generated.
8. It is of interest to note that the Breast Cancer Questionnaire (Levine et al. 1988) did not include weights when the assessments were summed to give a composite score.
9. The term happiness, more so than satisfaction, is likely to evoke an experienced emotion, although it is possible that comparisons are also made when happiness is reported (e.g., as when using attaching the word "happiness" to a feeling that was evaluated). In addition, satisfaction is traditionally thought of as a representative of a "cognitive" process in attitude research.
10. This statement is additional evidence for the presence of metaphoric arithmetization.
11. It would be interesting to know more about the characteristics of these persons. For example, were they healthier than the persons who indicated clearer domain preferences? If so, then this would be consistent with the data from subjects that were presumed to be "healthy" (see Table 12:2; Part A).
12. The standard deviations from Table 1 (Hagell and Westergren 2006) were actually uniformly lower than what was found for total QLI score and its four subdomains.
13. The kind of experiment that I think should be done would involve asking a person to rate their satisfaction of a particular domain, say along a 7 point Likert scale, and then having identified this rating, immediately ask the person to rate the importance of this particular satisfaction rating. I would also include a control group where the two assessments were done consecutively. There are several analyses that could be done. First, does presenting the satisfaction rating immediately followed by the importance rating increase the importance rating compared to rating the domain independently, and second whether multiplying the importance rating by the satisfaction ratings statistically significantly increases the domain score. Aggregating these domains into a global assessment might be more complex because some domain composite scores may go up or some may go down after weighting, so that demonstrating that the weighted satisfactions scores accounts for more variance may not be straightforward and would require detailed subgroup analyses.
14. As will become clear in this section, this is exactly what happens when such standard qualitative assessments as the HUI or the QWB are used. In both cases, what is summed is the value, or utilities, not the ratings of the evaluated descriptors.
15. Before I proceed I need to review and clarify the difference between a number of terms that I have been using. I have already clarified the difference between the phrase "evaluated descriptor" and "valuation" (Chap. 8, p. 270), discussed the contribution of evaluated descriptor to inductive theory building, how qualitative researchers are prone to deductive imposition (Spates 1983; p. 42) or metaphoric arithmetization, and how evaluation of descriptors involves a conscious act, while valuation is a continuous mostly unconscious process. Now I would like to differentiate the terms value and appraisal, since they appear to semantically and cognitively overlap. A value, as a noun, refers to an idealized outcome or state, as such it is a abstract cognitive entity whose meaning is clarified by its concrete examples. In contrast, the term appraisal, as a noun, refers to the fact that a process has occurred (e.g., an evaluation) involving an object or life lived, and therefore, would be cognitively represented as a more concrete entity. Both valuation and appraisal may involve an evaluation, but valuation as a metacognitive process (Chap. 3, p. 64) may occur either as a nonconscious or continuous process. Thus, the terms appraisal and valuation overlap, but are sufficiently distinct that they should not be used as synonyms.
16. The investigators who developed these methods may not acknowledge that their procedures reflected how a person thinks, but to the extent that a person can complete the tasks involved (e.g., using the VAS, TTO, or SG), then it is certainly reasonable to assume that a person is capable of thinking in these ways, and that some may actually think this way when making decisions or judgments.
17. Fryback's concern have a long history starting with a study by Patrick et al. (1973).
18. What I call "evaluated descriptors" Torrance et al. (1995) describes as a classification system, while what I describe as "valuation" they describe as preferences. The classification system consists of a variety of domains, and selection of levels within a domain results in a profile that characterizes the individual respondent. A preference implies that a choice has occurred, but a valuation refers to both a current and past process. Either can occur under conditions of certainty or uncertainty. Thus, the term valuation is a more comprehensive and inclusive term.

19. Bell et al. (1988) use their models to refer to decisions and choices under conditions of uncertainty, but I feel it can also be applied to decisions and choices where the outcomes are certain, such as a rating.
20. The TTO assumes that the utility of life duration is linear, so that it would be imperative that any correction supports this assumption.
21. An example of this type of study would be Sen's (2001) report that residents of more affluent areas of India have more expectations (and therefore loss aversion) than other poorer areas and as a result expect and receive better healthcare services.

References

Aaronson NK, Ahmedzai S, Bullinger M, Crabeels D, et al. (1991). The EORTC core quality of life questionnaire: Interim results of an international field study. In, (Ed.) D. Osoba, *The Effect of Cancer on Quality of life*. Boca Raton FL: CRC Press. (p. 185–203).

Abellan-Periñ JM, Bleichrodt H, Pinto-Prades JL. (2009). The predictive validity of prospect theory versus expected utility in health utility measurement. *J Health Econ*. 28, 1039–1047.

Allport GW. (1961). *Patterns and Growth in Personality*. New York NY: Holt Rinehart & Winston.

Anderson J. (1962). *Studies in Empirical Philosophy*. Sydney Australia: Angus and Robertson.

Balaban DJ, Sagi PC, Goldfarb NI, Nettler S. (1986). Weights for scoring Quality of Well-being instrument among rheumatoid arthritics. *Med Care*. 24, 973–980.

Bardi A, Schwartz SH. (2003). Values and behavior: Strength and structure of relations. *Personal Soc Psychol Bull*. 29, 1207–1220.

Barofsky I. (1996). Cognitive aspects of quality of life assessment. In, (Ed.) B. Spilker. *Quality of life and Pharmacoeconomics in Clinical Trials*, 2nd edition. New York NY: Raven Press. (pp. 107–115).

Barofsky I. (2000). The role of cognitive equivalence in studies of health-related quality of life assessments. *Med Care*. 38 (Supp II), 125–129.

Barsalou LW. (2008) Grounded cognition. *Annu Rev of Psychol*. 59, 617–645.

Bell DE, Raiffa H, Tversky A. (1988). Descriptive, normative and prescriptive interactions in decision making. In, (Eds.) DE Bell, H Raiffa, A Tversky. *Decision Making: Normative and Prescriptive Interactions*. New York NY: Cambridge University Press. (pp. 9–30).

Benyamini Y, Leventhal EA, Leventhal H. (2000). Gender differences in processing information for making self-assessments of health. *Psychosom Med*. 62, 354–364.

Bleichrodt H. (2002). A new explanation for the difference between time trade-off and standard gamble utilities. *Health Educ*. 11, 447–456.

Bleichrodt H, Abellan-Periñ JM, Pinto-Prades JL, et al. (2007). Revolving inconsistencies in utility measurement under risk: Tests of generalizations of expected utility. *Manag Sci*. 53, 469–482.

Bleichrodt H, Pinto L. (2005). The validity of QALYs under non-expected utility. *Econ J*. 115, 533–550.

Borg I, Lingoes JC. (1987). *Multidimensional Similarity Structure Analysis*. New York NY: Springer.

Borsboom D. (2005). *Measuring the Mind: Conceptual Issues in Contemporary Psychometrics*. Cambridge UK: Cambridge University Press.

Bowling A, Windsor J. (2001). Towards the good life: A population survey of dimensions of quality of life. *J Happiness Stud*. 2, 55–81.

Brazier J, Roberts J, Deverill M. (2002). The estimation of a preference-based measure of health from the SF-36. *J Health Econ*. 21, 271–292.

Brazier J, Roberts J, Tsuchiya A, Busschbach J. (2004). A comparison of EQ-5D and SF-6D across seven patient groups. *Health Econ*. 13, 873–884.

Bridgman PW. (1927). *The Logic of Modern Physics*. New York NY: Macmillan.

Bridgman PW. (1959). *The Way Things Are*. New York NY: Viking.

Byran S, Longworth L. (2005). Measuring health-related utility: Why the disparity between EQ-5D and SF-6D? *Eur J Health Econ*. 50, 253–260.

Calman KC. (1987). Definitions and dimensions of quality of life. In, (Eds.)NK Aaronson, JH Berkman. *The Quality of life of Cancer Patients*. New York NY: Raven Press. (pp. 1–9).

Campbell A, Converse PE, Rodgers WL. (1976). *The Quality of American Life: Perceptions, Evaluations and Satisfaction*. New York NY: Russell Sage Foundation.

Campbell NR. (1920). *Physics the Elements*. Cambridge, UK: Cambridge University Press.

Cella DF, Tulsky DS, Gray G, Sarafian B, et al. (1993).The Functional Assessment of Cancer Therapy (FACT) scale: Development and validation of the general measure. *J Clin Oncol*. 11, 570–579.

Cliff N. (1992). Abstract measurement of adult intelligence. *Psychol Bull*. 40, 153–193.

Connor-Spady B, Suarez-Alazor ME. (2003). Variation in estimation of quality-adjusted life years by different performance-based instruments. *Med Care*. 41, 791–801.

Cummins RA. (1997). *Comprehensive Quality of life Scale : Adult. Manuel*. Melbourne Australia: Deakin University .

Deleyiannis F W-B, Weymuller EA, Coltrera MD. (1997). Quality of life of disease-free survivors of advanced (Stage III or IV) Oropharyngeal cancer. *Head Neck*. 21, 466–473.

Derogatis LR, Derogatis MF. (1990). *The Psychological Adjustment to Illness Scale: Administration Scoring and Procedure Manuel*. Towson MD: Clinical Psychometric Research.

Diener E, Emmons RA, Larson RJ, Griffin S. (1985). The Satisfaction with Life Scale. *J Personal Assess*. 49, 71–75.

Dingle H. (1950). A theory of measurement. *Brit J Philos Sci*. 1, 5–26.

Dolan P. (2000). The effect of age on health state valuations. *J Health Serv Res Policy*. 5, 17–21.

Draycott SG, Kline P. (1994a). Further investigations into the nature of the BIP: a factor analysis of the BIP with primary abilities. *Personal Individ Individ*. 17, 201–209.

Draycott SG, Kline P. (1994b). Speed and ability: A research note. *Personal Individ Individ*. 17, 763–768.

Earles JLK, Connor LT, Smith AD, Park D. (1997).Interrelations of age, self-reported health, speed and memory. *Psychol Aging*. 12, 675–683.

Erickson P. (1998). Evaluation of a population-based measure of quality of life: the *Health and Activity Limitation Index* (HALex). *Qual Life Res*. 7,101-114.

Erickson P, Wilson R, Shannon I. (1995). *Years of Healthy Life*. Statistical Note #7. National Center for Health Statistics. Washington DC: Public Health Service.

Esper P, Mo F, Chodak G, Sinner M, et al. (1997). Measuring Quality of life in men with prostate cancer using the Functional Assessment of Cancer Therapy-Prostate (FACT-P) instrument. *Urology*. 50, 920–928.

Fanshel S, Bush, JW. (1970). A Health Status Index and its application to the health services outcomes. *Oper. Res*. 18, 1021–106.

Fechner GT. (1860). *Elemente der Psychophysik*. Leipzig: Breitkopf & Hartel.

Ferguson A, Myers CS, Bartlett RJ, Banister H, et al.(1940). Final report of the committee appointed to consider and report upon the possibility of quantitative estimates of sensory events. Rep Brit Assoc Adv Sci. 2, 331–349.

Ferrans CE, Frisch MB. (2005). Measuring quality of life: Is weighting with importance justified? Annual Meeting, International Society for Quality of life Research. San Francisco, CA. *Qual Life Res.* 14, A2012.

Ferrans CE, Powers MJ. (1985). Quality of life index: development and psychometric properties. *Adv Nurs Sci.* 8, 15–24.

Franks P, Hanner J, Fryback DG. (2006). Relative disutilities of 47 risk factors and conditions assessed with seven preference-based health status measures in a National U.S. sample: Toward consistency in cost-effectiveness analyses. *Med Care.* 44, 478–485.

Fraser CO. (1980). Measurement in psychology. *Br J Psychol.* 71. 23–34.

Freitas AL, Gollwitzer PM, Trope Y. (2004). The influence of abstract and concrete mindsets on anticipating and guiding other's self-regulatory efforts. *J Exp Soc Psychol.* 40, 739–752.

Fries JF, Spitz P, Kraines RG, Holman HR. (1980). Measurement of patient outcome in arthritis. *Arthritis Rheum.* 23,137–145.

Frisch MB. (1992). The quality of life inventory: A cognitive-behavioral tool for complete problem assessment, treatment planning, and outcome evaluation. *Behav Ther.* 16, 42–44.

Fryback DG, Dunham NC, Paita M, Hanner J, et al (2007). U.S. norms for six generic health-related quality of life indexes from the National Health Measurement study. *Med Care* 45, 1162–1170.

Fryback DG, Kim J-S, Palta M, Revicki DA. (2007). New perspectives on how preference-based indexes (EQ-5D, HUI2,HUI3, QWB-SA, and SF-6D) scale summary health-related quality of life. International Society for Quality of life Research Annual Meeting Abstracts *Qual of Life Res* A-5.

Fryback DG, Palta M, Cherepanov D, Bolt D, et al. (2009). Comparison of 5 health-related quality of life indexes using item response theory analysis. *Med Decis Mak.* First published online October 20, 2009 as doi:10.1177/0272989X09347016.

Fujita K, Trope Y, Liberman N, Levin-Sagi M. (2006). Construal levels and self-control. *J Personal Soc Psychol.* 90, 351–367.

Fulford KWM. (2000).Teleology without tears: Naturalism, neo-naturalism, and evaluationism in the analysis of function statements in biology (and a bet on the Twenty-first Century). *Philos Psychiatr Psychol,* 7,77–94.

Garster NC, Palta M, Sweitzer NK, Kaplan RM, et al (2009). Measuring health-related quality of life in population-based studies of coronary heart disease: Comparing six generic indexes and a disease-specific proxy score. *Qual Life Res.* 18,1239–1247.

Gollwitzer PM. (1996). The volitional benefits of planning. In, (Eds.) PM Gollwitzer, JA Bargh. *The Psychology of Action: Linking Cognition and Motivation to Behavior.* New York NY: Gilford Press. (p. 287–312).

Gorbatenko-Roth K, Levin I, Altmaier E, Doebbeling B. (2001). Accuracy of health-related quality of life assessment: What is the benefit of incorporating patients' preferences for domain functioning? *Health Psychol.* 20, 136–140.

Grice P. (1989). *Studies in the Way of Words.* Cambridge MA: Harvard University Press.

Grieve R, Grishchenko MK, Cairns J. (2009). SF-6D versus EQ-5D: Reasons for differences in utility scores and impact on reported cost-utility. *Eur J Health Econ.* 10, 15–23.

Guyatt GH, Nogradi S, Halcrow S, Singer J, et al. (1989). Development and testing of a new measure of health status for clinical trials in heart failure. *J Gen Intern Med.* 4, 101–107.

Guttman L. (1968). A general nonmetric technique for finding the smallest coordinate space for a configuration of points. *Psychom.* 33, 469–506.

Hagell P, Westergren A. (2006). The significance of importance: An evaluation of Ferrans and Powers' quality of life index. *Qual Life Res.* 15, 867–876.

Hamilton BB, Granger CV, Sheerwin FS, et al. (1987). *A Uniform National Data System for Medical Rehabilitation.* Baltimore MD: PH Brookes.

Hand DJ. (1996). Statistics and the theory of measurement. *J R Stat Soc A.* 159 (Part 3), 445–492.

Hawthorne G, Richardson J, Osborne R. (1999). The assessment of quality of life (AQoL) instrument: A psychometric measure of health-related quality of life. *Qual Life Res.* 8, 209–224.

Hawthorne G, Richardson J, Day NA. (2001). A comparison of the Assessment of Quality of life (AqoL) with four other generic utility instruments. *Ann Med.* 33, 358–370.

Hershey JC, Shoemaker PJH. (1985). Probability versus certainty equivalence methods in utility measurement: Are they equivalent? *Manag Sci.* 31, 1213–1231.

Hewlett S, Smith AP, Kirwan JR. (2001). Values for function in rheumatoid arthritis: patients, professionals, and public. *Ann Rheum Dis.* 60: 928–933.

Hewlett S, Smith AP, Kirwan JR. (2002). Measuring the meaning of disability in rheumatoid arthritis: Personal Impact Health Assessment Questionnaire (PI HAQ). *Ann Rheum Dis.* 61: 986–993.

Hickey A, O'Boyle CA, McGee HM, Joyce CRB. (1999). The Schedule for the Evaluation of Individual Quality of life. In, (Eds.) CRB Joyce, HM McGee, CA O'Boyle. *Individual Quality of life: Approaches to Conceptualization and Assessment.* Amsterdam The Netherlands: Harwood. (pp. 119–133).

Hsieh C-H. (2003). Counting importance: The case of life satisfaction and relative domain importance. *Soc Indic Res.* 61, 227–240.

Jostmann NB, D Lakens, TW Schubert. (2009). Weight as an embodiment of importance. *Psychol Sci.* 20, 1160–1174.

Kahneman D, Tversky A. (1979). Prospect theory: An analysis of decisions under risk. *Econometrica.* 47, 263–291.

Kant I. (1786). *Metaphysical Foundation of Natural Science.* (J. Ellington [Translator] 1970). Indianapolis IN: Bobbs-Merrill.

Kaplan RM, Anderson JP. (1990). The General Health Policy Model: An integrated approach. In, (Ed.) B. Spilker. *Quality of life Assessments in Clinical Trials.* New York NY: Ravens.

Kaplan R, Anderson JP. (1996). The General Health Policy Model: An integrated approach. In, (Ed.) B Spilker. *Quality of Life and Pharmacoeconomics in Clinical Trials.* New York NY: Ravens. (p. 309–322).

Kaplan RM, Bush JW, Berry CC. (1976). Health status: Types of validity and the index of well-being. *Health Serv Res.* 4, 478–507.

Kaplan RM, Groess EJ, Sengupta N, Sieder W, et al. (2005). Comparison of measured utilities scores and imputed scores from the SF-36, in patients with rheumatoid arthritis. *Med Care.* 43, 79–87.

Kaplan RM, Sieber WJ, Ganiats TG. (1997).The *Quality* of *Well-being Scale*: Comparison of the interviewer-administered version with a self-administered questionnaire. *Psychol Health.* 12, 783–791.

Keeney RL. (1988). Building models of values. *Eur J Oper Res.* 37, 149–157.

Kind P. (1996). The EuroQoL instrument: An index of health-related quality of life. In, (Ed.) B Spilker. *Quality of life and Pharmacoeconomics in Clinical Trials.* Philadelphia PA: Lippincott-Raven. (p. 191–201).

Kind P, Macran S. (2005). Eliciting social preference weights for Functional Assessment of Cancer Therapy-Lung health states. *Pharmecon.* 23, 1143–1153.

Kline P. (1998). *The New Psychometrics: Science, Psychology and Measurement.* London, UK: Routledge.

Knäuper B,Turner PA. (2003). Measuring health: Improving the validity of health assessments. *Qual Life Res.* 12(Suppl 1) 81–69.

Konerding U, Moock J, Kohlmann T. (2009). The classification of systems of the EQ-5D, the HUI II and the SF-6D: What do they have in common? *Qual Life Res.* 18,1249-1261.

Krantz DH, Luce RD, Suppes P, Tversky A. (1971). *Foundations of Measurement.* Vol 1. New York NY: Academic Press.

Kristiansen CH, Hotte AM. (1996). Mortality and the self: Implications for when and how of value-attitude-behavior relations. In, C Seligman, JM Olsen, MP Zanna (Eds.). *The Ontario Symposium.*

Vol 8. *The Psychology of Values*. Hillsdale NJ: Lawrence Erlbaum. (p. 77–106).

Lakoff G, Johnson M. (1980). *Metaphors We Live By*. Chicago Il: University of Chicago Press.

Laman H, Lankhorst GJ. (1994). Subjective weighting of disability: An approach to quality of life assessment in rehabilitation. *Disabil Rehabil*. 16, 198–204.

Levine MN, Guyatt GH, Gent M, De Pauw S, et al. (1988). Quality of life in Stage II breast cancer: An instrument for clinical trials. *J Clin Oncol*. 6, 1798–1810.

Locke EA. (1969). What is job satisfaction? *Organ Behav Hum Perform*. 4. 309–336.

Locke EA. (1970). Job satisfaction and job performance: A theoretical analysis. *Organ Behav Hum Perform*. 5, 484–500.

Luce RD, Krantz DH, Suppes P, Tversky A. (1990). *Foundations of Measurement*. Vol. 3. San Diego CA: Academic Press.

Luce RD, Tukey JW. (1964). Simultaneous conjoint measurement: A new type of fundamental measurement. *J Math Psychol*. 1, 1–27.

Marra CA, Woolcott JC, Kopec JA, Shojania K, et al (2005). A comparison of generic, indirect utility measures (the HUI2, HUI3,SF-6D and the EQ-5D) and disease-specific instruments in rheumatoid arthritis. *Soc Sci Med*, 60. 1571–1582.

Mastekaasa A. (1984). Multiplicative and additive models of job and life satisfaction. *Soc Indic Res*. 14, 141–163.

McCelland DC. (1985). *Human Motivation*. Glenview Il: Scott, Foreman.

McFarlin DB, Coster EA, Rice RW, Cooper AT. (1995). Facet importance and job satisfaction: Another look at the range-of-affect hypothesis. *Basic Appl Soc Psychol*. 16, 489–502.

McGee HM, O'Boyle CA, Hickey A, O'Malley K, et al. (1991). Assessing the quality of life of the individual: The SEIQoL with a healthy and a gastroenterology unit population. *Psychol Med*. 21, 749–759.

Merluzzi TV, Nairm RC, Hedge K, Martinez Sanchez MA, et al. (2001). Self-efficacy for coping with cancer: Revision of the Cancer Behavior Inventory (Version 2.0). *Psycho-Oncol*. 10, 206–217.

Michell J. (1997). Quantitative science and the definition of measurement in psychology. *Brit J Psychol*. 88, 355–383.

Michell J. (1999). *Measurement in Psychology: Critical history of a methodological concept*. Cambridge UK: Cambridge Press.

Michell J. (1986). Measurement scales and statistics: A class of paradigms. *Psychol Bull*. 100, 398–407.

Mobley WH, Locke EA. (1970). The relationship of value importance to satisfaction. *Organ Behav Hum Perform*. 5, 463–483.

Morita S, Ohashi Y, Kobayashi K, Matsumot T, et al. (2003). Individual different "weights" of quality of life assessment in patients with advanced nonsmall-cell lung cancer. *J Clin Epidemiol*. 56, 744–751.

Mozley CG, Huxley P, Sutcliffe C, Bagley H, et al (1999). "Not knowing where I am doesn't mean I don't know what I like": Cognitive impairment and quality of life responses in elderly people. *Int J Geriat Psychiatr*. 14, 776–783.

Neugarten BL, Havinghurst RJ, Tobin SS. (1961). Measurement of life satisfaction. *J Gerontol*. 16, 134–143.

O'Boyle, Brown J, Hickey A, McGee H, et al. (1995). *Schedule for the Evaluation of Individual Quality of life(SEIQoL): A direct weighting procedure for quality of life domains (SEIQoL-DW)*. Dublin UK: Administration Manuel; Department of Psychology, Royal College of Surgeons.

Osoba D. (1994). Lessons learned from measuring health-related quality of life in oncology. *J Clin Oncol*. 12, 508–516.

Osoba D, Hsu M, Copley-Merriman, Cooms J, et al. (2006). Stated preferences with cancer for health-related quality of life (HRQL) domains during treatment. *Qual Life Res*. 15, 273–283.

Park DC. (2000). The basic mechanisms accounting for age-related decline in cognitive function. In, (Eds.) DC Park, N Schwarz. *Cognitive Aging: A Primer*. Philadelphia PA: Taylor and Francis. (p. 3–21).

Patrick DL, Bush JW, Chen MM. (1973). Towards an operational definitional of health. *J Health Soc Behav*. 14, 6–23.

Phillip EJ, Merluzzi TV, Peterman A, Cronk LB. (2009). Measurement accuracy in assessing patient's quality of life: To weight or not to weight domains of quality of life. *Qual Life Res*. 18, 775–782.

Pickard AS, Shaw JW, Lin H-W, Trasj PC, et al. (2009). A Patient-based utility measure of health for clinical trials of cancer therapy based on the European Organization for the Research and Treatment of Cancer quality of life questionnaire. *Values Health* 12, 977–988.

Rand A. (1964). The objectivist ethics. In, (Ed.) A. Rand. *The Virtue of Selfishness*. New York NY: Signet. (pp. 13–35).

Reed Johnson F, Hauber AB, Osoba D, Hsu M-A, et al. (2006). Are chemotherapy patients' HRQOL importance weights consistent with linear scoring rule? A stated-choice approach. *Qual Life Res*. 15: 285–298.

Rice RW, Markus K, Moyer RP, McFarlin DB. (1991). Facet importance and job satisfaction: Two experimental tests of Locke's Range of Affect hypothesis. *J Appl Soc Psychol*. 24: 1977–1987.

Rogers SN, Laher SH, Overend L, Lowe D. (2002). Importance-rating using the University of Washington Quality of life questionnaire in patients treated by primary surgery for oral and oro-pharyngeal cancer. *J Cranio-Maxilofac Surg*. 30, 125–132.

Rokeach M. (1973). *The Nature of Human Values*. New York NY: Free Press.

Rosenberg, M. (1979). *Conceiving the self*. New York NY: Basic Books.

Ross M. (1989). Relation of implicit theories to the construction of personal histories. *Psychol Rev*. 96, 341–347.

Rosser R, Kind P. (1978). A scale of valuations of states of illness: Is there a social consensus? *Int J Epidemiol*. 7, 347–358.

Russell LB, Hubley AM. (2005). Importance ratings and weighting: Old concerns and new perspectives. *Int J Test*. 5, 105–130.

Russell LB, Hubley AM, Palepu A, Zumbo BD. (2006). Does weighting capture what's important? Revisiting subjective importance weighting with a quality of life measure. *Soc Indic Res*. 75, 141–167.

Ruta DA, Garratt AM, Leng M, Russell IT, et al. (1994). A new approach to the measurement of quality of life: The Patient Generated Inventory (PGI). *Med Care*. 32, 1109–1126.

Schwartz CE, Sprangers MAG. (2000). *Adaptation to Changing Health: Response Shift in Quality of life Research*. Washington DC, American Psychological Association.

Schwartz SH. (1992). Universals in the content and structure of values: Theoretical advances and empirical tests in 20 countries. In, (Ed.) MP Zunna. *Advances in Experimental Social Psychology*. (Vol. 25). New York NY: Academic Press. (pp. 1–65).

Schwarz N, Strack F. (1999). Reports of subjective well-being: Judgmental processes and their methodological implications. In, (Eds.) D Kahneman, E Diener, N Schwarz. *Well-being : The Foundations of Hedonic Psychology*. New York, Russell Sage Foundation. (p. 61–84).

Sen A. (2001). Objectivity and position: Assessment of health and well-being. In, (Eds.) J Drèze, A Sen. *India: Development and Participation*. Oxford UK: Oxford University Press. (p. 115–128).

Shipper H, Clinch J, McMurray A, Levitt M. (1984). Measuring the quality of life of cancer patients: The Functional Living Index-Cancer: Development and validation. *J Clin Oncol*. 2, 472–483.

Søgaard R, Christensen FB, Videba/ek TS, Bünger C, et al. (2009). Interchangeability of the EQ-5D and the SF-6D in long-lasting low back pain. *Value Health*. 12, 606–612.

Spates JL. (1983). The sociology of values. *Annu Rev Sociol*. 9, 27–49.

Stavem K, Frøland SS, Hellum KB. (2005). Comparison of preference-based utilities of the 15D, EQ-5D, and SF-6D in patients with HIV/AIDS. *Qual Life Res*. 14, 971–980.

References

Stevens SS. (1946). On the theory of scales of measurement. *Psychol Bull* 36, 221–263.

Strahan R, Gerbasi KC. (1972). Short homogenous versions of the Marlow-Crowne Social Desirability Scale. *J Clin Psychol.* 28: 191–193.

Stineman MG, Maislin G, Nosek M, Fiedler R, et al. (1998). Functional status; Application of a new feature trade-off consensus building tool. *Arch Phys Med Rehabil.* 79,1522–1529.

Stineman MG, Wechsler B, Ross R, Maislin G. (2003). A method for measuring quality of life through subjective weighting of functional status. *Arch Phys Med Rehabil.* 84 Suppl 2, S15-S22.

Stone PC, Murphy RF, Matar HE, Almerie MQ. (2009). Quality of life in patients with prostate cancer: Development and application of a hybrid assessment method. *Prostate Cancer Prostatic Dis.* 12, 72–78.

Stroop JR. (1935). Studies of interference in serial verbal reactions, *J Exp Psychol.* 18, 643–662.

Suppes P, Krantz DH, Luce RD, Tversky A. (1989). *Foundations of Measurement.* (Vol. 2). New York NY: Academic Press.

Todman J, Teunisse S, Phillips L. (2003). An "Innocuous Theoretical Indulgence"?: The use of weighting by importance. *Qual Life Res* 12: 825.

Torelli CJ, Kaikati AN. (2009). Values as predictors of judgments and behaviors: The role of abstract and concrete mindsets. *J Personal Soc Psychol.* 96, 231–247.

Torrance GW. (1976). Health status index models: A unified mathematical view. *Manag Sci.* 22, 990–1001.

Torrance GW. (1986). Measurement of health state utilities for economic appraisal: A review. *J Health Econ.* 5, 1–30.

Torrance GW, Furlong W, Feeny D, Boyle M. (1995).Multi-attribute preference functions: Health Utilities Index. *PharmEcon.* 7, 503–520.

Trauer T, Mackinnon A. (2001). Why are we weighting? The role of importance ratings in quality of life measurement. *Qual Life Res* 10, 579–585.

Tversky A, Kahneman D. (1992). Advances in Prospect Theory: Cumulative representation under uncertainty. *J Risk Uncertain.* 5, 297–323.

Tyas S, Snowdon DA, Desrosiers MF, Riley KP. (2007). Healthy aging in the Nun Study: Definition and neuropathologic correlates. *Age Aging.* 36, 650–655.

von Kries J. (1882). Über die Messung intensiver Grössen und über das sogenannte psychophysische Gesetz. *Vierteljahrsschr wiss Philos.* 6, 257–294.

von Neumann J, Morgenstern O. (1944). *Theory of Games and Economic Behavior.* Princeton, NJ: Princeton University Press.

von Osch SMC, Stiggelbout AE. (2008). The construction of the standard gamble utilities. *Health Econ.* 17, 31–40.

von Osch SMC, Wakker PP, van den Hout WB, Stiggelbout AE. (2004). Correcting biases in standard gamble and time tradeoff utilities. *Med Decis Mak.* 24, 511–517.

Wakefield JC. (1999). Evolutionary versus prototype analyses of the concept of disorder. *J Abnorm Psychol.* 108, 374–399.

Ware JE, Kosinski MA, Keller SD. (1994). *SF-36 Physical and Mental Health Summary Scales: A Users Manual.* Boston MA: The Health Institute, New England Medical Center.

Ware JE, Kosinski MA, Keller SD. (1995). *SF-12: How to Score the SF-12 Physical and Mental Health Summary Scales.* Boston MA: The Health Institute, New England Medical Center.

Ware JE, Snow KK, Kosinski MA, Gandek B. (1993). *SF-36 Health Survey: Manual and Interpretation Guide.* Boston MA: The Health Institute, New England Medical Center.

Wee H-L, Machin D, Loke W-C, Li S-C, et al. (2007). Assessing differences in utility scores: A comparison of four widely used preference-based instruments. *Value Health.* 10, 256–265.

Welham J, Haire M, Mercer D, Stedman T. (2001). A gap approach to exploring quality of life in mental health. *Qual Life Res.* 10: 421–429.

Wettergren L, Björkholm M, Langius-Eklöf A. (2005). Validation of an extended version of the SEIQoL-DW in a cohort of Hodgkin lymphoma' survivors. *Qual Life Res.* 14: 2329–2333.

The WHOQOL Group. (1998). The World Health Organization Quality of life Assessment (WHOQOL): Development and general psychometric properties. *Soc Sci Med.* 46, 1569–1585.

Williams B, Coyle J, Healy D. (1998). The meaning of patient satisfaction: An explanation of high reported levels. *Soc Sci Med.* 47, 1351–1359.

Williams PG, Wasserman MS, Lotto AJ. (2003). Individual differences in self-assessed health: An information-processing investigation of health and illness cognition. *Health Psychol* 22, 3–11.

Willis GB. (2005) Cognitive Interviewing: A Tool for Improving Questionnaire Design. Thousand Oaks CA: Sage.

Wu C-H. (2008). Can we weight satisfaction score with importance ranks across life domains? *Soc Indic Res.* 86: 469–480.

Wu C-H, Chen LH, Tsai Y-M. (2009). Investigating importance weights of satisfaction scores from a formative model with Partial Least Squares analysis. *Soc Indic Res.* 90, 351–363.

Wu C-H, Yeo G. (2006a). Do we need to weight satisfaction scores with importance ratings in measuring quality of life? *Soc Indic Res* 78: 305–326.

Wu C-H, Yeo G. (2006b). Do we need to weight item satisfaction scores by item importance? A perspective from Locke's range-of-affect hypothesis. *Soc Indic Res* 79, 485–502.

Wu C-H, Yeo G. (2007). Examining the relationship between global and domain measures of quality of life by three factor structure models. *Soc Indic Res* 84, 189–202.

Wundt W. (1874). *Grundzüge der physiologischen Psychologie.* Leipzig Germany: Wilhelm Engelmann.

Epilogue

It is highly likely that any quality-of-life researcher who ventures into this book will have to learn new material. I did and was well rewarded. Both looking at how language is used when a qualitative assessment occurs, and studying the cognitive processes that underlie a qualitative assessment are not normal activities for a quality-of-life researcher. Either or both of these perspectives permitted me to ask *how* a qualitative assessment occurred, and the more I learned the more I understood what a qualitative assessment was about. Still I only addressed a few of the many interesting questions that could have been asked. For example:

- *Time*, its perception and role in the assessment process, deserves an extensive review and examination. How does the brain perceive and process time, and how is this ability altered when adverse events occur (e.g., a stroke, loss of memory function, and so on). What is the ideal time period to maximize recall, for various groups of persons? What role does time perception play in the response shift or utility estimation?
- *Valuations*. I have only skimmed the surface of what is an extensive literature concerning the different valuations procedures. Each of the major traditions (e.g., hedonic approach, Aristotle's Eudaemonic approach, Sen's capacity model, etc.), as well as valuation procedures (e.g., standard gamble, time trade-off, willingness to pay, etc.) deserve extensive review, especially since each of the valuation procedures creates a different cognitive demand. What is known about these traditions and procedures should be critically reviewed from a language usage and cognitive science perspective. There is also the issue of how these procedures are used to elicit valuations or reflective processes. For example, is some of the observed variability in the use of the standard gamble procedure due to the cognitive demand of the instructions?
- *Defining endpoints* for a qualitative continuum is also a neglected issue. For example, certain utility models assume that conditions worse than death exist. Is this true? Alternatively, what does "very happy" or "very sad" mean? Quality-of-life researchers tend to ignore how investigators and respondents deal with these markers, yet they are critical in defining a continuum.
- *The role of metaphor* in qualitative research is generally ignored by psychometricians, yet its presence raises serious issues about what is being measured. For example, most statistical models include latent variables, yet do not address the issue that, linguistically, the inference from concrete data to an abstract construct (the latent variable) is the essence of how a metaphor is created. There is also the issue of arithmetic metaphorization, where numbers from diverse indicators are summed without regard to rationale for this summing. The argument that a number is a number and that I can apply any mathematical operation to it, does not free the investigator from having to interpret the number produced, especially when the number is to represent a subjective response.
- Metaphors, of course, are commonly used in the sciences, but usually also involve a unique language characteristic of a particular science. For example, both biology and physics have created special languages to characterize the phenomena that they study. In contrast, the language used in qualitative research is the common discourse; this makes it much more difficult to precisely define what the terms in the qualitative sciences mean. This prompts my next set of unanswered questions.
- The *meaning* any particular group of persons associate with common questions is an important area of investigation. For example, if I ask one group of persons who are intellectually compromised and another group who are not the same question, can I assume that the answers from the two groups have the same meaning? If they are not, how do I know this, and how do I determine if their answers have different meaning? Neuropsychologists will gladly help me answer this question, at least they can tell me the risk that I am in that the persons being assessed may be interpreting and answering the questions differently.

- What about the extreme cases; the severely demented, the terminal cancer patient, the person suffering "Job" like pain? People who, from the perspective of the reasonably well, might not seem to have any meaning in life, yet may indicate so if asked a common question (e.g., "How would you rate your Quality-of-Life; Excellent, Very Good, Good, Fair, or Poor?"). Do I as a qualitative researcher discard the extreme cases as aberrant and avoid including them in my sample, simply because they may increase the variance of my data or generate paradoxical findings? If I did, I would have a rather limited view of what the qualitative sciences should be concerned with, yet this, I believe, describes a good portion of the currently available qualitative research.
- The *need to infuse an ethical perspective* into the conduct of human affairs was a primary goal set forth in the introductory paragraph in Chap. 1. I stated; "My first and primary objective for this book is to contribute to what is known about how to infuse an ethical perspective into the conduct of human affairs, particularly in the conduct of medical care, by expanding the role that assessing quality plays in these activities" (Chap. 1, p. 2). Have I done this?
- I believe I have, if only by my proposal that a qualitative assessment is a central requirement in monitoring the unpleasant or debilitating events that befall every person, as they live their lives. I described two basic approaches to this problem; the prevention of adverse events or the enhancement of a person's well-being. I argued that the prevention of adverse events is probably the most efficient way to achieve the ethical objectives I described. Both prevention or enhancement interventions raise philosophical and practical policy issues, but I argued setting standards for enhancement (since they require a consensus on what should be achieved) are much more difficult to achieve than preventing certain adverse outcomes. This argument is particularly relevant to the medical care setting.

There are obviously many more questions that can be addressed, but this should be enough to wet the appetite of any qualitative researcher. However, my review and proposal concerning the role of a universal hybrid cognitive construct also deserves detailed investigation. For example, what evidence can be mounted that reflection occurs during a qualitative assessment, and how would an investigator measure this? I also claim that if two people engage in the same thought process, the fact that each may involve different content is irrelevant. Is this true, and if it is not then what role do differences in content play in how a qualitative assessment occurs? In addition, if I am right then much of what constitutes current quality-of-life research is only of limited importance.

An exciting aspect of this proposal is that persons from different cultures engage in the same thought processes when they assess quality. Is this true? Basically, I am arguing that demonstrating cognitive equivalence, not psychometric equivalence, take precedence when determining if two groups of respondents have provided an equivalent qualitative assessment. If this is so, then an investigator has a neurocognitive model that he or she can use to design qualitative assessments.

The potential that qualitative assessments offer for mitigating adversities attests to its potential contribution to ensuring the ethical character of human interactions. It does this by assuming that a common cognitive process underlies all qualitative assessments, and this offers a potential way whereby persons of diverse backgrounds, as defined by such characteristics as gender, ethnicity, culture, education, and cognitive states, as well as health and disease status, can be compared and given the opportunity to bring about improvements in quality-of-life that are both equitable and ethical.

Index

A

Abstract and concrete concepts, neurocognitive basis, 358
Accuracy of symptom reports, 129
    negative affect, 130
    Pennebaker, 129
    Verbrugge, 130
Ackoff, R.L., 7
Acquired modularity, 173, 224
    conditioned learning, 225
    IBS, 227
Adam in the Garden of Eden, 315
Adverse events
    conceptual metaphor, 145
    CTCAE, 143
    phase I trials, 145
Affect
    affective quality, 419
    circumplex, 419
    core affect, 419
    dimensional nature, 418
Affective expression
    Bramston, 406
    Clore, 398
    Colombe, 398
    emotional episodes, 406
    mood, 406
    qualitative assessment, 406
    Schwarz, 406
    Strack, 406
    subjective well-being, 406
Affect primacy hypothesis
    Zajonc, 422
Affordance, 67
    Gibsson, 67
    Norman, 67
Aging effects, cortical sites involved, 366
Alexandrova, A., 411
Alzheimer's disease
    cortical sites involved, 366
    decision making, 364
    lack of insight, 362
    natural history, 353
    reliability and validity of qualitative assessments, 359
American Society of Quality, 10
Analytical hierarchy process
    Thurstone, 343
    use in qualification, 343
Anaphora, 39

Andrews, F.M., 421
Appraisal, 69
    dual processing, 423
    Ellsworth, 426
    Gibson, 423
    Lazarus, 69, 423
    neurocognitive basis, 70
    Rapkin, 427
    Schwartz, 427
    Solomon, 427
    Strobeck, 423
    subjective well-being and quality-of-life assessment, 426
    Tong, 426
Aristotle, 33
Assessment language
    health opinion survey, 405
    Psychological General Well-being Index, 404
    satisfaction with life scale, 403
Automatic processing, 172
Averill
    on the nature of metaphors, 418
    quality-of-life assessment, 418

B

Baddeley, A.D., 181
Barofsky, I., 18, 50, 237
Barrett, L.F., 181, 424
    Izard, 416
Barsalou, L.W., 57, 440
Beckie, T.M., 278
Berlin, B., 57
Boorse, C., 251
Borsboom, D., 114, 474
Bowdle, B.F., 33
Bridgman, P.W., 19
Brown, H.I., 96
Brunswik, E., 72
    intellectual history, 72
    Lens model, 72
Bunge, M., 308

C

Calman, K.C., 462, 478
Capacity and performance models
    embodiment theories, 385
    neural network models, 378

Category formation, 162
    category learning, 165
    cultural determinants, 168
    kinds of categories, 167
    language usage, 163
    types, 167
Category learning
    Ashby, 165
    bilingualism, 170
    dual process theory, 165
    exemplar model, 168
    language, influence of, 170
    Maddox, 165
    rule-based, 170
    Sapir-Whorf hypothesis, 170
    types of learning, 165
Cattorini, P., 6
Chiang, Y.-P., 295
Chomsky, N., 26
Clark, W.C., 206, 222, 229
    multidimensional affect and pain survey, 229
Classification
    basic, 56
    Berlin, 57
    cognitive entities, 55
    comparison between basic and superordinate terms, 164
    cross-classification, 57
    DSM as a model of, 186
    hierarchical, 56
    hieroglyphics, 60
    models, 186
    Rosch, 56
    subordinate, 56
    superordinate, 56
    types, 186
Classifying pain, models, 228
Cleary, P.D., 185
Clinical management of symptoms, 148
    functional somatic syndromes, 148
    role of bodily distress, 149
Clinimetrics, 141
    cross-classification, 142
    and psychometric perspectives
        Streiner, 150
Codman, E.A., 98, 259
Cognitive algebra, 74, 75
Cognitive basis of emotions
    Barrett, 424
    Clore, 425
    Collins, 425
    Duncan, 424
    Ortony, 425
Cognitive degeneration
    aging effect, 363
    cortical sites, 366
    decision making, 364
    metaphors as a model, 369
Cognitive emotional processes
    affect primacy hypothesis, 422
    Andrews, 421
    cognitive primacy hypothesis, 423
    Cummins, 421
    De Haes, 421
    dimensional aspects, 421
    McKennell, 421
    Solomon, 421

Cognitive-emotional regulation
    can emotions be separated from its regulation, 430
    development perspective, 431
    Gross's model, 432
    hedonic treadmill, 429
    Lewis, 431
    neurocognitive basis, 431
    Phineas Gage, 428
    Stieben, 431
Cognitive entity, 50
    classification, 55
    formation, 54
    Qualia, 50
    role of metacognition, 51
Cognitive models of quality of life assessments, 72
    Anderson, 74
    Brunswik and SEIQoL, 72
    Bush, 74
    Kaplan, 74
Cognitive preservation
    cognitive reserve, 349
    longitudinal studies, 351
    nuns study, 351
    types, 349
    years of education, 350
Cognitive primacy hypothesis
    Lazarus, 423
Cognitive processes
    affordance, 19
    appraisal, 69
    automatic processing, 66
    classification, 55
    concept formation, 62
    developmental studies, 52
    formation of a cognitive entity, 50
    Hayek, 51
    metacognition, 64
    neural basis, 51
    neurocognitive theories, 51
    production rules, 54
    subjective probability, 77
    trait (facial) perception, 68
Combining NPTs and quality-of-life data
    common metric hypothesis, 344
    substitution hypothesis, 344
    summary statement, 384
Combining objective and subjective indicators, 97, 99
    aggregation and summary measures, 101
    Cummins, 101, 104
    economics, 100
    Fries, 103
    hierarchical models, 102
    history of the behavioral sciences, 97
    linear models, 104
    Michalos' rules for combining indicators, 99
    non-linear models, 107
    Ormel, 104
    Revicki, 103
    Spilker, 103
    Zapf, 98
Common terminology criteria for adverse events
        (CTCAE), 143
Complexity theory, 303
Complex regional pain syndrome (CRPS), 225, 226
    application of Fodor's criteria for modularity, 226
    Baron, 226

Jänig, 226
Oaklander, 226
type II-Causalgia, 226
type I-reflex sympathetic dystrophy, 226
Compromised person, 19, 203
assessment approaches, 203
pain assessment, 230
pain's impact on cognition, 230
role of care-givers, 231
Concept formation
classic model, 62
prototypes and exemplars, 62
Conceptual metaphors, 45
cognitive components, 36
THE DOCTOR-PATIENT RELATIONSHIP AS WAR, 126
empirical evidence against, 36
event structure, 145
EVENT STRUCTURE METAPHOR, 35
linguistic metaphors used, 126
A LOCATION EVENT STRUCTURE METAPHOR, 39
LOVE IS A JOURNEY, 35
the measuring stick metaphor, 44
role in a quality-of-life assessment, 370
SUBJECT-SELF METAPHOR, 370
Constructionism
empirical, 31
Gergen, 31
metaphysical, 31
Searle, 31
Zuriff, 31
Constructionist view, 31
Core affect, 419
Cross-classification, 57
Barsalou, 57
goal-directed and taxonomic-based categories, 60
Murphy, 57
Ross, 57
types, 57
CRPS. *See* Complex regional pain syndrome
Cultural factors in symptom assessment
Chinese, 204
Mexican, 204
Cultural factors, role of, 168
bilinguals, 170
Nisbett, 169
Cummins, R.A., 101, 421, 435

D

Daston, L., 131
Davidson, R.J., 438
Decision making, 364
Definitions of disorders, 255
cognitive basis, 258
Fulford, 257
Lilienfeld, 255
Marino, 255
Wakefield, 255
Definitions of normal
health status, 249
various definitions, 249
Depression, 201
Diener, E., 60
umbrella term, 43
Disability
definitions, 314, 315

difference from functional limitations, 316
Jette, 316
Verbrugge, 316
Disability assessment, 316
defining deviance and normalcy, 314
Fuhrer, 318
ICF model, 314, 317
Imrie, 317
Institute of Medicine's Models, 314, 317
Nagi's model, 314, 316
quality of life assessments, 318
role of values, 315
statistical modeling analyses, 319
Sullivan, 319
Disability paradox, 98
Distinguishing disease and illness, 254
Dolan, P., 412
Domain complexity, 174
Domain formation, 171
acquired modularity, 173
causal/indicator role, 185
classification models, 186
cognitive processes, 171
complexity of health status domains, 175
conceptual metaphors, role of, 171
domain complexity, 174
dual process theory, 172
Fodor, 176
Fodor's criteria of modular domain, 177
functional modules, 173
language usage, 171
modular basis, 175
modular domain, 173
objectification methods, 173
role of modules, 173
Donabedian, A., 262
Downward comparison, 60
Dual processing, 423
Dual process theory, 66, 165, 172
automatic processing, 172
controlled processing, 173
De Houwer, 66
Moors, 66
Dual usage terms, 38, 39
Duncan, S., 424

E

Einstein, A., 136
Ellsworth, P.C., 426
Embodiment, 36, 143
affect, 37
embodied truth, 36
Johnson, 36
Lakoff, 36
Meir, 37
Robinson, 37
Embodiment model of subjective well-being
Barsalou, 440
Niedenthal, 439
Embodiment theories
Gibbs, 382
Johnson, 382
Lakoff, 382

Emotions and cognition
    Andrews, 398
    McKennell, 398
    theoretical issues, 398
    Withey, 398
Emotion words
    Averill, 417
    happiness, 402
    Kövecses, 399
    Meier, 418
    metaphoric basis, 417
    Robinson, 418
    semantics, 399
    translation, 412
    Wierzbicka, 412
End-results and outcomes movement, 259
    Codman, 259
Enhancement models, 17
Eudaemonic, 394, 437, 441
EVENT STRUCTURE METAPOR, components, 35

F

Fatigue, 202, 207
    assessments, 208
        diversity of assessments, 208
        diversity of content, 208
        multi-dimensional fatigue inventory, 212
        predictors, 210
        single vs. multi-item assessment, 210
    cancer patients, 207
    definition of fatigue, 208
    depression and HRQOL, 211
        supporting data, 214
        theoretical model, 212
    metaphoric basis, 211
Feinstein, A.R., 4, 151
Ferrans, C.E., 4, 458
    quality of life index, 458
Figurative language, 26, 30
    metaphor, 30
Fisher, W.P. Jr., 111
Fodor, J., 26, 176, 225
Frank, A.W., 27
Frege, G., 28
    literal meaning hypothesis, 28
Fryback, D.G., 466
Fuhrer, M.J., 97
Fulford, K.W.M., 257, 471
Functionalism, 308
    models of-Mahner and Bunge, 308
    Parsons, 310
Functional modularity, 173, 181
Functional somatic syndromes, 148
Functional status
    Chiang, 296
    definitions, 295
    examples of items on assessments, 296
    Patrick, 296
Functional status modules, 180
Functions
    assessment, 314
        disabled, 314
        disablement process, 314
        ICF and quality of life assessment, 329
        independent observer's assessment and a quality of life assessment, 322
        quality of life assessment models, 320
        role of values-Hewlett, 326
        role of values-Stineman, 326
        self-report comparison to a quality of life assessment, 328
    definitions, 297
    statements, 297
        Barsalou, 299
        Boorse, 300
        Bunge, 320
        causal attributions, 300
        cognitive basis, 299
        Davis, 301
        figurative expression, 298
        health and disease, 300
        Mahner, 320
        mathematical expressions, 297
        quality of life assessments-difference from, 320
        teleological explanations, 302

G

Galison, P., 131
Gärdenfors, P., 12
Gentner, D., 33
    structure mapping theory, 34
Gibbs, R.W. Jr.
    literal meaning, 29
Gibson, J.J., 423
Gill, T.M., 151
Giora, R.
    saliency continuum, 35
Glucksberg, S., 29
    defining linguistic metaphors, 32
Goldwasser, O., 60
Goodwin, P.J., 235
Gorbatenko-Roth, K., 463
Grounded theory, 134
    Glaser, 134
    Strauss, 134

H

Halfon, N., 255
Hammond, K.R.
    cognitive continuum theory, 72
    Lens model, 73
    social judgment theory, 72
Happiness
    conceptual metaphors, 402
    linguistic conceptualization, 403
Hassin, R., 68
Hayduk, L.A., 278
Hayek, F.A., 5, 51
    complexity theory, 53
    early intellectual history, 52
    Hebb, 52
    qualitative research, 53
    Smith, A., 53
Headey, B., 434
Health, definitions, 250
    Boorse, 251
    developmental models, 255
    Dubos, 250
    genetic disorders, 254
    Johnson, 250
    Lakoff, 250

Larson, 253
Lay People, 252
Nordenfelt, 252
Offer, 253
Parsons, 251
Sabshin, 253
Health opinion survey, 405
Health-related quality-of-life (HRQOL)
  assessments
    Beckie, 278
    cognitive components, 270
    Cummins, 281
    current assessments, classified, 273
    examples of items, classified, 271
    Gotay, 280
    Hayduk, 278
    health status, difference from, 269
    list of criticisms, 281
    Michalos, 280
    Pagano, 280
    QALY, 282
    relationship to health status, 277
    types of items, classified, 270
  definitions, 30, 37
Health status, 178, 248
  applications, 269
  assessments, 259
  definitions of, 178
  Erickson, 178
  HRQOL, difference from, 269
  operations research, 260
  Patrick, 178
  relationship to HRQOL, 277
  Torrance's model, 178
  Wittenberger, 248
Hedonic treadmill, 429
Hedonic well-being
  Crisp, 395
  John Stuart Mill, 396
  model, 397
  Sidgwick, 396
Heinemann, A.W., 19
Helmholtz, H., 136
Hendry, F., 7
Heuristics, 55
Hieroglyphics, 60
  conceptual metaphors, 61
  Goldwasser, 60
Higgins, E.T.
  quality-of-life assessment, 437
  regulatory fit, 436
Hockstein, M., 255
HRQOL. *See* Health-related quality-of-life
Hubley, A.M., 465
Hybrid construct, 454
  advantages, 454
  Keeney, 456
  neurocognitive determinants, 351
  supporting evidence, 456

I
IBS. *See* Irritable bowel syndrome
Illich, I., 17
Importance ratings
  Campbell, 461
  metaphoric properties, 457

Introspection, 11
Invariance, 112
  Borsboom, 114
  Frege, 112
  Nozick, 112
  psychometric approaches-measurement invariance, 114
Irritable bowel syndrome (IBS), 126, 203, 227
  symptoms as metaphors, 127
Item overlap in HRQOL assessments
  Anderson, 234
  De Haes, 235
  Goodwin, 235
Item response theory (IRT), 173
Izard, C.E., 416

J
James, W., 422
Johnson, M., 143
Jostmann, N.B., 457
Joyce, C.R.B., 72

K
Kahneman, D., 19, 109
  Alxeandrova, 411
  Dolan, 412
  prospect theory, 471
  White, 412
Keil, F.C., 52
Kim-Prieto, C., 435
Kline, P., 475
Koriat, A., 65

L
Lack of insight, 362
Lakoff, G., 35, 143, 163
  conceptual metaphors, 35
Language usage
  Chomsky, 26
  Fodor's, 26
  grammar, 26
  principles, 26
  tree structure, 26
Larson, J.S., 253
Lascaratou, C., 217
Legendre, C., 379
Leventhal, H., 437
Levinas, E., 3
Levine, M.N., 462
Lewis, M.D., 431
Lilienfeld, S.O., 255
Lingenberg, S., 435
Linguistic metaphor, 32
Literal expression, 28
  Frege, 28
  Glucksberg, 29
Locke, E.A.
  range-of-affect, 462
Logical positivism, 31

M
Mackinnon, A., 465
Mahner, M., 308

Marino, L., 255
Maslow, A.H., 41
Mayr, E., 20
McGill pain questionnaire, relationship to hybrid construct, 223
McKennell, A.C., 421
McVitte, C., 7
Mechanical and structural objectivity, 135
    relationship to subjectivity, 136
Mehl, M., 27
Meier, B.P., 418
Melzack, R., 221, 224
    neuromatrix model, 224
Metacognition, 64
    Koriat, 65
    mechanism of action, 65
    psychopathology, 64
    role in valuation, 55
    Schwarz, 65
Metaphoric arithmetization
    conceptual blending, 298
    Lakoff, 298
    Núñez, 298
Metaphoric expression
    neurocognitive basis, 370
    visual analog scale, 372
Metaphors, 37
    Aristotle, 33
    comparison models, 33
    comprehension, 33
    conceptual, 35
    definition, 32
    embodiment, 36
    Glucksberg, 33
    HRQOL, 37
    linguistic, 32
    Miller, 32
    psychological symptoms, 125
    quality of life, 37
    saliency continuum, 35
    structure mapping theory, 34
    Susan Sontag, 124
    symptoms, 125
    use in doctor-patient interaction, 125
    use of in science, 30
    use so as to express a symptom, 126
Metonym, definition of, 400
Meyers, C.A., 346
Michalos, A.C., 99, 434
Michell, J., 474
Michotte, A., 11
Modular domain, 173
Modularity
    functional status, 180
    health status, 178
    neurocognitive indicators, 181
    subjective well-being, 182
    symptoms, 177
    visual HRQOL, 183
Momentary affective state
    Kahneman, 411
Mordacci, R., 6
Morreim, E.H., 7
Murdock, I., 353
Murphy, G.L., 57, 164, 205

N
Natural kinds, 167
    Barrett, 415
    Keil, 415
Neural network models, 378
Neurocognition
    cognitive capacity, 348
    degeneration, 352
    modeling the relationship between capacity and performance, 376
    Posner, 347
    preservation, 347, 348
    recovery of function, 368
    relationship to quality-of-life assessment, 341
    Stern, 348
Neurocognitive basis of subjective well-being
    Davidson, 438
Neurocognitive indicators, 181
    Baddeley, 181
Neurophenomenology, 11
Neuropsychological testing (NPT)
    combined with quality-of-life data, 341
    comparison to quality-of-life data, 341
    dementia rating scale (Mattis), 341
    neuropsychological testing, 341
Niedenthal, P.M., 439
Non-linear models, 224
    Catastrophe theory, 107
    history, 107
    Kahneman's model of objective happiness, 108
    McCulloch, 108
    Pitts, 108
    Smolensky and Legendre optimality theory, 108
    of subjective well-being
        Scherer's appraisal model, 439
Nord, E., 283
Nordenfelt, L., 252
Norman Anderson, 74
    cognitive algebra, 75
    information integration theory, 75
Nozick, R., 112
NPT. *See* Neuropsychological testing

O
Objectification methods, 173
Objectification process, 5, 17, 122, 473
    arithmetic metaphorization, 474
    Brown, 96
    classical model of measurement, 475
    fundamental or axiomatic measurement, 474
    item response theory, 173
    Kline, 476
    making subjective indicators objective, 92
    operational definitions, 473
Objective assessment of subjective indicators
    Kahneman's model of objective happiness, 109
    The Rasch model, 110
Objectives of Book, 17
    Levinas, 3
Objectivity
    conjoint measurement, 112
    Daston, 110, 131
    Galison, 110, 131
    history, 110
    mechanical objectivity, 132

Pythagoras, 111
    role of the expert, 132
    structural objectivity, 132
    truth-to-man, 132
    truth-to-nature, 131
Offer, D., 253
Operations research, 260
Optimality theory, 380
Ormel, J., 435
Ortony, A., 425
Osoba, D., 463
    definition of quality of life, 38

P

Pain, 202, 217
    acquired and functional modularity, 224
    assessments, 221
        Clark, 222
        Gracely, 222
        informational limitations, 223
        McGill pain questionnaire, 221
        Melzack, 221
        multidimensional approach, 222
        multidimensional pain survey, 229
        phenomenological approach, 221
        psychophysical approach, 222
        representative studies, 223
        Rudy, 228
        Turk, 228
    classification, 228
    compromised person, 203
    Cronje, 220
    CRPS, 225
    definitions, 220
    definitions of chronic pain, 225
    difference between sensory modalities, 219
    emotional pain, 220
    intellectual history, 224
    is pain a *Quale*, 221
    language of, 217
    Lascaratou, 217
    non-linear models, 224
    Williamson, 220
Parducci, A., 60
Parsons, T., 251, 310
    Erickson, 312
    Patrick, 312
    role in a HRQOL assessment, 312
    sick role, 310
    sick role-arguments against, 311
    structural functional analysis of medical practice, 310
Patient satisfaction, 264
    cognitive basis, 266
    Locke, 267
    major theories, 265
    Rapkin, 266
    Williams, 266
Patrick, D.L., 295
Pennebaker, J.W., 27, 129
Phase I trials, 145
    HRQOL assessment, 146
Philosophical functionalism, 306
    historical origin, 306
    Parsons, 310

Phineas Gage, 428
Pirsig, R.M., 14
    Phaedrus, 14
Positional objectivity, 95
Prevention models, 17
Price, C.I.M., 373
Production rules, 54
    heuristics, 55
PROMIS, 234
Prospect theory, 471
Psychological essence, 415
Psychological General Well-being Index, 404

Q

Quale
    Damasio, 12
    Dennett, 12
    experimental induction, 11
    Michotte, 11
    neurophenomenology, 11
    synaesthesia, 11
Qualia, 50, 77
    Quale, 11
Qualification
    definition, 91
    process, 92, 122
        objective indicators, 343
Qualitative assessments
    automaticity, 71
    cognitive processes, 19
    enhancement models, 17
    language usage, 19
    linear models, 19
    non-linear modeling, 19
    objective indicators, 91
    prevention models, 17
    subjective indicators, 91
    types, 14
Qualitative judgments
    Diener, 60
    models, 50
Quality
    aesthetic, 15
    conceptual space, 12
    functional, 15
    Gardenfors, 12
    Hayek, 14
    history of concept, 15, 97
    is quality a *Quale*, 10
    models, 12
    other dimensions, 16
    Shepard, 12
Quality adjusted life years (QALY), 282
    assets and limitations, 283
    Nord, 283
Quality-adjusted time without symptoms and toxicity (Q-TWiST), 146
    Gelber, 146
Quality control, 18
    American Society of Quality, 10
    operations research, 18
Quality of care
    Donabedian, 262
    limits of satisfaction indicators, 263
    process and outcome, 264

Quality of life
    assessments
        arguments against assessment, 6
        difficulties defining it, 4
        difficulties defining it-arguments for, 9
        difficulties defining it-Ferrans, 4
        difficulties defining it-Rapley, 4
        NPT items in, 345
        presence of metaphors, 369
        relationship to operations research, 18
        should it be assessed?, 6
    definitions, 30, 37
    HRQOL
        Barofsky's definition, 44
        causal definitions, 38
        list of definitions, 45
        matching models, 39
        well-being models, 41
    student definitions, 27
    value-based, 43
Quality of life index, 458
Quality of well-being scale, 74, 395
    Bush, 74
    expected utility, 75
    Fanshel, 74
    General Health Policy model, 74
    Kaplan, 74
    model of a hybrid construct, 75
Question understanding aid (QUAID), 380

R
Rapkin and Schwartz
    quality-of-life assessment, 428
    response shift, 428
    Tourangeau's cognitive model, 427
Rapkin, B., 266
Rapley, M., 4
    Wittgenstein, 7
Recovery of function
    metaphors as a model, 369
Reeves and Bednar, 10, 15
    industrial revolution, 15
Regulatory models of subjective well-being
    Cummins homeostatic model, 435
    Heady's set point model, 434
    Higgins self-management model, 436
    Kim-Prieto and Diener's model, 435
    Leventhal's self-regulation model, 437
    Lindenberg's self-management model, 436
    Michalos's multiple discrepancy model, 434
    social production function model, 435
Rehabilitation
    definitions, 315
    metaphoric properties, 315
Response shift, 428
Robinson, M.D., 418
Rokeach, M., 471
Rosch, E., 56
Ross, B.H., 57, 205
Rudy, T.E., 228
Russell, J.A., 419
Russell, L.B., 465
Ruta, D.A., 74
Ryff, C.D., 410

S
Saaty, T.L., 343
Sabshin, M., 253
Sackett, D.L., 249
Sacks, O., 133
Sagan, C., 339
Santayana, G.
    beauty, 15
Satisfaction with life scale, 403
Schedule for the evaluation of individual quality-of-life
        (SEIQoL), 72, 74
Schwartz, N., 65, 406, 472
SEIQoL. See Schedule for the evaluation of individual
        quality-of-life
Self-assessed health status, 9
Self-reports, 8
    Kahneman's concerns, 93
    Murphy and Salomon's concerns, 93
    Sen's concerns, 93
Semantic degeneration
    implications for quality-of-life assessments, 355
    Murdoch, Iris, 353
    neuroanatomical basis, 356
    progressive degeneration, 355
    semantic errors, 356
Semantics, 26
Sen, A.
    Bayes law, 95
    India study, 94
    objective illusions, 95
    positional objectivity, 95
    self-reports, 93
Sense of self
    impact on qualitative assessments, 375
Shepard, R.N., 12
Signs and symptoms, 131
    Daston, 131
    Galison, 131
    mechanical and structural objectivity, 135
    trained judgment-expertise, 139
    truth-to-nature, 132
Smets, E.M., 214
Smolensky, P., 379
Social indicator movement, 98
Social judgment theory
    Hammond, 72
Soft-tissue sarcoma, 18
Solomon, R.C., 421, 427
Sontag, S., 32, 43, 124
Speech acts, 27, 400
    Searle, 27
Sperber's theory of massive modularity, 225
Stability of subjective well-being, statistical models, 408
Stern, Y., 348
Stieben, J., 431
Stineman, M.G., 326, 457
Strack, F., 406
Strobeck, J., 423
Stroke
    metaphoric expression, 371
    visual analog scale, 373
    Winner and Gardner, 372
Structural objectivity, 136
    Einstein, 136
    Frege, 136

Helmholtz, 136
  psychometrics, 137
  qualitative assessment, 137
  symptom assessment, 137
Subjectification process, 5, 92
Subjective probability, 77
Subjective well-being
  artificial class, 416
  assessment language, 403
  classification issues, 415
  class of artifacts, 416
  cognitive basis, 395
  cognitive emotional processes, 421
  cognitive-emotional regulation, 428
  conceptual metaphors, 401
  data supporting stability, 407
  definition, 394
  developmental perspective, 410
  Diener's model, 417
  differences from mood, 407
  dimensional nature, 418
  discrete emotion, 413
  embodiment models, 439
  metaphoric basis, 401
  metonym, 400
  momentary affective state, 406
  neurocognitive basis, 438
  non-linear models, 439
  ordinary discourse, 400
  persistent state, 406
  philosophical perspectives, 395
  regulatory models, 433
  speech act, 400
  synonyms, 400
  Zajonc, 422
Subjective well-being (SWB), 182
  as an outcome of cognitive-emotional regulation, 428
  as a discrete emotion
    natural kind, 415
    semantic space, 414
  genetic factors, 182
  personality factors, 182
Subjectivity and clinical medicine, 142
  Sullivan, 142
Substitution hypothesis
  brain cancer, 345
  persons with schizophrenia, 345
Suffering
  Frank, 27
  Kleinman, 28
Sugerbaker, P.H., 18, 237
Sullivan, M., 142
SWB. *See* Subjective well-being
Symptoms, 123, 177
  accuracy of reports, 129
  assessment, 127, 128
    accuracy of reports, 129
    McCorkle, 128
    MD Anderson Symptom Inventory, 129
    objective assessment, 131
    Young, 128
  awareness of, 123
  clinical decision making and HRQOL, 232
    Anderson, 234
    Blazeby, 236

Efficace, 237
Goodwin, 235
Lenert, 236
domain formation, 200
  assessment item overlap, 234
  cognitive basis, 204
  cross-classification, 205, 233
  cultural factors, 204
  depression, 202
  fatigue, 207
  pain, 217
  PROMIS project, 234
  psychological determinants, 203
  Smets, 202
  Visser, 202
HRQOL assessment, 200
  symptom clustering, 205
  symptom management, 237
language usage, 124
Melzack, 125
metaphors in doctor-patient reporting, 125
metaphors in patient symptom reporting, 125
perception model, 130
qualification processes, 128
subjectivity and clinical medicine, 141
  clinical management of symptoms, 148
  clinimetric and psychometric perspective, 149
  defining adverse events, 143
Susan Sontag, 124
symptom perception model, 130
symptoms as metaphors, 126
Synaesthesia, 11
  Cytowic, 12

T
Teleological explanations, 302
  Aristotle, 302
  causes without teleology, 304
  complexity theory, 303
  Darwin, 303
  Fulford, 304
  Plato, 302
  teleology without values, 305
  teleology with values, 305
  types-Mayr, 303
Tong, E.M.W., 426
Tourangeau's cognitive model, 427
Trained judgment-expertise
  causal knowledge, 140
  cognitive basis, 140
Trait (facial) perception, 68
  Hassin, 68
  non-linear model, 68
  Trope, 68
Trauer, T., 465
Trope, Y., 68
Truth-to-nature, 133
  grounded theory, 134
  Sacks, 133
Turk, D.C., 228
Tversky, A.
  prospect theory, 471

V

Valuations, 457
  cognitive basis, 466
  cognitive differences in assessment procedures, 467
  comparison of summary statements methods, 470
  language-basis, 471
  propect theory, 471
  rational basis of decision making, 470
  variability in estimates-Fryback, 466
  violation of rational decision model, 470
  weighting procedures, 457
Value-based
  Erickson, 43
  Felce, 43
  Ferrans, 43
  Patrick, 43
  Perry, 43
Values, established, 471
  relationship to behavior, 472
  Rokeach, 471
  Schwartz, 472
Verbrugge, L., 130
Visser, M.R., 214
Visual analog scale, 372
Visual HRQOL, 183
  generic/disease-specific assessment, 183
  modular basis, 183

W

Wakefield, J.C., 255
Weighting procedures, 457
  Campbell, 464
  Gorbatenko-Roth, 463
  healthy subjects, 463
  Hubley, 465
  job satisfaction literature, 461
  Locke, 461
  Mackinnon, 465
  medically ill persons, 463
  metaphoric properties, 457
  methods, 458
  Osoba, 463
  Russell, 465
  Trauer, 465
  Wu, 462
Well-being, 41
  linguistic definitions, 43
White, M.P., 412
WHO definition of health, 6
  Bok, 394
Wierzbicka, A., 59, 412
Williams, B., 266
Wilson and Cleary model, 142
  Sullivan, 319
  supporting data, 105
Wilson, I.B., 185
Wisniewski, E.J., 164
Withey, S.B., 421
Wittenberg, E., 248
Wittgenstein, L., 8, 163
  language games, 8
  Michalos, 8
Wu, C-H., 462